T0155682

Lecture Notes in Mathematics

Volume 2320

This series reports on new developments in all areas of mathematics and their applications - quickly, informally and at a high level. Mathematical texts analysing new developments in modelling and numerical simulation are welcome. The type of material considered for publication includes:

1. Research monographs
2. Lectures on a new field or presentations of a new angle in a classical field
3. Summer schools and intensive courses on topics of current research.

Texts which are out of print but still in demand may also be considered if they fall within these categories. The timeliness of a manuscript is sometimes more important than its form, which may be preliminary or tentative.

Titles from this series are indexed by Scopus, Web of Science, Mathematical Reviews, and zbMATH.

Yinqin Li • Dachun Yang • Long Huang

Real-Variable Theory of Hardy Spaces Associated with Generalized Herz Spaces of Rafeiro and Samko

 Springer

Yinqin Li
Laboratory of Mathematics and Complex
Systems (Ministry of Education of China)
School of Mathematical Sciences
Beijing Normal University
Beijing, Beijing, China

Dachun Yang
(Corresponding Author)
Laboratory of Mathematics and Complex
Systems (Ministry of Education of China)
School of Mathematical Sciences
Beijing Normal University
Beijing, Beijing, China

Long Huang
School of Mathematics and Information
Science
Key Laboratory of Mathematics and
Interdisciplinary Sciences of the
Guangdong Higher Education Institute
Guangzhou University
Guangzhou, Guangdong, China

This work was supported by National Key Research and Development Program of China (2020YFA0712900), National Natural Science Foundation of China (11971058) (12071197), and Basic and Applied Basic Research Foundation of Guangdong Province (2021A1515110905)

ISSN 0075-8434   ISSN 1617-9692   (electronic)
Lecture Notes in Mathematics
ISBN 978-981-19-6787-0   ISBN 978-981-19-6788-7   (eBook)
https://doi.org/10.1007/978-981-19-6788-7

Mathematics Subject Classification: 42B35, 42B30, 42B25, 42B20, 42B10, 46E30, 47B47, 47G30

This Springer imprint is published by the registered company Springer Nature Singapore Pte Ltd.
The registered company address is: 152 Beach Road, #21-01/04 Gateway East, Singapore 189721, Singapore

Preface

It is well known that Herz spaces certainly play an important role in harmonic analysis and partial differential equations and have been systematically studied and developed so far; see, for instance, [79, 146, 150, 174, 275] for classical Herz spaces, [83, 239–241] for weighted Herz spaces, [132, 196, 202, 235–238, 276] for variable Herz spaces, [61, 68, 69, 93, 179, 262, 263] for Herz-type Hardy spaces, [59, 62, 64, 67, 257–259] for Herz-type Besov spaces, and [57, 60, 212, 256, 260, 261, 264] for Herz-type Triebel–Lizorkin spaces. Observe that the classical Herz space was originally introduced by Herz [104] in 1968 to study the Bernstein theorem on absolutely convergent Fourier transforms, while the research on Herz spaces can be traced back to the work of Beurling [13]. Indeed, in 1964, to study some convolution algebras, Beurling [13] first introduced a special Herz space $A^p(\mathbb{R}^n)$, with $p \in (1, \infty)$ [see Remark 7.1.2(iv) for its definition], which is also called the Beurling algebra. After that, great progress has been made on Herz spaces and their applications. For instance, in 1985, Baernstein and Sawyer [9] generalized these Herz spaces and gave many applications in both the embedding and the multiplier theorems for classical Hardy spaces on the n-dimensional Euclidean space; in 1984, to study the Wiener third Tauberian theorem for the n-dimensional Euclidean space, Feichtinger [79] introduced another norm of $A^p(\mathbb{R}^n)$, which is obviously equivalent to the norm defined by Beurling [13].

Moreover, Herz spaces play a crucial role in the convergence and the summability problems of both Fourier transforms and Fourier series. Recall that the study of summability means was originally motivated by the famous convergence problem of Dirichlet integral operators. As one of the deepest results in harmonic analysis, in the celebrated works of both Carleson [28] and Hunt [129], they showed that Dirichlet integral operators converge almost everywhere in one-dimensional case. In recent decades, via replacing Dirichlet integral operators by some other summability means, the summability of Fourier transforms was systematically studied by Butzer and Nessel [19], Trigub and Belinsky [230], and Feichtinger and Weisz [80–82] as well as Weisz [246–253]. Particularly, Weisz [254, 255] established the connection between the Herz spaces and the summability of both Fourier series and Fourier transforms. Let us give more details on the latter case. To this end, let

$$\theta \in L^1(\mathbb{R}^n) \cap C_0(\mathbb{R}^n),$$

where $C_0(\mathbb{R}^n)$ denotes the set of all the continuous functions f on $\mathbb{R}^n$ satisfying that

$$\lim_{|x| \to \infty} |f(x)| = 0.$$

Recall that, for any given $p \in [1, 2]$ and $\mathbf{T} := (T_1, \ldots, T_n) \in (0, \infty)^n$ and for any $f \in L^p(\mathbb{R}^n)$, the θ-mean of f is defined by setting, for any $x \in \mathbb{R}^n$,

$$\sigma_{\mathbf{T}}^\theta f(x) := \int_{\mathbb{R}^n} \theta\left(-\frac{\xi_1}{T_1}, \cdots, -\frac{\xi_n}{T_n}\right) \widehat{f}(\xi) e^{2\pi i x \cdot \xi} \, d\xi,$$

where $\widehat{f}$ denotes the Fourier transform of f, $i := \sqrt{-1}$, and

$$x \cdot \xi := \sum_{j=1}^n x_j \xi_j$$

for any $x := (x_1, \cdots, x_n), \xi := (\xi_1, \ldots, \xi_n) \in \mathbb{R}^n$ (see, for instance, [82, (3.1)]). Then, in the recent book [254], Weisz showed that the θ-means of some functions converge to these functions themselves at all their Lebesgue points of these functions under consideration if and only if the Fourier transform of θ belongs to a suitable Herz space. This means that those Herz spaces are the best choice in the study of the summability of Fourier transforms. Furthermore, Herz spaces also prove important in the recent article [207] of Sawano et al. Indeed, in 2017, Sawano et al. [207] introduced both the ball quasi-Banach function space X and the associated Hardy space $H_X(\mathbb{R}^n)$ via the grand maximal function. As was pointed out in [207], Sawano et al. used a certain inhomogeneous Herz space to overcome the essential difficulty appearing in the proof of the convergence of the atomic decomposition of $H_X(\mathbb{R}^n)$. Also, Herz-type spaces prove useful in the study related to partial differential equations. For instance, Scapellato [210] showed that the variable Herz spaces are the key tools in the study of the regularity of solutions to elliptic equations. Drihem [65, 66] studied semilinear parabolic equations with initial data in Herz spaces or Herz-type Triebel–Lizorkin spaces. In addition, the Fourier–Herz space has been proved to be one of the most suitable spaces to investigate the global stability for fractional Navier–Stokes equations; see, for instance, [31, 41, 159, 190]. For more progress on applications of Herz spaces, we refer the reader to [74, 112, 114, 116, 126, 180, 184, 189, 195, 274].

On the other hand, as a good substitute of Herz spaces, Herz-type Hardy spaces are also useful in many mathematical fields such as harmonic analysis and partial differential equations, which have been systematically studied and developed so far; see, for instance, [32, 40, 63, 86, 120, 158, 279]. Recall that the study of Hardy spaces can be traced back to the works of Hardy and Littlewood via tools from complex analysis (see [97, 98, 152]). Then, based on the real-variable methods introduced by Calderón and Zygmund in the 1950s (see [25, 26, 281]), the classical real Hardy space $H^p(\mathbb{R}^n)$ was originally initiated by Stein and Weiss [215] and then systematically developed by Fefferman and Stein [77]. Later on, Coifman [44] established the atomic characterization of the Hardy space $H^p(\mathbb{R})$; Strömberg [216] showed that an orthonormal wavelet basis is always an unconditional basis of the Hardy space $H^1(\mathbb{R}^n)$, and this remarkable discovery preluded the wavelet analysis. Moreover, Calderón et al. [21–23] investigated the Cauchy integral on Lipschitz curves using weighted Hardy spaces, and Kenig [143, 144] studied the (weighted) Hardy spaces on Lipschitz domains. For more developments of Hardy spaces, we refer the reader to [105, 106, 148, 160, 186, 187] for the real-variable theory of Hardy spaces as well as [46, 71, 183, 211, 213, 243, 244] for the applications of Hardy spaces to both harmonic analysis and partial differential equations. Furthermore, as a variant of the classical real Hardy spaces, the Hardy spaces associated with the Beurling algebras on the real line were first introduced by Chen and Lau [40] in 1989, in which they studied the dual spaces and the maximal function characterizations of these Herz–Hardy spaces. Later, García-Cuerva [86] in 1989 generalized the results of Chen and Lau [40] to higher-dimensional case, and García-Cuerva and Herrero [87] in 1994 further studied the maximal function, the atomic, and the Littlewood–Paley function characterizations of these Herz–Hardy spaces. Also, since 1990, Lu and Yang made a series of studies on the Hardy spaces associated with the Beurling algebras or with the Herz spaces (see, for instance, the monograph [175] and its references). In particular, in 1992, Lu and Yang [162] first studied both the Littlewood–Paley function and the ϕ-transform characterizations of the Herz–Hardy space $HK_2(\mathbb{R}^n)$. Moreover, in 1995, Lu and Yang [163] established the atomic and the molecular characterizations of the Herz–Hardy space with general indices and then, in 1997, they further gave various maximal function characterizations of weighted Herz–Hardy spaces in [172].

Meanwhile, Lu and Yang [164] proved that some oscillatory singular integral operators are bounded from Herz–Hardy spaces to Herz spaces and showed that their boundedness fails on Herz–Hardy spaces via a counterexample. Recall that, as was pointed out by Pan [193], the oscillatory singular integral operators may not be bounded from the classical Hardy space $H^1(\mathbb{R}^n)$ to the Lebesgue space $L^1(\mathbb{R}^n)$. Thus, in some sense, the results of [164] showed that the Herz–Hardy space is a proper substitution of $H^1(\mathbb{R}^n)$ in the study on oscillatory singular integral operators. In addition, the commutators and the multiplier theorems on the Herz–Hardy spaces were investigated by Lu and Yang, respectively, in [170] and [173]; both the interpolation of generalized Herz spaces and its applications were given in Hernández and Yang [102, 103]; weighted Herz spaces and their applications were considered by Lu and Yang in [166]; embedding theorems for Herz spaces were studied in [167]. Furthermore, Hernández et al. [101] established the ϕ-transform and the wavelet characterizations for some Herz and Herz-type Hardy spaces by

means of a local version of the discrete tent spaces at the origin. For more progress
on the Herz–Hardy spaces on $\mathbb{R}^n$ and their applications, we refer the reader to
[111, 119, 134, 165, 169, 171, 175, 198, 245]. Moreover, Herz-type spaces on local
fields and Vilenkin groups were also investigated by Fan, Lu, and Yang, respectively,
in [168] and [73, 174, 268].

Nowadays, more and more new function spaces continually spring up to meet
the increasing requirements arising in harmonic analysis and partial differential
equations; see, for instance, [1, 15, 65, 185, 197, 199, 270]. Particularly, extending
the classical Herz spaces to some more general settings has also attracted consider-
able attention recently. For instance, Rafeiro and Samko [197] creatively introduced
local and global generalized Herz spaces recently, which are the generalization
of classical homogeneous Herz spaces and connect with generalized Morrey type
spaces. To be precise, Rafeiro and Samko [197] showed that the scale of these
Herz spaces include Morrey type spaces and complementary Morrey type spaces
and, as applications, they also obtained the boundedness of a class of sublinear
operators on these generalized Herz spaces. Note that Morrey type spaces have
been studied in [2, 3, 191, 203, 204]. Moreover, as a generalization of classical
variable Herz spaces introduced by Izuki [130], Rafeiro and Samko [199] further
extended the generalized Herz spaces to the variable exponent setting. Furthermore,
in [199], Rafeiro and Samko showed that the generalized variable exponent Herz
spaces coincide with the generalized variable exponent Morrey type spaces, and
also established some boundedness of sublinear operators on these spaces.

Observe that Herz-type Hardy spaces have found lots of applications in many
branches of mathematics and that both the real-variable theory of function spaces
and their applications are always one of the central topics of harmonic analysis.
Then it is a natural and meaningful topic to introduce and develop the real-variable
theory of Hardy spaces associated with local and global generalized Herz spaces
of Rafeiro and Samko [197], which is the subject of the present book. To achieve
this, we first investigate some basic properties of these generalized Herz spaces and
realize that the Hardy spaces associated with generalized Herz spaces completely
fall into the framework of Hardy spaces associated with ball quasi-Banach function
spaces which were first studied by Sawano et al. [207] as was mentioned above.
Indeed, about the real-variable theory of Hardy spaces associated with ball quasi-
Banach function spaces, Sawano et al. [207] tried to answer the following *important
issue*: Characterize the quasi-Banach space X for which the associated Hardy space
$H_X(\mathbb{R}^n)$ can be described by various maximal functions, atomic decompositions,
molecular decompositions, Littlewood–Paley functions, and so on. To be precise, let
X be a ball quasi-Banach function space on $\mathbb{R}^n$ (see [207] or Definition 1.2.13 below
for its definition). Recall that, if the powered Hardy–Littlewood maximal operators
are bounded on both X and its associate space, and if X supports a Fefferman–Stein
vector-valued inequality, then Sawano et al. [207] established various real-variable
characterizations of the Hardy space $H_X(\mathbb{R}^n)$ associated with X, respectively, in
terms of atoms, molecules, and the Lusin area function. After the work of Sawano
et al. [207], the real-variable theory of function spaces associated with ball quasi-

Banach function spaces was well developed by Ho in [108, 113] and also by others in [29, 37–39, 122, 233, 242, 266, 277, 278].

This book is devoted to exploring further properties of the local and the global generalized Herz spaces and establishing a complete real-variable theory of Hardy spaces associated with local and global generalized Herz spaces via a totally fresh perspective, which means that we view these generalized Herz spaces as special cases of ball quasi-Banach function spaces. In this perspective, the real-variable theory of Hardy spaces associated with local generalized Herz spaces can be deduced directly from the general framework of the real-variable theory of $H_X(\mathbb{R}^n)$ because the local generalized Herz spaces satisfy all the assumptions of the results about Hardy-type spaces associated with ball quasi-Banach function spaces. However, due to the deficiency of the associate space of the global generalized Herz space, the known real-variable characterizations about Hardy-type spaces associated with ball quasi-Banach function spaces are not applicable to Hardy spaces associated with global generalized Herz spaces [see Remark 1.2.19(vi) below for details]. Therefore, the study of the real-variable theory of Hardy spaces associated with global generalized Herz spaces is more difficult. To overcome this obstacle, via replacing the assumptions on associate spaces by some weaker assumptions about the integral representations of quasi-norms of ball quasi-Banach function spaces [see Theorem 4.3.18(ii) below], we first develop some new real-variable characterizations of Hardy-type spaces associated with ball quasi-Banach function spaces, which even improve the known results on Hardy-type spaces associated with ball quasi-Banach function spaces and can be regarded as another way to handle the aforementioned important issue proposed in [207]; moreover, they surely have additional anticipating applications. In particular, applying these improved conclusions, we further obtain a complete real-variable theory of Hardy spaces associated with global generalized Herz spaces.

Precisely, in this book, we first give some basic properties of these generalized Herz spaces and obtain the boundedness and the compactness characterizations of commutators on them. Then, based on these local and global generalized Herz spaces, we introduce associated Herz–Hardy spaces, localized Herz–Hardy spaces, and weak Herz–Hardy spaces and develop a complete real-variable theory of these Herz–Hardy spaces, including various maximal function, atomic, molecular as well as various Littlewood–Paley function characterizations. As applications, we establish the boundedness of some important operators arising from harmonic analysis on these Herz–Hardy spaces. Finally, the inhomogeneous Herz–Hardy spaces and their complete real-variable theory are also investigated. We should point out that, with the aforementioned fresh perspective and the aforementioned improved conclusions on the real-variable theory of Hardy spaces associated with ball quasi-Banach function spaces, the exponents in all the obtained conclusions of this book are *sharp*. Moreover, all of these results in the book are *new* and have never been published before.

To be precise, this book is organized as follows.

In Chap. 1, we first recall the concepts of both the function class $M(\mathbb{R}_+)$ and the Matuszewska–Orlicz indices, and the definitions of the local generalized Herz space $\dot{\mathcal{K}}^{p,q}_{\omega,\mathbf{0}}(\mathbb{R}^n)$ as well as the global generalized Herz space $\dot{\mathcal{K}}^{p,q}_{\omega}(\mathbb{R}^n)$ introduced by Rafeiro and Samko [197], where $p, q \in (0, \infty]$, $\omega \in M(\mathbb{R}_+)$, and $\mathbf{0}$ denotes the *origin* of $\mathbb{R}^n$. In addition, under some reasonable and sharp assumptions, we show that these generalized Herz spaces are special ball quasi-Banach function spaces. Then we give some basic properties about these generalized Herz spaces, which include their convexity, the absolute continuity of the quasi-norm of $\dot{\mathcal{K}}^{p,q}_{\omega,\mathbf{0}}(\mathbb{R}^n)$, the boundedness criterion of sublinear operators on them which was essentially obtained by Rafeiro and Samko [197, Theorem 4.3], and Fefferman–Stein vector-valued inequalities. Furthermore, we find the dual space and the associate space of $\dot{\mathcal{K}}^{p,q}_{\omega,\mathbf{0}}(\mathbb{R}^n)$. Finally, we establish the extrapolation theorems of local and global generalized Herz spaces.

In Chap. 2, we first introduce the block space $\dot{\mathcal{B}}^{p,q}_{\omega}(\mathbb{R}^n)$ based on the concepts of (ω, p)-blocks. Then we investigate the properties of block spaces in two aspects. On the one hand, by establishing an equivalent characterization of block spaces via local generalized Herz spaces $\dot{\mathcal{K}}^{p,q}_{\omega,\mathbf{0}}(\mathbb{R}^n)$, and borrowing some ideas from the proof of [88, Theorem 6.1], we show that the global generalized Herz space $\dot{\mathcal{K}}^{p',q'}_{1/\omega}(\mathbb{R}^n)$ is just the dual space of the block space $\dot{\mathcal{B}}^{p,q}_{\omega}(\mathbb{R}^n)$. This dual theorem plays an essential role in the study of the real-variable theory of Hardy spaces associated with global generalized Herz spaces in the subsequent chapters. On the other hand, we establish the boundedness of some sublinear operators on block spaces. In particular, the boundedness of powered Hardy–Littlewood maximal operators on block spaces is obtained, which also plays an important role in the subsequent chapters.

The main target of Chap. 3 is to study the boundedness and the compactness characterizations of commutators on generalized Herz spaces. Recall that Tao et al. [228] established the boundedness and the compactness characterizations of commutators on ball Banach function spaces. Combining these and the aforementioned fact that the generalized Herz spaces are special ball Banach function spaces, we first show that local generalized Herz spaces satisfy all the assumptions of the results obtained in [228], and then obtain the boundedness and the compactness characterizations of commutators on the local generalized Herz space $\dot{\mathcal{K}}^{p,q}_{\omega,\mathbf{0}}(\mathbb{R}^n)$. However, the conclusions obtained in [228] are not applicable to show the boundedness and the compactness characterizations of commutators on $\dot{\mathcal{K}}^{p,q}_{\omega}(\mathbb{R}^n)$ due to the deficiency of the associate space of the global generalized Herz space $\dot{\mathcal{K}}^{p,q}_{\omega}(\mathbb{R}^n)$ [see Remark 1.2.19(vi) below for details]. Notice that the most important usage of associate spaces in the proof of [228] is that, under the assumption of the boundedness of the Hardy–Littlewood maximal operator on both ball Banach function spaces and their associate spaces, Tao et al. obtained the extrapolation theorem of ball Banach function spaces. Via this extrapolation theorem and some other technical lemmas independent of associate spaces, Tao et al. [228] then established the boundedness and the compactness characterizations of commutators on ball Banach function spaces. Therefore, to overcome the difficulty caused by the deficiency of associate spaces of $\dot{\mathcal{K}}^{p,q}_{\omega}(\mathbb{R}^n)$, we establish the new boundedness and the new compactness characterizations of commutators on ball Banach function spaces under the assumption that the extrapolation theorem holds true for ball

Banach function spaces instead of the assumption about associate spaces, which improves the corresponding results of [228]. Finally, applying these improved boundedness and compactness characterizations of commutators on ball Banach function spaces and the extrapolation theorem of $\dot{\mathcal{K}}_\omega^{p,q}(\mathbb{R}^n)$ obtained in Chap. 1, we then obtain the boundedness and the compactness characterizations of commutators on global generalized Herz spaces.

Chapter 4 is devoted to introducing the generalized Herz–Hardy space and then establishing its complete real-variable theory. To be precise, we first introduce the generalized Herz–Hardy spaces, $H\dot{\mathcal{K}}_{\omega,\mathbf{0}}^{p,q}(\mathbb{R}^n)$ and $H\dot{\mathcal{K}}_\omega^{p,q}(\mathbb{R}^n)$, associated, respectively, with the local generalized Herz space $\dot{\mathcal{K}}_{\omega,\mathbf{0}}^{p,q}(\mathbb{R}^n)$ and the global generalized Herz space $\dot{\mathcal{K}}_\omega^{p,q}(\mathbb{R}^n)$. Then, using the known real-variable characterizations of Hardy spaces associated with ball quasi-Banach function spaces, we establish various maximal function, atomic, molecular, and Littlewood–Paley function characterizations of the Herz–Hardy space $H\dot{\mathcal{K}}_{\omega,\mathbf{0}}^{p,q}(\mathbb{R}^n)$. Moreover, the duality and the Fourier transform properties of $H\dot{\mathcal{K}}_{\omega,\mathbf{0}}^{p,q}(\mathbb{R}^n)$ are also obtained based on the corresponding results about Hardy spaces associated with ball quasi-Banach function spaces. However, the study of $H\dot{\mathcal{K}}_\omega^{p,q}(\mathbb{R}^n)$ is more difficult than that of $H\dot{\mathcal{K}}_{\omega,\mathbf{0}}^{p,q}(\mathbb{R}^n)$ due to the deficiency of associate spaces of global generalized Herz spaces. To overcome this obstacle, via replacing the assumptions of the boundedness of powered Hardy–Littlewood maximal operators on associate spaces (see Assumption 1.2.33 below) used in [207, Theorems 3.6, 3,7, and 3.9] by some weaker assumptions about the integral representations of quasi-norms of ball quasi-Banach function spaces as well as some boundedness of powered Hardy–Littlewood maximal operators [see both (ii) and (iii) of Theorem 4.3.18 below], we establish the new atomic and the new molecular characterizations of the Hardy space $H_X(\mathbb{R}^n)$ associated with the ball quasi-Banach function space X, which improve the corresponding results obtained by Sawano et al. [207]. Using these improved characterizations and making full use of the obtained duality between block spaces and global generalized Herz spaces in Chap. 2 as well as the construction of the quasi-norm $\|\cdot\|_{\dot{\mathcal{K}}_\omega^{p,q}(\mathbb{R}^n)}$, we then obtain the maximal function, the (finite) atomic, the molecular, the various Littlewood–Paley function characterizations of $H\dot{\mathcal{K}}_\omega^{p,q}(\mathbb{R}^n)$ and also give some properties about Fourier transforms of distributions in $H\dot{\mathcal{K}}_\omega^{p,q}(\mathbb{R}^n)$. Finally, as applications, via first establishing two boundedness criteria of Calderón–Zygmund operators on Hardy spaces associated with ball quasi-Banach function spaces, we obtain the boundedness of Calderón–Zygmund operators on generalized Herz–Hardy spaces.

In Chap. 5, we first introduce the localized generalized Herz–Hardy space and then establish its complete real-variable theory. To achieve this, we begin with showing its various maximal function characterizations via the known maximal function characterizations of the local Hardy space $h_X(\mathbb{R}^n)$ associated with the ball quasi-Banach function space X. Then, via establishing the new atomic and the new molecular characterizations of $h_X(\mathbb{R}^n)$ as well as the boundedness of pseudo-differential operators on $h_X(\mathbb{R}^n)$ without using any assumptions about associate spaces (see Theorems 5.3.14, 5.4.11, and 5.6.9 below), we show the atomic and the molecular characterizations of localized generalized Herz–Hardy spaces and the boundedness of pseudo-differential operators on localized generalized Herz–Hardy spaces. In addition, to clarify the relation between localized generalized Herz–Hardy spaces and generalized Herz–Hardy spaces, we first establish the relation between $h_X(\mathbb{R}^n)$ and the Hardy space $H_X(\mathbb{R}^n)$ associated with the ball quasi-Banach function space X. This extends the results obtained by Goldberg [89, Lemma 4] for classical Hardy spaces and also Nakai and Sawano [186, Lemma 9.1] for variable Hardy spaces. Applying this and some auxiliary lemmas about generalized Herz spaces, we then establish the relation between localized generalized Herz–Hardy spaces and generalized Herz–Hardy spaces. As applications, we also establish various Littlewood–Paley function characterizations of $h_X(\mathbb{R}^n)$, which, together with the construction of the quasi-norm $\|\cdot\|_{\mathcal{K}^{p,q}_\omega(\mathbb{R}^n)}$, further imply the Littlewood–Paley function characterizations of localized generalized Herz–Hardy spaces.

The main target of Chap. 6 is to introduce weak generalized Herz–Hardy spaces and establish their complete real-variable theory. For this purpose, recall that Zhang et al. [278] and Wang et al. [242] investigated the real-variable theory of the weak Hardy space $WH_X(\mathbb{R}^n)$ associated with the ball quasi-Banach function space X. Via removing the assumption about associate spaces, we establish the new atomic and the new molecular characterizations of $WH_X(\mathbb{R}^n)$ as well as the new real interpolation between the Hardy space $H_X(\mathbb{R}^n)$ associated with the ball quasi-Banach function space X and the Lebesgue space $L^\infty(\mathbb{R}^n)$, which improve the corresponding results obtained in [278] and [242]. Then, using these improved real-variable characterizations of $WH_X(\mathbb{R}^n)$, we obtain various maximal function, atomic, and molecular characterizations of weak generalized Herz–Hardy spaces and also show that the real interpolation spaces between generalized Herz–Hardy spaces and the Lebesgue space $L^\infty(\mathbb{R}^n)$ are just the new introduced weak generalized Herz–Hardy spaces. In addition, by establishing a technique lemma about the quasi-norm $\|\cdot\|_{W\mathcal{K}^{p,q}_\omega(\mathbb{R}^n)}$ and the Littlewood–Paley function characterizations of $WH_X(\mathbb{R}^n)$ obtained in [242], we establish various Littlewood–Paley function characterizations of weak generalized Herz–Hardy spaces. Furthermore, we establish two boundedness criteria of Calderón–Zygmund operators from the Hardy space $H_X(\mathbb{R}^n)$ to the weak Hardy space $WH_X(\mathbb{R}^n)$ and, as a consequence, we finally deduce the boundedness of Calderón–Zygmund operators from generalized Herz–Hardy spaces to weak generalized Herz–Hardy spaces even in the critical case.

In Chap. 7, we first introduce the inhomogeneous generalized Herz spaces and then establish their corresponding conclusions obtained in Chaps. 1 through 3.

Furthermore, in Chap. 8, based on the inhomogeneous generalized Herz spaces studied in Chap. 7, we introduce the inhomogeneous generalized Herz–Hardy spaces, the inhomogeneous localized generalized Herz–Hardy spaces, and the inhomogeneous weak generalized Herz–Hardy spaces. Then we establish their various real-variable characterizations and also give some applications, which are the corresponding inhomogeneous variants obtained, respectively, in Chaps. 4 through 6.

Throughout this book, we always let $\mathbb{N} := \{1, 2, \ldots\}$, $\mathbb{Z}_+ := \mathbb{N} \cup \{0\}$, $\mathbb{R}_+ := (0, \infty)$, and

$$\mathbb{R}^{n+1}_+ := \{(x, t) : x \in \mathbb{R}^n, \ t \in (0, \infty)\}.$$

We also use $\mathbf{0} := (0, \ldots, 0)$ to denote the *origin* of $\mathbb{R}^n$. For any $x := (x_1, \ldots, x_n) \in \mathbb{R}^n$ and $\theta := (\theta_1, \ldots, \theta_n) \in (\mathbb{Z}_+)^n =: \mathbb{Z}^n_+$, let $|\theta| := \theta_1 + \cdots + \theta_n$,

$$x^\theta := x_1^{\theta_1} \cdots x_n^{\theta_n},$$

and

$$\partial^\gamma := \left(\frac{\partial}{\partial x_1}\right)^{\gamma_1} \cdots \left(\frac{\partial}{\partial x_n}\right)^{\gamma_n}.$$

We always denote by C a *positive constant* which is independent of the main parameters, but it may vary from line to line. We use $C_{(\alpha, \beta, \ldots)}$ to denote a positive constant depending on the indicated parameters α, β, $\ldots$. The notation $f \lesssim g$ means $f \leq Cg$ and, if $f \lesssim g \lesssim f$, then we write $f \sim g$. If $f \lesssim g$ and $g = h$ or $g \leq h$, we then write $f \lesssim g \sim h$ or $f \lesssim g \lesssim h$, rather than $f \lesssim g = h$ or $f \lesssim g \leq h$. For any $s \in \mathbb{R}$, the symbol $\lceil s \rceil$ denotes the smallest integer not less than s, and the symbol $\lfloor s \rfloor$ denotes the largest integer not greater than s. For any set $E \subset \mathbb{R}^n$, we denote the set $\mathbb{R}^n \setminus E$ by $E^\complement$, its *characteristic function* by $\mathbf{1}_E$, and its *n-dimensional Lebesgue measure* by $|E|$. For any $q \in [1, \infty]$, we denote by q' its *conjugate exponent*, that is, $1/q + 1/q' = 1$. In addition, we use

$$\mathbb{S}^{n-1} := \{x \in \mathbb{R}^n : |x| = 1\}$$

to denote the *unit sphere* in $\mathbb{R}^n$ and $d\sigma$ the *area measure* on $\mathbb{S}^{n-1}$. Furthermore, we always use the *symbol* $\mathscr{M}(\mathbb{R}^n)$ to denote the set of all measurable functions on $\mathbb{R}^n$. The symbol $\mathcal{Q}$ denotes the set of all cubes with edges parallel to the coordinate axes. Finally, for any cube $Q \in \mathcal{Q}$, rQ means a cube with the same center as Q and r times the edge length of Q.

This research of both Yinqin Li and Dachun Yang is supported by the National Key Research and Development Program of China (Grant No. 2020YFA0712900) and the National Natural Science Foundation of China (Grant Nos. 11971058 and 12071197). Long Huang is supported by Guangdong Basic and Applied Basic Research Foundation (Grant No. 2021A1515110905). Yinqin Li would like to express his deep gratitude to Dr. Hongchao Jia, Dr. Yangyang Zhang, and Dr. Yirui Zhao for some helpful discussions on the subject of this book. We would also like to thank the four referees of this book for their careful reading and for their many enlightening and useful comments which have definitely improved the presentation and the readability of the book. In particular, we are very grateful to one referee of the aforementioned four referees who gave us both a clear history on the development of the real-variable theory of Hardy spaces (which is now included in this preface) and also several deep insights on Herz spaces and their predual spaces (namely block spaces). In essence, this referee motivated us to improve the related chapters and encouraged us to illustrate both the main definitions and theorems with interesting examples that made the book more readily comprehensible. The improved contents that benefitted from this referee include: Proposition 1.2.3, Corollary 1.2.4, Remark 1.2.5, Theorems 1.2.6 and 1.2.9, Propositions 1.2.10, 1.2.12, 1.2.20, 1.4.6, 1.4.7, and 1.4.8, Theorem 2.1.5, Corollary 2.1.7, Theorems 2.2.4 and 2.2.5, Example 2.2.6, Theorem 2.2.8, Proposition 4.2.4, and Theorems 7.2.3, 7.2.7, 7.2.8, and 7.2.9.

Beijing, The People's Republic of China   Yinqin Li
Beijing, The People's Republic of China   Dachun Yang
Guangzhou, The People's Republic of China   Long Huang
August 2022

Abstract

This book is devoted to exploring properties of generalized Herz spaces and establishing a complete real-variable theory of Hardy spaces associated with local and global generalized Herz spaces via a totally fresh perspective which means that the authors view these generalized Herz spaces as special cases of ball quasi-Banach function spaces. To be precise, in this book, the authors first study some basic properties of generalized Herz spaces and obtain boundedness and compactness characterizations of commutators on them. Then the authors introduce the associated Herz–Hardy spaces, localized Herz–Hardy spaces, and weak Herz–Hardy spaces, and develop a complete real-variable theory of these Herz–Hardy spaces, including their various maximal function, atomic, molecular as well as various Littlewood–Paley function characterizations. As applications, the authors establish the boundedness of some important operators arising from harmonic analysis on these Herz–Hardy spaces. Finally, the inhomogeneous Herz–Hardy spaces and their complete real-variable theory are also investigated. Due to the deficiency of the associate space of the global Herz space, the known real-variable characterizations about Hardy-type spaces associated with ball quasi-Banach function spaces are not applicable to Hardy spaces associated with global generalized Herz spaces which need an improved generalization of the existing one, done by the authors also in this book and having more additional anticipating applications. The authors should also point out that, with the fresh perspective and the improved conclusions on the real-variable theory of Hardy spaces associated with ball quasi-Banach function spaces, the exponents in all the obtained results of this book are sharp. Moreover, all of these results in this book are new and have never been published before.

Keywords and Phrases

Generalized Herz space, Ball quasi-Banach function space, Block space, Hardy space, Localized Hardy space, Weak Hardy space, Atom, Molecule, Duality,

Maximal function, Littlewood–Paley function, Hardy–Littlewood maximal operator, Fefferman–Stein vector-valued inequality, Fourier transform, Interpolation, Calderón–Zygmund operator, Commutator, Pseudo-differential operator.

Contents

Chapter 1
Generalized Herz Spaces of Rafeiro and Samko

In this chapter, we first recall the concepts of both the function class $M(\mathbb{R}_+)$ and the Matuszewska–Orlicz indices, and the definitions of the local generalized Herz space

$$\dot{\mathcal{K}}^{p,q}_{\omega,\mathbf{0}}(\mathbb{R}^n)$$

as well as the global generalized Herz space

$$\dot{\mathcal{K}}^{p,q}_{\omega}(\mathbb{R}^n)$$

creatively introduced by Rafeiro and Samko [197], where $p, q \in (0, \infty]$ and $\omega \in M(\mathbb{R}_+)$. Next, we recall some basic concepts about ball quasi-Banach function spaces introduced by Sawano et al. [207]. Moreover, under some sharp assumptions, we show that these generalized Herz spaces are special ball quasi-Banach function spaces. Then we establish some basic properties about these generalized Herz spaces, which include their convexity, the absolutely continuity of the quasi-norm of $\dot{\mathcal{K}}^{p,q}_{\omega,\mathbf{0}}(\mathbb{R}^n)$, the boundedness criterion of sublinear operators on them which was essentially obtained by Rafeiro and Samko [197, Theorem 4.3], and Fefferman–Stein vector-valued inequalities. We should point out that the global generalized Herz space $\dot{\mathcal{K}}^{p,q}_{\omega}(\mathbb{R}^n)$ may not have the absolutely continuous quasi-norm. Indeed, using the quasi-norm of $\dot{\mathcal{K}}^{p,q}_{\omega}(\mathbb{R})$ and borrowing some ideas from [209, Example 5.1], we construct a special set E and show that its characteristic function $\mathbf{1}_E$ belongs to certain global generalized Herz space. But, $\mathbf{1}_E$ does not have an absolutely continuous quasi-norm in this global generalized Herz space. Finally, by introducing the local generalized Herz space $\dot{\mathcal{K}}^{p,q}_{\omega,\xi}(\mathbb{R}^n)$ for any given $\xi \in \mathbb{R}^n$ and establishing a dual result of it when $p \in [1, \infty)$ and $q \in (0, \infty)$, we find the associate space of $\dot{\mathcal{K}}^{p,q}_{\omega,\mathbf{0}}(\mathbb{R}^n)$ under some reasonable and sharp assumptions. As an

Y. Li et al., *Real-Variable Theory of Hardy Spaces Associated with Generalized Herz Spaces of Rafeiro and Samko*, Lecture Notes in Mathematics 2320, https://doi.org/10.1007/978-981-19-6788-7_1

application, we establish the extrapolation theorem of both $\dot{\mathcal{K}}_{\omega,\mathbf{0}}^{p,q}(\mathbb{R}^n)$ and $\dot{\mathcal{K}}_{\omega}^{p,q}(\mathbb{R}^n)$ at the end of this chapter.

1.1 Matuszewska–Orlicz Indices

In this section, we first recall the concept of a function class $M(\mathbb{R}_+)$ given in [197] and the concept of Matuszewska–Orlicz indices originally introduced by Matuszewska and Orlicz in [177, 178]. Then we present some fundamental properties related to $M(\mathbb{R}_+)$ and Matuszewska–Orlicz indices, which are widely used throughout this book.

To begin with, let ω be a nonnegative function on $\mathbb{R}_+$. Then the function ω is said to be *almost increasing* (resp., *almost decreasing*) on $\mathbb{R}_+$ if there exists a constant $C \in [1, \infty)$ such that, for any $t, \tau \in (0, \infty)$ satisfying $t \leq \tau$ (resp., $t \geq \tau$),

$$\omega(t) \leq C\omega(\tau)$$

(see, for instance, [145, p. 30]). Now, we recall the concept of the function class $M(\mathbb{R}_+)$ given in [197, Definition 2.1] as follows.

Definition 1.1.1 The *function class* $M(\mathbb{R}_+)$ is defined to be the set of all the positive functions ω on $\mathbb{R}_+$ such that, for any $0 < \delta < N < \infty$,

$$0 < \inf_{t \in (\delta, N)} \omega(t) \leq \sup_{t \in (\delta, N)} \omega(t) < \infty$$

and there exist four constants $\alpha_0, \beta_0, \alpha_\infty, \beta_\infty \in \mathbb{R}$ such that

(i) for any $t \in (0, 1]$, $\omega(t)t^{-\alpha_0}$ is almost increasing and $\omega(t)t^{-\beta_0}$ is almost decreasing;
(ii) for any $t \in [1, \infty)$, $\omega(t)t^{-\alpha_\infty}$ is almost increasing and $\omega(t)t^{-\beta_\infty}$ is almost decreasing.

Remark 1.1.2 Let $\omega \in M(\mathbb{R}_+)$, with α_0, β_0, α_∞, and β_∞, be as in Definition 1.1.1.

(i) Applying Definition 1.1.1(i), we conclude that, for any $t \in (0, 1]$,

$$t^{\beta_0} \lesssim \omega(t) \lesssim t^{\alpha_0}, \tag{1.1}$$

which further implies that $\alpha_0 \leq \beta_0$. However, we should point out that, for any positive function ω on $\mathbb{R}_+$, the condition (1.1) may not imply Definition 1.1.1(i), and hence both (1.1) and Definition 1.1.1(i) may not be equivalent.

Indeed, for any given $-\infty < \alpha_0 < \beta_0 < \infty$, let

$$\omega(t) := 2n(2n-1)\left[(2n-1)^{-\alpha_0} - (2n)^{-\beta_0}\right]\left(t - \frac{1}{2n-1}\right)$$
$$+ (2n-1)^{-\alpha_0}$$

for any $t \in (\frac{1}{2n}, \frac{1}{2n-1}]$ with $n \in \mathbb{N}$,

$$\omega(t) := 2n(2n+1)\left[(2n)^{-\beta_0} - (2n+1)^{-\alpha_0}\right]\left(t - \frac{1}{2n}\right) + (2n)^{-\beta_0}$$

for any $t \in (\frac{1}{2n+1}, \frac{1}{2n}]$ with $n \in \mathbb{N}$, and $\omega(t) := 1$ for any $t \in (1, \infty)$. Then, obviously, for any $t \in (0, 1]$, we have

$$t^{\beta_0} \le \omega(t) \le t^{\alpha_0}.$$

However, it is easy to show that

$$\frac{\omega(\frac{1}{2n+1})}{\omega(\frac{1}{2n})}\left(\frac{\frac{1}{2n+1}}{\frac{1}{2n}}\right)^{-\alpha_0} = (2n)^{\beta_0-\alpha_0} \to \infty$$

as $n \to \infty$, which implies that $\omega(t)t^{-\alpha_0}$ is not almost increasing when $t \in (0, 1]$. Similarly, it holds true that

$$\frac{\omega(\frac{1}{2n-1})}{\omega(\frac{1}{2n})}\left(\frac{\frac{1}{2n-1}}{\frac{1}{2n}}\right)^{-\beta_0} = (2n-1)^{\beta_0-\alpha_0} \to \infty$$

as $n \to \infty$, which further implies that $\omega(t)t^{-\beta_0}$ is not almost decreasing when $t \in (0, 1]$.

(ii) From Definition 1.1.1(ii), we deduce that, for any $t \in [1, \infty)$,

$$t^{\alpha_\infty} \lesssim \omega(t) \lesssim t^{\beta_\infty}. \tag{1.2}$$

By this, we further find that $\alpha_\infty \le \beta_\infty$. However, we should point out that, for any positive function ω on $\mathbb{R}_+$, the condition (1.2) may not imply Definition 1.1.1(ii), and hence both (1.2) and Definition 1.1.1(ii) may not be equivalent. Indeed, for any given $-\infty < \alpha_\infty < \beta_\infty < \infty$, let $\omega(t) := 1$ for any $t \in (0, 1]$,

$$\omega(t) := \left[(2n)^{\beta_\infty} - (2n-1)^{\alpha_\infty}\right](t - 2n) + (2n)^{\beta_\infty}$$

for any $t \in (2n - 1, 2n]$ with $n \in \mathbb{N}$, and

$$\omega(t) := \left[(2n + 1)^{\alpha_\infty} - (2n)^{\beta_\infty}\right](t - 2n - 1) + (2n + 1)^{\alpha_\infty}$$

for any $t \in (2n, 2n + 1]$ with $n \in \mathbb{N}$. Then, in this case, it holds true that, for any $t \in [1, \infty)$,

$$t^{\alpha_\infty} \leq \omega(t) \leq t^{\beta_\infty}.$$

However, we can easily find that

$$\frac{\omega(2n)}{\omega(2n + 1)}\left(\frac{2n}{2n + 1}\right)^{-\alpha_\infty} = (2n)^{\beta_\infty - \alpha_\infty} \to \infty$$

as $n \to \infty$. Thus, $\omega(t)t^{-\alpha_\infty}$ is not almost increasing when $t \in [1, \infty)$. On another hand, we have

$$\frac{\omega(2n)}{\omega(2n - 1)}\left(\frac{2n}{2n - 1}\right)^{-\beta_\infty} = (2n - 1)^{\beta_\infty - \alpha_\infty} \to \infty$$

as $n \to \infty$, which further implies that $\omega(t)t^{-\beta_\infty}$ is not almost decreasing when $t \in [1, \infty)$.

The following property of the function class $M(\mathbb{R}_+)$ was stated in [197, (7)] without the proof. For the convenience of the reader, we present it as follows and give its detailed proof.

Lemma 1.1.3 *Let $\omega \in M(\mathbb{R}_+)$. Then, for any given $\lambda \in (0, \infty)$, there exists a positive constant $C_{(\omega, \lambda)}$, depending only on ω and λ, such that, for any $t \in (0, \infty)$,*

$$\omega(\lambda t) \leq C_{(\omega, \lambda)}\omega(t).$$

Proof Let $\omega \in M(\mathbb{R}_+)$, with $\alpha_0, \beta_0, \alpha_\infty, \beta_\infty \in \mathbb{R}$, be as in Definition 1.1.1. Then, there exists a constant $C \in [1, \infty)$ such that, for any $h, t \in (0, 1]$,

$$\omega(ht)(ht)^{-\alpha_0} \leq C\omega(h)h^{-\alpha_0} \tag{1.3}$$

and

$$\omega(h)h^{-\beta_0} \leq C\omega(ht)(ht)^{-\beta_0} \tag{1.4}$$

and, for any $h, t \in [1, \infty)$,

$$\omega(h)h^{-\alpha_\infty} \leq C\omega(ht)(ht)^{-\alpha_\infty} \tag{1.5}$$

and

$$\omega(ht)(ht)^{-\beta_\infty} \le C\omega(h)h^{-\beta_\infty}. \tag{1.6}$$

Now, we show the desired inequality by considering the following two cases on λ.

Case (1) $\lambda \in (0, 1]$. In this case, using (1.3), we conclude that, for any $t \in (0, 1)$,

$$\omega(\lambda t)(\lambda t)^{-\alpha_0} \le C\omega(t)t^{-\alpha_0},$$

which further implies that

$$\omega(\lambda t) \le C\lambda^{\alpha_0}\omega(t) \le C^2\lambda^{\min\{\alpha_0,\alpha_\infty\}}\omega(t). \tag{1.7}$$

In addition, combining (1.3) and (1.5), we find that, for any $t \in [1, \frac{1}{\lambda}]$,

$$\omega(\lambda t)(\lambda t)^{-\min\{\alpha_0,\alpha_\infty\}}$$
$$\le \omega(\lambda t)(\lambda t)^{-\alpha_0} \le C\omega(1) \le C^2\omega(t)t^{-\alpha_\infty}$$
$$\le C^2\omega(t)t^{-\min\{\alpha_0,\alpha_\infty\}}.$$

This implies that, for any $t \in [1, \frac{1}{\lambda}]$,

$$\omega(\lambda t) \le C^2\lambda^{\min\{\alpha_0,\alpha_\infty\}}\omega(t). \tag{1.8}$$

On the other hand, applying (1.5), we know that, for any $t \in (\frac{1}{\lambda}, \infty)$,

$$\omega(\lambda t)(\lambda t)^{-\alpha_\infty} \le C\omega(t)t^{-\alpha_\infty},$$

which further implies that

$$\omega(\lambda t) \le C\lambda^{\alpha_\infty}\omega(t) \le C^2\lambda^{\min\{\alpha_0,\alpha_\infty\}}\omega(t).$$

This, together with (1.7) and (1.8), further implies that, in this case, Lemma 1.1.3 holds true with $C_{(\omega,\lambda)} := C^2\lambda^{\min\{\alpha_0,\alpha_\infty\}}$.

Case (2) $\lambda \in (1, \infty)$. In this case, applying (1.4), (1.6), and an argument similar to that used in the proof of Case (1), we conclude that Lemma 1.1.3 holds true with $C_{(\omega,\lambda)} := C^2\lambda^{\max\{\beta_0,\beta_\infty\}}$, which completes the proof of Lemma 1.1.3. $\qquad\square$

To describe the properties of positive functions at origin and infinity, we present the following Matuszewska–Orlicz indices from [177, 178] (see also [197]). Throughout this book, the *symbol* $h \to 0^+$ means that $h \in (0, \infty)$ and $h \to 0$.

Definition 1.1.4 Let ω be a positive function on $\mathbb{R}_+$. Then the *Matuszewska–Orlicz indices* $m_0(\omega)$, $M_0(\omega)$, $m_\infty(\omega)$, and $M_\infty(\omega)$ of ω are defined, respectively, by setting, for any $h \in (0, \infty)$,

$$m_0(\omega) := \sup_{t \in (0,1)} \frac{\ln(\overline{\lim_{h \to 0^+}} \frac{\omega(ht)}{\omega(h)})}{\ln t},$$

$$M_0(\omega) := \inf_{t \in (0,1)} \frac{\ln(\underline{\lim_{h \to 0^+}} \frac{\omega(ht)}{\omega(h)})}{\ln t},$$

$$m_\infty(\omega) := \sup_{t \in (1,\infty)} \frac{\ln(\underline{\lim_{h \to \infty}} \frac{\omega(ht)}{\omega(h)})}{\ln t},$$

and

$$M_\infty(\omega) := \inf_{t \in (1,\infty)} \frac{\ln(\overline{\lim_{h \to \infty}} \frac{\omega(ht)}{\omega(h)})}{\ln t}.$$

Remark 1.1.5

(i) By Rafeiro and Samko [197, p. 12], we find that, for any $\omega \in M(\mathbb{R}_+)$ and $h \in (0, \infty)$,

$$m_0(\omega) = \lim_{t \to 0^+} \frac{\ln(\overline{\lim_{h \to 0^+}} \frac{\omega(ht)}{\omega(h)})}{\ln t},$$

$$M_0(\omega) = \lim_{t \to 0^+} \frac{\ln(\underline{\lim_{h \to 0^+}} \frac{\omega(ht)}{\omega(h)})}{\ln t},$$

$$m_\infty(\omega) = \lim_{t \to \infty} \frac{\ln(\underline{\lim_{h \to \infty}} \frac{\omega(ht)}{\omega(h)})}{\ln t},$$

and

$$M_\infty(\omega) = \lim_{t \to \infty} \frac{\ln(\overline{\lim_{h \to \infty}} \frac{\omega(ht)}{\omega(h)})}{\ln t},$$

where, for any positive function ω on $\mathbb{R}_+$, $m_0(\omega)$, $M_0(\omega)$, $m_\infty(\omega)$, and $M_\infty(\omega)$ are as in Definition 1.1.4.

(ii) We point out that, for any positive function ω, its Matuszewska–Orlicz indices $m_0(\omega)$, $M_0(\omega)$, $m_\infty(\omega)$, and $M_\infty(\omega)$ may not be finite. Indeed, let $\omega(t) := e^{1/t}$ for any $t \in (0, \infty)$. Then, in this case, we have $m_0(\omega) = -\infty$. However, for any positive function $\omega \in M(\mathbb{R}_+)$, applying [176, Theorem 11.3] and [201, (6.14)], we conclude that

$$-\infty < m_0(\omega) \le M_0(\omega) < \infty$$

and

$$-\infty < m_\infty(\omega) \le M_\infty(\omega) < \infty.$$

(iii) For any given positive function ω on $\mathbb{R}_+$ and for any $t \in (0, \infty)$, let $\omega_*(t) := \omega(\frac{1}{t})$. Then, from [201, (6.14)], it follows that

$$m_0(\omega_*) = -M_\infty(\omega)$$

and

$$M_0(\omega_*) = -m_\infty(\omega),$$

where, for any positive function ϕ, $m_0(\phi)$, $M_0(\phi)$, $m_\infty(\phi)$, and $M_\infty(\phi)$ denote its Matuszewska–Orlicz indices.

For both the function class $M(\mathbb{R}_+)$ and the Matuszewska–Orlicz indices, we have the following conclusions which are just [201, (6.4), (6.5), and (6.14)].

Lemma 1.1.6 *Let $\omega \in M(\mathbb{R}_+)$. Then, for any given $s \in (0, \infty)$, it holds true that $1/\omega$, $\omega^s \in M(\mathbb{R}_+)$ and*

(i) $m_0(\frac{1}{\omega}) = -M_0(\omega)$ *and* $M_0(\frac{1}{\omega}) = -m_0(\omega)$;
(ii) $m_\infty(\frac{1}{\omega}) = -M_\infty(\omega)$ *and* $M_\infty(\frac{1}{\omega}) = -m_\infty(\omega)$;
(iii) $m_0(\omega^s) = s m_0(\omega)$ *and* $M_0(\omega^s) = s M_0(\omega)$;
(iv) $m_\infty(\omega^s) = s m_\infty(\omega)$ *and* $M_\infty(\omega^s) = s M_\infty(\omega)$.

Here, for any positive function ϕ on $\mathbb{R}_+$, $m_0(\phi)$, $M_0(\phi)$, $m_\infty(\phi)$, and $M_\infty(\phi)$ denote its Matuszewska–Orlicz indices.

Now, we give three typical examples of functions in $M(\mathbb{R}_+)$ and their concrete Matuszewska–Orlicz indices.

Example 1.1.7 For any given $\alpha \in \mathbb{R}$ and for any $t \in (0, \infty)$, let $\omega(t) := t^\alpha$. Then, using the fact that, for any $t \in (0, \infty)$, $\omega(t)t^{-\alpha} = 1$, we find that $\omega \in M(\mathbb{R}_+)$. In addition, for any $h, t \in (0, \infty)$, we have

$$\frac{\omega(ht)}{\omega(h)} = \frac{(ht)^\alpha}{h^\alpha} = t^\alpha.$$

By this and Definition 1.1.4, we conclude that

$$m_0(\omega) = \sup_{t \in (0,1)} \frac{\ln\left(\varlimsup_{h \to 0^+} \frac{\omega(ht)}{\omega(h)}\right)}{\ln t} = \sup_{t \in (0,1)} \frac{\ln(t^\alpha)}{\ln t} = \alpha.$$

Similarly, we easily obtain

$$m_\infty(\omega) = M_0(\omega) = M_\infty(\omega) = \alpha.$$

Example 1.1.8 For any $t \in (0, \infty)$, let

$$\omega(t) := \frac{t}{\ln(e+t)}.$$

Then notice that, for any $t \in (0, \infty)$, we have

$$\omega'(t) = \frac{\ln(e+t) - \frac{t}{e+t}}{[\ln(e+t)]^2} = \frac{(e+t)\ln(e+t) - t}{(e+t)[\ln(e+t)]^2}$$

$$> \frac{t \ln e - t}{(e+t)[\ln(e+t)]^2} = 0,$$

which implies that $\omega(t)$ is increasing when $t \in (0, \infty)$. On the other hand, it is easy to show that $\omega(t)t^{-1}$ is decreasing when $t \in (0, \infty)$. Therefore, we conclude that $\omega \in M(\mathbb{R}_+)$. Moreover, for any $t \in (0, \infty)$, we have

$$\lim_{h \to 0^+} \frac{\omega(ht)}{\omega(h)} = \lim_{h \to 0^+} \frac{\frac{ht}{\ln(e+ht)}}{\frac{h}{\ln(e+h)}} = t \lim_{h \to 0^+} \frac{\ln(e+h)}{\ln(e+ht)} = t.$$

From this and Definition 1.1.4, we deduce that

$$m_0(\omega) = \sup_{t \in (0,1)} \frac{\ln\left(\varlimsup_{h \to 0^+} \frac{\omega(ht)}{\omega(h)}\right)}{\ln t} = \sup_{t \in (0,1)} \frac{\ln t}{\ln t} = 1$$

and

$$M_0(\omega) = \inf_{t \in (0,1)} \frac{\ln\left(\varliminf_{h \to 0^+} \frac{\omega(ht)}{\omega(h)}\right)}{\ln t} = \inf_{t \in (0,1)} \frac{\ln t}{\ln t} = 1.$$

In addition, for any $t \in (0, \infty)$, we have

$$\lim_{h \to \infty} \frac{\omega(ht)}{\omega(h)} = \lim_{h \to \infty} \frac{\frac{ht}{\ln(e+ht)}}{\frac{h}{\ln(e+h)}} = t \lim_{h \to \infty} \frac{\ln(e+h)}{\ln(e+ht)}$$

$$= t \lim_{h \to \infty} \frac{1 + \frac{\ln(\frac{e}{h}+1)}{\ln h}}{1 + \frac{\ln(\frac{e}{h}+t)}{\ln h}} = t.$$

This, together with Definition 1.1.4, further implies that

$$m_\infty(\omega) = \sup_{t \in (1,\infty)} \frac{\ln(\lim_{h \to \infty} \frac{\omega(ht)}{\omega(h)})}{\ln t} = \sup_{t \in (1,\infty)} \frac{\ln t}{\ln t} = 1$$

and

$$M_\infty(\omega) = \inf_{t \in (1,\infty)} \frac{\ln(\overline{\lim_{h \to \infty}} \frac{\omega(ht)}{\omega(h)})}{\ln t} = \inf_{t \in (1,\infty)} \frac{\ln t}{\ln t} = 1.$$

Example 1.1.9 For any given $\alpha_1, \alpha_2 \in \mathbb{R}$ and for any $t \in (0, \infty)$, let

$$\omega(t) := \begin{cases} t^{\alpha_1}(1 - \ln t) & \text{when } t \in (0, 1], \\ t^{\alpha_2}(1 + \ln t) & \text{when } t \in (1, \infty). \end{cases}$$

Then, we find that, for any $t \in (0, 1)$,

$$\left(\omega(t)t^{-\alpha_1+1}\right)' = (t(1 - \ln t))' = -\ln t > 0$$

and, for any $t \in (1, \infty)$,

$$\left(\omega(t)t^{-\alpha_2-1}\right)' = \left(\frac{1 + \ln t}{t}\right)' = -\frac{\ln t}{t^2} < 0.$$

This further implies that $\omega(t)t^{-\alpha_1+1}$ is increasing when $t \in (0, 1]$, and $\omega(t)t^{-\alpha_2-1}$ is decreasing when $t \in [1, \infty)$. On the other hand, we can easily know that $\omega(t)t^{-\alpha_1}$ is decreasing when $t \in (0, 1]$, and $\omega(t)t^{-\alpha_2}$ is increasing when $t \in [1, \infty)$. Thus, the positive function ω belongs to $M(\mathbb{R}_+)$. In addition, it holds true that, for any $h, t \in (0, 1)$,

$$\lim_{h \to 0^+} \frac{\omega(ht)}{\omega(h)} = \lim_{h \to 0^+} \frac{(ht)^{\alpha_1}[1 - \ln(ht)]}{h^{\alpha_1}(1 - \ln h)} = t^{\alpha_1} \lim_{h \to 0^+} \left(1 - \frac{\ln t}{1 - \ln h}\right) = t^{\alpha_1}.$$

Combining this and Definition 1.1.4, we conclude that

$$m_0(\omega) = \sup_{t \in (0,1)} \frac{\ln(\varlimsup\limits_{h \to 0^+} \frac{\omega(ht)}{\omega(h)})}{\ln t} = \sup_{t \in (0,1)} \frac{\ln(t^{\alpha_1})}{\ln t} = \alpha_1$$

and

$$M_0(\omega) = \inf_{t \in (0,1)} \frac{\ln(\varliminf\limits_{h \to 0^+} \frac{\omega(ht)}{\omega(h)})}{\ln t} = \inf_{t \in (0,1)} \frac{\ln(t^{\alpha_1})}{\ln t} = \alpha_1.$$

Similarly, from Definition 1.1.4, we deduce that

$$m_\infty(\omega) = M_\infty(\omega) = \alpha_2.$$

The following equivalent formulae of Matuszewska–Orlicz indices are just [176, Theorem 11.13].

Lemma 1.1.10 *Let $\omega \in M(\mathbb{R}_+)$, and $m_0(\omega)$, $M_0(\omega)$, $m_\infty(\omega)$, and $M_\infty(\omega)$ denote its Matuszewska–Orlicz indices. Then*

$$m_0(\omega) = \sup\left\{\alpha_0 \in \mathbb{R} : \omega(t)t^{-\alpha_0} \text{ is almost increasing for any } t \in (0,1]\right\},$$
$$(1.9)$$
$$M_0(\omega) = \inf\left\{\beta_0 \in \mathbb{R} : \omega(t)t^{-\beta_0} \text{ is almost decreasing for any } t \in (0,1]\right\},$$

$$m_\infty(\omega) = \sup\left\{\alpha_\infty \in \mathbb{R} : \omega(t)t^{-\alpha_\infty} \text{ is almost increasing for any } t \in [1,\infty)\right\},$$

and

$$M_\infty(\omega) = \inf\left\{\beta_\infty \in \mathbb{R} : \omega(t)t^{-\beta_\infty} \text{ is almost decreasing for any } t \in [1,\infty)\right\}.$$

Remark 1.1.11 We should point out that the suprema and the infima in Lemma 1.1.10 may not be achieved. Indeed, for any $t \in (0,\infty)$, let

$$\omega(t) := \begin{cases} t(1 - \ln t) & \text{when } t \in (0,1], \\ t(1 + \ln t) & \text{when } t \in (1,\infty). \end{cases}$$

Then, by Example 1.1.9, we conclude that $m_0(\omega) = 1$. However, we have

$$\frac{\omega(\frac{1}{n})(\frac{1}{n})^{-1}}{\omega(\frac{1}{2})(\frac{1}{2})^{-1}} = \frac{1 + \ln n}{1 + \ln 2} \to \infty$$

as $n \to \infty$. This implies that $\omega(t)t^{-m_0(\omega)}$ is not almost increasing when $t \in (0, 1]$. Thus, in this case, the supremum in (1.9) of the Matuszewska–Orlicz index m_0 can not be achieved.

Applying Lemma 1.1.10, we immediately obtain the following estimates of the positive function $\omega \in M(\mathbb{R}_+)$, which were also stated in [197, (25) and (26)]; we omit the details.

Lemma 1.1.12 *Let $\omega \in M(\mathbb{R}_+)$, and $m_0(\omega)$, $M_0(\omega)$, $m_\infty(\omega)$, and $M_\infty(\omega)$ denote its Matuszewska–Orlicz indices. Then, for any given $\varepsilon \in (0, \infty)$, there exists a constant $C_{(\varepsilon)} \in [1, \infty)$, depending on ε, such that, for any $t \in (0, 1]$,*

$$C_{(\varepsilon)}^{-1} t^{M_0(\omega)+\varepsilon} \leq \inf_{\tau \in (0,1]} \frac{\omega(t\tau)}{\omega(\tau)} \leq \sup_{\tau \in (0,1]} \frac{\omega(t\tau)}{\omega(\tau)} \leq C_{(\varepsilon)} t^{m_0(\omega)-\varepsilon}$$

and, for any $t \in [1, \infty)$,

$$C_{(\varepsilon)}^{-1} t^{m_\infty(\omega)-\varepsilon} \leq \inf_{\tau \in [1,\infty)} \frac{\omega(t\tau)}{\omega(\tau)} \leq \sup_{\tau \in [1,\infty)} \frac{\omega(t\tau)}{\omega(\tau)} \leq C_{(\varepsilon)} t^{M_\infty(\omega)+\varepsilon}.$$

Remark 1.1.13 We should point out that, in Lemma 1.1.12, the constant $C_{(\varepsilon)}$ may not be uniform about ε. Indeed, for any $t \in (0, \infty)$, let

$$\omega(t) := \begin{cases} t(1 - \ln t) & \text{when } t \in (0, 1], \\ t(1 + \ln t) & \text{when } t \in (1, \infty). \end{cases}$$

Then, from Example 1.1.9, it follows that $m_0(\omega) = 1$. By this, we conclude that

$$\sup_{t,\,\tau \in (0,1]} \left\{ \frac{\omega(t\tau)}{\omega(\tau)t^{m_0(\omega)-\varepsilon}} \right\}$$

$$= \sup_{t,\,\tau \in (0,1]} \left\{ t^\varepsilon \frac{1 - \ln \tau - \ln t}{1 - \ln \tau} \right\} = \sup_{t \in (0,1]} \left\{ t^\varepsilon (1 - \ln t) \right\}$$

$$= \frac{e^{\varepsilon-1}}{\varepsilon} \to \infty$$

as $\varepsilon \to 0^+$, which further implies that $C_{(\varepsilon)} \to \infty$ as $\varepsilon \to 0^+$, where $C_{(\varepsilon)}$ is as in Lemma 1.1.12. Here and thereafter, $\varepsilon \to 0^+$ means $\varepsilon \in (0, \infty)$ and $\varepsilon \to 0$.

1.2   Generalized Herz Spaces

The targets of this section are threefold. The first one is to recall the concept of generalized Herz spaces of Rafeiro and Samko given in [197, Definition 2.2], as well as investigate some basic properties of them, which include the mapping property of the geometrical transformation called inversion [see (1.15) below for its definition] and the embedding property or the triviality of some special global generalized Herz spaces. The second one is to recall some concepts related to ball quasi-Banach function spaces. The last one is to show that generalized Herz spaces of Rafeiro and Samko are ball Banach function spaces or ball quasi-Banach function spaces under some reasonable and sharp assumptions.

First, we present the concept of generalized Herz spaces which were originally introduced by Rafeiro and Samko in [197, Definition 2.2] under the assumptions $p, q \in (0, \infty)$. In what follows, for any $x \in \mathbb{R}^n$ and $r \in (0, \infty)$, let

$$B(x, r) := \left\{ y \in \mathbb{R}^n : |x - y| < r \right\}$$

and

$$\mathbb{B} := \left\{ B(x, r) : x \in \mathbb{R}^n \text{ and } r \in (0, \infty) \right\}. \tag{1.10}$$

Moreover, for any ball $B := B(x, r)$ with $x \in \mathbb{R}^n$ and $r \in (0, \infty)$, r is called the *radius* of B, which is denoted by $r(B)$. Let $p \in (0, \infty)$. We always use $L^p_{\mathrm{loc}}(\mathbb{R}^n)$ and $L^p_{\mathrm{loc}}(\mathbb{R}^n \setminus \{\mathbf{0}\})$ to denote the set of all p-order locally integrable functions, respectively, on $\mathbb{R}^n$ and $\mathbb{R}^n \setminus \{\mathbf{0}\}$, and use $L^\infty_{\mathrm{loc}}(\mathbb{R}^n)$ and $L^\infty_{\mathrm{loc}}(\mathbb{R}^n \setminus \{\mathbf{0}\})$ to denote the set of all locally essential bounded functions, respectively, on $\mathbb{R}^n$ and $\mathbb{R}^n \setminus \{\mathbf{0}\}$,

Definition 1.2.1 Let $p, q \in (0, \infty]$ and $\omega \in M(\mathbb{R}_+)$.

(i) The *local generalized Herz space* $\dot{\mathcal{K}}^{p,q}_{\omega,\mathbf{0}}(\mathbb{R}^n)$ is defined to be the set of all the $f \in L^p_{\mathrm{loc}}(\mathbb{R}^n \setminus \{\mathbf{0}\})$ such that

$$\|f\|_{\dot{\mathcal{K}}^{p,q}_{\omega,\mathbf{0}}(\mathbb{R}^n)} := \left\{ \sum_{k \in \mathbb{Z}} \left[\omega(2^k) \right]^q \left\| f \mathbf{1}_{B(\mathbf{0},2^k) \setminus B(\mathbf{0},2^{k-1})} \right\|_{L^p(\mathbb{R}^n)}^q \right\}^{\frac{1}{q}} \tag{1.11}$$

is finite.

(ii) The *global generalized Herz space* $\dot{\mathcal{K}}^{p,q}_{\omega}(\mathbb{R}^n)$ is defined to be the set of all the $f \in L^p_{\mathrm{loc}}(\mathbb{R}^n)$ such that

$$\|f\|_{\dot{\mathcal{K}}^{p,q}_{\omega}(\mathbb{R}^n)} := \sup_{\xi \in \mathbb{R}^n} \left\{ \sum_{k \in \mathbb{Z}} \left[\omega(2^k) \right]^q \left\| f \mathbf{1}_{B(\xi,2^k) \setminus B(\xi,2^{k-1})} \right\|_{L^p(\mathbb{R}^n)}^q \right\}^{\frac{1}{q}} \tag{1.12}$$

is finite.

Remark 1.2.2

(i) Recall that, in [197, Definition 2.2], Rafeiro and Samko introduced both local generalized Herz space $\dot{\mathcal{K}}^{p,q}_{\omega,0}(\mathbb{R}^n)$ and global generalized Herz space $\dot{\mathcal{K}}^{p,q}_{\omega}(\mathbb{R}^n)$ with $p,q \in (0,\infty)$ and $\omega \in M(\mathbb{R}_+)$. In this book, we extend their definitions of generalized Herz spaces to $p = \infty$ or $q = \infty$. Moreover, in [197], Rafeiro and Samko also introduced the following *continuous versions* of both (1.11) and (1.12), respectively, by setting, for any measurable function f on $\mathbb{R}^n$,

$$\|f\|_{H^{p,q}_{\omega,0}(\mathbb{R}^n)} := \left\{ \int_0^\infty [\omega(t)]^q \left[\int_{t<|y|<2t} |f(y)|^p \, dy \right]^{\frac{q}{p}} \frac{dt}{t} \right\}^{\frac{1}{q}} \tag{1.13}$$

and

$$\|f\|_{H^{p,q}_{\omega}(\mathbb{R}^n)} := \sup_{\xi \in \mathbb{R}^n} \left\{ \int_0^\infty [\omega(t)]^q \left[\int_{t<|y-\xi|<2t} |f(y)|^p \, dy \right]^{\frac{q}{p}} \frac{dt}{t} \right\}^{\frac{1}{q}}, \tag{1.14}$$

where $p,q \in (0,\infty)$ and $\omega \in M(\mathbb{R}_+)$. We point out that, from [197, Lemma 2.3], it follows that, when $p,q \in [1,\infty)$ and $\omega \in M(\mathbb{R}_+)$, the continuous version (1.13) is equivalent to the discrete version (1.11) and, similarly, (1.14) is equivalent to (1.12).

(ii) Obviously, by Definition 1.2.1, we conclude that, for any measurable function f on $\mathbb{R}^n$,

$$\|f\|_{\dot{\mathcal{K}}^{p,q}_{\omega}(\mathbb{R}^n)} = \sup_{\xi \in \mathbb{R}^n} \|f(\cdot + \xi)\|_{\dot{\mathcal{K}}^{p,q}_{\omega,0}(\mathbb{R}^n)}.$$

(iii) Notice that the local generalized Herz space $\dot{\mathcal{K}}^{p,q}_{\omega,0}(\mathbb{R}^n)$ is always nontrivial. Indeed, from Definition 1.2.1(i), we immediately infer that $\mathbf{1}_{B(0,2)\setminus B(0,1)} \in \dot{\mathcal{K}}^{p,q}_{\omega,0}(\mathbb{R}^n)$. However, as is proved in both Theorems 1.2.6 and 1.2.7 later, when $\omega \in M(\mathbb{R}_+)$ satisfies that $M_0(\omega) \in (-\infty, -\frac{n}{p})$ or $m_\infty(\omega) \in (0,\infty)$, then

$$\dot{\mathcal{K}}^{p,q}_{\omega}(\mathbb{R}^n) = \left\{ f \in \mathcal{M}(\mathbb{R}^n) : \ f = 0 \text{ almost everywhere in } \mathbb{R}^n \right\},$$

and hence the global generalized Herz space $\dot{\mathcal{K}}^{p,q}_{\omega}(\mathbb{R}^n)$ is trivial in this case. Moreover, as is also pointed out in Remark 1.2.8 below, the above assumptions are sharp for the triviality of global generalized Herz spaces.

(iv) In Definition 1.2.1, let $p = q$ and $\omega \equiv 1$. Then, in this case, it is easy to show that

$$\dot{\mathcal{K}}^{p,q}_{\omega,0}(\mathbb{R}^n) = \dot{\mathcal{K}}^{p,q}_{\omega}(\mathbb{R}^n) = L^p(\mathbb{R}^n).$$

(v) Observe that, in Definition 1.2.1(i), for any given $\alpha \in \mathbb{R}$ and for any $t \in (0, \infty)$, let $\omega(t) := t^\alpha$. Then, in this case, the local generalized Herz space $\dot{\mathcal{K}}^{p,q}_{\omega,0}(\mathbb{R}^n)$ goes back to the classical *homogeneous Herz space* $\dot{K}^{\alpha,q}_p(\mathbb{R}^n)$, which was originally introduced in [163, Definition 1.1(a)] (see also [175, Chapter 1]), with the same quasi-norms.

(vi) Let $p, q \in [1, \infty)$ and $\omega \in M(\mathbb{R}_+)$ with $\max\{M_0(\omega), M_\infty(\omega)\} \in (-\infty, 0)$. Then, applying [197, Theorem 3.2 and Remark 3.3], we conclude that, in this case, the generalized Herz spaces $\dot{\mathcal{K}}^{p,q}_\omega(\mathbb{R}^n)$ and $\dot{\mathcal{K}}^{p,q}_{\omega,0}(\mathbb{R}^n)$, in the sense of equivalent norms, coincide, respectively, with the *local generalized Morrey space* $\boldsymbol{M}^{p,q}_{\omega,0}(\mathbb{R}^n)$ and the *global generalized Morrey space* $\boldsymbol{M}^{p,q}_\omega(\mathbb{R}^n)$, originally introduced by Guliev and Mustafaev in [94] (see also [197]), which are defined, respectively, to be the sets of all the measurable functions f on $\mathbb{R}^n$ such that

$$\|f\|_{\boldsymbol{M}^{p,q}_{\omega,0}(\mathbb{R}^n)} := \left\{ \int_0^\infty [\omega(t)]^q \left[\int_{|y|<t} |f(y)|^p \, dy \right]^{\frac{q}{p}} \frac{dt}{t} \right\}^{\frac{1}{q}} < \infty$$

and

$$\|f\|_{\boldsymbol{M}^{p,q}_\omega(\mathbb{R}^n)} := \sup_{\xi \in \mathbb{R}^n} \left\{ \int_0^\infty [\omega(t)]^q \left[\int_{|y-\xi|<t} |f(y)|^p \, dy \right]^{\frac{q}{p}} \frac{dt}{t} \right\}^{\frac{1}{q}} < \infty.$$

(vii) By (iv) of this remark, we know that local generalized Herz spaces in Definition 1.2.1(i) include the classical homogeneous Herz spaces. Recall that function spaces based on classical homogeneous Herz spaces prove useful in partial differential equations; see, for instance, [65, 66] for the applications of Herz spaces or Herz-type Triebel–Lizorkin spaces in the study of semilinear parabolic equations, and [31, 41, 159, 190] for the applications of Fourier–Herz spaces in the investigation of the global stability for fractional Navier–Stokes equations. However, it is still an interesting problem to be studied to find applications of these generalized Herz spaces and their related Hardy-type spaces studied in this book in partial differential equations.

We first give some basic properties of generalized Herz spaces. To this end, let f be a measurable function on $\mathbb{R}^n \setminus \{\boldsymbol{0}\}$ and recall that the *inversion* of f is denoted by $\widetilde{f}$, which is defined by setting, for any $x \in \mathbb{R}^n \setminus \{\boldsymbol{0}\}$,

$$\widetilde{f}(x) := f\left(\frac{x}{|x|^2} \right). \tag{1.15}$$

Then the following conclusion gives the mapping property of the inversion on local generalized Herz spaces.

Proposition 1.2.3 *Let* $p, q \in (0, \infty]$ *and* $\omega \in M(\mathbb{R}_+)$. *For any* $t \in (0, \infty)$, *let*

$$\widetilde{\omega}(t) := \omega\left(\frac{1}{t}\right) t^{-\frac{2n}{p}}.$$

Then the inversion (1.15) *maps the local generalized Herz space* $\dot{\mathcal{K}}^{p,q}_{\omega,0}(\mathbb{R}^n)$ *to the local generalized Herz space* $\dot{\mathcal{K}}^{p,q}_{\widetilde{\omega},0}(\mathbb{R}^n)$. *Moreover, there exists a constant* $C_{(n,\omega)} \in [1, \infty)$, *depends only on both* n *and* ω, *such that, for any* $f \in L^p_{\mathrm{loc}}(\mathbb{R}^n \setminus \{0\})$,

$$C_{(n,\omega)}^{-1} \|f\|_{\dot{\mathcal{K}}^{p,q}_{\omega,0}(\mathbb{R}^n)} \leq \|\widetilde{f}\|_{\dot{\mathcal{K}}^{p,q}_{\widetilde{\omega},0}(\mathbb{R}^n)} \leq C_{(n,\omega)} \|f\|_{\dot{\mathcal{K}}^{p,q}_{\omega,0}(\mathbb{R}^n)}.$$

Proof Let all the symbols be the same as in the present proposition, and $f \in L^p_{\mathrm{loc}}(\mathbb{R}^n \setminus \{0\})$. We show the present proposition via considering the following two cases on p.

Case (1) $p \in (0, \infty)$. In this case, by both the polar coordinate formula and the change of variables, we have, for any $k \in \mathbb{Z}$,

$$\left\|\widetilde{f} \mathbf{1}_{B(0,2^k)\setminus B(0,2^{k-1})}\right\|^p_{L^p(\mathbb{R}^n)}$$

$$= \int_{2^{k-1} \leq |x| < 2^k} \left|f\left(\frac{x}{|x|^2}\right)\right|^p dx$$

$$= \int_{2^{k-1}}^{2^k} \int_{\mathbb{S}^{n-1}} \left|f\left(\frac{\xi}{r}\right)\right|^p r^{n-1} d\sigma(\xi) dr$$

$$= \int_{2^{-k}}^{2^{-k+1}} \int_{\mathbb{S}^{n-1}} |f(r\xi)|^p r^{-n-1} d\sigma(\xi) dr$$

$$\sim 2^{2nk} \int_{2^{-k}}^{2^{-k+1}} \int_{\mathbb{S}^{n-1}} |f(r\xi)|^p r^{n-1} d\sigma(\xi) dr$$

$$\sim 2^{2nk} \left\|f \mathbf{1}_{B(0,2^{-k+1})\setminus B(0,2^{-k})}\right\|^p_{L^p(\mathbb{R}^n)}, \tag{1.16}$$

where the implicit positive constants depend only on n. Applying this, the change of variables, and Lemma 1.1.3, we further obtain

$$\|\widetilde{f}\|_{\dot{\mathcal{K}}^{p,q}_{\widetilde{\omega},0}(\mathbb{R}^n)} = \left\{\sum_{k \in \mathbb{Z}} \left[\omega(2^{-k})\right]^q 2^{-\frac{2nkq}{p}} \left\|\widetilde{f} \mathbf{1}_{B(0,2^k)\setminus B(0,2^{k-1})}\right\|^q_{L^p(\mathbb{R}^n)}\right\}^{\frac{1}{q}}$$

$$\sim \left\{\sum_{k \in \mathbb{Z}} \left[\omega(2^{-k})\right]^q \left\|f \mathbf{1}_{B(0,2^{-k+1})\setminus B(0,2^{-k})}\right\|^q_{L^p(\mathbb{R}^n)}\right\}^{\frac{1}{q}}$$

$$\sim \left\{ \sum_{k\in\mathbb{Z}} \left[\omega(2^{k-1})\right]^q \left\| f \mathbf{1}_{B(\mathbf{0},2^k)\setminus B(\mathbf{0},2^{k-1})} \right\|_{L^p(\mathbb{R}^n)}^q \right\}^{\frac{1}{q}}$$

$$\sim \left\{ \sum_{k\in\mathbb{Z}} \left[\omega(2^{k})\right]^q \left\| f \mathbf{1}_{B(\mathbf{0},2^k)\setminus B(\mathbf{0},2^{k-1})} \right\|_{L^p(\mathbb{R}^n)}^q \right\}^{\frac{1}{q}}$$

$$\sim \|f\|_{\dot{\mathcal{K}}_{\omega,\mathbf{0}}^{p,q}(\mathbb{R}^n)}, \tag{1.17}$$

where the implicit positive constants depend only on both n and ω. This then finishes the proof of Proposition 1.2.3 in this case.

Case (2) $p = \infty$. In this case, notice that, for any $k \in \mathbb{Z}$,

$$\left\| \widetilde{f} \mathbf{1}_{B(\mathbf{0},2^k)\setminus B(\mathbf{0},2^{k-1})} \right\|_{L^\infty(\mathbb{R}^n)} = \left\| f \mathbf{1}_{B(\mathbf{0},2^{-k+1})\setminus B(\mathbf{0},2^{-k})} \right\|_{L^\infty(\mathbb{R}^n)}. \tag{1.18}$$

Then, repeating an argument similar to that used in the estimation of (1.17) with (1.16) therein replaced by (1.18), we have

$$\left\| \widetilde{f} \right\|_{\dot{\mathcal{K}}_{\widetilde{\omega},\mathbf{0}}^{\infty,q}(\mathbb{R}^n)} \sim \|f\|_{\dot{\mathcal{K}}_{\omega,\mathbf{0}}^{\infty,q}(\mathbb{R}^n)},$$

which completes the proof of the present proposition in this case. Thus, Proposition 1.2.3 holds true. □

Observe that the inversion (1.15) exchanges $\mathbf{0}$ and ∞. Moreover, there is also a symmetry between $\mathbf{0}$ and ∞ in the definition of local generalized Herz spaces. Thus, the essential reason why the above proposition holds true is that the symmetry between $\mathbf{0}$ and ∞ under the inversion (1.15) meets the symmetry between $\mathbf{0}$ and ∞ in the definition of local generalized Herz spaces. In particular, using Proposition 1.2.3 with $\omega(t) := t^{-\frac{n}{p}}$ for any $t \in (0,\infty)$, then $\widetilde{\omega}(t) = t^{-\frac{n}{p}} = \omega(t)$ for any $t \in (0,\infty)$. By this, we immediately obtain the following property which shows that the inversion preserves a special local Herz space; we omit the details.

Corollary 1.2.4 *Let $p, q \in (0,\infty]$ and $\omega_p(t) := t^{-\frac{n}{p}}$ for any $t \in (0,\infty)$. Then the local generalized Herz space $\dot{\mathcal{K}}_{\omega_p,\mathbf{0}}^{p,q}(\mathbb{R}^n)$ is invariant under the inversion (1.15).*

Remark 1.2.5 Observe that the underlying space of local generalized Herz spaces in [197, Definition 2.2] is $\mathbb{R}^n$. However, in the definition of the local generalized Herz space $\dot{\mathcal{K}}_{\omega,\mathbf{0}}^{p,q}(\mathbb{R}^n)$, if we change $\mathbb{R}^n$ into $\mathbb{R}^n \setminus \{\mathbf{0}\}$, then we have the following three advantages:

(i) If a measurable function f is such that its quasi-norm in (1.11) is finite, then f automatically belongs to $L_{\text{loc}}^p(\mathbb{R}^n \setminus \{\mathbf{0}\})$ which is obvious by the definition of (1.11). But, f is not necessary to be locally integrable on $\mathbb{R}^n$. For instance, for $\beta \in (n, \infty)$, let

$$f_\beta(x) := \frac{1}{|x|^\beta} \mathbf{1}_{B(\mathbf{0},1)\setminus\{\mathbf{0}\}}(x)$$

for any $x \in \mathbb{R}^n$. Then $f_\beta \in \dot{\mathcal{K}}^{p,q}_{\omega,\mathbf{0}}(\mathbb{R}^n)$ with $p, q \in (0, \infty]$ and $\omega \in M(\mathbb{R}_+)$ satisfying $m_0(\omega) \in (\beta - \frac{n}{p}, \infty)$. Indeed, let β, p, q, and ω be the same as above, and $\varepsilon \in (0, m_0(\omega) + \frac{n}{p} - \beta)$ be a fixed positive constant. Then, using the fact that, for any $k \in \mathbb{Z}$ and $x \in \mathbb{R}^n$, $|x| \sim 2^k$, Lemma 1.1.12 with $t := 2^k$ for any $k \in \mathbb{Z} \setminus \mathbb{N}$, and the assumption $\varepsilon \in (0, m_0(\omega) + \frac{n}{p} - \beta)$, we conclude that

$$\|f_\beta\|_{\dot{\mathcal{K}}^{p,q}_\omega(\mathbb{R}^n)} = \left\{ \sum_{k \in \mathbb{Z}\setminus\mathbb{N}} \left[\omega(2^k)\right]^q \left\| |\cdot|^{-\beta} \mathbf{1}_{B(\mathbf{0},2^k)\setminus B(\mathbf{0},2^{k-1})} \right\|^q_{L^p(\mathbb{R}^n)} \right\}^{\frac{1}{q}}$$

$$\sim \left\{ \sum_{k \in \mathbb{Z}\setminus\mathbb{N}} 2^{-k\beta q} \left[\omega(2^k)\right]^q \left| B(\mathbf{0}, 2^k) \setminus B(\mathbf{0}, 2^{k-1}) \right|^{\frac{q}{p}} \right\}^{\frac{1}{q}}$$

$$\lesssim \left\{ \sum_{k \in \mathbb{Z}\setminus\mathbb{N}} 2^{kq[m_0(\omega)+\frac{n}{p}-\beta-\varepsilon]} \right\}^{\frac{1}{q}} < \infty.$$

But, obviously, $f_\beta \notin L^p_{\text{loc}}(\mathbb{R}^n)$ with $p \in (0, \infty)$ when $\beta \in [\frac{n}{p}, \infty)$, and $f_\beta \notin L^\infty_{\text{loc}}(\mathbb{R}^n)$ when $\beta \in (0, \infty)$.

(ii) It is also easy to show that any bounded function with compact support strictly contained in $\mathbb{R}^n \setminus \{\mathbf{0}\}$ belongs to all local generalized Herz spaces; however, a bounded function with compact support in $\mathbb{R}^n$ is not necessary to belong to some local generalized Herz spaces; see, for instance, Example 1.2.40 below for the details.

(iii) Only after we change the underlying space of local generalized Herz spaces into $\mathbb{R}^n \setminus \{\mathbf{0}\}$, we then have the interesting properties of these spaces in both Proposition 1.2.3 and Corollary 1.2.4. The essential reason why the above two conclusions hold true is that the symmetry between $\mathbf{0}$ and ∞ under the inversion (1.15) meets the symmetry between $\mathbf{0}$ and ∞ in the definition of local generalized Herz spaces.

We now study the triviality of generalized Herz spaces. Indeed, the following two theorems show that, under some assumptions, the global generalized Herz space $\dot{\mathcal{K}}^{p,q}_\omega(\mathbb{R}^n)$ is trivial.

Theorem 1.2.6 *Let $p, q \in (0, \infty]$ and $\omega \in M(\mathbb{R}_+)$ satisfy $M_0(\omega) \in (-\infty, -\frac{n}{p})$. Then*

$$\dot{\mathcal{K}}_\omega^{p,q}(\mathbb{R}^n) = \left\{ f \in \mathscr{M}(\mathbb{R}^n) : f(x) = 0 \text{ for almost every } x \in \mathbb{R}^n \right\}.$$

Proof Let all the symbols be the same as in the present theorem. Then, from Definition 1.2.1(ii), it follows that

$$\left\{ f \in \mathscr{M}(\mathbb{R}^n) : f(x) = 0 \text{ for almost every } x \in \mathbb{R}^n \right\} \subset \dot{\mathcal{K}}_\omega^{p,q}(\mathbb{R}^n).$$

Conversely, we now show

$$\dot{\mathcal{K}}_\omega^{p,q}(\mathbb{R}^n) \subset \left\{ f \in \mathscr{M}(\mathbb{R}^n) : f(x) = 0 \text{ for almost every } x \in \mathbb{R}^n \right\} \tag{1.19}$$

by considering the following two cases on p.

Case (1) $p \in (0, \infty)$. In this case, fixed an $f \in \dot{\mathcal{K}}_\omega^{p,q}(\mathbb{R}^n)$. Then we have $|f|^p \in L^1_{\text{loc}}(\mathbb{R}^n)$. We now claim that, for almost every $\xi \in \mathbb{R}^n$,

$$\lim_{k \to -\infty} 2^{-\frac{nk}{p}} \left\| f \mathbf{1}_{B(\xi, 2^k) \setminus B(\xi, 2^{k-1})} \right\|_{L^p(\mathbb{R}^n)} = |f(\xi)|. \tag{1.20}$$

To do this, let $\xi \in \mathbb{R}^n$ be the Lebesgue point of $|f|^p$. Then, from both the assumption $|f|^p \in L^1_{\text{loc}}(\mathbb{R}^n)$ and the Lebesgue differentiation theorem, we deduce that

$$\left| 2^{-nk} \left\| f \mathbf{1}_{B(\xi, 2^k) \setminus B(\xi, 2^{k-1})} \right\|_{L^p(\mathbb{R}^n)}^p - |f(\xi)|^p \right|$$

$$\lesssim 2^{-nk} \int_{B(\xi, 2^k) \setminus B(\xi, 2^{k-1})} \left| |f(x)|^p - |f(\xi)|^p \right| dx$$

$$\lesssim \frac{1}{|B(\xi, 2^k)|} \int_{B(\xi, 2^k)} \left| |f(x)|^p - |f(\xi)|^p \right| dx \to 0$$

as $k \to -\infty$, which further implies that

$$\lim_{k \to -\infty} 2^{-\frac{nk}{p}} \left\| f \mathbf{1}_{B(\xi, 2^k) \setminus B(\xi, 2^{k-1})} \right\|_{L^p(\mathbb{R}^n)} = |f(\xi)|,$$

and hence completes the proof of the above claim.

In addition, by Lemma 1.1.12, we find that, for any $t \in (0, 1]$,

$$t^{M_0(\omega)+\varepsilon} \lesssim \omega(t), \tag{1.21}$$

where $\varepsilon \in (0, -M_0(\omega) - \frac{n}{p})$ is a fixed positive constant and the implicit positive constant is independent of t but may depend on ε. Then, for almost every $\xi \in \mathbb{R}^n$

and for any $k \in \mathbb{Z} \setminus \mathbb{N}$, using (1.21) with $t := 2^k$, (1.12), and the assumption $f \in \dot{\mathcal{K}}^{p,q}_\omega(\mathbb{R}^n)$, we obtain

$$2^{-\frac{nk}{p}} \left\| f \mathbf{1}_{B(\xi,2^k) \setminus B(\xi,2^{k-1})} \right\|_{L^p(\mathbb{R}^n)}$$

$$\lesssim 2^{-k[M_0(\omega)+\frac{n}{p}+\varepsilon]} \omega(2^k) \left\| f \mathbf{1}_{B(\xi,2^k) \setminus B(\xi,2^{k-1})} \right\|_{L^p(\mathbb{R}^n)}$$

$$\lesssim 2^{-k[M_0(\omega)+\frac{n}{p}+\varepsilon]} \| f \|_{\dot{\mathcal{K}}^{p,q}_\omega(\mathbb{R}^n)}. \qquad (1.22)$$

Letting $k \to -\infty$ in (1.22), and applying both the above claim and the assumption $\varepsilon \in (0, -M_0(\omega) - \frac{n}{p})$, we conclude that, for almost every $\xi \in \mathbb{R}^n$, $f(\xi) = 0$. This finishes the proof of (1.19) in this case.

Case (2) $p = \infty$. In this case, fixed an $f \in \mathcal{M}(\mathbb{R}^n)$ satisfying that $\| f \|_{L^\infty(\mathbb{R}^n)} \in (0, \infty]$, and an $\alpha \in (0, \| f \|_{L^\infty(\mathbb{R}^n)})$. We first prove that, for any $k \in \mathbb{Z}$, there exists an $x_k \in \mathbb{R}^n$ such that

$$\left\| f \mathbf{1}_{B(x_k,2^k)} \right\|_{L^\infty(\mathbb{R}^n)} > \alpha. \qquad (1.23)$$

Otherwise, there exists a $k_0 \in \mathbb{Z}$ such that, for any $x \in \mathbb{R}^n$,

$$\left\| f \mathbf{1}_{B(x,2^{k_0})} \right\|_{L^\infty(\mathbb{R}^n)} \leq \alpha.$$

This implies that, for any given $x \in \mathbb{R}^n$ and for almost every $y \in B(x, 2^{k_0})$, $|f(y)| \leq \alpha$, and hence $|f| \leq \alpha$ almost everywhere in $\mathbb{R}^n$. Therefore, we have

$$\| f \|_{L^\infty(\mathbb{R}^n)} \leq \alpha,$$

which contradicts the assumption $\alpha \in (0, \| f \|_{L^\infty(\mathbb{R}^n)})$. This then finishes the proof that, for any $k \in \mathbb{Z}$, (1.23) holds true with some $x_k \in \mathbb{R}^n$.

Now, for any $k \in \mathbb{Z} \setminus \mathbb{N}$, let $\xi_k \in \mathbb{R}^n$ be such that $|\xi_k| = 3 \cdot 2^k$. Then, for any given $k \in \mathbb{Z} \setminus \mathbb{N}$ and for any $y \in B(x_k, 2^k)$, we have

$$|y - \xi_k| \leq |y - x_k| + |x_k - \xi_k| < 2^k + 3 \cdot 2^k = 2^{k+2}$$

and

$$|y - \xi_k| \geq |x_k - \xi_k| - |y - x_k| > 3 \cdot 2^k - 2^k = 2^{k+1},$$

which imply that $y \in B(\xi_k, 2^{k+2}) \setminus B(\xi_k, 2^{k+1})$, and hence

$$B(x_k, 2^k) \subset B(\xi_k, 2^{k+2}) \setminus B(\xi_k, 2^{k+1}). \qquad (1.24)$$

Let $\varepsilon \in (0, -M_0(\omega))$ be a fixed positive constant. Then, applying (1.12), (1.24), Lemmas 1.1.3 and 1.1.12, (1.23), and the assumption $\varepsilon \in (0, -M_0(\omega))$, we obtain

$$\|f\|_{\dot{\mathcal{K}}_\omega^{\infty,q}(\mathbb{R}^n)} \geq \sup_{\xi \in \mathbb{R}^n, \, k \in \mathbb{Z}} \left\{ \omega(2^k) \left\| f \mathbf{1}_{B(\xi,2^k) \setminus B(\xi,2^{k-1})} \right\|_{L^\infty(\mathbb{R}^n)} \right\}$$

$$\geq \omega(2^{k+2}) \left\| f \mathbf{1}_{B(\xi_k,2^{k+2}) \setminus B(\xi,2^{k+1})} \right\|_{L^\infty(\mathbb{R}^n)}$$

$$\gtrsim \omega(2^k) \left\| f \mathbf{1}_{B(x_k,2^k)} \right\|_{L^\infty(\mathbb{R}^n)} \gtrsim 2^{k[M_0(\omega)+\varepsilon]} \to \infty$$

as $k \to -\infty$, where the implicit positive constants depend only on ω and ε, which then implies that $f \notin \dot{\mathcal{K}}_\omega^{\infty,q}(\mathbb{R}^n)$, and hence (1.19) holds true in this case. This further implies that

$$\dot{\mathcal{K}}_\omega^{p,q}(\mathbb{R}^n) = \left\{ f \in \mathscr{M}(\mathbb{R}^n) : \; f(x) = 0 \text{ for almost every } x \in \mathbb{R}^n \right\},$$

which completes the proof of Theorem 1.2.6.   □

Theorem 1.2.7 *Let $p, q \in (0, \infty]$ and $\omega \in M(\mathbb{R}_+)$ satisfy $m_\infty(\omega) \in (0, \infty)$. Then*

$$\dot{\mathcal{K}}_\omega^{p,q}(\mathbb{R}^n) = \left\{ f \in \mathscr{M}(\mathbb{R}^n) : \; f(x) = 0 \text{ for almost every } x \in \mathbb{R}^n \right\}.$$

Proof Let all the symbols be as in the present theorem and f be a measurable function on $\mathbb{R}^n$ such that

$$\left| \{x \in \mathbb{R}^n : \; |f(x)| > 0\} \right| \neq 0. \tag{1.25}$$

We first show that there exists a $k_0 \in \mathbb{N}$ such that

$$\left| \{x \in B(\mathbf{0}, 2^{k_0}) : \; |f(x)| > 0\} \right| \neq 0.$$

Indeed, if, for any $k \in \mathbb{N}$, $|\{x \in B(\mathbf{0}, 2^k) : \; |f(x)| > 0\}| = 0$, from this and the fact that

$$\{x \in \mathbb{R}^n : \; |f(x)| > 0\} = \bigcup_{k \in \mathbb{N}} \{x \in B(\mathbf{0}, 2^k) : \; |f(x)| > 0\},$$

we deduce that

$$\left| \{x \in \mathbb{R}^n : \; |f(x)| > 0\} \right| \leq \sum_{k \in \mathbb{N}} \left| \{x \in B(\mathbf{0}, 2^k) : \; |f(x)| > 0\} \right| = 0,$$

which contradicts (1.25). Similarly to the argument above, we conclude that there exists a $k_1 \in \mathbb{N}$ such that $|E_{f,k_0,k_1}| \in (0, \infty)$, where

$$E_{f,k_0,k_1} := \left\{ x \in B(\mathbf{0}, 2^{k_0}) : |f(x)| > \frac{1}{k_1} \right\}.$$

Next, for any $i \in \mathbb{N} \cap [k_0 + 1, \infty)$, let $\xi_i \in \mathbb{R}^n$ satisfy $|\xi_i| = 2^i + 2^{k_0}$. Then, for any $x \in B(\mathbf{0}, 2^{k_0})$ and $i \in \mathbb{N} \cap [k_0 + 1, \infty)$, we have

$$|x - \xi_i| \leq |x| + |\xi_i| < 2^{k_0} + 2^i + 2^{k_0} \leq 2^{i+1}$$

and

$$|x - \xi_i| \geq |\xi_i| - |x| > 2^i + 2^{k_0} - 2^{k_0} = 2^i.$$

This further implies that, for any $i \in \mathbb{N} \cap [k_0 + 1, \infty)$,

$$B(\mathbf{0}, 2^{k_0}) \subset B(\xi_i, 2^{i+1}) \setminus B(\xi_i, 2^i).$$

Applying this and Definition 1.2.1, we find that, for any $i \in \mathbb{N} \cap [k_0 + 1, \infty)$,

$$\|f\|_{\mathcal{K}_\omega^{p,q}(\mathbb{R}^n)} \geq \left\| f \mathbf{1}_{B(\mathbf{0},2^{k_0})}(\cdot + \xi_i) \right\|_{\mathcal{K}_{\omega,0}^{p,q}(\mathbb{R}^n)} = \omega(2^{i+1}) \left\| f \mathbf{1}_{B(\mathbf{0},2^{k_0})} \right\|_{L^p(\mathbb{R}^n)}$$

$$\geq \omega(2^{i+1}) \frac{|E_{f,k_0,k_1}|^{1/p}}{k_1} \tag{1.26}$$

On the other hand, from Lemma 1.1.12, it follows that, for any $k \in \mathbb{N}$,

$$\omega(2^k) \gtrsim 2^{k[m_\infty(\omega) - \varepsilon]},$$

where $\varepsilon \in (0, m_\infty(\omega))$ is a fixed positive constant. This, together with (1.26) further implies that, for any $i \in \mathbb{N} \cap [k_0 + 1, \infty)$,

$$\|f\|_{\mathcal{K}_\omega^{p,q}(\mathbb{R}^n)} \gtrsim 2^{i[m_\infty(\omega) - \varepsilon]}$$

Using this and the assumption $m_\infty(\omega) \in (0, \infty)$, we further find that

$$\|f\|_{\dot{\mathcal{K}}_\omega^{p,q}(\mathbb{R}^n)} = \infty,$$

which completes the proof of Theorem 1.2.7. $\qquad\square$

Remark 1.2.8 We should point out that the assumptions of Theorems 1.2.6 and 1.2.7 are sharp. To be precise, if p, q, and ω do not satisfy the assumptions of Theorems 1.2.6 and 1.2.7, namely, $p, q \in (0, \infty]$ and $\omega \in M(\mathbb{R}_+)$ are such that

$M_0(\omega) \in [-\frac{n}{p}, \infty)$ and $m_\infty(\omega) \in (-\infty, 0]$, then global generalized Herz spaces may not be trivial. Indeed, let $p, q \in (0, \infty]$, $\alpha_1 \in [-\frac{n}{p}, \infty)$, $\alpha_2 \in (-\infty, 0]$, $\beta \in (\frac{1}{q}, \infty)$, and

$$\omega_{\alpha_1,\alpha_2,\beta}(t) := \begin{cases} t^{\alpha_1}(1 - \ln t)^{-\beta} & \text{when } t \in (0, 1], \\ t^{\alpha_2}(1 + \ln t)^{-\beta} & \text{when } t \in (1, \infty). \end{cases}$$

Then, similarly to Example 1.1.9, we find that $\omega_{\alpha_1,\alpha_2,\beta} \in M(\mathbb{R}_+)$,

$$m_0(\omega) = M_0(\omega) = \alpha_1 \in \left[-\frac{n}{p}, \infty\right),$$

and

$$m_\infty(\omega) = M_\infty(\omega) = \alpha_2 \in (-\infty, 0].$$

In addition, using the change of variables and the assumptions $\alpha_1 \in [-\frac{n}{p}, \infty)$, $\alpha_2 \in (-\infty, 0]$, and $\beta \in (\frac{1}{q}, \infty)$, we obtain

$$\left\| \mathbf{1}_{B(\mathbf{0},1)} \right\|_{\dot{\mathcal{K}}^{p,q}_{\omega_{\alpha_1,\alpha_2,\beta}}(\mathbb{R}^n)}$$

$$\lesssim \sup_{\xi \in \mathbb{R}^n} \left\{ \sum_{k \in \mathbb{Z} \setminus \mathbb{N}} 2^{k\alpha_1 q} (1 - k \ln 2)^{-\beta q} \left\| \mathbf{1}_{B(\mathbf{0},1)} \mathbf{1}_{B(\xi, 2^k) \setminus B(\xi, 2^{k-1})} \right\|_{L^p(\mathbb{R}^n)}^q \right\}^{\frac{1}{q}}$$

$$+ \sup_{\xi \in \mathbb{R}^n} \left\{ \sum_{k \in \mathbb{N}} 2^{k\alpha_2 q} (1 + k \ln 2)^{-\beta q} \left\| \mathbf{1}_{B(\mathbf{0},1)} \mathbf{1}_{B(\xi, 2^k) \setminus B(\xi, 2^{k-1})} \right\|_{L^p(\mathbb{R}^n)}^q \right\}^{\frac{1}{q}}$$

$$\lesssim \sup_{\xi \in \mathbb{R}^n} \left\{ \sum_{k \in \mathbb{N}} 2^{-k\alpha_1 q} (1 + k \ln 2)^{-\beta q} \left\| \mathbf{1}_{B(\xi, 2^{-k}) \setminus B(\xi, 2^{-k-1})} \right\|_{L^p(\mathbb{R}^n)}^q \right\}^{\frac{1}{q}}$$

$$+ \sup_{\xi \in \mathbb{R}^n} \left\{ \sum_{k \in \mathbb{N}} 2^{k\alpha_2 q} (1 + k \ln 2)^{-\beta q} \left\| \mathbf{1}_{B(\mathbf{0},1)} \right\|_{L^p(\mathbb{R}^n)}^q \right\}^{\frac{1}{q}}$$

$$\lesssim \left[\sum_{k \in \mathbb{N}} 2^{-k(\alpha_1 + \frac{n}{p})q} k^{-\beta q} \right]^{\frac{1}{q}} + \left[\sum_{k \in \mathbb{N}} 2^{k\alpha_2 q} k^{-\beta q} \right]^{\frac{1}{q}} < \infty,$$

which implies that $\mathbf{1}_{B(\mathbf{0},1)} \in \dot{\mathcal{K}}^{p,q}_{\omega_{\alpha_1,\alpha_2,\beta}}(\mathbb{R}^n)$, and hence the global generalized Herz space $\dot{\mathcal{K}}^{p,q}_{\omega_{\alpha_1,\alpha_2,\beta}}(\mathbb{R}^n)$ is nontrivial. Therefore, the assumptions of Theorems 1.2.6 and 1.2.7 are sharp.

Now, we consider the embedding relation between global generalized Herz spaces and Lebesgue spaces. Indeed, we have the following necessity result about exponents of the embedding.

Theorem 1.2.9 *Let $p, q, r \in (0, \infty]$ and $\omega \in M(\mathbb{R}_+)$ satisfy that there exists an $f_0 \in L^p_{\mathrm{loc}}(\mathbb{R}^n)$ such that $\|f_0\|_{\dot{\mathcal{K}}^{p,q}_\omega(\mathbb{R}^n)} \in (0, \infty)$. If the global generalized Herz space $\dot{\mathcal{K}}^{p,q}_\omega(\mathbb{R}^n)$ embeds continuously into the Lebesgue space $L^r(\mathbb{R}^n)$, namely, there exists a positive constant C such that, for any $f \in \dot{\mathcal{K}}^{p,q}_\omega(\mathbb{R}^n)$,*

$$\|f\|_{L^r(\mathbb{R}^n)} \leq C \|f\|_{\dot{\mathcal{K}}^{p,q}_\omega(\mathbb{R}^n)}, \tag{1.27}$$

then one has

$$\frac{n}{r} \in \left[\min\{m_0(\omega), m_\infty(\omega)\} + \frac{n}{p}, \max\{M_0(\omega), M_\infty(\omega)\} + \frac{n}{p} \right].$$

Proof Let all the symbols be the same as in the present theorem. Then, from the change of variables and (1.12), it follows that, for any $f \in \mathcal{M}(\mathbb{R}^n)$ and $j \in \mathbb{Z}$,

$$\left\| f(2^j \cdot) \right\|_{\dot{\mathcal{K}}^{p,q}_\omega(\mathbb{R}^n)}$$

$$= \sup_{\xi \in \mathbb{R}^n} \left\{ \sum_{k \in \mathbb{Z}} \left[\omega(2^k) \right]^q \left\| f\left(2^j \cdot\right) \mathbf{1}_{B(\xi, 2^k) \setminus B(\xi, 2^{k-1})} \right\|^q_{L^p(\mathbb{R}^n)} \right\}^{\frac{1}{q}}$$

$$= \sup_{\xi \in \mathbb{R}^n} \left\{ \sum_{k \in \mathbb{Z}} \left[\omega(2^k) \right]^q \left[2^{-\frac{jn}{p}} \left\| f \mathbf{1}_{B(\xi, 2^{k+j}) \setminus B(\xi, 2^{k+j-1})} \right\|_{L^p(\mathbb{R}^n)} \right]^q \right\}^{\frac{1}{q}}$$

$$= 2^{-\frac{jn}{p}} \sup_{\xi \in \mathbb{R}^n} \left\{ \sum_{k \in \mathbb{Z}} \left[\frac{\omega(2^k)}{\omega(2^{k+j})} \right]^q \left[\omega(2^{k+j}) \right]^q \left\| f \mathbf{1}_{B(\xi, 2^{k+j}) \setminus B(\xi, 2^{k+j-1})} \right\|^q_{L^p(\mathbb{R}^n)} \right\}^{\frac{1}{q}}$$

$$= 2^{-\frac{jn}{p}} \sup_{\xi \in \mathbb{R}^n} \left\{ \sum_{k \in \mathbb{Z}} \left[\frac{\omega(2^{-j+k})}{\omega(2^k)} \right]^q \left[\omega(2^k) \right]^q \left\| f \mathbf{1}_{B(\xi, 2^k) \setminus B(\xi, 2^{k-1})} \right\|^q_{L^p(\mathbb{R}^n)} \right\}^{\frac{1}{q}}$$

$$\lesssim 2^{-\frac{jn}{p}} \sup_{k \in \mathbb{Z}} \left\{ \frac{\omega(2^{-j+k})}{\omega(2^k)} \right\} \|f\|_{\dot{\mathcal{K}}^{p,q}_\omega(\mathbb{R}^n)}. \tag{1.28}$$

Let $\alpha_0, \alpha_\infty, \beta_0, \beta_\infty \in \mathbb{R}$ be such that both (i) and (ii) of Definition 1.1.1 hold true. Then, for any $j \in \mathbb{N}$, repeating an argument similar to that used in Case (1) of the proof of Lemma 1.1.3 with $\lambda := 2^{-j}$, we find that, for any $k \in \mathbb{Z}$,

$$\omega(2^{-j+k}) \lesssim 2^{-j \min\{\alpha_0, \alpha_\infty\}} \omega(2^k).$$

Applying this and (1.28) with $f := f_0$, we conclude that, for any $j \in \mathbb{N}$,

$$\left\| f_0 \left(2^j \cdot \right) \right\|_{\dot{\mathcal{K}}_{\omega}^{p,q}(\mathbb{R}^n)} \lesssim 2^{-j\left(\min\{\alpha_0, \alpha_\infty\} + \frac{n}{p}\right)} \|f_0\|_{\dot{\mathcal{K}}_{\omega}^{p,q}(\mathbb{R}^n)},$$

which, together with the change of variables and (1.27) with f therein replaced by f_0, further implies that

$$\|f_0\|_{L^r(\mathbb{R}^n)} = 2^{\frac{jn}{r}} \left\| f_0 \left(2^j \cdot \right) \right\|_{L^r(\mathbb{R}^n)} \lesssim 2^{\frac{jn}{r}} \left\| f_0 \left(2^j \cdot \right) \right\|_{\dot{\mathcal{K}}_{\omega}^{p,q}(\mathbb{R}^n)}$$

$$\lesssim 2^{j\left(\frac{n}{r} - \frac{n}{p} - \min\{\alpha_0, \alpha_\infty\}\right)} \|f_0\|_{\dot{\mathcal{K}}_{\omega}^{p,q}(\mathbb{R}^n)}. \tag{1.29}$$

If $\frac{n}{r} - \frac{n}{p} - \min\{\alpha_0, \alpha_\infty\} \in (-\infty, 0)$, then, using the assumption $\|f_0\|_{\dot{\mathcal{K}}_{\omega}^{p,q}(\mathbb{R}^n)} \in (0, \infty)$ and letting $j \to \infty$ in (1.29), we obtain

$$\|f_0\|_{L^r(\mathbb{R}^n)} = 0,$$

and hence $f_0 = 0$ almost everywhere in $\mathbb{R}^n$, which contradicts the assumption $\|f_0\|_{\dot{\mathcal{K}}_{\omega}^{p,q}(\mathbb{R}^n)} \in (0, \infty)$. From this and Lemma 1.1.10, we infer that

$$\frac{n}{r} \geq \sup \left\{ \min\{\alpha_0, \alpha_\infty\} + \frac{n}{p} : \alpha_0, \alpha_\infty \in \mathbb{R} \text{ satisfy Definition 1.1.1} \right\}$$

$$= \min\{m_0(\omega), m_\infty(\omega)\} + \frac{n}{p}. \tag{1.30}$$

On the other hand, for any $j \in \mathbb{Z} \setminus \mathbb{N}$, repeating an argument similar to that used in the estimation of (1.29) with Case (1) therein replaced by Case (2), we obtain

$$\|f_0\|_{L^r(\mathbb{R}^n)} \lesssim 2^{j\left(\frac{n}{r} - \frac{n}{p} - \max\{\beta_0, \beta_\infty\}\right)} \|f_0\|_{\dot{\mathcal{K}}_{\omega}^{p,q}(\mathbb{R}^n)}. \tag{1.31}$$

If $\frac{n}{r} - \frac{n}{p} - \max\{\beta_0, \beta_\infty\} \in (0, \infty)$, then, applying the assumption $\|f_0\|_{\dot{\mathcal{K}}_{\omega}^{p,q}(\mathbb{R}^n)} \in (0, \infty)$ again and letting $j \to -\infty$ in (1.31), we have

$$\|f_0\|_{L^r(\mathbb{R}^n)} = 0,$$

and hence $f_0 = 0$ almost everywhere in $\mathbb{R}^n$, which also contradicts the assumption $\|f_0\|_{\dot{\mathcal{K}}_{\omega}^{p,q}(\mathbb{R}^n)} \in (0, \infty)$. By this and Lemma 1.1.10 again, we find that

$$\frac{n}{r} \leq \inf \left\{ \max\{\beta_0, \beta_\infty\} + \frac{n}{p} : \beta_0, \beta_\infty \in \mathbb{R} \text{ satisfy Definition 1.1.1} \right\}$$

$$= \max\{M_0(\omega), M_\infty(\omega)\} + \frac{n}{p}.$$

Combining this and (1.30), we conclude that

$$\frac{n}{r} \in \left[\min\{m_0(\omega), m_\infty(\omega)\} + \frac{n}{p}, \max\{M_0(\omega), M_\infty(\omega)\} + \frac{n}{p} \right],$$

which then completes the proof of Theorem 1.2.9. □

Based on the above necessity result, we next investigate the embedding theorem of two special global generalized Herz spaces. We first study the embedding relation between the global generalized Herz space $\dot{\mathcal{K}}^{p,q}_{\omega_0}(\mathbb{R}^n)$ with $p, q \in (0, \infty]$ and $\omega_0 = 1$, and the Lebesgue space $L^r(\mathbb{R}^n)$ with $r \in (0, \infty]$. Indeed, let p, q, ω_0, and r be the same as above. Then, by Example 1.1.7 with $\alpha := 0$ and Theorem 1.2.9 with $\omega := \omega_0$, we have $\omega_0 \in M(\mathbb{R}_+)$ and

$$m_0(\omega_0) = M_0(\omega_0) = m_\infty(\omega_0) = M_\infty(\omega_0) = 0,$$

and hence, if $\dot{\mathcal{K}}^{p,q}_{\omega_0}(\mathbb{R}^n)$ embeds continuously into $L^r(\mathbb{R}^n)$, then $r = p$. Thus, we only need to consider the relation between $\dot{\mathcal{K}}^{p,q}_{\omega_0}(\mathbb{R}^n)$ and $L^p(\mathbb{R}^n)$. Indeed, we have the following conclusion.

Proposition 1.2.10 Let $p, q \in (0, \infty]$ and $\omega_0(t) := 1$ for any $t \in (0, \infty)$.

(i) Assume $0 < p \leq q \leq \infty$. Then

$$\dot{\mathcal{K}}^{p,q}_{\omega_0}(\mathbb{R}^n) = L^p(\mathbb{R}^n)$$

with the same quasi-norms.
(ii) Assume $0 < q < p \leq \infty$. Then the continuous embedding

$$\dot{\mathcal{K}}^{p,q}_{\omega_0}(\mathbb{R}^n) \hookrightarrow L^p(\mathbb{R}^n)$$

is strict.

To prove this proposition, we require the following basic inequality which is a part of [90, Exercise 1.1.4].

Lemma 1.2.11 Let $r \in (0, 1]$. Then, for any $\{a_j\}_{j \in \mathbb{N}} \subset \mathbb{C}$,

$$\left(\sum_{j \in \mathbb{N}} |a_j| \right)^r \leq \sum_{j \in \mathbb{N}} |a_j|^r.$$

Then we show Proposition 1.2.10 with the help of Lemma 1.2.11.

Proof of Proposition 1.2.10 Let all the symbols be the same as in the present proposition. Then we first prove (i) by considering the following two cases on p.

Case (1) $p \in (0, q] \cap (0, \infty)$. In this case, applying (1.12), the assumption $p \leq q$, and Lemma 1.2.11 with $r := \frac{p}{q}$, we find that, for any $f \in L^p(\mathbb{R}^n)$,

$$\|f\|_{\dot{\mathcal{K}}^{p,q}_{\omega_0}(\mathbb{R}^n)} \leq \sup_{\xi \in \mathbb{R}^n} \left\{ \sum_{k \in \mathbb{Z}} \left\| f \mathbf{1}_{B(\xi, 2^k) \setminus B(\xi, 2^{k-1})} \right\|^p_{L^p(\mathbb{R}^n)} \right\}^{\frac{1}{p}}$$

$$= \|f\|_{L^p(\mathbb{R}^n)} < \infty, \tag{1.32}$$

which implies that $L^p(\mathbb{R}^n) \subset \dot{\mathcal{K}}^{p,q}_{\omega_0}(\mathbb{R}^n)$. Conversely, we now show that $\dot{\mathcal{K}}^{p,q}_{\omega_0}(\mathbb{R}^n) \subset L^p(\mathbb{R}^n)$. To this end, we first claim that, for any $f \in \mathscr{M}(\mathbb{R}^n)$,

$$\sup_{\xi \in \mathbb{R}^n, k \in \mathbb{Z}} \left\{ \left\| f \mathbf{1}_{B(\xi, 2^k) \setminus B(\xi, 2^{k-1})} \right\|_{L^p(\mathbb{R}^n)} \right\} = \|f\|_{L^p(\mathbb{R}^n)}. \tag{1.33}$$

Indeed, for any $j \in \mathbb{N}$, we choose a $\xi_j \in \mathbb{R}^n$ satisfying that $|\xi_j| = 3 \cdot 2^j$. Then, for any given $j \in \mathbb{N}$ and for any $y \in B(\mathbf{0}, 2^j)$, we have

$$|y - \xi_j| \leq |y| + |\xi_j| < 2^{j+2}$$

and

$$|y - \xi_j| \geq |\xi_j| - |y - \xi_j| > 2^{j+1},$$

which further implies that $y \in B(\xi_j, 2^{j+2}) \setminus B(\xi_j, 2^{j+1})$, and hence

$$B(\mathbf{0}, 2^j) \subset B(\xi_j, 2^{j+2}) \setminus B(\xi_j, 2^{j+1}).$$

From this and the monotone convergence theorem, we deduce that, for any $f \in \mathscr{M}(\mathbb{R}^n)$,

$$\|f\|_{L^p(\mathbb{R}^n)} = \lim_{j \to \infty} \left\| f \mathbf{1}_{B(\mathbf{0}, 2^j)} \right\|_{L^p(\mathbb{R}^n)}$$

$$\leq \lim_{j \to \infty} \left\| f \mathbf{1}_{B(\xi_j, 2^{j+2}) \setminus B(\xi_j, 2^{j+1})} \right\|_{L^p(\mathbb{R}^n)}$$

$$\leq \sup_{\xi \in \mathbb{R}^n, k \in \mathbb{Z}} \left\{ \left\| f \mathbf{1}_{B(\xi, 2^k) \setminus B(\xi, 2^{k-1})} \right\|_{L^p(\mathbb{R}^n)} \right\}. \tag{1.34}$$

On the other hand, it is obvious that, for any $f \in \mathscr{M}(\mathbb{R}^n)$,

$$\sup_{\xi \in \mathbb{R}^n, k \in \mathbb{Z}} \left\{ \left\| f \mathbf{1}_{B(\xi, 2^k) \setminus B(\xi, 2^{k-1})} \right\|_{L^p(\mathbb{R}^n)} \right\} \leq \|f\|_{L^p(\mathbb{R}^n)},$$

which, together with (1.34), further implies that

$$\sup_{\xi\in\mathbb{R}^n,\,k\in\mathbb{Z}}\left\{\left\|f\mathbf{1}_{B(\xi,2^k)\setminus B(\xi,2^{k-1})}\right\|_{L^p(\mathbb{R}^n)}\right\}=\|f\|_{L^p(\mathbb{R}^n)}\,.$$

This then finishes the proof of the above claim. Combining this claim and (1.12) again, we conclude that, for any $f\in\dot{\mathcal{K}}^{p,q}_{\omega_0}(\mathbb{R}^n)$,

$$\begin{aligned}\|f\|_{L^p(\mathbb{R}^n)}&=\sup_{\xi\in\mathbb{R}^n,\,k\in\mathbb{Z}}\left\{\left\|f\mathbf{1}_{B(\xi,2^k)\setminus B(\xi,2^{k-1})}\right\|_{L^p(\mathbb{R}^n)}\right\}\\&\le\sup_{\xi\in\mathbb{R}^n}\left\{\sum_{k\in\mathbb{Z}}\left\|f\mathbf{1}_{B(\xi,2^k)\setminus B(\xi,2^{k-1})}\right\|_{L^p(\mathbb{R}^n)}^q\right\}^{\frac{1}{q}}\\&=\|f\|_{\dot{\mathcal{K}}^{p,q}_{\omega_0}(\mathbb{R}^n)}<\infty.\end{aligned}\qquad(1.35)$$

This then implies that

$$\dot{\mathcal{K}}^{p,q}_{\omega_0}(\mathbb{R}^n)\subset L^p(\mathbb{R}^n),$$

and hence finishes the proof that $\dot{\mathcal{K}}^{p,q}_{\omega_0}(\mathbb{R}^n)=L^p(\mathbb{R}^n)$. Moreover, using both (1.32) and (1.35), we find that, for any $f\in\dot{\mathcal{K}}^{p,q}_{\omega_0}(\mathbb{R}^n)$,

$$\|f\|_{\dot{\mathcal{K}}^{p,q}_{\omega_0}(\mathbb{R}^n)}=\|f\|_{L^p(\mathbb{R}^n)}\,,$$

which completes the proof of (i) in this case.

Case (2) $p=\infty$. In this case, we have $p=q=\infty$. Then, repeating an argument similar to that used in the proof of (1.33) with p therein replaced by ∞, we conclude that, for any $f\in\mathcal{M}(\mathbb{R}^n)$,

$$\|f\|_{\dot{\mathcal{K}}^{\infty,\infty}_{\omega_0}(\mathbb{R}^n)}=\sup_{\xi\in\mathbb{R}^n,\,k\in\mathbb{Z}}\left\{\left\|f\mathbf{1}_{B(\xi,2^k)\setminus B(\xi,2^{k-1})}\right\|_{L^\infty(\mathbb{R}^n)}\right\}=\|f\|_{L^\infty(\mathbb{R}^n)}\,.$$

This then implies that $\dot{\mathcal{K}}^{\infty,\infty}_{\omega_0}(\mathbb{R}^n)=L^\infty(\mathbb{R}^n)$ with the same norms, and hence finishes the proof of (i) in this case. Thus, we complete the proof of (i).

Next, we turn to prove (ii). Indeed, from (1.35), it follows that

$$\dot{\mathcal{K}}^{p,q}_{\omega_0}(\mathbb{R}^n)\hookrightarrow L^p(\mathbb{R}^n).$$

Therefore, to finish the proof of (ii), it suffices to show that

$$L^p(\mathbb{R}^n)\setminus\dot{\mathcal{K}}^{p,q}_{\omega_0}(\mathbb{R}^n)\neq\emptyset.\qquad(1.36)$$

To do this, for any $j \in \mathbb{N}$, let $\widetilde{\xi}_j \in \mathbb{R}^n$ be such that $|\widetilde{\xi}_j| = 2^j + 1$ and

$$f_0 := \sum_{j \in \mathbb{N}} j^{-\frac{1}{q}} \mathbf{1}_{B(\widetilde{\xi}_j, 1)}.$$

Then, for any given $j \in \mathbb{N}$ and for any $y \in B(\widetilde{\xi}_j, 1)$, we have

$$|y| \le |y - \widetilde{\xi}_j| + |\widetilde{\xi}_j| < 2 + 2^j \le 2^{j+1}$$

and

$$|y| \ge |\widetilde{\xi}_j| - |y - \widetilde{\xi}_j| > 2^j,$$

which implies that $y \in B(\mathbf{0}, 2^{j+1}) \setminus B(\mathbf{0}, 2^j)$. Thus, it holds true that, for any $j \in \mathbb{N}$,

$$B(\widetilde{\xi}_j, 1) \subset B(\mathbf{0}, 2^{j+1}) \setminus B(\mathbf{0}, 2^j).$$

Using this, (1.11) and (1.12), and the assumption $q < p$, we obtain

$$\|f_0\|_{\dot{\mathcal{K}}^{p,q}_{\omega_0}(\mathbb{R}^n)} \ge \|f_0\|_{\dot{\mathcal{K}}^{p,q}_{\omega_0,0}(\mathbb{R}^n)} = \left[\sum_{j \in \mathbb{N}} j^{-1} \left\| \mathbf{1}_{B(\widetilde{\xi}_j, 1)} \right\|^q_{L^p(\mathbb{R}^n)} \right]^{\frac{1}{q}}$$

$$\sim \left(\sum_{j \in \mathbb{N}} j^{-1} \right)^{\frac{1}{q}} = \infty$$

and

$$\|f_0\|_{L^p(\mathbb{R}^n)} = \left[\sum_{j \in \mathbb{N}} \int_{B(\widetilde{\xi}_j, 1)} j^{-\frac{p}{q}} \, dx \right]^{\frac{1}{p}} \sim \left(\sum_{j \in \mathbb{N}} j^{-\frac{p}{q}} \right)^{\frac{1}{q}} < \infty$$

when $p \in (0, \infty)$ or

$$\|f_0\|_{L^\infty(\mathbb{R}^n)} \le \sup_{j \in \mathbb{N}} \left\{ j^{-1} \right\} < \infty$$

when $p = \infty$. These then imply $f_0 \in L^p(\mathbb{R}^n) \setminus \dot{\mathcal{K}}^{p,q}_{\omega_0}(\mathbb{R}^n)$, and hence finish the proof of (1.36). Thus, the continuous embedding

$$\dot{\mathcal{K}}^{p,q}_{\omega_0}(\mathbb{R}^n) \hookrightarrow L^p(\mathbb{R}^n)$$

is strict, which completes the proof of (ii) and hence Proposition 1.2.10.   □

On the other hand, we turn to consider the embedding result of the global generalized Herz space $\dot{\mathcal{K}}^{p,q}_{\omega_0}(\mathbb{R}^n)$ with $p \in (0, \infty)$, $q \in (0, \infty]$, and $\omega_0 := t^{-\frac{n}{p}}$ for any $t \in (0, \infty)$. Indeed, let p, q, and ω_0 be the same as above. Then, from Example 1.1.7 with $\alpha := -\frac{n}{p}$ and Theorem 1.2.9 with $\omega := \omega_0$, we deduce that $\omega \in M(\mathbb{R}_+)$ and

$$m_0(\omega_0) = M_0(\omega_0) = m_\infty(\omega_0) = M_\infty(\omega_0) = -\frac{n}{p},$$

and hence, if $\dot{\mathcal{K}}^{p,q}_{\omega_0}(\mathbb{R}^n)$ embeds continuously into $L^r(\mathbb{R}^n)$ with $r \in (0, \infty]$, then $r = \infty$. Therefore, we only need to consider the relation between the global generalized Herz space $\dot{\mathcal{K}}^{p,q}_{\omega_0}(\mathbb{R}^n)$ and the Lebesgue space $L^\infty(\mathbb{R}^n)$ as follows.

Proposition 1.2.12 *Let* $p \in (0, \infty)$, $q \in (0, \infty]$, *and* $\omega_0(t) := t^{-\frac{n}{p}}$ *for any* $t \in (0, \infty)$. *Then the continuous embedding*

$$\dot{\mathcal{K}}^{p,q}_{\omega_0}(\mathbb{R}^n) \hookrightarrow L^\infty(\mathbb{R}^n)$$

is strict.

Proof Let all the symbols be the same as in the present proposition, and $f \in \dot{\mathcal{K}}^{p,q}_{\omega_0}(\mathbb{R}^n)$. Then, from (1.20) and (1.12), we infer that, for almost every $x \in \mathbb{R}^n$,

$$
\begin{aligned}
|f(x)| &= \lim_{k \to -\infty} 2^{-\frac{nk}{p}} \left\| f \mathbf{1}_{B(x,2^k) \setminus B(x,2^{k-1})} \right\|_{L^p(\mathbb{R}^n)} \\
&\leq \sup_{\xi \in \mathbb{R}, \, k \in \mathbb{Z}} \left\{ 2^{-\frac{nk}{p}} \left\| f \mathbf{1}_{B(\xi,2^k) \setminus B(\xi,2^{k-1})} \right\|_{L^p(\mathbb{R}^n)} \right\} \\
&\leq \sup_{\xi \in \mathbb{R}^n} \left\{ \sum_{k \in \mathbb{Z}} 2^{-\frac{nkq}{p}} \left\| f \mathbf{1}_{B(\xi,2^k) \setminus B(\xi,2^{k-1})} \right\|_{L^p(\mathbb{R}^n)}^q \right\}^{\frac{1}{q}} \\
&= \| f \|_{\dot{\mathcal{K}}^{p,q}_{\omega_0}(\mathbb{R}^n)} .
\end{aligned}
$$

This implies that

$$\| f \|_{L^\infty(\mathbb{R}^n)} \leq \| f \|_{\dot{\mathcal{K}}^{p,q}_{\omega_0}(\mathbb{R}^n)},$$

and hence

$$\dot{\mathcal{K}}^{p,q}_{\omega_0}(\mathbb{R}^n) \hookrightarrow L^\infty(\mathbb{R}^n). \tag{1.37}$$

In addition, by Example 1.2.40, we find that $\mathbf{1}_{B(0,1)} \in L^\infty(\mathbb{R}^n) \setminus \dot{\mathcal{K}}^{p,q}_{\omega_0}(\mathbb{R}^n)$. Thus, the continuous embedding (1.37) is strict, which then completes the proof of Proposition 1.2.12.   □

Now, we recall the definition of ball quasi-Banach function spaces and some other concepts which are used later. In what follows, we always use the *symbol* $\mathscr{M}(\mathbb{R}^n)$ to denote the set of all measurable functions on $\mathbb{R}^n$. Then the following concept of ball quasi-Banach function spaces was given in [207, Definition 2.2].

Definition 1.2.13 A quasi-normed linear space $X \subset \mathscr{M}(\mathbb{R}^n)$, equipped with a quasi-norm $\|\cdot\|_X$ which makes sense for all functions in $\mathscr{M}(\mathbb{R}^n)$, is called a *ball quasi-Banach function space* (for short, BQBF space) if it satisfies

(i) for any $f \in \mathscr{M}(\mathbb{R}^n)$, $\|f\|_X = 0$ implies that $f = 0$ almost everywhere;
(ii) for any $f, g \in \mathscr{M}(\mathbb{R}^n)$, $|g| \leq |f|$ almost everywhere implies that $\|g\|_X \leq \|f\|_X$;
(iii) for any $\{f_m\}_{m\in\mathbb{N}} \subset \mathscr{M}(\mathbb{R}^n)$ and $f \in \mathscr{M}(\mathbb{R}^n)$, $0 \leq f_m \uparrow f$ as $m \to \infty$ almost everywhere implies that $\|f_m\|_X \uparrow \|f\|_X$ as $m \to \infty$;
(iv) $B \in \mathbb{B}$ implies that $\mathbf{1}_B \in X$, where $\mathbb{B}$ is as in (1.10).

Moreover, a ball quasi-Banach function space X is called a *ball Banach function space* (for short, BBF space) if the quasi-norm $\|\cdot\|_X$ of X satisfies the triangle inequality: for any $f, g \in X$,

$$\|f + g\|_X \leq \|f\|_X + \|g\|_X \tag{1.38}$$

and, for any $B \in \mathbb{B}$, there exists a positive constant $C_{(B)}$, depending on B, such that, for any $f \in X$,

$$\int_B |f(x)|\, dx \leq C_{(B)}\|f\|_X. \tag{1.39}$$

Remark 1.2.14

(i) Observe that, in Definition 1.2.13(iv), if we replace any ball B by any bounded measurable set E, we obtain an equivalent formulation of ball quasi-Banach function spaces.
(ii) We should point out that, every ball quasi-Banach function space is complete (see Corollary 1.2.37 below).
(iii) Recall that a quasi-Banach space $X \subset \mathscr{M}(\mathbb{R}^n)$ is called a *quasi-Banach function space* (for short, QBF space) if it is a ball quasi-Banach function space and it satisfies Definition 1.2.13(iv) with ball therein replaced by any measurable set of *finite measure*. Furthermore, a *Banach function space* (for short, BF space) is a quasi-Banach function space satisfying (1.38) and (1.39) with ball therein replaced by any measurable set of *finite measure*, which was originally introduced in [11, Chapter 1, Definitions 1.1 and 1.3]. It is easy to see that every quasi-Banach function space (resp., Banach function space) is a ball quasi-Banach function space (resp., ball Banach function space), and the converse is not necessary to be true. Several examples about ball quasi-Banach function spaces are given in Remark 1.2.15 below.

Remark 1.2.15 We should point out that the family of (ball) quasi-Banach function spaces includes many important function spaces. Here are some examples.

(i) Let $p \in (0, \infty]$. Then it is easy to show that the *Lebesgue space* $L^p(\mathbb{R}^n)$ is a QBF space and, in particular, when $p \in [1, \infty]$, $L^p(\mathbb{R}^n)$ is a BF space.

(ii) Let $p \in (0, \infty)$ and $\upsilon \in A_\infty(\mathbb{R}^n)$ (see Definition 1.8.1 below for its definition). Then, as was pointed out in [207, p. 86], the *weighted Lebesgue space* $L^p_\upsilon(\mathbb{R}^n)$ [see Definition 2.3.7(i) below for its definition] is a BQBF space, but it may not be a QBF space. Moreover, when $p \in [1, \infty)$, we easily know that $L^p_\upsilon(\mathbb{R}^n)$ is a BBF space.

(iii) Let $\vec{p} := (p_1, \ldots, p_n) \in (0, \infty]^n$ and recall that the *mixed-norm Lebesgue space* $L^{\vec{p}}(\mathbb{R}^n)$, originally introduced in [10], is defined to be the set of all the measurable functions f on $\mathbb{R}^n$ such that

$$\|f\|_{L^{\vec{p}}(\mathbb{R}^n)} := \left\{ \int_{\mathbb{R}} \cdots \left[\int_{\mathbb{R}} |f(x_1, \ldots, x_n)|^{p_1} \, dx_1 \right]^{\frac{p_2}{p_1}} \cdots dx_n \right\}^{\frac{1}{p_n}} < \infty.$$

As was proved in [278, Subsection 7.2] (see also [279, Propositions 2.12, 2.14, and 2.22]), the mixed-norm Lebesgue space $L^{\vec{p}}(\mathbb{R}^n)$ is a BQBF space, but it may not be a QBF space. In particular, when $\vec{p} \in [1, \infty]^n$, $L^{\vec{p}}(\mathbb{R}^n)$ is a BBF space.

(iv) Let $0 < q \le p < \infty$ and recall that the *Morrey space* $M^p_q(\mathbb{R}^n)$, originally introduced by Morrey in [182], is defined to be the set of all the measurable functions f on $\mathbb{R}^n$ such that

$$\|f\|_{M^p_q(\mathbb{R}^n)} := \sup_{B \in \mathbb{B}} \left\{ |B|^{\frac{1}{p} - \frac{1}{q}} \|f \mathbf{1}_B\|_{L^q(\mathbb{R}^n)} \right\} < \infty,$$

where the supremum is taken over all balls $B \in \mathbb{B}$. It is easy to show that the Morrey space $M^p_q(\mathbb{R}^n)$ is a QBF space. In addition, Sawano and Tanaka [209, Example 3.3] constructed a counterexample to show that, when $1 < q < p < \infty$, $M^p_q(\mathbb{R}^n)$ does not satisfy (1.39), and hence the Morrey space $M^p_q(\mathbb{R}^n)$ may not be a BF space. However, applying the Hölder inequality, we can easily find that, when $1 \le q \le p < \infty$, the space $M^p_q(\mathbb{R}^n)$ is a BBF (see also [207, Subsection 7.4]).

(v) Let $p(\cdot) : \mathbb{R}^n \to [0, \infty)$ be a measurable function. Recall that the *variable Lebesgue space* $L^{p(\cdot)}(\mathbb{R}^n)$, originally introduced in [192] (see also [147]), is defined to be the set of all the measurable functions f on $\mathbb{R}^n$ such that

$$\|f\|_{L^{p(\cdot)}(\mathbb{R}^n)} := \inf \left\{ \lambda \in (0, \infty) : \int_{\mathbb{R}^n} \left[\frac{|f(x)|}{\lambda} \right]^{p(x)} \, dx \in [0, 1] \right\} < \infty.$$

Then, by Diening et al. [54, Lemma 3.2.6 and Theorem 3.2.13], we easily find that, when $p(\cdot) : \mathbb{R}^n \to (0, \infty)$, the variable Lebesgue space $L^{p(\cdot)}(\mathbb{R}^n)$ is a

QBF space and, when $p(x) \geq 1$ for almost every $x \in \mathbb{R}^n$, $L^{p(\cdot)}(\mathbb{R}^n)$ is a BF space (see also [207, Subsection 7.8]).

(vi) However, we should point out that the famous space $\mathrm{BMO}(\mathbb{R}^n)$ is not a BQBF space. Indeed, recall that the *space* $\mathrm{BMO}(\mathbb{R}^n)$, originally introduced by John and Nirenberg in [140], is defined to be the set of all the locally integrable functions f on $\mathbb{R}^n$ such that

$$\|f\|_{\mathrm{BMO}(\mathbb{R}^n)} := \sup_{B \in \mathbb{B}} \left\{ \frac{1}{|B|} \int_B |f(x) - f_B| \, dx \right\} < \infty,$$

where the supremum is taken over all balls $B \in \mathbb{B}$ and, for any given measurable set $E \subset \mathbb{R}^n$ with $|E| \in (0, \infty)$,

$$f_E := \frac{1}{|E|} \int_E f(x) \, dx. \tag{1.40}$$

Obviously, $\|f\|_{\mathrm{BMO}(\mathbb{R}^n)} = 0$ if and only if f is a constant almost everywhere in $\mathbb{R}^n$. In addition, for any $x \in \mathbb{R}$, let

$$f(x) := \frac{x}{|x|} \ln \left(\frac{1}{|x|} \right) \mathbf{1}_{(-1,1) \setminus \{0\}}(x).$$

Then, as was pointed out in [70, p. 118], $f \notin \mathrm{BMO}(\mathbb{R})$ but $|f| \in \mathrm{BMO}(\mathbb{R})$. Thus, the space $\mathrm{BMO}(\mathbb{R}^n)$ does not satisfy (i) or (ii) of Definition 1.2.13, and hence $\mathrm{BMO}(\mathbb{R}^n)$ is not a BQBF space.

(vii) We should also point out that the John–Nirenberg space $JN_p(\mathbb{R}^n)$ is not a BQBF space. Indeed, let $p \in (1, \infty)$. Recall that the *John–Nirenberg space* $JN_p(\mathbb{R}^n)$ is defined to be the set of all the locally integrable functions f on $\mathbb{R}^n$ such that

$$\|f\|_{JN_p(\mathbb{R}^n)} := \sup \left\{ \sum_j |Q_j| \left[\frac{1}{|Q_j|} \int_{Q_j} |f(x) - f_{Q_j}| \, dx \right]^p \right\}^{\frac{1}{p}} < \infty,$$

where the supremum is taken over all the collections $\{Q_j\}_j \subset \mathcal{Q}$ of interior pairwise disjoint cubes and, for any cube Q_j, f_{Q_j} is defined as in (1.40) with E replaced by Q_j, which is a special case of John–Nirenberg–Campanato spaces in [226, Definition 2.3] (see also [52, 136–138, 140, 227]). It is easy to show that $\|f\|_{JN_p(\mathbb{R}^n)} = 0$ if and only if f is a constant almost everywhere in $\mathbb{R}^n$. This implies that $JN_p(\mathbb{R}^n)$ does not satisfy Definition 1.2.13(i), and hence $JN_p(\mathbb{R}^n)$ is not a BQBF space.

The following basic properties of ball quasi-Banach function spaces can be deduced directly from Definition 1.2.13; we omit the details.

Proposition 1.2.16 *Let* $X \subset \mathscr{M}(\mathbb{R}^n)$ *be a ball quasi-Banach function space. Then*

(i) *for any* $f \in X$, $\|f\|_X = 0$ *if and only if* $f = 0$ *almost everywhere in* $\mathbb{R}^n$;
(ii) *for any* $f \in X$, $\|f\|_X = \||f|\|_X$.

Next, we recall the concept of the associate space of a ball Banach function space (see, for instance, [207, p. 9]).

Definition 1.2.17 For any Banach space $X \subset \mathscr{M}(\mathbb{R}^n)$, the *associate space (also called the* Köthe dual*)* X' is defined to be the set of all the $f \in \mathscr{M}(\mathbb{R}^n)$ such that

$$\|f\|_{X'} := \sup\left\{\|fg\|_{L^1(\mathbb{R}^n)} : g \in X, \ \|g\|_X = 1\right\} < \infty. \tag{1.41}$$

Remark 1.2.18 Let $X \subset \mathscr{M}(\mathbb{R}^n)$ be a Banach space.

(i) By Sawano et al. [207, Proposition 2.3], we know that, if X is a ball Banach function space, then its associate space X' is also a ball Banach function space.
(ii) The advantage of the associate space X' is that the norm of X' has an explicit integral expression which makes up for the deficiency of the explicit expression of the norm of X in some sense.
(iii) If X is a ball Banach function space, then, from [278, Lemma 2.6], we infer that

$$X = X''$$

with the same norms, where X'' denotes the second associate space of X. However, in general, the formula

$$X = X^{**}$$

does not hold true for a Banach space X and its second dual space X^{**}.
(iv) The explicit integral expression of the norm of the associate space makes it more convenient to find the associate space than the dual space of a given function space. Indeed, by Sawano and Tanaka [209, Theorem 4.1], we know that the associate space of the Morrey space is the block space, but the dual space of the Morrey space is still unknown so far; see also Remark 1.2.19(iv) below for more details.

Remark 1.2.19 There are several concrete examples on the associate spaces of ball Banach function spaces.

(i) Let $p \in [1, \infty]$. Then, from [11, Chapter 1, Theorem 2.5], it follows that

$$\left(L^p(\mathbb{R}^n)\right)' = L^{p'}(\mathbb{R}^n)$$

with the same norms, where $\frac{1}{p} + \frac{1}{p'} = 1$.

(ii) Let $p \in (1, \infty)$ and $\upsilon \in A_\infty(\mathbb{R}^n)$. Then, by Diening et al. [54, Theorem 2.7.4] (see also [29, Subsection 5.4]), we immediately find that

$$\left(L_\upsilon^p(\mathbb{R}^n)\right)' = L_{\upsilon^{1-p'}}^{p'}(\mathbb{R}^n)$$

with the same norms.

(iii) Let $\vec{p} := (p_1, \ldots, p_n) \in [1, \infty]^n$. Then, using [10, Theorem 2], we conclude that

$$\left(L^{\vec{p}}(\mathbb{R}^n)\right)' = L^{\vec{p'}}(\mathbb{R}^n)$$

with the same norms, where $\vec{p'} := (p_1', \ldots, p_n')$ with $\frac{1}{p_i} + \frac{1}{p_i'} = 1$ for any $i \in \{1, \ldots, n\}$.

(iv) Let $1 < q \le p < \infty$. A measurable function b on $\mathbb{R}^n$ is called a (p', q')-*block* if there exists a ball $B \in \mathbb{B}$ such that

$$\mathrm{supp}(b) := \{x \in \mathbb{R}^n : b(x) \ne 0\} \subset B$$

and

$$\left[\int_B |b(x)|^{q'}\, dx\right]^{\frac{1}{q'}} \le |B|^{\frac{1}{q'} - \frac{1}{p'}}.$$

Recall that the *block space* $B_{q'}^{p'}(\mathbb{R}^n)$, introduced in [14], is defined to be the set of all the measurable functions f on $\mathbb{R}^n$ such that there exists a sequence $\{\lambda_j\}_{j\in\mathbb{N}} \subset [0, \infty)$ and a sequence $\{b_j\}_{j\in\mathbb{N}} \subset L^{q'}(\mathbb{R}^n)$ of (p', q')-blocks supported, respectively, in the balls $\{B_j\}_{j\in\mathbb{N}} \subset \mathbb{B}$ such that both

$$f = \sum_{j\in\mathbb{N}} \lambda_j b_j$$

almost everywhere in $\mathbb{R}^n$ and

$$\sum_{j\in\mathbb{N}} \lambda_j < \infty.$$

Moreover, for any $f \in B_{q'}^{p'}(\mathbb{R}^n)$, let

$$\|f\|_{B_{q'}^{p'}(\mathbb{R}^n)} := \inf\left\{\sum_{j\in\mathbb{N}} \lambda_j\right\},$$

where the infimum is taken over all the decompositions of f as above. Then, as was proved in [209, Theorem 4.1],

$$\left(M_q^p(\mathbb{R}^n)\right)' = B_{q'}^{p'}(\mathbb{R}^n)$$

with the same norms.

(v) Let $p(\cdot) : \mathbb{R}^n \to (1, \infty)$ be a measurable function and, for any $x \in \mathbb{R}^n$, let $p'(x) \in (1, \infty)$ be such that

$$\frac{1}{p(x)} + \frac{1}{p'(x)} = 1.$$

Then, as was showed in [49, Proposition 2.37],

$$\left(L^{p(\cdot)}(\mathbb{R}^n)\right)' = L^{p'(\cdot)}(\mathbb{R}^n)$$

with equivalent norms.

(vi) Let $p, q \in (1, \infty)$ and $\omega \in M(\mathbb{R}_+)$ be such that

$$-\frac{n}{p} < m_0(\omega) \le M_0(\omega) < \frac{n}{p'}.$$

Then, by Theorem 1.7.9 below, we find that

$$\left(\dot{\mathcal{K}}_{\omega,0}^{p,q}(\mathbb{R}^n)\right)' = \dot{\mathcal{K}}_{1/\omega,0}^{p',q'}(\mathbb{R}^n)$$

with the same norms. However, we should point out that the associate space of the global generalized Herz space $\dot{\mathcal{K}}_{\omega}^{p,q}(\mathbb{R}^n)$ is still unknown. Indeed, as was proved in Theorem 2.2.1 later, the predual space of the global generalized Herz space $\dot{\mathcal{K}}_{\omega}^{p,q}(\mathbb{R}^n)$ is the block space $\dot{\mathcal{B}}_{1/\omega}^{p',q'}(\mathbb{R}^n)$ (see Definition 2.1.3 below for its definition). However, due to the unknown of the Fatou property of block spaces [namely, whether Definition 1.2.13(iii) holds true for block spaces is unknown], we can not show whether or not $\dot{\mathcal{B}}_{1/\omega}^{p',q'}(\mathbb{R}^n)$ is the associate space of $\dot{\mathcal{K}}_{\omega}^{p,q}(\mathbb{R}^n)$.

Now, we give a striking example about associate spaces. To this end, recall that the *Hardy space* $H^1(\mathbb{R}^n)$ is defined to be the set of all the $f \in L^1(\mathbb{R}^n)$ such that

$$\max_{t \in (0,\infty)} \left\{ \left\| \frac{1}{t^n} P\left(\frac{\cdot}{t}\right) * f \right\| \right\} \in L^1(\mathbb{R}^n), \tag{1.42}$$

where P is the *Poisson kernel* which is defined by setting, for any $x \in \mathbb{R}^n$,

$$P(x) := \frac{\Gamma(\frac{n+1}{2})}{\pi^{\frac{n+1}{2}}} \frac{1}{(1 + |x|^2)^{\frac{n+1}{2}}}$$

(see, for instance, [91, Section 2.1]). It was proved by Fefferman and Stein [77] that $H^1(\mathbb{R}^n)$ is a Banach space and the dual space of $H^1(\mathbb{R}^n)$ is the well-known space $\mathrm{BMO}(\mathbb{R}^n)$ (see also [214, Section 2.5]). However, from [214, p. 94, Exercise 24], we infer that there exists an $f_0 \in \mathrm{BMO}(\mathbb{R}^n)$ and a $g_0 \in H^1(\mathbb{R}^n)$ such that $f_0 g_0 \notin L^1(\mathbb{R}^n)$. This, together with (1.41), further implies that

$$\|f_0\|_{(H^1(\mathbb{R}^n))'} = \infty,$$

and hence the associate space of $H^1(\mathbb{R}^n)$ is not the space $\mathrm{BMO}(\mathbb{R}^n)$, which is quite surprising.

However, we can prove that the associate space of $H^1(\mathbb{R}^n)$ is just $L^\infty(\mathbb{R}^n)$ as follows.

Proposition 1.2.20 *The associate space of $H^1(\mathbb{R}^n)$ coincides with $L^\infty(\mathbb{R}^n)$ in the sense of equivalent norms.*

To prove the above proposition, we require some preliminary lemmas. First, the following result gives the embedding relation between $H^1(\mathbb{R}^n)$ and $L^1(\mathbb{R}^n)$, which is just [91, Theorem 2.1.2(b)].

Lemma 1.2.21 *The Hardy space $H^1(\mathbb{R}^n)$ embeds continuously into the Lebesgue space $L^1(\mathbb{R}^n)$. Namely, there exists a positive constant C such that, for any $f \in H^1(\mathbb{R}^n)$,*

$$\|f\|_{L^1(\mathbb{R}^n)} \le C \|f\|_{H^1(\mathbb{R}^n)}.$$

Moreover, we also need the following concept of atoms (see, for instance, [214, p. 81]).

Definition 1.2.22 A measurable function a on $\mathbb{R}^n$ is called an ∞-*atom* if there exists a ball $B \in \mathbb{B}$ such that

(i) $\mathrm{supp}(a) \subset B$;
(ii) $\|a\|_{L^\infty(\mathbb{R}^n)} \le \frac{1}{|B|}$;
(iii) $\int_{\mathbb{R}^n} a(x)\, dx = 0$.

We have the following technical estimate about the norm of ∞-atoms in $H^1(\mathbb{R}^n)$, which is just [214, p. 81, Corollary 5.3] and plays an important role in the proof of Proposition 1.2.20.

Lemma 1.2.23 *Let a be an ∞-atom. Then $a \in H^1(\mathbb{R}^n)$ and, moreover, there exists a positive constant C, independent of a, such that*

$$\|a\|_{H^1(\mathbb{R}^n)} \le C.$$

Furthermore, to show Proposition 1.2.20, we also need the following result about the approximate identity, which is just [84, Theorem 8.15].

Lemma 1.2.24 *Let $\phi \in L^1(\mathbb{R}^n)$ satisfy that there exist two positive constants C and ε such that, for any $x \in \mathbb{R}^n$,*

$$|\phi(x)| \le C\frac{1}{(1+|x|)^{n+\varepsilon}}.$$

Then, for any given $p \in [1, \infty]$ and for any $f \in L^p(\mathbb{R}^n)$ and any Lebesgue point x of f,

$$\lim_{t \to 0} f * \phi_t(x) = f(x) \int_{\mathbb{R}^n} \phi(y)\, dy,$$

where, for any $t \in (0, \infty)$,

$$\phi_t(\cdot) := \frac{1}{t^n}\phi\left(\frac{\cdot}{t}\right).$$

Via the above lemmas, we now establish an integral formula of the norm of $L^\infty(\mathbb{R}^n)$, which is a basic tool in the proof of Proposition 1.2.20.

Lemma 1.2.25 *Let $f \in L^\infty(\mathbb{R}^n)$. Then there exist two positive constants C_1 and C_2, independent of f, such that*

$$C_1 \|f\|_{L^\infty(\mathbb{R}^n)} \le \sup\left\{\|fg\|_{L^1(\mathbb{R}^n)} : \|g\|_{H^1(\mathbb{R}^n)} = 1\right\} \le C_2 \|f\|_{L^\infty(\mathbb{R}^n)}.$$

Proof Let $f \in L^\infty(\mathbb{R}^n)$. Then, for any $g \in H^1(\mathbb{R}^n)$ with $\|g\|_{H^1(\mathbb{R}^n)} = 1$, applying both the Hölder inequality and Lemma 1.2.21, we find that

$$\|fg\|_{L^1(\mathbb{R}^n)} \le \|f\|_{L^\infty(\mathbb{R}^n)} \|g\|_{L^1(\mathbb{R}^n)} \lesssim \|f\|_{L^\infty(\mathbb{R}^n)} \|g\|_{H^1(\mathbb{R}^n)} \sim \|f\|_{L^\infty(\mathbb{R}^n)},$$

which implies that

$$\sup\left\{\|fg\|_{L^1(\mathbb{R}^n)} : \|g\|_{H^1(\mathbb{R}^n)} = 1\right\} \lesssim \|f\|_{L^\infty(\mathbb{R}^n)}. \tag{1.43}$$

Conversely, we show that

$$\|f\|_{L^\infty(\mathbb{R}^n)} \lesssim \sup\left\{\|fg\|_{L^1(\mathbb{R}^n)} : \|g\|_{H^1(\mathbb{R}^n)} = 1\right\}. \tag{1.44}$$

To this end, we choose a function $a \in L^1(\mathbb{R}^n)$ satisfying that $\|a\|_{L^1(\mathbb{R}^n)} = 1$, $\mathrm{supp}(a) \subset B(\mathbf{0}, 1)$,

$$\|a\|_{L^\infty(\mathbb{R}^n)} \le \frac{1}{|B(\mathbf{0}, 1)|}, \tag{1.45}$$

and

$$\int_{\mathbb{R}^n} a(x)\, dx = 0, \tag{1.46}$$

For any given $x \in \mathbb{R}^n$ and $t \in (0, \infty)$ and for any $y \in \mathbb{R}^n$, let

$$a_{x,t}(y) := \frac{1}{t^n} a\left(\frac{x-y}{t}\right).$$

Then, from the assumption $\operatorname{supp}(a) \subset B(\mathbf{0}, 1)$, we infer that, for any $x \in \mathbb{R}^n$ and $t \in (0, \infty)$, $\operatorname{supp}(a_{x,t}) \subset B(x, t)$. In addition, using the change of variables, (1.45), and (1.46), we conclude that, for any $x \in \mathbb{R}^n$ and $t \in (0, \infty)$,

$$\int_{\mathbb{R}^n} a_{x,t}(y)\, dy = \int_{\mathbb{R}^n} a(y)\, dy = 0$$

and

$$\|a_{x,t}\|_{L^\infty(\mathbb{R}^n)} \leq \frac{1}{t^n} \frac{1}{|B(\mathbf{0}, 1)|} = \frac{1}{|B(x, t)|}.$$

Thus, for any $x \in \mathbb{R}^n$ and $t \in (0, \infty)$, $a_{x,t}$ is an ∞-atom. Combining this and Lemma 1.2.23, we have

$$\sup_{x\in\mathbb{R}^n,\, t\in(0,\infty)} \left\{ \|a_{x,t}\|_{H^1(\mathbb{R}^n)} \right\} \lesssim 1. \tag{1.47}$$

In addition, by (1.45), we find that, for any $y \in B(\mathbf{0}, 1)$,

$$|a(y)| \lesssim 1 \lesssim \frac{1}{(1+|y|)^{n+1}}. \tag{1.48}$$

Applying the assumption $\operatorname{supp}(a) \subset B(\mathbf{0}, 1)$, we also know that, for any $y \in [B(\mathbf{0}, 1)]^{\complement}$,

$$|a(y)| = 0 < \frac{1}{(1+|y|)^{n+1}},$$

From this, (1.48), and Lemma 1.2.24 with $\phi := |a|$, $\varepsilon := 1$, and $p := \infty$, and from the assumption $\|a\|_{L^1(\mathbb{R}^n)} = 1$, it follows that, for any Lebesgue point x of f,

$$\int_{\mathbb{R}^n} |f(y)| \, |a_{x,t}(y)|\, dy = |f| * |a|_t(x) \to |f(x)| \tag{1.49}$$

as $t \to 0$, where, for any $t \in (0, \infty)$,

$$|a|_t(\cdot) := \frac{1}{t^n}\left|a\left(\frac{\cdot}{t}\right)\right|.$$

Observe that, for any $x \in \mathbb{R}^n$ and $t \in (0, \infty)$, by Lemma 1.2.21 with a therein replaced by $a_{x,t}$, and by the assumption $\|a\|_{L^1(\mathbb{R}^n)} = 1$ again, we find that

$$\|a_{x,t}\|_{H^1(\mathbb{R}^n)} \gtrsim \|a_{x,t}\|_{L^1(\mathbb{R}^n)} \sim \|a\|_{L^1(\mathbb{R}^n)} \neq 0.$$

Using this, (1.49), and (1.47), we further conclude that, for any Lebesgue point x of f,

$$
\begin{aligned}
|f(x)| &\leq \sup_{t \in (0,\infty)} \left\{ \int_{\mathbb{R}^n} |f(y)| \, |a_{x,t}(y)| \, dy \right\} \\
&= \sup_{t \in (0,\infty)} \left\{ \|a_{x,t}\|_{H^1(\mathbb{R}^n)} \int_{\mathbb{R}^n} |f(y)| \frac{|a_{x,t}(y)|}{\|a_{x,t}\|_{H^1(\mathbb{R}^n)}} \, dy \right\} \\
&\lesssim \sup_{t \in (0,\infty)} \left\{ \int_{\mathbb{R}^n} |f(y)| \frac{|a_{x,t}(y)|}{\|a_{x,t}\|_{H^1(\mathbb{R}^n)}} \, dy \right\} \\
&\lesssim \sup \left\{ \|fg\|_{L^1(\mathbb{R}^n)} : \ \|g\|_{H^1(\mathbb{R}^n)} = 1 \right\},
\end{aligned}
$$

which then completes the proof of (1.44). This, combined with (1.43), further implies that

$$\|f\|_{L^\infty(\mathbb{R}^n)} \sim \sup \left\{ \|fg\|_{L^1(\mathbb{R}^n)} : \ \|g\|_{H^1(\mathbb{R}^n)} = 1 \right\}$$

with the positive equivalence constants independent of f, and hence finishes the proof of Lemma 1.2.25. $\square$

Next, we turn to show Proposition 1.2.20.

Proof of Proposition 1.2.20 Let all the symbols be the same as in the present proposition. We first prove that $L^\infty(\mathbb{R}^n) \subset (H^1(\mathbb{R}^n))'$. Indeed, for any $f \in L^\infty(\mathbb{R}^n)$, from (1.43), we infer that

$$\|f\|_{(H^1(\mathbb{R}^n))'} = \sup \left\{ \|fg\|_{L^1(\mathbb{R}^n)} : \ \|g\|_{H^1(\mathbb{R}^n)} = 1 \right\} \lesssim \|f\|_{L^\infty(\mathbb{R}^n)}, \qquad (1.50)$$

which implies that $f \in (H^1(\mathbb{R}^n))'$, and hence completes the proof that $L^\infty(\mathbb{R}^n) \subset (H^1(\mathbb{R}^n))'$.

Conversely, we next show that $(H^1(\mathbb{R}^n))' \subset L^\infty(\mathbb{R}^n)$. To do this, let $f \in (H^1(\mathbb{R}^n))'$ and, for any $k \in \mathbb{N}$,

$$f_k := f \mathbf{1}_{\{x \in \mathbb{R}^n : |f(x)| \leq k\}}.$$

Then, for any $k \in \mathbb{N}$, $0 \le |f_k| \le |f|$, $f_k \in L^\infty(\mathbb{R}^n)$, and $|f_k(x)| \uparrow |f(x)|$ as $k \to \infty$ for any $x \in \mathbb{R}^n$. Then, for any $k \in \mathbb{N}$, applying Lemma 1.2.25 with $f := f_k$, we find that

$$
\begin{aligned}
\|f_k\|_{L^\infty(\mathbb{R}^n)} &\sim \sup\left\{\|f_k g\|_{L^1(\mathbb{R}^n)} : \|g\|_{H^1(\mathbb{R}^n)} = 1\right\} \\
&\lesssim \sup\left\{\|f g\|_{L^1(\mathbb{R}^n)} : \|g\|_{H^1(\mathbb{R}^n)} = 1\right\} \\
&\sim \|f\|_{(H^1(\mathbb{R}^n))'}.
\end{aligned}
\tag{1.51}
$$

Now, we prove that $\|f_k\|_{L^\infty(\mathbb{R}^n)} \to \|f\|_{L^\infty(\mathbb{R}^n)}$ as $k \to \infty$. Indeed, it is obvious that

$$
\varlimsup_{k\to\infty} \|f_k\|_{L^\infty(\mathbb{R}^n)} \le \|f\|_{L^\infty(\mathbb{R}^n)}.
\tag{1.52}
$$

Conversely, for any $k \in \mathbb{N}$, there exists a measurable set $E_k \subset \mathbb{R}^n$ such that $|E_k| = 0$ and

$$
\|f_k\|_{L^\infty(\mathbb{R}^n)} = \sup_{x\in\mathbb{R}^n\setminus E_k} |f_k(x)|.
$$

From this, we deduce that

$$
\begin{aligned}
\|f\|_{L^\infty(\mathbb{R}^n)} &\le \sup_{x\in\mathbb{R}^n\setminus\bigcup_{j\in\mathbb{N}} E_j} |f(x)| = \lim_{k\to\infty} \sup_{\substack{x\in\mathbb{R}^n\setminus\bigcup_{j\in\mathbb{N}} E_j \\ |f(x)|\le k}} |f(x)| \\
&\le \lim_{k\to\infty} \sup_{x\in\mathbb{R}^n\setminus E_k} |f_k(x)| = \lim_{k\to\infty} \|f_k\|_{L^\infty(\mathbb{R}^n)}.
\end{aligned}
$$

This, combined with (1.52), further implies that $\|f_k\|_{L^\infty(\mathbb{R}^n)} \to \|f\|_{L^\infty(\mathbb{R}^n)}$ as $k \to \infty$. Using this and letting $k \to \infty$ in (1.51), we then conclude that

$$
\|f\|_{L^\infty(\mathbb{R}^n)} \lesssim \|f\|_{(H^1(\mathbb{R}^n))'},
\tag{1.53}
$$

which implies that $f \in L^\infty(\mathbb{R}^n)$. Therefore, we have $(H^1(\mathbb{R}^n))' \subset L^\infty(\mathbb{R}^n)$, and hence $(H^1(\mathbb{R}^n))' = L^\infty(\mathbb{R}^n)$. Moreover, by both (1.50) and (1.53), we further obtain

$$
\|f\|_{(H^1(\mathbb{R}^n))'} \sim \|f\|_{L^\infty(\mathbb{R}^n)},
$$

which implies that $(H^1(\mathbb{R}^n))' = L^\infty(\mathbb{R}^n)$ with equivalent norms, and hence completes the proof of Proposition 1.2.20.   □

The following concept on absolutely continuous quasi-norms of ball quasi-Banach function spaces given in [277, Definition 2.5] is similar to the absolute continuity of Banach function spaces (see, for instance, [11, Chapter 1, Definition 3.1]).

Definition 1.2.26 Let X be a ball quasi-Banach function space. A measurable function $f \in X$ is said to have an *absolutely continuous quasi-norm in X* if, for any sequence $\{E_i\}_{i \in \mathbb{N}}$ of measurable sets satisfying $\mathbf{1}_{E_i} \to 0$ almost everywhere as $i \to \infty$, $\|f\mathbf{1}_{E_i}\|_X \to 0$ as $i \to \infty$. Moreover, X is said to have an *absolutely continuous quasi-norm* if, for any $f \in X$, f has an absolutely continuous quasi-norm in X.

Remark 1.2.27 By Wang et al. [233, Remark 3.4], we find that Lebesgue spaces, weighted Lebesgue spaces, mixed-norm Lebesgue spaces, and variable Lebesgue spaces all have absolutely continuous quasi-norms, but Morrey spaces may not have (see also [209, Example 5.1]). In addition, from Theorem 1.4.1 below, we infer that local generalized Herz spaces have absolutely continuous quasi-norms. We should also point out that global generalized Herz spaces may not have absolutely continuous quasi-norms; see Example 1.4.4 below for more details.

Now, we recall the concepts of both the convexity and the concavity of ball quasi-Banach function spaces (see, for instance, [207, Definition 2.6]).

Definition 1.2.28 Let X be a ball quasi-Banach function space and $p \in (0, \infty)$.

(i) The *p-convexification* X^p of X is defined by setting

$$X^p := \left\{ f \in \mathscr{M}(\mathbb{R}^n) : |f|^p \in X \right\}$$

equipped with the quasi-norm $\|f\|_{X^p} := \||f|^p\|_X^{1/p}$.

(ii) The space X is said to be *p-convex* if there exists a positive constant C such that, for any $\{f_j\}_{j \in \mathbb{N}} \subset X^{1/p}$,

$$\left\| \sum_{j \in \mathbb{N}} |f_j| \right\|_{X^{1/p}} \leq C \sum_{j \in \mathbb{N}} \|f_j\|_{X^{1/p}}.$$

In particular, when $C = 1$, X is said to be *strictly p-convex*.

(iii) The space X is said to be *p-concave* if there exists a positive constant C such that, for any $\{f_j\}_{j \in \mathbb{N}} \subset X^{1/p}$,

$$\sum_{j \in \mathbb{N}} \|f_j\|_{X^{1/p}} \leq C \left\| \sum_{j \in \mathbb{N}} |f_j| \right\|_{X^{1/p}}.$$

In particular, when $C = 1$, X is said to be *strictly p-concave*.

Recall that the *Hardy–Littlewood maximal operator* $\mathcal{M}$ is defined by setting, for any $f \in L^1_{\mathrm{loc}}(\mathbb{R}^n)$ and $x \in \mathbb{R}^n$,

$$\mathcal{M}(f)(x) := \sup_{B \ni x} \frac{1}{|B|} \int_B |f(y)|\, dy, \tag{1.54}$$

where the supremum is taken over all the balls $B \in \mathbb{B}$ containing x. For any $\theta \in (0, \infty)$, the *powered Hardy–Littlewood maximal operator* $\mathcal{M}^{(\theta)}$ is defined by setting, for any $f \in L^1_{\mathrm{loc}}(\mathbb{R}^n)$ and $x \in \mathbb{R}^n$,

$$\mathcal{M}^{(\theta)}(f)(x) := \left[\mathcal{M} \left(|f|^\theta \right)(x) \right]^{\frac{1}{\theta}} . \tag{1.55}$$

We also need the following two assumptions about the *Fefferman–Stein vector-valued inequality* on ball quasi-Banach function spaces and the boundedness of powered Hardy–Littlewood maximal operators on associate spaces.

Assumption 1.2.29 *Let X be a ball quasi-Banach function space and θ, $s \in (0, 1]$. Assume that there exists a positive constant C such that, for any $\{f_j\}_{j \in \mathbb{N}} \subset L^1_{\mathrm{loc}}(\mathbb{R}^n)$,*

$$\left\| \left\{ \sum_{j \in \mathbb{N}} \left[\mathcal{M}^{(\theta)}(f_j) \right]^s \right\}^{1/s} \right\|_X \leq C \left\| \left(\sum_{j \in \mathbb{N}} |f_j|^s \right)^{1/s} \right\|_X . \tag{1.56}$$

Remark 1.2.30

(i) We claim that, for any given ball quasi-Banach function space X and any given $r \in (0, \infty)$, the powered Hardy–Littlewood maximal operator $\mathcal{M}^{(r)}$ is bounded on X if and only if the Hardy–Littlewood maximal operator $\mathcal{M}$ is bounded on $X^{\frac{1}{r}}$. Indeed, for any given $r \in (0, \infty)$, by the definition of the powered Hardy–Littlewood maximal operator $\mathcal{M}^{(r)}$, we conclude that, for any $f \in L^1_{\mathrm{loc}}(\mathbb{R}^n)$,

$$\left\| \mathcal{M}^{(r)}(f) \right\|_X = \left\| \left[\mathcal{M}(|f|^r) \right]^{\frac{1}{r}} \right\|_X = \left\| \mathcal{M}(|f|^r) \right\|_{X^{\frac{1}{r}}}^{\frac{1}{r}} . \tag{1.57}$$

In addition, from Definition 1.2.28(i), it follows that

$$\| f \|_X = \left\| |f|^r \right\|_{X^{\frac{1}{r}}}^{\frac{1}{r}} ,$$

which, combined with (1.57), implies that this claim holds true.

(ii) Similarly to Remark 1.2.30(i) above, we know that the inequality (1.56) holds true if and only if, for any given $\theta, s \in (0, 1]$ and any $\{f_j\}_{j \in \mathbb{N}} \subset L^1_{\mathrm{loc}}(\mathbb{R}^n)$,

$$\left\| \left\{ \sum_{j \in \mathbb{N}} \left[\mathcal{M}(f_j) \right]^{\frac{s}{\theta}} \right\}^{\frac{\theta}{s}} \right\|_{X^{\frac{1}{\theta}}} \lesssim \left\| \left(\sum_{j \in \mathbb{N}} |f_j|^{\frac{s}{\theta}} \right)^{\frac{\theta}{s}} \right\|_{X^{\frac{1}{\theta}}} ,$$

where the implicit positive constant is independent of $\{f_j\}_{j \in \mathbb{N}}$.

Remark 1.2.31 We point out that the Fefferman–Stein vector-valued inequality in Assumption 1.2.29 holds true for many important function spaces. Here are several examples.

(i) Let $p \in (0, \infty), \theta \in (0, \min\{1, p\})$, and $s \in (\theta, 1]$. Then, in this case, applying Remark 1.2.30(ii) and [76, Theorem 1], we conclude that Assumption 1.2.29 holds true with X therein replaced by $L^p(\mathbb{R}^n)$.

(ii) Let $\upsilon \in A_\infty(\mathbb{R}^n)$ and

$$q_\upsilon := \inf\{q \in [1, \infty) : \upsilon \in A_q(\mathbb{R}^n)\} \tag{1.58}$$

[see Definition 1.8.1 below for the definition of $A_q(\mathbb{R}^n)$]. Let $p \in (0, \infty)$, $\theta \in (0, \min\{1, p/q_\upsilon\})$, and $s \in (\theta, 1]$. Then, in this case, as was showed in [29, Subsection 5.4], Assumption 1.2.29 holds true with $X := L_\upsilon^p(\mathbb{R}^n)$.

(iii) Let $\vec{p} := (p_1, \ldots, p_n) \in (0, \infty)^n, \theta \in (0, \min\{1, p_1, \ldots, p_n\})$, and $s \in (\theta, 1]$. Then, in this case, from Remark 1.2.30(ii) and [123, Lemma 3.7], it follows that Assumption 1.2.29 holds true with X therein replaced by the mixed-norm Lebesgue space $L^{\vec{p}}(\mathbb{R}^n)$.

(iv) Let $0 < q \leq p < \infty, \theta \in (0, \min\{1, q\})$, and $s \in (\theta, 1]$. Then, in this case, using Remark 1.2.30(ii) and [224, Lemma 2.5], we find that the Morrey space $M_q^p(\mathbb{R}^n)$ satisfies Assumption 1.2.29.

(v) Recall that the measurable function $p(\cdot) : \mathbb{R}^n \to (0, \infty)$ is said to satisfy the *globally log-Hölder continuous condition*, denoted by $p(\cdot) \in C^{\log}(\mathbb{R}^n)$, if there exists a positive constant C and a constant $p_\infty \in \mathbb{R}$ such that, for any $x, y \in \mathbb{R}^n$,

$$|p(x) - p(y)| \leq \frac{C}{\ln(e + 1/|x - y|)}$$

and

$$|p(x) - p_\infty| \leq \frac{C}{\ln(e + |x|)}.$$

Let $p(\cdot) \in C^{\log}(\mathbb{R}^n)$,

$$p_- := \operatorname*{ess\,inf}_{x \in \mathbb{R}^n} p(x), \tag{1.59}$$

$\theta \in (0, \min\{1, p_-\})$, and $s \in (\theta, 1]$. Then, in this case, by Remark 1.2.30(ii), [54, Lemma 3.2.6], and [50, Corollary 2.1], we conclude that Assumption 1.2.29 holds true with $X := L^{p(\cdot)}(\mathbb{R}^n)$.

(vi) Let $p, q \in (0, \infty), \omega \in M(\mathbb{R}_+)$ satisfy both $m_0(\omega) \in (-\frac{n}{p}, \infty)$ and $m_\infty(\omega) \in (-\frac{n}{p}, \infty)$,

$$\theta \in \left(0, \min\left\{1, p, \frac{n}{\max\{M_0(\omega), M_\infty(\omega)\} + n/p}\right\}\right),$$

and $s \in (\theta, 1]$. Then, in this case, combining Remark 1.2.30(ii) and Lemmas 4.3.11 and 4.3.25 below, we find that both the local generalized Herz
space $\dot{\mathcal{K}}^{p,q}_{\omega,\mathbf{0}}(\mathbb{R}^n)$ and the global generalized Herz space $\mathcal{K}^{p,q}_{\omega}(\mathbb{R}^n)$ satisfy
Assumption 1.2.29.

Remark 1.2.32 We should point out that there exist some function spaces in which
Assumption 1.2.29 does not hold true or in which it is still unclear whether or not
Assumption 1.2.29 holds true. Here are two examples.

(i) Let $p(\cdot) : \mathbb{R}^n \to (0, \infty)$ and $q(\cdot) : \mathbb{R}^n \to (0, \infty)$ be two measurable
functions. Then, as was pointed out in [55, p. 1746], the variable exponents
space $l^{q(\cdot)}(L^{p(\cdot)}(\mathbb{R}^n))$ does not satisfy the vector-valued inequality. Fortunately,
there is an alternative tool for the vector-valued inequality established in [55,
Theorem 3.2].
(ii) As was proposed in [58, Question 9.3], it is still unclear whether or not the
Fefferman–Stein vector-valued inequality involving matrix-weights holds true.

Assumption 1.2.33 *Let X be a ball quasi-Banach function space, $s \in (0, 1]$, and
$r \in (1, \infty]$. Assume that $X^{1/s}$ is also a ball Banach function space and there exists
a positive constant C such that, for any $f \in L^1_{\mathrm{loc}}(\mathbb{R}^n)$,*

$$\left\| \mathcal{M}^{((r/s)')}(f) \right\|_{(X^{1/s})'} \le C \, \|f\|_{(X^{1/s})'}.$$

Remark 1.2.34 Here we give several concrete examples of function spaces satisfying Assumption 1.2.33.

(i) Let $p \in (0, \infty)$, $s \in (0, \min\{1, p\})$, and $r \in (\max\{1, p\}, \infty]$. Then, in this
case, as was pointed out in [233, Remark 2.7(a)], Assumption 1.2.33 holds
true with $X := L^p(\mathbb{R}^n)$.
(ii) Let $p \in (0, \infty)$, $\upsilon \in A_\infty(\mathbb{R}^n)$, and $s \in (0, \min\{1, p/q_\upsilon\})$, where q_υ is defined
in (1.58). Then, in this case, as was showed in [29, Subsection 5.4], there exists
an $r \in (1, \infty]$ such that Assumption 1.2.33 holds true with X therein replaced
by $L^p_\upsilon(\mathbb{R}^n)$.
(iii) Let $\vec{p} := (p_1, \ldots, p_n) \in (0, \infty)^n$,

$$s \in (0, \min\{1, p_1, \ldots, p_n\}),$$

and

$$r \in (\max\{1, p_1, \ldots, p_n\}, \infty].$$

Then, in this case, as was pointed out in [29, Subsection 5.2], the mixed-norm
Lebesgue space $L^{\vec{p}}(\mathbb{R}^n)$ satisfies Assumption 1.2.33.
(iv) Let $0 < q \le p < \infty$, $s \in (0, \min\{1, q\})$, and $r \in (\max\{1, p\}, \infty]$. Then, in
this case, as was mentioned in [233, Remark 2.7(e)], Assumption 1.2.33 holds
true with $X := M^p_q(\mathbb{R}^n)$.

(v) Let $p(\cdot) \in C^{\log}(\mathbb{R}^n)$, $s \in (0, \min\{1, p_-\})$, and $r \in (\max\{1, p_+\}, \infty]$, where p_- is the same as in (1.59) and

$$p_+ := \operatorname*{ess\,sup}_{x \in \mathbb{R}^n} p(x). \qquad (1.60)$$

Then, in this case, as was pointed out in [233, Remark 2.7(f)], the variable Lebesgue space $L^{p(\cdot)}(\mathbb{R}^n)$ satisfies Assumption 1.2.33.

(vi) Let p, q, ω, s, and r be the same as in Lemma 1.8.6 below. Then, in this case, as is proved in Lemma 1.8.6, Assumption 1.2.33 holds true with $X := \dot{\mathcal{K}}^{p,q}_{\omega,\mathbf{0}}(\mathbb{R}^n)$.

We point out that, for any $p, q \in (0, \infty]$ and $\omega \in M(\mathbb{R}_+)$, local and global generalized Herz spaces are both quasi-normed linear spaces and, in particular, for any $p, q \in [1, \infty]$ and $\omega \in M(\mathbb{R}_+)$, they are normed linear spaces. Therefore, we next investigate the completeness of these generalized Herz spaces. For this purpose, we need the following lemma which is a part of [53, Theorem 2].

Lemma 1.2.35 *Let $X \subset \mathcal{M}(\mathbb{R}^n)$ be a quasi-normed linear space such that Definition 1.2.13(ii) holds true. If, for any positive increasing Cauchy sequence $\{f_n\}_{n\in\mathbb{N}}$ of X, the function $f := \sup_{n\in\mathbb{N}} f_n$ belongs to X, then X is complete.*

Due to Lemma 1.2.35, we obtain the following conclusion about the completeness of quasi-normed linear spaces.

Proposition 1.2.36 *Let $X \subset \mathcal{M}(\mathbb{R}^n)$ be a quasi-normed linear space satisfying both (ii) and (iii) of Definition 1.2.13. Then X is complete.*

Proof Let $X \subset \mathcal{M}(\mathbb{R}^n)$ be a quasi-normed linear space satisfying both (ii) and (iii) of Definition 1.2.13, and $\{f_n\}_{n\in\mathbb{N}}$ be a positive increasing Cauchy sequence of X. Obviously, the sequence $\{f_n\}_{n\in\mathbb{N}}$ is bounded on X. This, together with Definition 1.2.13(iii), implies that

$$\left\| \sup_{n\in\mathbb{N}} f_n \right\|_X = \left\| \lim_{n\to\infty} f_n \right\|_X = \lim_{n\to\infty} \|f_n\|_X \leq \sup_{n\in\mathbb{N}} \|f_n\|_X < \infty.$$

Thus, the function $f := \sup_{n\in\mathbb{N}} f_n$ belongs to X. Using this and Lemma 1.2.35, we further conclude that X is complete. This finishes the proof of Proposition 1.2.36. $\qquad\square$

The following completeness of ball quasi-Banach function spaces can be deduced immediately from Proposition 1.2.36; we omit the details.

Corollary 1.2.37 *Let $X \subset \mathcal{M}(\mathbb{R}^n)$ be a ball quasi-Banach function space. Then X is complete.*

Via Proposition 1.2.36, we now show that local and global generalized Herz spaces are both complete and hence are quasi-Banach spaces.

Theorem 1.2.38 *Let $p, q \in (0, \infty]$ and $\omega \in M(\mathbb{R}_+)$. Then the local generalized Herz space $\dot{\mathcal{K}}^{p,q}_{\omega,0}(\mathbb{R}^n)$ is a quasi-Banach space.*

Proof Let $p, q \in (0, \infty]$ and $\omega \in M(\mathbb{R}_+)$. Obviously, $\dot{\mathcal{K}}^{p,q}_{\omega,0}(\mathbb{R}^n)$ is a quasi-normed linear space. Moreover, it is easy to show that the conditions (i) and (ii) of Definition 1.2.13 hold true and Definition 1.2.13(iii) is a simple corollary of the monotone convergence theorem. This, combined with Proposition 1.2.36, further implies that the local generalized Herz space $\dot{\mathcal{K}}^{p,q}_{\omega,0}(\mathbb{R}^n)$ is complete and hence finishes the proof of Theorem 1.2.38.   □

Theorem 1.2.39 *Let $p, q \in (0, \infty]$ and $\omega \in M(\mathbb{R}_+)$. Then the global generalized Herz space $\dot{\mathcal{K}}^{p,q}_{\omega}(\mathbb{R}^n)$ is a quasi-Banach space.*

Proof Let $p, q \in (0, \infty]$ and $\omega \in M(\mathbb{R}_+)$. We now show that the Herz space $\dot{\mathcal{K}}^{p,q}_{\omega}(\mathbb{R}^n)$ satisfies (i), (ii), and (iii) of Definition 1.2.13. Indeed, the global generalized Herz space $\dot{\mathcal{K}}^{p,q}_{\omega}(\mathbb{R}^n)$ satisfies both (i) and (ii) of Definition 1.2.13 obviously. Next, we prove that Definition 1.2.13(iii) holds true for $\dot{\mathcal{K}}^{p,q}_{\omega}(\mathbb{R}^n)$. To this end, assume that $f \in \mathscr{M}(\mathbb{R}^n)$ and $\{f_m\}_{m \in \mathbb{N}} \subset \mathscr{M}(\mathbb{R}^n)$ satisfies that $0 \leq f_m \uparrow f$ almost everywhere as $m \to \infty$. Then, for any given $\alpha \in (0, \|f\|_{\dot{\mathcal{K}}^{p,q}_{\omega}(\mathbb{R}^n)})$, by the definition of $\| \cdot \|_{\dot{\mathcal{K}}^{p,q}_{\omega}(\mathbb{R}^n)}$, we find that there exists a $\xi \in \mathbb{R}^n$ such that

$$\|f(\cdot + \xi)\|_{\dot{\mathcal{K}}^{p,q}_{\omega,0}(\mathbb{R}^n)} > \alpha.$$

Applying this and the monotone convergence theorem, we find that there exists an $N \in \mathbb{N}$ such that, for any $m \in \mathbb{N} \cap (N, \infty)$,

$$\|f_m(\cdot + \xi)\|_{\dot{\mathcal{K}}^{p,q}_{\omega,0}(\mathbb{R}^n)} > \alpha,$$

which, together with the definition of $\| \cdot \|_{\dot{\mathcal{K}}^{p,q}_{\omega}(\mathbb{R}^n)}$, further implies that, for any $m \in \mathbb{N} \cap (N, \infty)$,

$$\alpha < \sup_{\xi \in \mathbb{R}^n} \|f_m(\cdot + \xi)\|_{\dot{\mathcal{K}}^{p,q}_{\omega,0}(\mathbb{R}^n)} = \|f_m\|_{\dot{\mathcal{K}}^{p,q}_{\omega}(\mathbb{R}^n)}.$$

Therefore, we conclude that

$$\|f_m\|_{\dot{\mathcal{K}}^{p,q}_{\omega}(\mathbb{R}^n)} \uparrow \|f\|_{\dot{\mathcal{K}}^{p,q}_{\omega}(\mathbb{R}^n)}$$

as $m \to \infty$ and hence Definition 1.2.13(iii) holds true for $\dot{\mathcal{K}}^{p,q}_{\omega}(\mathbb{R}^n)$. From this and Proposition 1.2.36, we deduce that the global generalized Herz space $\dot{\mathcal{K}}^{p,q}_{\omega}(\mathbb{R}^n)$ is complete. This finishes the proof of Theorem 1.2.39.   □

However, the following two examples show that local and global generalized Herz spaces may not be a ball quasi-Banach function space. Namely, they may not satisfy Definition 1.2.13(iv).

Example 1.2.40 Let $p, q \in (0, \infty]$, $\alpha \in (-\infty, -\frac{n}{p}]$, and

$$\omega_\alpha(t) := t^\alpha \ln\left(e + \frac{1}{t}\right)$$

for any $t \in (0, \infty)$. Then $\mathbf{1}_{B(0,1)} \notin \dot{\mathcal{K}}^{p,q}_{\omega_\alpha,0}(\mathbb{R}^n)$ and $\mathbf{1}_{B(0,1)} \notin \dot{\mathcal{K}}^{p,q}_{\omega_\alpha}(\mathbb{R}^n)$, which imply that the local generalized Herz space $\dot{\mathcal{K}}^{p,q}_{\omega_\alpha,0}(\mathbb{R}^n)$ and the global generalized Herz space $\dot{\mathcal{K}}^{p,q}_{\omega_\alpha}(\mathbb{R}^n)$ are not ball quasi-Banach function spaces.

Proof Let all the symbols be the same as in the present example. Then, from the definition of $\|\cdot\|_{\dot{\mathcal{K}}^{p,q}_{\omega_\alpha,0}(\mathbb{R}^n)}$, it follows that

$$\left\|\mathbf{1}_{B(0,1)}\right\|_{\dot{\mathcal{K}}^{p,q}_{\omega_\alpha,0}(\mathbb{R}^n)}$$

$$= \left[\sum_{k\in\mathbb{Z}} 2^{k\alpha q}\left[\ln\left(e + 2^{-k}\right)\right]^q \left\|\mathbf{1}_{B(0,1)}\mathbf{1}_{B(0,2^k)\setminus B(0,2^{k-1})}\right\|^q_{L^p(\mathbb{R}^n)}\right]^{\frac{1}{q}}$$

$$= \left[\sum_{k\in\mathbb{Z}\setminus\mathbb{N}} 2^{k\alpha q}\left[\ln\left(e + 2^{-k}\right)\right]^q \left\|\mathbf{1}_{B(0,2^k)\setminus B(0,2^{k-1})}\right\|^q_{L^p(\mathbb{R}^n)}\right]^{\frac{1}{q}}$$

$$\sim \left\{\sum_{k\in\mathbb{Z}\setminus\mathbb{N}} 2^{kq(\alpha+\frac{n}{p})}\left[\ln\left(e + 2^{-k}\right)\right]^q\right\}^{\frac{1}{q}}.$$

Applying this and the assumption $\alpha \in (-\infty, -\frac{n}{p}]$, we conclude that

$$\left\|\mathbf{1}_{B(0,1)}\right\|_{\dot{\mathcal{K}}^{p,q}_{\omega_\alpha,0}(\mathbb{R}^n)} = \infty, \tag{1.61}$$

which implies that $\mathbf{1}_{B(0,1)} \notin \dot{\mathcal{K}}^{p,q}_{\omega_\alpha,0}(\mathbb{R}^n)$. Thus, the local generalized Herz space $\dot{\mathcal{K}}^{p,q}_{\omega_\alpha,0}(\mathbb{R}^n)$ is not a BQBF space.

In addition, by (1.61) and the definition of $\|\cdot\|_{\dot{\mathcal{K}}^{p,q}_{\omega_\alpha}(\mathbb{R}^n)}$, we find that

$$\left\|\mathbf{1}_{B(0,1)}\right\|_{\dot{\mathcal{K}}^{p,q}_{\omega_\alpha}(\mathbb{R}^n)} \geq \left\|\mathbf{1}_{B(0,1)}\right\|_{\dot{\mathcal{K}}^{p,q}_{\omega_\alpha,0}(\mathbb{R}^n)} = \infty.$$

Therefore, we conclude that $\mathbf{1}_B \notin \dot{\mathcal{K}}^{p,q}_{\omega_\alpha}(\mathbb{R}^n)$ and hence the global generalized Herz space $\dot{\mathcal{K}}^{p,q}_{\omega_\alpha}(\mathbb{R}^n)$ is also not a BQBF space. This finishes the proof of Example 1.2.40. $\qquad\square$

Example 1.2.41 Let $p, q \in (0, \infty], \alpha_1 \in \mathbb{R}, \alpha_2 \in [0, \infty)$, and

$$\omega(t) := \begin{cases} t^{\alpha_1}(1 - \ln t) & \text{when } t \in (0, 1], \\ t^{\alpha_2}(1 + \ln t) & \text{when } t \in (1, \infty). \end{cases}$$

Then $\mathbf{1}_{B(0,1)} \notin \dot{\mathcal{K}}_\omega^{p,q}(\mathbb{R}^n)$ and hence, in this case, the global generalized Herz space $\dot{\mathcal{K}}_\omega^{p,q}(\mathbb{R}^n)$ is not a ball quasi-Banach function space.

Proof Let all the symbols be as in the present example and, for any $k \in \mathbb{N}$, let $\xi_k \in \mathbb{R}^n$ satisfy $|\xi_k| = 2^k + 1$. Then we claim that, for any $k \in \mathbb{N}$,

$$B(0, 1) \subset B(\xi_k, 2^{k+1}) \setminus B(\xi_k, 2^k).$$

Indeed, for any $y \in B(0, 1)$, we have

$$|y - \xi_k| \leq |y| + |\xi_k| < 1 + |\xi_k| \leq 2^{k+1}$$

and

$$|y - \xi_k| \geq |\xi_k| - |y| > |\xi_k| - 1 = 2^k.$$

These imply that $y \in B(\xi_k, 2^{k+1}) \setminus B(\xi_k, 2^k)$ and hence the above claim holds true. From this and the definition of $\| \cdot \|_{\dot{\mathcal{K}}_\omega^{p,q}(\mathbb{R}^n)}$, it follows that

$$\left\| \mathbf{1}_{B(0,1)} \right\|_{\dot{\mathcal{K}}_\omega^{p,q}(\mathbb{R}^n)} \geq \left\| \mathbf{1}_{B(0,1)}(\cdot + \xi_k) \right\|_{\dot{\mathcal{K}}_{\omega,0}^{p,q}(\mathbb{R}^n)}$$
$$= \omega(2^{k+1}) \left\| \mathbf{1}_{B(0,1)} \right\|_{L^p(\mathbb{R}^n)}$$
$$\sim 2^{(k+1)\alpha_2} [1 + (k+1)\ln 2] \to \infty$$

as $k \to \infty$. Thus, $\mathbf{1}_{B(0,1)} \notin \dot{\mathcal{K}}_\omega^{p,q}(\mathbb{R}^n)$. This further implies that $\dot{\mathcal{K}}_\omega^{p,q}(\mathbb{R}^n)$ is not a BQBF space and hence finishes the proof of Example 1.2.41. □

The following two theorems show that local and global generalized Herz spaces are ball quasi-Banach function spaces under some additional and sharp assumptions of exponent $\omega \in M(\mathbb{R}_+)$.

Theorem 1.2.42 *Let $p, q \in (0, \infty]$ and $\omega \in M(\mathbb{R}_+)$ with $m_0(\omega) \in (-\frac{n}{p}, \infty)$. Then the local generalized Herz space $\dot{\mathcal{K}}_{\omega,0}^{p,q}(\mathbb{R}^n)$ is a ball quasi-Banach function space.*

Proof Let all the symbols be the same as in the present theorem. Then, by the proof of Theorem 1.2.38, we know that the local generalized Herz space $\dot{\mathcal{K}}^{p,q}_{\omega,0}(\mathbb{R}^n)$ is a quasi-Banach space satisfying (i), (ii), and (iii) of Definition 1.2.13. Therefore, to prove the present theorem, it suffices to show that Definition 1.2.13(iv) holds true for $\dot{\mathcal{K}}^{p,q}_{\omega,0}(\mathbb{R}^n)$.

For this purpose, let $B(x_0, r) \in \mathbb{B}$ with $x_0 \in \mathbb{R}^n$ and $r \in (0, \infty)$. Then we have

$$\left\| \mathbf{1}_{B(x_0,r)} \right\|_{\dot{\mathcal{K}}^{p,q}_{\omega,0}(\mathbb{R}^n)}$$

$$= \left\{ \sum_{k \in \mathbb{Z}} \left[\omega(2^k) \right]^q \left\| \mathbf{1}_{B(x_0,r)} \mathbf{1}_{B(0,2^k) \setminus B(0,2^{k-1})} \right\|^q_{L^p(\mathbb{R}^n)} \right\}^{\frac{1}{q}}$$

$$\sim \left\{ \sum_{k \in \mathbb{Z} \setminus \mathbb{N}} \left[\omega(2^k) \right]^q \left\| \mathbf{1}_{B(x_0,r)} \mathbf{1}_{B(0,2^k) \setminus B(0,2^{k-1})} \right\|^q_{L^p(\mathbb{R}^n)} \right\}^{\frac{1}{q}} + \left(\sum_{k \in \mathbb{N}} \cdots \right)^{\frac{1}{q}}$$

$$=: \mathrm{I}_1 + \mathrm{I}_2. \tag{1.62}$$

For I_1, from Lemma 1.1.12, it follows that, for any $k \in \mathbb{Z} \setminus \mathbb{N}$, $\omega(2^k) \lesssim 2^{k[m_0(\omega)-\varepsilon]}$, where $\varepsilon \in (0, m_0(\omega) + \frac{n}{p})$ is a fixed positive constant. By this and the assumption $m_0(\omega) + \frac{n}{p} - \varepsilon \in (0, \infty)$, we conclude that

$$\mathrm{I}_1 \lesssim \left\{ \sum_{k \in \mathbb{Z} \setminus \mathbb{N}} 2^{kq[m_0(\omega)-\varepsilon]} \left\| \mathbf{1}_{B(x_0,r)} \mathbf{1}_{B(0,2^k) \setminus B(0,2^{k-1})} \right\|^q_{L^p(\mathbb{R}^n)} \right\}^{\frac{1}{q}}$$

$$\lesssim \left\{ \sum_{k \in \mathbb{Z} \setminus \mathbb{N}} 2^{kq[m_0(\omega)+\frac{n}{p}-\varepsilon]} \right\} < \infty. \tag{1.63}$$

Now, we deal with I_2. To this end, we first claim that, for any $k \in \mathbb{N} \cap (\ln(r + |x_0|)/\ln 2 + 1, \infty)$,

$$B(x_0, r) \cap \left[B(0, 2^k) \setminus B(0, 2^{k-1}) \right] = \emptyset.$$

Indeed, from the assumption $k > \ln(r + |x_0|)/\ln 2 + 1$, we deduce that $2^{k-1} > r + |x_0|$. By this, we find that, for any $x \in B(x_0, r)$,

$$|x| \leq |x - x_0| + |x_0| < r + |x_0| < 2^{k-1},$$

which implies that $x \in B(\mathbf{0}, 2^{k-1})$ and hence $B(x_0, r) \subset B(\mathbf{0}, 2^{k-1})$. Therefore, in this case,

$$B(x_0, r) \cap \left[B(\mathbf{0}, 2^k) \setminus B(\mathbf{0}, 2^{k-1}) \right] = \emptyset,$$

which completes the proof of the above claim. Applying this claim, we conclude that

$$I_2 \sim \left\{ \sum_{k \in \mathbb{N} \cap [1, \frac{\ln(r+|x_0|)}{\ln 2}+1]} \left[\omega(2^k) \right]^q \left\| \mathbf{1}_{B(x_0,r)} \mathbf{1}_{B(\mathbf{0},2^k) \setminus B(\mathbf{0},2^{k-1})} \right\|^q_{L^p(\mathbb{R}^n)} \right\}^{\frac{1}{q}} < \infty,$$

which, together with (1.62) and (1.63), implies that $\mathbf{1}_{B(x_0,r)} \in \dot{\mathcal{K}}^{p,q}_{\omega,\mathbf{0}}(\mathbb{R}^n)$. Thus, Definition 1.2.13(iv) holds true for $\dot{\mathcal{K}}^{p,q}_{\omega,\mathbf{0}}(\mathbb{R}^n)$, which completes the proof of Theorem 1.2.42. □

Remark 1.2.43 We should point out that the assumptions of Theorem 1.2.42 are sharp for local generalized Herz spaces being ball quasi-Banach function spaces. Precisely, if p, q, and ω do not satisfy the assumptions of Theorem 1.2.42, namely, $p, q \in (0, \infty]$ and $\omega \in M(\mathbb{R}_+)$ are such that $m_0(\omega) \in (-\infty, -\frac{n}{p}]$, then the local generalized Herz space $\dot{\mathcal{K}}^{p,q}_{\omega,\mathbf{0}}(\mathbb{R}^n)$ may not be a ball quasi-Banach function space. Indeed, let $p, q \in (0, \infty]$, $\alpha \in (-\infty, -\frac{n}{p}]$, and ω_α be the same as in Example 1.2.40. Then, similarly to Example 1.1.8, we have $\omega_\alpha \in M(\mathbb{R}_+)$ and $m_0(\omega_\alpha) = \alpha \in (-\infty, -\frac{n}{p}]$. Moreover, as is proved in Example 1.2.40, the local generalized Herz space $\dot{\mathcal{K}}^{p,q}_{\omega_\alpha}(\mathbb{R}^n)$ is not a ball quasi-Banach function space. Therefore, in this sense, the assumptions of Theorem 1.2.42 are sharp.

Theorem 1.2.44 *Let $p, q \in (0, \infty]$ and $\omega \in M(\mathbb{R}_+)$ with $m_0(\omega) \in (-\frac{n}{p}, \infty)$ and $M_\infty(\omega) \in (-\infty, 0)$. Then the global generalized Herz space $\dot{\mathcal{K}}^{p,q}_{\omega}(\mathbb{R}^n)$ is a ball quasi-Banach function space.*

Proof Let all the symbols be the same as in the present theorem. Then, from the proof of Theorem 1.2.39, it follows that the global generalized Herz space $\dot{\mathcal{K}}^{p,q}_{\omega}(\mathbb{R}^n)$ is a quasi-Banach space satisfying (i), (ii), and (iii) of Definition 1.2.13.

Next, we prove that $\dot{\mathcal{K}}^{p,q}_{\omega}(\mathbb{R}^n)$ satisfies Definition 1.2.13(iv). For this purpose, let $B(x_0, r) \in \mathbb{B}$ with $x_0 \in \mathbb{R}^n$ and $r \in (0, \infty)$. In addition, by Lemma 1.1.12, we conclude that, for any $k \in \mathbb{Z} \setminus \mathbb{N}$,

$$\omega(2^k) \lesssim 2^{k[m_0(\omega)-\varepsilon]}$$

and, for any $k \in \mathbb{N}$,

$$\omega(2^k) \lesssim 2^{k[M_\infty(\omega)+\varepsilon]},$$

where $\varepsilon \in (0, \min\{m_0(\omega) + \frac{n}{p}, -M_\infty(\omega)\})$ is a fixed positive constant. Then, for any $\xi \in \mathbb{R}^n$, we have

$$\left\| \mathbf{1}_{B(x_0,r)}(\cdot + \xi) \right\|_{\dot{\mathcal{K}}^{p,q}_{\omega,\mathbf{0}}(\mathbb{R}^n)}$$

$$= \left\{ \sum_{k \in \mathbb{Z}} \left[\omega(2^k) \right]^q \left\| \mathbf{1}_{B(x_0,r)} \mathbf{1}_{B(\xi,2^k) \setminus B(\xi,2^{k-1})} \right\|^q_{L^p(\mathbb{R}^n)} \right\}^{\frac{1}{q}}$$

$$\sim \left\{ \sum_{k \in \mathbb{Z} \setminus \mathbb{N}} \left[\omega(2^k) \right]^q \left\| \mathbf{1}_{B(x_0,r)} \mathbf{1}_{B(\xi,2^k) \setminus B(\xi,2^{k-1})} \right\|^q_{L^p(\mathbb{R}^n)} \right\}^{\frac{1}{q}} + \left\{ \sum_{k \in \mathbb{N}} \cdots \right\}^{\frac{1}{q}}$$

$$=: \mathrm{I}_{\xi,1} + \mathrm{I}_{\xi,2}. \qquad (1.64)$$

We first deal with $\mathrm{I}_{\xi,1}$. Indeed, applying an argument similar to that used in the estimation of (1.63), we find that, for any $\xi \in \mathbb{R}^n$,

$$\mathrm{I}_{\xi,1} \lesssim \sum_{k \in \mathbb{Z} \setminus \mathbb{N}} 2^{kq[m_0(\omega)+\frac{n}{p}-\varepsilon]} < \infty. \qquad (1.65)$$

For $\mathrm{I}_{\xi,2}$, from the assumption $\omega(2^k) \lesssim 2^{k[M_\infty(\omega)+\varepsilon]}$ for any $k \in \mathbb{Z}$, and the assumption $M_\infty(\omega) + \varepsilon \in (-\infty, 0)$, it follows that, for any $\xi \in \mathbb{R}^n$,

$$\mathrm{I}_{\xi,2} \lesssim \left\{ \sum_{k \in \mathbb{N}} \left[\omega(2^k) \right]^q \left\| \mathbf{1}_{B(x_0,r)} \right\|^q_{L^p(\mathbb{R}^n)} \right\}^{\frac{1}{q}}$$

$$\lesssim \left\{ \sum_{k \in \mathbb{N}} 2^{kq[M_\infty(\omega)+\varepsilon]} \right\}^{\frac{1}{q}} < \infty, \qquad (1.66)$$

which, combined with (1.64) and (1.65), further implies that, for any $\xi \in \mathbb{R}^n$,

$$\left\| \mathbf{1}_{B(x_0,r)}(\cdot + \xi) \right\|_{\dot{\mathcal{K}}^{p,q}_{\omega,\mathbf{0}}(\mathbb{R}^n)}$$

$$\lesssim \left\{ \sum_{k \in \mathbb{Z} \setminus \mathbb{N}} 2^{kq[m_0(\omega)+\frac{n}{p}-\varepsilon]} \right\}^{\frac{1}{q}} + \left\{ \sum_{k \in \mathbb{N}} 2^{kq[M_\infty(\omega)+\varepsilon]} \right\}^{\frac{1}{q}} < \infty.$$

This implies that $\|\mathbf{1}_{B(x_0,r)}\|_{\dot{\mathcal{K}}_{\omega}^{p,q}(\mathbb{R}^n)} < \infty$ and hence $\mathbf{1}_{B(x_0,r)} \in \dot{\mathcal{K}}_{\omega}^{p,q}(\mathbb{R}^n)$. Thus, Definition 1.2.13(iv) also holds true, which completes the proof of Theorem 1.2.44. □

Remark 1.2.45 Applying Examples 1.1.9 and 1.2.41 and Remark 1.2.43, we conclude that the assumptions $m_0(\omega) \in (-\frac{n}{p}, \infty)$ and $M_\infty(\omega) \in (-\infty, 0)$ in Theorem 1.2.44 are sharp.

Moreover, by the following conclusion, we know that the local Herz spaces are ball Banach function spaces when $p, q \in [1, \infty]$ and $\omega \in M(\mathbb{R}_+)$ satisfy some reasonable and sharp assumptions.

Theorem 1.2.46 *Let $p, q \in [1, \infty]$ and $\omega \in M(\mathbb{R}_+)$ satisfy*

$$-\frac{n}{p} < m_0(\omega) \le M_0(\omega) < \frac{n}{p'},$$

where $\frac{1}{p} + \frac{1}{p'} = 1$. Then the local generalized Herz space $\dot{\mathcal{K}}_{\omega,\mathbf{0}}^{p,q}(\mathbb{R}^n)$ is a ball Banach function space.

Proof Let all the symbols be as in the present theorem. Then, applying Theorem 1.2.42, we know that the local generalized Herz space $\dot{\mathcal{K}}_{\omega,\mathbf{0}}^{p,q}(\mathbb{R}^n)$ is a BQBF space. In addition, notice that the quasi-norm $\|\cdot\|_{\dot{\mathcal{K}}_{\omega,\mathbf{0}}^{p,q}(\mathbb{R}^n)}$ satisfies the triangle inequality when $p, q \in [1, \infty]$. Thus, to complete the whole proof, it suffices to show that, for any $B \in \mathbb{B}$, (1.39) holds true with X replaced by $\dot{\mathcal{K}}_{\omega,\mathbf{0}}^{p,q}(\mathbb{R}^n)$. To this end, let $B(x_0, r) \in \mathbb{B}$ with $x_0 \in \mathbb{R}^n$ and $r \in (0, \infty)$. Then, from the Hölder inequality, it follows that, for any $f \in \dot{\mathcal{K}}_{\omega,\mathbf{0}}^{p,q}(\mathbb{R}^n)$,

$$\int_{B(x_0,r)} |f(y)|\, dy$$
$$= \int_{\mathbb{R}^n} |f(y)\mathbf{1}_{B(x_0,r)}(y)|\, dy$$
$$= \sum_{k\in\mathbb{Z}} \int_{B(0,2^k)\setminus B(0,2^{k-1})} |f(y)\mathbf{1}_{B(x_0,r)}(y)|\, dy$$
$$\le \sum_{k\in\mathbb{Z}} \|f\mathbf{1}_{B(0,2^k)\setminus B(0,2^{k-1})}\|_{L^p(\mathbb{R}^n)} \|\mathbf{1}_{B(x_0,r)}\mathbf{1}_{B(0,2^k)\setminus B(0,2^{k-1})}\|_{L^{p'}(\mathbb{R}^n)}$$
$$\le \left\{\sum_{k\in\mathbb{Z}} \left[\omega(2^k)\right]^q \|f\mathbf{1}_{B(0,2^k)\setminus B(0,2^{k-1})}\|_{L^p(\mathbb{R}^n)}^q\right\}^{\frac{1}{q}}$$

$$\times \left\{ \sum_{k \in \mathbb{Z}} \left[\omega(2^k) \right]^{-q'} \left\| \mathbf{1}_{B(x_0,r)} \mathbf{1}_{B(0,2^k) \setminus B(0,2^{k-1})} \right\|_{L^{p'}(\mathbb{R}^n)}^{q'} \right\}^{\frac{1}{q'}}$$

$$= \| f \|_{\dot{\mathcal{K}}^{p,q}_{\omega,0}(\mathbb{R}^n)} \| \mathbf{1}_{B(x_0,r)} \|_{\dot{\mathcal{K}}^{p',q'}_{1/\omega,0}(\mathbb{R}^n)}. \tag{1.67}$$

In addition, by Lemma 1.1.6, we find that

$$m_0 \left(\frac{1}{\omega} \right) = -M_0(\omega) > -\frac{n}{p'}.$$

This, together with Theorem 1.2.42, implies that the local generalized Herz space $\dot{\mathcal{K}}^{p',q'}_{1/\omega,0}(\mathbb{R}^n)$ is a BQBF space, and hence $\mathbf{1}_{B(x_0,r)} \in \dot{\mathcal{K}}^{p',q'}_{1/\omega,0}(\mathbb{R}^n)$. Using this and (1.67), we have, for any $f \in \dot{\mathcal{K}}^{p,q}_{\omega,0}(\mathbb{R}^n)$,

$$\int_{B(x_0,r)} |f(y)| \, dy \lesssim \| f \|_{\dot{\mathcal{K}}^{p,q}_{\omega,0}(\mathbb{R}^n)},$$

which completes the proof that (1.39) holds true with $X := \dot{\mathcal{K}}^{p,q}_{\omega,0}(\mathbb{R}^n)$ and $B := B(x_0, r)$ and hence Theorem 1.2.46. $\qquad\square$

The following examples show that the assumptions in Theorem 1.2.46 are sharp.

Example 1.2.47 Let $p, q \in [1, \infty]$ and $\alpha \in [\frac{n}{p'}, \infty)$, where $\frac{1}{p} + \frac{1}{p'} = 1$. For any $t \in (0, \infty)$, let $\omega(t) := t^\alpha$.

(i) If $\alpha \in (\frac{n}{p'}, \infty)$, then the local generalized Herz space $\dot{\mathcal{K}}^{p,q}_{\omega,0}(\mathbb{R}^n)$ is not a ball Banach function space.

(ii) If $p = q = 1$ and $\alpha = \frac{n}{p'} = 0$, then the local generalized Herz space $\dot{\mathcal{K}}^{p,q}_{\omega,0}(\mathbb{R}^n)$ coincides with $L^1(\mathbb{R}^n)$ and hence is a ball Banach function space.

(iii) If $p = q > 1$ and $\alpha = \frac{n}{p'}$, then the local generalized Herz space $\dot{\mathcal{K}}^{p,q}_{\omega,0}(\mathbb{R}^n)$ is not a ball Banach function space.

Proof Let $p, q \in [1, \infty]$ and $\omega(t) := t^\alpha$ for any $t \in (0, \infty)$ and any given $\alpha \in [\frac{n}{p'}, \infty)$, where $\frac{1}{p} + \frac{1}{p'} = 1$. We first show (i). Indeed, for any $x \in \mathbb{R}^n$, let

$$f(x) := \frac{1}{|x|^n} \mathbf{1}_{B(0,1) \setminus \{0\}}.$$

Then, for any $k \in \mathbb{Z} \setminus \mathbb{N}$ and $x \in B(\xi, 2^k) \setminus B(\xi, 2^{k-1})$, we have $f(x) \sim 2^{-nk}$. Using this and the assumption $\alpha \in (\frac{n}{p'}, \infty)$, we conclude that

$$\|f\|_{\dot{\mathcal{K}}^{p,q}_{\omega,\mathbf{0}}(\mathbb{R}^n)} = \left[\sum_{k \in \mathbb{Z}} 2^{k\alpha q} \left\| f \mathbf{1}_{B(\mathbf{0},2^k) \setminus B(\mathbf{0},2^{k-1})} \right\|^q_{L^p(\mathbb{R}^n)} \right]^{\frac{1}{q}}$$

$$\sim \left[\sum_{k \in \mathbb{Z} \setminus \mathbb{N}} 2^{k\alpha q} 2^{-nkq} \left| B(\mathbf{0}, 2^k) \setminus B(\mathbf{0}, 2^{k-1}) \right|^{\frac{q}{p}} \right]^{\frac{1}{q}}$$

$$\sim \left[\sum_{k \in \mathbb{Z} \setminus \mathbb{N}} 2^{kq(\alpha - \frac{n}{p'})} \right]^{\frac{1}{q}} \sim 1,$$

which implies that $f \in \dot{\mathcal{K}}^{p,q}_{\omega,\mathbf{0}}(\mathbb{R}^n)$. However, we have

$$\int_{B(\mathbf{0},1)} |f(y)| \, dy = \int_{B(\mathbf{0},1)} \frac{1}{|y|^n} \, dy = \int_0^1 r^{-1} \, dr \int_{\mathbb{S}^{n-1}} d\sigma(x) = \infty.$$

Therefore, in this case, the local generalized Herz space $\dot{\mathcal{K}}^{p,q}_{\omega,\mathbf{0}}(\mathbb{R}^n)$ does not satisfy (1.39) and hence $\dot{\mathcal{K}}^{p,q}_{\omega,\mathbf{0}}(\mathbb{R}^n)$ is not a BBF space. This implies that (i) holds true.

We next prove (ii). Indeed, from the assumption $p = 1$, it follows that $\alpha = \frac{n}{p'} = 0$. This further implies that, for any $f \in \mathscr{M}(\mathbb{R}^n)$,

$$\|f\|_{\dot{\mathcal{K}}^{p,q}_{\omega,\mathbf{0}}(\mathbb{R}^n)} = \sum_{k \in \mathbb{Z}} \left\| f \mathbf{1}_{B(\mathbf{0},2^k) \setminus B(\xi,2^{k-1})} \right\|_{L^1(\mathbb{R}^n)} = \|f\|_{L^1(\mathbb{R}^n)}.$$

Thus, in this case, it holds true that $\dot{\mathcal{K}}^{p,q}_{\omega,\mathbf{0}}(\mathbb{R}^n) = L^1(\mathbb{R}^n)$ and hence $\dot{\mathcal{K}}^{p,q}_{\omega,\mathbf{0}}(\mathbb{R}^n)$ is a BBF space, which completes the proof of (ii).

Now, we show (iii). To this end, for any given $\beta \in (0, n)$ and for any $x \in \mathbb{R}^n$, let

$$f_\beta(x) := \frac{1}{|x|^\beta} \mathbf{1}_{B(\mathbf{0},1) \setminus \{\mathbf{0}\}}.$$

Then, for any given $\beta \in (0, n)$ and for any $k \in \mathbb{Z} \setminus \mathbb{N}$ and $x \in B(0, 2^k) \setminus B(0, 2^{k-1})$, we have $|x| \sim 2^{\frac{np}{p'}}$. Using this, we find that, for any $\beta \in (0, n)$,

$$
\begin{aligned}
\left\| f_\beta \right\|_{\dot{\mathcal{K}}^{p,q}_{\omega,0}(\mathbb{R}^n)} &= \left[\sum_{k \in \mathbb{Z}} 2^{k\alpha p} \left\| f_\beta \mathbf{1}_{B(0,2^k)\setminus B(0,2^{k-1})} \right\|_{L^p}^p \right]^{\frac{1}{p}} \\
&\sim \left[\sum_{k \in \mathbb{Z}\setminus\mathbb{N}} |x|^{\frac{np}{p'}} \int_{B(0,2^k)\setminus B(0,2^{k-1})} |f_\beta(y)|^p \, dy \right]^{\frac{1}{p}} \\
&\sim \left[\int_{B(0,1)} |x|^{p(\frac{n}{p'}-\beta)} \, dy \right]^{\frac{1}{p}} \\
&\sim \left[\int_0^1 r^{p(n-\beta)-1} \, dr \int_{\mathbb{S}^{n-1}} d\sigma(x) \right]^{\frac{1}{p}} \\
&\sim \frac{1}{(n-\beta)^{\frac{1}{p}}}.
\end{aligned}
\tag{1.68}
$$

On another hand, for any $\beta \in (0, n)$, we have

$$
\begin{aligned}
\int_{B(0,1)} |f_\beta(y)| \, dy &= \int_{B(0,1)} \frac{1}{|x|^\beta} \, dy = \int_0^1 r^{n-\beta-1} \, dr \int_{\mathbb{S}^{n-1}} d\sigma(x) \\
&\sim \frac{1}{n-\beta}.
\end{aligned}
$$

Combining this and (1.68), we further conclude that

$$
\frac{\int_{B(0,1)} |f_\beta(y)| \, dy}{\left\| f_\beta \right\|_{\dot{\mathcal{K}}^{p,q}_{\omega,0}(\mathbb{R}^n)}} \sim (n-\beta)^{\frac{1}{p}-1} \to \infty
$$

as $\beta \to n$. This implies that, in this case, (1.39) does not hold true for the local generalized Herz space $\dot{\mathcal{K}}^{p,q}_{\omega,0}(\mathbb{R}^n)$, and hence $\dot{\mathcal{K}}^{p,q}_{\omega,0}(\mathbb{R}^n)$ is not a BBF space. Thus, we complete the proof of (iii) and hence Example 1.2.47. $\qquad \square$

Finally, we show that, under some reasonable and sharp assumptions, the global generalized Herz space $\dot{\mathcal{K}}^{p,q}_{\omega}(\mathbb{R}^n)$ is a ball Banach function space as follows.

Theorem 1.2.48 *Let $p, q \in [1, \infty]$ and $\omega \in M(\mathbb{R}_+)$ satisfy $m_0(\omega) \in (-\frac{n}{p}, \infty)$ and $M_\infty(\omega) \in (-\infty, 0)$. Then the global generalized Herz space $\dot{\mathcal{K}}^{p,q}_{\omega}(\mathbb{R}^n)$ is a ball Banach function space.*

Proof Let all the symbols be as in the present theorem. Then, by Theorem 1.2.44, we know that the global generalized Herz space $\dot{\mathcal{K}}_{\omega}^{p,q}(\mathbb{R}^n)$ is a BQBF space. Moreover, we can easily show that the quasi-norm $\|\cdot\|_{\dot{\mathcal{K}}_{\omega}^{p,q}(\mathbb{R}^n)}$ satisfies the triangle inequality because $p, q \in [1, \infty)$. Therefore, to finish the proof of the present theorem, we only need to prove that, for any $B \in \mathbb{B}$, (1.39) holds true with X replaced by $\dot{\mathcal{K}}_{\omega}^{p,q}(\mathbb{R}^n)$.

To this end, let $B(x_0, r) \in \mathbb{B}$ with $x_0 \in \mathbb{R}^n$, $k \in \mathbb{Z}$, and $r \in [2^{k-1}, 2^k)$. Assume $\xi \in \mathbb{R}^n$ satisfies that $|x_0 - \xi| = 3 \cdot 2^k$. Then we claim that

$$B(x_0, r) \subset B(\xi, 2^{k+2}) \setminus B(\xi, 2^{k+1}).$$

Indeed, applying the assumption $|x_0 - \xi| = 3 \cdot 2^k$, we find that, for any $y \in B(x_0, r)$,

$$|y - \xi| \leq |y - x_0| + |x_0 - \xi| < r + |x_0 - \xi| < 2^{k+2}$$

and

$$|y - \xi| \geq |x_0 - \xi| - |y - x_0| > |x_0 - \xi| - r > 2^{k+1}.$$

These imply that $B(x_0, r) \subset B(\xi, 2^{k+2}) \setminus B(\xi, 2^{k+1})$ and hence the above claim holds true. By this, the Hölder inequality, Definition 1.2.1(i), and Remark 1.2.2(ii), we find that, for any $f \in \dot{\mathcal{K}}_{\omega}^{p,q}(\mathbb{R}^n)$,

$$\int_{B(x_0,r)} |f(y)| \, dy \leq \int_{B(\xi, 2^{k+2}) \setminus B(\xi, 2^{k+1})} |f(y)| \, dy$$

$$\lesssim \left\| f \mathbf{1}_{B(\xi, 2^{k+2}) \setminus B(\xi, 2^{k+1})} \right\|_{L^p(\mathbb{R}^n)}$$

$$\lesssim \| f(\cdot + \xi) \|_{\dot{\mathcal{K}}_{\omega,\mathbf{0}}^{p,q}(\mathbb{R}^n)} \lesssim \| f \|_{\dot{\mathcal{K}}_{\omega}^{p,q}(\mathbb{R}^n)},$$

where the implicit positive constants depend only on p, ω, and k. Thus, for any $B \in \mathbb{B}$, (1.39) holds true with X replaced by $\dot{\mathcal{K}}_{\omega}^{p,q}(\mathbb{R}^n)$, which completes the proof of Theorem 1.2.48. $\qquad\square$

Remark 1.2.49 From Remark 1.2.45, it follows that Theorem 1.2.48 is sharp.

1.3 Convexities

In this section, we discuss the convexity of local and global generalized Herz spaces. To this end, we first investigate the relation between generalized Herz spaces and their convexification as follows.

Lemma 1.3.1 *Let $p, q, s \in (0, \infty)$ and $\omega \in M(\mathbb{R}_+)$. Then*

$$\left[\dot{\mathcal{K}}_{\omega,\mathbf{0}}^{p,q}(\mathbb{R}^n)\right]^{1/s} = \dot{\mathcal{K}}_{\omega^s,\mathbf{0}}^{p/s,q/s}(\mathbb{R}^n)$$

with the same quasi-norms.

Proof Let $p, q, s \in (0, \infty)$ and $\omega \in M(\mathbb{R}_+)$. By Definition 1.2.28(i) with $X := \dot{\mathcal{K}}_{\omega,\mathbf{0}}^{p,q}(\mathbb{R}^n)$, we conclude that, for any $f \in \mathscr{M}(\mathbb{R}^n)$,

$$\|f\|_{[\dot{\mathcal{K}}_{\omega,\mathbf{0}}^{p,q}(\mathbb{R}^n)]^{1/s}} = \left\||f|^{\frac{1}{s}}\right\|_{\dot{\mathcal{K}}_{\omega,\mathbf{0}}^{p,q}(\mathbb{R}^n)}^{s}$$

$$= \left\{\sum_{k\in\mathbb{Z}}\left[\omega(2^k)\right]^q\left[\int_{2^{k-1}<|y|<2^k}|f(y)|^{\frac{p}{s}}\,dy\right]^{\frac{q}{p}}\right\}^{\frac{s}{q}}$$

$$= \left\{\sum_{k\in\mathbb{Z}}\left[\omega^s(2^k)\right]^{\frac{q}{s}}\left[\int_{2^{k-1}<|y|<2^k}|f(y)|^{\frac{p}{s}}\,dy\right]^{\frac{q/s}{p/s}}\right\}^{\frac{s}{q}}$$

$$= \|f\|_{\dot{\mathcal{K}}_{\omega^s,\mathbf{0}}^{p/s,q/s}(\mathbb{R}^n)}. \tag{1.69}$$

This implies that $[\dot{\mathcal{K}}_{\omega,\mathbf{0}}^{p,q}(\mathbb{R}^n)]^{1/s} = \dot{\mathcal{K}}_{\omega^s,\mathbf{0}}^{p/s,q/s}(\mathbb{R}^n)$ with the same quasi-norms, and hence finishes the proof of Lemma 1.3.1. $\square$

Lemma 1.3.2 *Let $p, q, s \in (0, \infty)$ and $\omega \in M(\mathbb{R}_+)$. Then*

$$\left[\dot{\mathcal{K}}_{\omega}^{p,q}(\mathbb{R}^n)\right]^{1/s} = \dot{\mathcal{K}}_{\omega^s}^{p/s,q/s}(\mathbb{R}^n)$$

with the same quasi-norms.

Proof Let $p, q, s \in (0, \infty)$ and $\omega \in M(\mathbb{R}_+)$. Note that, for any $f \in \mathscr{M}(\mathbb{R}^n)$,

$$\left(\sup_{\xi\in\mathbb{R}^n}\left\{\sum_{k\in\mathbb{Z}}\left[\omega(2^k)\right]^q\left[\int_{2^{k-1}<|y-\xi|<2^k}|f(y)|^{\frac{p}{s}}\,dy\right]^{\frac{q}{p}}\right\}^{\frac{1}{q}}\right)^s$$

$$= \sup_{\xi\in\mathbb{R}^n}\left\{\sum_{k\in\mathbb{Z}}\left[\omega(2^k)\right]^q\left[\int_{2^{k-1}<|y-\xi|<2^k}|f(y)|^{\frac{p}{s}}\,dy\right]^{\frac{q}{p}}\right\}^{\frac{s}{q}}.$$

Applying this, Definition 1.2.28(i) with $X := \dot{\mathcal{K}}_{\omega}^{p,q}(\mathbb{R}^n)$, and some arguments similar to those used in the estimation of (1.69), we find that, for any $f \in \mathscr{M}(\mathbb{R}^n)$,

$$\|f\|_{[\dot{\mathcal{K}}_{\omega}^{p,q}(\mathbb{R}^n)]^{1/s}} = \|f\|_{\dot{\mathcal{K}}_{\omega^s}^{p/s,q/s}(\mathbb{R}^n)},$$

which completes the proof of Lemma 1.3.2. $\square$

The following two results show that local and global generalized Herz spaces are strictly convex under some assumptions of exponents p and q.

Theorem 1.3.3 *Let $p, q \in (0, \infty)$, $s \in (0, \min\{p, q\}]$, and $\omega \in M(\mathbb{R}_+)$. Then the local generalized Herz space $\dot{\mathcal{K}}_{\omega,0}^{p,q}(\mathbb{R}^n)$ is strictly s-convex.*

Proof Let $p, q, s \in (0, \infty)$ satisfy $s \in (0, \min\{p, q\}]$ and $\omega \in M(\mathbb{R}_+)$. Then, for any given sequence $\{f_j\}_{j \in \mathbb{N}}$ of measurable functions in $[\dot{\mathcal{K}}_{\omega,0}^{p,q}(\mathbb{R}^n)]^{1/s}$, from the assumption $s \in (0, \min\{p, q\}]$, we deduce that $p/s, q/s \in [1, \infty)$. This, together with Lemma 1.3.1 and the Minkowski inequality, further implies that, for any $N \in \mathbb{N}$,

$$\left\| \sum_{j=1}^{N} |f_j| \right\|_{[\dot{\mathcal{K}}_{\omega,0}^{p,q}(\mathbb{R}^n)]^{1/s}} \leq \sum_{j=1}^{N} \|f_j\|_{[\dot{\mathcal{K}}_{\omega,0}^{p,q}(\mathbb{R}^n)]^{1/s}} \leq \sum_{j \in \mathbb{N}} \|f_j\|_{[\dot{\mathcal{K}}_{\omega,0}^{p,q}(\mathbb{R}^n)]^{1/s}}.$$

Applying this and the monotone convergence theorem, we conclude that

$$\left\| \sum_{j \in \mathbb{N}} |f_j| \right\|_{[\dot{\mathcal{K}}_{\omega,0}^{p,q}(\mathbb{R}^n)]^{1/s}} \leq \sum_{j \in \mathbb{N}} \|f_j\|_{[\dot{\mathcal{K}}_{\omega,0}^{p,q}(\mathbb{R}^n)]^{1/s}},$$

which implies that the local generalized Herz space $\dot{\mathcal{K}}_{\omega,0}^{p,q}(\mathbb{R}^n)$ is strictly s-convex, and hence completes the proof of Theorem 1.3.3. $\qquad\qquad\square$

Theorem 1.3.4 *Let $p, q \in (0, \infty)$, $s \in (0, \min\{p, q\}]$, and $\omega \in M(\mathbb{R}_+)$. Then the global generalized Herz space $\dot{\mathcal{K}}_{\omega}^{p,q}(\mathbb{R}^n)$ is strictly s-convex.*

Proof Let $p, q, s \in (0, \infty)$ satisfy $s \in (0, \min\{p, q\}]$ and $\omega \in M(\mathbb{R}_+)$. Then, similarly to the proof of Theorem 1.3.3, by Lemma 1.3.2, we find that, for any given sequence $\{f_j\}_{j \in \mathbb{N}}$ of measurable functions in $[\dot{\mathcal{K}}_{\omega}^{p,q}(\mathbb{R}^n)]^{1/s}$,

$$\left\| \sum_{j \in \mathbb{N}} |f_j| \right\|_{[\dot{\mathcal{K}}_{\omega}^{p,q}(\mathbb{R}^n)]^{1/s}} \leq \sum_{j \in \mathbb{N}} \|f_j\|_{[\dot{\mathcal{K}}_{\omega}^{p,q}(\mathbb{R}^n)]^{1/s}}.$$

This implies that the global generalized Herz space $\dot{\mathcal{K}}_{\omega}^{p,q}(\mathbb{R}^n)$ is strictly s-convex, and hence finishes the proof of Theorem 1.3.4. $\qquad\qquad\square$

1.4 Absolutely Continuous Quasi-Norms

In this section, we investigate the absolute continuity of quasi-norms of local and global generalized Herz spaces. Indeed, we show that the local generalized Herz space $\dot{\mathcal{K}}_{\omega,0}^{p,q}(\mathbb{R}^n)$ has an absolutely continuous quasi-norm when $p, q \in (0, \infty)$ and otherwise, $\dot{\mathcal{K}}^{p,q}_{\omega,0}(\mathbb{R}^n)$ does not have one. Then, via constructing a special example under reasonable assumptions, we show that the global generalized Herz space may not have an absolutely continuous quasi-norm. Finally, based on the above example, we discuss the sharpness of embedding between global generalized Herz spaces and Lebesgue spaces.

First, the following conclusion shows that $\dot{\mathcal{K}}^{p,q}_{\omega,0}(\mathbb{R}^n)$ has an absolutely continuous quasi-norm when $p, q \in (0, \infty)$.

Theorem 1.4.1 *Let $p, q \in (0, \infty)$ and $\omega \in M(\mathbb{R}_+)$. Then the local generalized Herz space $\dot{\mathcal{K}}^{p,q}_{\omega,0}(\mathbb{R}^n)$ has an absolutely continuous quasi-norm.*

Proof Let $p, q \in (0, \infty)$ and $\omega \in M(\mathbb{R}_+)$. Assume that f is any given measurable function in $\dot{\mathcal{K}}^{p,q}_{\omega,0}(\mathbb{R}^n)$ and $\{E_i\}_{i \in \mathbb{N}}$ a sequence of measurable sets satisfying $\mathbf{1}_{E_i} \to 0$ almost everywhere as $i \to \infty$. For any $k \in \mathbb{Z}$ and $i \in \mathbb{N}$, let

$$a_{k,i} := \omega(2^k)\|f\mathbf{1}_{E_i}\mathbf{1}_{B(0,2^k)\setminus B(0,2^{k-1})}\|_{L^p(\mathbb{R}^n)}$$

and

$$a_k := \omega(2^k)\|f\mathbf{1}_{B(0,2^k)\setminus B(0,2^{k-1})}\|_{L^p(\mathbb{R}^n)}.$$

From $f \in \dot{\mathcal{K}}^{p,q}_{\omega,0}(\mathbb{R}^n)$ and the definition of $\dot{\mathcal{K}}^{p,q}_{\omega,0}(\mathbb{R}^n)$, it follows that

$$\left(\sum_{k \in \mathbb{Z}} |a_k|^q\right)^{\frac{1}{q}} < \infty. \tag{1.70}$$

Thus, for any $k \in \mathbb{Z}$,

$$\omega(2^k)\left\|f\mathbf{1}_{B(0,2^k)\setminus B(0,2^{k-1})}\right\|_{L^p(\mathbb{R}^n)} < \infty.$$

This implies that, for any $k \in \mathbb{Z}$,

$$f\mathbf{1}_{B(0,2^k)\setminus B(0,2^{k-1})} \in L^p(\mathbb{R}^n).$$

By this, the fact that, for any $i \in \mathbb{N}$,

$$|f\mathbf{1}_{E_i}\mathbf{1}_{B(0,2^k)\setminus B(0,2^{k-1})}| \le |f\mathbf{1}_{B(0,2^k)\setminus B(0,2^{k-1})}|,$$

and the dominated convergence theorem, we find that, for any $k \in \mathbb{Z}$, $a_{k,i} \to 0$ as $i \to \infty$. Applying this, the estimate that, for any $k \in \mathbb{Z}$ and $i \in \mathbb{N}$, $|a_{k,i}| \le |a_k|$, (1.70), and the dominated convergence theorem again, we conclude that

$$\lim_{i \to \infty}\|f\mathbf{1}_{E_i}\|_{\dot{\mathcal{K}}^{p,q}_{\omega,0}(\mathbb{R}^n)} = \lim_{i \to \infty}\left(\sum_{k \in \mathbb{Z}} |a_{k,i}|^q\right)^{\frac{1}{q}} = \left(\sum_{k \in \mathbb{Z}} \lim_{i \to \infty} |a_{k,i}|^q\right)^{\frac{1}{q}} = 0,$$

which implies that f has an absolutely continuous quasi-norm in $\dot{\mathcal{K}}^{p,q}_{\omega,0}(\mathbb{R}^n)$, and hence the local generalized Herz space $\dot{\mathcal{K}}^{p,q}_{\omega,0}(\mathbb{R}^n)$ has an absolutely quasi-norm. This then finishes the proof of Theorem 1.4.1. □

However, the following two results imply that, if $p = \infty$ or $q = \infty$, then local generalized Herz space $\dot{\mathcal{K}}^{p,q}_{\omega,0}(\mathbb{R}^n)$ does not have an absolutely continuous quasi-norm.

Theorem 1.4.2 *Let $q \in (0, \infty]$ and $\omega \in M(\mathbb{R}_+)$. Then the local generalized Herz space $\dot{\mathcal{K}}^{\infty,q}_{\omega,0}(\mathbb{R}^n)$ does not have an absolutely continuous quasi-norm.*

Proof Let all the symbols be the same as in the present theorem. Let

$$f_0 := \mathbf{1}_{B(0,2)\setminus B(0,1)}$$

and, for any $k \in \mathbb{N}$,

$$E_k := B\left(0, 1 + \frac{1}{k}\right) \setminus B(0,1).$$

Then, by Definition 1.2.1(i), we find that $f_0 \in \dot{\mathcal{K}}^{\infty,q}_{\omega,0}(\mathbb{R}^n)$. In addition, it is obvious that $\mathbf{1}_{E_k} \to 0$ almost everywhere as $k \to \infty$. However, for any $k \in \mathbb{N}$, we have

$$\left\| f_0 \mathbf{1}_{E_k} \right\|_{\dot{\mathcal{K}}^{\infty,q}_{\omega,0}(\mathbb{R}^n)} = \omega(2) \left\| \mathbf{1}_{B(0,\frac{1}{k})\setminus B(0,1)} \right\|_{L^\infty(\mathbb{R}^n)} = \omega(2) \nrightarrow 0$$

as $k \to \infty$. Therefore, the function f_0 does not have an absolutely continuous quasi-norm in $\dot{\mathcal{K}}^{\infty,q}_{\omega,0}(\mathbb{R}^n)$, and hence $\dot{\mathcal{K}}^{\infty,q}_{\omega,0}(\mathbb{R}^n)$ does not have an absolutely continuous quasi-norm. This then finishes the proof of Theorem 1.4.2. □

Theorem 1.4.3 *Let $p \in (0, \infty)$ and $\omega \in M(\mathbb{R}_+)$. Then the local generalized Herz space $\dot{\mathcal{K}}^{p,\infty}_{\omega,0}(\mathbb{R}^n)$ does not have an absolutely continuous quasi-norm.*

Proof Let all the symbols be the same as in the present theorem,

$$f_0 := \sum_{k \in \mathbb{N}} 2^{-\frac{nk}{p}} \left[\omega(2^k)\right]^{-1} \mathbf{1}_{B(0,2^k)\setminus B(0,2^{k-1})},$$

and, for any $k \in \mathbb{N}$,

$$E_j := \left[B(0, 2^{j-1})\right]^{\complement}.$$

Then, from Definition 1.2.1(i), we infer that

$$\|f_0\|_{\dot{\mathcal{K}}^{p,\infty}_{\omega,0}(\mathbb{R}^n)} = \sup_{k \in \mathbb{N}} \left\{ 2^{-\frac{nk}{p}} \left\| \mathbf{1}_{B(0,2^k)\setminus B(0,2^{k-1})} \right\|_{L^p(\mathbb{R}^n)} \right\} \sim 1,$$

which implies that $f_0 \in \dot{\mathcal{K}}^{p,\infty}_{\omega,\mathbf{0}}(\mathbb{R}^n)$. In addition, it is easy to show that $\mathbf{1}_{E_j} \to 0$ almost everywhere as $j \to \infty$. However, for any $j \in \mathbb{N}$, applying Definition 1.2.1(i) again, we have

$$
\left\| f_0 \mathbf{1}_{E_j} \right\|_{\dot{\mathcal{K}}^{p,\infty}_{\omega,\mathbf{0}}(\mathbb{R}^n)} = \sup_{k \in \mathbb{N} \cap [j,\infty)} \left\{ 2^{-\frac{nk}{p}} \left\| \mathbf{1}_{B(\mathbf{0},2^k) \setminus B(\mathbf{0},2^{k-1})} \right\|_{L^p(\mathbb{R}^n)} \right\}
$$

$$
\sim 1 \nrightarrow 0
$$

as $j \to \infty$. Thus, f_0 does not have an absolutely continuous quasi-norm in $\dot{\mathcal{K}}^{p,\infty}_{\omega,\mathbf{0}}(\mathbb{R}^n)$. This further implies that the global generalized Herz space $\dot{\mathcal{K}}^{p,\infty}_{\omega,\mathbf{0}}(\mathbb{R}^n)$ does not have an absolutely continuous quasi-norm, which then completes the proof of Theorem 1.4.3. $\qquad\square$

Via borrowing some ideas from the proof of [209, Example 5.1] and using the definition of global generalized Herz spaces, we construct a special set E and further show that the global generalized Herz space may not have an absolutely continuous quasi-norm.

Example 1.4.4 Let $p \in (0,\infty)$, $q \in (0,\infty]$, and $\alpha \in (-\frac{1}{p},0)$. For any $t \in (0,\infty)$, let $\omega(t) := t^\alpha$, and let

$$
E := \bigcup_{k \in \mathbb{N}} \left(k - 1 + k^{-\frac{2}{\alpha p}}, k + k^{-\frac{2}{\alpha p}} \right).
$$

Then the characteristic function $\mathbf{1}_E \in \dot{\mathcal{K}}^{p,q}_{\omega}(\mathbb{R})$, but $\mathbf{1}_E$ does not have an absolutely continuous quasi-norm in $\dot{\mathcal{K}}^{p,q}_{\omega}(\mathbb{R})$. This implies that the global generalized Herz space $\dot{\mathcal{K}}^{p,q}_{\omega}(\mathbb{R})$ may not have an absolutely continuous quasi-norm.

Proof Let $p,q \in (0,\infty)$ and $\alpha \in (-\frac{1}{p},0)$. For any $t \in (0,\infty)$, let $\omega(t) := t^\alpha$, and let

$$
E := \bigcup_{k \in \mathbb{N}} \left(k - 1 + k^{-\frac{2}{\alpha p}}, k + k^{-\frac{2}{\alpha p}} \right)
$$

and, for any $k \in \mathbb{N}$,

$$
E_k := \left(k - 1 + k^{-\frac{2}{\alpha p}}, k + k^{-\frac{2}{\alpha p}} \right).
$$

We now show that $\mathbf{1}_E \in \dot{\mathcal{K}}_\omega^{p,q}(\mathbb{R})$. Indeed, from the definition of $\|\cdot\|_{\dot{\mathcal{K}}_{\omega,0}^{p,q}(\mathbb{R})}$, it follows that, for any $\xi \in \mathbb{R}$,

$$\|\mathbf{1}_E(\cdot + \xi)\|_{\dot{\mathcal{K}}_{\omega,0}^{p,q}(\mathbb{R})}$$

$$= \left[\sum_{k \in \mathbb{Z}} 2^{k\alpha q} \left\|\mathbf{1}_E \mathbf{1}_{B(\xi,2^k)\setminus B(\xi,2^{k-1})}\right\|_{L^p(\mathbb{R})}^q\right]^{\frac{1}{q}}$$

$$\leq \left[\sum_{k \in \mathbb{Z}} 2^{k\alpha q} \left\|\mathbf{1}_E \mathbf{1}_{B(\xi,2^k)}\right\|_{L^p(\mathbb{R})}^q\right]^{\frac{1}{q}}$$

$$\lesssim \left[\sum_{k \in \mathbb{N}} 2^{k\alpha q} \left\|\mathbf{1}_E \mathbf{1}_{(\xi,\xi+2^k)}\right\|_{L^p(\mathbb{R})}^q\right]^{\frac{1}{q}} + \left[\sum_{k \in \mathbb{N}} 2^{k\alpha q} \left\|\mathbf{1}_E \mathbf{1}_{(\xi-2^k,\xi)}\right\|_{L^p(\mathbb{R})}^q\right]^{\frac{1}{q}}$$

$$+ \left[\sum_{k \in \mathbb{Z}\setminus\mathbb{N}} 2^{k\alpha q} \left\|\mathbf{1}_E \mathbf{1}_{B(\xi,2^k)}\right\|_{L^p(\mathbb{R})}^q\right]^{\frac{1}{q}}$$

$$=: J_{\xi,1} + J_{\xi,2} + J_{\xi,3}. \tag{1.71}$$

For $J_{\xi,1}$, we have

$$J_{\xi,1} \sim \left[\sum_{k \in \mathbb{N}} 2^{k\alpha q} \left|\bigcup_{\widetilde{k} \in \mathbb{N}} E_{\widetilde{k}} \cap (\xi, \xi + 2^k)\right|^{\frac{q}{p}}\right]^{\frac{1}{q}}. \tag{1.72}$$

Notice that, for any $k, \widetilde{k} \in \mathbb{N}$ and any $\xi \in \mathbb{R}$, $E_{\widetilde{k}} \cap (\xi, \xi + 2^k) \neq \emptyset$ if and only if

$$\xi \in \left(\widetilde{k} - 1 + \widetilde{k}^{-\frac{2}{\alpha p}} - 2^k, \widetilde{k} + \widetilde{k}^{-\frac{2}{\alpha p}}\right). \tag{1.73}$$

Now, we estimate $J_{\xi,1}$ by considering the following three cases on ξ.

Case (1) $\xi \in (-\infty, 1]$. In this case, by (1.73), we know that

$$\widetilde{k} - 1 + \widetilde{k}^{-\frac{2}{\alpha p}} < \xi + 2^k \leq 1 + 2^k.$$

It means that $\widetilde{k}^{-\frac{2}{\alpha p}} < \widetilde{k} + \widetilde{k}^{-\frac{2}{\alpha p}} < 2 + 2^k \le 2^{k+1}$ and hence $\widetilde{k} \in \mathbb{N} \cap (0, 2^{-\frac{(k+1)\alpha p}{2}})$.
This, together with (1.72) and the fact that, for any $\widetilde{k} \in \mathbb{N}$, $|E_{\widetilde{k}}| = 1$, further implies
that

$$
\mathrm{J}_{\xi,1} \lesssim \left\{ \sum_{k \in \mathbb{N}} 2^{k\alpha q} \left[\sum_{\widetilde{k} \in \mathbb{N} \cap (0, 2^{-\frac{(k+1)\alpha p}{2}})} |E_{\widetilde{k}}| \right]^{\frac{q}{p}} \right\}^{\frac{1}{q}}
$$

$$
\lesssim \left[\sum_{k \in \mathbb{N}} 2^{k\alpha q} 2^{-\frac{(k+1)\alpha q}{2}} \right]^{\frac{1}{q}} \sim 1. \tag{1.74}
$$

Case (2) $\xi \in (k_0 - 1 + k_0^{-\frac{2}{\alpha p}}, k_0 + k_0^{-\frac{2}{\alpha p}}]$ for some $k_0 \in \mathbb{N}$. In this case, from (1.73)
and the assumption $\xi > k_0 - 1 + k_0^{-\frac{2}{\alpha p}}$, it follows that

$$
k_0 - 1 + k_0^{-\frac{2}{\alpha p}} < \widetilde{k} + \widetilde{k}^{-\frac{2}{\alpha p}}
$$

and hence $\widetilde{k} \in \mathbb{N} \cap [k_0, \infty)$. On the other hand, by (1.73) and the assumption $\xi \le k_0 + k_0^{-\frac{2}{\alpha p}}$, we conclude that

$$
\widetilde{k} - 1 + \widetilde{k}^{-\frac{2}{\alpha p}} < k_0 + k_0^{-\frac{2}{\alpha p}} + 2^k.
$$

This, combined with $\widetilde{k} \in \mathbb{N} \cap [k_0, \infty)$, further implies that

$$
\widetilde{k}^{-\frac{2}{\alpha p}} - k_0^{-\frac{2}{\alpha p}} \le \widetilde{k} - k_0 + \widetilde{k}^{-\frac{2}{\alpha p}} - k_0^{-\frac{2}{\alpha p}} < 2^k + 1 < 2^{k+1}.
$$

Then, from $\widetilde{k}^{-\frac{2}{\alpha p}} - k_0^{-\frac{2}{\alpha p}} < 2^{k+1}$ and the assumption $\alpha \in (-\frac{1}{p}, 0)$, we deduce that

$$
\left(\widetilde{k} - k_0 \right)^{-\frac{2}{\alpha p}} \le \widetilde{k}^{-\frac{2}{\alpha p}} - k_0^{-\frac{2}{\alpha p}} < 2^{k+1}
$$

and hence $\widetilde{k} \in \mathbb{N} \cap [k_0, k_0 + 2^{-\frac{(k+1)\alpha p}{2}})$. Applying this, (1.72), and the fact that, for any $\widetilde{k} \in \mathbb{N}$, $|E_{\widetilde{k}}| = 1$, we find that

$$
\mathrm{J}_{\xi,1} \lesssim \left\{ \sum_{k \in \mathbb{N}} 2^{k\alpha q} \left[\sum_{\widetilde{k} \in \mathbb{N} \cap [k_0, k_0 + 2^{-\frac{(k+1)\alpha p}{2}})} |E_{\widetilde{k}}| \right]^{\frac{q}{p}} \right\}^{\frac{1}{q}}
$$

$$
\lesssim \left[\sum_{k \in \mathbb{N}} 2^{k\alpha q} 2^{-\frac{(k+1)\alpha q}{2}} \right]^{\frac{1}{q}} \sim 1. \tag{1.75}
$$

Case (3) $\xi \in (k_0 + k_0^{-\frac{2}{\alpha p}}, k_0 + (k_0+1)^{-\frac{2}{\alpha p}}]$ for some $k_0 \in \mathbb{N}$. In this case, by an argument similar to that used in Case (2), we conclude that, for any $k, \widetilde{k} \in \mathbb{N}$, if $E_{\widetilde{k}} \cap (\xi, \xi + 2^k) \neq \emptyset$, then

$$
\widetilde{k} \in \mathbb{N} \cap \left[k_0 + 1, k_0 + 1 + 2^{-\frac{k\alpha p}{2}} \right).
$$

From this, (1.72), and the fact that, for any $\widetilde{k} \in \mathbb{N}$, $|E_{\widetilde{k}}| = 1$, we deduce that

$$
\mathrm{J}_{\xi,1} \lesssim \left\{ \sum_{k \in \mathbb{N}} 2^{k\alpha q} \left[\sum_{\widetilde{k} \in \mathbb{N} \cap [k_0+1, k_0+1+2^{-\frac{k\alpha p}{2}})} |E_{\widetilde{k}}| \right]^{\frac{q}{p}} \right\}^{\frac{1}{q}}
$$

$$
\lesssim \left[\sum_{k \in \mathbb{N}} 2^{k\alpha q} 2^{-\frac{k\alpha q}{2}} \right]^{\frac{1}{q}} \sim 1,
$$

which, combined with (1.74) and (1.75), further implies that, for any $\xi \in \mathbb{R}$,

$$
\mathrm{J}_{\xi,1} \lesssim 1.
$$

Thus, we finish the estimation of $\mathrm{J}_{\xi,1}$.

Now, we deal with $\mathrm{J}_{\xi,2}$. Notice that

$$
\mathrm{J}_{\xi,2} \sim \left[\sum_{k \in \mathbb{N}} 2^{k\alpha q} \left| \bigcup_{\widetilde{k} \in \mathbb{N}} E_{\widetilde{k}} \cap (\xi - 2^k, \xi) \right|^{\frac{q}{p}} \right]^{\frac{1}{q}} \tag{1.76}
$$

and, for any $k, \widetilde{k} \in \mathbb{N}$ and any $\xi \in \mathbb{R}$, $E_{\widetilde{k}} \cap (\xi - 2^k, \xi) \neq \emptyset$ if and only if

$$\xi \in \left(\widetilde{k} - 1 + \widetilde{k}^{-\frac{2}{\alpha p}}, \widetilde{k} + \widetilde{k}^{-\frac{2}{\alpha p}} + 2^k \right). \tag{1.77}$$

We next estimate $J_{\xi,2}$ by considering the following three cases on ξ.

Case (i) $\xi \in (-\infty, 1]$. In this case, notice that, for any $\widetilde{k} \in \mathbb{N}$, $\xi \leq 1 \leq \widetilde{k} - 1 + \widetilde{k}^{-\frac{2}{\alpha p}}$, which contradicts (1.77). Thus, for any $\widetilde{k} \in \mathbb{N}$ and $\xi \in (-\infty, 1]$, we have

$$E_{\widetilde{k}} \cap (\xi - 2^k, \xi) = \emptyset$$

and hence, in this case,

$$J_{\xi,2} = 0. \tag{1.78}$$

Case (ii) $\xi \in (k_0 - 1 + k_0^{-\frac{2}{\alpha p}}, k_0 + k_0^{-\frac{2}{\alpha p}}]$ for some $k_0 \in \mathbb{N}$. In this case, applying (1.77) and the assumption $\xi \leq k_0 + k_0^{-\frac{2}{\alpha p}}$, we conclude that

$$\widetilde{k} - 1 + \widetilde{k}^{-\frac{2}{\alpha p}} < k_0 + k_0^{-\frac{2}{\alpha p}}$$

and hence $\widetilde{k} \in \mathbb{N} \cap (0, k_0]$. Moreover, from (1.77) and the assumption $\xi > k_0 - 1 + k_0^{-\frac{2}{\alpha p}}$, it follows that

$$k_0 - 1 + k_0^{-\frac{2}{\alpha p}} - 2^k < \widetilde{k} + \widetilde{k}^{-\frac{2}{\alpha p}}.$$

This, together with the fact that $\widetilde{k} \in \mathbb{N} \cap (0, k_0]$, further implies that

$$k_0^{-\frac{2}{\alpha p}} - \widetilde{k}^{-\frac{2}{\alpha p}} \leq k_0 - \widetilde{k} + k_0^{-\frac{2}{\alpha p}} - \widetilde{k}^{-\frac{2}{\alpha p}} < 1 + 2^k < 2^{k+1}.$$

By this and the assumption $\alpha \in (-\frac{1}{p}, 0)$, we find that

$$\left(k_0 - \widetilde{k} \right)^{-\frac{2}{\alpha p}} \leq k_0^{-\frac{2}{\alpha p}} - \widetilde{k}^{-\frac{2}{\alpha p}} < 2^{k+1}$$

and hence $\widetilde{k} > k_0 - 2^{-\frac{(k+1)\alpha p}{2}}$. Thus, from (1.76), $\widetilde{k} \in \mathbb{N} \cap (k_0 - 2^{-\frac{(k+1)\alpha p}{2}}, k_0]$, and the fact that, for any $\widetilde{k} \in \mathbb{N}$, $|E_{\widetilde{k}}| = 1$, we infer that

$$J_{\xi,2} \lesssim \left\{ \sum_{k \in \mathbb{N}} 2^{k\alpha q} \left[\sum_{\widetilde{k} \in \mathbb{N} \cap (k_0 - 2^{-\frac{(k+1)\alpha p}{2}}, k_0]} |E_{\widetilde{k}}| \right]^{\frac{q}{p}} \right\}^{\frac{1}{q}}$$

$$\lesssim \left[\sum_{k \in \mathbb{N}} 2^{k\alpha q} 2^{-\frac{(k+1)\alpha q}{2}} \right]^{\frac{1}{q}} \sim 1. \tag{1.79}$$

Case (iii) $\xi \in (k_0 + k_0^{-\frac{2}{\alpha p}}, k_0 + (k_0 + 1)^{-\frac{2}{\alpha p}}]$ for some $k_0 \in \mathbb{N}$. In this case, by an argument similar to that used in the proof of Case (ii), we conclude that, for any $k, \widetilde{k} \in \mathbb{N}$, if $E_{\widetilde{k}} \cap (\xi - 2^k, \xi) \neq \emptyset$, then

$$\widetilde{k} \in \mathbb{N} \cap \left(k_0 - 2^{-\frac{k\alpha p}{2}}, k_0 \right].$$

This, combined with (1.76) and the fact that, for any $\widetilde{k} \in \mathbb{N}$, $|E_{\widetilde{k}}| = 1$, further implies that

$$J_{\xi,2} \lesssim \left\{ \sum_{k \in \mathbb{N}} 2^{k\alpha q} \left[\sum_{\widetilde{k} \in \mathbb{N} \cap (k_0 - 2^{-\frac{k\alpha p}{2}}, k_0]} |E_{\widetilde{k}}| \right]^{\frac{q}{p}} \right\}^{\frac{1}{q}}$$

$$\lesssim \left[\sum_{k \in \mathbb{N}} 2^{k\alpha q} 2^{-\frac{k\alpha q}{2}} \right]^{\frac{1}{q}} \sim 1.$$

Combining this, (1.78), and (1.79), we further conclude that, for any $\xi \in \mathbb{R}$,

$$J_{\xi,2} \lesssim 1$$

and hence we finish the estimation of $J_{\xi,2}$.

Now, to deal with $J_{\xi,3}$, notice that, for any $\xi \in \mathbb{R}$,

$$J_{\xi,3} \lesssim \left[\sum_{k \in \mathbb{Z} \setminus \mathbb{N}} 2^{k\alpha q} \left\| \mathbf{1}_{B(\xi, 2^k)} \right\|_{L^p(\mathbb{R})}^q \right]^{\frac{1}{q}}$$

$$\sim \left[\sum_{k \in \mathbb{Z} \setminus \mathbb{N}} 2^{k\alpha q} \left| (\xi - 2^k, \xi + 2^k) \right|^{\frac{q}{p}} \right]^{\frac{1}{q}}$$

$$\sim \left[\sum_{k \in \mathbb{Z} \setminus \mathbb{N}} 2^{k\alpha q} 2^{\frac{(k+1)q}{p}} \right]^{\frac{1}{q}} \sim 1,$$

which, combined with (1.71) and the estimates of both $J_{\xi,1}$ and $J_{\xi,2}$, further implies that, for any $\xi \in \mathbb{R}$,

$$\left\| \mathbf{1}_E(\cdot + \xi) \right\|_{\dot{\mathcal{K}}_{\omega,0}^{p,q}(\mathbb{R})} \lesssim 1.$$

From this, the definition of $\| \cdot \|_{\dot{\mathcal{K}}_\omega^{p,q}(\mathbb{R})}$, and Remark 1.2.2(ii), it follows that $\mathbf{1}_E \in \dot{\mathcal{K}}_\omega^{p,q}(\mathbb{R})$.

Finally, we show that $\mathbf{1}_E$ does not have an absolutely continuous quasi-norm in the global generalized Herz space $\dot{\mathcal{K}}_\omega^{p,q}(\mathbb{R})$. To achieve this, for any $\widetilde{k} \in \mathbb{N}$, let $F_{\widetilde{k}} := (\widetilde{k}, \infty)$ and $\xi_{\widetilde{k}} := \widetilde{k} + \widetilde{k}^{-\frac{2}{\alpha p}} - \frac{1}{2}$. Then, obviously, for any $x \in \mathbb{R}$, we have $\mathbf{1}_{F_{\widetilde{k}}}(x) \to 0$ as $\widetilde{k} \to \infty$. In addition, notice that, for any $\widetilde{k} \in \mathbb{N}$,

$$\widetilde{k} - 1 + \widetilde{k}^{-\frac{2}{\alpha p}} \geq \widetilde{k}.$$

This implies that, for any $\widetilde{k} \in \mathbb{N}$, $E_{\widetilde{k}} \subset F_{\widetilde{k}}$. Thus, for any $\widetilde{k} \in \mathbb{N}$, we have

$$\left\| \mathbf{1}_E \mathbf{1}_{F_{\widetilde{k}}} \right\|_{\dot{\mathcal{K}}_\omega^{p,q}(\mathbb{R})}$$

$$\geq \left\| \mathbf{1}_{E_{\widetilde{k}}} \right\|_{\dot{\mathcal{K}}_\omega^{p,q}(\mathbb{R})}$$

$$\geq \left\| \mathbf{1}_{E_{\widetilde{k}}}(\cdot + \xi_{\widetilde{k}}) \right\|_{\dot{\mathcal{K}}_{\omega,0}^{p,q}(\mathbb{R})}$$

$$= \left[\sum_{k \in \mathbb{Z}} 2^{k\alpha q} \left\| \mathbf{1}_{E_{\widetilde{k}}} \mathbf{1}_{B(\xi_{\widetilde{k}}, 2^k) \setminus B(\xi_{\widetilde{k}}, 2^{k-1})} \right\|_{L^p(\mathbb{R})}^q \right]^{\frac{1}{q}}$$

$$\geq 2^{-\alpha} \left\| \mathbf{1}_{E_{\widetilde{k}}} \mathbf{1}_{(\xi_{\widetilde{k}} + \frac{1}{4}, \xi_{\widetilde{k}} + \frac{1}{2})} \right\|_{L^p(\mathbb{R})}$$

$$= 2^{-\alpha} \left| \left(\widetilde{k} - 1 + \widetilde{k}^{-\frac{2}{\alpha p}}, \widetilde{k} + \widetilde{k}^{-\frac{2}{\alpha p}} \right) \cap \left(\widetilde{k} - \frac{1}{4} + \widetilde{k}^{-\frac{2}{\alpha p}}, \widetilde{k} + \widetilde{k}^{-\frac{2}{\alpha p}} \right) \right|^{\frac{1}{p}}$$

$$= 2^{-\alpha - \frac{2}{p}}.$$

Therefore, $\mathbf{1}_E$ does not have an absolutely continuous quasi-norm in $\dot{\mathcal{K}}_\omega^{p,q}(\mathbb{R})$, which further implies that the global generalized Herz space $\dot{\mathcal{K}}_\omega^{p,q}(\mathbb{R})$ does not have an absolutely continuous quasi-norm. This finishes the proof of Example 1.4.4. □

Remark 1.4.5 We should point out that the assumptions in Example 1.4.4 are reasonable. Namely, the global generalized Herz space given in Example 1.4.4 is ball quasi-Banach function space. Indeed, combining Example 1.1.7 and Theorem 1.2.44, we find that, if the global generalized Herz space is the same as in Example 1.4.4, then all the assumptions of Theorem 1.2.44 are satisfied, and hence the global generalized Herz space under consideration is a ball quasi-Banach function space.

Now, based on the special element constructed in Example 1.4.4, we turn to study the sharpness of the embedding property between global generalized Herz spaces and Lebesgue spaces. Recall that, as is showed in both Theorems 1.2.6 and 1.2.7, if $p, q \in (0, \infty]$ and $\omega \in M(\mathbb{R}_+)$ are such that $M_0(\omega) \in (-\infty, -\frac{n}{p})$ or $m_\infty(\omega) \in (0, \infty)$, then the global generalized Herz space $\dot{\mathcal{K}}_\omega^{p,q}(\mathbb{R}^n)$ is trivial, namely,

$$\dot{\mathcal{K}}_\omega^{p,q}(\mathbb{R}^n) = \left\{ f \in \mathscr{M}(\mathbb{R}^n) : f(x) = 0 \text{ for almost every } x \in \mathbb{R}^n \right\},$$

which implies that, under the above assumptions on p, q, and ω, then, for any $r \in (0, \infty]$,

$$\dot{\mathcal{K}}_\omega^{p,q}(\mathbb{R}^n) \hookrightarrow L^r(\mathbb{R}^n). \tag{1.80}$$

We should point out that the above assumptions of ω,

$$M_0(\omega) \in \left(-\infty, -\frac{n}{p}\right) \text{ or } m_\infty(\omega) \in (0, \infty), \tag{1.81}$$

are sharp in the embedding (1.80). Namely, if ω does not satisfy (1.81), then the embedding (1.80) may not hold true. The following conclusion shows that, if $M_0(\omega) \in (-\frac{n}{p}, \infty)$ and $m_\infty(\omega) \in (-\infty, 0)$, then the global generalized Herz space $\dot{\mathcal{K}}_\omega^{p,q}(\mathbb{R}^n)$ may not embed continuously into any Lebesgue space.

Proposition 1.4.6 *Let $p \in (0, \infty)$, $q \in (0, \infty]$, $\alpha \in (-\frac{1}{p}, 0)$, and $\omega_\alpha(t) := t^\alpha$ for any $t \in (0, \infty)$. Then the global generalized Herz space $\dot{\mathcal{K}}_{\omega_\alpha}^{p,q}(\mathbb{R})$ can not embed into any Lebesgue space $L^r(\mathbb{R})$ with $r \in (0, \infty]$.*

Proof Let all the symbols be the same as in the present proposition. Then, by Example 1.1.7, we find that

$$m_0(\omega_\alpha) = M_0(\omega_\alpha) = m_\infty(\omega_\alpha)$$
$$= M_\infty(\omega_\alpha) \in \left(-\frac{1}{p}, 0\right).$$

From this and Theorem 1.2.42 with $n := 1$ and $\omega := \omega_\alpha$, we infer that the global generalized Herz space $\dot{\mathcal{K}}^{p,q}_{\omega_\alpha}(\mathbb{R})$ is a BQBF space. Thus, applying both (i) and (iv) of Definition 1.2.13, we obtain

$$\left\|\mathbf{1}_{B(0,1)}\right\|_{\dot{\mathcal{K}}^{p,q}_{\omega_\alpha}(\mathbb{R})} \in (0, \infty).$$

This, combined with Theorem 1.2.9 via replacing n, ω, and f_0 therein, respectively, by 1, ω_α, and $\mathbf{1}_{B(0,1)}$, further implies that, if the global generalized Herz space $\dot{\mathcal{K}}^{p,q}_{\omega_\alpha}(\mathbb{R})$ embeds continuously into the Lebesgue space $L^r(\mathbb{R})$ with $r \in (0, \infty]$, namely, for any $f \in \dot{\mathcal{K}}^{p,q}_{\omega_\alpha}(\mathbb{R})$,

$$\|f\|_{L^r(\mathbb{R})} \lesssim \|f\|_{\dot{\mathcal{K}}^{p,q}_{\omega_\alpha}(\mathbb{R})},$$

then $r = \frac{1}{\alpha+1/p}$.

Next, we prove that the embedding

$$\dot{\mathcal{K}}^{p,q}_{\omega_\alpha}(\mathbb{R}) \hookrightarrow L^{\frac{1}{\alpha+1/p}}(\mathbb{R}) \tag{1.82}$$

does not hold true. Indeed, let

$$E := \bigcup_{k\in\mathbb{N}} \left(k - 1 + k^{-\frac{2}{\alpha p}}, k + k^{-\frac{2}{\alpha p}}\right).$$

Obviously, $\mathbf{1}_E \notin L^{\frac{1}{\alpha+1/p}}(\mathbb{R})$. However, from Example 1.4.4, we infer that

$$\mathbf{1}_E \in \dot{\mathcal{K}}^{p,q}_{\omega_\alpha}(\mathbb{R}) \setminus L^{\frac{1}{\alpha+1/p}}(\mathbb{R}),$$

which implies that (1.82) does not hold true. Combining this and the above claim, we further find that the global generalized Herz space $\dot{\mathcal{K}}^{p,q}_{\omega_\alpha}(\mathbb{R})$ can not embed continuously into the Lebesgue space $L^r(\mathbb{R})$. This then finishes the proof of Proposition 1.4.6.   □

Moreover, the following result gives a example that, if $m_\infty(\omega) = 0$, then the global generalized Herz space $\dot{\mathcal{K}}^{p,q}_{\omega}(\mathbb{R}^n)$ can not embed continuously into any Lebesgue spaces.

Proposition 1.4.7 *Let $p \in (0, \infty)$, $q \in (0, \infty]$, and*

$$\omega_{p,q}(t) := \begin{cases} 1 & \text{when } t \in (0, 1], \\ (1 + \ln t)^{-\frac{1}{p} - \frac{2}{q}} & \text{when } t \in (1, \infty). \end{cases}$$

Then the global generalized Herz space $\dot{\mathcal{K}}^{p,q}_{\omega_{p,q}}(\mathbb{R})$ can not embed into any Lebesgue space $L^r(\mathbb{R})$ with $r \in (0, \infty]$.

Proof Let all the symbols be the same as in the present proposition. Then, similarly to Example 1.1.9, we have $\omega_{p,q} \in M(\mathbb{R}_+)$ and

$$m_0\left(\omega_{p,q}\right) = M_0\left(\omega_{p,q}\right) = m_\infty\left(\omega_{p,q}\right) = M_\infty\left(\omega_{p,q}\right) = 0. \tag{1.83}$$

On the other hand, repeating the arguments similar to those used in the estimations of (1.64), (1.65), and (1.66) with $B(x_0, r)$, $2^{k[m_0(\omega_{p,q})-\varepsilon]}$ for any $k \in \mathbb{Z} \setminus \mathbb{N}$, and $2^{k[M_\infty(\omega_{p,q})+\varepsilon]}\}$ for any $k \in \mathbb{N}$ therein replaced, respectively, by $B(0, 1)$, 1, and $(1 + k \ln 2)^{-\frac{1}{p} - \frac{2}{q}}$ for any $k \in \mathbb{N}$, we conclude that $\|\mathbf{1}_{B(0,1)}\|_{\dot{\mathcal{K}}^{p,q}_{\omega_{p,q}}(\mathbb{R})} \in (0, \infty)$. Using this, Theorem 1.2.9 with n, ω, and f_0 therein replaced, respectively, by 1, $\omega_{p,q}$, and $\mathbf{1}_B(0, 1)$, and (1.83), we find that, if $\dot{\mathcal{K}}^{p,q}_{\omega_{p,q}}(\mathbb{R})$ embeds continuously into $L^r(\mathbb{R})$ with $r \in (0, \infty]$, namely, for any $f \in \dot{\mathcal{K}}^{p,q}_{\omega_{p,q}}(\mathbb{R})$,

$$\|f\|_{L^r(\mathbb{R})} \lesssim \|f\|_{\dot{\mathcal{K}}^{p,q}_{\omega_{p,q}}(\mathbb{R})}, \tag{1.84}$$

then $\frac{1}{r} = \frac{1}{p}$, and hence $r = p$.

Next, we prove that the embedding (1.84) does not hold true with $r := p$. For this purpose, let

$$E := \bigcup_{k \in \mathbb{N}} \left(2^k - 1, 2^k\right).$$

Obviously, $\mathbf{1}_E \notin L^p(\mathbb{R})$. We now show that $\mathbf{1}_E \in \dot{\mathcal{K}}^{p,q}_{\omega_{p,q}}(\mathbb{R})$. Indeed, for any $\xi \in \mathbb{R}$, we have

$$\|\mathbf{1}_E(\cdot + \xi)\|_{\dot{\mathcal{K}}^{p,q}_{\omega_{p,q},0}(\mathbb{R})}$$

$$\sim \left\{\sum_{k \in \mathbb{N}} (1 + k \ln 2)^{-2 - \frac{q}{p}} \left\|\mathbf{1}_E \mathbf{1}_{B(\xi, 2^k) \setminus B(\xi, 2^{k-1})}\right\|_{L^p(\mathbb{R})}^q\right\}^{\frac{1}{q}}$$

$$+ \left\{\sum_{k \in \mathbb{Z} \setminus \mathbb{N}} \left\|\mathbf{1}_E \mathbf{1}_{B(\xi, 2^k) \setminus B(\xi, 2^{k-1})}\right\|_{L^p(\mathbb{R})}^q\right\}^{\frac{1}{q}}$$

$$\lesssim \left[\sum_{k \in \mathbb{N}} (1 + k \ln 2)^{-2 - \frac{q}{p}} \left\| \mathbf{1}_E \mathbf{1}_{(\xi, \xi + 2^k)} \right\|_{L^p(\mathbb{R})}^q \right]^{\frac{1}{q}}$$

$$+ \left[\sum_{k \in \mathbb{N}} (1 + k \ln 2)^{-2 - \frac{q}{p}} \left\| \mathbf{1}_E \mathbf{1}_{(\xi - 2^k, \xi)} \right\|_{L^p(\mathbb{R})}^q \right]^{\frac{1}{q}}$$

$$+ \left[\sum_{k \in \mathbb{Z} \setminus \mathbb{N}} \left\| \mathbf{1}_E \mathbf{1}_{B(\xi, 2^k)} \right\|_{L^p(\mathbb{R})}^q \right]^{\frac{1}{q}}$$

$$=: \mathrm{L}_{\xi,1} + \mathrm{L}_{\xi,2} + \mathrm{L}_{\xi,3}. \tag{1.85}$$

For any $\xi \in \mathbb{R}$, repeating the arguments similar to those used in the estimations of both $\mathrm{J}_{\xi,1}$ and $\mathrm{J}_{\xi,2}$ of the proof of Example 1.4.4 with $k + k^{-\frac{2}{\alpha p}}$ and $2^{k\alpha}$ therein replaced, respectively, by 2^k and $(1 + k \ln 2)^{-\frac{1}{p} - \frac{2}{q}}$, we obtain

$$\mathrm{L}_{\xi,1} \lesssim \left[\sum_{k \in \mathbb{N}} (1 + k \ln 2)^{-2 - \frac{q}{p}} (k + 2)^{\frac{q}{p}} \right]^{\frac{1}{q}}$$

$$\lesssim \left(\sum_{k \in \mathbb{N}} \frac{1}{k^2} \right)^{\frac{1}{q}} \sim 1 \tag{1.86}$$

and

$$\mathrm{L}_{\xi,2} \lesssim \left[\sum_{k \in \mathbb{N}} (1 + k \ln 2)^{-2 - \frac{q}{p}} (k + 2)^{\frac{q}{p}} \right]^{\frac{1}{q}}$$

$$\lesssim \left(\sum_{k \in \mathbb{N}} \frac{1}{k^2} \right)^{\frac{1}{q}} \sim 1. \tag{1.87}$$

These finish the estimation of both $\mathrm{L}_{\xi,1}$ and $\mathrm{L}_{\xi,2}$. In addition, for any $\xi \in \mathbb{R}$, repeating an argument similar to that used in the estimation of $\mathrm{J}_{\xi,3}$ of the proof of Example 1.4.4 with $2^{k\alpha}$ therein replaced by 1 for any $k \in \mathbb{Z} \setminus \mathbb{N}$, we have

$$\mathrm{L}_{\xi,3} \lesssim \left(\sum_{k \in \mathbb{Z} \setminus \mathbb{N}} 2^{\frac{(k+1)q}{p}} \right)^{\frac{1}{q}} \sim 1,$$

which completes the estimation of $L_{\xi,3}$. Combining this, Remark 1.2.2(ii), (1.85), (1.86), and (1.87), we find that

$$\|\mathbf{1}_E\|_{\dot{\mathcal{K}}^{p,q}_{\omega_{p,q}}(\mathbb{R})} = \sup_{\xi\in\mathbb{R}}\left\{\|\mathbf{1}_E(\cdot+\xi)\|_{\dot{\mathcal{K}}^{p,q}_{\omega_{p,q},0}(\mathbb{R})}\right\} \lesssim 1.$$

This implies that

$$\mathbf{1}_E \in \dot{\mathcal{K}}^{p,q}_{\omega_{p,q}}(\mathbb{R}) \setminus L^p(\mathbb{R}),$$

and hence the embedding (1.84) does not hold true with $r := p$, which completes the proof of Proposition 1.4.7.   □

Finally, we now show that, if $M_0(\omega) = -\frac{n}{p}$, then the global generalized Herz space $\dot{\mathcal{K}}^{p,q}_{\omega}(\mathbb{R}^n)$ may not embed continuously into any Lebesgue spaces as follows.

Proposition 1.4.8 *Let* $p \in (0, \infty)$ *and, for any* $t \in (0, \infty)$

$$\omega_p(t) := \frac{t^{-\frac{1}{p}}}{\ln(e+\frac{1}{t})}.$$

Then $\dot{\mathcal{K}}^{p,\infty}_{\omega_p}(\mathbb{R})$ *can not embed continuously into any* $L^r(\mathbb{R})$ *with* $r \in (0, \infty]$.

Proof Let all the symbols be the same as in the present proposition. Then, similarly to Example 1.1.8, we have $\omega_p \in M(\mathbb{R}_+)$ and

$$m_0\left(\omega_p\right) = M_0\left(\omega_p\right) = m_\infty\left(\omega_p\right) = M_\infty\left(\omega_p\right) = -\frac{1}{p}.$$

This, combined with Theorem 1.2.9 with $n := 1$, $q := \infty$, and $\omega := \omega_p$, further implies that, if $\dot{\mathcal{K}}^{p,\infty}_{\omega_p}(\mathbb{R})$ embeds continuously into the Lebesgue space $L^r(\mathbb{R})$ with $r \in (0, \infty]$, then $\frac{1}{r} = 0$, and hence $r = \infty$.

Next, we prove that $\dot{\mathcal{K}}^{p,\infty}_{\omega_p}(\mathbb{R})$ can not embed continuously into $L^\infty(\mathbb{R})$. Indeed, for any $x \in \mathbb{R}$, let

$$f_0(x) := \ln\left(e + \frac{1}{|x|}\right).$$

Obviously, $f_0 \notin L^\infty(\mathbb{R})$. On the other hand, repeating an argument similar to that used in the estimation of [206, (1.11)] with $r^{n(\frac{1}{p}-\frac{1}{q})}$ and q therein replaced, respectively, by ω_p and p, we obtain

$$
\begin{aligned}
\|f_0\|_{\dot{\mathcal{K}}^{p,\infty}_{\omega_p}(\mathbb{R})} &= \sup_{\xi \in \mathbb{R},\, k \in \mathbb{Z}} \left\{ \omega_p(2^k) \left\| f_0 \mathbf{1}_{B(\xi,2^k) \setminus B(\xi,2^{k-1})} \right\|_{L^p(\mathbb{R})} \right\} \\
&\sim \sup_{r \in (0,\infty)} \left\{ \omega_p(r) \left\| f_0 \mathbf{1}_{B(0,r)} \right\|_{L^p(\mathbb{R})} \right\} \\
&\sim \sup_{r \in (0,\infty)} \left\{ \frac{r^{-\frac{1}{p}}}{\ln(e+\frac{1}{r})} \left(\int_0^r \left[\ln\left(e+\frac{1}{t}\right) \right]^p dt \right)^{\frac{1}{p}} \right\} \sim 1,
\end{aligned}
$$

which implies that

$$
f_0 \in \dot{\mathcal{K}}^{p,\infty}_{\omega_p}(\mathbb{R}) \setminus L^\infty(\mathbb{R}),
$$

and hence $\dot{\mathcal{K}}^{p,\infty}_{\omega_p}(\mathbb{R})$ can not embed continuously into any $L^r(\mathbb{R})$ with $r \in (0,\infty]$. This then finishes the proof of Proposition 1.4.8. $\square$

1.5 Boundedness of Sublinear Operators

In this section, we establish a criterion about the boundedness of sublinear operators on local and global generalized Herz spaces, which was essentially obtained by Rafeiro and Samko in [197, Theorem 4.3]. As applications, we obtain the boundedness of both the Hardy–Littlewood maximal operator and the Calderón–Zygmund operator on these Herz spaces.

Recall that an operator T defined on $\mathscr{M}(\mathbb{R}^n)$ is called a *sublinear operator* if, for any $f, g \in \mathscr{M}(\mathbb{R}^n)$ and $\lambda \in \mathbb{C}$,

$$
|T(f+g)| \le |T(f)| + |T(g)|
$$

and

$$
|T(\lambda f)| = |\lambda||T(f)|.
$$

In addition, for any normed linear space X and any operator T on X, the *operator norm* $\|T\|_{X \to X}$ of T is defined by setting

$$
\|T\|_{X \to X} := \sup_{\{x \in X:\, \|x\|_X = 1\}} \|Tx\|_X . \tag{1.88}
$$

Then we present the boundedness criterion on generalized Herz spaces as follows.

Theorem 1.5.1 *Let $p \in (1, \infty]$, $q \in (0, \infty]$, and $\omega \in M(\mathbb{R}_+)$ satisfy*

$$-\frac{n}{p} < m_0(\omega) \leq M_0(\omega) < \frac{n}{p'} \tag{1.89}$$

and

$$-\frac{n}{p} < m_\infty(\omega) \leq M_\infty(\omega) < \frac{n}{p'}, \tag{1.90}$$

where $\frac{1}{p} + \frac{1}{p'} = 1$. Assume that T is a sublinear operator satisfying that T is bounded on $L^p(\mathbb{R}^n)$ and that there exists a positive constant $\widetilde{C}$ such that, for any $f \in \dot{\mathcal{K}}^{p,q}_{\omega,0}(\mathbb{R}^n)$ and $x \notin \overline{\mathrm{supp}(f)} := \{x \in \mathbb{R}^n : f(x) \neq 0\}$,

$$|T(f)(x)| \leq \widetilde{C} \int_{\mathbb{R}^n} \frac{|f(y)|}{|x - y|^n} \, dy. \tag{1.91}$$

Then there exists a positive constant C, independent of T, such that

(i) *for any $f \in \dot{\mathcal{K}}^{p,q}_{\omega,0}(\mathbb{R}^n)$,*

$$\|T(f)\|_{\dot{\mathcal{K}}^{p,q}_{\omega,0}(\mathbb{R}^n)} \leq C \left[\widetilde{C} + \|T\|_{L^p(\mathbb{R}^n) \to L^p(\mathbb{R}^n)}\right] \|f\|_{\dot{\mathcal{K}}^{p,q}_{\omega,0}(\mathbb{R}^n)};$$

(ii) *for any $f \in \dot{\mathcal{K}}^{p,q}_{\omega}(\mathbb{R}^n)$,*

$$\|T(f)\|_{\dot{\mathcal{K}}^{p,q}_{\omega}(\mathbb{R}^n)} \leq C \left[\widetilde{C} + \|T\|_{L^p(\mathbb{R}^n) \to L^p(\mathbb{R}^n)}\right] \|f\|_{\dot{\mathcal{K}}^{p,q}_{\omega}(\mathbb{R}^n)}.$$

To prove Theorem 1.5.1, we need the following auxiliary estimates of the positive function $\omega \in M(\mathbb{R}_+)$, which is just [197, Lemma 3.1].

Lemma 1.5.2 *Let $\omega \in M(\mathbb{R}_+)$ and $m_0(\omega)$, $M_0(\omega)$, $m_\infty(\omega)$, and $M_\infty(\omega)$ denote its Matuszewska–Orlicz indices. Then, for any given $\varepsilon \in (0, \infty)$, there exists a positive constant $C_{(\varepsilon)}$, depending on ε, such that, for any $0 < t < \tau < \infty$,*

$$\frac{\omega(t)}{\omega(\tau)} \leq C_{(\varepsilon)} \left(\frac{t}{\tau}\right)^{\min\{m_0(\omega), m_\infty(\omega)\} - \varepsilon}$$

and, for any $0 < \tau < t < \infty$,

$$\frac{\omega(t)}{\omega(\tau)} \leq C_{(\varepsilon)} \left(\frac{t}{\tau}\right)^{\max\{M_0(\omega), M_\infty(\omega)\} + \varepsilon}.$$

Proof of Theorem 1.5.1 Let all the symbols be as in the present theorem. We first
show (i). To this end, for any given $f \in \dot{\mathcal{K}}_{\omega,0}^{p,q}(\mathbb{R}^n)$ and for any $k \in \mathbb{Z}$, let

$$f_{k,1} := f \mathbf{1}_{B(0,2^{k-2})}, \quad f_{k,2} := f \mathbf{1}_{B(0,2^{k+1}) \backslash B(0,2^{k-2})},$$

and

$$f_{k,3} := f \mathbf{1}_{[B(0,2^{k+1})]^\complement}.$$

Then, obviously, for any $k \in \mathbb{Z}$, we have

$$f = f_{k,1} + f_{k,2} + f_{k,3}.$$

From this and the sublinearity of T, we deduce that

$$
\|T(f)\|_{\dot{\mathcal{K}}_{\omega,0}^{p,q}(\mathbb{R}^n)} \lesssim \left\{ \sum_{k \in \mathbb{Z}} \left[\omega(2^k)\right]^q \left\| T(f_{k,1}) \mathbf{1}_{B(0,2^k) \backslash B(0,2^{k-1})} \right\|_{L^p(\mathbb{R}^n)}^q \right\}^{\frac{1}{q}}
$$

$$
+ \left\{ \sum_{k \in \mathbb{Z}} \left[\omega(2^k)\right]^q \left\| T(f_{k,2}) \mathbf{1}_{B(0,2^k) \backslash B(0,2^{k-1})} \right\|_{L^p(\mathbb{R}^n)}^q \right\}^{\frac{1}{q}}
$$

$$
+ \left\{ \sum_{k \in \mathbb{Z}} \left[\omega(2^k)\right]^q \left\| T(f_{k,3}) \mathbf{1}_{B(0,2^k) \backslash B(0,2^{k-1})} \right\|_{L^p(\mathbb{R}^n)}^q \right\}^{\frac{1}{q}}
$$

$$
=: \mathrm{J}_1 + \mathrm{J}_2 + \mathrm{J}_3, \tag{1.92}
$$

where the implicit positive constant is independent of both T and f. We now deal
with J_1, J_2, and J_3, separately. For this purpose, let

$$
\varepsilon \in \left(0, \min\left\{ \min\{m_0(\omega), m_\infty(\omega)\} + \frac{n}{p}, \frac{n}{p'} - \max\{M_0(\omega), M_\infty(\omega)\} \right\}\right)
$$

be a fixed positive constant. Then, from Lemma 1.5.2, it follows that, for any $0 <
\tau < t < \infty$,

$$
\frac{\omega(t)}{\omega(\tau)} \lesssim \left(\frac{t}{\tau}\right)^{\max\{M_0(\omega), M_\infty(\omega)\} + \varepsilon} \tag{1.93}
$$

and, for any $0 < t < \tau < \infty$,

$$
\frac{\omega(t)}{\omega(\tau)} \lesssim \left(\frac{t}{\tau}\right)^{\min\{m_0(\omega), m_\infty(\omega)\} - \varepsilon}. \tag{1.94}
$$

For the simplicity of the presentation, let

$$m := \min\{m_0(\omega), m_\infty(\omega)\} - \varepsilon$$

and

$$M := \max\{M_0(\omega), M_\infty(\omega)\} + \varepsilon.$$

Obviously, we have $m \in (-\frac{n}{p}, \infty)$ and $M \in (-\infty, \frac{n}{p'})$.

We first estimate J_1. Notice that, for any $k, i \in \mathbb{Z}$ satisfying $i \in (-\infty, k-2]$, and for any $x, y \in \mathbb{R}^n$ satisfying $2^{k-1} \leq |x| < 2^k$ and $2^{i-1} \leq |y| < 2^i$, we have $|y| < 2^i \leq 2^{k-2}$. This implies that $|x - y| \geq |x| - |y| \geq 2^{k-2}$. Applying this, (1.91), and the Hölder inequality, we find that, for any $k \in \mathbb{Z}$ and $x \in \mathbb{R}^n$ with $2^{k-1} \leq |x| < 2^k$,

$$\left| T\left(f_{k,1} \right)(x) \right| \leq \widetilde{C} \int_{\mathbb{R}^n} \frac{|f_{k,1}(y)|}{|x-y|^n} \, dy$$

$$\leq \widetilde{C} \sum_{i=-\infty}^{k-2} 2^{-nk} \int_{B(0,2^i)\backslash B(0,2^{i-1})} |f(y)| \, dy$$

$$\lesssim \widetilde{C} \sum_{i=-\infty}^{k-2} 2^{-nk+\frac{ni}{p'}} \left\| f \mathbf{1}_{B(0,2^i)\backslash B(0,2^{i-1})} \right\|_{L^p(\mathbb{R}^n)}, \tag{1.95}$$

where the implicit positive constant is independent of both T and f. On the other hand, from both Lemma 1.1.3 and (1.93), it follows that, for any $k, i \in \mathbb{Z}$ satisfying $i \in (-\infty, k-2]$,

$$\frac{\omega(2^k)}{\omega(2^i)} \lesssim \frac{\omega(2^{k-2})}{\omega(2^i)} \lesssim \left(\frac{2^{k-2}}{2^i} \right)^{\max\{M_0(\omega), M_\infty(\omega)\}+\varepsilon} \sim 2^{(k-i)M},$$

which, combined with (1.95), further implies that, for any $k \in \mathbb{Z}$,

$$\omega(2^k) \left\| T(f_{k,1}) \mathbf{1}_{B(0,2^k)\backslash B(0,2^{k-1})} \right\|_{L^p(\mathbb{R}^n)}$$

$$\lesssim \widetilde{C}\omega(2^k) \sum_{i=-\infty}^{k-2} 2^{-nk+\frac{ni}{p'}} \left\| f \mathbf{1}_{B(0,2^i)\backslash B(0,2^{i-1})} \right\|_{L^p(\mathbb{R}^n)}$$

$$\times \left\| \mathbf{1}_{B(0,2^k)\backslash B(0,2^{k-1})} \right\|_{L^p(\mathbb{R}^n)}$$

$$\sim \widetilde{C} \sum_{i=-\infty}^{k-2} 2^{(i-k)\frac{n}{p'}} \frac{\omega(2^k)}{\omega(2^i)} \omega(2^i) \left\| f \mathbf{1}_{B(\mathbf{0},2^i)\setminus B(\mathbf{0},2^{i-1})} \right\|_{L^p(\mathbb{R}^n)}$$

$$\lesssim \widetilde{C} \sum_{i=-\infty}^{k-2} 2^{(k-i)(M-\frac{n}{p'})} \omega(2^i) \left\| f \mathbf{1}_{B(\mathbf{0},2^i)\setminus B(\mathbf{0},2^{i-1})} \right\|_{L^p(\mathbb{R}^n)}, \tag{1.96}$$

where the implicit positive constants are independent of both T and f. By this and Lemma 1.2.11, we conclude that, for any $q \in (0,1]$,

$$J_1 \lesssim \widetilde{C} \left\{ \sum_{k\in\mathbb{Z}} \left[\sum_{i=-\infty}^{k-2} 2^{(k-i)(M-\frac{n}{p'})} \omega(2^i) \left\| f \mathbf{1}_{B(\mathbf{0},2^i)\setminus B(\mathbf{0},2^{i-1})} \right\|_{L^p(\mathbb{R}^n)} \right]^q \right\}^{\frac{1}{q}}$$

$$\lesssim \widetilde{C} \left\{ \sum_{k\in\mathbb{Z}} \sum_{i=-\infty}^{k-2} 2^{(k-i)(M-\frac{n}{p'})q} \left[\omega(2^i) \right]^q \left\| f \mathbf{1}_{B(\mathbf{0},2^i)\setminus B(\mathbf{0},2^{i-1})} \right\|_{L^p(\mathbb{R}^n)}^q \right\}^{\frac{1}{q}}$$

$$\sim \widetilde{C} \left\{ \sum_{i\in\mathbb{Z}} \left[\omega(2^i) \right]^q \left\| f \mathbf{1}_{B(\mathbf{0},2^i)\setminus B(\mathbf{0},2^{i-1})} \right\|_{L^p(\mathbb{R}^n)}^q \sum_{k=i+2}^{\infty} 2^{(k-i)(M-\frac{n}{p'})q} \right\}^{\frac{1}{q}}$$

$$\sim \widetilde{C} \|f\|_{\dot{\mathcal{K}}_{\omega,\mathbf{0}}^{p,q}(\mathbb{R}^n)} \lesssim \left[\widetilde{C} + \|T\|_{L^p(\mathbb{R}^n)\to L^p(\mathbb{R}^n)} \right] \|f\|_{\dot{\mathcal{K}}_{\omega,\mathbf{0}}^{p,q}(\mathbb{R}^n)}, \tag{1.97}$$

where the implicit positive constants are independent of both T and f. Moreover, if $q \in (1,\infty]$, from both (1.96) and the Hölder inequality, we deduce that

$$\omega(2^k) \left\| T(f_{k,1}) \mathbf{1}_{B(\mathbf{0},2^k)\setminus B(\mathbf{0},2^{k-1})} \right\|_{L^p(\mathbb{R}^n)}$$

$$\lesssim \widetilde{C} \left[\sum_{i=-\infty}^{k-2} 2^{\frac{(k-i)}{2}(M-\frac{n}{p'})q'} \right]^{\frac{1}{q'}}$$

$$\times \left\{ \sum_{i=-\infty}^{k-2} 2^{\frac{(k-i)}{2}(M-\frac{n}{p'})q} \left[\omega(2^k) \right]^q \left\| f \mathbf{1}_{B(\mathbf{0},2^k)\setminus B(\mathbf{0},2^{k-1})} \right\|_{L^p(\mathbb{R}^n)}^q \right\}^{\frac{1}{q}}$$

$$\sim \widetilde{C} \left\{ \sum_{i=-\infty}^{k-2} 2^{\frac{(k-i)}{2}(M-\frac{n}{p'})q} \left[\omega(2^k) \right]^q \left\| f \mathbf{1}_{B(\mathbf{0},2^k)\setminus B(\mathbf{0},2^{k-1})} \right\|_{L^p(\mathbb{R}^n)}^q \right\}^{\frac{1}{q}},$$

where the implicit positive constants are independent of both T and f. Therefore,

$$
\begin{aligned}
J_1 &\lesssim \widetilde{C} \left\{ \sum_{k\in\mathbb{Z}} \sum_{i=-\infty}^{k-2} 2^{\frac{(k-i)}{2}(M-\frac{n}{p'})q} \left[\omega(2^i)\right]^q \left\| f\mathbf{1}_{B(0,2^i)\setminus B(0,2^{i-1})} \right\|_{L^p(\mathbb{R}^n)}^q \right\}^{\frac{1}{q}} \\
&\sim \widetilde{C} \left\{ \sum_{i\in\mathbb{Z}} \left[\omega(2^i)\right]^q \left\| f\mathbf{1}_{B(0,2^i)\setminus B(0,2^{i-1})} \right\|_{L^p(\mathbb{R}^n)}^q \sum_{k=i+2}^{\infty} 2^{\frac{(k-i)}{2}(M-\frac{n}{p'})q} \right\}^{\frac{1}{q}} \\
&\sim \widetilde{C} \|f\|_{\dot{\mathcal{K}}_{\omega,0}^{p,q}(\mathbb{R}^n)} \lesssim \left[\widetilde{C} + \|T\|_{L^p(\mathbb{R}^n)\to L^p(\mathbb{R}^n)}\right] \|f\|_{\dot{\mathcal{K}}_{\omega,0}^{p,q}(\mathbb{R}^n)},
\end{aligned}
\tag{1.98}
$$

where the implicit positive constants are independent of both T and f, which completes the estimation of J_1.

Next, for J_2, using the boundedness of T on $L^p(\mathbb{R}^n)$, we find that, for any $k\in\mathbb{Z}$,

$$
\begin{aligned}
&\left\| T(f_{k,2})\mathbf{1}_{B(0,2^k)\setminus B(0,2^{k-1})} \right\|_{L^p(\mathbb{R}^n)} \\
&\leq \left\| T(f_{k,2}) \right\|_{L^p(\mathbb{R}^n)} \leq \|T\|_{L^p(\mathbb{R}^n)\to L^p(\mathbb{R}^n)} \|f_{k,2}\|_{L^p(\mathbb{R}^n)} \\
&\leq \|T\|_{L^p(\mathbb{R}^n)\to L^p(\mathbb{R}^n)} \left[\left\| f\mathbf{1}_{B(0,2^{k+1})\setminus B(0,2^k)} \right\|_{L^p(\mathbb{R}^n)} + \left\| f\mathbf{1}_{B(0,2^k)\setminus B(0,2^{k-1})} \right\|_{L^p(\mathbb{R}^n)} \right. \\
&\quad \left. + \left\| f\mathbf{1}_{B(0,2^{k-1})\setminus B(0,2^{k-2})} \right\|_{L^p(\mathbb{R}^n)} \right].
\end{aligned}
\tag{1.99}
$$

In addition, by Lemma 1.1.3, we conclude that, for any $k\in\mathbb{Z}$,

$$
\omega(2^{k-1}) \sim \omega(2^k) \sim \omega(2^{k+1}).
$$

This, together with (1.99), further implies that

$$
\begin{aligned}
J_2 &\lesssim \|T\|_{L^p(\mathbb{R}^n)\to L^p(\mathbb{R}^n)} \left(\left\{ \sum_{k\in\mathbb{Z}} \left[\omega(2^{k+1})\right]^q \left\| f\mathbf{1}_{B(0,2^{k+1})\setminus B(0,2^k)} \right\|_{L^p(\mathbb{R}^n)}^q \right\}^{\frac{1}{q}} \right. \\
&\quad + \left\{ \sum_{k\in\mathbb{Z}} \left[\omega(2^k)\right]^q \left\| f\mathbf{1}_{B(0,2^k)\setminus B(0,2^{k-1})} \right\|_{L^p(\mathbb{R}^n)}^q \right\}^{\frac{1}{q}} \\
&\quad \left. + \left\{ \sum_{k\in\mathbb{Z}} \left[\omega(2^{k-1})\right]^q \left\| f\mathbf{1}_{B(0,2^{k-1})\setminus B(0,2^{k-2})} \right\|_{L^p(\mathbb{R}^n)}^q \right\}^{\frac{1}{q}} \right) \\
&\sim \|T\|_{L^p(\mathbb{R}^n)\to L^p(\mathbb{R}^n)} \|f\|_{\dot{\mathcal{K}}_{\omega,0}^{p,q}(\mathbb{R}^n)} \\
&\lesssim \left[\widetilde{C} + \|T\|_{L^p(\mathbb{R}^n)\to L^p(\mathbb{R}^n)}\right] \|f\|_{\dot{\mathcal{K}}_{\omega,0}^{p,q}(\mathbb{R}^n)},
\end{aligned}
\tag{1.100}
$$

where the implicit positive constants are independent of both T and f. Therefore, we complete the estimation of J_2.

Finally, we deal with J_3. Notice that, for any $k, i \in \mathbb{Z}$ satisfying $i \in [k+2, \infty)$, and for any $x, y \in \mathbb{R}^n$ satisfying $2^{k-1} \leq |x| < 2^k$ and $2^{i-1} \leq |y| < 2^i$, we have $|x| < 2^k \leq 2^{i-2}$, which implies that $|x - y| \geq |x| - |y| \geq 2^{i-2}$. Applying this, (1.91), and the Hölder inequality, we find that, for any $k \in \mathbb{Z}$ and $x \in \mathbb{R}^n$ with $2^{k-1} \leq |x| < 2^k$,

$$
\begin{aligned}
|T(f_{k,3})(x)| &\leq \widetilde{C} \int_{\mathbb{R}^n} \frac{|f_{k,3}(y)|}{|x-y|^n}\, dy \\
&\leq \widetilde{C} \sum_{i=k+2}^{\infty} 2^{-ni} \int_{B(0,2^i) \setminus B(0,2^{i-1})} |f(y)|\, dy \\
&\lesssim \widetilde{C} \sum_{i=k+2}^{\infty} 2^{-\frac{ni}{p}} \left\| f \mathbf{1}_{B(0,2^i) \setminus B(0,2^{i-1})} \right\|_{L^p(\mathbb{R}^n)},
\end{aligned}
\tag{1.101}
$$

where the implicit positive constant is independent of both T and f. On the other hand, from Lemma 1.1.3 and (1.94), it follows that, for any $k, i \in \mathbb{Z}$ satisfying $i \in [k+2, \infty)$,

$$
\frac{\omega(2^k)}{\omega(2^i)} \lesssim \frac{\omega(2^{k+2})}{\omega(2^i)} \lesssim \left(\frac{2^{k+2}}{2^i} \right)^{\min\{m_0(\omega), m_\infty(\omega)\} - \varepsilon} \sim 2^{(k-i)m},
$$

which, together with (1.101), further implies that, for any $k \in \mathbb{Z}$,

$$
\begin{aligned}
&\omega(2^k) \left\| T(f_{k,3}) \mathbf{1}_{B(0,2^k) \setminus B(0,2^{k-1})} \right\|_{L^p(\mathbb{R}^n)} \\
&\lesssim \widetilde{C} \omega(2^k) \sum_{i=k+2}^{\infty} 2^{-\frac{ni}{p}} \left\| f \mathbf{1}_{B(0,2^i) \setminus B(0,2^{i-1})} \right\|_{L^p(\mathbb{R}^n)} \left\| \mathbf{1}_{B(0,2^k) \setminus B(0,2^{k-1})} \right\|_{L^p(\mathbb{R}^n)} \\
&\sim \widetilde{C} \sum_{i=k+2}^{\infty} 2^{(k-i)\frac{n}{p}} \frac{\omega(2^k)}{\omega(2^i)} \omega(2^i) \left\| f \mathbf{1}_{B(0,2^i) \setminus B(0,2^{i-1})} \right\|_{L^p(\mathbb{R}^n)} \\
&\lesssim \widetilde{C} \sum_{i=k+2}^{\infty} 2^{(k-i)(m+\frac{n}{p})} \omega(2^i) \left\| f \mathbf{1}_{B(0,2^i) \setminus B(0,2^{i-1})} \right\|_{L^p(\mathbb{R}^n)},
\end{aligned}
$$

where the implicit positive constants are independent of both T and f. By this and an argument similar to that used in the estimations of both (1.97) and (1.98), we find that

$$
J_3 \lesssim \widetilde{C} \| f \|_{\dot{\mathcal{K}}_{\omega,0}^{p,q}(\mathbb{R}^n)} \lesssim \left[\widetilde{C} + \| T \|_{L^p(\mathbb{R}^n) \to L^p(\mathbb{R}^n)} \right] \| f \|_{\dot{\mathcal{K}}_{\omega,0}^{p,q}(\mathbb{R}^n)},
$$

80 Generalized Herz Spaces of Rafeiro and Samko

where the implicit positive constants are independent of both T and f, which completes the estimation of J_3. Combining this, (1.92), (1.97), (1.98), and (1.100), we conclude that there exists a positive constant C, independent of T, such that, for any $f \in \dot{\mathcal{K}}_{\omega,\mathbf{0}}^{p,q}(\mathbb{R}^n)$,

$$\|T(f)\|_{\dot{\mathcal{K}}_{\omega,\mathbf{0}}^{p,q}(\mathbb{R}^n)} \leq C\left[\widetilde{C} + \|T\|_{L^p(\mathbb{R}^n) \to L^p(\mathbb{R}^n)}\right] \|f\|_{\dot{\mathcal{K}}_{\omega,\mathbf{0}}^{p,q}(\mathbb{R}^n)}, \qquad (1.102)$$

which completes the proof of (i).

Next, we turn to prove (ii). For this purpose, let $f \in \dot{\mathcal{K}}_{\omega}^{p,q}(\mathbb{R}^n)$. By the definition of $\|\cdot\|_{\dot{\mathcal{K}}_{\omega}^{p,q}(\mathbb{R}^n)}$, we find that, for any $\xi \in \mathbb{R}^n$,

$$\|f(\cdot + \xi)\|_{\dot{\mathcal{K}}_{\omega,\mathbf{0}}^{p,q}(\mathbb{R}^n)} \leq \|f\|_{\dot{\mathcal{K}}_{\omega}^{p,q}(\mathbb{R}^n)} < \infty.$$

Thus, for any $\xi \in \mathbb{R}^n$, $f(\cdot + \xi) \in \dot{\mathcal{K}}_{\omega,\mathbf{0}}^{p,q}(\mathbb{R}^n)$. From this and (1.102), we deduce that

$$\|T(f)(\cdot + \xi)\|_{\dot{\mathcal{K}}_{\omega,\mathbf{0}}^{p,q}(\mathbb{R}^n)} \leq C\left[\widetilde{C} + \|T\|_{L^p(\mathbb{R}^n) \to L^p(\mathbb{R}^n)}\right] \|f(\cdot + \xi)\|_{\dot{\mathcal{K}}_{\omega,\mathbf{0}}^{p,q}(\mathbb{R}^n)}$$

$$\leq C\left[\widetilde{C} + \|T\|_{L^p(\mathbb{R}^n) \to L^p(\mathbb{R}^n)}\right] \|f\|_{\dot{\mathcal{K}}_{\omega}^{p,q}(\mathbb{R}^n)},$$

which, combined with the definition of $\|\cdot\|_{\dot{\mathcal{K}}_{\omega}^{p,q}(\mathbb{R}^n)}$ and the arbitrariness of $\xi \in \mathbb{R}^n$, further implies that, for any $f \in \dot{\mathcal{K}}_{\omega}^{p,q}(\mathbb{R}^n)$,

$$\|T(f)\|_{\dot{\mathcal{K}}_{\omega}^{p,q}(\mathbb{R}^n)} \leq C\left[\widetilde{C} + \|T\|_{L^p(\mathbb{R}^n) \to L^p(\mathbb{R}^n)}\right] \|f\|_{\dot{\mathcal{K}}_{\omega}^{p,q}(\mathbb{R}^n)},$$

where C is a positive constant independent of both T and f. This finishes the proof of (ii) and hence Theorem 1.5.1. $\qquad\square$

Remark 1.5.3

(i) We point out that, in Theorem 1.5.1, both the assumptions (1.89) and (1.90) on ω are reasonable and sharp. Indeed, for any given $\alpha \in \mathbb{R}$ and for any $t \in (0, \infty)$, let $\omega(t) := t^\alpha$. Then, from Example 1.1.7, it follows that both the assumptions (1.89) and (1.90) imply that $\alpha \in (-\frac{n}{p}, \frac{n}{p'})$. In this case, Theorem 1.5.1(i) coincides with the conclusions on the classical Herz spaces obtained by Li and Yang [150, Corollary 2.1]. Moreover, by [175, Examples 5.1.1 and 5.1.2], we find that the assumption $\alpha \in (-\frac{n}{p}, \frac{n}{p'})$ on [150, Corollary 2.1] is sharp. Therefore, in this sense, both the assumptions (1.89) and (1.90) in Theorem 1.5.1 are sharp.

(ii) Recall that, when $q \in (1, \infty)$, Rafeiro and Samko established the boundedness of sublinear operators on the global generalized Herz spaces $\dot{\mathcal{K}}_{\omega}^{p,q}(\mathbb{R}^n)$ in [197, Theorem 4.3], which is just a special case of Theorem 1.5.1(ii).

The following two conclusions show the boundedness of the Hardy–Littlewood maximal operator, respectively, on local and global generalized Herz spaces. Indeed, they are simple corollaries of Theorem 1.5.1 because the Hardy–Littlewood maximal operator $\mathcal{M}$ satisfies the size condition (1.91) and is also bounded on the Lebesgue space $L^p(\mathbb{R}^n)$ when $p \in (1, \infty]$; we omit the details.

Corollary 1.5.4 *Let p, q, and ω be as in Theorem 1.5.1 and $\mathcal{M}$ the Hardy–Littlewood maximal operator as in (1.54). Then there exists a positive constant C such that, for any $f \in L^1_{\mathrm{loc}}(\mathbb{R}^n)$,*

$$\|\mathcal{M}(f)\|_{\dot{\mathcal{K}}^{p,q}_{\omega,0}(\mathbb{R}^n)} \leq C \|f\|_{\dot{\mathcal{K}}^{p,q}_{\omega,0}(\mathbb{R}^n)}.$$

Remark 1.5.5 Let $\omega(t) := t^\alpha$ for any $t \in (0, \infty)$ and for any given $\alpha \in \mathbb{R}$ in Corollary 1.5.4. Then, in this case, Corollary 1.5.4 goes back to the boundedness of the Hardy–Littlewood maximal operator on the classical homogeneous Herz spaces obtained in [130, Theorem 4.1].

Corollary 1.5.6 *Let p, q, and ω be as in Theorem 1.5.1 and $\mathcal{M}$ the Hardy–Littlewood maximal operator as in (1.54). Then there exists a positive constant C such that, for any $f \in L^1_{\mathrm{loc}}(\mathbb{R}^n)$,*

$$\|\mathcal{M}(f)\|_{\dot{\mathcal{K}}^{p,q}_{\omega}(\mathbb{R}^n)} \leq C \|f\|_{\dot{\mathcal{K}}^{p,q}_{\omega}(\mathbb{R}^n)}.$$

Next, we establish the boundedness of Calderón–Zygmund operators on local and global generalized Herz spaces. For this purpose, we first recall some basic concepts (see, for instance, [213, Chapter III]). In what follows, for any $\gamma = (\gamma_1, \ldots, \gamma_n) \in \mathbb{Z}^n_+$, any γ-order differentiable function $F(\cdot, \cdot)$ on $\mathbb{R}^n \times \mathbb{R}^n$, and any $(x, y) \in \mathbb{R}^n \times \mathbb{R}^n$, let

$$\partial^\gamma_{(1)} F(x, y) := \frac{\partial^{|\gamma|}}{\partial x_1^{\gamma_1} \cdots \partial x_n^{\gamma_n}} F(x, y)$$

and

$$\partial^\gamma_{(2)} F(x, y) := \frac{\partial^{|\gamma|}}{\partial y_1^{\gamma_1} \cdots \partial y_n^{\gamma_n}} F(x, y).$$

Moreover, let

$$\Delta := \left\{ (x, y) \in \mathbb{R}^n \times \mathbb{R}^n : x = y \right\}$$

be the *diagonal* of $\mathbb{R}^n \times \mathbb{R}^n$. Then the definition of d-order standard kernels is as follows.

Definition 1.5.7 Let $d \in \mathbb{Z}_+$ and $\delta \in (0, 1]$. A measurable function K on $(\mathbb{R}^n \times \mathbb{R}^n) \setminus \Delta$ is called a *d-order standard kernel* if there exists a positive constant C such that the following two statements hold true:

(i) for any $\gamma \in \mathbb{Z}_+^n$ with $|\gamma| \leq d$, and for any $x, y \in \mathbb{R}^n$ with $x \neq y$,

$$\left| \partial_{(1)}^\gamma K(x, y) \right| + \left| \partial_{(2)}^\gamma K(x, y) \right| \leq \frac{C}{|x - y|^{n+|\gamma|}}; \qquad (1.103)$$

(ii) for any $\gamma \in \mathbb{Z}_+^n$ with $|\gamma| = d$, and for any $x, y, z \in \mathbb{R}^n$ with $x \neq y$ and $|x - y| > 2|y - z|$,

$$\left| \partial_{(2)}^\gamma K(x, y) - \partial_{(2)}^\gamma K(x, z) \right| \leq C \frac{|y - z|^\delta}{|x - y|^{n+d+\delta}} \qquad (1.104)$$

and

$$\left| \partial_{(1)}^\gamma K(y, x) - \partial_{(1)}^\gamma K(z, x) \right| \leq C \frac{|y - z|^\delta}{|x - y|^{n+d+\delta}}.$$

The following definition of Calderón–Zygmund operators was given in [91, Definition 4.1.8] (see also [70, Definition 5.11]).

Definition 1.5.8 Let $d \in \mathbb{Z}_+$ and K be a d-order standard kernel as in Definition 1.5.7. A linear operator T is called a *d-order Calderón–Zygmund operator* with kernel K if

(i) T is bounded on $L^2(\mathbb{R}^n)$;
(ii) for any given $f \in L^2(\mathbb{R}^n)$ with compact support, and for any $x \notin \overline{\mathrm{supp}(f)} := \overline{\{x \in \mathbb{R}^n : f(x) \neq 0\}}$,

$$T(f)(x) = \int_{\mathbb{R}^n} K(x, y) f(y) \, dy. \qquad (1.105)$$

Recall that the following $L^p(\mathbb{R}^n)$ boundedness of Calderón–Zygmund operators is just [70, Theorem 5.10].

Lemma 1.5.9 *Let* $p \in (1, \infty)$, $d \in \mathbb{Z}_+$, *K be a d-order standard kernel, and T a d-order Calderón–Zygmund operator with kernel K. Then T is well defined on $L^p(\mathbb{R}^n)$ and there exists a positive constant C such that, for any $f \in L^p(\mathbb{R}^n)$,*

$$\|T(f)\|_{L^p(\mathbb{R}^n)} \leq C \|f\|_{L^p(\mathbb{R}^n)}.$$

Now, we show the boundedness of the Calderón–Zygmund operator on local generalized Herz spaces as follows. Recall that $C_{\mathrm{c}}(\mathbb{R}^n)$ is defined to be the set of all the continuous functions on $\mathbb{R}^n$ with compact support.

Corollary 1.5.10 *Let $p \in (1, \infty)$, $q \in (0, \infty)$, and $\omega \in M(\mathbb{R}_+)$ satisfy both (1.89) and (1.90). Assume that $d \in \mathbb{Z}_+$ and T is a d-order Calderón–Zygmund operator with a d-order standard kernel K. Then T has a unique extension on $\dot{\mathcal{K}}_{\omega,\mathbf{0}}^{p,q}(\mathbb{R}^n)$ and there exists a positive constant C such that, for any $f \in \dot{\mathcal{K}}_{\omega,\mathbf{0}}^{p,q}(\mathbb{R}^n)$,*

$$\|T(f)\|_{\dot{\mathcal{K}}_{\omega,\mathbf{0}}^{p,q}(\mathbb{R}^n)} \leq C \|f\|_{\dot{\mathcal{K}}_{\omega,\mathbf{0}}^{p,q}(\mathbb{R}^n)}.$$

Proof Let all the symbols be as in the present corollary. We first assume $f \in C_{\mathrm{c}}(\mathbb{R}^n)$. Then, applying both (1.105) and (1.103) with $\gamma := \mathbf{0}$, we find that, for any $x \notin \mathrm{supp}(f)$,

$$|T(f)(x)| = \left| \int_{\mathbb{R}^n} K(x,y) f(y)\, dy \right| \lesssim \int_{\mathbb{R}^n} \frac{|f(y)|}{|x-y|^n}\, dy.$$

This, together with both Lemma 1.5.9 and Theorem 1.5.1, further implies that

$$\|T(f)\|_{\dot{\mathcal{K}}_{\omega,\mathbf{0}}^{p,q}(\mathbb{R}^n)} \lesssim \|f\|_{\dot{\mathcal{K}}_{\omega,\mathbf{0}}^{p,q}(\mathbb{R}^n)},$$

where the implicit positive constant is independent of f.

In addition, by the assumption $m_0(\omega) \in (-\frac{n}{p}, \infty)$ and Theorems 1.2.42 and 1.4.1, we conclude that the local generalized Herz space $\dot{\mathcal{K}}_{\omega,\mathbf{0}}^{p,q}(\mathbb{R}^n)$ is a BQBF space having an absolutely continuous quasi-norm. From this and [228, Proposition 3.8] with X therein replaced by $\dot{\mathcal{K}}_{\omega,\mathbf{0}}^{p,q}(\mathbb{R}^n)$, we deduce that $C_{\mathrm{c}}(\mathbb{R}^n)$ is dense in $\dot{\mathcal{K}}_{\omega,\mathbf{0}}^{p,q}(\mathbb{R}^n)$. This, combined with a standard density argument (see, for instance, [161, p. 45]), further implies that, for any $f \in \dot{\mathcal{K}}_{\omega,\mathbf{0}}^{p,q}(\mathbb{R}^n)$, $T(f)$ is well defined and

$$\|T(f)\|_{\dot{\mathcal{K}}_{\omega,\mathbf{0}}^{p,q}(\mathbb{R}^n)} \lesssim \|f\|_{\dot{\mathcal{K}}_{\omega,\mathbf{0}}^{p,q}(\mathbb{R}^n)}$$

with the implicit positive constant independent of f, which completes the proof of Corollary 1.5.10.   □

Remark 1.5.11 We should point out that, in Corollary 1.5.10, when $d = 0$ and $\omega(t) := t^\alpha$ for any $t \in (0, \infty)$ and for any given $\alpha \in \mathbb{R}$, then this corollary goes back to [175, Remark 5.1.1].

Applying Corollary 1.5.10 and an argument similar to that used in the proof of Theorem 1.6.1(ii), we conclude the following boundedness of the Calderón–Zygmund operator on global generalized Herz spaces immediately; we omit the details.

Corollary 1.5.12 *Let* $p \in (1, \infty)$, $q \in (0, \infty)$, *and* $\omega \in M(\mathbb{R}_+)$ *satisfy* (1.89) *and* (1.90). *Assume that* $d \in \mathbb{Z}_+$ *and* T *is a* d-*order Calderón–Zygmund operator with a* d-*order standard kernel* K. *Then* T *has a unique extension on* $\dot{\mathcal{K}}^{p,q}_\omega(\mathbb{R}^n)$ *and there exists a positive constant* C *such that, for any* $f \in \dot{\mathcal{K}}^{p,q}_\omega(\mathbb{R}^n)$,

$$\|T(f)\|_{\dot{\mathcal{K}}^{p,q}_\omega(\mathbb{R}^n)} \leq C \|f\|_{\dot{\mathcal{K}}^{p,q}_\omega(\mathbb{R}^n)}.$$

1.6 Fefferman–Stein Vector-Valued Inequalities

In this section, we show the Fefferman–Stein vector-valued inequalities on local and global generalized Herz spaces, which are important in the study of Hardy spaces associated with these generalized Herz spaces. For instance, in order to establish the atomic characterizations of generalized Herz–Hardy spaces, we need to prove some auxiliary lemmas about the Fefferman–Stein vector-valued inequality in Chap. 4 (see Lemmas 4.3.10 and 4.3.24 below).

We first have the following Fefferman–Stein vector-valued inequality on the local generalized Herz space $\dot{\mathcal{K}}^{p,q}_{\omega,\mathbf{0}}(\mathbb{R}^n)$.

Theorem 1.6.1 *Let* $p, r \in (1, \infty]$, $q \in (0, \infty]$, *and* $\omega \in M(\mathbb{R}_+)$ *satisfy*

$$-\frac{n}{p} < m_0(\omega) \leq M_0(\omega) < \frac{n}{p'}$$

and

$$-\frac{n}{p} < m_\infty(\omega) \leq M_\infty(\omega) < \frac{n}{p'},$$

where $\frac{1}{p} + \frac{1}{p'} = 1$. *Then there exists a positive constant* C *such that, for any* $\{f_j\}_{j \in \mathbb{N}} \subset L^1_{\text{loc}}(\mathbb{R}^n)$,

$$\left\| \left\{ \sum_{j \in \mathbb{N}} \left[\mathcal{M}(f_j) \right]^r \right\}^{\frac{1}{r}} \right\|_{\dot{\mathcal{K}}^{p,q}_{\omega,\mathbf{0}}(\mathbb{R}^n)} \leq C \left\| \left(\sum_{j \in \mathbb{N}} |f_j|^r \right)^{\frac{1}{r}} \right\|_{\dot{\mathcal{K}}^{p,q}_{\omega,\mathbf{0}}(\mathbb{R}^n)}.$$

To prove Theorem 1.6.1, we need the following Fefferman–Stein vector-valued inequality on the Lebesgue space $L^p(\mathbb{R}^n)$ with $p \in (1, \infty]$, which is given in [90, Theorem 5.6.6 and Remark 5.6.7].

Lemma 1.6.2 *Let* $p, r \in (1, \infty]$. *Then there exists a positive constant C such that, for any* $\{f_j\}_{j\in\mathbb{N}} \subset L^1_{\mathrm{loc}}(\mathbb{R}^n)$,

$$\left\| \left\{ \sum_{j\in\mathbb{N}} [\mathcal{M}(f_j)]^r \right\}^{\frac{1}{r}} \right\|_{L^p(\mathbb{R}^n)} \leq C \left\| \left(\sum_{j\in\mathbb{N}} |f_j|^r \right)^{\frac{1}{r}} \right\|_{L^p(\mathbb{R}^n)}.$$

We now show Theorem 1.6.1 via Lemma 1.6.2.

Proof of Theorem 1.6.1 Let all the symbols be as in the present theorem. For any given sequence $\{f_j\}_{j\in\mathbb{N}}$ of locally integrable functions on $\mathbb{R}^n$, and for any $g \in \mathcal{M}(\mathbb{R}^n)$ and any $x \in \mathbb{R}^n$, let

$$A(g)(x) := \left\{ \sum_{j\in\mathbb{N}} [\mathcal{M}(g\eta_j)(x)]^r \right\}^{\frac{1}{r}},$$

where, for any $j \in \mathbb{N}$ and $y \in \mathbb{R}^n$,

$$\eta_i(y) := \begin{cases} \dfrac{f_i(y)}{[\sum_{j\in\mathbb{N}} |f_j(y)|^r]^{\frac{1}{r}}}, & \text{when } \left[\sum_{j\in\mathbb{N}} |f_j(y)|^r \right]^{\frac{1}{r}} \neq 0, \\[4mm] 0, & \text{when } \left[\sum_{j\in\mathbb{N}} |f_j(y)|^r \right]^{\frac{1}{r}} = 0. \end{cases}$$

Then it is easy to show that, for any $\lambda \in \mathbb{C}$ and $g \in \mathcal{M}(\mathbb{R}^n)$,

$$A(\lambda g) = |\lambda| A(g). \tag{1.106}$$

Moreover, by the Minkowski inequality, we conclude that, for any $g_1, g_2 \in \mathcal{M}(\mathbb{R}^n)$,

$$A(g_1 + g_2) \leq A(g_1) + A(g_2),$$

which, combined with (1.106), further implies that A is a sublinear operator. Now, we show that A satisfies (1.91). Indeed, observe that, for any $y \in \mathbb{R}^n$,

$$\left[\sum_{j\in\mathbb{N}} |\eta_j(y)|^r \right]^{\frac{1}{r}} \leq 1. \tag{1.107}$$

This, together with both the fact that $\mathcal{M}$ satisfies (1.91) (see [197, Remark 4.4]), and the Minkowski integral inequality, implies that, for any $x \notin \operatorname{supp}(g)$,

$$|A(g)(x)| = \left\{\sum_{j\in\mathbb{N}}\left[\mathcal{M}(g\eta_j)(x)\right]^r\right\}^{\frac{1}{r}} \lesssim \left\{\sum_{j\in\mathbb{N}}\left[\int_{\mathbb{R}^n}\frac{|g(y)\eta_j(y)|}{|x-y|^n}\,dy\right]^r\right\}^{\frac{1}{r}}$$

$$\lesssim \int_{\mathbb{R}^n}\left\{\sum_{j\in\mathbb{N}}\left[\frac{|g(y)\eta_j(y)|}{|x-y|^n}\right]^r\right\}^{\frac{1}{r}}dy \lesssim \int_{\mathbb{R}^n}\frac{|g(y)|}{|x-y|^n}\,dy, \qquad (1.108)$$

where the implicit positive constants are independent of $\{f_j\}_{j\in\mathbb{N}}$, which further implies that the sublinear operator A satisfies (1.91). Then we show that A is bounded on $L^p(\mathbb{R}^n)$. Indeed, from both Lemma 1.6.2 and (1.107), we deduce that, for any $g \in \mathcal{M}(\mathbb{R}^n)$,

$$\|A(g)\|_{L^p(\mathbb{R}^n)} = \left\|\left\{\sum_{j\in\mathbb{N}}\left[\mathcal{M}(g\eta_j)\right]^r\right\}^{\frac{1}{r}}\right\|_{L^p(\mathbb{R}^n)}$$

$$\lesssim \left\|\left(\sum_{j\in\mathbb{N}}|g\eta_j|^r\right)^{\frac{1}{r}}\right\|_{L^p(\mathbb{R}^n)} \lesssim \|g\|_{L^p(\mathbb{R}^n)}, \qquad (1.109)$$

where the implicit positive constants are independent of $\{f_j\}_{j\in\mathbb{N}}$. Thus, the sublinear operator A is bounded on $L^p(\mathbb{R}^n)$ and hence A satisfies all the assumptions of Theorem 1.5.1. This, together with the fact that the implicit positive constants in both (1.108) and (1.109) are independent of $\{f_j\}_{j\in\mathbb{N}}$, further implies that, for any $g \in \mathcal{M}(\mathbb{R}^n)$,

$$\|A(g)\|_{\dot{\mathcal{K}}^{p,q}_{\omega,0}(\mathbb{R}^n)} \lesssim \|g\|_{\dot{\mathcal{K}}^{p,q}_{\omega,0}(\mathbb{R}^n)},$$

where the implicit positive constant is independent of $\{f_j\}_{j\in\mathbb{N}}$. Using this with $g := \{\sum_{j\in\mathbb{N}}|f_j|^r\}^{1/r}$, we find that

$$\left\|\left\{\sum_{j\in\mathbb{N}}\left[\mathcal{M}(f_j)\right]^r\right\}^{\frac{1}{r}}\right\|_{\dot{\mathcal{K}}^{p,q}_{\omega,0}(\mathbb{R}^n)}$$

$$= \|A(g)\|_{\dot{\mathcal{K}}^{p,q}_{\omega,0}(\mathbb{R}^n)} \lesssim \|g\|_{\dot{\mathcal{K}}^{p,q}_{\omega,0}(\mathbb{R}^n)} \sim \left\|\left(\sum_{j\in\mathbb{N}}|f_j|^r\right)^{\frac{1}{r}}\right\|_{\dot{\mathcal{K}}^{p,q}_{\omega,0}(\mathbb{R}^n)},$$

where the implicit positive constants are independent of $\{f_j\}_{j\in\mathbb{N}}$, which completes the proof of Theorem 1.6.1.   □

Remark 1.6.3 In Theorem 1.6.1, when $\omega(t) := t^\alpha$ for any $t \in (0, \infty)$ and for any given $\alpha \in \mathbb{R}$, then, in this case, the local generalized Herz space $\dot{\mathcal{K}}^{p,q}_{\omega,0}(\mathbb{R}^n)$ coincides with the classical homogeneous Herz space $\dot{K}^{\alpha,q}_p(\mathbb{R}^n)$. In this case, Theorem 1.6.1 goes back to [131, Corollary 4.5] with $p(\cdot) := p \in (1, \infty)$.

For the global generalized Herz space $\dot{\mathcal{K}}^{p,q}_\omega(\mathbb{R}^n)$, applying an argument similar to that used in the proof of Theorem 1.6.1 with $\dot{\mathcal{K}}^{p,q}_{\omega,0}(\mathbb{R}^n)$ replaced by $\dot{\mathcal{K}}^{p,q}_\omega(\mathbb{R}^n)$, we obtain the following Fefferman–Stein vector-valued inequality on $\dot{\mathcal{K}}^{p,q}_\omega(\mathbb{R}^n)$; we omit the details.

Theorem 1.6.4 *Let p, q, r, and ω be as in Theorem 1.6.1. Then there exists a positive constant C such that, for any $\{f_j\}_{j\in\mathbb{N}} \subset L^1_{\mathrm{loc}}(\mathbb{R}^n)$,*

$$\left\|\left\{\sum_{j\in\mathbb{N}}\left[\mathcal{M}(f_j)\right]^r\right\}^{\frac{1}{r}}\right\|_{\dot{\mathcal{K}}^{p,q}_\omega(\mathbb{R}^n)} \leq C \left\|\left(\sum_{j\in\mathbb{N}}|f_j|^r\right)^{\frac{1}{r}}\right\|_{\dot{\mathcal{K}}^{p,q}_\omega(\mathbb{R}^n)}.$$

1.7 Dual and Associate Spaces of Local Generalized Herz Spaces

In this section, we first investigate the dual spaces of local generalized Herz spaces. Then, using this dual result and the relation between the dual spaces and the associate spaces of ball Banach function spaces, we obtain the associate space of $\dot{\mathcal{K}}^{p,q}_{\omega,0}(\mathbb{R}^n)$ under some reasonable and sharp assumptions.

To begin with, we study the dual space of the local generalized Herz space $\dot{\mathcal{K}}^{p,q}_{\omega,0}(\mathbb{R}^n)$. Indeed, we can establish a more general conclusion, namely, the duality of the local generalized Herz space $\dot{\mathcal{K}}^{p,q}_{\omega,\xi}(\mathbb{R}^n)$ with given $\xi \in \mathbb{R}^n$. For this purpose, we first introduce the local generalized Herz space $\dot{\mathcal{K}}^{p,q}_{\omega,\xi}(\mathbb{R}^n)$ as follows.

Definition 1.7.1 Let $p, q \in (0, \infty]$, $\omega \in M(\mathbb{R}_+)$, and $\xi \in \mathbb{R}^n$. Then the *local generalized Herz space* $\dot{\mathcal{K}}^{p,q}_{\omega,\xi}(\mathbb{R}^n)$ is defined to be the set of all the measurable functions f on $\mathbb{R}^n$ such that

$$\|f\|_{\dot{\mathcal{K}}^{p,q}_{\omega,\xi}(\mathbb{R}^n)} := \left\{\sum_{k\in\mathbb{Z}}\left[\omega(2^k)\right]^q \left\|f\mathbf{1}_{B(\xi,2^k)\setminus B(\xi,2^{k-1})}\right\|^q_{L^p(\mathbb{R}^n)}\right\}^{\frac{1}{q}} < \infty.$$

Remark 1.7.2

(i) Obviously, for any given $\xi \in \mathbb{R}^n$ and for any $f \in \mathscr{M}(\mathbb{R}^n)$, we have

$$\|f\|_{\dot{\mathcal{K}}_{\omega,\xi}^{p,q}(\mathbb{R}^n)} = \|f(\cdot + \xi)\|_{\dot{\mathcal{K}}_{\omega,0}^{p,q}(\mathbb{R}^n)}$$

and

$$\|f\|_{\dot{\mathcal{K}}_{\omega}^{p,q}(\mathbb{R}^n)} = \sup_{\xi \in \mathbb{R}^n} \|f\|_{\dot{\mathcal{K}}_{\omega,\xi}^{p,q}(\mathbb{R}^n)}.$$

(ii) In Definition 1.7.1, let $q := 1$ and $w(t) := t^{n(1-1/p)}$ for any $t \in (0, \infty)$. Then, in this case, the local generalized Herz space $\dot{\mathcal{K}}_{\omega,\xi}^{p,q}(\mathbb{R}^n)$ coincides with the ξ-*concentrated homogeneous Herz space* $E_{p,\xi}^{\gamma}(\mathbb{R}^n)$ which was introduced in [220, p. 45] to study the convergence of multi-dimensional integral operators.

Then we have the following dual theorem on the Herz space $\dot{\mathcal{K}}_{\omega,\xi}^{p,q}(\mathbb{R}^n)$. To be precise, we show that, when $p \in [1, \infty)$ and $q \in (0, \infty)$, the dual space of $\dot{\mathcal{K}}_{\omega,\xi}^{p,q}(\mathbb{R}^n)$ is just $\dot{\mathcal{K}}_{1/\omega,\xi}^{p',q'}(\mathbb{R}^n)$.

Theorem 1.7.3 *Let* $p \in [1, \infty)$, $q \in (0, \infty)$, $\omega \in M(\mathbb{R}_+)$, *and* $\xi \in \mathbb{R}^n$. *Then the dual space of* $\dot{\mathcal{K}}_{\omega,\xi}^{p,q}(\mathbb{R}^n)$, *denoted by* $(\dot{\mathcal{K}}_{\omega,\xi}^{p,q}(\mathbb{R}^n))^*$, *is* $\dot{\mathcal{K}}_{1/\omega,\xi}^{p',q'}(\mathbb{R}^n)$ *in the following sense:*

(i) *Let* $g \in \dot{\mathcal{K}}_{1/\omega,\xi}^{p',q'}(\mathbb{R}^n)$. *Then the linear functional*

$$\phi_g : f \mapsto \phi_g(f) := \int_{\mathbb{R}^n} f(y)g(y)\,dy, \qquad (1.110)$$

defined for any $f \in \dot{\mathcal{K}}_{\omega,\xi}^{p,q}(\mathbb{R}^n)$, *is bounded on* $\dot{\mathcal{K}}_{\omega,\xi}^{p,q}(\mathbb{R}^n)$.

(ii) *Conversely, any continuous linear functional on* $\dot{\mathcal{K}}_{\omega,\xi}^{p,q}(\mathbb{R}^n)$ *arises as in* (1.110) *with a unique* $g \in \dot{\mathcal{K}}_{1/\omega,\xi}^{p',q'}(\mathbb{R}^n)$.

Here $\frac{1}{p} + \frac{1}{p'} = 1$ *and*

$$q' := \begin{cases} \dfrac{q}{q-1} & \text{when } q \in [1, \infty), \\ \infty & \text{when } q \in (0, 1). \end{cases}$$

Moreover, for any $g \in \dot{\mathcal{K}}_{1/\omega,\xi}^{p',q'}(\mathbb{R}^n)$,

$$\|g\|_{\dot{\mathcal{K}}_{1/\omega,\xi}^{p',q'}(\mathbb{R}^n)} = \|\phi_g\|_{(\dot{\mathcal{K}}_{\omega,\xi}^{p,q}(\mathbb{R}^n))^*}.$$

Proof Let all the symbols be as in the present theorem. We first show (i). Indeed, if $q \in [1, \infty)$, then, by the monotone convergence theorem and the Hölder inequality, we find that, for any $f \in \dot{\mathcal{K}}_{\omega,\xi}^{p,q}(\mathbb{R}^n)$ and $g \in \dot{\mathcal{K}}_{1/\omega,\xi}^{p',q'}(\mathbb{R}^n)$,

$$
\left| \int_{\mathbb{R}^n} f(y)g(y)\,dy \right|
$$

$$
\leq \int_{\mathbb{R}^n} |f(y)g(y)|\,dy
$$

$$
= \sum_{k \in \mathbb{Z}} \int_{B(\xi,2^k) \setminus B(\xi,2^{k-1})} |f(y)g(y)|\,dy
$$

$$
\leq \sum_{k \in \mathbb{Z}} \left\| f \mathbf{1}_{B(\xi,2^k) \setminus B(\xi,2^{k-1})} \right\|_{L^p(\mathbb{R}^n)} \left\| g \mathbf{1}_{B(\xi,2^k) \setminus B(\xi,2^{k-1})} \right\|_{L^{p'}(\mathbb{R}^n)}
$$

$$
\leq \left\{ \sum_{k \in \mathbb{Z}} \left[\omega(2^k) \right]^q \left\| f \mathbf{1}_{B(\xi,2^k) \setminus B(\xi,2^{k-1})} \right\|_{L^p(\mathbb{R}^n)}^q \right\}^{\frac{1}{q}}
$$

$$
\times \left\{ \sum_{k \in \mathbb{Z}} \left[\omega(2^k) \right]^{-q'} \left\| f \mathbf{1}_{B(\xi,2^k) \setminus B(\xi,2^{k-1})} \right\|_{L^{p'}(\mathbb{R}^n)}^{q'} \right\}^{\frac{1}{q'}}
$$

$$
= \| f \|_{\dot{\mathcal{K}}_{\omega,\xi}^{p,q}(\mathbb{R}^n)} \| g \|_{\dot{\mathcal{K}}_{1/\omega,\xi}^{p',q'}(\mathbb{R}^n)}. \tag{1.111}
$$

In addition, if $q \in (0, 1)$, then, using Lemma 1.2.11 and repeating an argument similar to that used in the estimation of (1.111), we conclude that, for any $f \in \dot{\mathcal{K}}_{\omega,\xi}^{p,q}(\mathbb{R}^n)$ and $g \in \dot{\mathcal{K}}_{1/\omega,\xi}^{p',\infty}(\mathbb{R}^n)$,

$$
\left| \int_{\mathbb{R}^n} f(y)g(y)\,dy \right|
$$

$$
\leq \int_{\mathbb{R}^n} |f(y)g(y)|\,dy
$$

$$
\leq \left\{ \sum_{k \in \mathbb{Z}} \left[\omega(2^k) \right]^q \left\| f \mathbf{1}_{B(\xi,2^k) \setminus B(\xi,2^{k-1})} \right\|_{L^p(\mathbb{R}^n)}^q \right\}^{\frac{1}{q}}
$$

$$
\times \sup_{k \in \mathbb{Z}} \left\{ \left[\omega(2^k) \right]^{-1} \left\| f \mathbf{1}_{B(\xi,2^k) \setminus B(\xi,2^{k-1})} \right\|_{L^{p'}(\mathbb{R}^n)} \right\}
$$

$$
= \| f \|_{\dot{\mathcal{K}}_{\omega,\xi}^{p,q}(\mathbb{R}^n)} \| g \|_{\dot{\mathcal{K}}_{1/\omega,\xi}^{p',\infty}(\mathbb{R}^n)}. \tag{1.112}
$$

This, together with (1.111), further implies that the linear functional ϕ_g as in (1.110) is bounded on $\dot{\mathcal{K}}_{\omega,\xi}^{p,q}(\mathbb{R}^n)$ and

$$\|\phi_g\|_{(\dot{\mathcal{K}}_{\omega,\xi}^{p,q}(\mathbb{R}^n))^*} \le \|g\|_{\dot{\mathcal{K}}_{1/\omega,\xi}^{p',q'}(\mathbb{R}^n)}. \qquad (1.113)$$

This finishes the proof of (i).

Conversely, we show (ii). To this end, let $\phi \in (\dot{\mathcal{K}}_{\omega,\xi}^{p,q}(\mathbb{R}^n))^*$. We first claim that, for any $k \in \mathbb{Z}$,

$$\phi \in \left(L^p(B(\xi, 2^k) \setminus B(\xi, 2^{k-1})) \right)^*, \qquad (1.114)$$

here and thereafter, for any measurable set E in $\mathbb{R}^n$, $L^p(E)$ is defined to be the set of all the measurable functions f on $\mathbb{R}^n$ such that

$$\text{supp}(f) := \left\{ x \in \mathbb{R}^n : f(x) \ne 0 \right\} \subset E$$

and

$$\|f\|_{L^p(E)} := \|f\|_{L^p(\mathbb{R}^n)} < \infty.$$

Indeed, for any $f \in L^p(B(\xi, 2^k) \setminus B(\xi, 2^{k-1}))$ with $k \in \mathbb{Z}$, we have

$$\|f\|_{\dot{\mathcal{K}}_{\omega,\xi}^{p,q}(\mathbb{R}^n)} = \omega(2^k) \|f\mathbf{1}_{B(\xi,2^k)\setminus B(\xi,2^{k-1})}\|_{L^p(\mathbb{R}^n)}$$

$$= \omega(2^k) \|f\|_{L^p(B(\xi,2^k)\setminus B(\xi,2^{k-1}))} < \infty,$$

which implies that $f \in \dot{\mathcal{K}}_{\omega,\xi}^{p,q}(\mathbb{R}^n)$ and hence

$$|\phi(f)| \le \|\phi\|_{(\dot{\mathcal{K}}_{\omega,\xi}^{p,q}(\mathbb{R}^n))^*} \|f\|_{\dot{\mathcal{K}}_{\omega,\xi}^{p,q}(\mathbb{R}^n)}$$

$$= \omega(2^k) \|\phi\|_{(\dot{\mathcal{K}}_{\omega,\xi}^{p,q}(\mathbb{R}^n))^*} \|f\|_{L^p(B(\xi,2^k)\setminus B(\xi,2^{k-1}))}.$$

Therefore, for any $k \in \mathbb{Z}$,

$$\phi \in \left(L^p(B(\xi, 2^k) \setminus B(\xi, 2^{k-1})) \right)^*.$$

From this and the Riesz representation theorem (see, for instance, [17, Theorem 4.11]), it follows that, for any $k \in \mathbb{Z}$, there exists a $g_k \in L^{p'}(B(\xi, 2^k) \setminus B(\xi, 2^{k-1}))$ such that, for any $f \in L^p(B(\xi, 2^k) \setminus B(\xi, 2^{k-1}))$,

$$\phi(f) = \int_{B(\xi,2^k)\setminus B(\xi,2^{k-1})} f(y)g_k(y)\, dy. \qquad (1.115)$$

We now show that

$$g := \sum_{k \in \mathbb{Z}} g_k \in \dot{\mathcal{K}}_{1/\omega,\xi}^{p',q'}(\mathbb{R}^n) \tag{1.116}$$

by considering the following two cases on the operator ϕ.

Case (1) $\phi = 0$. In this case, applying (1.115), we conclude that, for any $k \in \mathbb{Z}$, $g_k = 0$ almost everywhere in $B(\xi, 2^k) \setminus B(\xi, 2^{k-1})$, which further implies that $g_k = 0$ almost everywhere in $\mathbb{R}^n$. Using this, we find that $g = 0$ almost everywhere in $\mathbb{R}^n$ and hence $g \in \dot{\mathcal{K}}_{1/\omega,\xi}^{p',q'}(\mathbb{R}^n)$. Moreover, we have

$$\|g\|_{\dot{\mathcal{K}}_{1/\omega,\xi}^{p',q'}(\mathbb{R}^n)} = 0 = \|\phi\|_{(\dot{\mathcal{K}}_{\omega,\xi}^{p,q}(\mathbb{R}^n))^*}. \tag{1.117}$$

Case (2) $\phi \neq 0$. In this case, we first claim that there exists a $k_0 \in \mathbb{Z}$ such that

$$\left| \left\{ x \in B(\xi, 2^{k_0}) \setminus B(\xi, 2^{k_0-1}) : |g_{k_0}(x)| \neq 0 \right\} \right| \in (0, \infty). \tag{1.118}$$

Indeed, by the assumption $\phi \neq 0$, we find that there exists an $f_0 \in \dot{\mathcal{K}}_{\omega,\xi}^{p,q}(\mathbb{R}^n)$ such that $\phi(f_0) \neq 0$. In addition, for any $f \in \dot{\mathcal{K}}_{\omega,\xi}^{p,q}(\mathbb{R}^n)$, we have

$$\left\| f - \sum_{k=-N}^{N} f \mathbf{1}_{B(\xi,2^k) \setminus B(\xi,2^{k-1})} \right\|_{\dot{\mathcal{K}}_{\omega,\xi}^{p,q}(\mathbb{R}^n)}$$

$$= \left\{ \sum_{j \in \mathbb{Z}} \left[\omega(2^j) \right]^q \left\| \sum_{k \in \{k \in \mathbb{Z}: |k| > N\}} f \mathbf{1}_{B(\xi,2^k) \setminus B(\xi,2^{k-1})} \mathbf{1}_{B(\xi,2^j) \setminus B(\xi,2^{j-1})} \right\|_{L^p(\mathbb{R}^n)}^q \right\}^{\frac{1}{q}}$$

$$= \left\{ \sum_{j \in \{j \in \mathbb{Z}: |j| > N\}} \left[\omega(2^j) \right]^q \left\| f \mathbf{1}_{B(\xi,2^j) \setminus B(\xi,2^{j-1})} \right\|_{L^p(\mathbb{R}^n)}^q \right\}^{\frac{1}{q}} \to 0 \tag{1.119}$$

as $N \to \infty$. From this and the continuity of ϕ on $\dot{\mathcal{K}}_{\omega,\xi}^{p,q}(\mathbb{R}^n)$, it follows that, for any $f \in \dot{\mathcal{K}}_{\omega,\xi}^{p,q}(\mathbb{R}^n)$,

$$\lim_{N \to \infty} \phi \left(\sum_{k=-N}^{N} f \mathbf{1}_{B(\xi,2^k) \setminus B(\xi,2^{k-1})} \right) = \phi(f).$$

This further implies that there exists an $N_0 \in \mathbb{N}$ such that

$$\left| \phi \left(\sum_{k=-N_0}^{N_0} f_0 \mathbf{1}_{B(\xi,2^k) \setminus B(\xi,2^{k-1})} \right) \right| > \frac{|\phi(f_0)|}{2} > 0. \qquad (1.120)$$

Thus, there exists a $k_0 \in \mathbb{Z} \cap [-N_0, N_0]$ such that $\phi(f_0 \mathbf{1}_{B(\xi,2^{k_0}) \setminus B(\xi,2^{k_0-1})}) \neq 0$. Otherwise, for any $k \in \mathbb{Z} \cap [-N_0, N_0]$, it holds true that $\phi(f_0 \mathbf{1}_{B(\xi,2^k) \setminus B(\xi,2^{k-1})}) = 0$. This, together with the linearity of ϕ, further implies that

$$\phi \left(\sum_{k=-N_0}^{N_0} f_0 \mathbf{1}_{B(\xi,2^k) \setminus B(\xi,2^{k-1})} \right) = 0,$$

which contradicts (1.120). Therefore, for this $k_0 \in \mathbb{Z}$ such that

$$\phi \left(f_0 \mathbf{1}_{B(\xi,2^{k_0}) \setminus B(\xi,2^{k_0-1})} \right) \neq 0,$$

we have

$$\left\| f_0 \mathbf{1}_{B(\xi,2^{k_0}) \setminus B(\xi,2^{k_0-1})} \right\|_{L^p(\mathbb{R}^n)} \leq \left[\omega(2^{k_0}) \right]^{-1} \| f_0 \|_{\dot{\mathcal{K}}_{\omega,\xi}^{p,q}(\mathbb{R}^n)} < \infty.$$

Using this, the assumption $\phi(f_0 \mathbf{1}_{B(\xi,2^{k_0}) \setminus B(\xi,2^{k_0-1})}) \neq 0$, and (1.115), we further conclude that

$$\left| \left\{ x \in B(\xi, 2^{k_0}) \setminus B(\xi, 2^{k_0-1}) : |g_{k_0}(x)| \neq 0 \right\} \right| \in (0, \infty),$$

which completes the proof of the above claim.

Now, in order to finish the proof of Case (2), we consider the following four cases on p and q.

Case (2.1) $p, q \in (1, \infty)$. In this case, for any given $k \in \mathbb{Z}$, when $g_k = 0$ almost everywhere in $B(\xi, 2^k) \setminus B(\xi, 2^{k-1})$, let $f_k := 0$ and, otherwise, let

$$f_k := \left[\omega(2^k) \right]^{-q'} \left\| g_k \mathbf{1}_{B(\xi,2^k) \setminus B(\xi,2^{k-1})} \right\|_{L^{p'}(\mathbb{R}^n)}^{q'-p'} |g_k|^{p'-1} \operatorname{sgn}\overline{g_k},$$

here and thereafter, the function sgn is defined by setting

$$\operatorname{sgn} z := \begin{cases} \dfrac{z}{|z|}, & \text{when } z \in \mathbb{C} \setminus \{0\}, \\[2mm] 0, & \text{when } z = 0. \end{cases}$$

Then we show that, for any $k \in \mathbb{Z}$,

$$\left\| f_k \mathbf{1}_{B(\xi,2^k) \setminus B(\xi,2^{k-1})} \right\|_{L^p(\mathbb{R}^n)} = \left[\omega(2^k) \right]^{-q'} \left\| g_k \mathbf{1}_{B(\xi,2^k) \setminus B(\xi,2^{k-1})} \right\|_{L^{p'}(\mathbb{R}^n)}^{q'-1}$$

(1.121)

and

$$\int_{B(\xi,2^k) \setminus B(\xi,2^{k-1})} f_k(y) g_k(y)\, dy = \left[\omega(2^k) \right]^{-q'} \left\| g_k \mathbf{1}_{B(\xi,2^k) \setminus B(\xi,2^{k-1})} \right\|_{L^{p'}(\mathbb{R}^n)}^{q'}.$$

(1.122)

Indeed, for any $k \in \mathbb{Z}$ satisfying $g_k = 0$ almost everywhere in $B(\xi, 2^k) \setminus B(\xi, 2^{k-1})$, it is obvious that both (1.121) and (1.122) hold true. On the other hand, for any given $k \in \mathbb{Z}$ such that

$$\left| \left\{ x \in B(\xi, 2^k) \setminus B(\xi, 2^{k-1}) : |g_k(x)| \neq 0 \right\} \right| \in (0, \infty),$$

we have $\left\| g_k \mathbf{1}_{B(\xi,2^k) \setminus B(\xi,2^{k-1})} \right\|_{L^{p'}(\mathbb{R}^n)} \in (0, \infty)$. Thus,

$$\left\| f_k \mathbf{1}_{B(\xi,2^k) \setminus B(\xi,2^{k-1})} \right\|_{L^p(\mathbb{R}^n)}$$

$$= \left[\omega(2^k) \right]^{-q'} \left\| g_k \mathbf{1}_{B(\xi,2^k) \setminus B(\xi,2^{k-1})} \right\|_{L^{p'}(\mathbb{R}^n)}^{q'-p'} \left[\int_{B(\xi,2^k) \setminus B(\xi,2^{k-1})} |g_k(y)|^{p'}\, dy \right]^{\frac{1}{p}}$$

$$= \left[\omega(2^k) \right]^{-q'} \left\| g_k \mathbf{1}_{B(\xi,2^k) \setminus B(\xi,2^{k-1})} \right\|_{L^{p'}(\mathbb{R}^n)}^{q'-p'} \left\| g_k \mathbf{1}_{B(\xi,2^k) \setminus B(\xi,2^{k-1})} \right\|_{L^{p'}(\mathbb{R}^n)}^{\frac{p'}{p}}$$

$$= \left[\omega(2^k) \right]^{-q'} \left\| g_k \mathbf{1}_{B(\xi,2^k) \setminus B(\xi,2^{k-1})} \right\|_{L^{p'}(\mathbb{R}^n)}^{q'-1}$$

and

$$\int_{B(\xi,2^k) \setminus B(\xi,2^{k-1})} f_k(y) g_k(y)\, dy$$

$$= \left[\omega(2^k) \right]^{-q'} \left\| g_k \mathbf{1}_{B(\xi,2^k) \setminus B(\xi,2^{k-1})} \right\|_{L^{p'}(\mathbb{R}^n)}^{q'-p'} \int_{B(\xi,2^k) \setminus B(\xi,2^{k-1})} |g_k(y)|^{p'}\, dy$$

$$= \left[\omega(2^k) \right]^{-q'} \left\| g_k \mathbf{1}_{B(\xi,2^k) \setminus B(\xi,2^{k-1})} \right\|_{L^{p'}(\mathbb{R}^n)}^{q'-p'} \left\| g_k \mathbf{1}_{B(\xi,2^k) \setminus B(\xi,2^{k-1})} \right\|_{L^{p'}(\mathbb{R}^n)}^{p'}$$

$$= \left[\omega(2^k) \right]^{-q'} \left\| g_k \mathbf{1}_{B(\xi,2^k) \setminus B(\xi,2^{k-1})} \right\|_{L^{p'}(\mathbb{R}^n)}^{q'}.$$

These further imply that both (1.121) and (1.122) hold true. Moreover, apply-
ing (1.121) and the assumption that, for any $j \in \mathbb{N}$, $g_j \in L^{p'}(B(\xi, 2^j) \setminus B(\xi, 2^{j-1}))$,
we conclude that, for any $N \in \mathbb{N}$,

$$
\left\| \sum_{k=-N}^{N} f_k \right\|_{\dot{\mathcal{K}}^{p,q}_{\omega,\xi}(\mathbb{R}^n)}
$$

$$
= \left\{ \sum_{j \in \mathbb{Z}} \left[\omega(2^j) \right]^q \left\| \sum_{k=-N}^{N} f_k \mathbf{1}_{B(\xi, 2^j) \setminus B(\xi, 2^{j-1})} \right\|_{L^p(\mathbb{R}^n)}^q \right\}^{\frac{1}{q}}
$$

$$
= \left\{ \sum_{j=-N}^{N} \left[\omega(2^j) \right]^q \left\| f_j \mathbf{1}_{B(\xi, 2^j) \setminus B(\xi, 2^{j-1})} \right\|_{L^p(\mathbb{R}^n)}^q \right\}^{\frac{1}{q}}
$$

$$
= \left\{ \sum_{j=-N}^{N} \left[\omega(2^j) \right]^{-q'} \left\| g_j \mathbf{1}_{B(\xi, 2^j) \setminus B(\xi, 2^{j-1})} \right\|_{L^{p'}(\mathbb{R}^n)}^{q'} \right\}^{\frac{1}{q}} < \infty, \qquad (1.123)
$$

which implies that

$$
\sum_{k=-N}^{N} f_k \in \dot{\mathcal{K}}^{p,q}_{\omega,\xi}(\mathbb{R}^n).
$$

By this, the linearity of ϕ, (1.121), (1.115), and (1.122), we find that, for any $N \in \mathbb{N}$,

$$
\phi\left(\sum_{k=-N}^{N} f_k \right) = \sum_{k=-N}^{N} \phi(f_k)
$$

$$
= \sum_{k=-N}^{N} \int_{B(\xi, 2^k) \setminus B(\xi, 2^{k-1})} f_k(y) g_k(y) \, dy
$$

$$
= \sum_{k=-N}^{N} \left[\omega(2^k) \right]^{-q'} \left\| g_k \mathbf{1}_{B(\xi, 2^k) \setminus B(\xi, 2^{k-1})} \right\|_{L^{p'}(\mathbb{R}^n)}^{q'},
$$

which, together with (1.123), implies that

$$\sum_{k=-N}^{N} \left[\omega(2^k)\right]^{-q'} \left\| g_k \mathbf{1}_{B(\xi,2^k)\setminus B(\xi,2^{k-1})} \right\|_{L^{p'}(\mathbb{R}^n)}^{q'}$$

$$\leq \|\phi\|_{(\dot{\mathcal{K}}_{\omega,\xi}^{p,q}(\mathbb{R}^n))^*} \left\| \sum_{k=-N}^{N} f_k \right\|_{\dot{\mathcal{K}}_{\omega,\xi}^{p,q}(\mathbb{R}^n)}$$

$$= \|\phi\|_{(\dot{\mathcal{K}}_{\omega,\xi}^{p,q}(\mathbb{R}^n))^*} \left\{ \sum_{k=-N}^{N} \left[\omega(2^k)\right]^{-q'} \left\| g_k \mathbf{1}_{B(\xi,2^k)\setminus B(\xi,2^{k-1})} \right\|_{L^{p'}(\mathbb{R}^n)}^{q'} \right\}^{\frac{1}{q}}.$$

Then, by this, (1.116), and (1.118), we conclude that, for any $N \in \mathbb{N} \cap [|k_0|, \infty)$,

$$\left\{ \sum_{k=-N}^{N} \left[\omega(2^k)\right]^{-q'} \left\| g \mathbf{1}_{B(\xi,2^k)\setminus B(\xi,2^{k-1})} \right\|_{L^{p'}(\mathbb{R}^n)}^{q'} \right\}^{\frac{1}{q'}}$$

$$= \left\{ \sum_{k=-N}^{N} \left[\omega(2^k)\right]^{-q'} \left\| g_k \mathbf{1}_{B(\xi,2^k)\setminus B(\xi,2^{k-1})} \right\|_{L^{p'}(\mathbb{R}^n)}^{q'} \right\}^{\frac{1}{q'}}$$

$$\leq \|\phi\|_{(\dot{\mathcal{K}}_{\omega,\xi}^{p,q}(\mathbb{R}^n))^*}.$$

Letting $N \to \infty$, we obtain

$$\|g\|_{\dot{\mathcal{K}}_{1/\omega,\xi}^{p',q'}(\mathbb{R}^n)} \leq \|\phi\|_{(\dot{\mathcal{K}}_{\omega,\xi}^{p,q}(\mathbb{R}^n))^*}, \tag{1.124}$$

which further implies that $g \in \dot{\mathcal{K}}_{1/\omega,\xi}^{p',q'}(\mathbb{R}^n)$ in this case.

   Case (2.2) $p \in (1,\infty)$ and $q \in (0,1]$. In this case, for any given $k \in \mathbb{Z}$, when $g_k = 0$ almost everywhere in $B(\xi, 2^k) \setminus B(\xi, 2^{k-1})$, let $\widetilde{f}_k := 0$ and, otherwise, let

$$\widetilde{f}_k := \left[\omega(2^k)\right]^{-1} \left\| g_k \mathbf{1}_{B(\xi,2^k)\setminus B(\xi,2^{k-1})} \right\|_{L^{p'}(\mathbb{R}^n)}^{1-p'} |g_k|^{p'-1} \operatorname{sgn}\overline{g_k}.$$

Then, repeating an argument similar to that used in the proof of Case (2.1) via replacing $\{f_k\}_{k\in\mathbb{Z}}$ and q' therein, respectively, by $\{\widetilde{f}_k\}_{k\in\mathbb{Z}}$ and ∞, we obtain

$$\|g\|_{\dot{\mathcal{K}}_{1/\omega,\xi}^{p',\infty}(\mathbb{R}^n)} \leq \|\phi\|_{(\dot{\mathcal{K}}_{\omega,\xi}^{p,q}(\mathbb{R}^n))^*}, \tag{1.125}$$

which then completes the proof of (1.116) in this case.

Case (2.3) $p = 1$ and $q \in (1, \infty)$. In this case, for any given $k \in \mathbb{Z}$, when $g_k = 0$ almost everywhere in $B(\xi, 2^k) \setminus B(\xi, 2^{k-1})$, let $f_k^\star := 0$ and, otherwise, for any $\varepsilon \in (0, \|g_k\|_{L^\infty(\mathbb{R}^n)})$, let

$$E_{k,\varepsilon} := \left\{ x \in B(\xi, 2^k) \setminus B(\xi, 2^{k-1}) : |g_k(x)| > \|g_k\|_{L^\infty(\mathbb{R}^n)} - \varepsilon \right\} \qquad (1.126)$$

and

$$f_{k,\varepsilon}^\star := \frac{1}{|E_{k,\varepsilon}|} \left[\omega(2^k) \right]^{-q'} \|g_k\|_{L^\infty(\mathbb{R}^n)}^{q'-1} \operatorname{sgn}\overline{g_k} \mathbf{1}_{E_{k,\varepsilon}}.$$

Then, repeating an argument similar to that used in the proof of Case (2.1) with f_k and p' therein replaced, respectively, by $f_k^\star$ or $f_{k,\varepsilon}^\star$ and ∞, we conclude that

$$\|g\|_{\dot{\mathcal{K}}_{1/\omega,\xi}^{\infty,q'}(\mathbb{R}^n)} \lesssim \|\phi\|_{(\dot{\mathcal{K}}_{\omega,\xi}^{1,q}(\mathbb{R}^n))^*}. \qquad (1.127)$$

This then implies that $g \in \dot{\mathcal{K}}_{1/\omega,\xi}^{\infty,q'}(\mathbb{R}^n)$, and hence finishes the proof of (1.116) in this case.

Case (2.4) $p = 1$ and $q \in (0, 1]$. In this case, for any given $k \in \mathbb{Z}$, when $g_k = 0$ almost everywhere in $B(\xi, 2^k) \setminus B(\xi, 2^{k-1})$, let $f_k^\star := 0$ and, otherwise, for any $\varepsilon \in (0, \|g_k\|_{L^\infty(\mathbb{R}^n)})$, let

$$f_{k,\varepsilon}^\clubsuit := \frac{1}{|E_{k,\varepsilon}|} \left[\omega(2^k) \right]^{-1} \operatorname{sgn}\overline{g_k} \mathbf{1}_{E_{k,\varepsilon}},$$

where $E_{k,\varepsilon}$ is the same as in (1.126). Then, repeating an argument similar to that used in the proof of Case (2.1) with f_k, p', and q' therein replaced, respectively, by $f_k^\clubsuit$ or $f_{k,\varepsilon}^\star$, ∞, and ∞, we obtain

$$\|g\|_{\dot{\mathcal{K}}_{1/\omega,\xi}^{\infty,\infty}(\mathbb{R}^n)} \lesssim \|\phi\|_{(\dot{\mathcal{K}}_{\omega,\xi}^{1,q}(\mathbb{R}^n))^*}, \qquad (1.128)$$

which completes the proof of (1.116) in this case. Thus, (1.116) holds true. From this and (i), it follows that the linear functional ϕ_g as in (1.110) is bounded on $\dot{\mathcal{K}}_{\omega,\xi}^{p,q}(\mathbb{R}^n)$.

Next, we show that $\phi = \phi_g$. Indeed, applying (1.111), we find that, for any $f \in \dot{\mathcal{K}}_{\omega,\xi}^{p,q}(\mathbb{R}^n)$,

$$|fg| \in L^1(\mathbb{R}^n).$$

Combining this, (1.119), both the continuity and the linearity of ϕ on $\dot{\mathcal{K}}^{p,q}_{\omega,\xi}(\mathbb{R}^n)$, (1.115), (1.116), and the dominated convergence theorem, we further conclude that, for any $f \in \dot{\mathcal{K}}^{p,q}_{\omega,\xi}(\mathbb{R}^n)$,

$$
\begin{aligned}
\phi(f) &= \lim_{N\to\infty} \sum_{k=-N}^{N} \phi\left(f\mathbf{1}_{B(\xi,2^k)\setminus B(\xi,2^{k-1})}\right) \\
&= \lim_{N\to\infty} \sum_{k=-N}^{N} \int_{B(\xi,2^k)\setminus B(\xi,2^{k-1})} f(y)\mathbf{1}_{B(\xi,2^k)\setminus B(\xi,2^{k-1})}(y)g_k(y)\,dy \\
&= \lim_{N\to\infty} \int_{\mathbb{R}^n} f(y)\mathbf{1}_{B(\xi,2^N)\setminus B(\xi,2^{-N-1})}(y)g(y)\,dy \\
&= \int_{\mathbb{R}^n} f(y)g(y)\,dy = \phi_g(f).
\end{aligned}
$$

This finishes the proof that $\phi = \phi_g$. Moreover, from this, (1.113), (1.117), (1.124), (1.125), (1.127), and (1.128), we infer that

$$
\|\phi\|_{(\dot{\mathcal{K}}^{p,q}_{\omega,\xi}(\mathbb{R}^n))^*} = \|\phi_g\|_{(\dot{\mathcal{K}}^{p,q}_{\omega,\xi}(\mathbb{R}^n))^*} \leq \|g\|_{\dot{\mathcal{K}}^{p',q'}_{1/\omega,\xi}(\mathbb{R}^n)} \leq \|\phi\|_{(\dot{\mathcal{K}}^{p,q}_{\omega,\xi}(\mathbb{R}^n))^*}.
$$

Then we have

$$
\|\phi_g\|_{(\dot{\mathcal{K}}^{p,q}_{\omega,\xi}(\mathbb{R}^n))^*} = \|g\|_{\dot{\mathcal{K}}^{p,q}_{\omega,\xi}(\mathbb{R}^n)}.
$$

Using this and the linearity of ϕ_g about g, we further find that the g is unique. This finishes the proof of (ii) and hence Theorem 1.7.3.   □

Remark 1.7.4 We should point out that, in Theorem 1.7.3, when $\xi = \mathbf{0}$ and $\omega(t) := t^\alpha$ for any $t \in (0, \infty)$ and for any given $\alpha \in \mathbb{R}$, then the conclusion obtained in this theorem goes back to that of [175, Corollary 1.2.1].

On the other hand, we now show that the dual space of $\dot{\mathcal{K}}^{p,q}_{\omega,\xi}(\mathbb{R}^n)$ with $p \in (0, 1)$, $q \in (0, \infty)$, $\omega \in M(\mathbb{R}_+)$, and $\xi \in \mathbb{R}^n$ is trivial as follows, which plays an important role in the study of block spaces in the next chapter and also of independent interest.

Theorem 1.7.5 *Let* $p \in (0, 1)$, $q \in (0, \infty)$, $\omega \in M(\mathbb{R}_+)$, *and* $\xi \in \mathbb{R}^n$. *Then*

$$
\left(\dot{\mathcal{K}}^{p,q}_{\omega,\xi}(\mathbb{R}^n)\right)^* = \{0\}.
$$

To show the above theorem, we need the following triviality of the dual space of Lebesgue spaces which is a part of [90, Theorem 1.4.16].

Lemma 1.7.6 *Let $p \in (0, 1)$ and $E \subset \mathbb{R}^n$ be a measurable set. Then*

$$\left(L^p(E)\right)^* = \{0\}.$$

Via Lemma 1.7.6, we now prove Theorem 1.7.5.

Proof of Theorem 1.7.5 Let all the symbols be the same as in the present theorem. Obviously, we have

$$\{0\} \subset \left(\dot{\mathcal{K}}_{\omega,\xi}^{p,q}(\mathbb{R}^n)\right)^*.$$

Conversely, we prove that

$$\left(\dot{\mathcal{K}}_{\omega,\xi}^{p,q}(\mathbb{R}^n)\right)^* \subset \{0\}. \tag{1.129}$$

To do this, let $\phi \in (\dot{\mathcal{K}}_{\omega,\xi}^{p,q}(\mathbb{R}^n))^*$. Then, applying (1.114), we find that, for any $k \in \mathbb{Z}$,

$$\phi \in \left(L^p(B(\xi, 2^k) \setminus B(\xi, 2^{k-1}))\right)^*,$$

which, together with Lemma 1.7.6 with $E := B(\xi, 2^k) \setminus B(\xi, 2^{k-1})$, further implies that

$$\phi = 0 \text{ on } L^p(B(\xi, 2^k) \setminus B(\xi, 2^{k-1})). \tag{1.130}$$

Let $f \in \dot{\mathcal{K}}_{\omega,\xi}^{p,q}(\mathbb{R}^n)$ and, for any $N \in \mathbb{N}$,

$$f_N := \sum_{k=-N}^{N} f \mathbf{1}_{B(\xi,2^k) \setminus B(\xi,2^{k-1})}.$$

Then, from (1.119), we infer that, for any $N \in \mathbb{N}$, $f_N \in \dot{\mathcal{K}}_{\omega,\xi}^{p,q}(\mathbb{R}^n)$ and

$$\lim_{N \to \infty} \|f_N - f\|_{\dot{\mathcal{K}}_{\omega,\xi}^{p,q}(\mathbb{R}^n)} = 0. \tag{1.131}$$

On the other hand, by Definition 1.2.1(i), we conclude that, for any $k \in \mathbb{Z}$,

$$\left\|f \mathbf{1}_{B(\xi,2^k) \setminus B(\xi,2^{k-1})}\right\|_{L^p} \leq \left[\omega(2^k)\right]^{-1} \|f\|_{\dot{\mathcal{K}}_{\omega,\xi}^{p,q}(\mathbb{R}^n)} < \infty,$$

which then implies that

$$f \mathbf{1}_{B(\xi,2^k) \setminus B(\xi,2^{k-1})} \in L^p(B(\xi, 2^k) \setminus B(\xi, 2^{k-1})).$$

Using this, (1.131), the continuity and the linearity of ϕ on $\dot{\mathcal{K}}^{p,q}_{\omega,\xi}(\mathbb{R}^n)$, and (1.130), we obtain

$$\phi(f) = \lim_{N \to \infty} \phi(f_N) = \lim_{N \to \infty} \sum_{k=-N}^{N} \phi\left(f \mathbf{1}_{B(\xi,2^k) \setminus B(\xi,2^{k-1})}\right)$$

$$= \lim_{N \to \infty} \sum_{k=-N}^{N} 0 = 0.$$

This implies that $\phi = 0$ on $\dot{\mathcal{K}}^{p,q}_{\omega,\xi}(\mathbb{R}^n)$, and hence (1.129) holds true, which then completes the proof of Theorem 1.7.5. $\square$

Now, we turn to find the associate space of $\dot{\mathcal{K}}^{p,q}_{\omega,\mathbf{0}}(\mathbb{R}^n)$. Indeed, we have the following conclusion which is an essential tool to study the associate space of $\dot{\mathcal{K}}^{p,q}_{\omega,\mathbf{0}}(\mathbb{R}^n)$.

Lemma 1.7.7 *Let X be a ball Banach function space, and let X' and X^* denote, respectively, its associate space and dual space. Then*

(i) *for any $f \in X'$,*

$$\|f\|_{X'} = \sup\left\{\left|\int_{\mathbb{R}^n} f(x)g(x)\,dx\right| : g \in X,\ \|g\|_X = 1\right\};$$

(ii) $X' \subset X^*$;

(iii) $X' = X^*$ *with the same norms if and only if X has an absolutely continuous norm.*

By a slight modification, respectively, on the proofs of Bennett and Sharpley [11, Chapter 1, Lemma 2.8, Theorem 2.9, and Corollary 4.3], we obtain (i), (ii), and (iii) of Lemma 1.7.7; we omit the details.

Remark 1.7.8 We should point out that, in Lemma 1.7.7(iii), if X is not a ball Banach function space, then the conclusion of Lemma 1.7.7(iii) may not hold true. For instance, it was proved by Fefferman and Stein [77] that the dual space of the Hardy space $H^1(\mathbb{R}^n)$ is $\mathrm{BMO}(\mathbb{R}^n)$, but, by Proposition 1.2.20, the associate space of $H^1(\mathbb{R}^n)$ is $L^\infty(\mathbb{R}^n)$ which is a proper subset of $\mathrm{BMO}(\mathbb{R}^n)$.

Via using Lemma 1.7.7(iii), we now show that the associate space of $\dot{\mathcal{K}}^{p,q}_{\omega,\mathbf{0}}(\mathbb{R}^n)$ is $\dot{\mathcal{K}}^{p',q'}_{1/\omega,\mathbf{0}}(\mathbb{R}^n)$ under some reasonable and sharp assumptions. Concretely, we have the following conclusion.

Theorem 1.7.9 *Let $p, q \in (1, \infty)$ and $\omega \in M(\mathbb{R}_+)$ satisfy*

$$-\frac{n}{p} < m_0(\omega) \le M_0(\omega) < \frac{n}{p'}.$$

Then

$$\left(\dot{\mathcal{K}}_{\omega,\mathbf{0}}^{p,q}(\mathbb{R}^n)\right)' = \dot{\mathcal{K}}_{1/\omega,\mathbf{0}}^{p',q'}(\mathbb{R}^n)$$

with the same norms, where $(\dot{\mathcal{K}}_{\omega,\mathbf{0}}^{p,q}(\mathbb{R}^n))'$ *denotes the associate space of the local generalized Herz space* $\dot{\mathcal{K}}_{\omega,\mathbf{0}}^{p,q}(\mathbb{R}^n)$.

Proof Let all the symbols be as in the present theorem. Then, by Theorem 1.2.46 and the assumptions of the present theorem, we find that the local generalized Herz space $\dot{\mathcal{K}}_{\omega,\mathbf{0}}^{p,q}(\mathbb{R}^n)$ is a BBF space. In addition, from Theorem 1.4.1, it follows that $\dot{\mathcal{K}}_{\omega,\mathbf{0}}^{p,q}(\mathbb{R}^n)$ has an absolutely continuous norm. Combining this, Lemma 1.7.7(iii), and Theorem 1.7.3, we further conclude that

$$\left(\dot{\mathcal{K}}_{\omega,\mathbf{0}}^{p,q}(\mathbb{R}^n)\right)' = \left(\dot{\mathcal{K}}_{\omega,\mathbf{0}}^{p,q}(\mathbb{R}^n)\right)^* = \dot{\mathcal{K}}_{1/\omega,\mathbf{0}}^{p',q'}(\mathbb{R}^n)$$

with the same norms, which completes the proof of Theorem 1.7.9. $\square$

1.8 Extrapolation Theorems

In this section, we establish the extrapolation theorems of local and global generalized Herz spaces via Theorem 1.7.9 obtained in last section. To this end, we first recall the following definition of the Muckenhoupt weights (see, for instance, [90, Chapter 7]).

Definition 1.8.1 An $A_p(\mathbb{R}^n)$-*weight* υ, with $p \in [1, \infty)$, is a locally integrable and nonnegative function on $\mathbb{R}^n$ satisfying that, when $p \in (1, \infty)$,

$$[\upsilon]_{A_p(\mathbb{R}^n)} := \sup_{B \in \mathbb{B}} \left[\frac{1}{|B|} \int_B \upsilon(x)\,dx \right] \left\{ \frac{1}{|B|} \int_B [\upsilon(x)]^{\frac{1}{1-p}}\,dx \right\}^{p-1} < \infty$$

and

$$[\upsilon]_{A_1(\mathbb{R}^n)} := \sup_{B \in \mathbb{B}} \frac{1}{|B|} \int_B \upsilon(x)\,dx \left[\left\| \upsilon^{-1} \right\|_{L^\infty(B)} \right] < \infty.$$

Moreover, let

$$A_\infty(\mathbb{R}^n) := \bigcup_{p \in [1,\infty)} A_p(\mathbb{R}^n).$$

Then we have the following extrapolation theorem about the local generalized Herz space $\dot{\mathcal{K}}_{\omega,\mathbf{0}}^{p,q}(\mathbb{R}^n)$.

Theorem 1.8.2 *Let $p, q \in (1, \infty)$, $r_0 \in [1, \infty)$, and $\omega \in M(\mathbb{R}_+)$ satisfy*

$$-\frac{n}{p} < m_0(\omega) \le M_0(\omega) < \frac{n}{p'}$$

and

$$-\frac{n}{p} < m_\infty(\omega) \le M_\infty(\omega) < \frac{n}{p'}.$$

Assume $\mathcal{F}$ is a set of all pairs of nonnegative measurable functions (F, G) such that, for any given $\upsilon \in A_{r_0}(\mathbb{R}^n)$,

$$\int_{\mathbb{R}^n} [F(x)]^{r_0} \upsilon(x)\,dx \le C_{(r,[\upsilon]_{A_{r_0}(\mathbb{R}^n)})} \int_{\mathbb{R}^n} [G(x)]^{r_0} \upsilon(x)\,dx,$$

where the positive constant $C_{(r_0,[\upsilon]_{A_{r_0}(\mathbb{R}^n)})}$ is independent of (F, G), but depending on both r_0 and $[\upsilon]_{A_{r_0}(\mathbb{R}^n)}$. Then there exists a positive constant C such that, for any $(F, G) \in \mathcal{F}$ with $\|F\|_{\dot{\mathcal{K}}^{p,q}_{\omega,0}(\mathbb{R}^n)} < \infty$,

$$\|F\|_{\dot{\mathcal{K}}^{p,q}_{\omega,0}(\mathbb{R}^n)} \le C \|G\|_{\dot{\mathcal{K}}^{p,q}_{\omega,0}(\mathbb{R}^n)}.$$

Recall that Tao et al. [228, Lemma 2.13] obtained the following extrapolation theorem of general ball Banach function spaces.

Lemma 1.8.3 *Let X be a ball Banach function space satisfying that the Hardy–Littlewood maximal operator $\mathcal{M}$ is bounded on both X and its associate space X', and let $r_0 \in [1, \infty)$. Assume $\mathcal{F}$ is a set of all pairs of nonnegative measurable functions (F, G) such that, for any given $\upsilon \in A_{r_0}(\mathbb{R}^n)$,*

$$\int_{\mathbb{R}^n} [F(x)]^{r_0} \upsilon(x)\,dx \le C_{(r,[\upsilon]_{A_{r_0}(\mathbb{R}^n)})} \int_{\mathbb{R}^n} [G(x)]^{r_0} \upsilon(x)\,dx,$$

where the positive constant $C_{(r_0,[\upsilon]_{A_{r_0}(\mathbb{R}^n)})}$ is independent of (F, G), but depending on both r_0 and $[\upsilon]_{A_{r_0}(\mathbb{R}^n)}$. Then there exists a positive constant C such that, for any $(F, G) \in \mathcal{F}$ with $\|F\|_X < \infty$,

$$\|F\|_X \le C \|G\|_X.$$

Remark 1.8.4 We should point out that Lemma 1.8.3 has a wide range of applications. Here we give several function spaces to which Lemma 1.8.3 can be applied.

(i) Let $p \in (1, \infty)$ and $\upsilon \in A_p(\mathbb{R}^n)$. Then, in this case, as was mentioned in [228, Subsection 4.4], the weighted Lebesgue space $L^p_\upsilon(\mathbb{R}^n)$ satisfies all the assumptions of Lemma 1.8.3. Therefore, Lemma 1.8.3 with $X := L^p_\upsilon(\mathbb{R}^n)$

holds true, which is just the classical Rubio de Francia extrapolation theorem for weighted Lebesgue spaces (see, for instance, [49, Theorem 5.23]).

(ii) Let $\vec{p} := (p_1, \ldots, p_n) \in (1, \infty)^n$. Then, in this case, as was pointed out in [228, Subsection 4.2], the mixed-norm Lebesgue space $L^{\vec{p}}(\mathbb{R}^n)$ satisfies all the assumptions of Lemma 1.8.3. This further implies that Lemma 1.8.3 with X therein replaced by $L^{\vec{p}}(\mathbb{R}^n)$ holds true, which was also obtained in [126, Lemma 2.15].

(iii) Let $1 < q \leq p < \infty$. Then, in this case, as was mentioned in [29, Subsection 4.1], the Morrey space $M_q^p(\mathbb{R}^n)$ satisfies all the assumptions of Lemma 1.8.3. Thus, Lemma 1.8.3 with $X := M_q^p(\mathbb{R}^n)$ holds true, which is just [110, Theorem 3.1] with $u(x, r)$ and X therein replaced, respectively, by $r^{n(\frac{1}{q}-\frac{1}{p})}$ and $L^q(\mathbb{R}^n)$.

(iv) Let $p(\cdot) \in C^{\log}(\mathbb{R}^n)$ with $1 < p_- \leq p_+ < \infty$, where p_- and p_+ are defined, respectively, in (1.59) and (1.60). Then, in this case, as was pointed out in [228, Subsection 4.3], the variable Lebesgue space $L^{p(\cdot)}(\mathbb{R}^n)$ satisfies all the assumptions of Lemma 1.8.3. This further implies that Lemma 1.8.3 with X therein replaced by $L^{p(\cdot)}(\mathbb{R}^n)$ holds true, which is just the extrapolation theorem for variable Lebesgue spaces (see, for instance, [49, Corollary 5.32]).

Therefore, in order to prove Theorem 1.8.2, it suffices to show that, under the assumptions of Theorem 1.8.2, the local generalized Herz space $\dot{\mathcal{K}}_{\omega,\mathbf{0}}^{p,q}(\mathbb{R}^n)$ satisfies all the assumptions of Lemma 1.8.3. To achieve this, we need the following two auxiliary lemmas about the boundedness of the Hardy–Littlewood maximal operator on local generalized Herz spaces and their associate spaces.

Lemma 1.8.5 *Let* $p, q \in (0, \infty)$ *and* $\omega \in M(\mathbb{R}_+)$ *satisfy* $m_0(\omega) \in (-\frac{n}{p}, \infty)$ *and* $m_\infty(\omega) \in (-\frac{n}{p}, \infty)$. *Then, for any given* $r \in (0, \min\{p, \frac{n}{\max\{M_0(\omega), M_\infty(\omega)\}+n/p}\})$, *there exists a positive constant* C *such that, for any* $f \in L_{\mathrm{loc}}^1(\mathbb{R}^n)$,

$$\|\mathcal{M}(f)\|_{[\dot{\mathcal{K}}_{\omega,\mathbf{0}}^{p,q}(\mathbb{R}^n)]^{1/r}} \leq C \|f\|_{[\dot{\mathcal{K}}_{\omega,\mathbf{0}}^{p,q}(\mathbb{R}^n)]^{1/r}}.$$

Proof Let all the symbols be as in the present lemma. Then, applying Lemma 1.1.6 and the assumptions $r \in (0, \frac{n}{\max\{M_0(\omega), M_\infty(\omega)\}+n/p})$ and $\min\{m_0(\omega), m_\infty(\omega)\} \in (-\frac{n}{p}, \infty)$, we conclude that

$$\max\{M_0(\omega^r), M_\infty(\omega^r)\} = r \max\{M_0(\omega), M_\infty(\omega)\}$$
$$< r\left(\frac{n}{r} - \frac{n}{p}\right) = \frac{n}{(p/r)'} \tag{1.132}$$

and

$$\min\{m_0(\omega^r), m_\infty(\omega^r)\} = r \min\{m_0(\omega), m_\infty(\omega)\} > -\frac{n}{p/r}. \tag{1.133}$$

Using this, the fact that $\frac{p}{r} \in (1, \infty)$, and Corollary 1.5.4, we find that the Hardy–Littlewood maximal operator $\mathcal{M}$ is bounded on the local generalized Herz space $\dot{\mathcal{K}}^{p/r,q/r}_{\omega^r,\mathbf{0}}(\mathbb{R}^n)$. This, combined with Lemma 1.3.1, further implies that $\mathcal{M}$ is bounded on $[\dot{\mathcal{K}}^{p,q}_{\omega,\mathbf{0}}(\mathbb{R}^n)]^{1/r}$, which completes the proof of Lemma 1.8.5. $\qquad\square$

Lemma 1.8.6 *Let $p, q \in (0, \infty)$ and $\omega \in M(\mathbb{R}_+)$ satisfy $m_0(\omega) \in (-\frac{n}{p}, \infty)$ and $m_\infty(\omega) \in (-\frac{n}{p}, \infty)$. Then, for any given*

$$s \in \left(0, \min\left\{p, q, \frac{n}{\max\{M_0(\omega), M_\infty(\omega)\} + n/p}\right\}\right)$$

and

$$r \in \left(\max\left\{p, \frac{n}{\min\{m_0(\omega), m_\infty(\omega)\} + n/p}\right\}, \infty\right],$$

the Herz space $[\dot{\mathcal{K}}^{p,q}_{\omega,\mathbf{0}}(\mathbb{R}^n)]^{1/s}$ is a ball Banach function space and there exists a positive constant C such that, for any $f \in L^1_{\mathrm{loc}}(\mathbb{R}^n)$,

$$\left\|\mathcal{M}^{((r/s)')}(f)\right\|_{([\dot{\mathcal{K}}^{p,q}_{\omega,\mathbf{0}}(\mathbb{R}^n)]^{1/s})'} \leq C \|f\|_{([\dot{\mathcal{K}}^{p,q}_{\omega,\mathbf{0}}(\mathbb{R}^n)]^{1/s})'}. \tag{1.134}$$

Proof Let all the symbols be as in the present lemma. Then, by Lemma 1.1.6 and the assumptions that

$$s < \frac{n}{\max\{M_0(\omega), M_\infty(\omega)\} + n/p} \leq \frac{n}{M_0(\omega) + n/p}$$

and $m_0(\omega) \in (-n/p, \infty)$, we find that

$$M_0\left(\omega^s\right) = s M_0(\omega) < s\left(\frac{n}{s} - \frac{n}{p}\right) = \frac{n}{(p/s)'} \tag{1.135}$$

and

$$m_0\left(\omega^s\right) = s m_0(\omega) > -\frac{n}{p/s}.$$

These, together with the assumptions $p/s, q/s \in (1, \infty)$, and Theorem 1.2.46, imply that the local generalized Herz space $\dot{\mathcal{K}}^{p/s,q/s}_{\omega^s,\mathbf{0}}(\mathbb{R}^n)$ is a BBF space. Combining this and Lemma 1.3.1, we further conclude that the Herz space $[\dot{\mathcal{K}}^{p,q}_{\omega,\mathbf{0}}(\mathbb{R}^n)]^{1/s}$ is a BBF space.

Next, we show (1.134). Indeed, from Lemma 1.3.1, (1.135), and Theorem 1.7.9, we deduce that

$$\left([\dot{\mathcal{K}}^{p,q}_{\omega,0}(\mathbb{R}^n)]^{1/s}\right)' = \left(\dot{\mathcal{K}}^{p/s,q/s}_{\omega^s,0}(\mathbb{R}^n)\right)' = \dot{\mathcal{K}}^{(p/s)',(q/s)'}_{1/\omega^s,0}(\mathbb{R}^n). \qquad (1.136)$$

In addition, by Lemma 1.1.6 and the assumption $s \in (0, \frac{n}{\max\{M_0(\omega),M_\infty(\omega)\}+n/p})$, we conclude that

$$\min\left\{m_0(1/\omega^s), m_\infty(1/\omega^s)\right\}$$
$$= -s \max\{M_0(\omega), M_\infty(\omega)\} > -s\left(\frac{n}{s} - \frac{n}{p}\right) = -\frac{n}{(p/s)'}. \qquad (1.137)$$

On the other hand, from Lemma 1.1.6 again and the assumption

$$r \in \left(\frac{n}{\min\{m_0(\omega),m_\infty(\omega)\}+n/p}, \infty\right),$$

it follows that

$$\max\left\{M_0(1/\omega^s), M_\infty(1/\omega^s)\right\}$$
$$= -s \min\{m_0(\omega), m_\infty(\omega)\} < -s\left(\frac{n}{r} - \frac{n}{p}\right) = \frac{n}{(r/s)'} - \frac{n}{(p/s)'},$$

which, combined with $r \in (p, \infty)$, further implies that

$$\left(\frac{r}{s}\right)' < \frac{n}{\max\left\{M_0(1/\omega^s), M_\infty(1/\omega^s)\right\} + \frac{n}{(p/s)'}}$$

and

$$\left(\frac{r}{s}\right)' < \left(\frac{p}{s}\right)'.$$

Applying these, (1.137), and Lemma 1.8.5, we find that the operator $\mathcal{M}$ is bounded on $[\dot{\mathcal{K}}^{(p/s)',(q/s)'}_{1/\omega^s,0}(\mathbb{R}^n)]^{1/(r/s)'}$. Thus, from both Remark 1.2.30 and (1.136), it follows that, for any $f \in L^1_{\text{loc}}(\mathbb{R}^n)$,

$$\left\|\mathcal{M}^{((r/s)')}(f)\right\|_{([\dot{\mathcal{K}}^{p,q}_{\omega,0}(\mathbb{R}^n)]^{1/s})'} \lesssim \|f\|_{([\dot{\mathcal{K}}^{p,q}_{\omega,0}(\mathbb{R}^n)]^{1/s})'}.$$

This finishes the proof of Lemma 1.8.6.   □

Via Lemmas 1.8.3 and 1.8.6, we now show Theorem 1.8.2.

Proof of Theorem 1.8.2 Let all the symbols be as in the present theorem. Then, from the assumptions of the present theorem and Theorem 1.2.46, we deduce that the local generalized Herz space $\dot{\mathcal{K}}_{\omega,\mathbf{0}}^{p,q}(\mathbb{R}^n)$ is a BBF space.

Thus, to finish the proof of the present theorem, we only need to show that $\dot{\mathcal{K}}_{\omega,\mathbf{0}}^{p,q}(\mathbb{R}^n)$ satisfies all the assumptions of Lemma 1.8.3. On the one hand, applying Corollary 1.5.4, we find that the Hardy–Littlewood maximal operator $\mathcal{M}$ is bounded on $\dot{\mathcal{K}}_{\omega,\mathbf{0}}^{p,q}(\mathbb{R}^n)$, namely, for any $f \in L_{\mathrm{loc}}^1(\mathbb{R}^n)$,

$$\|\mathcal{M}(f)\|_{\dot{\mathcal{K}}_{\omega,\mathbf{0}}^{p,q}(\mathbb{R}^n)} \lesssim \|f\|_{\dot{\mathcal{K}}_{\omega,\mathbf{0}}^{p,q}(\mathbb{R}^n)}. \tag{1.138}$$

On the other hand, by the assumption

$$\max\{M_0(\omega), M_\infty(\omega)\} \in \left(-\frac{n}{p}, \frac{n}{p'}\right),$$

we conclude that

$$\frac{n}{\max\{M_0(\omega), M_\infty(\omega)\} + n/p} > \frac{n}{n(1/p' + 1/p)} = 1.$$

From this and the assumptions $p, q \in (1, \infty)$, it follows that

$$\min\left\{p, q, \frac{n}{\max\{M_0(\omega), M_\infty(\omega)\} + n/p}\right\} \in (1, \infty).$$

This, combined with Lemma 1.8.6, further implies that the Hardy–Littlewood maximal operator $\mathcal{M}$ is bounded on $(\dot{\mathcal{K}}_{\omega,\mathbf{0}}^{p,q}(\mathbb{R}^n))'$, namely, for any $f \in L_{\mathrm{loc}}^1(\mathbb{R}^n)$,

$$\|\mathcal{M}(f)\|_{(\dot{\mathcal{K}}_{\omega,\mathbf{0}}^{p,q}(\mathbb{R}^n))'} \lesssim \|f\|_{(\dot{\mathcal{K}}_{\omega,\mathbf{0}}^{p,q}(\mathbb{R}^n))'}. \tag{1.139}$$

Using this and (1.138), we conclude that $\dot{\mathcal{K}}_{\omega,\mathbf{0}}^{p,q}(\mathbb{R}^n)$ satisfies all the assumptions of Lemma 1.8.3, which completes the proof of Theorem 1.8.2. □

Next, we show the following extrapolation theorem of global generalized Herz spaces.

Theorem 1.8.7 *Let $p, q \in (1, \infty)$, $r_0 \in [1, \infty)$, and $\omega \in M(\mathbb{R}_+)$ satisfy*

$$-\frac{n}{p} < m_0(\omega) \leq M_0(\omega) < \frac{n}{p'}$$

and

$$-\frac{n}{p} < m_\infty(\omega) \leq M_\infty(\omega) < \frac{n}{p'}.$$

Assume $\mathcal{F}$ is a set of all pairs of nonnegative measurable functions (F, G) such that, for any given $\upsilon \in A_{r_0}(\mathbb{R}^n)$,

$$\int_{\mathbb{R}^n} [F(x)]^{r_0} \upsilon(x)\, dx \leq C_{(r, [\upsilon]_{A_{r_0}(\mathbb{R}^n)})} \int_{\mathbb{R}^n} [G(x)]^{r_0} \upsilon(x)\, dx, \qquad (1.140)$$

where the positive constant $C_{(r_0, [\upsilon]_{A_{r_0}(\mathbb{R}^n)})}$ is independent of (F, G), but depending on both r_0 and $[\upsilon]_{A_{r_0}(\mathbb{R}^n)}$. Then there exists a positive constant C such that, for any $(F, G) \in \mathcal{F}$ with $\|F\|_{\dot{\mathcal{K}}_\omega^{p,q}(\mathbb{R}^n)} < \infty$,

$$\|F\|_{\dot{\mathcal{K}}_\omega^{p,q}(\mathbb{R}^n)} \leq C \|G\|_{\dot{\mathcal{K}}_\omega^{p,q}(\mathbb{R}^n)}.$$

To prove this theorem, we require the following property of Muckenhoupt weights, which can be immediately inferred from Definition 1.8.1; we omit the details.

Lemma 1.8.8 *Let $p \in [1, \infty)$, $\xi \in \mathbb{R}^n$, and υ be a locally integrable and nonnegative function on $\mathbb{R}^n$. Then $\upsilon \in A_p(\mathbb{R}^n)$ if and only if $\upsilon(\cdot + \xi) \in A_p(\mathbb{R}^n)$. Moreover, for any locally integrable and nonnegative function υ,*

$$[\upsilon(\cdot + \xi)]_{A_p(\mathbb{R}^n)} = [\upsilon]_{A_p(\mathbb{R}^n)}.$$

Proof of Theorem 1.8.7 Let all the symbols be as in the present theorem. Then, for any given $\upsilon \in A_{r_0}(\mathbb{R}^n)$ and any given $\xi \in \mathbb{R}^n$, we claim that

$$\int_{\mathbb{R}^n} [F(x + \xi)]^{r_0} \upsilon(x)\, dx \lesssim \int_{\mathbb{R}^n} [G(x + \xi)]^{r_0} \upsilon(x)\, dx,$$

where the implicit positive constant is independent of both (F, G) and ξ. Indeed, from Lemma 1.8.8, we deduce that $\upsilon(\cdot - \xi) \in A_{r_0}(\mathbb{R}^n)$ and

$$[\upsilon(\cdot - \xi)]_{A_{r_0}(\mathbb{R}^n)} = [\upsilon]_{A_{r_0}(\mathbb{R}^n)}. \qquad (1.141)$$

By this and (1.140), we find that

$$\int_{\mathbb{R}^n} [F(x)]^{r_0} \upsilon(x - \xi)\, dx \lesssim \int_{\mathbb{R}^n} [G(x)]^{r_0} \upsilon(x - \xi)\, dx,$$

where the implicit positive constant is independent of (F, G). This, together with (1.141) again, further implies that

$$\int_{\mathbb{R}^n} [F(x + \xi)]^{r_0} \upsilon(x)\, dx \lesssim \int_{\mathbb{R}^n} [G(x + \xi)]^{r_0} \upsilon(x)\, dx,$$

where the implicit positive constant is independent of both (F, G) and ξ, which completes the proof of the above claim.

On the other hand, applying the assumption $\|F\|_{\dot{\mathcal{K}}_\omega^{p,q}(\mathbb{R}^n)} < \infty$ and Remark 1.2.2(ii), we conclude that, for any $\xi \in \mathbb{R}^n$,

$$\|F(\cdot + \xi)\|_{\dot{\mathcal{K}}_{\omega,\mathbf{0}}^{p,q}(\mathbb{R}^n)} < \infty.$$

Combining this, the above claim, Theorem 1.8.2, and Remark 1.2.2(ii), we further find that, for any $\xi \in \mathbb{R}^n$,

$$\|F(\cdot + \xi)\|_{\dot{\mathcal{K}}_{\omega,\mathbf{0}}^{p,q}(\mathbb{R}^n)} \lesssim \|G(\cdot + \xi)\|_{\dot{\mathcal{K}}_{\omega,\mathbf{0}}^{p,q}(\mathbb{R}^n)} \lesssim \|G\|_{\dot{\mathcal{K}}_\omega^{p,q}(\mathbb{R}^n)},$$

where the implicit positive constants are independent of (F, G) and ξ. From this and Remark 1.2.2(ii) again, it follows that

$$\|F\|_{\dot{\mathcal{K}}_\omega^{p,q}(\mathbb{R}^n)} \lesssim \|G\|_{\dot{\mathcal{K}}_\omega^{p,q}(\mathbb{R}^n)},$$

where the implicit positive constant is independent of (F, G), which completes the proof of Theorem 1.8.7. $\qquad\square$

Chapter 2
Block Spaces and Their Applications

In this chapter, we first introduce the concepts of both the (ω, p)-block and the block space $\dot{\mathcal{B}}_{\omega}^{p,q}(\mathbb{R}^n)$. Then we establish the duality between block spaces and global generalized Herz spaces, which plays a key role in the study of the real-variable theory of Hardy spaces associated with global generalized Herz spaces in the subsequent chapters. To this end, we first characterize local generalized Herz spaces via (ω, p)-blocks and then establish an equivalent characterization of block spaces via local generalized Herz spaces. Using this equivalent characterization and borrowing some ideas from the proof of [88, Theorem 6.1], we show that the global generalized Herz space $\dot{\mathcal{K}}_{1/\omega}^{p',q'}(\mathbb{R}^n)$ is just the dual space of the block space $\dot{\mathcal{B}}_{\omega}^{p,q}(\mathbb{R}^n)$. As applications, we obtain the boundedness of some sublinear operators on block spaces. To be precise, the boundedness criteria of both sublinear operators and Calderón–Zygmund operators are established. In particular, the boundedness of powered Hardy–Littlewood maximal operators, which can be deduced directly from that of sublinear operators, plays an important role in the subsequent chapters.

2.1 Block Spaces

In this section, we introduce block spaces and investigate some basic properties of them. To begin with, we introduce the following concept of (ω, p)-blocks.

Definition 2.1.1 Let $p \in (0, \infty)$ and $\omega \in M(\mathbb{R}_+)$. Then a measurable function b on $\mathbb{R}^n$ is called an (ω, p)-*block* if there exists a cube $Q \in \mathcal{Q}$ such that

$$\operatorname{supp}(b) := \left\{ x \in \mathbb{R}^n : b(x) \neq 0 \right\} \subset Q$$

and

$$\|b\|_{L^p(\mathbb{R}^n)} \le \left[\omega\left(|Q|^{\frac{1}{n}}\right)\right]^{-1}.$$

Before presenting the definition of block spaces, we need some notation. Recall that, for any $k \in \mathbb{Z}$, $\mathcal{D}_k$ is defined to be the set of all the standard dyadic cubes on $\mathbb{R}^n$ of level $-k$, namely,

$$\mathcal{D}_k := \left\{2^k \left\{j + [0, 1)^n\right\} : j \in \mathbb{Z}^n\right\}. \tag{2.1}$$

Moreover, let

$$\mathcal{D} := \bigcup_{k \in \mathbb{Z}} \mathcal{D}_k.$$

In what follows, for any cube $Q \in \mathcal{Q}$, we always use $l(Q)$ to denote its edge length. Furthermore, for any $x \in \mathbb{R}^n$ and $l \in (0, \infty)$, the *symbol* $Q(x, l)$ denotes the cube with center x and edge length l. Then, for any $k \in \mathbb{Z}$ and $\xi \in \mathbb{R}^n$, let

$$\mathcal{Q}_k^{(0)} := \mathcal{D}_{k-1} \cap \left[Q(\mathbf{0}, 2^{k+1}) \setminus Q(\mathbf{0}, 2^k)\right]$$

and

$$\mathcal{Q}_k^{(\xi)} := \left\{Q \in \mathcal{Q} : Q - \{\xi\} \in \mathcal{Q}_k^{(0)}\right\}. \tag{2.2}$$

Here and thereafter, $Q - \{\xi\} := \{x - \xi : x \in Q\}$.

Remark 2.1.2 We can easily show that, for any $\xi \in \mathbb{R}^n$ and $k \in \mathbb{Z}$, $\sharp \mathcal{Q}_k^{(\xi)} = 2^{2n} - 2^n$. Here and thereafter, for any set E, $\sharp E$ denotes its *cardinality*.

Now, we introduce the block space $\dot{\mathcal{B}}_\omega^{p,q}(\mathbb{R}^n)$ as follows.

Definition 2.1.3 Let $p, q \in (0, \infty)$ and $\omega \in M(\mathbb{R}_+)$. Then the *block space* $\dot{\mathcal{B}}_\omega^{p,q}(\mathbb{R}^n)$ is defined to be the set of all the measurable functions f on $\mathbb{R}^n$ such that both

$$f = \sum_{l \in \mathbb{N}} \sum_{k \in \mathbb{Z}} \sum_{Q \in \mathcal{Q}_k^{(\xi_l)}} \lambda_{\xi_l, k, Q} b_{\xi_l, k, Q} \tag{2.3}$$

almost everywhere in $\mathbb{R}^n$ and

$$\left\{ \sum_{l\in\mathbb{N}} \left[\sum_{k\in\mathbb{Z}} \sum_{Q\in\mathcal{Q}_k^{(\xi_l)}} \lambda_{\xi_l,k,Q}^q \right]^{\frac{1}{q}} \right\} < \infty,$$

where $\{\xi_l\}_{l\in\mathbb{N}} \subset \mathbb{R}^n$, $\{\lambda_{\xi_l,k,Q}\}_{l\in\mathbb{N},\,k\in\mathbb{Z},\,Q\in\mathcal{Q}_k^{(\xi_l)}} \subset [0, \infty)$, and, for any $l \in \mathbb{N}$, $k \in \mathbb{Z}$, and $Q \in \mathcal{Q}_k^{(\xi_l)}$, $b_{\xi_l,k,Q}$ is an (ω, p)-block supported in the cube Q. Moreover, for any $f \in \mathcal{M}(\mathbb{R}^n)$,

$$\|f\|_{\dot{\mathcal{B}}_\omega^{p,q}(\mathbb{R}^n)} := \inf \left\{ \sum_{l\in\mathbb{N}} \left[\sum_{k\in\mathbb{Z}} \sum_{Q\in\mathcal{Q}_k^{(\xi_l)}} \lambda_{\xi_l,k,Q}^q \right]^{\frac{1}{q}} \right\},$$

where the infimum is taken over all the decompositions of f as above.

Remark 2.1.4 It is easy to show that, for any $p, q \in (0, \infty)$ and $\omega \in M(\mathbb{R}_+)$, the block space $\dot{\mathcal{B}}_\omega^{p,q}(\mathbb{R}^n)$ is a linear space equipped with the seminorm $\|\cdot\|_{\dot{\mathcal{B}}_\omega^{p,q}(\mathbb{R}^n)}$.

Next, we turn to study the convergence of the series (2.3). For this purpose, let $p \in (0, \infty)$. Recall that the *space $L^1(\mathbb{R}^n) + L^p(\mathbb{R}^n)$* is defined to be the set of all the $f \in \mathcal{M}(\mathbb{R}^n)$ such that there exists an $f_1 \in L^1(\mathbb{R}^n)$ and an $f_2 \in L^p(\mathbb{R}^n)$ such that $f = f_1 + f_2$. Moreover, for any $f \in L^1(\mathbb{R}^n) + L^p(\mathbb{R}^n)$,

$$\|f\|_{L^1(\mathbb{R}^n)+L^p(\mathbb{R}^n)} := \inf \left\{ \|f_1\|_{L^1(\mathbb{R}^n)} + \|f_2\|_{L^p(\mathbb{R}^n)} \right\}, \qquad (2.4)$$

where the infimum is taken over all the decompositions of f as above. Then we have the following conclusion which shows that, under some assumptions, the summation (2.3) converges in $L^1(\mathbb{R}^n) + L^p(\mathbb{R}^n)$.

Theorem 2.1.5 *Let $p \in [1, \infty)$, $q \in (0, \infty)$, and $\omega \in M(\mathbb{R}_+)$ satisfy $M_0(\omega) \in (-\infty, \frac{n}{p'})$ and $m_\infty(\omega) \in (0, \infty)$, where $\frac{1}{p} + \frac{1}{p'} = 1$. Then the block space $\dot{\mathcal{B}}_\omega^{p,q}(\mathbb{R}^n)$ embeds continuously into $L^1(\mathbb{R}^n) + L^p(\mathbb{R}^n)$. Namely, there exists a positive constant C such that, for any $f \in \dot{\mathcal{B}}_\omega^{p,q}(\mathbb{R}^n)$,*

$$\|f\|_{L^1(\mathbb{R}^n)+L^p(\mathbb{R}^n)} \le C \|f\|_{\dot{\mathcal{B}}_\omega^{p,q}(\mathbb{R}^n)}.$$

Proof Let all the symbols be the same as in the present theorem, and let $f \in \dot{\mathcal{B}}_\omega^{p,q}(\mathbb{R}^n)$. Assume $\{\xi_l\}_{l\in\mathbb{N}} \subset \mathbb{R}^n$, $\{\lambda_{\xi_l,k,Q}\}_{l\in\mathbb{N},\,k\in\mathbb{Z},\,Q\in\mathcal{Q}_k^{(\xi_l)}} \subset [0, \infty)$, and, for any $l \in \mathbb{N}$, $k \in \mathbb{Z}$, and $Q \in \mathcal{Q}_k^{(\xi_l)}$, $b_{\xi_l,k,Q}$ is an (ω, p)-block supported in the cube Q

satisfying that both

$$f = \sum_{l \in \mathbb{N}} \sum_{k \in \mathbb{Z}} \sum_{Q \in \mathcal{Q}_k^{(\xi_l)}} \lambda_{\xi_l, k, Q} b_{\xi_l, k, Q} \tag{2.5}$$

almost everywhere in $\mathbb{R}^n$ and

$$\sum_{l \in \mathbb{N}} \left[\sum_{k \in \mathbb{Z}} \sum_{Q \in \mathcal{Q}_k^{(\xi_l)}} \lambda_{\xi_l, k, Q}^q \right]^{\frac{1}{q}} < \infty. \tag{2.6}$$

Let

$$f_- := \sum_{l \in \mathbb{N}} \sum_{k \in \mathbb{Z} \setminus \mathbb{N}} \sum_{Q \in \mathcal{Q}_k^{(\xi_l)}} \lambda_{\xi_l, k, Q} b_{\xi_l, k, Q}$$

and

$$f_+ := \sum_{l \in \mathbb{N}} \sum_{k \in \mathbb{N}} \sum_{Q \in \mathcal{Q}_k^{(\xi_l)}} \lambda_{\xi_l, k, Q} b_{\xi_l, k, Q}.$$

We next prove that $f_- \in L^1(\mathbb{R}^n)$ and $f_+ \in L^p(\mathbb{R}^n)$. Indeed, we can show a stronger conclusion, namely, the series

$$\sum_{l \in \mathbb{N}} \sum_{k \in \mathbb{Z} \setminus \mathbb{N}} \sum_{Q \in \mathcal{Q}_k^{(\xi_l)}} \lambda_{\xi_l, k, Q} \left| b_{\xi_l, k, Q} \right| \tag{2.7}$$

converges in $L^1(\mathbb{R}^n)$ and the series

$$\sum_{l \in \mathbb{N}} \sum_{k \in \mathbb{N}} \sum_{Q \in \mathcal{Q}_k^{(\xi_l)}} \lambda_{\xi_l, k, Q} \left| b_{\xi_l, k, Q} \right| \tag{2.8}$$

converges in $L^p(\mathbb{R}^n)$. For this purpose, we choose an

$$\varepsilon \in \left(0, \min \left\{ \frac{n}{p'} - M_0(\omega), m_\infty(\omega) \right\} \right).$$

Then, by Lemma 1.1.12, we find that, for any $t \in (0, 1]$,

$$t^{M_0(\omega)+\varepsilon} \lesssim \omega(t) \tag{2.9}$$

and, for any $t \in (1, \infty)$,

$$t^{m_\infty(\omega) - \varepsilon} \lesssim \omega(t). \tag{2.10}$$

We now estimate $\|f_-\|_{L^1(\mathbb{R}^n)}$. Indeed, from (2.5), the Fatou lemma, the Minkowski inequality, the Hölder inequality, Definition 2.1.1, and (2.9) with $t := 2^k$ and $k \in \mathbb{Z} \setminus \mathbb{N}$, we infer that

$$
\begin{aligned}
\|f_-\|_{L^1(\mathbb{R}^n)} &\leq \left\| \sum_{l \in \mathbb{N}} \sum_{k \in \mathbb{Z} \setminus \mathbb{N}} \sum_{Q \in \mathcal{Q}_k^{(\xi_l)}} \lambda_{\xi_l, k, Q} \left| b_{\xi_l, k, Q} \right| \right\|_{L^1(\mathbb{R}^n)} \\
&\leq \sum_{l \in \mathbb{N}} \sum_{k \in \mathbb{Z} \setminus \mathbb{N}} \sum_{Q \in \mathcal{Q}_k^{(\xi_l)}} \lambda_{\xi_l, k, Q} \left\| b_{\xi_l, k, Q} \right\|_{L^1(\mathbb{R}^n)} \\
&\leq \sum_{l \in \mathbb{N}} \sum_{k \in \mathbb{Z} \setminus \mathbb{N}} \sum_{Q \in \mathcal{Q}_k^{(\xi_l)}} \lambda_{\xi_l, k, Q} |Q|^{1 - \frac{1}{p}} \left\| b_{\xi_l, k, Q} \right\|_{L^p(\mathbb{R}^n)} \\
&\lesssim \sum_{l \in \mathbb{N}} \sum_{k \in \mathbb{Z} \setminus \mathbb{N}} \sum_{Q \in \mathcal{Q}_k^{(\xi_l)}} \lambda_{\xi_l, k, Q} |Q|^{1 - \frac{1}{p}} \left[\omega(2^k) \right]^{-1} \\
&\lesssim \sum_{l \in \mathbb{N}} \sum_{k \in \mathbb{Z} \setminus \mathbb{N}} \sum_{Q \in \mathcal{Q}_k^{(\xi_l)}} 2^{k[\frac{n}{p'} - M_0(\omega) - \varepsilon]} \lambda_{\xi_l, k, Q}.
\end{aligned} \tag{2.11}
$$

Using this, Lemma 1.2.11, the assumption $\varepsilon \in (0, \frac{n}{p'} - M_0(\omega))$, and (2.6), we conclude that, when $q \in (0, 1]$,

$$
\begin{aligned}
\|f_-\|_{L^1(\mathbb{R}^n)} &\lesssim \sum_{l \in \mathbb{N}} \left\{ \sum_{k \in \mathbb{Z} \setminus \mathbb{N}} \sum_{Q \in \mathcal{Q}_k^{(\xi_l)}} 2^{k[\frac{n}{p'} - M_0(\omega) - \varepsilon]q} \lambda_{\xi_l, k, Q}^q \right\}^{\frac{1}{q}} \\
&\lesssim \sum_{l \in \mathbb{N}} \left[\sum_{k \in \mathbb{Z} \setminus \mathbb{N}} \sum_{Q \in \mathcal{Q}_k^{(\xi_l)}} \lambda_{\xi_l, k, Q}^q \right]^{\frac{1}{q}} \\
&\lesssim \sum_{l \in \mathbb{N}} \left[\sum_{k \in \mathbb{Z}} \sum_{Q \in \mathcal{Q}_k^{(\xi_l)}} \lambda_{\xi_l, k, Q}^q \right]^{\frac{1}{q}} < \infty,
\end{aligned} \tag{2.12}
$$

which implies that the summation (2.7) converges in $L^1(\mathbb{R}^n)$ and $f_- \in L^1(\mathbb{R}^n)$ when $q \in (0, 1]$. In addition, applying (2.11), the Hölder inequality, the assumption $\varepsilon \in (0, \frac{n}{p'} - M_0(\omega))$, and (2.6), we further find that, when $q \in (1, \infty)$,

$$
\|f_-\|_{L^1(\mathbb{R}^n)} \lesssim \sum_{l \in \mathbb{N}} \left\{ \sum_{k \in \mathbb{Z} \backslash \mathbb{N}} \sum_{Q \in \mathcal{Q}_k^{(\xi_l)}} 2^{k[\frac{n}{p'} - M_0(\omega) - \varepsilon]q'} \right\}^{\frac{1}{q'}}
$$

$$
\times \sum_{l \in \mathbb{N}} \left[\sum_{k \in \mathbb{Z} \backslash \mathbb{N}} \sum_{Q \in \mathcal{Q}_k^{(\xi_l)}} \lambda_{\xi_l, k, Q}^q \right]^{\frac{1}{q}}
$$

$$
\lesssim \sum_{l \in \mathbb{N}} \left[\sum_{k \in \mathbb{Z}} \sum_{Q \in \mathcal{Q}_k^{(\xi_l)}} \lambda_{\xi_l, k, Q}^q \right]^{\frac{1}{q}} < \infty. \tag{2.13}
$$

This then implies that the summation (2.7) converges in $L^1(\mathbb{R}^n)$ and $f_- \in L^1(\mathbb{R}^n)$ when $q \in (1, \infty)$, and hence the summation (2.7) converges in $L^1(\mathbb{R}^n)$ and $f_- \in L^1(\mathbb{R}^n)$.

Now, we show that the summation (2.8) converges in $L^p(\mathbb{R}^n)$ and $f_+ \in L^p(\mathbb{R}^n)$. Indeed, from (2.5), the Fatou lemma, the Minkowski inequality, Definition 2.1.1, and (2.10) with $t := 2^k$ and $k \in \mathbb{N}$, it follows that

$$
\|f_+\|_{L^p(\mathbb{R}^n)} \leq \left\| \sum_{l \in \mathbb{N}} \sum_{k \in \mathbb{N}} \sum_{Q \in \mathcal{Q}_k^{(\xi_l)}} \lambda_{\xi_l, k, Q} \left| b_{\xi_l, k, Q} \right| \right\|_{L^p(\mathbb{R}^n)}
$$

$$
\leq \sum_{l \in \mathbb{N}} \sum_{k \in \mathbb{N}} \sum_{Q \in \mathcal{Q}_k^{(\xi_l)}} \lambda_{\xi_l, k, Q} \left\| b_{\xi_l, k, Q} \right\|_{L^p(\mathbb{R}^n)}
$$

$$
\lesssim \sum_{l \in \mathbb{N}} \sum_{k \in \mathbb{N}} \sum_{Q \in \mathcal{Q}_k^{(\xi_l)}} \lambda_{\xi_l, k, Q} \left[\omega(2^k) \right]^{-1}
$$

$$
\lesssim \sum_{l \in \mathbb{N}} \sum_{k \in \mathbb{N}} \sum_{Q \in \mathcal{Q}_k^{(\xi_l)}} 2^{-k[m_\infty(\omega) - \varepsilon]} \lambda_{\xi_l, k, Q}.
$$

Using this, the assumption $\varepsilon \in (0, m_\infty(\omega))$, and (2.6), and repeating arguments similar to those used in the estimations of both (2.12) and (2.13), we obtain

$$\|f_+\|_{L^p(\mathbb{R}^n)} \lesssim \sum_{l\in\mathbb{N}}\left[\sum_{k\in\mathbb{N}}\sum_{Q\in\mathcal{Q}_k^{(\xi_l)}}\lambda_{\xi_l,k,Q}^q\right]^{\frac{1}{q}}$$

$$\lesssim \sum_{l\in\mathbb{N}}\left[\sum_{k\in\mathbb{Z}}\sum_{Q\in\mathcal{Q}_k^{(\xi_l)}}\lambda_{\xi_l,k,Q}^q\right]^{\frac{1}{q}} < \infty, \tag{2.14}$$

which then implies that the summation (2.8) converges in $L^p(\mathbb{R}^n)$. Thus, we have $f_+ \in L^p(\mathbb{R}^n)$. This, together with the fact that $f_- \in L^1(\mathbb{R}^n)$, further implies that $f = f_- + f_+ \in L^1(\mathbb{R}^n) + L^p(\mathbb{R}^n)$. Moreover, from (2.12), (2.13), (2.14), and (2.4), we deduce that

$$\|f\|_{L^1(\mathbb{R}^n)+L^p(\mathbb{R}^n)} \leq \|f_-\|_{L^1(\mathbb{R}^n)} + \|f_+\|_{L^p(\mathbb{R}^n)}$$

$$\lesssim \sum_{l\in\mathbb{N}}\left[\sum_{k\in\mathbb{Z}}\sum_{Q\in\mathcal{Q}_k^{(\xi_l)}}\lambda_{\xi_l,k,Q}^q\right]^{\frac{1}{q}}.$$

Combining this, the choice of $\{\lambda_{\xi_l,k,Q}\}_{l\in\mathbb{N},\,k\in\mathbb{Z},\,Q\in\mathcal{Q}_k^{(\xi_l)}}$, and Definition 2.1.3, we then find that

$$\|f\|_{L^1(\mathbb{R}^n)+L^p(\mathbb{R}^n)} \lesssim \|f\|_{\dot{\mathcal{B}}_\omega^{p,q}(\mathbb{R}^n)},$$

which completes the proof of Theorem 2.1.5. $\square$

Remark 2.1.6 We point out that, in Definition 2.1.3, the assumption that the summation (2.3) converges almost everywhere in $\mathbb{R}^n$ is reasonable. Indeed, let p, q, and ω be the same as in Theorem 2.1.5 and assume that $\{\xi_l\}_{l\in\mathbb{N}} \subset \mathbb{R}^n$, $\{\lambda_{\xi_l,k,Q}\}_{l\in\mathbb{N},\,k\in\mathbb{Z},\,Q\in\mathcal{Q}_k^{(\xi_l)}} \subset [0,\infty)$, and, for any $l \in \mathbb{N}$, $k \in \mathbb{Z}$, and $Q \in \mathcal{Q}_k^{(\xi_l)}$, $b_{\xi_l,k,Q}$ is an (ω, p)-block supported in the cube Q satisfying that

$$\sum_{l\in\mathbb{N}}\left[\sum_{k\in\mathbb{Z}}\sum_{Q\in\mathcal{Q}_k^{(\xi_l)}}\lambda_{\xi_l,k,Q}^q\right]^{\frac{1}{q}} < \infty.$$

Then the summation $\sum_{l\in\mathbb{N}}\sum_{k\in\mathbb{Z}}\sum_{Q\in\mathcal{Q}_k^{\xi_l}}\lambda_{\xi_l,k,Q}b_{\xi_l,k,Q}$ converges almost everywhere in $\mathbb{R}^n$ automatically. To prove this, repeating an argument similar to that used in the estimation of f_- in the proof of Theorem 2.1.5, we find that

$$\sum_{l\in\mathbb{N}}\sum_{k\in\mathbb{Z}\setminus\mathbb{N}}\sum_{Q\in\mathcal{Q}_k^{\xi l}}\lambda_{\xi l,k,Q}\left|b_{\xi l,k,Q}\right|\in L^1(\mathbb{R}^n).$$

This further implies that the summation $\sum_{l\in\mathbb{N}}\sum_{k\in\mathbb{Z}\setminus\mathbb{N}}\sum_{Q\in\mathcal{Q}_k^{\xi l}}\lambda_{\xi l,k,Q}b_{\xi l,k,Q}$ converges almost everywhere in $\mathbb{R}^n$. In addition, repeating an argument similar to that used in the estimation of f_+ in the proof of Theorem 2.1.5, we obtain

$$\sum_{l\in\mathbb{N}}\sum_{k\in\mathbb{N}}\sum_{Q\in\mathcal{Q}_k^{\xi l}}\lambda_{\xi l,k,Q}\left|b_{\xi l,k,Q}\right|\in L^p(\mathbb{R}^n),$$

and hence the summation $\sum_{l\in\mathbb{N}}\sum_{k\in\mathbb{N}}\sum_{Q\in\mathcal{Q}_k^{\xi l}}\lambda_{\xi l,k,Q}b_{\xi l,k,Q}$ converges almost everywhere in $\mathbb{R}^n$. Thus, the summation $\sum_{l\in\mathbb{N}}\sum_{k\in\mathbb{Z}}\sum_{Q\in\mathcal{Q}_k^{\xi l}}\lambda_{\xi l,k,Q}b_{\xi l,k,Q}$ converges almost everywhere in $\mathbb{R}^n$. This then implies that the assumption on the convergence (namely, almost everywhere in $\mathbb{R}^n$) of (2.3) is reasonable.

Using Theorem 2.1.5, we now show that, under the same assumptions of Theorem 2.1.5, $\|\cdot\|_{\dot{\mathcal{B}}_\omega^{p,q}(\mathbb{R}^n)}$ is a true norm.

Corollary 2.1.7 *Let p, q, and ω be the same as in Theorem 2.1.5. Assume $f\in\mathscr{M}(\mathbb{R}^n)$. Then $\|f\|_{\dot{\mathcal{B}}_\omega^{p,q}(\mathbb{R}^n)}=0$ if and only if $f=0$ almost everywhere in $\mathbb{R}^n$.*

Proof Let all the symbols be the same as in the present corollary. Then, by Definition 2.1.3, we conclude that, if $f\in\mathscr{M}(\mathbb{R}^n)$ satisfies that $f=0$ almost everywhere in $\mathbb{R}^n$, then $\|f\|_{\dot{\mathcal{B}}_\omega^{p,q}(\mathbb{R}^n)}=0$. This finishes the proof of the sufficiency.

Conversely, we next show the necessity. To this end, let $f\in\mathscr{M}(\mathbb{R}^n)$ be such that $\|f\|_{\dot{\mathcal{B}}_\omega^{p,q}(\mathbb{R}^n)}=0$. Then, from Theorem 2.1.5, we deduce that

$$\|f\|_{L^1(\mathbb{R}^n)+L^p(\mathbb{R}^n)}\lesssim\|f\|_{\dot{\mathcal{B}}_\omega^{p,q}(\mathbb{R}^n)}=0,$$

which, together with (2.4), further implies that, for any $k\in\mathbb{N}$, there exists an $f_{1,k}\in L^1(\mathbb{R}^n)$ and an $f_{2,k}\in L^p(\mathbb{R}^n)$ such that

$$f=f_{1,k}+f_{2,k} \tag{2.15}$$

almost everywhere in $\mathbb{R}^n$,

$$\left\|f_{1,k}\right\|_{L^1(\mathbb{R}^n)}<\frac{1}{k},\text{ and }\left\|f_{2,k}\right\|_{L^p(\mathbb{R}^n)}<\frac{1}{k}.$$

Using these and the Riesz theorem, we find that there exists a sequence $\{f_{1,k_j}\}_{j\in\mathbb{N}}\subset L^1(\mathbb{R}^n)$ and a sequence $\{f_{2,k_j}\}_{j\in\mathbb{N}}\subset L^p(\mathbb{R}^n)$ such that

$$f_{1,k_j}\to 0\text{ and }f_{2,k_j}\to 0$$

almost everywhere in $\mathbb{R}^n$ as $j \to \infty$. Combining this and (2.15), we further obtain $f = 0$ almost everywhere in $\mathbb{R}^n$. This finishes the proof of the necessity and hence Corollary 2.1.7. □

2.2 Duality

In this section, we show that the dual space of the block space $\dot{\mathcal{B}}_\omega^{p,q}(\mathbb{R}^n)$ is the global generalized Herz space $\dot{\mathcal{K}}_{1/\omega}^{p',q'}(\mathbb{R}^n)$ with $p \in [1, \infty)$, $q \in (0, \infty)$, and $\omega \in M(\mathbb{R}_+)$ as well as prove the triviality of the dual space of $\dot{\mathcal{B}}_\omega^{p,q}(\mathbb{R}^n)$ with $p \in (0, 1)$, $q \in (0, \infty)$, and $\omega \in M(\mathbb{R}_+)$. Then, via the above two dual results, we investigate the triviality of the seminorm $\|\cdot\|_{\dot{\mathcal{B}}_\omega^{p,q}(\mathbb{R}^n)}$. First, we consider the dual space of $\dot{\mathcal{B}}_\omega^{p,q}(\mathbb{R}^n)$ with $p \in [1, \infty)$ and $q \in (0, \infty)$ as follows.

Theorem 2.2.1 *Let $p \in [1, \infty)$, $q \in (0, \infty)$, and $\omega \in M(\mathbb{R}_+)$. Then the dual space of $\dot{\mathcal{B}}_\omega^{p,q}(\mathbb{R}^n)$, denoted by $(\dot{\mathcal{B}}_\omega^{p,q}(\mathbb{R}^n))^*$, is $\dot{\mathcal{K}}_{1/\omega}^{p',q'}(\mathbb{R}^n)$ in the following sense:*

(i) *Let $g \in \dot{\mathcal{K}}_{1/\omega}^{p',q'}(\mathbb{R}^n)$. Then the linear functional*

$$\phi_g : f \mapsto \phi_g(f) := \int_{\mathbb{R}^n} f(y)g(y)\,dy, \qquad (2.16)$$

defined for any $f \in \dot{\mathcal{B}}_\omega^{p,q}(\mathbb{R}^n)$, is bounded on $\dot{\mathcal{B}}_\omega^{p,q}(\mathbb{R}^n)$.
(ii) *Conversely, any continuous linear functional on $\dot{\mathcal{B}}_\omega^{p,q}(\mathbb{R}^n)$ arises as in (2.16) with a unique $g \in \dot{\mathcal{K}}_{1/\omega}^{p',q'}(\mathbb{R}^n)$.*

Here $\frac{1}{p} + \frac{1}{p'} = 1$ and

$$q' := \begin{cases} \dfrac{q}{q-1} & \text{when } q \in [1, \infty), \\[2mm] \infty & \text{when } q \in (0, 1). \end{cases}$$

Moreover, there exists two positive constants C_1 and C_2 such that, for any $g \in \dot{\mathcal{K}}_{1/\omega}^{p',q'}(\mathbb{R}^n)$,

$$C_1 \|g\|_{\dot{\mathcal{K}}_{1/\omega}^{p',q'}(\mathbb{R}^n)} \leq \|\phi_g\|_{(\dot{\mathcal{B}}_\omega^{p,q}(\mathbb{R}^n))^*} \leq C_2 \|g\|_{\dot{\mathcal{K}}_{1/\omega}^{p',q'}(\mathbb{R}^n)}.$$

To show this dual theorem, we first establish the following equivalent characterizations of both the local generalized Herz spaces $\dot{\mathcal{K}}_{\omega,\xi}^{p,q}(\mathbb{R}^n)$ and the block space $\dot{\mathcal{B}}_\omega^{p,q}(\mathbb{R}^n)$, respectively, via (ω, p)-blocks and $\dot{\mathcal{K}}_{\omega,\xi}^{p,q}(\mathbb{R}^n)$.

Lemma 2.2.2 *Let* $p, q \in (0, \infty)$ *and* $\omega \in M(\mathbb{R}_+)$. *Then*

(i) *for any given* $\xi \in \mathbb{R}^n$, *a measurable function* f *belongs to* $\dot{\mathcal{K}}^{p,q}_{\omega,\xi}(\mathbb{R}^n)$ *if and only if*

$$f = \sum_{k\in\mathbb{Z}} \sum_{Q\in\mathcal{Q}_k^{(\xi)}} \lambda_{k,Q} b_{k,Q}$$

almost everywhere in $\mathbb{R}^n$ *and*

$$\left[\sum_{k\in\mathbb{Z}} \sum_{Q\in\mathcal{Q}_k^{(\xi)}} \lambda_{k,Q}^q \right]^{\frac{1}{q}} < \infty,$$

where $\{\lambda_{k,Q}\}_{k\in\mathbb{Z}, Q\in\mathcal{Q}_k^{(\xi)}} \subset [0,\infty)$ *and, for any* $k \in \mathbb{Z}$ *and* $Q \in \mathcal{Q}_k^{(\xi)}$, $b_{k,Q}$ *is an* (ω, p)-*block supported in the cube* Q. *Moreover, for any* $f \in \dot{\mathcal{K}}^{p,q}_{\omega,\xi}(\mathbb{R}^n)$,

$$\|f\|_{\dot{\mathcal{K}}^{p,q}_{\omega,\xi}(\mathbb{R}^n)} \sim \left[\sum_{k\in\mathbb{Z}} \sum_{Q\in\mathcal{Q}_k^{(\xi)}} \lambda_{k,Q}^q \right]^{\frac{1}{q}},$$

where the positive equivalence constants are independent of both ξ *and* f;

(ii) *a measurable function* f *belongs to* $\dot{\mathcal{B}}^{p,q}_{\omega}(\mathbb{R}^n)$ *if and only if*

$$\|f\|^\star_{\dot{\mathcal{B}}^{p,q}_{\omega}(\mathbb{R}^n)} := \inf \left\{ \sum_{l\in\mathbb{N}} \|f_{\xi_l}\|_{\dot{\mathcal{K}}^{p,q}_{\omega,\xi_l}(\mathbb{R}^n)} \right\} < \infty,$$

where the infimum is taken over all the sequences $\{\xi_l\}_{l\in\mathbb{N}} \subset \mathbb{R}^n$ *and* $\{f_{\xi_l}\}_{l\in\mathbb{N}} \subset \mathscr{M}(\mathbb{R}^n)$ *such that, for any* $l \in \mathbb{N}$, $f_{\xi_l} \in \dot{\mathcal{K}}^{p,q}_{\omega,\xi_l}(\mathbb{R}^n)$ *and*

$$f = \sum_{l\in\mathbb{N}} f_{\xi_l}$$

almost everywhere in $\mathbb{R}^n$. *Moreover, for any* $f \in \dot{\mathcal{B}}^{p,q}_{\omega}(\mathbb{R}^n)$,

$$\|f\|_{\dot{\mathcal{B}}^{p,q}_{\omega}(\mathbb{R}^n)} \sim \|f\|^\star_{\dot{\mathcal{B}}^{p,q}_{\omega}(\mathbb{R}^n)},$$

where the positive equivalence constants are independent of f.

Proof Let $p, q \in (0, \infty)$ and $\omega \in M(\mathbb{R}_+)$. We first show (i). Indeed, to prove the necessity of (i), let $\xi \in \mathbb{R}^n$ be a fixed point and $f \in \dot{\mathcal{K}}_{\omega,\xi}^{p,q}(\mathbb{R}^n)$. Then, for any $k \in \mathbb{Z}$ and $Q \in \mathcal{Q}_k^{(\xi)}$, let

$$\lambda_{k,Q} := \omega\left(|Q|^{\frac{1}{n}}\right)\left\|f\mathbf{1}_Q\right\|_{L^p(\mathbb{R}^n)}.$$

Moreover, for any $k \in \mathbb{Z}$ and $Q \in \mathcal{Q}_k^{(\xi)}$, let $b_{k,Q} := 0$ when $f = 0$ almost everywhere in Q; otherwise, let

$$b_{k,Q} := \left[\omega\left(|Q|^{\frac{1}{n}}\right)\right]^{-1}\left[\left\|f\mathbf{1}_Q\right\|_{L^p(\mathbb{R}^n)}\right]^{-1}f\mathbf{1}_Q.$$

Therefore, from the definition of $\mathcal{Q}_k^{(\xi)}$, it follows that

$$f = \sum_{k\in\mathbb{Z}} f\mathbf{1}_{Q(\xi,2^{k+1})\setminus Q(\xi,2^k)} = \sum_{k\in\mathbb{Z}}\sum_{Q\in\mathcal{Q}_k^{(\xi)}} f\mathbf{1}_Q = \sum_{k\in\mathbb{Z}}\sum_{Q\in\mathcal{Q}_k^{(\xi)}} \lambda_{k,Q}b_{k,Q}.$$

On the other hand, applying Definition 2.1.1, Remark 2.1.2, the fact that, for any $k \in \mathbb{Z}$,

$$\bigcup_{Q\in\mathcal{Q}_k^{(\xi)}} Q = Q(\xi, 2^{k+1}) \setminus Q(\xi, 2^k) \subset B(\xi, 2^{k+1}) \setminus B(\xi, 2^{k-1}),$$

and Lemma 1.1.3, we conclude that, for any $k \in \mathbb{Z}$ and $Q \in \mathcal{Q}_k^{(\xi)}$, $b_{k,Q}$ is an (ω, p)-block supported in Q, and

$$\left[\sum_{k\in\mathbb{Z}}\sum_{Q\in\mathcal{Q}_k^{(\xi)}} \lambda_{k,Q}^q\right]^{\frac{1}{q}} = \left\{\sum_{k\in\mathbb{Z}}\left[\omega(2^{k-1})\right]^q \sum_{Q\in\mathcal{Q}_k^{(\xi)}} \left\|f\mathbf{1}_Q\right\|_{L^p(\mathbb{R}^n)}^q\right\}^{\frac{1}{q}}$$

$$\sim \left\{\sum_{k\in\mathbb{Z}}\left[\omega(2^{k-1})\right]^q \left\|f\mathbf{1}_{Q(\xi,2^{k+1})\setminus Q(\xi,2^k)}\right\|_{L^p(\mathbb{R}^n)}^q\right\}^{\frac{1}{q}}$$

$$\lesssim \left\{\sum_{k\in\mathbb{Z}}\left[\omega(2^{k+1})\right]^q \left\|f\mathbf{1}_{B(\xi,2^{k+1})\setminus B(\xi,2^k)}\right\|_{L^p(\mathbb{R}^n)}^q\right\}^{\frac{1}{q}}$$

$$+ \left\{\sum_{k\in\mathbb{Z}}\left[\omega(2^k)\right]^q \left\|f\mathbf{1}_{B(\xi,2^k)\setminus B(\xi,2^{k-1})}\right\|_{L^p(\mathbb{R}^n)}^q\right\}^{\frac{1}{q}}$$

$$\sim \|f\|_{\dot{\mathcal{K}}_{\omega,\xi}^{p,q}(\mathbb{R}^n)} < \infty, \tag{2.17}$$

where the implicit positive constants are independent of both ξ and f. This finishes the proof of the necessity of (i).

Conversely, we deal with the sufficiency of (i). To this end, assume $f \in \mathcal{M}(\mathbb{R}^n)$ satisfying that both

$$f = \sum_{k \in \mathbb{Z}} \sum_{Q \in \mathcal{Q}_k^{(\xi)}} \lambda_{k,Q} b_{k,Q}$$

almost everywhere in $\mathbb{R}^n$ and

$$\left[\sum_{k \in \mathbb{Z}} \sum_{Q \in \mathcal{Q}_k^{(\xi)}} \lambda_{k,Q}^q \right]^{\frac{1}{q}} < \infty,$$

where $\{\lambda_{k,Q}\}_{k \in \mathbb{Z},\, Q \in \mathcal{Q}_k^{(\xi)}} \subset [0, \infty)$ and, for any $k \in \mathbb{Z}$ and $Q \in \mathcal{Q}_k^{(\xi)}$, $b_{k,Q}$ is an (ω, p)-block supported in the cube Q. Then, from the fact that, for any $j \in \mathbb{Z}$,

$$B(\xi, 2^j) \setminus B(\xi, 2^{j-1}) \subset Q(\xi, 2^{j+1}) \setminus Q(\xi, 2^{j-1}),$$

Remark 2.1.2, Definition 2.1.1, and Lemma 1.1.3, it follows that, for any $j \in \mathbb{Z}$,

$$\left\| f \mathbf{1}_{B(\xi, 2^j) \setminus B(\xi, 2^{j-1})} \right\|_{L^p(\mathbb{R}^n)}^q$$

$$\leq \left\| \sum_{k=j}^{j+1} \sum_{Q \in \mathcal{Q}_k^{(\xi)}} \lambda_{k,Q} b_{k,Q} \right\|_{L^p(\mathbb{R}^n)}^q$$

$$\lesssim \sum_{k=j}^{j+1} \sum_{Q \in \mathcal{Q}_k^{(\xi)}} \lambda_{k,Q}^q \left\| b_{k,Q} \right\|_{L^p(\mathbb{R}^n)}^q$$

$$\lesssim \left[\omega(2^{j-2}) \right]^{-q} \sum_{Q \in \mathcal{Q}_{j-1}^{(\xi)}} \lambda_{j-1,Q}^q + \left[\omega(2^{j-1}) \right]^{-q} \sum_{Q \in \mathcal{Q}_j^{(\xi)}} \lambda_{j,Q}^q$$

$$\sim \left[\omega(2^j) \right]^{-q} \sum_{k=j}^{j+1} \sum_{Q \in \mathcal{Q}_k^{(\xi)}} \lambda_{k,Q}^q. \qquad (2.18)$$

This further implies that

$$\|f\|_{\dot{\mathcal{K}}^{p,q}_{\omega,\xi}(\mathbb{R}^n)} \lesssim \left[\sum_{j\in\mathbb{Z}}\sum_{k=j}^{j+1}\sum_{Q\in\mathcal{Q}^{(\xi)}_k}\lambda^q_{k,Q}\right]^{\frac{1}{q}} \lesssim \left[\sum_{j\in\mathbb{Z}}\sum_{Q\in\mathcal{Q}^{(\xi)}_j}\lambda^q_{j,Q}\right]^{\frac{1}{q}} < \infty, \quad (2.19)$$

where the implicit positive constants are independent of both ξ and f. Thus, $f \in \dot{\mathcal{K}}^{p,q}_{\omega,\xi}(\mathbb{R}^n)$ and hence the proof of the sufficiency of (i) is completed. Moreover, combining (2.17) and (2.19), we conclude that, for any $f \in \dot{\mathcal{K}}^{p,q}_{\omega,\xi}(\mathbb{R}^n)$,

$$\|f\|_{\dot{\mathcal{K}}^{p,q}_{\omega,\xi}(\mathbb{R}^n)} \sim \left[\sum_{k\in\mathbb{Z}}\sum_{Q\in\mathcal{Q}^{(\xi)}_k}\lambda^q_{k,Q}\right]^{\frac{1}{q}},$$

where the positive equivalence constants are independent of both ξ and f, which completes the proof of (i).

Next, we show (ii). We first consider the necessity of (ii). For this purpose, let $f \in \dot{\mathcal{B}}^{p,q}_{\omega}(\mathbb{R}^n)$ satisfy

$$f = \sum_{l\in\mathbb{N}}\sum_{k\in\mathbb{Z}}\sum_{Q\in\mathcal{Q}^{(\xi_l)}_k}\lambda_{\xi_l,k,Q}b_{\xi_l,k,Q}$$

almost everywhere in $\mathbb{R}^n$, where $\{\xi_l\}_{l\in\mathbb{N}} \subset \mathbb{R}^n$, $\{\lambda_{\xi_l,k,Q}\}_{l\in\mathbb{N},\,k\in\mathbb{Z},\,Q\in\mathcal{Q}^{(\xi_l)}_k} \subset [0,\infty)$, and, for any $l \in \mathbb{N}$, $k \in \mathbb{Z}$, and $Q \in \mathcal{Q}^{(\xi_l)}_k$, $b_{\xi_l,k,Q}$ is an (ω, p)-block supported in the cube Q, and

$$\sum_{l\in\mathbb{N}}\left[\sum_{k\in\mathbb{Z}}\sum_{Q\in\mathcal{Q}^{(\xi_l)}_k}\lambda^q_{\xi_l,k,Q}\right]^{\frac{1}{q}} < \infty.$$

This implies that, for any $l \in \mathbb{N}$,

$$\left[\sum_{k\in\mathbb{Z}}\sum_{Q\in\mathcal{Q}^{(\xi_l)}_k}\lambda^q_{\xi_l,k,Q}\right]^{\frac{1}{q}} < \infty.$$

By this and (i), we find that, for any $l \in \mathbb{N}$,

$$f_{\xi_l} := \sum_{k \in \mathbb{Z}} \sum_{Q \in \mathcal{Q}_k^{(\xi_l)}} \lambda_{\xi_l, k, Q} b_{\xi_l, k, Q} \in \dot{\mathcal{K}}_{\omega, \xi_l}^{p, q}(\mathbb{R}^n)$$

and

$$\|f_{\xi_l}\|_{\dot{\mathcal{K}}_{\omega, \xi_l}^{p, q}(\mathbb{R}^n)} \sim \left[\sum_{k \in \mathbb{Z}} \sum_{Q \in \mathcal{Q}_k^{(\xi_l)}} \lambda_{\xi_l, k, Q}^q \right]^{\frac{1}{q}},$$

where the positive equivalence constants are independent of both l and f. Therefore, we have

$$\|f\|_{\dot{\mathcal{B}}_{\omega}^{p, q}(\mathbb{R}^n)}^{\star} \leq \sum_{l \in \mathbb{N}} \|f_{\xi_l}\|_{\dot{\mathcal{K}}_{\omega, \xi_l}^{p, q}(\mathbb{R}^n)} \sim \sum_{l \in \mathbb{N}} \left[\sum_{k \in \mathbb{Z}} \sum_{Q \in \mathcal{Q}_k^{(\xi_l)}} \lambda_{\xi_l, k, Q}^q \right]^{\frac{1}{q}} < \infty, \qquad (2.20)$$

where the positive equivalence constants are independent of f. This finishes the proof of the necessity of (ii). Moreover, applying (2.20) and the choice of the sequence $\{\lambda_{\xi_l, k, Q}\}_{l \in \mathbb{N}, k \in \mathbb{Z}, Q \in \mathcal{Q}_k^{(\xi_l)}}$, we further conclude that

$$\|f\|_{\dot{\mathcal{B}}_{\omega}^{p, q}(\mathbb{R}^n)}^{\star} \lesssim \|f\|_{\dot{\mathcal{B}}_{\omega}^{p, q}(\mathbb{R}^n)}, \qquad (2.21)$$

where the implicit positive constant is independent of f.

Conversely, we show the sufficiency of (ii). Indeed, assume that $f \in \mathscr{M}(\mathbb{R}^n)$ satisfies that there exists a sequence $\{\xi_l\}_{l \in \mathbb{N}} \subset \mathbb{R}^n$ and a sequence $\{f_{\xi_l}\}_{l \in \mathbb{N}} \subset \mathscr{M}(\mathbb{R}^n)$ such that both $f = \sum_{l \in \mathbb{N}} f_{\xi_l}$ almost everywhere in $\mathbb{R}^n$ and

$$\sum_{l \in \mathbb{N}} \|f_{\xi_l}\|_{\dot{\mathcal{K}}_{\omega, \xi_l}^{p, q}(\mathbb{R}^n)} < \infty.$$

Then, for any $l \in \mathbb{N}$, $f_{\xi_l} \in \dot{\mathcal{K}}_{\omega, \xi_l}^{p, q}(\mathbb{R}^n)$. This, together with (i), further implies that, for any $l \in \mathbb{N}$,

$$f_{\xi_l} = \sum_{k \in \mathbb{Z}} \sum_{Q \in \mathcal{Q}_k^{(\xi_l)}} \lambda_{\xi_l, k, Q} b_{\xi_l, k, Q}$$

almost everywhere in $\mathbb{R}^n$ and

$$\|f_{\xi_l}\|_{\dot{\mathcal{K}}_{\omega,\xi_l}^{p,q}(\mathbb{R}^n)} \sim \left[\sum_{k\in\mathbb{Z}}\sum_{Q\in\mathcal{Q}_k^{(\xi_l)}} \lambda_{\xi_l,k,Q}^q\right]^{\frac{1}{q}},$$

where the positive equivalence constants are independent of both l and f, and for any $k \in \mathbb{Z}$ and $Q \in \mathcal{Q}_k^{(\xi_l)}$, $\lambda_{\xi_l,k,Q} \in [0,\infty)$ and $b_{\xi_l,k,Q}$ is an (ω, p)-block supported in Q. From this and Definition 2.1.3, we deduce that both

$$f = \sum_{l\in\mathbb{N}}\sum_{k\in\mathbb{Z}}\sum_{Q\in\mathcal{Q}_k^{(\xi_l)}} \lambda_{\xi_l,k,Q} b_{\xi_l,k,Q}$$

almost everywhere in $\mathbb{R}^n$ and

$$\|f\|_{\dot{\mathcal{B}}_\omega^{p,q}(\mathbb{R}^n)} \le \sum_{l\in\mathbb{N}}\left[\sum_{k\in\mathbb{Z}}\sum_{Q\in\mathcal{Q}_k^{(\xi_l)}} \lambda_{\xi_l,k,Q}^q\right]^{\frac{1}{q}} \sim \sum_{l\in\mathbb{N}}\|f_{\xi_l}\|_{\dot{\mathcal{K}}_{\omega,\xi_l}^{p,q}(\mathbb{R}^n)} < \infty, \qquad (2.22)$$

where the positive equivalence constants are independent of f. Thus, we conclude that $f \in \dot{\mathcal{B}}_\omega^{p,q}(\mathbb{R}^n)$, and hence complete the proof of the sufficiency of (ii). In addition, using (2.22) and the choice of $\{f_{\xi_l}\}_{l\in\mathbb{N}}$, we further find that

$$\|f\|_{\dot{\mathcal{B}}_\omega^{p,q}(\mathbb{R}^n)} \lesssim \|f\|_{\dot{\mathcal{B}}_\omega^{p,q}(\mathbb{R}^n)}^\star,$$

where the implicit positive constant is independent of f. Combining this and (2.21), we obtain

$$\|f\|_{\dot{\mathcal{B}}_\omega^{p,q}(\mathbb{R}^n)} \sim \|f\|_{\dot{\mathcal{B}}_\omega^{p,q}(\mathbb{R}^n)}^\star,$$

where the positive equivalence constants are independent of f. This finishes the proof of (ii) and hence Lemma 2.2.2. $\qquad\square$

By Lemma 2.2.2(ii), we immediately obtain the following relation between local generalized Herz spaces and block spaces, which is also important in the proof of Theorem 2.2.1; we omit the details.

Lemma 2.2.3 *Let $p, q \in (0,\infty)$ and $\omega \in M(\mathbb{R}_+)$. Then, for any given $\xi \in \mathbb{R}^n$, $\dot{\mathcal{K}}_{\omega,\xi}^{p,q}(\mathbb{R}^n) \subset \dot{\mathcal{B}}_\omega^{p,q}(\mathbb{R}^n)$. Moreover, there exists a positive constant C, independent of ξ, such that, for any $f \in \dot{\mathcal{K}}_{\omega,\xi}^{p,q}(\mathbb{R}^n)$,*

$$\|f\|_{\dot{\mathcal{B}}_\omega^{p,q}(\mathbb{R}^n)} \le C\|f\|_{\dot{\mathcal{K}}_{\omega,\xi}^{p,q}(\mathbb{R}^n)}.$$

Via Lemmas 2.2.2 and 2.2.3, we now show Theorem 2.2.1.

Proof of Theorem 2.2.1 Let p, q, and ω be as in the present theorem. We first show (i). To this end, let $g \in \dot{\mathcal{K}}_{1/\omega}^{p',q'}(\mathbb{R}^n)$. Then, for any $f \in \dot{\mathcal{B}}_{\omega}^{p,q}(\mathbb{R}^n)$, from Lemma 2.2.2(ii), it follows that

$$\|f\|_{\dot{\mathcal{B}}_{\omega}^{p,q}(\mathbb{R}^n)}^{\star} = \inf\left\{\sum_{l\in\mathbb{N}} \|f_{\xi_l}\|_{\dot{\mathcal{K}}_{\omega,\xi_l}^{p,q}(\mathbb{R}^n)}\right\} < \infty, \qquad (2.23)$$

where the infimum is taken over all the sequences $\{\xi_l\}_{l\in\mathbb{N}} \subset \mathbb{R}^n$ and $\{f_{\xi_l}\}_{l\in\mathbb{N}} \subset \mathscr{M}(\mathbb{R}^n)$ such that, for any $l \in \mathbb{N}$, $f_{\xi_l} \in \dot{\mathcal{K}}_{\omega,\xi_l}^{p,q}(\mathbb{R}^n)$ and

$$f = \sum_{l\in\mathbb{N}} f_{\xi_l}$$

almost everywhere in $\mathbb{R}^n$. Therefore, applying (1.111), (1.112), and Remark 1.7.2, we conclude that

$$\left|\int_{\mathbb{R}^n} f(y)g(y)\,dy\right| \leq \sum_{l\in\mathbb{N}} \int_{\mathbb{R}^n} |f_{\xi_l}(y)g(y)|\,dy$$

$$\leq \sum_{l\in\mathbb{N}} \|f_{\xi_l}\|_{\dot{\mathcal{K}}_{\omega,\xi_l}^{p,q}(\mathbb{R}^n)} \|g\|_{\dot{\mathcal{K}}_{1/\omega,\xi_l}^{p',q'}(\mathbb{R}^n)}$$

$$\leq \|g\|_{\dot{\mathcal{K}}_{1/\omega}^{p',q'}(\mathbb{R}^n)} \sum_{l\in\mathbb{N}} \|f_{\xi_l}\|_{\dot{\mathcal{K}}_{\omega,\xi_l}^{p,q}(\mathbb{R}^n)}, \qquad (2.24)$$

which, together with (2.23) and Lemma 2.2.2(ii), further implies that

$$\left|\int_{\mathbb{R}^n} f(y)g(y)\,dy\right| \leq \|f\|_{\dot{\mathcal{B}}_{\omega}^{p,q}(\mathbb{R}^n)}^{\star} \|g\|_{\dot{\mathcal{K}}_{1/\omega}^{p',q'}(\mathbb{R}^n)} \sim \|f\|_{\dot{\mathcal{B}}_{\omega}^{p,q}(\mathbb{R}^n)} \|g\|_{\dot{\mathcal{K}}_{1/\omega}^{p',q'}(\mathbb{R}^n)}, \qquad (2.25)$$

where the positive equivalence constants are independent of both f and g. Thus, the linear functional ϕ_g defined as in (2.16) is bounded on $\dot{\mathcal{B}}_{\omega}^{p,q}(\mathbb{R}^n)$ and

$$\|\phi_g\|_{(\dot{\mathcal{B}}_{\omega}^{p,q}(\mathbb{R}^n))^*} \lesssim \|g\|_{\dot{\mathcal{K}}_{1/\omega}^{p',q'}(\mathbb{R}^n)}, \qquad (2.26)$$

where the implicit positive constant is independent of g. This finishes the proof of (i).

Next, we show (ii). For this purpose, let $\phi \in (\dot{\mathcal{B}}_{\omega}^{p,q}(\mathbb{R}^n))^*$. Then, using Lemma 2.2.2(ii) and Lemma 2.2.3, we find that, for any given $\xi \in \mathbb{R}^n$ and for any $f \in \dot{\mathcal{K}}_{\omega,\xi}^{p,q}(\mathbb{R}^n)$, $f \in \dot{\mathcal{B}}_{\omega}^{p,q}(\mathbb{R}^n)$ and

$$|\phi(f)| \leq \|\phi\|_{(\dot{\mathcal{B}}_{\omega}^{p,q}(\mathbb{R}^n))^*} \|f\|_{\dot{\mathcal{B}}_{\omega}^{p,q}(\mathbb{R}^n)} \lesssim \|\phi\|_{(\dot{\mathcal{B}}_{\omega}^{p,q}(\mathbb{R}^n))^*} \|f\|_{\dot{\mathcal{K}}_{\omega,\xi}^{p,q}(\mathbb{R}^n)}, \qquad (2.27)$$

where the implicit positive constant is independent of f. This implies that, for any $\xi \in \mathbb{R}^n$,

$$\phi \in \left(\dot{\mathcal{K}}^{p,q}_{\omega,\xi}(\mathbb{R}^n)\right)^*.$$

By this and Theorem 1.7.3, we conclude that, for any $\xi \in \mathbb{R}^n$, there exists a $g_\xi \in \dot{\mathcal{K}}^{p',q'}_{1/\omega,\xi}(\mathbb{R}^n)$ such that, for any $f \in \dot{\mathcal{K}}^{p,q}_{\omega,\xi}(\mathbb{R}^n)$,

$$\phi(f) = \int_{\mathbb{R}^n} f(y) g_\xi(y) \, dy. \tag{2.28}$$

Now, we show that, for any $\xi_1, \, \xi_2 \in \mathbb{R}^n$, $g_{\xi_1} = g_{\xi_2}$ almost everywhere in $\mathbb{R}^n$. Indeed, when $\xi_1 = \xi_2$, from Theorem 1.7.3(ii), we deduce that $g_{\xi_1} = g_{\xi_2}$ almost everywhere in $\mathbb{R}^n$. On the other hand, when $\xi_1 \neq \xi_2$, applying (2.28), we conclude that, for any $f \in \dot{\mathcal{K}}^{p,q}_{\omega,\xi_1}(\mathbb{R}^n) \cap \dot{\mathcal{K}}^{p,q}_{\omega,\xi_2}(\mathbb{R}^n)$,

$$\int_{\mathbb{R}^n} f(y) g_{\xi_1}(y) \, dy = \phi(f) = \int_{\mathbb{R}^n} f(y) g_{\xi_2}(y) \, dy,$$

which further implies that

$$\int_{\mathbb{R}^n} f(y) \left[g_{\xi_1}(y) - g_{\xi_2}(y) \right] dy = 0. \tag{2.29}$$

For any $k \in \mathbb{N}$, let

$$f_k := \mathrm{sgn}\left(\overline{g_{\xi_1} - g_{\xi_2}} \right) \mathbf{1}_{B(\xi_1, 2^k) \setminus B(\xi_1, 2^{-k})} \mathbf{1}_{[B(\xi_2, \frac{|\xi_1 - \xi_2|}{2})]^\complement}.$$

Then we have, for any $k \in \mathbb{N}$,

$$\| f_k \|_{\dot{\mathcal{K}}^{p,q}_{\omega,\xi_1}(\mathbb{R}^n)}$$

$$\leq \left\{ \sum_{j \in \mathbb{Z}} \left[\omega(2^j) \right]^q \left\| \mathbf{1}_{B(\xi_1, 2^k) \setminus B(\xi_1, 2^{-k})} \mathbf{1}_{B(\xi_1, 2^j) \setminus B(\xi_1, 2^{j-1})} \right\|^q_{L^p(\mathbb{R}^n)} \right\}^{\frac{1}{q}}$$

$$= \left\{ \sum_{j=-k+1}^{k} \left[\omega(2^j) \right]^q \left\| \mathbf{1}_{B(\xi_1, 2^j) \setminus B(\xi_1, 2^{j-1})} \right\|^q_{L^p(\mathbb{R}^n)} \right\}^{\frac{1}{q}} < \infty,$$

which implies that $f_k \in \dot{\mathcal{K}}^{p,q}_{\omega,\xi_1}(\mathbb{R}^n)$. Next, we prove that, for any $k \in \mathbb{N}$, $f_k \in \dot{\mathcal{K}}^{p,q}_{\omega,\xi_2}(\mathbb{R}^n)$. To this end, for any $k \in \mathbb{N}$, let

$$j_{k,1} := \left\lfloor \frac{\ln(|\xi_1 - \xi_2|)}{\ln 2} \right\rfloor$$

and

$$j_{k,2} := \left\lceil \frac{\ln(|\xi_1 - \xi_2| + 2^k)}{\ln 2} \right\rceil.$$

We claim that, for any given $k \in \mathbb{N}$ and for any $j \in \mathbb{Z} \cap (-\infty, j_{k,1})$,

$$B(\xi_2, 2^j) \setminus B(\xi_2, 2^{j-1}) \subset B\left(\xi_2, \frac{|\xi_1 - \xi_2|}{2}\right)$$

and, for any $j \in \mathbb{Z} \cap (j_{k,2}, \infty)$,

$$B(\xi_2, 2^j) \setminus B(\xi_2, 2^{j-1}) \subset \left[B(\xi_1, 2^k)\right]^{\complement}.$$

Indeed, for any given $k \in \mathbb{N}$ and for any $j \in \mathbb{Z} \cap (-\infty, j_{k,1})$, we have

$$j \le j_{k,1} - 1 \le \frac{\ln(|\xi_1 - \xi_2|)}{\ln 2} - 1,$$

which implies that $2^j \le \frac{|\xi_1 - \xi_2|}{2}$. This further implies that, for any given $k \in \mathbb{N}$ and for any $j \in \mathbb{Z} \cap (-\infty, j_{k,1})$,

$$B(\xi_2, 2^j) \setminus B(\xi_2, 2^{j-1}) \subset B(\xi_2, 2^j) \subset B\left(\xi_2, \frac{|\xi_1 - \xi_2|}{2}\right).$$

In addition, for any given $k \in \mathbb{N}$ and for any $j \in \mathbb{Z} \cap (j_{k,2}, \infty)$, we have

$$j \ge j_{k,2} + 1 \ge \frac{\ln(|\xi_1 - \xi_2| + 2^k)}{\ln 2} + 1,$$

which further implies that $2^{j-1} \ge |\xi_1 - \xi_2| + 2^k$. Using this, we conclude that, for any given $k \in \mathbb{N}$ and for any $j \in \mathbb{Z} \cap (j_{k,2}, \infty)$ and $x \in B(\xi_2, 2^j) \setminus B(\xi_2, 2^{j-1})$,

$$|x - \xi_1| \ge |x - \xi_2| - |\xi_1 - \xi_2| \ge 2^{j-1} - |\xi_1 - \xi_2| \ge 2^k.$$

From this, we further deduce that, for any given $k \in \mathbb{N}$ and for any $j \in \mathbb{Z} \cap (j_{k,2}, \infty)$,

$$B(\xi_2, 2^j) \setminus B(\xi_2, 2^{j-1}) \subset \left[B(\xi_1, 2^k) \right]^\complement$$

and hence the above claim holds true. Thus, we find that, for any $k \in \mathbb{N}$,

$$\|f_k\|_{\dot{\mathcal{K}}^{p,q}_{\omega,\xi_2}(\mathbb{R}^n)} \leq \left\{ \sum_{j=j_{k,1}}^{j_{k,2}} \left[\omega(2^j) \right]^q \left\| \mathbf{1}_{B(\xi_2, 2^j) \setminus B(\xi_2, 2^{j-1})} \right\|^q_{L^p(\mathbb{R}^n)} \right\}^{\frac{1}{q}} < \infty,$$

which implies that $f_k \in \dot{\mathcal{K}}^{p,q}_{\omega,\xi_2}(\mathbb{R}^n)$. Then, applying the fact that, for any $k \in \mathbb{N}$, $f_k \in \dot{\mathcal{K}}^{p,q}_{\omega,\xi_1}(\mathbb{R}^n) \cap \dot{\mathcal{K}}^{p,q}_{\omega,\xi_2}(\mathbb{R}^n)$ and (2.29) with f therein replaced by f_k, we conclude that, for any $k \in \mathbb{N}$,

$$\int_{[B(\xi_1, 2^k) \setminus B(\xi_1, 2^{-k})] \cap [B(\xi_2, \frac{|\xi_1 - \xi_2|}{2})]^\complement} \left| g_{\xi_1}(y) - g_{\xi_2}(y) \right| \, dy = 0.$$

This, together with the arbitrariness of k, further implies that $g_{\xi_1} = g_{\xi_2}$ almost everywhere in $\mathbb{R}^n \setminus B(\xi_2, \frac{|\xi_1 - \xi_2|}{2})$. Similarly, we can obtain $g_{\xi_1} = g_{\xi_2}$ almost everywhere in $\mathbb{R}^n \setminus B(\xi_1, \frac{|\xi_1 - \xi_2|}{2})$. Therefore, we further find that $g_{\xi_1} = g_{\xi_2}$ almost everywhere in $\mathbb{R}^n$.

Next, we show that

$$g := g_0 \in \dot{\mathcal{K}}^{p',q'}_{1/\omega}(\mathbb{R}^n).$$

Indeed, by Theorem 1.7.3 and (2.27), we conclude that, for any $\xi \in \mathbb{R}^n$,

$$\|g\|_{\dot{\mathcal{K}}^{p',q'}_{1/\omega,\xi}(\mathbb{R}^n)} = \|g_\xi\|_{\dot{\mathcal{K}}^{p',q'}_{1/\omega,\xi}(\mathbb{R}^n)} = \|\phi\|_{(\dot{\mathcal{K}}^{p,q}_{\omega,\xi}(\mathbb{R}^n))^*}$$

$$= \sup_{\|f\|_{\dot{\mathcal{K}}^{p,q}_{\omega,\xi}(\mathbb{R}^n)}=1} |\phi(f)| \lesssim \|\phi\|_{(\dot{\mathcal{B}}^{p,q}_\omega(\mathbb{R}^n))^*},$$

where the implicit positive constant is independent of both ξ and g. This, together with the arbitrariness of ξ, further implies that $g \in \dot{\mathcal{K}}^{p',q'}_{1/\omega}(\mathbb{R}^n)$ and

$$\|g\|_{\dot{\mathcal{K}}^{p',q'}_{1/\omega}(\mathbb{R}^n)} \lesssim \|\phi\|_{(\dot{\mathcal{B}}^{p,q}_\omega(\mathbb{R}^n))^*}, \tag{2.30}$$

where the implicit positive constant is independent of g. Thus, the function g can induce a ϕ_g as in (2.16).

Next, we show that $\phi = \phi_g$. Assume that $f \in \dot{\mathcal{B}}^{p,q}_\omega(\mathbb{R}^n)$. Then, from Lemma 2.2.2(ii), we infer that there exist sequences $\{\xi_l\}_{l \in \mathbb{N}} \subset \mathbb{R}^n$ and $\{f_{\xi_l}\}_{l \in \mathbb{N}} \subset$

$\mathscr{M}(\mathbb{R}^n)$ such that, for any $l \in \mathbb{N}$, $f_{\xi_l} \in \dot{\mathcal{K}}^{p,q}_{\omega,\xi_l}(\mathbb{R}^n)$, $f = \sum_{l \in \mathbb{N}} f_{\xi_l}$ almost everywhere in $\mathbb{R}^n$ and

$$\sum_{l \in \mathbb{N}} \left\| f_{\xi_l} \right\|_{\dot{\mathcal{K}}^{p,q}_{\omega,\xi_l}(\mathbb{R}^n)} \lesssim \|f\|_{\dot{\mathcal{B}}^{p,q}_{\omega}(\mathbb{R}^n)} + 1. \tag{2.31}$$

In addition, for any $N \in \mathbb{N}$, let

$$f^{(N)} := \sum_{l=1}^{N} f_{\xi_l}.$$

Then, applying Lemma 2.2.2(ii) and (2.31), we find that, for any $N \in \mathbb{N}$,

$$\left\| f^{(N)} \right\|_{\dot{\mathcal{B}}^{p,q}_{\omega}(\mathbb{R}^n)} \sim \left\| f^{(N)} \right\|^{\star}_{\dot{\mathcal{B}}^{p,q}_{\omega}(\mathbb{R}^n)} \lesssim \sum_{l=1}^{N} \left\| f_{\xi_l} \right\|_{\dot{\mathcal{K}}^{p,q}_{\omega,\xi_l}(\mathbb{R}^n)}$$

$$\lesssim \|f\|_{\dot{\mathcal{B}}^{p,q}_{\omega}(\mathbb{R}^n)} + 1 < \infty, \tag{2.32}$$

which implies that $f^{(N)} \in \dot{\mathcal{B}}^{p,q}_{\omega}(\mathbb{R}^n)$. Therefore, we have

$$\left\| f - f^{(N)} \right\|_{\dot{\mathcal{B}}^{p,q}_{\omega}(\mathbb{R}^n)} \sim \left\| f - f^{(N)} \right\|^{\star}_{\dot{\mathcal{B}}^{p,q}_{\omega}(\mathbb{R}^n)}$$

$$\lesssim \sum_{l=N+1}^{\infty} \left\| f_{\xi_l} \right\|_{\dot{\mathcal{K}}^{p,q}_{\omega,\xi_l}(\mathbb{R}^n)} \to 0 \tag{2.33}$$

as $N \to \infty$. This, combined with the continuity of ϕ on $\dot{\mathcal{B}}^{p,q}_{\omega}(\mathbb{R}^n)$, (2.28), (2.24), and the dominated convergence theorem, further implies that

$$\phi(f) = \lim_{N \to \infty} \phi\left(f^{(N)}\right) = \lim_{N \to \infty} \int_{\mathbb{R}^n} \sum_{l=1}^{N} f_{\xi_l}(y) g(y) \, dy$$

$$= \int_{\mathbb{R}^n} f(y) g(y) \, dy = \phi_g(f),$$

which completes the proof that $\phi = \phi_g$. Moreover, from (2.26) and (2.30), it follows that

$$\|\phi_g\|_{(\dot{\mathcal{B}}^{p,q}_{\omega}(\mathbb{R}^n))^*} \sim \|g\|_{\dot{\mathcal{K}}^{p',q'}_{1/\omega}(\mathbb{R}^n)}$$

with the positive equivalence constants independent of g. By this and the linearity of ϕ_g about g, we conclude that g is unique. This finishes the proof of (ii) and hence Theorem 2.2.1. $\qquad\qquad\qquad\square$

On the other hand, if $p \in (0, 1)$, $q \in (0, \infty)$, and $\omega \in M(\mathbb{R}_+)$, we then prove that the dual space of $\dot{\mathcal{B}}_\omega^{p,q}(\mathbb{R}^n)$ is trivial as follows.

Theorem 2.2.4 *Let $p \in (0, 1)$, $q \in (0, \infty)$, and $\omega \in M(\mathbb{R}_+)$. Then*

$$\left(\dot{\mathcal{B}}_\omega^{p,q}(\mathbb{R}^n)\right)^* = \{0\}.$$

Proof Let all the symbols be the same as in the present theorem. Then it is obvious that

$$\{0\} \subset \left(\dot{\mathcal{B}}_\omega^{p,q}(\mathbb{R}^n)\right)^*.$$

Conversely, we now show that

$$\left(\dot{\mathcal{B}}_\omega^{p,q}(\mathbb{R}^n)\right)^* \subset \{0\}. \tag{2.34}$$

Indeed, let $\phi \in (\dot{\mathcal{B}}_\omega^{p,q}(\mathbb{R}^n))^*$. Then, from (2.27), we deduce that, for any $\xi \in \mathbb{R}^n$, $\phi \in (\dot{\mathcal{K}}_{\omega,\xi}^{p,q}(\mathbb{R}^n))^*$. Thus, by Theorem 1.7.5, we find that, for any $\xi \in \mathbb{R}^n$,

$$\phi = 0 \text{ on } \dot{\mathcal{K}}_{\omega,\xi}^{p,q}(\mathbb{R}^n). \tag{2.35}$$

Assume $f \in \dot{\mathcal{B}}_\omega^{p,q}(\mathbb{R}^n)$. Then, using Lemma 2.2.2(ii), we conclude that there exist sequences $\{\xi_l\}_{l\in\mathbb{N}} \subset \mathbb{R}^n$ and $\{f_{\xi_l}\}_{l\in\mathbb{N}} \subset \mathscr{M}(\mathbb{R}^n)$ such that, for any $l \in \mathbb{N}$, $f_{\xi_l} \in \dot{\mathcal{K}}_{\omega,\xi_l}^{p,q}(\mathbb{R}^n)$, $f = \sum_{l\in\mathbb{N}} f_{\xi_l}$ almost everywhere in $\mathbb{R}^n$ and

$$\sum_{l\in\mathbb{N}} \left\| f_{\xi_l} \right\|_{\dot{\mathcal{K}}_{\omega,\xi_l}^{p,q}(\mathbb{R}^n)} \lesssim \|f\|_{\dot{\mathcal{B}}_\omega^{p,q}(\mathbb{R}^n)} + 1.$$

For any $N \in \mathbb{N}$, let

$$f^{(N)} := \sum_{l=1}^N f_{\xi_l}.$$

Then, combining both (2.32) and (2.33), we have, for any $N \in \mathbb{N}$, $f^{(N)} \in \dot{\mathcal{B}}_\omega^{p,q}(\mathbb{R}^n)$ and

$$\lim_{N\to\infty} \left\| f^{(N)} - f \right\|_{\dot{\mathcal{B}}_\omega^{p,q}(\mathbb{R}^n)} = 0.$$

From this, both the continuity and the linearity of ϕ on $\dot{\mathcal{B}}_\omega^{p,q}(\mathbb{R}^n)$, the assumption that, for any $l \in \mathbb{N}$, $f_{\xi_l} \in \dot{\mathcal{K}}_{\omega,\xi_l}^{p,q}(\mathbb{R}^n)$, and (2.35), we further infer that

$$\phi(f) = \lim_{N\to\infty} \phi\left(f^{(N)}\right) = \lim_{N\to\infty} \sum_{l=1}^{N} \phi\left(f_{\xi_l}\right)$$

$$= \lim_{N\to\infty} \sum_{l=1}^{N} 0 = 0,$$

which implies that $\phi = 0$ on $\dot{\mathcal{B}}_\omega^{p,q}(\mathbb{R}^n)$, and hence (2.34) holds true. This then finishes the proof of Theorem 2.2.4. $\square$

As an application of the above two dual theorems, we now investigate the triviality of the seminorm $\|\cdot\|_{\dot{\mathcal{B}}_\omega^{p,q}(\mathbb{R}^n)}$. First, we show that, under some sharp assumptions of ω, $\|\cdot\|_{\dot{\mathcal{B}}_\omega^{p,q}(\mathbb{R}^n)}$ is a trivial seminorm.

Theorem 2.2.5 *Let* $p \in [1,\infty)$, $q \in (0,\infty)$, *and* $\omega \in M(\mathbb{R}_+)$ *satisfy* $m_0(\omega) \in (\frac{n}{p'}, \infty)$ *or* $M_\infty(\omega) \in (-\infty, 0)$. *Then* $f \in \dot{\mathcal{B}}_\omega^{p,q}(\mathbb{R}^n)$ *if and only if* $f \in \mathscr{M}(\mathbb{R}^n)$ *and* $\|f\|_{\dot{\mathcal{B}}_\omega^{p,q}(\mathbb{R}^n)} = 0$.

Proof Let all the symbols be the same as in the present theorem. We first prove the sufficiency. To this end, let $f \in \mathscr{M}(\mathbb{R}^n)$ be such that $\|f\|_{\dot{\mathcal{B}}_\omega^{p,q}(\mathbb{R}^n)} = 0$. Then, by Definition 2.1.3, we have $f \in \dot{\mathcal{B}}_\omega^{p,q}(\mathbb{R}^n)$. This finishes the proof of the sufficiency.

Conversely, we next show the necessity. To do this, let $f \in \dot{\mathcal{B}}_\omega^{p,q}(\mathbb{R}^n)$,

$$E := \{\lambda f : \lambda \in \mathbb{C}\},$$

and, for any $\lambda \in \mathbb{C}$,

$$\mathcal{P}(\lambda f) := \lambda \|f\|_{\dot{\mathcal{B}}_\omega^{p,q}(\mathbb{R}^n)}^2. \qquad (2.36)$$

Then E is a linear subspace of $\dot{\mathcal{B}}_\omega^{p,q}(\mathbb{R}^n)$ and $\mathcal{P}$ is a linear functional on E. Moreover, observe that, for any $\lambda \in \mathbb{C}$,

$$|\mathcal{P}(\lambda f)| = \|f\|_{\dot{\mathcal{B}}_\omega^{p,q}(\mathbb{R}^n)} \|\lambda f\|_{\dot{\mathcal{B}}_\omega^{p,q}(\mathbb{R}^n)}.$$

This, combined with the fact that $\|\cdot\|_{\dot{\mathcal{B}}_\omega^{p,q}(\mathbb{R}^n)}$ is a seminorm on $\dot{\mathcal{B}}_\omega^{p,q}(\mathbb{R}^n)$, and the complex Hahn–Banach theorem, further implies that there exists a linear functional ϕ, defined on $\dot{\mathcal{B}}_\omega^{p,q}(\mathbb{R}^n)$, such that $\phi|_E = \mathcal{P}$ and, for any $g \in \dot{\mathcal{B}}_\omega^{p,q}(\mathbb{R}^n)$,

$$|\phi(g)| \leq \|f\|_{\dot{\mathcal{B}}_\omega^{p,q}(\mathbb{R}^n)} \|g\|_{\dot{\mathcal{B}}_\omega^{p,q}(\mathbb{R}^n)}. \qquad (2.37)$$

Thus, we have $\phi \in (\dot{\mathcal{B}}_\omega^{p,q}(\mathbb{R}^n))^*$. Using both (2.36) with $\lambda := 1$ and (2.37), we further obtain

$$\|\phi\|_{(\dot{\mathcal{B}}_\omega^{p,q}(\mathbb{R}^n))^*} = \|f\|_{\dot{\mathcal{B}}_\omega^{p,q}(\mathbb{R}^n)}.$$

From this and Theorem 2.2.1, we infer that there exists an $h \in \dot{\mathcal{K}}_{1/\omega}^{p',q'}(\mathbb{R}^n)$ such that

$$\|f\|_{\dot{\mathcal{B}}_\omega^{p,q}(\mathbb{R}^n)} = \|\phi\|_{(\dot{\mathcal{B}}_\omega^{p,q}(\mathbb{R}^n))^*} \sim \|h\|_{\dot{\mathcal{K}}_{1/\omega}^{p',q'}(\mathbb{R}^n)}. \tag{2.38}$$

In addition, using Lemma 1.1.6 and the assumptions $m_0(\omega) \in (\frac{n}{p'}, \infty)$ or $M_\infty(\omega) \in (-\infty, 0)$, we obtain

$$M_0\left(\frac{1}{\omega}\right) = -m_0(\omega) < -\frac{n}{p'}$$

or

$$m_\infty\left(\frac{1}{\omega}\right) = -M_\infty(\omega) > 0.$$

These, together with (2.38) and Theorems 1.2.6 and 1.2.7, further imply that

$$\|f\|_{\dot{\mathcal{B}}_\omega^{p,q}(\mathbb{R}^n)} \sim \|h\|_{\dot{\mathcal{K}}_{1/\omega}^{p',q'}(\mathbb{R}^n)} = 0,$$

which then completes the proof of the necessity and hence Theorem 2.2.5. $\quad\square$

Moreover, the following special example shows that both Corollary 2.1.7 and Theorem 2.2.5 are sharp.

Example 2.2.6 Let $p \in [1, \infty)$, $q \in (0, \infty)$, $\alpha_1, \alpha_2, \beta_1, \beta_2 \in \mathbb{R}$, and

$$\omega_{\alpha_1,\alpha_2,\beta_1,\beta_2}(t) := \begin{cases} t^{\alpha_1}(1 - \ln t)^{\beta_1} & \text{when } t \in (0, 1], \\ t^{\alpha_2}(1 + \ln t)^{\beta_2} & \text{when } t \in (1, \infty). \end{cases}$$

Then $\omega_{\alpha_1,\alpha_2,\beta_1,\beta_2} \in M(\mathbb{R}_+)$ and

$$m_0\left(\omega_{\alpha_1,\alpha_2,\beta_1,\beta_2}\right) = M_0\left(\omega_{\alpha_1,\alpha_2,\beta_1,\beta_2}\right) = \alpha_1 \tag{2.39}$$

and

$$m_\infty\left(\omega_{\alpha_1,\alpha_2,\beta_1,\beta_2}\right) = M_\infty\left(\omega_{\alpha_1,\alpha_2,\beta_1,\beta_2}\right) = \alpha_2. \tag{2.40}$$

Moreover,

(i) if $\alpha_1 \in [\frac{n}{p'}, \infty)$, $\alpha_2 \in (-\infty, 0]$, and $\beta_2 \in (-\infty, 0)$, then, for any ball $B \in \mathbb{B}$,

$$\left\| \mathbf{1}_B \right\|_{\dot{\mathcal{B}}^{p,q}_{\omega_{\alpha_1,\alpha_2,\beta_1,\beta_2}}(\mathbb{R}^n)} = 0;$$

(ii) if $\alpha_1 \in (-\infty, \frac{n}{p'}]$, $\alpha_2 \in [0, \infty)$, and $\beta_1, \beta_2 \in (\frac{1}{q}, \infty)$, then $f \in \dot{\mathcal{B}}^{p,q}_{\omega}(\mathbb{R}^n)$ if and only if $f \in \mathscr{M}(\mathbb{R}^n)$ and

$$\left\| f \right\|_{\dot{\mathcal{B}}^{p,q}_{\omega_{\alpha_1,\alpha_2,\beta_1,\beta_2}}(\mathbb{R}^n)} = 0.$$

Here $\frac{1}{p} + \frac{1}{p'} = 1$ and

$$q' := \begin{cases} \dfrac{q}{q-1} & \text{when } q \in [1, \infty), \\[2mm] \infty & \text{when } q \in (0, 1). \end{cases}$$

Proof Let all the symbols be the same as in the present example. Similarly to Example 1.1.9, we have both (2.39) and (2.40).

We now prove (i). To this end, let $B(x_0, r) \in \mathbb{B}$ with $x_0 \in \mathbb{R}^n$ and $r \in (0, \infty)$ and, in the remainder of this proof, for any $x := (x_1, \ldots, x_n) \in \mathbb{R}^n$, let

$$|x|_\infty := \max \left\{ |x_j| : \ j = 1, \ldots, n \right\}.$$

Then, for any $j \in \mathbb{N} \cap [2 + \frac{\ln r}{\ln 2}, \infty)$, we choose a $\xi_j \in \mathbb{R}^n$ satisfying that

$$\left| \xi_j - x_0 \right|_\infty \in [2^{j-1} + r, 2^j - r].$$

For any given $j \in \mathbb{N} \cap [2 + \frac{\ln r}{\ln 2}, \infty)$ and for any $y \in B(x_0, r)$, we have

$$\begin{aligned} \left| y - \xi_j \right|_\infty &\leq |y - x_0|_\infty + \left| x_0 - \xi_j \right|_\infty \\ &\leq |y - x_0| + \left| x_0 - \xi_j \right|_\infty < 2^j \end{aligned}$$

and

$$\begin{aligned} \left| y - \xi_j \right|_\infty &\geq \left| x_0 - \xi_j \right|_\infty - |y - x_0|_\infty \\ &\geq \left| x_0 - \xi_j \right|_\infty - |y - x_0| > 2^{j-1}, \end{aligned}$$

which imply that $y \in Q(\xi_j, 2^{j+1}) \setminus Q(\xi_j, 2^j)$. By this and the arbitrariness of y, we find that, for any $j \in \mathbb{N} \cap [2 + \frac{\ln r}{\ln 2}, \infty)$,

$$B(x_0, r) \subset Q(\xi_j, 2^{j+1}) \setminus Q(\xi_j, 2^j). \tag{2.41}$$

In addition, for any $j \in \mathbb{N} \cap [2 + \frac{\ln r}{\ln 2}, \infty)$ and $Q \in \mathcal{Q}_j^{(\xi_j)}$, let

$$b_{j,Q} := \left[\omega_{\alpha_1, \alpha_2, \beta_1, \beta_2}(2^{j-1}) \right]^{-1} |Q \cap B(x_0, r)|^{-\frac{1}{p}} \mathbf{1}_{Q \cap B(x_0,r)}.$$

Then, from Definition 2.1.1, we deduce that, for any $j \in \mathbb{N} \cap [2 + \frac{\ln r}{\ln 2}, \infty)$ and $Q \in \mathcal{Q}_j^{(\xi_j)}$, $b_{j,Q}$ is an $(\omega_{\alpha_1, \alpha_2, \beta_1, \beta_2}, p)$-block supported in the cube Q. Applying (2.41), we have, for any $j \in \mathbb{N} \cap [2 + \frac{\ln r}{\ln 2}, \infty)$

$$\mathbf{1}_{B(x_0,r)} = \sum_{Q \in \mathcal{Q}_j^{(\xi_j)}} \left[\omega_{\alpha_1, \alpha_2, \beta_1, \beta_2}(2^{j-1}) \right] |Q \cap B(x_0, r)|^{\frac{1}{p}} b_{j,Q}.$$

Combining this, the fact that $\{b_{j,Q}\}_{Q \in \mathcal{Q}_j^{(\xi_j)}}$ is a sequence of $(\omega_{\alpha_1, \alpha_2, \beta_1, \beta_2}, p)$-blocks, Definition 2.1.3, Remark 2.1.2, and the assumptions $\alpha_2 \in (-\infty, 0]$ and $\beta_2 \in (-\infty, 0)$, we conclude that, for any $j \in \mathbb{N} \cap [2 + \frac{\ln r}{\ln 2}, \infty)$,

$$\left\| \mathbf{1}_{B(x_0,r)} \right\|_{\dot{\mathcal{B}}^{p,q}_{\omega_{\alpha_1, \alpha_2, \beta_1, \beta_2}}(\mathbb{R}^n)}$$

$$\leq \left\{ \sum_{Q \in \mathcal{Q}_j^{(\xi_j)}} \left[\omega_{\alpha_1, \alpha_2, \beta_1, \beta_2}(2^{j-1}) \right]^q |Q \cap B(x_0, r)|^{\frac{q}{p}} \right\}^{\frac{1}{q}}$$

$$= \left\{ \sum_{Q \in \mathcal{Q}_j^{(\xi_j)}} 2^{(j-1)\alpha_2 q} [1 + (j-1)\ln 2]^{\beta_2 q} |Q \cap B(x_0, r)|^{\frac{q}{p}} \right\}^{\frac{1}{q}}$$

$$\leq 2^{(j-1)\alpha_2} [1 + (j-1)\ln 2]^{\beta_2} |B(x_0, r)|^{\frac{1}{p}} \left[\sharp \mathcal{Q}_j^{(\xi_j)} \right]^{\frac{1}{q}}$$

$$\sim 2^{j\alpha_2} (1 + j \ln 2)^{\beta_2} \to 0$$

as $j \to \infty$. Therefore, we have

$$\left\| \mathbf{1}_{B(x_0,r)} \right\|_{\dot{\mathcal{B}}^{p,q}_{\omega_{\alpha_1, \alpha_2, \beta_1, \beta_2}}(\mathbb{R}^n)} = 0,$$

which then completes the proof of (i).

Next, we show (ii). Indeed, using the assumptions $\alpha_1 \in (-\infty, \frac{n}{p'}]$, $\alpha_2 \in [0, \infty)$, and $\beta_1, \beta_2 \in (\frac{1}{q'}, \infty)$, and repeating an argument similar to that used in the proof of Theorem 2.1.5 with ω therein replaced by $\omega_{\alpha_1,\alpha_2,\beta_1,\beta_2}$, we find that, for any $f \in \dot{\mathcal{B}}^{p,q}_{\omega_{\alpha_1,\alpha_2,\beta_1,\beta_2}}(\mathbb{R}^n)$,

$$\|f\|_{L^1(\mathbb{R}^n)+L^p(\mathbb{R}^n)} \lesssim \|f\|_{\dot{\mathcal{B}}^{p,q}_{\omega_{\alpha_1,\alpha_2,\beta_1,\beta_2}}(\mathbb{R}^n)}.$$

This, together with Definition 2.1.3, further implies that, for any $f \in \mathcal{M}(\mathbb{R}^n)$, $\|f\|_{\dot{\mathcal{B}}^{p,q}_{\omega_{\alpha_1,\alpha_2,\beta_1,\beta_2}}(\mathbb{R}^n)} = 0$ if and only if $f = 0$ almost everywhere in $\mathbb{R}^n$. This then finishes the proof of (ii) and hence Example 2.2.6. □

Remark 2.2.7 Let $p \in [1, \infty)$ and $q \in (0, \infty)$.

(i) Example 2.2.6(i) shows that all the assumptions on the indices of ω in Corollary 2.1.7 are sharp. Namely, if ω does not satisfy all the assumptions of Corollary 2.1.7, then $\|\cdot\|_{\dot{\mathcal{B}}^{p,q}_{\omega}(\mathbb{R}^n)}$ may not be a norm. Indeed, let $\omega_{\alpha_1,\alpha_2,\beta_1,\beta_2}$ be the same as in Example 2.2.6(i). Then, applying both (2.39) and (2.40), we find that, in this case, $\omega_{\alpha_1,\alpha_2,\beta_1,\beta_2} \in M(\mathbb{R}_+)$, $m_0(\omega_{\alpha_1,\alpha_2,\beta_1,\beta_2}) \in [\frac{n}{p'}, \infty)$, and $M_\infty(\omega_{\alpha_1,\alpha_2,\beta_1,\beta_2}) \in (-\infty, 0]$. These imply that, in this case, $\omega_{\alpha_1,\alpha_2,\beta_1,\beta_2}$ fails the assumptions of Corollary 2.1.7 and $\|\cdot\|_{\dot{\mathcal{B}}^{p,q}_{\omega_{\alpha_1,\alpha_2,\beta_1,\beta_2}}(\mathbb{R}^n)}$ is not a norm because $\|\cdot\|_{\dot{\mathcal{B}}^{p,q}_{\omega_{\alpha_1,\alpha_2,\beta_1,\beta_2}}(\mathbb{R}^n)}$ does not satisfy the positive definiteness [namely, $\|f\|_{\dot{\mathcal{B}}^{p,q}_{\omega_{\alpha_1,\alpha_2,\beta_1,\beta_2}}(\mathbb{R}^n)} = 0$ can not imply $f = 0$ almost everywhere in $\mathbb{R}^n$]. Thus, in this sense, all the assumptions on the indices of ω in Corollary 2.1.7 are sharp.

(ii) Example 2.2.6(ii) shows that all the assumptions on the indices of ω in Theorem 2.2.5 are sharp. Precisely, if $\omega \in M(\mathbb{R}_+)$ fails the assumptions of Theorem 2.2.5, then $\|\cdot\|_{\dot{\mathcal{B}}^{p,q}_{\omega}(\mathbb{R}^n)}$ may be a true norm. Indeed, let $\omega_{\alpha_1,\alpha_2,\beta_1,\beta_2}$ be the same as in Example 2.2.6(ii). Then, from both (2.39) and (2.40), we infer that, in this case, $\omega_{\alpha_1,\alpha_2,\beta_1,\beta_2} \in M(\mathbb{R}_+)$, $m_0(\omega_{\alpha_1,\alpha_2,\beta_1,\beta_2}) \in (-\infty, \frac{n}{p'}]$, and $M_\infty(\omega_{\alpha_1,\alpha_2,\beta_1,\beta_2}) \in [0, \infty)$. Therefore, in this case, $\omega_{\alpha_1,\alpha_2,\beta_1,\beta_2}$ does not satisfy the assumptions of Theorem 2.2.5, but $\|\cdot\|_{\dot{\mathcal{B}}^{p,q}_{\omega_{\alpha_1,\alpha_2,\beta_1,\beta_2}}(\mathbb{R}^n)}$ is a true norm.

These further imply that, in this sense, all the assumptions on the indices of ω in Theorem 2.2.5 are sharp.

In addition, the following conclusion implies that, if $p \in (0, 1)$, then the seminorm $\|\cdot\|_{\dot{\mathcal{B}}^{p,q}_{\omega}(\mathbb{R}^n)}$ is always trivial.

Theorem 2.2.8 *Let $p \in (0, 1)$, $q \in (0, \infty)$, and $\omega \in M(\mathbb{R}_+)$. Then $f \in \dot{\mathcal{B}}^{p,q}_{\omega}(\mathbb{R}^n)$ if and only if $f \in \mathcal{M}(\mathbb{R}^n)$ and $\|f\|_{\dot{\mathcal{B}}^{p,q}_{\omega}(\mathbb{R}^n)} = 0$.*

Proof Let all the symbols be the same as in the present theorem. We first prove the sufficiency. Indeed, let $f \in \mathcal{M}(\mathbb{R}^n)$ be such that $\|f\|_{\dot{\mathcal{B}}^{p,q}_{\omega}(\mathbb{R}^n)} = 0$. Then, by Definition 2.1.3, we have $f \in \dot{\mathcal{B}}^{p,q}_{\omega}(\mathbb{R}^n)$. This finishes the proof of the sufficiency.

Conversely, we deal with the necessity. For this purpose, let $f \in \dot{\mathcal{B}}_\omega^{p,q}(\mathbb{R}^n)$. Then, repeating an argument similar to that used in the proof of Theorem 2.2.5 with Theorems 1.2.6 and 1.2.7 used therein replaced by Theorem 2.2.4, we find that there exists a $\phi \in (\dot{\mathcal{B}}_\omega^{p,q}(\mathbb{R}^n))^*$ such that

$$\|f\|_{\dot{\mathcal{B}}_\omega^{p,q}(\mathbb{R}^n)} = \|\phi\|_{(\dot{\mathcal{B}}_\omega^{p,q}(\mathbb{R}^n))^*} = \|0\|_{(\dot{\mathcal{B}}_\omega^{p,q}(\mathbb{R}^n))^*} = 0.$$

This then finishes the proof of the necessity and hence Theorem 2.2.8. $\qquad\square$

Remark 2.2.9 Theorems 2.2.5 and 2.2.8 show that, when $p \in [1, \infty)$, $q \in (0, \infty)$, and $\omega \in M(\mathbb{R}_+)$ satisfying $m_0(\omega) \in (\frac{n}{p'}, \infty)$ or $M_\infty(\omega) \in (-\infty, 0)$, or when $p \in (0, 1)$, $q \in (0, \infty)$, and $\omega \in M(\mathbb{R}_+)$, then the block space $\dot{\mathcal{B}}_\omega^{p,q}(\mathbb{R}^n)$ is meaningless. However, it does not affect applications of block spaces in this book. Indeed, in the study of generalized Herz–Hardy spaces in the subsequent chapters, we always use the block space $\dot{\mathcal{B}}_\omega^{p,q}(\mathbb{R}^n)$ under the assumptions $p, q \in (1, \infty)$ and $\omega \in M(\mathbb{R}_+)$ satisfying $M_0(\omega) \in (-\infty, \frac{n}{p'})$ and $m_\infty(\omega) \in (0, \infty)$ (see, for instance, Lemma 4.3.27 below). These block spaces used later do not satisfy the assumptions of both Theorems 2.2.5 and 2.2.8 and, as is proved in Corollary 2.1.7, their seminorms are true norms.

2.3 Boundedness of Sublinear Operators

The targets of this section are twofold. The first one is to establish a boundedness criterion of sublinear operators on the block space $\dot{\mathcal{B}}_\omega^{p,q}(\mathbb{R}^n)$ under some reasonable and sharp assumptions. To achieve this, we show the lattice property of block spaces (see Lemma 2.3.2 below). Using this lattice property and the characterization of block spaces via local generalized Herz spaces obtained in the last section, we conclude the boundedness of sublinear operators. As an application, we give the boundedness of powered Hardy–Littlewood maximal operators on block spaces, which plays an important role in the subsequent chapters. The second target is to investigate the boundedness of Calderón–Zygmund operators on block spaces. Indeed, by embedding $\dot{\mathcal{B}}_\omega^{p,q}(\mathbb{R}^n)$ into certain weighted Lebesgue space (see Lemma 2.3.8 below), the boundedness of Calderón–Zygmund operators can be concluded directly by that of sublinear operators just proved.

To begin with, we present the boundedness criterion of sublinear operators on $\dot{\mathcal{B}}_\omega^{p,q}(\mathbb{R}^n)$ as follows.

Theorem 2.3.1 *Let $p \in (1, \infty)$, $q \in (0, \infty)$, and $\omega \in M(\mathbb{R}_+)$ satisfy*

$$-\frac{n}{p} < m_0(\omega) \le M_0(\omega) < \frac{n}{p'}$$

and

$$-\frac{n}{p} < m_\infty(\omega) \le M_\infty(\omega) < \frac{n}{p'},$$

where $\frac{1}{p} + \frac{1}{p'} = 1$. Let T be a bounded sublinear operator on $L^p(\mathbb{R}^n)$ satisfying that there exists a positive constant C such that, for any $f \in \dot{\mathcal{K}}_{\omega,0}^{p,q}(\mathbb{R}^n)$ and $x \notin$ $\overline{\text{supp}(f)} := \overline{\{x \in \mathbb{R}^n : f(x) \ne 0\}}$,

$$|T(f)(x)| \le C \int_{\mathbb{R}^n} \frac{|f(y)|}{|x-y|^n}\, dy$$

and, for any $\{f_j\}_{j\in\mathbb{N}} \subset \mathcal{M}(\mathbb{R}^n)$ and almost every $x \in \mathbb{R}^n$,

$$\left| T\left(\sum_{j\in\mathbb{N}} f_j\right)(x) \right| \le \sum_{j\in\mathbb{N}} |T(f_j)(x)|. \tag{2.42}$$

If T is well defined on $\dot{\mathcal{B}}_\omega^{p,q}(\mathbb{R}^n)$, then there exists a positive constant C such that, for any $f \in \dot{\mathcal{B}}_\omega^{p,q}(\mathbb{R}^n)$,

$$\|T(f)\|_{\dot{\mathcal{B}}_\omega^{p,q}(\mathbb{R}^n)} \le C \|f\|_{\dot{\mathcal{B}}_\omega^{p,q}(\mathbb{R}^n)}.$$

In order to show Theorem 2.3.1, we require the following lattice property of block spaces.

Lemma 2.3.2 *Let $p, q \in (0, \infty)$ and $\omega \in M(\mathbb{R}_+)$. Then a measurable function f on $\mathbb{R}^n$ belongs to the block space $\dot{\mathcal{B}}_\omega^{p,q}(\mathbb{R}^n)$ if and only if there exists a measurable function $g \in \dot{\mathcal{B}}_\omega^{p,q}(\mathbb{R}^n)$ such that $|f| \le g$ almost everywhere in $\mathbb{R}^n$. Moreover, for these f and g,*

$$\|f\|_{\dot{\mathcal{B}}_\omega^{p,q}(\mathbb{R}^n)} \le \|g\|_{\dot{\mathcal{B}}_\omega^{p,q}(\mathbb{R}^n)}.$$

Proof Let $p, q \in (0, \infty)$ and $\omega \in M(\mathbb{R}_+)$. We first show the necessity. To this end, let $f \in \dot{\mathcal{B}}_\omega^{p,q}(\mathbb{R}^n)$. Then, by Definition 2.1.3, we find that

$$f = \sum_{l\in\mathbb{N}} \sum_{k\in\mathbb{Z}} \sum_{Q\in\mathcal{Q}_k^{(\xi_l)}} \lambda_{\xi_l,k,Q} b_{\xi_l,k,Q}$$

almost everywhere in $\mathbb{R}^n$ and

$$\sum_{l \in \mathbb{N}} \left[\sum_{k \in \mathbb{Z}} \sum_{Q \in \mathcal{Q}_k^{(\xi_l)}} \lambda_{\xi_l, k, Q}^q \right]^{\frac{1}{q}} \in \left[\|f\|_{\dot{\mathcal{B}}_\omega^{p,q}(\mathbb{R}^n)}, \|f\|_{\dot{\mathcal{B}}_\omega^{p,q}(\mathbb{R}^n)} + 1 \right),$$

where $\{\xi_l\}_{l \in \mathbb{N}} \subset \mathbb{R}^n$, $\{\lambda_{\xi_l, k, Q}\}_{l \in \mathbb{N}, k \in \mathbb{Z}, Q \in \mathcal{Q}_k^{(\xi_l)}} \subset [0, \infty)$, and, for any $l \in \mathbb{N}$, $k \in \mathbb{Z}$, and $Q \in \mathcal{Q}_k^{(\xi_l)}$, $b_{\xi_l, k, Q}$ is an (ω, p)-block supported in the cube Q. Let

$$g := \sum_{l \in \mathbb{N}} \sum_{k \in \mathbb{Z}} \sum_{Q \in \mathcal{Q}_k^{(\xi_l)}} \lambda_{\xi_l, k, Q} |b_{\xi_l, k, Q}|.$$

Obviously, $|f| \leq g$ and, for any $l \in \mathbb{N}$, $k \in \mathbb{Z}$, and $Q \in \mathcal{Q}_k^{(\xi_l)}$, the function $|b_{\xi_l, k, Q}|$ is also an (ω, p)-block supported in Q. This further implies that

$$\|g\|_{\dot{\mathcal{B}}_\omega^{p,q}(\mathbb{R}^n)} \leq \sum_{l \in \mathbb{N}} \left[\sum_{k \in \mathbb{Z}} \sum_{Q \in \mathcal{Q}_k^{(\xi_l)}} \lambda_{\xi_l, k, Q}^q \right]^{\frac{1}{q}} < \|f\|_{\dot{\mathcal{B}}_\omega^{p,q}(\mathbb{R}^n)} + 1 < \infty.$$

Thus, $g \in \dot{\mathcal{B}}_\omega^{p,q}(\mathbb{R}^n)$ and hence we finish the proof of the necessity.

Now, we show the sufficiency. To achieve this, let $f \in \mathcal{M}(\mathbb{R}^n)$ and $g \in \dot{\mathcal{B}}_\omega^{p,q}(\mathbb{R}^n)$ be such that $|f| \leq g$ almost everywhere in $\mathbb{R}^n$. Assume that

$$g = \sum_{l \in \mathbb{N}} \sum_{k \in \mathbb{Z}} \sum_{Q \in \mathcal{Q}_k^{(\xi_l)}} \lambda_{\xi_l, k, Q} b_{\xi_l, k, Q}$$

almost everywhere in $\mathbb{R}^n$, where $\{\xi_l\}_{l \in \mathbb{N}} \subset \mathbb{R}^n$, $\{\lambda_{\xi_l, k, Q}\}_{l \in \mathbb{N}, k \in \mathbb{Z}, Q \in \mathcal{Q}_k^{(\xi_l)}} \subset [0, \infty)$, and, for any $l \in \mathbb{N}$, $k \in \mathbb{Z}$, and $Q \in \mathcal{Q}_k^{(\xi_l)}$, $b_{\xi_l, k, Q}$ is an (ω, p)-block supported in the cube Q. Then we have

$$\mathbf{1}_{\{y \in \mathbb{R}^n: \, g(y) \neq 0\}} = \sum_{l \in \mathbb{N}} \sum_{k \in \mathbb{Z}} \sum_{Q \in \mathcal{Q}_k^{(\xi_l)}} \lambda_{\xi_l, k, Q} \frac{1}{g} b_{\xi_l, k, Q} \mathbf{1}_{\{y \in \mathbb{R}^n: \, g(y) \neq 0\}},$$

which, together with the assumption $|f| \leq g$ almost everywhere in $\mathbb{R}^n$, implies that

$$f = f \mathbf{1}_{\{y \in \mathbb{R}^n: \, g(y) \neq 0\}} = \sum_{l \in \mathbb{N}} \sum_{k \in \mathbb{Z}} \sum_{Q \in \mathcal{Q}_k^{(\xi_l)}} \lambda_{\xi_l, k, Q} \frac{f}{g} b_{\xi_l, k, Q} \mathbf{1}_{\{y \in \mathbb{R}^n: \, g(y) \neq 0\}}.$$

Now, for any $l \in \mathbb{N}$, $k \in \mathbb{Z}$, and $Q \in \mathcal{Q}_k^{(\xi_l)}$, let

$$b_{\xi_l,k,Q}^{\star} := \frac{f}{g} b_{\xi_l,k,Q} \mathbf{1}_{\{y \in \mathbb{R}^n : \, g(y) \neq 0\}}.$$

Obviously, for any $l \in \mathbb{N}$, $k \in \mathbb{Z}$, and $Q \in \mathcal{Q}_k^{(\xi_l)}$, $\operatorname{supp}(b_{\xi_l,k,Q}^{\star}) \subset Q$. In addition, from the assumption $|f| \leq g$ almost everywhere in $\mathbb{R}^n$, it follows that, for any $l \in \mathbb{N}$, $k \in \mathbb{Z}$, and $Q \in \mathcal{Q}_k^{(\xi_l)}$,

$$\left\| b_{\xi_l,k,Q}^{\star} \right\|_{L^p(\mathbb{R}^n)} \leq \left\| b_{\xi_l,k,Q} \right\|_{L^p(\mathbb{R}^n)} \leq \left[\omega \left(|Q|^{\frac{1}{n}} \right) \right]^{-1},$$

which further implies that, for any $l \in \mathbb{N}$, $k \in \mathbb{Z}$, and $Q \in \mathcal{Q}_k^{(\xi_l)}$, the function $b_{\xi_l,k,Q}^{\star}$ is an (ω, p)-block supported in Q, and hence

$$\| f \|_{\dot{\mathcal{B}}_\omega^{p,q}(\mathbb{R}^n)} \leq \sum_{l \in \mathbb{N}} \left[\sum_{k \in \mathbb{Z}} \sum_{Q \in \mathcal{Q}_k^{(\xi_l)}} \lambda_{\xi_l,k,Q}^q \right]^{\frac{1}{q}}.$$

This, combined with the choice of $\{\lambda_{\xi_l,k,Q}\}_{l \in \mathbb{N}, \, k \in \mathbb{N}, \, Q \in \mathcal{Q}_k^{(\xi_l)}}$, and the assumption $g \in \dot{\mathcal{B}}_\omega^{p,q}(\mathbb{R}^n)$, further implies that

$$\| f \|_{\dot{\mathcal{B}}_\omega^{p,q}(\mathbb{R}^n)} \leq \| g \|_{\dot{\mathcal{B}}_\omega^{p,q}(\mathbb{R}^n)} < \infty,$$

which completes the proof of the sufficiency and hence Lemma 2.3.2. $\qquad \square$

Remark 2.3.3 Let $p, q \in (0, \infty)$ and $\omega \in M(\mathbb{R}_+)$. We point out that a measurable function f on $\mathbb{R}^n$ belongs to the block space $\dot{\mathcal{B}}_\omega^{p,q}(\mathbb{R}^n)$ if and only if the function $|f|$ belongs to $\dot{\mathcal{B}}_\omega^{p,q}(\mathbb{R}^n)$. Indeed, the sufficiency of this conclusion is deduced from Lemma 2.3.2 directly. On the other hand, let $f \in \dot{\mathcal{B}}_\omega^{p,q}(\mathbb{R}^n)$ and g be as in the proof of the necessity of Lemma 2.3.2. Then, from the proof of the necessity of Lemma 2.3.2, it follows that $g \in \dot{\mathcal{B}}_\omega^{p,q}(\mathbb{R}^n)$. Applying this, the fact that $|f| \leq g$, and Lemma 2.3.2 again, we conclude that $|f| \in \dot{\mathcal{B}}_\omega^{p,q}(\mathbb{R}^n)$, which completes the proof of the necessity of the above claim. Moreover, repeating an argument similar to that used in the proof of Lemma 2.3.2, we easily find that, for any $f \in \dot{\mathcal{B}}_\omega^{p,q}(\mathbb{R}^n)$,

$$\| f \|_{\dot{\mathcal{B}}_\omega^{p,q}(\mathbb{R}^n)} = \| |f| \|_{\dot{\mathcal{B}}_\omega^{p,q}(\mathbb{R}^n)}.$$

Via the equivalent characterization of block spaces established in Lemma 2.2.2(ii), we obtain the following boundedness of sublinear operators on block spaces.

Proposition 2.3.4 *Let* $p, q \in (0, \infty)$, $\omega \in M(\mathbb{R}_+)$, *and* T *be a bounded sublinear operator on* $\dot{\mathcal{K}}_{\omega,0}^{p,q}(\mathbb{R}^n)$ *such that (2.42) holds true. If* T *is well defined on* $\dot{\mathcal{B}}_{\omega}^{p,q}(\mathbb{R}^n)$, *then there exists a positive constant* C *such that, for any* $f \in \dot{\mathcal{B}}_{\omega}^{p,q}(\mathbb{R}^n)$,

$$\|T(f)\|_{\dot{\mathcal{B}}_{\omega}^{p,q}(\mathbb{R}^n)} \leq C\|f\|_{\dot{\mathcal{B}}_{\omega}^{p,q}(\mathbb{R}^n)}.$$

Proof Let $p, q \in (0, \infty)$, $\omega \in M(\mathbb{R}_+)$, and $f \in \dot{\mathcal{B}}_{\omega}^{p,q}(\mathbb{R}^n)$. Then, from Lemma 2.2.2(ii) and Remark 1.7.2, it follows that, for any $k \in \mathbb{N}$, there exist sequences $\{\xi_l^{(k)}\}_{l\in\mathbb{N}} \subset \mathbb{R}^n$ and $\{f_{\xi_l^{(k)}}\}_{l\in\mathbb{N}} \subset \mathcal{M}(\mathbb{R}^n)$ such that both $f = \sum_{l\in\mathbb{N}} f_{\xi_l^{(k)}}$ almost everywhere in $\mathbb{R}^n$ and

$$\sum_{l\in\mathbb{N}} \left\| f_{\xi_l^{(k)}}\left(\cdot + \xi_l^{(k)}\right) \right\|_{\dot{\mathcal{K}}_{\omega,0}^{p,q}(\mathbb{R}^n)} \lesssim \|f\|_{\dot{\mathcal{B}}_{\omega}^{p,q}(\mathbb{R}^n)} + \frac{1}{k}. \tag{2.43}$$

Moreover, by (2.42), we find that, for any $k \in \mathbb{N}$ and almost every $x \in \mathbb{R}^n$,

$$|T(f)(x)| \leq \sum_{l\in\mathbb{N}} \left| T\left(f_{\xi_l^{(k)}}\right)(x) \right|. \tag{2.44}$$

Applying the boundedness of the operator T on $\dot{\mathcal{K}}_{\omega,0}^{p,q}(\mathbb{R}^n)$, we conclude that, for any $k, l \in \mathbb{N}$,

$$T\left(f_{\xi_l^{(k)}}\right)\left(\cdot + \xi_l^{(k)}\right) \in \dot{\mathcal{K}}_{\omega,0}^{p,q}(\mathbb{R}^n)$$

and

$$\left\| T\left(f_{\xi_l^{(k)}}\right)\left(\cdot + \xi_l^{(k)}\right) \right\|_{\dot{\mathcal{K}}_{\omega,0}^{p,q}(\mathbb{R}^n)} \lesssim \left\| f_{\xi_l^{(k)}}\left(\cdot + \xi_l^{(k)}\right) \right\|_{\dot{\mathcal{K}}_{\omega,0}^{p,q}(\mathbb{R}^n)},$$

where the implicit positive constant is independent of k, l, and f. This, combined with Lemma 2.3.2, Lemma 2.2.2(ii), (2.44), and (2.43), further implies that, for any $k \in \mathbb{N}$,

$$\begin{aligned}
\|T(f)\|_{\dot{\mathcal{B}}_{\omega}^{p,q}(\mathbb{R}^n)} &\leq \left\| \sum_{l\in\mathbb{N}} T\left(f_{\xi_l^{(k)}}\right) \right\|_{\dot{\mathcal{B}}_{\omega}^{p,q}(\mathbb{R}^n)} \sim \left\| \sum_{l\in\mathbb{N}} T\left(f_{\xi_l^{(k)}}\right) \right\|_{\dot{\mathcal{B}}_{\omega}^{p,q}(\mathbb{R}^n)}^{\star} \\
&\lesssim \sum_{l\in\mathbb{N}} \left\| T\left(f_{\xi_l^{(k)}}\right)\left(\cdot + \xi_l^{(k)}\right) \right\|_{\dot{\mathcal{K}}_{\omega,0}^{p,q}(\mathbb{R}^n)} \\
&\lesssim \sum_{l\in\mathbb{N}} \left\| f_{\xi_l^{(k)}}\left(\cdot + \xi_l^{(k)}\right) \right\|_{\dot{\mathcal{K}}_{\omega,0}^{p,q}(\mathbb{R}^n)} \\
&\lesssim \|f\|_{\dot{\mathcal{B}}_{\omega}^{p,q}(\mathbb{R}^n)} + \frac{1}{k},
\end{aligned} \tag{2.45}$$

where $\|\sum_{l\in\mathbb{N}} T(f_{\xi_l^{(k)}})\|^{\star}_{\dot{\mathcal{B}}_\omega^{p,q}(\mathbb{R}^n)}$ is defined as in Lemma 2.2.2(ii) with f therein replaced by $\sum_{l\in\mathbb{N}} T(f_{\xi_l^{(k)}})$, and the implicit positive constants are independent of both f and k. Letting $k \to \infty$ in (2.45), we obtain

$$\|T(f)\|_{\dot{\mathcal{B}}_\omega^{p,q}(\mathbb{R}^n)} \lesssim \|f\|_{\dot{\mathcal{B}}_\omega^{p,q}(\mathbb{R}^n)},$$

where the implicit positive constant is independent of f. This finishes the proof of Proposition 2.3.4.   □

We now show Theorem 2.3.1.

Proof of Theorem 2.3.1 Let all the symbols be as in the present theorem. Then, combining Theorem 1.5.1 and Proposition 2.3.4, we find that there exists a positive constant C such that, for any $f \in \dot{\mathcal{B}}_\omega^{p,q}(\mathbb{R}^n)$,

$$\|T(f)\|_{\dot{\mathcal{B}}_\omega^{p,q}(\mathbb{R}^n)} \leq C\|f\|_{\dot{\mathcal{B}}_\omega^{p,q}(\mathbb{R}^n)}.$$

This finishes the proof of Theorem 2.3.1.   □

As a consequence, we now give the boundedness of powered Hardy–Littlewood maximal operators on block spaces as follows.

Corollary 2.3.5 *Let $p \in (1, \infty)$, $q \in (0, \infty)$, $r \in [1, p)$, and $\omega \in M(\mathbb{R}_+)$ satisfy*

$$-\frac{n}{p} < m_0(\omega) \leq M_0(\omega) < n\left(\frac{1}{r} - \frac{1}{p}\right)$$

and

$$-\frac{n}{p} < m_\infty(\omega) \leq M_\infty(\omega) < n\left(\frac{1}{r} - \frac{1}{p}\right).$$

Then there exists a positive constant C such that, for any $f \in \dot{\mathcal{B}}_\omega^{p,q}(\mathbb{R}^n)$,

$$\left\|\mathcal{M}^{(r)}(f)\right\|_{\dot{\mathcal{B}}_\omega^{p,q}(\mathbb{R}^n)} \leq C\|f\|_{\dot{\mathcal{B}}_\omega^{p,q}(\mathbb{R}^n)},$$

where $\mathcal{M}^{(r)}$ is as in (1.55) with θ therein replaced by r.

Proof Let all the symbols be as in the present corollary. We first show that the operator $\mathcal{M}^{(r)}$ satisfies (2.42). Indeed, applying the Minkowski inequality of $L^r(\mathbb{R}^n)$, we know that, for any $\{f_j\}_{j\in\mathbb{N}} \subset \mathscr{M}(\mathbb{R}^n)$, any $x \in \mathbb{R}^n$, and any ball $B \in \mathbb{B}$ containing x,

$$\left[\int_B \left|\sum_{j\in\mathbb{N}} f_j(y)\right|^r dy\right]^{\frac{1}{r}} \leq \sum_{j\in\mathbb{N}} \left[\int_B |f_j(y)|^r dy\right]^{\frac{1}{r}},$$

which implies that

$$\left[\frac{1}{|B|}\int_B \left|\sum_{j\in\mathbb{N}} f_j(y)\right|^r dy\right]^{\frac{1}{r}} \le \sum_{j\in\mathbb{N}} \left[\frac{1}{|B|}\int_B |f_j(y)|^r dy\right]^{\frac{1}{r}}$$
$$\le \sum_{j\in\mathbb{N}} \mathcal{M}^{(r)}(f_j)(x).$$

This further implies that, for any $\{f_j\}_{j\in\mathbb{N}} \subset \mathcal{M}(\mathbb{R}^n)$ and $x \in \mathbb{R}^n$,

$$\mathcal{M}^{(r)}\left(\sum_{j\in\mathbb{N}} f_j\right)(x) \le \sum_{j\in\mathbb{N}} \mathcal{M}^{(r)}(f_j)(x).$$

Thus, the powered Hardy–Littlewood maximal operator $\mathcal{M}^{(r)}$ is a sublinear operator which satisfies (2.42).

Next, we prove that $\mathcal{M}^{(r)}$ is bounded on the local generalized Herz space $\dot{\mathcal{K}}^{p,q}_{\omega,\mathbf{0}}(\mathbb{R}^n)$. Indeed, from Lemma 1.3.1, it follows that

$$\left[\dot{\mathcal{K}}^{p,q}_{\omega,\mathbf{0}}(\mathbb{R}^n)\right]^{1/r} = \dot{\mathcal{K}}^{p/r,q/r}_{\omega^r,\mathbf{0}}(\mathbb{R}^n).$$

In addition, applying Lemma 1.1.6 and $p/r \in (1,\infty)$, we conclude that

$$\min\left\{m_0(\omega^r), m_\infty(\omega^r)\right\} = r\min\{m_0(\omega), m_\infty(\omega)\} > -\frac{n}{p/r}$$

and

$$\max\left\{M_0(\omega^r), M_\infty(\omega^r)\right\} = r\max\{M_0(\omega), M_\infty(\omega)\}$$
$$< r\left(\frac{n}{r} - \frac{n}{p}\right) = \frac{n}{(p/r)'}.$$

This, together with $p/r \in (1,\infty)$ and Corollary 1.5.4, further implies that $\mathcal{M}$ is bounded on the local generalized Herz space $\dot{\mathcal{K}}^{p/r,q/r}_{\omega^r,\mathbf{0}}(\mathbb{R}^n)$. Therefore, the operator $\mathcal{M}$ is bounded on $[\dot{\mathcal{K}}^{p,q}_{\omega,\mathbf{0}}(\mathbb{R}^n)]^{1/r}$. From this and Remark 1.2.30(i), it follows that the powered Hardy–Littlewood maximal operator $\mathcal{M}^{(r)}$ is bounded on $\dot{\mathcal{K}}^{p,q}_{\omega,\mathbf{0}}(\mathbb{R}^n)$. Using this, the fact that $\mathcal{M}^{(r)}$ satisfies (2.42), and Proposition 2.3.4, we then conclude that $\mathcal{M}^{(r)}$ is bounded on $\dot{\mathcal{B}}^{p,q}_{\omega}(\mathbb{R}^n)$, which completes the proof of Corollary 2.3.5. □

Next, we devote to establishing the boundedness of Calderón–Zygmund operators on block spaces. To be precise, we have the following conclusion.

Theorem 2.3.6 *Let* $p, q \in (1, \infty)$ *and* $\omega \in M(\mathbb{R}_+)$ *with*

$$-\frac{n}{p} < m_0(\omega) \le M_0(\omega) < \frac{n}{p'}$$

and

$$-\frac{n}{p} < m_\infty(\omega) \le M_\infty(\omega) < \frac{n}{p'},$$

where $\frac{1}{p} + \frac{1}{p'} = 1$. *Assume* $d \in \mathbb{Z}_+$ *and* T *is a* d-*order Calderón–Zygmund operator as in Definition 1.5.8. Then* T *is well defined on* $\dot{\mathcal{B}}_\omega^{p,q}(\mathbb{R}^n)$ *and there exists a positive constant* C *such that, for any* $f \in \dot{\mathcal{B}}_\omega^{p,q}(\mathbb{R}^n)$,

$$\|T(f)\|_{\dot{\mathcal{B}}_\omega^{p,q}(\mathbb{R}^n)} \le C \|f\|_{\dot{\mathcal{B}}_\omega^{p,q}(\mathbb{R}^n)}.$$

To prove Theorem 2.3.6, we need an embedding lemma about block spaces. First, we present the concepts of weighted (weak) Lebesgue spaces as follows.

Definition 2.3.7 Let $p \in (0, \infty)$ and $\upsilon \in A_\infty(\mathbb{R}^n)$. Then

(i) the *weighted Lebesgue space* $L_\upsilon^p(\mathbb{R}^n)$ is defined to be the set of all the measurable functions f on $\mathbb{R}^n$ such that

$$\|f\|_{L_\upsilon^p(\mathbb{R}^n)} := \left[\int_{\mathbb{R}^n} |f(x)|^p \, \upsilon(x) \, dx \right]^{\frac{1}{p}} < \infty;$$

(ii) the *weighted weak Lebesgue space* $WL_\upsilon^p(\mathbb{R}^n)$ is defined to be the set of all the measurable functions f on $\mathbb{R}^n$ such that

$$\|f\|_{WL_\upsilon^p(\mathbb{R}^n)} := \sup_{\lambda \in (0,\infty)} \left\{ \lambda \left[\int_{\{x \in \mathbb{R}^n : |f(x)| > \lambda\}} \upsilon(y) \, dy \right]^{\frac{1}{p}} \right\} < \infty.$$

Note that Chang et al. [29, Lemma 4.7] gave an embedding lemma about ball quasi-Banach function spaces. However, due to the deficiency of the Fatou property [namely, whether Definition 1.2.13(iii) holds true for block spaces is unknown], the block space $\dot{\mathcal{B}}_\omega^{p,q}(\mathbb{R}^n)$ may not be a ball quasi-Banach function space. Fortunately, we find that the proof of [29, Lemma 4.7] is still valid for Lemma 2.3.8 below; we omit the details.

Lemma 2.3.8 *Let* p, q, *and* ω *be as in Theorem 2.3.6. Then there exists an* $\varepsilon \in (0, 1)$ *and a positive constant* C *such that, for any* $f \in \mathcal{M}(\mathbb{R}^n)$,

$$\|f\|_{L_\upsilon^1(\mathbb{R}^n)} \le C \|f\|_{\dot{\mathcal{B}}_\omega^{p,q}(\mathbb{R}^n)},$$

where $\upsilon := [\mathcal{M}(\mathbf{1}_{B(0,1)})]^\varepsilon$ with $\mathcal{M}$ being the Hardy–Littlewood maximal operator as in (1.54).

In order to prove Theorem 2.3.6, we also require the following property of $A_1(\mathbb{R}^n)$-weights, which is just [70, Theorem 7.7(1)].

Lemma 2.3.9 *Let $f \in L^1_{\mathrm{loc}}(\mathbb{R}^n)$ satisfy $\mathcal{M}(f) < \infty$ almost everywhere in $\mathbb{R}^n$ and let $\delta \in [0, 1)$. Then $\upsilon := [\mathcal{M}(f)]^\delta \in A_1(\mathbb{R}^n)$.*

The following conclusion shows the weak boundedness of Calderón–Zygmund operators on weighted Lebesgue spaces, which is just [70, Theorem 7.12].

Lemma 2.3.10 *Let $\upsilon \in A_1(\mathbb{R}^n)$, $d \in \mathbb{Z}_+$, and T be a d-order Calderón–Zygmund operator as in Definition 1.5.8. Then there exists a positive constant C such that, for any $f \in L^1_\upsilon(\mathbb{R}^n)$ and $\lambda \in (0, \infty)$,*

$$\int_{\{x \in \mathbb{R}^n : |T(f)| > \lambda\}} \upsilon(y)\, dy \leq \frac{C}{\lambda} \|f\|_{L^1_\upsilon(\mathbb{R}^n)}$$

and hence

$$\|T(f)\|_{WL^1_\upsilon(\mathbb{R}^n)} \leq C\|f\|_{L^1_\upsilon(\mathbb{R}^n)}.$$

To prove Theorem 2.3.6, we also need the following technical estimate about the Hardy–Littlewood maximal operator of characteristic functions of balls (see, for instance, [90, Example 2.1.8]).

Lemma 2.3.11 *Let $r \in (0, \infty)$ and $\mathcal{M}$ be the Hardy–Littlewood maximal operator as in* (1.54). *Then, for any $x \in \mathbb{R}^n$,*

$$\frac{r^n}{(r + |x|)^n} \leq \mathcal{M}\left(\mathbf{1}_{B(0,r)}\right)(x) \leq \frac{6^n r^n}{(r + |x|)^n}.$$

Via above preparations, we now show Theorem 2.3.6.

Proof of Theorem 2.3.6 Let all the symbols be as in the present theorem and $f \in \dot{\mathcal{B}}^{p,q}_\omega(\mathbb{R}^n)$. Then, from Lemma 2.3.8, we deduce that $f \in L^1_\upsilon(\mathbb{R}^n)$, where $\upsilon := [\mathcal{M}(\mathbf{1}_{B(0,1)})]^\varepsilon$ with an $\varepsilon \in (0, 1)$. Applying Lemma 2.3.9, we find that $\upsilon \in A_1(\mathbb{R}^n)$. This, together with the fact that $f \in L^1_\upsilon(\mathbb{R}^n)$ and Lemma 2.3.10, further implies that $T(f) \in WL^1_\upsilon(\mathbb{R}^n)$, and hence T is well defined on $\dot{\mathcal{B}}^{p,q}_\omega(\mathbb{R}^n)$.

In addition, using Lemma 2.2.2(ii), we conclude that there exists a sequence $\{\xi_l\}_{l \in \mathbb{N}} \subset \mathbb{R}^n$ and a sequence $\{f_{\xi_l}\}_{l \in \mathbb{N}} \subset \mathcal{M}(\mathbb{R}^n)$ such that, for any $l \in \mathbb{N}$, $f_{\xi_l} \in \dot{\mathcal{K}}^{p,q}_{\omega,\xi_l}(\mathbb{R}^n)$, $f = \sum_{l \in \mathbb{N}} f_{\xi_l}$ almost everywhere in $\mathbb{R}^n$ and

$$\sum_{l \in \mathbb{N}} \|f_{\xi_l}\|_{\dot{\mathcal{K}}^{p,q}_{\omega,\xi_l}(\mathbb{R}^n)} < \infty. \tag{2.46}$$

We next prove that, for almost every $x \in \mathbb{R}^n$,

$$|T(f)(x)| \leq \sum_{l \in \mathbb{N}} |T(f_{\xi_l})(x)|. \tag{2.47}$$

To achieve this, for any $N \in \mathbb{N}$, let

$$f^{(N)} := \sum_{l=1}^{N} f_{\xi_l}.$$

Then, from (2.46), it follows that, for any $N \in \mathbb{N}$,

$$\sum_{l=1}^{N} \|f_{\xi_l}\|_{\dot{\mathcal{K}}_{\omega,\xi_l}^{p,q}(\mathbb{R}^n)} \leq \sum_{l \in \mathbb{N}} \|f_{\xi_l}\|_{\dot{\mathcal{K}}_{\omega,\xi_l}^{p,q}(\mathbb{R}^n)} < \infty,$$

which, combined with Lemma 2.2.2(ii) again, implies that $f^{(N)} \in \dot{\mathcal{B}}_\omega^{p,q}(\mathbb{R}^n)$. By this, Lemma 2.3.8, the linearity of T, Lemmas 2.3.10 and 2.2.2(ii), and (2.46), we conclude that, for any $N \in \mathbb{N}$,

$$f - f^{(N)} \in L_\upsilon^1(\mathbb{R}^n)$$

and, for any $\lambda \in (0, \infty)$,

$$\int_{\{x \in \mathbb{R}^n : |T(f)(x) - T(f^{(N)})(x)| > \lambda\}} \left[\mathcal{M}\left(\mathbf{1}_{B(0,1)}\right)(y)\right]^\varepsilon dy$$

$$\lesssim \frac{1}{\lambda} \left\|f - f^{(N)}\right\|_{L_\upsilon^1(\mathbb{R}^n)} \lesssim \frac{1}{\lambda} \left\|f - f^{(N)}\right\|_{\dot{\mathcal{B}}_\omega^{p,q}(\mathbb{R}^n)}$$

$$\lesssim \frac{1}{\lambda} \sum_{l=N+1}^{\infty} \|f_{\xi_l}\|_{\dot{\mathcal{K}}_{\omega,\xi_l}^{p,q}(\mathbb{R}^n)} \to 0 \tag{2.48}$$

as $N \to \infty$.

Now, we claim that there exists a subsequence $\{f^{(N_j)}\}_{j \in \mathbb{N}} \subset \{f^{(N)}\}_{N \in \mathbb{N}}$ such that $T(f^{(N_j)}) \to T(f)$ almost everywhere in $\mathbb{R}^n$ as $j \to \infty$. Indeed, applying Lemma 2.3.11 with $r := 1$, we find that, for any $x \in \mathbb{R}^n$,

$$\left[\mathcal{M}\left(\mathbf{1}_{B(0,1)}\right)(x)\right]^\varepsilon \sim (1 + |x|)^{-\varepsilon n}.$$

By this and (2.48), we conclude that, for any $\lambda \in (0, \infty)$,

$$\left| \left\{ x \in B(\mathbf{0}, 1) : \left| T(f)(x) - T\left(f^{(N)}\right)(x) \right| > \lambda \right\} \right|$$

$$\lesssim \int_{\{x \in \mathbb{R}^n : |T(f)(x) - T(f^{(N)})(x)| > \lambda\}} \left[\mathcal{M} \left(\mathbf{1}_{B(\mathbf{0},1)}\right)(y) \right]^{\varepsilon} dy \to 0$$

as $N \to \infty$ and, for any $k \in \mathbb{N}$,

$$\left| \left\{ x \in B(\mathbf{0}, 2^k) \setminus B(\mathbf{0}, 2^{k-1}) : \left| T(f)(x) - T\left(f^{(N)}\right)(x) \right| > \lambda \right\} \right|$$

$$\lesssim 2^{k\varepsilon n} \int_{\{x \in \mathbb{R}^n : |T(f)(x) - T(f^{(N)})(x)| > \lambda\}} \left[\mathcal{M} \left(\mathbf{1}_{B(\mathbf{0},1)}\right)(y) \right]^{\varepsilon} dy \to 0$$

as $N \to \infty$. These further imply that the sequence $\{T(f^{(N)})\}_{N \in \mathbb{N}}$ converges in measure to $T(f)$ both in $B(\mathbf{0}, 1)$ and $B(\mathbf{0}, 2^k) \setminus B(\mathbf{0}, 2^{k-1})$ for any $k \in \mathbb{N}$. From this, the Riesz theorem, and a diagonalization argument, we deduce that there exists a subsequence $\{f^{(N_j)}\}_{j \in \mathbb{N}} \subset \{f^{(N)}\}_{N \in \mathbb{N}}$ such that $T(f^{(N_j)}) \to T(f)$ almost everywhere in $\mathbb{R}^n$ as $j \to \infty$, which completes the proof of the above claim.

On the other hand, using the linearity of T, we conclude that, for any $j \in \mathbb{N}$,

$$\left| T\left(f^{(N_j)}\right) \right| = \left| \sum_{l=1}^{N_j} T\left(f_{\xi_l}\right)(x) \right| \leq \sum_{l=1}^{N_j} \left| T\left(f_{\xi_l}\right) \right| \leq \sum_{l \in \mathbb{N}} \left| T\left(f_{\xi_l}\right) \right|. \qquad (2.49)$$

By the above claim and letting $j \to \infty$ in (2.49), we further find that, for almost every $x \in \mathbb{R}^n$,

$$|T(f)(x)| \leq \sum_{l \in \mathbb{N}} \left| T\left(f_{\xi_l}\right)(x) \right|.$$

This finishes the proof of (2.47). Combining this, Corollary 1.5.10, and an argument similar to that used in the proof of (2.45), we conclude that

$$\|T(f)\|_{\dot{\mathcal{B}}_{\omega}^{p,q}(\mathbb{R}^n)} \lesssim \|f\|_{\dot{\mathcal{B}}_{\omega}^{p,q}(\mathbb{R}^n)},$$

where the implicit positive constant is independent of f, which completes the proof of Theorem 2.3.6. $\square$

Chapter 3
Boundedness and Compactness Characterizations of Commutators on Generalized Herz Spaces

In this chapter, we investigate the boundedness and the compactness characterizations of commutators on local and global generalized Herz spaces. Recall that the commutator plays an important role in various branch of mathematics, such as harmonic analysis (see, for instance, [7, 8, 30, 34, 95, 188, 208]) and partial differential equations (see, for instance, [33, 35, 225]). In 1976, Coifman et al. [47] first obtained a boundedness characterization of commutators on the Lebesgue space $L^p(\mathbb{R}^n)$ with $p \in (1, \infty)$. To be precise, let $b \in L^1_{\mathrm{loc}}(\mathbb{R}^n)$ and T_Ω be a singular integral operator with homogeneous kernel Ω. Coifman et al. [47] proved that, for any given $p \in (1, \infty)$ and any given $b \in \mathrm{BMO}(\mathbb{R}^n)$, the commutator $[b, T_\Omega]$ is bounded on $L^p(\mathbb{R}^n)$ and also that, if, for any given $b \in L^1_{\mathrm{loc}}(\mathbb{R}^n)$ and any Riesz transform R_j with $j \in \{1, \ldots, n\}$, $[b, R_j]$ is bounded on $L^p(\mathbb{R}^n)$, then $b \in \mathrm{BMO}(\mathbb{R}^n)$. Furthermore, in 1978, Uchiyama [231] showed that, for any given $p \in (1, \infty)$, the commutator $[b, T_\Omega]$ is bounded on $L^p(\mathbb{R}^n)$ if and only if $b \in \mathrm{BMO}(\mathbb{R}^n)$. Later on, as extensions of the results in Lebesgue spaces, such boundedness characterizations were established on various function spaces (see [56] for Morrey spaces, [161] for weighted Lebesgue spaces, and [141] for variable Lebesgue spaces).

On the other hand, the compactness characterizations of commutators were also studied. In 1978, Uchiyama [231] first showed that, for any given $p \in (1, \infty)$, the commutator $[b, T_\Omega]$ is compact on $L^p(\mathbb{R}^n)$ if and only if $b \in \mathrm{CMO}(\mathbb{R}^n)$. After that, this compactness characterization was extended to Morrey spaces (see [36]) and to weighted Lebesgue spaces (see [43, 96]). Moreover, very recently, Tao et al. [228] studied the boundedness and the compactness characterizations of commutators on ball Banach function spaces.

The main target of this chapter is to study the boundedness and the compactness characterizations of commutators on generalized Herz spaces. As was proved in Chap. 1, the generalized Herz spaces are special ball Banach function spaces under some reasonable and sharp assumptions of exponents. Thus, we can obtain the boundedness and the compactness characterizations of commutators on generalized

Y. Li et al., *Real-Variable Theory of Hardy Spaces Associated with Generalized Herz Spaces of Rafeiro and Samko*, Lecture Notes in Mathematics 2320,
https://doi.org/10.1007/978-981-19-6788-7_3

Herz spaces via proving that generalized Herz spaces satisfy all the assumptions of the results obtained in [228]. This approach is feasible for the boundedness and the compactness characterizations of commutators on local generalized Herz spaces. However, since the associate space of the global generalized Herz space $\dot{\mathcal{K}}_{\omega}^{p,q}(\mathbb{R}^n)$ is still unknown, we can not establish the boundedness and the compactness characterizations of commutators on $\dot{\mathcal{K}}_{\omega}^{p,q}(\mathbb{R}^n)$ by the aforementioned method. To overcome this difficulty, we replace the assumptions of conclusions in [228] about associate spaces by other assumptions about ball Banach function spaces which are more convenient to check for $\dot{\mathcal{K}}_{\omega}^{p,q}(\mathbb{R}^n)$. In particular, we point out that the extrapolation theorem of $\dot{\mathcal{K}}_{\omega}^{p,q}(\mathbb{R}^n)$ established in Chap. 1 is a key tool in the proof of the boundedness and the compactness characterizations of commutators on $\dot{\mathcal{K}}_{\omega}^{p,q}(\mathbb{R}^n)$.

We now recall some basic concepts. Let Ω be a Lipschitz function on the unit sphere of $\mathbb{R}^n$, which is homogeneous of degree zero and has mean value zero, namely, for any $x, y \in \mathbb{S}^{n-1}$ and $\mu \in (0, \infty)$,

$$|\Omega(x) - \Omega(y)| \le |x - y|, \tag{3.1}$$

$$\Omega(\mu x) = \Omega(x), \tag{3.2}$$

and

$$\int_{\mathbb{S}^{n-1}} \Omega(x) \, d\sigma(x) = 0, \tag{3.3}$$

here and thereafter, $\mathbb{S}^{n-1} := \{x \in \mathbb{R}^n : |x| = 1\}$ denotes the *unit sphere* of $\mathbb{R}^n$ and $d\sigma$ the *area measure* on $\mathbb{S}^{n-1}$. Furthermore, we state the following L^∞-*Dini* condition (see, for instance, [70, p. 93]).

Definition 3.0.12 A function $\Omega \in L^\infty(\mathbb{S}^{n-1})$ is said to satisfy the L^∞-*Dini condition* if

$$\int_0^1 \frac{\omega_\infty(\tau)}{\tau} \, d\tau < \infty,$$

where ω_∞ is defined by setting, for any $\tau \in (0, 1)$,

$$\omega_\infty(\tau) := \sup_{\{x, \, y \in \mathbb{S}^{n-1}: \, |x-y| < \tau\}} |\Omega(x) - \Omega(y)|.$$

Assume that Ω satisfies (3.2) and (3.3), and the L^∞-Dini condition, a linear operator T_Ω is called a *singular integral operator with homogeneous kernel* Ω (see, for instance, [161, p. 53, Corollary 2.1.1]) if, for any $f \in L^p(\mathbb{R}^n)$ with $p \in [1, \infty)$, and for any $x \in \mathbb{R}^n$,

$$T_\Omega(f)(x) := \text{p. v.} \int_{\mathbb{R}^n} \frac{\Omega(x-y)}{|x-y|^n} f(y)\,dy$$

$$:= \lim_{\varepsilon \to 0^+} \int_{\varepsilon < |x-y| < 1/\varepsilon} \frac{\Omega(x-y)}{|x-y|^n} f(y)\,dy. \qquad (3.4)$$

Now, we recall the concept of the commutator $[b, T_\Omega]$. For any given $b \in L^1_{\text{loc}}(\mathbb{R}^n)$, the commutator $[b, T_\Omega]$ is defined by setting, for any bounded function f with compact support, and for any $x \in \mathbb{R}^n$,

$$[b, T_\Omega](f)(x) := b(x)T_\Omega(f)(x) - T_\Omega(bf)(x). \qquad (3.5)$$

Finally, recall that the *space* BMO $(\mathbb{R}^n)$, introduced by John and Nirenberg [140], is defined to be the set of all the $f \in L^1_{\text{loc}}(\mathbb{R}^n)$ such that

$$\|f\|_{\text{BMO}(\mathbb{R}^n)} := \sup_{B \in \mathbb{B}} \left\{ \frac{1}{|B|} \int_B |f(x) - f_B|\,dx \right\} < \infty, \qquad (3.6)$$

where the supremum is taken over all balls $B \in \mathbb{B}$ and, for any given $B \in \mathbb{B}$,

$$f_B := \frac{1}{|B|} \int_B f(x)\,dx. \qquad (3.7)$$

3.1 Boundedness Characterizations

The target of this section is to establish the boundedness characterization of commutators on local and global generalized Herz spaces. We first consider commutators on local generalized Herz spaces. Namely, we show the following conclusion.

Theorem 3.1.1 *Let* $p, q \in (1, \infty)$ *and* $\omega \in M(\mathbb{R}_+)$ *satisfy*

$$-\frac{n}{p} < m_0(\omega) \leq M_0(\omega) < \frac{n}{p'}$$

and

$$-\frac{n}{p} < m_\infty(\omega) \leq M_\infty(\omega) < \frac{n}{p'},$$

where $\frac{1}{p} + \frac{1}{p'} = 1$. *Assume that* Ω *is a homogeneous Lipschitz function of degree zero on* $\mathbb{S}^{n-1}$ *satisfying* (3.3), T_Ω *a singular integral operator with homogeneous kernel* Ω, *and* $b \in L^1_{\text{loc}}(\mathbb{R}^n)$. *Then the commutator* $[b, T_\Omega]$ *is bounded on the local generalized Herz space* $\dot{\mathcal{K}}_{\omega,\mathbf{0}}^{p,q}(\mathbb{R}^n)$ *if and only if* $b \in \mathrm{BMO}\,(\mathbb{R}^n)$. *Moreover, there exist two positive constants* C_1 *and* C_2, *independent of* b, *such that*

$$C_1\|b\|_{\mathrm{BMO}\,(\mathbb{R}^n)} \leq \|[b, T_\Omega]\|_{\dot{\mathcal{K}}_{\omega,\mathbf{0}}^{p,q}(\mathbb{R}^n)\to\dot{\mathcal{K}}_{\omega,\mathbf{0}}^{p,q}(\mathbb{R}^n)} \leq C_2\|b\|_{\mathrm{BMO}\,(\mathbb{R}^n)},$$

where $\|[b, T_\Omega]\|_{\dot{\mathcal{K}}_{\omega,\mathbf{0}}^{p,q}(\mathbb{R}^n)\to\dot{\mathcal{K}}_{\omega,\mathbf{0}}^{p,q}(\mathbb{R}^n)}$ *is as in* (1.88) *with* T *and* X *therein replaced, respectively, by* $[b, T_\Omega]$ *and* $\dot{\mathcal{K}}_{\omega,\mathbf{0}}^{p,q}(\mathbb{R}^n)$.

First, we show the sufficiency of Theorem 3.1.1. Indeed, we have the following conclusion for the sufficiency of the boundedness of commutators on local generalized Herz spaces, where the assumption about Ω is weaker than that of Theorem 3.1.1.

Proposition 3.1.2 *Let* $p, q \in (1, \infty)$, $\omega \in M(\mathbb{R}_+)$ *be as in Theorem 3.1.1, and* $r \in (1, \infty]$. *Assume that* $b \in \mathrm{BMO}\,(\mathbb{R}^n)$, $\Omega \in L^r(\mathbb{S}^{n-1})$ *satisfies both* (3.2) *and* (3.3), *and* T_Ω *is a singular integral operator with homogeneous kernel* Ω. *Then there exists a positive constant* C *such that, for any* $f \in \dot{\mathcal{K}}_{\omega,\mathbf{0}}^{p,q}(\mathbb{R}^n)$,

$$\|[b, T_\Omega](f)\|_{\dot{\mathcal{K}}_{\omega,\mathbf{0}}^{p,q}(\mathbb{R}^n)} \leq C\|b\|_{\mathrm{BMO}\,(\mathbb{R}^n)}\|f\|_{\dot{\mathcal{K}}_{\omega,\mathbf{0}}^{p,q}(\mathbb{R}^n)}.$$

To prove Proposition 3.1.2, we require the following boundedness of commutators on ball Banach function spaces, which was obtained by Tao et al. in [228, Theorem 2.17].

Lemma 3.1.3 *Let* X *be a ball Banach function space satisfying that the Hardy–Littlewood maximal operator* $\mathcal{M}$ *is bounded on both* X *and its associate space* X', *and* $r \in (1, \infty]$. *Assume that* $b \in \mathrm{BMO}\,(\mathbb{R}^n)$, $\Omega \in L^r(\mathbb{S}^{n-1})$ *satisfies both* (3.2) *and* (3.3), *and* T_Ω *is a singular integral operator with homogeneous kernel* Ω. *Then there exists a positive constant* C *such that, for any* $f \in X$,

$$\|[b, T_\Omega](f)\|_X \leq C\|b\|_{\mathrm{BMO}\,(\mathbb{R}^n)}\|f\|_X.$$

Remark 3.1.4 We point out that Lemma 3.1.3 has a wide range of applications. Here we present several function spaces to which Lemma 3.1.3 can be applied (see also [228, Section 4]).

(i) Let $p \in (1, \infty)$. Then, in this case, by the boundedness of the Hardy–Littlewood maximal operator $\mathcal{M}$ on both $L^p(\mathbb{R}^n)$ and $L^{p'}(\mathbb{R}^n)$, and Remark 1.2.19(i), we find that the Lebesgue space $L^p(\mathbb{R}^n)$ satisfies all the assumptions of Lemma 3.1.3. Therefore, Lemma 3.1.3 with $X := L^p(\mathbb{R}^n)$ holds true. This result extends the known boundedness of commutators on Lebesgue spaces obtained in [47, Theorem I] (see also [161, Corollary 2.4.1]). Indeed, when Ω satisfies (3.1), (3.2), and (3.3), then the result obtained in this remark goes back to [47, Theorem I].

(ii) Let $p \in (1, \infty)$ and $\upsilon \in A_p(\mathbb{R}^n)$. Then, in this case, as was pointed out in [228, Subsection 4.4], the weighted Lebesgue space $L_\upsilon^p(\mathbb{R}^n)$ satisfies all the assumptions of Lemma 3.1.3. This implies that Lemma 3.1.3 with $X :=$ $L_\upsilon^p(\mathbb{R}^n)$ holds true. If further assume that $\Omega \in L^r(\mathbb{S}^{n-1})$ with $r' \in [1, p]$ and $\upsilon \in A_{p/r'}(\mathbb{R}^n)$, or $\Omega \in L^r(\mathbb{S}^{n-1})$ with $r \in [p, \infty]$ and $\upsilon^{1-p'} \in A_{p'/r'}(\mathbb{R}^n)$, then, in this case, the aforementioned result coincides with [161, (i) and (ii) of Theorem 2.4.4].

(iii) Let $\vec{p} := (p_1, \ldots, p_n) \in (1, \infty)^n$. Then, in this case, as was mentioned in [228, Theorem 4.6], the mixed-norm Lebesgue space $L^{\vec{p}}(\mathbb{R}^n)$ satisfies all the assumptions of Lemma 3.1.3. Thus, Lemma 3.1.3 with X therein replaced by $L^{\vec{p}}(\mathbb{R}^n)$ holds true. This result was also obtained in [228, Theorem 4.6].

(iv) Let $1 < q \leq p < \infty$. Then, in this case, as was pointed out in [228, Subsection 4.1], the Morrey space $M_q^p(\mathbb{R}^n)$ satisfies all the assumptions of Lemma 3.1.3. This implies that Lemma 3.1.3 with $X := M_q^p(\mathbb{R}^n)$ holds true. This result extends the known boundedness of commutators on Morrey spaces originally studied in [56, Theorem 1] (see also [228, Theorem 4.3]). Indeed, when Ω satisfies (3.1), (3.2), and (3.3), then, in this case, the aforementioned result coincides with [56, Theorem 1].

(v) Let $p(\cdot) \in C^{\log}(\mathbb{R}^n)$ with $1 < p_- \leq p_+ < \infty$, where p_- and p_+ are defined, respectively, in (1.59) and (1.60). Then, in this case, as was mentioned in [228, Subsection 4.3], the variable Lebesgue space $L^{p(\cdot)}(\mathbb{R}^n)$ satisfies all the assumptions of Lemma 3.1.3. This further implies that Lemma 3.1.3 with X therein replaced by $L^{p(\cdot)}(\mathbb{R}^n)$ holds true. This result extends the known boundedness of commutators on variable Lebesgue spaces given in [141, Theorem 1.1(a)] (see also [228, Theorem 4.8]). Indeed, when Ω is infinitely differentiable and satisfies both (3.2) and (3.3), then, in this case, the aforementioned result goes back to [141, Theorem 1.1(a)].

Via this general conclusion for ball Banach function spaces, we now show Proposition 3.1.2.

Proof of Proposition 3.1.2 Let all the symbols be as in the present proposition. Then, from Theorem 1.2.46, we deduce that, under the assumptions of the present proposition, the local generalized Herz space $\dot{\mathcal{K}}_{\omega,\mathbf{0}}^{p,q}(\mathbb{R}^n)$ is a BBF space.

Therefore, in order to complete the proof of the present proposition, it suffices to show that $\dot{\mathcal{K}}_{\omega,\mathbf{0}}^{p,q}(\mathbb{R}^n)$ satisfies all the assumptions of Lemma 3.1.3. Indeed, applying the assumptions of the present proposition, (1.138), and (1.139), we find that the Hardy–Littlewood maximal operator $\mathcal{M}$ is bounded on $\dot{\mathcal{K}}_{\omega,\mathbf{0}}^{p,q}(\mathbb{R}^n)$ and $(\dot{\mathcal{K}}_{\omega,\mathbf{0}}^{p,q}(\mathbb{R}^n))'$. This further implies that $\dot{\mathcal{K}}_{\omega,\mathbf{0}}^{p,q}(\mathbb{R}^n)$ satisfies all the assumptions of Lemma 3.1.3, which completes the proof of Proposition 3.1.2. $\square$

Next, we deal with the necessity of the boundedness of commutators on local generalized Herz spaces. Notice that the following proposition gives a more general conclusion than the necessity of Theorem 3.1.1.

Proposition 3.1.5 *Let* $p \in (1, \infty)$, $q \in [1, \infty)$, *and* $\omega \in M(\mathbb{R}_+)$ *be as in Theorem 3.1.1. Assume that* $b \in L^1_{\mathrm{loc}}(\mathbb{R}^n)$ *and* $\Omega \in L^\infty(\mathbb{S}^{n-1})$ *satisfies that there exists an open set* $\Lambda \subset \mathbb{S}^{n-1}$ *such that* Ω *never vanishes and never changes sign on* Λ. *If the commutator* $[b, T_\Omega]$ *is bounded on the local generalized Herz space* $\dot{\mathcal{K}}^{p,q}_{\omega,0}(\mathbb{R}^n)$ *and, for any bounded measurable set* $F \subset \mathbb{R}^n$ *and almost every* $x \in \mathbb{R}^n \setminus \overline{F}$,

$$[b, T_\Omega] (\mathbf{1}_F) (x) = \int_F [b(x) - b(y)] \frac{\Omega(x - y)}{|x - y|^n} \, dy, \qquad (3.8)$$

then $b \in \mathrm{BMO}(\mathbb{R}^n)$ *and there exists a positive constant* C, *independent of* b, *such that*

$$\|b\|_{\mathrm{BMO}(\mathbb{R}^n)} \le C \|[b, T_\Omega]\|_{\dot{\mathcal{K}}^{p,q}_{\omega,0}(\mathbb{R}^n) \to \dot{\mathcal{K}}^{p,q}_{\omega,0}(\mathbb{R}^n)},$$

where $\|[b, T_\Omega]\|_{\dot{\mathcal{K}}^{p,q}_{\omega,0}(\mathbb{R}^n) \to \dot{\mathcal{K}}^{p,q}_{\omega,0}(\mathbb{R}^n)}$ *is as in* (1.88) *with* T *and* X *therein replaced, respectively, by* $[b, T_\Omega]$ *and* $\dot{\mathcal{K}}^{p,q}_{\omega,0}(\mathbb{R}^n)$.

Recall that the following necessity of commutators on ball Banach function spaces was obtained in [228, Theorem 2.22].

Lemma 3.1.6 *Let* X *be a ball Banach function space satisfying that the Hardy–Littlewood maximal operator* $\mathcal{M}$ *is bounded on* X, *and let* $b \in L^1_{\mathrm{loc}}(\mathbb{R}^n)$. *Assume* $\Omega \in L^\infty(\mathbb{S}^{n-1})$ *satisfies that there exists an open set* $\Lambda \subset \mathbb{S}^{n-1}$ *such that* Ω *never vanishes and never changes sign on* Λ. *If* $[b, T_\Omega]$ *is bounded on* X *and satisfies* (3.8), *then* $b \in \mathrm{BMO}(\mathbb{R}^n)$ *and there exists a positive constant* C, *independent of* b, *such that*

$$\|b\|_{\mathrm{BMO}(\mathbb{R}^n)} \le C \|[b, T_\Omega]\|_{X \to X},$$

where $\|[b, T_\Omega]\|_{X \to X}$ *is as in* (1.88) *with* T *therein replaced by* $[b, T_\Omega]$.

Remark 3.1.7 We should point out that Lemma 3.1.6 has a wide range of applications. Here we present several function spaces to which Lemma 3.1.6 can be applied (see also [228, Section 4]).

(i) Let $p \in (1, \infty]$. Then, in this case, from the boundedness of the Hardy–Littlewood maximal operator $\mathcal{M}$ on $L^p(\mathbb{R}^n)$, we deduce that the Lebesgue space $L^p(\mathbb{R}^n)$ satisfies all the assumptions of Lemma 3.1.6. This then implies that Lemma 3.1.6 with X therein replaced by $L^p(\mathbb{R}^n)$ holds true. This result extends the work of Uchiyama [231, Theorem 1]. Indeed, when $p \in (1, \infty)$ and Ω satisfies (3.1), (3.2), and (3.3), then, in this case, the aforementioned result coincides with [231, Theorem 1].

(ii) Let $p \in (1, \infty)$ and $\upsilon \in A_p(\mathbb{R}^n)$. Then, in this case, as was pointed out in [228, Subsection 4.4], the weighted Lebesgue space $L^p_\upsilon(\mathbb{R}^n)$ satisfies all the assumptions of Lemma 3.1.6. Thus, Lemma 3.1.6 with $X := L_\upsilon^p(\mathbb{R}^n)$ holds true, which was also obtained in [149, Theorem 1.1(ii)].

(iii) Let $\vec{p} := (p_1, \ldots, p_n) \in (1, \infty)^n$. Then, in this case, as was mentioned in [228, Subsection 4.2], the mixed-norm Lebesgue space $L^{\vec{p}}(\mathbb{R}^n)$ satisfies all the assumptions of Lemma 3.1.6. This further implies that Lemma 3.1.6 with X replaced by $L^{\vec{p}}(\mathbb{R}^n)$ holds true, which is just a part of [228, Theorem 4.6].

(iv) Let $1 < q \le p < \infty$. Then, in this case, as was pointed out in [228, Subsection 4.1], the Morrey space $M_q^p(\mathbb{R}^n)$ satisfies all the assumptions of Lemma 3.1.6. Therefore, Lemma 3.1.6 with $X := M_q^p(\mathbb{R}^n)$ holds true. This result extends the boundedness characterization of commutators on Morrey spaces given in [56, Theorem 1]. Indeed, [56, Theorem 1] showed that, if, for any given $b \in L_{\mathrm{loc}}^1(\mathbb{R}^n)$ and for any Riesz transform R_j [namely, let $\Omega(x) := x_j/|x|$ for any $x \in \mathbb{R}^n \setminus \{0\}$ in (3.4)] with $j \in \{1, \ldots, n\}$, $[b, R_j]$ is bounded on the Morrey space $M_q^p(\mathbb{R}^n)$, then $b \in \mathrm{BMO}\,(\mathbb{R}^n)$, which is a special case of the aforementioned result.

(v) Let $p(\cdot) \in C^{\log}(\mathbb{R}^n)$ with $1 < p_- \le p_+ < \infty$, where p_- and p_+ are defined, respectively, in (1.59) and (1.60). Then, in this case, as was mentioned in [228, Subsection 4.3], the variable Lebesgue space $L^{p(\cdot)}(\mathbb{R}^n)$ satisfies all the assumptions of Lemma 3.1.6. Thus, Lemma 3.1.6 with X therein replaced by $L^{p(\cdot)}(\mathbb{R}^n)$ holds true. If further assume that Ω is odd, then, in this case, the aforementioned result goes back to [141, Theorem 1.1(b)].

We now show Proposition 3.1.5 via using Lemma 3.1.6. To this end, we only need to show that, under the assumptions of Proposition 3.1.5, the local generalized Herz space $\dot{\mathcal{K}}_{\omega,0}^{p,q}(\mathbb{R}^n)$ satisfies all the assumptions of Lemma 3.1.6.

Proof of Proposition 3.1.5 Let all the symbols be as in the present proposition. Then, combining the assumptions of the present proposition, and Theorem 1.2.46, we conclude that the local generalized Herz space $\dot{\mathcal{K}}_{\omega,0}^{p,q}(\mathbb{R}^n)$ is a BBF space. On the other hand, from Corollary 1.5.4, we deduce that the Hardy–Littlewood maximal operator $\mathcal{M}$ is bounded on the Herz space $\dot{\mathcal{K}}_{\omega,0}^{p,q}(\mathbb{R}^n)$. This implies that $\dot{\mathcal{K}}_{\omega,0}^{p,q}(\mathbb{R}^n)$ satisfies all the assumptions of Lemma 3.1.6 and hence finishes the proof of Proposition 3.1.5. $\qquad\square$

Via the sufficiency and the necessity of the boundedness of commutators on local generalized Herz spaces established above, we now prove Theorem 3.1.1. We first need the following integral representation of $[b, T_\Omega](\mathbf{1}_F)$ for any bounded measurable set F of $\mathbb{R}^n$, which might be well known. However, for the convenience of the reader, we give its detailed proof here.

Lemma 3.1.8 *Let Ω be a homogeneous Lipschitz function of degree zero on $\mathbb{S}^{n-1}$ satisfying (3.3), T_Ω a singular integral operator with homogeneous kernel Ω, and $b \in L_{\mathrm{loc}}^1(\mathbb{R}^n)$. Then, for any bounded measurable set F and almost every $x \in \mathbb{R}^n \setminus \overline{F}$,*

$$[b, T_\Omega]\,(\mathbf{1}_F)\,(x) = \int_F [b(x) - b(y)] \frac{\Omega(x-y)}{|x-y|^n}\,dy.$$

Proof Let Ω, T_Ω, b, and F be as in the present lemma. Then, applying (3.1), we conclude that, for any $\tau \in (0, 1)$,

$$\omega_\infty(\tau) \leq \sup_{\{x, y \in \mathbb{S}^{n-1}, |x-y|<\tau\}} |x - y| = \tau.$$

Thus,

$$\int_0^1 \frac{\omega_\infty(\tau)}{\tau} d\tau \leq 1. \tag{3.9}$$

In addition, for any $x \in \mathbb{R}^n \setminus \overline{F}$ and $\varepsilon \in (0, d(x, F))$, we have $[B(x, \varepsilon)]^\complement \cap F = F$, where, for any $x \in \mathbb{R}^n$ and $F \subset \mathbb{R}^n$,

$$d(x, F) := \inf\{|x - y| : y \in F\}.$$

From this, (3.9), and [161, Corollary 2.1.1], it follows that, for almost every $x \in \mathbb{R}^n \setminus \overline{F}$,

$$\begin{aligned}
T_\Omega(\mathbf{1}_F)(x) &= \lim_{\varepsilon \to 0^+} \int_{|x-y|\geq\varepsilon} \frac{\Omega(x - y)}{|x - y|^n} \mathbf{1}_F(y) \, dy \\
&= \lim_{\varepsilon \to 0^+} \int_{[B(x,\varepsilon)]^\complement \cap F} \frac{\Omega(x - y)}{|x - y|^n} \, dy \\
&= \int_F \frac{\Omega(x - y)}{|x - y|^n} \, dy.
\end{aligned} \tag{3.10}$$

Similarly, we find that, for almost every $x \in \mathbb{R}^n \setminus \overline{F}$,

$$T_\Omega(b\mathbf{1}_F)(x) = \int_F b(y) \frac{\Omega(x - y)}{|x - y|^n} \, dy.$$

Combining this and (3.10), we conclude that, for almost every $x \in \mathbb{R}^n \setminus \overline{F}$,

$$[b, T_\Omega](\mathbf{1}_F)(x) = \int_F [b(x) - b(y)] \frac{\Omega(x - y)}{|x - y|^n} \, dy,$$

which completes the proof of Lemma 3.1.8. $\qquad\square$

Via both Propositions 3.1.2 and 3.1.5, and Lemma 3.1.8, we next show Theorem 3.1.1.

Proof of Theorem 3.1.1 Let all the symbols be as in the present theorem. Notice that the assumption (3.1) on Ω implies that, for any given $r \in (0, \infty]$, $\Omega \in L^r(\mathbb{S}^{n-1})$ and the function Ω is continuous on $\mathbb{S}^{n-1}$. This further implies that there exists an open set $\Lambda \subset \mathbb{S}^{n-1}$ such that Ω never vanishes and never changes sign on Λ. By this, Lemma 3.1.8, and both Propositions 3.1.2 and 3.1.5, we obtain Theorem 3.1.1. □

Next, we establish the boundedness characterization of commutators on global generalized Herz spaces as follows.

Theorem 3.1.9 *Let* $p, q \in (1, \infty)$ *and* $\omega \in M(\mathbb{R}_+)$ *satisfy*

$$-\frac{n}{p} < m_0(\omega) \leq M_0(\omega) < \frac{n}{p'}$$

and

$$-\frac{n}{p} < m_\infty(\omega) \leq M_\infty(\omega) < 0,$$

where $\frac{1}{p} + \frac{1}{p'} = 1$. *Assume that* Ω *is a homogeneous function of degree zero on* $\mathbb{S}^{n-1}$ *satisfying both* (3.1) *and* (3.3), *and* T_Ω *a singular integral operator with homogeneous kernel* Ω. *Then, for any* $b \in L^1_{\mathrm{loc}}(\mathbb{R}^n)$, *the commutator* $[b, T_\Omega]$ *is bounded on the global generalized Herz space* $\dot{\mathcal{K}}^{p,q}_\omega(\mathbb{R}^n)$ *if and only if* $b \in$ BMO $(\mathbb{R}^n)$. *Moreover, there exist two positive constants* C_1 *and* C_2, *independent of* b, *such that*

$$C_1 \|b\|_{\mathrm{BMO}(\mathbb{R}^n)} \leq \|[b, T_\Omega]\|_{\dot{\mathcal{K}}^{p,q}_\omega(\mathbb{R}^n) \to \dot{\mathcal{K}}^{p,q}_\omega(\mathbb{R}^n)} \leq C_2 \|b\|_{\mathrm{BMO}(\mathbb{R}^n)},$$

where $\|[b, T_\Omega]\|_{\dot{\mathcal{K}}^{p,q}_\omega(\mathbb{R}^n) \to \dot{\mathcal{K}}^{p,q}_\omega(\mathbb{R}^n)}$ *is as in* (1.88) *with* T *and* X *therein replaced, respectively, by* $[b, T_\Omega]$ *and* $\dot{\mathcal{K}}^{p,q}_\omega(\mathbb{R}^n)$.

First, we consider the sufficiency of the above theorem. To this end, we establish the following boundedness of commutators on global generalized Herz spaces, where the assumption about Ω is weaker than that of Theorem 3.1.9.

Proposition 3.1.10 *Let* p, q, *and* ω *be as in Theorem 3.1.9 and* $r \in (1, \infty]$. *Assume that* $\Omega \in L^r(\mathbb{S}^{n-1})$ *satisfies both* (3.2) *and* (3.3), *and* T_Ω *is a singular integral operator with homogeneous kernel* Ω. *Then there exists a positive constant* C *such that, for any* $b \in$ BMO $(\mathbb{R}^n)$ *and* $f \in \dot{\mathcal{K}}^{p,q}_\omega(\mathbb{R}^n)$,

$$\|[b, T_\Omega](f)\|_{\dot{\mathcal{K}}^{p,q}_\omega(\mathbb{R}^n)} \leq C \|b\|_{\mathrm{BMO}(\mathbb{R}^n)} \|f\|_{\dot{\mathcal{K}}^{p,q}_\omega(\mathbb{R}^n)}.$$

To show Proposition 3.1.10, recall that Tao et al. [228, Theorem 2.17] proved the boundedness of commutators on the ball Banach function space X. However, due to the deficiency of the associate space of the global generalized Herz space $\dot{\mathcal{K}}^{p,q}_\omega(\mathbb{R}^n)$, we can not prove the above boundedness of commutators on $\dot{\mathcal{K}}^{p,q}_\omega(\mathbb{R}^n)$ by using [228, Theorem 2.17] directly [see Remark 1.2.19(vi) for the details]. To overcome this difficulty, via replacing the assumption about associate spaces by an assumption about the extrapolation inequality [see Proposition 3.1.11(ii) below], we now establish the following boundedness of commutators on ball Banach function spaces, which plays a key role in the proof of Proposition 3.1.10.

Proposition 3.1.11 *Let $b \in \mathrm{BMO}\,(\mathbb{R}^n)$, $r \in (1, \infty]$, $\Omega \in L^r(\mathbb{S}^{n-1})$ satisfy both (3.2) and (3.3), and T_Ω be a singular integral operator with homogeneous kernel Ω. Let X be a ball Banach function space satisfying that*

(i) *the commutator $[b, T_\Omega]$ is well defined on X;*
(ii) *there exists an $r_0 \in [r', \infty)$ and a positive constant $\widetilde{C}$ such that, for any $(F, G) \in \mathcal{F}$ with $\|F\|_X < \infty$,*

$$\|F\|_X \le \widetilde{C}\|G\|_X,$$

where $\mathcal{F}$ is the same as in Theorem 1.8.7.

Then there exists a positive constant C, independent of b, such that, for any $f \in X$,

$$\|[b, T_\Omega]\,(f)\|_X \le C\|b\|_{\mathrm{BMO}\,(\mathbb{R}^n)}\|f\|_X.$$

Proof Let all the symbols be as in the present proposition. Then, using (i) and repeating the proof of [228, Proposition 2.14] with $\mathcal{T}$ and [228, Lemma 2.13] therein replaced, respectively, by $[b, T_\Omega]$ and (ii), we conclude that, for any $f \in X$,

$$\|[b, T_\Omega(f)]\|_X \lesssim \|b\|_{\mathrm{BMO}\,(\mathbb{R}^n)}\|f\|_X.$$

This finishes the proof of Proposition 3.1.11.   □

We now prove Proposition 3.1.10.

Proof of Proposition 3.1.10 Let all the symbols be as in the present proposition. Then, combining the assumptions of both the present proposition and Theorem 1.2.48, we find that the global generalized Herz space $\dot{\mathcal{K}}_\omega^{p,q}(\mathbb{R}^n)$ is a BBF space. Thus, to complete the proof of the present proposition, it suffices for us to show that $\dot{\mathcal{K}}_\omega^{p,q}(\mathbb{R}^n)$ satisfies both (i) and (ii) of Proposition 3.1.11.

Indeed, from Definition 1.2.1, it follows that, for any $f \in \dot{\mathcal{K}}_\omega^{p,q}(\mathbb{R}^n)$,

$$\|f\|_{\dot{\mathcal{K}}_{\omega,\mathbf{0}}^{p,q}(\mathbb{R}^n)} \le \|f\|_{\dot{\mathcal{K}}_\omega^{p,q}(\mathbb{R}^n)} < \infty. \tag{3.11}$$

By both this and Proposition 3.1.2, we conclude that, for any $f \in \dot{\mathcal{K}}_\omega^{p,q}(\mathbb{R}^n)$, $[b, T_\Omega](f) \in \dot{\mathcal{K}}_{\omega,\mathbf{0}}^{p,q}(\mathbb{R}^n)$. Therefore, $[b, T_\Omega]$ is well defined on $\dot{\mathcal{K}}_\omega^{p,q}(\mathbb{R}^n)$. This further implies that $\dot{\mathcal{K}}_\omega^{p,q}(\mathbb{R}^n)$ satisfies Proposition 3.1.11(i).

On the other hand, using Theorem 1.8.7, we find that Proposition 3.1.11(ii) holds true for $\dot{\mathcal{K}}_\omega^{p,q}(\mathbb{R}^n)$. Then we conclude that all the assumptions of Proposition 3.1.11 are satisfied for the global generalized Herz space $\dot{\mathcal{K}}_\omega^{p,q}(\mathbb{R}^n)$, which completes the proof of Proposition 3.1.10.   □

Next, we focus on the necessity of Theorem 3.1.9. Indeed, via the necessity of the boundedness of commutators on ball Banach function spaces established in [228, Theorem 2.17] (see also Lemma 3.1.6 above), we obtain the following conclusion which is more general than the necessity of Theorem 3.1.9.

Proposition 3.1.12 *Let $p \in (1, \infty)$, $q \in [1, \infty)$, and $\omega \in M(\mathbb{R}_+)$ be as in Theorem 3.1.9. Assume that $b \in L^1_{loc}(\mathbb{R}^n)$ and $\Omega \in L^\infty(\mathbb{S}^{n-1})$ satisfying that there exists an open set $\Lambda \subset \mathbb{S}^{n-1}$ such that Ω never vanishes and never changes sign on Λ. If the commutator $[b, T_\Omega]$ is bounded on the global generalized Herz space $\dot{\mathcal{K}}^{p,q}_\omega(\mathbb{R}^n)$ and, for any bounded measurable set $F \subset \mathbb{R}^n$ and almost every $x \in \mathbb{R}^n \setminus \overline{F}$,*

$$[b, T_\Omega] (\mathbf{1}_F) (x) = \int_F [b(x) - b(y)] \frac{\Omega(x - y)}{|x - y|^n} \, dy,$$

then $b \in \mathrm{BMO}(\mathbb{R}^n)$ and there exists a positive constant C, independent of b, such that

$$\|b\|_{\mathrm{BMO}(\mathbb{R}^n)} \leq C \|[b, T_\Omega]\|_{\dot{\mathcal{K}}^{p,q}_\omega(\mathbb{R}^n) \to \dot{\mathcal{K}}^{p,q}_\omega(\mathbb{R}^n)},$$

where $\|[b, T_\Omega]\|_{\dot{\mathcal{K}}^{p,q}_\omega(\mathbb{R}^n) \to \dot{\mathcal{K}}^{p,q}_\omega(\mathbb{R}^n)}$ is as in (1.88) with T and X therein replaced, respectively, by $[b, T_\Omega]$ and $\dot{\mathcal{K}}^{p,q}_\omega(\mathbb{R}^n)$.

Proof Let all the symbols be as in the present proposition. Then, combining both Theorem 1.2.48 and Corollary 1.5.6, we conclude that $\dot{\mathcal{K}}^{p,q}_\omega(\mathbb{R}^n)$ is a BBF space and the Hardy–Littlewood maximal operator $\mathcal{M}$ is bounded on $\dot{\mathcal{K}}^{p,q}_\omega(\mathbb{R}^n)$. Using this and Lemma 3.1.6, we find that $b \in \mathrm{BMO}(\mathbb{R}^n)$ and

$$\|b\|_{\mathrm{BMO}(\mathbb{R}^n)} \lesssim \|[b, T_\Omega]\|_{\dot{\mathcal{K}}^{p,q}_\omega(\mathbb{R}^n) \to \dot{\mathcal{K}}^{p,q}_\omega(\mathbb{R}^n)},$$

which completes the proof of Proposition 3.1.12. $\qquad\square$

We finally show Theorem 3.1.9.

Proof of Theorem 3.1.9 Let all the symbols be as in the present theorem. Then, by (3.1), we conclude that, for any given $r \in (0, \infty]$, $\Omega \in L^r(\mathbb{S}^{n-1})$ and the function Ω is continuous on $\mathbb{S}^{n-1}$. This further implies that there exists an open set $\Lambda \subset \mathbb{S}^{n-1}$ such that Ω never vanishes and never changes sign on Λ. Using this, Lemma 3.1.8, and Propositions 3.1.10 and 3.1.12, we then complete the proof of Theorem 3.1.9. $\qquad\square$

3.2 Compactness Characterizations

In this section, we establish the compactness characterization of commutators on local and global generalized Herz spaces. We first present the compactness characterization on local generalized Herz spaces as follows. Recall that CMO $(\mathbb{R}^n)$ is defined to be the closure of infinitely differentiable functions with compact support in BMO $(\mathbb{R}^n)$.

Theorem 3.2.1 *Let $p, q \in (1, \infty)$ and $\omega \in M(\mathbb{R}_+)$ satisfy*

$$-\frac{n}{p} < m_0(\omega) \le M_0(\omega) < \frac{n}{p'}$$

and

$$-\frac{n}{p} < m_\infty(\omega) \le M_\infty(\omega) < \frac{n}{p'},$$

where $\frac{1}{p} + \frac{1}{p'} = 1$. Assume that Ω is a homogeneous function satisfying (3.1), (3.2), *and* (3.3), *and T_Ω a singular integral operator with homogeneous kernel Ω. Then, for any $b \in L^1_{\mathrm{loc}}(\mathbb{R}^n)$, the commutator $[b, T_\Omega]$ is compact on the local generalized Herz space $\dot{\mathcal{K}}^{p,q}_{\omega,\mathbf{0}}(\mathbb{R}^n)$ if and only if $b \in$ CMO $(\mathbb{R}^n)$.*

We prove this theorem by considering, respectively, the sufficiency and the necessity of the compactness of commutators on $\dot{\mathcal{K}}^{p,q}_{\omega,\mathbf{0}}(\mathbb{R}^n)$. First, for the sufficiency, we have the following more general conclusion.

Proposition 3.2.2 *Let p, q, and ω be as in Theorem 3.2.1. Assume that $b \in L^1_{\mathrm{loc}}(\mathbb{R}^n)$, $\Omega \in L^\infty(\mathbb{S}^{n-1})$ satisfies* (3.2), (3.3), *and the L^∞-Dini condition, and T_Ω is a singular integral operator with homogeneous kernel Ω. If $b \in$ CMO $(\mathbb{R}^n)$, then the commutator $[b, T_\Omega]$ is compact on the local generalized Herz space $\dot{\mathcal{K}}^{p,q}_{\omega,\mathbf{0}}(\mathbb{R}^n)$.*

Indeed, in [228, Theorem 3.1], Tao et al. established the following compactness of commutators on ball Banach function spaces.

Lemma 3.2.3 *Let X be a ball Banach function space satisfying that the Hardy–Littlewood maximal operator $\mathcal{M}$ is bounded on X and its associate space X'. Assume that $b \in L^1_{\mathrm{loc}}(\mathbb{R}^n)$, $\Omega \in L^\infty(\mathbb{S}^{n-1})$ satisfies* (3.2), (3.3), *and the L^∞-Dini condition, and T_Ω is a singular integral operator with homogeneous kernel Ω. If $b \in$ CMO $(\mathbb{R}^n)$, then the commutator $[b, T_\Omega]$ is compact on X.*

Remark 3.2.4 We point out that Lemma 3.2.3 has a wide range of applications. Here we give several function spaces to which Lemma 3.2.3 can be applied (see also [228, Section 4]).

(i) Let $p \in (1, \infty)$. Then, in this case, using the boundedness of the Hardy–Littlewood maximal operator $\mathcal{M}$ on both $L^p(\mathbb{R}^n)$ and $L^{p'}(\mathbb{R}^n)$, and Remark 1.2.19(i), we conclude that the Lebesgue space $L^p(\mathbb{R}^n)$ satisfies all the assumptions of Lemma 3.2.3. This further implies that Lemma 3.2.3 with $X := L^p(\mathbb{R}^n)$ holds true. This result extends the known compactness of commutators on Lebesgue spaces originally investigated in [231, Theorem 2]. Indeed, when Ω satisfies (3.1), (3.2), and (3.3), then, in this case, the aforementioned result coincides with the sufficiency of [231, Theorem 2].

(ii) Let $p \in (1, \infty)$ and $\upsilon \in A_p(\mathbb{R}^n)$. Then, in this case, as was mentioned in [228, Subsection 4.4], the weighted Lebesgue space $L_\upsilon^p(\mathbb{R}^n)$ satisfies all the assumptions of Lemma 3.2.3. Therefore, Lemma 3.2.3 with X replaced by $L_\upsilon^p(\mathbb{R}^n)$ holds true, which goes back to [96, Theorem 1.5].

(iii) Let $\vec{p} := (p_1, \ldots, p_n) \in (1, \infty)^n$. Then, in this case, as was pointed out in [228, Subsection 4.2], the mixed-norm Lebesgue space $L^{\vec{p}}(\mathbb{R}^n)$ satisfies all the assumptions of Lemma 3.2.3. This further implies that Lemma 3.2.3 with $X := L^{\vec{p}}(\mathbb{R}^n)$ holds true, which was also obtained in [228, Theorem 4.6].

(iv) Let $1 < q \le p < \infty$. Then, in this case, as was mentioned in [228, Subsection 4.1], the Morrey space $M_q^p(\mathbb{R}^n)$ satisfies all the assumptions of Lemma 3.2.3. Thus, Lemma 3.2.3 with X therein replaced by $M_q^p(\mathbb{R}^n)$ holds true. This result is just the known compactness of commutators on Morrey spaces proved in [36, Theorem 1.1].

(v) Let $p(\cdot) \in C^{\log}(\mathbb{R}^n)$ with $1 < p_- \le p_+ < \infty$, where p_- and p_+ are defined, respectively, in (1.59) and (1.60). Then, in this case, as was pointed out in [228, Subsection 4.3], the variable Lebesgue space $L^{p(\cdot)}(\mathbb{R}^n)$ satisfies all the assumptions of Lemma 3.2.3. Therefore, Lemma 3.2.3 with $X := L^{p(\cdot)}(\mathbb{R}^n)$ holds true, which was also obtained in [228, Theorem 4.8].

We now show Proposition 3.2.2 by using Lemma 3.2.3. To this end, we only need to check that all the assumptions of Lemma 3.2.3 are satisfied for $\dot{\mathcal{K}}_{\omega,\mathbf{0}}^{p,q}(\mathbb{R}^n)$ under the assumptions of Proposition 3.2.2.

Proof of Proposition 3.2.2 Let all the symbols be as in the present proposition. Then, from the assumptions $p, q \in (1, \infty)$ and $-\frac{n}{p} < m_0(\omega) \le M_0(\omega) < \frac{n}{p'}$, and Theorem 1.2.46, we deduce that the local generalized Herz space $\dot{\mathcal{K}}_{\omega,\mathbf{0}}^{p,q}(\mathbb{R}^n)$ is a BBF space. In addition, by the assumptions of the present proposition, (1.138), and (1.139), we find that $\mathcal{M}$ is bounded on both $\dot{\mathcal{K}}_{\omega,\mathbf{0}}^{p,q}(\mathbb{R}^n)$ and its associate space $(\dot{\mathcal{K}}_{\omega,\mathbf{0}}^{p,q}(\mathbb{R}^n))'$. This further implies that $\dot{\mathcal{K}}_{\omega,\mathbf{0}}^{p,q}(\mathbb{R}^n)$ satisfies all the assumptions of Lemma 3.2.3 and hence finishes the proof of Proposition 3.2.2. $\square$

Next, we consider the necessity of Theorem 3.2.1. Indeed, we have the following more general conclusion for the necessity of compactness of commutators on local generalized Herz spaces.

Proposition 3.2.5 *Let p, q, and ω be as in Theorem 3.2.1. Assume that $b \in L_{\text{loc}}^1(\mathbb{R}^n)$ and $\Omega \in L^\infty(\mathbb{S}^{n-1})$ satisfies that there exists an open set $\Lambda \subset \mathbb{S}^{n-1}$ such that Ω never vanishes and never changes sign on Λ. If the commutator $[b, T_\Omega]$ is compact on the local generalized Herz space $\dot{\mathcal{K}}_{\omega,\mathbf{0}}^{p,q}(\mathbb{R}^n)$ and (3.8) holds true, then $b \in \text{CMO}(\mathbb{R}^n)$.*

To prove this proposition, we require the following necessity of the compactness of commutators on ball Banach function spaces, which is just [228, Theorem 3.2].

Lemma 3.2.6 *Let X be a ball Banach function space satisfying that the Hardy–Littlewood $\mathcal{M}$ is bounded on both X and its associate space X'. Assume that $b \in L^1_{\mathrm{loc}}(\mathbb{R}^n)$ and $\Omega \in L^\infty(\mathbb{S}^{n-1})$ satisfies that there exists an open set $\Lambda \subset \mathbb{S}^{n-1}$ such that Ω never vanishes and never changes sign on Λ. If the commutator $[b, T_\Omega]$ is compact on X and (3.8) holds true, then $b \in \mathrm{CMO}(\mathbb{R}^n)$.*

Remark 3.2.7 We point out that Lemma 3.2.6 has a wide range of applications. Here we present several function spaces to which Lemma 3.2.6 can be applied (see also [228, Section 4]).

(i) Let $p \in (1, \infty)$. Then, in this case, from the boundedness of the Hardy–Littlewood maximal operator $\mathcal{M}$ on both $L^p(\mathbb{R}^n)$ and $L^{p'}(\mathbb{R}^n)$, and Remark 1.2.19(i), it follows that the Lebesgue space $L^p(\mathbb{R}^n)$ satisfies all the assumptions of Lemma 3.2.6. Therefore, Lemma 3.2.6 with $X := L^p(\mathbb{R}^n)$ holds true. This result extends the known compactness characterization of commutators on Lebesgue spaces originally studied by Uchiyama in [231, Theorem 2]. Indeed, when Ω satisfies (3.1), (3.2), and (3.3), then, in this case, the aforementioned result coincides with the necessity of [231, Theorem 2].

(ii) Let $p \in (1, \infty)$ and $\upsilon \in A_p(\mathbb{R}^n)$. Then, in this case, as was mentioned in [228, Subsection 4.4], the weighted Lebesgue space $L^p_\upsilon(\mathbb{R}^n)$ satisfies all the assumptions of Lemma 3.2.6. Therefore, Lemma 3.2.6 with X therein replaced by $L^p_\upsilon(\mathbb{R}^n)$ holds true. This result was also obtained in [96, Theorem 1.4].

(iii) Let $\vec{p} := (p_1, \ldots, p_n) \in (1, \infty)^n$. Then, in this case, as was pointed out in [228, Subsection 4.2], the mixed-norm Lebesgue space $L^{\vec{p}}(\mathbb{R}^n)$ satisfies all the assumptions of Lemma 3.2.6. This further implies that Lemma 3.2.6 with $X := L^{\vec{p}}(\mathbb{R}^n)$ holds true, which was also proved in [228, Theorem 4.6].

(iv) Let $1 < q \le p < \infty$. Then, in this case, as was mentioned in [228, Subsection 4.1], the Morrey space $M^p_q(\mathbb{R}^n)$ satisfies all the assumptions of Lemma 3.2.6. Therefore, Lemma 3.2.6 with X therein replaced by $M^p_q(\mathbb{R}^n)$ holds true. This result extends the known compactness characterization of commutators on Morrey spaces showed in [36, Theorem 1.2]. Indeed, when Ω satisfies (3.1), (3.2), and (3.3), then, in this case, the aforementioned result goes back to [36, Theorem 1.2].

(v) Let $p(\cdot) \in C^{\log}(\mathbb{R}^n)$ with $1 < p_- \le p_+ < \infty$, where p_- and p_+ are defined, respectively, in (1.59) and (1.60). Then, in this case, as was pointed out in [228, Subsection 4.3], the variable Lebesgue space $L^{p(\cdot)}(\mathbb{R}^n)$ satisfies all the assumptions of Lemma 3.2.6. This then implies that Lemma 3.2.6 with $X := L^{p(\cdot)}(\mathbb{R}^n)$ holds true, which was also obtained in [228, Theorem 4.8].

Via Lemma 3.2.6, we now show Proposition 3.2.5.

Proof of Proposition 3.2.5 Let all the symbols be as in the present proposition. Then, from the assumptions of the present proposition, and Theorem 1.2.46, we deduce that the local generalized Herz space $\dot{\mathcal{K}}^{p,q}_{\omega,\mathbf{0}}(\mathbb{R}^n)$ is a BBF space.

Therefore, to complete the proof of the present proposition, it suffices to show that $\dot{\mathcal{K}}_{\omega,\mathbf{0}}^{p,q}(\mathbb{R}^n)$ satisfies all the assumptions of Lemma 3.2.6. To be precise, we only need to prove that the Hardy–Littlewood maximal operator $\mathcal{M}$ is bounded on both $\dot{\mathcal{K}}_{\omega,\mathbf{0}}^{p,q}(\mathbb{R}^n)$ and $(\dot{\mathcal{K}}_{\omega,\mathbf{0}}^{p,q}(\mathbb{R}^n))'$. Indeed, applying (1.138) and (1.139), we find that, for any $f \in L_{\mathrm{loc}}^1(\mathbb{R}^n)$,

$$\|\mathcal{M}(f)\|_{\dot{\mathcal{K}}_{\omega,\mathbf{0}}^{p,q}(\mathbb{R}^n)} \lesssim \|f\|_{\dot{\mathcal{K}}_{\omega,\mathbf{0}}^{p,q}(\mathbb{R}^n)}$$

and

$$\|\mathcal{M}(f)\|_{(\dot{\mathcal{K}}_{\omega,\mathbf{0}}^{p,q}(\mathbb{R}^n))'} \lesssim \|f\|_{(\dot{\mathcal{K}}_{\omega,\mathbf{0}}^{p,q}(\mathbb{R}^n))'},$$

which imply that $\mathcal{M}$ is bounded on both $\dot{\mathcal{K}}_{\omega,\mathbf{0}}^{p,q}(\mathbb{R}^n)$ and $(\dot{\mathcal{K}}_{\omega,\mathbf{0}}^{p,q}(\mathbb{R}^n))'$, and hence complete the proof of Proposition 3.2.5. $\square$

We now give the proof of Theorem 3.2.1.

Proof of Theorem 3.2.1 Let all the symbols be as in the present theorem. Then, from (3.1), it follows that $\Omega \in L^\infty(\mathbb{S}^{n-1})$ and Ω is continuous on $\mathbb{S}^{n-1}$. This further implies that there exists an open set $\Lambda \subset \mathbb{S}^{n-1}$ such that Ω never vanishes and never changes sign on Λ. In addition, by (3.1) again, we conclude that, for any $\tau \in (0, 1)$, $\omega_\infty(\tau) \le \tau$, where $\omega_\infty(\tau)$ is as in Definition 3.0.12. This further implies that

$$\int_0^1 \frac{\omega_\infty(\tau)}{\tau} \, d\tau \in [0, 1].$$

Thus, the function Ω satisfies the L^∞-Dini condition. Then, applying Lemma 3.1.8 and both Propositions 3.2.2 and 3.2.5, we then complete the proof of Theorem 3.2.1.
                                    $\square$

Next, we establish the compactness characterization of commutators on global generalized Herz spaces. Namely, the following conclusion holds true.

Theorem 3.2.8 *Let $p, q \in (1, \infty)$ and $\omega \in M(\mathbb{R}_+)$ satisfy*

$$-\frac{n}{p} < m_0(\omega) \le M_0(\omega) < \frac{n}{p'}$$

and

$$-\frac{n}{p} < m_\infty(\omega) \le M_\infty(\omega) < 0,$$

where $\frac{1}{p} + \frac{1}{p'} = 1$. Assume that Ω is a homogeneous function of degree zero on $\mathbb{S}^{n-1}$ satisfying both (3.1) and (3.3), and T_Ω a singular integral operator with homogeneous kernel Ω. Then, for any $b \in L_{\mathrm{loc}}^1(\mathbb{R}^n)$, the commutator $[b, T_\Omega]$

*is compact on the global generalized Herz space $\dot{\mathcal{K}}_\omega^{p,q}(\mathbb{R}^n)$ if and only if $b \in$
CMO $(\mathbb{R}^n)$.*

To obtain this theorem, we first show the following compactness of commutators
on global generalized Herz spaces, which is more general than the sufficiency of
Theorem 3.2.8.

Proposition 3.2.9 *Let p, q, and ω be as in Theorem 3.2.8. Assume that $b \in$
$L_{\mathrm{loc}}^1(\mathbb{R}^n)$, $\Omega \in L^\infty(\mathbb{S}^{n-1})$ satisfies (3.2), (3.3), and the L^∞-Dini condition, and
T_Ω is a singular integral operator with homogeneous kernel Ω. If $b \in$ CMO $(\mathbb{R}^n)$,
then the commutator $[b, T_\Omega]$ is compact on $\dot{\mathcal{K}}_\omega^{p,q}(\mathbb{R}^n)$.*

To show Proposition 3.2.9, let $\Omega \in L^\infty(\mathbb{S}^{n-1})$ satisfy (3.2), (3.3), and the L^∞-
Dini condition, and T_Ω be a singular integral operator with homogeneous kernel Ω.
Recall that the *maximal operator* T_Ω^* is defined by setting, for any $f \in \mathscr{M}(\mathbb{R}^n)$ and
$x \in \mathbb{R}^n$,

$$T_\Omega^*(f)(x) := \sup_{\varepsilon \in (0,\infty)} \left| \int_{\{y \in \mathbb{R}^n: \, |x-y|>\varepsilon\}} \frac{\Omega(x-y)}{|x-y|^n} f(y) \, dy \right|. \tag{3.12}$$

This operator plays a vital role in the proof of Proposition 3.2.9. Indeed, we have
the following boundedness of T_Ω^* on X, which is just [228, Proposition 3.10].

Lemma 3.2.10 *Let X be a ball Banach function space satisfying that the Hardy–
Littlewood maximal operator $\mathcal{M}$ is bounded on both X and X'. Assume $\Omega \in$
$L^\infty(\mathbb{S}^{n-1})$ satisfies (3.2), (3.3), and the L^∞-Dini condition and T_Ω^* is as in (3.12).
Then there exists a positive constant C such that, for any $f \in X$,*

$$\left\| T_\Omega^*(f) \right\|_X \leq C \|f\|_X.$$

In addition, to show Proposition 3.2.9, due to the deficiency of the associate
space of the global generalized Herz space $\dot{\mathcal{K}}_\omega^{p,q}(\mathbb{R}^n)$, we can not obtain the above
compactness of commutators on $\dot{\mathcal{K}}_\omega^{p,q}(\mathbb{R}^n)$ by using [228, Theorem 3.1] directly
[see Remark 1.2.19(vi) for the details]. To overcome this obstacle, via replacing
the assumption about associate spaces by an assumption about the extrapolation
inequality [see Proposition 3.2.11(ii) below], we now establish the following
boundedness of commutators on ball Banach function spaces, which is important in
the proof of the compactness characterization of commutators on global generalized
Herz spaces.

Proposition 3.2.11 *Let $b \in$ CMO $(\mathbb{R}^n)$, $\Omega \in L^\infty(\mathbb{S}^{n-1})$ satisfy (3.2), (3.3), and the
L^∞-Dini condition, T_Ω be a singular integral operator with homogeneous kernel Ω,
and T_Ω^* as in (3.12). Assume that X is a ball Banach function space satisfying the
following three conditions:*

(i) *$[b, T_\Omega]$ and T_Ω^* are well defined on X;*
(ii) *the Hardy–Littlewood maximal operator $\mathcal{M}$ is bounded on X;*

(iii) *there exists an $r_0 \in (1, \infty)$ and a positive constant C such that, for any $(F, G) \in \mathcal{F}$ with $\|F\|_X < \infty$,*

$$\|F\|_X \leq C \|G\|_X,$$

where $\mathcal{F}$ is defined as in Theorem 1.8.7.

Then the commutator $[b, T_\Omega]$ is compact on X.

Proof Let all the symbols be as in the present proposition. Then, using (i) and (ii), and repeating an argument similar to that used in the proof of [228, Proposition 3.14] with $\mathcal{T}$ and [228, Lemma 2.13] therein replaced, respectively, by T_Ω^* and (iii) here, we conclude that, for any $f \in X$,

$$\left\| T_\Omega^*(f) \right\|_X \lesssim \|f\|_X.$$

From this and the proof of [228, Theorem 3.1] via replacing [228, Theorem 2.17] therein by Lemma 3.1.11 here, we infer that the commutator $[b, T_\Omega]$ is compact on X. This then finishes the proof of Proposition 3.2.11. $\square$

Now, we can show Proposition 3.2.9 via checking that the global generalized Herz space $\dot{\mathcal{K}}_\omega^{p,q}(\mathbb{R}^n)$ satisfies all the assumptions of Proposition 3.2.11. For this purpose, we first establish an auxiliary lemma about the maximal operator T_Ω^* as follows.

Lemma 3.2.12 *Let p, q, and ω be as in Theorem 3.2.8, $\Omega \in L^\infty(\mathbb{S}^{n-1})$ satisfy (3.2), (3.3), and the L^∞-Dini condition, and T_Ω^* be as in (3.12). Then T_Ω^* is well defined on $\dot{\mathcal{K}}_\omega^{p,q}(\mathbb{R}^n)$.*

Proof Let all the symbols be as in the present lemma. Then, using the assumptions of the present lemma and Theorem 1.2.46, we conclude that the local generalized Herz space $\dot{\mathcal{K}}_{\omega,\mathbf{0}}^{p,q}(\mathbb{R}^n)$ is a BBF space. In addition, from (1.138) and (1.139), we deduce that $\mathcal{M}$ is bounded on both $\dot{\mathcal{K}}_{\omega,\mathbf{0}}^{p,q}(\mathbb{R}^n)$ and $(\dot{\mathcal{K}}_{\omega,\mathbf{0}}^{p,q}(\mathbb{R}^n))'$. Therefore, by Lemma 3.2.10, we find that, for any $f \in \dot{\mathcal{K}}_{\omega,\mathbf{0}}^{p,q}(\mathbb{R}^n)$,

$$\left\| T_\Omega^*(f) \right\|_{\dot{\mathcal{K}}_{\omega,\mathbf{0}}^{p,q}(\mathbb{R}^n)} \lesssim \|f\|_{\dot{\mathcal{K}}_{\omega,\mathbf{0}}^{p,q}(\mathbb{R}^n)}. \tag{3.13}$$

On the other hand, applying (3.11), we conclude that $\dot{\mathcal{K}}_\omega^{p,q}(\mathbb{R}^n) \subset \dot{\mathcal{K}}_{\omega,\mathbf{0}}^{p,q}(\mathbb{R}^n)$. This, combined with (3.13), further implies that, for any $f \in \dot{\mathcal{K}}_\omega^{p,q}(\mathbb{R}^n)$, $T_\Omega^*(f) \in \dot{\mathcal{K}}_{\omega,\mathbf{0}}^{p,q}(\mathbb{R}^n)$, which shows that T_Ω^* is well defined on $\dot{\mathcal{K}}_\omega^{p,q}(\mathbb{R}^n)$, and hence completes the proof of Lemma 3.2.12. $\square$

Via the above two conclusions, we now show Proposition 3.2.9.

Proof of Proposition 3.2.9 Let all the symbols be as in the present proposition. Then, by the assumptions of the present propositions, and Theorem 1.2.48, we find that the global generalized Herz space $\dot{\mathcal{K}}_{\omega}^{p,q}(\mathbb{R}^n)$ is a BBF space. Thus, to finish the proof of the present proposition, we only need to show that $\dot{\mathcal{K}}_{\omega}^{p,q}(\mathbb{R}^n)$ satisfies (i), (ii), and (iii) of Proposition 3.2.11.

We first prove that Proposition 3.2.11(i) holds true for $\dot{\mathcal{K}}_{\omega}^{p,q}(\mathbb{R}^n)$. Indeed, from (3.11), it follows that $\dot{\mathcal{K}}_{\omega}^{p,q}(\mathbb{R}^n) \subset \dot{\mathcal{K}}_{\omega,\mathbf{0}}^{p,q}(\mathbb{R}^n)$. This, together with Proposition 3.1.2, implies that, for any $f \in \dot{\mathcal{K}}_{\omega}^{p,q}(\mathbb{R}^n)$, $[b, T_{\Omega}](f)$ is well defined. On the other hand, applying Lemma 3.2.12, we conclude that T_{Ω}^* is also well defined on $\dot{\mathcal{K}}_{\omega}^{p,q}(\mathbb{R}^n)$. Therefore, $\dot{\mathcal{K}}_{\omega}^{p,q}(\mathbb{R}^n)$ satisfies Proposition 3.2.11(i) under the assumptions of the present proposition.

We next show that $\dot{\mathcal{K}}_{\omega}^{p,q}(\mathbb{R}^n)$ satisfies Proposition 3.2.11(ii), namely, $\mathcal{M}$ is bounded on $\dot{\mathcal{K}}_{\omega}^{p,q}(\mathbb{R}^n)$. Indeed, from the assumptions of the present proposition, and Corollary 1.5.6, we deduce that, for any $f \in L_{\mathrm{loc}}^1(\mathbb{R}^n)$,

$$\|\mathcal{M}(f)\|_{\dot{\mathcal{K}}_{\omega}^{p,q}(\mathbb{R}^n)} \lesssim \|f\|_{\dot{\mathcal{K}}_{\omega}^{p,q}(\mathbb{R}^n)}.$$

This implies that $\mathcal{M}$ is bounded on $\dot{\mathcal{K}}_{\omega}^{p,q}(\mathbb{R}^n)$, and hence Proposition 3.2.11(ii) holds true for $\dot{\mathcal{K}}_{\omega}^{p,q}(\mathbb{R}^n)$.

Finally, by Theorem 1.8.7, we find that $\dot{\mathcal{K}}_{\omega}^{p,q}(\mathbb{R}^n)$ satisfies Proposition 3.2.11(iii). Thus, under the assumptions of the present proposition, the global generalized Herz space $\dot{\mathcal{K}}_{\omega}^{p,q}(\mathbb{R}^n)$ satisfies all the assumptions of Proposition 3.2.11. This implies that the commutator $[b, T_{\Omega}]$ is compact on $\dot{\mathcal{K}}_{\omega}^{p,q}(\mathbb{R}^n)$, and hence finishes the proof of Proposition 3.2.9.   □

Then we focus on the necessity of Theorem 3.2.8. Indeed, we have the following more general conclusion.

Proposition 3.2.13 *Let p, q, and ω be as in Theorem 3.2.8. Assume that $b \in L_{\mathrm{loc}}^1(\mathbb{R}^n)$ and $\Omega \in L^{\infty}(\mathbb{S}^{n-1})$ satisfies that there exists an open set $\Lambda \subset \mathbb{S}^{n-1}$ such that Ω never vanishes and never changes sign on Λ. If the commutator $[b, T_{\Omega}]$ is compact on the global generalized Herz space $\dot{\mathcal{K}}_{\omega}^{p,q}(\mathbb{R}^n)$, and satisfies (3.8), then $b \in \mathrm{CMO}(\mathbb{R}^n)$.*

To prove this proposition, Lemma 3.2.6 is unavailable due to the deficiency of associate spaces of global generalized Herz spaces [see Remark 1.2.19(vi) for the details]. To overcome this difficulty, via replacing the assumption about associate spaces by an assumption about the extrapolation inequality [see Proposition 3.2.14(ii) below], we now show the necessity of the compactness of commutators on ball Banach function spaces as follows.

Proposition 3.2.14 *Let $b \in L^1_{\mathrm{loc}}(\mathbb{R}^n)$, $\Omega \in L^\infty(\mathbb{S}^{n-1})$ satisfy that there exists an open set $\Lambda \subset \mathbb{S}^{n-1}$ such that Ω never vanishes and never changes sign on Λ, and X be a ball Banach function space satisfy the following two conditions:*

(i) *the Hardy–Littlewood maximal operator $\mathcal{M}$ is bounded on X;*
(ii) *there exists a positive constant C such that, for any $(F, G) \in \mathcal{F}$ with $\|F\|_X < \infty$,*

$$\|F\|_X \leq C\|G\|_X,$$

where $\mathcal{F}$ is defined as in Theorem 1.8.7 with $r_0 := 1$.

If the commutator $[b, T_\Omega]$ is compact on X, and the assumption (3.8) holds true, then $b \in \mathrm{CMO}(\mathbb{R}^n)$.

To prove this proposition, we need the following technical lemma which is a part of [135, Lemma 2.2 and Remark 2.3].

Lemma 3.2.15 *Let X be a ball Banach function space such that $\mathcal{M}$ is bounded on X. Then there exists a positive constant C such that, for any ball $B \in \mathbb{B}$,*

$$\frac{1}{|B|} \|\mathbf{1}_B\|_X \|\mathbf{1}_B\|_{X'} \leq C.$$

Via this technical lemma, we now show the following equivalent characterization of BMO $(\mathbb{R}^n)$ via ball Banach function spaces, which is an essential tool in the proof of Proposition 3.2.14 above.

Lemma 3.2.16 *Let X be a ball Banach function space satisfying the following two statements:*

(i) *the Hardy–Littlewood maximal operator $\mathcal{M}$ is bounded on X;*
(ii) *there exists a positive constant C such that, for any $(F, G) \in \mathcal{F}$ with $\|F\|_X < \infty$,*

$$\|F\|_X \leq C\|G\|_X,$$

where $\mathcal{F}$ is defined as in Theorem 1.8.7 with $r_0 := 1$.

Let $f \in L^1_{\mathrm{loc}}(\mathbb{R}^n)$. Then $f \in \mathrm{BMO}(\mathbb{R}^n)$ if and only if

$$\|f\|_{\mathrm{BMO}_X(\mathbb{R}^n)} := \sup_{B \in \mathbb{B}} \left\{ \frac{1}{\|\mathbf{1}_B\|_X} \||f - f_B| \mathbf{1}_B\|_X \right\} < \infty,$$

where f_B is defined as in (3.7). Moreover, there exist two positive constants C_1 and C_2 such that, for any $f \in \mathrm{BMO}(\mathbb{R}^n)$,

$$C_1\|f\|_{\mathrm{BMO}(\mathbb{R}^n)} \leq \|f\|_{\mathrm{BMO}_X(\mathbb{R}^n)} \leq C_2\|f\|_{\mathrm{BMO}(\mathbb{R}^n)}.$$

Proof Let all the symbols be as in the present lemma and $f \in L^1_{\mathrm{loc}}(\mathbb{R}^n)$. We first prove the sufficiency. To this end, assume that

$$\|f\|_{\mathrm{BMO}_X(\mathbb{R}^n)} < \infty.$$

By this, Definition 1.2.17, the assumption (i) of the present lemma, and Lemma 3.2.15, we conclude that, for any ball $B \in \mathbb{B}$,

$$\frac{1}{|B|} \int_B |f(x) - f_B| \, dx$$

$$\leq \frac{1}{|B|} \| |f - f_B| \mathbf{1}_B \|_X \| \mathbf{1}_B \|_{X'} \lesssim \frac{1}{\| \mathbf{1}_B \|_X} \| |f - f_B| \mathbf{1}_B \|_X$$

$$\lesssim \|f\|_{\mathrm{BMO}_X(\mathbb{R}^n)} < \infty,$$

which, combined with (3.6), further implies that

$$\|f\|_{\mathrm{BMO}(\mathbb{R}^n)} \lesssim \|f\|_{\mathrm{BMO}_X(\mathbb{R}^n)} < \infty. \tag{3.14}$$

This finishes the proof of the sufficiency.

Conversely, we next show the necessity. To achieve this, assume $f \in \mathrm{BMO}(\mathbb{R}^n)$. Then, for any given $\upsilon \in A_1(\mathbb{R}^n)$ and for any ball $B \in \mathbb{B}$, repeating the proof of [133, Theorem 1.2] with Q therein replaced by B, we find that

$$\int_B |f(x) - f_B| \upsilon(x) \, dx \lesssim [\upsilon]_{A_1(\mathbb{R}^n)} \|f\|_{\mathrm{BMO}(\mathbb{R}^n)} \int_B \upsilon(x) \, dx,$$

where the implicit positive constant depends only on n. From this and (ii), we infer that, for any $B \in \mathbb{B}$,

$$\| |f - f_B| \mathbf{1}_B \|_X \lesssim \| \mathbf{1}_B \|_X \|f\|_{\mathrm{BMO}(\mathbb{R}^n)}.$$

This further implies that

$$\|f\|_{\mathrm{BMO}_X(\mathbb{R}^n)} \lesssim \|f\|_{\mathrm{BMO}(\mathbb{R}^n)} < \infty,$$

which completes the proof of the necessity. Combining this and (3.14), we find that, for any $f \in \mathrm{BMO}(\mathbb{R}^n)$,

$$\|f\|_{\mathrm{BMO}(\mathbb{R}^n)} \sim \|f\|_{\mathrm{BMO}_X(\mathbb{R}^n)}$$

with the positive equivalence constants independent of f. Thus, the proof of Lemma 3.2.16 is then completed. $\square$

Applying Lemma 3.2.16, we now show Proposition 3.2.14.

Proof of Proposition 3.2.14 Let all the symbols be as in the present proposition and the commutator $[b, T_\Omega]$ is compact on the BBF space X under consideration. Then, repeating the proof of [228, Theorem 3.2] via replacing [228, Lemma 3.15] therein by Lemma 3.2.16 here, we conclude that $b \in \mathrm{CMO}\,(\mathbb{R}^n)$. This then finishes the proof of Proposition 3.2.14. $\square$

Via Proposition 3.2.14, we next prove Proposition 3.2.13.

Proof of Proposition 3.2.13 Let all the symbols be as in the present proposition. Then, from Theorem 1.2.48, we deduce that the global generalized Herz space $\dot{\mathcal{K}}_\omega^{p,q}(\mathbb{R}^n)$ is a BBF space under the assumptions of the present proposition. This implies that, to finish the proof of the present proposition, we only need to show that $\dot{\mathcal{K}}_\omega^{p,q}(\mathbb{R}^n)$ satisfies all the assumptions of Proposition 3.2.14.

Indeed, applying Corollary 1.5.6, we find that $\mathcal{M}$ is bounded on $\dot{\mathcal{K}}_\omega^{p,q}(\mathbb{R}^n)$, which implies that Proposition 3.2.14(i) holds true for $\dot{\mathcal{K}}_\omega^{p,q}(\mathbb{R}^n)$. On the other hand, using Theorem 1.8.7 with $r_0 := 1$, we conclude that $\dot{\mathcal{K}}_\omega^{p,q}(\mathbb{R}^n)$ satisfies Proposition 3.2.14(ii). This further implies that all the assumptions of Proposition 3.2.14 are satisfied for $\dot{\mathcal{K}}_\omega^{p,q}(\mathbb{R}^n)$. Thus, $b \in \mathrm{CMO}\,(\mathbb{R}^n)$ and we then complete the proof of Proposition 3.2.13. $\square$

We then show Theorem 3.2.8 via both Propositions 3.2.9 and 3.2.13.

Proof of Theorem 3.2.8 Let all the symbols be as in the present theorem. Then, by (3.1), we find that Ω is continuous and hence bounded on $\mathbb{S}^{n-1}$. This further implies that $\Omega \in L^\infty(\mathbb{S}^{n-1})$ and there exists an open set $\Lambda \subset \mathbb{S}^{n-1}$ such that Ω never vanishes and never changes sign on Λ. On the other hand, combining (3.1) and Definition 3.0.12, we conclude that, for any $\tau \in (0, 1)$, $\omega_\infty(\tau) \leq \tau$. This further implies that

$$\int_0^1 \frac{\omega_\infty(\tau)}{\tau}\, d\tau \in [0, 1].$$

Then the function Ω satisfies the L^∞-Dini condition. Thus, applying Lemma 3.1.8 and Propositions 3.2.9 and 3.2.13, we then complete the proof of Theorem 3.2.8. $\square$

Chapter 4
Generalized Herz–Hardy Spaces

Let $p, q \in (0, \infty)$ and $\omega \in M(\mathbb{R}_+)$. In this chapter, we first introduce the generalized Herz–Hardy spaces, both $H\dot{\mathcal{K}}^{p,q}_{\omega,0}(\mathbb{R}^n)$ and $H\dot{\mathcal{K}}^{p,q}_{\omega}(\mathbb{R}^n)$, by the grand maximal function. Then, via a totally fresh perspective, we establish their complete real-variable theory. To be precise, via viewing the local generalized Herz space $\dot{\mathcal{K}}^{p,q}_{\omega,0}(\mathbb{R}^n)$ as a special case of ball quasi-Banach function spaces and applying the known real-variable characterizations of Hardy spaces $H_X(\mathbb{R}^n)$ associated with ball quasi-Banach function spaces X, we establish various maximal function, atomic, molecular, and Littlewood–Paley function characterizations of the Herz–Hardy space $H\dot{\mathcal{K}}^{p,q}_{\omega,0}(\mathbb{R}^n)$. Moreover, both the duality and the Fourier transform properties of $H\dot{\mathcal{K}}^{p,q}_{\omega,0}(\mathbb{R}^n)$ are also obtained based on the corresponding results about $H_X(\mathbb{R}^n)$. However, the study of $H\dot{\mathcal{K}}^{p,q}_{\omega}(\mathbb{R}^n)$ is more difficult than that of $H\dot{\mathcal{K}}^{p,q}_{\omega,0}(\mathbb{R}^n)$ due to the deficiency of associate spaces of global generalized Herz spaces [see Remark 1.2.19(vi) for the details]. To overcome this obstacle, via replacing the assumptions of the boundedness of powered Hardy–Littlewood maximal operators on associate spaces (see Assumption 1.2.33 above) by some weaker assumptions about the integral representations of quasi-norms of ball quasi-Banach function spaces as well as some boundedness of powered Hardy–Littlewood maximal operators [see both (ii) and (iii) of Theorem 4.3.18 below], we establish some improved atomic and molecular characterizations of the Hardy space $H_X(\mathbb{R}^n)$, which include the corresponding results obtained by Sawano et al. [207]. Using these improved characterizations and making full use of the obtained duality between block spaces and global generalized Herz spaces in Chap. 2 as well as the construction of the quasi-norm $\| \cdot \|_{\dot{\mathcal{K}}^{p,q}_{\omega}(\mathbb{R}^n)}$, we then obtain the maximal function, atomic, molecular, the various Littlewood–Paley function characterizations of $H\dot{\mathcal{K}}^{p,q}_{\omega}(\mathbb{R}^n)$ and also give some properties about Fourier transforms of distributions in $H\dot{\mathcal{K}}^{p,q}_{\omega}(\mathbb{R}^n)$. As applications, via first showing two boundedness criteria of Calderón–Zygmund operators on $H_X(\mathbb{R}^n)$, we establish the boundedness of Calderón–Zygmund operators on generalized Herz–Hardy spaces. In addition, we also introduce the generalized Hardy–Morrey spaces. Using the fact that, under some reasonable assumptions on the exponents, the generalized Morrey spaces coincide with the generalized Herz spaces in the sense of equivalent norms, we obtain the complete real-variable characterizations of generalized Hardy–Morrey spaces immediately.

To begin with, we introduce the generalized Herz–Hardy spaces. For this purpose, we recall some basic concepts. Throughout this book, we use $\mathcal{S}(\mathbb{R}^n)$ to denote the *space of all Schwartz functions* on $\mathbb{R}^n$ equipped with the well-known topology determined by a countable family of norms, and $\mathcal{S}'(\mathbb{R}^n)$ its *topological dual space* equipped with the weak-$*$ topology. For any $N \in \mathbb{N}$ and $\phi \in \mathcal{S}(\mathbb{R}^n)$, let

$$p_N(\phi) := \sum_{\alpha \in \mathbb{Z}_+^n, |\alpha| \leq N} \sup_{x \in \mathbb{R}^n} \left\{ (1 + |x|)^{N+n} |\partial^\alpha \phi(x)| \right\} \qquad (4.1)$$

and

$$\mathcal{F}_N(\mathbb{R}^n) := \left\{ \phi \in \mathcal{S}(\mathbb{R}^n) : \ p_N(\phi) \in [0, 1] \right\}. \qquad (4.2)$$

Moreover, the *non-tangential grand maximal function* $\mathcal{M}_N(f)$ of $f \in \mathcal{S}'(\mathbb{R}^n)$ is defined by setting, for any $x \in \mathbb{R}^n$,

$$\mathcal{M}_N(f)(x) := \sup \left\{ |f * \phi_t(y)| : \ \phi \in \mathcal{F}_N(\mathbb{R}^n), \ t \in (0, \infty), \ |x - y| < t \right\}. \qquad (4.3)$$

We now introduce the generalized Herz–Hardy spaces via the grand maximal function as follows.

Definition 4.0.17 Let $p, q \in (0, \infty)$, $\omega \in M(\mathbb{R}_+)$, and $N \in \mathbb{N}$. Then

(i) the *generalized Herz–Hardy space* $H\dot{\mathcal{K}}_{\omega,\mathbf{0}}^{p,q}(\mathbb{R}^n)$, associated with the local generalized Herz space $\dot{\mathcal{K}}_{\omega,\mathbf{0}}^{p,q}(\mathbb{R}^n)$, is defined to be the set of all the $f \in \mathcal{S}'(\mathbb{R}^n)$ such that

$$\|f\|_{H\dot{\mathcal{K}}_{\omega,\mathbf{0}}^{p,q}(\mathbb{R}^n)} := \|\mathcal{M}_N(f)\|_{\dot{\mathcal{K}}_{\omega,\mathbf{0}}^{p,q}(\mathbb{R}^n)} < \infty;$$

(ii) the *generalized Herz–Hardy space* $H\dot{\mathcal{K}}_{\omega}^{p,q}(\mathbb{R}^n)$, associated with the global generalized Herz space $\dot{\mathcal{K}}_{\omega}^{p,q}(\mathbb{R}^n)$, is defined to be the set of all the $f \in \mathcal{S}'(\mathbb{R}^n)$ such that

$$\|f\|_{H\dot{\mathcal{K}}_{\omega}^{p,q}(\mathbb{R}^n)} := \|\mathcal{M}_N(f)\|_{\dot{\mathcal{K}}_{\omega}^{p,q}(\mathbb{R}^n)} < \infty.$$

Remark 4.0.18 If, in Definition 4.0.17, let $\omega(t) := t^\alpha$ for any $t \in (0, \infty)$ and for any given $\alpha \in \mathbb{R}$, then, in this case, the generalized Herz–Hardy space $H\dot{\mathcal{K}}_{\omega,\mathbf{0}}^{p,q}(\mathbb{R}^n)$ goes back to the classical *homogeneous Herz-type Hardy space* $H\dot{K}_p^{\alpha,q}(\mathbb{R}^n)$ which was originally introduced in [163, Definition 2.1] (see also [175, Chapter 2]). However, we point out that, to the best of our knowledge, even in this case, the generalized Herz–Hardy space $H\dot{\mathcal{K}}_\omega^{p,q}(\mathbb{R}^n)$ is also new.

Recall that the local and the global generalized Morrey spaces are defined as in Remark 1.2.2(vi). Then we introduce the generalized Hardy–Morrey spaces as follows.

Definition 4.0.19 Let $p, q \in (0, \infty)$, $\omega \in M(\mathbb{R}_+)$, and $N \in \mathbb{N}$. Then

(i) the *generalized Hardy–Morrey space* $HM_{\omega,\mathbf{0}}^{p,q}(\mathbb{R}^n)$, associated with the local generalized Morrey space $M_{\omega,\mathbf{0}}^{p,q}(\mathbb{R}^n)$, is defined to be the set of all the $f \in \mathcal{S}'(\mathbb{R}^n)$ such that

$$\|f\|_{HM_{\omega,\mathbf{0}}^{p,q}(\mathbb{R}^n)} := \|\mathcal{M}_N(f)\|_{M_{\omega,\mathbf{0}}^{p,q}(\mathbb{R}^n)} < \infty;$$

(ii) the *generalized Hardy–Morrey space* $HM_\omega^{p,q}(\mathbb{R}^n)$, associated with the global generalized Morrey space $M_\omega^{p,q}(\mathbb{R}^n)$, is defined to be the set of all the $f \in \mathcal{S}'(\mathbb{R}^n)$ such that

$$\|f\|_{HM_\omega^{p,q}(\mathbb{R}^n)} := \|\mathcal{M}_N(f)\|_{M_\omega^{p,q}(\mathbb{R}^n)} < \infty.$$

Remark 4.0.20

(i) To the best of our knowledge, in Definition 4.0.19, even when $\omega(t) := t^\alpha$ for any $t \in (0, \infty)$ and for any given $\alpha \in \mathbb{R}$, the generalized Hardy–Morrey spaces, both $HM_{\omega,\mathbf{0}}^{p,q}(\mathbb{R}^n)$ and $HM_\omega^{p,q}(\mathbb{R}^n)$, are also new.

(ii) In Definition 4.0.19, let $p, q \in [1, \infty)$ and $\omega \in M(\mathbb{R}_+)$ satisfy

$$\max\{M_0(\omega), M_\infty(\omega)\} \in (-\infty, 0).$$

Then, by Remark 1.2.2(vi), we conclude that, in this case, the generalized Hardy–Morrey spaces $HM_{\omega,\mathbf{0}}^{p,q}(\mathbb{R}^n)$ and $HM_\omega^{p,q}(\mathbb{R}^n)$ coincide, respectively, with the generalized Herz–Hardy spaces $H\dot{\mathcal{K}}_{\omega,\mathbf{0}}^{p,q}(\mathbb{R}^n)$ and $H\dot{\mathcal{K}}_\omega^{p,q}(\mathbb{R}^n)$ with equivalent norms.

4.1 Maximal Function Characterizations

In this section, we establish various maximal function characterizations of generalized Herz–Hardy spaces. We first present the concepts of various maximal functions as follows (see [207, Definition 2.12]).

Definition 4.1.1 Let $N \in \mathbb{N}$, $a, b \in (0, \infty)$, $\phi \in \mathcal{S}(\mathbb{R}^n)$, and $f \in \mathcal{S}'(\mathbb{R}^n)$.

(i) The *radial maximal function* $M(f, \phi)$ is defined by setting, for any $x \in \mathbb{R}^n$,

$$M(f, \phi)(x) := \sup_{t \in (0, \infty)} |f * \phi_t(x)|,$$

here and thereafter, for any $\phi \in \mathcal{S}(\mathbb{R}^n)$ and $x \in \mathbb{R}^n$,

$$\phi_t(x) := \frac{1}{t^n} \phi \left(\frac{x}{t} \right). \tag{4.4}$$

(ii) The *non-tangential maximal function* $M_a^*(f, \phi)$, with aperture $a \in (0, \infty)$, is defined by setting, for any $x \in \mathbb{R}^n$,

$$M_a^*(f, \phi)(x) := \sup_{(y,t) \in \Gamma_a(x)} |f * \phi_t(y)|,$$

here and thereafter, for any $x \in \mathbb{R}^n$ and $a \in (0, \infty)$,

$$\Gamma_a(x) := \{(y, t) \in \mathbb{R}_+^{n+1} : |y - x| < at\}. \tag{4.5}$$

(iii) The *maximal function* $M_b^{**}(f, \phi)$ *of Peetre type* is defined by setting, for any $x \in \mathbb{R}^n$,

$$M_b^{**}(f, \phi)(x) := \sup_{(y,t) \in \mathbb{R}_+^{n+1}} \frac{|f * \phi_t(x - y)|}{(1 + t^{-1}|y|)^b}.$$

(iv) The *grand maximal function* $\mathcal{M}_{b,N}^{**}(f)$ *of Peetre type* is defined by setting, for any $x \in \mathbb{R}^n$,

$$\mathcal{M}_{b,N}^{**}(f)(x) := \sup_{\phi \in \mathcal{F}_N(\mathbb{R}^n)} \sup_{(y,t) \in \mathbb{R}_+^{n+1}} \frac{|f * \phi_t(x - y)|}{(1 + t^{-1}|y|)^b},$$

where $\mathcal{F}_N(\mathbb{R}^n)$ is as in (4.2).

Via these maximal functions, we characterize the generalized Herz–Hardy space $H\dot{\mathcal{K}}_{\omega,0}^{p,q}(\mathbb{R}^n)$ as follows.

Theorem 4.1.2 *Let* $p, q, a, b \in (0, \infty)$, $\omega \in M(\mathbb{R}_+)$, $N \in \mathbb{N}$, *and* $\phi \in \mathcal{S}(\mathbb{R}^n)$ *satisfy* $\int_{\mathbb{R}^n} \phi(x)\, dx \neq 0$.

(i) *Let $N \in \mathbb{N} \cap [\lfloor b + 1 \rfloor, \infty)$ and ω satisfy $m_0(\omega) \in (-\frac{n}{p}, \infty)$. Then, for any $f \in \mathcal{S}'(\mathbb{R}^n)$,*

$$\|M(f, \phi)\|_{\dot{K}^{p,q}_{\omega,0}(\mathbb{R}^n)} \lesssim \|M_a^*(f, \phi)\|_{\dot{K}^{p,q}_{\omega,0}(\mathbb{R}^n)} \lesssim \|M_b^{**}(f, \phi)\|_{\dot{K}^{p,q}_{\omega,0}(\mathbb{R}^n)},$$

$$\|M(f, \phi)\|_{\dot{K}^{p,q}_{\omega,0}(\mathbb{R}^n)} \lesssim \|\mathcal{M}_N(f)\|_{\dot{K}^{p,q}_{\omega,0}(\mathbb{R}^n)} \lesssim \|\mathcal{M}_{\lfloor b+1 \rfloor}(f)\|_{\dot{K}^{p,q}_{\omega,0}(\mathbb{R}^n)}$$

$$\lesssim \|M_b^{**}(f, \phi)\|_{\dot{K}^{p,q}_{\omega,0}(\mathbb{R}^n)},$$

and

$$\|M_b^{**}(f, \phi)\|_{\dot{K}^{p,q}_{\omega,0}(\mathbb{R}^n)} \sim \|\mathcal{M}_{b,N}^{**}(f)\|_{\dot{K}^{p,q}_{\omega,0}(\mathbb{R}^n)},$$

where the implicit positive constants are independent of f.

(ii) *Let $\omega \in M(\mathbb{R}_+)$ satisfy $m_0(\omega) \in (-\frac{n}{p}, \infty)$ and $m_\infty(\omega) \in (-\frac{n}{p}, \infty)$. Assume $b \in (\max\{\frac{n}{p}, \max\{M_0(\omega), M_\infty(\omega)\} + \frac{n}{p}\}, \infty)$. Then, for any $f \in \mathcal{S}'(\mathbb{R}^n)$,*

$$\|M_b^{**}(f, \phi)\|_{\dot{K}^{p,q}_{\omega,0}(\mathbb{R}^n)} \lesssim \|M(f, \phi)\|_{\dot{K}^{p,q}_{\omega,0}(\mathbb{R}^n)},$$

where the implicit positive constant is independent of f. In particular, when $N \in \mathbb{N} \cap [\lfloor b + 1 \rfloor, \infty)$, if one of the quantities

$$\|M(f, \phi)\|_{\dot{K}^{p,q}_{\omega,0}(\mathbb{R}^n)}, \ \|M_a^*(f, \phi)\|_{\dot{K}^{p,q}_{\omega,0}(\mathbb{R}^n)}, \ \|\mathcal{M}_N(f)\|_{\dot{K}^{p,q}_{\omega,0}(\mathbb{R}^n)},$$

$$\|M_b^{**}(f, \phi)\|_{\dot{K}^{p,q}_{\omega,0}(\mathbb{R}^n)}, \ and \ \|\mathcal{M}_{b,N}^{**}(f)\|_{\dot{K}^{p,q}_{\omega,0}(\mathbb{R}^n)}$$

is finite, then the others are also finite and mutually equivalent with the positive equivalence constants independent of f.

Remark 4.1.3

(i) Let $p, q \in (0, \infty)$ and $\omega \in M(\mathbb{R}_+)$. Then the quasi-norm of $H\dot{K}^{p,q}_{\omega,0}(\mathbb{R}^n)$ in Definition 4.0.17(i) depends on N. However, by Theorem 4.1.2, we find that the Hardy space $H\dot{K}^{p,q}_{\omega,0}(\mathbb{R}^n)$ is independent of the choice of N whenever $\omega \in M(\mathbb{R}_+)$ satisfies

$$m_0(\omega) \in \left(-\frac{n}{p}, \infty\right) \text{ and } m_\infty(\omega) \in \left(-\frac{n}{p}, \infty\right),$$

and N satisfies

$$N \in \mathbb{N} \cap \left(1 + \max\left\{\frac{n}{p}, \max\{M_0(\omega), M_\infty(\omega)\} + \frac{n}{p}\right\}, \infty\right).$$

(ii) Observe that, if $p = q \in (0, \infty)$ and $\omega(t) := 1$ for any $t \in (0, \infty)$, then, in this case,

$$m_0(\omega) = M_0(\omega) = m_\infty(\omega) = M_\infty(\omega) = 0,$$

$H\dot{\mathcal{K}}^{p,q}_{\omega,\mathbf{0}}(\mathbb{R}^n)$ coincides with the classical Hardy space $H^p(\mathbb{R}^n)$ in the sense of equivalent quasi-norms, and all the conclusions in Theorem 4.1.2 coincide with the corresponding classical results on the Hardy space $H^p(\mathbb{R}^n)$; see, for instance [91, 160, 213].

In order to prove this theorem, we need the following maximal function characterizations of Hardy spaces associated with ball quasi-Banach function spaces, which is just [207, Theorem 3.1].

Lemma 4.1.4 *Let* $a, b \in (0, \infty)$, X *be a ball quasi-Banach function space, and* $\phi \in \mathcal{S}(\mathbb{R}^n)$ *satisfy* $\int_{\mathbb{R}^n} \phi(x) \, dx \neq 0$.

(i) *Let* $N \in \mathbb{N} \cap [\lfloor b + 1 \rfloor, \infty)$. *Then, for any* $f \in \mathcal{S}'(\mathbb{R}^n)$,

$$\|M(f, \phi)\|_X \lesssim \|M_a^*(f, \phi)\|_X \lesssim \|M_b^{**}(f, \phi)\|_X,$$

$$\|M(f, \phi)\|_X \lesssim \|\mathcal{M}_N(f)\|_X \lesssim \|\mathcal{M}_{\lfloor b+1 \rfloor}(f)\|_X \lesssim \|M_b^{**}(f, \phi)\|_X,$$

and

$$\|M_b^{**}(f, \phi)\|_X \sim \|\mathcal{M}_{b,N}^{**}(f)\|_X,$$

where the implicit positive constants are independent of f.

(ii) *Let* $r \in (0, \infty)$. *Assume* $b \in (\frac{n}{r}, \infty)$ *and* $\mathcal{M}$ *is bounded on* $X^{1/r}$. *Then, for any* $f \in \mathcal{S}'(\mathbb{R}^n)$,

$$\|M_b^{**}(f, \phi)\|_X \lesssim \|M(f, \phi)\|_X,$$

where the implicit positive constant is independent of f. *In particular, when* $N \in \mathbb{N} \cap [\lfloor b + 1 \rfloor, \infty)$, *if one of the quantities*

$$\|M(f, \phi)\|_X, \ \|M_a^*(f, \phi)\|_X, \ \|\mathcal{M}_N(f)\|_X,$$

$$\|M_b^{**}(f, \phi)\|_X, \ and \ \|\mathcal{M}_{b,N}^{**}(f)\|_X$$

is finite, then the others are also finite and mutually equivalent with the positive equivalence constants independent of f.

Remark 4.1.5 We point out that Lemma 4.1.4 has a wide range of applications. Here we present several function spaces to which Lemma 4.1.4 can be applied.

(i) Let $p \in (0, \infty)$ and $b \in (\frac{n}{p}, \infty)$. Then, in this case, by Remark 1.2.31(i), we can easily find that the Lebesgue space $L^p(\mathbb{R}^n)$ satisfies all the assumptions of Lemma 4.1.4. This then implies that Lemma 4.1.4 with $X := L^p(\mathbb{R}^n)$ holds true. This result coincides with the famous maximal function characterizations of the classical Hardy space $H^p(\mathbb{R}^n)$ (see, for instance, [77] or [213, Chapter III, Section 1]).

(ii) Let $p \in (0, \infty)$, $\upsilon \in A_\infty(\mathbb{R}^n)$, and $b \in (\frac{nq_\upsilon}{p}, \infty)$, where q_υ is the same as in (1.58). Then, in this case, from Remark 1.2.31(ii), we can easily infer that the weighted Lebesgue space $L_\upsilon^p(\mathbb{R}^n)$ satisfies all the assumptions of Lemma 4.1.4. Therefore, Lemma 4.1.4 with X therein replaced by $L_\upsilon^p(\mathbb{R}^n)$ holds true (see, for instance, [218, Chapter IV, Theorems 10, 11, and 12]).

(iii) Let $\vec{p} := (p_1, \ldots, p_n) \in (0, \infty)^n$ and $b \in (\frac{n}{\min\{p_1, \ldots, p_n\}}, \infty)$. Then, in this case, using Remark 1.2.31(iii), we can easily conclude that the mixed-norm Lebesgue space $L^{\vec{p}}(\mathbb{R}^n)$ satisfies all the assumptions of Lemma 4.1.4. Thus, Lemma 4.1.4 with $X := L^{\vec{p}}(\mathbb{R}^n)$ holds true. This result goes back to the maximal function characterizations of mixed-norm Hardy spaces given in [42, Theorem 3.4] (see also [127, Theorem 4.2]).

(iv) Let $0 < q \leq p < \infty$ and $b \in (\frac{n}{q}, \infty)$. Then, in this case, by Remark 1.2.31(iv), we easily find that the Morrey space $M_q^p(\mathbb{R}^n)$ satisfies all the assumptions of Lemma 4.1.4. This further implies that Lemma 4.1.4 with X replaced by $M_q^p(\mathbb{R}^n)$ holds true. This result coincides with the maximal function characterizations of Hardy–Morrey spaces obtained in [139, Section 2].

(v) Let $p(\cdot) \in C^{\log}(\mathbb{R}^n)$ and $b \in (\frac{n}{p_-}, \infty)$, where p_- is the same as in (1.59). Then, in this case, from Remark 1.2.31(v), we can easily deduce that the variable Lebesgue space $L^{p(\cdot)}(\mathbb{R}^n)$ satisfies all the assumptions of Lemma 4.1.4. Thus, Lemma 4.1.4 with $X := L^{p(\cdot)}(\mathbb{R}^n)$ holds true. This result goes back to the maximal function characterizations of variable Hardy spaces obtained in [186, Theorem 3.3].

As was pointed out in Chap. 1, under some reasonable and sharp assumptions, the local generalized Herz space $\dot{\mathcal{K}}_{\omega,\mathbf{0}}^{p,q}(\mathbb{R}^n)$ is a ball quasi-Banach function space. This implies that, to show Theorem 4.1.2, we only need to prove that $\dot{\mathcal{K}}_{\omega,\mathbf{0}}^{p,q}(\mathbb{R}^n)$ satisfies all the assumptions of Lemma 4.1.4.

Proof of Theorem 4.1.2 Let all the symbols be as in the present theorem. Then, from the assumption $m_0(\omega) \in (-\frac{n}{p}, \infty)$ and Theorem 1.2.42, it follows that the local generalized Herz space $\dot{\mathcal{K}}_{\omega,\mathbf{0}}^{p,q}(\mathbb{R}^n)$ is a BQBF space. Applying this and Lemma 4.1.4(i), we conclude that (i) holds true.

Next, we show (ii). To achieve this, we only need to prove that, under the assumptions of the present theorem, $\dot{\mathcal{K}}_{\omega,\mathbf{0}}^{p,q}(\mathbb{R}^n)$ satisfies all the assumptions of Lemma 4.1.4(ii), namely, there exists an $r \in (0, \infty)$ such that $b \in (\frac{n}{r}, \infty)$ and the Hardy–Littlewood maximal operator $\mathcal{M}$ is bounded on $[\dot{\mathcal{K}}_{\omega,\mathbf{0}}^{p,q}(\mathbb{R}^n)]^{1/r}$. Indeed, from the assumptions $\min\{m_0(\omega), m_\infty(\omega)\} \in (-\frac{n}{p}, \infty)$ and

$$b \in \left(\max\left\{ \frac{n}{p}, \max\{M_0(\omega), M_\infty(\omega)\} + \frac{n}{p} \right\}, \infty \right),$$

we deduce that

$$\frac{n}{b} \in \left(0, \min\left\{ p, \frac{n}{\max\{M_0(\omega), M_\infty(\omega)\} + n/p} \right\} \right).$$

Thus, we can choose an

$$r \in \left(\frac{n}{b}, \min\left\{ p, \frac{n}{\max\{M_0(\omega), M_\infty(\omega)\} + n/p} \right\} \right). \tag{4.6}$$

By this and Lemma 1.8.5, we find that $b \in (\frac{n}{r}, \infty)$ and $\mathcal{M}$ is bounded on the Herz space $[\dot{\mathcal{K}}^{p,q}_{\omega,\mathbf{0}}(\mathbb{R}^n)]^{1/r}$. This implies that all the assumptions of Lemma 4.1.4(ii) hold true for $\dot{\mathcal{K}}^{p,q}_{\omega,\mathbf{0}}(\mathbb{R}^n)$, which completes the proof of (ii) and hence Theorem 4.1.2. □

Remark 4.1.6 In Theorem 4.1.2, let $\omega(t) := t^\alpha$ for any $t \in (0, \infty)$ and for any given $\alpha \in \mathbb{R}$. Then, in this case, when $p \in (1, \infty)$, the maximal function characterizations established in Theorem 4.1.2 widen the range of $\alpha \in (0, \infty)$ in [175, Theorem 2.1] into $\alpha \in (-\frac{n}{p}, \infty)$; when $p \in (0, 1]$, to the best of out knowledge, the result of Theorem 4.1.2 is totally new.

As an application, we obtain the following maximal function characterizations of the generalized Hardy–Morrey space $HM^{p,q}_{\omega,\mathbf{0}}(\mathbb{R}^n)$, which is immediately deduced from Theorem 4.1.2 and Remark 1.2.2(vi); we omit the details.

Corollary 4.1.7 *Let $a, b \in (0, \infty)$, $p, q \in [1, \infty)$, $\omega \in M(\mathbb{R}_+)$, $N \in \mathbb{N}$, and $\phi \in \mathcal{S}(\mathbb{R}^n)$ satisfy $\int_{\mathbb{R}^n} \phi(x)\, dx \neq 0$.*

(i) *Let $N \in \mathbb{N} \cap [\lfloor b + 1 \rfloor, \infty)$ and ω satisfy $M_\infty(\omega) \in (-\infty, 0)$ and*

$$-\frac{n}{p} < m_0(\omega) \leq M_0(\omega) < 0.$$

Then, for any $f \in \mathcal{S}'(\mathbb{R}^n)$,

$$\|M(f, \phi)\|_{M^{p,q}_{\omega,\mathbf{0}}(\mathbb{R}^n)} \lesssim \|M^*_a(f, \phi)\|_{M^{p,q}_{\omega,\mathbf{0}}(\mathbb{R}^n)} \lesssim \|M^{**}_b(f, \phi)\|_{M^{p,q}_{\omega,\mathbf{0}}(\mathbb{R}^n)},$$

$$\|M(f, \phi)\|_{M^{p,q}_{\omega,\mathbf{0}}(\mathbb{R}^n)} \lesssim \|\mathcal{M}_N(f)\|_{M^{p,q}_{\omega,\mathbf{0}}(\mathbb{R}^n)} \lesssim \|\mathcal{M}_{\lfloor b+1 \rfloor}(f)\|_{M^{p,q}_{\omega,\mathbf{0}}(\mathbb{R}^n)}$$

$$\lesssim \|M^{**}_b(f, \phi)\|_{M^{p,q}_{\omega,\mathbf{0}}(\mathbb{R}^n)},$$

and

$$\|M_b^{**}(f,\phi)\|_{M_{\omega,0}^{p,q}(\mathbb{R}^n)} \sim \|\mathcal{M}_{b,N}^{**}(f)\|_{M_{\omega,0}^{p,q}(\mathbb{R}^n)},$$

where the implicit positive constants are independent of f.
(ii) *Let $\omega \in M(\mathbb{R}_+)$ satisfy*

$$-\frac{n}{p} < m_0(\omega) \le M_0(\omega) < 0$$

and

$$-\frac{n}{p} < m_0(\omega) \le M_0(\omega) < 0.$$

Assume $b \in (\frac{n}{p}, \infty)$. Then, for any $f \in \mathcal{S}'(\mathbb{R}^n)$,

$$\|M_b^{**}(f,\phi)\|_{M_{\omega,0}^{p,q}(\mathbb{R}^n)} \lesssim \|M(f,\phi)\|_{M_{\omega,0}^{p,q}(\mathbb{R}^n)},$$

where the implicit positive constant is independent of f. In particular, when $N \in \mathbb{N} \cap [\lfloor b+1 \rfloor, \infty)$, if one of the quantities

$$\|M(f,\phi)\|_{M_{\omega,0}^{p,q}(\mathbb{R}^n)}, \ \|M_a^*(f,\phi)\|_{M_{\omega,0}^{p,q}(\mathbb{R}^n)}, \ \|\mathcal{M}_N(f)\|_{M_{\omega,0}^{p,q}(\mathbb{R}^n)},$$

$$\|M_b^{**}(f,\phi)\|_{M_{\omega,0}^{p,q}(\mathbb{R}^n)}, \ and \ \|\mathcal{M}_{b,N}^{**}(f)\|_{M_{\omega,0}^{p,q}(\mathbb{R}^n)}$$

is finite, then the others are also finite and mutually equivalent with the positive equivalence constants independent of f.

Next, we show the following maximal function characterizations of the Hardy space $H\dot{\mathcal{K}}_\omega^{p,q}(\mathbb{R}^n)$ associated with the global generalized Herz space $\dot{\mathcal{K}}_\omega^{p,q}(\mathbb{R}^n)$.

Theorem 4.1.8 *Let $p,q,a,b \in (0,\infty)$, $\omega \in M(\mathbb{R}_+)$, $N \in \mathbb{N}$, and $\phi \in \mathcal{S}(\mathbb{R}^n)$ satisfy*

$$\int_{\mathbb{R}^n} \phi(x)\,dx \neq 0.$$

(i) *Let $N \in \mathbb{N} \cap [\lfloor b+1 \rfloor, \infty)$ and ω satisfy $m_0(\omega) \in (-\frac{n}{p}, \infty)$ and $M_\infty(\omega) \in (-\infty, 0)$. Then, for any $f \in \mathcal{S}'(\mathbb{R}^n)$,*

$$\|M(f,\phi)\|_{\dot{\mathcal{K}}_\omega^{p,q}(\mathbb{R}^n)} \lesssim \|M_a^*(f,\phi)\|_{\dot{\mathcal{K}}_\omega^{p,q}(\mathbb{R}^n)} \lesssim \|M_b^{**}(f,\phi)\|_{\dot{\mathcal{K}}_\omega^{p,q}(\mathbb{R}^n)},$$

$$\|M(f,\phi)\|_{\dot{\mathcal{K}}_\omega^{p,q}(\mathbb{R}^n)} \lesssim \|\mathcal{M}_N(f)\|_{\dot{\mathcal{K}}_\omega^{p,q}(\mathbb{R}^n)} \lesssim \|\mathcal{M}_{\lfloor b+1 \rfloor}(f)\|_{\dot{\mathcal{K}}_\omega^{p,q}(\mathbb{R}^n)}$$

$$\lesssim \|M_b^{**}(f,\phi)\|_{\dot{\mathcal{K}}_\omega^{p,q}(\mathbb{R}^n)},$$

and

$$\|M_b^{**}(f, \phi)\|_{\dot{\mathcal{K}}_\omega^{p,q}(\mathbb{R}^n)} \sim \|\mathcal{M}_{b,N}^{**}(f)\|_{\dot{\mathcal{K}}_\omega^{p,q}(\mathbb{R}^n)},$$

where the implicit positive constants are independent of f.

(ii) *Let* $\omega \in M(\mathbb{R}_+)$ *satisfy* $m_0(\omega) \in (-\frac{n}{p}, \infty)$ *and*

$$-\frac{n}{p} < m_\infty(\omega) \le M_\infty(\omega) < 0.$$

Assume $b \in (\max\{\frac{n}{p}, \max\{M_0(\omega), M_\infty(\omega)\} + \frac{n}{p}\}, \infty)$. *Then, for any* $f \in \mathcal{S}'(\mathbb{R}^n)$,

$$\|M_b^{**}(f, \phi)\|_{\dot{\mathcal{K}}_\omega^{p,q}(\mathbb{R}^n)} \lesssim \|M(f, \phi)\|_{\dot{\mathcal{K}}_\omega^{p,q}(\mathbb{R}^n)},$$

where the implicit positive constant is independent of f. *In particular, when* $N \in \mathbb{N} \cap [\lfloor b + 1 \rfloor, \infty)$, *if one of the quantities*

$$\|M(f, \phi)\|_{\dot{\mathcal{K}}_\omega^{p,q}(\mathbb{R}^n)}, \; \|M_a^*(f, \phi)\|_{\dot{\mathcal{K}}_\omega^{p,q}(\mathbb{R}^n)}, \; \|\mathcal{M}_N(f)\|_{\dot{\mathcal{K}}_\omega^{p,q}(\mathbb{R}^n)},$$

$$\|M_b^{**}(f, \phi)\|_{\dot{\mathcal{K}}_\omega^{p,q}(\mathbb{R}^n)}, \; \text{and} \; \|\mathcal{M}_{b,N}^{**}(f)\|_{\dot{\mathcal{K}}_\omega^{p,q}(\mathbb{R}^n)}$$

is finite, then the others are also finite and mutually equivalent with the positive equivalence constants independent of f.

Remark 4.1.9

(i) Let $p, q \in (0, \infty)$ and $\omega \in M(\mathbb{R}_+)$. Then the quasi-norm of $H\dot{\mathcal{K}}_\omega^{p,q}(\mathbb{R}^n)$ in Definition 4.0.17(ii) depends on N. However, from Theorem 4.1.8, we infer that the Hardy space $H\dot{\mathcal{K}}_\omega^{p,q}(\mathbb{R}^n)$ is independent of the choice of N whenever $\omega \in M(\mathbb{R}_+)$ satisfies $m_0(\omega) \in (-\frac{n}{p}, \infty)$ and

$$-\frac{n}{p} < m_\infty(\omega) \le M_\infty(\omega) < 0,$$

and N satisfies

$$N \in \mathbb{N} \cap \left(1 + \max\left\{\frac{n}{p}, \max\{M_0(\omega), M_\infty(\omega)\} + \frac{n}{p}\right\}, \infty\right).$$

(ii) Notice that, if $p = q \in (0, \infty)$ and $\omega(t) := 1$ for any $t \in (0, \infty)$, then, in this case,

$$m_0(\omega) = M_0(\omega) = m_\infty(\omega) = M_\infty(\omega) = 0$$

and $H\dot{\mathcal{K}}_{\omega}^{p,q}(\mathbb{R}^n)$ coincides with the classical Hardy space $H^p(\mathbb{R}^n)$ in the sense of equivalent quasi-norms. Thus, Theorem 4.1.8 completely excludes the classical Hardy space $H^p(\mathbb{R}^n)$ and, based on Remark 1.2.45, we find that the classical Hardy space $H^p(\mathbb{R}^n)$ is the critical case of $H\dot{\mathcal{K}}_{\omega}^{p,q}(\mathbb{R}^n)$ considered in Theorem 4.1.8.

To prove this theorem, we first present a lemma about the boundedness of the Hardy–Littlewood maximal operator as follows.

Lemma 4.1.10 *Let* $p, q \in (0, \infty)$ *and* $\omega \in M(\mathbb{R}_+)$ *satisfy* $m_0(\omega) \in (-\frac{n}{p}, \infty)$ *and* $m_\infty(\omega) \in (-\frac{n}{p}, \infty)$. *Then, for any given* $r \in (0, \min\{p, \frac{n}{\max\{M_0(\omega), M_\infty(\omega)\}+n/p}\})$, *there exists a positive constant C such that, for any $f \in L^1_{\mathrm{loc}}(\mathbb{R}^n)$,*

$$\|\mathcal{M}(f)\|_{[\dot{\mathcal{K}}_{\omega}^{p,q}(\mathbb{R}^n)]^{1/r}} \leq C\|f\|_{[\dot{\mathcal{K}}_{\omega}^{p,q}(\mathbb{R}^n)]^{1/r}}.$$

Proof Let all the symbols be as in the present lemma. From (1.132) and (1.133), it follows that

$$\max\left\{M_0(\omega^r), M_\infty(\omega^r)\right\} < \frac{n}{(p/r)'}$$

and

$$\min\left\{m_0(\omega^r), m_\infty(\omega^r)\right\} > -\frac{n}{p/r}.$$

This, together with the fact that $\frac{p}{r} \in (1, \infty)$ and Corollary 1.5.6, further implies that, for any $f \in L^1_{\mathrm{loc}}(\mathbb{R}^n)$,

$$\|\mathcal{M}(f)\|_{\dot{\mathcal{K}}_{\omega^r}^{p/r,q/r}(\mathbb{R}^n)} \lesssim \|f\|_{\dot{\mathcal{K}}_{\omega^r}^{p/r,q/r}(\mathbb{R}^n)}.$$

Therefore, by Lemma 1.3.2, we further find that the Hardy–Littlewood maximal operator $\mathcal{M}$ is bounded on $[\dot{\mathcal{K}}_{\omega}^{p,q}(\mathbb{R}^n)]^{1/r}$. This then finishes the proof of Lemma 4.1.10. $\qquad\square$

Via Lemma 4.1.10 and the maximal function characterizations of Hardy spaces associated with ball quasi-Banach function spaces presented in Lemma 4.1.4, we now show Theorem 4.1.8.

Proof of Theorem 4.1.8 Let all the symbols be as in the present theorem. Then, using the assumptions $m_0(\omega) \in (-\frac{n}{p}, \infty)$ and $M_\infty(\omega) \in (-\infty, 0)$, and Theorem 1.2.44, we conclude that the global generalized Herz space $\dot{\mathcal{K}}_{\omega}^{p,q}(\mathbb{R}^n)$ is a BQBF space. This, together with Lemma 4.1.4(i), then finishes the proof of (i).

Next, we show (ii). Indeed, let r be as in (4.6). Then, by Lemma 4.1.10, we find that the Hardy–Littlewood maximal operator $\mathcal{M}$ is bounded on $[\dot{\mathcal{K}}_{\omega}^{p,q}(\mathbb{R}^n)]^{1/r}$.

From this and Lemma 4.1.4(ii), it follows that (ii) holds true, which then completes the proof of Theorem 4.1.8.   □

Moreover, using the theorem above and Remark 1.2.2(vi), we conclude the following maximal function characterizations of the generalized Hardy–Morrey space $HM_\omega^{p,q}(\mathbb{R}^n)$; we omit the details.

Corollary 4.1.11 *Let $a, b \in (0, \infty)$, $p, q \in [1, \infty)$, $\omega \in M(\mathbb{R}_+)$, $N \in \mathbb{N}$, and $\phi \in \mathcal{S}(\mathbb{R}^n)$ satisfy $\int_{\mathbb{R}^n} \phi(x)\, dx \neq 0$.*

(i) *Let $N \in \mathbb{N} \cap [\lfloor b+1 \rfloor, \infty)$ and ω satisfy $M_\infty(\omega) \in (-\infty, 0)$ and*

$$-\frac{n}{p} < m_0(\omega) \leq M_0(\omega) < 0.$$

Then, for any $f \in \mathcal{S}'(\mathbb{R}^n)$,

$$\|M(f, \phi)\|_{M_\omega^{p,q}(\mathbb{R}^n)} \lesssim \|M_a^*(f, \phi)\|_{M_\omega^{p,q}(\mathbb{R}^n)} \lesssim \|M_b^{**}(f, \phi)\|_{M_\omega^{p,q}(\mathbb{R}^n)},$$

$$\|M(f, \phi)\|_{M_\omega^{p,q}(\mathbb{R}^n)} \lesssim \|\mathcal{M}_N(f)\|_{M_\omega^{p,q}(\mathbb{R}^n)} \lesssim \|\mathcal{M}_{\lfloor b+1 \rfloor}(f)\|_{M_\omega^{p,q}(\mathbb{R}^n)}$$

$$\lesssim \|M_b^{**}(f, \phi)\|_{M_\omega^{p,q}(\mathbb{R}^n)},$$

and

$$\|M_b^{**}(f, \phi)\|_{M_\omega^{p,q}(\mathbb{R}^n)} \sim \|\mathcal{M}_{b,N}^{**}(f)\|_{M_\omega^{p,q}(\mathbb{R}^n)},$$

where the implicit positive constants are independent of f.

(ii) *Let $\omega \in M(\mathbb{R}_+)$ satisfy*

$$-\frac{n}{p} < m_0(\omega) \leq M_0(\omega) < 0$$

and

$$-\frac{n}{p} < m_0(\omega) \leq M_0(\omega) < 0.$$

Assume $b \in (\frac{n}{p}, \infty)$. Then, for any $f \in \mathcal{S}'(\mathbb{R}^n)$,

$$\|M_b^{**}(f, \phi)\|_{M_\omega^{p,q}(\mathbb{R}^n)} \lesssim \|M(f, \phi)\|_{M_\omega^{p,q}(\mathbb{R}^n)},$$

where the implicit positive constant is independent of f. In particular, when $N \in \mathbb{N} \cap [\lfloor b+1 \rfloor, \infty)$, if one of the quantities

$$\|M(f, \phi)\|_{M_\omega^{p,q}(\mathbb{R}^n)}, \quad \|M_a^*(f, \phi)\|_{M_\omega^{p,q}(\mathbb{R}^n)}, \quad \|\mathcal{M}_N(f)\|_{M_\omega^{p,q}(\mathbb{R}^n)},$$

$$\|M_b^{**}(f, \phi)\|_{M_\omega^{p,q}(\mathbb{R}^n)}, \quad and \quad \|\mathcal{M}_{b,N}^{**}(f)\|_{M_\omega^{p,q}(\mathbb{R}^n)}$$

is finite, then the others are also finite and mutually equivalent with the positive equivalence constants independent of f.

Remark 4.1.12 Let $p, q \in (0, \infty)$ and $\omega \in M(\mathbb{R}_+)$. Then the quasi-norms of $HM_{\omega,0}^{p,q}(\mathbb{R}^n)$ and $HM_\omega^{p,q}(\mathbb{R}^n)$ in Definition 4.0.19 depend on N. However, applying Corollaries 4.1.7 and 4.1.11, we conclude that the Hardy spaces $HM_{\omega,0}^{p,q}(\mathbb{R}^n)$ and $HM_\omega^{p,q}(\mathbb{R}^n)$ are both independent of the choice of N whenever $p, q \in [1, \infty)$, $\omega \in M(\mathbb{R}_+)$ satisfies

$$-\frac{n}{p} < m_0(\omega) \le M_0(\omega) < 0$$

and

$$-\frac{n}{p} < m_\infty(\omega) \le M_\infty(\omega) < 0,$$

and N satisfies

$$N \in \mathbb{N} \cap \left(1 + \frac{n}{p}, \infty\right).$$

4.2 Relations with Generalized Herz Spaces

In this section, we investigate the relations between generalized Herz spaces and the associated Hardy spaces. The following conclusion shows that, under some assumptions on the exponents, the local generalized Herz space $\dot{\mathcal{K}}_{\omega,0}^{p,q}(\mathbb{R}^n)$ coincides with the generalized Herz–Hardy space $H\dot{\mathcal{K}}_{\omega,0}^{p,q}(\mathbb{R}^n)$ in the sense of equivalent quasi-norms.

Theorem 4.2.1 *Let $p \in (1, \infty)$, $q \in (0, \infty)$, and $\omega \in M(\mathbb{R}_+)$ satisfy*

$$-\frac{n}{p} < m_0(\omega) \le M_0(\omega) < \frac{n}{p'}$$

and

$$-\frac{n}{p} < m_\infty(\omega) \le M_\infty(\omega) < \frac{n}{p'},$$

where $\frac{1}{p} + \frac{1}{p'} = 1$. Then

(i) $\dot{\mathcal{K}}_{\omega,0}^{p,q}(\mathbb{R}^n) \hookrightarrow \mathcal{S}'(\mathbb{R}^n)$.

(ii) *If $f \in \dot{\mathcal{K}}_{\omega,0}^{p,q}(\mathbb{R}^n)$, then $f \in H\dot{\mathcal{K}}_{\omega,0}^{p,q}(\mathbb{R}^n)$ and there exists a positive constant C, independent of f, such that*

$$\|f\|_{H\dot{\mathcal{K}}_{\omega,0}^{p,q}(\mathbb{R}^n)} \leq C\|f\|_{\dot{\mathcal{K}}_{\omega,0}^{p,q}(\mathbb{R}^n)}.$$

(iii) *If $f \in H\dot{\mathcal{K}}_{\omega,0}^{p,q}(\mathbb{R}^n)$, then there exists a locally integrable function $g \in \dot{\mathcal{K}}_{\omega,0}^{p,q}(\mathbb{R}^n)$ such that g represents f, which means that $f = g$ in $\mathcal{S}'(\mathbb{R}^n)$,*

$$\|f\|_{H\dot{\mathcal{K}}_{\omega,0}^{p,q}(\mathbb{R}^n)} = \|g\|_{H\dot{\mathcal{K}}_{\omega,0}^{p,q}(\mathbb{R}^n)},$$

and there exists a positive constant C, independent of f, such that

$$\|g\|_{\dot{\mathcal{K}}_{\omega,0}^{p,q}(\mathbb{R}^n)} \leq C\|f\|_{H\dot{\mathcal{K}}_{\omega,0}^{p,q}(\mathbb{R}^n)}.$$

To prove this theorem, recall that Sawano et al. [207, Theorem 3.4] obtained the relation between the ball quasi-Banach function space X and the associated Hardy space $H_X(\mathbb{R}^n)$ (see also Lemma 4.2.6 below). Thus, to prove Theorem 4.2.1, it suffices to show that all the assumptions of [207, Theorem 3.4] hold true for local generalized Herz spaces. For this purpose, we first present the definition of the Hardy space $H_X(\mathbb{R}^n)$ as follows, which was given in [207, Definition 2.22].

Definition 4.2.2 Let X be a ball quasi-Banach function space and $N \in \mathbb{N}$. Then the *Hardy space $H_X(\mathbb{R}^n)$ is defined to be the set of all the $f \in \mathcal{S}'(\mathbb{R}^n)$ such that*

$$\|f\|_{H_X(\mathbb{R}^n)} := \|\mathcal{M}_N(f)\|_X < \infty.$$

Remark 4.2.3

(i) Let $p \in (0, \infty)$ and $X := L^p(\mathbb{R}^n)$. Then, in this case, $H_X(\mathbb{R}^n)$ is just the *Hardy space $H^p(\mathbb{R}^n)$ which was first studied in [77, 215]. In particular, Let $p = 1$. Then, in this case, by Grafakos [91, Theorem 2.1.4], we find that the two Hardy spaces defined, respectively, in (1.42) and this remark are the same.*

(ii) Let $p \in (0, \infty)$, $\upsilon \in A_\infty(\mathbb{R}^n)$, and $X := L_\upsilon^p(\mathbb{R}^n)$. Then, in this case, $H_X(\mathbb{R}^n)$ *is just the weighted Hardy space $H_\upsilon^p(\mathbb{R}^n)$ which was originally introduced in [217, p. 1054].*

(iii) Let $\vec{p} := (p_1, \ldots, p_n) \in (0, \infty)^n$ and $X := L^{\vec{p}}(\mathbb{R}^n)$. Then, in this case, $H_X(\mathbb{R}^n)$ *is just the mixed-norm Hardy space $H^{\vec{p}}(\mathbb{R}^n)$ which is a special case of [42, Definition 3.3] (see also [99, 123]).*

(iv) Let $0 < q \leq p < \infty$ and $X := M_q^p(\mathbb{R}^n)$. Then, in this case, $H_X(\mathbb{R}^n)$ is just the *Hardy–Morrey space $HM_q^p(\mathbb{R}^n)$ which was first introduced in [139, Definition 1.3].*

(v) Let $p(\cdot) : \mathbb{R}^n \to (0, \infty)$ and $X := L^{p(\cdot)}(\mathbb{R}^n)$. Then, in this case, $H_X(\mathbb{R}^n)$ is just the *variable Hardy space* $H^{p(\cdot)}(\mathbb{R}^n)$ which was first introduced in [186, Subsection 3.1].

Let δ_0 denote the *Dirac measure* at $\mathbf{0}$. Here we present a special element belonging to Hardy spaces, which shows that, for the Hardy–Morrey space $HM_p^{\frac{1}{m+1}}(\mathbb{R})$, when $m \to \infty$, and hence $p \to 0$ because $p \in (0, \frac{1}{m+1})$, then $N \to \infty$, where N is the same as in Definition 4.2.2 with $X := M_p^{\frac{1}{m+1}}(\mathbb{R})$. This indicates that the critical value of N in Definition 4.2.2 depends on X under consideration.

Proposition 4.2.4 *Let $m \in \mathbb{N}$ and $p \in (0, \frac{1}{m+1})$. Then $\frac{d^m \delta_0}{dx^m} \in HM_p^{\frac{1}{m+1}}(\mathbb{R})$, where the space $HM_p^{\frac{1}{m+1}}(\mathbb{R})$ is the Hardy–Morrey space defined as in Definition 4.2.2 with $X := M_p^{\frac{1}{m+1}}(\mathbb{R})$ and $N \in \mathbb{N} \cap [2m + 2, \infty)$.*

To prove this proposition, we require the following auxiliary result about Morrey spaces, which is just [206, Subsection 1.2.2, Example 6].

Lemma 4.2.5 *Let $0 < q < p < \infty$ and, for any $x \in \mathbb{R}^n \setminus \{\mathbf{0}\}$, let $f(x) := |x|^{-\frac{n}{p}}$. Then $f \in M_q^p(\mathbb{R}^n)$.*

Via Lemma 4.2.5, we now show Proposition 4.2.4.

Proof of Proposition 4.2.4 Let all the symbols be the same as in the present proposition and $\phi \in \mathcal{F}_N(\mathbb{R})$ in (4.2) with $n = 1$. Then, using both (4.1) and (4.2), we find that, for any $x \in \mathbb{R}$, $t \in (0, \infty)$, and $y \in \mathbb{R}$ with $|y - x| < t$,

$$
\left| \frac{d^m \delta_0}{dx^m} * \phi_t(y) \right| = \left| \left\langle \delta_0, \frac{d^m}{dx^m}(\phi_t)(y - \cdot) \right\rangle \right|
$$

$$
= \frac{1}{t^{m+1}} \left| \frac{d^m \phi}{dx^m} \left(\frac{y}{t} \right) \right| \le \frac{t^{N-m}}{(t + |y|)^{N+1}}.
$$

This, combined with (4.3), further implies that, for any $x \in \mathbb{R}$,

$$
\mathcal{M}_N \left(\frac{d^m \delta_0}{dx^m} \right)(x) \le \sup_{t \in (0,\infty), \, |y-x|<t} \left\{ \frac{t^{N-m}}{(t + |y|)^{N+1}} \right\}. \tag{4.7}
$$

Now, for any given $x, y \in \mathbb{R}$ and for any $t \in (|y - x|, \infty)$, let

$$
u_{x,y}(t) := \frac{t^{N-m}}{(t + |y|)^{N+1}}.
$$

Then we have

$$u'_{x,y}(t) = \frac{(N-m)|y| - (m+1)t}{(t+|y|)^{N+2}}. \tag{4.8}$$

We next prove that, for any x, $y \in \mathbb{R}$ and $t \in (|y-x|, \infty)$, $u_{x,y}(t) \lesssim |x|^{-m-1}$ by considering the following two cases on both x and y.

Case (1) $\frac{N-m}{m+1}|y| > |y-x|$. In this case, it holds true that

$$|x| \leq |y-x| + |y| < \frac{N+1}{m+1}|y|. \tag{4.9}$$

In addition, from (4.8), we infer that, for any $t \in (|y-x|, \frac{N-m}{m+1}|y|)$, $u'_{x,y}(t) > 0$ and, for any $t \in (\frac{N-m}{m+1}|y|, \infty)$, $u'_{x,y}(t) < 0$. These further imply that $u_{x,y}(t)$ is increasing when $t \in (|y-x|, \frac{N-m}{m+1}|y|)$ and decreasing when $t \in (\frac{N-m}{m+1}|y|, \infty)$. By this and (4.9), we then conclude that, for any $t \in (|y-x|, \infty)$,

$$u_{x,y}(t) \leq u_{x,y}\left(\frac{N-m}{m+1}|y|\right)$$
$$= \frac{(N-m)^{N-m}(m+1)^{m+1}}{(N+1)^{N+1}}|y|^{-m-1}$$
$$\leq (N-m)^{N-m}|x|^{-m-1}, \tag{4.10}$$

which is the desired estimate of $u_{x,y}$ in this case.

Case (2) $\frac{N-m}{m+1}|y| \leq |y-x|$. In this case, we have

$$|y-x| \leq |y| + |x| \leq \frac{m+1}{N-m}|y-x| + |x|,$$

which, together with the assumption $N \in (2m+1, \infty)$, further implies that

$$|y-x| \leq \frac{N-m}{N-2m-1}|x|. \tag{4.11}$$

In addition, applying both (4.8) and the assumption $\frac{N-m}{m+1}|y| \leq |y-x|$, we find that, for any $t \in (|y-x|, \infty)$, $u'_{x,y}(t) < 0$. Therefore, $u_{x,y}(t)$ is decreasing when $t \in (|y-x|, \infty)$. From this and (4.11), it follows that, for any $t \in (|y-x|, \infty)$

$$u_{x,y}(t) \leq u_{x,y}(|y-x|) = \frac{|y-x|^{N-m}}{(|y-x|+|y|)^{N+1}}$$
$$\leq \frac{|y-x|^{N-m}}{|x|^{N+1}} \leq \frac{(N-m)^{N-m}}{(N-2m-1)^{N-m}}|x|^{-m-1}. \tag{4.12}$$

This then finishes the estimation of $u_{x,y}$ in this case.

Combining (4.7), (4.10), and (4.12), we conclude that, for any $x \in \mathbb{R}$,

$$\mathcal{M}_N \left(\frac{d^m \delta_0}{dx^m} \right)(x) \leq \sup_{t \in (0,\infty), |y-x| < t} \{u_{x,y}(t)\} \lesssim |x|^{-m-1},$$

where the implicit positive constant depends only on both m and N. By this, Definition 4.2.2 with $X := M_p^{\frac{1}{m+1}}(\mathbb{R})$, and Lemma 4.2.5 with n, p, and q therein replaced, respectively, by 1, $\frac{1}{m+1}$, and p, we further obtain

$$\left\| \frac{d^m \delta_0}{dx^m} \right\|_{HM_p^{\frac{1}{m+1}}(\mathbb{R})} = \left\| \mathcal{M}_N \left(\frac{d^m \delta_0}{dx^m} \right) \right\|_{M_p^{\frac{1}{m+1}}(\mathbb{R})} \lesssim \left\| |\cdot|^{-m-1} \right\|_{M_p^{\frac{1}{m+1}}(\mathbb{R})} < \infty,$$

which implies that $\frac{d^m \delta_0}{dx^m} \in HM_p^{\frac{1}{m+1}}(\mathbb{R})$, and hence completes the proof of Proposition 4.2.4. $\quad\square$

The following lemma is just [207, Theorem 3.4], which is vital in the proof of Theorem 4.2.1.

Lemma 4.2.6 *Let $r \in (1, \infty)$ and X be a ball quasi-Banach function space satisfy that $\mathcal{M}$ is bounded on $X^{1/r}$. Then*

(i) $X \hookrightarrow \mathcal{S}'(\mathbb{R}^n)$.

(ii) *If $f \in X$, then $f \in H_X(\mathbb{R}^n)$ and there exists a positive constant C, independent of f, such that*

$$\|f\|_{H_X(\mathbb{R}^n)} \leq C \|f\|_X.$$

(iii) *If $f \in H_X(\mathbb{R}^n)$, then there exists a locally integrable function $g \in X$ such that g represents f, which means that $f = g$ in $\mathcal{S}'(\mathbb{R}^n)$,*

$$\|f\|_{H_X(\mathbb{R}^n)} = \|g\|_{H_X(\mathbb{R}^n)},$$

and there exists a positive constant C, independent of f, such that

$$\|g\|_X \leq C \|f\|_{H_X(\mathbb{R}^n)}.$$

Remark 4.2.7 We should point out that Lemma 4.2.6 has a wide range of applications. Here we present several function spaces to which Lemma 4.2.6 can be applied.

(i) Let $p \in (1, \infty)$. Then, in this case, from Remark 1.2.31(i), we can easily deduce that the Lebesgue space $L^p(\mathbb{R}^n)$ satisfies all the assumptions of Lemma 4.2.6. Thus, Lemma 4.2.6 with $X := L^p(\mathbb{R}^n)$ holds true (see, for instance, [213, p. 91, Subsection 1.2.1]).

(ii) Let $p \in (1, \infty)$ and $\upsilon \in A_p(\mathbb{R}^n)$. Then, in this case, using Remark 1.2.31(ii), we can easily conclude that the weighted Lebesgue space $L_\upsilon^p(\mathbb{R}^n)$ satisfies all the assumptions of Lemma 4.2.6. This then implies that Lemma 4.2.6 with X replaced by $L_\upsilon^p(\mathbb{R}^n)$ holds true (see, for instance, [218, Chapter VI, Theorem 1]).

(iii) Let $\vec{p} := (p_1, \ldots, p_n) \in (1, \infty)^n$. Then, in this case, by Remark 1.2.31(iii), we can easily find that the mixed-norm Lebesgue space $L^{\vec{p}}(\mathbb{R}^n)$ satisfies all the assumptions of Lemma 4.2.6. Therefore, Lemma 4.2.6 with $X := L^{\vec{p}}(\mathbb{R}^n)$ holds true. This result is just a special case of [42, Theorem 6.1] (see also [127, Proposition 4.1]).

(iv) Let $1 < q \leq p < \infty$. Then, in this case, from Remark 1.2.31(iv), we easily deduce that the Morrey space $M_q^p(\mathbb{R}^n)$ satisfies all the assumptions of Lemma 4.2.6. This further implies that Lemma 4.2.6 with X replaced by $M_q^p(\mathbb{R}^n)$ holds true. This result was also pointed out in [139, p. 100].

(v) Let $p(\cdot) \in C^{\log}(\mathbb{R}^n)$ satisfy $p_- \in (1, \infty)$, where p_- is the same as in (1.59). Then, in this case, applying Remark 1.2.31(v), we can easily conclude that the variable Lebesgue space $L^{p(\cdot)}(\mathbb{R}^n)$ satisfies all the assumptions of Lemma 4.2.6. Therefore, Lemma 4.2.6 with $X := L^{p(\cdot)}(\mathbb{R}^n)$ holds true. This result coincides with [186, Lemma 3.1].

Now, we show Theorem 4.2.1.

Proof of Theorem 4.2.1 Let all the symbols be as in the present theorem. Then, combining the assumption $m_0(\omega) \in (-\frac{n}{p}, \infty)$ and Theorem 1.2.42, we find that the local generalized Herz space $\dot{\mathcal{K}}_{\omega,0}^{p,q}(\mathbb{R}^n)$ is a BQBF space. This implies that, to complete the proof of the present theorem, we only need to show that $\dot{\mathcal{K}}_{\omega,0}^{p,q}(\mathbb{R}^n)$ satisfies all the assumptions of Lemma 4.2.6 under the assumptions of the present theorem. Indeed, from the assumptions $p \in (1, \infty)$ and $\max\{M_0(\omega), M_\infty(\omega)\} \in (-\frac{n}{p}, \frac{n}{p'})$, we deduce that

$$\min\left\{p, \frac{n}{\max\{M_0(\omega), M_\infty(\omega)\} + n/p}\right\} \in (1, \infty).$$

Therefore, we can choose an $r \in (1, \min\{p, \frac{n}{\max\{M_0(\omega), M_\infty(\omega)\}+n/p}\})$. For this r, by Lemma 1.8.5, we conclude that, for any $f \in L_{\text{loc}}^1(\mathbb{R}^n)$,

$$\|\mathcal{M}(f)\|_{[\dot{\mathcal{K}}_{\omega,0}^{p,q}(\mathbb{R}^n)]^{1/r}} \lesssim \|f\|_{[\dot{\mathcal{K}}_{\omega,0}^{p,q}(\mathbb{R}^n)]^{1/r}},$$

which further implies that $\mathcal{M}$ is bounded on $[\dot{\mathcal{K}}_{\omega,0}^{p,q}(\mathbb{R}^n)]^{1/r}$, and hence all the assumptions of Lemma 4.2.6 are satisfied for $\dot{\mathcal{K}}_{\omega,0}^{p,q}(\mathbb{R}^n)$. This finishes the proof of Theorem 4.2.1.   □

Remark 4.2.8 We should point out that, in Theorem 4.2.1, if $\omega(t) := t^\alpha$ for any $t \in (0, \infty)$ and for any given $\alpha \in \mathbb{R}$, then Theorem 4.2.1 goes back to [175, Proposition 2.1.1(1)].

Via Theorem 4.2.1 and Remark 4.0.20(ii), we immediately obtain the following corollary which shows that, under some assumptions, $HM_{\omega,0}^{p,q}(\mathbb{R}^n) = M_{\omega,0}^{p,q}(\mathbb{R}^n)$ with equivalent norms; we omit the details.

Corollary 4.2.9 *Let $p \in (1, \infty)$, $q \in [1, \infty)$, and $\omega \in M(\mathbb{R}_+)$ satisfy*

$$-\frac{n}{p} < m_0(\omega) \leq M_0(\omega) < 0$$

and

$$-\frac{n}{p} < m_\infty(\omega) \leq M_\infty(\omega) < 0.$$

Then

(i) $M_{\omega,0}^{p,q}(\mathbb{R}^n) \hookrightarrow \mathcal{S}'(\mathbb{R}^n)$.

(ii) *If $f \in M_{\omega,0}^{p,q}(\mathbb{R}^n)$, then $f \in HM_{\omega,0}^{p,q}(\mathbb{R}^n)$ and there exists a positive constant C, independent of f, such that*

$$\|f\|_{HM_{\omega,0}^{p,q}(\mathbb{R}^n)} \leq C\|f\|_{M_{\omega,0}^{p,q}(\mathbb{R}^n)}.$$

(iii) *If $f \in HM_{\omega,0}^{p,q}(\mathbb{R}^n)$, then there exists a locally integrable function $g \in M_{\omega,0}^{p,q}(\mathbb{R}^n)$ such that g represents f, which means that $f = g$ in $\mathcal{S}'(\mathbb{R}^n)$,*

$$\|f\|_{HM_{\omega,0}^{p,q}(\mathbb{R}^n)} = \|g\|_{HM_{\omega,0}^{p,q}(\mathbb{R}^n)},$$

and there exists a positive constant C, independent of f, such that

$$\|g\|_{M_{\omega,0}^{p,q}(\mathbb{R}^n)} \leq C\|f\|_{HM_{\omega,0}^{p,q}(\mathbb{R}^n)}.$$

Applying the known result of ball quasi-Banach function spaces mentioned in Lemma 4.2.6, we also have the following conclusion which shows that $H\dot{\mathcal{K}}_\omega^{p,q}(\mathbb{R}^n) = \dot{\mathcal{K}}_\omega^{p,q}(\mathbb{R}^n)$ with equivalent quasi-norms under some reasonable and sharp assumptions.

Theorem 4.2.10 *Let $p \in (1, \infty)$, $q \in (0, \infty)$, and $\omega \in M(\mathbb{R}_+)$ satisfy*

$$-\frac{n}{p} < m_0(\omega) \leq M_0(\omega) < \frac{n}{p'}$$

and

$$-\frac{n}{p} < m_\infty(\omega) \leq M_\infty(\omega) < 0,$$

where $\frac{1}{p} + \frac{1}{p'} = 1$. Then

(i) $\dot{\mathcal{K}}_\omega^{p,q}(\mathbb{R}^n) \hookrightarrow \mathcal{S}'(\mathbb{R}^n)$.

(ii) *If $f \in \dot{\mathcal{K}}_\omega^{p,q}(\mathbb{R}^n)$, then $f \in H\dot{\mathcal{K}}_\omega^{p,q}(\mathbb{R}^n)$ and there exists a positive constant C, independent of f, such that*

$$\|f\|_{H\dot{\mathcal{K}}_\omega^{p,q}(\mathbb{R}^n)} \le C\|f\|_{\dot{\mathcal{K}}_\omega^{p,q}(\mathbb{R}^n)}.$$

(iii) *If $f \in H\dot{\mathcal{K}}_\omega^{p,q}(\mathbb{R}^n)$, then there exists a locally integrable function $g \in \dot{\mathcal{K}}_\omega^{p,q}(\mathbb{R}^n)$ such that g represents f, which means that $f = g$ in $\mathcal{S}'(\mathbb{R}^n)$,*

$$\|f\|_{H\dot{\mathcal{K}}_\omega^{p,q}(\mathbb{R}^n)} = \|g\|_{H\dot{\mathcal{K}}_\omega^{p,q}(\mathbb{R}^n)},$$

and there exists a positive constant C, independent of f, such that

$$\|g\|_{\dot{\mathcal{K}}_\omega^{p,q}(\mathbb{R}^n)} \le C\|f\|_{H\dot{\mathcal{K}}_\omega^{p,q}(\mathbb{R}^n)}.$$

Proof Let all the symbols be as in the present theorem. Then, from the assumptions of the present theorem and Theorem 1.2.44, we deduce that the global generalized Herz space $\dot{\mathcal{K}}_\omega^{p,q}(\mathbb{R}^n)$ is a BQBF space. In addition, combining the assumptions $p \in (1, \infty)$ and $\max\{M_0(\omega), M_\infty(\omega)\} \in (-\frac{n}{p}, \frac{n}{p'})$, we conclude that

$$\min\left\{p, \frac{n}{\max\{M_0(\omega), M_\infty(\omega)\} + n/p}\right\} \in (1, \infty).$$

Let $r \in (1, \min\{p, \frac{n}{\max\{M_0(\omega), M_\infty(\omega)\} + n/p}\})$. Then, by Lemma 4.1.10, we find that $\mathcal{M}$ is bounded on $[\dot{\mathcal{K}}_\omega^{p,q}(\mathbb{R}^n)]^{1/r}$. Using this and Lemma 4.2.6, we then complete the proof of Theorem 4.2.10. $\qquad\qquad\square$

Moreover, from both Theorem 4.2.10 and Remark 4.0.20(ii), we immediately deduce the following relation between the global generalized Morrey space $M_\omega^{p,q}(\mathbb{R}^n)$ and the associated Hardy space $HM_\omega^{p,q}(\mathbb{R}^n)$; we omit the details.

Corollary 4.2.11 *Let p, q, and ω be as in Corollary 4.2.9. Then*

(i) $M_\omega^{p,q}(\mathbb{R}^n) \hookrightarrow \mathcal{S}'(\mathbb{R}^n)$.

(ii) *If $f \in M_\omega^{p,q}(\mathbb{R}^n)$, then $f \in HM_\omega^{p,q}(\mathbb{R}^n)$ and there exists a positive constant C, independent of f, such that*

$$\|f\|_{HM_\omega^{p,q}(\mathbb{R}^n)} \le C\|f\|_{M_\omega^{p,q}(\mathbb{R}^n)}.$$

(iii) *If $f \in HM_\omega^{p,q}(\mathbb{R}^n)$, then there exists a locally integrable function $g \in M_\omega^{p,q}(\mathbb{R}^n)$ such that g represents f, which means that $f = g$ in $\mathcal{S}'(\mathbb{R}^n)$,*

$$\|f\|_{HM_\omega^{p,q}(\mathbb{R}^n)} = \|g\|_{HM_\omega^{p,q}(\mathbb{R}^n)},$$

and there exists a positive constant C, independent of f, such that

$$\|g\|_{M^{p,q}_\omega(\mathbb{R}^n)} \le C\|f\|_{HM^{p,q}_\omega(\mathbb{R}^n)}.$$

4.3 Atomic Characterizations

The main target of this section is to characterize generalized Herz–Hardy spaces via atoms. By the known atomic characterization of Hardy spaces associated with ball quasi-Banach function spaces, we can directly obtain the atomic characterization of the Hardy space $H\dot{\mathcal{K}}^{p,q}_{\omega,0}(\mathbb{R}^n)$. However, the proof of the atomic characterization of $H\dot{\mathcal{K}}^{p,q}_\omega(\mathbb{R}^n)$ is more complex because of the deficiency of the associate space of $\dot{\mathcal{K}}^{p,q}_\omega(\mathbb{R}^n)$ [see Remark 1.2.19(vi) for the details]. Indeed, in order to show the atomic characterization of $H\dot{\mathcal{K}}^{p,q}_\omega(\mathbb{R}^n)$, we establish an improved atomic characterization of Hardy spaces associated with ball quasi-Banach function spaces under no assumption about associate spaces. Via this conclusion and some other auxiliary lemmas about global generalized Herz spaces and block spaces, we obtain the atomic characterization of $H\dot{\mathcal{K}}^{p,q}_\omega(\mathbb{R}^n)$ and get rid of the dependence on associate spaces.

First, we establish the atomic characterization of the generalized Herz–Hardy space $H\dot{\mathcal{K}}^{p,q}_{\omega,0}(\mathbb{R}^n)$. To achieve this, we introduce the definition of $(\dot{\mathcal{K}}^{p,q}_{\omega,0}(\mathbb{R}^n), r, d)$-atoms as follows.

Definition 4.3.1 Let $p, q \in (0, \infty)$, $\omega \in M(\mathbb{R}_+)$ with $m_0(\omega) \in (-\frac{n}{p}, \infty)$, $r \in [1, \infty]$, and $d \in \mathbb{Z}_+$. Then a measurable function a on $\mathbb{R}^n$ is called a $(\dot{\mathcal{K}}^{p,q}_{\omega,0}(\mathbb{R}^n), r, d)$-*atom* if there exists a ball $B \in \mathbb{B}$ such that

(i) $\mathrm{supp}\,(a) := \{x \in \mathbb{R}^n : a(x) \ne 0\} \subset B$;

(ii) $\|a\|_{L^r(\mathbb{R}^n)} \le \dfrac{|B|^{1/r}}{\|\mathbf{1}_B\|_{\dot{\mathcal{K}}^{p,q}_{\omega,0}(\mathbb{R}^n)}}$;

(iii) for any $\alpha \in \mathbb{Z}^n_+$ with $|\alpha| \le d$,

$$\int_{\mathbb{R}^n} a(x) x^\alpha \, dx = 0.$$

Via $(\dot{\mathcal{K}}^{p,q}_{\omega,0}(\mathbb{R}^n), r, d)$-atoms, we now introduce the following generalized atomic Herz–Hardy spaces associated with local generalized Herz spaces.

Definition 4.3.2 Let $p, q \in (0, \infty)$, $\omega \in M(\mathbb{R}_+)$ with $m_0(\omega) \in (-\frac{n}{p}, \infty)$ and $m_\infty(\omega) \in (-\frac{n}{p}, \infty)$, $r \in (\max\{1, p, \frac{n}{\min\{m_0(\omega), m_\infty(\omega)\} + n/p}\}, \infty]$,

$$s \in \left(0, \min\left\{1, p, q, \frac{n}{\max\{M_0(\omega), M_\infty(\omega)\} + n/p}\right\}\right),$$

and $d \geq \lfloor n(1/s - 1) \rfloor$ be a fixed integer. Then the *generalized atomic Herz–Hardy space* $H\dot{\mathcal{K}}^{p,q,r,d,s}_{\omega,0}(\mathbb{R}^n)$ is defined to be the set of all the $f \in \mathcal{S}'(\mathbb{R}^n)$ such that there exists $\{\lambda_j\}_{j\in\mathbb{N}} \subset [0, \infty)$ and a sequence $\{a_j\}_{j\in\mathbb{N}}$ of $(\dot{\mathcal{K}}^{p,q}_{\omega,0}(\mathbb{R}^n), r, d)$-atoms supported, respectively, in the balls $\{B_j\}_{j\in\mathbb{N}} \subset \mathbb{B}$ such that

$$f = \sum_{j\in\mathbb{N}} \lambda_j a_j$$

in $\mathcal{S}'(\mathbb{R}^n)$ and

$$\left\| \left\{ \sum_{j\in\mathbb{N}} \left[\frac{\lambda_j}{\|\mathbf{1}_{B_j}\|_{\dot{\mathcal{K}}^{p,q}_{\omega,0}(\mathbb{R}^n)}} \right]^s \mathbf{1}_{B_j} \right\}^{\frac{1}{s}} \right\|_{\dot{\mathcal{K}}^{p,q}_{\omega,0}(\mathbb{R}^n)} < \infty.$$

Moreover, for any $f \in H\dot{\mathcal{K}}^{p,q,r,d,s}_{\omega,0}(\mathbb{R}^n)$, let

$$\|f\|_{H\dot{\mathcal{K}}^{p,q,r,d,s}_{\omega,0}(\mathbb{R}^n)} := \inf \left\{ \left\| \left\{ \sum_{j\in\mathbb{N}} \left[\frac{\lambda_j}{\|\mathbf{1}_{B_j}\|_{\dot{\mathcal{K}}^{p,q}_{\omega,0}(\mathbb{R}^n)}} \right]^s \mathbf{1}_{B_j} \right\}^{\frac{1}{s}} \right\|_{\dot{\mathcal{K}}^{p,q}_{\omega,0}(\mathbb{R}^n)} \right\},$$

where the infimum is taken over all the decompositions of f as above.

Then we have the following atomic characterization of the generalized Herz–Hardy space $H\dot{\mathcal{K}}^{p,q}_{\omega,0}(\mathbb{R}^n)$.

Theorem 4.3.3 *Let p, q, ω, d, s, and r be as in Definition 4.3.2. Then*

$$H\dot{\mathcal{K}}^{p,q}_{\omega,0}(\mathbb{R}^n) = H\dot{\mathcal{K}}^{p,q,r,d,s}_{\omega,0}(\mathbb{R}^n)$$

with equivalent quasi-norms.

To show this theorem, we first recall the definition of atoms of general ball quasi-Banach function spaces as follows, which is just [207, Definition 3.5].

Definition 4.3.4 Let X be a ball quasi-Banach function space, $r \in [1, \infty]$, and $d \in \mathbb{Z}_+$. Then a measurable function a on $\mathbb{R}^n$ is called an (X, r, d)-*atom* if there exists a ball $B \in \mathbb{B}$ such that

(i) $\operatorname{supp}(a) := \{x \in \mathbb{R}^n : a(x) \neq 0\} \subset B$;

(ii) $\|a\|_{L^r(\mathbb{R}^n)} \leq \frac{|B|^{1/r}}{\|\mathbf{1}_B\|_X}$;

(iii) for any $\alpha \in \mathbb{Z}^n_+$ with $|\alpha| \leq d$,

$$\int_{\mathbb{R}^n} a(x) x^\alpha \, dx = 0.$$

The following technical lemma about atoms can be concluded by both Definition 4.3.4 and the Hölder inequality immediately; we omit the details.

Lemma 4.3.5 *Let X be a ball quasi-Banach function space, $r, t \in [1, \infty]$ with $r < t$, and $d \in \mathbb{Z}_+$. Assume a is an (X, t, d)-atom supported in the ball $B \in \mathbb{B}$. Then a is an (X, r, d)-atom supported in B.*

Note that Sawano et al. [207] established the atomic characterization of Hardy spaces associated with ball quasi-Banach function spaces. Indeed, the following conclusion is just the atomic reconstruction theorem from [207, Theorem 3.6] (see also [265, Proposition 4.2]). This conclusion plays a key role in the proof of the atomic characterization of $H\dot{\mathcal{K}}_{\omega,\mathbf{0}}^{p,q}(\mathbb{R}^n)$.

Lemma 4.3.6 *Let X be a ball quasi-Banach function space satisfying Assumption 1.2.29 with some $\theta, s \in (0, 1]$, $d \geq \lfloor n(1/\theta - 1) \rfloor$ be a fixed integer, and $r \in (1, \infty]$. Assume that $X^{1/s}$ is a ball Banach function space and there exists a positive constant C such that, for any $f \in L^1_{\mathrm{loc}}(\mathbb{R}^n)$,*

$$\left\| \mathcal{M}^{((r/s)')}(f) \right\|_{(X^{1/s})'} \leq C \, \|f\|_{(X^{1/s})'}. \tag{4.13}$$

Let $\{a_j\}_{j \in \mathbb{N}}$ be a sequence of (X, r, d)-atoms supported, respectively, in the balls $\{B_j\}_{j \in \mathbb{N}} \subset \mathbb{B}$, and $\{\lambda_j\}_{j \in \mathbb{N}} \subset [0, \infty)$ be such that

$$\left\| \left[\sum_{j \in \mathbb{N}} \left(\frac{\lambda_j}{\|\mathbf{1}_{B_j}\|_X} \right)^s \mathbf{1}_{B_j} \right]^{\frac{1}{s}} \right\|_X < \infty.$$

Then $f := \sum_{j \in \mathbb{N}} \lambda_j a_j$ converges in $\mathcal{S}'(\mathbb{R}^n)$, $f \in H_X(\mathbb{R}^n)$, and

$$\|f\|_{H_X(\mathbb{R}^n)} \lesssim \left\| \left[\sum_{j \in \mathbb{N}} \left(\frac{\lambda_j}{\|\mathbf{1}_{B_j}\|_X} \right)^s \mathbf{1}_{B_j} \right]^{\frac{1}{s}} \right\|_X,$$

where the implicit positive constant is independent of f.

Remark 4.3.7 Let X, $\{\lambda_j\}_{j \in \mathbb{N}}$, and $\{a_j\}_{j \in \mathbb{N}}$ be the same as in Lemma 4.3.6. If X has an absolutely continuous quasi-norm, then, by Sawano et al. [207, Corollary 3.11(ii)], we find that the summation $\sum_{j \in \mathbb{N}} \lambda_j a_j$ converges in $H_X(\mathbb{R}^n)$ (see also Lemma 4.8.20 below). Otherwise, the convergence of the summation $\sum_{j \in \mathbb{N}} \lambda_j a_j$ is only known to be in $\mathcal{S}'(\mathbb{R}^n)$ and the convergence in other sense is hard to judge because the (quasi-)norm of X does not have an explicit expression.

Moreover, the following atomic decomposition theorem of $H_X(\mathbb{R}^n)$ was obtained in [207, Theorem 3.7], which is also a vital tool in the proof of the atomic characterization of $H\dot{\mathcal{K}}_{\omega,\mathbf{0}}^{p,q}(\mathbb{R}^n)$.

Lemma 4.3.8 *Let X be a ball quasi-Banach function space satisfying Assumption 1.2.29 with some $\theta, s \in (0, 1]$, and $d \geq \lfloor n(1/\theta - 1) \rfloor$ be a fixed integer. Then, for any $f \in H_X(\mathbb{R}^n)$, there exists $\{\lambda_j\}_{j\in\mathbb{N}} \subset [0, \infty)$ and a sequence $\{a_j\}_{j\in\mathbb{N}}$ of (X, ∞, d)-atoms supported, respectively, in the balls $\{B_j\}_{j\in\mathbb{N}} \subset \mathbb{B}$ such that $f = \sum_{j\in\mathbb{N}} \lambda_j a_j$ in $\mathcal{S}'(\mathbb{R}^n)$ and*

$$\left\| \left[\sum_{j\in\mathbb{N}} \left(\frac{\lambda_j}{\|\mathbf{1}_{B_j}\|_X} \right)^s \mathbf{1}_{B_j} \right]^{\frac{1}{s}} \right\|_X \lesssim \|f\|_{H_X(\mathbb{R}^n)},$$

where the implicit positive constant is independent of f.

Remark 4.3.9 We point out that both Lemmas 4.3.6 and 4.3.8 have a wide range of applications. Here we give several function spaces to which Lemmas 4.3.6 and 4.3.8 can be applied.

(i) Let $\vec{p} := (p_1, \ldots, p_n) \in (0, \infty)^n$,

$$d \geq \left\lfloor n \left(\frac{1}{\min\{p_1, \ldots, p_n\}} - 1 \right) \right\rfloor$$

be a fixed nonnegative integer, $r \in (\max\{1, p_1, \ldots, p_n\}, \infty]$, and

$$s \in (0, \min\{1, p_1, \ldots, p_n\}).$$

Then, in this case, combining both Remarks 1.2.31(iii) and 1.2.34(iii), we can easily find that the mixed-norm Lebesgue space $L^{\vec{p}}(\mathbb{R}^n)$ satisfies all the assumptions of both Lemmas 4.3.6 and 4.3.8. Thus, both Lemmas 4.3.6 and 4.3.8 with $X := L^{\vec{p}}(\mathbb{R}^n)$ hold true. This result is just the atomic characterization of mixed-norm Hardy spaces established in [123, Theorem 3.16] (see also [127, Theorem 4.3]).

(ii) Let $0 < q \leq p < \infty$,

$$d \geq \left\lfloor n \left(\frac{1}{q} - 1 \right) \right\rfloor$$

be a fixed nonnegative integer, $r \in (\max\{1, p\}, \infty]$, and $s \in (0, \min\{1, q\})$. Then, in this case, from both Remarks 1.2.31(iv) and 1.2.34(iv), we easily infer that the Morrey space $M_q^p(\mathbb{R}^n)$ satisfies all the assumptions of both Lemmas 4.3.6 and 4.3.8. This then implies that both Lemmas 4.3.6 and 4.3.8 with $X := M_q^p(\mathbb{R}^n)$ hold true. This result goes back to the atomic characterization of Hardy–Morrey spaces obtained in [107, Theorems 5.1 and 5.2].

(iii) Let $p(\cdot) \in C^{\log}(\mathbb{R}^n)$ satisfy $0 < p_- \leq p_+ < \infty$, where p_- and p_+ are defined, respectively, in (1.59) and (1.60). Let

$$d \geq \left\lfloor n\left(\frac{1}{p_-} - 1\right)\right\rfloor$$

be a fixed nonnegative integer, $r \in (\max\{1, p_+\}, \infty]$, and $s := \min\{1, p_-\}$. Then, in this case, applying both Remarks 1.2.31(v) and 1.2.34(v), we can easily conclude that the variable Lebesgue space $L^{p(\cdot)}(\mathbb{R}^n)$ satisfies all the assumptions of both Lemmas 4.3.6 and 4.3.8. Therefore, both Lemmas 4.3.6 and 4.3.8 with $X := L^{p(\cdot)}(\mathbb{R}^n)$ hold true. This result coincides with the atomic characterization of variable Hardy spaces established in [186, Theorem 4.6].

In order to prove Theorem 4.3.3, we also require two auxiliary lemmas about the Fefferman–Stein vector-valued inequality on local generalized Herz spaces as follows.

Lemma 4.3.10 *Let $p, q \in (0, \infty)$ and $\omega \in M(\mathbb{R}_+)$ satisfy $m_0(\omega) \in (-\frac{n}{p}, \infty)$ and $m_\infty(\omega) \in (-\frac{n}{p}, \infty)$. Then, for any given $u \in (1, \infty)$ and*

$$r \in \left(0, \min\left\{p, \frac{n}{\max\{M_0(\omega), M_\infty(\omega)\} + n/p}\right\}\right),$$

there exists a positive constant C such that, for any $\{f_j\}_{j \in \mathbb{N}} \subset L^1_{\mathrm{loc}}(\mathbb{R}^n)$,

$$\left\| \left\{\sum_{j \in \mathbb{N}} [\mathcal{M}(f_j)]^u\right\}^{\frac{1}{u}} \right\|_{[\dot{\mathcal{K}}^{p,q}_{\omega,0}(\mathbb{R}^n)]^{1/r}} \leq C \left\| \left\{\sum_{j \in \mathbb{N}} |f_j|^u\right\}^{\frac{1}{u}} \right\|_{[\dot{\mathcal{K}}^{p,q}_{\omega,0}(\mathbb{R}^n)]^{1/r}}.$$

Proof Let all the symbols be as in the present lemma. Then, from (1.132) and (1.133), it follows that

$$\max\left\{M_0(\omega^r), M_\infty(\omega^r)\right\} < \frac{n}{(p/r)'}$$

and

$$\min\left\{m_0(\omega^r), m_\infty(\omega^r)\right\} > -\frac{n}{p/r}.$$

By this, the assumption $\frac{p}{r} \in (1, \infty)$, Theorem 1.6.1, and Lemma 1.3.1, we conclude that, for any $\{f_j\}_{j\in\mathbb{N}} \subset L^1_{\mathrm{loc}}(\mathbb{R}^n)$,

$$\left\| \left\{ \sum_{j\in\mathbb{N}} [\mathcal{M}(f_j)]^u \right\}^{\frac{1}{u}} \right\|_{[\dot{\mathcal{K}}^{p,q}_{\omega,0}(\mathbb{R}^n)]^{1/r}} \lesssim \left\| \left\{ \sum_{j\in\mathbb{N}} |f_j|^u \right\}^{\frac{1}{u}} \right\|_{[\dot{\mathcal{K}}^{p,q}_{\omega,0}(\mathbb{R}^n)]^{1/r}},$$

which completes the proof of Lemma 4.3.10.   □

Lemma 4.3.11 *Let* $p, q \in (0, \infty)$, $\omega \in M(\mathbb{R}_+)$ *satisfy* $m_0(\omega) \in (-\frac{n}{p}, \infty)$ *and* $m_\infty(\omega) \in (-\frac{n}{p}, \infty)$, $s \in (0, \infty)$, *and*

$$\theta \in \left(0, \min\left\{ s, p, \frac{n}{\max\{M_0(\omega), M_\infty(\omega)\} + n/p} \right\} \right).$$

Then there exists a positive constant C *such that, for any* $\{f_j\}_{j\in\mathbb{N}} \subset L^1_{\mathrm{loc}}(\mathbb{R}^n)$,

$$\left\| \left\{ \sum_{j\in\mathbb{N}} \left[\mathcal{M}^{(\theta)}(f_j) \right]^s \right\}^{1/s} \right\|_{\dot{\mathcal{K}}^{p,q}_{\omega,0}(\mathbb{R}^n)} \leq C \left\| \left(\sum_{j\in\mathbb{N}} |f_j|^s \right)^{1/s} \right\|_{\dot{\mathcal{K}}^{p,q}_{\omega,0}(\mathbb{R}^n)}.$$

Proof Let all the symbols be as in the present lemma. By the assumption $\theta \in (0, s)$, we find that $\frac{s}{\theta} \in (1, \infty)$. Then, using Lemma 4.3.10 with $u = \frac{s}{\theta}$ and $r = \frac{1}{\theta}$, we conclude that, for any $\{f_j\}_{j\subset\mathbb{N}} \subset L^1_{\mathrm{loc}}(\mathbb{R}^n)$,

$$\left\| \left\{ \sum_{j\in\mathbb{N}} [\mathcal{M}(f_j)]^{s/\theta} \right\}^{\theta/s} \right\|_{[\dot{\mathcal{K}}^{p,q}_{\omega,0}(\mathbb{R}^n)]^{1/\theta}} \lesssim \left\| \left(\sum_{j\in\mathbb{N}} |f_j|^{s/\theta} \right)^{\theta/s} \right\|_{[\dot{\mathcal{K}}^{p,q}_{\omega,0}(\mathbb{R}^n)]^{1/\theta}},$$

which, combined with Remark 1.2.30(ii) with X therein replaced by $\dot{\mathcal{K}}^{p,q}_{\omega,0}(\mathbb{R}^n)$, further implies that

$$\left\| \left\{ \sum_{j\in\mathbb{N}} \left[\mathcal{M}^{(\theta)}(f_j) \right]^s \right\}^{1/s} \right\|_{\dot{\mathcal{K}}^{p,q}_{\omega,0}(\mathbb{R}^n)} \lesssim \left\| \left(\sum_{j\in\mathbb{N}} |f_j|^s \right)^{1/s} \right\|_{\dot{\mathcal{K}}^{p,q}_{\omega,0}(\mathbb{R}^n)}.$$

This finishes the proof of Lemma 4.3.11.   □

Via above lemmas, we now show the atomic characterization of the generalized Herz–Hardy space $H\dot{\mathcal{K}}^{p,q}_{\omega,0}(\mathbb{R}^n)$.

Proof of Theorem 4.3.3 Let p, q, ω, d, s, and r be as in the present theorem. By the definition of $\lfloor n(1/s - 1) \rfloor$, we find that

$$\left\lfloor n\left(\frac{1}{s} - 1\right)\right\rfloor \leq n\left(\frac{1}{s} - 1\right) < \left\lfloor n\left(\frac{1}{s} - 1\right)\right\rfloor + 1.$$

Then we can choose a $\theta \in (0, s)$ such that

$$\left\lfloor n\left(\frac{1}{s} - 1\right)\right\rfloor \leq n\left(\frac{1}{\theta} - 1\right) < \left\lfloor n\left(\frac{1}{s} - 1\right)\right\rfloor + 1,$$

which further implies that $d \geq \lfloor n(1/s - 1)\rfloor = \lfloor n(1/\theta - 1)\rfloor$. Now, we claim that $\dot{\mathcal{K}}^{p,q}_{\omega,\mathbf{0}}(\mathbb{R}^n)$ satisfies Assumption 1.2.29 for this θ and s, namely, $\dot{\mathcal{K}}^{p,q}_{\omega,\mathbf{0}}(\mathbb{R}^n)$ is a BQBF space and, for any $\{f_j\}_{j\in\mathbb{N}} \subset L^1_{\mathrm{loc}}(\mathbb{R}^n)$,

$$\left\|\left\{\sum_{j\in\mathbb{N}}\left[\mathcal{M}^{(\theta)}(f_j)\right]^s\right\}^{1/s}\right\|_{\dot{\mathcal{K}}^{p,q}_{\omega,\mathbf{0}}(\mathbb{R}^n)} \lesssim \left\|\left(\sum_{j\in\mathbb{N}}|f_j|^s\right)^{1/s}\right\|_{\dot{\mathcal{K}}^{p,q}_{\omega,\mathbf{0}}(\mathbb{R}^n)}. \tag{4.14}$$

Indeed, from the assumption $m_0(\omega) \in (-\frac{n}{p}, \infty)$ and Theorem 1.2.42, it follows that the local generalized Herz space $\dot{\mathcal{K}}^{p,q}_{\omega,\mathbf{0}}(\mathbb{R}^n)$ is a BQBF space. In addition, using Lemma 4.3.11, we conclude that (4.14) holds true. This finishes the proof of the above claim.

Next, we show that $H\dot{\mathcal{K}}^{p,q,r,d,s}_{\omega,\mathbf{0}}(\mathbb{R}^n) \subset H\dot{\mathcal{K}}^{p,q}_{\omega,\mathbf{0}}(\mathbb{R}^n)$. To this end, we first prove that, under the assumptions of the present theorem, $\dot{\mathcal{K}}^{p,q}_{\omega,\mathbf{0}}(\mathbb{R}^n)$ satisfies all the assumptions of Lemma 4.3.6. Indeed, applying Lemma 1.8.6, we find that $[\dot{\mathcal{K}}^{p,q}_{\omega,\mathbf{0}}(\mathbb{R}^n)]^{1/s}$ is a BBF space and, for any $f \in L^1_{\mathrm{loc}}(\mathbb{R}^n)$,

$$\left\|\mathcal{M}^{((r/s)')}(f)\right\|_{([\dot{\mathcal{K}}^{p,q}_{\omega,\mathbf{0}}(\mathbb{R}^n)]^{1/s})'} \lesssim \|f\|_{([\dot{\mathcal{K}}^{p,q}_{\omega,\mathbf{0}}(\mathbb{R}^n)]^{1/s})'}.$$

Combining this and the above claim, we conclude that all the assumptions of Lemma 4.3.6 hold true for $\dot{\mathcal{K}}^{p,q}_{\omega,\mathbf{0}}(\mathbb{R}^n)$. Therefore, for any sequence $\{a_j\}_{j\in\mathbb{N}}$ of $(\dot{\mathcal{K}}^{p,q}_{\omega,\mathbf{0}}(\mathbb{R}^n), r, d)$-atoms supported, respectively, in the balls $\{B_j\}_{j\in\mathbb{N}} \subset \mathbb{B}$, and sequence $\{\lambda_j\}_{j\in\mathbb{N}} \subset [0, \infty)$ such that both $f := \sum_{j\in\mathbb{N}} \lambda_j a_j$ in $\mathcal{S}'(\mathbb{R}^n)$ and

$$\left\|\left\{\sum_{j\in\mathbb{N}}\left[\frac{\lambda_j}{\|\mathbf{1}_{B_j}\|_{\dot{\mathcal{K}}^{p,q}_{\omega,\mathbf{0}}(\mathbb{R}^n)}}\right]^s \mathbf{1}_{B_j}\right\}^{\frac{1}{s}}\right\|_{\dot{\mathcal{K}}^{p,q}_{\omega,\mathbf{0}}(\mathbb{R}^n)} < \infty,$$

it holds true that $f \in H\dot{\mathcal{K}}^{p,q}_{\omega,0}(\mathbb{R}^n)$ and

$$\|f\|_{H\dot{\mathcal{K}}^{p,q}_{\omega,0}(\mathbb{R}^n)} \lesssim \left\| \left\{ \sum_{j\in\mathbb{N}} \left[\frac{\lambda_j}{\|\mathbf{1}_{B_j}\|_{\dot{\mathcal{K}}^{p,q}_{\omega,0}(\mathbb{R}^n)}} \right]^s \mathbf{1}_{B_j} \right\}^{\frac{1}{s}} \right\|_{\dot{\mathcal{K}}^{p,q}_{\omega,0}(\mathbb{R}^n)}$$

with the implicit positive constant independent of f. From this and Definition 4.3.2, we deduce that $H\dot{\mathcal{K}}^{p,q,r,d,s}_{\omega,0}(\mathbb{R}^n) \subset H\dot{\mathcal{K}}^{p,q}_{\omega,0}(\mathbb{R}^n)$ and, for any $f \in H\dot{\mathcal{K}}^{p,q,r,d,s}_{\omega,0}(\mathbb{R}^n)$,

$$\|f\|_{H\dot{\mathcal{K}}^{p,q}_{\omega,0}(\mathbb{R}^n)} \lesssim \|f\|_{H\dot{\mathcal{K}}^{p,q,r,d,s}_{\omega,0}(\mathbb{R}^n)}, \tag{4.15}$$

where the implicit positive constant is independent of f.

We now prove that $H\dot{\mathcal{K}}^{p,q}_{\omega,0}(\mathbb{R}^n) \subset H\dot{\mathcal{K}}^{p,q,r,d,s}_{\omega,0}(\mathbb{R}^n)$. For this purpose, let $f \in H\dot{\mathcal{K}}^{p,q}_{\omega,0}(\mathbb{R}^n)$. Notice that the local generalized Herz space $\dot{\mathcal{K}}^{p,q}_{\omega,0}(\mathbb{R}^n)$ satisfies Assumptions 1.2.29 for the above θ and s. By this, the assumption $d \geq \lfloor n(1/s - 1) \rfloor = \lfloor n(1/\theta - 1) \rfloor$, and Lemma 4.3.8, we find that there exists $\{\lambda_j\}_{j\in\mathbb{N}} \subset [0,\infty)$ and a sequence $\{a_j\}_{j\in\mathbb{N}}$ of $(\dot{\mathcal{K}}^{p,q}_{\omega,0}(\mathbb{R}^n), \infty, d)$-atoms supported, respectively, in the balls $\{B_j\}_{j\in\mathbb{N}} \subset \mathbb{B}$ such that

$$f = \sum_{j\in\mathbb{N}} \lambda_j a_j \tag{4.16}$$

in $\mathcal{S}'(\mathbb{R}^n)$ and

$$\left\| \left\{ \sum_{j\in\mathbb{N}} \left[\frac{\lambda_j}{\|\mathbf{1}_{B_j}\|_{\dot{\mathcal{K}}^{p,q}_{\omega,0}(\mathbb{R}^n)}} \right]^s \mathbf{1}_{B_j} \right\}^{\frac{1}{s}} \right\|_{\dot{\mathcal{K}}^{p,q}_{\omega,0}(\mathbb{R}^n)} \lesssim \|f\|_{H\dot{\mathcal{K}}^{p,q}_{\omega,0}(\mathbb{R}^n)}, \tag{4.17}$$

where the implicit positive constant is independent of f. In addition, using Lemma 4.3.5 with X, t, and a therein replaced, respectively, by $\dot{\mathcal{K}}^{p,q}_{\omega,0}(\mathbb{R}^n)$, ∞, and a_j, we conclude that, for any $j \in \mathbb{N}$, a_j is a $(\dot{\mathcal{K}}^{p,q}_{\omega,0}(\mathbb{R}^n), r, d)$-atom supported in the ball B_j. From this, (4.16), (4.17), and Definition 4.3.2, it follows that $f \in H\dot{\mathcal{K}}^{p,q,r,d,s}_{\omega,0}(\mathbb{R}^n)$ and

$$\|f\|_{H\dot{\mathcal{K}}^{p,q,r,d,s}_{\omega,0}(\mathbb{R}^n)}$$

$$\leq \left\| \left\{ \sum_{j\in\mathbb{N}} \left[\frac{\lambda_j}{\|\mathbf{1}_{B_j}\|_{\dot{\mathcal{K}}^{p,q}_{\omega,0}(\mathbb{R}^n)}} \right]^s \mathbf{1}_{B_j} \right\}^{\frac{1}{s}} \right\|_{\dot{\mathcal{K}}^{p,q}_{\omega,0}(\mathbb{R}^n)} \lesssim \|f\|_{H\dot{\mathcal{K}}^{p,q}_{\omega,0}(\mathbb{R}^n)}, \tag{4.18}$$

where the implicit positive constant is independent of f. Thus, we have $H\dot{\mathcal{K}}_\omega^{p,q}(\mathbb{R}^n) \subset H\dot{\mathcal{K}}_\omega^{p,q,r,s}(\mathbb{R}^n)$. This further implies that

$$H\dot{\mathcal{K}}_{\omega,\mathbf{0}}^{p,q}(\mathbb{R}^n) = H\dot{\mathcal{K}}_{\omega,\mathbf{0}}^{p,q,r,s}(\mathbb{R}^n).$$

Moreover, combining (4.15) and (4.18), we conclude that, for any $f \in H\dot{\mathcal{K}}_{\omega,\mathbf{0}}^{p,q}(\mathbb{R}^n)$,

$$\|f\|_{H\dot{\mathcal{K}}_{\omega,\mathbf{0}}^{p,q}(\mathbb{R}^n)} \sim \|f\|_{H\dot{\mathcal{K}}_{\omega,\mathbf{0}}^{p,q,r,s}(\mathbb{R}^n)}$$

with the positive equivalence constants independent of f, which completes the proof of Theorem 4.3.3. $\qquad\square$

As an application, we next establish the atomic characterization of the generalized Hardy–Morrey space $HM_{\omega,\mathbf{0}}^{p,q}(\mathbb{R}^n)$. We first introduce the $(M_{\omega,\mathbf{0}}^{p,q}(\mathbb{R}^n), r, d)$-atoms as follows.

Definition 4.3.12 Let $p, q \in [1, \infty)$, $\omega \in M(\mathbb{R}_+)$ with $M_\infty(\omega) \in (-\infty, 0)$ and

$$-\frac{n}{p} < m_0(\omega) \le M_0(\omega) < 0,$$

$r \in [1, \infty]$, and $d \in \mathbb{Z}_+$. A measurable function a is called an $(M_{\omega,\mathbf{0}}^{p,q}(\mathbb{R}^n), r, d)$-atom if there exists a ball $B \in \mathbb{B}$ such that

(i) supp $(a) := \{x \in \mathbb{R}^n : a(x) \neq 0\} \subset B$;
(ii) $\|a\|_{L^r(\mathbb{R}^n)} \le \dfrac{|B|^{1/r}}{\|\mathbf{1}_B\|_{M_{\omega,\mathbf{0}}^{p,q}(\mathbb{R}^n)}}$;
(iii) for any $\alpha \in \mathbb{Z}_+^n$ with $|\alpha| \le d$,

$$\int_{\mathbb{R}^n} a(x)x^\alpha \, dx = 0.$$

Then, using Theorem 4.3.3 and Remarks 1.2.2(iv) and 4.0.20(ii), we immediately obtain the following atomic characterization of the generalized Hardy–Morrey space $HM_{\omega,\mathbf{0}}^{p,q}(\mathbb{R}^n)$; we omit the details.

Corollary 4.3.13 Let $p, q \in [1, \infty)$, $\omega \in M(\mathbb{R}_+)$ with

$$-\frac{n}{p} < m_0(\omega) \le M_0(\omega) < 0$$

and

$$-\frac{n}{p} < m_\infty(\omega) \le M_\infty(\omega) < 0,$$

$s \in (0, 1)$, $d \geq \lfloor n(1/s - 1) \rfloor$ *be a fixed integer, and* $r \in (\frac{n}{\min\{m_0(\omega),\, m_\infty(\omega)\}+n/p},\, \infty]$. *Then the generalized atomic Hardy–Morrey space* $HM_{\omega,0}^{p,q,r,d,s}(\mathbb{R}^n)$, *associated with the local generalized Morrey space* $M_{\omega,0}^{p,q}(\mathbb{R}^n)$, *is defined to be the set of all the* $f \in \mathcal{S}'(\mathbb{R}^n)$ *such that there exists a sequence* $\{a_j\}_{j\in\mathbb{N}}$ *of* $(M_{\omega,0}^{p,q}(\mathbb{R}^n),\, r,\, d)$- *atoms supported, respectively, in the balls* $\{B_j\}_{j\in\mathbb{N}} \subset \mathbb{B}$, *and a sequence* $\{\lambda_j\}_{j\in\mathbb{N}} \subset [0, \infty)$ *such that*

$$f = \sum_{j\in\mathbb{N}} \lambda_j a_j$$

in $\mathcal{S}'(\mathbb{R}^n)$ *and*

$$\left\| \left\{ \sum_{j\in\mathbb{N}} \left[\frac{\lambda_j}{\|\mathbf{1}_{B_j}\|_{M_{\omega,0}^{p,q}(\mathbb{R}^n)}} \right]^s \mathbf{1}_{B_j} \right\}^{\frac{1}{s}} \right\|_{M_{\omega,0}^{p,q}(\mathbb{R}^n)} < \infty.$$

Moreover, for any $f \in HM_{\omega,0}^{p,q,r,d,s}(\mathbb{R}^n)$,

$$\|f\|_{HM_{\omega,0}^{p,q,r,d,s}(\mathbb{R}^n)} := \inf \left\{ \left\| \left\{ \sum_{j\in\mathbb{N}} \left[\frac{\lambda_j}{\|\mathbf{1}_{B_j}\|_{M_{\omega,0}^{p,q}(\mathbb{R}^n)}} \right]^s \mathbf{1}_{B_j} \right\}^{\frac{1}{s}} \right\|_{M_{\omega,0}^{p,q}(\mathbb{R}^n)} \right\},$$

where the infimum is taken over all the decompositions of f *as above. Then*

$$HM_{\omega,0}^{p,q}(\mathbb{R}^n) = HM_{\omega,0}^{p,q,r,d,s}(\mathbb{R}^n)$$

with equivalent quasi-norms.

The remainder of this section is devoted to establishing the atomic characterization of the generalized Herz–Hardy space $H\dot{\mathcal{K}}_\omega^{p,q}(\mathbb{R}^n)$. To this end, we first introduce the following definition of $(\dot{\mathcal{K}}_\omega^{p,q}(\mathbb{R}^n),\, r,\, d)$-atoms.

Definition 4.3.14 Let $p, q \in (0, \infty)$, $\omega \in M(\mathbb{R}_+)$ with $m_0(\omega) \in (\frac{n}{p}, \infty)$ and $M_\infty(\omega) \in (-\infty, 0)$, $r \in [1, \infty]$, and $d \in \mathbb{Z}_+$. Then a measurable function a on $\mathbb{R}^n$ is called a $(\dot{\mathcal{K}}_\omega^{p,q}(\mathbb{R}^n),\, r,\, d)$-atom if there exists a ball $B \in \mathbb{B}$ such that

(i) $\mathrm{supp}\,(a) := \{x \in \mathbb{R}^n : a(x) \neq 0\} \subset B$;

(ii) $\|a\|_{L^r(\mathbb{R}^n)} \leq \dfrac{|B|^{1/r}}{\|\mathbf{1}_B\|_{\dot{\mathcal{K}}_\omega^{p,q}(\mathbb{R}^n)}}$;

(iii) for any $\alpha \in \mathbb{Z}_+^n$ with $|\alpha| \leq d$,

$$\int_{\mathbb{R}^n} a(x) x^\alpha \, dx = 0.$$

Then we introduce the following concept of the generalized atomic Herz–Hardy space $H\dot{\mathcal{K}}_{\omega}^{p,q,r,d,s}(\mathbb{R}^n)$.

Definition 4.3.15 Let $p, q \in (0, \infty)$, $\omega \in M(\mathbb{R}_+)$ with $m_0(\omega) \in (-\frac{n}{p}, \infty)$ and

$$-\frac{n}{p} < m_\infty(\omega) \leq M_\infty(\omega) < 0,$$

$r \in (\max\{1, p, \frac{n}{\min\{m_0(\omega), m_\infty(\omega)\}+n/p}\}, \infty]$, $s \in (0, \min\{1, p, q, \frac{n}{\max\{M_0(\omega), M_\infty(\omega)\}+n/p}\})$, and $d \geq \lfloor n(1/s - 1) \rfloor$ be a fixed integer. Then the *generalized atomic Herz–Hardy space* $H\dot{\mathcal{K}}_{\omega}^{p,q,r,d,s}(\mathbb{R}^n)$ is defined to be the set of all the $f \in \mathcal{S}'(\mathbb{R}^n)$ such that there exists $\{\lambda_j\}_{j\in\mathbb{N}} \subset [0, \infty)$ and a sequence $\{a_j\}_{j\in\mathbb{N}}$ of $(\dot{\mathcal{K}}_{\omega}^{p,q}(\mathbb{R}^n), r, d)$-atoms supported, respectively, in the balls $\{B_j\}_{j\in\mathbb{N}} \subset \mathbb{B}$ such that

$$f = \sum_{j\in\mathbb{N}} \lambda_j a_j$$

in $\mathcal{S}'(\mathbb{R}^n)$ and

$$\left\| \left\{ \sum_{j\in\mathbb{N}} \left[\frac{\lambda_j}{\|\mathbf{1}_{B_j}\|_{\dot{\mathcal{K}}_{\omega}^{p,q}(\mathbb{R}^n)}} \right]^s \mathbf{1}_{B_j} \right\}^{\frac{1}{s}} \right\|_{\dot{\mathcal{K}}_{\omega}^{p,q}(\mathbb{R}^n)} < \infty.$$

Moreover, for any $f \in H\dot{\mathcal{K}}_{\omega}^{p,q,r,d,s}(\mathbb{R}^n)$,

$$\|f\|_{H\dot{\mathcal{K}}_{\omega}^{p,q,r,d,s}(\mathbb{R}^n)} := \inf \left\{ \left\| \left\{ \sum_{j\in\mathbb{N}} \left[\frac{\lambda_j}{\|\mathbf{1}_{B_j}\|_{\dot{\mathcal{K}}_{\omega}^{p,q}(\mathbb{R}^n)}} \right]^s \mathbf{1}_{B_j} \right\}^{\frac{1}{s}} \right\|_{\dot{\mathcal{K}}_{\omega}^{p,q}(\mathbb{R}^n)} \right\},$$

where the infimum is taken over all the decompositions of f as above.

We now state the atomic characterization of the generalized Herz–Hardy space $H\dot{\mathcal{K}}_{\omega}^{p,q}(\mathbb{R}^n)$ as follows.

Theorem 4.3.16 *Let p, q, ω, r, s, and d be as in Definition 4.3.15. Then*

$$H\dot{\mathcal{K}}_{\omega}^{p,q}(\mathbb{R}^n) = H\dot{\mathcal{K}}_{\omega}^{p,q,r,d,s}(\mathbb{R}^n)$$

with equivalent quasi-norms.

To prove this theorem, observe that the associate spaces of global generalized Herz spaces are still unknown. Thus, we can not show Theorem 4.3.16 directly via using the known atomic characterization of Hardy spaces associated with ball quasi-Banach function spaces (see Lemmas 4.3.6 and 4.3.8 above). In order to overcome

this difficulty, we first establish an improved atomic characterization of Hardy
spaces associated with ball quasi-Banach function spaces under no assumption
about associate spaces. To achieve this, we first present the following definition
of atomic Hardy spaces introduced in [207] (see also [122, Definition 3.2])

Definition 4.3.17 Let X be a ball quasi-Banach function space, $r \in (1, \infty]$,

$$0 < \theta < s \leq 1,$$

and $d \geq \lfloor n(1/\theta - 1) \rfloor$ be a fixed integer. Then the *atomic Hardy space* $H^{X,r,d,s}(\mathbb{R}^n)$,
associated with X, is defined to be the set of all the $f \in \mathcal{S}'(\mathbb{R}^n)$ such that there
exists $\{\lambda_j\}_{j\in\mathbb{N}} \subset [0, \infty)$ and a sequence $\{a_j\}_{j\in\mathbb{N}}$ of (X, r, d)-atoms supported,
respectively, in the balls $\{B_j\}_{j\in\mathbb{N}} \subset \mathbb{B}$ such that

$$f = \sum_{j\in\mathbb{N}} \lambda_j a_j$$

in $\mathcal{S}'(\mathbb{R}^n)$ and

$$\left\| \left[\sum_{j\in\mathbb{N}} \left(\frac{\lambda_j}{\|\mathbf{1}_{B_j}\|_X} \right)^s \mathbf{1}_{B_j} \right]^{\frac{1}{s}} \right\|_X < \infty.$$

Moreover, for any $f \in H^{X,r,d,s}(\mathbb{R}^n)$,

$$\|f\|_{H^{X,r,d,s}(\mathbb{R}^n)} := \inf \left\{ \left\| \left[\sum_{j\in\mathbb{N}} \left(\frac{\lambda_j}{\|\mathbf{1}_{B_j}\|_X} \right)^s \mathbf{1}_{B_j} \right]^{\frac{1}{s}} \right\|_X \right\}, \tag{4.19}$$

where the infimum is taken over all the decompositions of f as above.

Then we have the following atomic characterization of the Hardy space $H_X(\mathbb{R}^n)$
associated with the ball quasi-Banach function space X, which is valid even when
the associate space of X is not clear.

Theorem 4.3.18 *Let X be a ball quasi-Banach function space satisfying:*

(i) *there exist $0 < \theta < s \leq 1$ such that Assumption 1.2.29 holds true;*
(ii) *for s in (i), there exists a linear space $Y \subset \mathcal{M}(\mathbb{R}^n)$ equipped with a seminorm
$\| \cdot \|_Y$ such that, for any $f \in \mathcal{M}(\mathbb{R}^n)$,*

$$\|f\|_{X^{1/s}} \sim \sup \left\{ \|fg\|_{L^1(\mathbb{R}^n)} : \|g\|_Y = 1 \right\},$$

where the positive equivalence constants are independent of f;

(iii) *for s in (i) and Y in (ii), there exists an* $r \in (1, \infty]$ *and a positive constant C such that, for any* $f \in L^1_{\text{loc}}(\mathbb{R}^n)$,

$$\left\| \mathcal{M}^{((r/s)')}(f) \right\|_Y \leq C \|f\|_Y .$$

Then

$$H_X(\mathbb{R}^n) = H^{X,r,d,s}(\mathbb{R}^n)$$

with equivalent quasi-norms.

Remark 4.3.19 We should point out that Theorem 4.3.18 is an improved version of the known atomic characterization established by Sawano et al. in [207, Theorems 3.6 and 3.7]. Indeed, if $Y \equiv (X^{1/s})'$ in Theorem 4.3.18, then this theorem goes back to [207, Theorems 3.6 and 3.7].

To show this atomic characterization of $H_X(\mathbb{R}^n)$, we need some preliminary lemmas. First, from the assumption (ii) of Theorem 4.3.18, the following Hölder inequality about X can be deduced directly, which is useful in the proof of Theorem 4.3.18; we omit the details.

Lemma 4.3.20 *Let X, s, and Y be as in Theorem 4.3.18. Then there exists a positive constant C such that, for any* $f, g \in \mathscr{M}(\mathbb{R}^n)$,

$$\|fg\|_{L^1(\mathbb{R}^n)} \leq C \|f\|_{X^{1/s}} \|g\|_Y .$$

To prove Theorem 4.3.18, we also need the following estimate about convolution, which is just [90, Corollary 2.1.12].

Lemma 4.3.21 *Let* $\phi \in L^1(\mathbb{R}^n)$ *and* Φ *be a nonnegative radial decreasing function on* $\mathbb{R}^n$ *such that* $|\phi| \leq \Phi$. *If* $\Phi \in L^1(\mathbb{R}^n)$, *then, for any* $f \in L^1_{\text{loc}}(\mathbb{R}^n)$ *and* $x \in \mathbb{R}^n$,

$$\sup_{t \in (0, \infty)} |f * \phi_t(x)| \leq \|\Phi\|_{L^1(\mathbb{R}^n)} \mathcal{M}(f)(x),$$

where $\mathcal{M}$ *is the Hardy–Littlewood maximal operator defined as in* (1.54).

Via the above estimate, we now establish a technical estimate about (X, r, d)-atoms as follows, which is an essential tool in the proof of Theorem 4.3.18.

Lemma 4.3.22 *Let X be a ball quasi-Banach function space,* $r \in (1, \infty]$, $\theta \in (0, 1]$, *and* $d \geq n(1/\theta - 1) - 1$ *be a nonnegative integer. Assume that* $\phi \in \mathcal{S}(\mathbb{R}^n)$ *satisfies* $\text{supp}\,(\phi) \subset B(\mathbf{0}, 1)$. *Then there exists a positive constant C such that, for* *any (X, r, d)-atom a supported in the ball B,*

$$M(a, \phi) \leq C \left[\mathcal{M}(a) \mathbf{1}_{2B} + \frac{1}{\|\mathbf{1}_B\|_X} \mathcal{M}^{(\theta)}(\mathbf{1}_B) \mathbf{1}_{(2B)^{\complement}} \right],$$

where M denotes the radial maximal function as in Definition 4.1.1(i).

Proof Let all the symbols be as in the present lemma. Then, using Lemma 4.3.21 with $\Phi := \|\phi\|_{L^\infty(\mathbb{R}^n)} \mathbf{1}_{B(\mathbf{0}, 1)}$ and $f := a$, we find that, for any $x \in \mathbb{R}^n$,

$$M(a, \phi)(x) = \sup_{t \in (0, \infty)} |a * \phi_t(x)| \lesssim \mathcal{M}(a)(x), \tag{4.20}$$

where the implicit positive constant is independent of both a and x.

In addition, let $x \in (2B)^{\complement}$ and $t \in (0, \infty)$. Then, from the assumption that $\operatorname{supp}(\phi) \subset B(\mathbf{0}, 1)$, it follows that $\operatorname{supp}(\phi_t(x - \cdot)) \subset B(x, t)$. Therefore, we have

$$|a * \phi_t(x)| = \left| \int_{B \cap B(x, t)} a(y) \phi_t(x - y) \, dy \right| = 0 \tag{4.21}$$

when $B \cap B(x, t) = \emptyset$. On the other hand, when $B \cap B(x, t) \neq \emptyset$, assume $B := B(x_B, r_B)$ with $x_B \in \mathbb{R}^n$ and $r_B \in (0, \infty)$. Then we claim that $|x - x_B| < t + r_B$. Otherwise, assume that $|x - x_B| \geq t + r_B$. Applying this, we find that, for any $y \in B(x, t)$,

$$|y - x_B| \geq |x - x_B| - |y - x| > r_B,$$

which implies that $B(x, t) \subset [B(x_B, r_B)]^{\complement}$ and hence contradicts to the assumption $B(x_B, r_B) \cap B(x, t) \neq \emptyset$. Thus, we obtain $|x - x_B| < t + r_B$ and finish the proof of the above claim. Combining this and the assumption $x \in (2B)^{\complement}$, we further conclude that

$$t > |x - x_B| - r_B \geq \frac{|x - x_B|}{2}.$$

From this, Definition 4.3.4(iii), and the Taylor remainder theorem, we deduce that, for any $y \in B = B(x_B, r_B)$, there exists a $t_y \in (0, 1)$ such that

$$|a * \phi_t(x)|$$

$$= \left| \int_{\mathbb{R}^n} a(y) \left[\phi_t(x - y) - \sum_{\substack{\gamma \in \mathbb{Z}_+^n \\ |\gamma| \leq d}} \frac{\partial^\gamma \phi_t(x - x_B)}{\gamma!} (y - x_B)^\gamma \right] dy \right|$$

$$
= \left| \int_{B(x_B, r_B)} a(y) \left[\sum_{\substack{\gamma \in \mathbb{Z}_+^n \\ |\gamma| = d+1}} \frac{\partial^\gamma \phi_t(x - t_y y - (1 - t_y)x_B)}{\gamma!} (y - x_B)^\gamma \right] dy \right|
$$

$$
\lesssim t^{-n-d-1} \int_{B(x_B, r_B)} |a(y)| |y - x_B|^{d+1} \, dy
$$

$$
\lesssim \frac{r_B^{d+1}}{|x - x_B|^{n+d+1}} \|a\|_{L^1(\mathbb{R}^n)}. \tag{4.22}
$$

Next, we claim that

$$
B\left(t_0 x + (1 - t_0)x_B, \frac{r_B}{2}\right) \subset [B(x_B, r_B) \cap B(x, |x - x_B|)],
$$

where $t_0 := \frac{r_B}{2|x - x_B|}$. Indeed, for any $y \in B(t_0 x + (1 - t_0)x_B, \frac{r_B}{2})$, we have

$$
|y - x_B| \leq |y - t_0 x - (1 - t_0)x_B| + t_0 |x - x_B| < \frac{r_B}{2} + \frac{r_B}{2} = r_B
$$

and

$$
|y - x| \leq |y - t_0 x - (1 - t_0)x_B| + (1 - t_0)|x - x_B|
$$
$$
< \frac{r_B}{2} + |x - x_B| - \frac{r_B}{2} = |x - x_B|.
$$

These imply $y \in B(x_B, r_B) \cap B(x, |x - x_B|)$ and finish the proof of the above claim. By this, (4.22), the Hölder inequality, Definition 4.3.4(ii), and the assumptions $x \in (2B)^\complement$ and $d \geq n(1/\theta - 1) - 1$, we conclude that

$$
|a * \phi_t(x)|
$$

$$
\lesssim \frac{r_B^{d+1}}{|x - x_B|^{n+d+1}} |B(x_B, r_B)|^{1 - \frac{1}{r}} \|a\|_{L^r(\mathbb{R}^n)}
$$

$$
\lesssim \frac{r_B^{d+1}}{|x - x_B|^{n+d+1}} \frac{|B(x_B, r_B)|}{\|\mathbf{1}_{B(x_B, r_B)}\|_X}
$$

$$
\sim \frac{1}{\|\mathbf{1}_{B(x_B, r_B)}\|_X} \left(\frac{r_B}{|x - x_B|}\right)^{n+d+1}
$$

$$
\lesssim \frac{1}{\|\mathbf{1}_{B(x_B, r_B)}\|_X} \left(\frac{r_B}{|x - x_B|}\right)^{\frac{n}{\theta}}
$$

$$\sim \frac{1}{\|\mathbf{1}_{B(x_B, r_B)}\|_X} \left[\frac{|B(t_0 x + (1 - t_0)x_B, \frac{r_B}{2})|}{|B(x, |x - x_B|)|} \right]^{\frac{1}{\theta}}$$

$$\lesssim \frac{1}{\|\mathbf{1}_{B(x_B, r_B)}\|_X} \left[\frac{1}{|B(x, |x - x_B|)|} \int_{B(x, |x - x_B|)} \left| \mathbf{1}_{B(x_B, r_B)}(y) \right|^{\theta} dy \right]^{\frac{1}{\theta}}$$

$$\lesssim \frac{1}{\|\mathbf{1}_B\|_X} \mathcal{M}^{(\theta)}(\mathbf{1}_B)(x). \tag{4.23}$$

This, combined with (4.20) and (4.21), further implies that, for any $x \in \mathbb{R}^n$,

$$M(a, \phi)(x) \lesssim \mathcal{M}(a)(x)\mathbf{1}_{2B}(x) + \frac{1}{\|\mathbf{1}_B\|_X} \mathcal{M}^{(\theta)}(x)\mathbf{1}_{(2B)^\complement}(x)$$

with the implicit positive constant independent of both a and x, which completes the proof of Lemma 4.3.22. $\square$

We now show Theorem 4.3.18.

Proof of Theorem 4.3.18 Let all the symbols be as in the present theorem. We first show

$$H^{X,r,d,s}(\mathbb{R}^n) \subset H_X(\mathbb{R}^n).$$

To this end, let $f \in H^{X,r,d,s}(\mathbb{R}^n)$ satisfy $f = \sum_{j \in \mathbb{N}} \lambda_j a_j$ in $\mathcal{S}'(\mathbb{R}^n)$ and $\phi \in \mathcal{S}(\mathbb{R}^n)$ satisfy $\mathrm{supp}\,(\phi) \subset B(\mathbf{0}, 1)$ and

$$\int_{\mathbb{R}^n} \phi(y)\, dy \neq 0,$$

where $\{\lambda_j\}_{j \in \mathbb{N}} \subset [0, \infty)$ and, for any $j \in \mathbb{N}$, a_j is a (X, r, d)-atom supported in the ball $B_j \in \mathbb{B}$. Then we find that, for any $t \in (0, \infty)$ and $x \in \mathbb{R}^n$,

$$|f * \phi_t(x)| = |\langle f, \phi_t(x - \cdot)\rangle| = \left| \sum_{j \in \mathbb{N}} \lambda_j \langle a_j, \phi_t(x - \cdot)\rangle \right|$$

$$= \left| \sum_{j \in \mathbb{N}} \lambda_j a_j * \phi_t(x) \right| \leq \sum_{j \in \mathbb{N}} \lambda_j |a_j * \phi_t(x)|. \tag{4.24}$$

This implies that

$$M(f, \phi) \leq \sum_{j \in \mathbb{N}} \lambda_j M(a_j, \phi). \tag{4.25}$$

In addition, by the definition of $\lfloor n(1/\theta - 1)\rfloor$, we conclude that

$$d > n\left(\frac{1}{\theta} - 1\right) - 1.$$

From this, (4.25), and Lemma 4.3.22 with a therein replaced by a_j, we deduce that

$$
\|M(f, \phi)\|_X \leq \left\| \sum_{j\in\mathbb{N}} \lambda_j M(a_j, \phi) \right\|_X
$$
$$
\lesssim \left\| \sum_{j\in\mathbb{N}} \lambda_j \mathcal{M}(a_j) \mathbf{1}_{2B_j} \right\|_X + \left\| \sum_{j\in\mathbb{N}} \frac{\lambda_j}{\|\mathbf{1}_{B_j}\|_X} \mathcal{M}^{(\theta)}\left(\mathbf{1}_{B_j}\right) \right\|_X
$$
$$
=: \mathrm{II}_1 + \mathrm{II}_2. \tag{4.26}
$$

Then we deal with II_1 and II_2, respectively. For II_1, let $g \in \mathcal{M}(\mathbb{R}^n)$ satisfy $\|g\|_Y = 1$. Then, using the Tonelli theorem, the Hölder inequality, the boundedness of the Hardy–Littlewood maximal operator $\mathcal{M}$ on $L^r(\mathbb{R}^n)$, and Definition 4.3.14(ii), we find that

$$
\int_{\mathbb{R}^n} \sum_{j\in\mathbb{N}} \lambda_j^s \left[\mathcal{M}(a_j)(y)\right]^s \mathbf{1}_{2B_j}(y) g(y)\, dy
$$
$$
\leq \sum_{j\in\mathbb{N}} \lambda_j^s \left\| \left[\mathcal{M}(a_j)\right]^s \right\|_{L^{r/s}(\mathbb{R}^n)} \left\| g\mathbf{1}_{2B_j} \right\|_{L^{(r/s)'}(\mathbb{R}^n)}
$$
$$
= \sum_{j\in\mathbb{N}} \lambda_j^s \left\| \mathcal{M}(a_j) \right\|_{L^r(\mathbb{R}^n)}^s \left\| g\mathbf{1}_{2B_j} \right\|_{L^{(r/s)'}(\mathbb{R}^n)}
$$
$$
\lesssim \sum_{j\in\mathbb{N}} \lambda_j^s \left\| a_j \right\|_{L^r(\mathbb{R}^n)}^s \left\| g\mathbf{1}_{2B_j} \right\|_{L^{(r/s)'}(\mathbb{R}^n)}
$$
$$
\lesssim \sum_{j\in\mathbb{N}} \lambda_j^s \frac{|B_j|^{\frac{s}{r}}}{\|\mathbf{1}_{B_j}\|_X^s} \left\| g\mathbf{1}_{2B_j} \right\|_{L^{(r/s)'}(\mathbb{R}^n)}. \tag{4.27}
$$

In addition, for any $j \in \mathbb{N}$ and $x \in B_j$, we have

$$
\mathcal{M}^{((r/s)')}(g)(x) \geq \left[\frac{1}{|2B_j|} \int_{2B_j} |g(y)|^{(r/s)'}\, dy \right]^{1/(r/s)'}
$$
$$
\sim |B_j|^{-1/(r/s)'} \left\| g\mathbf{1}_{2B_j} \right\|_{L^{(r/s)'}(\mathbb{R}^n)}.
$$

By this, (4.27), the Tonelli theorem, Lemma 4.3.20, and Definition 4.3.17(iii) together with the assumption $\|g\|_Y = 1$, we conclude that

$$
\int_{\mathbb{R}^n} \sum_{j \in \mathbb{N}} \lambda_j^s \left[\mathcal{M}(a_j)(y) \right]^s \mathbf{1}_{2B_j}(y) g(y) \, dy
$$

$$
\lesssim \sum_{j \in \mathbb{N}} \frac{\lambda_j^s}{\|\mathbf{1}_{B_j}\|_X^s} \int_{B_j} |B_j|^{-1/(r/s)'} \left\| g \mathbf{1}_{2B_j} \right\|_{L^{(r/s)'}(\mathbb{R}^n)} \, dx
$$

$$
\lesssim \sum_{j \in \mathbb{N}} \frac{\lambda_j^s}{\|\mathbf{1}_{B_j}\|_X^s} \int_{B_j} \mathcal{M}^{((r/s)')}(g)(x) \, dx
$$

$$
\sim \int_{\mathbb{R}^n} \sum_{j \in \mathbb{N}} \left(\frac{\lambda_j}{\|\mathbf{1}_{B_j}\|_X} \right)^s \mathbf{1}_{B_j}(x) \mathcal{M}^{((r/s)')}(g)(x) \, dx
$$

$$
\lesssim \left\| \sum_{j \in \mathbb{N}} \left(\frac{\lambda_j}{\|\mathbf{1}_{B_j}\|_X} \right)^s \mathbf{1}_{B_j} \right\|_{X^{1/s}} \left\| \mathcal{M}^{((r/s)')}(g) \right\|_Y
$$

$$
\lesssim \left\| \left[\sum_{j \in \mathbb{N}} \left(\frac{\lambda_j}{\|\mathbf{1}_{B_j}\|_X} \right)^s \mathbf{1}_{B_j} \right]^{\frac{1}{s}} \right\|_X^s .
$$

From this and Definition 4.3.17(ii), it follows that

$$
\left\| \sum_{j \in \mathbb{N}} \lambda_j^s \left[\mathcal{M}(a_j) \right]^s \mathbf{1}_{2B_j} \right\|_{X^{1/s}} \lesssim \left\| \left[\sum_{j \in \mathbb{N}} \left(\frac{\lambda_j}{\|\mathbf{1}_{B_j}\|_X} \right)^s \mathbf{1}_{B_j} \right]^{\frac{1}{s}} \right\|_X^s .
$$

This, together with Lemma 1.2.11 with r and $\{a_j\}_{j \in \mathbb{N}}$ therein replaced, respectively, by s and $\{\lambda_j \mathcal{M}(a_j) \mathbf{1}_{2B_j}\}_{j \in \mathbb{N}}$, further implies that

$$
\mathrm{II}_1 \lesssim \left\| \left\{ \sum_{j \in \mathbb{N}} \lambda_j^s \left[\mathcal{M}(a_j) \right]^s \mathbf{1}_{2B_j} \right\}^{\frac{1}{s}} \right\|_X
$$

$$
\sim \left\| \sum_{j \in \mathbb{N}} \lambda_j^s \left[\mathcal{M}(a_j) \right]^s \mathbf{1}_{2B_j} \right\|_{X^{1/s}}^{\frac{1}{s}}
$$

$$\lesssim \left\| \left[\sum_{j \in \mathbb{N}} \left(\frac{\lambda_j}{\|\mathbf{1}_{B_j}\|_X} \right)^s \mathbf{1}_{B_j} \right]^{\frac{1}{s}} \right\|_X, \tag{4.28}$$

which completes the estimation of II_1.

Next, we deal with II_2. Indeed, applying Definition 4.3.17(i), we find that Assumption 1.2.29 holds true for X, namely, for any $\{f_j\}_{j \in \mathbb{N}} \subset L^1_{\mathrm{loc}}(\mathbb{R}^n)$,

$$\left\| \left\{ \sum_{j \in \mathbb{N}} \left[\mathcal{M}^{(\theta)}(f_j) \right]^s \right\}^{1/s} \right\|_X \lesssim \left\| \left(\sum_{j \in \mathbb{N}} |f_j|^s \right)^{1/s} \right\|_X. \tag{4.29}$$

From this and Lemma 1.2.11 again, we deduce that

$$\mathrm{II}_2 \lesssim \left\| \left\{ \sum_{j \in \mathbb{N}} \left[\frac{\lambda_j}{\|\mathbf{1}_{B_j}\|_X} \mathcal{M}^{(\theta)}(\mathbf{1}_{B_j}) \right]^s \right\}^{\frac{1}{s}} \right\|_X$$

$$\sim \left\| \left\{ \sum_{j \in \mathbb{N}} \left[\mathcal{M}^{(\theta)} \left(\frac{\lambda_j}{\|\mathbf{1}_{B_j}\|_X} \mathbf{1}_{B_j} \right) \right]^s \right\}^{\frac{1}{s}} \right\|_X$$

$$\lesssim \left\| \left[\sum_{j \in \mathbb{N}} \left(\frac{\lambda_j}{\|\mathbf{1}_{B_j}\|_X} \right)^s \mathbf{1}_{B_j} \right]^{\frac{1}{s}} \right\|_X. \tag{4.30}$$

This finishes the estimation of II_2. Combining Lemma 4.1.4, (4.26), (4.28), (4.30), Definition 4.3.17, and the choice of $\{\lambda_j\}_{j \in \mathbb{N}}$, we further conclude that

$$\|f\|_{H_X(\mathbb{R}^n)} \sim \|M(f, \phi)\|_X \lesssim \|f\|_{H^{X,r,d,s}(\mathbb{R}^n)} < \infty, \tag{4.31}$$

where the implicit positive constants are independent of f. This implies that $f \in H_X(\mathbb{R}^n)$ and hence $H^{X,r,d,s}(\mathbb{R}^n) \subset H_X(\mathbb{R}^n)$.

Conversely, we prove that

$$H_X(\mathbb{R}^n) \subset H^{X,r,d,s}(\mathbb{R}^n).$$

To achieve this, let $f \in H_X(\mathbb{R}^n)$. Notice that X is a BQBF space. From this, (4.29), the assumption $d \geq \lfloor n(1/\theta - 1) \rfloor$, and Lemma 4.3.8, it follows that there exists $\{\lambda_j\}_{j \in \mathbb{N}} \subset [0, \infty)$ and a sequence $\{a_j\}_{j \in \mathbb{N}}$ of (X, ∞, d)-atoms supported, respectively, in the balls $\{B_j\}_{j\in\mathbb{N}} \subset \mathbb{B}$ such that

$$f = \sum_{j\in\mathbb{N}} \lambda_j a_j \tag{4.32}$$

in $\mathcal{S}'(\mathbb{R}^n)$ and

$$\left\| \left[\sum_{j\in\mathbb{N}} \left(\frac{\lambda_j}{\|\mathbf{1}_{B_j}\|_X} \right)^s \mathbf{1}_{B_j} \right]^{\frac{1}{s}} \right\|_X \lesssim \|f\|_{H_X(\mathbb{R}^n)} \tag{4.33}$$

with the implicit positive constant independent of f. In addition, applying Lemma 4.3.5 with $t = \infty$ and $a = a_j$ for any $j \in \mathbb{N}$, we conclude that, for any $j \in \mathbb{N}$, a_j is a (X, r, d)-atom supported in the ball B_j. This, together with (4.32), (4.33), and Definition 4.3.17, further implies that $f \in H^{X,r,d,s}(\mathbb{R}^n)$ and

$$\|f\|_{H^{X,r,d,s}(\mathbb{R}^n)} \le \left\| \left[\sum_{j\in\mathbb{N}} \left(\frac{\lambda_j}{\|\mathbf{1}_{B_j}\|_X} \right)^s \mathbf{1}_{B_j} \right]^{\frac{1}{s}} \right\|_X \lesssim \|f\|_{H_X(\mathbb{R}^n)}, \tag{4.34}$$

where the implicit positive constant is independent of f. Therefore, we have

$$H_X(\mathbb{R}^n) \subset H^{X,r,d,s}(\mathbb{R}^n).$$

This further implies that $H_X(\mathbb{R}^n) = H^{X,r,d,s}(\mathbb{R}^n)$. Moreover, by (4.31) and (4.34), we find that, for any $f \in H_X(\mathbb{R}^n)$,

$$\|f\|_{H_X(\mathbb{R}^n)} \sim \|f\|_{H^{X,r,d,s}(\mathbb{R}^n)}$$

with the positive equivalence constants independent of f, which completes the proof of Theorem 4.3.18. $\square$

Now, in order to establish the atomic characterization of the generalized Herz–Hardy space $H\dot{\mathcal{K}}_{\omega}^{p,q}(\mathbb{R}^n)$, we only need to prove that all the assumptions of Theorem 4.3.18 are satisfied for the Herz space $\dot{\mathcal{K}}_{\omega}^{p,q}(\mathbb{R}^n)$ under consideration. To this end, we first show several auxiliary lemmas about block spaces and global generalized Herz spaces. The following one shows the boundedness of powered Hardy–Littlewood maximal operators on block spaces.

Lemma 4.3.23 *Let $p, q \in (0, \infty)$ and $\omega \in M(\mathbb{R}_+)$ satisfy $m_0(\omega) \in (-\frac{n}{p}, \infty)$ and $m_\infty(\omega) \in (-\frac{n}{p}, \infty)$. Then, for any given $s \in (0, \min\{p, q, \frac{n}{\max\{M_0(\omega), M_\infty(\omega)\}+n/p}\})$ and*

$$r \in \left(\max\left\{ p, \frac{n}{\min\{m_0(\omega), m_\infty(\omega)\} + n/p} \right\}, \infty \right],$$

there exists a positive constant C such that, for any $f \in L^1_{\mathrm{loc}}(\mathbb{R}^n)$,

$$\left\| \mathcal{M}^{((r/s)')}(f) \right\|_{\dot{\mathcal{B}}^{(p/s)',(q/s)'}_{1/\omega^s}(\mathbb{R}^n)} \leq C \|f\|_{\dot{\mathcal{B}}^{(p/s)',(q/s)'}_{1/\omega^s}(\mathbb{R}^n)}.$$

Proof Let all the symbols be as in the present lemma. Then, from Lemma 1.1.6 and the assumption

$$s \in \left(0, \frac{n}{\max\{M_0(\omega), M_\infty(\omega)\} + n/p} \right),$$

it follows that

$$\min\left\{ m_0\left(\frac{1}{\omega^s}\right), m_\infty\left(\frac{1}{\omega^s}\right) \right\}$$
$$= -s \max\{M_0(\omega), M_\infty(\omega)\} > -s\left(\frac{n}{s} - \frac{n}{p}\right) = -\frac{n}{(p/s)'}. \tag{4.35}$$

In addition, using Lemma 1.1.6 and the assumption

$$r \in \left(\frac{n}{\min\{m_0(\omega), m_\infty(\omega)\} + n/p}, \infty \right),$$

we conclude that

$$\max\left\{ M_0\left(\frac{1}{\omega^s}\right), M_\infty\left(\frac{1}{\omega^s}\right) \right\}$$
$$= -s \min\{m_0(\omega), m_\infty(\omega)\} < -s\left(\frac{n}{r} - \frac{n}{p}\right) = \frac{n}{(r/s)'} - \frac{n}{(p/s)'}. \tag{4.36}$$

By the assumptions $r \in (p, \infty)$ and $s \in (0, p)$, we find that

$$\left(\frac{r}{s}\right)' \in \left(1, \left(\frac{p}{s}\right)'\right).$$

Applying this, (4.35), (4.36), and Corollary 2.3.5 with p, q, ω, and r therein replaced, respectively, by $(p/s)'$, $(q/s)'$, $1/\omega^s$, and $(r/s)'$, we conclude that, for any $f \in L^1_{\mathrm{loc}}(\mathbb{R}^n)$,

$$\left\| \mathcal{M}^{((r/s)')}(f) \right\|_{\dot{\mathcal{B}}^{(p/s)',(q/s)'}_{1/\omega^s}(\mathbb{R}^n)} \lesssim \|f\|_{\dot{\mathcal{B}}^{(p/s)',(q/s)'}_{1/\omega^s}(\mathbb{R}^n)},$$

which completes the proof of Lemma 4.3.23. $\square$

We also need the following two technical lemmas about the Fefferman–Stein vector-valued inequality on global generalized Herz spaces.

Lemma 4.3.24 *Let* $p, q \in (0, \infty)$ *and* $\omega \in M(\mathbb{R}_+)$ *satisfy* $m_0(\omega) \in (-\frac{n}{p}, \infty)$ *and* $m_\infty(\omega) \in (-\frac{n}{p}, \infty)$. *Then, for any given* $u \in (1, \infty)$ *and*

$$r \in \left(0, \min\left\{ p, \frac{n}{\max\{M_0(\omega), M_\infty(\omega)\} + n/p} \right\} \right),$$

there exists a positive constant C *such that, for any* $\{f_j\}_{j \in \mathbb{N}} \subset L^1_{\mathrm{loc}}(\mathbb{R}^n)$,

$$\left\| \left\{ \sum_{j \in \mathbb{N}} [\mathcal{M}(f_j)]^u \right\}^{\frac{1}{u}} \right\|_{[\dot{\mathcal{K}}^{p,q}_\omega(\mathbb{R}^n)]^{1/r}} \leq C \left\| \left(\sum_{j \in \mathbb{N}} |f_j|^u \right)^{\frac{1}{u}} \right\|_{[\dot{\mathcal{K}}^{p,q}_\omega(\mathbb{R}^n)]^{1/r}}.$$

Proof Let all the symbols be as in the present lemma. Then, by (1.132) and (1.133), we find that

$$\max\left\{ M_0(\omega^r), M_\infty(\omega^r) \right\} < \frac{n}{(p/r)'}$$

and

$$\min\left\{ m_0(\omega^r), m_\infty(\omega^r) \right\} > -\frac{n}{p/r}.$$

This, together with the assumption $\frac{p}{r} \in (1, \infty)$, Theorem 1.6.4, and Lemma 1.3.2, further implies that, for any $\{f_j\}_{j \in \mathbb{N}} \subset L^1_{\mathrm{loc}}(\mathbb{R}^n)$,

$$\left\| \left\{ \sum_{j \in \mathbb{N}} [\mathcal{M}(f_j)]^u \right\}^{\frac{1}{u}} \right\|_{[\dot{\mathcal{K}}^{p,q}_\omega(\mathbb{R}^n)]^{1/r}} \lesssim \left\| \left(\sum_{j \in \mathbb{N}} |f_j|^u \right)^{\frac{1}{u}} \right\|_{[\dot{\mathcal{K}}^{p,q}_\omega(\mathbb{R}^n)]^{1/r}},$$

which completes the proof of Lemma 4.3.24. $\square$

Lemma 4.3.25 *Let $p, q \in (0, \infty)$, $\omega \in M(\mathbb{R}_+)$ satisfy $m_0(\omega) \in (-\frac{n}{p}, \infty)$ and $m_\infty(\omega) \in (-\frac{n}{p}, \infty)$, $s \in (0, \infty)$, and*

$$\theta \in \left(0, \min\left\{s, p, \frac{n}{\max\{M_0(\omega), M_\infty(\omega)\} + n/p}\right\}\right).$$

Then there exists a positive constant C such that, for any $\{f_j\}_{j \in \mathbb{N}} \subset L^1_{\mathrm{loc}}(\mathbb{R}^n)$,

$$\left\| \left\{ \sum_{j \in \mathbb{N}} \left[\mathcal{M}^{(\theta)}(f_j) \right]^s \right\}^{1/s} \right\|_{\dot{\mathcal{K}}^{p,q}_\omega(\mathbb{R}^n)} \leq C \left\| \left(\sum_{j \in \mathbb{N}} |f_j|^s \right)^{1/s} \right\|_{\dot{\mathcal{K}}^{p,q}_\omega(\mathbb{R}^n)}.$$

Proof Let all the symbols be as in the present lemma. Then, from the assumption $\theta \in (0, s)$, we deduce that $\frac{s}{\theta} \in (1, \infty)$. Combining this and Lemma 4.3.24 with u and r therein replaced, respectively, by $\frac{s}{\theta}$ and $\frac{1}{\theta}$, we conclude that, for any $\{f_j\}_{j \in \mathbb{N}} \subset L^1_{\mathrm{loc}}(\mathbb{R}^n)$,

$$\left\| \left\{ \sum_{j \in \mathbb{N}} \left[\mathcal{M}(f_j) \right]^{s/\theta} \right\}^{\theta/s} \right\|_{[\dot{\mathcal{K}}^{p,q}_\omega(\mathbb{R}^n)]^{1/\theta}} \lesssim \left\| \left(\sum_{j \in \mathbb{N}} |f_j|^{s/\theta} \right)^{\theta/s} \right\|_{[\dot{\mathcal{K}}^{p,q}_\omega(\mathbb{R}^n)]^{1/\theta}}.$$

By this and Remark 1.2.30(ii) with $X = \dot{\mathcal{K}}^{p,q}_{\omega,\mathbf{0}}(\mathbb{R}^n)$, we find that, for any $\{f_j\}_{j \in \mathbb{N}} \subset L^1_{\mathrm{loc}}(\mathbb{R}^n)$,

$$\left\| \left\{ \sum_{j \in \mathbb{N}} \left[\mathcal{M}^{(\theta)}(f_j) \right]^s \right\}^{1/s} \right\|_{\dot{\mathcal{K}}^{p,q}_\omega(\mathbb{R}^n)} \lesssim \left\| \left(\sum_{j \in \mathbb{N}} |f_j|^s \right)^{1/s} \right\|_{\dot{\mathcal{K}}^{p,q}_\omega(\mathbb{R}^n)}.$$

This finishes the proof of Lemma 4.3.25.   □

Moreover, we require the following variant of the Hölder inequality of global generalized Herz spaces, which can be deduced directly from (2.25); we omit the details.

Lemma 4.3.26 *Let $p, q \in (1, \infty)$ and $\omega \in M(\mathbb{R}_+)$. Then there exists a positive constant C such that, for any $f, g \in \mathscr{M}(\mathbb{R}^n)$,*

$$\|fg\|_{L^1(\mathbb{R}^n)} \leq C \|f\|_{\dot{\mathcal{K}}^{p,q}_\omega(\mathbb{R}^n)} \|g\|_{\dot{\mathcal{B}}^{p',q'}_{1/\omega}(\mathbb{R}^n)}.$$

We point out that the following equivalent characterization of the global generalized Herz space $\dot{\mathcal{K}}^{p,q}_\omega(\mathbb{R}^n)$ is an essential tool used to overcome the deficiency of the associate space of $\dot{\mathcal{K}}_{\omega}^{p,q}(\mathbb{R}^n)$, which plays a key role throughout this book [see Remark 1.2.19(vi) for the details].

Lemma 4.3.27 *Let* $p, q \in (1, \infty)$ *and* $\omega \in M(\mathbb{R}_+)$ *satisfy* $m_0(\omega) \in (-\frac{n}{p}, \infty)$ *and* $M_\infty(\omega) \in (-\infty, 0)$. *Then a measurable function* f *belongs to the global generalized Herz space* $\dot{\mathcal{K}}_{\omega}^{p,q}(\mathbb{R}^n)$ *if and only if*

$$\|f\|^{\star}_{\dot{\mathcal{K}}_{\omega}^{p,q}(\mathbb{R}^n)} := \sup\left\{ \|fg\|_{L^1(\mathbb{R}^n)} : \|g\|_{\dot{\mathcal{B}}_{1/\omega}^{p',q'}(\mathbb{R}^n)} = 1 \right\} < \infty.$$

Moreover, there exist two positive constants C_1 *and* C_2 *such that, for any* $f \in \dot{\mathcal{K}}_{\omega}^{p,q}(\mathbb{R}^n)$,

$$C_1 \|f\|_{\dot{\mathcal{K}}_{\omega}^{p,q}(\mathbb{R}^n)} \leq \|f\|^{\star}_{\dot{\mathcal{K}}_{\omega}^{p,q}(\mathbb{R}^n)} \leq C_2 \|f\|_{\dot{\mathcal{K}}_{\omega}^{p,q}(\mathbb{R}^n)}.$$

Proof Let all the symbols be as in the present lemma. We first show the necessity. Indeed, let $f \in \dot{\mathcal{K}}_{\omega}^{p,q}(\mathbb{R}^n)$. Then, from Lemma 4.3.26, we deduce that, for any $g \in \mathcal{M}(\mathbb{R}^n)$ with $\|g\|_{\dot{\mathcal{B}}_{1/\omega}^{p',q'}(\mathbb{R}^n)} = 1$,

$$\|fg\|_{L^1(\mathbb{R}^n)} \lesssim \|f\|_{\dot{\mathcal{K}}_{\omega}^{p,q}(\mathbb{R}^n)} \|g\|_{\dot{\mathcal{B}}_{1/\omega}^{p',q'}(\mathbb{R}^n)} \sim \|f\|_{\dot{\mathcal{K}}_{\omega}^{p,q}(\mathbb{R}^n)},$$

which further implies that

$$\|f\|^{\star}_{\dot{\mathcal{K}}_{\omega}^{p,q}(\mathbb{R}^n)} \lesssim \|f\|_{\dot{\mathcal{K}}_{\omega}^{p,q}(\mathbb{R}^n)} < \infty. \tag{4.37}$$

Thus, we complete the proof of the necessity.

Conversely, we show the sufficiency. To this end, assume $f \in \mathcal{M}(\mathbb{R}^n)$ satisfying $\|f\|^{\star}_{\dot{\mathcal{K}}_{\omega}^{p,q}(\mathbb{R}^n)} < \infty$, and, for any $k \in \mathbb{N}$, let

$$f_k := \min\{|f|, k\}\mathbf{1}_{B(0,2^k)}.$$

Obviously, for any $k \in \mathbb{N}$, $|f_k| \leq k\mathbf{1}_{B(0,2^k)}$ and $0 \leq f_k \uparrow |f|$ as $k \to \infty$. In addition, applying Theorem 1.2.44, we conclude that the global generalized Herz space $\dot{\mathcal{K}}_{\omega}^{p,q}(\mathbb{R}^n)$ is a BQBF space. Therefore, for any $k \in \mathbb{N}$, we have $f_k \in \dot{\mathcal{K}}_{\omega}^{p,q}(\mathbb{R}^n)$. By this and Theorem 2.2.1 with $\dot{\mathcal{B}}_{\omega}^{p,q}(\mathbb{R}^n)$ replaced by $\dot{\mathcal{B}}_{1/\omega}^{p',q'}(\mathbb{R}^n)$, we find that, for any $k \in \mathbb{N}$,

$$\|f_k\|_{\dot{\mathcal{K}}_{\omega}^{p,q}(\mathbb{R}^n)} \sim \sup\left\{ \left| \int_{\mathbb{R}^n} f_k(y)g(y)\,dy \right| : \|g\|_{\dot{\mathcal{B}}_{1/\omega}^{p',q'}(\mathbb{R}^n)} = 1 \right\}$$

$$\lesssim \sup\left\{ \|fg\|_{L^1(\mathbb{R}^n)} : \|g\|_{\dot{\mathcal{B}}_{1/\omega}^{p',q'}(\mathbb{R}^n)} = 1 \right\} \sim \|f\|^{\star}_{\dot{\mathcal{K}}_{\omega}^{p,q}(\mathbb{R}^n)}. \tag{4.38}$$

On the other hand, from Definition 1.2.13(iii) and the fact that $0 \leq f_k \uparrow |f|$ as $k \to \infty$, it follows that

$$\lim_{k \to \infty} \|f_k\|_{\dot{\mathcal{K}}_{\omega}^{p,q}(\mathbb{R}^n)} = \|f\|_{\dot{\mathcal{K}}_{\omega}^{p,q}(\mathbb{R}^n)}.$$

Combining this and (4.38), we further find that

$$\|f\|_{\dot{\mathcal{K}}_{\omega}^{p,q}(\mathbb{R}^n)} \lesssim \|f\|^{\star}_{\dot{\mathcal{K}}_{\omega}^{p,q}(\mathbb{R}^n)} < \infty. \tag{4.39}$$

Then we conclude that $f \in \dot{\mathcal{K}}_{\omega}^{p,q}(\mathbb{R}^n)$ and hence complete the proof of the sufficiency. Moreover, by both (4.37) and (4.39), we find that

$$\|f\|_{\dot{\mathcal{K}}_{\omega}^{p,q}(\mathbb{R}^n)} \sim \|f\|^{\star}_{\dot{\mathcal{K}}_{\omega}^{p,q}(\mathbb{R}^n)},$$

where the positive equivalence constants are independent of f. This finishes the proof of Lemma 4.3.27. $\qquad \square$

Via above lemmas, we now show Theorem 4.3.16.

Proof of Theorem 4.3.16 Let p, q, ω, r, s, and d be as in the present theorem. Then, applying the assumptions $m_0(\omega) \in (-\frac{n}{p}, \infty)$ and $M_{\infty}(\omega) \in (-\infty, 0)$, and Theorem 1.2.44, we find that the global generalized Herz space $\dot{\mathcal{K}}_{\omega}^{p,q}(\mathbb{R}^n)$ under consideration is a BQBF space. Thus, by Theorem 4.3.18, we conclude that, to complete the proof of the present theorem, it suffices to show that the assumptions (i), (ii), and (iii) of Theorem 4.3.18 are satisfied for $\dot{\mathcal{K}}_{\omega}^{p,q}(\mathbb{R}^n)$.

Indeed, from the definition of $\lfloor n(1/s - 1) \rfloor$, we deduce that

$$\left\lfloor n\left(\frac{1}{s} - 1\right) \right\rfloor \leq n\left(\frac{1}{s} - 1\right) < \left\lfloor n\left(\frac{1}{s} - 1\right) \right\rfloor + 1.$$

Then we can choose a $\theta \in (0, s)$ such that

$$\left\lfloor n\left(\frac{1}{s} - 1\right) \right\rfloor \leq n\left(\frac{1}{\theta} - 1\right) < \left\lfloor n\left(\frac{1}{s} - 1\right) \right\rfloor + 1,$$

which further implies that $d \geq \lfloor n(1/s - 1) \rfloor = \lfloor n(1/\theta - 1) \rfloor$. Now, we show that Theorem 4.3.18(i) holds true for the above θ and s. Namely, for the above θ and s, $\dot{\mathcal{K}}_{\omega}^{p,q}(\mathbb{R}^n)$ satisfies Assumption 1.2.29. Indeed, applying Lemma 4.3.25, we find that, for any $\{f_j\}_{j \in \mathbb{N}} \subset L^1_{\text{loc}}(\mathbb{R}^n)$,

$$\left\| \left\{ \sum_{j \in \mathbb{N}} \left[\mathcal{M}^{(\theta)}(f_j) \right]^s \right\}^{1/s} \right\|_{\dot{\mathcal{K}}_{\omega}^{p,q}(\mathbb{R}^n)} \lesssim \left\| \left(\sum_{j \in \mathbb{N}} |f_j|^s \right)^{1/s} \right\|_{\dot{\mathcal{K}}_{\omega}^{p,q}(\mathbb{R}^n)}. \tag{4.40}$$

This implies that Assumption 1.2.29 is satisfied for $\dot{\mathcal{K}}_\omega^{p,q}(\mathbb{R}^n)$ and hence finishes the proof that Theorem 4.3.18(i) holds true.

Next, we prove that $\dot{\mathcal{K}}_\omega^{p,q}(\mathbb{R}^n)$ satisfies Theorem 4.3.18(ii). Indeed, from the assumptions $m_0(\omega) \in (-\frac{n}{p}, \infty)$ and $M_\infty(\omega) \in (-\infty, 0)$, and Lemma 1.1.6, it follows that

$$m_0\left(\omega^s\right) = sm_0(\omega) > -\frac{n}{p/s}$$

and

$$M_\infty\left(\omega^s\right) = sM_\infty(\omega) < 0.$$

These, combined with the assumptions $\frac{p}{s}, \frac{q}{s} \in (1, \infty)$ and Lemma 4.3.27 with p, q, and ω therein replaced, respectively, by $\frac{p}{s}, \frac{q}{s}$, and ω^s, further imply that, for any $f \in \mathscr{M}(\mathbb{R}^n)$,

$$\|f\|_{\dot{\mathcal{K}}_{\omega^s}^{p/s,q/s}(\mathbb{R}^n)} \sim \sup\left\{\|fg\|_{L^1(\mathbb{R}^n)} : \|g\|_{\dot{\mathcal{B}}_{1/\omega^s}^{(p/s)',(q/s)'}(\mathbb{R}^n)} = 1\right\} \tag{4.41}$$

with the positive equivalence constants independent of f. On the other hand, by Theorem 1.3.2, we find that $\dot{\mathcal{K}}_{\omega^s}^{p/s,q/s}(\mathbb{R}^n) = [\dot{\mathcal{K}}_\omega^{p,q}(\mathbb{R}^n)]^{1/s}$. Combining this and (4.41), we conclude that, for any $f \in \mathscr{M}(\mathbb{R}^n)$,

$$\|f\|_{[\dot{\mathcal{K}}_\omega^{p,q}(\mathbb{R}^n)]^{1/s}} \sim \sup\left\{\|fg\|_{L^1(\mathbb{R}^n)} : \|g\|_{\dot{\mathcal{B}}_{1/\omega^s}^{(p/s)',(q/s)'}(\mathbb{R}^n)} = 1\right\}. \tag{4.42}$$

Therefore, $\dot{\mathcal{K}}_\omega^{p,q}(\mathbb{R}^n)$ satisfies Theorem 4.3.18(ii) with $Y := \dot{\mathcal{B}}_{1/\omega^s}^{(p/s)',(q/s)'}(\mathbb{R}^n)$.

Finally, we show that Theorem 4.3.18(iii) is satisfied for $\dot{\mathcal{K}}_\omega^{p,q}(\mathbb{R}^n)$. Indeed, from Lemma 4.3.23, it follows that, for any $f \in L_{\mathrm{loc}}^1(\mathbb{R}^n)$,

$$\left\|\mathcal{M}^{((r/s)')}(f)\right\|_{\dot{\mathcal{B}}_{1/\omega^s}^{(p/s)',(q/s)'}(\mathbb{R}^n)} \lesssim \|f\|_{\dot{\mathcal{B}}_{1/\omega^s}^{(p/s)',(q/s)'}(\mathbb{R}^n)}. \tag{4.43}$$

This implies that Theorem 4.3.18(iii) holds true for $\dot{\mathcal{K}}_\omega^{p,q}(\mathbb{R}^n)$ with

$$Y := \dot{\mathcal{B}}_{1/\omega^s}^{(p/s)',(q/s)'}(\mathbb{R}^n).$$

Moreover, combining (4.40), (4.42), and (4.43), we further find that the Herz space $\dot{\mathcal{K}}_\omega^{p,q}(\mathbb{R}^n)$ under consideration satisfies all the assumptions of Theorem 4.3.18 and then complete the proof of Theorem 4.3.16. □

As an application of this atomic characterization, we give the atomic characterization of the generalized Hardy–Morrey space $HM_\omega^{p,q}(\mathbb{R}^n)$. First, we introduce the following $(M_\omega^{p,q}(\mathbb{R}^n), r, d)$-atoms.

Definition 4.3.28 Let p, q, and ω be as in Definition 4.3.12, $r \in [1, \infty]$, and $d \in \mathbb{Z}_+$. Then a measurable function a on $\mathbb{R}^n$ is called an $(M_\omega^{p,q}(\mathbb{R}^n), r, d)$-atom if there exists a ball $B \in \mathbb{B}$ such that

(i) $\operatorname{supp}(a) := \{x \in \mathbb{R}^n : a(x) \neq 0\} \subset B$;

(ii) $\|a\|_{L^r(\mathbb{R}^n)} \leq \dfrac{|B|^{1/r}}{\|\mathbf{1}_B\|_{M_\omega^{p,q}(\mathbb{R}^n)}}$;

(iii) for any $\alpha \in \mathbb{Z}_+^n$ with $|\alpha| \leq d$,

$$\int_{\mathbb{R}^n} a(x) x^\alpha \, dx = 0.$$

By Theorem 4.3.16 and Remarks 1.2.2(iv) and 4.0.20(ii), we immediately obtain the atomic characterization of the generalized Hardy–Morrey space $HM_\omega^{p,q}(\mathbb{R}^n)$ as follows; we omit the details.

Corollary 4.3.29 *Let p, q, ω, r, d, and s be as in Corollary 4.3.13. The generalized atomic Hardy–Morrey space $HM_\omega^{p,q,r,d,s}(\mathbb{R}^n)$, associated with the global generalized Morrey space $M_\omega^{p,q}(\mathbb{R}^n)$, is defined to be the set of all the $f \in \mathcal{S}'(\mathbb{R}^n)$ such that there exists a sequence $\{a_j\}_{j\in\mathbb{N}}$ of $(M_\omega^{p,q}(\mathbb{R}^n), r, d)$-atoms supported, respectively, in the balls $\{B_j\}_{j\in\mathbb{N}} \subset \mathbb{B}$, and a sequence $\{\lambda_j\}_{j\in\mathbb{N}} \subset [0, \infty)$ such that*

$$f = \sum_{j\in\mathbb{N}} \lambda_j a_j$$

in $\mathcal{S}'(\mathbb{R}^n)$ and

$$\left\| \left\{ \sum_{j\in\mathbb{N}} \left[\frac{\lambda_j}{\|\mathbf{1}_{B_j}\|_{M_\omega^{p,q}(\mathbb{R}^n)}} \right]^s \mathbf{1}_{B_j} \right\}^{\frac{1}{s}} \right\|_{M_\omega^{p,q}(\mathbb{R}^n)} < \infty.$$

Moreover, for any $f \in HM_\omega^{p,q,r,d,s}(\mathbb{R}^n)$,

$$\|f\|_{HM_\omega^{p,q,r,d,s}(\mathbb{R}^n)} := \inf \left\{ \left\| \left\{ \sum_{j\in\mathbb{N}} \left[\frac{\lambda_j}{\|\mathbf{1}_{B_j}\|_{M_\omega^{p,q}(\mathbb{R}^n)}} \right]^s \mathbf{1}_{B_j} \right\}^{\frac{1}{s}} \right\|_{M_\omega^{p,q}(\mathbb{R}^n)} \right\},$$

where the infimum is taken over all the decompositions of f as above. Then

$$HM_\omega^{p,q}(\mathbb{R}^n) = HM_\omega^{p,q,r,d,s}(\mathbb{R}^n)$$

with equivalent quasi-norms.

4.4 Generalized Finite Atomic Herz–Hardy Spaces

In this section, we first introduce the generalized finite atomic Herz–Hardy spaces, which are equipped with a finite atomic quasi-norm, and we then prove that, on these generalized finite atomic Herz–Hardy spaces, this finite atomic quasi-norm is equivalent to the quasi-norm of the corresponding generalized Herz–Hardy spaces.

We begin with the following concept of finite atomic Hardy spaces associated with $\dot{\mathcal{K}}^{p,q}_{\omega,0}(\mathbb{R}^n)$.

Definition 4.4.1 Let $p, q \in (0, \infty)$, $\omega \in M(\mathbb{R}_+)$ with $m_0(\omega) \in (-\frac{n}{p}, \infty)$ and $m_\infty(\omega) \in (-\frac{n}{p}, \infty)$, $r \in (\max\{1, p, \frac{n}{\min\{m_0(\omega), m_\infty(\omega)\}+n/p}\}, \infty]$,

$$
s \in \left(0, \min\left\{1, p, q, \frac{n}{\max\{M_0(\omega), M_\infty(\omega)\} + n/p}\right\}\right),
$$

and $d \geq \lfloor n(1/s - 1) \rfloor$ be a fixed integer. Then the *generalized finite atomic Herz–Hardy space* $H\dot{\mathcal{K}}^{p,q,r,s}_{\omega,0,\mathrm{fin}}(\mathbb{R}^n)$, associated with $\dot{\mathcal{K}}^{p,q}_{\omega,0}(\mathbb{R}^n)$, is defined to be the set of all finite linear combinations of $(\dot{\mathcal{K}}^{p,q}_{\omega,0}(\mathbb{R}^n), r, d)$-atoms. Moreover, for any $f \in H\dot{\mathcal{K}}^{p,q,r,d,s}_{\omega,0,\mathrm{fin}}(\mathbb{R}^n)$,

$$
\|f\|_{H\dot{\mathcal{K}}^{p,q,r,d,s}_{\omega,0,\mathrm{fin}}(\mathbb{R}^n)} := \left\{ \inf \left\| \left\{ \sum_{j=1}^{N} \left[\frac{\lambda_j}{\|\mathbf{1}_{B_j}\|_{\dot{\mathcal{K}}^{p,q}_{\omega,0}(\mathbb{R}^n)}} \right]^s \mathbf{1}_{B_j} \right\}^{\frac{1}{s}} \right\|_{\dot{\mathcal{K}}^{p,q}_{\omega,0}(\mathbb{R}^n)} \right\},
$$

where the infimum is taken over all finite linear combinations of f, namely, $N \in \mathbb{N}$,

$$
f = \sum_{j=1}^{N} \lambda_j a_j,
$$

$\{\lambda_j\}_{j=1}^{N} \subset [0, \infty)$, and $\{a_j\}_{j=1}^{N}$ being $(\dot{\mathcal{K}}^{p,q}_{\omega,0}(\mathbb{R}^n), r, d)$-atoms supported, respectively, in the balls $\{B_j\}_{j=1}^{N} \subset \mathbb{B}$.

Based on the known finite atomic characterization of the Hardy space $H_X(\mathbb{R}^n)$ associated with the ball quasi-Banach function space X (see [266, Theorem 1.10] or Lemma 4.4.3 below), we have the following equivalence between the quasi-norms $\|\cdot\|_{H\dot{\mathcal{K}}^{p,q}_{\omega,0}(\mathbb{R}^n)}$ and $\|\cdot\|_{H\dot{\mathcal{K}}^{p,q,r,d,s}_{\omega,0,\mathrm{fin}}(\mathbb{R}^n)}$ on $H\dot{\mathcal{K}}^{p,q,r,d,s}_{\omega,0,\mathrm{fin}}(\mathbb{R}^n)$. Throughout this book, $C(\mathbb{R}^n)$ is defined to be the set of all continuous functions on $\mathbb{R}^n$.

Theorem 4.4.2 *Let p, q, ω, d, s, and r be as in Definition 4.4.1.*

(i) *If*

$$r \in \left(\max\left\{ 1, p, \frac{n}{\min\{m_0(\omega), m_\infty(\omega)\} + n/p} \right\}, \infty \right),$$

then $\| \cdot \|_{H\dot{\mathcal{K}}^{p,q,r,d,s}_{\omega,0,\mathrm{fin}}(\mathbb{R}^n)}$ and $\| \cdot \|_{H\dot{\mathcal{K}}^{p,q}_{\omega,0}(\mathbb{R}^n)}$ are equivalent quasi-norms on the generalized finite atomic Herz–Hardy space $H\dot{\mathcal{K}}^{p,q,r,d,s}_{\omega,0,\mathrm{fin}}(\mathbb{R}^n)$.

(ii) *If $r = \infty$, then $\| \cdot \|_{H\dot{\mathcal{K}}^{p,q,\infty,d,s}_{\omega,0,\mathrm{fin}}(\mathbb{R}^n)}$ and $\| \cdot \|_{H\dot{\mathcal{K}}^{p,q}_{\omega,0}(\mathbb{R}^n)}$ are equivalent quasi-norms on $H\dot{\mathcal{K}}^{p,q,\infty,d,s}_{\omega,0,\mathrm{fin}}(\mathbb{R}^n) \cap C(\mathbb{R}^n)$.*

To prove this theorem, recall that, on Hardy spaces associated with ball quasi-Banach function spaces, Yan et al. [266, Theorem 1.10] introduced the following finite atomic Hardy spaces, which are equipped with a finite atomic quasi-norm, and then showed that, on these finite atomic Hardy spaces, this finite atomic quasi-norm is equivalent to the quasi-norm of the corresponding Hardy spaces.

Lemma 4.4.3 *Let X, r, d, and s be as in Lemma 4.3.6. Then the finite atomic Hardy space $H^{X,r,d,s}_{\mathrm{fin}}(\mathbb{R}^n)$, associated with X, is defined to be the set of all finite linear combinations of (X, r, d)-atoms. Moreover, for any $f \in H^{X,r,d,s}_{\mathrm{fin}}(\mathbb{R}^n)$,*

$$\|f\|_{H^{X,r,d,s}_{\mathrm{fin}}(\mathbb{R}^n)} := \left\{ \inf \left\| \left\{ \sum_{j=1}^{N} \left(\frac{\lambda_j}{\|\mathbf{1}_{B_j}\|_X} \right)^s \mathbf{1}_{B_j} \right\}^{\frac{1}{s}} \right\|_X \right\}, \tag{4.44}$$

where the infimum is taken over all finite linear combinations of f, namely, $N \in \mathbb{N}$,

$$f = \sum_{j=1}^{N} \lambda_j a_j,$$

$\{\lambda_j\}_{j=1}^{N} \subset [0, \infty)$, and $\{a_j\}_{j=1}^{N}$ being (X, r, d)-atoms supported, respectively, in the balls $\{B_j\}_{j=1}^{N} \subset \mathbb{B}$. Then

(i) *if $r \in (1, \infty)$, $\| \cdot \|_{H^{X,r,d,s}_{\mathrm{fin}}(\mathbb{R}^n)}$ and $\| \cdot \|_{H_X(\mathbb{R}^n)}$ are equivalent quasi-norms on $H^{X,r,d,s}_{\mathrm{fin}}(\mathbb{R}^n)$;*

(ii) *if $r = \infty$, $\| \cdot \|_{H^{X,\infty,d,s}_{\mathrm{fin}}(\mathbb{R}^n)}$ and $\| \cdot \|_{H_X(\mathbb{R}^n)}$ are equivalent quasi-norms on $H^{X,\infty,d,s}_{\mathrm{fin}}(\mathbb{R}^n) \cap C(\mathbb{R}^n)$.*

Remark 4.4.4

(i) We point out that the finite atomic Hardy space $H_{\mathrm{fin}}^{X,r,d,s}(\mathbb{R}^n)$ may not be equal to the Hardy space $H_X(\mathbb{R}^n)$. Indeed, let $X := L^1(\mathbb{R})$ and, for any $x \in \mathbb{R}$, let

$$f(x) := \frac{x}{|x|(1+x^2)}.$$

Then $H_X(\mathbb{R}^n)$ goes back to the Hardy space $H^1(\mathbb{R})$ and, as was mentioned in [213, p. 178, Subsection 6.2], $f \in H^1(\mathbb{R})$. But, f does not have compact support and hence f does not belong to any finite atomic Hardy space associated with $L^1(\mathbb{R})$. However, as was showed in [207, Remark 3.12], when the ball quasi-Banach function space X has an absolutely continuous quasi-norm, then $H_{\mathrm{fin}}^{X,r,d,s}(\mathbb{R}^n)$ is dense in $H_X(\mathbb{R}^n)$.

(ii) To simplify the presentation, in what follows, if a Hardy space has properties similar to those in Lemma 4.4.3 (or Theorem 4.4.2), we simply call that this Hardy space has the *finite atomic characterization*.

Remark 4.4.5 We should point out that Lemma 4.4.3 has a wide range of applications. Here we present several function spaces to which Lemma 4.4.3 can be applied.

(i) Let $\vec{p} := (p_1, \ldots, p_n) \in (0, \infty)^n$,

$$d \geq \left\lfloor n \left(\frac{1}{\min\{p_1, \ldots, p_n\}} - 1 \right) \right\rfloor$$

be a fixed nonnegative integer, $r \in (\max\{1, p_1, \ldots, p_n\}, \infty]$, and

$$s \in (0, \min\{1, p_1, \ldots, p_n\}).$$

Then, in this case, by both Remarks 1.2.31(iii) and 1.2.34(iii), we easily conclude that the mixed-norm Lebesgue space $L^{\vec{p}}(\mathbb{R}^n)$ satisfies all the assumptions of Lemma 4.4.3. This implies that Lemma 4.4.3 with X therein replaced by $L^{\vec{p}}(\mathbb{R}^n)$ holds true. This result is just the finite atomic characterization of mixed-norm Hardy spaces showed in [123, Theorem 5.9] (see also [127, Theorem 4.5]).

(ii) Let $0 < q \leq p < \infty$,

$$d \geq \left\lfloor n \left(\frac{1}{q} - 1 \right) \right\rfloor$$

be a fixed nonnegative integer, $r \in (\max\{1, p\}, \infty]$, and $s \in (0, \min\{1, q\})$. Then, in this case, from both Remarks 1.2.31(iv) and 1.2.34(iv), we can easily deduce that the Morrey space $M_q^p(\mathbb{R}^n)$ satisfies all the assumptions of Lemma 4.4.3. This then implies that Lemma 4.4.3 with $X := M_q^p(\mathbb{R}^n)$ holds true. To the best of our knowledge, this finite atomic characterization of Hardy–Morrey spaces is totally new.

(iii) Let $p(\cdot) \in C^{\log}(\mathbb{R}^n)$ satisfy $0 < p_- \le p_+ < \infty$, where p_- and p_+ are defined, respectively, in (1.59) and (1.60). Let

$$d \ge \left\lfloor n \left(\frac{1}{p_-} - 1 \right) \right\rfloor$$

be a fixed nonnegative integer, $r \in (\max\{1, p_+\}, \infty]$, and $s := \min\{1, p_-\}$. Then, in this case, using both Remarks 1.2.31(v) and 1.2.34(v), we can easily find that the variable Lebesgue space $L^{p(\cdot)}(\mathbb{R}^n)$ satisfies all the assumptions of Lemma 4.4.3. Thus, Lemma 4.4.3 with $X := L^{p(\cdot)}(\mathbb{R}^n)$ holds true. This result is just the finite atomic characterization of variable Hardy spaces established in [156, Theorem 5.4].

Lemma 4.4.3 implies that, to show Theorem 4.4.2, we only need to prove that, under the assumptions of Theorem 4.4.2, $\dot{\mathcal{K}}^{p,q}_{\omega,\mathbf{0}}(\mathbb{R}^n)$ satisfies all the assumptions of Lemma 4.4.3. Applying this idea and some technical lemmas obtained in the last section, we now show Theorem 4.4.2.

Proof of Theorem 4.4.2 Let p, q, ω, d, s, and r be as in the present theorem. Then, using the assumption $m_0(\omega) \in (-\frac{n}{p}, \infty)$ and Theorem 1.2.42, we conclude that the local generalized Herz space $\dot{\mathcal{K}}^{p,q}_{\omega,\mathbf{0}}(\mathbb{R}^n)$ is a BQBF space. Thus, in order to finish the proof of the present theorem, it suffices to show that all the assumptions of Lemma 4.4.3 hold true for $\dot{\mathcal{K}}^{p,q}_{\omega,\mathbf{0}}(\mathbb{R}^n)$.

First, let $\theta \in (0, s)$ be such that

$$\left\lfloor n \left(\frac{1}{s} - 1 \right) \right\rfloor \le n \left(\frac{1}{\theta} - 1 \right) < \left\lfloor n \left(\frac{1}{s} - 1 \right) \right\rfloor + 1.$$

This implies that $d \ge \lfloor n(1/s - 1) \rfloor = \lfloor n(1/\theta - 1) \rfloor$. We now prove that $\dot{\mathcal{K}}^{p,q}_{\omega,\mathbf{0}}(\mathbb{R}^n)$ satisfies Assumption 1.2.29 for the above θ and s. Indeed, from Lemma 4.3.11, it follows that, for any $\{f_j\}_{j\in\mathbb{N}} \subset L^1_{\mathrm{loc}}(\mathbb{R}^n)$,

$$\left\| \left\{ \sum_{j\in\mathbb{N}} \left[\mathcal{M}^{(\theta)}(f_j) \right]^s \right\}^{1/s} \right\|_{\dot{\mathcal{K}}^{p,q}_{\omega,\mathbf{0}}(\mathbb{R}^n)} \lesssim \left\| \left(\sum_{j\in\mathbb{N}} |f_j|^s \right)^{1/s} \right\|_{\dot{\mathcal{K}}^{p,q}_{\omega,\mathbf{0}}(\mathbb{R}^n)},$$

which implies that $\dot{\mathcal{K}}^{p,q}_{\omega,\mathbf{0}}(\mathbb{R}^n)$ satisfies Assumption 1.2.29 for the above θ and s.

On the other hand, by Lemma 1.8.6, we find that $[\dot{\mathcal{K}}^{p,q}_{\omega,\mathbf{0}}(\mathbb{R}^n)]^{1/s}$ is a BBF space and, for any $f \in L^1_{\mathrm{loc}}(\mathbb{R}^n)$,

$$\left\| \mathcal{M}^{((r/s)')}(f) \right\|_{([\dot{\mathcal{K}}^{p,q}_{\omega,\mathbf{0}}(\mathbb{R}^n)]^{1/s})'} \lesssim \|f\|_{([\dot{\mathcal{K}}^{p,q}_{\omega,\mathbf{0}}(\mathbb{R}^n)]^{1/s})'}.$$

This further implies that, under the assumptions of the present theorem, $\dot{\mathcal{K}}^{p,q}_{\omega,\mathbf{0}}(\mathbb{R}^n)$ satisfies all the assumptions of Lemma 4.4.3, which completes the proof of Theorem 4.4.2.   □

Then, as an application of Theorem 4.4.2, we have the following finite atomic characterization of generalized Hardy–Morrey spaces, which is a simple corollary of Theorem 4.4.2 and Remarks 1.2.2(iv) and 4.0.20(ii); we omit the details.

Corollary 4.4.6 *Let* p, q, ω, r, d, *and* s *be as in Corollary 4.3.13. Then the generalized finite atomic Hardy–Morrey space* $HM^{p,q,r,d,s}_{\omega,\mathbf{0},\mathrm{fin}}(\mathbb{R}^n)$, *associated with the local generalized Morrey space* $M^{p,q}_{\omega,\mathbf{0}}(\mathbb{R}^n)$, *is defined to be the set of all finite linear combinations of* $(M^{p,q}_{\omega,\mathbf{0}}(\mathbb{R}^n),\ r,\ d)$-*atoms. Moreover, for any* $f \in HM^{p,q,r,d,s}_{\omega,\mathbf{0},\mathrm{fin}}(\mathbb{R}^n)$,

$$\|f\|_{HM^{p,q,r,d,s}_{\omega,\mathbf{0},\mathrm{fin}}(\mathbb{R}^n)} := \left\{ \inf \left\| \left\{ \sum_{j=1}^{N} \left[\frac{\lambda_j}{\|\mathbf{1}_{B_j}\|_{M^{p,q}_{\omega,\mathbf{0}}(\mathbb{R}^n)}} \right]^s \mathbf{1}_{B_j} \right\}^{\frac{1}{s}} \right\|_{M^{p,q}_{\omega,\mathbf{0}}(\mathbb{R}^n)} \right\},$$

where the infimum is taken over all finite linear combinations of f, *namely,* $N \in \mathbb{N}$,

$$f = \sum_{j=1}^{N} \lambda_j a_j,$$

$\{\lambda_j\}_{j=1}^{N} \subset [0, \infty)$, *and* $\{a_j\}_{j=1}^{N}$ *being* $(M^{p,q}_{\omega,\mathbf{0}}(\mathbb{R}^n),\ r,\ d)$-*atoms supported, respectively, in the balls* $\{B_j\}_{j=1}^{N} \subset \mathbb{B}$. *Then*

(i) *if*

$$r \in \left(\frac{n}{\min\{m_0(\omega), m_\infty(\omega)\} + n/p}, \infty \right),$$

$\|\cdot\|_{HM^{p,q,r,d,s}_{\omega,\mathbf{0},\mathrm{fin}}(\mathbb{R}^n)}$ *and* $\|\cdot\|_{HM^{p,q}_{\omega,\mathbf{0}}(\mathbb{R}^n)}$ *are equivalent quasi-norms on the generalized finite atomic Hardy–Morrey space* $HM^{p,q,r,d,s}_{\omega,\mathbf{0},\mathrm{fin}}(\mathbb{R}^n)$;

(ii) *if* $r = \infty$, $\|\cdot\|_{HM^{p,q,\infty,d,s}_{\omega,\mathbf{0},\mathrm{fin}}(\mathbb{R}^n)}$ *and* $\|\cdot\|_{HM^{p,q}_{\omega,\mathbf{0}}(\mathbb{R}^n)}$ *are equivalent quasi-norms on* $HM^{p,q,\infty,d,s}_{\omega,\mathbf{0},\mathrm{fin}}(\mathbb{R}^n) \cap C(\mathbb{R}^n)$.

Next, we are devoted to introducing and investigating the finite atomic Hardy space associated with the global generalized Herz space $\dot{\mathcal{K}}^{p,q}_{\omega}(\mathbb{R}^n)$. For this purpose, we first introduce the following generalized finite atomic Herz–Hardy space.

Definition 4.4.7 Let $p, q \in (0, \infty)$, $\omega \in M(\mathbb{R}_+)$ with $m_0(\omega) \in (-\frac{n}{p}, \infty)$ and

$$-\frac{n}{p} < m_\infty(\omega) \leq M_\infty(\omega) < 0,$$

$r \in (\max\{1, p, \frac{n}{\min\{m_0(\omega), m_\infty(\omega)\}+n/p}\}, \infty]$, $s \in (0, \min\{1, p, q, \frac{n}{\max\{M_0(\omega), M_\infty(\omega)\}+n/p}\})$, and $d \geq \lfloor n(1/s - 1) \rfloor$ be a fixed integer. Then the *generalized finite atomic Herz–Hardy space* $H\dot{\mathcal{K}}^{p,q,r,d,s}_{\omega,\mathrm{fin}}(\mathbb{R}^n)$, associated with $\dot{\mathcal{K}}^{p,q}_\omega(\mathbb{R}^n)$, is defined to be the set of all finite linear combinations of $(\dot{\mathcal{K}}^{p,q}_\omega(\mathbb{R}^n), r, d)$-atoms. Moreover, for any $f \in H\dot{\mathcal{K}}^{p,q,r,d,s}_{\omega,\mathrm{fin}}(\mathbb{R}^n)$,

$$\|f\|_{H\dot{\mathcal{K}}^{p,q,r,d,s}_{\omega,\mathrm{fin}}(\mathbb{R}^n)} := \left\{ \inf \left\| \left\{ \sum_{j=1}^{N} \left[\frac{\lambda_j}{\|\mathbf{1}_{B_j}\|_{\dot{\mathcal{K}}^{p,q}_\omega(\mathbb{R}^n)}} \right]^s \mathbf{1}_{B_j} \right\}^{\frac{1}{s}} \right\|_{\dot{\mathcal{K}}^{p,q}_\omega(\mathbb{R}^n)} \right\},$$

where the infimum is taken over all finite linear combinations of f, namely, $N \in \mathbb{N}$,

$$f = \sum_{j=1}^{N} \lambda_j a_j,$$

$\{\lambda_j\}_{j=1}^{N} \subset [0, \infty)$, and $\{a_j\}_{j=1}^{N}$ being $(\dot{\mathcal{K}}^{p,q}_\omega(\mathbb{R}^n), r, d)$-atoms supported, respectively, in the balls $\{B_j\}_{j=1}^{N} \subset \mathbb{B}$.

Then the following conclusion implies that the quasi-norms of both the generalized Herz–Hardy space $H\dot{\mathcal{K}}^{p,q}_\omega(\mathbb{R}^n)$ and the generalized finite atomic Herz–Hardy space $H\dot{\mathcal{K}}^{p,q,r,d,s}_{\omega,\mathrm{fin}}(\mathbb{R}^n)$ are equivalent on $H\dot{\mathcal{K}}^{p,q,r,d,s}_{\omega,\mathrm{fin}}(\mathbb{R}^n)$.

Theorem 4.4.8 *Let* $p, q, \omega, d, s,$ *and* r *be as in Definition 4.4.7.*

(i) *If*

$$r \in \left(\max\left\{1, p, \frac{n}{\min\{m_0(\omega), m_\infty(\omega)\} + n/p}\right\}, \infty \right),$$

then $\| \cdot \|_{H\dot{\mathcal{K}}^{p,q,r,d,s}_{\omega,\mathrm{fin}}(\mathbb{R}^n)}$ *and* $\| \cdot \|_{H\dot{\mathcal{K}}^{p,q}_\omega(\mathbb{R}^n)}$ *are equivalent quasi-norms on the generalized finite atomic Herz–Hardy space* $H\dot{\mathcal{K}}^{p,q,r,d,s}_{\omega,\mathrm{fin}}(\mathbb{R}^n)$.

(ii) *If* $r = \infty$*, then* $\| \cdot \|_{H\dot{\mathcal{K}}^{p,q,\infty,d,s}_{\omega,\mathrm{fin}}(\mathbb{R}^n)}$ *and* $\| \cdot \|_{H\dot{\mathcal{K}}^{p,q}_\omega(\mathbb{R}^n)}$ *are equivalent quasi-norms on* $H\dot{\mathcal{K}}^{p,q,\infty,d,s}_{\omega,\mathrm{fin}}(\mathbb{R}^n) \cap C(\mathbb{R}^n)$.

To show Theorem 4.4.8, note that the associate space of the global generalized Herz space $\dot{\mathcal{K}}^{p,q}_\omega(\mathbb{R}^n)$ is still unknown. This means that we can not prove Theorem 4.4.8 via applying Lemma 4.4.3 directly. Fortunately, by the proof of [266, Theorem 1.10], we find that, if the atomic characterization of $H_X(\mathbb{R}^n)$ holds true,

222 4 Generalized Herz–Hardy Spaces

Lemma 4.4.3 still holds true even when there is not any assumption about the associate space of X. Namely, we have the following finite atomic characterization of $H_X(\mathbb{R}^n)$.

Lemma 4.4.9 *Let X be a ball quasi-Banach function space satisfy Assumption 1.2.29 with some $\theta, s \in (0, 1]$, $r \in (1, \infty]$, and $d \geq \lfloor n(1/\theta - 1) \rfloor$ be a fixed integer such that*

$$H_X(\mathbb{R}^n) = H^{X,r,d,s}(\mathbb{R}^n)$$

with equivalent quasi-norms, where the atomic Hardy space $H^{X,r,d,s}(\mathbb{R}^n)$ is defined as in Definition 4.3.17. Then

(i) *if $r \in (1, \infty)$, $\| \cdot \|_{H^{X,r,d,s}_{\mathrm{fin}}(\mathbb{R}^n)}$ and $\| \cdot \|_{H_X(\mathbb{R}^n)}$ are equivalent quasi-norms on $H^{X,r,d,s}_{\mathrm{fin}}(\mathbb{R}^n)$;*

(ii) *if $r = \infty$, $\| \cdot \|_{H^{X,\infty,d,s}_{\mathrm{fin}}(\mathbb{R}^n)}$ and $\| \cdot \|_{H_X(\mathbb{R}^n)}$ are equivalent quasi-norms on $H^{X,\infty,d,s}_{\mathrm{fin}}(\mathbb{R}^n) \cap C(\mathbb{R}^n)$.*

Proof Let all the symbols be as in the present lemma and $f \in H^{X,r,d,s}_{\mathrm{fin}}(\mathbb{R}^n)$. Then, applying Lemma 4.1.4(ii), (4.44), (4.19), and the assumption that

$$H_X(\mathbb{R}^n) = H^{X,r,d,s}(\mathbb{R}^n)$$

with equivalent quasi-norms, we find that

$$\|f\|_{H_X(\mathbb{R}^n)} \sim \|f\|_{H^{X,r,d,s}(\mathbb{R}^n)} \lesssim \|f\|_{H^{X,r,d,s}_{\mathrm{fin}}(\mathbb{R}^n)}. \tag{4.45}$$

Conversely, using Assumption 1.2.29 and repeating the proof of [266, Theorem 1.10] with q therein replaced by r, we conclude that, if $r \in (1, \infty)$, then

$$\|f\|_{H^{X,r,d,s}_{\mathrm{fin}}(\mathbb{R}^n)} \lesssim \|f\|_{H_X(\mathbb{R}^n)}$$

and, if $r = \infty$ and $f \in C(\mathbb{R}^n)$, then

$$\|f\|_{H^{X,\infty,d,s}_{\mathrm{fin}}(\mathbb{R}^n)} \lesssim \|f\|_{H_X(\mathbb{R}^n)}.$$

These, combined with (4.45), further imply that both (i) and (ii) of the present lemma hold true. Thus, the proof of Lemma 4.4.9 is completed. $\qquad \square$

Via this finite atomic characterization of $H_X(\mathbb{R}^n)$, we now show Theorem 4.4.8.

Proof of Theorem 4.4.8 Let p, q, ω, d, s, and r be as in the present theorem. Then, applying the assumptions $m_0(\omega) \in (-\frac{n}{p}, \infty)$ and $M_\infty(\omega) \in (-\infty, 0)$, and Theorem 1.2.44, we find that the global generalized Herz space $\dot{\mathcal{K}}^{p,q}_\omega(\mathbb{R}^n)$ is a BQBF space. Therefore, to complete the proof of the present theorem, we only need to show that all the assumptions of Lemma 4.4.9 are satisfied for $\dot{\mathcal{K}}_\omega^{p,q}(\mathbb{R}^n)$.

First, we prove that there exists a $\theta \in (0, 1]$ such that $d \geq \lfloor n(1/\theta - 1) \rfloor$ and, for this θ and the s same as in the present theorem, $\dot{\mathcal{K}}_\omega^{p,q}(\mathbb{R}^n)$ satisfies Assumption 1.2.29, namely, for any $\{f_j\}_{j \in \mathbb{N}} \subset L_{\mathrm{loc}}^1(\mathbb{R}^n)$,

$$
\left\| \left\{ \sum_{j \in \mathbb{N}} \left[\mathcal{M}^{(\theta)}(f_j) \right]^s \right\}^{1/s} \right\|_{\dot{\mathcal{K}}_{\omega,0}^{p,q}(\mathbb{R}^n)} \lesssim \left\| \left(\sum_{j \in \mathbb{N}} |f_j|^s \right)^{1/s} \right\|_{\dot{\mathcal{K}}_{\omega,0}^{p,q}(\mathbb{R}^n)}. \tag{4.46}
$$

Indeed, from the definition of $\lfloor n(1/s - 1) \rfloor$, it follows that

$$
\left\lfloor n \left(\frac{1}{s} - 1 \right) \right\rfloor \leq n \left(\frac{1}{s} - 1 \right) < \left\lfloor n \left(\frac{1}{s} - 1 \right) \right\rfloor + 1.
$$

Thus, we can choose a $\theta \in (0, s)$ such that

$$
\left\lfloor n \left(\frac{1}{s} - 1 \right) \right\rfloor \leq n \left(\frac{1}{\theta} - 1 \right) < \left\lfloor n \left(\frac{1}{s} - 1 \right) \right\rfloor + 1,
$$

which further implies that $d \geq \lfloor n(1/s - 1) \rfloor = \lfloor n(1/\theta - 1) \rfloor$. On the other hand, by Lemma 4.3.25, we conclude that (4.46) holds true. This implies that Assumption 1.2.29 holds true for this θ and s.

Next, using Theorem 4.3.16, we find that

$$
H\dot{\mathcal{K}}_\omega^{p,q}(\mathbb{R}^n) = H\dot{\mathcal{K}}_\omega^{p,q,r,d,s}(\mathbb{R}^n)
$$

with equivalent quasi-norms. This further implies that $\dot{\mathcal{K}}_\omega^{p,q}(\mathbb{R}^n)$ satisfies all the assumptions of Lemma 4.4.9, which completes the proof of Theorem 4.4.8. $\quad\square$

Combining Theorem 4.4.8 and Remarks 1.2.2(iv) and 4.0.20(ii), we immediately obtain the following finite atomic characterization of the generalized Hardy–Morrey space $HM_\omega^{p,q}(\mathbb{R}^n)$; we omit the details.

Corollary 4.4.10 *Let* p, q, ω, r, d, *and* s *be as in Corollary 4.3.13. Then the generalized finite atomic Hardy–Morrey space* $HM_{\omega,\mathrm{fin}}^{p,q,r,d,s}(\mathbb{R}^n)$, *associated with the global generalized Morrey space* $M_\omega^{p,q}(\mathbb{R}^n)$, *is defined to be the set of all finite linear combinations of* $(M_\omega^{p,q}(\mathbb{R}^n), r, d)$-*atoms. Moreover, for any* $f \in HM_{\omega,\mathrm{fin}}^{p,q,r,d,s}(\mathbb{R}^n)$,

$$
\|f\|_{HM_{\omega,\mathrm{fin}}^{p,q,r,d,s}(\mathbb{R}^n)} := \left\{ \inf \left\| \left\{ \sum_{j=1}^N \left[\frac{\lambda_j}{\|\mathbf{1}_{B_j}\|_{M_\omega^{p,q}(\mathbb{R}^n)}} \right]^s \mathbf{1}_{B_j} \right\}^{\frac{1}{s}} \right\|_{M_\omega^{p,q}(\mathbb{R}^n)} \right\},
$$

where the infimum is taken over all finite linear combinations of f, namely, $N \in \mathbb{N}$,

$$f = \sum_{j=1}^{N} \lambda_j a_j,$$

$\{\lambda_j\}_{j=1}^{N} \subset [0, \infty)$, *and* $\{a_j\}_{j=1}^{N}$ *being* $(M_\omega^{p,q}(\mathbb{R}^n), r, d)$-*atoms supported, respectively, in the balls* $\{B_j\}_{j=1}^{N} \subset \mathbb{B}$. *Then*

(i) *if*

$$r \in \left(\frac{n}{\min\{m_0(\omega), m_\infty(\omega)\} + n/p}, \infty \right),$$

then $\| \cdot \|_{HM_{\omega,\mathrm{fin}}^{p,q,r,d,s}(\mathbb{R}^n)}$ *and* $\| \cdot \|_{HM_\omega^{p,q}(\mathbb{R}^n)}$ *are equivalent quasi-norms on the generalized finite atomic Hardy–Morrey space* $HM_{\omega,\mathrm{fin}}^{p,q,r,d,s}(\mathbb{R}^n)$;

(ii) *if* $r = \infty$, *then* $\| \cdot \|_{HM_{\omega,\mathrm{fin}}^{p,q,\infty,d,s}(\mathbb{R}^n)}$ *and* $\| \cdot \|_{HM_\omega^{p,q}(\mathbb{R}^n)}$ *are equivalent quasi-norms on* $HM_{\omega,\mathrm{fin}}^{p,q,\infty,d,s}(\mathbb{R}^n) \cap C(\mathbb{R}^n)$.

4.5 Molecular Characterizations

The target of this section is to establish the molecular characterization of generalized Herz–Hardy spaces. Indeed, we first show the molecular characterization of the generalized Herz–Hardy space $H\dot{\mathcal{K}}_{\omega,0}^{p,q}(\mathbb{R}^n)$ via the known molecular characterization of the Hardy space $H_X(\mathbb{R}^n)$ associated with the ball quasi-Banach function space X (see Lemma 4.5.5 below). However, we should point out that, due to the deficiency of the associate space of $\dot{\mathcal{K}}_\omega^{p,q}(\mathbb{R}^n)$, the molecular characterization of the generalized Herz–Hardy space $H\dot{\mathcal{K}}_\omega^{p,q}(\mathbb{R}^n)$ can not be obtained by applying Lemma 4.5.5 [see Remark 1.2.19(vi) for the details]. To overcome this obstacle, we first establish an improved molecular characterization of $H_X(\mathbb{R}^n)$ (see Theorem 4.5.11 below) via borrowing some ideas from the proof of [207, Theorem 3.9] and get the rid of associate spaces. Then, by this molecular characterization, we obtain the molecular characterization of $H\dot{\mathcal{K}}_\omega^{p,q}(\mathbb{R}^n)$.

We first use molecules to characterize the Hardy space $H\dot{\mathcal{K}}_{\omega,0}^{p,q}(\mathbb{R}^n)$ associated with the local generalized Herz space $\dot{\mathcal{K}}_{\omega,0}^{p,q}(\mathbb{R}^n)$. To this end, we now introduce the following concept of $(\dot{\mathcal{K}}_{\omega,0}^{p,q}(\mathbb{R}^n), r, d, \tau)$-molecules. In what follows, for any $j \in \mathbb{N}$ and $B \in \mathbb{B}$, let

$$S_j(B) := \left(2^j B \right) \setminus \left(2^{j-1} B \right) \text{ and } S_0(B) := B.$$

Definition 4.5.1 Let $p, q \in (0, \infty)$, $\omega \in M(\mathbb{R}_+)$ with $m_0(\omega) \in (-\frac{n}{p}, \infty)$, $\tau \in (0, \infty)$, $r \in [1, \infty]$, and $d \in \mathbb{Z}_+$. Then a measurable function m on $\mathbb{R}^n$ is called a $(\dot{\mathcal{K}}_{\omega,0}^{p,q}(\mathbb{R}^n), r, d, \tau)$-*molecule* centered at a ball $B \in \mathbb{B}$ if

(i) for any $j \in \mathbb{Z}_+$,

$$\left\| m \mathbf{1}_{S_j(B)} \right\|_{L^r(\mathbb{R}^n)} \leq 2^{-\tau j} \frac{|B|^{1/r}}{\|\mathbf{1}_B\|_{\dot{\mathcal{K}}_{\omega,0}^{p,q}(\mathbb{R}^n)}};$$

(ii) for any $\alpha \in \mathbb{Z}_+^n$ with $|\alpha| \leq d$,

$$\int_{\mathbb{R}^n} m(x) x^\alpha \, dx = 0.$$

Then we establish the molecular characterization of the generalized Herz–Hardy space $H\dot{\mathcal{K}}_{\omega,0}^{p,q}(\mathbb{R}^n)$ as follows.

Theorem 4.5.2 Let $p, q \in (0, \infty)$, $\omega \in M(\mathbb{R}_+)$ with $m_0(\omega) \in (-\frac{n}{p}, \infty)$ and $m_\infty(\omega) \in (-\frac{n}{p}, \infty)$,

$$s \in \left(0, \min\left\{ 1, p, q, \frac{n}{\max\{M_0(\omega), M_\infty(\omega)\} + n/p} \right\} \right),$$

$d \geq \lfloor n(1/s - 1) \rfloor$ be a fixed integer,

$$r \in \left(\max\left\{ 1, p, \frac{n}{\min\{m_0(\omega), m_\infty(\omega)\} + n/p} \right\}, \infty \right],$$

and $\tau \in (0, \infty)$ with $\tau > n(1/s - 1/r)$. Then $f \in H\dot{\mathcal{K}}_{\omega,0}^{p,q}(\mathbb{R}^n)$ if and only if $f \in \mathcal{S}'(\mathbb{R}^n)$ and there exists a sequence $\{m_j\}_{j\in\mathbb{N}}$ of $(\dot{\mathcal{K}}_{\omega,0}^{p,q}(\mathbb{R}^n), r, d, \tau)$-molecules centered, respectively, at the balls $\{B_j\}_{j\in\mathbb{N}} \subset \mathbb{B}$ and a sequence $\{\lambda_j\}_{j\in\mathbb{N}} \subset [0, \infty)$ such that $f = \sum_{j\in\mathbb{N}} \lambda_j m_j$ in $\mathcal{S}'(\mathbb{R}^n)$ and

$$\left\| \left\{ \sum_{j\in\mathbb{N}} \left[\frac{\lambda_j}{\|\mathbf{1}_{B_j}\|_{\dot{\mathcal{K}}_{\omega,0}^{p,q}(\mathbb{R}^n)}} \right]^s \mathbf{1}_{B_j} \right\}^{\frac{1}{s}} \right\|_{\dot{\mathcal{K}}_{\omega,0}^{p,q}(\mathbb{R}^n)} < \infty.$$

Moreover, there exist two positive constants C_1 and C_2 such that, for any $f \in H\dot{\mathcal{K}}^{p,q}_{\omega,0}(\mathbb{R}^n)$,

$$
C_1 \|f\|_{H\dot{\mathcal{K}}^{p,q}_{\omega,0}(\mathbb{R}^n)} \leq \inf \left\{ \left\| \left\{ \sum_{j \in \mathbb{N}} \left[\frac{\lambda_i}{\|\mathbf{1}_{B_j}\|_{\dot{\mathcal{K}}^{p,q}_{\omega,0}(\mathbb{R}^n)}} \right]^s \mathbf{1}_{B_j} \right\}^{\frac{1}{s}} \right\|_{\dot{\mathcal{K}}^{p,q}_{\omega,0}(\mathbb{R}^n)} \right\}
$$

$$
\leq C_2 \|f\|_{H\dot{\mathcal{K}}^{p,q}_{\omega,0}(\mathbb{R}^n)},
$$

where the infimum is taken over all the decompositions of f as above.

To show this theorem, we need the molecular characterization of Hardy spaces associated with ball quasi-Banach function spaces. First, we present the definition of (X, r, d, τ)-molecules introduced in [207, Definition 3.8].

Definition 4.5.3 Let X be a ball quasi-Banach function space, $r \in [1, \infty]$, $d \in \mathbb{Z}_+$, and $\tau \in (0, \infty)$. Then a measurable function m on $\mathbb{R}^n$ is called an (X, r, d, τ)-*molecule* centered at a ball $B \in \mathbb{B}$ if

(i) for any $j \in \mathbb{Z}_+$,

$$
\left\| m \mathbf{1}_{S_j(B)} \right\|_{L^r(\mathbb{R}^n)} \leq 2^{-\tau j} \frac{|B|^{1/r}}{\|\mathbf{1}_B\|_X};
$$

(ii) for any $\alpha \in \mathbb{Z}^n_+$ with $|\alpha| \leq d$,

$$
\int_{\mathbb{R}^n} m(x) x^\alpha \, dx = 0.
$$

Remark 4.5.4 Let X, r, d, and τ be as in Definition 4.5.3. Then it is easy to see that, for any (X, r, d)-atom a supported in the ball $B \in \mathbb{B}$, a is also an (X, r, d, τ)-molecule centered at B.

Via these molecules, Sawano et al. [207, Theorem 3.9] established the molecular characterization of the Hardy space $H_X(\mathbb{R}^n)$, which plays a key role in the proof of the molecular characterization of $H\dot{\mathcal{K}}^{p,q}_{\omega,0}(\mathbb{R}^n)$. Namely, the following conclusion holds true.

Lemma 4.5.5 *Let X, r, d, θ, and s be as in Lemma 4.3.6 and $\tau \in (0, \infty)$ with $\tau > n(1/\theta - 1/r)$. Then $f \in H_X(\mathbb{R}^n)$ if and only if $f \in \mathcal{S}'(\mathbb{R}^n)$ and there exists a sequence $\{m_j\}_{j \in \mathbb{N}}$ of (X, r, d, τ)-molecules centered, respectively, at the balls $\{B_j\}_{j \in \mathbb{N}} \subset \mathbb{B}$ and a sequence $\{\lambda_j\}_{j \in \mathbb{N}} \subset [0, \infty)$ such that*

$$
f = \sum_{j \in \mathbb{N}} \lambda_j m_j
$$

in $\mathcal{S}'(\mathbb{R}^n)$ and

$$\left\| \left[\sum_{j \in \mathbb{N}} \left(\frac{\lambda_j}{\|\mathbf{1}_{B_j}\|_X} \right)^s \mathbf{1}_{B_j} \right]^{\frac{1}{s}} \right\|_X < \infty.$$

Moreover, for any $f \in H_X(\mathbb{R}^n)$,

$$\|f\|_{H_X(\mathbb{R}^n)} \sim \inf \left\{ \left\| \left[\sum_{j \in \mathbb{N}} \left(\frac{\lambda_i}{\|\mathbf{1}_{B_j}\|_X} \right)^s \mathbf{1}_{B_j} \right]^{\frac{1}{s}} \right\|_X \right\}$$

with the positive equivalence constants independent of f, where the infimum is taken over all the decompositions of f as above.

Remark 4.5.6 We should point out that Lemma 4.5.5 has a wide range of applications. Here we present several function spaces to which Lemma 4.5.5 can be applied.

(i) Let $\vec{p} := (p_1, \ldots, p_n) \in (0, \infty)^n$,

$$d \geq \left\lfloor n \left(\frac{1}{\min\{p_1, \ldots, p_n\}} - 1 \right) \right\rfloor$$

be a fixed nonnegative integer, $r \in (\max\{1, p_1, \ldots, p_n\}, \infty]$,

$$s \in (0, \min\{1, p_1, \ldots, p_n\}),$$

and

$$\tau \in \left(n \left[\frac{1}{\min\{1, p_1, \ldots, p_n\}} - \frac{1}{r} \right], \infty \right).$$

Then, in this case, from both Remarks 1.2.31(iii) and 1.2.34(iii), we easily infer that the mixed-norm Lebesgue space $L^{\vec{p}}(\mathbb{R}^n)$ satisfies all the assumptions of Lemma 4.5.5. Therefore, Lemma 4.5.5 with $X := L^{\vec{p}}(\mathbb{R}^n)$ holds true. This result goes back to the molecular characterization of mixed-norm Hardy spaces established in [155, Theorem 1].

(ii) Let $0 < q \leq p < \infty$,

$$d \geq \left\lfloor n \left(\frac{1}{q} - 1 \right) \right\rfloor$$

be a fixed nonnegative integer, $r \in (\max\{1, p\}, \infty]$, $s \in (0, \min\{1, q\})$, and

$$\tau \in \left(n \left[\frac{1}{\min\{1, q\}} - \frac{1}{r} \right], \infty \right).$$

Then, in this case, combining both Remarks 1.2.31(iv) and 1.2.34(iv), we can easily find that the Morrey space $M_q^p(\mathbb{R}^n)$ satisfies all the assumptions of Lemma 4.5.5. This then implies that Lemma 4.5.5 with $X := M_q^p(\mathbb{R}^n)$ holds true. Recall that the molecular characterization of Hardy–Morrey spaces was also given in [107, Theorem 5.10] via replacing Definition 4.5.3(i) by some pointwise estimates (namely, [107, Definition 5.3]) which is stronger than Definition 4.5.3(i). Therefore, this aforementioned result extends [107, Theorem 5.10].

(iii) Let $p(\cdot) \in C^{\log}(\mathbb{R}^n)$ satisfy $0 < p_- \le p_+ < \infty$, where p_- and p_+ are defined, respectively, in (1.59) and (1.60). Let

$$d \ge \left\lfloor n \left(\frac{1}{p_-} - 1 \right) \right\rfloor$$

be a fixed nonnegative integer, $r \in (\max\{1, p_+\}, \infty]$, $s := \min\{1, p_-\}$, and

$$\tau \in \left(n \left[\frac{1}{\min\{1, p_-\}} - \frac{1}{r} \right], \infty \right).$$

Then, in this case, using both Remarks 1.2.31(v) and 1.2.34(v), we easily conclude that the variable Lebesgue space $L^{p(\cdot)}(\mathbb{R}^n)$ satisfies all the assumptions of Lemma 4.5.5. Thus, Lemma 4.5.5 with $X := L^{p(\cdot)}(\mathbb{R}^n)$ holds true. This result coincides with the molecular characterization of variable Hardy spaces given in [154, Theorem 3.1].

Applying Lemma 4.5.5, we next show Theorem 4.5.2.

Proof of Theorem 4.5.2 Let p, q, ω, r, d, s, and τ be as in the present theorem. Then, from the assumption $m_0(\omega) \in (-\frac{n}{p}, \infty)$ and Theorem 1.2.42, it follows that the local generalized Herz space $\dot{\mathcal{K}}_{\omega,0}^{p,q}(\mathbb{R}^n)$ is a BQBF space. Therefore, to complete the proof of the present theorem, we only need to show that all the assumptions of Lemma 4.5.5 are satisfied for $\dot{\mathcal{K}}_{\omega,0}^{p,q}(\mathbb{R}^n)$.

First, let $\theta \in (0, s)$ satisfy

$$\left\lfloor n \left(\frac{1}{s} - 1 \right) \right\rfloor \le n \left(\frac{1}{\theta} - 1 \right) < \left\lfloor n \left(\frac{1}{s} - 1 \right) \right\rfloor + 1$$

and

$$\tau > n \left(\frac{1}{\theta} - \frac{1}{r} \right).$$

This implies that $d \geq \lfloor n(1/s - 1) \rfloor = \lfloor n(1/\theta - 1) \rfloor$. We now show that $\dot{\mathcal{K}}_{\omega,0}^{p,q}(\mathbb{R}^n)$ satisfies Assumption 1.2.29 for the above θ and s. Indeed, applying Lemma 4.3.11, we find that, for any $\{f_j\}_{j \in \mathbb{N}} \subset L_{loc}^1(\mathbb{R}^n)$,

$$\left\| \left\{ \sum_{j \in \mathbb{N}} \left[\mathcal{M}^{(\theta)}(f_j) \right]^s \right\}^{1/s} \right\|_{\dot{\mathcal{K}}_{\omega,0}^{p,q}(\mathbb{R}^n)} \lesssim \left\| \left(\sum_{j \in \mathbb{N}} |f_j|^s \right)^{1/s} \right\|_{\dot{\mathcal{K}}_{\omega,0}^{p,q}(\mathbb{R}^n)}, \tag{4.47}$$

which implies that Assumption 1.2.29 holds true for $\dot{\mathcal{K}}_{\omega,0}^{p,q}(\mathbb{R}^n)$ with the above θ and s.

On the other hand, by Lemma 1.8.6, we conclude that $[\dot{\mathcal{K}}_{\omega,0}^{p,q}(\mathbb{R}^n)]^{1/s}$ is a BBF space and, for any $f \in L_{loc}^1(\mathbb{R}^n)$,

$$\left\| \mathcal{M}^{((r/s)')}(f) \right\|_{([\dot{\mathcal{K}}_{\omega,0}^{p,q}(\mathbb{R}^n)]^{1/s})'} \lesssim \|f\|_{([\dot{\mathcal{K}}_{\omega,0}^{p,q}(\mathbb{R}^n)]^{1/s})'}.$$

This, together with (4.47), further implies that, under the assumptions of the present theorem, $\dot{\mathcal{K}}_{\omega,0}^{p,q}(\mathbb{R}^n)$ satisfies all the assumptions of Lemma 4.5.5 and hence finishes the proof of Theorem 4.5.2. $\qquad \square$

As an application of Theorem 4.5.2, we now consider the molecular characterization of the generalized Hardy–Morrey space $HM_{\omega,0}^{p,q}(\mathbb{R}^n)$. For this purpose, we first introduce $(M_{\omega,0}^{p,q}(\mathbb{R}^n), r, d, \tau)$-molecules as follows.

Definition 4.5.7 Let $p, q \in [1, \infty)$, $\omega \in M(\mathbb{R}_+)$ with $M_\infty(\omega) \in (-\infty, 0)$ and

$$-\frac{n}{p} < m_0(\omega) \leq M_0(\omega) < 0,$$

$\tau \in (0, \infty)$, $r \in [1, \infty]$, and $d \in \mathbb{Z}_+$. Then a measurable function m on $\mathbb{R}^n$ is called an $(M_{\omega,0}^{p,q}(\mathbb{R}^n), r, d, \tau)$-*molecule* centered at a ball $B \in \mathbb{B}$ if

(i) for any $j \in \mathbb{Z}_+$,

$$\left\| m \mathbf{1}_{S_j(B)} \right\|_{L^r(\mathbb{R}^n)} \leq 2^{-\tau j} \frac{|B|^{1/r}}{\|\mathbf{1}_B\|_{M_{\omega,0}^{p,q}(\mathbb{R}^n)}};$$

(ii) for any $\alpha \in \mathbb{Z}_+^n$ with $|\alpha| \leq d$, $\int_{\mathbb{R}^n} m(x) x^\alpha \, dx = 0$.

Then we have the following molecular characterization of $HM_{\omega,0}^{p,q}(\mathbb{R}^n)$, which can be deduced directly from Theorem 4.5.2 and Remarks 1.2.2(iv) and 4.0.20(ii); we omit the details.

Corollary 4.5.8 *Let p, q, ω, r, d, and s be as in Corollary 4.3.13 and $\tau \in (0, \infty)$ with $\tau > n(1/s - 1/r)$. Then $f \in HM_{\omega,0}^{p,q}(\mathbb{R}^n)$ if and only if $f \in \mathcal{S}'(\mathbb{R}^n)$*

and there exists a sequence $\{m_j\}_{j\in\mathbb{N}}$ *of* $(\boldsymbol{M}_{\omega,\mathbf{0}}^{p,q}(\mathbb{R}^n),\ r,\ d,\ \tau)$-*molecules centered, respectively, at the balls* $\{B_j\}_{j\in\mathbb{N}} \subset \mathbb{B}$ *and a sequence* $\{\lambda_j\}_{j\in\mathbb{N}} \subset [0,\infty)$ *such that*

$$f = \sum_{j\in\mathbb{N}} \lambda_j m_j$$

in $\mathcal{S}'(\mathbb{R}^n)$ *and*

$$\left\| \left\{ \sum_{j\in\mathbb{N}} \left[\frac{\lambda_j}{\|\mathbf{1}_{B_j}\|_{\boldsymbol{M}_{\omega,\mathbf{0}}^{p,q}(\mathbb{R}^n)}} \right]^s \mathbf{1}_{B_j} \right\}^{\frac{1}{s}} \right\|_{\boldsymbol{M}_{\omega,\mathbf{0}}^{p,q}(\mathbb{R}^n)} < \infty.$$

Moreover, for any $f \in H\boldsymbol{M}_{\omega,\mathbf{0}}^{p,q}(\mathbb{R}^n)$,

$$\|f\|_{H\boldsymbol{M}_{\omega,\mathbf{0}}^{p,q}(\mathbb{R}^n)} \sim \inf \left\{ \left\| \left\{ \sum_{j\in\mathbb{N}} \left[\frac{\lambda_i}{\|\mathbf{1}_{B_j}\|_{\boldsymbol{M}_{\omega,\mathbf{0}}^{p,q}(\mathbb{R}^n)}} \right]^s \mathbf{1}_{B_j} \right\}^{\frac{1}{s}} \right\|_{\boldsymbol{M}_{\omega,\mathbf{0}}^{p,q}(\mathbb{R}^n)} \right\}$$

with the positive equivalence constants independent of f, *where the infimum is taken over all the decompositions of* f *as above.*

Next, we establish the molecular characterization of the generalized Herz–Hardy space $H\dot{\mathcal{K}}_{\omega}^{p,q}(\mathbb{R}^n)$. To achieve this, we first introduce the following concept of $(\dot{\mathcal{K}}_{\omega}^{p,q}(\mathbb{R}^n),\ r,\ d,\ \tau)$-molecules.

Definition 4.5.9 Let $p,q \in (0,\infty)$, $\omega \in M(\mathbb{R}_+)$ with $m_0(\omega) \in (-\frac{n}{p},\infty)$ and $M_\infty(\omega) \in (-\infty,0)$, $\tau \in (0,\infty)$, $r \in [1,\infty]$, and $d \in \mathbb{Z}_+$. Then a measurable function m on $\mathbb{R}^n$ is called a $(\dot{\mathcal{K}}_{\omega}^{p,q}(\mathbb{R}^n),\ r,\ d,\ \tau)$-*molecule* centered at a ball $B \in \mathbb{B}$ if

(i) for any $j \in \mathbb{Z}_+$,

$$\left\| m\mathbf{1}_{S_j(B)} \right\|_{L^r(\mathbb{R}^n)} \le 2^{-j\tau} \frac{|B|^{1/r}}{\|\mathbf{1}_B\|_{\dot{\mathcal{K}}_{\omega}^{p,q}(\mathbb{R}^n)}};$$

(ii) for any $\alpha \in \mathbb{Z}_+^n$ with $|\alpha| \le d$,

$$\int_{\mathbb{R}^n} m(x)x^\alpha \, dx = 0.$$

Via these molecules, we characterize the generalized Herz–Hardy space $H\dot{\mathcal{K}}_{\omega}^{p,q}(\mathbb{R}^n)$ associated with the global generalized Herz space $\dot{\mathcal{K}}_{\omega}^{p,q}(\mathbb{R}^n)$ as follows.

Theorem 4.5.10 *Let* $p, q \in (0, \infty)$, $\omega \in M(\mathbb{R}_+)$ *with* $m_0(\omega) \in (-\frac{n}{p}, \infty)$ *and*

$$-\frac{n}{p} < m_\infty(\omega) \leq M_\infty(\omega) < 0,$$

$r \in (\max\{1, p, \frac{n}{\min\{m_0(\omega), m_\infty(\omega)\} + n/p}\}, \infty]$, $s \in (0, \min\{1, p, q, \frac{n}{\max\{M_0(\omega), M_\infty(\omega)\} + n/p}\})$, $d \geq \lfloor n(\frac{1}{s} - 1) \rfloor$ *be a fixed integer, and* $\tau \in (n(\frac{1}{s} - \frac{1}{r}), \infty)$. *Then* $f \in H\dot{\mathcal{K}}_\omega^{p,q}(\mathbb{R}^n)$ *if and only if* $f \in \mathcal{S}'(\mathbb{R}^n)$ *and there exists* $\{\lambda_j\}_{j\in\mathbb{N}} \subset [0, \infty)$ *and a sequence* $\{m_j\}_{j\in\mathbb{N}}$ *of* $(\dot{\mathcal{K}}_\omega^{p,q}(\mathbb{R}^n), r, d, \tau)$-*molecules centered, respectively, at the balls* $\{B_j\}_{j\in\mathbb{N}} \subset \mathbb{B}$ *such that*

$$f = \sum_{j\in\mathbb{N}} \lambda_j m_j$$

in $\mathcal{S}'(\mathbb{R}^n)$ *and*

$$\left\| \left\{ \sum_{j\in\mathbb{N}} \left[\frac{\lambda_j}{\|\mathbf{1}_{B_j}\|_{\dot{\mathcal{K}}_\omega^{p,q}(\mathbb{R}^n)}} \right]^s \mathbf{1}_{B_j} \right\}^{\frac{1}{s}} \right\|_{\dot{\mathcal{K}}_\omega^{p,q}(\mathbb{R}^n)} < \infty.$$

Moreover, there exist two positive constants C_1 *and* C_2 *such that, for any* $f \in H\dot{\mathcal{K}}_\omega^{p,q}(\mathbb{R}^n)$,

$$C_1 \|f\|_{H\dot{\mathcal{K}}_\omega^{p,q}(\mathbb{R}^n)} \leq \inf \left\{ \left\| \left\{ \sum_{j\in\mathbb{N}} \left[\frac{\lambda_j}{\|\mathbf{1}_{B_j}\|_{\dot{\mathcal{K}}_\omega^{p,q}(\mathbb{R}^n)}} \right]^s \mathbf{1}_{B_j} \right\}^{\frac{1}{s}} \right\|_{\dot{\mathcal{K}}_\omega^{p,q}(\mathbb{R}^n)} \right\}$$

$$\leq C_2 \|f\|_{H\dot{\mathcal{K}}_\omega^{p,q}(\mathbb{R}^n)},$$

where the infimum is taken over all the decompositions of f *as above.*

Due to the deficiency of associate spaces, we can not show Theorem 4.5.10 via using the known molecular characterization of the Hardy space $H_X(\mathbb{R}^n)$ associated with the ball quasi-Banach function space X (see Lemma 4.5.5 above) directly [see Remark 1.2.19(vi) for the details]. To overcome this obstacle, we first give the following molecular characterization of $H_X(\mathbb{R}^n)$ when there is no assumption about the associate space X'.

Theorem 4.5.11 *Let* X, r, d, s, *and* θ *be as in Theorem 4.3.18 and* $\tau \in (0, \infty)$ *with* $\tau > n(\frac{1}{\theta} - \frac{1}{r})$. *Then* $f \in H_X(\mathbb{R}^n)$ *if and only if* $f \in \mathcal{S}'(\mathbb{R}^n)$ *and there exists* $\{\lambda_j\}_{j\in\mathbb{N}} \subset [0, \infty)$ *and a sequence* $\{m_j\}_{j\in\mathbb{N}}$ *of* (X, r, d, τ)-*molecules centered,* *respectively, at the balls* $\{B_j\}_{j\in\mathbb{N}} \subset \mathbb{B}$ *such that*

$$f = \sum_{j\in\mathbb{N}} \lambda_j m_j \tag{4.48}$$

in $\mathcal{S}'(\mathbb{R}^n)$ *and*

$$\left\| \left[\sum_{j\in\mathbb{N}} \left(\frac{\lambda_j}{\|\mathbf{1}_{B_j}\|_X} \right)^s \mathbf{1}_{B_j} \right]^{\frac{1}{s}} \right\|_X < \infty. \tag{4.49}$$

Moreover, there exist two positive constants C_1 *and* C_2 *such that, for any* $f \in H_X(\mathbb{R}^n)$,

$$C_1 \|f\|_{H_X(\mathbb{R}^n)} \leq \inf \left\{ \left\| \left[\sum_{j\in\mathbb{N}} \left(\frac{\lambda_i}{\|\mathbf{1}_{B_j}\|_X} \right)^s \mathbf{1}_{B_j} \right]^{\frac{1}{s}} \right\|_X \right\} \leq C_2 \|f\|_{H_X(\mathbb{R}^n)}, \tag{4.50}$$

where the infimum is taken over all the decompositions of f *as above.*

Remark 4.5.12 We should point out that Theorem 4.5.11 is an improved version of the known molecular characterization obtained in [207, Theorem 3.9]. Indeed, if $Y \equiv (X^{1/s})'$ in Theorem 4.5.11, then this theorem goes back to [207, Theorem 3.9].

In order to show the above theorem, we require the following pointwise estimate on the radial maximal function of (X, r, d, τ)-molecules.

Lemma 4.5.13 *Let* X *be a ball quasi-Banach function space,* $r \in (1, \infty]$, $\theta \in (0, 1]$, $d \geq \lfloor n(1/\theta - 1) \rfloor$ *be a fixed integer, and* $\tau \in (n(\frac{1}{\theta} - \frac{1}{r}), \infty)$. *Assume that* $\phi \in \mathcal{S}(\mathbb{R}^n)$ *satisfies* $\operatorname{supp}(\phi) \subset B(\mathbf{0}, 1)$. *Then there exists a positive constant* C *such that, for any* (X, r, d, τ)-*molecule* m *centered at the ball* $B \in \mathbb{B}$,

$$M(m, \phi) \leq C \left[\mathcal{M}(m)\mathbf{1}_{4B} + \sum_{j=3}^{\infty} \mathcal{M}\left(m\mathbf{1}_{(2^{j+1}B)\backslash(2^{j-2}B)} \right) \mathbf{1}_{(2^j B)\backslash(2^{j-1}B)} \right.$$
$$\left. + \frac{1}{\|\mathbf{1}_B\|_X} \mathcal{M}^{(\theta)}\left(\mathbf{1}_B \right) \right],$$

where M *is the radial maximal function defined as in Definition 4.1.1(i).*

Proof Let all the symbols be as in the present lemma and $B := B(x_0, r_0)$ with $x_0 \in \mathbb{R}^n$ and $r_0 \in (0, \infty)$. Then, from Lemma 4.3.21 with Φ and f therein replaced, respectively, by $\|\phi\|_{L^\infty(\mathbb{R}^n)}\mathbf{1}_{B(0,1)}$ and m, we deduce that, for any $x \in 4B$,

$$M(m, \phi)(x) = \sup_{t\in(0,\infty)} |m * \phi_t(x)| \lesssim \mathcal{M}(m)(x), \qquad (4.51)$$

which is the desired estimate of $M(m, \phi)(x)$ when $x \in 4B$.

Next, we estimate $M(m, \phi)(x)$ for any $x \in S_j(B)$ with $j \in \mathbb{N} \cap [3, \infty)$. To this end, let $j \in \mathbb{N} \cap [3, \infty)$, $x \in S_j(B)$, $t \in (0, \infty)$, and

$$d_\theta := \left\lfloor n\left(\frac{1}{\theta} - 1\right)\right\rfloor.$$

Then, by Definition 4.5.3(ii), we find that

$$|m * \phi_t(x)|$$

$$= \left| \int_{\mathbb{R}^n} \phi_t(x - y)m(y)\,dy \right|$$

$$= \left| \frac{1}{t^n} \int_{\mathbb{R}^n} \left[\phi\left(\frac{x-y}{t}\right) - \sum_{\substack{\gamma\in\mathbb{Z}_+^n \\ |\gamma|\le d_\theta}} \frac{\partial^\gamma(\phi(\frac{x-\cdot}{t}))(x_0)}{\gamma!}(y - x_0)^\gamma \right] m(y)\,dy \right|$$

$$\le \frac{1}{t^n} \left[\left| \int_{2^{j-2}B} \left[\phi\left(\frac{x-y}{t}\right) \right. \right. \right.$$

$$\left. \left. \left. - \sum_{\substack{\gamma\in\mathbb{Z}_+^n \\ |\gamma|\le d_\theta}} \frac{\partial^\gamma(\phi(\frac{x-\cdot}{t}))(x_0)}{\gamma!}(y - x_0)^\gamma \right] m(y)\,dy \right| \right.$$

$$\left. + \left| \int_{(2^{j+1}B)\backslash(2^{j-2}B)} \cdots \right| + \left| \int_{(2^{j-2}B)^\complement} \cdots \right| \right]$$

$$\le \int_{2^{j-2}B} \left| \phi\left(\frac{x-y}{t}\right) - \sum_{\substack{\gamma\in\mathbb{Z}_+^n \\ |\gamma|\le d_\theta}} \frac{\partial^\gamma(\phi(\frac{x-\cdot}{t}))(x_0)}{\gamma!}(y - x_0)^\gamma \right| |m(y)|\,dy$$

$$+ \left| \int_{(2^{j+1}B)\backslash(2^{j-2}B)} \phi_t(x - y)m(y)\,dy \right|$$

$$+ \frac{1}{t^n} \int_{(2^{j+1}B)^\complement} \left| \phi\left(\frac{x-y}{t}\right) \right| |m(y)|\,dy$$

$$+ \frac{1}{t^n} \int_{(2^{j-2}B)^{\complement}} \left| \sum_{\substack{\gamma \in \mathbb{Z}_+^n \\ |\gamma| \le d_\theta}} \frac{\partial^\gamma (\phi(\frac{x-\cdot}{t}))(x_0)}{\gamma!} (y - x_0)^\gamma \right| |m(y)| \, dy$$

$$=: A_1 + A_2 + A_3 + A_4. \tag{4.52}$$

We next estimate A_1, A_2, A_3, and A_4, respectively.

First, we deal with A_1. Indeed, from the Taylor remainder theorem and the Tonelli theorem, it follows that, for any $y \in 2^{j-2}B$, there exists a $t_y \in (0, 1)$ such that

$$A_1 = \sum_{k=0}^{j-2} \frac{1}{t^n} \int_{S_k(B)} \left| \sum_{\substack{\gamma \in \mathbb{Z}_+^n \\ |\gamma| = d_\theta + 1}} \frac{\partial^\gamma (\phi(\frac{x-\cdot}{t}))(t_y + (1 - t_y)x_0)}{\gamma!} (y - x_0)^\gamma \right|$$

$$\times |m(y)| \, dy$$

$$\lesssim \sum_{k=0}^{j-2} \sum_{\substack{\gamma \in \mathbb{Z}_+^n \\ |\gamma| = d_\theta + 1}} \frac{1}{t^{n+d_\theta+1}}$$

$$\times \int_{S_k(B)} \left| \partial^\gamma \phi \left(\frac{x - t_y - (1 - t_y)x_0}{t} \right) \right| |y - x_0|^{d_\theta+1} |m(y)| \, dy$$

$$\lesssim \sum_{k=0}^{j-2} \int_{S_k(B)} \frac{|y - x_0|^{d_\theta+1}}{|x - t_y - (1 - t_y)x_0|^{n+d_\theta+1}} |m(y)| \, dy.$$

Observe that, for any $y \in 2^{j-2}B$, we have

$$|y - x_0| < 2^{j-2}r_0 \le \frac{1}{2}|x - x_0|,$$

which, combined with the assumption $t_y \in (0, 1)$, further implies that

$$|x - t_y - (1 - t_y)x_0| \ge |x - x_0| - |y - y_0| > \frac{1}{2}|x - x_0|.$$

Using this and the assumption $n(1/\theta - 1) < d_\theta + 1$, we conclude that

$$A_1 \lesssim \sum_{k=0}^{j-2} \int_{S_k(B)} \frac{|y - x_0|^{d_\theta+1}}{|x - x_0|^{n+d_\theta+1}} |m(y)| \, dy$$

$$\lesssim \sum_{k=0}^{j-2} \int_{S_k(B)} \frac{|y - x_0|^{\frac{n}{\theta}-n}}{|x - x_0|^{\frac{n}{\theta}}} \, dy \lesssim \sum_{k=0}^{j-2} \frac{(2^k r_0)^{\frac{n}{\theta}-n}}{|x - x_0|^{\frac{n}{\theta}}} \left\| m \mathbf{1}_{S_k(B)} \right\|_{L^1(\mathbb{R}^n)}. \tag{4.53}$$

Notice that, for any $k \in \mathbb{Z}_+$, by the Hölder inequality and Definition 4.5.3(i), we find that

$$\left\| m \mathbf{1}_{S_k(B)} \right\|_{L^1(\mathbb{R}^n)} \leq |S_k(B)|^{1-\frac{1}{r}} \left\| m \mathbf{1}_{S_k(B)} \right\|_{L^r(\mathbb{R}^n)}$$

$$\leq |S_k(B)|^{1-\frac{1}{r}} 2^{-k\tau} \frac{|B|^{\frac{1}{r}}}{\|\mathbf{1}_B\|_X} \sim 2^{-k(\tau-n+\frac{n}{r})} \frac{r_0^n}{\|\mathbf{1}_B\|_X}. \qquad (4.54)$$

From this, (4.53), and the assumption $\tau \in (n(\frac{1}{\theta} - \frac{1}{r}), \infty)$, we deduce that

$$\mathrm{A}_1 \lesssim \frac{1}{\|\mathbf{1}_B\|_X} \frac{r_0^{\frac{n}{\theta}}}{|x - x_0|^{\frac{n}{\theta}}} \sum_{k=0}^{j-2} 2^{-k(\tau-\frac{n}{\theta}+\frac{n}{r})} \sim \frac{1}{\|\mathbf{1}_B\|_X} \frac{r_0^{\frac{n}{\theta}}}{|x - x_0|^{\frac{n}{\theta}}}, \qquad (4.55)$$

which is the desired estimate of A_1.

Next, we estimate A_2. Indeed, applying Lemma 4.3.21 with $f := m\mathbf{1}_{(2^{j+1}B)\setminus(2^{j-2}B)}$ and $\Phi := \|\phi\|_{L^\infty(\mathbb{R}^n)} \mathbf{1}_{B(0,1)}$, we conclude that

$$\mathrm{A}_2 = \left| \left[m\mathbf{1}_{(2^{j+1}B)\setminus(2^{j-2}B)} \right] * \phi(x) \right| \lesssim \mathcal{M}\left(m\mathbf{1}_{(2^{j+1}B)\setminus(2^{j-2}B)} \right)(x). \qquad (4.56)$$

This finishes the estimate of A_2.

We now deal with A_3. Notice that, for any $y \in (2^{j+1}B)^\complement$, we have

$$|y - x_0| \geq 2^{j+1} r_0 > 2|x - x_0|,$$

which further implies that

$$|x - y| \geq |y - x_0| - |x - x_0| > |x - x_0|.$$

From this, the Tonelli theorem, and (4.54), it then follows that

$$\mathrm{A}_3 \lesssim \sum_{k=j+2}^{\infty} \int_{S_k(B)} \frac{1}{|x - y|^n} |m(y)| \, dy \lesssim \sum_{k=j+2}^{\infty} \frac{1}{|x - x_0|^n} \left\| m\mathbf{1}_{S_k(B)} \right\|_{L^1(\mathbb{R}^n)}$$

$$\lesssim \sum_{k=j+2}^{\infty} 2^{-k(\tau+\frac{n}{r})} \frac{1}{\|\mathbf{1}_B\|_X} \frac{(2^k r_0)^n}{|x - x_0|^n}.$$

In addition, for any $k \in \mathbb{N} \cap [j+2, \infty)$, we have

$$|x - x_0| < 2^j r_0 \leq 2^{k-2} r_0,$$

which implies that $\frac{2^k r_0}{|x - x_0|} \in [4, \infty)$. This, together with the assumptions $\theta \in (0, 1]$ and $\tau \in (n(\frac{1}{\theta} - \frac{1}{r}), \infty)$, further implies that

$$
\begin{aligned}
A_3 &\lesssim \sum_{k=j+2}^{\infty} 2^{-k(\tau + \frac{n}{r})} \frac{1}{\|\mathbf{1}_B\|_X} \left(\frac{2^k r_0}{|x - x_0|} \right)^{\frac{n}{\theta}} \\
&\sim \frac{1}{\|\mathbf{1}_B\|_X} \frac{r_0^{\frac{n}{\theta}}}{|x - x_0|^{\frac{n}{\theta}}} \sum_{k=j+2}^{\infty} 2^{-k(\tau - \frac{n}{\theta} + \frac{n}{r})} \sim \frac{1}{\|\mathbf{1}_B\|_X} \frac{r_0^{\frac{n}{\theta}}}{|x - x_0|^{\frac{n}{\theta}}},
\end{aligned}
\tag{4.57}
$$

which is the desired estimate of A_3.

Finally, we turn to estimate A_4. Indeed, for any $y \in (2^{j-2}B)^{\complement}$, it holds true that

$$
|y - x_0| \geq 2^{j-2} r_0 > \frac{1}{4}|x - x_0|,
$$

which further implies that $\frac{4|y - x_0|}{|x - x_0|} \in (1, \infty)$. Applying this, the assumption $d_\theta \leq n(1/\theta - 1)$, (4.54), and the assumption $\tau \in (n(\frac{1}{\theta} - \frac{1}{r}), \infty)$, we conclude that

$$
\begin{aligned}
A_4 &\lesssim \sum_{k=j-1}^{\infty} \sum_{\substack{\gamma \in \mathbb{Z}_+^n \\ |\gamma| \leq d_\theta}} \frac{1}{t^{n+|\gamma|}} \int_{S_k(B)} \left| \partial^\gamma \phi \left(\frac{x - x_0}{t} \right) \right| |y - x_0|^{|\gamma|} |m(y)| \, dy \\
&\lesssim \sum_{k=j-1}^{\infty} \sum_{\substack{\gamma \in \mathbb{Z}_+^n \\ |\gamma| \leq d_\theta}} \int_{S_k(B)} \frac{|y - x_0|^{|\gamma|}}{|x - x_0|^{n+|\gamma|}} \, dy \\
&\lesssim \sum_{k=j-1}^{\infty} \int_{S_k(B)} \frac{|y - x_0|^{\frac{n}{\theta} - n}}{|x - x_0|^{\frac{n}{\theta}}} |m(y)| \, dy \\
&\lesssim \sum_{k=j-1}^{\infty} \frac{(2^k r_0)^{\frac{n}{\theta} - n}}{|x - x_0|^{\frac{n}{\theta}}} \left\| m \mathbf{1}_{S_k(B)} \right\|_{L^1(\mathbb{R}^n)} \\
&\lesssim \frac{1}{\|\mathbf{1}_B\|_X} \frac{r_0^{\frac{n}{\theta}}}{|x - x_0|^{\frac{n}{\theta}}} \sum_{k=j-1}^{\infty} 2^{-k(\tau - \frac{n}{\theta} + \frac{n}{r})} \sim \frac{1}{\|\mathbf{1}_B\|_X} \frac{r_0^{\frac{n}{\theta}}}{|x - x_0|^{\frac{n}{\theta}}},
\end{aligned}
$$

which is the desired estimate of A_4. Combining this, (4.52), (4.55), (4.56), (4.57), and an argument similar to that used in the estimation of (4.23) with r_B and x_B therein replaced, respectively, by r_0 and x_0, we find that, for any $x \in S_j(B)$ with

$j \in \mathbb{N} \cap [3, \infty)$ and $t \in (0, \infty)$,

$$|m * \phi_t(x)| \lesssim \mathcal{M} \left(m \mathbf{1}_{(2^{j+1}B)\setminus(2^{j-2}B)} \right)(x) + \frac{1}{\|\mathbf{1}_B\|_X} \frac{r_0^{\frac{n}{\theta}}}{|x - x_0|^{\frac{n}{\theta}}}$$

$$\lesssim \mathcal{M} \left(m \mathbf{1}_{(2^{j+1}B)\setminus(2^{j-2}B)} \right)(x) + \frac{1}{\|\mathbf{1}_B\|_X} \mathcal{M}^{(\theta)}(\mathbf{1}_B)(x),$$

which, together with the arbitrariness of t and (4.51), further implies that

$$M(m, \phi) \lesssim \mathcal{M}(m) \mathbf{1}_{4B} + \sum_{j=3}^{\infty} \mathcal{M} \left(m \mathbf{1}_{(2^{j+1}B)\setminus(2^{j-2}B)} \right) \mathbf{1}_{(2^j B)\setminus(2^{j-1}B)}$$

$$+ \frac{1}{\|\mathbf{1}_B\|_X} \mathcal{M}^{(\theta)}(\mathbf{1}_B).$$

This then finishes the proof of Lemma 4.5.13. $\qquad \square$

We now show Theorem 4.5.11.

Proof of Theorem 4.5.11 Let all the symbols be as in the present theorem. We first prove the necessity. Indeed, let $f \in H_X(\mathbb{R}^n)$. Then, applying Theorem 4.3.18, we find that, under the assumptions of the present theorem, $H_X(\mathbb{R}^n) = H^{X,r,d,s}(\mathbb{R}^n)$ with equivalent quasi-norms. This implies that $f \in H^{X,r,d,s}(\mathbb{R}^n)$. Therefore, there exists a sequence $\{\lambda_j\}_{j\in\mathbb{N}} \subset [0, \infty)$ and $\{a_j\}_{j\in\mathbb{N}}$ of (X, r, d)-atoms supported, respectively, in the balls $\{B_j\}_{j\in\mathbb{N}} \subset \mathbb{B}$ such that

$$f = \sum_{j\in\mathbb{N}} \lambda_j a_j \tag{4.58}$$

in $\mathcal{S}'(\mathbb{R}^n)$ and

$$\left\| \left[\sum_{j\in\mathbb{N}} \left(\frac{\lambda_j}{\|\mathbf{1}_{B_j}\|_X} \right)^s \mathbf{1}_{B_j} \right]^{\frac{1}{s}} \right\|_X < \infty. \tag{4.59}$$

Moreover, by Remark 4.5.4, we conclude that, for any $j \in \mathbb{N}$, a_j is a (X, r, d, τ)-molecule centered at the ball B_j. This, combined with (4.58) and (4.59), then finishes the proof of the necessity. In addition, from the choice of $\{\lambda_j\}_{j\in\mathbb{N}}$, Definition 4.3.17, and Theorem 4.3.18, we deduce that

$$\inf \left\{ \left\| \left[\sum_{j\in\mathbb{N}} \left(\frac{\lambda_j}{\|\mathbf{1}_{B_j}\|_X} \right)^s \mathbf{1}_{B_j} \right]^{\frac{1}{s}} \right\|_X \right\} \leq \|f\|_{H^{X,r,d,s}(\mathbb{R}^n)} \sim \|f\|_{H_X(\mathbb{R}^n)}, \tag{4.60}$$

where the infimum is taken over all the sequences $\{\lambda_j\}_{j\in\mathbb{N}} \subset [0, \infty)$ and $\{m_j\}_{j\in\mathbb{N}}$ of (X, r, d, τ)-molecules centered, respectively, at the balls $\{B_j\}_{j\in\mathbb{N}} \subset \mathbb{B}$ such that (4.48) and (4.49) hold true.

Next, we show the sufficiency. To this end, let $f \in \mathcal{S}'(\mathbb{R}^n)$ satisfy $f = \sum_{j\in\mathbb{N}} \lambda_j m_j$ in $\mathcal{S}'(\mathbb{R}^n)$, where $\{\lambda_j\}_{j\in\mathbb{N}} \subset [0, \infty)$ and $\{m_j\}_{j\in\mathbb{N}}$ is a sequence of (X, r, d, τ)-molecules centered, respectively, at the balls $\{B_j\}_{j\in\mathbb{N}} \subset \mathbb{B}$ such that

$$\left\| \left[\sum_{j\in\mathbb{N}} \left(\frac{\lambda_j}{\|\mathbf{1}_{B_j}\|_X} \right)^s \mathbf{1}_{B_j} \right]^{\frac{1}{s}} \right\|_X < \infty. \tag{4.61}$$

To show the sufficiency, we only need to prove that $f \in H_X(\mathbb{R}^n)$. To achieve this, we choose a $\phi \in \mathcal{S}(\mathbb{R}^n)$ satisfying that $\operatorname{supp}(\phi) \subset B(\mathbf{0}, 1)$ and $\int_{\mathbb{R}^n} \phi(x)\, dx \neq 0$. Then, by Lemma 4.1.4, we find that, to prove $f \in H_X(\mathbb{R}^n)$, it suffices to show $\|M(f, \phi)\|_X < \infty$. Indeed, applying both the assumption that $f = \sum_{j\in\mathbb{N}} \lambda_j m_j$ in $\mathcal{S}'(\mathbb{R}^n)$ and an argument similar to that used in the proof of (4.25) with a_j therein replaced by m_j for any $j \in \mathbb{N}$, we conclude that

$$M(f, \phi) \leq \sum_{j\in\mathbb{N}} \lambda_j M(m_j, \phi).$$

From this, the assumption $d \geq \lfloor n(1/\theta - 1) \rfloor$, and Lemma 4.5.13 with m replaced by m_j for any $j \in \mathbb{N}$, we deduce that

$$\|M(f, \phi)\|_X \lesssim \left\| \sum_{j\in\mathbb{N}} \lambda_j \mathcal{M}(m_j) \mathbf{1}_{4B_j} \right\|_X$$

$$+ \left\| \sum_{j\in\mathbb{N}} \sum_{k=3}^{\infty} \lambda_j \mathcal{M}\left(m_j \mathbf{1}_{(2^{k+1}B_j)\setminus(2^{k-2}B_j)} \right) \mathbf{1}_{(2^k B_j)\setminus(2^{k-1}B_j)} \right\|_X$$

$$+ \left\| \sum_{j\in\mathbb{N}} \frac{\lambda_j}{\|\mathbf{1}_{B_j}\|_X} \mathcal{M}^{(\theta)}(\mathbf{1}_{B_j}) \right\|_X$$

$$=: \mathrm{III}_1 + \mathrm{III}_2 + \mathrm{III}_3. \tag{4.62}$$

We first estimate III_1. Indeed, applying Definition 4.5.3(i) and an argument similar to that used in the estimation of II_1 in the proof of Theorem 4.3.18 with

$\mathcal{M}(a_j)\mathbf{1}_{2B_j}$ therein replaced by $\mathcal{M}(m_j)\mathbf{1}_{4B_j}$ for any $j \in \mathbb{N}$, we find that

$$\mathrm{III}_1 \lesssim \left\| \left[\sum_{j \in \mathbb{N}} \left(\frac{\lambda_j}{\|\mathbf{1}_{B_j}\|_X} \right)^s \mathbf{1}_{B_j} \right]^{\frac{1}{s}} \right\|_X. \tag{4.63}$$

This is the desired estimate of III_1.

Next, we deal with III_2. Indeed, from the $L^r(\mathbb{R}^n)$ boundedness of $\mathcal{M}$ and Definition 4.5.3(i), it follows that, for any $j \in \mathbb{N}$ and $k \in \mathbb{N} \cap [3, \infty)$,

$$\left\| \mathcal{M}\left(m_j \mathbf{1}_{(2^{k+1}B_j) \setminus (2^{k-2}B_j)} \right) \mathbf{1}_{(2^k B_j) \setminus (2^{k-1}B_j)} \right\|_{L^r(\mathbb{R}^n)}$$

$$\lesssim \left\| m_j \mathbf{1}_{(2^{k+1}B_j) \setminus (2^{k-2}B_j)} \right\|_{L^r(\mathbb{R}^n)} \lesssim \sum_{l=k-1}^{k+1} \left\| m_j \mathbf{1}_{(2^l B_j) \setminus (2^{l-1}B_j)} \right\|_{L^r(\mathbb{R}^n)}$$

$$\lesssim 2^{-k\tau} \frac{|B_j|^{1/r}}{\|\mathbf{1}_{B_j}\|_X} \sim \left[2^{-k(\tau+\frac{n}{r})} \frac{\|\mathbf{1}_{2^k B_j}\|_X}{\|\mathbf{1}_{B_j}\|_X} \right] \frac{|2^k B_j|^{1/r}}{\|\mathbf{1}_{2^k B_j}\|_X},$$

where the implicit positive constants are independent of both j and k. This implies that there exists a positive constant C such that, for any $j \in \mathbb{N}$ and $k \in \mathbb{N} \cap [3, \infty)$,

$$\left\| \mathcal{M}\left(m_j \mathbf{1}_{(2^{k+1}B_j) \setminus (2^{k-2}B_j)} \right) \mathbf{1}_{(2^k B_j) \setminus (2^{k-1}B_j)} \right\|_{L^r(\mathbb{R}^n)}$$

$$\leq C \left[2^{-k(\tau+\frac{n}{r})} \frac{\|\mathbf{1}_{2^k B_j}\|_X}{\|\mathbf{1}_{B_j}\|_X} \right] \frac{|2^k B_j|^{1/r}}{\|\mathbf{1}_{2^k B_j}\|_X}. \tag{4.64}$$

For any $j \in \mathbb{N}$ and $k \in \mathbb{N} \cap [3, \infty)$, let

$$\mu_{j,k} := C 2^{-k(\tau+\frac{n}{r})} \frac{\|\mathbf{1}_{2^k B_j}\|_X}{\|\mathbf{1}_{B_j}\|_X}$$

and

$$a_{j,k} := C^{-1} 2^{k(\tau+\frac{n}{r})} \frac{\|\mathbf{1}_{B_j}\|_X}{\|\mathbf{1}_{2^k B_j}\|_X} \mathcal{M}\left(m_j \mathbf{1}_{(2^{k+1}B_j) \setminus (2^{k-2}B_j)} \right) \mathbf{1}_{(2^k B_j) \setminus (2^{k-1}B_j)}.$$

Then, by (4.64), we conclude that, for any $j \in \mathbb{N}$ and $k \in \mathbb{N} \cap [3, \infty)$, $\mathrm{supp}\,(a_{j,k}) \subset 2^k B_j$ and

$$\|a_{j,k}\|_{L^r(\mathbb{R}^n)} \leq \frac{|2^k B_j|^{1/r}}{\|2^k B_j\|_X}.$$

These, together with the definitions of both $\mu_{j,k}$ and $a_{j,k}$, and an argument similar to that used in the estimation of II_1 in the proof of Theorem 4.3.18 with $\{\lambda_j\}_{j\in\mathbb{N}}$ and $\{\mathcal{M}(a_j)\mathbf{1}_{2B_j}\}_{j\in\mathbb{N}}$ therein replaced, respectively, by

$$\left\{\lambda_j\mu_{j,k}\right\}_{j\in\mathbb{N},k\in\mathbb{N}\cap[3,\infty)}$$

and

$$\left\{a_{j,k}\mathbf{1}_{2^kB_j}\right\}_{j\in\mathbb{N},k\in\mathbb{N}\cap[3,\infty)},$$

further imply that

$$\mathrm{III}_2 \sim \left\|\sum_{j\in\mathbb{N}}\sum_{k=3}^\infty \lambda_j\mu_{j,k}a_{j,k}\right\|_X$$

$$\lesssim \left\|\left[\sum_{j\in\mathbb{N}}\sum_{k=3}^\infty \left(\frac{\lambda_j\mu_{j,k}}{\|\mathbf{1}_{2^kB_j}\|_X}\right)^s \mathbf{1}_{2^kB_j}\right]^{\frac{1}{s}}\right\|_X$$

$$\sim \left\|\left[\sum_{j\in\mathbb{N}}\left(\frac{\lambda_j}{\|\mathbf{1}_{B_j}\|_X}\right)^s \sum_{k=3}^\infty 2^{-ks(\tau+\frac{n}{r})}\mathbf{1}_{2^kB_j}\right]^{\frac{1}{s}}\right\|_X. \qquad (4.65)$$

In order to complete the estimation of III_2, we now estimate the characteristic function $\mathbf{1}_{2^kB_j}$ with $j\in\mathbb{N}$ and $k\in\mathbb{N}\cap[3,\infty)$. Indeed, from (1.55), we deduce that, for any $j\in\mathbb{N}$, $k\in\mathbb{N}\cap[3,\infty)$, and $x\in2^kB_j$,

$$\left[\mathcal{M}^{(\theta)}\left(\mathbf{1}_{B_j}\right)(x)\right]^s \geq \left\{\frac{1}{|2^kB_j|}\int_{2^kB_j}\left[\mathbf{1}_{B_j}(y)\right]^\theta dy\right\}^{\frac{s}{\theta}}$$

$$\sim \left(\frac{|B_j|}{|2^kB_j|}\right)^{\frac{s}{\theta}} \sim 2^{-\frac{nks}{\theta}},$$

which implies that

$$\mathbf{1}_{2^kB_j} \lesssim 2^{\frac{nks}{\theta}}\left[\mathcal{M}^{(\theta)}\left(\mathbf{1}_{B_j}\right)\right]^s.$$

Combining this and the assumption $\tau\in(n(\frac{1}{\theta}-\frac{1}{r}),\infty)$, we further conclude that

$$\sum_{k=3}^\infty 2^{-ks(\tau+\frac{n}{r})}\mathbf{1}_{2^kB_j} \lesssim \sum_{k=3}^\infty 2^{-ks(\tau-\frac{n}{\theta}+\frac{n}{r})}\left[\mathcal{M}^{(\theta)}\left(\mathbf{1}_{B_j}\right)\right]^s \sim \left[\mathcal{M}^{(\theta)}\left(\mathbf{1}_{B_j}\right)\right]^s.$$

Using this, (4.65), and Definition 4.3.17(i) with f_j therein replaced by $\frac{\lambda_j}{\|\mathbf{1}_{B_j}\|_X}\mathbf{1}_{B_j}$ for any $j \in \mathbb{N}$, we find that

$$\mathrm{III}_2 \lesssim \left\| \left\{ \sum_{j\in\mathbb{N}} \left[\mathcal{M}^{(\theta)}\left(\frac{\lambda_j}{\|\mathbf{1}_{B_j}\|_X}\mathbf{1}_{B_j} \right) \right]^s \right\}^{\frac{1}{s}} \right\|_X$$

$$\lesssim \left\| \left[\sum_{j\in\mathbb{N}} \left(\frac{\lambda_j}{\|\mathbf{1}_{B_j}\|_X} \right)^s \mathbf{1}_{B_j} \right]^{\frac{1}{s}} \right\|_X, \tag{4.66}$$

which completes the estimation of III_2.

Finally, for the term III_3, from (4.30), we infer that

$$\mathrm{III}_3 \lesssim \left\| \left[\sum_{j\in\mathbb{N}} \left(\frac{\lambda_j}{\|\mathbf{1}_{B_j}\|_X} \right)^s \mathbf{1}_{B_j} \right]^{\frac{1}{s}} \right\|_X. \tag{4.67}$$

This is the desired estimate of III_3. Thus, combining (4.62), (4.63), (4.66), (4.67), and (4.61), we conclude that

$$\|M(f,\phi)\|_X \lesssim \left\| \left[\sum_{j\in\mathbb{N}} \left(\frac{\lambda_j}{\|\mathbf{1}_{B_j}\|_X} \right)^s \mathbf{1}_{B_j} \right]^{\frac{1}{s}} \right\|_X < \infty, \tag{4.68}$$

which further implies that $f \in H_X(\mathbb{R}^n)$ and hence completes the proof of the sufficiency. Moreover, from Lemma 4.1.4, (4.68), and the choice of $\{\lambda_j\}_{j\in\mathbb{N}}$, it follows that

$$\|f\|_{H_X(\mathbb{R}^n)} \sim \|M(f,\phi)\|_X$$

$$\lesssim \inf \left\{ \left\| \left[\sum_{j\in\mathbb{N}} \left(\frac{\lambda_j}{\|\mathbf{1}_{B_j}\|_X} \right)^s \mathbf{1}_{B_j} \right]^{\frac{1}{s}} \right\|_X \right\},$$

where the infimum is taken over all the sequences $\{\lambda_j\}_{j\in\mathbb{N}} \subset [0,\infty)$ and $\{m_j\}_{j\in\mathbb{N}}$ of (X, r, d, τ)-molecules centered, respectively, at the balls $\{B_j\}_{j\in\mathbb{N}} \subset \mathbb{B}$ such that (4.48) and (4.49) hold true. This, together with (4.60), implies that (4.50) holds true, which completes the proof of Theorem 4.5.11. □

Via the above molecular characterization of $H_X(\mathbb{R}^n)$, we next prove Theorem 4.5.10.

Proof of Theorem 4.5.10 Let p, q, ω, r, d, s, and τ be as in the present theorem. Then, combining the assumptions, $m_0(\omega) \in (-\frac{n}{p}, \infty)$ and $M_\infty(\omega) \in (-\infty, 0)$, and Theorem 1.2.44, we find that the global generalized Herz space $\dot{\mathcal{K}}_\omega^{p,q}(\mathbb{R}^n)$ under consideration is a BQBF space. Therefore, in order to show the present theorem, we only need to prove that all the assumptions of Theorem 4.5.11 are satisfied.

Indeed, let $\theta \in (0, s)$ be such that

$$\left\lfloor n\left(\frac{1}{s} - 1\right)\right\rfloor \le n\left(\frac{1}{\theta} - 1\right) < \left\lfloor n\left(\frac{1}{s} - 1\right)\right\rfloor + 1$$

and

$$\tau > n\left(\frac{1}{\theta} - \frac{1}{r}\right).$$

Thus, we have $d \ge \lfloor n(1/s - 1)\rfloor = \lfloor n(1/\theta - 1)\rfloor$. Then, from (4.40), (4.42), and (4.43), it follows that the following three statements hold true:

(i) for any $\{f_j\}_{j\in\mathbb{N}} \subset L^1_{\mathrm{loc}}(\mathbb{R}^n)$,

$$\left\|\left\{\sum_{j\in\mathbb{N}}\left[\mathcal{M}^{(\theta)}(f_j)\right]^s\right\}^{1/s}\right\|_{\dot{\mathcal{K}}_\omega^{p,q}(\mathbb{R}^n)} \lesssim \left\|\left(\sum_{j\in\mathbb{N}}|f_j|^s\right)^{1/s}\right\|_{\dot{\mathcal{K}}_\omega^{p,q}(\mathbb{R}^n)};$$

(ii) for any $f \in \mathcal{M}(\mathbb{R}^n)$,

$$\|f\|_{[\dot{\mathcal{K}}_\omega^{p,q}(\mathbb{R}^n)]^{1/s}} \sim \sup\left\{\|fg\|_{L^1(\mathbb{R}^n)} : \|g\|_{\dot{\mathcal{B}}_{1/\omega^s}^{(p/s)',(q/s)'}(\mathbb{R}^n)} = 1\right\};$$

(iii) for any $f \in L^1_{\mathrm{loc}}(\mathbb{R}^n)$,

$$\left\|\mathcal{M}^{((r/s)')}(f)\right\|_{\dot{\mathcal{B}}_{1/\omega^s}^{(p/s)',(q/s)'}(\mathbb{R}^n)} \lesssim \|f\|_{\dot{\mathcal{B}}_{1/\omega^s}^{(p/s)',(q/s)'}(\mathbb{R}^n)}.$$

These further imply that, under the assumptions of the present theorem, $\dot{\mathcal{K}}_\omega^{p,q}(\mathbb{R}^n)$ satisfies all the assumptions of Theorem 4.5.11, which completes the proof of Theorem 4.5.10. $\square$

Similarly, we now characterize the generalized Hardy–Morrey space $HM_\omega^{p,q}(\mathbb{R}^n)$ via molecules. To this end, we first introduce the following $(M_\omega^{p,q}(\mathbb{R}^n), r, d, \tau)$-molecules.

Definition 4.5.14 Let p, q, ω, r, d, and τ be as in Definition 4.5.7. Then a measurable function m on $\mathbb{R}^n$ is called an $(M_\omega^{p,q}(\mathbb{R}^n), r, d, \tau)$-*molecule* centered at a ball $B \in \mathbb{B}$ if

(i) for any $j \in \mathbb{Z}_+$,

$$\left\| m \mathbf{1}_{S_j(B)} \right\|_{L^r(\mathbb{R}^n)} \le 2^{-\tau j} \frac{|B|^{1/r}}{\|\mathbf{1}_B\|_{M_\omega^{p,q}(\mathbb{R}^n)}};$$

(ii) for any $\alpha \in \mathbb{Z}_+^n$ with $|\alpha| \le d$,

$$\int_{\mathbb{R}^n} m(x) x^\alpha \, dx = 0.$$

Then we immediately obtain the following molecular characterization of the generalized Hardy–Morrey space $HM_\omega^{p,q}(\mathbb{R}^n)$ via Theorem 4.5.10 and Remarks 1.2.2(iv) and 4.0.20(ii); we omit the details.

Corollary 4.5.15 *Let p, q, ω, r, d, and s be as in Corollary 4.3.13 and $\tau \in (0, \infty)$ with $\tau > n(1/s - 1/r)$. Then $f \in HM_\omega^{p,q}(\mathbb{R}^n)$ if and only if $f \in \mathcal{S}'(\mathbb{R}^n)$ and there exists a sequence $\{m_j\}_{j \in \mathbb{N}}$ of $(M_\omega^{p,q}(\mathbb{R}^n), r, d, \tau)$-molecules centered, respectively, at the balls $\{B_j\}_{j \in \mathbb{N}} \subset \mathbb{B}$ and a sequence $\{\lambda_j\}_{j \in \mathbb{N}} \subset [0, \infty)$ such that*

$$f = \sum_{j \in \mathbb{N}} \lambda_j m_j$$

in $\mathcal{S}'(\mathbb{R}^n)$ and

$$\left\| \left\{ \sum_{j \in \mathbb{N}} \left[\frac{\lambda_j}{\|\mathbf{1}_{B_j}\|_{M_\omega^{p,q}(\mathbb{R}^n)}} \right]^s \mathbf{1}_{B_j} \right\}^{\frac{1}{s}} \right\|_{M_\omega^{p,q}(\mathbb{R}^n)} < \infty.$$

Moreover, for any $f \in HM_\omega^{p,q}(\mathbb{R}^n)$,

$$\|f\|_{HM_\omega^{p,q}(\mathbb{R}^n)} \sim \inf \left\{ \left\| \left\{ \sum_{j \in \mathbb{N}} \left[\frac{\lambda_i}{\|\mathbf{1}_{B_j}\|_{M_\omega^{p,q}(\mathbb{R}^n)}} \right]^s \mathbf{1}_{B_j} \right\}^{\frac{1}{s}} \right\|_{M_\omega^{p,q}(\mathbb{R}^n)} \right\}$$

with the positive equivalence constants independent of f, where the infimum is taken over all the decompositions of f as above.

4.6 Littlewood–Paley Function Characterizations

In this section, we establish various Littlewood–Paley function characterizations of
generalized Herz–Hardy spaces. To be precise, we characterize generalized Herz–
Hardy spaces via the Lusin area function, the Littlewood–Paley g-function, and the
Littlewood–Paley g_λ^*-function.

Throughout this book, for any $\varphi \in \mathcal{S}(\mathbb{R}^n)$, $\mathcal{F}\varphi$ or $\widehat{\varphi}$ always denotes its *Fourier
transform*, namely, for any $x \in \mathbb{R}^n$,

$$\mathcal{F}\varphi(x) := \widehat{\varphi}(x) := \int_{\mathbb{R}^n} \varphi(\xi) e^{-2\pi i x \cdot \xi} \, d\xi,$$

where $i := \sqrt{-1}$ and $x \cdot \xi := \sum_{j=1}^n x_j \xi_j$ for any $x := (x_1, \ldots, x_n), \xi :=
(\xi_1, \ldots, \xi_n) \in \mathbb{R}^n$. Moreover, for any $f \in \mathcal{S}'(\mathbb{R}^n)$, its *Fourier transform*, also
denoted by $\mathcal{F}f$ or $\widehat{f}$, is defined by setting, for any $\phi \in \mathcal{S}(\mathbb{R}^n)$,

$$\langle \mathcal{F}f, \phi \rangle := \langle \widehat{f}, \phi \rangle := \langle f, \widehat{\phi} \rangle.$$

Then we state the definitions of the Lusin area function, the Littlewood–Paley g-
function, and the Littlewood–Paley g_λ^*-function as follows (see, for instance, [29,
Definitions 4.1 and 4.2]).

Definition 4.6.1 Let $\varphi \in \mathcal{S}(\mathbb{R}^n)$ satisfy $\widehat{\varphi}(0) = 0$ and, for any $\xi \in \mathbb{R}^n \setminus \{0\}$, there
exists a $t \in (0, \infty)$ such that $\widehat{\varphi}(t\xi) \neq 0$. Then, for any $f \in \mathcal{S}'(\mathbb{R}^n)$, the *Lusin
area function* $S(f)$ and the *Littlewood–Paley g_λ^*-function* $g_\lambda^*(f)$ with $\lambda \in (0, \infty)$
are defined, respectively, by setting, for any $x \in \mathbb{R}^n$,

$$S(f)(x) := \left[\int_{\Gamma(x)} |f * \varphi_t(y)|^2 \frac{dy \, dt}{t^{n+1}} \right]^{\frac{1}{2}}$$

and

$$g_\lambda^*(f)(x) := \left[\int_0^\infty \int_{\mathbb{R}^n} \left(\frac{t}{t + |x - y|} \right)^{\lambda n} |f * \varphi_t(y)|^2 \frac{dy \, dt}{t^{n+1}} \right]^{\frac{1}{2}},$$

here and thereafter, for any $x \in \mathbb{R}^n$,

$$\Gamma(x) := \{(y, t) \in \mathbb{R}_+^{n+1} : |y - x| < t\}.$$

Definition 4.6.2 Let $\varphi \in \mathcal{S}(\mathbb{R}^n)$ satisfy $\widehat{\varphi}(0) = 0$ and, for any $x \in \mathbb{R}^n \setminus \{0\}$, there
exists a $j \in \mathbb{Z}$ such that $\widehat{\varphi}(2^j x) \neq 0$. Then, for any $f \in \mathcal{S}'(\mathbb{R}^n)$, the *Littlewood–*

Paley g-function $g(f)$ is defined by setting, for any $x \in \mathbb{R}^n$,

$$g(f)(x) := \left[\int_0^\infty |f * \varphi_t(x)|^2 \frac{dt}{t} \right]^{\frac{1}{2}}.$$

Recall that $f \in \mathcal{S}'(\mathbb{R}^n)$ is said to *vanish weakly at infinity* if, for any $\phi \in \mathcal{S}(\mathbb{R}^n)$, $f * \phi_t \to 0$ in $\mathcal{S}'(\mathbb{R}^n)$ as $t \to \infty$ with $\phi_t(\cdot) = t^{-n}\phi(\cdot/t)$. Now, we establish various Littlewood–Paley function characterizations of the generalized Herz–Hardy spaces $H\dot{\mathcal{K}}^{p,q}_{\omega,\mathbf{0}}(\mathbb{R}^n)$ and $H\dot{\mathcal{K}}^{p,q}_{\omega}(\mathbb{R}^n)$ as follows. To begin with, we give the Littlewood–Paley function characterizations of $H\dot{\mathcal{K}}^{p,q}_{\omega,\mathbf{0}}(\mathbb{R}^n)$.

Theorem 4.6.3 *Let* $p, q \in (0, \infty)$, $\omega \in M(\mathbb{R}_+)$ *satisfy* $m_0(\omega) \in (-\frac{n}{p}, \infty)$ *and* $m_\infty(\omega) \in (-\frac{n}{p}, \infty)$,

$$s_0 := \min\left\{ 1, p, q, \frac{n}{\max\{M_0(\omega), M_\infty(\omega)\} + n/p} \right\},$$

and $\lambda \in (\max\{1, 2/s_0\}, \infty)$. *Then the following four statements are mutually equivalent:*

(i) $f \in H\dot{\mathcal{K}}^{p,q}_{\omega,\mathbf{0}}(\mathbb{R}^n)$;
(ii) $f \in \mathcal{S}'(\mathbb{R}^n)$, f *vanishes weakly at infinity, and* $S(f) \in \dot{\mathcal{K}}^{p,q}_{\omega,\mathbf{0}}(\mathbb{R}^n)$;
(iii) $f \in \mathcal{S}'(\mathbb{R}^n)$, f *vanishes weakly at infinity, and* $g(f) \in \dot{\mathcal{K}}^{p,q}_{\omega,\mathbf{0}}(\mathbb{R}^n)$;
(iv) $f \in \mathcal{S}'(\mathbb{R}^n)$, f *vanishes weakly at infinity, and* $g^*_\lambda(f) \in \dot{\mathcal{K}}^{p,q}_{\omega,\mathbf{0}}(\mathbb{R}^n)$.

Moreover, for any $f \in H\dot{\mathcal{K}}^{p,q}_{\omega,\mathbf{0}}(\mathbb{R}^n)$,

$$\|f\|_{H\dot{\mathcal{K}}^{p,q}_{\omega,\mathbf{0}}(\mathbb{R}^n)} \sim \|S(f)\|_{\dot{\mathcal{K}}^{p,q}_{\omega,\mathbf{0}}(\mathbb{R}^n)} \sim \|g(f)\|_{\dot{\mathcal{K}}^{p,q}_{\omega,\mathbf{0}}(\mathbb{R}^n)} \sim \|g^*_\lambda(f)\|_{\dot{\mathcal{K}}^{p,q}_{\omega,\mathbf{0}}(\mathbb{R}^n)},$$

where the positive equivalence constants are independent of f.

To show Theorem 4.6.3, recall that Chang et al. [29, Theorems 4.9, 4.11, and 4.13] investigated the Lusin area function, the g-function, and the g^*_λ-function characterizations of the Hardy space $H_X(\mathbb{R}^n)$ as follows, which is vital in the proof of the Littlewood–Paley function characterizations of $H\dot{\mathcal{K}}^{p,q}_{\omega,\mathbf{0}}(\mathbb{R}^n)$.

Lemma 4.6.4 *Let* $s \in (0, 1]$, $\theta \in (0, s)$, $\lambda \in (\max\{1, \frac{2}{s}\}, \infty)$, *and* X *be a ball quasi-Banach function space. Assume that Assumption 1.2.29 holds true for both* X *and* $X^{s/2}$ *with* θ *and* s *as above and* X *satisfies Assumption 1.2.33 with* s *as above. Then the following four statements are mutually equivalent:*

(i) $f \in H_X(\mathbb{R}^n)$;
(ii) $f \in \mathcal{S}'(\mathbb{R}^n)$, f *vanishes weakly at infinity, and* $S(f) \in X$;
(iii) $f \in \mathcal{S}'(\mathbb{R}^n)$, f *vanishes weakly at infinity, and* $g(f) \in X$;
(iv) $f \in \mathcal{S}'(\mathbb{R}^n)$, f *vanishes weakly at infinity, and* $g^*_\lambda(f) \in X$.

Moreover, for any $f \in H_X(\mathbb{R}^n)$,

$$\|f\|_{H_X(\mathbb{R}^n)} \sim \|S(f)\|_X \sim \|g(f)\|_X \sim \|g_\lambda^*(f)\|_X$$

with the positive equivalence constants independent of f.

Remark 4.6.5 We point out that Lemma 4.6.4 has a wide range of applications. Here we give several function spaces to which Lemma 4.6.4 can be applied (see also [29, Section 5]).

(i) Let $p \in (0, \infty)$ and $\lambda \in (\max\{1, \frac{2}{p}\}, \infty)$. Then, in this case, by both Remarks 1.2.31(i) and 1.2.34(i), we can easily conclude that the Lebesgue space $L^p(\mathbb{R}^n)$ satisfies all the assumptions of Lemma 4.6.4. Therefore, Lemma 4.6.4 with $X := L^p(\mathbb{R}^n)$ holds true. This result is the known Littlewood–Paley function characterizations of the Hardy space $H^p(\mathbb{R}^n)$ (see, for instance, [85, Chapter 7]).

(ii) Let $p \in (0, \infty)$, $\upsilon \in A_\infty(\mathbb{R}^n)$, and $\lambda \in (\max\{1, \frac{2q_\upsilon}{p}\}, \infty)$, where q_υ is the same as in (1.58). Then, in this case, as was pointed out in [29, Subsection 5.4], the weighted Lebesgue space $L_\upsilon^p(\mathbb{R}^n)$ satisfies all the assumptions of Lemma 4.6.4. This then implies that Lemma 4.6.4 with $X := L_\upsilon^p(\mathbb{R}^n)$ holds true (see, for instance, [29, Theorems 5.16 and 5.18]).

(iii) Let $\vec{p} := (p_1, \ldots, p_n) \in (0, \infty)^n$ and

$$\lambda \in \left(\max\left\{1, \frac{2}{\min\{p_1, \ldots, p_n\}}\right\}, \infty \right).$$

Then, in this case, as was mentioned in [29, Subsection 5.2], the mixed-norm Lebesgue space $L^{\vec{p}}(\mathbb{R}^n)$ satisfies all the assumptions of Lemma 4.6.4. Thus, Lemma 4.6.4 with $X := L^{\vec{p}}(\mathbb{R}^n)$ holds true. These Littlewood–Paley function characterizations of mixed-norm Hardy spaces were also established in [29, Theorems 5.8 and 5.10] (see also [123, 127]).

(iv) Let $0 < q \leq p < \infty$ and $\lambda \in (\max\{1, \frac{2}{q}\}, \infty)$. Then, in this case, as was pointed out in [29, Subsection 5.1], the Morrey space $M_q^p(\mathbb{R}^n)$ satisfies all the assumptions of Lemma 4.6.4. Therefore, Lemma 4.6.4 with $X := M_q^p(\mathbb{R}^n)$ holds true (see, for instance, [29, Theorems 5.3 and 5.5]).

(v) Let $p(\cdot) \in C^{\log}(\mathbb{R}^n)$ satisfy $0 < p_- \leq p_+ < \infty$, and $\lambda \in (\max\{1, \frac{2}{p_-}\}, \infty)$, where p_- and p_+ are defined, respectively, in (1.59) and (1.60). Then, in this case, as was mentioned in [29, Subsection 5.3], the variable Lebesgue space $L^{p(\cdot)}(\mathbb{R}^n)$ satisfies all the assumptions of Lemma 4.6.4. This further implies that Lemma 4.6.4 with $X := L^{p(\cdot)}(\mathbb{R}^n)$ holds true (see, for instance, [29, Theorems 5.12 and 5.14]).

Via this lemma, we now show Theorem 4.6.3.

Proof of Theorem 4.6.3 Let all the symbols be as in the present theorem. Then, combining the assumption $m_0(\omega) \in (-\frac{n}{p}, \infty)$ and Theorem 1.2.42, we find that the local generalized Herz space $\dot{\mathcal{K}}^{p,q}_{\omega,\mathbf{0}}(\mathbb{R}^n)$ is a BQBF space. This implies that, to complete the proof of the present theorem, we only need to show that, under the assumptions of the present theorem, $\dot{\mathcal{K}}^{p,q}_{\omega,\mathbf{0}}(\mathbb{R}^n)$ satisfies all the assumptions of Lemma 4.6.4.

For this purpose, let $s \in (\frac{2}{\lambda}, s_0)$ and $\theta \in (0, \min\{s, \frac{s^2}{2}\})$. We now prove that both $\dot{\mathcal{K}}^{p,q}_{\omega,\mathbf{0}}(\mathbb{R}^n)$ and $[\dot{\mathcal{K}}^{p,q}_{\omega,\mathbf{0}}(\mathbb{R}^n)]^{s/2}$ satisfy Assumption 1.2.29 with these θ and s. Indeed, applying Lemma 4.3.11, we conclude that, for any $\{f_j\}_{j\in\mathbb{N}} \subset L^1_{\mathrm{loc}}(\mathbb{R}^n)$,

$$\left\| \left\{ \sum_{j\in\mathbb{N}} \left[\mathcal{M}^{(\theta)}(f_j) \right]^s \right\}^{1/s} \right\|_{\dot{\mathcal{K}}^{p,q}_{\omega,\mathbf{0}}(\mathbb{R}^n)} \lesssim \left\| \left(\sum_{j\in\mathbb{N}} |f_j|^s \right)^{1/s} \right\|_{\dot{\mathcal{K}}^{p,q}_{\omega,\mathbf{0}}(\mathbb{R}^n)}.$$

This implies that, for the above θ and s, Assumption 1.2.29 holds true for $\dot{\mathcal{K}}^{p,q}_{\omega,\mathbf{0}}(\mathbb{R}^n)$. On the other hand, from the assumptions $\theta \in (0, \min\{s, \frac{s^2}{2}\})$ and

$$s \in \left(0, \min\left\{ p, \frac{n}{\max\{M_0(\omega), M_\infty(\omega)\} + n/p} \right\} \right),$$

we deduce that $\frac{s}{\theta} \in (1, \infty)$ and

$$\min\left\{ p, \frac{n}{\max\{M_0(\omega), M_\infty(\omega)\} + n/p} \right\} > s > \frac{2\theta}{s}.$$

This, together with Lemma 4.3.11 with $\dot{\mathcal{K}}^{p,q}_{\omega,\mathbf{0}}(\mathbb{R}^n)$ therein replaced by $[\dot{\mathcal{K}}^{p,q}_{\omega,\mathbf{0}}(\mathbb{R}^n)]^{s/2}$, further implies that, for any $\{f_j\}_{j\in\mathbb{N}} \subset L^1_{\mathrm{loc}}(\mathbb{R}^n)$,

$$\left\| \left\{ \sum_{j\in\mathbb{N}} \left[\mathcal{M}^{(\theta)}(f_j) \right]^s \right\}^{1/s} \right\|_{[\dot{\mathcal{K}}^{p,q}_{\omega,\mathbf{0}}(\mathbb{R}^n)]^{s/2}} \lesssim \left\| \left(\sum_{j\in\mathbb{N}} |f_j|^s \right)^{1/s} \right\|_{[\dot{\mathcal{K}}^{p,q}_{\omega,\mathbf{0}}(\mathbb{R}^n)]^{s/2}},$$

which completes the proof that Assumption 1.2.29 holds true for both $\dot{\mathcal{K}}^{p,q}_{\omega,\mathbf{0}}(\mathbb{R}^n)$ and $[\dot{\mathcal{K}}^{p,q}_{\omega,\mathbf{0}}(\mathbb{R}^n)]^{s/2}$.

In addition, notice that $\lambda \in (\max\{1, \frac{2}{s}\}, \infty)$ and we can choose an

$$r \in \left(\max\left\{ 1, p, \frac{n}{\min\{m_0(\omega), m_\infty(\omega)\} + n/p} \right\}, \infty \right].$$

Then, by Lemma 1.8.6, we find that $[\dot{\mathcal{K}}_{\omega,\mathbf{0}}^{p,q}(\mathbb{R}^n)]^{1/s}$ is a BBF space and, for any $f \in L^1_{\mathrm{loc}}(\mathbb{R}^n)$,

$$\left\| \mathcal{M}^{((r/s)')}(f) \right\|_{([\dot{\mathcal{K}}_{\omega,\mathbf{0}}^{p,q}(\mathbb{R}^n)]^{1/s})'} \lesssim \|f\|_{([\dot{\mathcal{K}}_{\omega,\mathbf{0}}^{p,q}(\mathbb{R}^n)]^{1/s})'},$$

which further implies that all the assumptions of Lemma 4.6.4 are satisfied for $\dot{\mathcal{K}}_{\omega,\mathbf{0}}^{p,q}(\mathbb{R}^n)$. This finishes the proof of Theorem 4.6.3. $\qquad\square$

Remark 4.6.6 We point out that, in Theorem 4.6.3, when $p \in (1,\infty)$, $q \in [1,\infty)$, and $\omega(t) := t^\alpha$ for any $t \in (0,\infty)$ and for any given $\alpha \in (-\frac{n}{p}, \frac{n}{p})$, from Remark 4.0.18 and Theorem 4.2.1, it follows that the generalized Herz–Hardy space $H\dot{\mathcal{K}}_{\omega,\mathbf{0}}^{p,q}(\mathbb{R}^n)$ coincides with the classical homogeneous Herz space $\dot{K}_p^{\alpha,q}(\mathbb{R}^n)$ and hence the conclusion obtained in this theorem goes back to [175, Theorem 1.1.1].

Using Theorem 4.6.3 and Remarks 1.2.2(iv) and 4.0.20(ii), we immediately obtain the following Littlewood–Paley function characterizations of the generalized Hardy–Morrey space $HM_{\omega,\mathbf{0}}^{p,q}(\mathbb{R}^n)$; we omit the details.

Corollary 4.6.7 *Let $p, q \in [1,\infty)$, $\omega \in M(\mathbb{R}_+)$ satisfy*

$$-\frac{n}{p} < m_0(\omega) \le M_0(\omega) < 0$$

and

$$-\frac{n}{p} < m_\infty(\omega) \le M_\infty(\omega) < 0,$$

and $\lambda \in (2,\infty)$. Then the following four statements are mutually equivalent:

(i) $f \in HM_{\omega,\mathbf{0}}^{p,q}(\mathbb{R}^n)$;
(ii) $f \in \mathcal{S}'(\mathbb{R}^n)$, f vanishes weakly at infinity, and $S(f) \in M_{\omega,\mathbf{0}}^{p,q}(\mathbb{R}^n)$;
(iii) $f \in \mathcal{S}'(\mathbb{R}^n)$, f vanishes weakly at infinity, and $g(f) \in M_{\omega,\mathbf{0}}^{p,q}(\mathbb{R}^n)$;
(iv) $f \in \mathcal{S}'(\mathbb{R}^n)$, f vanishes weakly at infinity, and $g_\lambda^*(f) \in M_{\omega,\mathbf{0}}^{p,q}(\mathbb{R}^n)$.

Moreover, for any $f \in HM_{\omega,\mathbf{0}}^{p,q}(\mathbb{R}^n)$,

$$\|f\|_{HM_{\omega,\mathbf{0}}^{p,q}(\mathbb{R}^n)} \sim \|S(f)\|_{M_{\omega,\mathbf{0}}^{p,q}(\mathbb{R}^n)} \sim \|g(f)\|_{M_{\omega,\mathbf{0}}^{p,q}(\mathbb{R}^n)} \sim \|g_\lambda^*(f)\|_{M_{\omega,\mathbf{0}}^{p,q}(\mathbb{R}^n)},$$

where the positive equivalence constants are independent of f.

On the other hand, we now show the following Littlewood–Paley function characterizations of the generalized Herz–Hardy space $H\dot{\mathcal{K}}_\omega^{p,q}(\mathbb{R}^n)$.

Theorem 4.6.8 *Let* $p, q \in (0, \infty)$, $\omega \in M(\mathbb{R}_+)$ *satisfy* $m_0(\omega) \in (-\frac{n}{p}, \infty)$ *and*

$$-\frac{n}{p} < m_\infty(\omega) \le M_\infty(\omega) < 0,$$

$$s_0 := \min\left\{1, p, q, \frac{n}{\max\{M_0(\omega), M_\infty(\omega)\} + n/p}\right\},$$

and $\lambda \in (\max\{1, 2/s_0\}, \infty)$. *Then the following four statements are mutually equivalent:*

(i) $f \in H\dot{\mathcal{K}}_\omega^{p,q}(\mathbb{R}^n)$;
(ii) $f \in \mathcal{S}'(\mathbb{R}^n)$, f *vanishes weakly at infinity, and* $S(f) \in \dot{\mathcal{K}}_\omega^{p,q}(\mathbb{R}^n)$;
(iii) $f \in \mathcal{S}'(\mathbb{R}^n)$, f *vanishes weakly at infinity, and* $g(f) \in \dot{\mathcal{K}}_\omega^{p,q}(\mathbb{R}^n)$;
(iv) $f \in \mathcal{S}'(\mathbb{R}^n)$, f *vanishes weakly at infinity, and* $g_\lambda^*(f) \in \dot{\mathcal{K}}_\omega^{p,q}(\mathbb{R}^n)$.

Moreover, for any $f \in H\dot{\mathcal{K}}_\omega^{p,q}(\mathbb{R}^n)$,

$$\|f\|_{H\dot{\mathcal{K}}_\omega^{p,q}(\mathbb{R}^n)} \sim \|S(f)\|_{\dot{\mathcal{K}}_\omega^{p,q}(\mathbb{R}^n)} \sim \|g(f)\|_{\dot{\mathcal{K}}_\omega^{p,q}(\mathbb{R}^n)} \sim \|g_\lambda^*(f)\|_{\dot{\mathcal{K}}_\omega^{p,q}(\mathbb{R}^n)},$$

where the positive equivalence constants are independent of f.

To prove Theorem 4.6.8, we first give some symbols. Recall that, for any given $\xi \in \mathbb{R}^n$, the *translation operator* τ_ξ is defined by setting, for any $f \in \mathcal{M}(\mathbb{R}^n)$ and $x \in \mathbb{R}^n$,

$$\tau_\xi(f)(x) := f(x - \xi). \tag{4.69}$$

Furthermore, the *translation operator* τ_ξ of distributions with $\xi \in \mathbb{R}^n$ is defined by setting, for any $f \in \mathcal{S}'(\mathbb{R}^n)$ and $\phi \in \mathcal{S}(\mathbb{R}^n)$,

$$\langle \tau_\xi(f), \phi \rangle := \langle f, \tau_{-\xi}(\phi) \rangle = \langle f, \phi(\cdot + \xi) \rangle.$$

The following technical lemma establishes the relations among translations, convolutions, and various Littlewood–Paley functions, which plays a key role in the proof of Theorem 4.6.8.

Lemma 4.6.9 *Let* $f \in \mathcal{S}'(\mathbb{R}^n)$, $\phi \in \mathcal{S}(\mathbb{R}^n)$, *and* $\xi \in \mathbb{R}^n$. *Then*

(i) $[\tau_\xi(f)] * \phi = \tau_\xi(f * \phi)$;
(ii) $M(\tau_\xi(f), \phi) = \tau_\xi(M(f, \phi))$, *where the radial maximal function M is defined as in Definition 4.1.1(i)*;
(iii) $A(\tau_\xi(f)) = \tau_\xi(A(f))$, *where* $A \in \{S, g, g_\lambda^*\}$ *with* $\lambda \in (0, \infty)$;
(iv) *if f vanishes weakly at infinity, then $\tau_\xi(f)$ vanishes weakly at infinity.*

Proof Let all the symbols be as in the present lemma. We first prove (i). Indeed, for any $x \in \mathbb{R}^n$, we have

$$\left[\tau_\xi(f)\right] * \phi(x) = \left\langle \tau_\xi(f), \phi(x - \cdot) \right\rangle = \left\langle f, \phi(x - (\cdot + \xi)) \right\rangle$$

$$= \left\langle f, \phi(x - \xi - \cdot) \right\rangle = (f * \phi)(x - \xi)$$

$$= \tau_\xi(f * \phi)(x),$$

which implies that (i) holds true.

Next, we show (ii). Applying (i) with ϕ therein replaced by ϕ_t for any $t \in (0, \infty)$, we find that

$$M\left(\tau_\xi(f), \phi\right) = \sup_{t \in (0,\infty)} \left\{ \left| \left[\tau_\xi(f)\right] * \phi_t \right| \right\} = \sup_{t \in (0,\infty)} \left\{ \left| \tau_\xi(f * \phi_t) \right| \right\}$$

$$= \tau_\xi \left(\sup_{t \in (0,\infty)} \left\{ |f * \phi_t| \right\} \right) = \tau_\xi \left(M(f, \phi) \right).$$

This finishes the proof of (ii).

We then prove (iii). First, let $A := S$. Then, from Definition 4.6.1 and (i) with $\phi = \varphi_t$ for any $t \in (0, \infty)$, it follows that, for any $x \in \mathbb{R}^n$,

$$\left[S\left(\tau_\xi(f) \right)(x) \right]^2 = \int_{\Gamma(x)} \left| \left[\tau_\xi(f)\right] * \varphi_t(y) \right|^2 \frac{dy\,dt}{t^{n+1}}$$

$$= \int_{\Gamma(x)} |(f * \varphi_t)(y - \xi)|^2 \frac{dy\,dt}{t^{n+1}}$$

$$= \int_{\Gamma(x-\xi)} |f * \varphi_t(y)|^2 \frac{dy\,dt}{t^{n+1}} = \left[\tau_\xi(S(f))(x) \right]^2.$$

Thus, $S(\tau_\xi(f)) = \tau_\xi(S(f))$ holds true. Similarly, we can obtain $g(\tau_\xi(f)) = \tau_\xi(g(f))$ and $g_\lambda^*(\tau_\xi(f)) = \tau_\xi(g_\lambda^*(f))$ and hence (iii) holds true.

Finally, we prove (iv). Indeed, for any ψ, using (i) with ϕ therein replaced by ψ_t for any $t \in (0, \infty)$, and the assumption that f vanishes weakly at infinity, we find that, for any $\eta \in \mathcal{S}(\mathbb{R}^n)$,

$$\int_{\mathbb{R}^n} \left[\tau_\xi(f)\right] * \psi_t(y)\eta(y)\,dy = \int_{\mathbb{R}^n} (f * \psi_t)(y - \xi)\eta(y)\,dy$$

$$= \int_{\mathbb{R}^n} (f * \psi_t)(y)\eta(y + \xi)\,dy \to 0$$

as $t \to \infty$. This further implies that, for any $\psi \in \mathcal{S}(\mathbb{R}^n)$, $\left[\tau_\xi(f)\right] * \psi_t \to 0$ in $\mathcal{S}'(\mathbb{R}^n)$ as $t \to \infty$, which completes the proof of (iv) and hence Lemma 4.6.9. $\square$

Now, we show Theorem 4.6.8 via Lemma 4.6.9.

Proof of Theorem 4.6.8 Let p, q, ω, and λ be as in the present theorem, and let $\phi \in \mathcal{S}(\mathbb{R}^n)$ satisfy that $\int_{\mathbb{R}^n} \phi(x)\,dx \neq 0$. We now first show that (i) implies (ii). To this end, let $f \in H\dot{\mathcal{K}}_{\omega}^{p,q}(\mathbb{R}^n)$. Then, applying this and Theorem 4.1.8, we find that $M(f, \phi) \in \dot{\mathcal{K}}_{\omega}^{p,q}(\mathbb{R}^n)$. This, together with Lemma 4.6.9(ii), further implies that, for any $\xi \in \mathbb{R}^n$,

$$\left\| M\left(\tau_\xi(f), \phi\right) \right\|_{\dot{\mathcal{K}}_{\omega,\mathbf{0}}^{p,q}(\mathbb{R}^n)} = \left\| \tau_\xi\left(M(f, \phi)\right) \right\|_{\dot{\mathcal{K}}_{\omega,\mathbf{0}}^{p,q}(\mathbb{R}^n)}$$

$$\leq \|M(f, \phi)\|_{\dot{\mathcal{K}}_{\omega}^{p,q}(\mathbb{R}^n)} \sim \|f\|_{H\dot{\mathcal{K}}_{\omega}^{p,q}(\mathbb{R}^n)} < \infty.$$

Therefore, for any $\xi \in \mathbb{R}^n$, $M(\tau_\xi(f), \phi) \in \dot{\mathcal{K}}_{\omega,\mathbf{0}}^{p,q}(\mathbb{R}^n)$. From this and Theorem 4.1.2, we deduce that, for any $\xi \in \mathbb{R}^n$, $\tau_\xi(f) \in H\dot{\mathcal{K}}_{\omega,\mathbf{0}}^{p,q}(\mathbb{R}^n)$. Combining this and Theorem 4.6.3, we conclude that, for any $\xi \in \mathbb{R}^n$, $\tau_\xi(f)$ vanishes weakly at infinity, $S(\tau_\xi(f)) \in \dot{\mathcal{K}}_{\omega,\mathbf{0}}^{p,q}(\mathbb{R}^n)$, and

$$\left\| S\left(\tau_\xi(f)\right) \right\|_{\dot{\mathcal{K}}_{\omega,\mathbf{0}}^{p,q}(\mathbb{R}^n)} \sim \left\| \tau_\xi(f) \right\|_{H\dot{\mathcal{K}}_{\omega,\mathbf{0}}^{p,q}(\mathbb{R}^n)}. \qquad (4.70)$$

In particular, letting $\xi := \mathbf{0}$, we have f vanishes weakly at infinity. Then, from Lemma 4.6.9(iii), (4.70), Theorem 4.1.2, and Lemma 4.6.9(ii), we further deduce that, for any $\xi \in \mathbb{R}^n$,

$$\left\| \tau_\xi\left(S(f)\right) \right\|_{\dot{\mathcal{K}}_{\omega,\mathbf{0}}^{p,q}(\mathbb{R}^n)} = \left\| S\left(\tau_\xi(f)\right) \right\|_{\dot{\mathcal{K}}_{\omega,\mathbf{0}}^{p,q}(\mathbb{R}^n)}$$

$$\sim \left\| \tau_\xi(f) \right\|_{H\dot{\mathcal{K}}_{\omega,\mathbf{0}}^{p,q}(\mathbb{R}^n)} \sim \left\| M\left(\tau_\xi(f), \phi\right) \right\|_{\dot{\mathcal{K}}_{\omega,\mathbf{0}}^{p,q}(\mathbb{R}^n)}$$

$$\sim \left\| \tau_\xi\left(M(f, \phi)\right) \right\|_{\dot{\mathcal{K}}_{\omega,\mathbf{0}}^{p,q}(\mathbb{R}^n)}.$$

By this, the definition of $\|\cdot\|_{\dot{\mathcal{K}}_{\omega}^{p,q}(\mathbb{R}^n)}$, and Theorem 4.1.8, we find that

$$\|S(f)\|_{\dot{\mathcal{K}}_{\omega}^{p,q}(\mathbb{R}^n)} \sim \|M(f, \phi)\|_{\dot{\mathcal{K}}_{\omega}^{p,q}(\mathbb{R}^n)} \sim \|f\|_{H\dot{\mathcal{K}}_{\omega}^{p,q}(\mathbb{R}^n)} < \infty, \qquad (4.71)$$

which further implies that $S(f) \in \dot{\mathcal{K}}_{\omega}^{p,q}(\mathbb{R}^n)$, and hence (i) implies (ii).

Conversely, we show that (ii) implies (i), namely, assume $f \in \mathcal{S}'(\mathbb{R}^n)$ vanishes weakly at infinity and $S(f) \in \dot{\mathcal{K}}_{\omega}^{p,q}(\mathbb{R}^n)$, we need to prove that $f \in H\dot{\mathcal{K}}_{\omega}^{p,q}(\mathbb{R}^n)$. Indeed, from Lemma 4.6.9(iii), we deduce that, for any $\xi \in \mathbb{R}^n$,

$$\left\| S\left(\tau_\xi(f)\right) \right\|_{\dot{\mathcal{K}}_{\omega,\mathbf{0}}^{p,q}(\mathbb{R}^n)} = \left\| \tau_\xi\left(S(f)\right) \right\|_{\dot{\mathcal{K}}_{\omega,\mathbf{0}}^{p,q}(\mathbb{R}^n)} \leq \|S(f)\|_{\dot{\mathcal{K}}_{\omega}^{p,q}(\mathbb{R}^n)} < \infty, \qquad (4.72)$$

which implies that $S(\tau_\xi(f)) \in \dot{\mathcal{K}}_{\omega,\mathbf{0}}^{p,q}(\mathbb{R}^n)$. On the other hand, by Lemma 4.6.9(iv), we conclude that, for any $\xi \in \mathbb{R}^n$, $\tau_\xi(f)$ vanishes weakly at infinity. Combining this, the fact that $S(\tau_\xi(f)) \in \dot{\mathcal{K}}_{\omega,\mathbf{0}}^{p,q}(\mathbb{R}^n)$ for any $\xi \in \mathbb{R}^n$, and Theorem 4.6.3, we further conclude that, for any $\xi \in \mathbb{R}^n$, $\tau_\xi(f) \in H\dot{\mathcal{K}}_{\omega,\mathbf{0}}^{p,q}(\mathbb{R}^n)$. Then, applying Theorem 4.1.8,

Lemma 4.6.9(ii), Theorems 4.1.2 and 4.6.3, and (4.72), we find that

$$
\|f\|_{H\dot{\mathcal{K}}_{\omega}^{p,q}(\mathbb{R}^n)} \sim \sup_{\xi\in\mathbb{R}^n} \left\|\tau_\xi\left(M(f,\phi)\right)\right\|_{\dot{\mathcal{K}}_{\omega,\mathbf{0}}^{p,q}(\mathbb{R}^n)}
$$
$$
\sim \sup_{\xi\in\mathbb{R}^n} \left\|M\left(\tau_\xi(f),\phi\right)\right\|_{\dot{\mathcal{K}}_{\omega,\mathbf{0}}^{p,q}(\mathbb{R}^n)} \sim \sup_{\xi\in\mathbb{R}^n} \left\|\tau_\xi(f)\right\|_{H\dot{\mathcal{K}}_{\omega,\mathbf{0}}^{p,q}(\mathbb{R}^n)}
$$
$$
\sim \sup_{\xi\in\mathbb{R}^n} \left\|S\left(\tau_\xi(f)\right)\right\|_{\dot{\mathcal{K}}_{\omega,\mathbf{0}}^{p,q}(\mathbb{R}^n)} \lesssim \|S(f)\|_{\dot{\mathcal{K}}_{\omega}^{p,q}(\mathbb{R}^n)} < \infty,
$$

which further implies that $f \in H\dot{\mathcal{K}}_{\omega}^{p,q}(\mathbb{R}^n)$, and hence (ii) implies (i). Moreover, from (4.71), it follows that, for any $f \in H\dot{\mathcal{K}}_{\omega}^{p,q}(\mathbb{R}^n)$,

$$
\|f\|_{H\dot{\mathcal{K}}_{\omega}^{p,q}(\mathbb{R}^n)} \sim \|S(f)\|_{\dot{\mathcal{K}}_{\omega}^{p,q}(\mathbb{R}^n)}.
$$

Similarly, we can obtain (i) is equivalent to both (iii) and (iv) and, for any $f \in H\dot{\mathcal{K}}_{\omega}^{p,q}(\mathbb{R}^n)$,

$$
\|f\|_{H\dot{\mathcal{K}}_{\omega}^{p,q}(\mathbb{R}^n)} \sim \|A(f)\|_{\dot{\mathcal{K}}_{\omega}^{p,q}(\mathbb{R}^n)},
$$

where $A \in \{g, g_\lambda^*\}$. This then finishes the proof of Theorem 4.6.8. $\square$

As an application, we then characterize the generalized Hardy–Morrey space $HM_{\omega}^{p,q}(\mathbb{R}^n)$ via the Lusin area function, the g-function, and the g_λ^*-function. Namely, the following conclusion holds true, which is a direct corollary of Theorem 4.6.8 and Remarks 1.2.2(iv) and 4.0.20(ii); we omit the details.

Corollary 4.6.10 *Let p, q, ω, and λ be as in Corollary 4.6.7. Then the following four statements are mutually equivalent:*

(i) $f \in HM_{\omega}^{p,q}(\mathbb{R}^n)$;
(ii) $f \in \mathcal{S}'(\mathbb{R}^n)$, f *vanishes weakly at infinity, and* $S(f) \in M_{\omega}^{p,q}(\mathbb{R}^n)$;
(iii) $f \in \mathcal{S}'(\mathbb{R}^n)$, f *vanishes weakly at infinity, and* $g(f) \in M_{\omega}^{p,q}(\mathbb{R}^n)$;
(iv) $f \in \mathcal{S}'(\mathbb{R}^n)$, f *vanishes weakly at infinity, and* $g_\lambda^*(f) \in M_{\omega}^{p,q}(\mathbb{R}^n)$.

Moreover, for any $f \in HM_{\omega}^{p,q}(\mathbb{R}^n)$,

$$
\|f\|_{HM_{\omega}^{p,q}(\mathbb{R}^n)} \sim \|S(f)\|_{M_{\omega}^{p,q}(\mathbb{R}^n)} \sim \|g(f)\|_{M_{\omega}^{p,q}(\mathbb{R}^n)} \sim \|g_\lambda^*(f)\|_{M_{\omega}^{p,q}(\mathbb{R}^n)}
$$

with the positive equivalence constants independent of f.

4.7 Dual Space of $H\dot{\mathcal{K}}_{\omega,\mathbf{0}}^{p,q}(\mathbb{R}^n)$

In this section, we investigate the dual space of the generalized Herz–Hardy space $H\dot{\mathcal{K}}_{\omega,\mathbf{0}}^{p,q}(\mathbb{R}^n)$. Throughout this book, for any $d \in \mathbb{Z}_+$, the *symbol* $\mathcal{P}_d(\mathbb{R}^n)$ denotes the set of all polynomials on $\mathbb{R}^n$ with degree not greater than d. Moreover, for any ball $B \in \mathbb{B}$ and any $g \in L_{\text{loc}}^1(\mathbb{R}^n)$, $P_B^d g$ denotes the *minimizing polynomial* of g with degree not greater than d, which means that $P_B^d g$ is the unique polynomial $f \in \mathcal{P}_d(\mathbb{R}^n)$ such that, for any $h \in \mathcal{P}_d(\mathbb{R}^n)$,

$$\int_B [g(x) - f(x)]h(x)\,dx = 0.$$

Now, we introduce the following Campanato-type function spaces associated with local generalized Herz spaces, which were originally introduced in [277, Definition 3.2] for any given general ball quasi-Banach function space X.

Definition 4.7.1 Let $p, q, s \in (0, \infty)$, $r \in [1, \infty)$, $d \in \mathbb{Z}_+$, and $\omega \in M(\mathbb{R}_+)$ with $m_0(\omega) \in (-\frac{n}{p}, \infty)$. Then the *Campanato-type function space* $\dot{\mathcal{L}}_{\omega,\mathbf{0}}^{p,q,r,d,s}(\mathbb{R}^n)$, associated with the local generalized Herz space $\dot{\mathcal{K}}_{\omega,\mathbf{0}}^{p,q}(\mathbb{R}^n)$, is defined to be the set of all the $f \in L_{\text{loc}}^r(\mathbb{R}^n)$ such that

$$\|f\|_{\dot{\mathcal{L}}_{\omega,\mathbf{0}}^{p,q,r,d,s}(\mathbb{R}^n)}$$

$$:= \sup \left\| \left\{ \sum_{i=1}^m \left[\frac{\lambda_i}{\|\mathbf{1}_{B_i}\|_{\dot{\mathcal{K}}_{\omega,\mathbf{0}}^{p,q}(\mathbb{R}^n)}} \right]^s \mathbf{1}_{B_i} \right\}^{\frac{1}{s}} \right\|_{\dot{\mathcal{K}}_{\omega,\mathbf{0}}^{p,q}(\mathbb{R}^n)}^{-1}$$

$$\times \sum_{j=1}^m \left\{ \frac{\lambda_j |B_j|}{\|\mathbf{1}_{B_j}\|_{\dot{\mathcal{K}}_{\omega,\mathbf{0}}^{p,q}(\mathbb{R}^n)}} \left[\frac{1}{|B_j|} \int_{B_j} \left| f(x) - P_{B_j}^d f(x) \right|^r dx \right]^{\frac{1}{r}} \right\}$$

is finite, where the supremum is taken over all $m \in \mathbb{N}$, $\{B_j\}_{j=1}^m \subset \mathbb{B}$, and $\{\lambda_j\}_{j=1}^m \subset [0, \infty)$ with $\sum_{j=1}^m \lambda_j \neq 0$.

Remark 4.7.2 By Definition 4.7.1, we can easily show that $\mathcal{P}_d(\mathbb{R}^n) \subset \dot{\mathcal{L}}_{\omega,\mathbf{0}}^{p,q,r,d,s}(\mathbb{R}^n)$ and, for any $f \in \dot{\mathcal{L}}_{\omega,\mathbf{0}}^{p,q,r,d,s}(\mathbb{R}^n)$, $\|f\|_{\dot{\mathcal{L}}_{\omega,\mathbf{0}}^{p,q,r,d,s}(\mathbb{R}^n)} = 0$ if and only if $f \in \mathcal{P}_d(\mathbb{R}^n)$. Therefore, in what follows, we always identify $f \in \dot{\mathcal{L}}_{\omega,\mathbf{0}}^{p,q,r,d,s}(\mathbb{R}^n)$ with $\{f + P : P \in \mathcal{P}_d(\mathbb{R}^n)\}$.

Applying [277, Remark 3.3(iii) and Proposition 3.4] with X therein replaced by $\dot{\mathcal{K}}_{\omega,\mathbf{0}}^{p,q}(\mathbb{R}^n)$, we immediately obtain the following equivalent characterizations of these Campanato-type function spaces; we omit the details.

Proposition 4.7.3 *Let* p, q, ω, r, d, *and* s *be as in Definition 4.7.1. Then the following three statements are equivalent:*

(i) $f \in \dot{\mathcal{L}}_{\omega,0}^{p,q,r,d,s}(\mathbb{R}^n)$;

(ii) $f \in L_{\mathrm{loc}}^r(\mathbb{R}^n)$ *and*

$$\|f\|^{\star}_{\dot{\mathcal{L}}_{\omega,0}^{p,q,r,d,s}(\mathbb{R}^n)}$$

$$:= \sup\inf \left\| \left\{ \sum_{i=1}^m \left[\frac{\lambda_i}{\|\mathbf{1}_{B_i}\|_{\dot{\mathcal{K}}_{\omega,0}^{p,q}(\mathbb{R}^n)}} \right]^s \mathbf{1}_{B_i} \right\}^{\frac{1}{s}} \right\|^{-1}_{\dot{\mathcal{K}}_{\omega,0}^{p,q}(\mathbb{R}^n)}$$

$$\times \sum_{j=1}^m \left\{ \frac{\lambda_j |B_j|}{\|\mathbf{1}_{B_j}\|_{\dot{\mathcal{K}}_{\omega,0}^{p,q}(\mathbb{R}^n)}} \left[\frac{1}{|B_j|} \int_{B_j} |f(x) - P(x)|^r \, dx \right]^{\frac{1}{r}} \right\}$$

is finite, where the supremum is the same as in Definition 4.7.1 and the infimum is taken over all $P \in \mathcal{P}_d(\mathbb{R}^n)$;

(iii) $f \in L_{\mathrm{loc}}^r(\mathbb{R}^n)$ *and*

$$\|\widetilde{f}\|_{\dot{\mathcal{L}}_{\omega,0}^{p,q,r,d,s}(\mathbb{R}^n)}$$

$$:= \sup \left\| \left\{ \sum_{i\in\mathbb{N}} \left[\frac{\lambda_i}{\|\mathbf{1}_{B_i}\|_{\dot{\mathcal{K}}_{\omega,0}^{p,q}(\mathbb{R}^n)}} \right]^s \mathbf{1}_{B_i} \right\}^{\frac{1}{s}} \right\|^{-1}_{\dot{\mathcal{K}}_{\omega,0}^{p,q}(\mathbb{R}^n)}$$

$$\times \sum_{j\in\mathbb{N}} \left\{ \frac{\lambda_j |B_j|}{\|\mathbf{1}_{B_j}\|_{\dot{\mathcal{K}}_{\omega,0}^{p,q}(\mathbb{R}^n)}} \left[\frac{1}{|B_j|} \int_{B_j} \left| f(x) - P_{B_j}^d f(x) \right|^r \, dx \right]^{\frac{1}{r}} \right\}$$

is finite, where the supremum is taken over all $\{B_j\}_{j\in\mathbb{N}} \subset \mathbb{B}$ *and* $\{\lambda_j\}_{j\in\mathbb{N}} \subset [0,\infty)$ *satisfying*

$$\left\| \left\{ \sum_{i\in\mathbb{N}} \left[\frac{\lambda_i}{\|\mathbf{1}_{B_i}\|_{\dot{\mathcal{K}}_{\omega,0}^{p,q}(\mathbb{R}^n)}} \right]^s \mathbf{1}_{B_i} \right\}^{\frac{1}{s}} \right\|_{\dot{\mathcal{K}}_{\omega,0}^{p,q}(\mathbb{R}^n)} \in (0,\infty).$$

Moreover, there exist two positive constants C_1 *and* C_2 *such that, for any* $f \in L_{\mathrm{loc}}^r(\mathbb{R}^n)$,

$$C_1 \|f\|_{\dot{\mathcal{L}}_{\omega,0}^{p,q,r,d,s}(\mathbb{R}^n)} \leq \|f\|^{\star}_{\dot{\mathcal{L}}_{\omega,0}^{p,q,r,d,s}(\mathbb{R}^n)} \leq C_2 \|f\|_{\dot{\mathcal{L}}_{\omega,0}^{p,q,r,d,s}(\mathbb{R}^n)}.$$

and

$$\|\widetilde{f}\|_{\dot{\mathcal{L}}^{p,q,r,s}_{\omega,0}(\mathbb{R}^n)} = \|f\|_{\dot{\mathcal{L}}^{p,q,r,s}_{\omega,0}(\mathbb{R}^n)}.$$

Via the known dual theorem of the Hardy space $H_X(\mathbb{R}^n)$ associated with the ball quasi-Banach function space X, we now show that the dual space of the generalized Herz–Hardy space $H\dot{\mathcal{K}}^{p,q}_{\omega,0}(\mathbb{R}^n)$ is just the Campanato-type function space $\dot{\mathcal{L}}^{p,q,r',d,s}_{\omega,0}(\mathbb{R}^n)$. Namely, we have the following theorem.

Theorem 4.7.4 *Let* $p, q \in (0, \infty)$, $\omega \in M(\mathbb{R}_+)$ *with* $m_0(\omega) \in (-\frac{n}{p}, \infty)$ *and* $m_\infty(\omega) \in (-\frac{n}{p}, \infty)$,

$$p_- := \min\left\{1, p, \frac{n}{\max\{M_0(\omega), M_\infty(\omega)\} + n/p}\right\},$$

$d \geq \lfloor n(1/p_- - 1)\rfloor$ *be a fixed integer,* $s \in (0, \min\{p_-, q\})$, *and*

$$r \in \left(\max\left\{1, p, \frac{n}{\min\{m_0(\omega), m_\infty(\omega)\} + n/p}\right\}, \infty\right].$$

Then $\dot{\mathcal{L}}^{p,q,r',d,s}_{\omega,0}(\mathbb{R}^n)$ *is the dual space of* $H\dot{\mathcal{K}}^{p,q}_{\omega,0}(\mathbb{R}^n)$ *in the following sense:*

(i) *Let* $g \in \dot{\mathcal{L}}^{p,q,r',d,s}_{\omega,0}(\mathbb{R}^n)$. *Then the linear functional*

$$L_g : f \mapsto L_g(f) := \int_{\mathbb{R}^n} f(x)g(x)\,dx, \tag{4.73}$$

initially defined for any $f \in H\dot{\mathcal{K}}^{p,q,r,d,s}_{\omega,0,\mathrm{fin}}(\mathbb{R}^n)$, *has a bounded extension to the generalized Herz–Hardy space* $H\dot{\mathcal{K}}^{p,q}_{\omega,0}(\mathbb{R}^n)$.

(ii) *Conversely, any continuous linear functional on* $H\dot{\mathcal{K}}^{p,q}_{\omega,0}(\mathbb{R}^n)$ *arises as in* (4.73) *with a unique* $g \in \dot{\mathcal{L}}^{p,q,r',d,s}_{\omega,0}(\mathbb{R}^n)$.

Moreover, there exist two positive constants C_1 *and* C_2 *such that, for any* $g \in \dot{\mathcal{L}}^{p,q,r',d,s}_{\omega,0}(\mathbb{R}^n)$,

$$C_1\|g\|_{\dot{\mathcal{L}}^{p,q,r',d,s}_{\omega,0}(\mathbb{R}^n)} \leq \|L_g\|_{(H\dot{\mathcal{K}}^{p,q}_{\omega,0}(\mathbb{R}^n))^*} \leq C_2\|g\|_{\dot{\mathcal{L}}^{p,q,r',d,s}_{\omega,0}(\mathbb{R}^n)},$$

where $(H\dot{\mathcal{K}}^{p,q}_{\omega,0}(\mathbb{R}^n))^*$ *denotes the dual space of* $H\dot{\mathcal{K}}^{p,q}_{\omega,0}(\mathbb{R}^n)$.

To show this theorem, we first recall the definition of the ball Campanato-type function space $\mathcal{L}_{X,r,d,s}(\mathbb{R}^n)$ associated with the general ball quasi-Banach function space X as follows, which is just [277, Definition 3.2].

Definition 4.7.5 Let X be a ball quasi-Banach function space, $r \in [1,\infty)$, $s \in (0,\infty)$, and $d \in \mathbb{Z}_+$. Then the *ball Campanato-type function space* $\mathcal{L}_{X,r,d,s}(\mathbb{R}^n)$, associated with X, is defined to be the set of all the $f \in L^r_{\mathrm{loc}}(\mathbb{R}^n)$ such that

$$
\|f\|_{\mathcal{L}_{X,r,d,s}(\mathbb{R}^n)}
$$

$$
:= \sup \left\{ \left\| \left[\sum_{i=1}^m \left(\frac{\lambda_i}{\|\mathbf{1}_{B_i}\|_X} \right)^s \mathbf{1}_{B_i} \right]^{\frac{1}{s}} \right\|_X^{-1} \right.
$$

$$
\left. \times \sum_{j=1}^m \left\{ \frac{\lambda_j |B_j|}{\|\mathbf{1}_{B_j}\|_X} \left[\frac{1}{|B_j|} \int_{B_j} \left| f(x) - P^d_{B_j} f(x) \right|^r dx \right]^{\frac{1}{r}} \right\} \right\}
$$

is finite, where the supremum is taken over all $m \in \mathbb{N}$, $\{B_j\}_{j=1}^m \subset \mathbb{B}$, and $\{\lambda_j\}_{j=1}^m \subset [0,\infty)$ with $\sum_{j=1}^m \lambda_j \neq 0$.

Notice that Zhang et al. [277, Theorem 3.14] established the following duality between the Hardy space $H_X(\mathbb{R}^n)$ and the ball Campanato-type function space $\mathcal{L}_{X,r',d,s}(\mathbb{R}^n)$, which is an essential tool for us to show Theorem 4.7.4.

Lemma 4.7.6 *Let X be a ball quasi-Banach function space satisfy:*

(i) *there exists a $p_- \in (0,\infty)$ such that, for any given $\theta \in (0, p_-)$ and $u \in (1,\infty)$, and for any $\{f_j\}_{j\in\mathbb{N}} \subset L^1_{\mathrm{loc}}(\mathbb{R}^n)$,*

$$
\left\| \left\{ \sum_{j\in\mathbb{N}} [\mathcal{M}(f_j)]^u \right\}^{\frac{1}{u}} \right\|_{X^{1/\theta}} \lesssim \left\| \left\{ \sum_{j\in\mathbb{N}} |f_j|^u \right\}^{\frac{1}{u}} \right\|_{X^{1/\theta}},
$$

where the implicit positive constant is independent of $\{f_j\}_{j\in\mathbb{N}}$;
(ii) *for the above p_-, there exists an $s_0 \in (0, \min\{1, p_-\})$ and an $r_0 \in (s_0, \infty)$ such that X^{1/s_0} is a ball Banach function space and, for any $f \in L^1_{\mathrm{loc}}(\mathbb{R}^n)$,*

$$
\left\| \mathcal{M}^{((r_0/s_0)')}(f) \right\|_{(X^{1/s_0})'} \lesssim \|f\|_{(X^{1/s_0})'},
$$

where the implicit positive constant is independent of f;
(iii) *X has an absolutely continuous quasi-norm.*

Assume $d \geq \lfloor n(1/\min\{1, p_-\} - 1)\rfloor$, $s \in (0, s_0]$, *and* $r \in (\max\{1, r_0\}, \infty]$. *Then the dual space of* $H_X(\mathbb{R}^n)$, *denoted by* $(H_X(\mathbb{R}^n))^*$, *is* $\mathcal{L}_{X,r',d,s}(\mathbb{R}^n)$ *in the following sense:*

(a) *Let* $g \in \mathcal{L}_{X,r',d,s}(\mathbb{R}^n)$. *Then the linear functional*

$$L_g : f \mapsto L_g(f) := \int_{\mathbb{R}^n} f(x)g(x)\, dx, \qquad (4.74)$$

initially defined for any $f \in H_{\text{fin}}^{X,r,d,s}(\mathbb{R}^n)$, *has a bounded extension to* $H_X(\mathbb{R}^n)$.
(b) *Conversely, any continuous linear functional on* $H_X(\mathbb{R}^n)$ *arises as in* (4.74) *with a unique* $g \in \mathcal{L}_{X,r',d,s}(\mathbb{R}^n)$.

Moreover, $\|g\|_{\mathcal{L}_{X,r',d,s}(\mathbb{R}^n)} \sim \|L_g\|_{(H_X(\mathbb{R}^n))^*}$, *where the positive equivalence constants are independent of g.*

Remark 4.7.7 We point out that the dual theorem, Lemma 4.7.6, has a wide range of applications. Here we present two function spaces to which Lemma 4.7.6 can be applied (see also [277, Section 6]).

 (i) Let $\vec{p} := (p_1, \ldots, p_n) \in (0, \infty)^n$,

$$d \geq \left\lfloor n\left(\frac{1}{\min\{p_1, \ldots, p_n\}} - 1\right)\right\rfloor$$

be a fixed nonnegative integer, $s \in (0, \min\{1, p_1, \ldots, p_n\})$, and

$$r \in (\max\{1, p_1, \ldots, p_n\}, \infty].$$

Then, in this case, as was pointed out in [277, Subsection 6.1], the mixed-norm Lebesgue space $L^{\vec{p}}(\mathbb{R}^n)$ satisfies all the assumptions of Lemma 4.7.6. Therefore, Lemma 4.7.6 with $X := L^{\vec{p}}(\mathbb{R}^n)$ holds true. This result coincides with [277, Theorem 6.2] (see also [128]). Moreover, the dual theorem of the mixed-norm Hardy space $H^{\vec{p}}(\mathbb{R}^n)$, with $\vec{p} \in (0, 1]^n$, was first established in [124, Theorem 3.10] (see also [127]), which is a special case of the aforementioned result.
(ii) Let $p(\cdot) \in C^{\log}(\mathbb{R}^n)$ satisfy $0 < p_- \leq p_+ < \infty$, where p_- and p_+ are defined, respectively, in (1.59) and (1.60). Let

$$d \geq \left\lfloor n\left(\frac{1}{p_-} - 1\right)\right\rfloor$$

be a fixed nonnegative integer, $s \in (0, \min\{1, p_-\})$, and $r \in (\max\{1, p_+\}, \infty]$. Then, in this case, as was mentioned in [277, Subsection 6.2], the variable Lebesgue space $L^{p(\cdot)}(\mathbb{R}^n)$ satisfies all the assumptions of Lemma 4.7.6. Thus, Lemma 4.7.6 with $X := L^{p(\cdot)}(\mathbb{R}^n)$ holds true. This dual result is just [277, Theorem 6.7] (see also [125]) and extends [186, Theorem 7.5]. Indeed, when

$0 < p_- \le p_+ \le 1$, then, in this case, the aforementioned result coincides with [186, Theorem 7.5].

With the help of the above lemma, we then prove Theorem 4.7.4.

Proof of Theorem 4.7.4 Let all the symbols be as in the present theorem. Then, from the assumption $m_0(\omega) \in (-\frac{n}{p}, \infty)$ and Theorem 1.2.42, we deduce that the local generalized Herz space $\dot{\mathcal{K}}_{\omega,0}^{p,q}(\mathbb{R}^n)$ is a BQBF space. This implies that, to finish the proof of the present theorem, we only need to show that $\dot{\mathcal{K}}_{\omega,0}^{p,q}(\mathbb{R}^n)$ satisfies (i), (ii), and (iii) of Lemma 4.7.6.

First, we prove that Lemma 4.7.6(i) holds true for $\dot{\mathcal{K}}_{\omega,0}^{p,q}(\mathbb{R}^n)$. Indeed, for any given $\theta \in (0, p_-)$ and $u \in (1, \infty)$, by Lemma 4.3.10 with $r := \theta$, we conclude that, for any $\{f_j\}_{j \in \mathbb{N}} \subset L^1_{\mathrm{loc}}(\mathbb{R}^n)$,

$$\left\| \left\{ \sum_{j \in \mathbb{N}} \left[\mathcal{M}(f_j)\right]^u \right\}^{\frac{1}{u}} \right\|_{[\dot{\mathcal{K}}_{\omega,0}^{p,q}(\mathbb{R}^n)]^{1/\theta}} \lesssim \left\| \left(\sum_{j \in \mathbb{N}} |f_j|^u \right)^{\frac{1}{u}} \right\|_{[\dot{\mathcal{K}}_{\omega,0}^{p,q}(\mathbb{R}^n)]^{1/\theta}},$$

which implies that Lemma 4.7.6(i) holds true for $\dot{\mathcal{K}}_{\omega,0}^{p,q}(\mathbb{R}^n)$.

Next, we show that $\dot{\mathcal{K}}_{\omega,0}^{p,q}(\mathbb{R}^n)$ satisfies Lemma 4.7.6(ii). To this end, let $s_0 \in (s, \min\{p_-, q\})$ and

$$r_0 \in \left(\max \left\{ 1, p, \frac{n}{\min\{m_0(\omega), m_\infty(\omega)\} + n/p} \right\}, r \right).$$

Then, from Lemma 1.8.6 with s and r therein replaced, respectively, by s_0 and r_0, it follows that $[\dot{\mathcal{K}}_{\omega,0}^{p,q}(\mathbb{R}^n)]^{1/s_0}$ is a BBF space and, for any $f \in L^1_{\mathrm{loc}}(\mathbb{R}^n)$,

$$\left\| \mathcal{M}^{((r_0/s_0)')}(f) \right\|_{([\dot{\mathcal{K}}_{\omega,0}^{p,q}(\mathbb{R}^n)]^{1/s_0})'} \lesssim \|f\|_{([\dot{\mathcal{K}}_{\omega,0}^{p,q}(\mathbb{R}^n)]^{1/s_0})'}.$$

This implies that, under the assumptions of the present theorem, $\dot{\mathcal{K}}_{\omega,0}^{p,q}(\mathbb{R}^n)$ satisfies Lemma 4.7.6(ii).

Finally, applying Theorem 1.4.1, we find that the local generalized Herz space $\dot{\mathcal{K}}_{\omega,0}^{p,q}(\mathbb{R}^n)$ has an absolutely continuous quasi-norm. Therefore, all the assumptions of Lemma 4.7.6 hold true for $\dot{\mathcal{K}}_{\omega,0}^{p,q}(\mathbb{R}^n)$. Thus, the proof of Theorem 4.7.4 is then completed. $\square$

Remark 4.7.8 We point out that the absolutely continuous quasi-norm of the ball quasi-Banach function space X plays a key role in the dual theorem of the associated Hardy space $H_X(\mathbb{R}^n)$. However, by Example 1.4.4, we find that the global generalized Herz space does not have an absolutely continuous quasi-norm. Thus, the dual space of the generalized Herz–Hardy space $H\dot{\mathcal{K}}_{\omega}^{p,q}(\mathbb{R}^n)$ is still unknown.

From the above dual theorem, we immediately deduce the following equivalence of the Campanato-type function space $\dot{\mathcal{L}}_{\omega,0}^{p,q,r,d,s}(\mathbb{R}^n)$; we omit the details.

Corollary 4.7.9 *Let* p, q, ω, p_-, d, *and* s *be as in Theorem 4.7.4,*

$$p_+ := \max\left\{1, p, \frac{n}{\min\{m_0(\omega), m_\infty(\omega)\} + n/p}\right\},$$

$r \in [1, p_+')$, $d_0 := \lfloor n(1/p_- - 1) \rfloor$, *and* $s_0 \in (0, \min\{p_-, q\})$. *Then*

$$\dot{\mathcal{L}}_{\omega,0}^{p,q,r,d,s}(\mathbb{R}^n) = \dot{\mathcal{L}}_{\omega,0}^{p,q,1,d_0,s_0}(\mathbb{R}^n)$$

with equivalent quasi-norms.

As an application of Theorem 4.7.4, we next investigate the dual space of the generalized Hardy–Morrey space $HM_{\omega,0}^{p,q}(\mathbb{R}^n)$ via introducing the following Campanato-type function space $\mathscr{L}_{\omega,0}^{p,q,r,d,s}(\mathbb{R}^n)$ associated with the local generalized Morrey space.

Definition 4.7.10 Let p, q, $r \in [1, \infty)$, $s \in (0, \infty)$, $d \in \mathbb{Z}_+$, and $\omega \in M(\mathbb{R}_+)$ satisfy $M_\infty(\omega) \in (-\infty, 0)$ and

$$-\frac{n}{p} < m_0(\omega) \leq M_0(\omega) < 0.$$

Then the *Campanato-type function space* $\mathscr{L}_{\omega,0}^{p,q,r,d,s}(\mathbb{R}^n)$, associated with the local generalized Morrey space $M_{\omega,0}^{p,q}(\mathbb{R}^n)$, is defined to be the set of all the $f \in L_{\mathrm{loc}}^r(\mathbb{R}^n)$ such that

$$\|f\|_{\mathscr{L}_{\omega,0}^{p,q,r,d,s}(\mathbb{R}^n)}$$

$$:= \sup \left\| \left\{ \sum_{i=1}^m \left[\frac{\lambda_i}{\|\mathbf{1}_{B_i}\|_{M_{\omega,0}^{p,q}(\mathbb{R}^n)}} \right]^s \mathbf{1}_{B_i} \right\}^{\frac{1}{s}} \right\|_{M_{\omega,0}^{p,q}(\mathbb{R}^n)}^{-1}$$

$$\times \sum_{j=1}^m \left\{ \frac{\lambda_j |B_j|}{\|\mathbf{1}_{B_j}\|_{M_{\omega,0}^{p,q}(\mathbb{R}^n)}} \left[\frac{1}{|B_j|} \int_{B_j} \left| f(x) - P_{B_j}^d f(x) \right|^r dx \right]^{\frac{1}{r}} \right\}$$

is finite, where the supremum is taken over all $m \in \mathbb{N}$, $\{B_j\}_{j=1}^m \subset \mathbb{B}$, and $\{\lambda_j\}_{j=1}^m \subset [0, \infty)$ with $\sum_{j=1}^m \lambda_j \neq 0$.

Using Theorem 4.7.4 and Remarks 1.2.2(iv) and 4.0.20(ii), we immediately obtain the following conclusion, which shows that the dual space of the generalized Hardy–Morrey space $HM_{\omega,0}^{p,q}(\mathbb{R}^n)$ is just the Campanato-type function space $\mathscr{L}_{\omega,0}^{p,q,r',d,s}(\mathbb{R}^n)$; we omit the details.

Corollary 4.7.11 *Let* $p, q \in [1, \infty)$, $\omega \in M(\mathbb{R}_+)$ *with*

$$-\frac{n}{p} < m_0(\omega) \le M_0(\omega) < 0$$

and

$$-\frac{n}{p} < m_\infty(\omega) \le M_\infty(\omega) < 0,$$

$d \in \mathbb{Z}_+$, $s \in (0, 1)$, *and*

$$r \in \left(\frac{n}{\min\{m_0(\omega), m_\infty(\omega)\} + n/p}, \infty \right].$$

Then $\mathscr{L}_{\omega,0}^{p,q,r',d,s}(\mathbb{R}^n)$ *is the dual space of* $HM_{\omega,0}^{p,q}(\mathbb{R}^n)$ *in the following sense:*

(i) *Let* $g \in \mathscr{L}_{\omega,0}^{p,q,r',d,s}(\mathbb{R}^n)$. *Then the linear functional*

$$L_g : f \mapsto L_g(f) := \int_{\mathbb{R}^n} f(x)g(x)\,dx, \qquad (4.75)$$

initially defined for any $f \in HM_{\omega,0,\mathrm{fin}}^{p,q,r,d,s}(\mathbb{R}^n)$, *has a bounded extension to the generalized Hardy–Morrey space* $HM_{\omega,0}^{p,q}(\mathbb{R}^n)$.

(ii) *Conversely, any continuous linear functional* $L \in (HM_{\omega,0}^{p,q}(\mathbb{R}^n))^*$ *arises as in* (4.75) *with a unique* $g \in \mathscr{L}_{\omega,0}^{p,q,r',d,s}(\mathbb{R}^n)$.

Moreover, there exist two positive constants C_1 *and* C_2 *such that, for any* $g \in \mathscr{L}_{\omega,0}^{p,q,r',d,s}(\mathbb{R}^n)$,

$$C_1 \|g\|_{\mathscr{L}_{\omega,0}^{p,q,r',d,s}(\mathbb{R}^n)} \le \|L_g\|_{(HM_{\omega,0}^{p,q}(\mathbb{R}^n))^*} \le C_2 \|g\|_{\mathscr{L}_{\omega,0}^{p,q,r',d,s}(\mathbb{R}^n)},$$

where $(HM_{\omega,0}^{p,q}(\mathbb{R}^n))^*$ *denotes the dual space of* $HM_{\omega,0}^{p,q}(\mathbb{R}^n)$.

4.8 Boundedness of Calderón–Zygmund Operators

The main target of this section is to investigate the boundedness of Calderón–Zygmund operators on generalized Herz–Hardy spaces. To this end, we first establish two general boundedness criteria of Calderón–Zygmund operators on Hardy spaces associated with ball quasi-Banach function spaces (see Proposition 4.8.12 and Theorem 4.8.17 below) under some reasonable assumptions. Via these results and the facts that both local and global generalized Herz spaces are ball quasi-Banach function spaces, we then obtain the boundedness of Calderón–Zygmund operators on generalized Herz–Hardy spaces.

Let $d \in \mathbb{Z}_+$ and T be a d-order Calderón–Zygmund operator defined as in Definition 1.5.8. Recall the well-known assumption on T that, for any $\gamma \in \mathbb{Z}_+^n$ with $|\gamma| \le d$, $T^*(x^\gamma) = 0$, namely, for any $a \in L^2(\mathbb{R}^n)$ having compact support and satisfying that, for any $\gamma \in \mathbb{Z}_+^n$ with $|\gamma| \le d$, $\int_{\mathbb{R}^n} a(x)x^\gamma \, dx = 0$, it holds true that

$$\int_{\mathbb{R}^n} T(a)(x)x^\gamma \, dx = 0$$

(see, for instance, [160, p. 119]).

Definition 4.8.1 Let $d \in \mathbb{Z}_+$. A d-order Calderón–Zygmund operator T is said to have the *vanishing moments up to order d* if, for any $\gamma \in \mathbb{Z}_+^n$ with $|\gamma| \le d$, $T^*(x^\gamma) = 0$.

We should point out that the assumption that the d-order Calderón–Zygmund operator T has the vanishing moments up to order d is reasonable. Indeed, this assumption holds true automatically when T is a Calderón–Zygmund operator with kernel $K(x, y) := K_1(x - y)$ for a locally integrable function K_1 on $\mathbb{R}^n \setminus \{0\}$. To be precise, we have the following interesting proposition about convolutional type Calderón–Zygmund operators, which might be well known. But, we do not find its detailed proof in the literature. Thus, for the convenience of the reader, we present its detailed proof as follows.

Proposition 4.8.2 *Let $d \in \mathbb{Z}_+$, $K \in \mathcal{S}'(\mathbb{R}^n)$, and the operator T be defined by setting, for any $f \in \mathcal{S}(\mathbb{R}^n)$, $T(f) := K * f$. Assume that the following three statements hold true:*

(i) $\widehat{K} \in L^\infty(\mathbb{R}^n)$;

(ii) *K coincides with a function belonging to $C^d(\mathbb{R}^n \setminus \{0\})$ in the sense that, for any given $a \in L^2(\mathbb{R}^n)$ with compact support and for any $x \notin \overline{\mathrm{supp}\,(a)} := \overline{\{x \in \mathbb{R}^n : a(x) \ne 0\}}$,*

$$T(a)(x) = \int_{\mathbb{R}^n} K(x - y)a(y) \, dy;$$

(iii) *there exists a positive constant C and a $\delta \in (0, 1]$ such that, for any $\gamma \in \mathbb{Z}_+^n$ with $|\gamma| \le d$ and for any $x \in \mathbb{R}^n \setminus \{0\}$,*

$$\left| \partial^\gamma K(x) \right| \le \frac{C}{|x|^{n+|\gamma|}} \tag{4.76}$$

and, for any $\gamma \in \mathbb{Z}_+^n$ *with* $|\gamma| = d$ *and* $x, y \in \mathbb{R}^n$ *with* $|x| > 2|y|$,

$$\left|\partial^\gamma K(x - y) - \partial^\gamma K(x)\right| \leq C \frac{|y|^\delta}{|x|^{n+d+\delta}}. \tag{4.77}$$

Then T has a vanishing moments up to order d.

To show Proposition 4.8.2, we need some preliminary lemmas. Recall that the following conclusion gives the $L^p(\mathbb{R}^n)$ boundedness of convolutional type Calderón–Zygmund operators, which is just [70, Theorem 5.1].

Lemma 4.8.3 *Let* $p \in (1, \infty)$ *and T be as in Proposition 4.8.2. Then T is well defined on* $L^p(\mathbb{R}^n)$ *and there exists a positive constant C such that, for any* $f \in L^p(\mathbb{R}^n)$,

$$\|T(f)\|_{L^p(\mathbb{R}^n)} \leq C \|f\|_{L^p(\mathbb{R}^n)}.$$

Via borrowing some ideas from the proof of [213, p. 117, Lemma], we next establish the following technical estimates about the kernel K, which plays an important role in the proof of Proposition 4.8.2.

Lemma 4.8.4 *Let d, K, and* δ *be as in Proposition 4.8.2. For any* $t \in (0, \infty)$, *let* $K^{(t)} := K * \phi_t$, *where* $\phi \in \mathcal{S}(\mathbb{R}^n)$ *with* $\operatorname{supp}(\phi) \subset B(0, 1)$. *Then there exists a positive constant C such that*

(i) *for any* $t \in (0, \infty)$, $\gamma \in \mathbb{Z}_+^n$ *with* $|\gamma| \leq d$, *and* $x \in \mathbb{R}^n \setminus \{0\}$,

$$\left|\partial^\gamma K^{(t)}(x)\right| \leq \frac{C}{|x|^{n+|\gamma|}};$$

(ii) *for any* $t \in (0, \infty)$, $\gamma \in \mathbb{Z}_+^n$ *with* $|\gamma| = d$, *and* $x, y \in \mathbb{R}^n$ *with* $|x| > 4|y|$,

$$\left|\partial^\gamma K^{(t)}(x - y) - \partial^\gamma K^{(t)}(x)\right| \leq C \frac{|y|^\delta}{|x|^{n+d+\delta}}.$$

Proof Let all the symbols be as in the present lemma and $t \in (0, \infty)$. We first prove (i). Indeed, notice that, for any $x \in \mathbb{R}^n$,

$$K^{(t)}(x) = \int_{\mathbb{R}^n} e^{2\pi i x \cdot \xi} \widehat{K}(\xi) \widehat{\phi}(t\xi) \, d\xi.$$

From this, we further deduce that, for any $\gamma \in \mathbb{Z}_+^n$ with $|\gamma| \leq d$ and for any $x \in \mathbb{R}^n$,

$$\partial^\gamma K^{(t)}(x) = \int_{\mathbb{R}^n} (2\pi i \xi)^\gamma \, e^{2\pi i x \cdot \xi} \widehat{K}(\xi) \widehat{\phi}(t\xi) \, d\xi. \tag{4.78}$$

which, together with the assumption $\widehat{K} \in L^\infty(\mathbb{R}^n)$, further implies that, for any $x \in \mathbb{R}^n$ with $0 < |x| < 2t$,

$$\left| \partial^\gamma K^{(t)}(x) \right| \lesssim \int_{\mathbb{R}^n} |\xi|^\gamma \left| \widehat{\phi}(t\xi) \right| d\xi \sim \frac{1}{t^{n+|\gamma|}} \int_{\mathbb{R}^n} \left| \widehat{\phi}(\xi) \right| d\xi$$

$$\sim \frac{1}{t^{n+|\gamma|}} \lesssim \frac{1}{|x|^{n+|\gamma|}}, \tag{4.79}$$

where the implicit positive constants are independent of t.

On the other hand, applying the both assumptions Proposition 4.8.2(ii) and

$$\mathrm{supp}\,(\phi_t) \subset B(\mathbf{0}, t),$$

we find that, for any $x \in \mathbb{R}^n$ with $|x| \geq 2t$,

$$K^{(t)}(x) = \int_{\mathbb{R}^n} K(x - y)\phi_t(y)\, dy.$$

This implies that, for any $\gamma \in \mathbb{Z}_+^n$ with $|\gamma| \leq d$ and for any $x \in \mathbb{R}^n$ with $|x| \geq 2t$,

$$\partial^\gamma K^{(t)}(x) = \int_{\mathbb{R}^n} \partial^\gamma K(x - y)\phi_t(y)\, dy$$

$$= \int_{B(\mathbf{0},t)} \partial^\gamma K(x - y)\phi_t(y)\, dy. \tag{4.80}$$

In addition, for any $x \in \mathbb{R}^n$ with $|x| \geq 2t$ and for any $y \in B(\mathbf{0}, t)$, we have

$$|x - y| \geq |x| - |y| > \frac{1}{2}|x|.$$

Combining this, (4.80), and (4.76) with x therein replaced by $x - y$, we further conclude that, for any $\gamma \in \mathbb{Z}_+^n$ with $|\gamma| \leq d$ and for any $x \in \mathbb{R}^n$ with $|x| \geq 2t$,

$$\left| \partial^\gamma K^{(t)}(x) \right| \lesssim \frac{1}{|x - y|^{n+|\gamma|}} \int_{\mathbb{R}^n} |\phi(y)|\, dy \lesssim \frac{1}{|x|^{n+|\gamma|}}, \tag{4.81}$$

where the implicit positive constants are independent of t. Moreover, from (4.79) and (4.81), it follows that (i) holds true.

Next, we show (ii). To this end, fix $x, y \in \mathbb{R}^n$ with $|x| > 4|y|$. We now prove (ii) by considering the following two cases on x.

Case (1) $x \in \mathbb{R}^n$ with $|x| < 4t$. In this case, we have $|y| < \frac{1}{4}|x| < t$. Then, using (4.78), the Lagrange mean value theorem, and the assumptions $\widehat{K} \in L^\infty(\mathbb{R}^n)$ and $\delta \in (0, 1]$, we find that there exists a $y_1 \in \mathbb{R}^n$ such that, for any $\gamma \in \mathbb{Z}_+^n$ with $|\gamma| = d$,

$$
\begin{aligned}
&\left| \partial^\gamma K^{(t)}(x - y) - \partial^\gamma K^{(t)}(x) \right| \\
&= \left| \int_{\mathbb{R}^n} (2\pi i \xi)^\gamma \left[e^{2\pi i (x-y)\cdot\xi} - e^{2\pi i x \cdot \xi} \right] \widehat{K}(\xi) \widehat{\phi}(t\xi) \, d\xi \right| \\
&= \left| \int_{\mathbb{R}^n} (2\pi i \xi)^\gamma (2\pi i \xi) \cdot y e^{2\pi i (x-y_1)\cdot\xi} \widehat{K}(\xi) \widehat{\phi}(t\xi) \, d\xi \right| \\
&\lesssim |y| \int_{\mathbb{R}^n} |\xi|^{d+1} \left| \widehat{\phi}(t\xi) \right| d\xi \sim \frac{|y|}{t^{n+d+1}} \int_{\mathbb{R}^n} \left| \widehat{\phi}(\xi) \right| d\xi \\
&\lesssim \frac{1}{t^{n+d}} \frac{|y|^\delta}{t^\delta} \lesssim \frac{|y|^\delta}{|x|^{n+d+\delta}},
\end{aligned}
\tag{4.82}
$$

where the implicit positive constants are independent of t. This finishes the proof of (ii) under the assumption $|x| < 4t$.

Case (2) $|x| \geq 4t$. In this case, we first claim that $|x - y| > t$. Indeed, when $|y| < t$, we have

$$
|x - y| \geq |x| - |y| > 3t > t.
$$

On the other hand, if $|y| \geq t$, from the assumption $|x| > 4|y|$, we deduce that

$$
|x - y| \geq |x| - |y| > 3|y| > t.
$$

Therefore, the above claim holds true. Using this claim and the assumptions Proposition 4.8.2(ii), $|x| \geq 4t$, and $\operatorname{supp}(\phi_t) \subset B(\mathbf{0}, t)$, we conclude that

$$
K^{(t)}(x) = \int_{|z| < t} K(x - z) \phi_t(z) \, dz
$$

and

$$
K^{(t)}(x - y) = \int_{|z| < t} K(x - y - z) \phi_t(z) \, dz,
$$

which further imply that, for any $\gamma \in \mathbb{Z}_+^n$ with $|\gamma| = d$,

$$
\partial^\gamma K^{(t)}(x) = \int_{|z| < t} \partial^\gamma K(x - z) \phi_t(z) \, dz
\tag{4.83}
$$

and

$$\partial^\gamma K^{(t)}(x - y) = \int_{|z|<t} \partial^\gamma K(x - y - z)\phi_t(z)\,dz. \tag{4.84}$$

We now estimate $|x - z|$ for any $|z| < t$. Indeed, for any $z \in \mathbb{R}^n$ with $|z| < t$, we have $|x| \geq 4t > 4|z|$. This implies that, for any $z \in \mathbb{R}^n$ with $|z| < t$,

$$|x - z| \geq |x| - |z| > \frac{3}{4}|x| > 3|y|.$$

Applying this, (4.83), (4.84), and the assumption (4.77) with x and y therein replaced, respectively, by $x - z$ and y, we conclude that, for any $\gamma \in \mathbb{Z}_+^n$ with $|\gamma| = d$,

$$\left|\partial^\gamma K^{(t)}(x - y) - \partial^\gamma K^{(t)}(x)\right|$$

$$\leq \int_{|z|<1} \left|\partial^\gamma K(x - y - z) - \partial^\gamma K(x - z)\right| |\phi_t(z)|\,dz$$

$$\lesssim \int_{|z|<1} \frac{|y|^\delta}{|x - z|^{n+d+\delta}} |\phi_t(z)|\,dz \lesssim \|\phi\|_{L^1(\mathbb{R}^n)} \frac{|y|^\delta}{|x|^{n+d+\delta}}, \tag{4.85}$$

where the implicit positive constants are independent of t. This finishes the proof of (ii) under the assumption $|x| \geq 4t$. Combining both (4.82) and (4.85), we then complete the proof of (ii) and hence Lemma 4.8.4. $\square$

In addition, via borrowing some ideas from the arguments used in [213, p. 118], we obtain the following auxiliary lemma about convolutional type Calderón–Zygmund operators and radial maximal functions, which plays a key role in the proof of Proposition 4.8.2.

Lemma 4.8.5 *Let $\phi \in \mathcal{S}(\mathbb{R}^n)$ with $\mathrm{supp}\,(\phi) \subset B(\mathbf{0}, 1)$, and d, K, δ, and T be as in Proposition 4.8.2. Assume $a \in L^2(\mathbb{R}^n)$, with compact support, satisfying that, for any $\gamma \in \mathbb{Z}_+^n$ with $|\gamma| \leq d$,*

$$\int_{\mathbb{R}^n} a(x)x^\gamma\,dx = 0.$$

Then $M(T(a), \phi)| \cdot |^{|\gamma|} \in L^1(\mathbb{R}^n)$ for any $\gamma \in \mathbb{Z}_+^n$ with $|\gamma| \leq d$, where, for any $f \in \mathcal{S}'(\mathbb{R}^n)$ and $\psi \in \mathcal{S}(\mathbb{R}^n)$, $M(f, \psi)$ is defined as in Definition 4.1.1(i).

Proof Let all the symbols be as in the present lemma and $\mathrm{supp}\,(a) \subset B(x_0, r_0)$ with $x_0 \in \mathbb{R}^n$ and $r_0 \in (0, \infty)$. We first estimate $M(T(a), \phi)$. Indeed, from the Hölder inequality, (4.20), the $L^2(\mathbb{R}^n)$ boundedness of the Hardy–Littlewood maximal operator $\mathcal{M}$, and Lemma 4.8.3 with $p := 2$, it follows that

$$\int_{B(x_0,4r_0)} M\,(T(a),\phi)\,(x)\,dx$$

$$\lesssim \left\| M\,(T(a),\phi)\,\mathbf{1}_{B(x_0,4r_0)} \right\|_{L^2(\mathbb{R}^n)} \lesssim \left\| \mathcal{M}\,(T(a)) \right\|_{L^2(\mathbb{R}^n)}$$

$$\lesssim \|T(a)\|_{L^2(\mathbb{R}^n)} \lesssim \|a\|_{L^2(\mathbb{R}^n)} < \infty. \tag{4.86}$$

This is the desired estimate of $M(T(a),\phi)(x)$ with $x \in B(x_0, 4r_0)$.

On the other hand, we estimate $M(T(a),\phi)(x)$ when $x \in [B(x_0, 4r_0)]^{\complement}$. To achieve this, fix a $t \in (0, \infty)$. Then, by the assumption that, for any $\gamma \in \mathbb{Z}_+^n$ with $|\gamma| \le d$, $\int_{\mathbb{R}^n} a(x)x^\gamma\,dx = 0$ and the Taylor remainder theorem, we find that, for any $y \in B(x_0, r_0)$, there exists a $t_y \in [0, 1]$ such that, for any $x \in \mathbb{R}^n$,

$$T(a) * \phi_t(x)$$

$$= a * K^{(t)}(x) = \int_{B(x_0,r_0)} K^{(t)}(x-y)a(y)\,dy$$

$$= \int_{B(x_0,r_0)} \left[K^{(t)}(x-y) - \sum_{\substack{\gamma \in \mathbb{Z}_+^n \\ |\gamma| \le d}} \frac{\partial^\gamma K^{(t)}(x-x_0)}{\gamma!}(y-x_0)^\gamma \right] a(y)\,dy$$

$$= \int_{B(x_0,r_0)} \sum_{\substack{\gamma \in \mathbb{Z}_+^n \\ |\gamma| = d}} \frac{\partial^\gamma K^{(t)}(x - t_y y - (1-t_y)x_0) - \partial^\gamma K^{(t)}(x-x_0)}{\gamma!}$$

$$\times (y - x_0)^\gamma a(y)\,dy. \tag{4.87}$$

Notice that, for any $x \in [B(x_0, 4r_0)]^{\complement}$ and $y \in B(x_0, r_0)$, we have

$$|x - x_0| \ge 4r_0 > 4|t_y(y - x_0)|,$$

where t_y is as in (4.87). This, together with (4.87), Lemma 4.8.4(ii) with x and y replaced, respectively, by $x - x_0$ and $t_y(y - x_0)$ for any $y \in B(x_0, r_0)$, and the Hölder inequality, further implies that, for any $x \in [B(x_0, 4r_0)]^{\complement}$,

$$|T(a) * \phi_t(x)| \lesssim \int_{B(x_0,r_0)} \frac{|y - x_0|^{d+\delta}}{|x - x_0|^{n+d+\delta}}|a(y)|\,dy \lesssim \frac{\|a\|_{L^2(\mathbb{R}^n)}}{|x - x_0|^{n+d+\delta}},$$

where the implicit positive constants are independent of t. Therefore, for any $x \in [B(x_0, 4r_0)]^{\complement}$, it holds true that

$$M(T(a), \phi)(x) \lesssim \frac{\|a\|_{L^2(\mathbb{R}^n)}}{|x - x_0|^{n+d+\delta}}, \tag{4.88}$$

which is the desired estimate of $M(T(a), \phi)(x)$ with $x \in [B(x_0, 4r_0)]^{\complement}$. Then, for any $\gamma \in \mathbb{Z}_+^n$ with $|\gamma| \leq d$, applying (4.86), (4.88), and the facts $n + d + \delta > n$ and $n + d - |\gamma| + \delta > n$, we further conclude that

$$\int_{\mathbb{R}^n} M(T(a), \phi)(x)|x|^{|\gamma|} \, dx$$

$$= \int_{B(x_0, 4r_0)} M(T(a), \phi)(x)|x|^{|\gamma|} \, dx + \int_{[B(x_0, 4r_0)]^{\complement}} \cdots$$

$$\lesssim (x_0 + 4r_0)^{|\gamma|} \int_{B(x_0, 4r_0)} M(T(a), \phi)(x) \, dx$$

$$+ \|a\|_{L^2(\mathbb{R}^n)} \left[\int_{[B(x_0, 4r_0)]^{\complement}} \left(\frac{|x_0|^{|\gamma|}}{|x - x_0|^{n+d+\delta}} + \frac{1}{|x - x_0|^{n+d-|\gamma|+\delta}} \right) dx \right]$$

$$\lesssim \|a\|_{L^2(\mathbb{R}^n)} \left[1 + \int_{[B(x_0, 4r_0)]^{\complement}} \left(\frac{1}{|x - x_0|^{n+d+\delta}} + \frac{1}{|x - x_0|^{n+d-|\gamma|+\delta}} \right) dx \right]$$

$$< \infty.$$

This implies that, for any $\gamma \in \mathbb{Z}_+^n$ with $|\gamma| \leq d$, $M(T(a), \phi)| \cdot |^{|\gamma|} \in L^1(\mathbb{R}^n)$ and hence finishes the proof of Lemma 4.8.5. □

Recall that the *Hardy space* $H^1(\mathbb{R}^n)$ is defined as in Definition 4.2.2 with $X := L^1(\mathbb{R}^n)$. The following atomic decomposition of $H^1(\mathbb{R}^n)$ can be deduced from the proof of [160, Chapter 2, Proposition 3.3] immediately, which is an essential tool in the proof of Proposition 4.8.2; we omit the details.

Lemma 4.8.6 *Let* $N \in \mathbb{N}$, $d \in \mathbb{Z}_+$, *and* $f \in H^1(\mathbb{R}^n) \cap L^2(\mathbb{R}^n)$. *Then there exists* $\{\lambda_{i,j}\}_{i \in \mathbb{Z}, j \in \mathbb{N}} \subset [0, \infty)$ *and a sequence* $\{a_{i,j}\}_{i \in \mathbb{Z}, j \in \mathbb{N}}$ *of* $(L^1(\mathbb{R}^n), \infty, d)$-*atoms supported, respectively, in the balls* $\{B_{i,j}\}_{i \in \mathbb{Z}, j \in \mathbb{N}} \subset \mathbb{B}$ *such that*

$$f = \sum_{i \in \mathbb{Z}} \sum_{j \in \mathbb{N}} \lambda_{i,j} a_{i,j}$$

almost everywhere in $\mathbb{R}^n$ *and there exists a positive constant* C *such that, for any* $i \in \mathbb{Z}$ *and* $j \in \mathbb{N}$, *the following three statements hold true:*

(i) $\lambda_{i,j} |a_{i,j}| \leq C2^i$;
(ii) $\bigcup_{j \in \mathbb{N}} B_{i,j} = \Omega_i := \{x \in \mathbb{R}^n : \mathcal{M}_N(f)(x) > 2^i\}$ *with* $\mathcal{M}_N$ *as in* (4.3);
(iii) $\sum_{j \in \mathbb{N}} \mathbf{1}_{B_{i,j}} \leq C$.

In order to show Proposition 4.8.2, we also require some auxiliary conclusions about weighted function spaces. The following one about powered weights is a part of [90, Example 7.1.7].

Lemma 4.8.7 *Let $a \in \mathbb{R}$ and $p \in (1, \infty)$. Then $|\cdot|^a \in A_p(\mathbb{R}^n)$ if and only if $-n < a < n(p-1)$.*

Moreover, the following conclusion gives the strong type inequality characterization of $A_p(\mathbb{R}^n)$-weights with $p \in (1, \infty)$, which can be found in [70, Theorem 7.3].

Lemma 4.8.8 *Let $p \in (1, \infty)$ and $\upsilon \in A_\infty(\mathbb{R}^n)$. Then the Hardy–Littlewood maximal operator $\mathcal{M}$ is bounded on $L_\upsilon^p(\mathbb{R}^n)$ if and only if $\upsilon \in A_p(\mathbb{R}^n)$.*

Via the above two lemmas, we show the following technical conclusion about grand maximal functions and radial maximal functions, which is vital in the proof of Proposition 4.8.2.

Lemma 4.8.9 *Let $d \in \mathbb{Z}_+$, $N \in \mathbb{N} \cap [n + d + 1, \infty)$, $f \in \mathcal{S}'(\mathbb{R}^n)$, and $\phi \in \mathcal{S}(\mathbb{R}^n)$ with $\mathrm{supp}\,(\phi) \subset B(\mathbf{0}, 1)$ and*

$$\int_{\mathbb{R}^n} \phi(x)\, dx \neq 0.$$

Assume that $M(f, \phi)|\cdot|^d \in L^1(\mathbb{R}^n)$, then $\mathcal{M}_N(f)|\cdot|^d \in L^1(\mathbb{R}^n)$ and

$$\left\| \mathcal{M}_N(f)|\cdot|^d \right\|_{L^1(\mathbb{R}^n)} \sim \left\| M(f, \phi)|\cdot|^d \right\|_{L^1(\mathbb{R}^n)}$$

with the positive equivalence constants independent of f.

Proof Let all the symbols be as in the present lemma and $r \in (0, \frac{n}{n+d})$. Then we have $d < n(\frac{1}{r} - 1)$. This, together with Lemma 4.8.7 with $a := d$ and $p := \frac{1}{r}$, implies that $|\cdot|^d \in A_{1/r}(\mathbb{R}^n)$. Applying this and Lemma 4.8.8 with $p := \frac{1}{r}$ and $\upsilon := |\cdot|^d$, we conclude that, for any $g \in L^1_{\mathrm{loc}}(\mathbb{R}^n)$,

$$\|\mathcal{M}(g)\|_{L^{1/r}_{|\cdot|^d}(\mathbb{R}^n)} \lesssim \|g\|_{L^{1/r}_{|\cdot|^d}(\mathbb{R}^n)}. \tag{4.89}$$

In addition, observe that $L^{1/r}_{|\cdot|^d}(\mathbb{R}^n) = [L^1_{|\cdot|^d}(\mathbb{R}^n)]^{1/r}$. Thus, from this, (4.89), and Lemma 4.1.4 with $X := L^1_{|\cdot|^d}(\mathbb{R}^n)$, it follows that

$$\left\| \mathcal{M}_N(f)|\cdot|^d \right\|_{L^1(\mathbb{R}^n)} \sim \left\| M(f, \phi)|\cdot|^d \right\|_{L^1(\mathbb{R}^n)},$$

where the positive equivalence constants are independent of f. This finishes the proof of Lemma 4.8.9. $\qquad\square$

Based on above preparations, we now prove Proposition 4.8.2.

Proof of Proposition 4.8.2 Let all the symbols be as in the present proposition and $a \in L^2(\mathbb{R}^n)$, with compact support, satisfying that, for any $\gamma \in \mathbb{Z}_+^n$ with $|\gamma| \le d$,

$$\int_{\mathbb{R}^n} a(x) x^\gamma \, dx = 0.$$

Then, from Lemma 4.8.3 with $p = 2$, we deduce that $T(a) \in L^2(\mathbb{R}^n)$. In addition, let $\phi \in \mathcal{S}(\mathbb{R}^n)$ with $\operatorname{supp}(\phi) \subset B(\mathbf{0}, 1)$ and

$$\int_{\mathbb{R}^n} \phi(x) \, dx \ne 0.$$

By Lemma 4.8.5 with $\gamma = \mathbf{0}$, we find that $M(T(a), \phi) \in L^1(\mathbb{R}^n)$ and hence $T(a) \in H^1(\mathbb{R}^n)$. Thus, we have $T(a) \in H^1(\mathbb{R}^n) \cap L^2(\mathbb{R}^n)$. Choose an $N \in \mathbb{N} \cap [n + d + 1, \infty)$. Then, applying Lemma 4.8.6 with f therein replaced by $T(a)$, we conclude that there exists $\{\lambda_{i,j}\}_{i \in \mathbb{Z}, \, j \in \mathbb{N}} \subset [0, \infty)$ and a sequence $\{a_{i,j}\}_{i \in \mathbb{Z}, \, j \in \mathbb{N}}$ of $(L^1(\mathbb{R}^n), \infty, d)$-atoms such that

$$T(a) = \sum_{i \in \mathbb{Z}} \sum_{j \in \mathbb{N}} \lambda_{i,j} a_{i,j} \tag{4.90}$$

almost everywhere in $\mathbb{R}^n$ and (i), (ii), and (iii) of Lemma 4.8.6 hold true. Assume $E \subset \mathbb{R}^n$ such that $|E| = 0$ and, for any $x \in \mathbb{R}^n \setminus E$,

$$T(a)(x) = \sum_{i \in \mathbb{Z}} \sum_{j \in \mathbb{N}} \lambda_{i,j} a_{i,j}(x).$$

Now, we show that

$$\sum_{i \in \mathbb{Z}, \, j \in \mathbb{N}} \lambda_{i,j} \left| a_{i,j} \right| \lesssim \mathcal{M}_N(T(a))$$

almost everywhere in $\mathbb{R}^n$ by considering the following two cases on x.

Case (1) $x \in \left(\bigcup_{i \in \mathbb{Z}} \Omega_i \right)^{\complement} \setminus E$. Here and thereafter, for any $i \in \mathbb{Z}$, Ω_i is defined as in Lemma 4.8.6(ii) with f replaced by $T(a)$. In this case, we have $\mathcal{M}_N(T(a))(x) = 0$. In addition, for any $i \in \mathbb{Z}$ and $j \in \mathbb{N}$, by the assumption $\operatorname{supp}(a_{i,j}) \subset B_{i,j}$ and Lemma 4.8.6(ii), we conclude that $a_{i,j}(x) = 0$. This, together with $\mathcal{M}_N(T(a))(x) = 0$, further implies that

$$\sum_{i \in \mathbb{Z}, \, j \in \mathbb{N}} \lambda_{i,j} \left| a_{i,j}(x) \right| = 0 = \mathcal{M}_N(T(a))(x). \tag{4.91}$$

Thus,

$$\sum_{i\in\mathbb{Z},\, j\in\mathbb{N}} \lambda_{i,j}\left|a_{i,j}\right| \lesssim \mathcal{M}_N\left(T\left(a\right)\right)$$

almost everywhere in $\left(\bigcup_{i\in\mathbb{Z}}\Omega_i\right)^{\complement}$.

Case (2) $x \in (\Omega_{i_0} \setminus \Omega_{i_0+1}) \setminus E$ for some $i_0 \in \mathbb{Z}$. In this case, for any $i \in \mathbb{N} \cap [i_0 + 1, \infty)$ and $j \in \mathbb{N}$, we have $a_{i,j}(x) = 0$. Applying this, both (i) and (iii) of Lemma 4.8.6, the assumption $\operatorname{supp}(a_{i,j}) \subset B_{i,j}$ for any $i \in \mathbb{Z}$ and $j \in \mathbb{N}$, and the definition of Ω_i, we find that, for any $x \in (\Omega_{i_0} \setminus \Omega_{i_0+1}) \setminus E$,

$$
\sum_{i\in\mathbb{Z},\, j\in\mathbb{N}} \lambda_{i,j}\left|a_{i,j}(x)\right| = \sum_{i=-\infty}^{i_0}\sum_{j\in\mathbb{N}} \lambda_{i,j}\left|a_{i,j}(x)\right|
$$

$$
\lesssim \sum_{i=-\infty}^{i_0} 2^i \sum_{j\in\mathbb{N}} \mathbf{1}_{B_{i,j}} \lesssim \sum_{i=-\infty}^{i_0} 2^i \sim 2^{i_0}
$$

$$
\sim \mathcal{M}_N\left(T(a)\right)(x), \tag{4.92}
$$

which further implies that

$$\sum_{i\in\mathbb{Z},\, j\in\mathbb{N}} \lambda_{i,j}\left|a_{i,j}\right| \lesssim \mathcal{M}_N\left(T\left(a\right)\right)$$

almost everywhere in $\Omega_{i_0} \setminus \Omega_{i_0+1}$. Combining (4.91) and (4.92), we then conclude that

$$\sum_{i\in\mathbb{Z},\, j\in\mathbb{N}} \lambda_{i,j}\left|a_{i,j}\right| \lesssim \mathcal{M}_N(T(a))$$

almost everywhere in $\mathbb{R}^n$.

Therefore, for any $\gamma \in \mathbb{Z}_+^n$ with $|\gamma| \le d$, from both Lemmas 4.8.5 and 4.8.9 with d therein replaced by $|\gamma|$, we deduce that

$$\sum_{i\in\mathbb{Z},\, j\in\mathbb{N}} \lambda_{i,j}\left|a_{i,j}\right| |\cdot|^{|\gamma|} \lesssim \mathcal{M}_N(T(a))|\cdot|^{|\gamma|} \in L^1(\mathbb{R}^n).$$

Using this, (4.90), the dominated convergence theorem, and the assumption that $\{a_{i,j}\}_{i\in\mathbb{Z},\,j\in\mathbb{N}}$ is a sequence of $(L^1(\mathbb{R}^n),\,\infty,\,d)$-atoms, we further find that, for any $\gamma\in\mathbb{Z}_+^n$ with $|\gamma|\le d$,

$$\int_{\mathbb{R}^n} T(a)(x)x^\gamma\,dx = \sum_{i\in\mathbb{Z}}\sum_{j\in\mathbb{N}}\int_{\mathbb{R}^n} a_{i,j}(x)x^\gamma\,dx = 0,$$

which completes the proof of Proposition 4.8.2.   □

Under the reasonable assumption that the Calderón–Zygmund operator T has vanishing moments, we then have the boundedness of Calderón–Zygmund operators on the generalized Herz–Hardy space $H\dot{\mathcal{K}}^{p,q}_{\omega,\mathbf{0}}(\mathbb{R}^n)$ as follows.

Theorem 4.8.10 *Let* $d\in\mathbb{Z}_+$, $\delta\in(0,1]$, $p,q\in(\frac{n}{n+d+\delta},\infty)$, K *be a d-order standard kernel defined as in Definition 1.5.7, T a d-order Calderón–Zygmund operator with kernel K having the vanishing moments up to order d, and $\omega\in M(\mathbb{R}_+)$ with*

$$-\frac{n}{p} < m_0(\omega) \le M_0(\omega) < n - \frac{n}{p} + d + \delta$$

and

$$-\frac{n}{p} < m_\infty(\omega) \le M_\infty(\omega) < n - \frac{n}{p} + d + \delta.$$

Then T has a unique extension on $H\dot{\mathcal{K}}^{p,q}_{\omega,\mathbf{0}}(\mathbb{R}^n)$ and there exists a positive constant C such that, for any $f\in H\dot{\mathcal{K}}^{p,q}_{\omega,\mathbf{0}}(\mathbb{R}^n)$,

$$\|T(f)\|_{H\dot{\mathcal{K}}^{p,q}_{\omega,\mathbf{0}}(\mathbb{R}^n)} \le C\|f\|_{H\dot{\mathcal{K}}^{p,q}_{\omega,\mathbf{0}}(\mathbb{R}^n)}.$$

Remark 4.8.11 We should point out that, in Theorem 4.8.10, when $d=0$, $p\in(1,\infty)$, and $\omega(t):=t^\alpha$ for any $t\in(0,\infty)$ and for any given $\alpha\in[n(1-\frac{1}{p}),n(1-\frac{1}{p})+\delta)$, then Theorem 4.8.10 goes back to [157, Theorem 1].

To show Theorem 4.8.10, we first establish the boundedness of Calderón–Zygmund operators on Hardy spaces associated with ball quasi-Banach function spaces as follows.

Proposition 4.8.12 *Let X be a ball quasi-Banach function space satisfying Assumption 1.2.29 for some $0 < \theta < s \le 1$. Assume that $X^{1/s}$ is a ball Banach function space and there exists an $r_0\in(1,\infty)$ and a positive constant C such that, for any $f\in L^1_{\mathrm{loc}}(\mathbb{R}^n)$,*

$$\left\|\mathcal{M}^{((r_0/s)')}(f)\right\|_{(X^{1/s})'} \le C\|f\|_{(X^{1/s})'}. \tag{4.93}$$

Let $d \in \mathbb{Z}_+$, $\delta \in (0, 1]$, K be a d-order standard kernel defined as in Definition 1.5.7, and T a d-order Calderón–Zygmund operator with kernel K having the vanishing moments up to order d. If $\theta \in (\frac{n}{n+d+\delta}, \frac{n}{n+d}]$ and X has an absolutely continuous quasi-norm, then T has a unique extension on $H_X(\mathbb{R}^n)$ and there exists a positive constant C such that, for any $f \in H_X(\mathbb{R}^n)$,

$$\|T(f)\|_{H_X(\mathbb{R}^n)} \leq C\|f\|_{H_X(\mathbb{R}^n)}.$$

Remark 4.8.13 We point out that Proposition 4.8.12 has a wide range of applications. Here we present several function spaces to which Proposition 4.8.12 can be applied. In what follows, let d and δ be the same as in Proposition 4.8.12.

(i) Let $p \in (\frac{n}{n+d+\delta}, \infty)$. Then, in this case, combining Remarks 1.2.27, 1.2.31(i), and 1.2.34(i), we can easily conclude that the Lebesgue space $L^p(\mathbb{R}^n)$ satisfies all the assumptions of Proposition 4.8.12. This then implies that Proposition 4.8.12 with $X := L^p(\mathbb{R}^n)$ holds true. If further assume that $p \in (0, 1]$ and T is a convolutional type operators defined as in Proposition 4.8.2, then, in this case, the aforementioned result is just [213, p. 115, Theorem 4].

(ii) Let $\upsilon \in A_\infty(\mathbb{R}^n)$, q_υ be the same as in (1.58), and $p \in (\frac{nq_\upsilon}{n+d+\delta}, \infty)$. Then, in this case, from Remarks 1.2.27, 1.2.31(ii), and 1.2.34(ii), we can easily deduce that the weighted Lebesgue space $L^p_\upsilon(\mathbb{R}^n)$ satisfies all the assumptions of Proposition 4.8.12. Therefore, Proposition 4.8.12 with $X := L^p_\upsilon(\mathbb{R}^n)$ holds true. If further assume that $p \in (0, 1]$ and $d := 0$ in Proposition 4.8.12, then, in this case, the aforementioned result goes back to [194, Theorem 3].

(iii) Let $\vec{p} := (p_1, \dots, p_n) \in (\frac{n}{n+d+\delta}, \infty)^n$. Then, in this case, by Remarks 1.2.27, 1.2.31(iii), and 1.2.34(iii), we can easily find that the mixed-norm Lebesgue space $L^{\vec{p}}(\mathbb{R}^n)$ satisfies all the assumptions of Proposition 4.8.12. Thus, Proposition 4.8.12 with $X := L^{\vec{p}}(\mathbb{R}^n)$ holds true. If further assume that K is a convolutional type operator defined as in Proposition 4.8.2, and $d := 0$, then, in this case, the aforementioned result was established in [123, Theorem 6.4] (see also [127, 155]).

(iv) Let $p(\cdot) \in C^{\log}(\mathbb{R}^n)$ satisfy $\frac{n}{n+d+\delta} < p_- \leq p_+ < \infty$, where p_- and p_+ are defined, respectively, in (1.59) and (1.60). Then, in this case, from Remarks 1.2.27, 1.2.31(v), and 1.2.34(v), we can easily infer that the variable Lebesgue space $L^{p(\cdot)}(\mathbb{R}^n)$ satisfies all the assumptions of Proposition 4.8.12. This then implies that Proposition 4.8.12 with $X := L^{p(\cdot)}(\mathbb{R}^n)$ holds true. If further assume that $p_+ \in (0, 1]$ and $d := 0$ in Proposition 4.8.12, then, in this case, the aforementioned result was also given in [280, Theorem 6.6].

To show Proposition 4.8.12, we need the following auxiliary lemma which established the relations among Calderón–Zygmund operators, atoms, and molecules.

Lemma 4.8.14 *Let X be a ball quasi-Banach function space, $r \in [2, \infty)$, and $d \in \mathbb{Z}_+$. Assume that K is a d-order standard kernel defined as in Definition 1.5.7 with some $\delta \in (0, 1]$, and T a d-order Calderón–Zygmund operator with kernel K having the vanishing moments up to order d. Then, for any (X, r, d)-*

atom a supported in the ball $B \in \mathbb{B}$, $T(a)$ is a harmless constant multiple of an $(X, r, d, d + \delta + \frac{n}{r'})$*-molecule centered at B.*

Proof Let all the symbols be as in the present lemma and a an (X, r, d)-atom supported in the ball $B := B(x_0, r_0)$ with $x_0 \in \mathbb{R}^n$ and $r_0 \in (0, \infty)$. Then, combining Definitions 4.3.4(iii) and 4.8.1, we find that, for any $\gamma \in \mathbb{Z}_+^n$ with $|\gamma| \leq d$,

$$\int_{\mathbb{R}^n} T(a)(x) x^\gamma \, dx = 0.$$

This implies that $T(a)$ satisfies Definition 4.5.3(ii).

Next, we show that Definition 4.5.3(i) holds true for a harmless constant multiple of $T(a)$. Indeed, from Lemma 1.5.9 with $p := r$ and Definition 4.3.4(ii), it follows that

$$\|T(a)\mathbf{1}_B\|_{L^r(\mathbb{R}^n)} \leq \|T(a)\|_{L^r(\mathbb{R}^n)} \lesssim \|a\|_{L^r(\mathbb{R}^n)} \lesssim \frac{|B|^{1/r}}{\|\mathbf{1}_B\|_X} \tag{4.94}$$

and

$$\left\| T(a)\mathbf{1}_{(2B)\setminus B} \right\|_{L^r(\mathbb{R}^n)}$$

$$\leq \|T(a)\|_{L^r(\mathbb{R}^n)} \lesssim \|a\|_{L^r(\mathbb{R}^n)} \lesssim \frac{|B|^{1/r}}{\|\mathbf{1}_B\|_X} \sim 2^{-(d+\delta+\frac{n}{r'})} \frac{|B|^{1/r}}{\|\mathbf{1}_B\|_X}. \tag{4.95}$$

These are the desired estimates of $\|T(a)\mathbf{1}_B\|_{L^r(\mathbb{R}^n)}$ and $\|T(a)\mathbf{1}_{(2B)\setminus B}\|_{L^r(\mathbb{R}^n)}$, respectively.

In addition, recall that, for any $j \in \mathbb{N}$, $S_j(B) := (2^j B) \setminus (2^{j-1} B)$. We next estimate $\|T(a)\mathbf{1}_{S_{j+1}(B)}\|_{L^r(\mathbb{R}^n)}$. Indeed, by Definition 1.5.8(ii), we find that, for any $j \in \mathbb{N}$ and $x \in S_{j+1}(B)$,

$$T(a)(x) = \int_{\mathbb{R}^n} K(x, y) a(y).$$

This, together with Definition 4.3.4(iii) and the Taylor remainder theorem, further implies that, for any $j \in \mathbb{N}$ and $y \in B$, there exists a $t_y \in [0, 1]$ such that

$$\left\| T(a)\mathbf{1}_{S_{j+1}(B)} \right\|_{L^r(\mathbb{R}^n)}$$

$$= \left[\int_{S_{j+1}(B)} \left| \int_B K(x, y) a(y) \, dy \right|^r dx \right]^{\frac{1}{r}}$$

$$
= \left\{ \int_{S_{j+1}(B)} \left| \int_B \left[K(x, y) - \sum_{\substack{\gamma \in \mathbb{Z}_+^n \\ |\gamma| \le d}} \frac{\partial_{(2)}^{\gamma} K(x, x_0)}{\gamma!} (y - x_0)^{\gamma} \right] a(y) \, dy \right|^r dx \right\}^{\frac{1}{r}}
$$

$$
= \left\{ \int_{S_{j+1}(B)} \left| \int_B \sum_{\substack{\gamma \in \mathbb{Z}_+^n \\ |\gamma| = d}} \frac{\partial_{(2)}^{\gamma} K(x, t_y y + (1 - t_y)x_0) - \partial_{(2)}^{\gamma} K(x, x_0)}{\gamma!} \right. \right.
$$

$$
\left. \left. \times (y - x_0)^{\gamma} a(y) \, dy \right|^r dx \right\}^{\frac{1}{r}}
$$

$$
\lesssim \int_B |y - x_0|^d a(y) \left[\int_{S_{j+1}(B)} \right.
$$

$$
\left. \times \sum_{\substack{\gamma \in \mathbb{Z}_+^n \\ |\gamma| = d}} \left| \partial_{(2)}^{\gamma} K(x, t_y y + (1 - t_y)x_0) - \partial_{(2)}^{\gamma} K(x, x_0) \right|^r dx \right]^{\frac{1}{r}} dy. \qquad (4.96)
$$

On the other hand, for any $j \in \mathbb{N}$, $x \in S_{j+1}(B)$, and $y \in B$, we have

$$
|x - x_0| \ge 2^j r_0 > 2|y - x_0| \ge 2|t_y(y - x_0)|.
$$

Using this, (4.96), (1.104) with y and z therein replaced, respectively, by x_0 and $t_y y + (1 - t_y)x_0$ for any $y \in B$, the Hölder inequality, and Definition 4.3.4(ii), we conclude that, for any $j \in \mathbb{N}$,

$$
\| T(a) \mathbf{1}_{S_{j+1}(B)} \|_{L^r(\mathbb{R}^n)}
$$

$$
\lesssim \int_B r_0^d |a(y)| \left[\int_{S_{j+1}(B)} \frac{|y - x_0|^{\delta r}}{|x - x_0|^{(n+d+\delta)r}} \, dx \right]^{\frac{1}{r}} dy
$$

$$
\lesssim r_0^{d+\delta} (2^j r_0)^{-n+d+\delta} |S_{j+1}(B)|^{\frac{1}{r}} \int_B |a(y)| \, dy
$$

$$
\lesssim 2^{-j(d+\delta+\frac{n}{r})} \|a\|_{L^r(\mathbb{R}^n)} \lesssim 2^{-j(d+\delta+\frac{n}{r})} \frac{|B|^{1/r}}{\|\mathbf{1}_B\|_X}
$$

$$
\sim 2^{-(j+1)(d+\delta+\frac{n}{r})} \frac{|B|^{1/r}}{\|\mathbf{1}_B\|_X}, \qquad (4.97)
$$

which is the desired estimate of $\| T(a) \mathbf{1}_{S_{j+1}(B)} \|_{L^r(\mathbb{R}^n)}$ with $j \in \mathbb{N}$. Applying (4.94), (4.95), and (4.97), we find that there exists a positive constant C, independent of a, such that, for any $j \in \mathbb{Z}_+$,

$$\| CT(a) \mathbf{1}_{S_j(B)} \|_{L^r(\mathbb{R}^n)} \leq 2^{-j(d+\delta+\frac{n}{r'})} \frac{|B|^{1/r}}{\|\mathbf{1}_B\|_X}.$$

This further implies that $CT(a)$ satisfies Definition 4.5.3(ii) and hence $CT(a)$ is an $(X, r, d, d + \delta + \frac{n}{r'})$-molecule centered at B, which then completes the proof of Lemma 4.8.14. □

Via Lemma 4.8.14, we next show Proposition 4.8.12.

Proof of Proposition 4.8.12 Let all the symbols be as in the present proposition and $r := \max\{2, r_0\}$. Then, from the Hölder inequality, it follows that, for any $f \in L^1_{\mathrm{loc}}(\mathbb{R}^n)$, $x \in \mathbb{R}^n$, and $B \in \mathbb{B}$ satisfying $x \in B$,

$$\left[\frac{1}{|B|} \int_B |f(y)|^{(r/s)'} \, dy \right]^{\frac{1}{(r/s)'}} \leq \left[\frac{1}{|B|} \int_B |f(y)|^{(r_0/s)'} \, dy \right]^{\frac{1}{(r_0/s)'}}$$
$$\leq \mathcal{M}^{((r_0/s)')}(f)(x),$$

which further implies that

$$\mathcal{M}^{((r/s)')}(f)(x) \leq \mathcal{M}^{((r_0/s)')}(f)(x).$$

Using this, (4.93), the assumption that $X^{1/s}$ is a BBF space, Remark 1.2.18, and Definition 1.2.13(ii), we find that, for any $f \in L^1_{\mathrm{loc}}(\mathbb{R}^n)$,

$$\left\| \mathcal{M}^{((r/s)')}(f) \right\|_{(X^{1/s})'} \leq \left\| \mathcal{M}^{((r_0/s)')}(f) \right\|_{(X^{1/s})'} \lesssim \| f \|_{(X^{1/s})'}. \tag{4.98}$$

On the other hand, by the assumptions $\theta \in (\frac{n}{n+d+\delta}, \frac{n}{n+d}]$ and $\delta \in (0, 1]$, we conclude that

$$d > n \left(\frac{1}{\theta} - 1 \right) - \delta \geq n \left(\frac{1}{\theta} - 1 \right) - 1$$

and

$$d \leq n \left(\frac{1}{\theta} - 1 \right).$$

These imply that $d = \lfloor n(1/\theta - 1) \rfloor$. From this, (4.98), the assumption that X satisfies Assumption 1.2.29 for the above θ and s, and Lemma 4.4.3, we deduce that, for any $g \in H_{\mathrm{fin}}^{X,r,d,s}(\mathbb{R}^n)$,

$$\|g\|_{H_{\mathrm{fin}}^{X,r,d,s}(\mathbb{R}^n)} \sim \|g\|_{H_X(\mathbb{R}^n)} \qquad (4.99)$$

with the positive equivalence constants independent of g.

Now, let $f \in H_{\mathrm{fin}}^{X,r,d,s}(\mathbb{R}^n)$, $m \in \mathbb{N}$, $\{\lambda_j\}_{j=1}^m \subset [0, \infty)$, and $\{a_j\}_{j=1}^m$ of (X, r, d)-atoms supported, respectively, in the balls $\{B_j\}_{j=1}^m \subset \mathbb{B}$ such that

$$f = \sum_{j=1}^m \lambda_j a_j.$$

This, combined with the linearity of T, implies that

$$T(f) = \sum_{j=1}^m T(a_j). \qquad (4.100)$$

On the other hand, from Lemma 4.8.14, it follows that, for any $j \in \{1, \dots, m\}$, $T(a_j)$ is a harmless constant multiple of an $(X, r, d, d + \delta + \frac{n}{r'})$-molecule centered at B_j. Moreover, by the assumption $\theta > \frac{n}{n+d+\delta}$, we find that

$$d + \delta + \frac{n}{r'} > n\left(\frac{1}{\theta} - \frac{1}{r}\right).$$

Combining this, the assumption that X satisfies Assumption 1.2.29 for the above θ and s, (4.98), (4.100), the fact that, for any $j \in \{1, \dots, m\}$, $T(a_j)$ is a harmless constant multiple of an $(X, r, d, d + \delta + \frac{n}{r'})$-molecule centered at B_j, and Lemma 4.5.5, we further conclude that $T(f) \in H_X(\mathbb{R}^n)$ and

$$\|T(f)\|_{H_X(\mathbb{R}^n)} \lesssim \left\| \left[\sum_{j=1}^m \left(\frac{\lambda_j}{\|\mathbf{1}_{B_j}\|_X} \right)^s \mathbf{1}_{B_j} \right]^{\frac{1}{s}} \right\|_X.$$

This, together with the choice of $\{\lambda_j\}_{j=1}^m$, (4.44), and (4.99), implies that

$$\|T(f)\|_{H_X(\mathbb{R}^n)} \lesssim \|f\|_{H_{\mathrm{fin}}^{X,r,d,s}(\mathbb{R}^n)} \sim \|f\|_{H_X(\mathbb{R}^n)}.$$

Therefore, T is bounded on the finite atomic Hardy space $H_{\mathrm{fin}}^{X,r,d,s}(\mathbb{R}^n)$.

Finally, from the assumption that X has an absolutely continuous quasi-norm and [207, Remark 3.12], we deduce that the finite atomic Hardy space $H_{\mathrm{fin}}^{X,r,d,s}(\mathbb{R}^n)$ is dense in the Hardy space $H_X(\mathbb{R}^n)$. Thus, by a standard density argument, we find that T has a unique extension on $H_X(\mathbb{R}^n)$ and, for any $f \in H_X(\mathbb{R}^n)$,

$$\|T(f)\|_{H_X(\mathbb{R}^n)} \lesssim \|f\|_{H_X(\mathbb{R}^n)},$$

which completes the proof of Proposition 4.8.12. □

Via the above boundedness of Calderón–Zygmund operators on the Hardy space $H_X(\mathbb{R}^n)$, we next prove Theorem 4.8.10.

Proof of Theorem 4.8.10 Let all the symbols be as in the present theorem. Then, from the assumption $m_0(\omega) \in (-\frac{n}{p}, \infty)$ and Theorem 1.2.42, it follows that the local generalized Herz space $\dot{\mathcal{K}}^{p,q}_{\omega,0}(\mathbb{R}^n)$ is a BQBF space. We now show Theorem 4.8.10 via proving that all the assumptions of Proposition 4.8.12 hold true for $\dot{\mathcal{K}}^{p,q}_{\omega,0}(\mathbb{R}^n)$.

First, we show that there exist $\theta, s \in (0, 1]$ such that Assumption 1.2.29 holds true for $\dot{\mathcal{K}}^{p,q}_{\omega,0}(\mathbb{R}^n)$. Indeed, by the assumption $\max\{M_0(\omega), M_\infty(\omega)\} \in (-\frac{n}{p}, n - \frac{n}{p} + d + \delta)$, we conclude that

$$\frac{n}{\max\{M_0(\omega), M_\infty(\omega)\} + n/p} \in \left(\frac{n}{n+d+\delta}, \infty\right).$$

This, combined with the assumptions $p, q \in (\frac{n}{n+d+\delta}, \infty)$, further implies that

$$\min\left\{1, p, q, \frac{n}{\max\{M_0(\omega), M_\infty(\omega)\} + n/p}\right\} \in \left(\frac{n}{n+d+\delta}, \infty\right).$$

Therefore, we can choose an

$$s \in \left(\frac{n}{n+d+\delta}, \min\left\{1, p, q, \frac{n}{\max\{M_0(\omega), M_\infty(\omega)\} + n/p}\right\}\right)$$

and a

$$\theta \in \left(\frac{n}{n+d+\delta}, \min\left\{s, \frac{n}{n+d}\right\}\right).$$

Then, applying Lemma 4.3.11, we find that, for any $\{f_j\}_{j\in\mathbb{N}} \subset L^1_{\mathrm{loc}}(\mathbb{R}^n)$,

$$\left\|\left\{\sum_{j\in\mathbb{N}}\left[\mathcal{M}^{(\theta)}(f_j)\right]^s\right\}^{1/s}\right\|_{\dot{\mathcal{K}}^{p,q}_{\omega,0}(\mathbb{R}^n)} \lesssim \left\|\left(\sum_{j\in\mathbb{N}}|f_j|^s\right)^{1/s}\right\|_{\dot{\mathcal{K}}^{p,q}_{\omega,0}(\mathbb{R}^n)}. \tag{4.101}$$

This implies that, for the above θ and s, $\dot{\mathcal{K}}^{p,q}_{\omega,0}(\mathbb{R}^n)$ satisfies Assumption 1.2.29.

Next, from Lemma 1.8.6, we deduce that the Herz space $[\dot{\mathcal{K}}^{p,q}_{\omega,0}(\mathbb{R}^n)]^{1/s}$ is a BBF space. In addition, let

$$r_0 \in \left(\max\left\{ 1, p, \frac{n}{\min\{m_0(\omega), m_\infty(\omega)\} + n/p} \right\}, \infty \right).$$

We now show that, for this r_0 and the above s, (4.13) holds true. Indeed, using Lemma 1.8.6 with $r := r_0$, we conclude that, for any $f \in L^1_{\text{loc}}(\mathbb{R}^n)$,

$$\left\| \mathcal{M}^{((r_0/s)')}(f) \right\|_{([\dot{\mathcal{K}}^{p,q}_{\omega,0}(\mathbb{R}^n)]^{1/s})'} \lesssim \|f\|_{([\dot{\mathcal{K}}^{p,q}_{\omega,0}(\mathbb{R}^n)]^{1/s})'}. \tag{4.102}$$

Finally, from Theorem 1.4.1, it follows that the local generalized Herz space $\dot{\mathcal{K}}^{p,q}_{\omega,0}(\mathbb{R}^n)$ has an absolutely continuous quasi-norm. Combining this, (4.101), the fact that $[\dot{\mathcal{K}}^{p,q}_{\omega,0}(\mathbb{R}^n)]^{1/s}$ is a BBF space, and (4.102), we find that all the assumptions of Lemma 4.8.12 hold true for $\dot{\mathcal{K}}^{p,q}_{\omega,0}(\mathbb{R}^n)$ under consideration and then complete the proof of Theorem 4.8.10. $\square$

As an application, we have the following boundedness of Calderón–Zygmund operators on the generalized Hardy–Morrey space $HM^{p,q}_{\omega,0}(\mathbb{R}^n)$, which can be deduced from Theorem 4.8.10 and Remarks 1.2.2(iv) and 4.0.20(ii) immediately; we omit the details.

Corollary 4.8.15 *Let $d \in \mathbb{Z}_+$, $p, q \in [1, \infty)$, K be a d-order standard kernel defined as in Definition 1.5.7 with some $\delta \in (0, 1]$, T a d-order Calderón–Zygmund operator with kernel K having the vanishing moments up to order d, and $\omega \in M(\mathbb{R}_+)$ with*

$$-\frac{n}{p} < m_0(\omega) \leq M_0(\omega) < 0$$

and

$$-\frac{n}{p} < m_\infty(\omega) \leq M_\infty(\omega) < 0.$$

Then T has a unique extension on $HM^{p,q}_{\omega,0}(\mathbb{R}^n)$ and there exists a positive constant C such that, for any $f \in HM^{p,q}_{\omega,0}(\mathbb{R}^n)$,

$$\|T(f)\|_{HM^{p,q}_{\omega,0}(\mathbb{R}^n)} \leq C\|f\|_{HM^{p,q}_{\omega,0}(\mathbb{R}^n)}.$$

The remainder of this section is devoted to showing the boundedness of Calderón–Zygmund operators on the generalized Herz–Hardy space $H\dot{\mathcal{K}}^{p,q}_{\omega}(\mathbb{R}^n)$. Precisely, we turn to prove the following theorem.

Theorem 4.8.16 *Let $d \in \mathbb{Z}_+$, $\delta \in (0, 1]$, $p, q \in (\frac{n}{n+d+\delta}, \infty)$, K be a d-order standard kernel with defined as in Definition 1.5.7, T a d-order Calderón–Zygmund operator with kernel K having the vanishing moments up to order d, and $\omega \in M(\mathbb{R}_+)$ with*

$$-\frac{n}{p} < m_0(\omega) \leq M_0(\omega) < n - \frac{n}{p} + d + \delta$$

and

$$-\frac{n}{p} < m_\infty(\omega) \leq M_\infty(\omega) < 0.$$

Then T can be extended into a bounded linear operator on $H\dot{\mathcal{K}}^{p,q}_\omega(\mathbb{R}^n)$ and there exists a positive constant C such that, for any $f \in H\dot{\mathcal{K}}^{p,q}_\omega(\mathbb{R}^n)$,

$$\|T(f)\|_{H\dot{\mathcal{K}}^{p,q}_\omega(\mathbb{R}^n)} \leq C\|f\|_{H\dot{\mathcal{K}}^{p,q}_\omega(\mathbb{R}^n)}.$$

To show this theorem, we first establish a general result about the boundedness of Calderón–Zygmund operators on Hardy spaces associated with ball quasi-Banach function spaces as follows.

Theorem 4.8.17 *Let X be a ball quasi-Banach function space, Y a linear space equipped with a quasi-seminorm $\|\cdot\|_Y$, Y_0 a linear space equipped with a quasi-seminorm $\|\cdot\|_{Y_0}$, $\eta \in (1, \infty)$, and $0 < \theta < s < s_0 \leq 1$ such that*

(i) *for the above θ and s, Assumption 1.2.29 holds true;*
(ii) *both $\|\cdot\|_Y$ and $\|\cdot\|_{Y_0}$ satisfy Definition 1.2.13(ii);*
(iii) *$\mathbf{1}_{B(0,1)} \in Y_0$;*
(iv) *for any $f \in \mathscr{M}(\mathbb{R}^n)$,*

$$\|f\|_{X^{1/s}} \sim \sup\{\|fg\|_{L^1(\mathbb{R}^n)} : \|g\|_Y = 1\}$$

and

$$\|f\|_{X^{1/s_0}} \sim \sup\{\|fg\|_{L^1(\mathbb{R}^n)} : \|g\|_{Y_0} = 1\}$$

with the positive equivalence constants independent of f;
(v) *$\mathcal{M}^{(\eta)}$ is bounded on both Y and Y_0.*

Assume that $d \in \mathbb{Z}_+$, $\delta \in (0, 1]$, K is a d-order standard kernel defined as in Definition 1.5.7, and T a d-order Calderón–Zygmund operator with kernel K having the vanishing moments up to d. If $\theta \in (\frac{n}{n+d+\delta}, \frac{n}{n+d}]$, then T can be extended into a bounded linear operator on $H_X(\mathbb{R}^n)$, namely, there exists a positive constant

C such that, for any $f \in H_X(\mathbb{R}^n)$,

$$\|T(f)\|_{H_X(\mathbb{R}^n)} \leq C\|f\|_{H_X(\mathbb{R}^n)}.$$

To show Theorem 4.8.17, the following two embedding theorems about X and $H_X(\mathbb{R}^n)$ are the essential tools.

Lemma 4.8.18 *Let* X *be a ball quasi-Banach function space,* $Y \subset \mathscr{M}(\mathbb{R}^n)$ *a linear space equipped with a quasi-seminorm* $\|\cdot\|_Y$, $\theta \in (1, \infty)$, *and* $s \in (0, \infty)$ *satisfy the following four statements:*

(i) $\|\cdot\|_Y$ *satisfies Definition 1.2.13(ii);*
(ii) $\mathbf{1}_{B(0,1)} \in Y$;
(iii) *for any* $f \in \mathscr{M}(\mathbb{R}^n)$,

$$\|f\|_{X^{1/s}} \sim \sup \left\{ \|fg\|_{L^1(\mathbb{R}^n)} : \|g\|_Y = 1 \right\},$$

where the positive equivalence constants are independent of f;
(iv) $\mathcal{M}^{(\theta)}$ *is bounded on* Y.

Assume $\varepsilon \in (\frac{1}{\theta}, 1)$ *and* $\upsilon := [\mathcal{M}(\mathbf{1}_{B(0,1)})]^\varepsilon$. *Then there exists a positive constant* C *such that, for any* $f \in X$,

$$\|f\|_{L^s_\upsilon(\mathbb{R}^n)} \leq C\|f\|_X.$$

Proof Let all the symbols be as in the present lemma and $f \in X$. Then, by Lemma 2.3.11 with $r := 1$, we find that, for any $x \in \mathbb{R}^n$,

$$\upsilon(x) := \left[\mathcal{M}\left(\mathbf{1}_{B(0,1)}\right) \right]^\varepsilon \sim (1 + |x|)^{-\varepsilon n}.$$

This implies that, for any $x \in B(0, 1)$, $\upsilon(x) \sim 1$ and, for any $k \in \mathbb{N}$ and $x \in B(0, 2^k) \setminus B(0, 2^{k-1})$, $\upsilon(x) \sim 2^{-k\varepsilon n}$. From these, we further deduce that

$$\int_{\mathbb{R}^n} |f(y)|^s \upsilon(y)\, dy$$

$$= \int_{B(0,1)} |f(y)|^s \upsilon(y)\, dy + \sum_{k \in \mathbb{N}} \int_{B(0,2^k) \setminus B(0,2^{k-1})} |f(y)|^s \upsilon(y)\, dy$$

$$\sim \int_{B(0,1)} |f(y)|^s\, dy + \sum_{k \in \mathbb{N}} 2^{-k\varepsilon n} \int_{B(0,2^k) \setminus B(0,2^{k-1})} |f(y)|^s\, dy$$

$$\lesssim \int_{\mathbb{R}^n} |f(y)|^s \mathbf{1}_{B(0,1)}(y)\, dy + \sum_{k \in \mathbb{N}} 2^{-k\varepsilon n} \int_{\mathbb{R}^n} |f(y)|^s \mathbf{1}_{B(0,2^k)}(y)\, dy. \qquad (4.103)$$

Next, applying (1.55), we conclude that, for any $k \in \mathbb{N}$ and $x \in B(\mathbf{0}, 2^k)$,

$$\mathcal{M}^{(\theta)}\left(\mathbf{1}_{B(0,1)}\right)(x) \geq \left\{\frac{1}{|B(\mathbf{0}, 2^k)|} \int_{B(0,2^k)} \left[\mathbf{1}_{B(0,1)}(y)\right]^\theta \, dy\right\}^{\frac{1}{\theta}}$$

$$\sim \left[\frac{|B(\mathbf{0}, 1)|}{|B(\mathbf{0}, 2^k)|}\right]^{\frac{1}{\theta}} \sim 2^{-\frac{nk}{\theta}},$$

which implies that

$$\mathbf{1}_{B(0,2^k)} \lesssim 2^{\frac{nk}{\theta}} \mathcal{M}^{(\theta)}\left(\mathbf{1}_{B(0,1)}\right),$$

where the implicit positive constant is independent of k. Applying this and both the assumptions (i) and (iv) of the present lemma, we find that, for any $k \in \mathbb{N}$,

$$\left\|\mathbf{1}_{B(0,2^k)}\right\|_Y \lesssim 2^{\frac{nk}{\theta}} \left\|\mathcal{M}^{(\theta)}\left(\mathbf{1}_{B(0,1)}\right)\right\|_Y \lesssim 2^{\frac{nk}{\theta}} \left\|\mathbf{1}_{B(0,1)}\right\|_Y, \qquad (4.104)$$

which is the desired estimate of $\mathbf{1}_{B(0,2^k)}$. Thus, using (4.103), the assumption (iii) of the present lemma, (4.104), and the assumption $\varepsilon - \frac{1}{\theta} \in (0, \infty)$, we find that

$$\int_{\mathbb{R}^n} |f(y)|^s \upsilon(y) \, dy$$

$$\lesssim \left\| |f|^s \right\|_{X^{1/s}} \left[\left\|\mathbf{1}_{B(0,1)}\right\|_Y + \sum_{k\in\mathbb{N}} 2^{-k\varepsilon n} \left\|\mathbf{1}_{B(0,2^k)}\right\|_Y\right]$$

$$\lesssim \|f\|_X^s \left\|\mathbf{1}_{B(0,1)}\right\|_Y \left[1 + \sum_{k\in\mathbb{N}} 2^{-kn(\varepsilon-\frac{1}{\theta})}\right]$$

$$\sim \|f\|_X^s \left\|\mathbf{1}_{B(0,1)}\right\|_Y. \qquad (4.105)$$

On the other hand, by the assumption (ii) of the present lemma, we conclude that

$$\left\|\mathbf{1}_{B(0,1)}\right\|_Y < \infty.$$

Therefore, from (4.105), we infer that

$$\|f\|_{L_\upsilon^s(\mathbb{R}^n)}^s = \int_{\mathbb{R}^n} |f(y)|^s \upsilon(y) \, dy \lesssim \|f\|_X^s,$$

which completes the proof of Lemma 4.8.18. $\qquad\qquad\qquad\qquad\qquad\qquad\square$

Lemma 4.8.19 *Let X be a ball quasi-Banach function space and $N \in \mathbb{N}$. Then the Hardy space $H_X(\mathbb{R}^n)$ embeds continuously into $\mathcal{S}'(\mathbb{R}^n)$. Namely, there exists a* *positive constant C such that, for any $f \in H_X(\mathbb{R}^n)$ and $\phi \in \mathcal{S}(\mathbb{R}^n)$,*

$$|\langle f, \phi \rangle| \leq C p_N(\phi) \| f \|_{H_X(\mathbb{R}^n)},$$

where p_N is defined as in (4.1).

Proof Let all the symbols be as in the present lemma, $f \in H_X(\mathbb{R}^n)$, and $\phi \in \mathcal{S}(\mathbb{R}^n)$. Then we prove the present lemma by considering the following two cases on ϕ.

Case (1) $p_N(\phi) = 0$. In this case, we have $\phi = 0$. This further implies that

$$|\langle f, \phi \rangle| = 0 = p_N(\phi) \| f \|_{H_X(\mathbb{R}^n)}, \tag{4.106}$$

which completes the proof of Lemma 4.8.19 in this case.

Case (2) $p_N(\phi) \neq 0$. In this case, from (4.1), we deduce that $\frac{\phi(-\cdot)}{p_N(\phi)} \in \mathcal{S}(\mathbb{R}^n)$ and

$$p_N\left(\frac{\phi(-\cdot)}{p_N(\phi)} \right) = \frac{p_N(\phi)}{p_N(\phi)} = 1.$$

Therefore, $\frac{\phi(-\cdot)}{p_N(\phi)} \in \mathcal{F}_N(\mathbb{R}^n)$ with $\mathcal{F}_N(\mathbb{R}^n)$ as in (4.2). By this and (4.3), we conclude that, for any $x \in B(\mathbf{0}, 1)$,

$$|\langle f, \phi \rangle| = |f * \phi(-\cdot)(\mathbf{0})|$$

$$= p_N(\phi) \left| f * \left[\frac{\phi(-\cdot)}{p_N(\phi)} \right] (\mathbf{0}) \right| \leq p_N(\phi) \mathcal{M}_N(f)(x).$$

This further implies that

$$|\langle f, \phi \rangle| \mathbf{1}_{B(\mathbf{0},1)} \leq p_N(\phi) \mathcal{M}_N(f). \tag{4.107}$$

In addition, since X is a BQBF space, from both (i) and (iv) of Definition 1.2.13, we deduce that $\| \mathbf{1}_{B(\mathbf{0},1)} \|_X \in (0, \infty)$. This, together with (4.107) and Definitions 1.2.13(ii) and 4.2.2, further implies that

$$|\langle f, \phi \rangle| \leq \left\| \mathbf{1}_{B(\mathbf{0},1)} \right\|_X^{-1} p_N(\phi) \| \mathcal{M}_N(f) \|_X$$

$$= \left\| \mathbf{1}_{B(\mathbf{0},1)} \right\|_X^{-1} p_N(\phi) \| f \|_{H_X(\mathbb{R}^n)}. \tag{4.108}$$

Using this, we find that the present lemma holds true in this case. Combining this and (4.106), we then complete the proof of Lemma 4.8.19. $\square$

Furthermore, the following auxiliary lemma about function spaces having absolutely continuous quasi-norms is given in [207, Corollary 3.11(ii)], which is important in the proof of Theorem 4.8.17.

Lemma 4.8.20 *Let* X, r, d, s, $\{\lambda_j\}_{j\in\mathbb{N}}$, *and* $\{a_j\}_{j\in\mathbb{N}}$ *be as in Lemma 4.3.6. If* X *has an absolutely continuous quasi-norm, then* $f \in H_X(\mathbb{R}^n)$ *and* $f := \sum_{j\in\mathbb{N}} \lambda_j a_j$ *converges in* $H_X(\mathbb{R}^n)$.

To prove Theorem 4.8.17, we also need weighted function spaces. In what follows, the *weighted Hardy space* $H_\upsilon^p(\mathbb{R}^n)$, with $p \in (0, \infty)$ and $\upsilon \in A_\infty(\mathbb{R}^n)$, is defined as in Definition 4.2.2 with $X := L_\upsilon^p(\mathbb{R}^n)$ [see also Remark 4.2.3(ii)]. Then, via Lemma 4.8.20 above, we establish the following atomic reconstruction theorem about weighted Hardy spaces, which is one of the key tools used in the proof of Theorem 4.8.17.

Lemma 4.8.21 *Let* $0 < \theta < s < s_0 \leq 1$, $\upsilon \in A_1(\mathbb{R}^n)$, *and* $d \geq \lfloor n(1/\theta - 1) \rfloor$ *be a fixed integer. Assume that* $\{a_j\}_{j\in\mathbb{N}}$ *is a sequence of* $(L_\upsilon^{s_0}(\mathbb{R}^n), \infty, d)$*-atoms supported, respectively, in the balls* $\{B_j\}_{j\in\mathbb{N}} \subset \mathbb{B}$ *and* $\{\lambda_j\}_{j\in\mathbb{N}} \subset [0, \infty)$ *such that* $f := \sum_{j\in\mathbb{N}} \lambda_j a_j$ *in* $\mathcal{S}'(\mathbb{R}^n)$ *and*

$$\left\| \left\{ \sum_{j\in\mathbb{N}} \left[\frac{\lambda_j}{\|\mathbf{1}_{B_j}\|_{L_\upsilon^{s_0}(\mathbb{R}^n)}} \right]^s \mathbf{1}_{B_j} \right\}^{\frac{1}{s}} \right\|_{L_\upsilon^{s_0}(\mathbb{R}^n)} < \infty.$$

Then $f \in H_\upsilon^{s_0}(\mathbb{R}^n)$ *and* $f = \sum_{j\in\mathbb{N}} \lambda_j a_j$ *holds true in* $H_\upsilon^{s_0}(\mathbb{R}^n)$.

Proof Let all the symbols be as in the present lemma. Then, applying [233, Remarks 2.4(b), 2.7(b), and 3.4(i)], we find that the following four statements hold true:

(i) $L_\upsilon^{s_0}(\mathbb{R}^n)$ is a BQBF space;
(ii) for any $\{f_j\}_{j\in\mathbb{N}} \subset L_{\mathrm{loc}}^1(\mathbb{R}^n)$,

$$\left\| \left\{ \sum_{j\in\mathbb{N}} \left[\mathcal{M}^{(\theta)}(f_j) \right]^s \right\}^{1/s} \right\|_{L_\upsilon^{s_0}(\mathbb{R}^n)} \lesssim \left\| \left(\sum_{j\in\mathbb{N}} |f_j|^s \right)^{1/s} \right\|_{L_\upsilon^{s_0}(\mathbb{R}^n)};$$

(iii) $[L_\upsilon^{s_0}(\mathbb{R}^n)]^{1/s}$ is a BBF space and, for any $f \in L_{\mathrm{loc}}^1(\mathbb{R}^n)$,

$$\|\mathcal{M}(f)\|_{([L_\upsilon^{s_0}(\mathbb{R}^n)]^{1/s})'} \lesssim \|f\|_{([L_\upsilon^{s_0}(\mathbb{R}^n)]^{1/s})'};$$

(iv) $L_\upsilon^{s_0}(\mathbb{R}^n)$ has an absolutely continuous quasi-norm.

Thus, the weighted Lebesgue space $L_\upsilon^{s_0}(\mathbb{R}^n)$ under consideration satisfies all the assumptions of Lemma 4.8.20 with r therein replaced by ∞. This finishes the proof of Lemma 4.8.21. $\square$

We now show Theorem 4.8.17.

Proof of Theorem 4.8.17 Let all the symbols be as in the present theorem and $f \in H_X(\mathbb{R}^n)$. Then, by the assumptions $\theta \in (\frac{n}{n+d+\delta}, \frac{n}{n+d}]$ and $\delta \in (0, 1]$, we conclude that

$$d > n\left(\frac{1}{\theta} - 1\right) - \delta \geq n\left(\frac{1}{\theta} - 1\right) - 1$$

and

$$d \leq n\left(\frac{1}{\theta} - 1\right).$$

Thus, $d = \lfloor n(1/\theta - 1)\rfloor$. From this, the assumption (i) of the present theorem, and Lemma 4.3.8, it follows that there exists $\{\lambda_j\}_{j\in\mathbb{N}} \subset [0, \infty)$ and $\{a_j\}_{j\in\mathbb{N}}$ of (X, ∞, d)-atoms supported, respectively, in the balls $\{B_j\}_{j\in\mathbb{N}} \subset \mathbb{B}$ such that

$$f = \sum_{j\in\mathbb{N}} \lambda_j a_j \tag{4.109}$$

in $\mathcal{S}'(\mathbb{R}^n)$ and

$$\left\|\left[\sum_{j\in\mathbb{N}}\left(\frac{\lambda_j}{\|\mathbf{1}_{B_j}\|_X}\right)^s \mathbf{1}_{B_j}\right]^{\frac{1}{s}}\right\|_X \lesssim \|f\|_{H_X(\mathbb{R}^n)}. \tag{4.110}$$

In addition, by the assumptions (ii) through (v) of the present theorem and Lemma 4.8.18 with $Y := Y_0$, $\theta := \eta$, and $s := s_0$, we find that there exists an $\varepsilon \in (0, 1)$ such that, for any $h \in \mathcal{M}(\mathbb{R}^n)$,

$$\|h\|_{L_\upsilon^{s_0}(\mathbb{R}^n)} \lesssim \|h\|_X,$$

where $\upsilon := [\mathcal{M}(\mathbf{1}_{B(0,1)})]^\varepsilon$. Combining this, (4.109), and (4.110), we conclude that

$$f = \sum_{j\in\mathbb{N}} \lambda_j a_j = \sum_{j\in\mathbb{N}}\left[\lambda_j \frac{\|\mathbf{1}_{B_j}\|_{L_\upsilon^{s_0}(\mathbb{R}^n)}}{\|\mathbf{1}_{B_j}\|_X}\right]\left[\frac{\|\mathbf{1}_{B_j}\|_X}{\|\mathbf{1}_{B_j}\|_{L_\upsilon^{s_0}(\mathbb{R}^n)}} a_j\right] \tag{4.111}$$

in $\mathcal{S}'(\mathbb{R}^n)$ and

$$
\left\| \left\{ \sum_{j \in \mathbb{N}} \left[\frac{\lambda_j \frac{\|\mathbf{1}_{B_j}\|_{L_v^{s_0}(\mathbb{R}^n)}}{\|\mathbf{1}_{B_j}\|_X}}{\|\mathbf{1}_{B_j}\|_{L_v^{s_0}(\mathbb{R}^n)}} \right]^s \mathbf{1}_{B_j} \right\}^{\frac{1}{s}} \right\|_{L_v^{s_0}(\mathbb{R}^n)}
$$

$$
\approx \left\| \left[\sum_{j \in \mathbb{N}} \left(\frac{\lambda_j}{\|\mathbf{1}_{B_j}\|_X} \right)^s \mathbf{1}_{B_j} \right]^{\frac{1}{s}} \right\|_X \lesssim \|f\|_{H_X(\mathbb{R}^n)} < \infty. \tag{4.112}
$$

Observe that, for any $j \in \mathbb{N}$, $\frac{\|\mathbf{1}_{B_j}\|_X}{\|\mathbf{1}_{B_j}\|_{L_v^{s_0}(\mathbb{R}^n)}} a_j$ is an $(L_v^{s_0}(\mathbb{R}^n), \infty, d)$-atom supported in B_j. This, together with (4.111), (4.112), and Lemma 4.8.21, further implies that $f \in H_v^{s_0}(\mathbb{R}^n)$ and $f = \sum_{j \in \mathbb{N}} \lambda_j a_j$ in $H_v^{s_0}(\mathbb{R}^n)$. Applying this and Remark 4.8.13(ii) with $p := s_0$, we find that $T(f) = \sum_{j \in \mathbb{N}} \lambda_j T(a_j)$ holds true in $H_v^{s_0}(\mathbb{R}^n)$. Therefore, from Lemma 4.8.19 with X replaced by $L_v^{s_0}(\mathbb{R}^n)$, we further infer that

$$
T(f) = \sum_{j \in \mathbb{N}} \lambda_j T(a_j) \tag{4.113}
$$

in $\mathcal{S}'(\mathbb{R}^n)$.

Next, we prove that $T(f) \in H_X(\mathbb{R}^n)$. To this end, choose an $r \in (\max\{2, s\eta'\}, \infty)$ with $\frac{1}{\eta} + \frac{1}{\eta'} = 1$. Then, by Lemma 4.3.5 with $t = \infty$, we conclude that, for any $j \in \mathbb{N}$, a_j is an (X, r, d)-atom supported in the ball B_j. This, together with Lemma 4.8.14, implies that, for any $j \in \mathbb{N}$, $T(a_j)$ is a harmless multiple of an $(X, r, d, d + \delta + \frac{n}{r'})$-molecule centered at B_j. Moreover, since $\theta > \frac{n}{n+d+\delta}$, it follows that

$$
d + \delta + \frac{n}{r'} > n \left(\frac{1}{\theta} - \frac{1}{r} \right).
$$

On the other hand, using the assumption (ii) of the present theorem and an argument similar to that used in the proof of (4.98) with $(r_0/s)'$ and $(X^{1/s})'$ therein replaced, respectively, by η and Y, we find that, for any $f \in L_{\mathrm{loc}}^1(\mathbb{R}^n)$,

$$
\left\| \mathcal{M}^{((r/s)')}(f) \right\|_Y \lesssim \|f\|_Y.
$$

From this, both the assumptions (i) and (iv) of the present theorem, Theorem 4.5.11, (4.113), and (4.110), we deduce that $T(f) \in H_X(\mathbb{R}^n)$ and

$$\|T(f)\|_{H_X(\mathbb{R}^n)} \lesssim \left\| \left[\sum_{j \in \mathbb{N}} \left(\frac{\lambda_j}{\|\mathbf{1}_{B_j}\|_X} \right)^s \mathbf{1}_{B_j} \right]^{\frac{1}{s}} \right\|_X \lesssim \|f\|_{H_X(\mathbb{R}^n)}.$$

This finishes the proof that $T(f) \in H_X(\mathbb{R}^n)$ and further implies that T is bounded on $H_X(\mathbb{R}^n)$. Thus, the proof of Theorem 4.8.17 is completed. $\qquad\square$

Now, we prove Theorem 4.8.16.

Proof of Theorem 4.8.16 Let all the symbols be as in the present theorem. Then, using the assumptions $m_0(\omega) \in (-\frac{n}{p}, \infty)$ and $M_\infty(\omega) \in (-\infty, 0)$, and Theorem 1.2.44, we find that, under the assumptions of the present theorem, the global generalized Herz space $\dot{\mathcal{K}}_\omega^{p,q}(\mathbb{R}^n)$ is a BQBF space. Therefore, to prove the present theorem, it suffices to show that all the assumptions of Theorem 4.8.17 hold true for $\dot{\mathcal{K}}_\omega^{p,q}(\mathbb{R}^n)$.

First, by the assumptions $p, q \in (\frac{n}{n+d+\delta}, \infty)$ and

$$\max\{M_0(\omega), M_\infty(\omega)\} \in \left(-\frac{n}{p}, n - \frac{n}{p} + d + \delta \right),$$

we conclude that

$$\frac{n}{n+d+\delta} < \min\left\{ 1, p, q, \frac{n}{\max\{M_0(\omega), M_\infty(\omega)\} + n/p} \right\}.$$

Let

$$s \in \left(\frac{n}{n+d+\delta}, \min\left\{ 1, p, q, \frac{n}{\max\{M_0(\omega), M_\infty(\omega)\} + n/p} \right\} \right)$$

and

$$\theta \in \left(\frac{n}{n+d+\delta}, \min\left\{ s, \frac{n}{n+d} \right\} \right).$$

We now prove that $\dot{\mathcal{K}}_\omega^{p,q}(\mathbb{R}^n)$ satisfies Theorem 4.8.17(i) for the above θ and s. Indeed, from Lemma 4.3.25, it follows that, for any $\{f_j\}_{j \in \mathbb{N}} \subset L_{\text{loc}}^1(\mathbb{R}^n)$,

$$\left\| \left\{ \sum_{j \in \mathbb{N}} \left[\mathcal{M}^{(\theta)}(f_j) \right]^s \right\}^{1/s} \right\|_{\dot{\mathcal{K}}_\omega^{p,q}(\mathbb{R}^n)} \lesssim \left\| \left(\sum_{j \in \mathbb{N}} |f_j|^s \right)^{1/s} \right\|_{\dot{\mathcal{K}}_\omega^{p,q}(\mathbb{R}^n)}.$$

This implies that, for the above θ and s, Assumption 1.2.29 holds true for $\dot{\mathcal{K}}_\omega^{p,q}(\mathbb{R}^n)$ and hence Theorem 4.8.17(i) is satisfied for $\dot{\mathcal{K}}_\omega^{p,q}(\mathbb{R}^n)$.

We next show that there exist two linear spaces Y and Y_0 and $\eta, s \in (0, \infty)$ such that the assumptions (ii) through (v) of Theorem 4.8.17 hold true. To this end, let $Y := \dot{\mathcal{B}}_{1/\omega^s}^{(p/s)',(q/s)'}(\mathbb{R}^n)$, $s_0 \in (0, 1]$ be such that

$$s_0 \in \left(s, \min\left\{ 1, p, q, \frac{n}{\max\{M_0(\omega), M_\infty(\omega)\} + n/p} \right\} \right),$$

$Y_0 := \dot{\mathcal{B}}_{1/\omega^{s_0}}^{(p/s_0)',(q/s_0)'}(\mathbb{R}^n)$, and $\eta \in (1, \infty)$ be such that

$$\eta < \min\left\{ \frac{n}{n(1 - s/p) - s\min\{m_0(\omega), m_\infty(\omega)\}}, \left(\frac{p}{s}\right)' \right\}. \tag{4.114}$$

Then, applying Lemma 2.3.2, we find that both

$$\dot{\mathcal{B}}_{1/\omega^s}^{(p/s)',(q/s)'}(\mathbb{R}^n) \text{ and } \dot{\mathcal{B}}_{1/\omega^{s_0}}^{(p/s_0)',(q/s_0)'}(\mathbb{R}^n)$$

satisfy Definition 1.2.13(ii). This further implies that Theorem 4.8.17(ii) holds true.

Now, we show that the assumption (iii) of Theorem 4.8.17 holds true for the above Y_0. Namely, $\mathbf{1}_{B(0,1)} \in \dot{\mathcal{B}}_{1/\omega^{s_0}}^{(p/s_0)',(q/s_0)'}(\mathbb{R}^n)$. Indeed, combining Lemma 1.1.6 and the assumptions $s < \frac{n}{\max\{M_0(\omega), M_\infty(\omega)\}+n/p}$ and $M_0(\omega) > -\frac{n}{p}$, we conclude that

$$m_0\left(\omega^{-s_0}\right) = -s_0 M_0(\omega) > -s_0 \max\{M_0(\omega), M_\infty(\omega)\}$$

$$> -n\left(1 - \frac{s}{p}\right) = -\frac{n}{(p/s)'}.$$

Applying this and Theorem 1.2.42 with p, q, and ω therein replaced, respectively, by $(p/s_0)'$, $(q/s_0)'$, and $1/\omega^{s_0}$, we find that the Herz space $\dot{\mathcal{K}}_{1/\omega^{s_0},0}^{(p/s_0)',(q/s_0)'}(\mathbb{R}^n)$ is a BQBF space. This, together with Lemma 2.2.3 with p, q, ω, and ξ therein replaced, respectively, by $(p/s_0)'$, $(q/s_0)'$, $1/\omega^{s_0}$, and $\mathbf{0}$, and Definition 1.2.13(iv), further implies that

$$\left\| \mathbf{1}_{B(0,1)} \right\|_{\dot{\mathcal{B}}_{1/\omega^{s_0}}^{(p/s_0)',(q/s_0)'}(\mathbb{R}^n)} \lesssim \left\| \mathbf{1}_{B(0,1)} \right\|_{\dot{\mathcal{K}}_{1/\omega^{s_0},0}^{(p/s_0)',(q/s_0)'}(\mathbb{R}^n)} < \infty,$$

which completes the proof that $\mathbf{1}_{B(0,1)} \in \dot{\mathcal{B}}_{1/\omega^{s_0}}^{(p/s_0)',(q/s_0)'}(\mathbb{R}^n)$. Therefore, Theorem 4.8.17(iii) holds true.

Next, we prove that both $\dot{\mathcal{B}}_{1/\omega^s}^{(p/s)',(q/s)'}(\mathbb{R}^n)$ and $\dot{\mathcal{B}}_{1/\omega^{s_0}}^{(p/s_0)',(q/s_0)'}(\mathbb{R}^n)$ satisfy Theorem 4.8.17(iv). Indeed, using Lemma 1.1.6 and the assumption $m_0(\omega) > -\frac{n}{p}$, we find that

$$m_0\left(\omega^s\right) = s m_0(\omega) > -\frac{n}{p/s}. \tag{4.115}$$

On the other hand, from Lemma 1.1.6 again and the assumption $M_\infty(\omega) < 0$, we deduce that

$$M_\infty\left(\omega^s\right) = s M_\infty(\omega) < 0.$$

Combining this, (4.115), and Lemma 4.3.27 with p, q, and ω therein replaced, respectively, by p/s, q/s, and ω^s, we conclude that, for any $f \in \mathcal{M}(\mathbb{R}^n)$,

$$\|f\|_{\dot{\mathcal{K}}_{\omega^s}^{p/s,q/s}(\mathbb{R}^n)} \sim \sup\left\{\|fg\|_{L^1(\mathbb{R}^n)} : \|g\|_{\dot{\mathcal{B}}_{1/\omega^s}^{(p/s)',(q/s)'}(\mathbb{R}^n)} = 1\right\}, \tag{4.116}$$

where the positive equivalence constants are independent of f. Similarly, repeating an argument used in the proof of (4.116) with s replaced by s_0, we find that, for any $f \in \mathcal{M}(\mathbb{R}^n)$,

$$\|f\|_{\dot{\mathcal{K}}_{\omega^{s_0}}^{p/s_0,q/s_0}(\mathbb{R}^n)} \sim \sup\left\{\|fg\|_{L^1(\mathbb{R}^n)} : \|g\|_{\dot{\mathcal{B}}_{1/\omega^{s_0}}^{(p/s_0)',(q/s_0)'}(\mathbb{R}^n)} = 1\right\},$$

which, together with (4.116), further implies that Theorem 4.8.16(iv) holds true with $Y = \dot{\mathcal{B}}_{1/\omega^s}^{(p/s)',(q/s)'}(\mathbb{R}^n)$ and $Y_0 = \dot{\mathcal{B}}_{1/\omega^{s_0}}^{(p/s_0)',(q/s_0)'}(\mathbb{R}^n)$.

Finally, we prove that the powered Hardy–Littlewood maximal operator $\mathcal{M}^{(\eta)}$ is bounded on both $\dot{\mathcal{B}}_{1/\omega^s}^{(p/s)',(q/s)'}(\mathbb{R}^n)$ and $\dot{\mathcal{B}}_{1/\omega^{s_0}}^{(p/s_0)',(q/s_0)'}(\mathbb{R}^n)$. Indeed, from the assumption $s < \frac{n}{\max\{M_0(\omega), M_\infty(\omega)\} + n/p}$, we deduce that

$$n\left(1 - \frac{s}{p}\right) - s\min\{m_0(\omega), m_\infty(\omega)\}$$

$$= n - s\left[\min\{m_0(\omega), m_\infty(\omega)\} + \frac{n}{p}\right]$$

$$\geq n - s\left[\max\{M_0(\omega), M_\infty(\omega)\} + \frac{n}{p}\right] > 0,$$

which, combined with Lemma 1.1.6 and (4.114), further implies that

$$\max\left\{M_0\left(\omega^{-s}\right),\, M_\infty\left(\omega^{-s}\right)\right\}$$

$$= -s\,\min\{m_0(\omega),\, m_\infty(\omega)\} < n\left[\frac{1}{\eta} - \frac{1}{(p/s)'}\right]. \tag{4.117}$$

In addition, applying Lemma 1.1.6 again and the assumptions

$$s < \frac{n}{\max\{M_0(\omega),\, M_\infty(\omega)\} + n/p}$$

and $\max\{M_0(\omega),\, M_\infty(\omega)\} > -\frac{n}{p}$, we conclude that

$$\min\left\{m_0\left(\omega^{-s}\right),\, m_\infty\left(\omega^{-s}\right)\right\}$$

$$= -s\,\max\{M_0(\omega),\, M_\infty(\omega)\} > -n\left(1 - \frac{s}{p}\right) = -\frac{n}{(p/s)'}.$$

From this, (4.117), the assumption $\eta < (p/s)'$, and Corollary 2.3.5 with p, q, ω, and r therein replaced, respectively, by $(p/s)'$, $(q/s)'$, $1/\omega^s$, and η, it follows that, for any $f \in L^1_{\mathrm{loc}}(\mathbb{R}^n)$,

$$\left\|\mathcal{M}^{(\eta)}(f)\right\|_{\dot{\mathcal{B}}^{(p/s)',(q/s)'}_{1/\omega^s}(\mathbb{R}^n)} \lesssim \|f\|_{\dot{\mathcal{B}}^{(p/s)',(q/s)'}_{1/\omega^s}(\mathbb{R}^n)}. \tag{4.118}$$

This further implies that $\mathcal{M}^{(\eta)}$ is bounded on $Y = \dot{\mathcal{B}}^{(p/s)',(q/s)'}_{1/\omega^s}(\mathbb{R}^n)$. Moreover, repeating an argument used in the proof of (4.118) with s replaced by s_0, and using the assumption $s_0 > s$, we find that, for any $f \in L^1_{\mathrm{loc}}(\mathbb{R}^n)$,

$$\left\|\mathcal{M}^{(\eta)}(f)\right\|_{\dot{\mathcal{B}}^{(p/s_0)',(q/s_0)'}_{1/\omega^{s_0}}(\mathbb{R}^n)} \lesssim \|f\|_{\dot{\mathcal{B}}^{(p/s_0)',(q/s_0)'}_{1/\omega^{s_0}}(\mathbb{R}^n)},$$

which implies that $\mathcal{M}^{(\eta)}$ is bounded on $Y_0 = \dot{\mathcal{B}}^{(p/s_0)',(q/s_0)'}_{1/\omega^{s_0}}(\mathbb{R}^n)$ and hence Theorem 4.8.17(v) holds true. Therefore, all the assumptions of Theorem 4.8.17 hold true for $\dot{\mathcal{K}}^{p,q}_\omega(\mathbb{R}^n)$. Therefore, T can be extended into a bounded linear operator on $H\dot{\mathcal{K}}^{p,q}_\omega(\mathbb{R}^n)$ and, for any $f \in H\dot{\mathcal{K}}^{p,q}_\omega(\mathbb{R}^n)$,

$$\|T(f)\|_{H\dot{\mathcal{K}}^{p,q}_\omega(\mathbb{R}^n)} \lesssim \|f\|_{H\dot{\mathcal{K}}^{p,q}_\omega(\mathbb{R}^n)}.$$

This finishes the proof of Theorem 4.8.16.   □

Via Theorem 4.8.16 and Remarks 1.2.2(iv) and 4.0.20(ii), we immediately obtain the following boundedness of Calderón–Zygmund operators on the generalized Hardy–Morrey space $HM^{p,q}_\omega(\mathbb{R}^n)$; we omit the details.

Corollary 4.8.22 *Let* $d \in \mathbb{Z}_+$, *K* *be a* *d-order standard kernel defined as in Definition 1.5.7 with some* $\delta \in (0, 1]$, *T* *a* *d-order Calderón–Zygmund operator with kernel K having the vanishing moments up to order d, and p, q, and* ω *as in Corollary 4.8.15. Then T can be extended into a bounded linear operator on* $HM_{\omega}^{p,q}(\mathbb{R}^n)$ *and there exists a positive constant C such that, for any* $f \in HM_{\omega}^{p,q}(\mathbb{R}^n)$,

$$\|T(f)\|_{HM_{\omega}^{p,q}(\mathbb{R}^n)} \leq C\|f\|_{HM_{\omega}^{p,q}(\mathbb{R}^n)}.$$

4.9   Fourier Transform

The target of this section is to investigate the Fourier transform of a distribution in the generalized Herz–Hardy space $H\dot{\mathcal{K}}_{\omega,0}^{p,q}(\mathbb{R}^n)$ or $H\dot{\mathcal{K}}_{\omega}^{p,q}(\mathbb{R}^n)$. Recall that, in 1974, Coifman [45] characterized all $\widehat{f}$ via entire functions of exponential type for $n = 1$, where $f \in H^p(\mathbb{R})$ with $p \in (0, 1]$. Later, a number of authors investigated the characterization of $\widehat{f}$ with distribution f belonging to Hardy spaces in higher dimensions (see, for instance, [16, 48, 121, 122, 221]). In particular, Huang et al. [122] studied the Fourier transform of the distribution belonging to the Hardy space associated with the ball quasi-Banach function space, which plays an important role in this section.

We first consider the Fourier transform properties of the Hardy spaces associated with the local generalized Herz spaces. Namely, we have the following theorem.

Theorem 4.9.1 *Let* $p, q \in (0, 1]$, $\omega \in M(\mathbb{R}_+)$ *with* $m_0(\omega) \in (0, \infty)$ *and* $m_\infty(\omega) \in (0, \infty)$, *and* $p_- \in (0, \frac{n}{\max\{M_0(\omega), M_\infty(\omega)\} + n/p})$. *Then, for any* $f \in H\dot{\mathcal{K}}_{\omega,0}^{p,q}(\mathbb{R}^n)$, *there exists a continuous function g on* $\mathbb{R}^n$ *such that* $\widehat{f} = g$ *in* $\mathcal{S}'(\mathbb{R}^n)$ *and*

$$\lim_{|x| \to 0^+} \frac{|g(x)|}{|x|^{n(\frac{1}{p_-} - 1)}} = 0.$$

Moreover, there exists a positive constant C, independent of both f and g, such that, for any $x \in \mathbb{R}^n$,

$$|g(x)| \leq C\|f\|_{H\dot{\mathcal{K}}_{\omega,0}^{p,q}(\mathbb{R}^n)} \max\left\{1, |x|^{n(\frac{1}{p_-} - 1)}\right\}$$

and

$$\int_{\mathbb{R}^n} |g(x)| \min\left\{|x|^{-\frac{n}{p_-}}, |x|^{-n}\right\} dx \leq C\|f\|_{H\dot{\mathcal{K}}_{\omega,0}^{p,q}(\mathbb{R}^n)}.$$

To show this theorem, we first investigate the properties of the Fourier transform of the distribution belonging to the Hardy space $H_X(\mathbb{R}^n)$ associated with the general ball quasi-Banach function space X. Indeed, we have the following technical lemma, which is essential obtained by Huang et al. [122, Theorems 2.1, 2.2, and 2.3].

Lemma 4.9.2 *Let X be a ball quasi-Banach function space and $p_- \in (0, 1]$ such that, for any given $\theta \in (0, p_-)$ and $u \in (1, \infty)$, there exists a positive constant C such that, for any $\{f_j\}_{j \in \mathbb{N}} \subset L^1_{\mathrm{loc}}(\mathbb{R}^n)$,*

$$\left\| \left\{ \sum_{j \in \mathbb{N}} [\mathcal{M}(f_j)]^u \right\}^{\frac{1}{u}} \right\|_{X^{1/\theta}} \leq C \left\| \left(\sum_{j \in \mathbb{N}} |f_j|^u \right)^{\frac{1}{u}} \right\|_{X^{1/\theta}}. \tag{4.119}$$

Assume that there exists a $p_0 \in [p_-, 1]$ such that X is p_0-concave and there exists a positive constant C such that, for any $B \in \mathbb{B}$,

$$\|\mathbf{1}_B\|_X \geq C \min \left\{ |B|^{\frac{1}{p_0}}, |B|^{\frac{1}{p_-}} \right\}. \tag{4.120}$$

Then, for any $f \in H_X(\mathbb{R}^n)$, there exists a continuous function g on $\mathbb{R}^n$ such that $\widehat{f} = g$ in $\mathcal{S}'(\mathbb{R}^n)$ and

$$\lim_{|x| \to 0^+} \frac{|g(x)|}{|x|^{n(\frac{1}{p_-}-1)}} = 0.$$

Moreover, there exists a positive constant C, independent of both f and g, such that, for any $x \in \mathbb{R}^n$,

$$|g(x)| \leq C \|f\|_{H_X(\mathbb{R}^n)} \max \left\{ |x|^{n(\frac{1}{p_0}-1)}, |x|^{n(\frac{1}{p_-}-1)} \right\}$$

and

$$\left[\int_{\mathbb{R}^n} |g(x)|^{p_0} \min \left\{ |x|^{n(p_0-1-\frac{p_0}{p_-})}, |x|^{n(p_0-2)} \right\} dx \right]^{1/p_0} \leq C \|f\|_{H_X(\mathbb{R}^n)}.$$

Proof Let all the symbols be as in the present lemma, $d \geq \lfloor n(1/p_- - 1) \rfloor$ a fixed integer, and $f \in \mathcal{S}'(\mathbb{R}^n)$. Then, by (4.119) and the known atomic decomposition of $H_X(\mathbb{R}^n)$ (see [207, Theorem 3.7] or Lemma 4.3.8), we find that there exists $\{\lambda_j\}_{j \in \mathbb{N}} \subset [0, \infty)$ and a sequence $\{a_j\}_{j \in \mathbb{N}}$ of (X, ∞, d)-atoms supported, respectively, in the balls $\{B_j\}_{j \in \mathbb{N}} \subset \mathbb{B}$ such that

$$f = \sum_{j \in \mathbb{N}} \lambda_j a_j$$

in $\mathcal{S}'(\mathbb{R}^n)$ and

$$\left\|\left[\sum_{j\in\mathbb{N}}\left(\frac{\lambda_j}{\|\mathbf{1}_{B_j}\|_X}\right)^s \mathbf{1}_{B_j}\right]^{\frac{1}{s}}\right\|_X \lesssim \|f\|_{H_X(\mathbb{R}^n)}.$$

Using this and repeating the proof of [122, Theorems 2.1, 2.2, and 2.3], we conclude that there exists a continuous function g on $\mathbb{R}^n$ such that $\widehat{f} = g$ in $\mathcal{S}'(\mathbb{R}^n)$,

$$\lim_{|x|\to 0^+}\frac{|g(x)|}{|x|^{n(\frac{1}{p_-}-1)}} = 0,$$

and, for any $x \in \mathbb{R}^n$,

$$|g(x)| \lesssim \|f\|_{H_X(\mathbb{R}^n)}\max\left\{|x|^{n(\frac{1}{p_0}-1)}, |x|^{n(\frac{1}{p_-}-1)}\right\}$$

and

$$\left[\int_{\mathbb{R}^n}|g(x)|^{p_0}\min\left\{|x|^{n(p_0-1-\frac{p_0}{p_-})}, |x|^{n(p_0-2)}\right\}dx\right]^{1/p_0} \lesssim \|f\|_{H_X(\mathbb{R}^n)}$$

with the implicit positive constants independent of both f and g. This then finishes the proof of Lemma 4.9.2. $\square$

Remark 4.9.3 We point out that Lemma 4.9.2 has a wide range of applications. Here we give several function spaces to which Lemma 4.9.2 can be applied.

(i) Let $p \in (0, 1]$ and $p_- = p_0 = p$. Then, in this case, using Remark 1.2.31(i) and the fact that, for any ball $B \in \mathbb{B}$, $\|\mathbf{1}_B\|_{L^p(\mathbb{R}^n)} = |B|^{\frac{1}{p}}$, we can easily find that the Lebesgue space $L^p(\mathbb{R}^n)$ satisfies all the assumptions of Lemma 4.9.2. Therefore, Lemma 4.9.2 with $X := L^p(\mathbb{R}^n)$ holds true. This result was originally established in [221].

(ii) Let $\vec{p} := (p_1, \ldots, p_n) \in (0, 1]^n$,

$$p_- := \min\{p_1, \ldots, p_n\},$$

and

$$p_0 := \max\{p_1, \ldots, p_n\}.$$

Then, in this case, as was mentioned in [122, Subsection 4.1], the mixed-norm Lebesgue space $L^{\vec{p}}(\mathbb{R}^n)$ satisfies all the assumptions of Lemma 4.9.2. Thus, Lemma 4.9.2 with X replaced by $L^{\vec{p}}(\mathbb{R}^n)$ holds true. This result was also obtained in [122, Theorem 4.1].

(iii) Let $p(\cdot) \in C^{\log}(\mathbb{R}^n)$ satisfy $0 < p_- \leq p_+ \leq 1$, where p_- and p_+ are defined, respectively, in (1.59) and (1.60). Let $p_0 = p_+$. Then, in this case, as was pointed out in [122, Subsection 4.2], the variable Lebesgue space $L^{p(\cdot)}(\mathbb{R}^n)$ satisfies all the assumptions of Lemma 4.9.2. This then implies that Lemma 4.9.2 with $X := L^{p(\cdot)}(\mathbb{R}^n)$ holds true, which coincides with [122, Theorem 4.2].

In order to prove Theorem 4.9.1, we also require the following estimate about the quasi-norm of the characteristic functions of balls on the local generalized Herz space $\dot{\mathcal{K}}^{p,q}_{\omega,\mathbf{0}}(\mathbb{R}^n)$, which plays an important role in the proof of Theorem 4.9.1 and is also of independent interest.

Lemma 4.9.4 *Let p, q, ω, and p_- be as in Theorem 4.9.1. Then there exists a positive constant C such that, for any $B \in \mathbb{B}$,*

$$\left\| \mathbf{1}_B \right\|_{\dot{\mathcal{K}}^{p,q}_{\omega,\mathbf{0}}(\mathbb{R}^n)} \geq C \min \left\{ |B|, |B|^{\frac{1}{p_-}} \right\}. \tag{4.121}$$

Proof Let all the symbols be as in the present lemma. We first show that, for any $B(x_0, 2^{k_0}) \in \mathbb{B}$ with $x_0 \in \mathbb{R}^n$ and $k_0 \in \mathbb{Z}$,

$$2^{\frac{nk_0}{p}} \omega(2^{k_0}) \lesssim \left\| \mathbf{1}_{B(x_0, 2^{k_0})} \right\|_{\dot{\mathcal{K}}^{p,q}_{\omega,\mathbf{0}}(\mathbb{R}^n)}. \tag{4.122}$$

To achieve this, we consider the following four cases on x_0.

Case (1) $x_0 = \mathbf{0}$. In this case, from the fact that $B(\mathbf{0}, 2^{k_0}) \setminus B(\mathbf{0}, 2^{k_0-1}) \subset B(\mathbf{0}, 2^{k_0})$ and the definition of $\| \cdot \|_{\dot{\mathcal{K}}^{p,q}_{\omega,\mathbf{0}}(\mathbb{R}^n)}$, we deduce that

$$2^{\frac{nk_0}{p}} \omega(2^{k_0}) \sim \omega(2^{k_0}) \left\| \mathbf{1}_{B(\mathbf{0},2^{k_0})\setminus B(\mathbf{0},2^{k_0-1})} \right\|_{L^p(\mathbb{R}^n)}$$

$$\lesssim \left\| \mathbf{1}_{B(\mathbf{0},2^{k_0})} \right\|_{\dot{\mathcal{K}}^{p,q}_{\omega,\mathbf{0}}(\mathbb{R}^n)}, \tag{4.123}$$

which completes the proof of (4.122) in this case.

Case (2) $|x_0| \in (0, 3 \cdot 2^{k_0-1})$. In this case, we first claim that

$$B\left(\frac{3 \cdot 2^{k_0-2}}{|x_0|} x_0, 2^{k_0-2} \right) \subset \left[B(x_0, 2^{k_0}) \cap \left(B(\mathbf{0}, 2^{k_0}) \setminus B(\mathbf{0}, 2^{k_0-1}) \right) \right]. \tag{4.124}$$

Indeed, for any $y \in B((3 \cdot 2^{k_0-2}/|x_0|)x_0, 2^{k_0-2})$, we have

$$|y| \leq \left| y - \frac{3 \cdot 2^{k_0-2}}{|x_0|} x_0 \right| + \left| \frac{3 \cdot 2^{k_0-2}}{|x_0|} x_0 \right| < 2^{k_0-2} + 3 \cdot 2^{k_0-2} = 2^{k_0}$$

and

$$|y| \geq \left| \frac{3 \cdot 2^{k_0-2}}{|x_0|} x_0 \right| - \left| y - \frac{3 \cdot 2^{k_0-2}}{|x_0|} x_0 \right| > 3 \cdot 2^{k_0-2} - 2^{k_0-2} = 2^{k_0-1}.$$

This implies that

$$B\left(\frac{3 \cdot 2^{k_0-2}}{|x_0|} x_0, 2^{k_0-2} \right) \subset \left[B(0, 2^{k_0}) \setminus B(0, 2^{k_0-1}) \right]. \tag{4.125}$$

On the other hand, for any $y \in B((3 \cdot 2^{k_0-2}/|x_0|)x_0, 2^{k_0-2})$, we have

$$|y - x_0| \leq \left| y - \frac{3 \cdot 2^{k_0-2}}{|x_0|} x_0 \right| + \left| \frac{3 \cdot 2^{k_0-2}}{|x_0|} x_0 - x_0 \right|$$

$$< 2^{k_0-2} + \left| 3 \cdot 2^{k_0-2} - |x_0| \right|.$$

This further implies that, for any $y \in B((3 \cdot 2^{k_0-2}/|x_0|)x_0, 2^{k_0-2})$ with $x_0 \in [3 \cdot 2^{k_0-2}, 3 \cdot 2^{k_0-1})$,

$$|y - x_0| < |x_0| - 2^{k_0-1} < 2^{k_0}$$

and, for any $y \in B((3 \cdot 2^{k_0-2}/|x_0|)x_0, 2^{k_0-2})$ with $|x_0| \in (0, 3 \cdot 2^{k_0-2})$,

$$|y - x_0| < 2^{k_0} - |x_0| < 2^{k_0}.$$

Therefore, $B((3 \cdot 2^{k_0-2}/|x_0|)x_0, 2^{k_0-2}) \subset B(x_0, 2^{k_0})$. By this and (4.125), we find that

$$B\left(\frac{3 \cdot 2^{k_0-2}}{|x_0|} x_0, 2^{k_0-2} \right) \subset \left[B(x_0, 2^{k_0}) \cap \left\{ B(0, 2^{k_0}) \setminus B(0, 2^{k_0-1}) \right\} \right].$$

This finishes the proof of the above claim. Thus, from the definition of the quasi-norm $\| \cdot \|_{\dot{\mathcal{K}}^{p,q}_{\omega,0}(\mathbb{R}^n)}$, it follows that

$$2^{\frac{nk_0}{p}} \omega(2^{k_0}) \sim \omega(2^{k_0}) \left| B\left(\frac{3 \cdot 2^{k_0-2}}{|x_0|} x_0, 2^{k_0-2} \right) \right|^{\frac{1}{p}}$$

$$\lesssim \omega(2^{k_0}) \left| B(x_0, 2^{k_0}) \cap \left[B(0, 2^{k_0}) \setminus B(0, 2^{k_0-1}) \right] \right|^{\frac{1}{p}}$$

$$\lesssim \left\| \mathbf{1}_{B(x_0, 2^{k_0})} \right\|_{\dot{\mathcal{K}}^{p,q}_{\omega,0}(\mathbb{R}^n)} \tag{4.126}$$

and hence (4.122) holds true in this case.

Case (3) $|x_0| \in [3 \cdot 2^{k_0-1}, 2^{k_0+1})$. In this case, for any $y \in B(x_0, 2^{k_0})$, we have

$$|y| \leq |y - x_0| + |x_0| < 2^{k_0} + 2^{k_0+1} < 2^{k_0+2}$$

and

$$|y| \geq |x_0| - |y - x_0| > 3 \cdot 2^{k_0-1} - 2^{k_0} = 2^{k_0-1}.$$

This implies that $B(x_0, 2^{k_0}) \subset B(\mathbf{0}, 2^{k_0+2}) \setminus B(\mathbf{0}, 2^{k_0-1})$. Applying this and Lemma 1.1.3, we conclude that

$$2^{\frac{nk_0q}{p}} \left[\omega(2^{k_0})\right]^q \sim \left[\omega(2^{k_0})\right]^q \left|B(x_0, 2^{k_0})\right|^{\frac{q}{p}}$$

$$\lesssim \left[\omega(2^{k_0+2})\right]^q \left|B(x_0, 2^{k_0}) \cap \left[B(\mathbf{0}, 2^{k_0+2}) \setminus B(\mathbf{0}, 2^{k_0+1})\right]\right|^{\frac{q}{p}}$$

$$+ \left[\omega(2^{k_0+1})\right]^q \left|B(x_0, 2^{k_0}) \cap \left[B(\mathbf{0}, 2^{k_0+1}) \setminus B(\mathbf{0}, 2^{k_0})\right]\right|^{\frac{q}{p}}$$

$$+ \left[\omega(2^{k_0})\right]^q \left|B(x_0, 2^{k_0}) \cap \left[B(\mathbf{0}, 2^{k_0}) \setminus B(\mathbf{0}, 2^{k_0-1})\right]\right|^{\frac{q}{p}}$$

$$\lesssim \left\|\mathbf{1}_{B(x_0, 2^{k_0})}\right\|_{\dot{\mathcal{K}}_{\omega,0}^{p,q}(\mathbb{R}^n)}^q, \tag{4.127}$$

which completes the estimation of (4.122) in this case.

Case (4) $|x_0| \in [2^k, 2^{k+1})$ with $k \in \mathbb{Z} \cap [k_0 + 1, \infty)$. In this case, from Lemma 1.5.2, it follows that, for any $0 < t < \tau < \infty$,

$$\frac{\omega(t)}{\omega(\tau)} \lesssim \left(\frac{t}{\tau}\right)^{\min\{m_0(\omega), m_\infty(\omega)\} - \varepsilon},$$

where $\varepsilon \in (0, \min\{m_0(\omega), m_\infty(\omega)\})$ is a fixed positive constant. Using this, Lemma 1.1.3, and the assumptions $k \in [k_0 + 1, \infty)$ and

$$\varepsilon \in (0, \min\{m_0(\omega), m_\infty(\omega)\}),$$

we further find that

$$\frac{\omega(2^{k_0})}{\omega(2^k)} \sim \frac{\omega(2^{k_0+1})}{\omega(2^k)} \lesssim \left(\frac{2^{k_0+1}}{2^k}\right)^{\min\{m_0(\omega), m_\infty(\omega)\} - \varepsilon} \lesssim 1. \tag{4.128}$$

On the other hand, applying the assumption $|x_0| \in [2^k, 2^{k+1})$ with $k \in \mathbb{Z} \cap [k_0 + 1, \infty)$, we conclude that, for any $y \in B(x_0, 2^{k_0})$,

$$|y| \leq |y - x_0| + |x_0| < 2^{k_0} + 2^{k+1} \leq 2^{k-1} + 2^{k+1} < 2^{k+2}$$

and

$$|y| \geq |x_0| - |y - x_0| > 2^k - 2^{k_0} \geq 2^k - 2^{k-1} = 2^{k-1}.$$

This implies that

$$B(x_0, 2^{k_0}) \subset B(\mathbf{0}, 2^{k+2}) \setminus B(\mathbf{0}, 2^{k-1}). \tag{4.129}$$

Thus, from (4.128) and an argument similar to that used in the estimation of (4.127), we deduce that

$$2^{-\frac{nk_0}{p}} \omega(2^{k_0}) \lesssim \omega(2^k) \left| B(x_0, 2^{k_0}) \right|^{\frac{n}{p}} \lesssim \left\| \mathbf{1}_{B(x_0, 2^{k_0})} \right\|_{\dot{\mathcal{K}}_{\omega,\mathbf{0}}^{p,q}(\mathbb{R}^n)},$$

which, together with (4.123), (4.126), and (4.127), implies that (4.122) holds true.

Next, we show that, for any $B(x_0, r) \in \mathbb{B}$ with $x_0 \in \mathbb{R}^n$ and $r \in (0, \infty)$,

$$r^{\frac{n}{p}} \omega(r) \lesssim \left\| \mathbf{1}_{B(x_0,r)} \right\|_{\dot{\mathcal{K}}_{\omega,\mathbf{0}}^{p,q}(\mathbb{R}^n)}. \tag{4.130}$$

Indeed, for any $r \in (0, \infty)$, there exists a $k \in \mathbb{Z}$ such that $r \in [2^k, 2^{k+1})$. Thus, applying Lemma 1.1.3 and (4.122), we find that

$$r^{\frac{n}{p}} \omega(r) \sim 2^{\frac{nk}{p}} \omega(2^k) \lesssim \left\| \mathbf{1}_{B(x_0, 2^k)} \right\|_{\dot{\mathcal{K}}_{\omega,\mathbf{0}}^{p,q}(\mathbb{R}^n)} \lesssim \left\| \mathbf{1}_{B(x_0,r)} \right\|_{\dot{\mathcal{K}}_{\omega,\mathbf{0}}^{p,q}(\mathbb{R}^n)}.$$

This finishes the proof of (4.130).

Finally, from the definition of p_- and Lemma 1.1.12, we infer that, for any $r \in (0, 1]$,

$$\omega(r) \gtrsim r^{M_0(\omega)+\varepsilon}$$

and, for any $r \in (1, \infty)$,

$$\omega(r) \gtrsim r^{m_\infty(\omega)-\varepsilon}, \tag{4.131}$$

where

$$\varepsilon \in \left(0, \min\left\{ \frac{n}{p_-} - M_0(\omega) - \frac{n}{p}, -n + m_\infty(\omega) + \frac{n}{p} \right\} \right)$$

is a fixed positive constant. Therefore, using the assumption $\varepsilon \in (0, \frac{n}{p_-} - M_0(\omega) - \frac{n}{p})$ and (4.130), we conclude that, for any $x_0 \in \mathbb{R}^n$ and $r \in (0, 1]$,

$$|B(x_0, r)|^{\frac{1}{p_-}} \sim r^{\frac{n}{p_-}} \lesssim r^{M_0(\omega) + \frac{n}{p} + \varepsilon} \lesssim r^{\frac{n}{p}} \omega(r) \lesssim \left\| \mathbf{1}_{B(x_0,r)} \right\|_{\dot{\mathcal{K}}^{p,q}_{\omega,\mathbf{0}}(\mathbb{R}^n)}. \tag{4.132}$$

Moreover, applying the fact that $\varepsilon \in (0, -n + m_\infty(\omega) + \frac{n}{p})$, (4.131), and (4.130), we find that, for any $x_0 \in \mathbb{R}^n$ and $r \in (1, \infty)$,

$$|B(x_0, r)| \sim r^n \lesssim r^{m_\infty(\omega) + \frac{n}{p} - \varepsilon} \lesssim r^{\frac{n}{p}} \omega(r) \lesssim \left\| \mathbf{1}_{B(x_0,r)} \right\|_{\dot{\mathcal{K}}^{p,q}_{\omega,\mathbf{0}}(\mathbb{R}^n)}, \tag{4.133}$$

which, together with (4.132), further implies that (4.121) holds true and hence completes the proof of Lemma 4.9.4. □

Via both Lemmas 4.9.2 and 4.9.4, we now show Theorem 4.9.1.

Proof of Theorem 4.9.1 Let all the symbols be as in the present theorem. Then, from the assumption $m_0(\omega) \in (0, \infty)$ and Theorem 1.2.42, it follows that the local generalized Herz space $\dot{\mathcal{K}}^{p,q}_{\omega,\mathbf{0}}(\mathbb{R}^n)$ is a BQBF space. This implies that, in order to complete the proof of the present theorem, we only need to show that the Herz space $\dot{\mathcal{K}}^{p,q}_{\omega,\mathbf{0}}(\mathbb{R}^n)$ under consideration satisfies all the assumptions of Lemma 4.9.2. Namely, $\dot{\mathcal{K}}^{p,q}_{\omega,\mathbf{0}}(\mathbb{R}^n)$ satisfies (4.119), the p_0-concavity for some $p_0 \in [p_-, 1]$, and (4.120).

We first prove that (4.119) holds true for $\dot{\mathcal{K}}^{p,q}_{\omega,\mathbf{0}}(\mathbb{R}^n)$. Indeed, by the fact that $\min\{m_0(\omega), m_\infty(\omega)\} \in (0, \infty)$ and Remark 1.1.5(iii), we find that

$$\max\{M_0(\omega), M_\infty(\omega)\} \in (0, \infty)$$

and hence

$$\frac{n}{\max\{M_0(\omega), M_\infty(\omega)\} + n/p} \in (0, p).$$

Applying this and Lemma 4.3.10, we conclude that, for any given $\theta \in (0, p_-)$ and $u \in (1, \infty)$, and for any $\{f_j\}_{j \in \mathbb{N}} \subset L^1_{\mathrm{loc}}(\mathbb{R}^n)$,

$$\left\| \left\{ \sum_{j \in \mathbb{N}} \left[\mathcal{M}(f_j) \right]^u \right\}^{\frac{1}{u}} \right\|_{[\dot{\mathcal{K}}^{p,q}_{\omega,\mathbf{0}}(\mathbb{R}^n)]^{1/\theta}} \lesssim \left\| \left(\sum_{j \in \mathbb{N}} |f_j|^u \right)^{\frac{1}{u}} \right\|_{[\dot{\mathcal{K}}^{p,q}_{\omega,\mathbf{0}}(\mathbb{R}^n)]^{1/\theta}}. \tag{4.134}$$

This further implies that $\dot{\mathcal{K}}^{p,q}_{\omega,\mathbf{0}}(\mathbb{R}^n)$ satisfies (4.119).

In addition, from $p, q \in (0, 1]$, it follows that, for any $\{f_j\}_{j\in\mathbb{N}} \subset \dot{\mathcal{K}}^{p,q}_{\omega,0}(\mathbb{R}^n)$,

$$\sum_{j\in\mathbb{N}} \|f_j\|_{\dot{\mathcal{K}}^{p,q}_{\omega,0}(\mathbb{R}^n)} \leq \left\| \sum_{j\in\mathbb{N}} |f_j| \right\|_{\dot{\mathcal{K}}^{p,q}_{\omega,0}(\mathbb{R}^n)}, \tag{4.135}$$

which implies that $\dot{\mathcal{K}}^{p,q}_{\omega,0}(\mathbb{R}^n)$ is strictly 1-concave and hence $\dot{\mathcal{K}}^{p,q}_{\omega,0}(\mathbb{R}^n)$ satisfies the p_0-concavity with $p_0 = 1$.

Finally, applying Lemma 4.9.4, we conclude that, for any $B \in \mathbb{B}$,

$$\|\mathbf{1}_B\|_{\dot{\mathcal{K}}^{p,q}_{\omega,0}(\mathbb{R}^n)} \gtrsim \min\left\{ |B|, |B|^{\frac{1}{p_-}} \right\}.$$

Thus, (4.120) also holds true for $\dot{\mathcal{K}}^{p,q}_{\omega,0}(\mathbb{R}^n)$ with $p_0 = 1$. This, combined with both (4.134) and (4.135), further implies that all the assumptions of Lemma 4.9.2 hold true for $\dot{\mathcal{K}}^{p,q}_{\omega,0}(\mathbb{R}^n)$, and hence finishes the proof of Theorem 4.9.1. $\qquad\square$

Next, we investigate the Fourier transform properties of the generalized Herz–Hardy space $H\dot{\mathcal{K}}^{p,q}_\omega(\mathbb{R}^n)$. Indeed, we have the following conclusion.

Theorem 4.9.5 *Let* $p \in (0, 1)$, $q \in (0, 1]$, $\omega \in M(\mathbb{R}_+)$ *with* $m_0(\omega) \in (-\frac{n}{p}, \infty)$ *and*

$$n\left(1 - \frac{1}{p}\right) < m_\infty(\omega) \leq M_\infty(\omega) < 0,$$

and $p_- \in (0, \min\{p, \frac{n}{\max\{M_0(\omega), M_\infty(\omega)\}+n/p}\})$. *Then, for any* $f \in H\dot{\mathcal{K}}^{p,q}_\omega(\mathbb{R}^n)$, *there exists a continuous function g on* $\mathbb{R}^n$ *such that* $\widehat{f} = g$ *in* $\mathcal{S}'(\mathbb{R}^n)$ *and*

$$\lim_{|x|\to 0^+} \frac{|g(x)|}{|x|^{n(\frac{1}{p_-}-1)}} = 0.$$

Moreover, there exists a positive constant C, independent of both f and g, such that, for any $x \in \mathbb{R}^n$,

$$|g(x)| \leq C\|f\|_{H\dot{\mathcal{K}}^{p,q}_\omega(\mathbb{R}^n)} \max\left\{ 1, |x|^{n(\frac{1}{p_-}-1)} \right\}$$

and

$$\int_{\mathbb{R}^n} |g(x)| \min\left\{ |x|^{-\frac{n}{p_-}}, |x|^{-n} \right\} dx \leq C\|f\|_{H\dot{\mathcal{K}}^{p,q}_\omega(\mathbb{R}^n)}.$$

To show Theorem 4.9.5, we need the following technical estimate for the quasi-norm of the characteristic function of balls on the global generalized Herz space $\dot{\mathcal{K}}_{\omega}^{p,q}(\mathbb{R}^n)$.

Lemma 4.9.6 *Let p, q, ω, and p_- be as in Theorem 4.9.5. Then there exists a positive constant C such that, for any $B \in \mathbb{B}$,*

$$\left\| \mathbf{1}_B \right\|_{\dot{\mathcal{K}}_{\omega}^{p,q}(\mathbb{R}^n)} \geq C \min \left\{ |B|, |B|^{\frac{1}{p_-}} \right\}. \tag{4.136}$$

Proof Let all the symbols be as in the present lemma and $B := B(x_0, r) \in \mathbb{B}$ with $x_0 \in \mathbb{R}^n$ and $r \in (0, \infty)$. Then, there exists a $k \in \mathbb{Z}$ such that $r \in [2^k, 2^{k+1})$. This implies that $B(x_0, 2^k) \setminus B(x_0, 2^{k-1}) \subset B(x_0, r)$. Therefore, by Lemma 1.1.3 and the definition of the quasi-norm $\| \cdot \|_{\dot{\mathcal{K}}_{\omega}^{p,q}(\mathbb{R}^n)}$, we find that

$$r^{\frac{n}{p}} \omega(r) \sim 2^{\frac{nk}{p}} \omega(2^k) \sim \omega(2^k) \left\| \mathbf{1}_{B(x_0, 2^k) \setminus B(x_0, 2^{k-1})} \right\|_{L^p(\mathbb{R}^n)}$$

$$\lesssim \left\| \mathbf{1}_{B(x_0, r)} \right\|_{\dot{\mathcal{K}}_{\omega}^{p,q}(\mathbb{R}^n)}. \tag{4.137}$$

Moreover, from both the definition of p_- and Lemma 1.1.12, it follows that, for any $r \in (0, 1]$,

$$\omega(r) \gtrsim r^{M_0(\omega) + \varepsilon} \tag{4.138}$$

and, for any $r \in (1, \infty)$,

$$\omega(r) \gtrsim r^{m_\infty(\omega) - \varepsilon}, \tag{4.139}$$

where

$$\varepsilon \in \left(0, \min \left\{ \frac{n}{p_-} - M_0(\omega) - \frac{n}{p}, -n + m_\infty(\omega) + \frac{n}{p} \right\} \right)$$

is a fixed positive constant. We now show (4.136) by considering the following two cases on r.

Case (1) $r \in (0, 1]$. In this case, using the assumption $\varepsilon \in (0, \frac{n}{p_-} - M_0(\omega) - \frac{n}{p})$, (4.138), and (4.137), we conclude that

$$|B(x_0, r)|^{\frac{1}{p_-}} \sim r^{\frac{n}{p_-}} \lesssim r^{M_0(\omega) + \frac{n}{p} + \varepsilon} \lesssim r^{\frac{n}{p}} \omega(r) \lesssim \left\| \mathbf{1}_{B(x_0, r)} \right\|_{\dot{\mathcal{K}}_{\omega}^{p,q}(\mathbb{R}^n)}. \tag{4.140}$$

This finishes the estimation of (4.136) in this case.

Case (2) r $\in (1, \infty)$. In this case, applying the fact that $\varepsilon \in (0, -n+m_\infty(\omega)+\frac{n}{p})$, (4.139), and (4.137), we find that

$$|B(x_0, r)| \sim r^n \lesssim r^{m_\infty(\omega)+\frac{n}{p}-\varepsilon} \lesssim r^{\frac{n}{p}} \omega(r) \lesssim \left\|\mathbf{1}_{B(x_0,r)}\right\|_{\dot{\mathcal{K}}_\omega^{p,q}(\mathbb{R}^n)},$$

which completes the proof of (4.136) in this case. Combining this and (4.140), we conclude that Lemma 4.9.6 holds true.   □

Based on the above lemma and the Fourier transform properties of Hardy spaces associated with ball quasi-Banach function spaces established in Lemma 4.9.2, we next prove Theorem 4.9.5.

Proof of Theorem 4.9.5 Let all the symbols be as in the present theorem. Then, by the assumptions $m_0(\omega) \in (-\frac{n}{p}, \infty)$ and $M_\infty(\omega) \in (-\infty, 0)$, and Theorem 1.2.44, we find that the global generalized Herz space $\dot{\mathcal{K}}_\omega^{p,q}(\mathbb{R}^n)$ under consideration is a BQBF space. Thus, to finish the proof of the present theorem, it suffices to prove that the Herz space $\dot{\mathcal{K}}_\omega^{p,q}(\mathbb{R}^n)$ satisfies all the assumptions of Lemma 4.9.2.

First, we show that (4.119) holds true. Indeed, from Lemma 4.3.24 with $r = \theta$, it follows that, for any given $\theta \in (0, p_-)$ and $u \in (1, \infty)$, and for any $\{f_j\}_{j\in\mathbb{N}} \subset L^1_{\text{loc}}(\mathbb{R}^n)$,

$$\left\|\left\{\sum_{j\in\mathbb{N}} [\mathcal{M}(f_j)]^u\right\}^{\frac{1}{u}}\right\|_{[\dot{\mathcal{K}}_\omega^{p,q}(\mathbb{R}^n)]^{1/\theta}} \leq C \left\|\left(\sum_{j\in\mathbb{N}} |f_j|^u\right)^{\frac{1}{u}}\right\|_{[\dot{\mathcal{K}}_\omega^{p,q}(\mathbb{R}^n)]^{1/\theta}},$$

which implies that $\dot{\mathcal{K}}_\omega^{p,q}(\mathbb{R}^n)$ satisfies (4.119).

Next, we prove that $\dot{\mathcal{K}}_\omega^{p,q}(\mathbb{R}^n)$ is strictly 1-concave. Indeed, applying the assumptions $p, q \in (0, 1]$, we conclude that, for any $\{f_j\}_{j\in\mathbb{N}} \subset \dot{\mathcal{K}}_\omega^{p,q}(\mathbb{R}^n)$,

$$\sum_{j\in\mathbb{N}} \|f_j\|_{\dot{\mathcal{K}}_\omega^{p,q}(\mathbb{R}^n)} \leq \left\|\sum_{j\in\mathbb{N}} |f_j|\right\|_{\dot{\mathcal{K}}_\omega^{p,q}(\mathbb{R}^n)}.$$

This implies that $\dot{\mathcal{K}}_\omega^{p,q}(\mathbb{R}^n)$ is strictly p_0-concave with $p_0 := 1$.

Finally, using Lemma 4.9.6, we find that, for any $B \in \mathbb{B}$,

$$\|\mathbf{1}_B\|_{\dot{\mathcal{K}}_\omega^{p,q}(\mathbb{R}^n)} \gtrsim \min\left\{|B|, |B|^{\frac{1}{p_-}}\right\},$$

which implies that (4.120) holds true for $\dot{\mathcal{K}}_\omega^{p,q}(\mathbb{R}^n)$ with $p_0 := 1$. Thus, the Herz space $\dot{\mathcal{K}}_\omega^{p,q}(\mathbb{R}^n)$ under consideration satisfies (4.119), 1-concavity, and (4.120) with $p_0 := 1$. By this and Lemma 4.9.2, we then obtain Theorem 4.9.5.   □

Chapter 5
Localized Generalized Herz–Hardy Spaces

Recall that, in 1979, Goldberg [89] introduced the local Hardy space $h^p(\mathbb{R}^n)$, with $p \in (0, \infty)$, as a localization of the classical Hardy spaces $H^p(\mathbb{R}^n)$ and showed some properties which are different from the classical Hardy spaces. Indeed, Goldberg proved that the local Hardy space $h^p(\mathbb{R}^n)$ contains $\mathcal{S}(\mathbb{R}^n)$ as a dense subspace and established the boundedness of pseudo-differential operators of order zero on $h^p(\mathbb{R}^n)$. Moreover, due to [89], the localized Hardy space is well defined on manifolds. After that, lots of nice works have been done in the study of localized Hardy spaces and their applications; see, for instance, [27, 51, 181, 273]. Specially, some variants of the local Hardy spaces $h^p(\mathbb{R}^n)$ have also been studied; see, for instance, [223] for weighted localized Hardy spaces, [234] for localized Herz–Hardy spaces, [271] for (weighted) localized Orlicz–Hardy spaces, and [272] for localized Musielak–Orlicz–Hardy spaces. Very recently, Sawano et al. [207] introduced the local Hardy space $h_X(\mathbb{R}^n)$ associated with the ball quasi-Banach function space X and gave various maximal function characterizations of $h_X(\mathbb{R}^n)$. Wang et al. [233] then established atomic, molecular, and various Littlewood–Paley function characterizations of $h_X(\mathbb{R}^n)$ as well as showed the boundedness of pseudo-differential operators on these localized Hardy spaces.

The main target of this chapter is devoted to introducing the localized generalized Herz–Hardy spaces and then establishing their complete real-variable theory. To achieve this, we begin with showing their various maximal function characterizations via the known maximal function characterizations of the local Hardy space $h_X(\mathbb{R}^n)$ associated with the ball quasi-Banach function space X. Then, to overcome the obstacle caused by the deficiency of associate spaces of global generalized Herz spaces, we establish improved atomic and molecular characterizations of $h_X(\mathbb{R}^n)$ as well as the boundedness of pseudo-differential operators on $h_X(\mathbb{R}^n)$ without recourse to associate spaces (see Theorems 5.3.14, 5.4.11, and 5.6.9 below). Via these improved conclusions, we establish the atomic and molecular characterizations of localized generalized Herz–Hardy spaces and the boundedness of pseudo-differential operators on localized generalized Herz–Hardy spaces. In

© The Author(s), under exclusive license to Springer Nature Singapore Pte Ltd. 2022
Y. Li et al., *Real-Variable Theory of Hardy Spaces Associated with Generalized Herz Spaces of Rafeiro and Samko*, Lecture Notes in Mathematics 2320,
https://doi.org/10.1007/978-981-19-6788-7_5

addition, to clarify the relation between localized generalized Herz–Hardy spaces
and generalized Herz–Hardy spaces, we find the relation between $h_X(\mathbb{R}^n)$ and
Hardy spaces $H_X(\mathbb{R}^n)$ associated with the ball quasi-Banach function space X
(see Theorem 5.2.3 below). Applying this and some auxiliary lemmas about
generalized Herz spaces, we then obtain the relation between localized generalized
Herz–Hardy spaces and generalized Herz–Hardy spaces. As applications, we also
establish various Littlewood–Paley function characterizations of $h_X(\mathbb{R}^n)$ and hence,
together with the construction of the quasi-norm $\|\cdot\|_{\dot{\mathcal{K}}_{\omega}^{p,q}(\mathbb{R}^n)}$, we further obtain the
Littlewood–Paley function characterizations of localized generalized Herz–Hardy
spaces. In addition, we introduce localized generalized Hardy–Morrey spaces.
Using the equivalence between generalized Herz spaces and generalized Morrey
spaces, we also conclude the corresponding real-variable characterizations and
applications of localized generalized Hardy–Morrey spaces in this chapter.

For any given $N \in \mathbb{N}$ and for any $f \in \mathcal{S}'(\mathbb{R}^n)$, the *local grand maximal function*
$m_N(f)$ is defined by setting, for any $x \in \mathbb{R}^n$,

$$m_N(f)(x) := \sup\left\{|f * \phi_t(y)| : t \in (0,1), |x - y| < t, \phi \in \mathcal{F}_N(\mathbb{R}^n)\right\},$$
(5.1)

where $\mathcal{F}_N(\mathbb{R}^n)$ is as in (4.2). Then we introduce localized generalized Herz–Hardy
spaces as follows.

Definition 5.0.1 Let $p, q \in (0, \infty)$, $\omega \in M(\mathbb{R}_+)$, and $N \in \mathbb{N}$. Then

(i) the *local generalized Herz–Hardy space* $h\dot{\mathcal{K}}_{\omega,\mathbf{0}}^{p,q}(\mathbb{R}^n)$, associated with the local
     generalized Herz space $\dot{\mathcal{K}}_{\omega,\mathbf{0}}^{p,q}(\mathbb{R}^n)$, is defined to be the set of all the $f \in \mathcal{S}'(\mathbb{R}^n)$
     such that

$$\|f\|_{h\dot{\mathcal{K}}_{\omega,\mathbf{0}}^{p,q}(\mathbb{R}^n)} := \|m_N(f)\|_{\dot{\mathcal{K}}_{\omega,\mathbf{0}}^{p,q}(\mathbb{R}^n)} < \infty;$$

(ii) the *local generalized Herz–Hardy space* $h\dot{\mathcal{K}}_{\omega}^{p,q}(\mathbb{R}^n)$, associated with the global
     generalized Herz space $\dot{\mathcal{K}}_{\omega}^{p,q}(\mathbb{R}^n)$, is defined to be the set of all the $f \in \mathcal{S}'(\mathbb{R}^n)$
     such that

$$\|f\|_{h\dot{\mathcal{K}}_{\omega}^{p,q}(\mathbb{R}^n)} := \|m_N(f)\|_{\dot{\mathcal{K}}_{\omega}^{p,q}(\mathbb{R}^n)} < \infty.$$

Remark 5.0.2 In Definition 5.0.1, for any given $\alpha \in \mathbb{R}$ and for any $t \in (0, \infty)$, let
$\omega(t) := t^{\alpha}$. Then, in this case, the local generalized Herz–Hardy space $h\dot{\mathcal{K}}_{\omega,\mathbf{0}}^{p,q}(\mathbb{R}^n)$
goes back to the classical *homogeneous local Herz-type Hardy space* $h\dot{K}_p^{\alpha,q}(\mathbb{R}^n)$
which was originally introduced in [72, Definition 1.2] (see also [175, Section 2.6]).
However, to the best of our knowledge, even in this case, the local generalized Herz–
Hardy space $h\dot{\mathcal{K}}_{\omega}^{p,q}(\mathbb{R}^n)$ is also new.

Recall that local and global generalized Morrey spaces are given in Remark 1.2.2(vi). We now introduce the following concepts of localized generalized Hardy–Morrey spaces.

Definition 5.0.3 Let $p, q \in (0, \infty)$, $\omega \in M(\mathbb{R}_+)$, and $N \in \mathbb{N}$.

(i) The *local generalized Hardy–Morrey space* $hM_{\omega,0}^{p,q}(\mathbb{R}^n)$, associated with the local generalized Morrey space $M_{\omega,0}^{p,q}(\mathbb{R}^n)$, is defined to be the set of all the $f \in \mathcal{S}'(\mathbb{R}^n)$ such that

$$\|f\|_{hM_{\omega,0}^{p,q}(\mathbb{R}^n)} := \|m_N(f)\|_{M_{\omega,0}^{p,q}(\mathbb{R}^n)} < \infty;$$

(ii) The *local generalized Hardy–Morrey space* $hM_\omega^{p,q}(\mathbb{R}^n)$, associated with the global generalized Morrey space $M_\omega^{p,q}(\mathbb{R}^n)$, is defined to be the set of all the $f \in \mathcal{S}'(\mathbb{R}^n)$ such that

$$\|f\|_{hM_\omega^{p,q}(\mathbb{R}^n)} := \|m_N(f)\|_{M_\omega^{p,q}(\mathbb{R}^n)} < \infty.$$

Remark 5.0.4

(i) We should point out that, to the best of our knowledge, in Definition 5.0.3, even when $\omega(t) := t^\alpha$ for any $t \in (0, \infty)$ and for any given $\alpha \in \mathbb{R}$, the local generalized Hardy–Morrey spaces $hM_{\omega,0}^{p,q}(\mathbb{R}^n)$ and $hM_\omega^{p,q}(\mathbb{R}^n)$ are also new.

(ii) In Definition 5.0.3, let $p, q \in [1, \infty)$ and ω satisfy

$$\max\{M_0(\omega), M_\infty(\omega)\} \in (-\infty, 0).$$

Then, in this case, by Remark 1.2.2(vi), we find that the local generalized Hardy–Morrey spaces $hM_{\omega,0}^{p,q}(\mathbb{R}^n)$ and $hM_\omega^{p,q}(\mathbb{R}^n)$ coincide, respectively, with the local generalized Herz–Hardy spaces $h\dot{\mathcal{K}}_{\omega,0}^{p,q}(\mathbb{R}^n)$ and $h\dot{\mathcal{K}}_\omega^{p,q}(\mathbb{R}^n)$ in the sense of equivalent norms.

5.1 Maximal Function Characterizations

In this section, we establish the maximal function characterizations of localized generalized Herz–Hardy spaces. To begin with, we recall the following definitions of various localized maximal functions (see, for instance, [207, Definition 5.1]).

Definition 5.1.1 Let $N \in \mathbb{N}$, $a, b \in (0, \infty)$, $\phi \in \mathcal{S}(\mathbb{R}^n)$, and $f \in \mathcal{S}'(\mathbb{R}^n)$.

(i) The *local radial maximal function* $m(f, \phi)$ is defined by setting, for any $x \in \mathbb{R}^n$,

$$m(f, \phi)(x) := \sup_{t \in (0,1)} |f * \phi_t(x)|.$$

(ii) The *local non-tangential maximal function* $m_a^*(f, \phi)$, with aperture $a \in (0, \infty)$, is defined by setting, for any $x \in \mathbb{R}^n$,

$$m_a^*(f, \phi)(x) := \sup_{t \in (0,1)} \sup_{\{y \in \mathbb{R}^n: \, |y-x|<at\}} |f * \phi_t(y)|.$$

(iii) The *local maximal function* $m_b^{**}(f, \phi)$ *of Peetre type* is defined by setting, for any $x \in \mathbb{R}^n$,

$$m_b^{**}(f, \phi)(x) := \sup_{(y,t) \in \mathbb{R}^n \times (0,1)} \frac{|f * \phi_t(x - y)|}{(1 + t^{-1}|y|)^b}.$$

(iv) The *local grand maximal function* $m_{b,N}^{**}(f)$ *of Peetre type* is defined by setting, for any $x \in \mathbb{R}^n$,

$$m_{b,N}^{**}(f)(x) := \sup_{\phi \in \mathcal{F}_N(\mathbb{R}^n)} \sup_{(y,t) \in \mathbb{R}^n \times (0,1)} \frac{|f * \phi_t(x - y)|}{(1 + t^{-1}|y|)^b},$$

where, for any $N \in \mathbb{N}$, $\mathcal{F}_N(\mathbb{R}^n)$ is defined as in (4.2).

Now, we establish the following maximal function characterizations of the local generalized Herz–Hardy space $h\dot{\mathcal{K}}_{\omega,\mathbf{0}}^{p,q}(\mathbb{R}^n)$.

Theorem 5.1.2 *Let* $p, q, a, b \in (0, \infty)$, $\omega \in M(\mathbb{R}_+)$, $N \in \mathbb{N}$, *and* $\phi \in \mathcal{S}(\mathbb{R}^n)$ *satisfy*

$$\int_{\mathbb{R}^n} \phi(x) \, dx \neq 0.$$

(i) *Let* $N \in \mathbb{N} \cap [\lfloor b + 1 \rfloor, \infty)$ *and* ω *satisfy* $m_0(\omega) \in (-\frac{n}{p}, \infty)$. *Then, for any* $f \in \mathcal{S}'(\mathbb{R}^n)$,

$$\|m(f, \phi)\|_{\dot{\mathcal{K}}_{\omega,\mathbf{0}}^{p,q}(\mathbb{R}^n)} \lesssim \|m_a^*(f, \phi)\|_{\dot{\mathcal{K}}_{\omega,\mathbf{0}}^{p,q}(\mathbb{R}^n)} \lesssim \|m_b^{**}(f, \phi)\|_{\dot{\mathcal{K}}_{\omega,\mathbf{0}}^{p,q}(\mathbb{R}^n)},$$

$$\|m(f, \phi)\|_{\dot{\mathcal{K}}_{\omega,\mathbf{0}}^{p,q}(\mathbb{R}^n)} \lesssim \|m_N(f)\|_{\dot{\mathcal{K}}_{\omega,\mathbf{0}}^{p,q}(\mathbb{R}^n)} \lesssim \|m_{\lfloor b+1 \rfloor}(f)\|_{\dot{\mathcal{K}}_{\omega,\mathbf{0}}^{p,q}(\mathbb{R}^n)}$$

$$\lesssim \|m_b^{**}(f, \phi)\|_{\dot{\mathcal{K}}_{\omega,\mathbf{0}}^{p,q}(\mathbb{R}^n)},$$

and

$$\|m_b^{**}(f, \phi)\|_{\dot{\mathcal{K}}_{\omega,\mathbf{0}}^{p,q}(\mathbb{R}^n)} \sim \|m_{b,N}^{**}(f)\|_{\dot{\mathcal{K}}_{\omega,\mathbf{0}}^{p,q}(\mathbb{R}^n)},$$

where the implicit positive constants are independent of f.

(ii) *Let ω satisfy $m_0(\omega) \in (-\frac{n}{p}, \infty)$ and $m_\infty(\omega) \in (-\frac{n}{p}, \infty)$, and*

$$b \in \left(2\max\left\{ \frac{n}{p}, \frac{n}{q}, \max\{M_0(\omega), M_\infty(\omega)\} + \frac{n}{p} \right\}, \infty \right).$$

Then, for any $f \in \mathcal{S}'(\mathbb{R}^n)$,

$$\|m_b^{**}(f, \phi)\|_{\dot{\mathcal{K}}_{\omega,0}^{p,q}(\mathbb{R}^n)} \lesssim \|m(f, \phi)\|_{\dot{\mathcal{K}}_{\omega,0}^{p,q}(\mathbb{R}^n)},$$

where the implicit positive constant is independent of f. In particular, when $N \in \mathbb{N} \cap [\lfloor b+1 \rfloor, \infty)$, if one of the quantities

$$\|m(f, \phi)\|_{\dot{\mathcal{K}}_{\omega,0}^{p,q}(\mathbb{R}^n)}, \quad \|m_a^*(f, \phi)\|_{\dot{\mathcal{K}}_{\omega,0}^{p,q}(\mathbb{R}^n)}, \quad \|m_N(f)\|_{\dot{\mathcal{K}}_{\omega,0}^{p,q}(\mathbb{R}^n)},$$

$$\|m_b^{**}(f, \phi)\|_{\dot{\mathcal{K}}_{\omega,0}^{p,q}(\mathbb{R}^n)}, \quad \text{and} \quad \|m_{b,N}^{**}(f)\|_{\dot{\mathcal{K}}_{\omega,0}^{p,q}(\mathbb{R}^n)}$$

is finite, then the others are also finite and mutually equivalent with the positive equivalence constants independent of f.

Remark 5.1.3

(i) Let $p, q \in (0, \infty)$ and $\omega \in M(\mathbb{R}_+)$. Then the quasi-norm of $h\dot{\mathcal{K}}_{\omega,0}^{p,q}(\mathbb{R}^n)$ in Definition 5.0.1(i) depends on N. However, from Theorem 5.1.2, we infer that the local Hardy space $h\dot{\mathcal{K}}_{\omega,0}^{p,q}(\mathbb{R}^n)$ is independent of the choice of N whenever $\omega \in M(\mathbb{R}_+)$ satisfies

$$m_0(\omega) \in \left(-\frac{n}{p}, \infty \right) \text{ and } m_\infty(\omega) \in \left(-\frac{n}{p}, \infty \right),$$

and N satisfies

$$N \in \mathbb{N} \cap \left(1 + 2\max\left\{ \frac{n}{p}, \frac{n}{q}, \max\{M_0(\omega), M_\infty(\omega)\} + \frac{n}{p} \right\}, \infty \right).$$

(ii) Observe that, if $p = q \in (0, \infty)$ and $\omega(t) := 1$ for any $t \in (0, \infty)$, then, in this case,

$$m_0(\omega) = M_0(\omega) = m_\infty(\omega) = M_\infty(\omega) = 0,$$

$h\dot{\mathcal{K}}_{\omega,0}^{p,q}(\mathbb{R}^n)$ coincides with the classical local Hardy space $h^p(\mathbb{R}^n)$ in the sense of equivalent quasi-norms, and all the conclusions in Theorem 5.1.2 coincide with the corresponding classical results on the local Hardy space $h^p(\mathbb{R}^n)$ established in [89].

To prove the above maximal function characterizations, we require the following auxiliary inequality which was obtained in [233, Remark 4.4].

Lemma 5.1.4 *Let X be a ball quasi-Banach function space and $r \in (0, \infty)$ such that $\mathcal{M}^{(r)}$ is bounded on X. Then there exists a positive constant C such that, for any $f \in X$ and $z \in \mathbb{R}^n$,*

$$\left\| \left\{ \int_{z+[0,1]^n} |f(\cdot - y)|^r \, dy \right\}^{\frac{1}{r}} \right\|_X \leq C(1 + |z|)^{\frac{n}{r}} \|f\|_X.$$

We also need the maximal function characterizations of localized Hardy spaces associated with ball quasi-Banach function spaces as follows, which is just [207, Theorem 5.3] and plays a key role in the proof of Theorem 5.1.2.

Lemma 5.1.5 *Let X be a ball quasi-Banach function space, $a, b \in (0, \infty)$, $N \in \mathbb{N}$, and $\phi \in \mathcal{S}(\mathbb{R}^n)$ satisfy*

$$\int_{\mathbb{R}^n} \phi(x) \, dx \neq 0.$$

(i) *Let $N \in \mathbb{N} \cap [\lfloor b + 1 \rfloor, \infty)$. Then, for any $f \in \mathcal{S}'(\mathbb{R}^n)$,*

$$\|m(f, \phi)\|_X \lesssim \|m_a^*(f, \phi)\|_X \lesssim \|m_b^{**}(f, \phi)\|_X,$$

$$\|m(f, \phi)\|_X \lesssim \|m_N(f)\|_X \lesssim \|m_{\lfloor b+1 \rfloor}(f)\|_X \lesssim \|m_b^{**}(f, \phi)\|_X,$$

and

$$\|m_b^{**}(f, \phi)\|_X \sim \|m_{b,N}^{**}(f)\|_X,$$

where the implicit positive constants are independent of f.

(ii) *Let $r, A \in (0, \infty)$ satisfy $(b - A)r > n$. Assume that X is strictly r-convex and there exists a positive constant C such that, for any $g \in X$ and $z \in \mathbb{R}^n$,*

$$\left\| \left\{ \int_{z+[0,1]^n} |g(\cdot - y)|^r \, dy \right\}^{\frac{1}{r}} \right\|_X \leq C(1 + |z|)^A \|g\|_X. \tag{5.2}$$

Then, for any $f \in \mathcal{S}'(\mathbb{R}^n)$,

$$\|m_b^{**}(f, \phi)\|_X \lesssim \|m(f, \phi)\|_X,$$

where the implicit positive constant is independent of f. In particular, when $N \in \mathbb{N} \cap [\lfloor b+1 \rfloor, \infty)$, if one of the quantities

$$\|m(f, \phi)\|_X, \quad \|m_a^*(f, \phi)\|_X, \quad \|m_N(f)\|_X,$$

$$\|m_b^{**}(f, \phi)\|_X, \quad \text{and} \quad \|m_{b,N}^{**}(f)\|_X$$

is finite, then the others are also finite and mutually equivalent with the positive equivalence constants independent of f.

Remark 5.1.6 Let X be a ball quasi-Banach function space satisfying Assumption 1.2.29 for some $0 < \theta < s \leq 1$ and Assumption 1.2.33 for the same s. Then we claim that X satisfies all the assumptions of Lemma 5.1.5(ii). Indeed, by Assumption 1.2.33 and [29, Lemma 2.6] with X and p therein replaced, respectively, by $X^{1/s}$ and $\frac{s}{\theta}$, we find that $X^{1/\theta}$ is a ball Banach function space and hence X is strictly θ-convex. On the other hand, from both Assumption 1.2.29 and Lemma 5.1.4 with $r := \theta$, it follows that (5.2) holds true with $r := \theta$ and $A \in (\frac{n}{\theta}, \infty)$. This then implies that all the assumptions of Lemma 5.1.5(ii) hold true for the above X and hence finishes the proof of the above claim.

Remark 5.1.7 We should point out that Lemma 5.1.5 has a wide range of applications. Here we present several function spaces to which Lemma 5.1.5 can be applied.

(i) Let $p \in (0, \infty)$ and $b \in (\frac{2n}{p}, \infty)$. Then, in this case, combining Remarks 1.2.31(i), 1.2.34(i), and 5.1.6, we can easily find that the Lebesgue space $L^p(\mathbb{R}^n)$ satisfies all the assumptions of Lemma 5.1.5. This then implies that Lemma 5.1.5 with $X := L^p(\mathbb{R}^n)$ holds true, which coincides with the known maximal function characterizations of localized Hardy spaces established in [89, Theorem 1].

(ii) Let $p \in (0, \infty)$, $\upsilon \in A_\infty(\mathbb{R}^n)$, and $b \in (\frac{2nq_\upsilon}{p}, \infty)$, where q_υ is the same as in (1.58). Then, in this case, from Remarks 1.2.31(ii), 1.2.34(ii), and 5.1.6, we can easily infer that the weighted Lebesgue space $L_\upsilon^p(\mathbb{R}^n)$ satisfies all the assumptions of Lemma 5.1.5. Therefore, Lemma 5.1.5 with X therein replaced by $L_\upsilon^p(\mathbb{R}^n)$ holds true. This result coincides with [18, Theorem 4.2].

(iii) Let $\vec{p} := (p_1, \ldots, p_n) \in (0, \infty)^n$ and $b \in (\frac{2n}{\min\{p_1, \ldots, p_n\}}, \infty)$. Then, in this case, using Remarks 1.2.31(iii), 1.2.34(iii), and 5.1.6, we can easily conclude that the mixed-norm Lebesgue space $L^{\vec{p}}(\mathbb{R}^n)$ satisfies all the assumptions of Lemma 5.1.5. Thus, Lemma 5.1.5 with $X := L^{\vec{p}}(\mathbb{R}^n)$ holds true. To the best of our knowledge, this result is totally new.

(iv) Let $0 < q \leq p < \infty$ and $b \in (\frac{2n}{q}, \infty)$. Then, in this case, by Remarks 1.2.31(iv), 1.2.34(iv), and 5.1.6, we easily find that the Morrey space $M_q^p(\mathbb{R}^n)$ satisfies all the assumptions of Lemma 5.1.5. This further implies that Lemma 5.1.5 with X replaced by $M_q^p(\mathbb{R}^n)$ holds true. To the best of our knowledge, this result is totally new.

(v) Let $p(\cdot) \in C^{\log}(\mathbb{R}^n)$ and $b \in (\frac{2n}{p_-}, \infty)$, where p_- is the same as in (1.59). Then, in this case, from Remarks 1.2.31(v), 1.2.34(v), and 5.1.6, we can easily deduce that the variable Lebesgue space $L^{p(\cdot)}(\mathbb{R}^n)$ satisfies all the assumptions of Lemma 5.1.5. Thus, Lemma 5.1.5 with $X := L^{p(\cdot)}(\mathbb{R}^n)$ holds true. This result was also pointed out in [186, p. 3747].

Via the above two lemmas, we now show Theorem 5.1.2.

Proof of Theorem 5.1.2 Let all the symbols be as in the present theorem. Then, from the assumption $m_0(\omega) \in (-\frac{n}{p}, \infty)$ and Theorem 1.2.42, it follows that the local generalized Herz space $\dot{\mathcal{K}}^{p,q}_{\omega,0}(\mathbb{R}^n)$ under consideration is a BQBF space. This, together with Lemma 5.1.5(i), then finishes the proof of (i).

Next, we show (ii). Indeed, by Lemma 5.1.5(ii), we conclude that it suffices to prove that all the assumptions of Lemma 5.1.5(ii) hold true for the Herz space $\dot{\mathcal{K}}^{p,q}_{\omega,0}(\mathbb{R}^n)$ under consideration. Namely, there exist r, $A \in (0, \infty)$ with $(b-A)r > n$ satisfying that $\dot{\mathcal{K}}^{p,q}_{\omega,0}(\mathbb{R}^n)$ is strictly r-convex and, for any $f \in \dot{\mathcal{K}}^{p,q}_{\omega,0}(\mathbb{R}^n)$ and $z \in \mathbb{R}^n$,

$$\left\| \left\{ \int_{z+[0,1]^n} |f(\cdot - y)|^r \, dy \right\}^{\frac{1}{r}} \right\|_{\dot{\mathcal{K}}^{p,q}_{\omega,0}(\mathbb{R}^n)} \lesssim (1 + |z|)^A \|f\|_{\dot{\mathcal{K}}^{p,q}_{\omega,0}(\mathbb{R}^n)}. \qquad (5.3)$$

To achieve this, let

$$r \in \left(\frac{2n}{b}, \min \left\{ p, q, \frac{n}{\max\{M_0(\omega), M_\infty(\omega)\} + n/p} \right\} \right).$$

Then, from Theorem 1.3.3 with $s := r$, we deduce that the local generalized Herz space $\dot{\mathcal{K}}^{p,q}_{\omega,0}(\mathbb{R}^n)$ is strictly r-convex. On the other hand, let $A \in (\frac{n}{r}, b - \frac{n}{r})$. We then prove that (5.3) holds true for this A. Indeed, by Lemma 1.8.5 and Remark 1.2.30(i), we find that, for any $f \in L^1_{\mathrm{loc}}(\mathbb{R}^n)$,

$$\left\| \mathcal{M}^{(r)}(f) \right\|_{\dot{\mathcal{K}}^{p,q}_{\omega,0}(\mathbb{R}^n)} \lesssim \|f\|_{\dot{\mathcal{K}}^{p,q}_{\omega,0}(\mathbb{R}^n)}.$$

Applying this, Lemma 5.1.4, and the assumption $A > \frac{n}{r}$, we further conclude that, for any $f \in \dot{\mathcal{K}}^{p,q}_{\omega,0}(\mathbb{R}^n)$ and $z \in \mathbb{R}^n$,

$$\left\| \left\{ \int_{z+[0,1]^n} |f(\cdot - y)|^r \, dy \right\}^{\frac{1}{r}} \right\|_{\dot{\mathcal{K}}^{p,q}_{\omega,0}(\mathbb{R}^n)}$$
$$\lesssim (1 + |z|)^{\frac{n}{r}} \|f\|_{\dot{\mathcal{K}}^{p,q}_{\omega,0}(\mathbb{R}^n)} \lesssim (1 + |z|)^A \|f\|_{\dot{\mathcal{K}}^{p,q}_{\omega,0}(\mathbb{R}^n)}.$$

This finishes the proof of (5.3). Thus, under the assumptions of the present theorem, $\dot{\mathcal{K}}_{\omega,0}^{p,q}(\mathbb{R}^n)$ satisfies all the assumptions of Lemma 5.1.5(ii), which completes the proof of (ii) and hence Theorem 5.1.2. □

Remark 5.1.8 We point out that, in Theorem 5.1.2, if $\omega(t) := t^\alpha$ for any $t \in (0, \infty)$ and for any given $\alpha \in [n(1 - \frac{1}{p}), \infty)$, then Theorem 5.1.2 goes back to [175, Theorem 2.6.1].

Via Theorem 5.1.2 and Remark 5.0.4(ii), we immediately obtain the maximal function characterizations of the local generalized Hardy–Morrey space $h\boldsymbol{M}_{\omega,0}^{p,q}(\mathbb{R}^n)$ as follows; we omit the details.

Corollary 5.1.9 *Let* $a, b \in (0, \infty)$, $p, q \in [1, \infty)$, $\omega \in M(\mathbb{R}_+)$, $N \in \mathbb{N}$, *and* $\phi \in \mathcal{S}(\mathbb{R}^n)$ *satisfy*

$$\int_{\mathbb{R}^n} \phi(x)\, dx \neq 0.$$

(i) *Let* $N \in \mathbb{N} \cap [\lfloor b + 1 \rfloor, \infty)$ *and* ω *satisfy* $M_\infty(\omega) \in (-\infty, 0)$ *and*

$$-\frac{n}{p} < m_0(\omega) \leq M_0(\omega) < 0.$$

Then, for any $f \in \mathcal{S}'(\mathbb{R}^n)$,

$$\|m(f, \phi)\|_{\boldsymbol{M}_{\omega,0}^{p,q}(\mathbb{R}^n)} \lesssim \|m_a^*(f, \phi)\|_{\boldsymbol{M}_{\omega,0}^{p,q}(\mathbb{R}^n)} \lesssim \|m_b^{**}(f, \phi)\|_{\boldsymbol{M}_{\omega,0}^{p,q}(\mathbb{R}^n)},$$

$$\|m(f, \phi)\|_{\boldsymbol{M}_{\omega,0}^{p,q}(\mathbb{R}^n)} \lesssim \|m_N(f)\|_{\boldsymbol{M}_{\omega,0}^{p,q}(\mathbb{R}^n)} \lesssim \|m_{\lfloor b+1 \rfloor}(f)\|_{\boldsymbol{M}_{\omega,0}^{p,q}(\mathbb{R}^n)}$$

$$\lesssim \|m_b^{**}(f, \phi)\|_{\boldsymbol{M}_{\omega,0}^{p,q}(\mathbb{R}^n)},$$

and

$$\|m_b^{**}(f, \phi)\|_{\boldsymbol{M}_{\omega,0}^{p,q}(\mathbb{R}^n)} \sim \|m_{b,N}^{**}(f)\|_{\boldsymbol{M}_{\omega,0}^{p,q}(\mathbb{R}^n)},$$

where the implicit positive constants are independent of f.
(ii) *Let* ω *satisfy*

$$-\frac{n}{p} < m_0(\omega) \leq M_0(\omega) < 0$$

and

$$-\frac{n}{p} < m_\infty(\omega) \leq M_\infty(\omega) < 0,$$

and $b \in (2n \max\{\frac{1}{p}, \frac{1}{q}\}, \infty)$. *Then, for any* $f \in \mathcal{S}'(\mathbb{R}^n)$,

$$\|m_b^{**}(f, \phi)\|_{M_{\omega,0}^{p,q}(\mathbb{R}^n)} \lesssim \|m(f, \phi)\|_{M_{\omega,0}^{p,q}(\mathbb{R}^n)},$$

where the implicit positive constant is independent of f. *In particular, when* $N \in \mathbb{N} \cap [\lfloor b + 1 \rfloor, \infty)$, *if one of the quantities*

$$\|m(f, \phi)\|_{M_{\omega,0}^{p,q}(\mathbb{R}^n)}, \quad \|m_a^*(f, \phi)\|_{M_{\omega,0}^{p,q}(\mathbb{R}^n)}, \quad \|m_N(f)\|_{M_{\omega,0}^{p,q}(\mathbb{R}^n)},$$

$$\|m_b^{**}(f, \phi)\|_{M_{\omega,0}^{p,q}(\mathbb{R}^n)}, \quad and \quad \|m_{b,N}^{**}(f)\|_{M_{\omega,0}^{p,q}(\mathbb{R}^n)}$$

is finite, then the others are also finite and mutually equivalent with the positive equivalence constants independent of f.

Next, we establish the following maximal function characterizations of the local generalized Herz–Hardy space $h\dot{\mathcal{K}}_\omega^{p,q}(\mathbb{R}^n)$ with the help of the known maximal function characterizations of the local Hardy space $h_X(\mathbb{R}^n)$ presented in Lemma 5.1.5.

Theorem 5.1.10 *Let* $p, q, a, b \in (0, \infty)$, $\omega \in M(\mathbb{R}_+)$, $N \in \mathbb{N}$, *and* $\phi \in \mathcal{S}(\mathbb{R}^n)$ *satisfy*

$$\int_{\mathbb{R}^n} \phi(x)\, dx \neq 0.$$

(i) *Let* $N \in \mathbb{N} \cap [\lfloor b + 1 \rfloor, \infty)$ *and* ω *satisfy* $m_0(\omega) \in (-\frac{n}{p}, \infty)$ *and* $M_\infty(\omega) \in (-\infty, 0)$. *Then, for any* $f \in \mathcal{S}'(\mathbb{R}^n)$,

$$\|m(f, \phi)\|_{\dot{\mathcal{K}}_\omega^{p,q}(\mathbb{R}^n)} \lesssim \|m_a^*(f, \phi)\|_{\dot{\mathcal{K}}_\omega^{p,q}(\mathbb{R}^n)} \lesssim \|m_b^{**}(f, \phi)\|_{\dot{\mathcal{K}}_\omega^{p,q}(\mathbb{R}^n)},$$

$$\|m(f, \phi)\|_{\dot{\mathcal{K}}_\omega^{p,q}(\mathbb{R}^n)} \lesssim \|m_N(f)\|_{\dot{\mathcal{K}}_\omega^{p,q}(\mathbb{R}^n)} \lesssim \|m_{\lfloor b+1 \rfloor}(f)\|_{\dot{\mathcal{K}}_\omega^{p,q}(\mathbb{R}^n)}$$

$$\lesssim \|m_b^{**}(f, \phi)\|_{\dot{\mathcal{K}}_\omega^{p,q}(\mathbb{R}^n)},$$

and

$$\|m_b^{**}(f, \phi)\|_{\dot{\mathcal{K}}_\omega^{p,q}(\mathbb{R}^n)} \sim \|m_{b,N}^{**}(f)\|_{\dot{\mathcal{K}}_\omega^{p,q}(\mathbb{R}^n)},$$

where the implicit positive constants are independent of f.

(ii) *Let* ω *satisfy* $m_0(\omega) \in (-\frac{n}{p}, \infty)$ *and*

$$-\frac{n}{p} < m_\infty(\omega) \leq M_\infty(\omega) < 0,$$

and

$$b \in \left(2 \max \left\{ \frac{n}{p}, \frac{n}{q}, \max\{M_0(\omega), M_\infty(\omega)\} + \frac{n}{p} \right\}, \infty \right).$$

Then, for any $f \in \mathcal{S}'(\mathbb{R}^n)$,

$$\left\| m_b^{**}(f, \phi) \right\|_{\dot{\mathcal{K}}_\omega^{p,q}(\mathbb{R}^n)} \lesssim \left\| m(f, \phi) \right\|_{\dot{\mathcal{K}}_\omega^{p,q}(\mathbb{R}^n)},$$

where the implicit positive constant is independent of f. *In particular, when* $N \in \mathbb{N} \cap [\lfloor b + 1 \rfloor, \infty)$, *if one of the quantities*

$$\left\| m(f, \phi) \right\|_{\dot{\mathcal{K}}_\omega^{p,q}(\mathbb{R}^n)}, \quad \left\| m_a^*(f, \phi) \right\|_{\dot{\mathcal{K}}_\omega^{p,q}(\mathbb{R}^n)}, \quad \left\| m_N(f) \right\|_{\dot{\mathcal{K}}_\omega^{p,q}(\mathbb{R}^n)},$$

$$\left\| m_b^{**}(f, \phi) \right\|_{\dot{\mathcal{K}}_\omega^{p,q}(\mathbb{R}^n)}, \quad \text{and} \quad \left\| m_{b,N}^{**}(f) \right\|_{\dot{\mathcal{K}}_\omega^{p,q}(\mathbb{R}^n)}$$

is finite, then the others are also finite and mutually equivalent with the positive equivalence constants independent of f.

Proof Let all the symbols be as in the present theorem. Note that ω satisfies $m_0(\omega) \in (-\frac{n}{p}, \infty)$ and $M_\infty(\omega) \in (-\infty, 0)$. From this and Theorem 1.2.44, it follows that the global generalized Herz space $\dot{\mathcal{K}}_\omega^{p,q}(\mathbb{R}^n)$ under consideration is a BQBF space. Thus, applying Lemma 5.1.5(i), we then complete the proof of (i).

Now, we prove (ii). Indeed, we only need to show that the Herz space $\dot{\mathcal{K}}_\omega^{p,q}(\mathbb{R}^n)$ under consideration satisfies all the assumptions of Lemma 5.1.5(ii). For this purpose, let $r, A \in (0, \infty)$ be such that

$$r \in \left(\frac{2n}{b}, \min \left\{ p, q, \frac{n}{\max\{M_0(\omega), M_\infty(\omega)\} + n/p} \right\} \right)$$

and $A \in (\frac{n}{r}, b - \frac{n}{r})$. Then, applying Theorem 1.3.4 with $s := r$, we conclude that the global generalized Herz space $\dot{\mathcal{K}}_\omega^{p,q}(\mathbb{R}^n)$ is strictly r-convex. On the other hand, from Lemma 4.1.10 and Remark 1.2.30(i), we deduce that, for any $f \in L_{\mathrm{loc}}^1(\mathbb{R}^n)$,

$$\left\| \mathcal{M}^{(r)}(f) \right\|_{\dot{\mathcal{K}}_\omega^{p,q}(\mathbb{R}^n)} \lesssim \|f\|_{\dot{\mathcal{K}}_\omega^{p,q}(\mathbb{R}^n)}.$$

By this, Lemma 5.1.4, and the assumption $A > \frac{n}{r}$, we further find that, for any $f \in \dot{\mathcal{K}}_\omega^{p,q}(\mathbb{R}^n)$ and $z \in \mathbb{R}^n$,

$$\left\| \left\{ \int_{z+[0,1]^n} |f(\cdot - y)|^r \, dy \right\}^{\frac{1}{r}} \right\|_{\dot{\mathcal{K}}_\omega^{p,q}(\mathbb{R}^n)}$$

$$\lesssim (1 + |z|)^{\frac{n}{r}} \|f\|_{\dot{\mathcal{K}}_\omega^{p,q}(\mathbb{R}^n)} \lesssim (1 + |z|)^A \|f\|_{\dot{\mathcal{K}}_\omega^{p,q}(\mathbb{R}^n)}.$$

Combining this and the fact that $\dot{\mathcal{K}}_\omega^{p,q}(\mathbb{R}^n)$ is strictly r-convex, we conclude that all
the assumptions of Lemma 5.1.4(ii) hold true for $\dot{\mathcal{K}}_\omega^{p,q}(\mathbb{R}^n)$. This finishes the proof
of (ii) and hence Theorem 5.1.10.   □

Remark 5.1.11

(i) Let $p, q \in (0, \infty)$ and $\omega \in M(\mathbb{R}_+)$. Then the quasi-norm of $h\dot{\mathcal{K}}_\omega^{p,q}(\mathbb{R}^n)$ in
Definition 5.0.1(ii) depends on N. However, by Theorem 5.1.10, we conclude
that the local Hardy space $h\dot{\mathcal{K}}_\omega^{p,q}(\mathbb{R}^n)$ is independent of the choice of N
whenever $\omega \in M(\mathbb{R}_+)$ satisfies $m_0(\omega) \in (-\frac{n}{p}, \infty)$ and

$$-\frac{n}{p} < m_\infty(\omega) \leq M_\infty(\omega) < 0,$$

and N satisfies

$$N \in \mathbb{N} \cap \left(1 + 2\max\left\{\frac{n}{p}, \frac{n}{q}, \max\{M_0(\omega), M_\infty(\omega)\} + \frac{n}{p}\right\}, \infty\right).$$

(ii) Notice that, if $p = q \in (0, \infty)$ and $\omega(t) := 1$ for any $t \in (0, \infty)$, then, in this
case,

$$m_0(\omega) = M_0(\omega) = m_\infty(\omega) = M_\infty(\omega) = 0$$

and $h\dot{\mathcal{K}}_\omega^{p,q}(\mathbb{R}^n)$ coincides with the classical local Hardy space $h^p(\mathbb{R}^n)$ in
the sense of equivalent quasi-norms. Therefore, Theorem 5.1.10 completely
excludes the classical local Hardy space $h^p(\mathbb{R}^n)$ and, by Remark 1.2.45, we
conclude that the classical local Hardy space $h^p(\mathbb{R}^n)$ is the critical case of
$h\dot{\mathcal{K}}_\omega^{p,q}(\mathbb{R}^n)$ considered in Theorem 5.1.10.

Using Theorem 5.1.10 and Remark 5.0.4(ii), we immediately obtain the following maximal function characterizations of the local generalized Hardy–Morrey
space $h M_\omega^{p,q}(\mathbb{R}^n)$; we omit the details.

Corollary 5.1.12 *Let $a, b \in (0, \infty)$, $p, q \in [1, \infty)$, $\omega \in M(\mathbb{R}_+)$, $N \in \mathbb{N}$, and
$\phi \in \mathcal{S}(\mathbb{R}^n)$ satisfy*

$$\int_{\mathbb{R}^n} \phi(x)\,dx \neq 0.$$

(i) *Let $N \in \mathbb{N} \cap [\lfloor b + 1 \rfloor, \infty)$ and ω satisfy $M_\infty(\omega) \in (-\infty, 0)$ and*

$$-\frac{n}{p} < m_0(\omega) \leq M_0(\omega) < 0.$$

Then, for any $f \in \mathcal{S}'(\mathbb{R}^n)$,

$$\|m(f,\phi)\|_{M_\omega^{p,q}(\mathbb{R}^n)} \lesssim \|m_a^*(f,\phi)\|_{M_\omega^{p,q}(\mathbb{R}^n)} \lesssim \|m_b^{**}(f,\phi)\|_{M_\omega^{p,q}(\mathbb{R}^n)},$$

$$\|m(f,\phi)\|_{M_\omega^{p,q}(\mathbb{R}^n)} \lesssim \|m_N(f)\|_{M_\omega^{p,q}(\mathbb{R}^n)} \lesssim \|m_{\lfloor b+1 \rfloor}(f)\|_{M_\omega^{p,q}(\mathbb{R}^n)}$$

$$\lesssim \|m_b^{**}(f,\phi)\|_{M_\omega^{p,q}(\mathbb{R}^n)},$$

and

$$\|m_b^{**}(f,\phi)\|_{M_\omega^{p,q}(\mathbb{R}^n)} \sim \|m_{b,N}^{**}(f)\|_{M_\omega^{p,q}(\mathbb{R}^n)},$$

where the implicit positive constants are independent of f.
(ii) *Let* ω *satisfy*

$$-\frac{n}{p} < m_0(\omega) \le M_0(\omega) < 0$$

and

$$-\frac{n}{p} < m_\infty(\omega) \le M_\infty(\omega) < 0,$$

and $b \in (2n \max\{\frac{1}{p}, \frac{1}{q}\}, \infty)$. *Then, for any* $f \in \mathcal{S}'(\mathbb{R}^n)$,

$$\|m_b^{**}(f,\phi)\|_{M_\omega^{p,q}(\mathbb{R}^n)} \lesssim \|m(f,\phi)\|_{M_\omega^{p,q}(\mathbb{R}^n)},$$

where the implicit positive constant is independent of f. *In particular, when* $N \in \mathbb{N} \cap [\lfloor b+1 \rfloor, \infty)$, *if one of the quantities*

$$\|m(f,\phi)\|_{M_\omega^{p,q}(\mathbb{R}^n)}, \quad \|m_a^*(f,\phi)\|_{M_\omega^{p,q}(\mathbb{R}^n)}, \quad \|m_N(f)\|_{M_\omega^{p,q}(\mathbb{R}^n)},$$

$$\|m_b^{**}(f,\phi)\|_{M_\omega^{p,q}(\mathbb{R}^n)}, \quad \text{and } \|m_{b,N}^{**}(f)\|_{M_\omega^{p,q}(\mathbb{R}^n)}$$

is finite, then the others are also finite and mutually equivalent with the positive equivalence constants independent of f.

Remark 5.1.13 Let $p, q \in (0, \infty)$ and $\omega \in M(\mathbb{R}_+)$. Then the quasi-norms of $hM_{\omega,0}^{p,q}(\mathbb{R}^n)$ and $hM_\omega^{p,q}(\mathbb{R}^n)$ in Definition 5.0.3 depend on N. However, from Corollaries 5.1.9 and 5.1.12, we deduce that the local Hardy spaces $hM_{\omega,0}^{p,q}(\mathbb{R}^n)$ and $hM_\omega^{p,q}(\mathbb{R}^n)$ are both independent of the choice of N whenever $p, q \in [1, \infty)$, $\omega \in M(\mathbb{R}_+)$ satisfies

$$-\frac{n}{p} < m_0(\omega) \le M_0(\omega) < 0$$

and

$$-\frac{n}{p} < m_\infty(\omega) \le M_\infty(\omega) < 0,$$

and N satisfies

$$N \in \mathbb{N} \cap \left(1 + 2n \max \left\{ \frac{1}{p}, \frac{1}{q} \right\}, \infty \right).$$

5.2   Relations with Generalized Herz–Hardy Spaces

The target of this section is to investigate the relation between localized generalized Herz–Hardy spaces and generalized Herz–Hardy spaces. For this purpose, we first establish a relation between localized Hardy spaces and Hardy spaces associated with ball quasi-Banach function spaces, which extends the results for various concrete function spaces (see Remark 5.2.4 below). Using this general conclusion, we then obtain the relation between localized generalized Herz–Hardy spaces and generalized Herz–Hardy spaces.

First, we give the relation between localized Hardy spaces and Hardy spaces associated with ball quasi-Banach function spaces. To this end, we first recall the definition of localized Hardy spaces associated with ball quasi-Banach function spaces as follows, which was introduced in [207, Definition 5.2].

Definition 5.2.1 Let X be a ball quasi-Banach function space and $N \in \mathbb{N}$. Then the *local Hardy space* $h_X(\mathbb{R}^n)$ is defined to be the set of all the $f \in \mathcal{S}'(\mathbb{R}^n)$ such that

$$\|f\|_{h_X(\mathbb{R}^n)} := \|m_N(f)\|_X < \infty.$$

Remark 5.2.2

(i) Let $p \in (0, \infty)$ and $X := L^p(\mathbb{R}^n)$. Then, in this case, $h_X(\mathbb{R}^n)$ is just the *local Hardy space* $h^p(\mathbb{R}^n)$ which was originally introduced in [89].
(ii) Let $p \in (0, \infty)$, $\upsilon \in A_\infty(\mathbb{R}^n)$, and $X := L^p_\upsilon(\mathbb{R}^n)$. Then, in this case, $h_X(\mathbb{R}^n)$ is just the *weighted local Hardy space* $h^p_\upsilon(\mathbb{R}^n)$ which was originally studied in [18].
(iii) Let $\vec{p} := (p_1, \ldots, p_n) \in (0, \infty)^n$ and $X := L^{\vec{p}}(\mathbb{R}^n)$. Then the *mixed-norm local Hardy space* $h^{\vec{p}}(\mathbb{R}^n)$ is defined as in Definition 5.2.1 with X therein replaced by $L^{\vec{p}}(\mathbb{R}^n)$. To the best of our knowledge, $h^{\vec{p}}(\mathbb{R}^n)$ is totally new.
(iv) Let $0 < q \le p < \infty$ and $X := M^p_q(\mathbb{R}^n)$. Then, in this case, $H_X(\mathbb{R}^n)$ is just the *local Hardy–Morrey space* $hM^p_q(\mathbb{R}^n)$ which was originally introduced in [205, Section 4].

(v) Let $p(\cdot) : \mathbb{R}^n \rightarrow (0, \infty)$ and $X := L^{p(\cdot)}(\mathbb{R}^n)$. Then, in this case, $h_X(\mathbb{R}^n)$ is just the *variable local Hardy space* $h^{p(\cdot)}(\mathbb{R}^n)$ which was originally introduced in [186, Section 9].

Then we have the following conclusion.

Theorem 5.2.3 *Let X be a ball quasi-Banach function space and let $r \in (0, \infty)$ be such that X is strictly r-convex and $\mathcal{M}$ is bounded on $X^{1/r}$. Assume $\varphi \in \mathcal{S}(\mathbb{R}^n)$ satisfying that*

$$\mathbf{1}_{B(0,1)} \leq \widehat{\varphi} \leq \mathbf{1}_{B(0,2)}.$$

Then there exist two positive constants C_1 and C_2 such that, for any $f \in \mathcal{S}'(\mathbb{R}^n)$,

$$C_1 \|f\|_{h_X(\mathbb{R}^n)} \leq \|f * \varphi\|_X + \|f - f * \varphi\|_{H_X(\mathbb{R}^n)} \leq C_2 \|f\|_{h_X(\mathbb{R}^n)}.$$

Remark 5.2.4 We should point out that Theorem 5.2.3 has a wide range of applications. Here we give several function spaces to which Theorem 5.2.3 can be applied.

(i) Let $p \in (0, \infty)$. Then, in this case, applying Remark 1.2.31(i) and the fact that, for any $r \in (0, p]$, $L^p(\mathbb{R}^n)$ is strictly r-convex, we can easily conclude that the Lebesgue space $L^p(\mathbb{R}^n)$ satisfies all the assumptions of Theorem 5.2.3. Thus, Theorem 5.2.3 with $X := L^p(\mathbb{R}^n)$ holds true. This result goes back to [89, Lemma 4].

(ii) Let $p \in (0, \infty)$ and $\upsilon \in A_\infty(\mathbb{R}^n)$. Then, in this case, from Remarks 1.2.31(i) and the fact that $L_\upsilon^p(\mathbb{R}^n)$ is strictly r-convex for any $r \in (0, p]$, we can easily deduce that the weighted Lebesgue space $L_\upsilon^p(\mathbb{R}^n)$ satisfies all the assumptions of Theorem 5.2.3. Therefore, Theorem 5.2.3 with X therein replaced by $L_\upsilon^p(\mathbb{R}^n)$ holds true. If further assume that $p := 1$, then, in this case, the aforementioned result goes back to [18, Proposition 4.1].

(iii) Let $\vec{p} := (p_1, \ldots, p_n) \in (0, \infty)^n$. Then, in this case, using Remark 1.2.31(iii) and the fact that, for any $r \in (0, \min\{p_1, \ldots, p_n\}]$, $L^{\vec{p}}(\mathbb{R}^n)$ is strictly r-convex, we can easily find that the mixed-norm Lebesgue space $L^{\vec{p}}(\mathbb{R}^n)$ satisfies all the assumptions of Theorem 5.2.3. This then implies that Theorem 5.2.3 with $X := L^{\vec{p}}(\mathbb{R}^n)$ holds true. To the best of our knowledge, this result is totally new.

(iv) Let $0 < q \leq p < \infty$. Then, in this case, by Remark 1.2.31(iv) and the fact that, for any $r \in (0, q]$, $M_q^p(\mathbb{R}^n)$ is strictly r-convex, we easily conclude that the Morrey space $M_q^p(\mathbb{R}^n)$ satisfies all the assumptions of Theorem 5.2.3. Thus, Theorem 5.2.3 with X replaced by $M_q^p(\mathbb{R}^n)$ holds true. This result was also obtained in [205, Proposition 4.8].

(v) Let $p(\cdot) \in C^{\log}(\mathbb{R}^n)$. Then, in this case, from Remarks 1.2.31(v), 1.2.34(v), and 5.1.6, we can easily infer that the variable Lebesgue space $L^{p(\cdot)}(\mathbb{R}^n)$ satisfies all the assumptions of Theorem 5.2.3. This further implies that

Theorem 5.2.3 with $X := L^{p(\cdot)}(\mathbb{R}^n)$ holds true. This result coincides with [186, Lemma 9.1].

To prove this theorem, we need the following Plancherel–Pólya–Nikol'skij inequality which is a consequence of [229, p. 16, Theorem] (see also [186, Lemma 2.6]).

Lemma 5.2.5 *Let $\varphi \in \mathcal{S}(\mathbb{R}^n)$ be such that $\widehat{\varphi}$ has compact support, and let $r \in (0, \infty)$. Then there exists a positive constant C, independent of φ, such that, for any $f \in \mathcal{S}'(\mathbb{R}^n)$ and $x, y \in \mathbb{R}^n$,*

$$|f * \varphi(y)| \le C \, (1 + |x - y|)^{\frac{n}{r}} \, \mathcal{M}^{(r)} \, (f * \varphi) \, (x).$$

Via Lemma 5.2.5, we next show Theorem 5.2.3.

Proof of Theorem 5.2.3 Let all the symbols be as in the present theorem and $f \in \mathcal{S}'(\mathbb{R}^n)$. We first prove that

$$\|f\|_{h_X(\mathbb{R}^n)} \lesssim \|f * \varphi\|_X + \|f - f * \varphi\|_{H_X(\mathbb{R}^n)}. \tag{5.4}$$

Indeed, from Definitions 4.2.2 and 5.2.1, (4.3), and (5.1), we deduce that

$$\|f - f * \varphi\|_{h_X(\mathbb{R}^n)} \le \|f - f * \varphi\|_{H_X(\mathbb{R}^n)}. \tag{5.5}$$

On the other hand, by the assumption that $\mathcal{M}$ is bounded on $X^{1/r}$ and Remark 1.2.30(i), we conclude that $\mathcal{M}^{(r)}$ is bounded on X. This, together with Lemma 5.1.4, further implies that, for any $g \in X$ and $z \in \mathbb{R}^n$,

$$\left\| \left\{ \int_{z+[0,1]^n} |g(\cdot - y)|^r \, dy \right\}^{\frac{1}{r}} \right\|_X \lesssim (1 + |z|)^{\frac{n}{r}} \|g\|_X.$$

Applying this, the assumption that X is strictly r-convex, and Lemma 5.1.5(ii), we find that

$$\|f * \varphi\|_{h_X(\mathbb{R}^n)} \sim \|m(f * \varphi, \varphi)\|_X,$$

which, combined with (5.5), implies that

$$\begin{aligned}
\|f\|_{h_X(\mathbb{R}^n)} &\lesssim \|f * \varphi\|_{h_X(\mathbb{R}^n)} + \|f - f * \varphi\|_{h_X(\mathbb{R}^n)} \\
&\lesssim \|m(f * \varphi, \varphi)\|_X + \|f - f * \varphi\|_{H_X(\mathbb{R}^n)}.
\end{aligned} \tag{5.6}$$

Now, we estimate $m(f * \varphi, \varphi)$. Indeed, using Lemma 5.2.5 with y replaced by $x - y$, we conclude that, for any $t \in (0, 1)$ and $x \in \mathbb{R}^n$,

$$
\begin{aligned}
&|(f * \varphi) * \varphi_t(x)| \\
&= \left| \int_{\mathbb{R}^n} f * \varphi(x - y) \varphi_t(y) \, dy \right| \\
&\lesssim \int_{\mathbb{R}^n} (1 + |y|)^{\frac{n}{r}} \, \mathcal{M}^{(r)} (f * \varphi)(x) \, |\varphi_t(y)| \, dy \\
&\lesssim \frac{1}{t^n} \mathcal{M}^{(r)} (f * \varphi)(x) \int_{\mathbb{R}^n} \left(1 + \left| \frac{y}{t} \right| \right)^{\frac{n}{r}} \left| \varphi \left(\frac{y}{t} \right) \right| \, dy \\
&\sim \mathcal{M}^{(r)} (f * \varphi)(x) \int_{\mathbb{R}^n} (1 + |y|)^{\frac{n}{r}} |\varphi(y)| \, dy \\
&\lesssim \mathcal{M}^{(r)} (f * \varphi)(x) \int_{\mathbb{R}^n} \frac{1}{(1 + |y|)^{n+1}} \, dy \sim \mathcal{M}^{(r)} (f * \varphi)(x).
\end{aligned}
$$

Therefore, from (5.1), it follows that

$$
m (f * \varphi, \varphi) \lesssim \mathcal{M}^{(r)} (f * \varphi).
$$

Then, combining this, Definition 1.2.13(ii), the assumption that $\mathcal{M}$ is bounded on $X^{1/r}$, and Remark 1.2.30(i), we conclude that

$$
\| m (f * \varphi, \varphi) \|_X \lesssim \left\| \mathcal{M}^{(r)} (f * \varphi) \right\|_X \lesssim \| f * \varphi \|_X .
$$

By this and (5.6), we further obtain

$$
\| f \|_{h_X(\mathbb{R}^n)} \lesssim \| f * \varphi \|_X + \| f - f * \varphi \|_{H_X(\mathbb{R}^n)} ,
$$

which completes the proof of (5.4).

Conversely, we show

$$
\| f * \varphi \|_X + \| f - f * \varphi \|_{H_X(\mathbb{R}^n)} \lesssim \| f \|_{h_X(\mathbb{R}^n)} . \tag{5.7}
$$

To achieve this, we first estimate the term $\| f - f * \varphi \|_{H_X(\mathbb{R}^n)}$. Indeed, applying Definition 4.1.1(i), we find that

$$
\begin{aligned}
M (f - f * \varphi, \varphi) &= \sup_{t \in (0, \infty)} |(f - f * \varphi) * \varphi_t| \\
&= \sup_{t \in (0, \infty)} |f * \varphi_t - f * (\varphi * \varphi_t)| . \tag{5.8}
\end{aligned}
$$

We next claim that, for any $t \in [2, \infty)$, $f * \varphi_t - f * (\varphi * \varphi_t) = 0$. Indeed, notice that, for any $t \in (0, \infty)$,

$$f * \varphi_t - f * (\varphi * \varphi_t) \in \mathcal{S}'(\mathbb{R}^n).$$

Thus, for any $t \in (0, \infty)$, we have

$$\mathcal{F}(f * \varphi_t - f * (\varphi * \varphi_t)) = \widehat{f}\widehat{\varphi}(t \cdot)(1 - \widehat{\varphi}). \tag{5.9}$$

Using the assumption $\widehat{\varphi}\mathbf{1}_{B(0,1)} = 1$, we conclude that, for any $t \in (0, \infty)$ and $x \in B(0, 1)$,

$$\widehat{\varphi}(tx)[1 - \widehat{\varphi}(x)] = 0.$$

On the other hand, for any $t \in [2, \infty)$ and $x \in [B(0, 1)]^{\complement}$, we have $|tx| \geq t \geq 2$. This, together with the assumption that $\widehat{\varphi}\mathbf{1}_{[B(0,2)]^{\complement}} = 0$, further implies that, for any $t \in [2, \infty)$ and $x \in [B(0, 1)]^{\complement}$,

$$\widehat{\varphi}(tx)[1 - \widehat{\varphi}(x)] = 0.$$

Therefore, for any $t \in [2, \infty)$, we have

$$\widehat{\varphi}(t \cdot)(1 - \widehat{\varphi}) = 0.$$

By this and (5.9), we find that, for any $t \in [2, \infty)$,

$$\mathcal{F}(f * \varphi_t - f * (\varphi * \varphi_t)) = 0$$

in $\mathcal{S}'(\mathbb{R}^n)$, and hence $f * \varphi_t - f * (\varphi * \varphi_t) = 0$, which completes the proof of the above claim. Now, we define $\psi \in \mathcal{S}(\mathbb{R}^n)$ by setting, for any $x \in \mathbb{R}^n$,

$$\psi(x) := \frac{1}{2^n} \varphi\left(\frac{x}{2}\right). \tag{5.10}$$

Then, from (5.8) and the above claim, we deduce that

$$
\begin{aligned}
M(f - f * \varphi, \varphi) &= \sup_{t \in (0,2)} |f * \varphi_t - f * (\varphi * \varphi_t)| \\
&\leq \sup_{t \in (0,1)} |f * \psi_t| + \sup_{t \in (0,2)} |f * (\varphi * \varphi_t)| \\
&= m(f, \psi) + \sup_{t \in (0,2)} |f * (\varphi * \varphi_t)|.
\end{aligned}
$$

Combining this and both Lemmas 4.1.4(ii) and 5.1.5(ii), we find that

$$\|f - f * \varphi\|_{H_X(\mathbb{R}^n)} \sim \|M(f - f * \varphi, \varphi)\|_X$$

$$\lesssim \|m(f, \psi)\|_X + \left\| \sup_{t \in (0,2)} |f * (\varphi * \varphi_t)| \right\|_X$$

$$\sim \|f\|_{h_X(\mathbb{R}^n)} + \left\| \sup_{t \in (0,2)} |f * (\varphi * \varphi_t)| \right\|_X. \qquad (5.11)$$

Next, we prove that

$$\left\| \sup_{t \in (0,2)} |f * (\varphi * \varphi_t)| \right\|_X \lesssim \|f\|_{h_X(\mathbb{R}^n)}. \qquad (5.12)$$

Indeed, for any $t \in (0, \infty)$, we have

$$\varphi * \varphi_t = \mathcal{F}(\widehat{\varphi}(-\cdot)\widehat{\varphi}(-t\cdot)).$$

Thus, for any $\alpha, \beta \in \mathbb{Z}_+^n$, $t \in (0, 2)$, and $x \in \mathbb{R}^n$, we obtain

$$\left| x^\beta \partial^\alpha (\varphi * \varphi_t)(x) \right|$$

$$\sim \left| \mathcal{F}\left(\partial^\beta \left((\cdot)^\alpha \widehat{\varphi}(-\cdot)\widehat{\varphi}(-t\cdot) \right) \right)(x) \right|$$

$$\lesssim \int_{\mathbb{R}^n} \left| \partial^\beta \left((\cdot)^\alpha \widehat{\varphi}(-\cdot)\widehat{\varphi}(-t\cdot) \right)(x) \right| dx$$

$$\lesssim \sum_{\substack{\gamma \in \mathbb{Z}_+^n \\ \gamma \leq \beta}} \int_{\mathbb{R}^n} \left[\left| x^\alpha \widehat{\varphi}(-x) \partial^\gamma (\widehat{\varphi}(-t\cdot))(x) \right| \right.$$

$$\left. + \left| \widehat{\varphi}(-tx) \partial^{\beta-\gamma} \left((\cdot)^\alpha \widehat{\varphi}(-\cdot) \right)(x) \right| \right] dx$$

$$\lesssim \sum_{\substack{\gamma \in \mathbb{Z}_+^n \\ \gamma \leq \beta}} \int_{\mathbb{R}^n} \left[t^{|\gamma|} \left| x^\alpha \widehat{\varphi}(-x) \right| + \left| \partial^{\beta-\gamma} \left((\cdot)^\alpha \widehat{\varphi}(-\cdot) \right)(x) \right| \right] dx$$

$$\lesssim \int_{\mathbb{R}^n} \frac{1}{(1+|x|)^{n+1}} \, dx \sim 1. \qquad (5.13)$$

Fix an $N \in \mathbb{N} \cap (\frac{n}{r}+1, \infty)$. Then, applying (5.13), we conclude that, for any $\alpha \in \mathbb{Z}_+^n$ with $|\alpha| \leq N$, any $t \in (0, 2)$, and $x \in \mathbb{R}^n$,

$$
(1 + |x|)^{N+n} \left| \partial^\alpha \left(\frac{1}{2^n} (\varphi * \varphi_t) \left(\frac{\cdot}{2} \right) \right) (x) \right|
$$

$$
\lesssim \sum_{\beta \in \mathbb{Z}_+^n, |\beta| \leq N} \left| x^\beta \partial^\alpha \left(\frac{1}{2^n} (\varphi * \varphi_t) \left(\frac{\cdot}{2} \right) \right) (x) \right|
$$

$$
\lesssim \sum_{\beta \in \mathbb{Z}_+^n, |\beta| \leq N} \left| x^\beta \partial^\alpha (\varphi * \varphi_t) \left(\frac{x}{2} \right) \right| \lesssim 1.
$$

This, together with (4.1), further implies that, for any $t \in (0, 2)$,

$$
p_N \left(\frac{1}{2^n} (\varphi * \varphi_t) \left(\frac{\cdot}{2} \right) \right) \lesssim 1,
$$

where the implicit positive constant is independent of t. From this, (4.2), and (5.1), it follows that

$$
\sup_{t \in (0, 2)} |f * (\varphi * \varphi_t)| \lesssim m_N(f).
$$

Combining this and Definition 1.2.13(ii), we further conclude that

$$
\left\| \sup_{t \in (0, 2)} |f * (\varphi * \varphi_t)| \right\|_X \lesssim \|m_N(f)\|_X \sim \|f\|_{h_X(\mathbb{R}^n)}.
$$

This finishes the estimation of (5.12). From this and (5.11), we further infer that

$$
\|f - f * \varphi\|_{H_X(\mathbb{R}^n)} \lesssim \|f\|_{h_X(\mathbb{R}^n)}. \tag{5.14}
$$

In addition, applying (5.10) and Definition 5.1.1(i), we find that

$$
|f * \varphi| = \left| f * \psi_{\frac{1}{2}} \right| \leq m(f, \psi).
$$

From this, Definition 1.2.13(ii), and Lemma 5.1.5(ii), it follows that

$$
\|f * \varphi\|_X \leq \|m(f, \psi)\|_X \sim \|f\|_{h_X(\mathbb{R}^n)}.
$$

This, together with (5.14), further implies that

$$
\|f * \varphi\|_X + \|f - f * \varphi\|_{H_X(\mathbb{R}^n)} \lesssim \|f\|_{h_X(\mathbb{R}^n)},
$$

which completes the proof of (5.7). Combining both (5.4) and (5.7), we obtain

$$\|f\|_{h_X(\mathbb{R}^n)} \sim \|f * \varphi\|_X + \|f - f * \varphi\|_{H_X(\mathbb{R}^n)}.$$

This finishes the proof of Theorem 5.2.3.   □

Based on the above relation between $h_X(\mathbb{R}^n)$ and $H_X(\mathbb{R}^n)$, we now establish the following relation between the local generalized Herz–Hardy space $h\dot{\mathcal{K}}_{\omega,\mathbf{0}}^{p,q}(\mathbb{R}^n)$ and the generalized Herz–Hardy space $H\dot{\mathcal{K}}_{\omega,\mathbf{0}}^{p,q}(\mathbb{R}^n)$.

Theorem 5.2.6 *Let* $p, q \in (0, \infty)$, $\omega \in M(\mathbb{R}_+)$ *with* $m_0(\omega) \in (-\frac{n}{p}, \infty)$ *and* $m_\infty(\omega) \in (-\frac{n}{p}, \infty)$, *and* $\varphi \in \mathcal{S}(\mathbb{R}^n)$ *with*

$$\mathbf{1}_{B(0,1)} \leq \widehat{\varphi} \leq \mathbf{1}_{B(0,2)}.$$

Then there exist two positive constants C_1 *and* C_2 *such that, for any* $f \in \mathcal{S}'(\mathbb{R}^n)$,

$$C_1 \|f\|_{h\dot{\mathcal{K}}_{\omega,\mathbf{0}}^{p,q}(\mathbb{R}^n)} \leq \|f * \varphi\|_{\dot{\mathcal{K}}_{\omega,\mathbf{0}}^{p,q}(\mathbb{R}^n)} + \|f - f * \varphi\|_{H\dot{\mathcal{K}}_{\omega,\mathbf{0}}^{p,q}(\mathbb{R}^n)}$$

$$\leq C_2 \|f\|_{h\dot{\mathcal{K}}_{\omega,\mathbf{0}}^{p,q}(\mathbb{R}^n)}.$$

Proof Let all the symbols be as in the present theorem. Then, combining the assumption $m_0(\omega) \in (-\frac{n}{p}, \infty)$ and Theorem 1.2.42, we find that, under the assumptions of the present theorem, the local generalized Herz space $\dot{\mathcal{K}}_{\omega,\mathbf{0}}^{p,q}(\mathbb{R}^n)$ is a BQBF space. This implies that, to complete the proof of the present theorem, we only need to show that the Herz space $\dot{\mathcal{K}}_{\omega,\mathbf{0}}^{p,q}(\mathbb{R}^n)$ under consideration satisfies all the assumptions of Theorem 5.2.3.

Indeed, let

$$r \in \left(0, \min\left\{p, q, \frac{n}{\max\{M_0(\omega), M_\infty(\omega)\} + n/p}\right\}\right).$$

Then, applying Theorem 1.3.3 with $s := r$, we conclude that $\dot{\mathcal{K}}_{\omega,\mathbf{0}}^{p,q}(\mathbb{R}^n)$ is strictly r-convex. On the other hand, from Lemma 1.8.5, it follows that, for any $f \in L_{\mathrm{loc}}^1(\mathbb{R}^n)$,

$$\|\mathcal{M}(f)\|_{[\dot{\mathcal{K}}_{\omega,\mathbf{0}}^{p,q}(\mathbb{R}^n)]^{1/r}} \lesssim \|f\|_{[\dot{\mathcal{K}}_{\omega,\mathbf{0}}^{p,q}(\mathbb{R}^n)]^{1/r}}.$$

Therefore, all the assumptions of Theorem 5.2.3 hold true for $\dot{\mathcal{K}}_{\omega,\mathbf{0}}^{p,q}(\mathbb{R}^n)$ and hence the proof of Theorem 5.2.6 is completed.   □

Theorem 5.2.6, together with Remark 5.0.4(ii), further implies that the following conclusion holds true for the local generalized Hardy–Morrey space $h\boldsymbol{M}_{\omega,\mathbf{0}}^{p,q}(\mathbb{R}^n)$; we omit the details.

Corollary 5.2.7 *Let* $p, q \in [1, \infty)$, $\omega \in M(\mathbb{R}_+)$ *with*

$$-\frac{n}{p} < m_0(\omega) \leq M_0(\omega) < 0$$

and

$$-\frac{n}{p} < m_\infty(\omega) \leq M_\infty(\omega) < 0,$$

and $\varphi \in \mathcal{S}(\mathbb{R}^n)$ *with* $\mathbf{1}_{B(0,1)} \leq \widehat{\varphi} \leq \mathbf{1}_{B(0,2)}$. *Then there exist two positive constants* C_1 *and* C_2 *such that, for any* $f \in \mathcal{S}'(\mathbb{R}^n)$,

$$C_1 \|f\|_{hM_{\omega,0}^{p,q}(\mathbb{R}^n)} \leq \|f * \varphi\|_{M_{\omega,0}^{p,q}(\mathbb{R}^n)} + \|f - f * \varphi\|_{HM_{\omega,0}^{p,q}(\mathbb{R}^n)}$$

$$\leq C_2 \|f\|_{hM_{\omega,0}^{p,q}(\mathbb{R}^n)}.$$

Next, we establish the relation between the local generalized Herz–Hardy space $h\dot{\mathcal{K}}_\omega^{p,q}(\mathbb{R}^n)$ and the generalized Herz–Hardy space $H\dot{\mathcal{K}}_\omega^{p,q}(\mathbb{R}^n)$ as follows.

Theorem 5.2.8 *Let* $p, q \in (0, \infty)$, $\omega \in M(\mathbb{R}_+)$ *with* $m_0(\omega) \in (-\frac{n}{p}, \infty)$ *and*

$$-\frac{n}{p} < m_\infty(\omega) \leq M_\infty(\omega) < 0,$$

and $\varphi \in \mathcal{S}(\mathbb{R}^n)$ *with*

$$\mathbf{1}_{B(0,1)} \leq \widehat{\varphi} \leq \mathbf{1}_{B(0,2)}.$$

Then there exist two positive constants C_1 *and* C_2 *such that, for any* $f \in \mathcal{S}'(\mathbb{R}^n)$,

$$C_1 \|f\|_{h\dot{\mathcal{K}}_\omega^{p,q}(\mathbb{R}^n)} \leq \|f * \varphi\|_{\dot{\mathcal{K}}_\omega^{p,q}(\mathbb{R}^n)} + \|f - f * \varphi\|_{H\dot{\mathcal{K}}_\omega^{p,q}(\mathbb{R}^n)}$$

$$\leq C_2 \|f\|_{h\dot{\mathcal{K}}_\omega^{p,q}(\mathbb{R}^n)}.$$

Proof Let all the symbols be as in the present theorem. Then, since $\omega \in M(\mathbb{R}_+)$ satisfies that $m_0(\omega) \in (-\frac{n}{p}, \infty)$ and $M_\infty(\omega) \in (-\infty, 0)$, from Theorem 1.2.44, it follows that the global generalized Herz space $\dot{\mathcal{K}}_\omega^{p,q}(\mathbb{R}^n)$ under consideration is a BQBF space. Thus, to finish the proof of the present theorem, it suffices to prove that all the assumptions of Theorem 5.2.3 hold true for $\dot{\mathcal{K}}_\omega^{p,q}(\mathbb{R}^n)$. Namely, there exists an $r \in (0, \infty)$ such that $\dot{\mathcal{K}}_\omega^{p,q}(\mathbb{R}^n)$ is strictly r-convex and the Hardy–Littlewood maximal operator $\mathcal{M}$ is bounded on $[\dot{\mathcal{K}}_\omega^{p,q}(\mathbb{R}^n)]^{1/r}$.

To this end, let

$$r \in \left(0, \min \left\{ p, q, \frac{n}{\max\{M_0(\omega), M_\infty(\omega)\} + n/p} \right\} \right).$$

Then, by Theorem 1.3.4 with s therein replaced by r, we find that $\dot{\mathcal{K}}_\omega^{p,q}(\mathbb{R}^n)$ is strictly r-convex. On the other hand, applying Lemma 4.1.10, we conclude that, for any $f \in L^1_{\text{loc}}(\mathbb{R}^n)$,

$$\|\mathcal{M}(f)\|_{[\dot{\mathcal{K}}_\omega^{p,q}(\mathbb{R}^n)]^{1/r}} \lesssim \|f\|_{[\dot{\mathcal{K}}_\omega^{p,q}(\mathbb{R}^n)]^{1/r}}.$$

This further implies that all the assumptions of Theorem 5.2.3 hold true for $\dot{\mathcal{K}}_\omega^{p,q}(\mathbb{R}^n)$ and then finishes the proof of Theorem 5.2.8.   □

Via Theorem 5.2.8 and Remark 5.0.4(ii), we immediately obtain the following relation between the local Hardy space $hM_\omega^{p,q}(\mathbb{R}^n)$ and the Hardy space $HM_\omega^{p,q}(\mathbb{R}^n)$ associated with the global generalized Morrey space $M_\omega^{p,q}(\mathbb{R}^n)$; we omit the details.

Corollary 5.2.9 *Let p, q, ω, and φ be as in Corollary 5.2.7. Then there exist two positive constants C_1 and C_2 such that, for any $f \in \mathcal{S}'(\mathbb{R}^n)$,*

$$C_1 \|f\|_{hM_\omega^{p,q}(\mathbb{R}^n)} \le \|f * \varphi\|_{M_\omega^{p,q}(\mathbb{R}^n)} + \|f - f * \varphi\|_{HM_\omega^{p,q}(\mathbb{R}^n)}$$

$$\le C_2 \|f\|_{hM_\omega^{p,q}(\mathbb{R}^n)}.$$

5.3 Atomic Characterizations

The main target of this section is to establish the atomic characterization of the local generalized Herz–Hardy spaces $h\dot{\mathcal{K}}_{\omega,0}^{p,q}(\mathbb{R}^n)$ and $h\dot{\mathcal{K}}_\omega^{p,q}(\mathbb{R}^n)$. Indeed, by the known atomic characterization of localized Hardy spaces associated with ball quasi-Banach function spaces, we obtain the atomic characterization of $h\dot{\mathcal{K}}_{\omega,0}^{p,q}(\mathbb{R}^n)$. On the other hand, in order to show the atomic characterization of $h\dot{\mathcal{K}}_\omega^{p,q}(\mathbb{R}^n)$, recall that the associate spaces of global generalized Herz spaces are still unknown. To overcome this difficulty, we first prove an improved atomic characterization of the local Hardy space $h_X(\mathbb{R}^n)$, with X being a ball quasi-Banach function space, without recourse to the associate space X'. From this improved conclusion of $h_X(\mathbb{R}^n)$, we deduce the desired atomic characterization of $h\dot{\mathcal{K}}_\omega^{p,q}(\mathbb{R}^n)$.

To begin with, we establish the atomic characterization of the local generalized Herz–Hardy space $h\dot{\mathcal{K}}_{\omega,0}^{p,q}(\mathbb{R}^n)$. For this purpose, we first introduce the definitions of local atoms and the local atomic Hardy space $h\dot{\mathcal{K}}_{\omega,0}^{p,q,d,s}(\mathbb{R}^n)$ associated with the local generalized Herz space $\dot{\mathcal{K}}_{\omega,0}^{p,q}(\mathbb{R}^n)$ as follows.

Definition 5.3.1 Let $p, q \in (0, \infty)$, $\omega \in M(\mathbb{R}_+)$ with $m_0(\omega) \in (-\frac{n}{p}, \infty)$, $r \in [1, \infty]$, and $d \in \mathbb{Z}_+$. Then a measurable function a is called a *local-* $(\dot{\mathcal{K}}_{\omega,0}^{p,q}(\mathbb{R}^n), r, d)$-*atom* if

(i) there exists a ball $B(x_0, r_0) \in \mathbb{B}$, with $x_0 \in \mathbb{R}^n$ and $r_0 \in (0, \infty)$, such that

$$\mathrm{supp}\,(a) := \{x \in \mathbb{R}^n : a(x) \neq 0\} \subset B(x_0, r_0);$$

(ii) $\|a\|_{L^r(\mathbb{R}^n)} \leq \dfrac{|B(x_0, r_0)|^{1/r}}{\|\mathbf{1}_{B(x_0, r_0)}\|_{\dot{\mathcal{K}}_{\omega,0}^{p,q}(\mathbb{R}^n)}};$

(iii) when $r_0 \in (0, 1)$, then, for any $\alpha \in \mathbb{Z}_+^n$ satisfying $|\alpha| \leq d$,

$$\int_{\mathbb{R}^n} a(x) x^\alpha \, dx = 0.$$

Definition 5.3.2 Let $p, q \in (0, \infty)$, $\omega \in M(\mathbb{R}_+)$ with $m_0(\omega) \in (-\frac{n}{p}, \infty)$ and $m_\infty(\omega) \in (-\frac{n}{p}, \infty)$,

$$s \in \left(0, \min\left\{1, p, q, \frac{n}{\max\{M_0(\omega), M_\infty(\omega)\} + n/p}\right\}\right),$$

$d \geq \lfloor n(1/s - 1) \rfloor$ be a fixed integer, and

$$r \in \left(\max\left\{1, p, \frac{n}{\min\{m_0(\omega), m_\infty(\omega)\} + n/p}\right\}, \infty\right].$$

Then the *local generalized atomic Herz–Hardy space* $h\dot{\mathcal{K}}_{\omega,0}^{p,q,r,d,s}(\mathbb{R}^n)$, associated with the local generalized Herz space $\dot{\mathcal{K}}_{\omega,0}^{p,q}(\mathbb{R}^n)$, is defined to be the set of all the $f \in \mathcal{S}'(\mathbb{R}^n)$ such that there exists a sequence $\{a_j\}_{j \in \mathbb{N}}$ of local-$(\dot{\mathcal{K}}_{\omega,0}^{p,q}(\mathbb{R}^n), r, d)$-atoms supported, respectively, in the balls $\{B_j\}_{j \in \mathbb{N}} \subset \mathbb{B}$ and a sequence $\{\lambda_j\}_{j \in \mathbb{N}} \subset [0, \infty)$ satisfying that

$$f = \sum_{j \in \mathbb{N}} \lambda_j a_j$$

in $\mathcal{S}'(\mathbb{R}^n)$ and

$$\left\| \left\{ \sum_{j \in \mathbb{N}} \left[\frac{\lambda_j}{\|\mathbf{1}_{B_j}\|_{\dot{\mathcal{K}}_{\omega,0}^{p,q}(\mathbb{R}^n)}} \right]^s \mathbf{1}_{B_j} \right\}^{\frac{1}{s}} \right\|_{\dot{\mathcal{K}}_{\omega,0}^{p,q}(\mathbb{R}^n)} < \infty.$$

Moreover, for any $f \in h\dot{\mathcal{K}}_{\omega,0}^{p,q,r,d,s}(\mathbb{R}^n)$,

$$
\|f\|_{h\dot{\mathcal{K}}_{\omega,0}^{p,q,r,d,s}(\mathbb{R}^n)} := \inf \left\{ \left\| \left\{ \sum_{j\in\mathbb{N}} \left[\frac{\lambda_j}{\|\mathbf{1}_{B_j}\|_{\dot{\mathcal{K}}_{\omega,0}^{p,q}(\mathbb{R}^n)}} \right]^s \mathbf{1}_{B_j} \right\}^{\frac{1}{s}} \right\|_{\dot{\mathcal{K}}_{\omega,0}^{p,q}(\mathbb{R}^n)} \right\},
$$

where the infimum is taken over all the decompositions of f as above.

Then we have the following atomic characterization of the local generalized Herz–Hardy space $h\dot{\mathcal{K}}_{\omega,0}^{p,q}(\mathbb{R}^n)$.

Theorem 5.3.3 *Let p, q, ω, d, s, and r be as in Definition 5.3.2. Then*

$$
h\dot{\mathcal{K}}_{\omega,0}^{p,q}(\mathbb{R}^n) = h\dot{\mathcal{K}}_{\omega,0}^{p,q,r,d,s}(\mathbb{R}^n)
$$

with equivalent quasi-norms.

To show the above atomic characterization, we require the known atomic characterization of the local Hardy space $h_X(\mathbb{R}^n)$ associated with ball quasi-Banach function space X. First, we recall the following definition of local-(X, r, d)-atoms, which is just [233, Definition 4.6].

Definition 5.3.4 Let X be a ball quasi-Banach function space, $r \in [1, \infty]$, and $d \in \mathbb{Z}_+$. Then a measurable function a is called a *local-(X, r, d)-atom* if

(i) there exists a ball $B(x_0, r_0) \in \mathbb{B}$, with $x_0 \in \mathbb{R}^n$ and $r_0 \in (0, \infty)$, such that
$\operatorname{supp}(a) := \{x \in \mathbb{R}^n : a(x) \neq 0\} \subset B(x_0, r_0)$;

(ii) $\|a\|_{L^r(\mathbb{R}^n)} \leq \frac{|B(x_0,r_0)|^{1/r}}{\|\mathbf{1}_{B(x_0,r_0)}\|_X}$;

(iii) when $r_0 \in (0, 1)$, then, for any $\alpha \in \mathbb{Z}_+^n$ satisfying $|\alpha| \leq d$,

$$
\int_{\mathbb{R}^n} a(x)x^\alpha \, dx = 0.
$$

Remark 5.3.5 Let X be a ball quasi-Banach function space, $r, t \in [1, \infty]$, and $d \in \mathbb{Z}_+$.

(i) Assume that a is a local-(X, t, d)-atom supported in a ball $B \in \mathbb{B}$ and $r \leq t$. Then it is obvious that a is also a local-(X, r, d)-atom supported in B.

(ii) Obviously, for any (X, r, d)-atom a supported in the ball $B \in \mathbb{B}$, a is also a local-(X, r, d)-atom supported in B.

Via local-(X, r, d)-atoms, we now present the following definition of local atomic Hardy spaces associated with ball quasi-Banach function spaces (see, for instance, [233]).

Definition 5.3.6 Let X be a ball quasi-Banach function space, $r \in (1, \infty]$, $d \in \mathbb{Z}_+$, and $s \in (0, 1]$. Then the *local atomic Hardy space* $h^{X,r,d,s}(\mathbb{R}^n)$, associated with X, is defined to be the set of all the $f \in \mathcal{S}'(\mathbb{R}^n)$ such that there exists a sequence $\{a_j\}_{j \in \mathbb{N}}$ of local-(X, r, d)-atoms supported, respectively, in the balls $\{B_j\}_{j \in \mathbb{N}} \subset \mathbb{B}$ and a sequence $\{\lambda_j\}_{j \in \mathbb{N}} \subset [0, \infty)$ satisfying that

$$f = \sum_{j \in \mathbb{N}} \lambda_j a_j$$

in $\mathcal{S}'(\mathbb{R}^n)$ and

$$\left\| \left[\sum_{j \in \mathbb{N}} \left(\frac{\lambda_j}{\|\mathbf{1}_{B_j}\|_X} \right)^s \mathbf{1}_{B_j} \right]^{\frac{1}{s}} \right\|_X < \infty.$$

Moreover, for any $f \in h^{X,r,d,s}(\mathbb{R}^n)$,

$$\|f\|_{h^{X,r,d,s}(\mathbb{R}^n)} := \inf \left\{ \left\| \left[\sum_{j \in \mathbb{N}} \left(\frac{\lambda_j}{\|\mathbf{1}_{B_j}\|_X} \right)^s \mathbf{1}_{B_j} \right]^{\frac{1}{s}} \right\|_X \right\},$$

where the infimum is taken over all the decompositions of f as above.

Then we state the following atomic characterization of the local Hardy space $h_X(\mathbb{R}^n)$, which was obtained in [233, Theorem 4.8] and plays an essential role in the proof of Theorem 5.3.3.

Lemma 5.3.7 *Let X be a ball quasi-Banach function space satisfying both Assumption 1.2.29 with $0 < \theta < s \le 1$ and Assumption 1.2.33 with the same s and $r \in (1, \infty]$, and let $d \ge \lfloor n(1/\theta - 1) \rfloor$ be a fixed integer. Then*

$$h_X(\mathbb{R}^n) = h^{X,r,d,s}(\mathbb{R}^n)$$

with equivalent quasi-norms.

Remark 5.3.8 We point out that Lemma 5.3.7 has a wide range of applications. Here we present several function spaces to which Lemma 5.3.7 can be applied.

(i) Let $\vec{p} := (p_1, \ldots, p_n) \in (0, \infty)^n$,

$$d \ge \left\lfloor n \left(\frac{1}{\min\{1, p_1, \ldots, p_n\}} - 1 \right) \right\rfloor$$

be a fixed integer, $r \in (\max\{1, p_1, \ldots, p_n\}, \infty]$, and $s \in (0, \min\{1, p_1, \ldots, p_n\})$. Then, in this case, combining both Remarks 1.2.31(iii) and 1.2.34(iii), we can easily find that the mixed-norm Lebesgue space $L^{\vec{p}}(\mathbb{R}^n)$ satisfies all the assumptions of Lemma 5.3.7. Thus, Lemma 5.3.7 with $X := L^{\vec{p}}(\mathbb{R}^n)$ holds true. To the best of our knowledge, this result is totally new.

(ii) Let $0 < q \leq p < \infty$,

$$d \geq \left\lfloor n\left(\frac{1}{\min\{1, q\}} - 1\right)\right\rfloor$$

be a fixed integer, $r \in (\max\{1, p\}, \infty]$, and $s \in (0, \min\{1, q\})$. Then, in this case, from both Remarks 1.2.31(iv) and 1.2.34(iv), we easily infer that the Morrey space $M_q^p(\mathbb{R}^n)$ satisfies all the assumptions of Lemma 5.3.7. This then implies that Lemma 5.3.7 with $X := M_q^p(\mathbb{R}^n)$ holds true. To the best of our knowledge, this result is totally new.

(iii) Let $p(\cdot) \in C^{\log}(\mathbb{R}^n)$ satisfy $0 < p_- \leq p_+ < \infty$, where p_- and p_+ are defined, respectively, in (1.59) and (1.60). Let

$$d \geq \left\lfloor n\left(\frac{1}{p_-} - 1\right)\right\rfloor$$

be a fixed nonnegative integer, $r \in (\max\{1, p_+\}, \infty]$, and $s := \min\{1, p_-\}$. Then, in this case, applying both Remarks 1.2.31(v) and 1.2.34(v), we can easily conclude that the variable Lebesgue space $L^{p(\cdot)}(\mathbb{R}^n)$ satisfies all the assumptions of Lemma 5.3.7. Therefore, Lemma 5.3.7 with $X := L^{p(\cdot)}(\mathbb{R}^n)$ holds true. This result coincides with [222, Corollary 4.1].

Via Lemma 5.3.7, we now show Theorem 5.3.3.

Proof of Theorem 5.3.3 Let p, q, ω, d, s, and r be as in the present theorem. Then, from the assumption $m_0(\omega) \in (-\frac{n}{p}, \infty)$ and Theorem 1.2.42, we deduce that the local generalized Herz space $\dot{\mathcal{K}}_{\omega,\mathbf{0}}^{p,q}(\mathbb{R}^n)$ under consideration is a BQBF space. This implies that, to complete the proof of the present theorem, we only need to prove that the Herz space $\dot{\mathcal{K}}_{\omega,\mathbf{0}}^{p,q}(\mathbb{R}^n)$ under consideration satisfies all the assumptions of Lemma 5.3.7.

First, let $\theta \in (0, s)$ satisfy

$$\left\lfloor n\left(\frac{1}{s} - 1\right)\right\rfloor \leq n\left(\frac{1}{\theta} - 1\right) < \left\lfloor n\left(\frac{1}{s} - 1\right)\right\rfloor + 1. \tag{5.15}$$

We now show that $\dot{\mathcal{K}}^{p,q}_{\omega,0}(\mathbb{R}^n)$ satisfies Assumption 1.2.29 with the above θ and s. Indeed, by Lemma 4.3.11, we conclude that, for any $\{f_j\}_{j\in\mathbb{N}} \subset L^1_{\mathrm{loc}}(\mathbb{R}^n)$,

$$\left\| \left\{ \sum_{j\in\mathbb{N}} \left[\mathcal{M}^{(\theta)}(f_j) \right]^s \right\}^{1/s} \right\|_{\dot{\mathcal{K}}^{p,q}_{\omega,0}(\mathbb{R}^n)} \lesssim \left\| \left(\sum_{j\in\mathbb{N}} |f_j|^s \right)^{1/s} \right\|_{\dot{\mathcal{K}}^{p,q}_{\omega,0}(\mathbb{R}^n)}, \tag{5.16}$$

which implies that Assumption 1.2.29 holds true for $\dot{\mathcal{K}}^{p,q}_{\omega,0}(\mathbb{R}^n)$ with the above θ and s.

Next, we prove that $\dot{\mathcal{K}}^{p,q}_{\omega,0}(\mathbb{R}^n)$ satisfies Assumption 1.2.33 with the above s and r. Indeed, from Lemma 1.8.6, we deduce that $[\dot{\mathcal{K}}^{p,q}_{\omega,0}(\mathbb{R}^n)]^{1/s}$ is a BBF space and, for any $f \in L^1_{\mathrm{loc}}(\mathbb{R}^n)$,

$$\left\| \mathcal{M}^{((r/s)')}(f) \right\|_{([\dot{\mathcal{K}}^{p,q}_{\omega,0}(\mathbb{R}^n)]^{1/s})'} \lesssim \|f\|_{([\dot{\mathcal{K}}^{p,q}_{\omega,0}(\mathbb{R}^n)]^{1/s})'}. \tag{5.17}$$

This implies that, for the above s and r, Assumption 1.2.33 holds true with $X := \dot{\mathcal{K}}^{p,q}_{\omega,0}(\mathbb{R}^n)$. Finally, using (5.15), we find that $d \geq \lfloor n(1/\theta - 1) \rfloor$. This, together with (5.16), the fact that $[\dot{\mathcal{K}}^{p,q}_{\omega,0}(\mathbb{R}^n)]^{1/s}$ is a BBF space, and (5.17), further implies that, under the assumptions of the present theorem, $\dot{\mathcal{K}}^{p,q}_{\omega,0}(\mathbb{R}^n)$ satisfies all the assumptions of Lemma 5.3.7. Therefore, we have

$$h\dot{\mathcal{K}}^{p,q}_{\omega,0}(\mathbb{R}^n) = h\dot{\mathcal{K}}^{p,q,r,d,s}_{\omega,0}(\mathbb{R}^n)$$

with equivalent quasi-norms, which completes the proof of Theorem 5.3.3. □

As an application, we now establish the atomic characterization of the local generalized Hardy–Morrey space $h M^{p,q}_{\omega,0}(\mathbb{R}^n)$. To achieve this, we first introduce the definition of the local atoms associated with the local generalized Morrey space $M^{p,q}_{\omega,0}(\mathbb{R}^n)$ as follows.

Definition 5.3.9 Let $p, q \in [1, \infty)$, $\omega \in M(\mathbb{R}_+)$ with $M_\infty(\omega) \in (-\infty, 0)$ and

$$-\frac{n}{p} < m_0(\omega) \leq M_0(\omega) < 0,$$

$r \in [1, \infty]$, and $d \in \mathbb{Z}_+$. Then a measurable function a on $\mathbb{R}^n$ is called a *local-$(M^{p,q}_{\omega,0}(\mathbb{R}^n), r, d)$-atom* if

(i) there exists a ball $B(x_0, r_0) \in \mathbb{B}$, with $x_0 \in \mathbb{R}^n$ and $r_0 \in (0, \infty)$, such that

$$\mathrm{supp}\,(a) := \{x \in \mathbb{R}^n : a(x) \neq 0\} \subset B(x_0, r_0);$$

(ii) $\|a\|_{L^r(\mathbb{R}^n)} \leq \dfrac{|B(x_0, r_0)|^{1/r}}{\|\mathbf{1}_{B(x_0, r_0)}\|_{M^{p,q}_{\omega,0}(\mathbb{R}^n)}};$

(iii) when $r_0 \in (0, 1)$, then, for any $\alpha \in \mathbb{Z}_+^n$ satisfying that $|\alpha| \le d$,

$$\int_{\mathbb{R}^n} a(x)x^\alpha \, dx = 0.$$

Now, we give the following atomic characterization of $h M_{\omega,0}^{p,q}(\mathbb{R}^n)$, which is a corollary of both Theorem 5.3.3 and Remark 1.2.2(vi); we omit the details.

Corollary 5.3.10 *Let* $p, q \in [1, \infty)$, $\omega \in M(\mathbb{R}_+)$ *with*

$$-\frac{n}{p} < m_0(\omega) \le M_0(\omega) < 0$$

and

$$-\frac{n}{p} < m_\infty(\omega) \le M_\infty(\omega) < 0,$$

$s \in (0, 1)$, $d \ge \lfloor n(1/s - 1)\rfloor$ *be a fixed integer, and*

$$r \in \left(\frac{n}{\min\{m_0(\omega), m_\infty(\omega)\} + n/p}, \infty\right].$$

Then the local generalized atomic Hardy–Morrey space $h M_{\omega,0}^{p,q,r,d,s}(\mathbb{R}^n)$, *associated with the local generalized Morrey space* $M_{\omega,0}^{p,q}(\mathbb{R}^n)$, *is defined to be the set of all the* $f \in \mathcal{S}'(\mathbb{R}^n)$ *such that there exists a sequence* $\{a_j\}_{j\in\mathbb{N}}$ *of local-*$(M_{\omega,0}^{p,q}(\mathbb{R}^n), r, d)$-*atoms supported, respectively, in the balls* $\{B_j\}_{j\in\mathbb{N}} \subset \mathbb{B}$ *and a sequence* $\{\lambda_j\}_{j\in\mathbb{N}} \subset [0, \infty)$ *satisfying that*

$$f = \sum_{j\in\mathbb{N}} \lambda_j a_j$$

in $\mathcal{S}'(\mathbb{R}^n)$ *and*

$$\left\| \left\{ \sum_{j\in\mathbb{N}} \left[\frac{\lambda_j}{\|\mathbf{1}_{B_j}\|_{M_{\omega,0}^{p,q}(\mathbb{R}^n)}} \right]^s \mathbf{1}_{B_j} \right\}^{\frac{1}{s}} \right\|_{M_{\omega,0}^{p,q}(\mathbb{R}^n)} < \infty.$$

Moreover, for any $f \in h M_{\omega,0}^{p,q,r,d,s}(\mathbb{R}^n)$,

$$\|f\|_{h M_{\omega,0}^{p,q,r,d,s}(\mathbb{R}^n)} := \inf \left\{ \left\| \left\{ \sum_{j\in\mathbb{N}} \left[\frac{\lambda_j}{\|\mathbf{1}_{B_j}\|_{M_{\omega,0}^{p,q}(\mathbb{R}^n)}} \right]^s \mathbf{1}_{B_j} \right\}^{\frac{1}{s}} \right\|_{M_{\omega,0}^{p,q}(\mathbb{R}^n)} \right\},$$

where the infimum is taken over all the decompositions of f as above. Then

$$hM^{p,q}_{\omega,0}(\mathbb{R}^n) = hM^{p,q,r,d,s}_{\omega,0}(\mathbb{R}^n)$$

with equivalent quasi-norms.

The remainder of this section is devoted to establishing the atomic characterization of the local generalized Herz–Hardy space $h\dot{\mathcal{K}}^{p,q}_{\omega}(\mathbb{R}^n)$. To this end, we first introduce the following local-$(\dot{\mathcal{K}}^{p,q}_{\omega}(\mathbb{R}^n)$, r, $d)$-atoms and the local atomic Hardy space associated with $\dot{\mathcal{K}}^{p,q}_{\omega}(\mathbb{R}^n)$.

Definition 5.3.11 Let $p, q \in (0, \infty)$, $\omega \in M(\mathbb{R}_+)$ with $m_0(\omega) \in (-\frac{n}{p}, \infty)$ and $M_\infty(\omega) \in (-\infty, 0)$, $r \in [1, \infty]$, and $d \in \mathbb{Z}_+$. Then a measurable function a is called a *local-$(\dot{\mathcal{K}}^{p,q}_{\omega}(\mathbb{R}^n)$, r, $d)$-atom* if

(i) there exists a ball $B(x_0, r_0) \in \mathbb{B}$, with $x_0 \in \mathbb{R}^n$ and $r_0 \in (0, \infty)$, such that

$$\mathrm{supp}\,(a) := \{x \in \mathbb{R}^n : a(x) \neq 0\} \subset B(x_0, r_0);$$

(ii) $\|a\|_{L^r(\mathbb{R}^n)} \leq \frac{|B(x_0,r_0)|^{1/r}}{\|\mathbf{1}_{B(x_0,r_0)}\|_{\dot{\mathcal{K}}^{p,q}_{\omega}(\mathbb{R}^n)}}$;

(iii) when $r_0 \in (0, 1)$, then, for any $\alpha \in \mathbb{Z}^n_+$ satisfying $|\alpha| \leq d$,

$$\int_{\mathbb{R}^n} a(x)x^\alpha \, dx = 0.$$

Definition 5.3.12 Let $p, q \in (0, \infty)$, $\omega \in M(\mathbb{R}_+)$ with $m_0(\omega) \in (-\frac{n}{p}, \infty)$ and

$$-\frac{n}{p} < m_\infty(\omega) \leq M_\infty(\omega) < 0,$$

$$s \in \left(0, \min\left\{1, p, q, \frac{n}{\max\{M_0(\omega), M_\infty(\omega)\} + n/p}\right\}\right),$$

$d \geq \lfloor n(1/s - 1) \rfloor$ be a fixed integer, and

$$r \in \left(\max\left\{1, p, \frac{n}{\min\{m_0(\omega), m_\infty(\omega)\} + n/p}\right\}, \infty\right].$$

Then the *local generalized atomic Herz–Hardy space* $h\dot{\mathcal{K}}^{p,q,r,d,s}_{\omega}(\mathbb{R}^n)$, associated with the global generalized Herz space $\dot{\mathcal{K}}^{p,q}_{\omega}(\mathbb{R}^n)$, is defined to be the set of all the $f \in \mathcal{S}'(\mathbb{R}^n)$ such that there exists a sequence $\{a_j\}_{j\in\mathbb{N}}$ of local-$(\dot{\mathcal{K}}^{p,q}_{\omega}(\mathbb{R}^n)$, r, $d)$-atoms supported, respectively, in the balls $\{B_j\}_{j\in\mathbb{N}} \subset \mathbb{B}$ and a sequence $\{\lambda_j\}_{j\in\mathbb{N}} \subset [0, \infty)$ satisfying that

$$f = \sum_{j\in\mathbb{N}} \lambda_j a_j$$

in $\mathcal{S}'(\mathbb{R}^n)$ and

$$\left\| \left\{ \sum_{j \in \mathbb{N}} \left[\frac{\lambda_j}{\|\mathbf{1}_{B_j}\|_{\dot{\mathcal{K}}_\omega^{p,q}(\mathbb{R}^n)}} \right]^s \mathbf{1}_{B_j} \right\}^{\frac{1}{s}} \right\|_{\dot{\mathcal{K}}_\omega^{p,q}(\mathbb{R}^n)} < \infty.$$

Moreover, for any $f \in h\dot{\mathcal{K}}_\omega^{p,q,r,d,s}(\mathbb{R}^n)$,

$$\|f\|_{h\dot{\mathcal{K}}_\omega^{p,q,r,d,s}(\mathbb{R}^n)} := \inf \left\{ \left\| \left\{ \sum_{j \in \mathbb{N}} \left[\frac{\lambda_j}{\|\mathbf{1}_{B_j}\|_{\dot{\mathcal{K}}_\omega^{p,q}(\mathbb{R}^n)}} \right]^s \mathbf{1}_{B_j} \right\}^{\frac{1}{s}} \right\|_{\dot{\mathcal{K}}_\omega^{p,q}(\mathbb{R}^n)} \right\},$$

where the infimum is taken over all the decompositions of f as above.

Then we have the following atomic characterization of the local generalized Herz–Hardy space $h\dot{\mathcal{K}}_\omega^{p,q}(\mathbb{R}^n)$.

Theorem 5.3.13 *Let p, q, ω, d, s, and r be as in Definition 5.3.12. Then*

$$h\dot{\mathcal{K}}_\omega^{p,q}(\mathbb{R}^n) = h\dot{\mathcal{K}}_\omega^{p,q,r,d,s}(\mathbb{R}^n)$$

with equivalent quasi-norms.

To prove this atomic characterization, we first establish an atomic characterization of the local Hardy space $h_X(\mathbb{R}^n)$ via borrowing some ideas from [233, Theorem 4.8] and get rid of the usage of associate spaces. Namely, we have the following conclusion.

Theorem 5.3.14 *Let X be a ball quasi-Banach function space satisfy:*

(i) *Assumption 1.2.29 holds true with $0 < \theta < s \le 1$;*
(ii) *for the above s, $X^{1/s}$ is a ball Banach function space and there exists a linear space $Y \subset \mathcal{M}(\mathbb{R}^n)$ equipped with a seminorm $\|\cdot\|_Y$ such that, for any $f \in \mathcal{M}(\mathbb{R}^n)$,*

$$\|f\|_{X^{1/s}} \sim \sup \left\{ \|fg\|_{L^1(\mathbb{R}^n)} : \|g\|_Y = 1 \right\}, \tag{5.18}$$

where the positive equivalence constants are independent of f;
(iii) *for the above s and Y, there exists an $r \in (1, \infty]$ and a positive constant C such that, for any $f \in L^1_{\mathrm{loc}}(\mathbb{R}^n)$,*

$$\left\| \mathcal{M}^{((r/s)')}(f) \right\|_Y \le C \|f\|_Y.$$

Then

$$h_X(\mathbb{R}^n) = h^{X,r,d,s}(\mathbb{R}^n)$$

with equivalent quasi-norms.

Remark 5.3.15 We should point out that Theorem 5.3.14 is an improved version of the known atomic characterization of $h_X(\mathbb{R}^n)$ obtained in [233, Theorem 4.8]. Indeed, if $Y \equiv (X^{1/s})'$ in Theorem 5.3.14, then this theorem goes back to [233, Theorem 4.8].

To show Theorem 5.3.14, we require the following atomic decomposition of the local Hardy space $h_X(\mathbb{R}^n)$, which was obtained in [233, pp. 37–39].

Lemma 5.3.16 *Let X be a ball quasi-Banach function space satisfying Assumption 1.2.29 with $0 < \theta < s \le 1$, $d \ge \lfloor n(1/\theta - 1) \rfloor$ be a fixed integer, and $f \in h_X(\mathbb{R}^n)$. Then there exists $\{\lambda_j\}_{j\in\mathbb{N}} \subset [0,\infty)$ and $\{a_j\}_{j\in\mathbb{N}}$ of local-(X, ∞, d)-atoms supported, respectively, in the balls $\{B_j\}_{j\in\mathbb{N}} \subset \mathbb{B}$ satisfying that*

$$f = \sum_{j\in\mathbb{N}} \lambda_j a_j$$

in $\mathcal{S}'(\mathbb{R}^n)$ and

$$\left\| \left[\sum_{j\in\mathbb{N}} \left(\frac{\lambda_j}{\|\mathbf{1}_{B_j}\|_X} \right)^s \mathbf{1}_{B_j} \right]^{\frac{1}{s}} \right\|_X \lesssim \|f\|_{h_X(\mathbb{R}^n)},$$

where the implicit positive constant is independent of f.

Now, we turn to prove Theorem 5.3.14.

Proof of Theorem 5.3.14 Let X, r, d, and s be as in the present theorem. We first prove that $h_X(\mathbb{R}^n) \subset h^{X,r,d,s}(\mathbb{R}^n)$. For this purpose, let $f \in h_X(\mathbb{R}^n)$. Then, applying the assumption (i) of the present theorem and Lemma 5.3.16, we find that there exists $\{\lambda_j\}_{j\in\mathbb{N}} \subset [0,\infty)$ and $\{a_j\}_{j\in\mathbb{N}}$ of local-(X, ∞, d)-atoms supported, respectively, in the balls $\{B_j\}_{j\in\mathbb{N}} \subset \mathbb{B}$ satisfying that

$$f = \sum_{j\in\mathbb{N}} \lambda_j a_j \tag{5.19}$$

in $\mathcal{S}'(\mathbb{R}^n)$ and

$$\left\| \left[\sum_{j\in\mathbb{N}} \left(\frac{\lambda_j}{\|\mathbf{1}_{B_j}\|_X} \right)^s \mathbf{1}_{B_j} \right]^{\frac{1}{s}} \right\|_X \lesssim \|f\|_{h_X(\mathbb{R}^n)}. \tag{5.20}$$

In addition, for any $j \in \mathbb{N}$, from Remark 5.3.5(i) with $a := a_j$ and $t := \infty$, it follows that a_j is a local-(X, r, d)-atom supported in the ball B_j. This, combined with (5.19), (5.20), and Definition 5.3.6, further implies that $f \in h^{X,r,d,s}(\mathbb{R}^n)$ and

$$\|f\|_{h^{X,r,d,s}(\mathbb{R}^n)} \leq \left\| \left[\sum_{j\in\mathbb{N}} \left(\frac{\lambda_j}{\|\mathbf{1}_{B_j}\|_X} \right)^s \mathbf{1}_{B_j} \right]^{\frac{1}{s}} \right\|_X \lesssim \|f\|_{h_X(\mathbb{R}^n)}, \tag{5.21}$$

which completes the proof that $h_X(\mathbb{R}^n) \subset h^{X,r,d,s}(\mathbb{R}^n)$.

Conversely, we next show that $h^{X,r,d,s}(\mathbb{R}^n) \subset h_X(\mathbb{R}^n)$. Indeed, let $f \in h^{X,r,d,s}(\mathbb{R}^n)$, $\{\lambda_{l,j}\}_{l\in\{1,2\}, j\in\mathbb{N}} \subset [0, \infty)$, and $\{a_{l,j}\}_{l\in\{1,2\}, j\in\mathbb{N}}$ be a sequence of local-(X, r, d)-atoms supported, respectively, in the balls $\{B_{l,j}\}_{l\in\{1,2\}, j\in\mathbb{N}} \subset \mathbb{B}$ satisfying that, for any $j \in \mathbb{N}$, $r(B_{1,j}) \in (0, 1)$ and $r(B_{2,j}) \in [1, \infty)$,

$$f = \sum_{l=1}^{2} \sum_{j\in\mathbb{N}} \lambda_{l,j} a_{l,j} \tag{5.22}$$

in $\mathcal{S}'(\mathbb{R}^n)$, and

$$\left\| \left[\sum_{l=1}^{2} \sum_{j\in\mathbb{N}} \left(\frac{\lambda_{l,j}}{\|\mathbf{1}_{B_{l,j}}\|_X} \right)^s \mathbf{1}_{B_{l,j}} \right]^{\frac{1}{s}} \right\|_X < \infty. \tag{5.23}$$

In what follows, let $\phi \in \mathcal{S}(\mathbb{R}^n)$ be such that $\mathrm{supp}\,(\phi) \subset B(\mathbf{0}, 1)$ and $\int_{\mathbb{R}^n} \phi(x)\, dx \neq 0$. Then, using (5.22) and repeating an argument similar to that used in the estimation of (4.24) with $\{\lambda_j\}_{j\in\mathbb{N}}$ and $\{a_j\}_{j\in\mathbb{N}}$ therein replaced, respectively, by $\{\lambda_{l,j}\}_{l\in\{1,2\}, j\in\mathbb{N}}$ and $\{a_{l,j}\}_{l\in\{1,2\}, j\in\mathbb{N}}$, we conclude that, for any $t \in (0, \infty)$,

$$|f * \phi_t| \leq \sum_{l=1}^{2} \sum_{j\in\mathbb{N}} \lambda_{l,j} \left| a_{l,j} * \phi_t \right|.$$

This, together with Definition 5.1.1(i), further implies that

$$m(f, \phi) \leq \sum_{l=1}^{2} \sum_{j\in\mathbb{N}} \lambda_{l,j} m(a_{l,j}, \phi).$$

Using this and Definition 1.2.13(ii), we find that

$$\|m(f,\phi)\|_X \lesssim \left\|\sum_{j\in\mathbb{N}}\lambda_{1,j}m(a_{1,j},\phi)\right\|_X + \left\|\sum_{j\in\mathbb{N}}\lambda_{2,j}m(a_{2,j},\phi)\right\|_X$$
$$=: \mathrm{IV}_1 + \mathrm{IV}_2. \tag{5.24}$$

We first estimate IV_1. Indeed, for any given $j \in \mathbb{N}$, by the assumption $r(B_{1,j}) \in (0,1)$ and both Definitions 5.3.4 and 4.3.1, we conclude that a_j is an (X, r, d)-atom supported in B_j. Thus, using both Definitions 5.1.1(i) and 4.1.1(i) and repeating the arguments similar to those used in the estimations of (4.26), (4.28), and (4.30) with $\{\lambda_j\}_{j\in\mathbb{N}}$, $\{a_j\}_{j\in\mathbb{N}}$, and $\{B_j\}_{j\in\mathbb{N}}$ therein replaced, respectively, by $\{\lambda_{1,j}\}_{j\in\mathbb{N}}$, $\{a_{1,j}\}_{j\in\mathbb{N}}$, and $\{B_{1,j}\}_{j\in\mathbb{N}}$, we find that

$$\mathrm{IV}_1 \lesssim \left\|\sum_{j\in\mathbb{N}}\lambda_{1,j}M(a_{1,j},\phi)\right\|_X \lesssim \left\|\left[\sum_{j\in\mathbb{N}}\left(\frac{\lambda_{1,j}}{\|\mathbf{1}_{B_{1,j}}\|_X}\right)^s\mathbf{1}_{B_{1,j}}\right]^{\frac{1}{s}}\right\|_X. \tag{5.25}$$

This is the desired estimate of IV_1.

On the other hand, we deal with IV_2. To this end, we first claim that, for any given $j \in \mathbb{N}$ and for any $x \in (2B_{2,j})^\complement$, $m(a_{2,j},\phi)(x) = 0$. Indeed, fix a $j \in \mathbb{N}$. Then, applying the assumption $r(B_{2,j}) \in [1,\infty)$, we find that, for any $t \in (0,1)$, $x \in (2B_{2,j})^\complement$, and $y \in B_{2,j}$,

$$|x - y| \geq r(B_{2,j}) \geq 1 > t,$$

which implies that $y \in [B(x,t)]^\complement$. This, together with the assumption $\mathrm{supp}\,(\phi) \subset B(\mathbf{0},1)$, further implies that, for any $t \in (0,1)$ and $x \in (2B_{2,j})^\complement$,

$$a_{2,j} * \phi_t(x) = \int_{\mathbb{R}^n}\phi_t(x-y)a_{2,j}(y)\,dy$$
$$= \int_{B_{2,j}\cap B(x,t)}\phi_t(x-y)a_{2,j}(y)\,dy = 0.$$

Therefore, by Definition 5.1.1(i), we further conclude that, for any given $j \in \mathbb{N}$ and for any $x \in (2B_{2,j})^\complement$, $m(a_{2,j},\phi)(x) = 0$, which completes the proof of the above claim. Combining this claim and both Definitions 5.1.1(i) and 4.1.1(i), we obtain

$$\mathrm{IV}_2 \sim \left\|\sum_{j\in\mathbb{N}}\lambda_{2,j}m(a_{2,j},\phi)\mathbf{1}_{2B_{2,j}}\right\|_X \lesssim \left\|\sum_{j\in\mathbb{N}}\lambda_{2,j}M(a_{2,j},\phi)\mathbf{1}_{2B_{2,j}}\right\|_X.$$

From this and an argument similar to that used in the estimation of II_1 in the proof of Theorem 4.3.18 with $\{\lambda_j\}_{j\in\mathbb{N}}$, $\{a_j\}_{j\in\mathbb{N}}$, and $\{B_j\}_{j\in\mathbb{N}}$ therein replaced, respectively, by $\{\lambda_{2,j}\}_{j\in\mathbb{N}}$, $\{a_{2,j}\}_{j\in\mathbb{N}}$, and $\{B_{2,j}\}_{j\in\mathbb{N}}$, it follows that

$$\mathrm{IV}_2 \lesssim \left\| \left[\sum_{j\in\mathbb{N}} \left(\frac{\lambda_{2,j}}{\|\mathbf{1}_{B_{2,j}}\|_X} \right)^s \mathbf{1}_{B_{2,j}} \right]^{\frac{1}{s}} \right\|_X, \tag{5.26}$$

which is the desired estimate of IV_2. Thus, by both the assumptions (i) and (ii) of the present theorem, Remark 5.1.6, Lemma 5.1.5(ii), (5.24), (5.25), (5.26), and (5.23), we find that

$$\|f\|_{h_X(\mathbb{R}^n)} \sim \|m(f,\phi)\|_X \lesssim \left\| \left[\sum_{l=1}^{2} \sum_{j\in\mathbb{N}} \left(\frac{\lambda_{l,j}}{\|\mathbf{1}_{B_{l,j}}\|_X} \right)^s \mathbf{1}_{B_{l,j}} \right]^{\frac{1}{s}} \right\|_X < \infty. \tag{5.27}$$

This further implies that $f \in h_X(\mathbb{R}^n)$ and hence $h_X(\mathbb{R}^n) \subset h^{X,r,d,s}(\mathbb{R}^n)$. Moreover, from (5.27), the choice of $\{\lambda_{l,j}\}_{l\in\{1,2\},j\in\mathbb{N}}$, and Definition 5.3.6, we deduce that

$$\|f\|_{h_X(\mathbb{R}^n)} \lesssim \|f\|_{h^{X,r,d,s}(\mathbb{R}^n)},$$

which, combined with (5.21), implies that

$$\|f\|_{h_X(\mathbb{R}^n)} \sim \|f\|_{h^{X,r,d,s}(\mathbb{R}^n)}.$$

Thus, we have

$$h_X(\mathbb{R}^n) = h^{X,r,d,s}(\mathbb{R}^n)$$

with equivalent quasi-norms, which completes the proof of Theorem 5.3.14. $\square$

Via the above atomic characterization of local Hardy spaces associated with ball quasi-Banach function spaces, we now prove Theorem 5.3.13.

Proof of Theorem 5.3.13 Let p, q, ω, r, s, and d be as in the present theorem. Then, combining the assumptions $m_0(\omega) \in (-\frac{n}{p}, \infty)$ and $M_\infty(\omega) \in (-\infty, 0)$, and Theorem 1.2.44, we conclude that the global generalized Herz space $\dot{\mathcal{K}}_\omega^{p,q}(\mathbb{R}^n)$ under consideration is a BQBF space. From this and Theorem 5.3.14, it follows that, to finish the proof of the present theorem, we only need to show that the assumptions (i) through (iii) of Theorem 5.3.14 hold true for $\dot{\mathcal{K}}_\omega^{p,q}(\mathbb{R}^n)$.

First, we show that Theorem 5.3.14(i) holds true for $\dot{\mathcal{K}}_\omega^{p,q}(\mathbb{R}^n)$. To this end, let $\theta \in (0, s)$ be such that

$$\left\lfloor n\left(\frac{1}{s}-1\right) \right\rfloor \le n\left(\frac{1}{\theta}-1\right) < \left\lfloor n\left(\frac{1}{s}-1\right) \right\rfloor + 1. \tag{5.28}$$

Then, applying Lemma 4.3.25, we find that, for any $\{f_j\}_{j\in\mathbb{N}} \subset L^1_{\mathrm{loc}}(\mathbb{R}^n)$,

$$\left\| \left\{ \sum_{j\in\mathbb{N}} \left[\mathcal{M}^{(\theta)}(f_j) \right]^s \right\}^{1/s} \right\|_{\dot{\mathcal{K}}^{p,q}_{\omega}(\mathbb{R}^n)} \lesssim \left\| \left(\sum_{j\in\mathbb{N}} |f_j|^s \right)^{1/s} \right\|_{\dot{\mathcal{K}}^{p,q}_{\omega}(\mathbb{R}^n)}.$$

This implies that, for the above θ and s, $\dot{\mathcal{K}}^{p,q}_{\omega}(\mathbb{R}^n)$ satisfies Assumption 1.2.29 and hence Theorem 5.3.14(i) holds true.

Next, we prove that $\dot{\mathcal{K}}^{p,q}_{\omega}(\mathbb{R}^n)$ satisfies Theorem 5.3.14(ii). Indeed, from the assumptions $m_0(\omega) \in (-\frac{n}{p}, \infty)$ and $M_\infty(\omega) \in (-\infty, 0)$, and Lemma 1.1.6, it follows that

$$m_0\left(\omega^s\right) = s m_0(\omega) > -\frac{n}{p/s}$$

and

$$M_\infty\left(\omega^s\right) = s M_\infty(\omega) < 0.$$

Applying these, the assumptions $p/s, q/s \in (1, \infty)$, and Theorem 1.2.48 with p, q, and ω replaced, respectively, by p/s, q/s, and ω^s, we find that $\dot{\mathcal{K}}^{p/s,q/s}_{\omega^s}(\mathbb{R}^n)$ is a BBF space. Moreover, by Lemma 1.3.2, we conclude that

$$\left[\dot{\mathcal{K}}^{p,q}_{\omega}(\mathbb{R}^n) \right]^{1/s} = \dot{\mathcal{K}}^{p/s,q/s}_{\omega^s}(\mathbb{R}^n).$$

Thus, $[\dot{\mathcal{K}}^{p,q}_{\omega}(\mathbb{R}^n)]^{1/s}$ is a BBF space. On the other hand, from (4.42), it follows that, for any $f \in \mathscr{M}(\mathbb{R}^n)$,

$$\|f\|_{[\dot{\mathcal{K}}^{p,q}_{\omega}(\mathbb{R}^n)]^{1/s}} \sim \sup \left\{ \|fg\|_{L^1(\mathbb{R}^n)} : \|g\|_{\dot{\mathcal{B}}^{(p/s)',(q/s)'}_{1/\omega^s}(\mathbb{R}^n)} = 1 \right\}.$$

Therefore, $[\dot{\mathcal{K}}^{p,q}_{\omega}(\mathbb{R}^n)]^{1/s}$ is a BBF space and (5.18) holds true with

$$Y := \dot{\mathcal{B}}^{(p/s)',(q/s)'}_{1/\omega^s}(\mathbb{R}^n).$$

These further imply that the Herz space $\dot{\mathcal{K}}^{p,q}_{\omega}(\mathbb{R}^n)$ under consideration satisfies Theorem 5.3.14(ii).

Finally, we show that Theorem 5.3.14(iii) holds true for $\dot{\mathcal{K}}^{p,q}_{\omega}(\mathbb{R}^n)$. Indeed, using Lemma 4.3.23, we conclude that, for any $f \in L^1_{\mathrm{loc}}(\mathbb{R}^n)$,

$$\left\| \mathcal{M}^{((r/s)')}(f) \right\|_{\dot{\mathcal{B}}^{(p/s)',(q/s)'}_{1/\omega^s}(\mathbb{R}^n)} \lesssim \|f\|_{\dot{\mathcal{B}}^{(p/s)',(q/s)'}_{1/\omega^s}(\mathbb{R}^n)},$$

which implies that Theorem 5.3.14(iii) holds true for $\dot{\mathcal{K}}_{\omega}^{p,q}(\mathbb{R}^n)$ with

$$Y := \dot{\mathcal{B}}_{1/\omega^s}^{(p/s)',(q/s)'}(\mathbb{R}^n).$$

Moreover, by (5.28), we find that $d \geq \lfloor n(1/\theta - 1) \rfloor$. This, together with the fact that the assumptions (i) through (iii) of Theorem 5.3.14 hold true for the Herz space $\dot{\mathcal{K}}_{\omega}^{p,q}(\mathbb{R}^n)$ under consideration, further implies that

$$h\dot{\mathcal{K}}_{\omega}^{p,q}(\mathbb{R}^n) = h\dot{\mathcal{K}}_{\omega}^{p,q,r,d,s}(\mathbb{R}^n)$$

with equivalent quasi-norms, which completes the proof of Theorem 5.3.13. □

As an application of Theorem 5.3.13, we now establish the atomic characterization of the local generalized Hardy–Morrey space $hM_{\omega}^{p,q}(\mathbb{R}^n)$. To this end, we first introduce the definition of local-$(M_{\omega}^{p,q}(\mathbb{R}^n), r, d)$-atoms as follows.

Definition 5.3.17 Let p, q, ω, r, and d be as in Definition 5.3.9. Then a measurable function a on $\mathbb{R}^n$ is called a *local-$(M_{\omega}^{p,q}(\mathbb{R}^n), r, d)$-atom* if

(i) there exists a ball $B(x_0, r_0) \in \mathbb{B}$, with $x_0 \in \mathbb{R}^n$ and $r_0 \in (0, \infty)$, such that

$$\operatorname{supp}(a) := \{x \in \mathbb{R}^n : a(x) \neq 0\} \subset B(x_0, r_0);$$

(ii) $\|a\|_{L^r(\mathbb{R}^n)} \leq \dfrac{|B(x_0, r_0)|^{1/r}}{\|\mathbf{1}_{B(x_0,r_0)}\|_{M_{\omega}^{p,q}(\mathbb{R}^n)}}$;

(iii) when $r_0 \in (0, 1)$, then, for any $\alpha \in \mathbb{Z}_+^n$ satisfying $|\alpha| \leq d$,

$$\int_{\mathbb{R}^n} a(x)x^\alpha \, dx = 0.$$

Then we have the following atomic characterization of $hM_{\omega}^{p,q}(\mathbb{R}^n)$, which can be deduced from Theorem 5.3.13 and Remark 1.2.2(vi) immediately; we omit the details.

Corollary 5.3.18 *Let p, q, ω, r, d, and s be as in Corollary 5.3.10. Then the local generalized atomic Hardy–Morrey space $hM_{\omega}^{p,q,r,d,s}(\mathbb{R}^n)$, associated with the global generalized Morrey space $M_{\omega}^{p,q}(\mathbb{R}^n)$, is defined to be the set of all the $f \in \mathcal{S}'(\mathbb{R}^n)$ such that there exists a sequence $\{a_j\}_{j\in\mathbb{N}}$ of local-$(M_{\omega}^{p,q}(\mathbb{R}^n), r, d)$-atoms supported, respectively, in the balls $\{B_j\}_{j\in\mathbb{N}} \subset \mathbb{B}$ and a sequence $\{\lambda_j\}_{j\in\mathbb{N}} \subset [0, \infty)$ satisfying that $f = \sum_{j\in\mathbb{N}} \lambda_j a_j$ in $\mathcal{S}'(\mathbb{R}^n)$ and*

$$\left\| \left\{ \sum_{j\in\mathbb{N}} \left[\frac{\lambda_j}{\|\mathbf{1}_{B_j}\|_{M_{\omega}^{p,q}(\mathbb{R}^n)}} \right]^s \mathbf{1}_{B_j} \right\}^{\frac{1}{s}} \right\|_{M_{\omega}^{p,q}(\mathbb{R}^n)} < \infty.$$

Moreover, for any $f \in hM_\omega^{p,q,r,d,s}(\mathbb{R}^n)$,

$$\|f\|_{hM_\omega^{p,q,r,d,s}(\mathbb{R}^n)} := \inf \left\{ \left\| \left\{ \sum_{j \in \mathbb{N}} \left[\frac{\lambda_j}{\|\mathbf{1}_{B_j}\|_{M_\omega^{p,q}(\mathbb{R}^n)}} \right]^s \mathbf{1}_{B_j} \right\}^{\frac{1}{s}} \right\|_{M_\omega^{p,q}(\mathbb{R}^n)} \right\},$$

where the infimum is taken over all the decompositions of f *as above. Then*

$$hM_\omega^{p,q}(\mathbb{R}^n) = hM_\omega^{p,q,r,d,s}(\mathbb{R}^n)$$

with equivalent quasi-norms.

5.4 Molecular Characterizations

In this section, we investigate the molecular characterization of localized general-ized Herz–Hardy spaces via viewing generalized Herz spaces as special cases of ball quasi-Banach function spaces. Precisely, we establish the molecular character-ization of the local generalized Herz–Hardy space $h\dot{\mathcal{K}}_{\omega,\mathbf{0}}^{p,q}(\mathbb{R}^n)$ via using the known molecular characterization of localized Hardy spaces associated with ball quasi-Banach function spaces obtained in [233, Theorem 5.2] (see also Lemma 5.4.5 below). On the other hand, to prove the molecular characterization of the local generalized Herz–Hardy space $h\dot{\mathcal{K}}_\omega^{p,q}(\mathbb{R}^n)$, recall that the associate spaces of the global generalized Herz spaces are still unknown. To overcome this obstacle, we first establish an improved molecular characterization of localized Hardy spaces associated with ball quasi-Banach function spaces (see Theorem 5.4.11 below) without recourse to associate spaces. Combining this molecular characterization and the fact that the global generalized Herz space $\dot{\mathcal{K}}_\omega^{p,q}(\mathbb{R}^n)$ is a special ball quasi-Banach function space, we then obtain the desired molecular characterization of $h\dot{\mathcal{K}}_\omega^{p,q}(\mathbb{R}^n)$.

We first investigate the molecular characterization of $h\dot{\mathcal{K}}_{\omega,\mathbf{0}}^{p,q}(\mathbb{R}^n)$. To begin with, we introduce the local molecules associated with the local generalized Herz space $\dot{\mathcal{K}}_{\omega,\mathbf{0}}^{p,q}(\mathbb{R}^n)$ as follows.

Definition 5.4.1 Let $p, q \in (0, \infty)$, $\omega \in M(\mathbb{R}_+)$ with $m_0(\omega) \in (-\frac{n}{p}, \infty)$, $r \in [1, \infty]$, $d \in \mathbb{Z}_+$, and $\tau \in (0, \infty)$. Then a measurable function m on $\mathbb{R}^n$ is called a *local-*$(\dot{\mathcal{K}}_{\omega,\mathbf{0}}^{p,q}(\mathbb{R}^n), r, d, \tau)$-*molecule* centered at a ball $B(x_0, r_0) \in \mathbb{B}$, with $x_0 \in \mathbb{R}^n$ and $r_0 \in (0, \infty)$, if

(i) for any $i \in \mathbb{Z}_+$,

$$\left\| m\mathbf{1}_{S_i(B(x_0,r_0))} \right\|_{L^r(\mathbb{R}^n)} \leq 2^{-\tau i} \frac{|B(x_0, r_0)|^{1/r}}{\|\mathbf{1}_{B(x_0,r_0)}\|_{\dot{\mathcal{K}}_{\omega,\mathbf{0}}^{p,q}(\mathbb{R}^n)}};$$

(ii) when $r_0 \in (0, 1)$, then, for any $\alpha \in \mathbb{Z}_+^n$ with $|\alpha| \le d$,

$$\int_{\mathbb{R}^n} m(x)x^\alpha \, dx = 0.$$

Then we establish the following molecular characterization of the local generalized Herz–Hardy space $h\dot{\mathcal{K}}_{\omega,\mathbf{0}}^{p,q}(\mathbb{R}^n)$.

Theorem 5.4.2 *Let* $p, q \in (0, \infty)$, $\omega \in M(\mathbb{R}_+)$ *with* $m_0(\omega) \in (-\frac{n}{p}, \infty)$ *and* $m_\infty(\omega) \in (-\frac{n}{p}, \infty)$,

$$s \in \left(0, \min\left\{1, p, q, \frac{n}{\max\{M_0(\omega), M_\infty(\omega)\} + n/p}\right\}\right),$$

$d \ge \lfloor n(1/s - 1) \rfloor$ *be a fixed integer,*

$$r \in \left(\max\left\{1, p, \frac{n}{\min\{m_0(\omega), m_\infty(\omega)\} + n/p}\right\}, \infty\right],$$

and $\tau \in (0, \infty)$ *with* $\tau > n(1/s - 1/r)$. *Then* $f \in h\dot{\mathcal{K}}_{\omega,\mathbf{0}}^{p,q}(\mathbb{R}^n)$ *if and only if* $f \in \mathcal{S}'(\mathbb{R}^n)$ *and there exists a sequence* $\{m_j\}_{j\in\mathbb{N}}$ *of local-$(\dot{\mathcal{K}}_{\omega,\mathbf{0}}^{p,q}(\mathbb{R}^n)$, r, d, $\tau)$-molecules centered, respectively, at the balls* $\{B_j\}_{j\in\mathbb{N}} \subset \mathbb{B}$ *and a sequence* $\{\lambda_j\}_{j\in\mathbb{N}} \subset [0, \infty)$ *such that*

$$f = \sum_{j\in\mathbb{N}} \lambda_j m_j$$

in $\mathcal{S}'(\mathbb{R}^n)$ *and*

$$\left\| \left\{ \sum_{j\in\mathbb{N}} \left[\frac{\lambda_j}{\|\mathbf{1}_{B_j}\|_{\dot{\mathcal{K}}_{\omega,\mathbf{0}}^{p,q}(\mathbb{R}^n)}} \right]^s \mathbf{1}_{B_j} \right\}^{\frac{1}{s}} \right\|_{\dot{\mathcal{K}}_{\omega,\mathbf{0}}^{p,q}(\mathbb{R}^n)} < \infty.$$

Moreover, there exist two positive constants C_1 *and* C_2 *such that, for any* $f \in h\dot{\mathcal{K}}_{\omega,\mathbf{0}}^{p,q}(\mathbb{R}^n)$,

$$C_1\|f\|_{h\dot{\mathcal{K}}_{\omega,\mathbf{0}}^{p,q}(\mathbb{R}^n)} \le \inf\left\{ \left\| \left\{ \sum_{j\in\mathbb{N}} \left[\frac{\lambda_i}{\|\mathbf{1}_{B_j}\|_{\dot{\mathcal{K}}_{\omega,\mathbf{0}}^{p,q}(\mathbb{R}^n)}} \right]^s \mathbf{1}_{B_j} \right\}^{\frac{1}{s}} \right\|_{\dot{\mathcal{K}}_{\omega,\mathbf{0}}^{p,q}(\mathbb{R}^n)} \right\}$$

$$\le C_2\|f\|_{h\dot{\mathcal{K}}_{\omega,\mathbf{0}}^{p,q}(\mathbb{R}^n)},$$

where the infimum is taken over all the decompositions of f *as above.*

To obtain this molecular characterization, we first recall the following definition of local-(X, r, d, τ)-molecules with X being a ball quasi-Banach function space, which is just [233, Definition 5.1].

Definition 5.4.3 Let X be a ball quasi-Banach function space, $r \in [1, \infty]$, $d \in \mathbb{Z}_+$, and $\tau \in (0, \infty)$. Then a measurable function m on $\mathbb{R}^n$ is called a *local-(X, r, d, τ)-molecule* centered at a ball $B(x_0, r_0) \in \mathbb{B}$, with $x_0 \in \mathbb{R}^n$ and $r_0 \in (0, \infty)$, if

(i) for any $i \in \mathbb{Z}_+$,

$$\left\| m \mathbf{1}_{S_i(B(x_0, r_0))} \right\|_{L^r(\mathbb{R}^n)} \leq 2^{-\tau i} \frac{|B(x_0, r_0)|^{1/r}}{\|\mathbf{1}_{B(x_0, r_0)}\|_X};$$

(ii) when $r_0 \in (0, 1)$, then, for any $\alpha \in \mathbb{Z}_+^n$ with $|\alpha| \leq d$,

$$\int_{\mathbb{R}^n} m(x) x^\alpha \, dx = 0.$$

Remark 5.4.4 Let X be a ball quasi-Banach function space, $r \in [1, \infty]$, $d \in \mathbb{Z}_+$, and $\tau \in (0, \infty)$. Then it is obvious that

(i) for any local-(X, r, d)-atom a supported in $B \in \mathbb{B}$, a is a local-(X, r, d, τ)-molecule centered at B;

(ii) for any (X, r, d, τ)-molecule m centered at a ball $B \in \mathbb{B}$, m is a local-(X, r, d, τ)-molecule centered at B.

The following molecular characterization of $h_X(\mathbb{R}^n)$ was established in [233, Theorem 5.2], which is an essential tool in the proof of Theorem 5.4.2.

Lemma 5.4.5 *Let X be a ball quasi-Banach function space satisfying both Assumption 1.2.29 with $0 < \theta < s \leq 1$ and Assumption 1.2.33 with the same s and $r \in (1, \infty]$, $\tau \in (n(1/\theta - 1/r), \infty)$, and let $d \geq \lfloor n(1/\theta - 1) \rfloor$ be a fixed integer. Then $f \in h_X(\mathbb{R}^n)$ if and only if $f \in \mathcal{S}'(\mathbb{R}^n)$ and there exists a sequence $\{m_j\}_{j \in \mathbb{N}}$ of local-(X, r, d, τ)-molecules centered, respectively, at the balls $\{B_j\}_{j \in \mathbb{N}} \subset \mathbb{B}$ and a sequence $\{\lambda_j\}_{j \in \mathbb{N}} \subset [0, \infty)$ such that both $f = \sum_{j \in \mathbb{N}} \lambda_j m_j$ in $\mathcal{S}'(\mathbb{R}^n)$ and*

$$\left\| \left[\sum_{j \in \mathbb{N}} \left(\frac{\lambda_j}{\|\mathbf{1}_{B_j}\|_X} \right)^s \mathbf{1}_{B_j} \right]^{\frac{1}{s}} \right\|_X < \infty.$$

Moreover, for any $f \in h_X(\mathbb{R}^n)$,

$$\|f\|_{h_X(\mathbb{R}^n)} \sim \inf \left\{ \left\| \left[\sum_{j \in \mathbb{N}} \left(\frac{\lambda_i}{\|\mathbf{1}_{B_j}\|_X} \right)^s \mathbf{1}_{B_j} \right]^{\frac{1}{s}} \right\|_X \right\}$$

with the positive equivalence constants independent of f, where the infimum is taken over all the decompositions of f as above.

Remark 5.4.6 We point out that Lemma 5.4.5 has a wide range of applications. Here we give several function spaces to which Lemma 5.4.5 can be applied.

(i) Let $\vec{p} := (p_1, \ldots, p_n) \in (0, \infty)^n$,

$$d \geq \left\lfloor n \left(\frac{1}{\min\{1, p_1, \ldots, p_n\}} - 1 \right) \right\rfloor$$

be a fixed integer, $r \in (\max\{1, p_1, \ldots, p_n\}, \infty]$, $s \in (0, \min\{1, p_1, \ldots, p_n\})$, and

$$\tau \in \left(n \left[\frac{1}{\min\{1, p_1, \ldots, p_n\}} - \frac{1}{r} \right], \infty \right).$$

Then, in this case, by both Remarks 1.2.31(iii) and 1.2.34(iii), we can easily find that the mixed-norm Lebesgue space $L^{\vec{p}}(\mathbb{R}^n)$ satisfies all the assumptions of Lemma 5.4.5. Thus, Lemma 5.4.5 with $X := L^{\vec{p}}(\mathbb{R}^n)$ holds true. To the best of our knowledge, this result is totally new.

(ii) Let $0 < q \leq p < \infty$,

$$d \geq \left\lfloor n \left(\frac{1}{\min\{1, q\}} - 1 \right) \right\rfloor$$

be a fixed integer, $r \in (\max\{1, p\}, \infty]$, $s \in (0, \min\{1, q\})$, and

$$\tau \in \left(n \left[\frac{1}{\min\{1, q\}} - \frac{1}{r} \right], \infty \right).$$

Then, in this case, from both Remarks 1.2.31(iv) and 1.2.34(iv), we can easily deduce that the Morrey space $M_q^p(\mathbb{R}^n)$ satisfies all the assumptions of Lemma 5.4.5. This then implies that Lemma 5.4.5 with $X := M_q^p(\mathbb{R}^n)$ holds true. To the best of our knowledge, this result is totally new.

(iii) Let $p(\cdot) \in C^{\log}(\mathbb{R}^n)$ satisfy $0 < p_- \leq p_+ < \infty$, where p_- and p_+ are defined, respectively, in (1.59) and (1.60). Let

$$d \geq \left\lfloor n \left(\frac{1}{\min\{1, p_-\}} - 1 \right) \right\rfloor$$

be a fixed integer, $r \in (\max\{1, p_+\}, \infty]$, $s := \min\{1, p_-\}$, and

$$\tau \in \left(n \left[\frac{1}{\min\{1, p_-\}} - \frac{1}{r} \right], \infty \right).$$

Then, in this case, using both Remarks 1.2.31(v) and 1.2.34(v), we can easily conclude that the variable Lebesgue space $L^{p(\cdot)}(\mathbb{R}^n)$ satisfies all the assumptions of Lemma 5.4.5. Therefore, Lemma 5.4.5 with $X := L^{p(\cdot)}(\mathbb{R}^n)$ holds true. To the best of our knowledge, this result is totally new.

Based on the above lemma, we next show Theorem 5.4.2.

Proof of Theorem 5.4.2 Let p, q, ω, r, d, and s be as in the present theorem. Since ω satisfies that $m_0(\omega) \in (-\frac{n}{p}, \infty)$, from Theorem 1.2.42, it follows that the local generalized Herz space $\dot{\mathcal{K}}^{p,q}_{\omega,0}(\mathbb{R}^n)$ under consideration is a BQBF space. Thus, to complete the present theorem, it suffices to prove that all the assumptions of Lemma 5.4.5 hold true for the Herz space $\dot{\mathcal{K}}^{p,q}_{\omega,0}(\mathbb{R}^n)$ under consideration. Namely, $\dot{\mathcal{K}}^{p,q}_{\omega,0}(\mathbb{R}^n)$ satisfies both Assumption 1.2.29 with the above s and some $\theta \in (0, s)$, and Assumption 1.2.33 with the above s and r.

Indeed, let $\theta \in (0, s)$ satisfy that

$$\left\lfloor n\left(\frac{1}{s} - 1\right)\right\rfloor \leq n\left(\frac{1}{\theta} - 1\right) < \left\lfloor n\left(\frac{1}{s} - 1\right)\right\rfloor + 1 \qquad (5.29)$$

and

$$\tau > n\left(\frac{1}{\theta} - \frac{1}{r}\right). \qquad (5.30)$$

Then, by Lemma 4.3.25, we find that, for any $\{f_j\}_{j\in\mathbb{N}} \subset L^1_{\mathrm{loc}}(\mathbb{R}^n)$,

$$\left\|\left\{\sum_{j\in\mathbb{N}}\left[\mathcal{M}^{(\theta)}(f_j)\right]^s\right\}^{1/s}\right\|_{\dot{\mathcal{K}}^{p,q}_{\omega,0}(\mathbb{R}^n)} \lesssim \left\|\left(\sum_{j\in\mathbb{N}}|f_j|^s\right)^{1/s}\right\|_{\dot{\mathcal{K}}^{p,q}_{\omega,0}(\mathbb{R}^n)},$$

which implies that, for the above θ and s, Assumption 1.2.29 holds true for $\dot{\mathcal{K}}^{p,q}_{\omega,0}(\mathbb{R}^n)$. On the other hand, applying Lemma 1.8.6, we conclude that $[\dot{\mathcal{K}}^{p,q}_{\omega,0}(\mathbb{R}^n)]^{1/s}$ is a BBF space and, for any $f \in L^1_{\mathrm{loc}}(\mathbb{R}^n)$,

$$\left\|\mathcal{M}^{((r/s)')}(f)\right\|_{([\dot{\mathcal{K}}^{p,q}_{\omega,0}(\mathbb{R}^n)]^{1/s})'} \lesssim \|f\|_{([\dot{\mathcal{K}}^{p,q}_{\omega,0}(\mathbb{R}^n)]^{1/s})'}.$$

This implies that $\dot{\mathcal{K}}^{p,q}_{\omega,0}(\mathbb{R}^n)$ satisfies Assumption 1.2.33 with the above s and r. Finally, from (5.29), we deduce that $d \geq \lfloor n(1/\theta - 1)\rfloor$. Combining this, (5.30), and the fact that $\dot{\mathcal{K}}^{p,q}_{\omega,0}(\mathbb{R}^n)$ satisfies both Assumptions 1.2.29 and 1.2.33, we find that the Herz space $\dot{\mathcal{K}}^{p,q}_{\omega,0}(\mathbb{R}^n)$ under consideration satisfies all the assumptions of Lemma 5.4.5 and then complete the proof of Theorem 5.4.2.   □

As an application of Theorem 5.4.2, we now establish the molecular characterization of the local generalized Hardy–Morrey space $hM_{\omega,0}^{p,q}(\mathbb{R}^n)$ via introducing the definition of local-$(M_{\omega,0}^{p,q}(\mathbb{R}^n)$, r, d, $\tau)$-molecules as follows.

Definition 5.4.7 Let $p, q \in [1, \infty)$, $\omega \in M(\mathbb{R}_+)$ with $M_\infty(\omega) \in (-\infty, 0)$ and

$$-\frac{n}{p} < m_0(\omega) \le M_0(\omega) < 0,$$

$r \in [1, \infty]$, $d \in \mathbb{Z}_+$, and $\tau \in (0, \infty)$. Then a measurable function m on $\mathbb{R}^n$ is called a *local-$(M_{\omega,0}^{p,q}(\mathbb{R}^n)$, r, d, $\tau)$-molecule* centered at a ball $B(x_0, r_0) \in \mathbb{B}$, with $x_0 \in \mathbb{R}^n$ and $r_0 \in (0, \infty)$, if

(i) for any $i \in \mathbb{Z}_+$,

$$\left\| m \mathbf{1}_{S_i(B(x_0,r_0))} \right\|_{L^r(\mathbb{R}^n)} \le 2^{-\tau i} \frac{|B(x_0, r_0)|^{1/r}}{\left\| \mathbf{1}_{B(x_0,r_0)} \right\|_{M_{\omega,0}^{p,q}(\mathbb{R}^n)}};$$

(ii) when $r_0 \in (0, 1)$, then, for any $\alpha \in \mathbb{Z}_+^n$ with $|\alpha| \le d$,

$$\int_{\mathbb{R}^n} m(x) x^\alpha \, dx = 0.$$

Then, combining Remark 5.0.4(ii) and Theorem 5.4.2, we immediately conclude the following molecular characterization of $hM_{\omega,0}^{p,q}(\mathbb{R}^n)$; we omit the details.

Corollary 5.4.8 *Let* $p, q \in [1, \infty)$, $\omega \in M(\mathbb{R}_+)$ *with*

$$-\frac{n}{p} < m_0(\omega) \le M_0(\omega) < 0$$

and

$$-\frac{n}{p} < m_\infty(\omega) \le M_\infty(\omega) < 0,$$

$s \in (0, 1)$, $d \ge \lfloor n(1/s - 1) \rfloor$ *be a fixed integer,*

$$r \in \left(\frac{n}{\min\{m_0(\omega), m_\infty(\omega)\} + n/p}, \infty \right],$$

and $\tau \in (0, \infty)$ *with* $\tau > n(1/s - 1/r)$. *Then* $f \in h\boldsymbol{M}_{\omega,0}^{p,q}(\mathbb{R}^n)$ *if and only if* $f \in \mathcal{S}'(\mathbb{R}^n)$ *and there exists a sequence* $\{m_j\}_{j\in\mathbb{N}}$ *of local-*$(\boldsymbol{M}_{\omega,0}^{p,q}(\mathbb{R}^n), r, d, \tau)$-*molecules centered, respectively, at cubes* $\{B_j\}_{j\in\mathbb{N}} \subset \mathbb{B}$ *and a sequence* $\{\lambda_j\}_{j\in\mathbb{N}} \subset [0, \infty)$ *such that both*

$$f = \sum_{j\in\mathbb{N}} \lambda_j m_j$$

in $\mathcal{S}'(\mathbb{R}^n)$ *and*

$$\left\| \left\{ \sum_{j\in\mathbb{N}} \left[\frac{\lambda_j}{\|\mathbf{1}_{B_j}\|_{\boldsymbol{M}_{\omega,0}^{p,q}(\mathbb{R}^n)}} \right]^s \mathbf{1}_{B_j} \right\}^{\frac{1}{s}} \right\|_{\boldsymbol{M}_{\omega,0}^{p,q}(\mathbb{R}^n)} < \infty.$$

Moreover, there exist two positive constants C_1 *and* C_2 *such that, for any* $f \in h\boldsymbol{M}_{\omega,0}^{p,q}(\mathbb{R}^n)$,

$$C_1 \|f\|_{h\boldsymbol{M}_{\omega,0}^{p,q}(\mathbb{R}^n)} \leq \inf \left\{ \left\| \left\{ \sum_{j\in\mathbb{N}} \left[\frac{\lambda_i}{\|\mathbf{1}_{B_j}\|_{\boldsymbol{M}_{\omega,0}^{p,q}(\mathbb{R}^n)}} \right]^s \mathbf{1}_{B_j} \right\}^{\frac{1}{s}} \right\|_{\boldsymbol{M}_{\omega,0}^{p,q}(\mathbb{R}^n)} \right\}$$

$$\leq C_2 \|f\|_{h\boldsymbol{M}_{\omega,0}^{p,q}(\mathbb{R}^n)},$$

where the infimum is taken over all the decompositions of f *as above.*

Next, we are devoted to establishing the molecular characterization of the local generalized Herz–Hardy space $h\dot{\mathcal{K}}_\omega^{p,q}(\mathbb{R}^n)$. For this purpose, we first introduce the definition of local-$(\dot{\mathcal{K}}_\omega^{p,q}(\mathbb{R}^n), r, d, \tau)$-molecules as follows.

Definition 5.4.9 Let $p, q \in (0, \infty)$, $\omega \in M(\mathbb{R}_+)$ with $m_0(\omega) \in (-\frac{n}{p}, \infty)$ and

$$-\frac{n}{p} < m_\infty(\omega) \leq M_\infty(\omega) < 0,$$

$r \in [1, \infty]$, $d \in \mathbb{Z}_+$, and $\tau \in (0, \infty)$. Then a measurable function m on $\mathbb{R}^n$ is called a *local-*$(\dot{\mathcal{K}}_\omega^{p,q}(\mathbb{R}^n), r, d, \tau)$-*molecule centered at a ball* $B(x_0, r_0) \in \mathbb{B}$, with $x_0 \in \mathbb{R}^n$ and $r_0 \in (0, \infty)$, if

(i) for any $i \in \mathbb{Z}_+$,

$$\left\| m\mathbf{1}_{S_i(B(x_0,r_0))} \right\|_{L^r(\mathbb{R}^n)} \leq 2^{-\tau i} \frac{|B(x_0, r_0)|^{1/r}}{\|\mathbf{1}_{B(x_0,r_0)}\|_{\dot{\mathcal{K}}_\omega^{p,q}(\mathbb{R}^n)}};$$

(ii) when $r_0 \in (0, 1)$, then, for any $\alpha \in \mathbb{Z}_+^n$ with $|\alpha| \le d$,

$$\int_{\mathbb{R}^n} m(x) x^\alpha \, dx = 0.$$

Via local-$(\dot{\mathcal{K}}_\omega^{p,q}(\mathbb{R}^n)$, r, d, $\tau)$-molecules, we have the following molecular characterization of the local generalized Herz–Hardy space $h\dot{\mathcal{K}}_\omega^{p,q}(\mathbb{R}^n)$.

Theorem 5.4.10 *Let $p, q \in (0, \infty)$, $\omega \in M(\mathbb{R}_+)$ with $m_0(\omega) \in (-\frac{n}{p}, \infty)$ and*

$$-\frac{n}{p} < m_\infty(\omega) \le M_\infty(\omega) < 0,$$

$$s \in \left(0, \min\left\{1, p, q, \frac{n}{\max\{M_0(\omega), M_\infty(\omega)\} + n/p}\right\}\right),$$

$d \ge \lfloor n(1/s - 1) \rfloor$ be a fixed integer,

$$r \in \left(\max\left\{1, p, \frac{n}{\min\{m_0(\omega), m_\infty(\omega)\} + n/p}\right\}, \infty\right],$$

and $\tau \in (0, \infty)$ with $\tau > n(1/s - 1/r)$. Then $f \in h\dot{\mathcal{K}}_\omega^{p,q}(\mathbb{R}^n)$ if and only if $f \in \mathcal{S}'(\mathbb{R}^n)$ and there exists a sequence $\{m_j\}_{j\in\mathbb{N}}$ of local-$(\dot{\mathcal{K}}_\omega^{p,q}(\mathbb{R}^n)$, r, d, $\tau)$-molecules centered, respectively, at the balls $\{B_j\}_{j\in\mathbb{N}} \subset \mathbb{B}$ and a sequence $\{\lambda_j\}_{j\in\mathbb{N}} \subset [0, \infty)$ such that both

$$f = \sum_{j\in\mathbb{N}} \lambda_j m_j$$

in $\mathcal{S}'(\mathbb{R}^n)$ and

$$\left\| \left\{ \sum_{j\in\mathbb{N}} \left[\frac{\lambda_j}{\|\mathbf{1}_{B_j}\|_{\dot{\mathcal{K}}_\omega^{p,q}(\mathbb{R}^n)}} \right]^s \mathbf{1}_{B_j} \right\}^{\frac{1}{s}} \right\|_{\dot{\mathcal{K}}_\omega^{p,q}(\mathbb{R}^n)} < \infty.$$

Moreover, there exist two positive constants C_1 and C_2 such that, for any $f \in h\dot{\mathcal{K}}_\omega^{p,q}(\mathbb{R}^n)$,

$$C_1 \|f\|_{h\dot{\mathcal{K}}_\omega^{p,q}(\mathbb{R}^n)} \le \inf\left\{ \left\| \left\{ \sum_{j\in\mathbb{N}} \left[\frac{\lambda_i}{\|\mathbf{1}_{B_j}\|_{\dot{\mathcal{K}}_\omega^{p,q}(\mathbb{R}^n)}} \right]^s \mathbf{1}_{B_j} \right\}^{\frac{1}{s}} \right\|_{\dot{\mathcal{K}}_\omega^{p,q}(\mathbb{R}^n)} \right\}$$

$$\le C_2 \|f\|_{h\dot{\mathcal{K}}_\omega^{p,q}(\mathbb{R}^n)},$$

where the infimum is taken over all the decompositions of f as above.

To establish the above molecular characterization, we first show the following improved molecular characterization of localized Hardy spaces associated with ball quasi-Banach function spaces.

Theorem 5.4.11 *Let X be a ball quasi-Banach function space satisfy:*

(i) *Assumption 1.2.29 holds true with $0 < \theta < s \leq 1$;*

(ii) *for the above s, $X^{1/s}$ is a ball Banach function space and there exists a linear space $Y \subset \mathscr{M}(\mathbb{R}^n)$ equipped with a seminorm $\|\cdot\|_Y$ such that, for any $f \in \mathscr{M}(\mathbb{R}^n)$,*

$$\|f\|_{X^{1/s}} \sim \sup\left\{\|fg\|_{L^1(\mathbb{R}^n)} : \|g\|_Y = 1\right\},$$

where the positive equivalence constants are independent of f;

(iii) *for the above s and Y, there exists an $r \in (1, \infty]$ and a positive constant C such that, for any $f \in L^1_{\mathrm{loc}}(\mathbb{R}^n)$,*

$$\left\|\mathcal{M}^{((r/s)')}(f)\right\|_Y \leq C\|f\|_Y.$$

Let $\tau \in (0, \infty)$ satisfy $\tau > n(1/\theta - 1/r)$, and $d \geq \lfloor n(1/\theta - 1) \rfloor$ be a fixed integer. Then $f \in h_X(\mathbb{R}^n)$ if and only if $f \in \mathcal{S}'(\mathbb{R}^n)$ and there exists a sequence $\{m_j\}_{j\in\mathbb{N}}$ of local-(X, r, d, τ)-molecules centered, respectively, at the balls $\{B_j\}_{j\in\mathbb{N}} \subset \mathbb{B}$ and a sequence $\{\lambda_j\}_{j\in\mathbb{N}} \subset [0, \infty)$ such that both

$$f = \sum_{j\in\mathbb{N}} \lambda_j m_j \tag{5.31}$$

in $\mathcal{S}'(\mathbb{R}^n)$ and

$$\left\|\left[\sum_{j\in\mathbb{N}}\left(\frac{\lambda_j}{\|\mathbf{1}_{B_j}\|_X}\right)^s \mathbf{1}_{B_j}\right]^{\frac{1}{s}}\right\|_X < \infty. \tag{5.32}$$

Moreover, there exist two positive constants C_1 and C_2 such that, for any $f \in h_X(\mathbb{R}^n)$,

$$C_1\|f\|_{h_X(\mathbb{R}^n)} \leq \inf\left\{\left\|\left[\sum_{j\in\mathbb{N}}\left(\frac{\lambda_i}{\|\mathbf{1}_{B_j}\|_X}\right)^s \mathbf{1}_{B_j}\right]^{\frac{1}{s}}\right\|_X\right\}$$

$$\leq C_2\|f\|_{h_X(\mathbb{R}^n)}, \tag{5.33}$$

where the infimum is taken over all the decompositions of f as above.

Proof Let all the symbols be as in the present theorem. We first prove the necessity. To achieve this, let $f \in h_X(\mathbb{R}^n)$. Then, from the assumptions (i) through (iii) of the present theorem, and Theorem 5.3.14, we deduce that $f \in h^{X,r,d,s}(\mathbb{R}^n)$. Thus, by Definition 5.3.6, we conclude that there exists a sequence $\{\lambda_j\}_{j \in \mathbb{N}} \subset [0, \infty)$ and a sequence $\{a_j\}_{j \in \mathbb{N}}$ of local-(X, r, d)-atoms supported, respectively, in the balls $\{B_j\}_{j \in \mathbb{N}}$ such that both

$$f = \sum_{j \in \mathbb{N}} \lambda_j a_j \tag{5.34}$$

in $\mathcal{S}'(\mathbb{R}^n)$ and

$$\left\| \left[\sum_{j \in \mathbb{N}} \left(\frac{\lambda_j}{\|\mathbf{1}_{B_j}\|_X} \right)^s \mathbf{1}_{B_j} \right]^{\frac{1}{s}} \right\|_X < \infty. \tag{5.35}$$

Using Remark 5.4.4(i), we find that, for any $j \in \mathbb{N}$, a_j is a local-(X, r, r, τ)-molecule centered at B_j. This, combined with both (5.34) and (5.35), implies that the necessity holds true. Furthermore, by the choice of $\{\lambda_j\}_{j \in \mathbb{N}}$, Definition 5.3.6, and Theorem 5.3.14 again, we conclude that

$$\inf \left\{ \left\| \left[\sum_{j \in \mathbb{N}} \left(\frac{\lambda_i}{\|\mathbf{1}_{B_j}\|_X} \right)^s \mathbf{1}_{B_j} \right]^{\frac{1}{s}} \right\|_X \right\} \le \|f\|_{h^{X,r,d,s}(\mathbb{R}^n)} \sim \|f\|_{h_X(\mathbb{R}^n)}, \tag{5.36}$$

where the infimum is taken over all the sequences $\{\lambda_j\}_{j \in \mathbb{N}} \subset [0, \infty)$ and $\{m_j\}_{j \in \mathbb{N}}$ of local-(X, r, d, τ)-molecules centered, respectively, at the balls $\{B_j\}_{j \in \mathbb{N}} \subset \mathbb{B}$ such that both (5.31) and (5.32) hold true.

Conversely, we show the sufficiency. Let $f \in \mathcal{S}'(\mathbb{R}^n)$, $\{\lambda_{l,j}\}_{l \in \{1,2\}, j \in \mathbb{N}} \subset [0, \infty)$, and $\{m_{l,j}\}_{l \in \{1,2\}, j \in \mathbb{N}}$ be a sequence of local-(X, r, d, τ)-molecules centered, respectively, at the balls $\{B_{l,j}\}_{l \in \{1,2\}, j \in \mathbb{N}} \in \mathbb{B}$ such that, for any $j \in \mathbb{N}$, $r(B_{1,j}) \in (0, 1)$ and $r(B_{2,j}) \in [1, \infty)$,

$$f = \sum_{l=1}^{2} \sum_{j \in \mathbb{N}} \lambda_{l,j} m_{l,j} \tag{5.37}$$

in $S'(\mathbb{R}^n)$, and

$$\left\| \left[\sum_{l=1}^{2} \sum_{j \in \mathbb{N}} \left(\frac{\lambda_{l,j}}{\|\mathbf{1}_{B_{l,j}}\|_X} \right)^s \mathbf{1}_{B_{l,j}} \right]^{\frac{1}{s}} \right\|_X < \infty. \tag{5.38}$$

In what follows, fix a $\phi \in \mathcal{S}(\mathbb{R}^n)$ satisfying that $\mathrm{supp}\,(\phi) \subset B(\mathbf{0}, 1)$ and

$$\int_{\mathbb{R}^n} \phi(x)\,dx \neq 0.$$

Then, applying (5.37) and an argument similar to that used in the estimation of (4.24) with $\{\lambda_j\}_{j\in\mathbb{N}}$ and $\{a_j\}_{j\in\mathbb{N}}$ therein replaced, respectively, by $\{\lambda_{l,j}\}_{l\in\{1,2\},\,j\in\mathbb{N}}$ and $\{m_{l,j}\}_{l\in\{1,2\},\,j\in\mathbb{N}}$, we find that, for any $t \in (0, \infty)$,

$$|f * \phi_t| \leq \sum_{l=1}^{2} \sum_{j\in\mathbb{N}} \lambda_{l,j}\, \left|m_{l,j} * \phi_t\right|.$$

Combining this and Definition 5.1.1(i), we further obtain

$$m(f, \phi) \leq \sum_{l=1}^{2} \sum_{j\in\mathbb{N}} \lambda_{l,j} m(m_{l,j}, \phi).$$

This, together with Definition 1.2.13(ii), implies that

$$\|m(f, \phi)\|_X \lesssim \left\| \sum_{j\in\mathbb{N}} \lambda_{1,j} m(m_{1,j}, \phi) \right\|_X + \left\| \sum_{j\in\mathbb{N}} \lambda_{2,j} m(m_{2,j}, \phi) \right\|_X$$
$$=: V_1 + V_2. \tag{5.39}$$

We next estimate V_1 and V_2, respectively.

First, we deal with V_1. Indeed, for any $j \in \mathbb{N}$, by Remark 5.4.4(ii) with $m := m_{1,j}$, we conclude that $m_{1,j}$ is an (X, r, d, τ)-molecule centered at $B_{1,j}$. Therefore, from Definitions 5.1.1(i), 4.1.1(i), and 1.2.13(ii), and some arguments similar to those used in the estimations of (4.62), (4.63), (4.66), and (4.67), it follows that

$$V_1 \lesssim \left\| \sum_{j\in\mathbb{N}} \lambda_j M(m_{1,j}, \phi) \right\|_X \lesssim \left\| \left[\sum_{j\in\mathbb{N}} \left(\frac{\lambda_{1,j}}{\|\mathbf{1}_{B_{1,j}}\|_X} \right)^s \mathbf{1}_{B_{1,j}} \right]^{1/s} \right\|_X, \tag{5.40}$$

which is the desired estimate of V_1.

Conversely, we next estimate V_2. To this end, for any $j \in \mathbb{N}$ and $k \in \mathbb{Z}_+$, let

$$\mu_{j,k} := 2^{-k(\tau+\frac{n}{r})} \frac{\|\mathbf{1}_{2^k B_{2,j}}\|_X}{\|\mathbf{1}_{B_{2,j}}\|_X}$$

and

$$a_{j,k} := 2^{k(\tau+\frac{n}{r})} \frac{\|\mathbf{1}_{B_{2,j}}\|_X}{\|\mathbf{1}_{2^k B_{2,j}}\|_X} m_{2,j} \mathbf{1}_{S_k(B_{2,j})}.$$

Then, for any $j \in \mathbb{N}$, we have

$$m_{2,j} = \sum_{k \in \mathbb{Z}_+} m_{2,j} \mathbf{1}_{S_k(B_{2,j})} = \sum_{k \in \mathbb{Z}_+} \mu_{j,k} a_{j,k} \qquad (5.41)$$

almost everywhere in $\mathbb{R}^n$. In addition, by the Tonelli theorem, the Hölder inequality, and Definition 5.4.3(i) with m therein replaced by $m_{2,j}$ with $j \in \mathbb{N}$, we conclude that, for any $t \in (0, \infty)$ and $x \in \mathbb{R}^n$,

$$\sum_{k \in \mathbb{Z}_+} \int_{\mathbb{R}^n} \mu_{j,k} \left| a_{j,k}(y) \right| |\phi_t(x - y)| \, dy$$

$$= \sum_{k \in \mathbb{Z}_+} \int_{S_k(B_{2,j})} \left| m_{2,j}(y) \right| |\phi_t(x - y)| \, dy$$

$$\leq \sum_{k \in \mathbb{Z}_+} \left\| m_{2,j} \mathbf{1}_{S_k(B_{2,j})} \right\|_{L^r(\mathbb{R}^n)} \|\phi_t\|_{L^{r'}(\mathbb{R}^n)}$$

$$\leq \frac{|B_{2,j}|^{1/r}}{\|\mathbf{1}_{B_{2,j}}\|_X} \|\phi_t\|_{L^{r'}(\mathbb{R}^n)} \sum_{k \in \mathbb{Z}_+} 2^{-k\tau} \sim 1.$$

From this, (5.41), and the Fubini theorem, we deduce that, for any $j \in \mathbb{N}$, $t \in (0, \infty)$, and $x \in \mathbb{R}^n$,

$$m_{2,j} * \phi_t(x) = \int_{\mathbb{R}^n} \sum_{k \in \mathbb{Z}_+} \mu_{j,k} a_{j,k}(y) \phi_t(x - y) \, dy$$

$$= \sum_{k \in \mathbb{Z}_+} \mu_{j,k} \int_{\mathbb{R}^n} a_{j,k} \phi_{j,k}(x - y) \, dy$$

$$= \sum_{k \in \mathbb{Z}_+} \mu_{j,k} a_{j,k} * \phi_t(x).$$

This, together with Definition 5.1.1, further implies that, for any $j \in \mathbb{N}$,

$$m(m_{2,j}, \phi) \leq \sum_{k \in \mathbb{Z}_+} \mu_{j,k} m(a_{j,k}, \phi).$$

By this and Definition 1.2.13(ii), we conclude that

$$
V_2 \lesssim \left\| \sum_{j \in \mathbb{N}} \sum_{k \in \mathbb{Z}_+} \lambda_{2,j} \mu_{j,k} m(a_{j,k}, \phi) \right\|_X . \tag{5.42}
$$

We now claim that, for any $j \in \mathbb{N}$ and $k \in \mathbb{Z}_+$, $a_{j,k}$ is a local-(X, r, d)-atom supported in $2^k B_{2,j}$. Indeed, applying Definition 5.4.3(i), we find that, for any $j \in \mathbb{N}$ and $k \in \mathbb{Z}_+$,

$$
\| a_{j,k} \|_{L^r(\mathbb{R}^n)} \leq 2^{\frac{nk}{r}} \frac{|B_{2,j}|^{1/r}}{\| \mathbf{1}_{2^k B_{2,j}} \|_X} = \frac{|2^k B_{2,j}|^{1/r}}{\| \mathbf{1}_{2^k B_{2,j}} \|}. \tag{5.43}
$$

On the other hand, for any $j \in \mathbb{N}$ and $k \in \mathbb{Z}_+$, observe that $\operatorname{supp}(a_{j,k}) \subset 2^k B_{2,j}$ and

$$
r(2^k B_{2,j}) = 2^k r(B_{2,j}) \geq 2^k \geq 1,
$$

which, combined with (5.43) and Definition 5.3.4, imply that $a_{j,k}$ is a local-(X, r, d)-atom supported in $2^k B_{2,j}$. This finishes the proof of the above claim. Combining this claim, the fact that, for any $j \in \mathbb{N}$ and $k \in \mathbb{Z}_+$, $r(2^k B_{2,j}) \geq 1$, the claim obtained in the proof of Theorem 5.3.14, (5.42), and Definitions 5.1.1(i) and 4.1.1(i), we conclude that

$$
V_2 \lesssim \left\| \sum_{j \in \mathbb{N}} \sum_{k \in \mathbb{Z}_+} \lambda_{2,j} \mu_{j,k} m(a_{j,k}, \phi) \mathbf{1}_{2^{k+1} B_{2,j}} \right\|_X
$$

$$
\lesssim \left\| \sum_{j \in \mathbb{N}} \sum_{k \in \mathbb{Z}_+} \lambda_{2,j} \mu_{j,k} M(a_{j,k}, \phi) \mathbf{1}_{2^{k+1} B_{2,j}} \right\|_X .
$$

Using this and an argument similar to that used in the estimation of II_1 in the proof of Theorem 4.3.18 with $\{\lambda_j\}_{j \in \mathbb{N}}$ and $\{a_j\}_{j \in \mathbb{N}}$ therein replaced, respectively, by $\{\lambda_{2,j} \mu_{j,k}\}_{j \in \mathbb{N}, k \in \mathbb{Z}_+}$ and $\{a_{j,k}\}_{j \in \mathbb{N}, k \in \mathbb{Z}_+}$, we find that

$$
V_2 \lesssim \left\| \left[\sum_{j \in \mathbb{N}} \sum_{k \in \mathbb{Z}_+} \left(\frac{\lambda_{2,j} \mu_{j,k}}{\| \mathbf{1}_{2^k B_{2,j}} \|_X} \right)^s \mathbf{1}_{2^k B_{2,j}} \right]^{\frac{1}{s}} \right\|_X
$$

$$
\sim \left\| \left[\sum_{j \in \mathbb{N}} \left(\frac{\lambda_{2,j}}{\| \mathbf{1}_{B_{2,j}} \|_X} \right)^s \sum_{k \in \mathbb{Z}_+} 2^{-ks(\tau + \frac{n}{r})} \mathbf{1}_{2^k B_{2,j}} \right]^{\frac{1}{s}} \right\|_X .
$$

This, together with an argument similar to that used in the estimation of (4.66) with $\{\lambda_j\}_{j\in\mathbb{N}}$ and $\{B_j\}_{j\in\mathbb{N}}$ therein replaced, respectively, by $\{\lambda_{2,j}\}_{j\in\mathbb{N}}$ and $\{B_{2,j}\}_{j\in\mathbb{N}}$, further implies that

$$\mathrm{V}_2 \lesssim \left\|\left[\sum_{j\in\mathbb{N}}\left(\frac{\lambda_{2,j}}{\|\mathbf{1}_{B_{2,j}}\|_X}\right)^s \mathbf{1}_{B_{1,j}}\right]^{1/s}\right\|_X, \tag{5.44}$$

which is the desired estimate of V_2. Then, from both the assumptions (i) and (ii) of the present theorem, Remark 5.1.6, Lemma 5.1.5(ii), (5.39), (5.40), (5.44), and (5.38), it follows that

$$\|f\|_{h_X(\mathbb{R}^n)} \sim \|m(f,\phi)\|_X \lesssim \left\|\left[\sum_{l=1}^{2}\sum_{j\in\mathbb{N}}\left(\frac{\lambda_{l,j}}{\|\mathbf{1}_{B_{l,j}}\|_X}\right)^s \mathbf{1}_{B_{l,j}}\right]^{\frac{1}{s}}\right\|_X < \infty. \tag{5.45}$$

This further implies that $f \in h_X(\mathbb{R}^n)$, and hence finishes the proof of the sufficiency. Moreover, by (5.45) and the choice of $\{\lambda_{l,j}\}_{l\in\{1,2\},\,j\in\mathbb{N}}$, we conclude that

$$\|f\|_{h_X(\mathbb{R}^n)} \lesssim \inf\left\{\left\|\left[\sum_{j\in\mathbb{N}}\left(\frac{\lambda_i}{\|\mathbf{1}_{B_j}\|_X}\right)^s \mathbf{1}_{B_j}\right]^{\frac{1}{s}}\right\|_X\right\},$$

where the infimum is taken over all the sequences $\{\lambda_j\}_{j\in\mathbb{N}} \subset [0,\infty)$ and $\{m_j\}_{j\in\mathbb{N}}$ of local-(X, r, d, τ)-molecules centered, respectively, at the balls $\{B_j\}_{j\in\mathbb{N}} \subset \mathbb{B}$ such that both (5.31) and (5.32) hold true. From this and (5.36), we deduce that (5.33) holds true, which completes the proof of Theorem 5.4.11. □

Remark 5.4.12 We should point out that Theorem 5.4.11 is an improved version of the known molecular characterization of $h_X(\mathbb{R}^n)$ established by Wang et al. in [233, Theorem 5.2]. Indeed, if $Y \equiv (X^{1/s})'$ in Theorem 5.4.11, then this theorem goes back to [233, Theorem 5.2].

With the help of the above improved molecular characterization of $h_X(\mathbb{R}^n)$, we now prove Theorem 5.4.10.

Proof of Theorem 5.4.10 Let p, q, ω, r, d, and s be as in the present theorem. Then, using the assumptions $m_0(\omega) \in (-\frac{n}{p},\infty)$ and $M_\infty(\omega) \in (-\infty,0)$, and Theorem 1.2.44, we find that the global generalized Herz space $\dot{\mathcal{K}}_\omega^{p,q}(\mathbb{R}^n)$ under consideration is a BQBF space. Therefore, to complete the proof of the present theorem, we only need to show that $\dot{\mathcal{K}}_\omega^{p,q}(\mathbb{R}^n)$ satisfies (i) through (iii) of

Theorem 5.4.11. Indeed, let $\theta \in (0, s)$ satisfy that

$$\left\lfloor n\left(\frac{1}{s} - 1\right) \right\rfloor \leq n\left(\frac{1}{\theta} - 1\right) < \left\lfloor n\left(\frac{1}{s} - 1\right) \right\rfloor + 1 \qquad (5.46)$$

and

$$\tau > n\left(\frac{1}{\theta} - \frac{1}{r}\right). \qquad (5.47)$$

For the above θ, s, and r, by the proof of Theorem 5.3.13, we conclude that the following three statements hold true:

(i) for any $\{f_j\}_{j \in \mathbb{N}} \subset L^1_{\mathrm{loc}}(\mathbb{R}^n)$,

$$\left\| \left\{ \sum_{j \in \mathbb{N}} \left[\mathcal{M}^{(\theta)}(f_j) \right]^s \right\}^{1/s} \right\|_{\dot{\mathcal{K}}^{p,q}_\omega(\mathbb{R}^n)} \lesssim \left\| \left(\sum_{j \in \mathbb{N}} |f_j|^s \right)^{1/s} \right\|_{\dot{\mathcal{K}}^{p,q}_\omega(\mathbb{R}^n)} ;$$

(ii) $[\dot{\mathcal{K}}^{p,q}_\omega(\mathbb{R}^n)]^{1/s}$ is a BBF space and, for any $f \in \mathscr{M}(\mathbb{R}^n)$,

$$\|f\|_{[\dot{\mathcal{K}}^{p,q}_\omega(\mathbb{R}^n)]^{1/s}} \sim \sup\left\{ \|fg\|_{L^1(\mathbb{R}^n)} : \|g\|_{\dot{\mathcal{B}}^{(p/s)',(q/s)'}_{1/\omega^s}(\mathbb{R}^n)} = 1 \right\}$$

with the positive equivalence constants independent of f;

(iii) for any $f \in L^1_{\mathrm{loc}}(\mathbb{R}^n)$,

$$\left\| \mathcal{M}^{((r/s)')}(f) \right\|_{\dot{\mathcal{B}}^{(p/s)',(q/s)'}_{1/\omega^s}(\mathbb{R}^n)} \lesssim \|f\|_{\dot{\mathcal{B}}^{(p/s)',(q/s)'}_{1/\omega^s}(\mathbb{R}^n)}.$$

These further imply that the Herz space $\dot{\mathcal{K}}^{p,q}_\omega(\mathbb{R}^n)$ under consideration satisfies the assumptions (i) through (iii) of Theorem 5.4.11. In addition, from (5.46) and the assumption $d \geq \lfloor n(1/s - 1) \rfloor$, it follows that $d \geq \lfloor n(1/s - 1) \rfloor = \lfloor n(1/\theta - 1) \rfloor$. This, together with (5.47) and the fact that (i) through (iii) of Theorem 5.4.11 hold true, then finishes the proof of Theorem 5.4.10. $\qquad \square$

As an application of Theorem 5.4.10, we next establish the molecular characterization of the local generalized Hardy–Morrey space $h M^{p,q}_\omega(\mathbb{R}^n)$. To begin with, we introduce the following concept of local molecules associated with the global generalized Morrey space $M^{p,q}_\omega(\mathbb{R}^n)$.

Definition 5.4.13 Let p, q, ω, r, d, and τ be as in Definition 5.4.7. Then a measurable function m on $\mathbb{R}^n$ is called a *local-$(M_\omega^{p,q}(\mathbb{R}^n)$, r, d, $\tau)$-molecule* centered at a ball $B(x_0, r_0) \in \mathbb{B}$, with $x_0 \in \mathbb{R}^n$ and $r_0 \in (0, \infty)$, if

(i) for any $i \in \mathbb{Z}_+$,

$$\left\| m \mathbf{1}_{S_i(B(x_0,r_0))} \right\|_{L^r(\mathbb{R}^n)} \leq 2^{-\tau i} \frac{|B(x_0,r_0)|^{1/r}}{\left\| \mathbf{1}_{B(x_0,r_0)} \right\|_{M_\omega^{p,q}(\mathbb{R}^n)}};$$

(ii) when $r_0 \in (0, 1)$, then, for any $\alpha \in \mathbb{Z}_+^n$ with $|\alpha| \leq d$,

$$\int_{\mathbb{R}^n} m(x) x^\alpha \, dx = 0.$$

Then, by Remark 5.0.4(ii) and Theorem 5.4.10, we conclude the molecular characterization of the local generalized Hardy–Morrey space $hM_\omega^{p,q}(\mathbb{R}^n)$ as follows; we omit the details.

Corollary 5.4.14 *Let p, q, ω, r, d, and s be as in Corollary 5.4.8. Then $f \in hM_\omega^{p,q}(\mathbb{R}^n)$ if and only if $f \in \mathcal{S}'(\mathbb{R}^n)$ and there exists a sequence $\{m_j\}_{j \in \mathbb{N}}$ of local-$(M_\omega^{p,q}(\mathbb{R}^n)$, r, d, $\tau)$-molecules centered, respectively, at cubes $\{B_j\}_{j \in \mathbb{N}} \subset \mathbb{B}$ and a sequence $\{\lambda_j\}_{j \in \mathbb{N}} \subset [0, \infty)$ such that both*

$$f = \sum_{j \in \mathbb{N}} \lambda_j m_j$$

in $\mathcal{S}'(\mathbb{R}^n)$ and

$$\left\| \left\{ \sum_{j \in \mathbb{N}} \left[\frac{\lambda_j}{\left\| \mathbf{1}_{B_j} \right\|_{M_\omega^{p,q}(\mathbb{R}^n)}} \right]^s \mathbf{1}_{B_j} \right\}^{\frac{1}{s}} \right\|_{M_\omega^{p,q}(\mathbb{R}^n)} < \infty.$$

Moreover, there exist two positive constants C_1 and C_2 such that, for any $f \in hM_\omega^{p,q}(\mathbb{R}^n)$,

$$C_1 \| f \|_{hM_\omega^{p,q}(\mathbb{R}^n)} \leq \inf \left\{ \left\| \left\{ \sum_{j \in \mathbb{N}} \left[\frac{\lambda_i}{\left\| \mathbf{1}_{B_j} \right\|_{M_\omega^{p,q}(\mathbb{R}^n)}} \right]^s \mathbf{1}_{B_j} \right\}^{\frac{1}{s}} \right\|_{M_\omega^{p,q}(\mathbb{R}^n)} \right\}$$

$$\leq C_2 \| f \|_{hM_\omega^{p,q}(\mathbb{R}^n)},$$

where the infimum is taken over all the decompositions of f as above.

5.5 Littlewood–Paley Function Characterizations

The main target of this section is to prove the Littlewood–Paley function characterizations of the local generalized Herz–Hardy spaces $h\dot{\mathcal{K}}_{\omega,\mathbf{0}}^{p,q}(\mathbb{R}^n)$ and $h\dot{\mathcal{K}}_{\omega}^{p,q}(\mathbb{R}^n)$. To achieve this, we first establish various Littlewood–Paley function characterizations of the local Hardy space $h_X(\mathbb{R}^n)$ with X being a ball quasi-Banach function space. To begin with, we present the following concepts of the local Littlewood–Paley functions.

Definition 5.5.1 Let $\varphi_0 \in \mathcal{S}(\mathbb{R}^n)$ satisfy

$$\mathbf{1}_{B(0,1)} \le \widehat{\varphi_0} \le \mathbf{1}_{B(0,2)},$$

and $\varphi \in \mathcal{S}(\mathbb{R}^n)$ satisfy

$$\mathbf{1}_{B(0,4)\setminus B(0,2)} \le \widehat{\varphi} \le \mathbf{1}_{B(0,8)\setminus B(0,1)}.$$

Then, for any $f \in \mathcal{S}'(\mathbb{R}^n)$, the *local Lusin area function* $S_{\mathrm{loc}}(f)$, the *local Littlewood–Paley g-function* $g_{\mathrm{loc}}(f)$, and the *local Littlewood–Paley g_λ^*-function* $(g_\lambda^*)_{\mathrm{loc}}(f)$ with $\lambda \in (0, \infty)$ are defined, respectively, by setting, for any $x \in \mathbb{R}^n$,

$$S_{\mathrm{loc}}(f)(x) := |f * \varphi_0(x)| + \left[\int_0^1 \int_{B(x,t)} |f * \varphi_t(y)|^2 \frac{dy\,dt}{t^{n+1}} \right]^{\frac{1}{2}},$$

$$g_{\mathrm{loc}}(f)(x) := |f * \varphi_0(x)| + \left[\int_0^1 |f * \varphi_t(x)|^2 \frac{dt}{t} \right]^{\frac{1}{2}},$$

and

$$\left(g_\lambda^*\right)_{\mathrm{loc}}(f)(x)$$

$$:= |f * \varphi_0(x)| + \left[\int_0^1 \int_{\mathbb{R}^n} \left(\frac{t}{t + |x - y|} \right)^{\lambda n} |f * \varphi_t(y)|^2 \frac{dy\,dt}{t^{n+1}} \right]^{\frac{1}{2}}.$$

We now establish the following Lusin area function characterization of the local Hardy space $h_X(\mathbb{R}^n)$.

Theorem 5.5.2 *Let X be a ball quasi-Banach function space satisfying both Assumptions 1.2.29 and 1.2.33 with the same s. Then $f \in h_X(\mathbb{R}^n)$ if and only if $f \in \mathcal{S}'(\mathbb{R}^n)$ and $S_{\mathrm{loc}}(f) \in X$. Moreover, there exist two positive constants C_1 and C_2 such that, for any $f \in h_X(\mathbb{R}^n)$,*

$$C_1 \|f\|_{h_X(\mathbb{R}^n)} \le \|S_{\mathrm{loc}}(f)\|_X \le C_2 \|f\|_{h_X(\mathbb{R}^n)}.$$

In order to show this theorem, we require some auxiliary lemmas and concepts. First, we need the following Calderón reproducing formula which was given in [233, (5.3)] (see also [91, Proposition 1.1.6]).

Lemma 5.5.3 *Let φ_0 and φ be as in Theorem 5.5.2. Then there exist $a, b, c \in (0, \infty)$ and $\psi_0, \psi \in \mathcal{S}(\mathbb{R}^n)$ such that*

$$\operatorname{supp}\left(\widehat{\psi_0}\right) \subset B(\mathbf{0}, a),$$

$$\operatorname{supp}\left(\widehat{\psi}\right) \subset B(\mathbf{0}, c) \setminus B(\mathbf{0}, b),$$

and, for any $f \in \mathcal{S}'(\mathbb{R}^n)$,

$$f = f * \varphi_0 * \psi_0 + \int_0^1 f * \varphi_t * \psi_t \, \frac{dt}{t}$$

in $\mathcal{S}'(\mathbb{R}^n)$, namely,

$$f = f * \varphi_0 * \psi_0 + \lim_{\varepsilon \to 0^+} \int_\varepsilon^1 f * \varphi_t * \psi_t \, \frac{dt}{t}$$

in $\mathcal{S}'(\mathbb{R}^n)$.

We also need the following auxiliary estimate about convolutions, which is just [233, (5.9)].

Lemma 5.5.4 *Let $f \in \mathcal{S}'(\mathbb{R}^n)$ and $\varphi \in \mathcal{S}(\mathbb{R}^n)$. Then there exists a positive integer m, depending only on f, and a positive constant C, independent of f, such that, for any $t \in (0, 1]$ and $x \in \mathbb{R}^n$,*

$$|f * \varphi_t(x)| \le C t^{-n-m} \left(1 + |x|\right)^m.$$

Let S be the Lusin area operator as in Definition 4.6.1. Then the following L^p boundedness of S plays an important role in the proof of Theorem 5.5.2, which was obtained in [85, Theorem 7.8].

Lemma 5.5.5 *Let $p \in (1, \infty)$ and S be as in Definition 4.6.1. Then there exists a positive constant C such that, for any $f \in L^p(\mathbb{R}^n)$,*

$$\|S(f)\|_{L^p(\mathbb{R}^n)} \le C \|f\|_{L^p(\mathbb{R}^n)}.$$

Moreover, the following conclusion is useful in the proof of Theorem 5.5.2, which characterizes the Hardy space $H_X(\mathbb{R}^n)$ via the Lusin area function S and was showed in [207, Theorem 3.21].

Lemma 5.5.6 *Let* X *be a ball quasi-Banach function space satisfying both Assumptions 1.2.29 and 1.2.33 with the same* $s \in (0, 1]$, φ *be as in Definition 5.5.1, and* S *be as in Definition 4.6.1 with the above* φ. *Then* $f \in H_X(\mathbb{R}^n)$ *if and only if* $f \in \mathcal{S}'(\mathbb{R}^n)$, f *vanishes weakly at infinity, and* $S(f) \in X$. *Moreover, for any* $f \in H_X(\mathbb{R}^n)$,

$$\|f\|_{H_X(\mathbb{R}^n)} \sim \|S(f)\|_X$$

with the positive equivalence constants independent of f.

In what follows, for any $f \in \mathcal{S}'(\mathbb{R}^n)$ and $x \in \mathbb{R}^n$, let

$$\widetilde{S}_{\mathrm{loc}}(f)(x) := \left[\int_0^1 \int_{B(x,t)} |f * \varphi_t(y)|^2 \, \frac{dy \, dt}{t^{n+1}} \right]^{\frac{1}{2}}. \tag{5.48}$$

Then, in order to prove Theorem 5.5.2, we also need a technical estimate about $\widetilde{S}_{\mathrm{loc}}$ as follows, which was obtained in [233, p. 53].

Lemma 5.5.7 *Let* X *be a ball quasi-Banach function space,* $d \in \mathbb{Z}_+$, *and* $\theta \in (0, \infty)$. *Then there exists a positive constant* C *such that, for any local-*(X, ∞, d)*-atom* a *supported in the ball* $B \in \mathbb{B}$ *with* $r(B) \in [1, \infty)$,

$$\widetilde{S}_{\mathrm{loc}}(a) \mathbf{1}_{(4B)^{\complement}} \leq C \frac{1}{\|\mathbf{1}_B\|_X} \mathcal{M}^{(\theta)}(\mathbf{1}_B).$$

Furthermore, to show Theorem 5.5.2, we require some conclusions about X-tent spaces. We first recall some basic concepts. Let $\alpha \in (0, \infty)$ and $f : \mathbb{R}^{n+1}_+ \to \mathbb{C}$ be a measurable function. Then the *Lusin area function* $\mathcal{A}_\alpha(f)$, with aperture α, is defined by setting, for any $x \in \mathbb{R}^n$,

$$\mathcal{A}^{(\alpha)}(f)(x) := \left\{ \int_{\Gamma_\alpha(x)} |f * \varphi_t(y)|^2 \, \frac{dy \, dt}{t^{n+1}} \right\}^{\frac{1}{2}},$$

where, for any $x \in \mathbb{R}^n$ and $\alpha \in (0, \infty)$, $\Gamma_\alpha(x)$ is defined as in (4.5). For any given ball quasi-Banach function space X, the X-*tent space* $T_X^\alpha(\mathbb{R}^{n+1}_+)$, with aperture α, is defined to be the set of all the measurable functions $f : \mathbb{R}^{n+1}_+ \to \mathbb{C}$ such that

$$\|f\|_{T_X^\alpha(\mathbb{R}^{n+1}_+)} := \|\mathcal{A}_\alpha(f)\|_X < \infty.$$

In addition, let $\alpha \in (0, \infty)$. Then, for any ball $B(x, r) \in \mathbb{B}$ with $x \in \mathbb{R}^n$ and $r \in (0, \infty)$, let

$$T_\alpha(B) := \left\{ (y, t) \in \mathbb{R}^{n+1}_+ : 0 < t < r/\alpha, \ |y - x| < r - \alpha t \right\}.$$

When $\alpha := 1$, we denote $T_\alpha(B)$ simply by $T(B)$. Then the following definition of atoms is just [207, Definition 3.17].

Definition 5.5.8 Let X be a ball quasi-Banach function space, $p \in (1, \infty)$, and $\alpha \in (0, \infty)$. A measurable function $a : \mathbb{R}_+^{n+1} \to \mathbb{C}$ is called a (T_X^α, p)-atom if there exists a ball $B \in \mathbb{B}$ such that

(i) $\operatorname{supp}(a) := \{(y, t) \in \mathbb{R}_+^{n+1} : a(y, t) \neq 0\} \subset T_\alpha(B)$;

(ii) $\|\mathcal{A}_\alpha(a)\|_{L^p(\mathbb{R}^n)} \leq \frac{|B|^{1/p}}{\|\mathbf{1}_B\|_X}$.

Moreover, if a is a (T_X^α, p)-atom for any $p \in (1, \infty)$, then a is called a (T_X^α, ∞)-atom.

Repeating an argument similar to that used in the proof of [115, Lemma 4.8] with $L^\varphi(\mathbb{R}^n)$ therein replaced by X, we obtain the following auxiliary conclusion about (T_X^1, ∞)-atoms, which plays a key role in the proof of the Lusin area function characterization of $h_X(\mathbb{R}^n)$; we omit the details.

Lemma 5.5.9 Let $d \in \mathbb{Z}_+$, $\psi \in \mathcal{S}(\mathbb{R}^n)$ satisfy that, for any $\gamma \in \mathbb{Z}_+^n$ with $|\gamma| \leq d$, $\int_{\mathbb{R}^n} \psi(x) x^\gamma \, dx = 0$, and X be a ball quasi-Banach function space. Assume that a is a (T_X^1, ∞)-atom supported in $T(B)$ with $B \in \mathbb{B}$. Then, for any $\tau \in (0, \infty)$, $\int_0^\infty a(\cdot, t) * \psi_t \frac{dt}{t}$ is a harmless constant multiple of an (X, ∞, d, τ)-molecule centered at the ball B.

Moreover, the following atomic characterization of X-tent spaces obtained in [207, Theorem 3.19] is also an essential tool in the proof of Theorem 5.5.2.

Lemma 5.5.10 Let $f : \mathbb{R}_+^{n+1} \to \mathbb{C}$ be a measurable function and X a ball quasi-Banach function space. Assume that X satisfies both Assumptions 1.2.29 and 1.2.33 with the same $s \in (0, 1]$. Then $f \in T_X^1(\mathbb{R}_+^{n+1})$ if and only if there exists a sequence $\{\lambda_j\}_{j \in \mathbb{N}} \subset [0, \infty)$ and a sequence $\{a_j\}_{j \in \mathbb{N}}$ of (T_X^1, ∞)-atoms supported, respectively, in $\{T(B_j)\}_{j \in \mathbb{N}}$ with $\{B_j\}_{j \in \mathbb{N}} \subset \mathbb{B}$ such that, for almost every $(x, t) \in \mathbb{R}_+^{n+1}$,

$$f(x, t) = \sum_{j \in \mathbb{N}} \lambda_j a_j(x, t)$$

and

$$|f(x, t)| = \sum_{j \in \mathbb{N}} \lambda_j |a_j(x, t)|.$$

Moreover,

$$\|f\|_{T_X^1(\mathbb{R}_+^{n+1})} \sim \left\| \left[\sum_{j\in\mathbb{N}} \left(\frac{\lambda_j}{\|\mathbf{1}_{B_j}\|_X} \right)^s \mathbf{1}_{B_j} \right]^{1/s} \right\|_X ,$$

where the positive equivalence constants are independent of f.

Via these preparations, we now show Theorem 5.5.2.

Proof of Theorem 5.5.2 Let all the symbols be as in the present theorem. We first prove the sufficiency. To this end, let $f \in \mathcal{S}'(\mathbb{R}^n)$ satisfy $S_{\mathrm{loc}}(f) \in X$. Then, applying Lemma 5.5.3, we find that there exist $a, b, c \in (0, \infty)$ and $\psi_0, \psi \in \mathcal{S}(\mathbb{R}^n)$ such that $\mathrm{supp}\,(\widehat{\psi_0}) \subset B(0, a)$, $\mathrm{supp}\,(\widehat{\psi}) \subset B(0, c) \setminus B(0, b)$, and

$$f = f * \varphi_0 * \psi_0 + \int_0^1 f * \varphi_t * \psi_t \, \frac{dt}{t} \tag{5.49}$$

in $\mathcal{S}'(\mathbb{R}^n)$. From both Definitions 5.5.1 and 1.2.13(ii), it follows that

$$\begin{aligned}
\left\| f * \varphi_t \mathbf{1}_{\{\tau\in(0,1)\}} \right\|_{T_X^1(\mathbb{R}_+^{n+1})} &= \left\| \left[\int_{\Gamma(\cdot)} \left| f * \varphi_t(y) \mathbf{1}_{\{\tau:\,\tau\in(0,1)\}}(t) \right|^2 \frac{dy\,dt}{t^{n+1}} \right]^{\frac{1}{2}} \right\|_X \\
&= \left\| \left[\int_0^1 \int_{B(\cdot,t)} |f * \varphi_t(y)|^2 \frac{dy\,dt}{t^{n+1}} \right]^{\frac{1}{2}} \right\|_X \\
&\leq \|S_{\mathrm{loc}}(f)\|_X < \infty.
\end{aligned}$$

This, combined with Lemma 5.5.10, further implies that there exists a sequence $\{\lambda_{1,j}\}_{j\in\mathbb{N}} \subset [0, \infty)$ and a sequence $\{a_{1,j}\}_{j\in\mathbb{N}}$ of (T_X^1, ∞)-atoms supported, respectively, in $\{T(B_{1,j})\}_{j\in\mathbb{N}}$ with $\{B_{1,j}\}_{j\in\mathbb{N}} \subset \mathbb{B}$ such that, for almost every $(x, t) \in \mathbb{R}_+^{n+1}$

$$f * \varphi_t(x)\mathbf{1}_{\{\tau:\,\tau\in(0,1)\}}(t) = \sum_{j\in\mathbb{N}} \lambda_{1,j} a_{1,j}(x, t), \tag{5.50}$$

$$|f * \varphi_t(x)|\,\mathbf{1}_{\{\tau:\,\tau\in(0,1)\}}(t) = \sum_{j\in\mathbb{N}} \lambda_{1,j} |a_{1,j}|(x, t)|, \tag{5.51}$$

and

$$\left\| \left[\sum_{j=1}^{\infty} \left(\frac{\lambda_{1,j}}{\|\mathbf{1}_{B_{1,j}}\|_X} \right)^s \mathbf{1}_{B_{1,j}} \right]^{1/s} \right\|_X \sim \left\| f * \varphi_t \mathbf{1}_{\{\tau \in (0,1)\}} \right\|_{T_X^1(\mathbb{R}_+^{n+1})}$$

$$\lesssim \|S_{\mathrm{loc}}(f)\|_X. \tag{5.52}$$

Next, we show that

$$\int_0^1 f * \varphi_t * \psi_t \frac{dt}{t} = \sum_{j \in \mathbb{N}} \lambda_{1,j} \int_0^1 a_j(\cdot, t) * \psi_t \frac{dt}{t} \tag{5.53}$$

in $\mathcal{S}'(\mathbb{R}^n)$. Indeed, by Lemma 5.5.4, we find that there exists an $m \in \mathbb{N}$, depending only on f, such that, for any $t \in (0, 1]$ and $x \in \mathbb{R}^n$,

$$|f * \varphi_t(x)| \lesssim t^{-n-m} (1 + |x|)^m. \tag{5.54}$$

This further implies that, for any $t \in (0, 1]$ and $x \in \mathbb{R}^n$,

$$|f * \varphi_t| * |\psi_t| (x)$$

$$= \int_{\mathbb{R}^n} |f * \varphi_t(y)| \, |\psi_t(x - y)| \, dy$$

$$\lesssim t^{-n-m} \int_{\mathbb{R}^n} (1 + |y|)^m \frac{1}{t^n} \left| \psi \left(\frac{x - y}{t} \right) \right| dy$$

$$\sim t^{-n-m} \int_{\mathbb{R}^n} (1 + |x - ty|)^m \, |\psi(y)| \, dy$$

$$\lesssim t^{-n-m} (1 + |x|)^m \left[\int_{\mathbb{R}^n} |\psi(y)| \, dy + \int_{\mathbb{R}^n} (1 + |y|)^m \, |\psi(y)| \, dy \right]$$

$$\lesssim t^{-n-m} (1 + |x|)^m \int_{\mathbb{R}^n} \frac{1}{(1 + |y|)^{n+1}} \, dy \sim t^{-n-m} (1 + |x|)^m.$$

From this and (5.51), we deduce that

$$\int_0^1 \int_{\mathbb{R}^n} \sum_{j \in \mathbb{N}} \lambda_{1,j} \, |a_{1,j}| * |\psi_t| (x) \, |\eta(x)| \, \frac{dx \, dt}{t}$$

$$= \int_0^1 \int_{\mathbb{R}^n} \sum_{j \in \mathbb{N}} \lambda_{1,j} \, |a_{1,j}(x, t)| \, |\psi_t(-\cdot)| * |\eta| (x) \, \frac{dx \, dt}{t}$$

$$= \int_0^1 \int_{\mathbb{R}^n} |f * \varphi_t(x)| \, |\psi_t(-\cdot)| * |\eta| \, (x) \, \frac{dx \, dt}{t}$$

$$= \int_0^1 \int_{\mathbb{R}^n} |f * \varphi_t| * |\psi_t| \, (x) \, |\eta(x)| \, \frac{dx \, dt}{t}$$

$$\lesssim \int_0^1 \int_{\mathbb{R}^n} t^{-n-m} \, (1 + |x|)^m \, (1 + |x|)^{-n-m-1} \, \frac{dx \, dt}{t}$$

$$\sim \int_0^1 \frac{1}{t^{n+m+1}} \, dt \int_{\mathbb{R}^n} \frac{1}{(1 + |x|)^{n+1}} \, dx < \infty.$$

Combining this, (5.49), (5.50), and the Fubini theorem, we further conclude that

$$\int_{\mathbb{R}^n} \left[\int_0^1 f * \varphi_t * \psi_t(x) \, \frac{dt}{t} \right] \eta(x) \, dx$$

$$= \int_0^1 \int_{\mathbb{R}^n} f * \varphi_t * \psi_t(x) \eta(x) \, \frac{dx \, dt}{t}$$

$$= \int_0^1 \int_{\mathbb{R}^n} f * \varphi_t(x) \psi_t(-\cdot) * \eta(x) \, \frac{dx \, dt}{t}$$

$$= \int_0^1 \int_{\mathbb{R}^n} \sum_{j \in \mathbb{N}} \lambda_{1,j} a_{1,j}(x, t) \psi_t(-\cdot) * \eta(x) \, \frac{dx \, dt}{t}$$

$$= \int_{\mathbb{R}^n} \left[\sum_{j \in \mathbb{N}} \lambda_{1,j} \int_0^1 a_{1,j} * \psi_t(x) \, \frac{dt}{t} \right] \eta(x) \, dx,$$

which implies that (5.53) holds true.

In addition, observe that $\operatorname{supp}(\widehat{\psi}) \subset B(\mathbf{0}, c) \setminus B(\mathbf{0}, b)$. This implies that, for any $\gamma \in \mathbb{Z}_+^n$,

$$\int_{\mathbb{R}^n} \psi(x) x^\gamma \, dx = (-2\pi i)^{-|\gamma|} \int_{\mathbb{R}^n} \psi(x) \, (2\pi i x)^\gamma \, dx$$

$$= (-2\pi i)^{-|\gamma|} \mathcal{F} \left(\psi \, [2\pi i \, (\cdot)]^\gamma \right) (\mathbf{0})$$

$$= (-2\pi i)^{-|\gamma|} \partial^\gamma \widehat{\psi}(\mathbf{0}) = 0.$$

In what follows, let $d \geq \lfloor n(1/\theta - 1) \rfloor$ be a fixed integer and $\tau \in (n(\frac{1}{\theta} - \frac{1}{r}), \infty)$ a fixed positive constant. Then, from Lemma 5.5.9, we deduce that, for any $j \in \mathbb{N}$, $\int_0^\infty a_{1,j}(\cdot, t) * \psi_t \, \frac{dt}{t}$ is a harmless constant multiple of an (X, ∞, d, τ)-molecule centered at $B_{1,j}$. On the other hand, by (5.51), we find that, for any $j \in \mathbb{N}$ and

$t \in [1, \infty)$ and for almost every $x \in \mathbb{R}^n$, $a_{1,j}(x, t) = 0$. This further implies that, for any $j \in \mathbb{N}$,

$$\int_0^1 a_{1,j}(\cdot, t) * \psi_t(y) \frac{dt}{t} = \int_0^\infty a_{1,j}(\cdot, t) * \psi_t(y) \frac{dt}{t}.$$

Therefore, for any $j \in \mathbb{N}$, $\int_0^1 a_{1,j}(\cdot, t) * \psi_t \frac{dt}{t}$ is a harmless constant multiple of an (X, ∞, d, τ)-molecule centered at $B_{1,j}$. Combining this, (5.53), Lemma 5.4.5, and (5.52), we further conclude that

$$\left\| \int_0^1 f * \varphi_t * \psi_t(\cdot) \frac{dt}{t} \right\|_{h_X(\mathbb{R}^n)}$$

$$\lesssim \left\| \left[\sum_{j=1}^\infty \left(\frac{\lambda_{1,j}}{\|\mathbf{1}_{B_{1,j}}\|_X} \right)^s \mathbf{1}_{B_{1,j}} \right]^{1/s} \right\|_X \lesssim \|S_{\text{loc}}(f)\|_X. \tag{5.55}$$

Next, we prove that

$$\|f * \varphi_0 * \psi_0\|_{h_X(\mathbb{R}^n)} \lesssim \|S_{\text{loc}}(f)\|_X. \tag{5.56}$$

To achieve this, for any $\alpha \in \mathbb{Z}^n$, let $Q_\alpha := 2\alpha + [0, 2)^n$, $\mathbf{e} := (1, \ldots, 1) \in \mathbb{R}^n$ denote the *unit* of $\mathbb{R}^n$, and $B_\alpha := B(\mathbf{e} + \alpha, \sqrt{n})$. Take arrangements of $\{Q_\alpha\}_{\alpha \in \mathbb{Z}^n}$ and $\{B_\alpha\}_{\alpha \in \mathbb{Z}^n}$, which are denoted, respectively, by $\{Q_{2,j}\}_{j \in \mathbb{N}}$ and $\{B_{2,j}\}_{j \in \mathbb{N}}$. Then it holds true that, for any $j \in \mathbb{N}$, $Q_{2,j} \subset B_{2,j}$, $\{Q_j\}_{j \in \mathbb{N}}$ are pairwise disjoint cubes,

$$\bigcup_{j \in \mathbb{N}} Q_{2,j} = \mathbb{R}^n, \text{ and } \sum_{j \in \mathbb{N}} \mathbf{1}_{B_{2,j}} \leq 3^n.$$

For any $j \in \mathbb{N}$, if $\|f * \varphi_0 * \psi_0 \mathbf{1}_{Q_{2,j}}\|_{L^\infty(\mathbb{R}^n)} = 0$, define $\lambda_{2,j} := 0$ and $a_{2,j} := 0$; if $\|f * \varphi_0 * \psi_0 \mathbf{1}_{Q_{2,j}}\|_{L^\infty(\mathbb{R}^n)} \neq 0$, define

$$\lambda_{2,j} := \left\| f * \varphi_0 * \psi_0 \mathbf{1}_{Q_{2,j}} \right\|_{L^\infty(\mathbb{R}^n)} \left\| \mathbf{1}_{B_{2,j}} \right\|_X$$

and

$$a_{2,j} := \frac{f * \varphi_0 * \psi_0 \mathbf{1}_{Q_{2,j}}}{\|f * \varphi_0 * \psi_0 \mathbf{1}_{Q_{2,j}}\|_{L^\infty(\mathbb{R}^n)} \|\mathbf{1}_{B_{2,j}}\|_X}.$$

Then, for any $x \in \mathbb{R}^n$,

$$f * \varphi_0 * \psi_0(x) = \sum_{j \in \mathbb{N}} \lambda_{2,j} a_{2,j}(x) \tag{5.57}$$

and

$$|f * \varphi_0 * \psi_0(x)| = \sum_{j\in\mathbb{N}} \lambda_{2,j} \left|a_{2,j}(x)\right|. \tag{5.58}$$

We now prove that $f * \varphi_0 * \psi_0 = \sum_{j\in\mathbb{N}} \lambda_{2,j}a_{2,j}$ in $\mathcal{S}'(\mathbb{R}^n)$. Indeed, applying (5.54) with φ and t therein replaced, respectively, by $\varphi_0 * \psi_0$ and 1, we find that, for any $x \in \mathbb{R}^n$,

$$|f * \varphi_0 * \psi_0(x)| \lesssim (1 + |x|)^m,$$

which, together with (5.58), further implies that, for any given $\eta \in \mathcal{S}(\mathbb{R}^n)$,

$$\int_{\mathbb{R}^n} \sum_{j\in\mathbb{N}} \lambda_{2,j} \left|a_{2,j}(x)\right| |\eta(x)| \, dx$$

$$= \int_{\mathbb{R}^n} |f * \varphi_0 * \psi_0(x)| |\eta(x)| \, dx$$

$$\lesssim \int_{\mathbb{R}^n} (1 + |x|)^m |\eta(x)| \, dx \sim \int_{\mathbb{R}^n} \frac{1}{(1 + |x|)^{n+1}} \, dx < \infty.$$

From this, (5.57), and the Fubini theorem, we deduce that, for any given $\eta \in \mathcal{S}(\mathbb{R}^n)$,

$$\int_{\mathbb{R}^n} f * \varphi_0 * \psi_0(x)\eta(x) \, dx$$

$$= \int_{\mathbb{R}^n} \sum_{j\in\mathbb{N}} \lambda_{2,j}a_{2,j}(x)\eta(x) \, dx = \sum_{j\in\mathbb{N}} \lambda_{2,j} \int_{\mathbb{R}^n} a_{2,j}(x)\eta(x) \, dx.$$

This further implies that $f * \varphi_0 * \psi_0 = \sum_{j\in\mathbb{N}} \lambda_{2,j}a_{2,j}$ holds true in $\mathcal{S}'(\mathbb{R}^n)$. In addition, notice that, for any $j \in \mathbb{N}$, $\mathrm{supp}\,(a_{2,j}) \subset B_{2,j}$ and

$$\left\|a_{2,j}\right\|_{L^\infty(\mathbb{R}^n)} \leq \frac{1}{\|\mathbf{1}_{B_{2,j}}\|_X}.$$

These further imply that, for any $j \in \mathbb{N}$,

$$\left\|a_{2,j}\right\|_{L^r(\mathbb{R}^n)} \leq \frac{|B_{2,j}|^{1/r}}{\|\mathbf{1}_{B_{2,j}}\|_X}.$$

Moreover, for any $j \in \mathbb{N}$, we have $r(B_{2,j}) = \sqrt{n} \geq 1$. Therefore, applying Definition 5.3.4, we find that, for any $j \in \mathbb{N}$, $a_{2,j}$ is a local-(X, r, d)-atom supported in the ball $B_{2,j}$. Combining this, the fact that $f * \varphi_0 * \psi_0 = \sum_{j\in\mathbb{N}} \lambda_{2,j}a_{2,j}$

holds true in $\mathcal{S}'(\mathbb{R}^n)$, Lemma 5.3.7, and Definition 5.3.6, we conclude that

$$\|f * \varphi_0 * \psi_0\|_{h_X(\mathbb{R}^n)} \sim \|f * \varphi_0 * \psi_0\|_{h^{X,r,d,s}(\mathbb{R}^n)}$$

$$\lesssim \left\|\left[\sum_{j \in \mathbb{N}} \left(\frac{\lambda_{2,j}}{\|\mathbf{1}_{B_{2,j}}\|_X}\right)^s \mathbf{1}_{B_{2,j}}\right]^{\frac{1}{s}}\right\|_X$$

$$\sim \left\|\left[\sum_{j \in \mathbb{N}} \left\|f * \varphi_0 * \psi_0 \mathbf{1}_{Q_{2,j}}\right\|_{L^\infty(\mathbb{R}^n)}^s \mathbf{1}_{B_{2,j}}\right]^{\frac{1}{s}}\right\|_X . \quad (5.59)$$

Next, we estimate $\|f * \varphi_0 * \psi_0 \mathbf{1}_{Q_{2,j}}\|_{L^\infty(\mathbb{R}^n)} \mathbf{1}_{B_{2,j}}$ for any $j \in \mathbb{N}$. Indeed, for any given $j \in \mathbb{N}$ and $x \in B_{2,j}$, and for any $y \in Q_{2,j}$, we have $|y - x| < 2r(B_{2,j}) = 2\sqrt{n}$. This, combined with Lemma 5.2.5 with φ, r, and y therein replaced, respectively, by φ_0, θ, and z, implies that, for any $j \in \mathbb{N}$ and $x \in B_{2,j}$,

$$\left\|f * \varphi_0 * \psi_0 \mathbf{1}_{Q_{2,j}}\right\|_{L^\infty(\mathbb{R}^n)}$$

$$\leq \sup_{y \in \mathbb{R}^n, |y-x|<2\sqrt{n}} |f * \varphi_0 * \psi_0(y)|$$

$$\leq \sup_{y \in \mathbb{R}^n, |y-x|<2\sqrt{n}} \int_{\mathbb{R}^n} |f * \varphi_0(z)| |\psi_0(y - z)| \, dz$$

$$\lesssim \mathcal{M}^{(\theta)} (f * \varphi_0)(x) \sup_{y \in \mathbb{R}^n, |y-x|<2\sqrt{n}} \int_{\mathbb{R}^n} (1 + |x - z|)^{\frac{n}{\theta}} |\psi_0(y - z)| \, dz.$$

$$(5.60)$$

Observe that, for any $x, y, z \in \mathbb{R}^n$ with $|y - x| < 2\sqrt{n}$,

$$|x - z| \leq |y - x| + |y - z| < 2\sqrt{n} + |y - z|.$$

From this and (5.60), we deduce that, for any $j \in \mathbb{N}$ and $x \in B_{2,j}$,

$$\left\|f * \varphi_0 * \psi_0 \mathbf{1}_{Q_{2,j}}\right\|_{L^\infty(\mathbb{R}^n)}$$

$$\lesssim \mathcal{M}^{(\theta)} (f * \varphi_0)(x) \sup_{y \in \mathbb{R}^n, |y-x|<2\sqrt{n}} \int_{\mathbb{R}^n} (1 + |y - z|)^{\frac{n}{\theta}} |\psi_0(y - z)| \, dz$$

$$\lesssim \mathcal{M}^{(\theta)} (f * \varphi_0)(x) \int_{\mathbb{R}^n} \frac{1}{(1 + |z|)^{n+1}} \, dz \sim \mathcal{M}^{(\theta)} (f * \varphi_0)(x).$$

This further implies that, for any $j \in \mathbb{N}$,

$$\left\| f * \varphi_0 * \psi_0 \mathbf{1}_{Q_{2,j}} \right\|_{L^\infty(\mathbb{R}^n)} \mathbf{1}_{B_{2,j}} \lesssim \mathcal{M}^{(\theta)} \left(f * \varphi_0 \right) \mathbf{1}_{B_{2,j}}.$$

Applying this, Definition 1.2.13(ii), (5.59), the fact that $\sum_{j \in \mathbb{N}} \mathbf{1}_{B_{2,j}} \leq 3^n$, Assumption 1.2.29 with $f_1 := f * \varphi_0$ and $f_j := 0$ for any $j \in \mathbb{N} \cap [2, \infty)$, and Definition 5.1.1, we find that

$$\| f * \varphi_0 * \psi_0 \|_{h_X(\mathbb{R}^n)} \lesssim \left\| \mathcal{M}^{(\theta)} \left(f * \varphi_0 \right) \left(\sum_{j \in \mathbb{N}} \mathbf{1}_{B_{2,j}} \right)^{\frac{1}{s}} \right\|_X$$

$$\lesssim \left\| \mathcal{M}^{(\theta)} \left(f * \varphi_0 \right) \right\|_X \lesssim \| f * \varphi_0 \|_X \lesssim \| S_{\mathrm{loc}} (f) \|_X,$$

which completes the proof of (5.56). Combining (5.49), (5.55), and (5.56), we have

$$\| f \|_{h_X(\mathbb{R}^n)} \lesssim \| f * \varphi_0 * \psi_0 \|_{h_X(\mathbb{R}^n)} + \left\| \int_0^1 f * \varphi_t * \psi_t(\cdot) \frac{dt}{t} \right\|_{h_X(\mathbb{R}^n)}$$

$$\lesssim \| S_{\mathrm{loc}} (f) \|_X < \infty. \tag{5.61}$$

This further implies that $f \in h_X(\mathbb{R}^n)$ and then finishes the proof of the sufficiency.

Next, we prove the necessity. For this purpose, let $f \in h_X(\mathbb{R}^n)$. Then, by Remark 5.1.6 and Theorem 5.2.3, we find that

$$\| f * \varphi_0 \|_X \lesssim \| f \|_{h_X(\mathbb{R}^n)}. \tag{5.62}$$

On the other hand, we estimate $\widetilde{S}_{\mathrm{loc}} (f)$ with $\widetilde{S}_{\mathrm{loc}}$ as in (5.48). Indeed, from both (5.48) and Definition 4.6.1, we infer that

$$\left\| \widetilde{S}_{\mathrm{loc}} (f) \right\|_X \lesssim \left\| \widetilde{S}_{\mathrm{loc}} (f * \varphi_0) \right\|_X + \left\| \widetilde{S}_{\mathrm{loc}} (f - f * \varphi_0) \right\|_X$$

$$\lesssim \left\| \widetilde{S}_{\mathrm{loc}} (f * \varphi_0) \right\|_X + \| S (f - f * \varphi_0) \|_X, \tag{5.63}$$

where $S(f - f * \varphi_0)$ is defined as in Definition 4.6.1 with f therein replaced by $f - f * \varphi_0$. By Remark 5.1.6 and Theorem 5.2.3, we conclude that

$$f - f * \varphi_0 \in H_X(\mathbb{R}^n).$$

From this, Lemma 5.5.6, and Theorem 5.2.3 again, we further deduce that

$$\| S (f - f * \varphi_0) \|_X \sim \| f - f * \varphi_0 \|_{H_X(\mathbb{R}^n)} \lesssim \| f \|_{h_X(\mathbb{R}^n)}. \tag{5.64}$$

Now, let $\{Q_{2,j}\}_{j\in\mathbb{N}}$ and $\{B_{2,j}\}_{j\in\mathbb{N}}$ be the same as in the proof of the sufficiency. For any $j \in \mathbb{N}$, if $\|f * \varphi_0 \mathbf{1}_{Q_{2,j}}\|_{L^\infty(\mathbb{R}^n)} = 0$, define

$$\mu_{2,j} := 0 \text{ and } b_{2,j} := 0;$$

if $\|f * \varphi_0 \mathbf{1}_{Q_{2,j}}\|_{L^\infty(\mathbb{R}^n)} \neq 0$, define

$$\mu_{2,j} := \left\|f * \varphi_0 \mathbf{1}_{Q_{2,j}}\right\|_{L^\infty(\mathbb{R}^n)} \left\|\mathbf{1}_{B_{2,j}}\right\|_X$$

and

$$b_{2,j} := \frac{f * \varphi_0 \mathbf{1}_{Q_{2,j}}}{\|f * \varphi_0 \mathbf{1}_{Q_{2,j}}\|_{L^\infty(\mathbb{R}^n)} \|\mathbf{1}_{B_{2,j}}\|_X}.$$

Then, repeating an argument similar to that used in the proof of the sufficiency of the present theorem with $\varphi_0 * \psi_0$ therein replaced by φ_0, we find that, for any $j \in \mathbb{N}$, $b_{2,j}$ is a local-(X, r, d)-atom supported in the ball $B_{2,j}$, and

$$f * \varphi_0 = \sum_{j\in\mathbb{N}} \mu_{2,j} b_{2,j}$$

in $\mathcal{S}'(\mathbb{R}^n)$. This further implies that, for any $t \in (0, \infty)$ and $x \in \mathbb{R}^n$,

$$f * \varphi_0 * \varphi_t(x) = \int_{\mathbb{R}^n} f * \varphi_0(y)\varphi_t(x - y)\, dy$$

$$= \sum_{j\in\mathbb{N}} \mu_{2,j} \int_{\mathbb{R}^n} b_{2,j}(y)\varphi_t(x - y)\, dy$$

$$= \sum_{j\in\mathbb{N}} \mu_{2,j} b_{2,j} * \varphi_t(x).$$

By this, (5.48), the Minkowski inequality, and Definition 1.2.13(ii), we obtain

$$\left\|\widetilde{S}_{\mathrm{loc}}\left(f * \varphi_0\right)\right\|_X$$

$$\leq \left\|\sum_{j\in\mathbb{N}} \mu_{2,j} \widetilde{S}_{\mathrm{loc}}\left(b_{2,j}\right)\right\|_X$$

$$\lesssim \left\|\sum_{j\in\mathbb{N}} \mu_{2,j} \widetilde{S}_{\mathrm{loc}}\left(b_{2,j}\right) \mathbf{1}_{2B_{2,j}}\right\|_X + \left\|\sum_{j\in\mathbb{N}} \mu_{2,j} \widetilde{S}_{\mathrm{loc}}\left(b_{2,j}\right) \mathbf{1}_{(2B_{2,j})^\complement}\right\|_X$$

$$=: \mathrm{VI}_1 + \mathrm{VI}_2. \tag{5.65}$$

For any $j \in \mathbb{N}$, from (5.48), Lemma 5.5.5 with $p := r$ and $f := b_{2,j}$, and the fact that $b_{2,j}$ is a local-(X, ∞, r)-atom supported in the ball $B_{2,j}$, it follows that

$$\left\| \widetilde{S}_{\mathrm{loc}}\left(b_{2,j}\right) \right\|_{L^r(\mathbb{R}^n)} \lesssim \left\| S\left(b_{2,j}\right) \right\|_{L^r(\mathbb{R}^n)} \lesssim \left\| b_{2,j} \right\|_{L^r(\mathbb{R}^n)} \lesssim \frac{|B_{2,j}|^{1/r}}{\left\| \mathbf{1}_{B_{2,j}} \right\|_X}.$$

This, together with an argument similar to that used in the estimation of II_1, further implies that

$$\mathrm{VI}_1 \lesssim \left\| \left[\sum_{j \in \mathbb{N}} \left(\frac{\mu_{2,j}}{\left\| \mathbf{1}_{B_{2,j}} \right\|_X} \right)^s \mathbf{1}_{B_{2,j}} \right]^{\frac{1}{s}} \right\|_X$$

$$\sim \left\| \left[\sum_{j \in \mathbb{N}} \left\| f * \varphi_0 \mathbf{1}_{Q_{2,j}} \right\|_{L^\infty(\mathbb{R}^n)}^s \mathbf{1}_{B_{2,j}} \right]^{\frac{1}{s}} \right\|_X. \tag{5.66}$$

On the other hand, applying Lemma 5.5.7 with $a := b_{2,j}$ for any $j \in \mathbb{N}$ and (4.30) with $\{\lambda_j\}_{j \in \mathbb{N}}$ and $\{B_j\}_{j \in \mathbb{N}}$ therein replaced, respectively, by $\{\mu_{2,j}\}_{j \in \mathbb{N}}$ and $\{B_{2,j}\}_{j \in \mathbb{N}}$, we find that

$$\mathrm{VI}_2 \lesssim \left\| \left\{ \sum_{j \in \mathbb{N}} \left[\frac{\mu_{2,j}}{\left\| \mathbf{1}_{B_{2,j}} \right\|_X} \mathcal{M}^{(\theta)}(\mathbf{1}_{B_{2,j}}) \right]^s \right\}^{\frac{1}{s}} \right\|_X$$

$$\lesssim \left\| \left[\sum_{j \in \mathbb{N}} \left(\frac{\mu_{2,j}}{\left\| \mathbf{1}_{B_{2,j}} \right\|_X} \right)^s \mathbf{1}_{B_{2,j}} \right]^{\frac{1}{s}} \right\|_X$$

$$\sim \left\| \left[\sum_{j \in \mathbb{N}} \left\| f * \varphi_0 \mathbf{1}_{Q_{2,j}} \right\|_{L^\infty(\mathbb{R}^n)}^s \mathbf{1}_{B_{2,j}} \right]^{\frac{1}{s}} \right\|_X.$$

Combining this, (5.65), and (5.66), we further obtain

$$\left\| \widetilde{S}_{\mathrm{loc}}\left(f * \varphi_0\right) \right\|_X \lesssim \left\| \left[\sum_{j \in \mathbb{N}} \left\| f * \varphi_0 \mathbf{1}_{Q_{2,j}} \right\|_{L^\infty(\mathbb{R}^n)}^s \mathbf{1}_{B_{2,j}} \right]^{\frac{1}{s}} \right\|_X. \tag{5.67}$$

Next, in order to complete the estimation of $\|\widetilde{S}_{\mathrm{loc}}(f*\varphi_0)\|_X$, we first deal with $\|f*\varphi_0\mathbf{1}_{Q_{2,j}}\|_{L^\infty(\mathbb{R}^n)}\mathbf{1}_{B_{2,j}}$ for any $j\in\mathbb{N}$. To this end, define a function ϕ_0 by setting, for any $x\in\mathbb{R}^n$,

$$\phi_0(x):=\frac{1}{2^n}\varphi_0\left(\frac{x}{2}\right).$$

Then, by Definition 5.1.1(ii) with $\phi:=\phi_0$ and $a:=4\sqrt{n}$, we conclude that, for any $j\in\mathbb{N}$ and $x\in B_{2,j}$,

$$\left\|f*\varphi_0\mathbf{1}_{Q_{2,j}}\right\|_{L^\infty(\mathbb{R}^n)}=\sup_{y\in Q_{2,j}}|f*\varphi_0(y)|\le\sup_{y\in\mathbb{R}^n,\,|y-x|<2\sqrt{n}}\left|f*(\phi_0)_{\frac{1}{2}}(y)\right|$$
$$\le m^*_{4\sqrt{n}}(f,\phi_0)(x),$$

where $(\phi_0)_{\frac{1}{2}}$ is defined as in (4.4) with $\phi:=\phi_0$ and $t:=\frac{1}{2}$. This implies that, for any $j\in\mathbb{N}$,

$$\left\|f*\varphi_0\mathbf{1}_{Q_{2,j}}\right\|_{L^\infty(\mathbb{R}^n)}\mathbf{1}_{B_{2,j}}\le m^*_{4\sqrt{n}}(f,\phi_0)\mathbf{1}_{B_{2,j}}.\qquad(5.68)$$

On the other hand, notice that

$$\int_{\mathbb{R}^n}\phi_0(x)\,dx=\int_{\mathbb{R}^n}\varphi_0(x)\,dx=\widehat{\varphi_0}(\mathbf{0})\ne0.$$

From this, (5.67), Definition 1.2.13(ii), (5.68), the fact that $\sum_{j\in\mathbb{N}}\mathbf{1}_{B_{2,j}}\le3^n$, Remark 5.1.6, and Lemma 5.1.5, it further follows that

$$\left\|\widetilde{S}_{\mathrm{loc}}(f*\varphi_0)\right\|_X\lesssim\left\|m^*_{4\sqrt{n}}(f,\phi_0)\left(\sum_{j\in\mathbb{N}}\mathbf{1}_{B_{2,j}}\right)^{\frac{1}{s}}\right\|_X$$
$$\lesssim\left\|m^*_{4\sqrt{n}}(f,\phi_0)\right\|_X\sim\|f\|_{h_X(\mathbb{R}^n)},$$

which then completes the estimation of $\|\widetilde{S}_{\mathrm{loc}}(f*\varphi_0)\|_X$. Combining this, (5.63), (5.64), and (5.62), we further conclude that

$$\|S_{\mathrm{loc}}(f)\|_X\lesssim\|f*\varphi_0\|_X+\left\|\widetilde{S}_{\mathrm{loc}}(f)\right\|_X\lesssim\|f\|_{h_X(\mathbb{R}^n)}<\infty.$$

This implies that $S_{\mathrm{loc}}(f)\in X$, and hence finishes the proof of the necessity. Moreover, by both (5.61) and (5.68), we find that

$$\|f\|_{h_X(\mathbb{R}^n)}\sim\|S_{\mathrm{loc}}(f)\|_X,$$

where the positive equivalence constants are independent of f, which then completes the proof of Theorem 5.5.2. $\square$

Next, we are devoted to establishing the Littlewood–Paley g-function characterization of the local Hardy space $h_X(\mathbb{R}^n)$ as follows.

Theorem 5.5.11 *Let X be a ball quasi-Banach function space satisfying Assumption 1.2.33 with some $s \in (0, 1]$. Assume that, for the above s, Assumption 1.2.29 holds true for both X and $X^{s/2}$. Then $f \in h_X(\mathbb{R}^n)$ if and only if $f \in S'(\mathbb{R}^n)$ and $g_{\mathrm{loc}}(f) \in X$. Moreover, there exist two positive constants C_1 and C_2 such that, for any $f \in h_X(\mathbb{R}^n)$,*

$$C_1 \|f\|_{h_X(\mathbb{R}^n)} \le \|g_{\mathrm{loc}}(f)\|_X \le C_2 \|f\|_{h_X(\mathbb{R}^n)} .$$

To show this theorem, we need a technical estimate about the local Littlewood–Paley g-function. For this purpose, let $\varphi \in S(\mathbb{R}^n)$ be as in Definition 5.5.1, b, $t \in (0, \infty)$, and, for any $f \in S'(\mathbb{R}^n)$ and $x \in \mathbb{R}^n$,

$$\left(\varphi_t^* f\right)_b (x) := \sup_{y \in \mathbb{R}^n} \frac{|f * \varphi_t(y)|}{(1 + t^{-1}|x - y|)^b}$$

and

$$g_{b,*}^{(l)}(f)(x) := \left\{ \int_0^1 \left[\left(\varphi_t^* f\right)_b (x) \right]^2 \frac{dt}{t} \right\}^{\frac{1}{2}} .$$

Then we have the following estimate about g_{loc} and $g_{b,*}^{(l)}$, which plays an essential role in the proof of the Littlewood–Paley g-function characterization of $h_X(\mathbb{R}^n)$.

Proposition 5.5.12 *Let $0 < \theta < s \le 1$, $b \in (\frac{ns}{2\theta}, \infty)$, and X be a ball quasi-Banach function space satisfying that there exists a positive constant C such that, for any $\{f_j\}_{j \in \mathbb{N}} \subset L_{\mathrm{loc}}^1(\mathbb{R}^n)$,*

$$\left\| \left\{ \sum_{j \in \mathbb{N}} \left[\mathcal{M}^{(\theta)}(f_j) \right]^s \right\}^{1/s} \right\|_{X^{s/2}} \le C \left\| \left(\sum_{j \in \mathbb{N}} |f_j|^s \right)^{1/s} \right\|_{X^{s/2}} .$$

Then there exists a positive constant C such that, for any $f \in S'(\mathbb{R}^n)$,

$$\left\| g_{b,*}^{(l)}(f) \right\|_X \le C \|g_{\mathrm{loc}}(f)\|_X .$$

To prove Proposition 5.5.12, we first present the following auxiliary lemma which was obtained in [242, Lemma 3.21].

Lemma 5.5.13 *Let* $\varphi \in \mathcal{S}(\mathbb{R}^n)$ *satisfy that, for any* $\xi \in \mathbb{R}^n \setminus \{0\}$, *there exists a* $j \in \mathbb{Z}$ *such that* $\widehat{\varphi}(2^j \xi) \neq 0$. *Then, for any given* $N_0 \in \mathbb{N}$ *and* $r \in (0, \infty)$, *there exists a positive constant* $C_{(n, N_0, r, \varphi)}$, *depending only on* n, N_0, r, *and* φ, *such that, for any* $t \in [1, 2]$, $b \in (0, N_0]$, $j \in \mathbb{Z}$, $f \in \mathcal{S}'(\mathbb{R}^n)$, *and* $x \in \mathbb{R}^n$,

$$\left[\left(\varphi_{2^{-j}t}^* f \right)_b (x) \right]^r$$

$$\leq C_{(n, N_0, r, \varphi)} \sum_{k=0}^{\infty} 2^{-kN_0 r} 2^{(k+j)n} \int_{\mathbb{R}^n} \frac{|\varphi_{2^{-k-j}t} * f(y)|^r}{(1 + 2^j t^{-1} |x - y|)^{br}} \, dy.$$

Furthermore, we need the following well-known Aoki–Rolewicz theorem (see [90, Exercise 1.46] and also [6, 200]).

Lemma 5.5.14 *Let* X *be a ball quasi-Banach function space. Then there exists a* $\nu \in (0, 1]$ *such that, for any* $\{f_j\}_{j \in \mathbb{N}} \subset \mathcal{M}(\mathbb{R}^n)$,

$$\left\| \sum_{j \in \mathbb{N}} f_j \right\|_X^\nu \leq 4 \sum_{j \in \mathbb{N}} \| f_j \|_X^\nu.$$

Based on the above two conclusions, we now show Proposition 5.5.12.

Proof of Proposition 5.5.12 Let all the symbols be as in the present proposition and $f \in \mathcal{S}'(\mathbb{R}^n)$. Fix an $N_0 \in \mathbb{N} \cap [b, \infty)$. Then, from Lemma 5.5.13 with $r := \frac{2\theta}{s}$ and the Minkowski inequality, it follows that, for any $x \in \mathbb{R}^n$,

$$\left[g_{b,*}(f)(x) \right]^2$$

$$= \sum_{j \in \mathbb{N}} \int_{2^{-j}}^{2^{-j+1}} \left| \left(\varphi_t^* f \right)(x) \right|^2 \frac{dt}{t}$$

$$= \sum_{j \in \mathbb{N}} \int_1^2 \left| \left(\varphi_{2^{-j}t}^* f \right)(x) \right|^2 \frac{dt}{t}$$

$$\lesssim \sum_{j \in \mathbb{N}} \left[\sum_{k=0}^{\infty} 2^{-\frac{2kN_0\theta}{s}} 2^{(k+j)n} \int_{\mathbb{R}^n} \frac{[\int_1^2 |\varphi_{2^{-k-j}t} * f(y)|^2 \frac{dt}{t}]^{\frac{\theta}{s}}}{(1 + 2^j |x - y|)^{\frac{2b\theta}{s}}} \, dy \right]^{\frac{s}{\theta}}. \tag{5.69}$$

Moreover, by the assumption $b > \frac{ns}{2\theta}$, we find that, for any $j \in \mathbb{N}$, $k \in \mathbb{Z}_+$, and $x \in \mathbb{R}^n$,

$$
\int_{\mathbb{R}^n} \frac{[\int_1^2 |\varphi_{2^{-k-j}t} * f(y)|^2 \frac{dt}{t}]^{\frac{\theta}{s}}}{(1 + 2^j |x - y|)^{\frac{2b\theta}{s}}}\, dy
$$

$$
= \int_{|y-x|<2^{-j}} \frac{[\int_1^2 |\varphi_{2^{-k-j}t} * f(y)|^2 \frac{dt}{t}]^{\frac{\theta}{s}}}{(1 + 2^j |x - y|)^{\frac{2b\theta}{s}}}\, dy + \sum_{j\in\mathbb{N}} \int_{2^{i-j-1} \le |y-x| < 2^{i-j}} \cdots
$$

$$
\lesssim \sum_{i=0}^{\infty} 2^{-\frac{2ib\theta}{s}} \int_{|y-x|<2^{i-j}} \left[\int_1^2 |\varphi_{2^{-k-j}t} * f(y)|^2 \frac{dt}{t} \right]^{\frac{\theta}{s}}\, dy
$$

$$
\lesssim 2^{-jn} \mathcal{M} \left(\left[\int_1^2 |\varphi_{2^{-k-j}t} * f(\cdot)|^2 \frac{dt}{t} \right]^{\frac{\theta}{s}} \right)(x) \sum_{i=0}^{\infty} 2^{-\frac{2i\theta}{s}(b - \frac{ns}{2\theta})}
$$

$$
\sim 2^{-jn} \mathcal{M} \left(\left[\int_1^2 |\varphi_{2^{-k-j}t} * f(\cdot)|^2 \frac{dt}{t} \right]^{\frac{\theta}{s}} \right)(x).
$$

This, combined with (5.69) and the Minkowski inequality again, further implies that, for any $x \in \mathbb{R}^n$,

$$
\left[g_{b,*}^{(l)}(f)(x) \right]^{\frac{2\theta}{s}}
$$

$$
\lesssim \left\{ \sum_{j\in\mathbb{N}} \left[\sum_{k=0}^{\infty} 2^{-\frac{2k\theta}{s}(N_0 - \frac{ns}{2\theta})} \mathcal{M} \left(\left[\int_1^2 |\varphi_{2^{-k-j}t} * f(\cdot)|^2 \frac{dt}{t} \right]^{\frac{\theta}{s}} \right)(x) \right]^{\frac{s}{\theta}} \right\}^{\frac{\theta}{s}}
$$

$$
\lesssim \sum_{k=0}^{\infty} 2^{-\frac{2k\theta}{s}(N_0 - \frac{ns}{2\theta})} \left\{ \sum_{j\in\mathbb{N}} \left[\mathcal{M} \left(\left[\int_1^2 |\varphi_{2^{-k-j}t} * f(\cdot)|^2 \frac{dt}{t} \right]^{\frac{\theta}{s}} \right)(x) \right]^{\frac{s}{\theta}} \right\}^{\frac{\theta}{s}}.
$$

Using this, Definition 1.2.13(ii), Lemma 5.5.14 with X replaced by $X^{\frac{s}{2\theta}}$, and Remark 1.2.30(ii) with X replaced by $X^{\frac{s}{2}}$, we find that there exists a constant $\nu \in (0, 1]$ such that, for any $x \in \mathbb{R}^n$,

$$
\left\| g_{b,*}^{(l)}(f) \right\|_X^{\frac{2\theta\nu}{s}}
$$

$$
= \left\| \left| g_{b,*}^{(l)}(f) \right|^{\frac{2\theta}{s}} \right\|_{X^{\frac{s}{2\theta}}}^{\nu}
$$

$$\lesssim \sum_{k=0}^{\infty} 2^{-\frac{2k\theta v}{s}(N_0-\frac{ns}{2\theta})} \left\| \left\{ \sum_{j\in\mathbb{N}} \left[\mathcal{M}\left(\left[\int_1^2 |\varphi_{2^{-k-j}t} * f(\cdot)|^2 \frac{dt}{t} \right]^{\frac{\theta}{s}} \right) \right]^{\frac{s}{\theta}} \right\}^{\frac{\theta}{s}} \right\|_{X^{\frac{s}{2\theta}}}^{v}$$

$$\lesssim \sum_{k=0}^{\infty} 2^{-\frac{2k\theta v}{s}(N_0-\frac{ns}{2\theta})} \left\| \left[\sum_{j\in\mathbb{N}} \int_1^2 |\varphi_{2^{-k-j}t} * f(\cdot)|^2 \frac{dt}{t} \right]^{\frac{\theta}{s}} \right\|_{X^{\frac{s}{2\theta}}}^{v}$$

$$\sim \sum_{k=0}^{\infty} 2^{-\frac{2k\theta v}{s}(N_0-\frac{ns}{2\theta})} \left\| \left[\sum_{j\in\mathbb{N}} \int_1^2 |\varphi_{2^{-k-j}t} * f(\cdot)|^2 \frac{dt}{t} \right]^{\frac{1}{2}} \right\|_X^{\frac{2\theta v}{s}}. \tag{5.70}$$

Observe that, for any $k \in \mathbb{Z}_+$ and $x \in \mathbb{R}^n$,

$$\sum_{j\in\mathbb{N}} \int_1^2 |\varphi_{2^{-k-j}t} * f(x)|^2 \frac{dt}{t}$$

$$= \sum_{j\in\mathbb{N}} \int_{2^{-k-j}}^{2^{-k-j+1}} |\varphi_t * f(x)|^2 \frac{dt}{t}$$

$$= \int_0^{2^{-k}} |\varphi_t * f(x)|^2 \frac{dt}{t} \le g_{\mathrm{loc}}(f)(x).$$

Thus, combining this, (5.70), Definition 1.2.13(ii), and the assumption $N_0 \ge b > \frac{ns}{2\theta}$, we further obtain

$$\left\| g_{b,*}^{(l)}(f) \right\|_X^{\frac{2\theta v}{s}} \lesssim \| g_{\mathrm{loc}} \|_X^{\frac{2\theta v}{s}} \sum_{k=0}^{\infty} 2^{-\frac{2k\theta v}{s}(N_0-\frac{ns}{2\theta})} \sim \| g_{\mathrm{loc}} \|_X^{\frac{2\theta v}{s}},$$

which then completes the proof of Proposition 5.5.12. $\square$

We then turn to show Theorem 5.5.11. To this end, we need the L^p boundedness of the Littlewood–Paley g-function as follows, which was obtained in [85, Theorem 7.7].

Lemma 5.5.15 *Let $p \in (1, \infty)$ and g be as in Definition 4.6.2. Then there exists a positive constant C such that, for any $f \in L^p(\mathbb{R}^n)$,*

$$\|g(f)\|_{L^p(\mathbb{R}^n)} \le C \|f\|_{L^p(\mathbb{R}^n)}.$$

Moreover, let φ be as in Definition 5.5.1. For any $f \in \mathcal{S}'(\mathbb{R}^n)$ and $x \in \mathbb{R}^n$, let

$$\widetilde{g}_{\mathrm{loc}}(f)(x) := \left[\int_0^1 |f * \varphi_t(x)|^2 \frac{dt}{t} \right]^{\frac{1}{2}}.$$

Then, repeating an argument similar to that used in the proof of [233, p. 53] with $\widetilde{S}_l$ therein replaced by $\widetilde{g}_{\mathrm{loc}}$, we then obtain the following technical lemma about $\widetilde{g}_{\mathrm{loc}}$, which is important in the proof of Theorem 5.5.11; we omit the details.

Lemma 5.5.16 *Let X be a ball quasi-Banach function space, $d \in \mathbb{Z}_+$ and $\theta \in (0, \infty)$. Then there exists a positive constant C such that, for any local-(X, ∞, d)-atom a supported in the ball $B \in \mathbb{B}$ with $r(B) \in [1, \infty)$,*

$$\widetilde{g}_{\mathrm{loc}}(a)\mathbf{1}_{(2B)^\complement} \leq C \frac{1}{\|\mathbf{1}_B\|_X} \mathcal{M}^{(\theta)}(\mathbf{1}_B).$$

Via the above preparations, we now prove Theorem 5.5.11.

Proof of Theorem 5.5.11 Let all the symbols be as in the present theorem. We first prove the necessity. For this purpose, let $f \in h_X(\mathbb{R}^n)$. Then, repeating an argument similar to that used in the proof of the necessity of Theorem 5.5.2 with Lemmas 5.5.6, 5.5.5, and 5.5.7, therein replaced, respectively, by Lemmas 5.48, 5.5.15, and 5.5.16, we find that $g_{\mathrm{loc}}(f) \in X$ and

$$\|g_{\mathrm{loc}}(f)\| \lesssim \|f\|_{h_X(\mathbb{R}^n)}, \tag{5.71}$$

which then completes the proof of the necessity.

Now, we show the sufficiency. To this end, let $f \in \mathcal{S}'(\mathbb{R}^n)$ with $g_{\mathrm{loc}}(f) \in X$ and $b \in (\frac{ns}{2\theta}, \infty)$. Notice that, for any given $t \in (0, \infty)$ and $x \in \mathbb{R}^n$, and for any $y \in B(x, t)$, we have $\frac{1}{1+t^{-1}|x-y|} \sim 1$. This further implies that, for any $x \in \mathbb{R}^n$,

$$\widetilde{S}_{\mathrm{loc}}(f)(x) \sim \left\{ \int_0^1 \int_{B(x,t)} \left[\frac{|f * \varphi_t(y)|}{(1+t^{-1}|x-y|)^b} \right]^2 \frac{dy\,dt}{t^{n+1}} \right\}^{\frac{1}{2}}$$

$$\lesssim \left\{ \int_0^1 \left[(\varphi_t^* f)_b(x) \right]^2 |B(x,t)| \frac{dt}{t^{n+1}} \right\}^{\frac{1}{2}} \sim g_{b,*}^{(l)}(f)(x).$$

From this, Definition 5.5.1, (5.48), Definition 1.2.13(ii), and Proposition 5.5.12, it follows that

$$\|S_{\mathrm{loc}}(f)\|_X \lesssim \|f * \varphi_0\|_X + \|\widetilde{S}_{\mathrm{loc}}(f)\|_X$$
$$\lesssim \|f * \varphi_0\|_X + \|g_{b,*}^{(l)}(f)\|_X$$
$$\lesssim \|g_{\mathrm{loc}}(f)\|_X < \infty,$$

which, combined with Theorem 5.5.2, further implies that $f \in h_X(\mathbb{R}^n)$ and

$$\|f\|_{h_X(\mathbb{R}^n)} \sim \|S_{\mathrm{loc}}(f)\|_X \lesssim \|g_{\mathrm{loc}}(f)\|_X. \tag{5.72}$$

This implies that the sufficiency holds true. Therefore, by both (5.71) and (5.72), we conclude that

$$\|f\|_{h_X(\mathbb{R}^n)} \sim \|g_{\mathrm{loc}}(f)\|_X$$

with the positive equivalence constants independent of f, which then completes the proof of Theorem 5.5.11. □

We next investigate the Littlewood–Paley g_λ^*-function characterization of the Hardy space $h_X(\mathbb{R}^n)$. Namely, we have the following conclusion.

Theorem 5.5.17 *Let X be a ball quasi-Banach function space satisfying both Assumptions 1.2.29 and 1.2.33 with the same $s \in (0, 1]$, and $\lambda \in (\max\{1, \frac{2}{s}\}, \infty)$. Then $f \in h_X(\mathbb{R}^n)$ if and only if $f \in \mathcal{S}'(\mathbb{R}^n)$ and $(g_\lambda^*)_{\mathrm{loc}}(f) \in X$. Moreover, there exist two positive constants C_1 and C_2 such that, for any $f \in h_X(\mathbb{R}^n)$,*

$$C_1 \|f\|_{h_X(\mathbb{R}^n)} \leq \|(g_\lambda^*)_{\mathrm{loc}}(f)\|_X \leq C_2 \|f\|_{h_X(\mathbb{R}^n)}.$$

In order to prove this theorem, we first show the following auxiliary estimate about $(g_\lambda^*)_{\mathrm{loc}}$ and S_{loc}.

Proposition 5.5.18 *Let X be a ball quasi-Banach function space, $s \in (0, 1]$, and $\lambda \in (\max\{1, \frac{2}{s}\}, \infty)$. Assume that $X^{1/s}$ is a ball Banach function space and the Hardy–Littlewood maximal operator $\mathcal{M}$ is bounded on $(X^{1/s})'$. Then there exists a positive constant C such that, for any $f \in \mathcal{S}'(\mathbb{R}^n)$,*

$$\|(g_\lambda^*)_{\mathrm{loc}}(f)\|_X \leq C \|S_{\mathrm{loc}}(f)\|_X.$$

To obtain this proposition, we need the following estimate on the change of angles in X-tent spaces, which is just [29, Theorem 3.3].

Lemma 5.5.19 *Let X be a ball quasi-Banach function space and $s \in (0, 1]$. Assume that $X^{1/s}$ is a ball Banach function space and the Hardy–Littlewood* *maximal operator $\mathcal{M}$ is bounded on $(X^{1/s})'$. Then there exists a positive constant C such that, for any $\alpha \in [1, \infty)$ and any measurable function F on $\mathbb{R}^{n+1}_+$,*

$$\left\| \mathcal{A}^{(\alpha)}(F) \right\|_X \le C \alpha^{n \max\{\frac{1}{2}, \frac{1}{s}\}} \left\| \mathcal{A}^{(1)}(F) \right\|_X.$$

Now, we show Proposition 5.5.18.

Proof of Proposition 5.5.18 Let all the symbols be as in the present proposition, $f \in \mathcal{S}'(\mathbb{R}^n)$, and, for any α and $x \in \mathbb{R}^n$,

$$\widetilde{S}^{(\alpha)}_{\mathrm{loc}}(f)(x) := \left[\int_0^1 \int_{B(x,\alpha t)} |f * \varphi_t(y)|^2 \, \frac{dy \, dt}{t^{n+1}} \right]^{\frac{1}{2}}.$$

Then, applying Lemma 5.5.19 with $F := f * \varphi_t \mathbf{1}_{\{\tau \in (0,1)\}}$, we find that, for any $\alpha \in [1, \infty)$,

$$
\begin{aligned}
\left\| \widetilde{S}^{(\alpha)}_{\mathrm{loc}}(f) \right\|_X &= \left\| \left\{ \int_{\Gamma(\cdot)} \left[|f * \varphi_t(y)| \, \mathbf{1}_{\{\tau: \tau \in (0,1)\}}(t) \right]^2 \frac{dy \, dt}{t^{n+1}} \right\}^{\frac{1}{2}} \right\|_X \\
&= \left\| \mathcal{A}^{(\alpha)} \left(f * \varphi_t \mathbf{1}_{\{\tau \in (0,1)\}} \right) \right\|_X \\
&\lesssim \alpha^{n \max\{\frac{1}{2}, \frac{1}{s}\}} \left\| \mathcal{A}^{(1)} \left(f * \varphi_t \mathbf{1}_{\{\tau \in (0,1)\}} \right) \right\|_X \\
&\sim \alpha^{n \max\{\frac{1}{2}, \frac{1}{s}\}} \left\| \widetilde{S}_{\mathrm{loc}}(f) \right\|_X.
\end{aligned}
$$
(5.73)

In addition, for any $x \in \mathbb{R}^n$, we have

$$
\begin{aligned}
& \left(g^*_\lambda \right)_{\mathrm{loc}}(f)(x) \\
& \le |f * \varphi_0(x)| + \left\{ \int_0^1 \int_{|y-x|<t} \left(\frac{t}{t+|x-y|} \right)^{\lambda n} |f * \varphi_t(x)|^2 \, \frac{dy \, dt}{t^{n+1}} \right. \\
& \left. + \sum_{j \in \mathbb{N}} \int_0^1 \int_{2^{j-1}t \le |y-x| < 2^j t} \cdots \right\}^{\frac{1}{2}} \\
& \lesssim |f * \varphi_0(x)| + \left\{ \int_0^1 \int_{|y-x|<t} \left(\frac{t}{t+|x-y|} \right)^{\lambda n} |f * \varphi_t(x)|^2 \, \frac{dy \, dt}{t^{n+1}} \right\}^{\frac{1}{2}}
\end{aligned}
$$

$$+ \sum_{j \in \mathbb{N}} 2^{-\frac{j\lambda n}{2}} \left\{ \left[\int_0^1 \int_{|y-x|<2^j t} \cdots \right] \right\}^{\frac{1}{2}}$$

$$\sim |f * \varphi_0(x)| + \sum_{j=0}^{\infty} 2^{-\frac{j\lambda n}{2}} \widetilde{S}_{\mathrm{loc}}^{(2^j)}(f)(x).$$

This, combined with Definition 1.2.13(ii), Lemma 5.5.14 with f_1 therein replaced by $|f * \varphi_0|$ and f_{j+2} therein replaced by $2^{-\frac{j\lambda n}{2}} \widetilde{S}_{\mathrm{loc}}^{(2^j)}(f)$ for any $j \in \mathbb{Z}_+$, (5.73) with α therein replaced by 2^j for any $j \in \mathbb{Z}_+$, and the assumption $\lambda > \max\{1, \frac{2}{s}\}$, further implies that there exists a $\nu \in (0, 1]$ such that

$$\left\| \left(g_\lambda^* \right)_{\mathrm{loc}}(f) \right\|_X^\nu \lesssim \|f * \varphi_0\|_X^\nu + \sum_{j=0}^{\infty} 2^{-\frac{j\lambda n \nu}{2}} \left\| \widetilde{S}_{\mathrm{loc}}^{(2^j)}(f) \right\|_X^\nu$$

$$\lesssim \|f * \varphi_0\|_X^\nu + \left\| \widetilde{S}_{\mathrm{loc}}(f) \right\|_X^\nu \sum_{j=0}^{\infty} 2^{-\frac{j n \nu}{2} \max\{\lambda-1, \lambda-\frac{2}{s}\}}$$

$$\sim \|f * \varphi_0\|_X^\nu + \left\| \widetilde{S}_{\mathrm{loc}}(f) \right\|_X^\nu \lesssim \|S_{\mathrm{loc}}(f)\|_X^\nu,$$

which then completes the proof of Proposition 5.5.18. $\qquad \square$

Via Proposition 5.5.18, we next prove Theorem 5.5.17.

Proof of Theorem 5.5.17 Let all the symbols be as in the present theorem. We first show the necessity. For this purpose, let $f \in \mathcal{S}'(\mathbb{R}^n)$. Then, by Proposition 5.5.18 and Theorem 5.5.2, we conclude that

$$\left\| \left(g_\lambda^* \right)_{\mathrm{loc}}(f) \right\|_X \lesssim \|S_{\mathrm{loc}}(f)\|_X \sim \|f\|_{h_X(\mathbb{R}^n)} < \infty, \tag{5.74}$$

which implies that $(g_\lambda^*)_{\mathrm{loc}}(f) \in X$ and then completes the proof of the necessity.

Next, we prove the sufficiency. To this end, let $f \in \mathcal{S}'(\mathbb{R}^n)$ satisfy that $(g_\lambda^*)_{\mathrm{loc}}(f) \in X$. Observe that, for any given $t \in (0, \infty)$ and $x \in \mathbb{R}^n$, and for any $y \in B(x, t)$,

$$\frac{t}{t + |x - y|} \sim 1.$$

This, together with Definition 5.5.1, further implies that, for any $x \in \mathbb{R}^n$,

$$S_{\text{loc}}(f)(x)$$

$$\sim |f * \varphi_0(x)| + \left[\int_0^1 \int_{B(x,t)} \left(\frac{t}{t + |x - y|} \right)^{\lambda n} |f * \varphi_t(y)|^2 \, \frac{dy \, dt}{t^{n+1}} \right]^{\frac{1}{2}}$$

$$\lesssim \left(g_\lambda^* \right)_{\text{loc}}(f)(x).$$

From this and Definition 1.2.13(ii), it follows that

$$\left\| S_{\text{loc}}(f) \right\|_X \lesssim \left\| \left(g_\lambda^* \right)_{\text{loc}}(f) \right\|_X < \infty.$$

This, combined with Theorem 5.5.2, further implies that $f \in h_X(\mathbb{R}^n)$ and

$$\|f\|_{h_X(\mathbb{R}^n)} \sim \left\| S_{\text{loc}}(f) \right\|_X \lesssim \left\| \left(g_\lambda^* \right)_{\text{loc}}(f) \right\|_X,$$

which completes the proof of the sufficiency. By this and (5.74), we further conclude that

$$\|f\|_{h_X(\mathbb{R}^n)} \sim \left\| \left(g_\lambda^* \right)_{\text{loc}}(f) \right\|_X,$$

where the positive equivalence constants are independent of f, which then completes the proof of Theorem 5.5.17.   □

Remark 5.5.20 We point out that Theorems 5.5.2, 5.5.11, and 5.5.17 have a wide range of applications. Here we present several function spaces to which Theorems 5.5.2, 5.5.11, and 5.5.17 can be applied.

(i) Let $p \in (0, \infty)$ and $\lambda \in (\max\{1, \frac{2}{p}\}, \infty)$. Then, in this case, by both Remarks 1.2.31(i) and 1.2.34(i), we can easily conclude that the Lebesgue space $L^p(\mathbb{R}^n)$ satisfies all the assumptions of Theorems 5.5.2, 5.5.11, and 5.5.17. Therefore, Theorems 5.5.2, 5.5.11, and 5.5.17 with $X := L^p(\mathbb{R}^n)$ hold true. These Littlewood–Paley function characterizations of the local Hardy space $h^p(\mathbb{R}^n)$ can also be deduced from both [18, (15)] and [232, Theorem 2.6].

(ii) Let $p \in (0, \infty)$, $\upsilon \in A_\infty(\mathbb{R}^n)$, and $\lambda \in (\max\{1, \frac{2q_\upsilon}{p}\}, \infty)$, where q_υ is the same as in (1.58). Then, in this case, as was pointed out in [29, Subsection 5.4], the weighted Lebesgue space $L_\upsilon^p(\mathbb{R}^n)$ satisfies all the assumptions of Theorems 5.5.2, 5.5.11, and 5.5.17. This then implies that Theorems 5.5.2, 5.5.11, and 5.5.17 with $X := L_\upsilon^p(\mathbb{R}^n)$ hold true. To the best of our knowledge, this result is totally new.

(iii) Let $\vec{p} := (p_1, \ldots, p_n) \in (0, \infty)^n$ and

$$\lambda \in \left(\max\left\{ 1, \frac{2}{\min\{p_1, \ldots, p_n\}} \right\}, \infty \right).$$

Then, in this case, as was mentioned in [29, Subsection 5.2], the mixed-norm Lebesgue space $L^{\vec{p}}(\mathbb{R}^n)$ satisfies all the assumptions of Theorems 5.5.2, 5.5.11, and 5.5.17. Thus, Theorems 5.5.2, 5.5.11, and 5.5.17 with $X := L^{\vec{p}}(\mathbb{R}^n)$ hold true. To the best of our knowledge, this result is totally new.

(iv) Let $0 < q \leq p < \infty$ and $\lambda \in (\max\{1, \frac{2}{q}\}, \infty)$. Then, in this case, as was pointed out in [29, Subsection 5.1], the Morrey space $M_q^p(\mathbb{R}^n)$ satisfies all the assumptions of Theorems 5.5.2, 5.5.11, and 5.5.17. Therefore, Theorems 5.5.2, 5.5.11, and 5.5.17 with $X := M_q^p(\mathbb{R}^n)$ hold true. To the best of our knowledge, this result is totally new.

(v) Let $p(\cdot) \in C^{\log}(\mathbb{R}^n)$ satisfy $0 < p_- \leq p_+ < \infty$, and $\lambda \in (\max\{1, \frac{2}{p_-}\}, \infty)$, where p_- and p_+ are defined, respectively, in (1.59) and (1.60). Then, in this case, as was mentioned in [29, Subsection 5.3], the variable Lebesgue space $L^{p(\cdot)}(\mathbb{R}^n)$ satisfies all the assumptions of Theorems 5.5.2, 5.5.11, and 5.5.17. This further implies that Theorems 5.5.2, 5.5.11, and 5.5.17 with $X := L^{p(\cdot)}(\mathbb{R}^n)$ hold true. To the best of our knowledge, this result is totally new.

Via the various Littlewood–Paley function characterizations of $h_X(\mathbb{R}^n)$ obtained above, in the remainder of this section, we turn to establish the Lusin area function, the Littlewood–Paley g-function, and the Littlewood–Paley g_λ^*-function characterization of local generalized Herz–Hardy spaces $h\dot{\mathcal{K}}_{\omega,0}^{p,q}(\mathbb{R}^n)$ and $h\dot{\mathcal{K}}_\omega^{p,q}(\mathbb{R}^n)$. Indeed, we have the following Littlewood–Paley function characterizations of the local generalized Herz–Hardy space $h\dot{\mathcal{K}}_{\omega,0}^{p,q}(\mathbb{R}^n)$.

Theorem 5.5.21 *Let* $p, q \in (0, \infty)$, $\omega \in M(\mathbb{R}_+)$ *satisfy* $m_0(\omega) \in (-\frac{n}{p}, \infty)$ *and* $m_\infty(\omega) \in (-\frac{n}{p}, \infty)$,

$$s_0 := \min\left\{1, p, q, \frac{n}{\max\{M_0(\omega), M_\infty(\omega)\} + n/p}\right\},$$

and $\lambda \in (\max\{1, 2/s_0\}, \infty)$. *Then the following four statements are mutually equivalent:*

(i) $f \in h\dot{\mathcal{K}}_{\omega,0}^{p,q}(\mathbb{R}^n)$;

(ii) $f \in \mathcal{S}'(\mathbb{R}^n)$ *and* $S_{\mathrm{loc}}(f) \in \dot{\mathcal{K}}_{\omega,0}^{p,q}(\mathbb{R}^n)$;

(iii) $f \in \mathcal{S}'(\mathbb{R}^n)$ *and* $g_{\mathrm{loc}}(f) \in \dot{\mathcal{K}}_{\omega,0}^{p,q}(\mathbb{R}^n)$;

(iv) $f \in \mathcal{S}'(\mathbb{R}^n)$ *and* $(g_\lambda^*)_{\mathrm{loc}}(f) \in \dot{\mathcal{K}}_{\omega,0}^{p,q}(\mathbb{R}^n)$.

Moreover, for any $f \in h\dot{\mathcal{K}}_{\omega,0}^{p,q}(\mathbb{R}^n)$,

$$\|f\|_{h\dot{\mathcal{K}}_{\omega,0}^{p,q}(\mathbb{R}^n)} \sim \|S_{\mathrm{loc}}(f)\|_{\dot{\mathcal{K}}_{\omega,0}^{p,q}(\mathbb{R}^n)} \sim \|g_{\mathrm{loc}}(f)\|_{\dot{\mathcal{K}}_{\omega,0}^{p,q}(\mathbb{R}^n)}$$

$$\sim \left\|\left(g_\lambda^*\right)_{\mathrm{loc}}(f)\right\|_{\dot{\mathcal{K}}_{\omega,0}^{p,q}(\mathbb{R}^n)},$$

where the positive equivalence constants are independent of f.

Proof Let all the symbols be as in the present theorem, $s \in (\frac{2}{\lambda}, s_0)$, and $\theta \in (0, \min\{s, \frac{s^2}{2}\})$. Then, by the proof of Theorem 4.6.3, we find that the following three statements hold true:

(i) $\dot{\mathcal{K}}^{p,q}_{\omega,0}(\mathbb{R}^n)$ is the BQBF space and $[\dot{\mathcal{K}}^{p,q}_{\omega,0}(\mathbb{R}^n)]^{1/s}$ is a BBF space;

(ii) for any $\{f_j\}_{j\in\mathbb{N}} \subset L^1_{\mathrm{loc}}(\mathbb{R}^n)$,

$$\left\| \left\{ \sum_{j\in\mathbb{N}} \left[\mathcal{M}^{(\theta)}(f_j) \right]^s \right\}^{1/s} \right\|_{\dot{\mathcal{K}}^{p,q}_{\omega,0}(\mathbb{R}^n)} \lesssim \left\| \left(\sum_{j\in\mathbb{N}} |f_j|^s \right)^{1/s} \right\|_{\dot{\mathcal{K}}^{p,q}_{\omega,0}(\mathbb{R}^n)}$$

and

$$\left\| \left\{ \sum_{j\in\mathbb{N}} \left[\mathcal{M}^{(\theta)}(f_j) \right]^s \right\}^{1/s} \right\|_{[\dot{\mathcal{K}}^{p,q}_{\omega,0}(\mathbb{R}^n)]^{s/2}} \lesssim \left\| \left(\sum_{j\in\mathbb{N}} |f_j|^s \right)^{1/s} \right\|_{[\dot{\mathcal{K}}^{p,q}_{\omega,0}(\mathbb{R}^n)]^{s/2}} ;$$

(iii) for any $f \in L^1_{\mathrm{loc}}(\mathbb{R}^n)$,

$$\left\| \mathcal{M}^{((r/s)')}(f) \right\|_{([\dot{\mathcal{K}}^{p,q}_{\omega,0}(\mathbb{R}^n)]^{1/s})'} \lesssim \|f\|_{([\dot{\mathcal{K}}^{p,q}_{\omega,0}(\mathbb{R}^n)]^{1/s})'}.$$

Combining these and Theorems 5.5.2, 5.5.11, and 5.5.17, we further conclude that (i), (ii), (iii), and (iv) are mutually equivalent and, for any $f \in h\dot{\mathcal{K}}^{p,q}_{\omega,0}(\mathbb{R}^n)$,

$$\|f\|_{h\dot{\mathcal{K}}^{p,q}_{\omega,0}(\mathbb{R}^n)} \sim \|S_{\mathrm{loc}}(f)\|_{\dot{\mathcal{K}}^{p,q}_{\omega,0}(\mathbb{R}^n)} \sim \|g_{\mathrm{loc}}(f)\|_{\dot{\mathcal{K}}^{p,q}_{\omega,0}(\mathbb{R}^n)}$$

$$\sim \left\| \left(g^*_\lambda\right)_{\mathrm{loc}}(f) \right\|_{\dot{\mathcal{K}}^{p,q}_{\omega,0}(\mathbb{R}^n)}$$

with the positive constants independent of f. This finishes the proof of Theorem 5.5.21. $\qquad\square$

From Theorem 5.5.21 and Remark 5.0.4(ii), we immediately deduce the following Littlewood–Paley function characterizations of the local generalized Hardy–Morrey space $hM^{p,q}_{\omega,0}(\mathbb{R}^n)$; we omit the details.

Corollary 5.5.22 *Let $p, q \in [1, \infty)$, $\omega \in M(\mathbb{R}_+)$ satisfy*

$$-\frac{n}{p} < m_0(\omega) \leq M_0(\omega) < 0$$

and

$$-\frac{n}{p} < m_\infty(\omega) \le M_\infty(\omega) < 0,$$

and $\lambda \in (2, \infty)$. *Then the following four statements are mutually equivalent:*

(i) $f \in h\boldsymbol{M}^{p,q}_{\omega,\mathbf{0}}(\mathbb{R}^n)$;

(ii) $f \in \mathcal{S}'(\mathbb{R}^n)$ *and* $S_{\mathrm{loc}}(f) \in \boldsymbol{M}^{p,q}_{\omega,\mathbf{0}}(\mathbb{R}^n)$;

(iii) $f \in \mathcal{S}'(\mathbb{R}^n)$ *and* $g_{\mathrm{loc}}(f) \in \boldsymbol{M}^{p,q}_{\omega,\mathbf{0}}(\mathbb{R}^n)$;

(iv) $f \in \mathcal{S}'(\mathbb{R}^n)$ *and* $(g^*_\lambda)_{\mathrm{loc}}(f) \in \boldsymbol{M}^{p,q}_{\omega,\mathbf{0}}(\mathbb{R}^n)$.

Moreover, for any $f \in h\boldsymbol{M}^{p,q}_{\omega,\mathbf{0}}(\mathbb{R}^n)$,

$$\|f\|_{h\boldsymbol{M}^{p,q}_{\omega,\mathbf{0}}(\mathbb{R}^n)} \sim \|S_{\mathrm{loc}}(f)\|_{\boldsymbol{M}^{p,q}_{\omega,\mathbf{0}}(\mathbb{R}^n)} \sim \|g_{\mathrm{loc}}(f)\|_{\boldsymbol{M}^{p,q}_{\omega,\mathbf{0}}(\mathbb{R}^n)} \sim \left\|(g^*_\lambda)_{\mathrm{loc}}(f)\right\|_{\boldsymbol{M}^{p,q}_{\omega,\mathbf{0}}(\mathbb{R}^n)},$$

where the positive equivalence constants are independent of f.

On the other hand, the following theorem gives the Littlewood–Paley function characterizations of the local generalized Herz–Hardy space $h\dot{\mathcal{K}}^{p,q}_\omega(\mathbb{R}^n)$.

Theorem 5.5.23 *Let* $p, q \in (0, \infty)$, $\omega \in M(\mathbb{R}_+)$ *satisfy* $m_0(\omega) \in (-\frac{n}{p}, \infty)$ *and*

$$-\frac{n}{p} < m_\infty(\omega) \le M_\infty(\omega) < 0,$$

$$s_0 := \min\left\{1, p, q, \frac{n}{\max\{M_0(\omega), M_\infty(\omega)\} + n/p}\right\},$$

and $\lambda \in (\max\{1, 2/s_0\}, \infty)$. *Then the following four statements are mutually equivalent:*

(i) $f \in h\dot{\mathcal{K}}^{p,q}_\omega(\mathbb{R}^n)$;

(ii) $f \in \mathcal{S}'(\mathbb{R}^n)$ *and* $S_{\mathrm{loc}}(f) \in \dot{\mathcal{K}}^{p,q}_\omega(\mathbb{R}^n)$;

(iii) $f \in \mathcal{S}'(\mathbb{R}^n)$ *and* $g_{\mathrm{loc}}(f) \in \dot{\mathcal{K}}^{p,q}_\omega(\mathbb{R}^n)$;

(iv) $f \in \mathcal{S}'(\mathbb{R}^n)$ *and* $(g^*_\lambda)_{\mathrm{loc}}(f) \in \dot{\mathcal{K}}^{p,q}_\omega(\mathbb{R}^n)$.

Moreover, for any $f \in h\dot{\mathcal{K}}^{p,q}_\omega(\mathbb{R}^n)$,

$$\|f\|_{h\dot{\mathcal{K}}^{p,q}_\omega(\mathbb{R}^n)} \sim \|S_{\mathrm{loc}}(f)\|_{\dot{\mathcal{K}}^{p,q}_\omega(\mathbb{R}^n)} \sim \|g_{\mathrm{loc}}(f)\|_{\dot{\mathcal{K}}^{p,q}_\omega(\mathbb{R}^n)}$$

$$\sim \left\|(g^*_\lambda)_{\mathrm{loc}}(f)\right\|_{\dot{\mathcal{K}}^{p,q}_\omega(\mathbb{R}^n)},$$

where the positive equivalence constants are independent of f.

To show this theorem, we need the following auxiliary lemma about translations, whose proof is just to repeat the proof of Lemma 4.6.9(iii) with $\{S, g, g^*_\lambda\}$ therein replaced by $\{S_{\mathrm{loc}}, g_{\mathrm{loc}}, (g^*_\lambda)_{\mathrm{loc}}\}$; we omit the details.

Lemma 5.5.24 *Let* $f \in \mathcal{S}'(\mathbb{R}^n)$, $\xi \in \mathbb{R}^n$, *and* $\lambda \in (0, \infty)$. *Then, for any* $A \in \{S_{\mathrm{loc}}, g_{\mathrm{loc}}, (g_\lambda^*)_{\mathrm{loc}}\}$,

$$A\left(\tau_\xi(f)\right) = \tau_\xi\left(A(f)\right).$$

With the help of Lemma 5.5.24, we now show Theorem 5.5.23.

Proof of Theorem 5.5.23 Let all the symbols be as in the present theorem. Then, repeating an argument similar to that used in the proof of Theorem 4.6.8 with Lemma 4.6.9(iii) replaced by Lemma 5.5.24, we find that (i), (ii), (iii), and (iv) are mutually equivalent and, for any $f \in h\dot{\mathcal{K}}_\omega^{p,q}(\mathbb{R}^n)$,

$$\|f\|_{h\dot{\mathcal{K}}_\omega^{p,q}(\mathbb{R}^n)} \sim \|S_{\mathrm{loc}}(f)\|_{\dot{\mathcal{K}}_\omega^{p,q}(\mathbb{R}^n)} \sim \|g_{\mathrm{loc}}(f)\|_{\dot{\mathcal{K}}_\omega^{p,q}(\mathbb{R}^n)}$$
$$\sim \left\|\left(g_\lambda^*\right)_{\mathrm{loc}}(f)\right\|_{\dot{\mathcal{K}}_\omega^{p,q}(\mathbb{R}^n)}$$

with the positive equivalence constants independent of f. This then finishes the proof of Theorem 5.5.23. $\qquad\qquad\Box$

As an application of Theorem 5.5.23, we have the Littlewood–Paley function characterizations of the local generalized Hardy–Morrey space $hM_\omega^{p,q}(\mathbb{R}^n)$ as follows, which can be deduced from Theorem 5.5.23 and Remark 5.0.4(ii) directly; we omit the details.

Corollary 5.5.25 *Let* p, q, ω, *and* λ *be as in Corollary 5.5.22. Then the following four statements are mutually equivalent:*

(i) $f \in hM_\omega^{p,q}(\mathbb{R}^n)$;
(ii) $f \in \mathcal{S}'(\mathbb{R}^n)$ *and* $S_{\mathrm{loc}}(f) \in M_\omega^{p,q}(\mathbb{R}^n)$;
(iii) $f \in \mathcal{S}'(\mathbb{R}^n)$ *and* $g_{\mathrm{loc}}(f) \in M_\omega^{p,q}(\mathbb{R}^n)$;
(iv) $f \in \mathcal{S}'(\mathbb{R}^n)$ *and* $(g_\lambda^*)_{\mathrm{loc}}(f) \in M_\omega^{p,q}(\mathbb{R}^n)$.

Moreover, for any $f \in hM_\omega^{p,q}(\mathbb{R}^n)$,

$$\|f\|_{hM_\omega^{p,q}(\mathbb{R}^n)} \sim \|S_{\mathrm{loc}}(f)\|_{M_\omega^{p,q}(\mathbb{R}^n)} \sim \|g_{\mathrm{loc}}(f)\|_{M_\omega^{p,q}(\mathbb{R}^n)}$$
$$\sim \left\|\left(g_\lambda^*\right)_{\mathrm{loc}}(f)\right\|_{M_\omega^{p,q}(\mathbb{R}^n)},$$

where the positive equivalence constants are independent of f.

5.6 Boundedness of Pseudo-Differential Operators

The target of this section is to show the boundedness of pseudo-differential operators on the local generalized Herz–Hardy spaces $h\dot{\mathcal{K}}_{\omega,0}^{p,q}(\mathbb{R}^n)$ and $h\dot{\mathcal{K}}_\omega^{p,q}(\mathbb{R}^n)$. Recall that the *Hörmander class* $S_{1,0}^0(\mathbb{R}^n)$ is defined to be the set of all the infinitely differentiable functions σ on $\mathbb{R}^n \times \mathbb{R}^n$ such that, for any $\alpha, \beta \in \mathbb{Z}_+^n$, there exists a positive constant $C_{(\alpha,\beta)}$, depending on both α and β, such that, for any $x, \xi \in \mathbb{R}^n$,

$$\left| \partial_{(1)}^\alpha \partial_{(2)}^\beta \sigma(x,\xi) \right| \leq C_{(\alpha,\beta)} (1 + |\xi|)^{-|\beta|}. \tag{5.75}$$

Via the Hörmander class $S_{1,0}^0(\mathbb{R}^n)$, we present the definition of pseudo-differential operators as follows (see, for instance, [91, Subsection 4.5.4]).

Definition 5.6.1 Let $\sigma \in S_{1,0}^0(\mathbb{R}^n)$. Then the *pseudo-differential operator* T_σ is defined by setting, for any $f \in \mathcal{S}(\mathbb{R}^n)$ and $x \in \mathbb{R}^n$,

$$T_\sigma(f)(x) := \int_{\mathbb{R}^n} \sigma(x,\xi) e^{2\pi i x \cdot \xi} \widehat{f}(\xi)\, d\xi,$$

where σ is called the *symbol* of T_σ.

Remark 5.6.2 Let T_σ be a pseudo-differential operator with symbol $\sigma \in S_{1,0}^0(\mathbb{R}^n)$. Then, applying [213, p. 250, (43)], we find that there exists a function K on $\mathbb{R}^n \times \mathbb{R}^n$ such that, for any $f \in L^2(\mathbb{R}^n)$ with compact support, and for almost every $x \notin \overline{\mathrm{supp}\,(f)} := \overline{\{x \in \mathbb{R}^n : f(x) \neq 0\}}$,

$$T_\sigma(f)(x) = \int_{\mathbb{R}^n} K(x, x - y) f(y)\, dy. \tag{5.76}$$

Moreover, from [213, p. 235, (9)], it follows that the above K satisfies that, for any given $M \in (0, \infty)$, there exists a positive constant C such that, for any $z \in \mathbb{R}^n$ with $|z| \geq 1$,

$$|K(x, z)| \leq C|z|^{-M}. \tag{5.77}$$

The function K satisfying both (5.76) and (5.77) is called the *kernel* of T_σ.

Then the following theorem shows that the pseudo-differential operator T_σ, with $\sigma \in S_{1,0}^0(\mathbb{R}^n)$, is bounded on the local generalized Herz–Hardy space $h\dot{\mathcal{K}}_{\omega,\mathbf{0}}^{p,q}(\mathbb{R}^n)$.

Theorem 5.6.3 *Let $p, q \in (0, \infty)$, $\omega \in M(\mathbb{R}_+)$ satisfy $m_0(\omega) \in (-\frac{n}{p}, \infty)$ and $m_\infty(\omega) \in (-\frac{n}{p}, \infty)$, and T_σ be a pseudo-differential operator with $\sigma \in S_{1,0}^0(\mathbb{R}^n)$. Then T_σ is well defined on $h\dot{\mathcal{K}}_{\omega,\mathbf{0}}^{p,q}(\mathbb{R}^n)$ and there exists a positive constant C such that, for any $f \in h\dot{\mathcal{K}}_{\omega,\mathbf{0}}^{p,q}(\mathbb{R}^n)$,*

$$\|T_\sigma(f)\|_{h\dot{\mathcal{K}}_{\omega,\mathbf{0}}^{p,q}(\mathbb{R}^n)} \leq C \|f\|_{h\dot{\mathcal{K}}_{\omega,\mathbf{0}}^{p,q}(\mathbb{R}^n)}.$$

To prove Theorem 5.6.3, we need the following boundedness of pseudo-differential operators on the local Hardy space $h_X(\mathbb{R}^n)$, which is just [233, Theorem 4.5].

Lemma 5.6.4 *Let X be a ball quasi-Banach function space satisfying both Assumptions 1.2.29 and 1.2.33 with the same $s \in (0, 1]$, and having an absolutely continuous quasi-norm. Assume that T_σ is a pseudo-differential operator with $\sigma \in S^0_{1,0}(\mathbb{R}^n)$. Then T_σ is well defined on $H_X(\mathbb{R}^n)$ and there exists a positive constant C such that, for any $f \in H_X(\mathbb{R}^n)$,*

$$\|T_\sigma(f)\|_{H_X(\mathbb{R}^n)} \leq C\|f\|_{H_X(\mathbb{R}^n)}.$$

Remark 5.6.5 We point out that Lemma 5.6.4 has a wide range of applications. Here we present several function spaces to which Lemma 5.6.4 can be applied.

(i) Let $p \in (0, \infty)$. Then, in this case, combining Remarks 1.2.27, 1.2.31(i), and 1.2.34(i), we can easily conclude that the Lebesgue space $L^p(\mathbb{R}^n)$ satisfies all the assumptions of Lemma 5.6.4. This then implies that Lemma 5.6.4 with $X := L^p(\mathbb{R}^n)$ holds true. If further assume that $p \in (0, 1]$, then, in this case, the aforementioned result is just [89, Theorem 4].

(ii) Let $p \in (0, \infty)$ and $\upsilon \in A_\infty(\mathbb{R}^n)$. Then, in this case, from 1.2.27, 1.2.31(ii), and 1.2.34(ii), we can easily deduce that the weighted Lebesgue space $L^p_\upsilon(\mathbb{R}^n)$ satisfies all the assumptions of Lemma 5.6.4. Therefore, Lemma 5.6.4 with $X := L^p_\upsilon(\mathbb{R}^n)$ holds true. If further assume that $p \in (0, 1]$, then, in this case, the aforementioned result was also given in [233, Corollary 4.14(b)].

(iii) Let $\vec{p} := (p_1, \dots, p_n) \in (0, \infty)^n$. Then, in this case, by Remarks 1.2.27, 1.2.31 (iii), and 1.2.34(iii), we can easily find that the mixed-norm Lebesgue space $L^{\vec{p}}(\mathbb{R}^n)$ satisfies all the assumptions of Lemma 5.6.4. Thus, Lemma 5.6.4 with $X := L^{\vec{p}}(\mathbb{R}^n)$ holds true. To the best of our knowledge, this result is totally new.

(iv) Let $p(\cdot) \in C^{\log}(\mathbb{R}^n)$ satisfy $0 < p_- \leq p_+ < \infty$, where p_- and p_+ are defined, respectively, in (1.59) and (1.60). Then, in this case, from Remarks 1.2.27, 1.2.31(v), and 1.2.34(v), we can easily infer that the variable Lebesgue space $L^{p(\cdot)}(\mathbb{R}^n)$ satisfies all the assumptions of Lemma 5.6.4. This then implies that Lemma 5.6.4 with $X := L^{p(\cdot)}(\mathbb{R}^n)$ holds true. If further assume that $p_- \in (1, \infty)$, then, in this case, the aforementioned result goes back to [142, Theorem 1.2].

Via the above lemma, we now show Theorem 5.6.3.

Proof of Theorem 5.6.3 Let all the symbols be as in the present theorem. Then, combining the assumption $m_0(\omega) \in (-\frac{n}{p}, \infty)$ and both Theorems 1.2.42 and 1.4.1, we conclude that the local generalized Herz space $\dot{\mathcal{K}}^{p,q}_{\omega,0}(\mathbb{R}^n)$ under consideration is a BQBF space having an absolutely continuous quasi-norm. This, combined with Lemma 5.6.4, implies that, to finish the proof of the present theorem, we only need to show that the Herz space $\dot{\mathcal{K}}^{p,q}_{\omega,0}(\mathbb{R}^n)$ satisfies both Assumptions 1.2.29 and 1.2.33 with the same $s \in (0, 1]$.

To achieve this, let

$$s \in \left(0, \min\left\{1, p, q, \frac{n}{\max\{M_0(\omega), M_\infty(\omega)\} + n/p}\right\}\right)$$

and $\theta \in (0, s)$. Then, applying Lemma 4.3.25, we find that, for any $\{f_j\}_{j \in \mathbb{N}} \subset L^1_{\mathrm{loc}}(\mathbb{R}^n)$,

$$\left\| \left\{ \sum_{j \in \mathbb{N}} \left[\mathcal{M}^{(\theta)}(f_j) \right]^s \right\}^{1/s} \right\|_{\dot{\mathcal{K}}^{p,q}_{\omega,\mathbf{0}}(\mathbb{R}^n)} \lesssim \left\| \left(\sum_{j \in \mathbb{N}} |f_j|^s \right)^{1/s} \right\|_{\dot{\mathcal{K}}^{p,q}_{\omega,\mathbf{0}}(\mathbb{R}^n)} ,$$

which further implies that Assumption 1.2.29 holds true for $\dot{\mathcal{K}}^{p,q}_{\omega,\mathbf{0}}(\mathbb{R}^n)$ with the above θ and s.

On the other hand, from Lemma 1.8.6 with $r := \infty$, it follows that $[\dot{\mathcal{K}}^{p,q}_{\omega,\mathbf{0}}(\mathbb{R}^n)]^{1/s}$ is a BBF space and, for any $f \in L^1_{\mathrm{loc}}(\mathbb{R}^n)$,

$$\|\mathcal{M}(f)\|_{([\dot{\mathcal{K}}^{p,q}_{\omega,\mathbf{0}}(\mathbb{R}^n)]^{1/s})'} \lesssim \|f\|_{([\dot{\mathcal{K}}^{p,q}_{\omega,\mathbf{0}}(\mathbb{R}^n)]^{1/s})'} .$$

This implies that the Herz space $\dot{\mathcal{K}}^{p,q}_{\omega,\mathbf{0}}(\mathbb{R}^n)$ under consideration satisfies Assumption 1.2.33 with the above s and $r := \infty$. Therefore, both Assumptions 1.2.29 and 1.2.33 hold true for $\dot{\mathcal{K}}^{p,q}_{\omega,\mathbf{0}}(\mathbb{R}^n)$ with the same s. Combining this, the assumptions that $\dot{\mathcal{K}}^{p,q}_{\omega,\mathbf{0}}(\mathbb{R}^n)$ under consideration is a BQBF space having an absolutely continuous quasi-norm, and Lemma 5.6.3 with $X := \dot{\mathcal{K}}^{p,q}_{\omega,\mathbf{0}}(\mathbb{R}^n)$, we further conclude that T_σ is well defined on $h\dot{\mathcal{K}}^{p,q}_{\omega,\mathbf{0}}(\mathbb{R}^n)$ and, for any $f \in h\dot{\mathcal{K}}^{p,q}_{\omega,\mathbf{0}}(\mathbb{R}^n)$,

$$\|T_\sigma(f)\|_{h\dot{\mathcal{K}}^{p,q}_{\omega,\mathbf{0}}(\mathbb{R}^n)} \lesssim \|f\|_{h\dot{\mathcal{K}}^{p,q}_{\omega,\mathbf{0}}(\mathbb{R}^n)},$$

which completes the proof of Theorem 5.6.3. $\square$

Remark 5.6.6 We should point out that, in Theorem 5.6.3, when $p \in (1, \infty)$ and $\omega(t) := t^\alpha$ for any $t \in (0, \infty)$ and for any given $\alpha \in [n(1 - \frac{1}{p}), \infty)$, then Theorem 5.6.3 goes back to [234, Theorem 2.6].

Via Theorem 5.6.3 and Remark 5.0.4(ii), we immediately obtain the boundedness of the pseudo-differential operators T_σ on the local generalized Hardy–Morrey space $h\mathbf{M}^{p,q}_{\omega,\mathbf{0}}(\mathbb{R}^n)$ as follows; we omit the details.

Corollary 5.6.7 *Let $p, q \in [1, \infty)$, $\omega \in M(\mathbb{R}_+)$ with*

$$-\frac{n}{p} < m_0(\omega) \leq M_0(\omega) < 0$$

and

$$-\frac{n}{p} < m_\infty(\omega) \le M_\infty(\omega) < 0,$$

and T_σ be a pseudo-differential operator with $\sigma \in S^0_{1,0}(\mathbb{R}^n)$. Then T_σ is well defined on $hM^{p,q}_{\omega,\mathbf{0}}(\mathbb{R}^n)$ and there exists a positive constant C such that, for any $f \in hM^{p,q}_{\omega,\mathbf{0}}(\mathbb{R}^n)$,

$$\|T_\sigma(f)\|_{hM^{p,q}_{\omega,\mathbf{0}}(\mathbb{R}^n)} \le C\|f\|_{hM^{p,q}_{\omega,\mathbf{0}}(\mathbb{R}^n)}.$$

We next turn to establish the boundedness of pseudo-differential operators on the local generalized Herz–Hardy space $h\dot{\mathcal{K}}^{p,q}_\omega(\mathbb{R}^n)$. Namely, we have the following conclusion.

Theorem 5.6.8 *Let $p, q \in (0, \infty)$, $\omega \in M(\mathbb{R}_+)$ satisfy $m_0(\omega) \in (-\frac{n}{p}, \infty)$ and*

$$-\frac{n}{p} < m_\infty(\omega) \le M_\infty(\omega) < 0,$$

and T_σ be a pseudo-differential operator with symbol $\sigma \in S^0_{1,0}(\mathbb{R}^n)$. Then T_σ is well defined on $h\dot{\mathcal{K}}^{p,q}_\omega(\mathbb{R}^n)$ and there exists a positive constant C such that, for any $f \in h\dot{\mathcal{K}}^{p,q}_\omega(\mathbb{R}^n)$,

$$\|T_\sigma(f)\|_{h\dot{\mathcal{K}}^{p,q}_\omega(\mathbb{R}^n)} \le C\|f\|_{h\dot{\mathcal{K}}^{p,q}_\omega(\mathbb{R}^n)}.$$

Due to the deficiency of associate spaces and the absolutely continuity of quasi-norms of global generalized Herz spaces, we can not show Theorem 5.6.8 using Lemma 5.6.4 directly [see Remark 1.2.19(vi) and Example 1.4.4 for the details]. To overcome these difficulties, we first establish a new boundedness criterion of pseudo-differential operators on the local Hardy space $h_X(\mathbb{R}^n)$ associated with the ball quasi-Banach function space X as follows.

Theorem 5.6.9 *Let X be a ball quasi-Banach function space, Y a linear space equipped with a quasi-seminorm $\|\cdot\|_Y$, and Y_0 a linear space equipped with a quasi-seminorm $\|\cdot\|_{Y_0}$, and let $\eta \in (1, \infty)$ and $0 < \theta < s < s_0 \le 1$ be such that*

(i) *for the above θ and s, Assumption 1.2.29 holds true;*
(ii) *both $\|\cdot\|_Y$ and $\|\cdot\|_{Y_0}$ satisfy Definition 1.2.13(ii);*
(iii) *$\mathbf{1}_{B(\mathbf{0},1)} \in Y_0$;*
(iv) *$X^{1/s}$ is a ball Banach function space and, for any $f \in \mathscr{M}(\mathbb{R}^n)$,*

$$\|f\|_{X^{1/s}} \sim \sup\left\{\|fg\|_{L^1(\mathbb{R}^n)} : \|g\|_Y = 1\right\}$$

and

$$\|f\|_{X^{1/s_0}} \sim \sup \left\{ \|fg\|_{L^1(\mathbb{R}^n)} : \|g\|_{Y_0} = 1 \right\}$$

with the positive equivalence constants independent of f;
(v) $\mathcal{M}^{(\eta)}$ *is bounded on Y and Y_0.*

Assume that T_σ is a pseudo-differential operator with symbol $\sigma \in S_{1,0}^0(\mathbb{R}^n)$. Then T_σ is well defined on $h_X(\mathbb{R}^n)$ and there exists a positive constant C such that, for any $f \in h_X(\mathbb{R}^n)$,

$$\|T_\sigma(f)\|_{h_X(\mathbb{R}^n)} \leq C \|f\|_{h_X(\mathbb{R}^n)} .$$

To prove this theorem, we require some auxiliary conclusions. First, the following lemma shows that the local Hardy space $h_X(\mathbb{R}^n)$ can be continuously embedded into $\mathcal{S}'(\mathbb{R}^n)$, which plays a vital role in the proof of Theorem 5.6.9.

Lemma 5.6.10 *Let X be a ball quasi-Banach function space and $N \in \mathbb{N}$. Then the local Hardy space $h_X(\mathbb{R}^n)$ embeds continuously into $\mathcal{S}'(\mathbb{R}^n)$. Namely, there exists a positive constant C such that, for any $f \in h_X(\mathbb{R}^n)$ and $\phi \in \mathcal{S}(\mathbb{R}^n)$,*

$$|\langle f, \phi \rangle| \leq C p_N(\phi) \|f\|_{h_X(\mathbb{R}^n)},$$

where p_N is defined as in (4.1).

Proof Let all the symbols be as in the present lemma, $f \in h_X(\mathbb{R}^n)$, and $\phi \in \mathcal{S}(\mathbb{R}^n)$. We define ψ by setting, for any $x \in \mathbb{R}^n$,

$$\psi(x) := \frac{1}{2^n} \phi \left(\frac{x}{2} \right) .$$

Now, we show the present lemma by considering the following two cases on ψ.
  Case (1) $p_N(\psi) = 0$. In this case, applying 4.1 with ϕ therein replaced by ψ, we find that $\psi = 0$. This further implies that $\phi = 0$. Thus, we have

$$|\langle f, \phi \rangle| = 0 = p_N(\phi) \|f\|_{h_X(\mathbb{R}^n)} , \tag{5.78}$$

which completes the proof of the present lemma in this case.
  Case (2) $p_N(\psi) \neq 0$. In this case, we have $\frac{\psi(-\cdot)}{p_N(\psi)} \in \mathcal{S}(\mathbb{R}^n)$,

$$\left(\frac{\psi(-\cdot)}{p_N(\psi)} \right)_{\frac{1}{2}} = \frac{\phi(-\cdot)}{p_N(\psi)}, \tag{5.79}$$

and

$$p_N \left(\frac{\psi(-\cdot)}{p_N(\psi)} \right) = \frac{p_N(\psi)}{p_N(\psi)} = 1, \qquad (5.80)$$

where $(\frac{\psi(-\cdot)}{p_N(\psi)})_{\frac{1}{2}}$ is defined as in (4.4) with ϕ and t therein replaced, respectively, by $\frac{\psi(-\cdot)}{p_N(\psi)}$ and $\frac{1}{2}$. From (5.80), it follows that $\frac{\psi(-\cdot)}{p_N(\psi)} \in \mathcal{F}_N(\mathbb{R}^n)$. This, combined with (5.79) and (5.1), further implies that, for any $x \in B(0, 1)$,

$$|\langle f, \phi \rangle| = |f * [\phi(-\cdot)] (\mathbf{0})|$$

$$= p_N(\psi) \left| f * \left[\left(\frac{\psi(-\cdot)}{p_N(\psi)} \right)_{\frac{1}{2}} \right] (\mathbf{0}) \right|$$

$$\leq p_N(\psi) m_N(f)(x). \qquad (5.81)$$

In addition, notice that

$$p_N(\psi) = \sum_{\alpha \in \mathbb{Z}_+^n, \, |\alpha| \leq N} \sup_{x \in \mathbb{R}^n} (1 + |x|)^{N+n} \left| \partial^\alpha \left(\frac{1}{2^n} \phi \left(\frac{\cdot}{2} \right) \right) (x) \right|$$

$$\sim \sum_{\alpha \in \mathbb{Z}_+^n, \, |\alpha| \leq N} \sup_{x \in \mathbb{R}^n} (1 + |x|)^{N+n} \left| \partial^\alpha \phi \left(\frac{x}{2} \right) \right|$$

$$\sim \sum_{\alpha \in \mathbb{Z}_+^n, \, |\alpha| \leq N} \sup_{x \in \mathbb{R}^n} (1 + |x|)^{N+n} \left| \partial^\alpha \phi(x) \right| \sim p_N(\phi).$$

By this and (5.81), we conclude that

$$|\langle f, \phi \rangle| \mathbf{1}_{B(0,1)} \lesssim p_N(\phi) m_N(f).$$

Combining this, an argument similar to that used in the estimation of (4.108) with $\mathcal{M}_N(f)$ therein replaced by $m_N(f)$, and Definition 5.2.1, we further obtain

$$|\langle f, \phi \rangle| \lesssim p_N(\phi) \|f\|_{h_X(\mathbb{R}^n)}, \qquad (5.82)$$

which completes the proof of the present lemma in this case. Thus, by (5.78) and (5.82), we then complete the proof of Lemma 5.6.10. □

To prove Theorem 5.6.9, we also need the following technical estimate about local atoms.

Proposition 5.6.11 *Let T_σ be a pseudo-differential operator with $\sigma \in S_{1,0}^0(\mathbb{R}^n)$, X a ball quasi-Banach function space, $\phi \in \mathcal{S}(\mathbb{R}^n)$, $\theta \in (0, 1]$, $d \geq n(\frac{1}{\theta} - 1) - 1$ be a* *fixed integer, and* $r \in [1, \infty]$. *Then there exists a positive constant C such that, for any local-(X, r, d)-atom a supported in the ball $B \in \mathbb{B}$,*

$$m\left(T_\sigma(a), \phi\right) \mathbf{1}_{(2B)^\complement} \le C \frac{1}{\|\mathbf{1}_B\|_X} \mathcal{M}^{(\theta)}(\mathbf{1}_B).$$

In order to show this estimate, we need the following auxiliary lemma about symbols and kernels of pseudo-differential operators, which is just [89, Lemma 6].

Lemma 5.6.12 *Let T_σ be a pseudo-differential operator with symbol $\sigma \in S^0_{1,0}(\mathbb{R}^n)$, $\phi \in \mathcal{S}(\mathbb{R}^n)$, and $t \in (0,1)$. The operator $T_\sigma^{(t)}$ is defined by setting, for any $f \in \mathcal{S}(\mathbb{R}^n)$,*

$$T_\sigma^{(t)}(f) := [T(f)] * \phi_t.$$

Then $T_\sigma^{(t)}$ is a pseudo-differential operator with symbol σ_t and kernel K_t satisfying that, for any $\alpha, \beta \in \mathbb{Z}^n_+$, there exists a positive constant $C_{(\alpha, \beta)}$, depending on α and β but independent of t, such that, for any $x, \xi \in \mathbb{R}^n$,

$$\left|\partial^\alpha_{(1)}\partial^\beta_{(2)}\sigma_t(x, \xi)\right| \le C_{(\alpha, \beta)}(1 + |\xi|)^{-|\beta|}$$

and, for any $z \in \mathbb{R}^n \setminus \{\mathbf{0}\}$,

$$\left|\partial^\alpha_{(1)}\partial^\beta_{(2)}K_t(x, z)\right| \le C_{(\alpha, \beta)}|z|^{-n-|\beta|}. \tag{5.83}$$

Applying the above lemma, we next prove Proposition 5.6.11.

Proof of Proposition 5.6.11 Let all the symbols be as in the present proposition and a a local-(X, r, d)-atom supported in the ball $B(x_0, r_0)$ with $x_0 \in \mathbb{R}^n$ and $r_0 \in (0, \infty)$. Then we claim that, for any $t \in (0, 1)$,

$$|T_\sigma(a) * \phi_t| \mathbf{1}_{[B(x_0, 2r_0)]^\complement} \lesssim \frac{r_0^{d+1}}{|x - x_0|^{n+d+1}} \|a\|_{L^1(\mathbb{R}^n)}, \tag{5.84}$$

where the implicit positive constant is independent of a. Assume that this claim holds true for the moment. Then, by Definition 5.1.1(i), (5.84), and an argument similar to that used in the estimation of (4.23), we conclude that

$$m\left(T_\sigma(a), \phi\right) \mathbf{1}_{[B(x_0, 2r_0)]^\complement} \lesssim \frac{1}{\|\mathbf{1}_{B(x_0, r_0)}\|_X} \mathcal{M}^{(\theta)}\left(\mathbf{1}_{B(x_0, r_0)}\right)$$

with the implicit positive constant independent of a, which completes the proof of the present proposition. Therefore, to complete the whole proof, it suffices to show the above claim. To achieve this, we now prove the above claim by considering the following two cases on r_0.

Case (1) $r_0 \in (0, 1)$. In this case, for any $t \in (0, \infty)$ and $x \in [B(x_0, 2r_0)]^\complement$, from Lemma 5.6.12, (5.76) with T_σ, K, and f therein replaced, respectively, by $T_\sigma^{(t)}$, K_t, and a, Definition 5.3.4(iii), and the Taylor remainder theorem, it follows that, for any $y \in B(x_0, r_0)$, there exists a $t_y \in (0, 1)$ such that

$$[T_\sigma(a)] * \phi_t(x)$$

$$= T_\sigma^{(t)}(a)(x) = \int_{B(x_0, r_0)} K_t(x, x - y)a(y)\, dy$$

$$= \int_{B(x_0, r_0)} \left[K_t(x, x - y) - \sum_{\substack{\gamma \in \mathbb{Z}_+^n \\ |\gamma| \leq d}} \frac{\partial_{(2)}^\gamma K(x, x - x_0)}{\gamma!}(y - x_0)^\gamma \right] a(y)\, dy$$

$$= \int_{B(x_0, r_0)} \sum_{\substack{\gamma \in \mathbb{Z}_+^n \\ |\gamma| = d+1}} \frac{\partial_{(2)}^\gamma K_t(x, x - t_y y - (1 - t_y)x_0)}{\gamma!} a(y)\, dy. \tag{5.85}$$

In addition, notice that, for any $x \in [B(x_0, 2r_0)]^\complement$ and $y \in B(x_0, r_0)$,

$$|x - x_0| \geq 2r_0 > 2|y - x_0|,$$

which further implies that

$$\left| x - t_y y - \left(1 - t_y\right) x_0 \right| = \left| x - x_0 - t_y \left(y - x_0 \right) \right|$$

$$\geq |x - x_0| - |y - x_0| > \frac{1}{2}|x - x_0|.$$

Combining this, (5.85), and (5.83) with α, β, and z therein replaced, respectively, by $\mathbf{0}$, γ for any $\gamma \in \mathbb{Z}_+^n$ satisfying that $|\gamma| = d + 1$, and $x - t_y y - (1 - t_y)x_0$ for any $x \in [B(x_0, 2r_0)]^\complement$ and $y \in B(x_0, r_0)$, we find that, for any $t \in (0, 1)$ and $x \in [B(x_0, 2r_0)]^\complement$,

$$|T_\sigma(a) * \phi_t(x)| \lesssim \int_{B(x_0, r_0)} \frac{|y - x_0|^{d+1}}{|x - t_y y - (1 - t_y)x_0|^{n+d+1}} |a(y)|\, dy$$

$$\lesssim \frac{r_0^{d+1}}{|x - x_0|^{n+d+1}} \|a\|_{L^1(\mathbb{R}^n)},$$

where the implicit positive constant is independent of a. This finishes the proof of the above claim in this case.

Case (2) $r_0 \in [1, \infty)$. In this case, for any $y \in B(x_0, r_0)$, we have

$$|x - y| \geq |x - x_0| - |y - y_0| > \frac{1}{2} |x - x_0| \geq r_0 \geq 1.$$

Using this, (5.76) with T_σ, K, and f therein replaced, respectively, by $T_\sigma^{(t)}$ and K_t for any $t \in (0, 1)$, and a, Lemma 5.6.12, and (5.77) with K, z, and M therein replaced, respectively, by K_t for any $t \in (0, 1)$, $x - y$, and $n + d + 1$, we conclude that, for any $t \in (0, 1)$ and for almost every $x \in [B(x_0, 2r_0)]^\complement$,

$$|T_\sigma(a) * \phi_t(x)|$$

$$= \left| T_\sigma^{(t)}(a)(x) \right| \leq \int_{B(x_0, r_0)} |K_t(x, x - y)| \, |a(y)| \, dy$$

$$\lesssim \int_{B(x_0, r_0)} \frac{|a(y)|}{|x - y|^{n+d+1}} \, dy \lesssim \frac{r_0^{d+1}}{|x - x_0|^{n+d+1}} \|a\|_{L^1(\mathbb{R}^n)}, \qquad (5.86)$$

where the implicit positive constants are independent of t and a. This then finishes the proof of the above claim in this case. Therefore, combining (5.85) and (5.86), we further find that (5.84) holds true and hence complete the proof of the above claim. This further finishes the proof of Proposition 5.6.11. $\qquad \square$

Furthermore, the following proposition gives the atomic decomposition of the local Hardy space $h_X(\mathbb{R}^n)$ with convergence in $h_X(\mathbb{R}^n)$, which is an essential tool in the proof of Theorem 5.6.9.

Proposition 5.6.13 *Let X be a ball quasi-Banach function space satisfying both Assumption 1.2.29 with $0 < \theta < s \leq 1$ and Assumption 1.2.33 with the same s, $d \geq \lfloor n(1/\theta - 1) \rfloor$ be a fixed integer, and $r \in (1, \infty]$. Assume that X has an absolutely continuous quasi-norm. Let $\{a_j\}_{j \in \mathbb{N}}$ be a sequence of local-(X, r, d)-atoms supported, respectively, in the balls $\{B_j\}_{j \in \mathbb{N}} \subset \mathbb{B}$ and $\{\lambda_j\}_{j \in \mathbb{N}} \subset [0, \infty)$ such that*

$$f := \sum_{j \in \mathbb{N}} \lambda_j a_j$$

in $\mathcal{S}'(\mathbb{R}^n)$ and

$$\left\| \left[\sum_{j \in \mathbb{N}} \left(\frac{\lambda_j}{\|\mathbf{1}_{B_j}\|_X} \right)^s \mathbf{1}_{B_j} \right]^{\frac{1}{s}} \right\|_X < \infty. \qquad (5.87)$$

Then $f \in h_X(\mathbb{R}^n)$ and $f = \sum_{j \in \mathbb{N}} \lambda_j a_j$ holds true in $h_X(\mathbb{R}^n)$.

In order to prove this proposition, we first present the following dominated convergence theorem of ball quasi-Banach function spaces, which was obtained in [265, Lemma 6.3] (see also [11, Chapter 1, Proposition 3.6]).

Lemma 5.6.14 *Let X be a ball quasi-Banach function space having an absolutely continuous quasi-norm. Assume that $g \in X$ and $\{f_k\}_{k\in\mathbb{N}}$ is a sequence of measurable functions satisfying that $|f_k| \leq |g|$ for any $k \in \mathbb{N}$ and $\lim_{k\to\infty} f_k = f$ almost everywhere in $\mathbb{R}^n$. Then*

$$\lim_{k\to\infty} \|f_k - f\|_X = 0.$$

We now prove Proposition 5.6.13.

Proof of Proposition 5.6.13 Let all the symbols be as in the present theorem. Then, using both Definition 5.3.6 and Lemma 5.3.7, we find that $f \in h_X(\mathbb{R}^n)$ and $f = \sum_{j\in\mathbb{N}} \lambda_j a_j$ holds true in $\mathcal{S}'(\mathbb{R}^n)$. Next, we show that $f = \sum_{j\in\mathbb{N}} \lambda_j a_j$ also holds true in $h_X(\mathbb{R}^n)$. To this end, for any $k \in \mathbb{N}$, let

$$f_k := \sum_{j=1}^{k} \lambda_j a_j.$$

Then, in order to show that $f = \sum_{j\in\mathbb{N}} \lambda_j a_j$ holds true in $h_X(\mathbb{R}^n)$, it suffices to prove that

$$\lim_{k\to\infty} \|f_k - f\|_{h_X(\mathbb{R}^n)} = 0. \tag{5.88}$$

Indeed, for any $k \in \mathbb{N}$, we have

$$f_k - f = \sum_{j=k+1}^{\infty} \lambda_j a_j$$

in $\mathcal{S}'(\mathbb{R}^n)$. Applying this, Lemma 5.3.7, Definition 5.3.6, (5.87), and Lemma 5.6.14 with $\{f_k\}_{k\in\mathbb{N}}$, f, and g therein replaced, respectively, by

$$\left\{ \left[\sum_{j=1}^{k} \left(\frac{\lambda_j}{\|\mathbf{1}_{B_j}\|_X} \right)^s \mathbf{1}_{B_j} \right]^{\frac{1}{s}} \right\}_{k\in\mathbb{N}},$$

$$\left[\sum_{j\in\mathbb{N}} \left(\frac{\lambda_j}{\|\mathbf{1}_{B_j}\|_X} \right)^s \mathbf{1}_{B_j} \right]^{\frac{1}{s}},$$

and

$$\left[\sum_{j\in\mathbb{N}}\left(\frac{\lambda_j}{\|\mathbf{1}_{B_j}\|_X}\right)^s \mathbf{1}_{B_j}\right]^{\frac{1}{s}},$$

we conclude that

$$\|f_k - f\|_{h_X(\mathbb{R}^n)} \sim \|f_k - f\|_{h^{X,r,d,s}(\mathbb{R}^n)}$$

$$\lesssim \left\|\left[\sum_{j=k+1}^{\infty}\left(\frac{\lambda_j}{\|\mathbf{1}_{B_j}\|_X}\right)^s \mathbf{1}_{B_j}\right]^{\frac{1}{s}}\right\|_X \to 0$$

as $k \to \infty$. This finishes the proof of (5.88), and further implies that $f = \sum_{j\in\mathbb{N}} \lambda_j a_j$ holds true in $h_X(\mathbb{R}^n)$, which then completes the proof of Proposition 5.6.13. □

In addition, to show Theorem 5.6.9, we also need the boundedness of pseudo-differential operators on Lebesgue spaces and the localized weighted Hardy spaces. Recall that the *local weighted Hardy space* $h_\upsilon^p(\mathbb{R}^n)$, with $p \in (0, \infty)$ and $\upsilon \in A_\infty(\mathbb{R}^n)$, is defined as in Definition 5.2.1 with $X := L_\upsilon^p(\mathbb{R}^n)$ [see also Remark 5.2.2(ii)]. Then we have the following two lemmas which can be found, respectively, in [213, p. 250, Proposition 4] and [233, Corollary 4.14(b)].

Lemma 5.6.15 *Let* $p \in (1, \infty)$ *and* T_σ *be a pseudo-differential operator with* $\sigma \in S_{1,0}^0(\mathbb{R}^n)$. *Then* T_σ *is well defined on* $L^p(\mathbb{R}^n)$ *and there exists a positive constant* C *such that, for any* $f \in L^p(\mathbb{R}^n)$,

$$\|T_\sigma(f)\|_{L^p(\mathbb{R}^n)} \leq C\|f\|_{L^p(\mathbb{R}^n)}.$$

Lemma 5.6.16 *Let* $p \in (0, 1]$, $\upsilon \in A_\infty(\mathbb{R}^n)$, *and* T_σ *be a pseudo-differential operator with symbol* $\sigma \in S_{1,0}^0(\mathbb{R}^n)$. *Then* T_σ *is well defined on* $h_\upsilon^p(\mathbb{R}^n)$ *and there exists a positive constant* C *such that, for any* $f \in h_\upsilon^p(\mathbb{R}^n)$,

$$\|T_\sigma(f)\|_{h_\upsilon^p(\mathbb{R}^n)} \leq C\|f\|_{h_\upsilon^p(\mathbb{R}^n)}.$$

Via above preparations, we next prove Theorem 5.6.9.

Proof of Theorem 5.6.9 Let all the symbols be as in the present theorem, $f \in h_X(\mathbb{R}^n)$, and $d \geq \lfloor n(1/\theta - 1) \rfloor$ be a fixed integer. Then, by the assumption (i) of the present theorem and Lemma 5.3.16, we find that there exists $\{\lambda_j\}_{j\in\mathbb{N}} \subset [0, \infty)$ and

$\{a_j\}_{j\in\mathbb{N}}$ of local-(X, ∞, d)-atoms supported, respectively, in the balls $\{B_j\}_{j\in\mathbb{N}} \subset \mathbb{B}$ such that

$$f = \sum_{j\in\mathbb{N}} \lambda_j a_j \tag{5.89}$$

in $\mathcal{S}'(\mathbb{R}^n)$ and

$$\left\| \left[\sum_{j\in\mathbb{N}} \left(\frac{\lambda_j}{\|\mathbf{1}_{B_j}\|_X} \right)^s \mathbf{1}_{B_j} \right]^{\frac{1}{s}} \right\|_X \lesssim \|f\|_{h_X(\mathbb{R}^n)}. \tag{5.90}$$

In addition, from the assumptions (ii) through (v) of the present theorem and Lemma 4.8.18 with $Y := Y_0$, $\theta := \eta$, and $s := s_0$, we deduce that there exists an $\varepsilon \in (0, 1)$ such that, for any $g \in \mathcal{M}(\mathbb{R}^n)$,

$$\|g\|_{L_\upsilon^{s_0}(\mathbb{R}^n)} \lesssim \|g\|_X,$$

where $\upsilon := [\mathcal{M}(\mathbf{1}_{B(0,1)})]^\varepsilon$. This, combined with both (5.89) and (5.90), further implies that

$$f = \sum_{j\in\mathbb{N}} \lambda_j a_j = \sum_{j\in\mathbb{N}} \left[\lambda_j \frac{\|\mathbf{1}_{B_j}\|_{L_\upsilon^{s_0}(\mathbb{R}^n)}}{\|\mathbf{1}_{B_j}\|_X} \right] \left[\frac{\|\mathbf{1}_{B_j}\|_X}{\|\mathbf{1}_{B_j}\|_{L_\upsilon^{s_0}(\mathbb{R}^n)}} a_j \right] \tag{5.91}$$

in $\mathcal{S}'(\mathbb{R}^n)$ and

$$\left\| \left\{ \sum_{j\in\mathbb{N}} \left[\frac{\lambda_j \frac{\|\mathbf{1}_{B_j}\|_{L_\upsilon^{s_0}(\mathbb{R}^n)}}{\|\mathbf{1}_{B_j}\|_X}}{\|\mathbf{1}_{B_j}\|_{L_\upsilon^{s_0}(\mathbb{R}^n)}} \right]^s \mathbf{1}_{B_j} \right\}^{\frac{1}{s}} \right\|_{L_\upsilon^{s_0}(\mathbb{R}^n)}$$

$$\sim \left\| \left[\sum_{j\in\mathbb{N}} \left(\frac{\lambda_j}{\|\mathbf{1}_{B_j}\|_X} \right)^s \mathbf{1}_{B_j} \right]^{\frac{1}{s}} \right\|_X \lesssim \|f\|_{h_X(\mathbb{R}^n)} < \infty. \tag{5.92}$$

Notice that, for any $j \in \mathbb{N}$, applying Definition 5.3.4 with $X := L_\upsilon^{s_0}(\mathbb{R}^n)$, we conclude that $\frac{\|\mathbf{1}_{B_j}\|_X}{\|\mathbf{1}_{B_j}\|_{L_\upsilon^{s_0}(\mathbb{R}^n)}} a_j$ is a local-$(L_\upsilon^{s_0}(\mathbb{R}^n), \infty, d)$-atom supported in B_j.

Now, we prove that $f \in h_v^{s_0}(\mathbb{R}^n)$ and $f = \sum_{j \in \mathbb{N}} \lambda_j a_j$ holds true in $h_v^{s_0}(\mathbb{R}^n)$. Indeed, using [233, Remarks 2.4(b), 2.7(b), and 3.4(i)], we find that the following four statements hold true:

(i) $L_v^{s_0}(\mathbb{R}^n)$ is a BQBF space;

(ii) for any $\{f_j\}_{j \in \mathbb{N}} \subset L_{\text{loc}}^1(\mathbb{R}^n)$,

$$\left\| \left\{ \sum_{j \in \mathbb{N}} \left[\mathcal{M}^{(\theta)}(f_j) \right]^s \right\}^{1/s} \right\|_{L_v^{s_0}(\mathbb{R}^n)} \lesssim \left\| \left(\sum_{j \in \mathbb{N}} |f_j|^s \right)^{1/s} \right\|_{L_v^{s_0}(\mathbb{R}^n)} ;$$

(iii) $[L_v^{s_0}(\mathbb{R}^n)]^{1/s}$ is a BBF space and, for any $f \in L_{\text{loc}}^1(\mathbb{R}^n)$,

$$\| \mathcal{M}(f) \|_{([L_v^{s_0}(\mathbb{R}^n)]^{1/s})'} \lesssim \| f \|_{([L_v^{s_0}(\mathbb{R}^n)]^{1/s})'} ;$$

(iv) $L_v^{s_0}(\mathbb{R}^n)$ has an absolutely continuous quasi-norm.

These, combined with Proposition 5.6.13 with $X := L_v^{s_0}(\mathbb{R}^n)$ and $r := \infty$, (5.91), and (5.92), further imply that $f \in h_v^{s_0}(\mathbb{R}^n)$ and $f = \sum_{j \in \mathbb{N}} \lambda_j a_j$ holds true in $h_v^{s_0}(\mathbb{R}^n)$. From this and Lemma 5.6.16 with $p := s_0$, it follows that $T_\sigma(f) = \sum_{j \in \mathbb{N}} \lambda_j T_\sigma(a_j)$ holds true in $h_v^{s_0}(\mathbb{R}^n)$. This, together with Lemma 5.6.10 with $X := L_v^{s_0}(\mathbb{R}^n)$, further implies that

$$T_\sigma(f) = \sum_{j \in \mathbb{N}} \lambda_j T_\sigma(a_j) \tag{5.93}$$

in $\mathcal{S}'(\mathbb{R}^n)$.

Now, we show that $T_\sigma(f) \in h_X(\mathbb{R}^n)$ and

$$\| T_\sigma(f) \|_{h_X(\mathbb{R}^n)} \lesssim \| f \|_{h_X(\mathbb{R}^n)} .$$

To achieve this, fix a $\phi \in \mathcal{S}(\mathbb{R}^n)$ satisfying that $\text{supp}(\phi) \subset B(\mathbf{0}, 1)$ and

$$\int_{\mathbb{R}^n} \phi(x) \, dx \neq 0.$$

Then, by (5.93) and repeating an argument similar to that used in the estimation of (4.24) with f and $\{a_j\}_{j \in \mathbb{N}}$ therein replaced, respectively, by $T_\sigma(f)$ and $\{T_\sigma(a_j)\}_{j \in \mathbb{N}}$, we find that, for any $t \in (0, \infty)$,

$$|T_\sigma(f) * \phi_t| \leq \sum_{j \in \mathbb{N}} \lambda_j \left| T_\sigma(a_j) * \phi_t \right|,$$

which, combined with Definition 5.1.1, further implies that

$$m\left(T_\sigma(f), \phi\right) \leq \sum_{j\in\mathbb{N}} \lambda_j m\left(T_\sigma(a_j), \phi\right).$$

Applying this and Definition 1.2.13(ii), we conclude that

$$\left\| m\left(T_\sigma(f), \phi\right)\right\|_X$$

$$\leq \left\| \sum_{j\in\mathbb{N}} \lambda_j m\left(T_\sigma(a_j), \phi\right)\right\|_X$$

$$\lesssim \left\| \sum_{j\in\mathbb{N}} \lambda_j m\left(T_\sigma(a_j), \phi\right)\mathbf{1}_{2B_j}\right\|_X + \left\| \sum_{j\in\mathbb{N}} \lambda_j m\left(T_\sigma(a_j), \phi\right)\mathbf{1}_{(2B_j)^\complement}\right\|_X$$

$$=: \mathrm{VII}_1 + \mathrm{VII}_2. \tag{5.94}$$

We then estimate VII_1 and VII_2 respectively.

First, we deal with VII_1. To this end, we first estimate $m(T_\sigma(a_j), \phi)$ for any $j \in \mathbb{N}$. Indeed, for any $j \in \mathbb{N}$, from Definition 5.1.1(i) and Lemma 4.3.21 with $f := T_\sigma(a_j)$ and $\Phi := \|\phi\|_{L^\infty(\mathbb{R}^n)}\mathbf{1}_{B(0,1)}$, we infer that

$$m\left(T_\sigma(a_j), \phi\right) \leq \sup_{t\in(0,\infty)} \left| T_\sigma(a_j) * \phi_t\right| \lesssim \mathcal{M}\left(T_\sigma(a_j)\right). \tag{5.95}$$

Let $r \in (\max\{1, s\eta'\}, \infty)$ with $\frac{1}{\eta} + \frac{1}{\eta'} = 1$. Then, using (5.95), the $L^r(\mathbb{R}^n)$ boundedness of the Hardy–littlewood maximal operator $\mathcal{M}$, Lemma 5.6.15 with $p := r$, and Definition 5.3.4(ii), we find that, for any $j \in \mathbb{N}$,

$$\left\| m\left(T_\sigma(a_j, \phi)\right)\right\|_{L^r(\mathbb{R}^n)} \lesssim \left\|\mathcal{M}\left(T_\sigma(a_j)\right)\right\|_{L^r(\mathbb{R}^n)} \lesssim \left\|T_\sigma(a_j)\right\|_{L^r(\mathbb{R}^n)}$$

$$\lesssim \left\|a_j\right\|_{L^r(\mathbb{R}^n)} \lesssim \frac{|B_j|^{1/r}}{\|\mathbf{1}_{B_j}\|_X}. \tag{5.96}$$

In addition, from the assumption (ii) of the present theorem and an argument similar to that used in the proof of (4.98) with $(r_0/s)'$ and $(X^{1/s})'$ therein replaced, respectively, by η and Y, it follows that, for any $f \in L^1_{\mathrm{loc}}(\mathbb{R}^n)$,

$$\left\|\mathcal{M}^{((r/s)')}(f)\right\|_Y \lesssim \|f\|_Y.$$

This, together with (5.96), an argument similar to that used in the estimation of II_1 in the proof of Theorem 4.3.18 with $\{\mathcal{M}(a_j)\}_{j\in\mathbb{N}}$ therein replaced by

$$\left\{ m\left(T_\sigma(a_j), \phi\right)\right\}_{j\in\mathbb{N}},$$

and (5.90), implies that

$$
\mathrm{VII}_1 \lesssim \left\| \left[\sum_{j\in\mathbb{N}} \left(\frac{\lambda_j}{\|\mathbf{1}_{B_j}\|_X} \right)^s \mathbf{1}_{B_j} \right]^{\frac{1}{s}} \right\|_X \lesssim \|f\|_{h_X(\mathbb{R}^n)}, \qquad (5.97)
$$

which is the desired estimate of VII_1.

On the other hand, we estimate VII_2. Indeed, applying Proposition 5.6.11 with $a := a_j$ for any $j \in \mathbb{N}$ and Definition 1.2.13(ii), we find that

$$
\mathrm{VII}_2 \lesssim \left\| \sum_{j\in\mathbb{N}} \frac{\lambda_j}{\|\mathbf{1}_{B_j}\|_X} \mathcal{M}^{(\theta)}(\mathbf{1}_{B_j}) \right\|_X .
$$

From this, (4.30), and (5.90), we further deduce that

$$
\mathrm{VII}_2 \lesssim \left\| \left[\sum_{j\in\mathbb{N}} \left(\frac{\lambda_j}{\|\mathbf{1}_{B_j}\|_X} \right)^s \mathbf{1}_{B_j} \right]^{\frac{1}{s}} \right\|_X \lesssim \|f\|_{h_X(\mathbb{R}^n)}, \qquad (5.98)
$$

which completes the estimation of VII_2. Moreover, combining both the assumptions (i) and (iv) of the present theorem, Remark 5.1.6, Lemma 5.1.5(ii), (5.94), (5.97), and (5.98), we obtain

$$
\|T_\sigma(f)\|_{h_X(\mathbb{R}^n)} \sim \|m\,(T_\sigma(f),\phi)\|_X \lesssim \|f\|_{h_X(\mathbb{R}^n)}.
$$

This finishes the proof of Theorem 5.6.9   □

Next, we show Theorem 5.6.8 with the help of Theorem 5.6.9.

Proof of Theorem 5.6.8 Let all the symbols be as in the present theorem. Then, applying the assumptions $m_0(\omega) \in (-\frac{n}{p}, \infty)$ and $M_\infty(\omega) \in (-\infty, 0)$, and Theorem 1.2.44, we find that the Herz space $\dot{\mathcal{K}}_\omega^{p,q}(\mathbb{R}^n)$ under consideration is a BQBF space. In addition, let

$$
s \in \left(0, \min\left\{ 1, p, q, \frac{n}{\max\{M_0(\omega), M_\infty(\omega)\} + n/p} \right\} \right) .
$$

Then, from the assumptions $m_0(\omega) \in (-\frac{n}{p}, \infty)$ and $M_\infty(\omega) \in (-\infty, 0)$, and Lemma 1.1.6, it follows that

$$
m_0\left(\omega^s\right) = s m_0(\omega) > -\frac{n}{p/s}.
$$

and

$$M_\infty\left(\omega^s\right) = s M_\infty(\omega) < 0.$$

Combining these, the assumptions $p/s, q/s \in (1, \infty)$, and Theorem 1.2.48 with p, q, and ω replaced, respectively, by p/s, q/s, and ω^s, we conclude that the Herz space $\dot{\mathcal{K}}_{\omega^s}^{p/s,q/s}(\mathbb{R}^n)$ is a BBF space. From this and Lemma 1.3.2, it follows that $[\dot{\mathcal{K}}_\omega^{p,q}(\mathbb{R}^n)]^{1/s}$ is a BBF space. Moreover, let $\theta \in (0, s)$,

$$s_0 \in \left(s, \min\left\{1, p, q, \frac{n}{\max\{M_0(\omega), M_\infty(\omega)\} + n/p}\right\}\right),$$

and $\eta \in (1, \infty)$ satisfy

$$\eta < \min\left\{\frac{n}{n(1 - s/p) - s \min\{m_0(\omega), m_\infty(\omega)\}}, \left(\frac{p}{s}\right)'\right\}.$$

Then, for the above θ, s, s_0, and η, repeating an argument similar to that used in the proof of Theorem 4.8.16, we find that the following five statements hold true:

(i) for any $\{f_j\}_{j\in\mathbb{N}} \subset L_{\mathrm{loc}}^1(\mathbb{R}^n)$,

$$\left\|\left\{\sum_{j\in\mathbb{N}}\left[\mathcal{M}^{(\theta)}(f_j)\right]^s\right\}^{1/s}\right\|_{\dot{\mathcal{K}}_\omega^{p,q}(\mathbb{R}^n)} \lesssim \left\|\left(\sum_{j\in\mathbb{N}}|f_j|^s\right)^{1/s}\right\|_{\dot{\mathcal{K}}_\omega^{p,q}(\mathbb{R}^n)};$$

(ii) for any $f \in \mathscr{M}(\mathbb{R}^n)$,

$$\|f\|_{[\dot{\mathcal{K}}_\omega^{p,q}(\mathbb{R}^n)]^{1/s}} \sim \sup\left\{\|fg\|_{L^1(\mathbb{R}^n)}: \|g\|_{\dot{\mathcal{B}}_{1/\omega^s}^{(p/s)',(q/s)'}(\mathbb{R}^n)} = 1\right\}$$

and

$$\|f\|_{[\dot{\mathcal{K}}_\omega^{p,q}(\mathbb{R}^n)]^{1/s_0}} \sim \sup\left\{\|fg\|_{L^1(\mathbb{R}^n)}: \|g\|_{\dot{\mathcal{B}}_{1/\omega^{s_0}}^{(p/s_0)',(q/s_0)'}(\mathbb{R}^n)} = 1\right\}$$

with the positive equivalence constants independent of f;

(iii) both $\|\cdot\|_{\dot{\mathcal{B}}_{1/\omega^s}^{(p/s)',(q/s)'}(\mathbb{R}^n)}$ and $\|\cdot\|_{\dot{\mathcal{B}}_{1/\omega^{s_0}}^{(p/s_0)',(q/s_0)'}(\mathbb{R}^n)}$ satisfy Definition 1.2.13(ii);

(iv) $\mathbf{1}_{B(0,1)} \in \dot{\mathcal{B}}_{1/\omega^{s_0}}^{(p/s_0)',(q/s_0)'}(\mathbb{R}^n)$;

(v) $\mathcal{M}^{(\eta)}$ is bounded on $\dot{\mathcal{B}}_{1/\omega^s}^{(p/s)',(q/s)'}(\mathbb{R}^n)$ and $\dot{\mathcal{B}}_{1/\omega^{s_0}}^{(p/s_0)',(q/s_0)'}(\mathbb{R}^n)$.

These, together with the facts that $\dot{\mathcal{K}}_\omega^{p,q}(\mathbb{R}^n)$ is a BQBF space and $[\dot{\mathcal{K}}_\omega^{p,q}(\mathbb{R}^n)]^{1/s}$ is a BBF space, and Theorem 5.6.9 with $X := \dot{\mathcal{K}}_\omega^{p,q}(\mathbb{R}^n)$, $Y := \dot{\mathcal{B}}_{1/\omega^s}^{(p/s)',(q/s)'}(\mathbb{R}^n)$, and

$Y_0 := \dot{\mathcal{B}}^{(p/s_0)',(q/s_0)'}_{1/\omega^{s_0}}(\mathbb{R}^n)$, further imply that T_σ is well defined on $h\dot{\mathcal{K}}^{p,q}_\omega(\mathbb{R}^n)$ and, for any $f \in h\dot{\mathcal{K}}^{p,q}_\omega(\mathbb{R}^n)$,

$$\|T_\sigma(f)\|_{h\dot{\mathcal{K}}^{p,q}_\omega(\mathbb{R}^n)} \lesssim \|f\|_{h\dot{\mathcal{K}}^{p,q}_\omega(\mathbb{R}^n)},$$

which then completes the proof of Theorem 5.6.8. □

By both Theorem 5.6.8 and Remark 5.0.4(ii), we immediately conclude that the pseudo-differential operator T_σ, with $\sigma \in S^0_{1,0}(\mathbb{R}^n)$, is bounded on the local generalized Hardy–Morrey space $hM^{p,q}_\omega(\mathbb{R}^n)$ as follows; we omit the details.

Corollary 5.6.17 *Let p, q, and ω be as in Corollary 5.6.7 and T_σ be a pseudo-differential operator with symbol $\sigma \in S^0_{1,0}(\mathbb{R}^n)$. Then T_σ is well defined on $hM^{p,q}_\omega(\mathbb{R}^n)$ and there exists a positive constant C such that, for any $f \in hM^{p,q}_\omega(\mathbb{R}^n)$,*

$$\|T_\sigma(f)\|_{hM^{p,q}_\omega(\mathbb{R}^n)} \le C\|f\|_{hM^{p,q}_\omega(\mathbb{R}^n)}.$$

Chapter 6
Weak Generalized Herz–Hardy Spaces

The main target of this chapter is to establish a complete real-variable theory of weak generalized Herz–Hardy spaces. Note that the classical weak Hardy space $WH^p(\mathbb{R}^n)$, with $p \in (0, 1]$, plays an important role in the study of the boundedness of operators in harmonic analysis. Indeed, in 1986, to find the biggest function space X such that Calderón–Zygmund operators are bounded from X to the weak Lebesgue space $WL^1(\mathbb{R}^n)$, Fefferman and Soria [78] introduced the weak Hardy space $WH^1(\mathbb{R}^n)$. Via establishing the ∞-atomic characterization of $WH^1(\mathbb{R}^n)$, Fefferman and Soria showed that the convolutional Calderón–Zygmund operators with kernels satisfying the Dini condition were bounded from $WH^1(\mathbb{R}^n)$ to $WL^1(\mathbb{R}^n)$. In addition, let $\delta \in (0, 1]$ and T be a convolutional δ-type Calderón–Zygmund operator. It is well known that, for any given $p \in (n/(n + \delta), 1]$, T is bounded on the classical Hardy space $H^p(\mathbb{R}^n)$ (see [4]). However, T is not bounded on $H^{n/(n+\delta)}(\mathbb{R}^n)$, which is called the *critical case* or the *endpoint case*. In 1988, to deal with this critical case, Liu [153] first introduced the weak Hardy space $WH^p(\mathbb{R}^n)$ with $p \in (0, 1]$, and showed that the convolutional δ-type Calderón–Zygmund operators are bounded from $H^{n/(n+\delta)}(\mathbb{R}^n)$ to $WH^{n/(n+\delta)}(\mathbb{R}^n)$ via establishing the ∞-atomic characterization of $WH^p(\mathbb{R}^n)$.

Furthermore, the classical weak Hardy space is the real interpolation space between the classical Hardy space and the Lebesgue space $L^\infty(\mathbb{R}^n)$ (see [75]), which is another motivation to develop the real-variable theory of weak Hardy spaces. Later on, a lot of works have been done in the study of weak Hardy spaces and their variants; see, for instance, [194] for weighted weak Hardy spaces, [267] for variable weak Hardy spaces, [109] for weak Hardy–Morrey spaces, [151, 269] for weak Musielak–Orlicz Hardy spaces, and [92, 100] for vector-valued weak Hardy spaces. Particularly, Zhang et al. [278] introduced the weak Hardy space $WH_X(\mathbb{R}^n)$ associated with the ball quasi-Banach function space X and characterized $WH_X(\mathbb{R}^n)$ via various maximal functions, atoms, and molecules. Moreover, they also established the boundedness of Calderón–Zygmund operators from the Hardy space $H_X(\mathbb{R}^n)$ to the weak Hardy space $WH_X(\mathbb{R}^n)$ in [278]. After

Y. Li et al., *Real-Variable Theory of Hardy Spaces Associated with Generalized Herz Spaces of Rafeiro and Samko*, Lecture Notes in Mathematics 2320, https://doi.org/10.1007/978-981-19-6788-7_6

that, various Littlewood–Paley function characterizations of $WH_X(\mathbb{R}^n)$ and the real interpolation theorems have been studied by Wang et al. in [242].

In this chapter, we first introduce weak generalized Herz–Hardy spaces and then establish their complete real-variable theory. For this purpose, recall that Zhang et al. [278] and Wang et al. [242] investigated the real-variable theory of the weak Hardy space $WH_X(\mathbb{R}^n)$ associated with the ball quasi-Banach function space X. Combining these known results and establishing some improved atomic and molecular characterizations of $WH_X(\mathbb{R}^n)$ (see Propositions 6.3.11, 6.3.15, and 6.4.5 below) as well as new real interpolation between the Hardy space $H_X(\mathbb{R}^n)$ and the Lebesgue space $L^\infty(\mathbb{R}^n)$ (see Proposition 6.7.7 below), we obtain various maximal function, atomic, and molecular characterizations of weak generalized Herz–Hardy spaces and also show that the real interpolation spaces between generalized Herz–Hardy spaces and the Lebesgue space $L^\infty(\mathbb{R}^n)$ are just the new introduced weak generalized Herz–Hardy spaces. In addition, by establishing a technical lemma about the quasi-norm $\|\cdot\|_{W\dot{\mathcal{K}}_{\omega}^{p,q}(\mathbb{R}^n)}$ and the Littlewood–Paley function characterizations of $WH_X(\mathbb{R}^n)$ obtained in [242], we show various Littlewood–Paley function characterizations of weak generalized Herz–Hardy spaces. Furthermore, we establish two boundedness criteria of Calderón–Zygmund operators from the Hardy space $H_X(\mathbb{R}^n)$ to the weak Hardy space $WH_X(\mathbb{R}^n)$ and, as a consequence, we finally deduce the boundedness of Calderón–Zygmund operators from generalized Herz–Hardy spaces to weak generalized Herz–Hardy spaces even in the critical case. In addition, we introduce weak generalized Hardy–Morrey spaces. From the fact that, under some reasonable and sharp assumptions on exponents, generalized Morrey spaces are included into the scale of generalized Herz spaces, we also obtain the corresponding real-variable characterizations and applications of weak generalized Hardy–Morrey spaces in this chapter.

We now introduce weak generalized Herz spaces as follows.

Definition 6.0.18 Let $p, q \in (0, \infty)$ and $\omega \in M(\mathbb{R}_+)$.

(i) The *weak local generalized Herz space* $W\dot{\mathcal{K}}_{\omega,\mathbf{0}}^{p,q}(\mathbb{R}^n)$ is defined to be the set of all the measurable functions f on $\mathbb{R}^n$ such that

$$\|f\|_{W\dot{\mathcal{K}}_{\omega,\mathbf{0}}^{p,q}(\mathbb{R}^n)} := \sup_{\alpha \in (0,\infty)} \left\{ \alpha \left\| \mathbf{1}_{\{x \in \mathbb{R}^n : |f(x)| > \alpha\}} \right\|_{\dot{\mathcal{K}}_{\omega,\mathbf{0}}^{p,q}(\mathbb{R}^n)} \right\} < \infty.$$

(ii) The *weak global generalized Herz space* $W\dot{\mathcal{K}}_{\omega}^{p,q}(\mathbb{R}^n)$ is defined to be the set of all the measurable functions f on $\mathbb{R}^n$ such that

$$\|f\|_{W\dot{\mathcal{K}}_{\omega}^{p,q}(\mathbb{R}^n)} := \sup_{\alpha \in (0,\infty)} \left\{ \alpha \left\| \mathbf{1}_{\{x \in \mathbb{R}^n : |f(x)| > \alpha\}} \right\|_{\dot{\mathcal{K}}_{\omega}^{p,q}(\mathbb{R}^n)} \right\} < \infty.$$

Remark 6.0.19 In Definition 6.0.18, when $\omega(t) := t^\alpha$ for any $t \in (0, \infty)$ and for any given $\alpha \in \mathbb{R}$, then the weak local generalized Herz space $W\dot{\mathcal{K}}_{\omega,\mathbf{0}}^{p,q}(\mathbb{R}^n)$ coincides with the classical *homogeneous weak Herz space* $W\dot{K}_p^{\alpha,q}(\mathbb{R}^n)$ which was originally introduced in [117] (see also [175, Section 1.4]). However, we should point out that, even in this case, the weak global generalized Herz space $W\dot{\mathcal{K}}_{\omega}^{p,q}(\mathbb{R}^n)$ is also new.

For any $N \in \mathbb{N}$ and $f \in \mathcal{S}'(\mathbb{R}^n)$, via the non-tangential grand maximal function $\mathcal{M}_N(f)$ of f as in (4.3), we then introduce the definitions of weak generalized Herz–Hardy spaces as follows.

Definition 6.0.20 Let $p, q \in (0, \infty)$, $\omega \in M(\mathbb{R}_+)$, and $N \in \mathbb{N}$.

(i) The *weak generalized Herz–Hardy space* $WH\dot{\mathcal{K}}_{\omega,\mathbf{0}}^{p,q}(\mathbb{R}^n)$, associated with the weak local generalized Herz space $W\dot{\mathcal{K}}_{\omega,\mathbf{0}}^{p,q}(\mathbb{R}^n)$, is defined to be the set of all the $f \in \mathcal{S}'(\mathbb{R}^n)$ such that

$$\|f\|_{WH\dot{\mathcal{K}}_{\omega,\mathbf{0}}^{p,q}(\mathbb{R}^n)} := \|\mathcal{M}_N(f)\|_{W\dot{\mathcal{K}}_{\omega,\mathbf{0}}^{p,q}(\mathbb{R}^n)} < \infty.$$

(ii) The *weak generalized Herz–Hardy space* $WH\dot{\mathcal{K}}_{\omega}^{p,q}(\mathbb{R}^n)$, associated with the weak global generalized Herz space $W\dot{\mathcal{K}}_{\omega}^{p,q}(\mathbb{R}^n)$, is defined to be the set of all the $f \in \mathcal{S}'(\mathbb{R}^n)$ such that

$$\|f\|_{WH\dot{\mathcal{K}}_{\omega}^{p,q}(\mathbb{R}^n)} := \|\mathcal{M}_N(f)\|_{W\dot{\mathcal{K}}_{\omega}^{p,q}(\mathbb{R}^n)} < \infty.$$

Remark 6.0.21 In Definition 6.0.20, if $\omega(t) := t^{\alpha}$ for any $t \in (0, \infty)$ and for any given $\alpha \in \mathbb{R}$, then the weak generalized Herz–Hardy space $WH\dot{\mathcal{K}}_{\omega,\mathbf{0}}^{p,q}(\mathbb{R}^n)$ coincides with the classical *homogeneous weak Herz-type Hardy space* $WH\dot{K}_p^{\alpha,q}(\mathbb{R}^n)$ which was originally introduced by Hu et al. in [118, Definition 3] (see also [175, Definition 2.1.2]). However, even in this case, the weak generalized Herz–Hardy space $WH\dot{\mathcal{K}}_{\omega}^{p,q}(\mathbb{R}^n)$ is also new.

We next introduce the following concepts of both weak generalized Morrey spaces and associated Hardy spaces.

Definition 6.0.22 Let $p, q \in (0, \infty)$ and $\omega \in M(\mathbb{R}_+)$.

(i) The *weak local generalized Morrey space* $WM_{\omega,\mathbf{0}}^{p,q}(\mathbb{R}^n)$ is defined to be the set of all the measurable functions f on $\mathbb{R}^n$ such that

$$\|f\|_{WM_{\omega,\mathbf{0}}^{p,q}(\mathbb{R}^n)} := \sup_{\alpha \in (0,\infty)} \left\{ \alpha \left\| \mathbf{1}_{\{x \in \mathbb{R}^n : |f(x)| > \alpha\}} \right\|_{M_{\omega,\mathbf{0}}^{p,q}(\mathbb{R}^n)} \right\} < \infty.$$

(ii) The *weak global generalized Morrey space* $WM_{\omega}^{p,q}(\mathbb{R}^n)$ is defined to be the set of all the measurable functions f on $\mathbb{R}^n$ such that

$$\|f\|_{WM_{\omega}^{p,q}(\mathbb{R}^n)} := \sup_{\alpha \in (0,\infty)} \left\{ \alpha \left\| \mathbf{1}_{\{x \in \mathbb{R}^n : |f(x)| > \alpha\}} \right\|_{M_{\omega}^{p,q}(\mathbb{R}^n)} \right\} < \infty.$$

Remark 6.0.23

(i) We should point out that, in Definition 6.0.22, even when $\omega(t) := t^\alpha$ for any
$t \in (0, \infty)$ and for any given $\alpha \in \mathbb{R}$, the weak generalized Morrey spaces
$WM_{\omega,\mathbf{0}}^{p,q}(\mathbb{R}^n)$ and $WM_\omega^{p,q}(\mathbb{R}^n)$ are also new.
(ii) In Definition 6.0.22, let $p, q \in [1, \infty)$ and ω satisfy

$$\max\{M_0(\omega), M_\infty(\omega)\} \in (-\infty, 0).$$

Then, in this case, by Remark 1.2.2(vi), we conclude that the weak generalized
Morrey spaces $WM_{\omega,\mathbf{0}}^{p,q}(\mathbb{R}^n)$ and $WM_\omega^{p,q}(\mathbb{R}^n)$ coincide, respectively, with the
weak generalized Herz spaces $W\dot{\mathcal{K}}_{\omega,\mathbf{0}}^{p,q}(\mathbb{R}^n)$ and $W\dot{\mathcal{K}}_\omega^{p,q}(\mathbb{R}^n)$ in the sense of
equivalent quasi-norms.

Definition 6.0.24 Let $p, q \in (0, \infty)$, $\omega \in M(\mathbb{R}_+)$, and $N \in \mathbb{N}$.

(i) The *weak generalized Hardy–Morrey space* $WHM_{\omega,\mathbf{0}}^{p,q}(\mathbb{R}^n)$, associated with the
weak local generalized Morrey space $WM_{\omega,\mathbf{0}}^{p,q}(\mathbb{R}^n)$, is defined to be the set of
all the $f \in \mathcal{S}'(\mathbb{R}^n)$ such that

$$\|f\|_{WHM_{\omega,\mathbf{0}}^{p,q}(\mathbb{R}^n)} := \|\mathcal{M}_N(f)\|_{WM_{\omega,\mathbf{0}}^{p,q}(\mathbb{R}^n)} < \infty.$$

(ii) The *weak generalized Hardy–Morrey space* $WHM_\omega^{p,q}(\mathbb{R}^n)$, associated with the
weak global generalized Morrey space $WM_{\omega,\mathbf{0}}^{p,q}(\mathbb{R}^n)$, is defined to be the set of
all the $f \in \mathcal{S}'(\mathbb{R}^n)$ such that

$$\|f\|_{WHM_\omega^{p,q}(\mathbb{R}^n)} := \|\mathcal{M}_N(f)\|_{WM_\omega^{p,q}(\mathbb{R}^n)} < \infty.$$

Remark 6.0.25 In Definition 6.0.24, let $p, q \in [1, \infty)$ and $\omega \in M(\mathbb{R}_+)$ satisfy

$$\max\{M_0(\omega), M_\infty(\omega)\} \in (-\infty, 0).$$

In this case, from Remark 6.0.23(ii), it follows that the weak generalized Hardy–
Morrey spaces $WHM_{\omega,\mathbf{0}}^{p,q}(\mathbb{R}^n)$ and $WHM_\omega^{p,q}(\mathbb{R}^n)$ coincide, respectively, with
the weak generalized Herz–Hardy spaces $WH\dot{\mathcal{K}}_{\omega,\mathbf{0}}^{p,q}(\mathbb{R}^n)$ and $WH\dot{\mathcal{K}}_\omega^{p,q}(\mathbb{R}^n)$ with
equivalent quasi-norms.

6.1 Maximal Function Characterizations

In this section, we establish the maximal function characterizations of the weak
generalized Herz–Hardy spaces $WH\dot{\mathcal{K}}_{\omega,\mathbf{0}}^{p,q}(\mathbb{R}^n)$ and $WH\dot{\mathcal{K}}_\omega^{p,q}(\mathbb{R}^n)$. Recall that various radial and non-tangential maximal functions are defined as in Definition 4.1.1.

Via these maximal functions, we then show the following maximal function characterizations of the weak generalized Herz–Hardy space $W H\dot{\mathcal{K}}^{p,q}_{\omega,0}(\mathbb{R}^n)$.

Theorem 6.1.1 *Let* $p, q, a, b \in (0, \infty)$, $\omega \in M(\mathbb{R}_+)$, $N \in \mathbb{N}$, *and* $\phi \in \mathcal{S}(\mathbb{R}^n)$ *satisfy* $\int_{\mathbb{R}^n} \phi(x)\, dx \neq 0$.

(i) *Let* $N \in \mathbb{N} \cap [\lfloor b + 1 \rfloor, \infty)$ *and* ω *satisfy* $m_0(\omega) \in (-\frac{n}{p}, \infty)$. *Then, for any* $f \in \mathcal{S}'(\mathbb{R}^n)$,

$$\|M(f, \phi)\|_{W\dot{\mathcal{K}}^{p,q}_{\omega,0}(\mathbb{R}^n)} \lesssim \|M_a^*(f, \phi)\|_{W\dot{\mathcal{K}}^{p,q}_{\omega,0}(\mathbb{R}^n)} \lesssim \|M_b^{**}(f, \phi)\|_{W\dot{\mathcal{K}}^{p,q}_{\omega,0}(\mathbb{R}^n)},$$

$$\|M(f, \phi)\|_{W\dot{\mathcal{K}}^{p,q}_{\omega,0}(\mathbb{R}^n)} \lesssim \|\mathcal{M}_N(f)\|_{W\dot{\mathcal{K}}^{p,q}_{\omega,0}(\mathbb{R}^n)} \lesssim \|\mathcal{M}_{\lfloor b+1 \rfloor}(f)\|_{W\dot{\mathcal{K}}^{p,q}_{\omega,0}(\mathbb{R}^n)}$$

$$\lesssim \|M_b^{**}(f, \phi)\|_{W\dot{\mathcal{K}}^{p,q}_{\omega,0}(\mathbb{R}^n)},$$

and

$$\|M_b^{**}(f, \phi)\|_{W\dot{\mathcal{K}}^{p,q}_{\omega,0}(\mathbb{R}^n)} \sim \|\mathcal{M}_{b,N}^{**}(f)\|_{W\dot{\mathcal{K}}^{p,q}_{\omega,0}(\mathbb{R}^n)},$$

where the implicit positive constants are independent of f.
(ii) *Let* $\omega \in M(\mathbb{R}_+)$ *satisfy* $m_0(\omega) \in (-\frac{n}{p}, \infty)$ *and* $m_\infty(\omega) \in (-\frac{n}{p}, \infty)$. *Assume*

$$b \in \left(\max\left\{ \frac{n}{p}, \max\{M_0(\omega), M_\infty(\omega)\} + \frac{n}{p} \right\}, \infty \right).$$

Then, for any $f \in \mathcal{S}'(\mathbb{R}^n)$,

$$\|M_b^{**}(f, \phi)\|_{W\dot{\mathcal{K}}^{p,q}_{\omega,0}(\mathbb{R}^n)} \lesssim \|M(f, \phi)\|_{W\dot{\mathcal{K}}^{p,q}_{\omega,0}(\mathbb{R}^n)},$$

where the implicit positive constant is independent of f. *In particular, when* $N \in \mathbb{N} \cap [\lfloor b + 1 \rfloor, \infty)$, *if one of the quantities*

$$\|M(f, \phi)\|_{W\dot{\mathcal{K}}^{p,q}_{\omega,0}(\mathbb{R}^n)}, \quad \|M_a^*(f, \phi)\|_{W\dot{\mathcal{K}}^{p,q}_{\omega,0}(\mathbb{R}^n)}, \quad \|\mathcal{M}_N(f)\|_{W\dot{\mathcal{K}}^{p,q}_{\omega,0}(\mathbb{R}^n)},$$

$$\|M_b^{**}(f, \phi)\|_{W\dot{\mathcal{K}}^{p,q}_{\omega,0}(\mathbb{R}^n)}, \quad and \quad \|\mathcal{M}_{b,N}^{**}(f)\|_{W\dot{\mathcal{K}}^{p,q}_{\omega,0}(\mathbb{R}^n)}$$

is finite, then the others are also finite and mutually equivalent with the positive equivalence constants independent of f.

Remark 6.1.2

(i) Let $p, q \in (0, \infty)$ and $\omega \in M(\mathbb{R}_+)$. Then the quasi-norm of $W H \dot{\mathcal{K}}_{\omega,0}^{p,q}(\mathbb{R}^n)$ in Definition 6.0.20(i) depends on N. However, by Theorem 6.1.1, we find that the weak Hardy space $W H \dot{\mathcal{K}}_{\omega,0}^{p,q}(\mathbb{R}^n)$ is independent of the choice of N whenever $\omega \in M(\mathbb{R}_+)$ satisfies

$$m_0(\omega) \in \left(-\frac{n}{p}, \infty \right) \text{ and } m_\infty(\omega) \in \left(-\frac{n}{p}, \infty \right),$$

and N satisfies

$$N \in \mathbb{N} \cap \left(1 + \max \left\{ \frac{n}{p}, \max\{M_0(\omega), M_\infty(\omega)\} + \frac{n}{p} \right\}, \infty \right).$$

(ii) Observe that, if $p = q \in (0, \infty)$ and $\omega(t) := 1$ for any $t \in (0, \infty)$, then, in this case,

$$m_0(\omega) = M_0(\omega) = m_\infty(\omega) = M_\infty(\omega) = 0,$$

$W H \dot{\mathcal{K}}_{\omega,0}^{p,q}(\mathbb{R}^n)$ coincides with the classical weak Hardy space $W H^p(\mathbb{R}^n)$ in the sense of equivalent quasi-norms, and all the conclusions in Theorem 6.1.1 coincide with the corresponding classical results on the weak Hardy space $W H^p(\mathbb{R}^n)$; see, for instance, [153].

To prove Theorem 6.1.1, we first recall the following concept of weak ball quasi-Banach function spaces introduced in [278, Definition 2.8].

Definition 6.1.3 Let X be a ball quasi-Banach function space. The *weak ball quasi-Banach function space* $W X$ is defined to be the set of all the measurable functions f on $\mathbb{R}^n$ such that

$$\|f\|_{WX} := \sup_{\lambda \in (0,\infty)} \left\{ \lambda \left\| \mathbf{1}_{\{x \in \mathbb{R}^n : |f(x)| > \lambda\}} \right\|_X \right\} < \infty.$$

Remark 6.1.4 Let X be a ball quasi-Banach function space and $E \subset \mathbb{R}^n$ a measurable set. Then it is easy to show that

$$\|\mathbf{1}_E\|_{WX} = \|\mathbf{1}_E\|_X.$$

The following conclusion obtained in [278, Lemma 2.13] shows that the weak ball quasi-Banach function space is also a ball quasi-Banach function space. This plays a key role in the proof of Theorem 6.1.1.

Lemma 6.1.5 *Let X be a ball quasi-Banach function space. Then the weak ball quasi-Banach function space $W X$ is also a ball quasi-Banach function space.*

To show the maximal function characterizations of $WH\dot{\mathcal{K}}_{\omega,\mathbf{0}}^{p,q}(\mathbb{R}^n)$, we also require a lemma about the boundedness of the Hardy–Littlewood maximal operator on weak local generalized Herz spaces as follows.

Lemma 6.1.6 *Let* $p, q \in (0, \infty)$ *and* $\omega \in M(\mathbb{R}_+)$ *satisfy* $m_0(\omega) \in (-\frac{n}{p}, \infty)$ *and* $m_\infty(\omega) \in (-\frac{n}{p}, \infty)$. *Then, for any given*

$$r \in \left(0, \min\left\{p, \frac{n}{\max\{M_0(\omega), M_\infty(\omega)\} + n/p}\right\}\right),$$

there exists a positive constant C *such that, for any* $f \in L_{\text{loc}}^1(\mathbb{R}^n)$,

$$\|\mathcal{M}(f)\|_{[W\dot{\mathcal{K}}_{\omega,\mathbf{0}}^{p,q}(\mathbb{R}^n)]^{1/r}} \leq C\|f\|_{[W\dot{\mathcal{K}}_{\omega,\mathbf{0}}^{p,q}(\mathbb{R}^n)]^{1/r}}.$$

To prove Lemma 6.1.6, we first state the following auxiliary conclusion in ball quasi-Banach function spaces, which was obtained in [219, Theorem 4.4].

Lemma 6.1.7 *Let* X *be a ball quasi-Banach function space and there exists a* $p_- \in (0, \infty)$ *such that, for any given* $r \in (0, p_-)$ *and* $u \in (1, \infty)$, *there exists a positive constant* C *such that, for any* $\{f_j\}_{j\in\mathbb{N}} \subset L_{\text{loc}}^1(\mathbb{R}^n)$,

$$\left\|\left\{\sum_{j\in\mathbb{N}}[\mathcal{M}(f_j)]^u\right\}^{\frac{1}{u}}\right\|_{X^{1/r}} \leq C\left\|\left(\sum_{j\in\mathbb{N}}|f_j|^u\right)^{\frac{1}{u}}\right\|_{X^{1/r}}.$$

Then, for any given $r \in (0, p_-)$ *and* $u \in (1, \infty)$, *there exists a positive constant* C *such that, for any* $\{f_j\}_{j\in\mathbb{N}} \subset L_{\text{loc}}^1(\mathbb{R}^n)$,

$$\left\|\left\{\sum_{j\in\mathbb{N}}[\mathcal{M}(f_j)]^u\right\}^{\frac{1}{u}}\right\|_{(WX)^{1/r}} \leq C\left\|\left(\sum_{j\in\mathbb{N}}|f_j|^u\right)^{\frac{1}{u}}\right\|_{(WX)^{1/r}}.$$

Via this conclusion, we now prove Lemma 6.1.6.

Proof of Lemma 6.1.6 Let all the symbols be as in the present lemma. Then, by the assumption $m_0(\omega) \in (-\frac{n}{p}, \infty)$ and Theorem 1.2.42, we conclude that the local generalized Herz space $\dot{\mathcal{K}}_{\omega,\mathbf{0}}^{p,q}(\mathbb{R}^n)$ under consideration is a BQBF space. Moreover, let

$$p_- := \min\left\{p, \frac{n}{\max\{M_0(\omega), M_\infty(\omega)\} + n/p}\right\}.$$

Then, for any given $r \in (0, p_-)$ and $u \in (1, \infty)$, applying Lemma 4.3.10, we find that, for any $\{f_j\}_{j\in\mathbb{N}} \subset L^1_{\text{loc}}(\mathbb{R}^n)$,

$$\left\| \left\{ \sum_{j\in\mathbb{N}} [\mathcal{M}(f_j)]^u \right\}^{\frac{1}{u}} \right\|_{[\dot{\mathcal{K}}^{p,q}_{\omega,0}(\mathbb{R}^n)]^{1/r}} \lesssim \left\| \left(\sum_{j\in\mathbb{N}} |f_j|^u \right)^{\frac{1}{u}} \right\|_{[\dot{\mathcal{K}}^{p,q}_{\omega,0}(\mathbb{R}^n)]^{1/r}}.$$

This, combined with the fact that $\dot{\mathcal{K}}^{p,q}_{\omega,0}(\mathbb{R}^n)$ is a BQBF space and Lemma 6.1.7 with $X := \dot{\mathcal{K}}^{p,q}_{\omega,0}(\mathbb{R}^n)$, $f_1 := f$, and $f_j := 0$ for any $j \in \mathbb{N} \cap [2, \infty)$, further implies that, for any given $r \in (0, p_-)$ and for any $f \in L^1_{\text{loc}}(\mathbb{R}^n)$,

$$\|\mathcal{M}(f)\|_{[W\dot{\mathcal{K}}^{p,q}_{\omega,0}(\mathbb{R}^n)]^{1/r}} \lesssim \|f\|_{[W\dot{\mathcal{K}}^{p,q}_{\omega,0}(\mathbb{R}^n)]^{1/r}},$$

which then completes the proof of Lemma 6.1.6.   □

Via the above lemmas and the known maximal function characterizations of Hardy spaces associated with ball quasi-Banach function spaces proved in [207, Theorem 3.1] (see also Lemma 4.1.4 above), we next show Theorem 6.1.1.

Proof of Theorem 6.1.1 Let all the symbols be as in the present theorem. Then, from the assumption $m_0(\omega) \in (-\frac{n}{p}, \infty)$ and Theorem 1.2.42, it follows that the local generalized Herz space $\dot{\mathcal{K}}^{p,q}_{\omega,0}(\mathbb{R}^n)$ under consideration is a BQBF space. By this and Lemma 6.1.5 with $X := \dot{\mathcal{K}}^{p,q}_{\omega,0}(\mathbb{R}^n)$, we conclude that the weak space $W\dot{\mathcal{K}}^{p,q}_{\omega,0}(\mathbb{R}^n)$ is also a BQBF space. This, combined with Lemma 4.1.4(i), finishes the proof of (i).

Next, we show (ii). Indeed, let

$$r \in \left(\frac{n}{b}, \min\left\{ p, \frac{n}{\max\{M_0(\omega), M_\infty(\omega)\} + n/p} \right\} \right).$$

Then, using Lemma 6.1.6, we find that, for any $f \in L^1_{\text{loc}}(\mathbb{R}^n)$,

$$\|\mathcal{M}(f)\|_{[W\dot{\mathcal{K}}^{p,q}_{\omega,0}(\mathbb{R}^n)]^{1/r}} \lesssim \|f\|_{[W\dot{\mathcal{K}}^{p,q}_{\omega,0}(\mathbb{R}^n)]^{1/r}}.$$

From this, the fact that $b \in (\frac{n}{r}, \infty)$, and Lemma 4.1.4(ii), we deduce that (ii) holds true and hence complete the proof of Theorem 6.1.1.   □

Via both Theorem 6.1.1 and Remark 6.0.25, we immediately obtain the following maximal function characterizations of the weak generalized Hardy–Morrey spaces $WHM^{p,q}_{\omega,0}(\mathbb{R}^n)$; we omit the details.

Corollary 6.1.8 *Let* $a, b \in (0, \infty)$, $p, q \in [1, \infty)$, $\omega \in M(\mathbb{R}_+)$, $N \in \mathbb{N}$, *and* $\phi \in \mathcal{S}(\mathbb{R}^n)$ *satisfy* $\int_{\mathbb{R}^n} \phi(x)\, dx \neq 0$.

(i) *Let $N \in \mathbb{N} \cap [\lfloor b + 1 \rfloor, \infty)$ and ω satisfy $M_\infty(\omega) \in (-\infty, 0)$ and*

$$-\frac{n}{p} < m_0(\omega) \leq M_0(\omega) < 0.$$

Then, for any $f \in \mathcal{S}'(\mathbb{R}^n)$,

$$\|M(f, \phi)\|_{WM_{\omega,0}^{p,q}(\mathbb{R}^n)} \lesssim \|M_a^*(f, \phi)\|_{WM_{\omega,0}^{p,q}(\mathbb{R}^n)} \lesssim \|M_b^{**}(f, \phi)\|_{WM_{\omega,0}^{p,q}(\mathbb{R}^n)},$$

$$\|M(f, \phi)\|_{WM_{\omega,0}^{p,q}(\mathbb{R}^n)} \lesssim \|\mathcal{M}_N(f)\|_{WM_{\omega,0}^{p,q}(\mathbb{R}^n)} \lesssim \|\mathcal{M}_{\lfloor b+1 \rfloor}(f)\|_{WM_{\omega,0}^{p,q}(\mathbb{R}^n)}$$

$$\lesssim \|M_b^{**}(f, \phi)\|_{WM_{\omega,0}^{p,q}(\mathbb{R}^n)},$$

and

$$\|M_b^{**}(f, \phi)\|_{WM_{\omega,0}^{p,q}(\mathbb{R}^n)} \sim \|\mathcal{M}_{b,N}^{**}(f)\|_{WM_{\omega,0}^{p,q}(\mathbb{R}^n)},$$

where the implicit positive constants are independent of f.
(ii) *Let $\omega \in M(\mathbb{R}_+)$ satisfy*

$$-\frac{n}{p} < m_0(\omega) \leq M_0(\omega) < 0$$

and

$$-\frac{n}{p} < m_0(\omega) \leq M_0(\omega) < 0.$$

Assume $b \in (\frac{n}{p}, \infty)$. Then, for any $f \in \mathcal{S}'(\mathbb{R}^n)$,

$$\|M_b^{**}(f, \phi)\|_{WM_{\omega,0}^{p,q}(\mathbb{R}^n)} \lesssim \|M(f, \phi)\|_{WM_{\omega,0}^{p,q}(\mathbb{R}^n)},$$

where the implicit positive constant is independent of f. In particular, when $N \in \mathbb{N} \cap [\lfloor b + 1 \rfloor, \infty)$, if one of the quantities

$$\|M(f, \phi)\|_{WM_{\omega,0}^{p,q}(\mathbb{R}^n)}, \quad \|M_a^*(f, \phi)\|_{WM_{\omega,0}^{p,q}(\mathbb{R}^n)}, \quad \|\mathcal{M}_N(f)\|_{WM_{\omega,0}^{p,q}(\mathbb{R}^n)},$$

$$\|M_b^{**}(f, \phi)\|_{WM_{\omega,0}^{p,q}(\mathbb{R}^n)}, \quad and \quad \|\mathcal{M}_{b,N}^{**}(f)\|_{WM_{\omega,0}^{p,q}(\mathbb{R}^n)}$$

is finite, then the others are also finite and mutually equivalent with the positive equivalence constants independent of f.

We now establish the maximal function characterizations of the weak generalized Herz–Hardy space $WH\dot{\mathcal{K}}_\omega^{p,q}(\mathbb{R}^n)$ as follows.

Theorem 6.1.9 *Let* $p, q, a, b \in (0, \infty)$, $\omega \in M(\mathbb{R}_+)$, $N \in \mathbb{N}$, *and* $\phi \in \mathcal{S}(\mathbb{R}^n)$ *satisfy* $\int_{\mathbb{R}^n} \phi(x)\,dx \neq 0$.

(i) *Let* $N \in \mathbb{N} \cap [\lfloor b + 1 \rfloor, \infty)$ *and* ω *satisfy* $m_0(\omega) \in (-\frac{n}{p}, \infty)$ *and* $M_\infty(\omega) \in (-\infty, 0)$. *Then, for any* $f \in \mathcal{S}'(\mathbb{R}^n)$,

$$\|M(f, \phi)\|_{W\dot{\mathcal{K}}_\omega^{p,q}(\mathbb{R}^n)} \lesssim \|M_a^*(f, \phi)\|_{W\dot{\mathcal{K}}_\omega^{p,q}(\mathbb{R}^n)} \lesssim \|M_b^{**}(f, \phi)\|_{W\dot{\mathcal{K}}_\omega^{p,q}(\mathbb{R}^n)},$$

$$\|M(f, \phi)\|_{W\dot{\mathcal{K}}_\omega^{p,q}(\mathbb{R}^n)} \lesssim \|\mathcal{M}_N(f)\|_{W\dot{\mathcal{K}}_\omega^{p,q}(\mathbb{R}^n)} \lesssim \|\mathcal{M}_{\lfloor b+1 \rfloor}(f)\|_{W\dot{\mathcal{K}}_\omega^{p,q}(\mathbb{R}^n)}$$
$$\lesssim \|M_b^{**}(f, \phi)\|_{W\dot{\mathcal{K}}_\omega^{p,q}(\mathbb{R}^n)},$$

and

$$\|M_b^{**}(f, \phi)\|_{W\dot{\mathcal{K}}_\omega^{p,q}(\mathbb{R}^n)} \sim \|\mathcal{M}_{b,N}^{**}(f)\|_{W\dot{\mathcal{K}}_\omega^{p,q}(\mathbb{R}^n)},$$

where the implicit positive constants are independent of f.

(ii) *Let* $\omega \in M(\mathbb{R}_+)$ *satisfy* $m_0(\omega) \in (-\frac{n}{p}, \infty)$ *and*

$$-\frac{n}{p} < m_\infty(\omega) \leq M_\infty(\omega) < 0.$$

Assume $b \in (\max\{\frac{n}{p}, \max\{M_0(\omega), M_\infty(\omega)\} + \frac{n}{p}\}, \infty)$. *Then, for any* $f \in \mathcal{S}'(\mathbb{R}^n)$,

$$\|M_b^{**}(f, \phi)\|_{W\dot{\mathcal{K}}_\omega^{p,q}(\mathbb{R}^n)} \lesssim \|M(f, \phi)\|_{W\dot{\mathcal{K}}_\omega^{p,q}(\mathbb{R}^n)},$$

where the implicit positive constant is independent of f. *In particular, when* $N \in \mathbb{N} \cap [\lfloor b + 1 \rfloor, \infty)$, *if one of the quantities*

$$\|M(f, \phi)\|_{W\dot{\mathcal{K}}_\omega^{p,q}(\mathbb{R}^n)}, \quad \|M_a^*(f, \phi)\|_{W\dot{\mathcal{K}}_\omega^{p,q}(\mathbb{R}^n)}, \quad \|\mathcal{M}_N(f)\|_{W\dot{\mathcal{K}}_\omega^{p,q}(\mathbb{R}^n)},$$

$$\|M_b^{**}(f, \phi)\|_{W\dot{\mathcal{K}}_\omega^{p,q}(\mathbb{R}^n)}, \quad and \quad \|\mathcal{M}_{b,N}^{**}(f)\|_{W\dot{\mathcal{K}}_\omega^{p,q}(\mathbb{R}^n)}$$

is finite, then the others are also finite and mutually equivalent with the positive equivalence constants independent of f.

Remark 6.1.10

(i) Let $p, q \in (0, \infty)$ and $\omega \in M(\mathbb{R}_+)$. Then the quasi-norm of $W H\dot{\mathcal{K}}_\omega^{p,q}(\mathbb{R}^n)$ in Definition 6.0.20(ii) depends on N. However, from Theorem 6.1.9, we infer that the weak Hardy space $W H\dot{\mathcal{K}}_\omega^{p,q}(\mathbb{R}^n)$ is independent of the choice of N whenever $\omega \in M(\mathbb{R}_+)$ satisfies $m_0(\omega) \in (-\frac{n}{p}, \infty)$ and

$$-\frac{n}{p} < m_\infty(\omega) \leq M_\infty(\omega) < 0,$$

and N satisfies

$$N \in \mathbb{N} \cap \left(1 + \max\left\{\frac{n}{p}, \max\{M_0(\omega), M_\infty(\omega)\} + \frac{n}{p}\right\}, \infty\right).$$

(ii) Notice that, if $p = q \in (0, \infty)$ and $\omega(t) := 1$ for any $t \in (0, \infty)$, then, in this case,

$$m_0(\omega) = M_0(\omega) = m_\infty(\omega) = M_\infty(\omega) = 0$$

and $W H \dot{\mathcal{K}}_\omega^{p,q}(\mathbb{R}^n)$ coincides with the classical weak Hardy space $W H^p(\mathbb{R}^n)$ in the sense of equivalent quasi-norms. Therefore, Theorem 6.1.9 completely excludes the classical weak Hardy space $W H^p(\mathbb{R}^n)$ and, based on Remark 1.2.45, we find that the classical weak Hardy space $W H^p(\mathbb{R}^n)$ is the critical case of $W H \dot{\mathcal{K}}_\omega^{p,q}(\mathbb{R}^n)$ considered in Theorem 6.1.9.

To prove this maximal function characterizations, we first show the boundedness of the Hardy–Littlewood maximal operator on weak global generalized Herz spaces as follows.

Lemma 6.1.11 *Let $p, q \in (0, \infty)$ and $\omega \in M(\mathbb{R}_+)$ satisfy $m_0(\omega) \in (-\frac{n}{p}, \infty)$ and $m_\infty(\omega) \in (-\frac{n}{p}, \infty)$. Then, for any given*

$$r \in \left(0, \min\left\{p, \frac{n}{\max\{M_0(\omega), M_\infty(\omega)\} + n/p}\right\}\right),$$

there exists a positive constant C such that, for any $f \in L^1_{\mathrm{loc}}(\mathbb{R}^n)$,

$$\|\mathcal{M}(f)\|_{[W\dot{\mathcal{K}}_\omega^{p,q}(\mathbb{R}^n)]^{1/r}} \le C \|f\|_{[W\dot{\mathcal{K}}_\omega^{p,q}(\mathbb{R}^n)]^{1/r}}.$$

Proof Let all the symbols be as in the present lemma. Then, from the assumptions $m_0(\omega) \in (-\frac{n}{p}, \infty)$ and $M_\infty(\omega) \in (-\infty, 0)$, and Theorem 1.2.44, it follows that the global generalized Herz space $\dot{\mathcal{K}}_\omega^{p,q}(\mathbb{R}^n)$ under consideration is a BQBF space. Let

$$p_- := \min\left\{p, \frac{n}{\max\{M_0(\omega), M_\infty(\omega)\} + n/p}\right\}.$$

Then, for any given $r \in (0, p_-)$ and $u \in (1, \infty)$, applying Lemma 4.3.24, we find that, for any $\{f_j\}_{j \in \mathbb{N}} \subset L^1_{\mathrm{loc}}(\mathbb{R}^n)$,

$$\left\|\left\{\sum_{j \in \mathbb{N}} [\mathcal{M}(f_j)]^u\right\}^{\frac{1}{u}}\right\|_{[\dot{\mathcal{K}}_\omega^{p,q}(\mathbb{R}^n)]^{1/r}} \lesssim \left\|\left(\sum_{j \in \mathbb{N}} |f_j|^u\right)^{\frac{1}{u}}\right\|_{[\dot{\mathcal{K}}_\omega^{p,q}(\mathbb{R}^n)]^{1/r}}.$$

This, combined with the fact that $\dot{\mathcal{K}}_\omega^{p,q}(\mathbb{R}^n)$ is a BQBF space and Lemma 6.1.7 with $X := \dot{\mathcal{K}}_\omega^{p,q}(\mathbb{R}^n)$, $f_1 := f$, and $f_j := 0$ for any $j \in \mathbb{N} \cap [2, \infty)$, further implies that, for any given $r \in (0, p_-)$ and for any $f \in L_{\mathrm{loc}}^1(\mathbb{R}^n)$,

$$\|\mathcal{M}(f)\|_{[W\dot{\mathcal{K}}_\omega^{p,q}(\mathbb{R}^n)]^{1/r}} \lesssim \|f\|_{[W\dot{\mathcal{K}}_\omega^{p,q}(\mathbb{R}^n)]^{1/r}},$$

which then completes the proof of Lemma 6.1.11.   □

Using Lemma 6.1.11, we next show Theorem 6.1.9.

Proof of Theorem 6.1.9 Let all the symbols be as in the present theorem. Then, applying the assumptions $m_0(\omega) \in (-\frac{n}{p}, \infty)$ and $M_\infty(\omega) \in (-\infty, 0)$, and Theorem 1.2.44, we conclude that the global generalized Herz space $\dot{\mathcal{K}}_\omega^{p,q}(\mathbb{R}^n)$ under consideration is a BQBF space. This, together with Lemma 6.1.5, further implies that the weak Herz space $W\dot{\mathcal{K}}_\omega^{p,q}(\mathbb{R}^n)$ is a BQBF space. Thus, by Lemma 4.1.4(i), we then complete the proof of (i).

Next, we prove (ii). Indeed, let

$$r \in \left(\frac{n}{b}, \min\left\{p, \frac{n}{\max\{M_0(\omega), M_\infty(\omega)\} + n/p}\right\}\right). \tag{6.1}$$

Then, from Lemma 6.1.11, it follows that, for any $f \in L_{\mathrm{loc}}^1(\mathbb{R}^n)$,

$$\|\mathcal{M}(f)\|_{[W\dot{\mathcal{K}}_\omega^{p,q}(\mathbb{R}^n)]^{1/r}} \lesssim \|f\|_{[W\dot{\mathcal{K}}_\omega^{p,q}(\mathbb{R}^n)]^{1/r}}. \tag{6.2}$$

In addition, by (6.1), we find that $b \in (\frac{n}{r}, \infty)$. Combining this, (6.2), and Lemma 4.1.4(ii), we then complete the proof of (ii) and hence Theorem 6.1.9.   □

As an application of Theorem 6.1.9, we next present the following result about the maximal function characterizations of the weak generalized Hardy–Morrey space $WHM_\omega^{p,q}(\mathbb{R}^n)$, which is just a immediately corollary of both Theorem 6.1.9 and Remark 6.0.25; we omit the details.

Corollary 6.1.12 *Let $a, b \in (0, \infty)$, $p, q \in [1, \infty)$, $\omega \in M(\mathbb{R}_+)$, $N \in \mathbb{N}$, and $\phi \in \mathcal{S}(\mathbb{R}^n)$ satisfy $\int_{\mathbb{R}^n} \phi(x)\, dx \neq 0$.*

(i) *Let $N \in \mathbb{N} \cap [\lfloor b+1 \rfloor, \infty)$ and ω satisfy $M_\infty(\omega) \in (-\infty, 0)$ and*

$$-\frac{n}{p} < m_0(\omega) \leq M_0(\omega) < 0.$$

Then, for any $f \in \mathcal{S}'(\mathbb{R}^n)$,

$$\|M(f, \phi)\|_{WM_\omega^{p,q}(\mathbb{R}^n)} \lesssim \|M_a^*(f, \phi)\|_{WM_\omega^{p,q}(\mathbb{R}^n)} \lesssim \|M_b^{**}(f, \phi)\|_{WM_\omega^{p,q}(\mathbb{R}^n)},$$

$$\|M(f, \phi)\|_{WM_\omega^{p,q}(\mathbb{R}^n)} \lesssim \|\mathcal{M}_N(f)\|_{WM_\omega^{p,q}(\mathbb{R}^n)} \lesssim \|\mathcal{M}_{\lfloor b+1 \rfloor}(f)\|_{WM_\omega^{p,q}(\mathbb{R}^n)}$$

$$\lesssim \|M_b^{**}(f, \phi)\|_{WM_\omega^{p,q}(\mathbb{R}^n)},$$

and

$$\|M_b^{**}(f,\phi)\|_{WM_\omega^{p,q}(\mathbb{R}^n)} \sim \|\mathcal{M}_{b,N}^{**}(f)\|_{WM_\omega^{p,q}(\mathbb{R}^n)},$$

where the implicit positive constants are independent of f.
(ii) *Let $\omega \in M(\mathbb{R}_+)$ satisfy*

$$-\frac{n}{p} < m_0(\omega) \le M_0(\omega) < 0$$

and

$$-\frac{n}{p} < m_0(\omega) \le M_0(\omega) < 0.$$

Assume $b \in (\frac{n}{p}, \infty)$. Then, for any $f \in \mathcal{S}'(\mathbb{R}^n)$,

$$\|M_b^{**}(f,\phi)\|_{WM_\omega^{p,q}(\mathbb{R}^n)} \lesssim \|M(f,\phi)\|_{WM_\omega^{p,q}(\mathbb{R}^n)},$$

where the implicit positive constant is independent of f. In particular, when $N \in \mathbb{N} \cap [\lfloor b + 1 \rfloor, \infty)$, if one of the quantities

$$\|M(f,\phi)\|_{WM_\omega^{p,q}(\mathbb{R}^n)}, \quad \|M_a^*(f,\phi)\|_{WM_\omega^{p,q}(\mathbb{R}^n)}, \quad \|\mathcal{M}_N(f)\|_{WM_\omega^{p,q}(\mathbb{R}^n)},$$

$$\|M_b^{**}(f,\phi)\|_{WM_\omega^{p,q}(\mathbb{R}^n)}, \quad \text{and} \quad \|\mathcal{M}_{b,N}^{**}(f)\|_{WM_\omega^{p,q}(\mathbb{R}^n)}$$

is finite, then the others are also finite and mutually equivalent with the positive equivalence constants independent of f.

Remark 6.1.13 Let $p, q \in (0, \infty)$ and $\omega \in M(\mathbb{R}_+)$. Then the quasi-norms of $WHM_{\omega,0}^{p,q}(\mathbb{R}^n)$ and $WHM_\omega^{p,q}(\mathbb{R}^n)$ in Definition 6.0.24 depend on N. However, from Corollaries 6.1.8 and 6.1.12, it follows that the spaces $WHM_{\omega,0}^{p,q}(\mathbb{R}^n)$ and $WHM_\omega^{p,q}(\mathbb{R}^n)$ are both independent of the choice of N whenever $p, q \in [1, \infty)$, $\omega \in M(\mathbb{R}_+)$ satisfies

$$-\frac{n}{p} < m_0(\omega) \le M_0(\omega) < 0$$

and

$$-\frac{n}{p} < m_\infty(\omega) \le M_\infty(\omega) < 0,$$

and N satisfies

$$N \in \mathbb{N} \cap \left(1 + \frac{n}{p}, \infty\right).$$

6.2 Relations with Weak Generalized Herz Spaces

In this section, via the relation between ball quasi-Banach function spaces and associated Hardy spaces obtained in [207, Theorem 3.4] (see also Lemma 4.2.6 above), we investigate the relations between weak generalized Herz spaces and associated Hardy spaces. We first establish the relation between $W\dot{\mathcal{K}}_{\omega,0}^{p,q}(\mathbb{R}^n)$ and the associated Hardy space $WH\dot{\mathcal{K}}_{\omega,0}^{p,q}(\mathbb{R}^n)$ as follows. Indeed, the following theorem shows that, under some reasonable and sharp assumptions, $W\dot{\mathcal{K}}_{\omega,0}^{p,q}(\mathbb{R}^n) = WH\dot{\mathcal{K}}_{\omega,0}^{p,q}(\mathbb{R}^n)$ with equivalent quasi-norms.

Theorem 6.2.1 *Let $p \in (1, \infty)$, $q \in (0, \infty)$, and $\omega \in M(\mathbb{R}_+)$ satisfy*

$$-\frac{n}{p} < m_0(\omega) \leq M_0(\omega) < \frac{n}{p'}$$

and

$$-\frac{n}{p} < m_\infty(\omega) \leq M_\infty(\omega) < \frac{n}{p'},$$

where $\frac{1}{p} + \frac{1}{p'} = 1$. Then

(i) $W\dot{\mathcal{K}}_{\omega,0}^{p,q}(\mathbb{R}^n) \hookrightarrow \mathcal{S}'(\mathbb{R}^n)$.

(ii) *If $f \in W\dot{\mathcal{K}}_{\omega,0}^{p,q}(\mathbb{R}^n)$, then $f \in WH\dot{\mathcal{K}}_{\omega,0}^{p,q}(\mathbb{R}^n)$ and there exists a positive constant C, independent of f, such that*

$$\|f\|_{WH\dot{\mathcal{K}}_{\omega,0}^{p,q}(\mathbb{R}^n)} \leq C \|f\|_{W\dot{\mathcal{K}}_{\omega,0}^{p,q}(\mathbb{R}^n)}.$$

(iii) *If $f \in WH\dot{\mathcal{K}}_{\omega,0}^{p,q}(\mathbb{R}^n)$, then there exists a locally integrable function g belonging to $W\dot{\mathcal{K}}_{\omega,0}^{p,q}(\mathbb{R}^n)$ such that g represents f, which means that $f = g$ in $\mathcal{S}'(\mathbb{R}^n)$,*

$$\|f\|_{WH\dot{\mathcal{K}}_{\omega,0}^{p,q}(\mathbb{R}^n)} = \|g\|_{WH\dot{\mathcal{K}}_{\omega,0}^{p,q}(\mathbb{R}^n)},$$

and there exists a positive constant C, independent of f, such that

$$\|g\|_{W\dot{\mathcal{K}}_{\omega,0}^{p,q}(\mathbb{R}^n)} \leq C \|f\|_{WH\dot{\mathcal{K}}_{\omega,0}^{p,q}(\mathbb{R}^n)}.$$

Proof Let all the symbols be as in the present theorem. Then, from the assumption $m_0(\omega) \in (-\frac{n}{p}, \infty)$ and Theorem 1.2.42, it follows that the local generalized Herz space $\dot{\mathcal{K}}_{\omega,0}^{p,q}(\mathbb{R}^n)$ under consideration is a BQBF space. This, together with Lemma 6.1.5, further implies that the weak local generalized Herz space $W\dot{\mathcal{K}}_{\omega,0}^{p,q}(\mathbb{R}^n)$ is a BQBF space. Therefore, to finish the proof of the present theorem, we only need to show that $W\dot{\mathcal{K}}_{\omega,0}^{p,q}(\mathbb{R}^n)$ satisfies all the assumptions of Lemma 4.2.6. Namely, there exists an $r \in (1, \infty)$ such that the Hardy–Littlewood maximal operator $\mathcal{M}$ is bounded on $[W\dot{\mathcal{K}}_{\omega,0}^{p,q}(\mathbb{R}^n)]^{1/r}$.

Indeed, applying the assumptions

$$\min\{m_0(\omega), m_\infty(\omega)\} \in \left(-\frac{n}{p}, \infty\right)$$

and

$$\max\{M_0(\omega), M_\infty(\omega)\} \in \left(-\infty, \frac{n}{p'}\right),$$

and Remark 1.1.5(iii), we conclude that

$$\frac{n}{\max\{M_0(\omega), M_\infty(\omega)\} + n/p} \in (1, \infty).$$

Combining this and the assumption $p \in (1, \infty)$, we further find that

$$\min\left\{p, \frac{n}{\max\{M_0(\omega), M_\infty(\omega)\} + n/p}\right\} \in (1, \infty).$$

Thus, we can choose an

$$r \in \left(1, \min\left\{p, \frac{n}{\max\{M_0(\omega), M_\infty(\omega)\} + n/p}\right\}\right).$$

For this r, from Lemma 6.1.6, we infer that, for any $f \in L_{\mathrm{loc}}^1(\mathbb{R}^n)$,

$$\|\mathcal{M}(f)\|_{[W\dot{\mathcal{K}}_{\omega,0}^{p,q}(\mathbb{R}^n)]^{1/r}} \lesssim \|f\|_{[W\dot{\mathcal{K}}_{\omega,0}^{p,q}(\mathbb{R}^n)]^{1/r}}.$$

This further implies that there exists an $r \in (1, \infty)$ such that $\mathcal{M}$ is bounded on $[W\dot{\mathcal{K}}_{\omega,0}^{p,q}(\mathbb{R}^n)]^{1/r}$. Thus, all the assumptions of Lemma 4.2.6 hold true for $W\dot{\mathcal{K}}_{\omega,0}^{p,q}(\mathbb{R}^n)$, and hence the proof of Theorem 6.2.1 is then completed. $\square$

Using Theorem 6.2.1 and Remark 6.0.25, we immediately obtain the following conclusion, which shows that $W M_{\omega,0}^{p,q}(\mathbb{R}^n) = W H M_{\omega,0}^{p,q}(\mathbb{R}^n)$ with equivalent quasi-norms under some reasonable and sharp assumptions; we omit the details.

Corollary 6.2.2 *Let $p \in (1, \infty)$, $q \in [1, \infty)$, and $\omega \in M(\mathbb{R}_+)$ satisfy*

$$-\frac{n}{p} < m_0(\omega) \leq M_0(\omega) < 0$$

and

$$-\frac{n}{p} < m_\infty(\omega) \le M_\infty(\omega) < 0.$$

Then

(i) $WM_{\omega,0}^{p,q}(\mathbb{R}^n) \hookrightarrow \mathcal{S}'(\mathbb{R}^n)$.

(ii) *If* $f \in WM_{\omega,0}^{p,q}(\mathbb{R}^n)$, *then* $f \in WHM_{\omega,0}^{p,q}(\mathbb{R}^n)$ *and there exists a positive constant C, independent of* f, *such that*

$$\|f\|_{WHM_{\omega,0}^{p,q}(\mathbb{R}^n)} \le C \|f\|_{WM_{\omega,0}^{p,q}(\mathbb{R}^n)}.$$

(iii) *If* $f \in WHM_{\omega,0}^{p,q}(\mathbb{R}^n)$, *then there exists a locally integrable function g belonging to* $WM_{\omega,0}^{p,q}(\mathbb{R}^n)$ *such that g represents* f, *which means that* $f = g$ *in* $\mathcal{S}'(\mathbb{R}^n)$,

$$\|f\|_{WHM_{\omega,0}^{p,q}(\mathbb{R}^n)} = \|g\|_{WHM_{\omega,0}^{p,q}(\mathbb{R}^n)},$$

and there exists a positive constant C, independent of f, *such that*

$$\|g\|_{WM_{\omega,0}^{p,q}(\mathbb{R}^n)} \le C \|f\|_{WHM_{\omega,0}^{p,q}(\mathbb{R}^n)}.$$

Similarly, we next prove that, under some reasonable and sharp assumptions, the weak generalized Herz–Hardy space $WH\dot{\mathcal{K}}_\omega^{p,q}(\mathbb{R}^n)$ coincides with the weak global generalized Herz space $W\dot{\mathcal{K}}_\omega^{p,q}(\mathbb{R}^n)$ with equivalent quasi-norms as follows.

Theorem 6.2.3 *Let* $p \in (1,\infty)$, $q \in (0,\infty)$, *and* $\omega \in M(\mathbb{R}_+)$ *with*

$$-\frac{n}{p} < m_0(\omega) \le M_0(\omega) < \frac{n}{p'}$$

and

$$-\frac{n}{p} < m_\infty(\omega) \le M_\infty(\omega) < 0.$$

Then

(i) $W\dot{\mathcal{K}}_\omega^{p,q}(\mathbb{R}^n) \hookrightarrow \mathcal{S}'(\mathbb{R}^n)$.

(ii) *If* $f \in W\dot{\mathcal{K}}_\omega^{p,q}(\mathbb{R}^n)$, *then* $f \in WH\dot{\mathcal{K}}_\omega^{p,q}(\mathbb{R}^n)$ *and there exists a positive constant C, independent of* f, *such that*

$$\|f\|_{WH\dot{\mathcal{K}}_\omega^{p,q}(\mathbb{R}^n)} \le C \|f\|_{W\dot{\mathcal{K}}_\omega^{p,q}(\mathbb{R}^n)}.$$

(iii) *If* $f \in WH\dot{\mathcal{K}}_{\omega}^{p,q}(\mathbb{R}^n)$, *then there exists a locally integrable function* g *belonging to* $W\dot{\mathcal{K}}_{\omega}^{p,q}(\mathbb{R}^n)$ *such that* g *represents* f, *which means that* $f = g$ *in* $\mathcal{S}'(\mathbb{R}^n)$,

$$\|f\|_{WH\dot{\mathcal{K}}_{\omega}^{p,q}(\mathbb{R}^n)} = \|g\|_{WH\dot{\mathcal{K}}_{\omega}^{p,q}(\mathbb{R}^n)},$$

and there exists a positive constant C, *independent of* f, *such that*

$$\|g\|_{W\dot{\mathcal{K}}_{\omega}^{p,q}(\mathbb{R}^n)} \leq C \|f\|_{WH\dot{\mathcal{K}}_{\omega}^{p,q}(\mathbb{R}^n)}.$$

Proof Let all the symbols be as in the present theorem. Notice that ω satisfies both $m_0(\omega) \in (-\frac{n}{p}, \infty)$ and $M_{\infty}(\omega) \in (-\infty, 0)$. From these and Theorem 1.2.44, it follows that the global generalized Herz space $\dot{\mathcal{K}}_{\omega}^{p,q}(\mathbb{R}^n)$ under consideration is a BQBF space. By this and Lemma 6.1.5, we conclude that the weak global generalized Herz space $W\dot{\mathcal{K}}_{\omega}^{p,q}(\mathbb{R}^n)$ is also a BQBF space. On the other hand, let

$$r \in \left(1, \min\left\{p, \frac{n}{\max\{M_0(\omega), M_{\infty}(\omega)\} + n/p}\right\}\right).$$

Then, applying Lemma 6.1.11, we find that, for any $f \in L_{\text{loc}}^1(\mathbb{R}^n)$,

$$\|\mathcal{M}(f)\|_{[W\dot{\mathcal{K}}_{\omega}^{p,q}(\mathbb{R}^n)]^{1/r}} \lesssim \|f\|_{[W\dot{\mathcal{K}}_{\omega}^{p,q}(\mathbb{R}^n)]^{1/r}}.$$

This, together with the facts that $r \in (1, \infty)$ and $W\dot{\mathcal{K}}_{\omega}^{p,q}(\mathbb{R}^n)$ is a BQBF space, and Lemma 4.2.6 with $X := W\dot{\mathcal{K}}_{\omega}^{p,q}(\mathbb{R}^n)$, finishes the proof of Theorem 6.2.1. $\square$

Via both Theorem 6.2.3 and Remark 6.0.25, we immediately obtain the following relation between the weak generalized Hardy–Morrey space $WHM_{\omega}^{p,q}(\mathbb{R}^n)$ and the weak global generalized Morrey space $WM_{\omega}^{p,q}(\mathbb{R}^n)$; we omit the details.

Corollary 6.2.4 *Let* p, q, *and* ω *be as in Corollary 6.2.2. Then*

(i) $WM_{\omega}^{p,q}(\mathbb{R}^n) \hookrightarrow \mathcal{S}'(\mathbb{R}^n)$.

(ii) *If* $f \in WM_{\omega}^{p,q}(\mathbb{R}^n)$, *then* $f \in WHM_{\omega}^{p,q}(\mathbb{R}^n)$ *and there exists a positive constant* C, *independent of* f, *such that*

$$\|f\|_{WHM_{\omega}^{p,q}(\mathbb{R}^n)} \leq C \|f\|_{WM_{\omega}^{p,q}(\mathbb{R}^n)}.$$

(iii) *If* $f \in WHM_{\omega}^{p,q}(\mathbb{R}^n)$, *then there exists a locally integrable function* g *belonging to* $WM_{\omega}^{p,q}(\mathbb{R}^n)$ *such that* g *represents* f, *which means that* $f = g$ *in* $\mathcal{S}'(\mathbb{R}^n)$,

$$\|f\|_{WHM_{\omega}^{p,q}(\mathbb{R}^n)} = \|g\|_{WHM_{\omega}^{p,q}(\mathbb{R}^n)},$$

and there exists a positive constant C, independent of f, such that

$$\|g\|_{WM_\omega^{p,q}(\mathbb{R}^n)} \leq C \|f\|_{WHM_\omega^{p,q}(\mathbb{R}^n)} .$$

6.3 Atomic Characterizations

The main target of this section is to characterize weak generalized Herz–Hardy spaces via atoms. For this purpose, we first establish the atomic characterization of the weak generalized Herz–Hardy space $WH\dot{\mathcal{K}}_{\omega,0}^{p,q}(\mathbb{R}^n)$ via the known atomic characterization of weak Hardy spaces associated with ball quasi-Banach function spaces directly. However, due to the deficiency of associate spaces of global generalized Herz spaces, the atomic characterization of Hardy spaces associated with ball quasi-Banach function spaces mentioned above is not applicable to establish the atomic characterization of the weak generalized Herz–Hardy space $WH\dot{\mathcal{K}}_\omega^{p,q}(\mathbb{R}^n)$ [see Remark 1.2.19(vi) for the details]. To overcome this obstacle, we first establish an improved atomic characterization of weak Hardy spaces associated with ball quasi-Banach function spaces (see both Propositions 6.3.11 and 6.3.15 below) without recourse to associate spaces. Then, using this improved conclusion, we obtain the desired atomic characterization of $WH\dot{\mathcal{K}}_\omega^{p,q}(\mathbb{R}^n)$.

Recall that the concept of $(\dot{\mathcal{K}}_{\omega,0}^{p,q}(\mathbb{R}^n), r, d)$-atoms is given in Definition 4.3.1. We now introduce the weak generalized atomic Herz–Hardy spaces associated with the local generalized Herz spaces as follows.

Definition 6.3.1 Let $p, q \in (0, \infty)$, $\omega \in M(\mathbb{R}_+)$ with $m_0(\omega) \in (-\frac{n}{p}, \infty)$ and $m_\infty(\omega) \in (-\frac{n}{p}, \infty)$,

$$p_- \in \left(0, \frac{\min\{p, q, \frac{n}{\max\{M_0(\omega), M_\infty(\omega)\}+n/p}\}}{\max\{1, p, q\}} \right),$$

$d \geq \lfloor n(1/p_- - 1) \rfloor$ be a fixed integer, and

$$r \in \left(\max\left\{ 1, p, \frac{n}{\min\{m_0(\omega), m_\infty(\omega)\} + n/p} \right\}, \infty \right].$$

Then the *weak generalized atomic Herz–Hardy space* $WH\dot{\mathcal{K}}_{\omega,0}^{p,q,r,d}(\mathbb{R}^n)$, associated with the weak local generalized Herz space $W\dot{\mathcal{K}}_{\omega,0}^{p,q}(\mathbb{R}^n)$, is defined to be the set of all the $f \in \mathcal{S}'(\mathbb{R}^n)$ such that there exists a sequence $\{a_{i,j}\}_{i \in \mathbb{Z}, j \in \mathbb{N}}$ of $(\dot{\mathcal{K}}_{\omega,0}^{p,q}(\mathbb{R}^n), r, d)$-atoms supported, respectively, in the balls $\{B_{i,j}\}_{i \in \mathbb{Z}, j \in \mathbb{N}} \subset \mathbb{B}$ and three positive constants $c \in (0, 1]$, A, and $\widetilde{A}$, independent of f, satisfying that, for any $i \in \mathbb{Z}$,

$$\sum_{j \in \mathbb{N}} \mathbf{1}_{cB_{i,j}} \leq A,$$

$$f = \sum_{i \in \mathbb{Z}} \sum_{j \in \mathbb{N}} \widetilde{A} 2^i \left\| \mathbf{1}_{B_{i,j}} \right\|_{\dot{\mathcal{K}}^{p,q}_{\omega,0}(\mathbb{R}^n)} a_{i,j}$$

in $\mathcal{S}'(\mathbb{R}^n)$, and

$$\sup_{i \in \mathbb{Z}} \left\{ 2^i \left\| \sum_{j \in \mathbb{N}} \mathbf{1}_{B_{i,j}} \right\|_{\dot{\mathcal{K}}^{p,q}_{\omega,0}(\mathbb{R}^n)} \right\} < \infty.$$

Moreover, for any $f \in WH\dot{\mathcal{K}}^{p,q,r,d}_{\omega,0}(\mathbb{R}^n)$,

$$\|f\|_{WH\dot{\mathcal{K}}^{p,q,r,d}_{\omega,0}(\mathbb{R}^n)} := \inf \sup_{i \in \mathbb{Z}} 2^i \left\{ \left\| \sum_{j \in \mathbb{N}} \mathbf{1}_{B_{i,j}} \right\|_{\dot{\mathcal{K}}^{p,q}_{\omega,0}(\mathbb{R}^n)} \right\},$$

where the infimum is taken over all the decompositions of f as above.

Then we have the following atomic characterization of the weak generalized Herz–Hardy space $WH\dot{\mathcal{K}}^{p,q}_{\omega,0}(\mathbb{R}^n)$.

Theorem 6.3.2 *Let p, q, ω, r, and d be as in Definition 6.3.1. Then*

$$WH\dot{\mathcal{K}}^{p,q}_{\omega,0}(\mathbb{R}^n) = WH\dot{\mathcal{K}}^{p,q,r,d}_{\omega,0}(\mathbb{R}^n)$$

with equivalent quasi-norms.

To show this theorem, we first recall the following definition of weak Hardy spaces associated with ball quasi-Banach function spaces, which is just [278, Definition 2.21].

Definition 6.3.3 Let X be a ball quasi-Banach function space and $N \in \mathbb{N}$. Then the *weak Hardy space* $WH_X(\mathbb{R}^n)$ is defined to be the set of all the $f \in \mathcal{S}'(\mathbb{R}^n)$ such that

$$\|f\|_{WH_X(\mathbb{R}^n)} := \|\mathcal{M}_N(f)\|_{WX} < \infty.$$

Remark 6.3.4

(i) Let $p \in (0, \infty)$ and $X := L^p(\mathbb{R}^n)$. Then, in this case, the space $WH_X(\mathbb{R}^n)$ is just the classical *weak Hardy space* $WH^p(\mathbb{R}^n)$ which was originally introduced in [153, p. 114].

(ii) Let $p \in (0, \infty)$, $\upsilon \in A_\infty(\mathbb{R}^n)$, and $X := L^p_\upsilon(\mathbb{R}^n)$. Then, in this case, the space $WH_X(\mathbb{R}^n)$ is just the *weighted weak Hardy space* $WH^p_\upsilon(\mathbb{R}^n)$ which was introduced in [194, Definition 3].

(iii) Let $\vec{p} := (p_1, \ldots, p_n) \in (0, \infty)^n$ and $X := L^{\vec{p}}(\mathbb{R}^n)$. Then, in this case, the space $WH_X(\mathbb{R}^n)$ is just the *weak mixed-norm Hardy space* $WH^{\vec{p}}(\mathbb{R}^n)$ which was introduced in [278, Definition 7.27].

(iv) Let $0 < q \leq p < \infty$ and $X := M^p_q(\mathbb{R}^n)$. Then, in this case, the space $WH_X(\mathbb{R}^n)$ is just the *weak Hardy–Morrey space* $WHM^p_q(\mathbb{R}^n)$ which was originally introduced in [109, Definition 4.1].

(v) Let $p(\cdot) : \mathbb{R}^n \to (0, \infty)$ and $X := L^{p(\cdot)}(\mathbb{R}^n)$. Then, in this case, the space $WH_X(\mathbb{R}^n)$ is just the *variable weak Hardy space* $WH^{p(\cdot)}(\mathbb{R}^n)$ which was originally introduced in [267, Definition 2.13].

The following conclusion shows that, under some assumptions of the ball quasi-Banach function space X, distributions in the weak Hardy space $WH_X(\mathbb{R}^n)$ can be decomposed into linear combinations of atoms, which was proved in [278, Theorem 4.2] and plays an essential role in the proof of Theorem 6.3.2 above.

Lemma 6.3.5 *Let X be a ball quasi-Banach function space and let $r \in (0, 1)$, $r_0, s, p_- \in (0, \infty)$, and $\theta \in (1, \infty)$ be such that the following four statements hold true:*

(i) *there exists a positive constant C such that, for any $\{f_j\}_{j \in \mathbb{N}} \subset L^1_{\mathrm{loc}}(\mathbb{R}^n)$,*

$$\left\| \left\{ \sum_{j \in \mathbb{N}} [\mathcal{M}(f_j)]^{1/r} \right\}^r \right\|_{X^{1/r}} \leq C \left\| \left(\sum_{j \in \mathbb{N}} |f_j|^{1/r} \right)^r \right\|_{X^{1/r}} ;$$

(ii) *X is θ-concave and $\mathcal{M}$ bounded on $X^{\frac{1}{\theta p_-}}$;*

(iii) *$X^{1/s}$ is a ball Banach function space and $\mathcal{M}$ bounded on $(X^{1/s})'$;*

(iv) *$\mathcal{M}$ is bounded on $(WX)^{1/r_0}$.*

Assume that $d \geq \lfloor n(1/p_- - 1) \rfloor$ is a fixed nonnegative integer and $f \in WH_X(\mathbb{R}^n)$. Then there exists $\{a_{i,j}\}_{i \in \mathbb{Z}, \, j \in \mathbb{N}}$ of (X, ∞, d)-atoms supported, respectively, in the balls $\{B_{i,j}\}_{i \in \mathbb{Z}, \, j \in \mathbb{N}} \subset \mathbb{B}$ and three positive constants $c \in (0, 1]$, A, and $\widetilde{A}$, independent of f, such that, for any $i \in \mathbb{Z}$,

$$\sum_{j \in \mathbb{N}} \mathbf{1}_{cB_{i,j}} \leq A,$$

$$f = \sum_{i \in \mathbb{Z}} \sum_{j \in \mathbb{N}} \widetilde{A} 2^i \left\| \mathbf{1}_{B_{i,j}} \right\|_X a_{i,j}$$

in $\mathcal{S}'(\mathbb{R}^n)$, and

$$\sup_{i \in \mathbb{Z}} \left\{ 2^i \left\| \sum_{j \in \mathbb{N}} \mathbf{1}_{B_{i,j}} \right\|_X \right\} \lesssim \|f\|_{WH_X(\mathbb{R}^n)},$$

where the implicit positive constant is independent of f.

The following atomic reconstruction theorem of the weak Hardy space $WH_X(\mathbb{R}^n)$ is just [278, Theorem 4.7], which also plays a vital role in the proof of Theorem 6.3.2.

Lemma 6.3.6 *Let X be a ball quasi-Banach function space and let $p_- \in (0, 1)$ be such that the following three statements hold true:*

(i) *for any given $\theta \in (0, p_-)$ and $u \in (1, \infty)$, there exists a positive constant C such that, for any $\{f_j\}_{j \in \mathbb{N}} \subset L^1_{\mathrm{loc}}(\mathbb{R}^n)$,*

$$\left\| \left\{ \sum_{j \in \mathbb{N}} \left[\mathcal{M}(f_j) \right]^u \right\}^{\frac{1}{u}} \right\|_{X^{1/\theta}} \leq C \left\| \left\{ \sum_{j \in \mathbb{N}} |f_j|^u \right\}^{\frac{1}{u}} \right\|_{X^{1/\theta}};$$

(ii) *for any $s \in (0, p_-)$, $X^{1/s}$ is a ball Banach function space;*
(iii) *there exists an $s_0 \in (0, p_-)$, an $r_0 \in (s_0, \infty)$, and a $C \in (0, \infty)$ such that, for any $f \in L^1_{\mathrm{loc}}(\mathbb{R}^n)$,*

$$\left\| \mathcal{M}^{((r_0/s_0)')} (f) \right\|_{(X^{1/s_0})'} \leq C \|f\|_{(X^{1/s_0})'}.$$

Let $d \geq \lfloor n(1/p_- - 1) \rfloor$ be a fixed integer, $c \in (0, 1]$, $r \in (\max\{1, r_0\}, \infty]$, and $A, \tilde{A} \in (0, \infty)$. Assume that $\{a_{i,j}\}_{i \in \mathbb{Z}, j \in \mathbb{N}}$ is a sequence of (X, r, d)-atoms supported, respectively, in the balls $\{B_{i,j}\}_{i \in \mathbb{Z}, j \in \mathbb{N}} \subset \mathbb{B}$ such that, for any $i \in \mathbb{Z}$,

$$\sum_{j \in \mathbb{N}} \mathbf{1}_{cB_{i,j}} \leq A,$$

$f := \sum_{i \in \mathbb{Z}} \sum_{j \in \mathbb{N}} \tilde{A} 2^i \|\mathbf{1}_{B_{i,j}}\|_X a_{i,j}$ converges in $\mathcal{S}'(\mathbb{R}^n)$, and

$$\sup_{i \in \mathbb{Z}} \left\{ 2^i \left\| \sum_{j \in \mathbb{N}} \mathbf{1}_{B_{i,j}} \right\|_X \right\} < \infty.$$

Then $f \in WH_X(\mathbb{R}^n)$ and

$$\|f\|_{WH_X(\mathbb{R}^n)} \lesssim \sup_{i\in\mathbb{Z}} 2^i \left\| \sum_{j\in\mathbb{N}} \mathbf{1}_{B_{i,j}} \right\|_X ,$$

where the implicit positive constant is independent of f.

Remark 6.3.7 We point out that both Lemmas 6.3.5 and 6.3.6 have a wide range of applications. Here we present two function spaces to which both Lemmas 6.3.5 and 6.3.6 can be applied (see also [278, Section 7]).

(i) Let $\vec{p} := (p_1, \ldots, p_n) \in (0, \infty)^n$,

$$d \geq \left\lfloor n \left(\frac{\max\{1, p_1, \ldots, p_n\}}{\min\{p_1, \ldots, p_n\}} - 1 \right) \right\rfloor$$

be a fixed integer, and $r \in (\max\{1, p_1, \ldots, p_n\}, \infty]$. Then, in this case, as was mentioned in [278, Subsection 7.2], the mixed-norm Lebesgue space $L^{\vec{p}}(\mathbb{R}^n)$ satisfies all the assumptions of both Lemmas 6.3.5 and 6.3.6. Thus, both Lemmas 6.3.5 and 6.3.6 with $X := L^{\vec{p}}(\mathbb{R}^n)$ hold true. These results coincide with [278, Theorem 7.30].

(ii) Let $0 < q \leq p < \infty$,

$$d \geq \left\lfloor n \left(\frac{1}{\min\{1, q\}} - 1 \right) \right\rfloor$$

be a fixed integer, and $r \in (\max\{1, p\}, \infty]$. Then, in this case, as was pointed out in [278, Subsection 7.1], the Morrey space $M_q^p(\mathbb{R}^n)$ satisfies all the assumptions of both Lemmas 6.3.5 and 6.3.6. This then implies that both Lemmas 6.3.5 and 6.3.6 with $X := M_q^p(\mathbb{R}^n)$ hold true. These results coincide with [278, Theorem 7.11].

Based on the above two lemmas, we now show the atomic characterization of the weak generalized Herz–Hardy space $WH\dot{\mathcal{K}}_{\omega,0}^{p,q}(\mathbb{R}^n)$.

Proof of Theorem 6.3.2 Let p, q, ω, r, p_-, and d be as in the present theorem. We first prove $WH\dot{\mathcal{K}}_{\omega,0}^{p,q}(\mathbb{R}^n) \subset WH\dot{\mathcal{K}}_{\omega,0}^{p,q,r,d}(\mathbb{R}^n)$. To this end, let $f \in WH\dot{\mathcal{K}}_{\omega,0}^{p,q}(\mathbb{R}^n)$. Now, we claim that, for the above r and p_-, all the assumptions of Lemma 6.3.5 hold true with $X := \dot{\mathcal{K}}_{\omega,0}^{p,q}(\mathbb{R}^n)$. Assume that this claim holds true for the moment. Then, from Lemma 6.3.5 with X therein replaced by $\dot{\mathcal{K}}_{\omega,0}^{p,q}(\mathbb{R}^n)$, it follows that there exists $\{a_{i,j}\}_{i\in\mathbb{Z}, j\in\mathbb{N}}$ of $(\dot{\mathcal{K}}_{\omega,0}^{p,q}(\mathbb{R}^n), \infty, d)$-atoms supported, respectively, in the balls $\{B_{i,j}\}_{i\in\mathbb{Z},\,j\in\mathbb{N}} \subset \mathbb{B}$ and three positive constants $c \in (0,1]$, A, and $\widetilde{A}$, independent of f, such that, for any $i \in \mathbb{Z}$,

$$\sum_{j\in\mathbb{N}} \mathbf{1}_{cB_{i,j}} \leq A,$$

$$f = \sum_{i\in\mathbb{Z}}\sum_{j\in\mathbb{N}} \widetilde{A} 2^i \left\| \mathbf{1}_{B_{i,j}} \right\|_{\dot{\mathcal{K}}^{p,q}_{\omega,\mathbf{0}}(\mathbb{R}^n)} a_{i,j} \tag{6.3}$$

in $\mathcal{S}'(\mathbb{R}^n)$, and

$$\sup_{i\in\mathbb{Z}} \left\{ 2^i \left\| \sum_{j\in\mathbb{N}} \mathbf{1}_{B_{i,j}} \right\|_{\dot{\mathcal{K}}^{p,q}_{\omega,\mathbf{0}}(\mathbb{R}^n)} \right\} \lesssim \|f\|_{WH\dot{\mathcal{K}}^{p,q}_{\omega,\mathbf{0}}(\mathbb{R}^n)} < \infty. \tag{6.4}$$

In addition, by Lemma 4.3.5 with $t := \infty$, we conclude that, for any $i \in \mathbb{Z}$ and $j \in \mathbb{N}$, $a_{i,j}$ is a $(\dot{\mathcal{K}}^{p,q}_{\omega,\mathbf{0}}(\mathbb{R}^n), r, d)$-atom supported in the ball $B_{i,j}$. This, combined with (6.3), (6.4), and Definition 6.3.1, further implies that $f \in WH\dot{\mathcal{K}}^{p,q,r,d}_{\omega,\mathbf{0}}(\mathbb{R}^n)$ and then finishes the proof that $WH\dot{\mathcal{K}}^{p,q}_{\omega,\mathbf{0}}(\mathbb{R}^n) \subset WH\dot{\mathcal{K}}^{p,q,r,d}_{\omega,\mathbf{0}}(\mathbb{R}^n)$.

Thus, to complete the proof that $WH\dot{\mathcal{K}}^{p,q}_{\omega,\mathbf{0}}(\mathbb{R}^n) \subset WH\dot{\mathcal{K}}^{p,q,r,d}_{\omega,\mathbf{0}}(\mathbb{R}^n)$, it remains to show the above claim. Indeed, applying the assumption $m_0 \in (-\frac{n}{p}, \infty)$ and Theorem 1.2.42, we find that the local generalized Herz space $\dot{\mathcal{K}}^{p,q}_{\omega,\mathbf{0}}(\mathbb{R}^n)$ under consideration is a BQBF space. Using Lemma 4.3.10 with $u := \frac{1}{r}$, we conclude that, for any $\{f_j\}_{j\in\mathbb{N}} \subset L^1_{\text{loc}}(\mathbb{R}^n)$,

$$\left\| \left\{ \sum_{j\in\mathbb{N}} [\mathcal{M}(f_j)]^{1/r} \right\}^r \right\|_{[\dot{\mathcal{K}}^{p,q}_{\omega,\mathbf{0}}(\mathbb{R}^n)]^{1/r}} \lesssim \left\| \left(\sum_{j\in\mathbb{N}} |f_j|^{1/r} \right)^r \right\|_{[\dot{\mathcal{K}}^{p,q}_{\omega,\mathbf{0}}(\mathbb{R}^n)]^{1/r}},$$

which implies that the assumption (i) of Lemma 6.3.5 holds true for $\dot{\mathcal{K}}^{p,q}_{\omega,\mathbf{0}}(\mathbb{R}^n)$. We next prove that there exists a θ, an s, and an r_0 such that (ii), (iii), and (iv) of Lemma 6.3.5 hold true. Indeed, let

$$\theta \in \left(\max\{1, p, q\}, \frac{1}{p_-} \min\left\{ p, q, \frac{n}{\max\{M_0(\omega), M_\infty(\omega)\} + n/p} \right\} \right).$$

Then, from the assumptions $p/\theta, q/\theta \in (0,1)$, the reverse Minkowski inequality, and Lemma 1.3.1, we deduce that, for any $\{f_j\}_{j\in\mathbb{N}} \subset \mathscr{M}(\mathbb{R}^n)$,

$$\sum_{j\in\mathbb{N}} \|f_j\|_{[\dot{\mathcal{K}}^{p,q}_{\omega,\mathbf{0}}(\mathbb{R}^n)]^{1/\theta}} \leq \left\| \sum_{j\in\mathbb{N}} |f_j| \right\|_{[\dot{\mathcal{K}}^{p,q}_{\omega,\mathbf{0}}(\mathbb{R}^n)]^{1/\theta}}, \tag{6.5}$$

which implies that $\dot{\mathcal{K}}_{\omega,0}^{p,q}(\mathbb{R}^n)$ is θ-concave. On the other hand, using Lemma 1.8.5 with $r := \theta p_-$, we conclude that, for any $f \in L_{\mathrm{loc}}^1(\mathbb{R}^n)$,

$$\|\mathcal{M}(f)\|_{[\dot{\mathcal{K}}_{\omega,0}^{p,q}(\mathbb{R}^n)]^{\frac{1}{\theta p_-}}} \lesssim \|f\|_{[\dot{\mathcal{K}}_{\omega,0}^{p,q}(\mathbb{R}^n)]^{\frac{1}{\theta p_-}}} .$$

From this and (6.5), we further infer that, for the above θ and p_-, Lemma 6.3.5(ii) holds true. Let

$$s \in \left(0, \min\left\{1, p, q, \frac{n}{\max\{M_0(\omega), M_\infty(\omega)\} + n/p}\right\}\right).$$

Then, by Lemma 1.8.6 with $r := \infty$, we find that $[\dot{\mathcal{K}}_{\omega,0}^{p,q}(\mathbb{R}^n)]^{1/s}$ is a BBF space and, for any $f \in L_{\mathrm{loc}}^1(\mathbb{R}^n)$,

$$\|\mathcal{M}(f)\|_{([\dot{\mathcal{K}}_{\omega,0}^{p,q}(\mathbb{R}^n)]^{1/s})'} \lesssim \|f\|_{([\dot{\mathcal{K}}_{\omega,0}^{p,q}(\mathbb{R}^n)]^{1/s})'} ,$$

which imply that Lemma 6.3.5(iii) holds true with the above s. Finally, let

$$r_0 \in \left(0, \min\left\{p, \frac{n}{\max\{M_0(\omega), M_\infty(\omega)\} + n/p}\right\}\right).$$

Then, from Lemma 6.1.6 with $r := r_0$, we deduce that, for any $f \in L_{\mathrm{loc}}^1(\mathbb{R}^n)$,

$$\|\mathcal{M}(f)\|_{[W\dot{\mathcal{K}}_{\omega,0}^{p,q}(\mathbb{R}^n)]^{1/r_0}} \lesssim \|f\|_{[W\dot{\mathcal{K}}_{\omega,0}^{p,q}(\mathbb{R}^n)]^{1/r_0}} .$$

This implies that Lemma 6.3.5(iv) holds true with the above r_0. Therefore, the Herz space $\dot{\mathcal{K}}_{\omega,0}^{p,q}(\mathbb{R}^n)$ under consideration satisfies all the assumptions of Lemma 6.3.5 with r and p_- in the present theorem. This then finishes the proof of the above claim and further implies that

$$W H\dot{\mathcal{K}}_{\omega,0}^{p,q}(\mathbb{R}^n) \subset W H\dot{\mathcal{K}}_{\omega,0}^{p,q,d}(\mathbb{R}^n).$$

Moreover, combining (6.3), (6.4), and Definition 6.3.1 again, we conclude that, for any $f \in W H\dot{\mathcal{K}}_{\omega,0}^{p,q}(\mathbb{R}^n)$,

$$\|f\|_{W H\dot{\mathcal{K}}_{\omega,0}^{p,q,d}(\mathbb{R}^n)} \lesssim \|f\|_{W H\dot{\mathcal{K}}_{\omega,0}^{p,q}(\mathbb{R}^n)} . \tag{6.6}$$

Conversely, we now prove that $W H\dot{\mathcal{K}}_{\omega,0}^{p,q,d}(\mathbb{R}^n) \subset W H\dot{\mathcal{K}}_{\omega,0}^{p,q}(\mathbb{R}^n)$. Indeed, from the proof that $W H\dot{\mathcal{K}}_{\omega,0}^{p,q}(\mathbb{R}^n) \subset W H\dot{\mathcal{K}}_{\omega,0}^{p,q,d}(\mathbb{R}^n)$, it follows that the local generalized Herz space $\dot{\mathcal{K}}_{\omega,0}^{p,q}(\mathbb{R}^n)$ under consideration is a BQBF space. Thus, in order to finish the proof that $W H\dot{\mathcal{K}}_{\omega,0}^{p,q,d}(\mathbb{R}^n) \subset W H\dot{\mathcal{K}}_{\omega,0}^{p,q}(\mathbb{R}^n)$, we only need to show that the assumptions (i) through (iii) of Lemma 6.3.6 hold true for $\dot{\mathcal{K}}_{\omega,\mathbf{0}}^{p,q}(\mathbb{R}^n)$. Indeed, for any given $\theta \in (0, p_-)$ and $u \in (1, \infty)$, applying Lemma 4.3.10 with $r := \theta$, we find that, for any $\{f_j\}_{j\in\mathbb{N}} \subset L_{\mathrm{loc}}^1(\mathbb{R}^n)$,

$$\left\| \left\{ \sum_{j\in\mathbb{N}} [\mathcal{M}(f_j)]^u \right\}^{\frac{1}{u}} \right\|_{[\dot{\mathcal{K}}_{\omega,\mathbf{0}}^{p,q}(\mathbb{R}^n)]^{1/\theta}} \lesssim \left\| \left(\sum_{j\in\mathbb{N}} |f_j|^u \right)^{\frac{1}{u}} \right\|_{[\dot{\mathcal{K}}_{\omega,\mathbf{0}}^{p,q}(\mathbb{R}^n)]^{1/\theta}}.$$

This implies that Lemma 6.3.6(i) holds true for $\dot{\mathcal{K}}_{\omega,\mathbf{0}}^{p,q}(\mathbb{R}^n)$. In addition, from Lemma 1.8.6, we deduce that, for any $s \in (0, p_-)$, $[\dot{\mathcal{K}}_{\omega,\mathbf{0}}^{p,q}(\mathbb{R}^n)]^{1/s}$ is a BBF space. This implies that Lemma 6.3.6(ii) holds true. Finally, we show that there exists an s_0 and an r_0 such that Lemma 6.3.6(iii) holds true with these s_0 and r_0. To do this, let $s_0 \in (0, p_-)$ and

$$r_0 \in \left(\max\left\{ 1, p, \frac{n}{\max\{M_0(\omega), M_\infty(\omega)\} + n/p} \right\}, r \right).$$

Then, using Lemma 1.8.6 again with s and r therein replaced, respectively, by s_0 and r_0, we conclude that, for any $f \in L_{\mathrm{loc}}^1(\mathbb{R}^n)$,

$$\left\| \mathcal{M}^{((r_0/s_0)')}(f) \right\|_{([\dot{\mathcal{K}}_{\omega,\mathbf{0}}^{p,q}(\mathbb{R}^n)]^{1/s_0})'} \lesssim \|f\|_{([\dot{\mathcal{K}}_{\omega,\mathbf{0}}^{p,q}(\mathbb{R}^n)]^{1/s_0})'},$$

which implies that, for the above s_0 and r_0, Lemma 6.3.6(iii) holds true for $\dot{\mathcal{K}}_{\omega,\mathbf{0}}^{p,q}(\mathbb{R}^n)$. Therefore, all the assumptions of Lemma 6.3.6 hold true for the Herz space $\dot{\mathcal{K}}_{\omega,\mathbf{0}}^{p,q}(\mathbb{R}^n)$ under consideration. This then implies that $WH\dot{\mathcal{K}}_{\omega,\mathbf{0}}^{p,q,r,d}(\mathbb{R}^n) \subset WH\dot{\mathcal{K}}_{\omega,\mathbf{0}}^{p,q}(\mathbb{R}^n)$. Moreover, using Definition 6.3.1 and Lemma 6.3.6 again, we further find that, for any $f \in WH\dot{\mathcal{K}}_{\omega,\mathbf{0}}^{p,q,r,d}(\mathbb{R}^n)$,

$$\|f\|_{WH\dot{\mathcal{K}}_{\omega,\mathbf{0}}^{p,q}(\mathbb{R}^n)} \lesssim \|f\|_{WH\dot{\mathcal{K}}_{\omega,\mathbf{0}}^{p,q,r,d}(\mathbb{R}^n)},$$

which, together with (6.6), further implies that

$$WH\dot{\mathcal{K}}_{\omega,\mathbf{0}}^{p,q}(\mathbb{R}^n) = WH\dot{\mathcal{K}}_{\omega,\mathbf{0}}^{p,q,r,d}(\mathbb{R}^n)$$

with equivalent quasi-norms. This finishes the proof of Theorem 6.3.2. $\quad\square$

As an application, we now establish the atomic characterization of the weak generalized Hardy–Morrey space $WHM_{\omega,\mathbf{0}}^{p,q}(\mathbb{R}^n)$. Recall that the definition of the $(M_{\omega,\mathbf{0}}^{p,q}(\mathbb{R}^n), r, d)$-atoms is as in Definition 4.3.12. Then, applying Theorem 6.3.2 and Remark 6.0.25, we obtain the following atomic characterization of the weak generalized Hardy–Morrey space $WHM_{\omega,\mathbf{0}}^{p,q}(\mathbb{R}^n)$ immediately; we omit the details.

Corollary 6.3.8 *Let* $p, q \in [1, \infty)$, $p_- \in (0, \min\{p, q\}/\max\{p, q\})$, $\omega \in M(\mathbb{R}_+)$ *with*

$$-\frac{n}{p} < m_0(\omega) \le M_0(\omega) < 0$$

and

$$-\frac{n}{p} < m_\infty(\omega) \le M_\infty(\omega) < 0,$$

$d \ge \lfloor n(1/p_- - 1) \rfloor$ *be a fixed integer, and*

$$r \in \left(\frac{n}{\min\{m_0(\omega), m_\infty(\omega)\} + n/p}, \infty \right].$$

Then the weak generalized atomic Hardy–Morrey space $WHM_{\omega,0}^{p,q,r,d}(\mathbb{R}^n)$, *associated with the weak local generalized Morrey space* $WM_{\omega,0}^{p,q}(\mathbb{R}^n)$, *is defined to be the set of all the* $f \in \mathcal{S}'(\mathbb{R}^n)$ *such that there exists a sequence* $\{a_{i,j}\}_{i \in \mathbb{Z}, j \in \mathbb{N}}$ *of* $(M_{\omega,0}^{p,q}(\mathbb{R}^n), r, d)$-*atoms supported, respectively, in the balls* $\{B_{i,j}\}_{i \in \mathbb{Z}, j \in \mathbb{N}} \subset \mathbb{B}$ *and three positive constants* $c \in (0, 1]$, A, *and* $\widetilde{A}$, *independent of* f, *satisfying that, for any* $i \in \mathbb{Z}$,

$$\sum_{j \in \mathbb{N}} \mathbf{1}_{cB_{i,j}} \le A,$$

$$f = \sum_{i \in \mathbb{Z}} \sum_{j \in \mathbb{N}} \widetilde{A} 2^i \left\| \mathbf{1}_{B_{i,j}} \right\|_{M_{\omega,0}^{p,q}(\mathbb{R}^n)} a_{i,j}$$

in $\mathcal{S}'(\mathbb{R}^n)$, *and*

$$\sup_{i \in \mathbb{Z}} \left\{ 2^i \left\| \sum_{j \in \mathbb{N}} \mathbf{1}_{B_{i,j}} \right\|_{M_{\omega,0}^{p,q}(\mathbb{R}^n)} \right\} < \infty.$$

Moreover, for any $f \in WHM_{\omega,0}^{p,q,r,d}(\mathbb{R}^n)$,

$$\|f\|_{WHM_{\omega,0}^{p,q,r,d}(\mathbb{R}^n)} := \inf \sup_{i \in \mathbb{Z}} \left\{ 2^i \left\| \sum_{j \in \mathbb{N}} \mathbf{1}_{B_{i,j}} \right\|_{M_{\omega,0}^{p,q}(\mathbb{R}^n)} \right\},$$

where the infimum is taken over all the decompositions of f as above. Then

$$WHM_{\omega,0}^{p,q}(\mathbb{R}^n) = WHM_{\omega,0}^{p,q,r,d}(\mathbb{R}^n)$$

with equivalent quasi-norms.

The remainder of this section is devoted to establishing the atomic characterization of the weak generalized Herz–Hardy space $WH\dot{\mathcal{K}}_{\omega}^{p,q}(\mathbb{R}^n)$. To this end, we first introduce the following definition of weak generalized atomic Herz–Hardy spaces via $(\dot{\mathcal{K}}_{\omega}^{p,q}(\mathbb{R}^n),\ r,\ d)$-atoms introduced in Definition 4.3.14.

Definition 6.3.9 Let $p, q \in (0, \infty)$, $\omega \in M(\mathbb{R}_+)$ with $m_0(\omega) \in (-\frac{n}{p}, \infty)$ and

$$-\frac{n}{p} < m_\infty(\omega) \le M_\infty(\omega) < 0,$$

$$p_- \in \left(0, \frac{\min\{p, q, \frac{n}{\max\{M_0(\omega), M_\infty(\omega)\}+n/p}\}}{\max\{1, p, q\}}\right),$$

$d \ge \lfloor n(1/p_- - 1) \rfloor$ be a fixed integer, and

$$r \in \left(\max\left\{1, p, \frac{n}{\min\{m_0(\omega), m_\infty(\omega)\}+n/p}\right\}, \infty\right].$$

Then the *weak generalized atomic Herz–Hardy space $WH\dot{\mathcal{K}}_{\omega}^{p,q,r,d}(\mathbb{R}^n)$*, associated with the weak global generalized Herz space $W\dot{\mathcal{K}}_{\omega}^{p,q}(\mathbb{R}^n)$, is defined to be the set of all the $f \in \mathcal{S}'(\mathbb{R}^n)$ such that there exists a sequence $\{a_{i,j}\}_{i\in\mathbb{Z},\ j\in\mathbb{N}}$ of $(\dot{\mathcal{K}}_{\omega}^{p,q}(\mathbb{R}^n),\ r,\ d)$-atoms supported, respectively, in the balls $\{B_{i,j}\}_{i\in\mathbb{Z},\ j\in\mathbb{N}} \subset \mathbb{B}$ and three positive constants $c \in (0, 1]$, A, and $\widetilde{A}$, independent of f, satisfying that, for any $i \in \mathbb{Z}$,

$$\sum_{j\in\mathbb{N}} \mathbf{1}_{cB_{i,j}} \le A,$$

$$f = \sum_{i\in\mathbb{Z}}\sum_{j\in\mathbb{N}} \widetilde{A}2^i \left\| \mathbf{1}_{B_{i,j}} \right\|_{\dot{\mathcal{K}}_{\omega}^{p,q}(\mathbb{R}^n)} a_{i,j}$$

in $\mathcal{S}'(\mathbb{R}^n)$, and

$$\sup_{i\in\mathbb{Z}}\left\{2^i \left\| \sum_{j\in\mathbb{N}} \mathbf{1}_{B_{i,j}} \right\|_{\dot{\mathcal{K}}_{\omega}^{p,q}(\mathbb{R}^n)}\right\} < \infty,$$

Moreover, for any $f \in WH\dot{\mathcal{K}}_{\omega}^{p,q,d}(\mathbb{R}^n)$,

$$\|f\|_{WH\dot{\mathcal{K}}_{\omega}^{p,q,d}(\mathbb{R}^n)} := \inf \sup_{i \in \mathbb{Z}} \left\{ 2^i \left\| \sum_{j \in \mathbb{N}} \mathbf{1}_{B_{i,j}} \right\|_{\dot{\mathcal{K}}_{\omega}^{p,q}(\mathbb{R}^n)} \right\},$$

where the infimum is taken over all the decompositions of f as above.

Then we have the following atomic characterization of the weak generalized Herz–Hardy space $WH\dot{\mathcal{K}}_{\omega}^{p,q}(\mathbb{R}^n)$.

Theorem 6.3.10 *Let p, q, ω, r, and d be as in Definition 6.3.9. Then*

$$WH\dot{\mathcal{K}}_{\omega}^{p,q}(\mathbb{R}^n) = WH\dot{\mathcal{K}}_{\omega}^{p,q,d}(\mathbb{R}^n)$$

with equivalent quasi-norms.

Due to the deficiency of associate spaces of global generalized Herz spaces, we can not prove this theorem via the atomic characterization of weak Hardy spaces associated with ball quasi-Banach function spaces obtained in [278, Theorems 4.2 and 4.7] (see also both Lemmas 6.3.5 and 6.3.6) directly. To overcome this difficulty, we establish an improved atomic characterization of $WH_X(\mathbb{R}^n)$ associated with the ball quasi-Banach function space X without recourse to the associate space X'. Indeed, we have the following atomic decomposition theorem.

Proposition 6.3.11 *Let X be a ball quasi-Banach function space and $Y \subset \mathcal{M}(\mathbb{R}^n)$ a linear space equipped with a quasi-seminorm $\|\cdot\|_Y$, and let $r \in (0, 1)$, r_0, s, $p_- \in (0, \infty)$, and θ, $\theta_0 \in (1, \infty)$ be such that the following seven statements hold true:*

(i) *there exists a positive constant C such that, for any $\{f_j\}_{j \in \mathbb{N}} \subset L^1_{\mathrm{loc}}(\mathbb{R}^n)$,*

$$\left\| \left\{ \sum_{j \in \mathbb{N}} [\mathcal{M}(f_j)]^{1/r} \right\}^r \right\|_{X^{1/r}} \leq C \left\| \left(\sum_{j \in \mathbb{N}} |f_j|^{1/r} \right)^r \right\|_{X^{1/r}};$$

(ii) *X is θ-concave and $\mathcal{M}$ bounded on $X^{\frac{1}{\theta p_-}}$;*
(iii) *$\mathcal{M}$ is bounded on $(WX)^{1/r_0}$;*
(iv) *$\|\cdot\|_Y$ satisfies Definition 1.2.13(ii);*
(v) *$\mathbf{1}_{B(0,1)} \in Y$;*
(vi) *for any $f \in \mathcal{M}(\mathbb{R}^n)$,*

$$\|f\|_{X^{1/s}} \sim \sup \left\{ \|fg\|_{L^1(\mathbb{R}^n)} : \|g\|_Y = 1 \right\},$$

where the positive equivalence constants are independent of f;
(vii) *$\mathcal{M}^{(\theta_0)}$ is bounded on Y.*

Assume that $d \geq \lfloor n(1/p_- - 1) \rfloor$ is a fixed nonnegative integer and $f \in WH_X(\mathbb{R}^n)$. Then there exists $\{a_{i,j}\}_{i \in \mathbb{Z}, j \in \mathbb{N}}$ of (X, ∞, d)-atoms supported, respectively, in the balls $\{B_{i,j}\}_{i \in \mathbb{Z}, j \in \mathbb{N}} \subset \mathbb{B}$ and three positive constants $c \in (0, 1]$, A, and $\tilde{A}$, independent of f, such that, for any $i \in \mathbb{Z}$,

$$\sum_{j \in \mathbb{N}} \mathbf{1}_{cB_{i,j}} \leq A,$$

$$f = \sum_{i \in \mathbb{Z}} \sum_{j \in \mathbb{N}} \tilde{A} 2^i \left\| \mathbf{1}_{B_{i,j}} \right\|_X a_{i,j}$$

in $\mathcal{S}'(\mathbb{R}^n)$, and

$$\sup_{i \in \mathbb{Z}} \left\{ 2^i \left\| \sum_{j \in \mathbb{N}} \mathbf{1}_{B_{i,j}} \right\|_X \right\} \lesssim \|f\|_{WH_X(\mathbb{R}^n)},$$

where the implicit positive constant is independent of f.

Remark 6.3.12 We should point out that Proposition 6.3.11 is an improved version of the known atomic decomposition of $WH_X(\mathbb{R}^n)$ obtained in [278, Theorem 4.2]. Indeed, if $Y \equiv (X^{1/s})'$ in Proposition 6.3.11, then this proposition goes back to [278, Theorem 4.2].

To show this proposition, we first prove the following technical lemma by borrowing some ideas from the proof of [278, Lemma 4.3], which shows a crucial fact that distributions in weak Hardy spaces vanish weakly at infinity.

Lemma 6.3.13 *Let X be a ball quasi-Banach function space, $Y \subset \mathcal{M}(\mathbb{R}^n)$ a linear space equipped with a quasi-seminorm $\| \cdot \|_Y$, $\theta \in (1, \infty)$, and $s \in (0, \infty)$ satisfy the following four statements:*

(i) *$\| \cdot \|_Y$ satisfies Definition 1.2.13(ii);*
(ii) *$\mathbf{1}_{B(0,1)} \in Y$;*
(iii) *for any $f \in \mathcal{M}(\mathbb{R}^n)$,*

$$\|f\|_{X^{1/s}} \sim \sup \left\{ \|fg\|_{L^1(\mathbb{R}^n)} : \|g\|_Y = 1 \right\},$$

where the positive equivalence constants are independent of f;
(iv) *$\mathcal{M}^{(\theta)}$ is bounded on Y.*

Assume $f \in WH_X(\mathbb{R}^n)$. Then f vanishes weakly at infinity.

In order to prove Lemma 6.3.13, we require the following auxiliary estimate about A_p-weights (see, for instance, [70, Corollary 7.6(3)]).

Lemma 6.3.14 *Let* $p \in [1, \infty)$ *and* $\upsilon \in A_p(\mathbb{R}^n)$. *Then there exist two positive constants* δ *and* C *such that, for any ball* $B_1 \in \mathbb{B}$ *and any ball* $B_2 \subset B_1$,

$$\frac{\upsilon(B_2)}{\upsilon(B_1)} \leq C \left(\frac{|B_2|}{|B_1|} \right)^{\delta}.$$

We next show Lemma 6.3.13.

Proof of Lemma 6.3.13 Let all the symbols be as in the present lemma. Then, from the assumptions (i) through (iv) of the present proposition and Lemma 4.8.18, we deduce that there exists an $\varepsilon \in (0, 1)$ such that, for any $g \in \mathcal{M}(\mathbb{R}^n)$,

$$\|g\|_{L_{\upsilon}^s(\mathbb{R}^n)} \lesssim \|g\|_X, \tag{6.7}$$

where $\upsilon := [\mathcal{M}(\mathbf{1}_{B(\mathbf{0},1)})]^{\varepsilon}$. Let $\varphi \in \mathcal{S}(\mathbb{R}^n)$ and $N \in \mathbb{N}$. Then, by (4.3), we conclude that, for any $t \in (0, \infty)$ and $x, y \in \mathbb{R}^n$ with $|y - x| < t$,

$$|f * \varphi_t(x)| \lesssim \mathcal{M}_N(f)(y),$$

which further implies that

$$|f * \varphi_t(x)| \, \mathbf{1}_{B(x,t)} \lesssim \mathcal{M}_N(f) \mathbf{1}_{B(x,t)}.$$

Applying this, Definitions 2.3.7(ii) and 6.1.3, (6.7), Remark 6.1.4 with $X := L_{\upsilon}^s(\mathbb{R}^n)$, and Definition 6.3.3, we find that, for any $t \in (0, \infty)$ and $x \in \mathbb{R}^n$,

$$|f * \varphi_t(x)| \lesssim \frac{\|\mathcal{M}_N(f) \mathbf{1}_{B(x,t)}\|_{WL_{\upsilon}^s(\mathbb{R}^n)}}{\|\mathbf{1}_{B(x,t)}\|_{WL_{\upsilon}^s(\mathbb{R}^n)}}$$

$$\lesssim \frac{\|\mathcal{M}_N(f)\|_{WX}}{\|\mathbf{1}_{B(x,t)}\|_{L_{\upsilon}^s(\mathbb{R}^n)}} \sim \frac{\|f\|_{WH_X(\mathbb{R}^n)}}{\|\mathbf{1}_{B(x,t)}\|_{L_{\upsilon}^s(\mathbb{R}^n)}}.$$

Therefore, for any given $\phi \in \mathcal{S}(\mathbb{R}^n)$ and for any $t \in (0, \infty)$, we have

$$\left| \int_{\mathbb{R}^n} f * \varphi_t(x) \phi(x) \, dx \right|$$

$$\lesssim \int_{\mathbb{R}^n} \frac{1}{\|\mathbf{1}_{B(x,t)}\|_{L_{\upsilon}^s(\mathbb{R}^n)}} |\phi(x)| \, dx$$

$$\sim \int_{\mathbb{R}^n} \frac{1}{\|\mathbf{1}_{B(x,1)}\|_{L_{\upsilon}^s(\mathbb{R}^n)}} \frac{\|\mathbf{1}_{B(x,1)}\|_{L_{\upsilon}^s(\mathbb{R}^n)}}{\|\mathbf{1}_{B(x,t)}\|_{L_{\upsilon}^s(\mathbb{R}^n)}} |\phi(x)| \, dx. \tag{6.8}$$

Moreover, using Lemma 2.3.11 with $r := 1$, we find that, for any $x \in \mathbb{R}^n$,

$$\left\|\mathbf{1}_{B(x,1)}\right\|_{L^s_\upsilon(\mathbb{R}^n)} = \left\{\int_{B(x,1)} \left[\mathcal{M}\left(\mathbf{1}_{B(0,1)}\right)(y)\right]^\varepsilon dy\right\}^{\frac{1}{s}}$$

$$\sim \left[\int_{B(x,1)} (1+|y|)^{-n\varepsilon} dy\right]^{\frac{1}{s}}$$

$$\gtrsim (1+|x|)^{-\frac{n\varepsilon}{s}} |B(x,1)|^{\frac{1}{s}} \sim (1+|x|)^{-\frac{n\varepsilon}{s}}. \tag{6.9}$$

On the other hand, from Lemma 2.3.9 with $f := \mathbf{1}_{B(0,1)}$ and $\delta := \varepsilon$, it follows that $\upsilon \in A_1(\mathbb{R}^n)$. Then, for any $t \in [1, \infty)$ and $x \in \mathbb{R}^n$, by Lemma 6.3.14 with $p := 1$, $B_1 := B(x,t)$, and $B_2 := B(x,1)$, we find that there exists a $\delta \in (0, \infty)$, independent of t and x, such that

$$\frac{\left\|\mathbf{1}_{B(x,1)}\right\|_{L^s_\upsilon(\mathbb{R}^n)}}{\left\|\mathbf{1}_{B(x,t)}\right\|_{L^s_\upsilon(\mathbb{R}^n)}} \sim \left[\frac{\upsilon(B(x,1))}{\upsilon(B(x,t))}\right]^{\frac{1}{s}} \lesssim \left[\frac{|B(x,1)|}{|B(x,t)|}\right]^{\frac{\delta}{s}} \sim t^{-\frac{n\delta}{s}}.$$

Combining this, (6.8), and (6.9), we further conclude that, for any given $\phi \in \mathcal{S}(\mathbb{R}^n)$,

$$\left|\int_{\mathbb{R}^n} f * \varphi_t(x)\phi(x) dx\right|$$

$$\lesssim t^{-\frac{n\delta}{s}} \int_{\mathbb{R}^n} (1+|x|)^{\frac{n\varepsilon}{s}} |\phi(x)| dx \lesssim t^{-\frac{n\delta}{s}} \to 0$$

as $t \to \infty$. This implies that $f * \varphi_t \to 0$ in $\mathcal{S}'(\mathbb{R}^n)$ as $t \to \infty$, and hence f vanishes weakly at infinity, which then completes the proof of Lemma 6.3.13. $\square$

Applying Lemma 6.3.13, we now prove the atomic decomposition of $WH_X(\mathbb{R}^n)$ as follows.

Proof of Proposition 6.3.11 Let all the symbols be as in the present proposition and $f \in WH_X(\mathbb{R}^n)$. Then, by the assumptions (iv) through (vii) of the present proposition and Lemma 6.3.13, we find that f vanishes weakly at infinity. Using this, (i) through (iii) of the present proposition, and repeating the proof of [278, Theorem 4.2], we conclude that there exists a sequence $\{a_{i,j}\}_{i\in\mathbb{Z}, j\in\mathbb{N}}$ of (X, ∞, d)-atoms supported, respectively, in the balls $\{B_{i,j}\}_{i\in\mathbb{Z}, j\in\mathbb{N}} \subset \mathbb{B}$ and three positive constants $c \in (0, 1]$, A, and $\widetilde{A}$, independent of f, such that, for any $i \in \mathbb{Z}$,

$$\sum_{j\in\mathbb{N}} \mathbf{1}_{cB_{i,j}} \leq A,$$

$$f = \sum_{i\in\mathbb{Z}} \sum_{j\in\mathbb{N}} \widetilde{A} 2^i \left\|\mathbf{1}_{B_{i,j}}\right\|_X a_{i,j}$$

in $\mathcal{S}'(\mathbb{R}^n)$, and

$$\sup_{i\in\mathbb{Z}}\left\{2^i\left\|\sum_{j\in\mathbb{N}}\mathbf{1}_{B_{i,j}}\right\|_X\right\}\lesssim \|f\|_{WH_X(\mathbb{R}^n)}.$$

This finishes the proof of Proposition 6.3.11. □

On the other hand, we establish the following atomic reconstruction theorem of $WH_X(\mathbb{R}^n)$ without recourse to associate spaces.

Proposition 6.3.15 *Let X be a ball quasi-Banach function space and let $p_- \in (0, 1)$ be such that the following three statements hold true:*

(i) *for any given $\theta \in (0, p_-)$ and $u \in (1, \infty)$, there exists a positive constant C such that, for any $\{f_j\}_{j\in\mathbb{N}} \subset L^1_{\mathrm{loc}}(\mathbb{R}^n)$,*

$$\left\|\left\{\sum_{j\in\mathbb{N}}\left[\mathcal{M}(f_j)\right]^u\right\}^{\frac{1}{u}}\right\|_{X^{1/\theta}} \le C \left\|\left\{\sum_{j\in\mathbb{N}}|f_j|^u\right\}^{\frac{1}{u}}\right\|_{X^{1/\theta}};$$

(ii) *for any $s \in (0, p_-)$, $X^{1/s}$ is a ball Banach function space;*
(iii) *there exists an $s_0 \in (0, p_-)$, an $r_0 \in (s_0, \infty)$, a $C \in (0, \infty)$, and a linear space $Y_0 \subset \mathcal{M}(\mathbb{R}^n)$ equipped with a quasi-seminorm $\|\cdot\|_{Y_0}$ such that, for any $f \in \mathcal{M}(\mathbb{R}^n)$,*

$$\|f\|_{X^{1/s_0}} \sim \sup\left\{\|fg\|_{L^1(\mathbb{R}^n)} : \|g\|_{Y_0} = 1\right\}$$

with the positive equivalence constants independent of f and, for any $f \in L^1_{\mathrm{loc}}(\mathbb{R}^n)$,

$$\left\|\mathcal{M}^{((r_0/s_0)')}(f)\right\|_{Y_0} \le C\|f\|_{Y_0}.$$

Let $d \ge \lfloor n(1/p_- - 1)\rfloor$ be a fixed integer, $c \in (0, 1]$, $r \in (\max\{1, r_0\}, \infty]$, and $A, \widetilde{A} \in (0, \infty)$. Assume that $\{a_{i,j}\}_{i\in\mathbb{Z}, j\in\mathbb{N}}$ is a sequence of (X, r, d)-atoms supported, respectively, in the balls $\{B_{i,j}\}_{i\in\mathbb{Z}, j\in\mathbb{N}} \subset \mathbb{B}$ such that, for any $i \in \mathbb{Z}$,

$$\sum_{j\in\mathbb{N}}\mathbf{1}_{cB_{i,j}} \le A,$$

$$f := \sum_{i\in\mathbb{Z}}\sum_{j\in\mathbb{N}}\widetilde{A}2^i\left\|\mathbf{1}_{B_{i,j}}\right\|_X a_{i,j}$$

converges in $\mathcal{S}'(\mathbb{R}^n)$, *and*

$$\sup_{i\in\mathbb{Z}}\left\{2^i\left\|\sum_{j\in\mathbb{N}}\mathbf{1}_{B_{i,j}}\right\|_X\right\}<\infty.$$

Then $f\in WH_X(\mathbb{R}^n)$ *and*

$$\|f\|_{WH_X(\mathbb{R}^n)}\lesssim\sup_{i\in\mathbb{Z}}2^i\left\|\sum_{j\in\mathbb{N}}\mathbf{1}_{B_{i,j}}\right\|_X,$$

where the implicit positive constant is independent of f.

Remark 6.3.16 We should point out that Proposition 6.3.15 is an improved version of the known atomic reconstruction of $WH_X(\mathbb{R}^n)$ established in [278, Theorem 4.8]. Indeed, if $Y_0\equiv(X^{1/s_0})'$ in Proposition 6.3.15, then this proposition goes back to [278, Theorem 4.8].

To prove Proposition 6.3.15, we first establish the following technical lemma via borrowing some ideas from [278, Lemma 4.8].

Lemma 6.3.17 *Let* $s\in(0,\infty)$, $r\in(s,\infty]$, X *be a ball quasi-Banach function space, and* $Y\subset\mathcal{M}(\mathbb{R}^n)$ *a linear space equipped with a quasi-seminorm* $\|\cdot\|_Y$. *Assume that the following two statements hold true:*

(i) *for any* $f\in\mathcal{M}(\mathbb{R}^n)$,

$$\|f\|_{X^{1/s}}\sim\sup\left\{\|fg\|_{L^1(\mathbb{R}^n)}:\ \|g\|_Y=1\right\},$$

where the positive equivalence constants are independent of f;
(ii) *there exists a positive constant* C *such that, for any* $f\in L^1_{\mathrm{loc}}(\mathbb{R}^n)$,

$$\left\|\mathcal{M}^{((r/s)')}(f)\right\|_Y\le C\|f\|_Y.$$

Let $\{\lambda_j\}_{j\in\mathbb{N}}\subset[0,\infty)$, $\{B_j\}_{j\in\mathbb{N}}\subset\mathbb{B}$, *and* $\{a_j\}_{j\in\mathbb{N}}\subset\mathcal{M}(\mathbb{R}^n)$ *satisfy that, for any* $j\in\mathbb{N}$, $\mathrm{supp}\,(a_j)\subset B_j$ *and* $\|a_j\|_{L^r(\mathbb{R}^n)}\le|B_j|^{1/r}$. *Then there exists a positive constant* C, *independent of* $\{\lambda_j\}_{j\in\mathbb{N}}$, $\{B_j\}_{j\in\mathbb{N}}$, *and* $\{a_j\}_{j\in\mathbb{N}}$, *such that*

$$\left\|\left(\sum_{j\in\mathbb{N}}|\lambda_j a_j|^s\right)^{\frac{1}{s}}\right\|_X\le C\left\|\left(\sum_{j\in\mathbb{N}}|\lambda_j\mathbf{1}_{B_j}|^s\right)^{\frac{1}{s}}\right\|_X$$

432   6 Weak Generalized Herz–Hardy Spaces

Proof Let all the symbols be as in the present lemma and $g \in \mathscr{M}(\mathbb{R}^n)$ with $\|g\|_Y = 1$. Then, by the Tonelli theorem, the assumption that, for any $j \in \mathbb{N}$, supp $(a_j) \subset B_j$, the Hölder inequality, and the assumption that, for any $j \in \mathbb{N}$, $\|a_j\|_{L^r(\mathbb{R}^n)} \le |B_j|^{1/r}$, we find that

$$
\left| \int_{\mathbb{R}^n} \sum_{j \in \mathbb{N}} |\lambda_j a_j(x)|^s g(x)\, dx \right|
$$

$$
\le \sum_{j \in \mathbb{N}} \lambda_j^s \int_{B_j} |a_j(x)|^s |g(x)|\, dx
$$

$$
\le \sum_{\mathbb{R}^n} \lambda_j^s \left\| |a_j|^s \right\|_{L^{r/s}(\mathbb{R}^n)} \left\| g \mathbf{1}_{B_j} \right\|_{L^{(r/s)'}(\mathbb{R}^n)}
$$

$$
= \sum_{j \in \mathbb{N}} \lambda_j^s \left\| a_j \right\|_{L^r(\mathbb{R}^n)}^s \left\| g \mathbf{1}_{B_j} \right\|_{L^{(r/s)'}(\mathbb{R}^n)}
$$

$$
\le \sum_{j \in \mathbb{N}} \lambda_j^s |B_j|^{\frac{s}{r}} \left\| g \mathbf{1}_{B_j} \right\|_{L^{(r/s)'}(\mathbb{R}^n)}. \qquad (6.10)
$$

Observe that, from (1.54) and (1.55) with $\theta := (r/s)'$ and $f := g$, it follows that, for any $j \in \mathbb{N}$ and $x \in B_j$,

$$
\mathcal{M}^{((r/s)')}(g)(x) \ge \left[\frac{1}{|B_j|} \int_{B_j} |g(y)|^{(r/s)'}\, dy \right]^{1/(r/s)'}
$$

$$
\sim |B_j|^{-1/(r/s)'} \left\| g \mathbf{1}_{B_j} \right\|_{L^{(r/s)'}(\mathbb{R}^n)}. \qquad (6.11)
$$

On the other hand, by the assumption (ii), we find that, for any $f, g \in \mathscr{M}(\mathbb{R}^n)$,

$$
\|fg\|_{L^1(\mathbb{R}^n)} \lesssim \|f\|_{X^{1/s}} \|g\|_Y.
$$

This, combined with (6.10), (6.11), the Tonelli theorem, the assumption (ii) of the present lemma, and the assumption that $\|g\|_Y = 1$, further implies that

$$
\left| \int_{\mathbb{R}^n} \sum_{j \in \mathbb{N}} |\lambda_j a_j(x)|^s g(x)\, dx \right|
$$

$$
\lesssim \sum_{j \in \mathbb{N}} \lambda_j^s \int_{B_j} |B_j|^{-1/(r/s)'} \left\| g \mathbf{1}_{B_j} \right\|_{L^{(r/s)'}(\mathbb{R}^n)}\, dx
$$

$$
\lesssim \sum_{j \in \mathbb{N}} \lambda_j^s \int_{B_j} \mathcal{M}^{((r/s)')}(g)(x)\, dx
$$

$$\sim \int_{\mathbb{R}^n} \sum_{j \in \mathbb{N}} \left| \lambda_j \mathbf{1}_{B_j}(x) \right|^s \, \mathcal{M}^{((r/s)')}(g)(x) \, dx$$

$$\lesssim \left\| \sum_{j \in \mathbb{N}} \left| \lambda_j \mathbf{1}_{B_j} \right|^s \right\|_{X^{1/s}} \left\| \mathcal{M}^{((r/s)')}(g) \right\|_Y \lesssim \left\| \left(\sum_{j \in \mathbb{N}} \left| \lambda_j \mathbf{1}_{B_j} \right|^s \right)^{\frac{1}{s}} \right\|_X^s.$$

Applying this, Definition 1.2.28(i) with $p := \frac{1}{s}$, the arbitrariness of g, and the assumption (i) of the present lemma, we conclude that

$$\left\| \left(\sum_{j \in \mathbb{N}} \left| \lambda_j a_j \right|^s \right)^{\frac{1}{s}} \right\|_X = \left\| \sum_{j \in \mathbb{N}} \left| \lambda_j a_j \right|^s \right\|_{X^{1/s}}^{\frac{1}{s}} \lesssim \left\| \left(\sum_{j \in \mathbb{N}} \left| \lambda_j \mathbf{1}_{B_j} \right|^s \right)^{\frac{1}{s}} \right\|_X.$$

This then finishes the proof of Lemma 6.3.17. □

Via Lemma 6.3.17, we now prove the atomic reconstruction theorem.

Proof of Proposition 6.3.15 Let X, r, d, c, A, and $\widetilde{A}$ be as in the present proposition and $\{a_{i,j}\}_{i \in \mathbb{Z}, \, j \in \mathbb{N}}$ a sequence of (X, r, d)-atoms supported, respectively, in the balls $\{B_{i,j}\}_{i \in \mathbb{Z}, \, j \in \mathbb{N}} \subset \mathbb{B}$ such that, for any $i \in \mathbb{Z}$,

$$\sum_{j \in \mathbb{N}} \mathbf{1}_{cB_{i,j}} \leq A,$$

$$\sum_{i \in \mathbb{Z}} \sum_{j \in \mathbb{N}} \widetilde{A} 2^i \left\| \mathbf{1}_{B_{i,j}} \right\|_X a_{i,j}$$

converges in $\mathcal{S}'(\mathbb{R}^n)$, and

$$\sup_{i \in \mathbb{Z}} \left\{ 2^i \left\| \sum_{j \in \mathbb{N}} \mathbf{1}_{B_{i,j}} \right\|_X \right\} < \infty.$$

Then, repeating the proof of [278, Theorem 4.7] via replacing [278, Lemma 4.8] therein by Lemma 6.3.17 here, we obtain $f \in WH_X(\mathbb{R}^n)$ and

$$\| f \|_{WH_X(\mathbb{R}^n)} \lesssim \sup_{i \in \mathbb{Z}} 2^i \left\| \sum_{j \in \mathbb{N}} \mathbf{1}_{B_{i,j}} \right\|_X,$$

which then completes the proof of Proposition 6.3.15. □

Via the improved atomic characterization of $WH_X(\mathbb{R}^n)$ established in both Propositions 6.3.11 and 6.3.15 above, we next show the atomic characterization of the weak generalized Herz–Hardy space $WH\dot{\mathcal{K}}_\omega^{p,q}(\mathbb{R}^n)$.

Proof of Theorem 6.3.10 Let p, q, ω, r, p_-, and d be as in the present theorem. We first prove that $WH\dot{\mathcal{K}}_\omega^{p,q}(\mathbb{R}^n) \subset WH\dot{\mathcal{K}}_\omega^{p,q,r,d}(\mathbb{R}^n)$. To this end, we show that, the global generalized Herz space $\dot{\mathcal{K}}_\omega^{p,q}(\mathbb{R}^n)$ under consideration satisfies all the assumptions of Proposition 6.3.11. Indeed, from the assumptions $m_0(\omega) \in (-\frac{n}{p}, \infty)$ and $M_\infty(\omega) \in (-\infty, 0)$, and Theorem 1.2.44, it follows that $\dot{\mathcal{K}}_\omega^{p,q}(\mathbb{R}^n)$ is a BQBF space. Applying Lemma 4.3.24 with $u := \frac{1}{r}$, we conclude that, for any $\{f_j\}_{j\in\mathbb{N}} \subset L_{\mathrm{loc}}^1(\mathbb{R}^n)$,

$$\left\| \left\{ \sum_{j\in\mathbb{N}} [\mathcal{M}(f_j)]^{1/r} \right\}^r \right\|_{[\dot{\mathcal{K}}_\omega^{p,q}(\mathbb{R}^n)]^{1/r}} \lesssim \left\| \left(\sum_{j\in\mathbb{N}} |f_j|^{1/r} \right)^r \right\|_{[\dot{\mathcal{K}}_\omega^{p,q}(\mathbb{R}^n)]^{1/r}},$$

which implies that Proposition 6.3.11(i) holds true. Now, we choose a

$$\theta \in \left(\max\{1, p, q\}, \frac{1}{p_-} \min\left\{ p, q, \frac{n}{\max\{M_0(\omega), M_\infty(\omega)\} + n/p} \right\} \right).$$

Then, by the assumptions $p/\theta, q/\theta \in (0, 1)$, the reverse Minkowski inequality, and Lemma 1.3.2, we find that, for any $\{f_j\}_{j\in\mathbb{N}} \subset \mathscr{M}(\mathbb{R}^n)$,

$$\sum_{j\in\mathbb{N}} \|f_j\|_{[\dot{\mathcal{K}}_\omega^{p,q}(\mathbb{R}^n)]^{1/\theta}} \le \left\| \sum_{j\in\mathbb{N}} |f_j| \right\|_{[\dot{\mathcal{K}}_\omega^{p,q}(\mathbb{R}^n)]^{1/\theta}}.$$

This implies that $\dot{\mathcal{K}}_\omega^{p,q}(\mathbb{R}^n)$ is θ-concave. Moreover, from Lemma 4.1.10 with $r := \theta p_-$, we deduce that, for any $f \in L_{\mathrm{loc}}^1(\mathbb{R}^n)$,

$$\|\mathcal{M}(f)\|_{[\dot{\mathcal{K}}_\omega^{p,q}(\mathbb{R}^n)]^{\frac{1}{\theta p_-}}} \lesssim \|f\|_{[\dot{\mathcal{K}}_\omega^{p,q}(\mathbb{R}^n)]^{\frac{1}{\theta p_-}}},$$

which, together with the fact that $\dot{\mathcal{K}}_\omega^{p,q}(\mathbb{R}^n)$ is θ-concave, further implies that Proposition 6.3.11(ii) holds true. In addition, let

$$r_0 \in \left(0, \min\left\{ p, \frac{n}{\max\{M_0(\omega), M_\infty(\omega)\} + n/p} \right\} \right).$$

Then, using Lemma 6.1.11 with $r := r_0$, we conclude that, for any $f \in L_{\mathrm{loc}}^1(\mathbb{R}^n)$,

$$\|\mathcal{M}(f)\|_{[W\dot{\mathcal{K}}_\omega^{p,q}(\mathbb{R}^n)]^{1/r_0}} \lesssim \|f\|_{[W\dot{\mathcal{K}}_\omega^{p,q}(\mathbb{R}^n)]^{1/r_0}},$$

which implies that Proposition 6.3.11(iii) holds true with the above r_0. Finally, let

$$s \in \left(0, \min \left\{ 1, p, q, \frac{n}{\max\{M_0(\omega), M_\infty(\omega)\} + n/p} \right\} \right)$$

and $\theta_0 \in (1, \infty)$ satisfy

$$\theta_0 < \min \left\{ \frac{n}{n(1 - s/p) - s \min\{m_0(\omega), m_\infty(\omega)\}}, \left(\frac{p}{s} \right)' \right\}.$$

Then, repeating an argument similar to that used in the proof of Theorem 4.8.16 with η therein replaced by θ_0, we find that

(i) for any $f \in \mathcal{M}(\mathbb{R}^n)$,

$$\|f\|_{[\dot{\mathcal{K}}_\omega^{p,q}(\mathbb{R}^n)]^{1/s}} \sim \sup \left\{ \|fg\|_{L^1(\mathbb{R}^n)} : \|g\|_{\dot{\mathcal{B}}_{1/\omega^s}^{(p/s)',(q/s)'}(\mathbb{R}^n)} = 1 \right\}$$

with positive equivalence constants independent of f;

(ii) $\| \cdot \|_{\dot{\mathcal{B}}_{1/\omega^s}^{(p/s)',(q/s)'}(\mathbb{R}^n)}$ satisfies Definition 1.2.13(ii);

(iii) $\mathbf{1}_{B(0,1)} \in \dot{\mathcal{B}}_{1/\omega^s}^{(p/s)',(q/s)'}(\mathbb{R}^n)$;

(iv) $\mathcal{M}^{(\theta_0)}$ is bounded on $\dot{\mathcal{B}}_{1/\omega^s}^{(p/s)',(q/s)'}(\mathbb{R}^n)$.

These imply that the assumptions (iv) through (vii) of Proposition 6.3.11 hold true with the above s and θ_0. Thus, all the assumptions of Proposition 6.3.11 hold true for the Herz space $\dot{\mathcal{K}}_\omega^{p,q}(\mathbb{R}^n)$ under consideration.

Let $f \in WH\dot{\mathcal{K}}_\omega^{p,q}(\mathbb{R}^n)$. Then, from Proposition 6.3.11 with $X := \dot{\mathcal{K}}_\omega^{p,q}(\mathbb{R}^n)$, we infer that there exists $\{a_{i,j}\}_{i \in \mathbb{Z}, j \in \mathbb{N}}$ of $(\dot{\mathcal{K}}_\omega^{p,q}(\mathbb{R}^n), \infty, d)$-atoms supported, respectively, in the balls $\{B_{i,j}\}_{i \in \mathbb{Z}, j \in \mathbb{N}} \subset \mathbb{B}$ and three positive constants $c \in (0, 1]$, A, and $\widetilde{A}$, independent of f, such that, for any $i \in \mathbb{Z}$,

$$\sum_{j \in \mathbb{N}} \mathbf{1}_{cB_{i,j}} \leq A,$$

$$f = \sum_{i \in \mathbb{Z}} \sum_{j \in \mathbb{N}} \widetilde{A} 2^i \left\| \mathbf{1}_{B_{i,j}} \right\|_{\dot{\mathcal{K}}_\omega^{p,q}(\mathbb{R}^n)} a_{i,j} \tag{6.12}$$

in $\mathcal{S}'(\mathbb{R}^n)$, and

$$\sup_{i \in \mathbb{Z}} \left\{ 2^i \left\| \sum_{j \in \mathbb{N}} \mathbf{1}_{B_{i,j}} \right\|_{\dot{\mathcal{K}}_\omega^{p,q}(\mathbb{R}^n)} \right\} \lesssim \|f\|_{WH\dot{\mathcal{K}}_\omega^{p,q}(\mathbb{R}^n)}. \tag{6.13}$$

In addition, using Lemma 4.3.5 with $X := \dot{\mathcal{K}}_\omega^{p,q}(\mathbb{R}^n)$ and $t := \infty$, we find that, for any $i \in \mathbb{Z}$ and $j \in \mathbb{N}$, $a_{i,j}$ is a $(\dot{\mathcal{K}}_\omega^{p,q}(\mathbb{R}^n), r, d)$-atom supported in the ball $B_{i,j}$. Combining this, (6.12), (6.13), and Definition 6.3.9, we conclude that $f \in WH\dot{\mathcal{K}}_\omega^{p,q,r,d}(\mathbb{R}^n)$ and

$$\|f\|_{WH\dot{\mathcal{K}}_\omega^{p,q,r,d}(\mathbb{R}^n)} \lesssim \|f\|_{WH\dot{\mathcal{K}}_\omega^{p,q}(\mathbb{R}^n)}. \tag{6.14}$$

This then finishes the proof that $WH\dot{\mathcal{K}}_\omega^{p,q}(\mathbb{R}^n) \subset WH\dot{\mathcal{K}}_\omega^{p,q,r,d}(\mathbb{R}^n)$.

Conversely, we next show that $WH\dot{\mathcal{K}}_\omega^{p,q,r,d}(\mathbb{R}^n) \subset WH\dot{\mathcal{K}}_\omega^{p,q}(\mathbb{R}^n)$. For this purpose, we first prove that the Herz space $\dot{\mathcal{K}}_\omega^{p,q}(\mathbb{R}^n)$ under consideration satisfies the assumptions (i), (ii), and (iii) of Proposition 6.3.15 with p_- as in the present theorem. Indeed, for any given $\theta \in (0, p_-)$ and $u \in (1, \infty)$, from Lemma 4.3.24 with $r := \theta$, it follows that, for any $\{f_j\}_{j\in\mathbb{N}} \subset \mathscr{M}(\mathbb{R}^n)$,

$$\left\| \left\{ \sum_{j\in\mathbb{N}} [\mathcal{M}(f_j)]^u \right\}^{\frac{1}{u}} \right\|_{[\dot{\mathcal{K}}_\omega^{p,q}(\mathbb{R}^n)]^{1/\theta}} \lesssim \left\| \left(\sum_{j\in\mathbb{N}} |f_j|^u \right)^{\frac{1}{u}} \right\|_{[\dot{\mathcal{K}}_\omega^{p,q}(\mathbb{R}^n)]^{1/\theta}},$$

which implies that Proposition 6.3.15(i) holds true. Next, we show that, for any $s \in (0, p_-)$, $[\dot{\mathcal{K}}_\omega^{p,q}(\mathbb{R}^n)]^{1/s}$ is a BBF. Indeed, fix an $s \in (0, p_-)$. Then, from the assumptions $m_0(\omega) \in (-\frac{n}{p}, \infty)$ and $M_\infty(\omega) \in (-\infty, 0)$, and Lemma 1.1.6, it follows that

$$m_0\left(\omega^s\right) = sm_0(\omega) > -\frac{n}{p/s}$$

and

$$M_\infty\left(\omega^s\right) = sM_\infty(\omega) < 0.$$

Applying these, the assumptions $p/s, q/s \in (1, \infty)$, and Theorem 1.2.48 with p, q, and ω replaced, respectively, by p/s, q/s, and ω^s, we find that $\dot{\mathcal{K}}_{\omega^s}^{p/s,q/s}(\mathbb{R}^n)$ is a BBF space. From this and Lemma 1.3.2, we further infer that $[\dot{\mathcal{K}}_\omega^{p,q}(\mathbb{R}^n)]^{1/s}$ is a BBF. This then implies that Proposition 6.3.15(ii) holds true.

Finally, let $s_0 \in (0, p_-)$ and

$$r_0 \in \left(\max\left\{ 1, p, \frac{n}{\max\{M_0(\omega), M_\infty(\omega)\} + n/p} \right\}, \infty \right).$$

Then, repeating an argument similar to that used in the proof of Theorem 4.3.16 with s and r therein replaced, respectively, by s_0 and r_0, we conclude that, for any $f \in \mathcal{M}(\mathbb{R}^n)$,

$$\|f\|_{[\dot{\mathcal{K}}_{\omega}^{p,q}(\mathbb{R}^n)]^{1/s_0}} \sim \sup\left\{\|fg\|_{L^1(\mathbb{R}^n)} : \|g\|_{\dot{\mathcal{B}}_{1/\omega^{s_0}}^{(p/s_0)',(q/s_0)'}(\mathbb{R}^n)} = 1\right\}$$

and, for any $f \in L^1_{\mathrm{loc}}(\mathbb{R}^n)$,

$$\left\|\mathcal{M}^{((r_0/s_0)')}(f)\right\|_{\dot{\mathcal{B}}_{1/\omega^{s_0}}^{(p/s_0)',(q/s_0)'}(\mathbb{R}^n)} \lesssim \|f\|_{\dot{\mathcal{B}}_{1/\omega^{s_0}}^{(p/s_0)',(q/s_0)'}(\mathbb{R}^n)}.$$

These imply that the global generalized Herz space $\dot{\mathcal{K}}_{\omega}^{p,q}(\mathbb{R}^n)$ under consideration satisfies Proposition 6.3.15(iii) with the above s_0, r_0, and $Y_0 := \dot{\mathcal{B}}_{1/\omega^{s_0}}^{(p/s_0)',(q/s_0)'}(\mathbb{R}^n)$. Therefore, the assumptions (i) through (iii) of Proposition 6.3.15 hold true for $\dot{\mathcal{K}}_{\omega}^{p,q}(\mathbb{R}^n)$.

In addition, by the proof that $WH\dot{\mathcal{K}}_{\omega}^{p,q}(\mathbb{R}^n) \subset WH\dot{\mathcal{K}}_{\omega}^{p,q,d}(\mathbb{R}^n)$, we find that $\dot{\mathcal{K}}_{\omega}^{p,q}(\mathbb{R}^n)$ is a BQBF space. Combining this, the fact that $\dot{\mathcal{K}}_{\omega}^{p,q}(\mathbb{R}^n)$ satisfies (i), (ii), and (iii) of Proposition 6.3.15, Proposition 6.3.15 with $X := \dot{\mathcal{K}}_{\omega}^{p,q}(\mathbb{R}^n)$, and Definition 6.3.9, we conclude that $WH\dot{\mathcal{K}}_{\omega}^{p,q,r,d}(\mathbb{R}^n) \subset WH\dot{\mathcal{K}}_{\omega}^{p,q}(\mathbb{R}^n)$ and, for any $f \in WH\dot{\mathcal{K}}_{\omega}^{p,q,r,d}(\mathbb{R}^n)$,

$$\|f\|_{WH\dot{\mathcal{K}}_{\omega}^{p,q}(\mathbb{R}^n)} \lesssim \|f\|_{WH\dot{\mathcal{K}}_{\omega}^{p,q,r,d}(\mathbb{R}^n)}.$$

This, combined with (6.14), further implies that

$$WH\dot{\mathcal{K}}_{\omega}^{p,q}(\mathbb{R}^n) = WH\dot{\mathcal{K}}_{\omega}^{p,q,r,d}(\mathbb{R}^n)$$

with equivalent quasi-norms, which then completes the proof of Theorem 6.3.10. $\square$

Finally, combining Theorem 6.3.10 and Remark 6.0.25, we have the following atomic characterization of $WHM_{\omega}^{p,q}(\mathbb{R}^n)$; we omit the details.

Corollary 6.3.18 *Let p, q, ω, r, and d be as in Corollary 6.3.8. Then the weak generalized atomic Hardy–Morrey space $WHM_{\omega}^{p,q,r,d}(\mathbb{R}^n)$, associated with the weak local generalized Morrey space $WM_{\omega}^{p,q}(\mathbb{R}^n)$, is defined to be the set of all the $f \in \mathcal{S}'(\mathbb{R}^n)$ such that there exists a sequence $\{a_{i,j}\}_{i\in\mathbb{Z}, j\in\mathbb{N}}$ of $(M_{\omega}^{p,q}(\mathbb{R}^n), r, d)$-atoms supported, respectively, in the balls $\{B_{i,j}\}_{i\in\mathbb{Z}, j\in\mathbb{N}} \subset \mathbb{B}$ and three positive* *constants* $c \in (0, 1]$, A, *and* $\widetilde{A}$, *independent of* f, *satisfying that, for any* $i \in \mathbb{Z}$,

$$\sum_{j \in \mathbb{N}} \mathbf{1}_{cB_{i,j}} \leq A,$$

$$f = \sum_{i \in \mathbb{Z}} \sum_{j \in \mathbb{N}} \widetilde{A} 2^i \left\| \mathbf{1}_{B_{i,j}} \right\|_{M_\omega^{p,q}(\mathbb{R}^n)} a_{i,j}$$

in $\mathcal{S}'(\mathbb{R}^n)$, *and*

$$\sup_{i \in \mathbb{Z}} \left\{ 2^i \left\| \sum_{j \in \mathbb{N}} \mathbf{1}_{B_{i,j}} \right\|_{M_\omega^{p,q}(\mathbb{R}^n)} \right\} < \infty,$$

Moreover, for any $f \in WHM_\omega^{p,q,r,d}(\mathbb{R}^n)$,

$$\|f\|_{WHM_\omega^{p,q,r,d}(\mathbb{R}^n)} := \inf \sup_{i \in \mathbb{Z}} \left\{ 2^i \left\| \sum_{j \in \mathbb{N}} \mathbf{1}_{B_{i,j}} \right\|_{M_\omega^{p,q}(\mathbb{R}^n)} \right\},$$

where the infimum is taken over all the decompositions of f *as above. Then*

$$WHM_\omega^{p,q}(\mathbb{R}^n) = WHM_\omega^{p,q,r,d}(\mathbb{R}^n)$$

with equivalent quasi-norms.

6.4 Molecular Characterizations

In this section, we establish the molecular characterization of weak generalized Herz–Hardy spaces. Indeed, applying the molecular characterization of weak Hardy spaces $WH_X(\mathbb{R}^n)$ associated with ball quasi-Banach function spaces X obtained in [278], we immediately obtain the molecular characterization of the weak generalized Herz–Hardy space $WH\dot{\mathcal{K}}_{\omega,\mathbf{0}}^{p,q}(\mathbb{R}^n)$. However, due to the deficiency of the associate space of the global generalized Herz space $\dot{\mathcal{K}}_\omega^{p,q}(\mathbb{R}^n)$, we can not prove the molecular characterization of the weak generalized Herz–Hardy space $WH\dot{\mathcal{K}}_\omega^{p,q}(\mathbb{R}^n)$ via the known molecular characterization of $WH_X(\mathbb{R}^n)$ directly [see Remark 1.2.19(vi) for the details]. To overcome this difficulty, we first prove an improved molecular reconstruction theorem of weak Hardy spaces associated with ball quasi-Banach function spaces (see Proposition 6.4.5 below) without recourse to associate spaces. Then, combining this improved conclusion and the atomic characterization of $WH\dot{\mathcal{K}}_\omega^{p,q}(\mathbb{R}^n)$ established in the last section, we obtain the desired molecular characterization of $WH\dot{\mathcal{K}}_\omega^{p,q}(\mathbb{R}^n)$.

Recall that the definition of $(\dot{\mathcal{K}}_{\omega,0}^{p,q}(\mathbb{R}^n))$, r, d, $\tau)$-molecules is given in Definition 4.5.1. We now characterize the weak generalized Herz–Hardy space $WH\dot{\mathcal{K}}_{\omega,0}^{p,q}(\mathbb{R}^n)$ via these molecules as follows.

Theorem 6.4.1 *Let* $p,q \in (0,\infty)$, $\omega \in M(\mathbb{R}_+)$ *satisfy* $m_0(\omega) \in (-\frac{n}{p},\infty)$ *and* $m_\infty(\omega) \in (-\frac{n}{p},\infty)$,

$$p_- \in \left(0, \frac{\min\{p, q, \overline{\frac{n}{\max\{M_0(\omega), M_\infty(\omega)\}+n/p}}\}}{\max\{1, p, q\}}\right),$$

$d \geq \lfloor n(1/p_- - 1)\rfloor$ *be a fixed integer,*

$$r \in \left(\max\left\{1, p, \frac{n}{\min\{m_0(\omega), m_\infty(\omega)\} + n/p}\right\}, \infty\right],$$

and $\tau \in (n(\frac{1}{p_-} - \frac{1}{r}), \infty)$. *Then* f *belongs to the weak generalized Herz–Hardy space* $WH\dot{\mathcal{K}}_{\omega,0}^{p,q}(\mathbb{R}^n)$ *if and only if* $f \in \mathcal{S}'(\mathbb{R}^n)$ *and there exists a sequence* $\{m_{i,j}\}_{i\in\mathbb{Z},\,j\in\mathbb{N}}$ *of* $(\dot{\mathcal{K}}_{\omega,0}^{p,q}(\mathbb{R}^n))$, r, d, $\tau)$-molecules centered, respectively, at the balls $\{B_{i,j}\}_{i\in\mathbb{Z},\,j\in\mathbb{N}} \subset \mathbb{B}$ *and three positive constants* $c \in (0,1]$, A, *and* $\tilde{A}$, *independent of* f, *such that, for any* $i \in \mathbb{Z}$,

$$\sum_{j\in\mathbb{N}} \mathbf{1}_{cB_{i,j}} \leq A,$$

$$f = \sum_{i\in\mathbb{Z}}\sum_{j\in\mathbb{N}} \tilde{A}2^i \left\|\mathbf{1}_{B_{i,j}}\right\|_{\dot{\mathcal{K}}_{\omega,0}^{p,q}(\mathbb{R}^n)} m_{i,j}$$

in $\mathcal{S}'(\mathbb{R}^n)$, *and*

$$\sup_{i\in\mathbb{Z}}\left\{2^i\left\|\sum_{j\in\mathbb{N}}\mathbf{1}_{B_{i,j}}\right\|_{\dot{\mathcal{K}}_{\omega,0}^{p,q}(\mathbb{R}^n)}\right\} < \infty.$$

Moreover, there exist two positive constants C_1 *and* C_2 *such that, for any* $f \in WH\dot{\mathcal{K}}_{\omega,0}^{p,q}(\mathbb{R}^n)$,

$$C_1\|f\|_{WH\dot{\mathcal{K}}_{\omega,0}^{p,q}(\mathbb{R}^n)} \leq \inf\left\{\sup_{i\in\mathbb{Z}}2^i\left\|\sum_{j\in\mathbb{N}}\mathbf{1}_{B_{i,j}}\right\|_{\dot{\mathcal{K}}_{\omega,0}^{p,q}(\mathbb{R}^n)}\right\}$$

$$\leq C_2\|f\|_{WH\dot{\mathcal{K}}_{\omega,0}^{p,q}(\mathbb{R}^n)},$$

where the infimum is taken over all the decompositions of f *as above.*

To show this molecular characterization, we first recall the following molecular characterization of weak Hardy spaces $WH_X(\mathbb{R}^n)$ associated with ball quasi-Banach function spaces X, which was obtained by Zhang et al. in [278, Theorem 5.3].

Lemma 6.4.2 *Let X be a ball quasi-Banach function space and let $p_- \in (0, 1)$ and $p_+ \in [p_-, \infty)$ be such that the following three statements hold true:*

(i) *for any given $\theta \in (0, p_-)$ and $u \in (1, \infty)$, there exists a positive constant C such that, for any $\{f_j\}_{j \in \mathbb{N}} \subset L^1_{\text{loc}}(\mathbb{R}^n)$,*

$$\left\| \left\{ \sum_{j \in \mathbb{N}} [\mathcal{M}(f_j)]^u \right\}^{\frac{1}{u}} \right\|_{X^{1/\theta}} \leq C \left\| \left\{ \sum_{j \in \mathbb{N}} |f_j|^u \right\}^{\frac{1}{u}} \right\|_{X^{1/\theta}};$$

(ii) *for any $s \in (0, p_-)$, $X^{1/s}$ is a ball Banach function space;*
(iii) *for any given $s \in (0, p_-)$ and $r \in (p_+, \infty)$, there exists a positive constant C such that, for any $f \in L^1_{\text{loc}}(\mathbb{R}^n)$,*

$$\left\| \mathcal{M}^{((r/s)')}(f) \right\|_{(X^{1/s})'} \leq C \|f\|_{(X^{1/s})'}.$$

Let $d \geq \lfloor n(1/p_- - 1) \rfloor$ be a fixed integer, $r \in (\max\{1, p_+\}, \infty]$, $\tau \in (n(\frac{1}{p_-} - \frac{n}{r}), \infty)$, $c \in (0, 1]$, and $A, \tilde{A} \in (0, \infty)$. Assume that $\{m_{i,j}\}_{i \in \mathbb{Z}, j \in \mathbb{N}}$ is a sequence of (X, r, d, τ)-molecules centered, respectively, at the balls $\{B_{i,j}\}_{i \in \mathbb{Z}, j \in \mathbb{N}} \subset \mathbb{B}$ such that, for any $i \in \mathbb{Z}$,

$$\sum_{j \in \mathbb{N}} \mathbf{1}_{cB_{i,j}} \leq A,$$

$$f := \sum_{i \in \mathbb{Z}} \sum_{j \in \mathbb{N}} \tilde{A} 2^i \left\| \mathbf{1}_{B_{i,j}} \right\|_X m_{i,j}$$

converges in $\mathcal{S}'(\mathbb{R}^n)$, and

$$\sup_{i \in \mathbb{Z}} \left\{ 2^i \left\| \sum_{j \in \mathbb{N}} \mathbf{1}_{B_{i,j}} \right\|_X \right\} < \infty.$$

Then $f \in WH_X(\mathbb{R}^n)$ and

$$\|f\|_{WH_X(\mathbb{R}^n)} \lesssim \sup_{i \in \mathbb{Z}} \left\{ 2^i \left\| \sum_{j \in \mathbb{N}} \mathbf{1}_{B_{i,j}} \right\|_X \right\},$$

where the implicit positive constant is independent of f.

With the help of the above lemma, we next show Theorem 6.4.1.

Proof of Theorem 6.4.1 Let all the symbols be as in the present theorem. We first prove the necessity. To achieve this, let $f \in WH\dot{\mathcal{K}}^{p,q}_{\omega,0}(\mathbb{R}^n)$. Then, applying Theorem 6.3.2, we find that $f \in WH\dot{\mathcal{K}}^{p,q,r,d}_{\omega,0}(\mathbb{R}^n)$. From this and Definition 6.3.1, it follows that there exists a sequence $\{a_{i,j}\}_{i \in \mathbb{Z},\, j \in \mathbb{N}}$ of $(\dot{\mathcal{K}}^{p,q}_{\omega,0}(\mathbb{R}^n),\, r,\, d)$-atoms supported, respectively, in the balls $\{B_{i,j}\}_{i \in \mathbb{Z},\, j \in \mathbb{N}} \subset \mathbb{B}$ and three positive constants $c \in (0,1]$, A, and $\widetilde{A}$, independent of f, such that, for any $i \in \mathbb{Z}$,

$$\sum_{j \in \mathbb{N}} \mathbf{1}_{cB_{i,j}} \le A,$$

$$f = \sum_{i \in \mathbb{Z}} \sum_{j \in \mathbb{N}} \widetilde{A} 2^i \left\| \mathbf{1}_{B_{i,j}} \right\|_{\dot{\mathcal{K}}^{p,q}_{\omega,0}(\mathbb{R}^n)} a_{i,j} \tag{6.15}$$

in $\mathcal{S}'(\mathbb{R}^n)$, and

$$\sup_{i \in \mathbb{Z}} \left\{ 2^i \left\| \sum_{j \in \mathbb{N}} \mathbf{1}_{B_{i,j}} \right\|_{\dot{\mathcal{K}}^{p,q}_{\omega,0}(\mathbb{R}^n)} \right\} < \infty. \tag{6.16}$$

On the other hand, by Remark 4.5.4 with $X := \dot{\mathcal{K}}^{p,q}_{\omega,0}(\mathbb{R}^n)$, we conclude that, for any $i \in \mathbb{Z}$ and $j \in \mathbb{N}$, $a_{i,j}$ is a $(\dot{\mathcal{K}}^{p,q}_{\omega,0}(\mathbb{R}^n),\, r,\, d,\, \tau)$-molecule centered at $B_{i,j}$. Combining this, (6.15), and (6.16), we then complete the proof the necessity. Moreover, using both Definition 6.3.1 and Theorem 6.3.2 again, we find that

$$\inf \left\{ \sup_{i \in \mathbb{Z}} 2^i \left\| \sum_{j \in \mathbb{N}} \mathbf{1}_{B_{i,j}} \right\|_{\dot{\mathcal{K}}^{p,q}_{\omega,0}(\mathbb{R}^n)} \right\} \lesssim \|f\|_{WH\dot{\mathcal{K}}^{p,q,r,d}_{\omega,0}(\mathbb{R}^n)}$$

$$\sim \|f\|_{WH\dot{\mathcal{K}}^{p,q}_{\omega,0}(\mathbb{R}^n)}, \tag{6.17}$$

where the infimum is taken over all the decompositions of f as in the present theorem.

Conversely, we next prove the sufficiency. For this purpose, assume that $f \in \mathcal{S}'(\mathbb{R}^n)$, $\{m_{i,j}\}_{i \in \mathbb{Z}, j \in \mathbb{N}}$ is a sequence of $(\dot{\mathcal{K}}_{\omega,0}^{p,q}(\mathbb{R}^n), r, d, \tau)$-molecules centered, respectively, at the balls $\{B_{i,j}\}_{i \in \mathbb{Z}, j \in \mathbb{N}} \subset \mathbb{B}$, $c \in (0, 1]$, and $A, \widetilde{A} \in (0, \infty)$ satisfying that, for any $i \in \mathbb{Z}$,

$$\sum_{j \in \mathbb{N}} \mathbf{1}_{cB_{i,j}} \leq A,$$

$$f = \sum_{i \in \mathbb{Z}} \sum_{j \in \mathbb{N}} \widetilde{A} 2^i \left\| \mathbf{1}_{B_{i,j}} \right\|_{\dot{\mathcal{K}}_{\omega,0}^{p,q}(\mathbb{R}^n)} m_{i,j}$$

in $\mathcal{S}'(\mathbb{R}^n)$, and

$$\sup_{i \in \mathbb{Z}} \left\{ 2^i \left\| \sum_{j \in \mathbb{N}} \mathbf{1}_{B_{i,j}} \right\|_{\dot{\mathcal{K}}_{\omega,0}^{p,q}(\mathbb{R}^n)} \right\} < \infty.$$

Let

$$p_+ := \max \left\{ 1, p, \frac{n}{\min\{m_0(\omega), m_\infty(\omega)\} + n/p} \right\}.$$

Then, repeating an argument similar to that used in the proof of Theorem 6.3.2 with s_0 and r_0 therein replaced, respectively, by s and r, we conclude that the local generalized Herz space $\dot{\mathcal{K}}_{\omega,0}^{p,q}(\mathbb{R}^n)$ under consideration satisfies the following four statements:

(i) $\dot{\mathcal{K}}_{\omega,0}^{p,q}(\mathbb{R}^n)$ is a BQBF space;
(ii) for any given $\theta \in (0, p_-)$ and $u \in (1, \infty)$, and for any $\{f_j\}_{j \in \mathbb{N}} \subset L_{\mathrm{loc}}^1(\mathbb{R}^n)$,

$$\left\| \left\{ \sum_{j \in \mathbb{N}} [\mathcal{M}(f_j)]^u \right\}^{\frac{1}{u}} \right\|_{[\dot{\mathcal{K}}_{\omega,0}^{p,q}(\mathbb{R}^n)]^{1/\theta}} \lesssim \left\| \left\{ \sum_{j \in \mathbb{N}} |f_j|^u \right\}^{\frac{1}{u}} \right\|_{[\dot{\mathcal{K}}_{\omega,0}^{p,q}(\mathbb{R}^n)]^{1/\theta}} ;$$

(iii) for any $s \in (0, p_-)$, $[\dot{\mathcal{K}}_{\omega,0}^{p,q}(\mathbb{R}^n)]^{1/s}$ is a BBF space;
(iv) for any given $s \in (0, p_-)$ and $r \in (p_+, \infty)$, and for any $f \in L_{\mathrm{loc}}^1(\mathbb{R}^n)$,

$$\left\| \mathcal{M}^{((r/s)')}(f) \right\|_{([\dot{\mathcal{K}}_{\omega,0}^{p,q}(\mathbb{R}^n)]^{1/s})'} \lesssim \|f\|_{([\dot{\mathcal{K}}_{\omega,0}^{p,q}(\mathbb{R}^n)]^{1/s})'} .$$

These, together with Lemma 6.4.2 with $X := \dot{\mathcal{K}}_{\omega,0}^{p,q}(\mathbb{R}^n)$, further imply that $f \in WH\dot{\mathcal{K}}_{\omega,0}^{p,q}(\mathbb{R}^n)$. This then finishes the proof of the sufficiency. Moreover, applying

Lemma 6.4.2 again with $X := \dot{\mathcal{K}}_{\omega,0}^{p,q}(\mathbb{R}^n)$ and the choice of $\{m_{i,j}\}_{i\in\mathbb{Z}, j\in\mathbb{N}}$, we find that

$$
\|f\|_{WH\dot{\mathcal{K}}_{\omega,0}^{p,q}(\mathbb{R}^n)} \lesssim \inf_{i\in\mathbb{Z}} \sup \left\{ 2^i \left\| \sum_{j\in\mathbb{N}} \mathbf{1}_{B_{i,j}} \right\|_{\dot{\mathcal{K}}_{\omega,0}^{p,q}(\mathbb{R}^n)} \right\},
$$

where the infimum is taken over all the decompositions of f as in the present theorem. From this and (6.17), we further infer that, for any $f \in WH\dot{\mathcal{K}}_{\omega,0}^{p,q}(\mathbb{R}^n)$,

$$
\|f\|_{WH\dot{\mathcal{K}}_{\omega,0}^{p,q}(\mathbb{R}^n)} \sim \inf_{i\in\mathbb{Z}} \sup \left\{ 2^i \left\| \sum_{j\in\mathbb{N}} \mathbf{1}_{B_{i,j}} \right\|_{\dot{\mathcal{K}}_{\omega,0}^{p,q}(\mathbb{R}^n)} \right\},
$$

where the infimum is taken over all the decompositions of f as in the present theorem. This finishes the proof of Theorem 6.4.1. □

From Theorem 6.4.1 and Remark 6.0.25 above, we deduce the following molecular characterization of the weak generalized Hardy–Morrey space $WHM_{\omega,0}^{p,q}(\mathbb{R}^n)$; we omit the details.

Corollary 6.4.3 *Let* $p, q \in [1, \infty)$, $p_- \in (0, \min\{p, q\}/\max\{p, q\})$, $\omega \in M(\mathbb{R}_+)$ *with*

$$
-\frac{n}{p} < m_0(\omega) \le M_0(\omega) < 0
$$

and

$$
-\frac{n}{p} < m_\infty(\omega) \le M_\infty(\omega) < 0,
$$

$d \ge \lfloor n(1/p_- - 1) \rfloor$ *be a fixed integer,*

$$
r \in \left(\frac{n}{\min\{m_0(\omega), m_\infty(\omega)\} + n/p}, \infty \right],
$$

and $\tau \in (n(\frac{1}{p_-} - \frac{1}{r}), \infty)$. *Then* f *belongs to the weak generalized Hardy–Morrey space* $WHM_{\omega,0}^{p,q}(\mathbb{R}^n)$ *if and only if* $f \in \mathcal{S}'(\mathbb{R}^n)$ *and there exists a sequence* $\{m_{i,j}\}_{i\in\mathbb{Z}, j\in\mathbb{N}}$ *of* $(M_{\omega,0}^{p,q}(\mathbb{R}^n), r, d, \tau)$-*molecules centered, respectively, at the balls* $\{B_{i,j}\}_{i\in\mathbb{Z}, j\in\mathbb{N}} \subset \mathbb{B}$ *and three positive constants* $c \in (0, 1]$, A, *and* $\tilde{A}$, *independent of f, such that, for any i ∈ ℤ,*

$$\sum_{j\in\mathbb{N}} \mathbf{1}_{cB_{i,j}} \le A,$$

$$f = \sum_{i\in\mathbb{Z}}\sum_{j\in\mathbb{N}} \tilde{A}2^i \left\| \mathbf{1}_{B_{i,j}} \right\|_{M^{p,q}_{\omega,0}(\mathbb{R}^n)} m_{i,j}$$

in $\mathcal{S}'(\mathbb{R}^n)$, and

$$\sup_{i\in\mathbb{Z}} \left\{ 2^i \left\| \sum_{j\in\mathbb{N}} \mathbf{1}_{B_{i,j}} \right\|_{M^{p,q}_{\omega,0}(\mathbb{R}^n)} \right\} < \infty.$$

Moreover, there exist two positive constants C_1 and C_2 such that, for any $f \in WHM^{p,q}_{\omega,0}(\mathbb{R}^n)$,

$$C_1 \|f\|_{WHM^{p,q}_{\omega,0}(\mathbb{R}^n)} \le \inf \sup_{i\in\mathbb{Z}} \left\{ 2^i \left\| \sum_{j\in\mathbb{N}} \mathbf{1}_{B_{i,j}} \right\|_{M^{p,q}_{\omega,0}(\mathbb{R}^n)} \right\}$$

$$\le C_2 \|f\|_{WHM^{p,q}_{\omega,0}(\mathbb{R}^n)},$$

where the infimum is taken over all the decompositions of f as above.

Now, we turn to establish the molecular characterization of the weak generalized Herz–Hardy space $WH\dot{\mathcal{K}}^{p,q}_\omega(\mathbb{R}^n)$. Recall that the concept of $(\dot{\mathcal{K}}^{p,q}_\omega(\mathbb{R}^n), r, d, \tau)$-molecules is introduced in Definition 4.5.9 above. Then we show the following conclusion.

Theorem 6.4.4 *Let $p, q \in (0,\infty)$, $\omega \in M(\mathbb{R}_+)$ satisfy $m_0(\omega) \in (-\frac{n}{p}, \infty)$ and*

$$-\frac{n}{p} < m_\infty(\omega) \le M_\infty(\omega) < 0,$$

$r \in (\max\{1, p, \frac{n}{\min\{m_0(\omega),m_\infty(\omega)\}+n/p}\}, \infty]$,

$$p_- \in \left(0, \frac{\min\{p, q, \frac{n}{\max\{M_0(\omega),M_\infty(\omega)\}+n/p}\}}{\max\{1,p,q\}}\right),$$

$d \geq \lfloor n(1/p_- - 1) \rfloor$ *be a fixed integer, and* $\tau \in (n(\frac{1}{p_-} - \frac{1}{r}), \infty)$. *Then* f *belongs to the weak generalized Herz–Hardy space* $WH\dot{\mathcal{K}}_{\omega}^{p,q}(\mathbb{R}^n)$ *if and only if* $f \in \mathcal{S}'(\mathbb{R}^n)$ *and there exists a sequence* $\{m_{i,j}\}_{i \in \mathbb{Z}, j \in \mathbb{N}}$ *of* $(\dot{\mathcal{K}}_{\omega}^{p,q}(\mathbb{R}^n), r, d, \tau)$- *molecules centered, respectively, at the balls* $\{B_{i,j}\}_{i \in \mathbb{Z}, j \in \mathbb{N}} \subset \mathbb{B}$ *and three positive constants* $c \in (0, 1]$, A, *and* $\widetilde{A}$, *independent of* f, *such that, for any* $i \in \mathbb{Z}$,

$$\sum_{j \in \mathbb{N}} \mathbf{1}_{cB_{i,j}} \leq A,$$

$$f = \sum_{i \in \mathbb{Z}} \sum_{j \in \mathbb{N}} \widetilde{A} 2^i \left\| \mathbf{1}_{B_{i,j}} \right\|_{\dot{\mathcal{K}}_{\omega}^{p,q}(\mathbb{R}^n)} m_{i,j}$$

in $\mathcal{S}'(\mathbb{R}^n)$, *and*

$$\sup_{i \in \mathbb{Z}} \left\{ 2^i \left\| \sum_{j \in \mathbb{N}} \mathbf{1}_{B_{i,j}} \right\|_{\dot{\mathcal{K}}_{\omega}^{p,q}(\mathbb{R}^n)} \right\} < \infty.$$

Moreover, there exist two positive constants C_1 *and* C_2 *such that, for any* $f \in WH\dot{\mathcal{K}}_{\omega}^{p,q}(\mathbb{R}^n)$,

$$C_1 \|f\|_{WH\dot{\mathcal{K}}_{\omega}^{p,q}(\mathbb{R}^n)} \leq \inf \sup_{i \in \mathbb{Z}} \left\{ 2^i \left\| \sum_{j \in \mathbb{N}} \mathbf{1}_{B_{i,j}} \right\|_{\dot{\mathcal{K}}_{\omega}^{p,q}(\mathbb{R}^n)} \right\}$$

$$\leq C_2 \|f\|_{WH\dot{\mathcal{K}}_{\omega}^{p,q}(\mathbb{R}^n)},$$

where the infimum is taken over all the decompositions of f *as above.*

In order to show this theorem, we first establish the following molecular reconstruction theorem of the weak Hardy space $WH_X(\mathbb{R}^n)$, which improves [278, Theorem 5.3] via removing the assumption about associate spaces.

Proposition 6.4.5 *Let* X *be a ball quasi-Banach function space and let* $p_- \in (0, 1)$ *and* $p_+ \in [p_-, \infty)$ *be such that the following three statements hold true:*

(i) *for any given* $\theta \in (0, p_-)$ *and* $u \in (1, \infty)$, *there exists a positive constant* C *such that, for any* $\{f_j\}_{j \in \mathbb{N}} \subset L_{\mathrm{loc}}^1(\mathbb{R}^n)$,

$$\left\| \left\{ \sum_{j \in \mathbb{N}} \left[\mathcal{M}(f_j) \right]^u \right\}^{\frac{1}{u}} \right\|_{X^{1/\theta}} \leq C \left\| \left\{ \sum_{j \in \mathbb{N}} |f_j|^u \right\}^{\frac{1}{u}} \right\|_{X^{1/\theta}};$$

(ii) *for any $s \in (0, p_-)$, $X^{1/s}$ is a ball Banach function space and there exists a linear space $Y_s \subset \mathscr{M}(\mathbb{R}^n)$ equipped with a quasi-seminorm $\|\cdot\|_{Y_s}$ such that, for any $f \in \mathscr{M}(\mathbb{R}^n)$,*

$$\|f\|_{X^{1/s}} \sim \sup\left\{\|fg\|_{L^1(\mathbb{R}^n)} : \|g\|_{Y_s} = 1\right\}$$

with the positive equivalence constants independent of f;

(iii) *for any given $s \in (0, p_-)$ and $r \in (p_+, \infty)$, there exists a positive constant C such that, for any $f \in L^1_{\mathrm{loc}}(\mathbb{R}^n)$,*

$$\left\|\mathcal{\vec{M}}^{((r/s)')}(f)\right\|_{Y_s} \le C \|f\|_{Y_s}.$$

Let $d \ge \lfloor n(1/p_- - 1) \rfloor$ be a fixed integer, $r \in (\max\{1, p_+\}, \infty]$, $\tau \in (n(\frac{1}{p_-} - \frac{n}{r}), \infty)$, $c \in (0, 1]$, and $A, \widetilde{A} \in (0, \infty)$. Assume that $\{m_{i,j}\}_{i\in\mathbb{Z}, j\in\mathbb{N}}$ is a sequence of (X, r, d, τ)-molecules centered, respectively, at the balls $\{B_{i,j}\}_{i\in\mathbb{Z}, j\in\mathbb{N}} \subset \mathbb{B}$ such that, for any $i \in \mathbb{Z}$,

$$\sum_{j\in\mathbb{N}} \mathbf{1}_{cB_{i,j}} \le A,$$

$$f := \sum_{i\in\mathbb{Z}}\sum_{j\in\mathbb{N}} \widetilde{A}2^i \left\|\mathbf{1}_{B_{i,j}}\right\|_X m_{i,j}$$

converges in $\mathcal{S}'(\mathbb{R}^n)$, and

$$\sup_{i\in\mathbb{Z}}\left\{2^i \left\|\sum_{j\in\mathbb{N}}\mathbf{1}_{B_{i,j}}\right\|_X\right\} < \infty.$$

Then $f \in WH_X(\mathbb{R}^n)$ and

$$\|f\|_{WH_X(\mathbb{R}^n)} \lesssim \sup_{i\in\mathbb{Z}} 2^i \left\|\sum_{j\in\mathbb{N}}\mathbf{1}_{B_{i,j}}\right\|_X,$$

where the implicit positive constant is independent of f.

Proof Let X, r, d, τ, c, A, and $\widetilde{A}$ be as in the present proposition and $\{m_{i,j}\}_{i\in\mathbb{Z}, j\in\mathbb{N}}$ a sequence of (X, r, d, τ)-molecules centered, respectively, at the balls $\{B_{i,j}\}_{i\in\mathbb{Z}, j\in\mathbb{N}} \subset \mathbb{B}$ such that, for any $i \in \mathbb{Z}$,

$$\sum_{j\in\mathbb{N}} \mathbf{1}_{cB_{i,j}} \le A,$$

the summation

$$\sum_{i\in\mathbb{Z}}\sum_{j\in\mathbb{N}}\widetilde{A}2^i\left\|\mathbf{1}_{B_{i,j}}\right\|_X m_{i,j}$$

converges in $\mathcal{S}'(\mathbb{R}^n)$, and

$$\sup_{i\in\mathbb{Z}}\left\{2^i\left\|\sum_{j\in\mathbb{N}}\mathbf{1}_{B_{i,j}}\right\|_X\right\}<\infty.$$

Then, repeating the proof of [278, Theorem 5.3] via replacing [278, Lemma 4.8] therein by Lemma 6.3.17 here, we conclude that

$$f:=\sum_{i\in\mathbb{Z}}\sum_{j\in\mathbb{N}}\widetilde{A}2^i\left\|\mathbf{1}_{B_{i,j}}\right\|_X m_{i,j}\in WH_X(\mathbb{R}^n)$$

and

$$\|f\|_{WH_X(\mathbb{R}^n)}\lesssim\sup_{i\in\mathbb{Z}}\left\{2^i\left\|\sum_{j\in\mathbb{N}}\mathbf{1}_{B_{i,j}}\right\|_X\right\},$$

which then completes the proof of Proposition 6.4.5. □

Remark 6.4.6 We should point out that Proposition 6.4.5 is an improved version of the known molecular reconstruction theorem obtained by Zhang et al. in [278, Theorem 5.3]. Indeed, if $Y_s\equiv(X^{1/s})'$ in Proposition 6.4.5, then this proposition is just [278, Theoreem 5.3].

Using this proposition and the atomic characterization of the weak generalized Herz–Hardy space $WH\dot{\mathcal{K}}_\omega^{p,q}(\mathbb{R}^n)$ obtained in the last section, we now prove the molecular characterization of $WH\dot{\mathcal{K}}_\omega^{p,q}(\mathbb{R}^n)$.

Proof of Theorem 6.4.4 Let all the symbols be as in the present theorem. We first show the necessity. Indeed, let $f\in WH\dot{\mathcal{K}}_\omega^{p,q}(\mathbb{R}^n)$. Then, applying Theorem 6.3.10, we find that $f\in WH\dot{\mathcal{K}}_\omega^{p,q,r,d}(\mathbb{R}^n)$. This, together with Definition 6.3.9, further implies that there exists a sequence $\{a_{i,j}\}_{i\in\mathbb{Z},\,j\in\mathbb{N}}$ of $(\dot{\mathcal{K}}_\omega^{p,q}(\mathbb{R}^n),r,d)$-atoms supported, respectively, in the balls $\{B_{i,j}\}_{i\in\mathbb{Z},\,j\in\mathbb{N}}\subset\mathbb{B}$ and three positive constants $c\in(0,1]$, A, and $\widetilde{A}$, independent of f, such that, for any $i\in\mathbb{Z}$,

$$\sum_{j\in\mathbb{N}}\mathbf{1}_{cB_{i,j}}\leq A,$$

$$f=\sum_{i\in\mathbb{Z}}\sum_{j\in\mathbb{N}}\widetilde{A}2^i\left\|\mathbf{1}_{B_{i,j}}\right\|_{\dot{\mathcal{K}}_\omega^{p,q}(\mathbb{R}^n)}a_{i,j}\tag{6.18}$$

in $\mathcal{S}'(\mathbb{R}^n)$, and

$$\sup_{i \in \mathbb{Z}} \left\{ 2^i \left\| \sum_{j \in \mathbb{N}} \mathbf{1}_{B_{i,j}} \right\|_{\dot{\mathcal{K}}_\omega^{p,q}(\mathbb{R}^n)} \right\} < \infty. \tag{6.19}$$

In addition, from Remark 4.5.4 with $X := \dot{\mathcal{K}}_\omega^{p,q}(\mathbb{R}^n)$, it follows that, for any $i \in \mathbb{Z}$ and $j \in \mathbb{N}$, $a_{i,j}$ is a $(\dot{\mathcal{K}}_\omega^{p,q}(\mathbb{R}^n), r, d, \tau)$-molecule centered at $B_{i,j}$. This, combined with (6.18) and (6.19), finishes the proof of the necessity. Moreover, using both Definition 6.3.9 and Theorem 6.3.10 again, we conclude that

$$\inf \left\{ \sup_{i \in \mathbb{Z}} 2^i \left\| \sum_{j \in \mathbb{N}} \mathbf{1}_{B_{i,j}} \right\|_{\dot{\mathcal{K}}_\omega^{p,q}(\mathbb{R}^n)} \right\} \lesssim \| f \|_{WH\dot{\mathcal{K}}_\omega^{p,q,r,d,}(\mathbb{R}^n)}$$

$$\sim \| f \|_{WH\dot{\mathcal{K}}_\omega^{p,q}(\mathbb{R}^n)}, \tag{6.20}$$

where the infimum is taken over all the decompositions of f as in the present theorem.

Conversely, we prove the sufficiency. To this end, let $f \in \mathcal{S}'(\mathbb{R}^n)$, $\{m_{i,j}\}_{i \in \mathbb{Z}, j \in \mathbb{N}}$ be a sequence of $(\dot{\mathcal{K}}_\omega^{p,q}(\mathbb{R}^n), r, d, \tau)$-molecules centered, respectively, at the balls $\{B_{i,j}\}_{i \in \mathbb{Z}, j \in \mathbb{N}} \subset \mathbb{B}$, $c \in (0, 1]$, and $A, \tilde{A} \in (0, \infty)$ such that, for any $i \in \mathbb{Z}$,

$$\sum_{j \in \mathbb{N}} \mathbf{1}_{cB_{i,j}} \leq A,$$

$$f = \sum_{i \in \mathbb{Z}} \sum_{j \in \mathbb{N}} \tilde{A} 2^i \left\| \mathbf{1}_{B_{i,j}} \right\|_{\dot{\mathcal{K}}_\omega^{p,q}(\mathbb{R}^n)} m_{i,j}$$

in $\mathcal{S}'(\mathbb{R}^n)$, and

$$\sup_{i \in \mathbb{Z}} \left\{ 2^i \left\| \sum_{j \in \mathbb{N}} \mathbf{1}_{B_{i,j}} \right\|_{\dot{\mathcal{K}}_\omega^{p,q}(\mathbb{R}^n)} \right\} < \infty.$$

We next claim that the global generalized Herz space $\dot{\mathcal{K}}_\omega^{p,q}(\mathbb{R}^n)$ under consideration satisfies all the assumptions of Proposition 6.4.5. Assume that this claim holds true for the moment. Then, applying Proposition 6.4.5 with $X := \dot{\mathcal{K}}_\omega^{p,q}(\mathbb{R}^n)$, we conclude that $f \in WH\dot{\mathcal{K}}_\omega^{p,q}(\mathbb{R}^n)$, which completes the proof of the sufficiency.

Thus, to finish the proof of the sufficiency, we only need to show the above claim. For this purpose, let

$$p_+ := \max \left\{ 1, p, \frac{n}{\min\{m_0(\omega), m_\infty(\omega)\} + n/p} \right\}.$$

Then, repeating an argument similar to that used in the proof of Theorem 6.3.10 with s_0 and r_0 therein replaced, respectively, by s and r, we find that the global generalized Herz space $\dot{\mathcal{K}}_\omega^{p,q}(\mathbb{R}^n)$ under consideration satisfies the following four statements:

(i) $\dot{\mathcal{K}}_\omega^{p,q}(\mathbb{R}^n)$ is a BQBF space;
(i) for any given $\theta \in (0, p_-)$ and $u \in (1, \infty)$, and for any $\{f_j\}_{j\in\mathbb{N}} \subset L_{\mathrm{loc}}^1(\mathbb{R}^n)$,

$$\left\|\left\{\sum_{j\in\mathbb{N}}[\mathcal{M}(f_j)]^u\right\}^{\frac{1}{u}}\right\|_{[\dot{\mathcal{K}}_\omega^{p,q}(\mathbb{R}^n)]^{1/\theta}} \lesssim \left\|\left\{\sum_{j\in\mathbb{N}}|f_j|^u\right\}^{\frac{1}{u}}\right\|_{[\dot{\mathcal{K}}_\omega^{p,q}(\mathbb{R}^n)]^{1/\theta}};$$

(ii) for any $s \in (0, p_-)$, $[\dot{\mathcal{K}}_\omega^{p,q}(\mathbb{R}^n)]^{1/s}$ is a BBF space and, for any $f \in \mathcal{M}(\mathbb{R}^n)$,

$$\|f\|_{[\dot{\mathcal{K}}_\omega^{p,q}(\mathbb{R}^n)]^{1/s}} \sim \sup\left\{\|fg\|_{L^1(\mathbb{R}^n)} : \|g\|_{\dot{\mathcal{B}}_{1/\omega^s}^{(p/s)',(q/s)'}(\mathbb{R}^n)}\right\}$$

with the positive equivalence constants independent of f;
(iii) for any given $s \in (0, p_-)$ and $r \in (p_+, \infty)$, and for any $f \in L_{\mathrm{loc}}^1(\mathbb{R}^n)$,

$$\left\|\mathcal{M}^{((r/s)')}(f)\right\|_{\dot{\mathcal{B}}_{1/\omega^s}^{(p/s)',(q/s)'}(\mathbb{R}^n)} \lesssim \|f\|_{\dot{\mathcal{B}}_{1/\omega^s}^{(p/s)',(q/s)'}(\mathbb{R}^n)}.$$

These further imply that the global generalized Herz space $\dot{\mathcal{K}}_\omega^{p,q}(\mathbb{R}^n)$ under consideration satisfies all the assumptions of Proposition 6.4.4 with the above p_-, p_+, and $Y_s := \dot{\mathcal{B}}_{1/\omega^s}^{(p/s)',(q/s)'}(\mathbb{R}^n)$ for any $s \in (0, p_-)$. This then finishes the above claim and further implies that the sufficiency holds true.

In addition, from Proposition 6.4.5 again with $X := \dot{\mathcal{K}}_\omega^{p,q}(\mathbb{R}^n)$ and the choice of $\{m_{i,j}\}_{i\in\mathbb{Z},\,j\in\mathbb{N}}$, we deduce that

$$\|f\|_{WH\dot{\mathcal{K}}_\omega^{p,q}(\mathbb{R}^n)} \lesssim \inf\sup_{i\in\mathbb{Z}}\left\{2^i\left\|\sum_{j\in\mathbb{N}}\mathbf{1}_{B_{i,j}}\right\|_{\dot{\mathcal{K}}_\omega^{p,q}(\mathbb{R}^n)}\right\},$$

where the infimum is taken over all the decompositions of f as in the present theorem. This, together with (6.20), further implies that, for any $f \in WH\dot{\mathcal{K}}_\omega^{p,q}(\mathbb{R}^n)$,

$$\|f\|_{WH\dot{\mathcal{K}}_\omega^{p,q}(\mathbb{R}^n)} \sim \inf\sup_{i\in\mathbb{Z}}\left\{2^i\left\|\sum_{j\in\mathbb{N}}\mathbf{1}_{B_{i,j}}\right\|_{\dot{\mathcal{K}}_\omega^{p,q}(\mathbb{R}^n)}\right\},$$

where the infimum is taken over all the decompositions of f as in the present theorem. This then finishes the proof of Theorem 6.4.4. $\qquad\square$

As an application, we now establish the following molecular characterization of weak Hardy spaces associated with global generalized Morrey spaces, which is just a simple corollary of Theorem 6.4.4 and Remark 6.0.25; we omit the details.

Corollary 6.4.7 *Let p, q, ω, r, d, and τ be as in Corollary 6.4.3. Then f belongs to the weak generalized Hardy–Morrey space $WHM_\omega^{p,q}(\mathbb{R}^n)$ if and only if $f \in \mathcal{S}'(\mathbb{R}^n)$ and there exists a sequence $\{m_{i,j}\}_{i\in\mathbb{Z},\,j\in\mathbb{N}}$ of $(M_\omega^{p,q}(\mathbb{R}^n), r, d, \tau)$-molecules centered, respectively, at the balls $\{B_{i,j}\}_{i\in\mathbb{Z},\,j\in\mathbb{N}} \subset \mathbb{B}$ and three positive constants $c \in (0, 1]$, A, and $\tilde{A}$, independent of f, such that, for any $i \in \mathbb{Z}$,*

$$\sum_{j\in\mathbb{N}} \mathbf{1}_{cB_{i,j}} \leq A,$$

$$f = \sum_{i\in\mathbb{Z}}\sum_{j\in\mathbb{N}} \tilde{A}2^i \left\|\mathbf{1}_{B_{i,j}}\right\|_{M_\omega^{p,q}(\mathbb{R}^n)} m_{i,j}$$

in $\mathcal{S}'(\mathbb{R}^n)$, and

$$\sup_{i\in\mathbb{Z}}\left\{2^i \left\|\sum_{j\in\mathbb{N}}\mathbf{1}_{B_{i,j}}\right\|_{M_\omega^{p,q}(\mathbb{R}^n)}\right\} < \infty.$$

Moreover, there exist two positive constants C_1 and C_2 such that, for any $f \in WHM_\omega^{p,q}(\mathbb{R}^n)$,

$$C_1\|f\|_{WHM_\omega^{p,q}(\mathbb{R}^n)} \leq \inf\left\{\sup_{i\in\mathbb{Z}}2^i\left\|\sum_{j\in\mathbb{N}}\mathbf{1}_{B_{i,j}}\right\|_{M_\omega^{p,q}(\mathbb{R}^n)}\right\}$$

$$\leq C_2\|f\|_{WHM_\omega^{p,q}(\mathbb{R}^n)},$$

where the infimum is taken over all the decompositions of f as above.

6.5 Littlewood–Paley Function Characterizations

The main target of this section is to characterize weak generalized Herz–Hardy spaces via various Littlewood–Paley functions. Precisely, using the Lusin area function S, the g-function g, and the g_λ^*-function g_λ^* presented in Definitions 4.6.1 and 4.6.2, we establish several equivalent characterizations of the weak generalized Herz–Hardy spaces $WH\dot{\mathcal{K}}_{\omega,0}^{p,q}(\mathbb{R}^n)$ and $WH\dot{\mathcal{K}}_\omega^{p,q}(\mathbb{R}^n)$. To begin with, we state the following Littlewood–Paley function characterizations of the Hardy space $WH\dot{\mathcal{K}}_{\omega,0}^{p,q}(\mathbb{R}^n)$.

Theorem 6.5.1 *Let* $p, q \in (0, \infty)$, $\omega \in M(\mathbb{R}_+)$ *with* $m_0(\omega) \in (-\frac{n}{p}, \infty)$ *and* $m_\infty(\omega) \in (-\frac{n}{p}, \infty)$,

$$p_- := \min \left\{ p, q, \frac{n}{\max\{M_0(\omega), M_\infty(\omega)\} + n/p} \right\},$$

$$p_+ := \max \left\{ p, \frac{n}{\min\{m_0(\omega), m_\infty(\omega)\} + n/p} \right\},$$

and

$$\lambda \in \left(\max \left\{ \frac{2}{\min\{1, p_-\}}, 1 - \frac{2}{\max\{1, p_+\}} + \frac{2}{\min\{1, p_-\}} \right\}, \infty \right).$$

Assume that, for any $f \in \mathcal{S}'(\mathbb{R}^n)$, $S(f)$ *and* $g_\lambda^*(f)$ *are as in Definition 4.6.1, and* $g(f)$ *is as in Definition 4.6.2. Then the following four statements are mutually equivalent:*

(i) $f \in WH\dot{\mathcal{K}}_{\omega,\mathbf{0}}^{p,q}(\mathbb{R}^n)$;
(ii) $f \in \mathcal{S}'(\mathbb{R}^n)$, f *vanishes weakly at infinity, and* $S(f) \in W\dot{\mathcal{K}}_{\omega,\mathbf{0}}^{p,q}(\mathbb{R}^n)$;
(iii) $f \in \mathcal{S}'(\mathbb{R}^n)$, f *vanishes weakly at infinity, and* $g(f) \in W\dot{\mathcal{K}}_{\omega,\mathbf{0}}^{p,q}(\mathbb{R}^n)$;
(iv) $f \in \mathcal{S}'(\mathbb{R}^n)$, f *vanishes weakly at infinity, and* $g_\lambda^*(f) \in W\dot{\mathcal{K}}_{\omega,\mathbf{0}}^{p,q}(\mathbb{R}^n)$.

Moreover, for any $f \in WH\dot{\mathcal{K}}_{\omega,\mathbf{0}}^{p,q}(\mathbb{R}^n)$,

$$\|f\|_{WH\dot{\mathcal{K}}_{\omega,\mathbf{0}}^{p,q}(\mathbb{R}^n)} \sim \|S(f)\|_{W\dot{\mathcal{K}}_{\omega,\mathbf{0}}^{p,q}(\mathbb{R}^n)} \sim \|g(f)\|_{W\dot{\mathcal{K}}_{\omega,\mathbf{0}}^{p,q}(\mathbb{R}^n)}$$

$$\sim \left\| g_\lambda^*(f) \right\|_{W\dot{\mathcal{K}}_{\omega,\mathbf{0}}^{p,q}(\mathbb{R}^n)},$$

where the positive equivalence constants are independent of f.

In order to prove this theorem, we need the following Littlewood–Paley function characterizations of weak Hardy spaces associated with ball quasi-Banach function spaces, which are just [242, Theorems 3.12, 3.16, and 3.21].

Lemma 6.5.2 *Let* X *be a ball quasi-Banach function space and let* $p_- \in (0, \infty)$, $p_+ \in [p_-, \infty)$, *and* $\theta_0 \in (1, \infty)$ *be such that the following four statements hold true:*

(i) *for any given* $\theta \in (0, p_-)$ *and* $u \in (1, \infty)$, *there exists a positive constant* C *such that, for any* $\{f_j\}_{j \in \mathbb{N}} \subset L_{\text{loc}}^1(\mathbb{R}^n)$,

$$\left\| \left\{ \sum_{j \in \mathbb{N}} \left[\mathcal{M}(f_j) \right]^u \right\}^{\frac{1}{u}} \right\|_{X^{1/\theta}} \leq C \left\| \left\{ \sum_{j \in \mathbb{N}} |f_j|^u \right\}^{\frac{1}{u}} \right\|_{X^{1/\theta}}$$

and

$$\left\| \left\{ \sum_{j\in\mathbb{N}} [\mathcal{M}(f_j)]^u \right\}^{\frac{1}{u}} \right\|_{(WX)^{1/\theta}} \le C \left\| \left\{ \sum_{j\in\mathbb{N}} |f_j|^u \right\}^{\frac{1}{u}} \right\|_{(WX)^{1/\theta}} ;$$

(ii) X *is* θ_0-*concave;*
(iii) *for any* $s \in (0, \min\{1, p_-\})$, $X^{1/s}$ *is a ball Banach function space;*
(iv) *for any given* $s \in (0, \min\{1, p_-\})$ *and* $r \in (p_+, \infty)$, *there exists a positive constant* C *such that, for any* $f \in L^1_{\mathrm{loc}}(\mathbb{R}^n)$,

$$\left\| \mathcal{M}^{((r/s)')}(f) \right\|_{(X^{1/s})'} \le C \|f\|_{(X^{1/s})'}.$$

Let

$$\lambda \in \left(\max\left\{ \frac{2}{\min\{1, p_-\}}, 1 - \frac{2}{\max\{1, p_+\}} + \frac{2}{\min\{1, p_-\}} \right\}, \infty \right).$$

Then the following four statements are mutually equivalent:

(i) $f \in WH_X(\mathbb{R}^n)$;
(ii) $f \in \mathcal{S}'(\mathbb{R}^n)$, f *vanishes weakly at infinity, and* $S(f) \in WX$;
(iii) $f \in \mathcal{S}'(\mathbb{R}^n)$, f *vanishes weakly at infinity, and* $g(f) \in WX$;
(iv) $f \in \mathcal{S}'(\mathbb{R}^n)$, f *vanishes weakly at infinity, and* $g_\lambda^*(f) \in WX$.

Moreover, for any $f \in WH_X(\mathbb{R}^n)$,

$$\|f\|_{WH_X(\mathbb{R}^n)} \sim \|S(f)\|_{WX} \sim \|g(f)\|_{WX} \sim \|g_\lambda^*(f)\|_{WX}$$

with the positive equivalence constants independent of f.

Remark 6.5.3 We point out that Lemma 6.5.2 has a wide range of applications. Here we give several function spaces to which Lemma 6.5.2 can be applied (see also [242, Section 5]).

(i) Let $\vec{p} := (p_1, \ldots, p_n) \in (0, \infty)^n$ and $\lambda \in (0, \infty)$ satisfy

$$\lambda > \max\left\{ \frac{2}{\min\{1, p_1, \ldots, p_n\}}, 1 - \frac{2}{\max\{1, p_1, \ldots, p_n\}} + \frac{2}{\min\{1, p_1, \ldots, p_n\}} \right\}.$$

Then, in this case, as was pointed out in [242, Subsection 5.2], the mixed-norm Lebesgue space $L^{\vec{p}}(\mathbb{R}^n)$ satisfies all the assumptions of Lemma 6.5.2. This implies that Lemma 6.5.2 with $X := L^{\vec{p}}(\mathbb{R}^n)$ holds true. This result coincides with [242, Theorems 5.15 and 5.16].

(ii) Let $0 < q \le p < \infty$ and

$$\lambda \in \left(\max\left\{ \frac{2}{\min\{1,q\}}, \; 1 - \frac{2}{\max\{1,p\}} + \frac{2}{\min\{1,q\}} \right\}, \infty \right).$$

Then, in this case, as was mentioned in [242, Subsection 5.1], the Morrey space $M_q^p(\mathbb{R}^n)$ satisfies all the assumptions of Lemma 6.5.2. Thus, Lemma 6.5.2 with $X := M_q^p(\mathbb{R}^n)$ holds true. This result coincides with [242, Theorems 5.6 and 5.7].

(iii) Let $p(\cdot) \in C^{\log}(\mathbb{R}^n)$ satisfy $0 < p_- \le p_+ < \infty$, and

$$\lambda \in \left(\max\left\{ \frac{2}{\min\{1,p_-\}}, \; 1 - \frac{2}{\max\{1,p_+\}} + \frac{2}{\min\{1,p_-\}} \right\}, \infty \right),$$

where p_- and p_+ are defined, respectively, in (1.59) and (1.60). In this case, as was pointed out in [242, Subsection 5.3], the variable Lebesgue space $L^{p(\cdot)}(\mathbb{R}^n)$ satisfies all the assumptions of Lemma 6.5.2. Therefore, Lemma 6.5.2 with $X := L^{p(\cdot)}(\mathbb{R}^n)$ holds true. This result coincides with [242, Theorems 5.18 and 5.19].

Moreover, to prove Theorem 6.5.1, we still require the following auxiliary lemma about the Fefferman–Stein vector-valued inequality on weak local generalized Herz spaces.

Lemma 6.5.4 *Let* $p, q \in (0, \infty)$ *and* $\omega \in M(\mathbb{R}_+)$ *satisfy* $m_0(\omega) \in (-\frac{n}{p}, \infty)$ *and* $m_\infty(\omega) \in (-\frac{n}{p}, \infty)$. *Then, for any given*

$$r \in \left(0, \min\left\{ p, \frac{n}{\max\{M_0(\omega), M_\infty(\omega)\} + n/p} \right\} \right)$$

and $u \in (1, \infty)$, *there exists a positive constant* C *such that, for any* $\{f_j\}_{j \in \mathbb{N}} \subset L^1_{\mathrm{loc}}(\mathbb{R}^n)$,

$$\left\| \left\{ \sum_{j \in \mathbb{N}} [\mathcal{M}(f_j)]^u \right\}^{\frac{1}{u}} \right\|_{[W\dot{\mathcal{K}}^{p,q}_{\omega,\mathbf{0}}(\mathbb{R}^n)]^{1/r}} \le C \left\| \left\{ \sum_{j \in \mathbb{N}} |f_j|^u \right\}^{\frac{1}{u}} \right\|_{[W\dot{\mathcal{K}}^{p,q}_{\omega,\mathbf{0}}(\mathbb{R}^n)]^{1/r}}.$$

Proof Let all the symbols be as in the present lemma. Then, by the assumption $m_0(\omega) \in (-\frac{n}{p}, \infty)$ and Theorem 1.2.42, we conclude that the local generalized Herz space $\dot{\mathcal{K}}^{p,q}_{\omega,\mathbf{0}}(\mathbb{R}^n)$ under consideration is a BQBF space. Moreover, let

$$p_- := \min\left\{ p, \frac{n}{\max\{M_0(\omega), M_\infty(\omega)\} + n/p} \right\}.$$

Then, for any given $r \in (0, p_-)$ and $u \in (1, \infty)$, applying Lemma 4.3.10, we find that, for any $\{f_j\}_{j\in\mathbb{N}} \subset L^1_{\mathrm{loc}}(\mathbb{R}^n)$,

$$\left\| \left\{ \sum_{j\in\mathbb{N}} [\mathcal{M}(f_j)]^u \right\}^{\frac{1}{u}} \right\|_{[\dot{\mathcal{K}}^{p,q}_{\omega,0}(\mathbb{R}^n)]^{1/r}} \lesssim \left\| \left(\sum_{j\in\mathbb{N}} |f_j|^u \right)^{\frac{1}{u}} \right\|_{[\dot{\mathcal{K}}^{p,q}_{\omega,0}(\mathbb{R}^n)]^{1/r}} .$$

This, combined with the fact that $\dot{\mathcal{K}}^{p,q}_{\omega,0}(\mathbb{R}^n)$ is a BQBF space and Lemma 6.1.7 with $X := \dot{\mathcal{K}}^{p,q}_{\omega,0}(\mathbb{R}^n)$, further implies that, for any given $r \in (0, p_-)$ and $u \in (1, \infty)$, and for any $\{f_j\}_{j\in\mathbb{N}} \subset L^1_{\mathrm{loc}}(\mathbb{R}^n)$,

$$\left\| \left\{ \sum_{j\in\mathbb{N}} [\mathcal{M}(f_j)]^u \right\}^{\frac{1}{u}} \right\|_{[W\dot{\mathcal{K}}^{p,q}_{\omega,0}(\mathbb{R}^n)]^{1/r}} \lesssim \left\| \left(\sum_{j\in\mathbb{N}} |f_j|^u \right)^{\frac{1}{u}} \right\|_{[W\dot{\mathcal{K}}^{p,q}_{\omega,0}(\mathbb{R}^n)]^{1/r}} ,$$

which then completes the proof of Lemma 6.5.4. $\qquad\square$

Via the above two lemmas, we next prove Theorem 6.5.1.

Proof of Theorem 6.5.1 Let all the symbols be as in the present theorem. Then, by the assumption $m_0(\omega) \in (-\frac{n}{p}, \infty)$ and Theorem 1.2.42, we conclude that the local generalized Herz space $\dot{\mathcal{K}}^{p,q}_{\omega,0}(\mathbb{R}^n)$ under consideration is a BQBF space. Therefore, from Lemma 6.5.2, it follows that, to finish the proof of the present theorem, we only need to show that $\dot{\mathcal{K}}^{p,q}_{\omega,0}(\mathbb{R}^n)$ satisfies all the assumptions of Lemma 6.5.2. Indeed, for any given $\theta \in (0, p_-)$ and $u \in (1, \infty)$, using Lemma 6.5.4 with $r := \theta$, we find that, for any $\{f_j\}_{j\in\mathbb{N}} \subset L^1_{\mathrm{loc}}(\mathbb{R}^n)$,

$$\left\| \left\{ \sum_{j\in\mathbb{N}} [\mathcal{M}(f_j)]^u \right\}^{\frac{1}{u}} \right\|_{[W\dot{\mathcal{K}}^{p,q}_{\omega,0}(\mathbb{R}^n)]^{1/\theta}} \lesssim \left\| \left\{ \sum_{j\in\mathbb{N}} |f_j|^u \right\}^{\frac{1}{u}} \right\|_{[W\dot{\mathcal{K}}^{p,q}_{\omega,0}(\mathbb{R}^n)]^{1/\theta}} . \tag{6.21}$$

In addition, let $\theta_0 \in (\max\{1, p, q\}, \infty)$. Then, repeating an argument similar to that used in the proof of Theorem 6.3.2 with s_0 and r_0 therein replaced, respectively, by s and r, we conclude that the following four statements hold true:

(i) for any given $\theta \in (0, p_-)$ and $u \in (1, \infty)$, and for any $\{f_j\}_{j\in\mathbb{N}} \subset L^1_{\mathrm{loc}}(\mathbb{R}^n)$,

$$\left\| \left\{ \sum_{j\in\mathbb{N}} [\mathcal{M}(f_j)]^u \right\}^{\frac{1}{u}} \right\|_{[\dot{\mathcal{K}}^{p,q}_{\omega,0}(\mathbb{R}^n)]^{1/\theta}} \lesssim \left\| \left\{ \sum_{j\in\mathbb{N}} |f_j|^u \right\}^{\frac{1}{u}} \right\|_{[\dot{\mathcal{K}}^{p,q}_{\omega,0}(\mathbb{R}^n)]^{1/\theta}} ;$$

(ii) $\dot{\mathcal{K}}^{p,q}_{\omega,\mathbf{0}}(\mathbb{R}^n)$ is θ_0-concave;

(iii) for any $s \in (0, \min\{1, p_-\})$, $[\dot{\mathcal{K}}^{p,q}_{\omega,\mathbf{0}}(\mathbb{R}^n)]^{1/s}$ is a BBF space;

(iv) for any given $s \in (0, \min\{1, p_-\})$ and $r \in (p_+, \infty)$, and for any $f \in L^1_{\mathrm{loc}}(\mathbb{R}^n)$,

$$\left\| \mathcal{M}^{((r/s)')}(f) \right\|_{([\dot{\mathcal{K}}^{p,q}_{\omega,\mathbf{0}}(\mathbb{R}^n)]^{1/s})'} \lesssim \|f\|_{([\dot{\mathcal{K}}^{p,q}_{\omega,\mathbf{0}}(\mathbb{R}^n)]^{1/s})'}.$$

These, combined with (6.21), further imply that the Herz space $\dot{\mathcal{K}}^{p,q}_{\omega,\mathbf{0}}(\mathbb{R}^n)$ under consideration satisfies all the assumptions of Lemma 6.5.2 with p_- and p_+ as in the present theorem. This then implies that (i), (ii), (iii), and (iv) of the present theorem are mutually equivalent and, for any $f \in WH\dot{\mathcal{K}}^{p,q}_{\omega,\mathbf{0}}(\mathbb{R}^n)$,

$$\|f\|_{WH\dot{\mathcal{K}}^{p,q}_{\omega,\mathbf{0}}(\mathbb{R}^n)} \sim \|S(f)\|_{W\dot{\mathcal{K}}^{p,q}_{\omega,\mathbf{0}}(\mathbb{R}^n)} \sim \|g(f)\|_{W\dot{\mathcal{K}}^{p,q}_{\omega,\mathbf{0}}(\mathbb{R}^n)}$$

$$\sim \left\| g^*_\lambda(f) \right\|_{W\dot{\mathcal{K}}^{p,q}_{\omega,\mathbf{0}}(\mathbb{R}^n)}$$

with the positive equivalence constants independent of f, which completes the proof of Theorem 6.5.1.   □

Combining Theorem 6.5.1 and Remark 6.0.25, we immediately obtain the following Littlewood–Paley function characterizations of the weak generalized Hardy–Morrey space $WH\boldsymbol{M}^{p,q}_{\omega,\mathbf{0}}(\mathbb{R}^n)$; we omit the details.

Corollary 6.5.5 *Let* $p, q \in [1, \infty)$, $\omega \in M(\mathbb{R}_+)$ *with*

$$-\frac{n}{p} < m_0(\omega) \le M_0(\omega) < 0$$

and

$$-\frac{n}{p} < m_\infty(\omega) \le M_\infty(\omega) < 0,$$

and

$$\lambda \in \left(\max\left\{ 2, 3 - \frac{2(\min\{m_0(\omega), m_\infty(\omega)\} + n/p)}{n} \right\}, \infty \right).$$

Then the following four statements are mutually equivalent:

(i) $f \in W\boldsymbol{M}^{p,q}_{\omega,\mathbf{0}}(\mathbb{R}^n)$;

(ii) $f \in \mathcal{S}'(\mathbb{R}^n)$, f *vanishes weakly at infinity, and* $S(f) \in W\boldsymbol{M}^{p,q}_{\omega,\mathbf{0}}(\mathbb{R}^n)$;

(iii) $f \in \mathcal{S}'(\mathbb{R}^n)$, f *vanishes weakly at infinity, and* $g(f) \in W\boldsymbol{M}^{p,q}_{\omega,\mathbf{0}}(\mathbb{R}^n)$;

(iv) $f \in \mathcal{S}'(\mathbb{R}^n)$, f *vanishes weakly at infinity, and* $g^*_\lambda(f) \in W\boldsymbol{M}^{p,q}_{\omega,\mathbf{0}}(\mathbb{R}^n)$.

Moreover, for any $f \in WM_{\omega,0}^{p,q}(\mathbb{R}^n)$,

$$\|f\|_{WM_{\omega,0}^{p,q}(\mathbb{R}^n)} \sim \|S(f)\|_{WM_{\omega,0}^{p,q}(\mathbb{R}^n)} \sim \|g(f)\|_{WM_{\omega,0}^{p,q}(\mathbb{R}^n)}$$

$$\sim \|g_\lambda^*(f)\|_{WM_{\omega,0}^{p,q}(\mathbb{R}^n)},$$

where the positive equivalence constants are independent of f.

Now, we turn to investigate the Littlewood–Paley function characterizations of the weak generalized Herz–Hardy space $WH\dot{\mathcal{K}}_\omega^{p,q}(\mathbb{R}^n)$. Indeed, we have the following conclusion.

Theorem 6.5.6 *Let* $p, q \in (0, \infty)$, $\omega \in M(\mathbb{R}_+)$ *with* $m_0(\omega) \in (-\frac{n}{p}, \infty)$ *and*

$$-\frac{n}{p} < m_\infty(\omega) \le M_\infty(\omega) < 0,$$

$$p_- := \min\left\{p, q, \frac{n}{\max\{M_0(\omega), M_\infty(\omega)\} + n/p}\right\},$$

$$p_+ := \max\left\{p, \frac{n}{\min\{m_0(\omega), m_\infty(\omega)\} + n/p}\right\},$$

and

$$\lambda \in \left(\max\left\{\frac{2}{\min\{1, p_-\}}, 1 - \frac{2}{\max\{1, p_+\}} + \frac{2}{\min\{1, p_-\}}\right\}, \infty\right).$$

Assume that, for any $f \in \mathcal{S}'(\mathbb{R}^n)$, $S(f)$ *and* $g_\lambda^*(f)$ *are as in Definition 4.6.1, and* $g(f)$ *is as in Definition 4.6.2. Then the following four statements are mutually equivalent:*

(i) $f \in WH\dot{\mathcal{K}}_\omega^{p,q}(\mathbb{R}^n)$;
(ii) $f \in \mathcal{S}'(\mathbb{R}^n)$, f *vanishes weakly at infinity, and* $S(f) \in W\dot{\mathcal{K}}_\omega^{p,q}(\mathbb{R}^n)$;
(iii) $f \in \mathcal{S}'(\mathbb{R}^n)$, f *vanishes weakly at infinity, and* $g(f) \in W\dot{\mathcal{K}}_\omega^{p,q}(\mathbb{R}^n)$;
(iv) $f \in \mathcal{S}'(\mathbb{R}^n)$, f *vanishes weakly at infinity, and* $g_\lambda^*(f) \in W\dot{\mathcal{K}}_\omega^{p,q}(\mathbb{R}^n)$.

Moreover, for any $f \in WH\dot{\mathcal{K}}_\omega^{p,q}(\mathbb{R}^n)$,

$$\|f\|_{WH\dot{\mathcal{K}}_\omega^{p,q}(\mathbb{R}^n)} \sim \|S(f)\|_{W\dot{\mathcal{K}}_\omega^{p,q}(\mathbb{R}^n)} \sim \|g(f)\|_{W\dot{\mathcal{K}}_\omega^{p,q}(\mathbb{R}^n)}$$

$$\sim \|g_\lambda^*(f)\|_{W\dot{\mathcal{K}}_\omega^{p,q}(\mathbb{R}^n)},$$

where the positive equivalence constants are independent of f.

To prove these Littlewood–Paley function characterizations, we first establish the following representation formula of the quasi-norm $\|\cdot\|_{W\dot{\mathcal{K}}_\omega^{p,q}(\mathbb{R}^n)}$.

Lemma 6.5.7 *Let* $p, q \in (0, \infty)$ *and* $\omega \in M(\mathbb{R}_+)$. *Then, for any* $f \in \mathscr{M}(\mathbb{R}^n)$,

$$\|f\|_{W\dot{\mathcal{K}}^{p,q}_\omega(\mathbb{R}^n)} = \sup_{\xi \in \mathbb{R}^n} \left\| \tau_\xi(f) \right\|_{W\dot{\mathcal{K}}^{p,q}_{\omega,0}(\mathbb{R}^n)},$$

where, for any $\xi \in \mathbb{R}^n$, *the operator* τ_ξ *is defined as in* (4.69).

Proof Let all the symbols be as in the present lemma. Then, using Definition 6.0.18(ii), Remark 1.2.2(ii), and Definition 6.0.18(i), we find that, for any $f \in \mathscr{M}(\mathbb{R}^n)$,

$$\|f\|_{W\dot{\mathcal{K}}^{p,q}_\omega(\mathbb{R}^n)} = \sup_{\lambda \in (0,\infty)} \left[\lambda \left\| \mathbf{1}_{\{y \in \mathbb{R}^n: \, |f(y)| > \lambda\}} \right\|_{\dot{\mathcal{K}}^{p,q}_\omega(\mathbb{R}^n)} \right]$$

$$= \sup_{\lambda \in (0,\infty)} \left\{ \lambda \sup_{\xi \in \mathbb{R}^n} \left[\left\| \mathbf{1}_{\{y \in \mathbb{R}^n: \, |f(y)| > \lambda\}} (\cdot - \xi) \right\|_{\dot{\mathcal{K}}^{p,q}_{\omega,0}(\mathbb{R}^n)} \right] \right\}$$

$$= \sup_{\lambda \in (0,\infty), \, \xi \in \mathbb{R}^n} \left[\lambda \left\| \mathbf{1}_{\{y \in \mathbb{R}^n: \, |\tau_\xi(f)(y)| > \lambda\}} \right\|_{\dot{\mathcal{K}}^{p,q}_{\omega,0}(\mathbb{R}^n)} \right]$$

$$= \sup_{\xi \in \mathbb{R}^n} \left\{ \sup_{\lambda \in (0,\infty)} \left[\lambda \left\| \mathbf{1}_{\{y \in \mathbb{R}^n: \, |\tau_\xi(f)(y)| > \lambda\}} \right\|_{\dot{\mathcal{K}}^{p,q}_{\omega,0}(\mathbb{R}^n)} \right] \right\}$$

$$= \sup_{\xi \in \mathbb{R}^n} \left\| \tau_\xi(f) \right\|_{\dot{\mathcal{K}}^{p,q}_{\omega,0}(\mathbb{R}^n)}.$$

This finishes the proof of Lemma 6.5.7.   □

Based on this formula and the Littlewood–Paley function characterizations of $W H \dot{\mathcal{K}}^{p,q}_{\omega,0}(\mathbb{R}^n)$ obtained in Theorem 6.5.1 above, we next prove Theorem 6.5.6.

Proof of Theorem 6.5.6 Let all the symbols be as in the present theorem, $f \in \mathcal{S}'(\mathbb{R}^n)$, and $\phi \in \mathcal{S}(\mathbb{R}^n)$ satisfy $\int_{\mathbb{R}^n} \phi(x)\, dx \neq 0$. We first show that (i) implies (ii). Indeed, assume $f \in W H \dot{\mathcal{K}}^{p,q}_\omega(\mathbb{R}^n)$. Let M denote the radial maximal function defined as in Definition 4.1.1(i). Then, for any $\xi \in \mathbb{R}^n$, using Theorem 6.1.1(ii) with f therein replaced by $\tau_\xi(f)$, Lemma 4.6.9(ii), Lemma 6.5.7 with f therein replaced by $M(f, \phi)$, and Theorem 6.1.9(ii), we conclude that

$$\left\| \tau_\xi(f) \right\|_{W H \dot{\mathcal{K}}^{p,q}_{\omega,0}(\mathbb{R}^n)}$$

$$\sim \left\| M\left(\tau_\xi(f), \phi\right) \right\|_{W\dot{\mathcal{K}}^{p,q}_{\omega,0}(\mathbb{R}^n)} \sim \left\| \tau_\xi(M(f, \phi)) \right\|_{W\dot{\mathcal{K}}^{p,q}_{\omega,0}(\mathbb{R}^n)}$$

$$\lesssim \left\| M(f, \phi) \right\|_{W\dot{\mathcal{K}}^{p,q}_\omega(\mathbb{R}^n)} \sim \|f\|_{W H \dot{\mathcal{K}}^{p,q}_\omega(\mathbb{R}^n)} < \infty, \qquad (6.22)$$

which further implies that $\tau_\xi(f) \in W H \dot{\mathcal{K}}^{p,q}_{\omega,0}(\mathbb{R}^n)$. Applying this with $\xi := \mathbf{0}$, we further find that $f \in W H \dot{\mathcal{K}}^{p,q}_{\omega,0}(\mathbb{R}^n)$. This, together with Theorem 6.5.1, implies that f vanishes weakly at infinity. On the other hand, from Lemma 4.6.9(iii) with

$A := S$, the fact that $\tau_\xi(f) \in WH\dot{\mathcal{K}}_{\omega,0}^{p,q}(\mathbb{R}^n)$ for any $\xi \in \mathbb{R}^n$, Theorem 6.5.1, and (6.22), it follows that

$$\left\| \tau_\xi\left(S\left(f\right)\right) \right\|_{W\dot{\mathcal{K}}_{\omega,0}^{p,q}(\mathbb{R}^n)} \sim \left\| S\left(\tau_\xi\left(f\right)\right) \right\|_{W\dot{\mathcal{K}}_{\omega,0}^{p,q}(\mathbb{R}^n)}$$

$$\sim \left\| \tau_\xi\left(f\right) \right\|_{WH\dot{\mathcal{K}}_{\omega,0}^{p,q}(\mathbb{R}^n)} \lesssim \|f\|_{WH\dot{\mathcal{K}}_{\omega}^{p,q}(\mathbb{R}^n)}.$$

Using this and Lemma 6.5.7 again with f replaced by $S(f)$, we find that

$$\|S(f)\|_{W\dot{\mathcal{K}}_{\omega}^{p,q}(\mathbb{R}^n)} \lesssim \|f\|_{WH\dot{\mathcal{K}}_{\omega}^{p,q}(\mathbb{R}^n)} < \infty, \tag{6.23}$$

which completes the proof that (i) implies (ii).

Conversely, we next prove that (ii) implies (i). Indeed, assume that f vanishes weakly at infinity and $S(f) \in W\dot{\mathcal{K}}_{\omega}^{p,q}(\mathbb{R}^n)$. Then, by Lemma 4.6.9(iv), we conclude that, for any $\xi \in \mathbb{R}^n$, $\tau_\xi(f)$ vanishes weakly at infinity. In addition, from Lemma 4.6.9(iii) with $A := S$ and Lemma 6.5.7 with f therein replaced by $S(f)$, we deduce that, for any $\xi \in \mathbb{R}^n$,

$$\left\| S\left(\tau_\xi\left(f\right)\right) \right\|_{W\dot{\mathcal{K}}_{\omega,0}^{p,q}(\mathbb{R}^n)} = \left\| \tau_\xi\left(S\left(f\right)\right) \right\|_{W\dot{\mathcal{K}}_{\omega,0}^{p,q}(\mathbb{R}^n)}$$

$$\leq \|S(f)\|_{W\dot{\mathcal{K}}_{\omega}^{p,q}(\mathbb{R}^n)} < \infty, \tag{6.24}$$

which, combined with the fact that $\tau_\xi(f)$ vanishes weakly at infinity and Theorem 6.5.1, further implies that $\tau_\xi(f) \in WH\dot{\mathcal{K}}_{\omega,0}^{p,q}(\mathbb{R}^n)$. Moreover, for any $\xi \in \mathbb{R}^n$, using Lemma 4.6.9(ii), Theorem 6.1.1(ii) with f replaced by $\tau_\xi(f)$, Theorem 6.5.1 again, and (6.24), we conclude that

$$\left\| \tau_\xi\left(M\left(f,\phi\right)\right) \right\|_{W\dot{\mathcal{K}}_{\omega,0}^{p,q}(\mathbb{R}^n)} = \left\| M\left(\tau_\xi\left(f\right),\phi\right) \right\|_{W\dot{\mathcal{K}}_{\omega,0}^{p,q}(\mathbb{R}^n)}$$

$$\sim \left\| \tau_\xi\left(f\right) \right\|_{WH\dot{\mathcal{K}}_{\omega,0}^{p,q}(\mathbb{R}^n)} \sim \left\| S\left(\tau_\xi\left(f\right)\right) \right\|_{W\dot{\mathcal{K}}_{\omega,0}^{p,q}(\mathbb{R}^n)}$$

$$\lesssim \|S(f)\|_{W\dot{\mathcal{K}}_{\omega}^{p,q}(\mathbb{R}^n)},$$

which, together with Theorem 6.1.9(ii) and Lemma 6.5.7 with f replaced by $M(f,\phi)$, further implies that

$$\|f\|_{WH\dot{\mathcal{K}}_{\omega}^{p,q}(\mathbb{R}^n)} \sim \|M(f,\phi)\|_{W\dot{\mathcal{K}}_{\omega}^{p,q}(\mathbb{R}^n)}$$

$$\sim \sup_{\xi \in \mathbb{R}^n} \left\{ \left\| \tau_\xi\left(M\left(f,\phi\right)\right) \right\|_{W\dot{\mathcal{K}}_{\omega,0}^{p,q}(\mathbb{R}^n)} \right\}$$

$$\lesssim \|S(f)\|_{W\dot{\mathcal{K}}_{\omega}^{p,q}(\mathbb{R}^n)} < \infty, \tag{6.25}$$

which completes the proof that (ii) implies (i). Therefore, (i) is equivalent to (ii). Moreover, combining (6.23) and (6.25), we find that

$$\|f\|_{W\dot{\mathcal{K}}_\omega^{p,q}(\mathbb{R}^n)} \sim \|S(f)\|_{WH\dot{\mathcal{K}}_\omega^{p,q}(\mathbb{R}^n)}.$$

Now, repeating an argument similar to that used in the estimations of both (6.23) and (6.25) above with S therein replaced by g or g_λ^*, we conclude that (i) is equivalent to (iii) and (i) is also equivalent to (iv). Moreover, for any $A \in \{g, g_\lambda^*\}$ and $f \in WH\dot{\mathcal{K}}_\omega^{p,q}(\mathbb{R}^n)$, it holds true that

$$\|f\|_{WH\dot{\mathcal{K}}_\omega^{p,q}(\mathbb{R}^n)} \sim \|A(f)\|_{W\dot{\mathcal{K}}_\omega^{p,q}(\mathbb{R}^n)}.$$

This then finishes the proof of Theorem 6.5.6. □

Using the above theorem and Remark 6.0.25, we immediately obtain the following Littlewood–Paley function characterizations of the weak generalized Hardy–Morrey space $WHM_\omega^{p,q}(\mathbb{R}^n)$; we omit the details.

Corollary 6.5.8 *Let p, q, ω, and λ be as in Corollary 6.5.5. Then the following four statements are mutually equivalent:*

(i) $f \in WM_\omega^{p,q}(\mathbb{R}^n)$;
(ii) $f \in \mathcal{S}'(\mathbb{R}^n)$, f vanishes weakly at infinity, and $S(f) \in WM_\omega^{p,q}(\mathbb{R}^n)$;
(iii) $f \in \mathcal{S}'(\mathbb{R}^n)$, f vanishes weakly at infinity, and $g(f) \in WM_\omega^{p,q}(\mathbb{R}^n)$;
(iv) $f \in \mathcal{S}'(\mathbb{R}^n)$, f vanishes weakly at infinity, and $g_\lambda^*(f) \in WM_\omega^{p,q}(\mathbb{R}^n)$.

Moreover, for any $f \in WM_\omega^{p,q}(\mathbb{R}^n)$,

$$\|f\|_{WM_\omega^{p,q}(\mathbb{R}^n)} \sim \|S(f)\|_{WM_\omega^{p,q}(\mathbb{R}^n)} \sim \|g(f)\|_{WM_\omega^{p,q}(\mathbb{R}^n)}$$

$$\sim \left\|g_\lambda^*(f)\right\|_{WM_\omega^{p,q}(\mathbb{R}^n)},$$

where the positive equivalence constants are independent of f.

6.6 Boundedness of Calderón–Zygmund Operators

Let $\delta \in (0,1)$, $d \in \mathbb{Z}_+$, K be a standard kernel as in Definition 1.5.7 with $\delta \in (0,1)$, and T a d-order Calderón–Zygmund operator as in Definition 1.5.8 with kernel K. The main target of this section is to investigate the boundedness of operator T from generalized Herz–Hardy spaces to weak generalized Herz–Hardy spaces. First, we establish two boundedness criteria of T from Hardy spaces $H_X(\mathbb{R}^n)$ to weak Hardy spaces $WH_X(\mathbb{R}^n)$, which improve the boundedness of convolutional type Calderón–Zygmund operators from $H_X(\mathbb{R}^n)$ to $WH_X(\mathbb{R}^n)$ obtained in [278, Theorem 6.3]. As applications, we obtain the boundedness of T from generalized Herz–Hardy spaces to weak generalized Herz–Hardy spaces even in the critical case $p = \frac{n}{n+d+\delta}$.

Next, we show the following boundedness of Calderón–Zygmund operators about Hardy-type spaces associated with ball quasi-Banach function spaces.

Theorem 6.6.1 *Let* $d \in \mathbb{Z}_+$, K *be a d-order standard kernel with* $\delta \in (0, 1)$, *and* T *a d-order Calderón–Zygmund operator with kernel* K *having the vanishing moments up to* d. *Let* X *be a ball quasi-Banach function space,* Y *a linear space equipped with a quasi-seminorm* $\|\cdot\|_Y$, *and* Y_0 *a linear space equipped with a quasi-seminorm* $\|\cdot\|_{Y_0}$, *and let* $r_1 \in (0, \infty)$, $\eta \in (1, \infty)$, *and* $0 < \theta < s < s_0 \leq 1$ *be such that*

(i) *for the above* θ *and* s, *Assumption 1.2.29 holds true;*
(ii) *both* $\|\cdot\|_Y$ *and* $\|\cdot\|_{Y_0}$ *satisfy Definition 1.2.13(ii);*
(iii) $\mathbf{1}_{B(\mathbf{0},1)} \in Y_0$;
(iv) *for any* $f \in \mathscr{M}(\mathbb{R}^n)$,

$$\|f\|_{X^{1/s}} \sim \sup\left\{\|fg\|_{L^1(\mathbb{R}^n)} : \|g\|_Y = 1\right\}$$

and

$$\|f\|_{X^{1/s_0}} \sim \sup\left\{\|fg\|_{L^1(\mathbb{R}^n)} : \|g\|_{Y_0} = 1\right\}$$

with the positive equivalence constants independent of f;
(v) $\mathcal{M}^{(\eta)}$ *is bounded on both* Y *and* Y_0;
(vi) $\mathcal{M}$ *is bounded on* $(WX)^{1/r_1}$;
(vii) *there exists a positive constant* C *such that, for any* $\{f_j\}_{j\in\mathbb{N}} \subset L^1_{\text{loc}}(\mathbb{R}^n)$,

$$\left\|\left\{\sum_{j\in\mathbb{N}}\left[\mathcal{M}\left(f_j\right)\right]^{\frac{n+d+\delta}{n}}\right\}^{\frac{n}{n+d+\delta}}\right\|_{(WX)^{\frac{n+d+\delta}{n}}} \leq C \left\|\left(\sum_{j\in\mathbb{N}}|f_j|^{\frac{n+d+\delta}{n}}\right)^{\frac{n}{n+d+\delta}}\right\|_{X^{\frac{n+d+\delta}{n}}}.$$

If $\theta \in (0, \frac{n}{n+d}]$, *then* T *is well defined on* $H_X(\mathbb{R}^n)$ *and there exists a positive constant* C *such that, for any* $f \in H_X(\mathbb{R}^n)$,

$$\|T(f)\|_{WH_X(\mathbb{R}^n)} \leq C\|f\|_{H_X(\mathbb{R}^n)}.$$

Remark 6.6.2 We should point out that Theorem 6.6.1 is an improved version of the known boundedness of convolutional type Calderón–Zygmund operators from $H_X(\mathbb{R}^n)$ to $WH_X(\mathbb{R}^n)$ established by Zhang et al. in [278, Theorem 6.3]. Indeed, in Theorem 6.6.1, if $d = 0$, $K(x, y) \equiv K_1(x - y)$ for some $K_1 \in L^1_{\text{loc}}(\mathbb{R}^n \setminus \{\mathbf{0}\})$, $Y \equiv (X^{1/s})'$, and $Y_0 \equiv (X^{1/s_0})'$, then T coincides with the *convolution* δ-*type Calderón–Zygmund operator* as in [278, Theorem 6.3], and this proposition goes back to [278, Theorem 6.3].

To establish this boundedness criterion, we first show the following boundedness of Calderón–Zygmund operators under stronger assumptions than Theorem 6.6.1.

Proposition 6.6.3 *Let* $d \in \mathbb{Z}_+$, K *be a d-order standard kernel with* $\delta \in (0, 1)$, *and* T *a d-order Calderón–Zygmund operator with kernel* K *having the vanishing moments up to order* d. *Let* X *be a ball quasi-Banach function space satisfying Assumption 1.2.29 with some* $0 < \theta < s \le 1$ *and Assumption 1.2.33 with the same* s *and some* $r_0 \in (1, \infty)$. *Assume that there exists an* $r_1 \in (0, \infty)$ *such that* $\mathcal{M}$ *is bounded on* $(WX)^{1/r_1}$, *and there exists a positive constant* C *such that, for any* $\{f_j\}_{j \in \mathbb{N}} \subset L^1_{loc}(\mathbb{R}^n)$,

$$\left\| \left\{ \sum_{j \in \mathbb{N}} \left[\mathcal{M}(f_j) \right]^{\frac{n+d+\delta}{n}} \right\}^{\frac{n}{n+d+\delta}} \right\|_{(WX)^{\frac{n+d+\delta}{n}}}$$

$$\le C \left\| \left(\sum_{j \in \mathbb{N}} |f_j|^{\frac{n+d+\delta}{n}} \right)^{\frac{n}{n+d+\delta}} \right\|_{X^{\frac{n+d+\delta}{n}}}. \qquad (6.26)$$

If $\theta \in (0, \frac{n}{n+d}]$ *and* X *has an absolutely continuous quasi-norm, then* T *has a unique extension on* $H_X(\mathbb{R}^n)$ *and there exists a positive constant* C *such that, for any* $f \in H_X(\mathbb{R}^n)$,

$$\|T(f)\|_{WH_X(\mathbb{R}^n)} \le C \|f\|_{H_X(\mathbb{R}^n)}.$$

Remark 6.6.4 We point out that Proposition 6.6.3 has a wide range of applications. Here we present two function spaces to which Proposition 6.6.3 can be applied. In what follows, let d and δ be the same as in Proposition 6.6.3.

(i) Let $\vec{p} := (p_1, \ldots, p_n) \in [\frac{n}{n+d+\delta}, \infty)^n$. Then, in this case, as was pointed out in both Remark 1.2.27 and [278, Subsection 7.2], the mixed-norm Lebesgue space $L^{\vec{p}}(\mathbb{R}^n)$ satisfies all the assumptions of Proposition 6.6.3. Thus, Proposition 6.6.3 with $X := L^{\vec{p}}(\mathbb{R}^n)$ holds true. If further assume that K is a convolutional type operator defined as in Proposition 4.8.2, and $d := 0$, then, in this case, the aforementioned result coincides with [278, Theorem 7.36].

(ii) Let $p(\cdot) \in C^{\log}(\mathbb{R}^n)$ satisfy $\frac{n}{n+d+\delta} \le p_- \le p_+ < \infty$, where p_- and p_+ are defined, respectively, in (1.59) and (1.60). Then, in this case, combining Remarks 1.2.27, 1.2.31(v), and 1.2.34(v) and [267, Proposition 7.8], we can easily find that the variable Lebesgue space $L^{p(\cdot)}(\mathbb{R}^n)$ satisfies all the assumptions of Proposition 6.6.3. This then implies that Proposition 6.6.3 with $X := L^{p(\cdot)}(\mathbb{R}^n)$ holds true. If further assume that $p_+ \in (0, 1]$, K is a convolutional type operator defined as in Proposition 4.8.2, and $d := 0$, then, in this case, the aforementioned result is just [267, Theorem 7.4].

To show this proposition, we need the following technical estimate about radial maximal functions.

Lemma 6.6.5 *Let $d \in \mathbb{Z}_+$, K be a d-order standard kernel with $\delta \in (0, 1)$, and T a d-order Calderón–Zygmund with kernel K having the vanishing moments up to order d. Assume that X is a ball quasi-Banach function space, $\phi \in \mathcal{S}(\mathbb{R}^n)$, and $r \in [2, \infty)$. Then there exists a positive constant C such that, for any (X, r, d)-atom a supported in the ball $B \in \mathbb{B}$,*

$$M\,(T\,(a)\,,\phi)\,\mathbf{1}_{(4B)^{\complement}} \leq C\,\frac{1}{\|\mathbf{1}_B\|_X}\,[\mathcal{M}\,(\mathbf{1}_B)]^{\frac{n+d+\delta}{n}},$$

where the radial maximal function M is defined as in Definition 4.1.1(i).

Proof Let all the symbols be as in the present lemma and a an (X, r, d)-atom supported in the ball $B := B(x_0, r_0)$ with $x_0 \in \mathbb{R}^n$ and $r_0 \in (0, \infty)$. Then, combining Definition 4.3.4(iii), the fact that T has the vanishing moments up to order d, and Definition 4.8.1, we conclude that, for any $\gamma \in \mathbb{Z}_+^n$ with $|\gamma| \leq d$,

$$\int_{\mathbb{R}^n} T(a)(x)x^\gamma \, dx = 0. \tag{6.27}$$

Fix a $t \in (0, \infty)$ and an $x \in [B(x_0, 4r_0)]^{\complement}$. Then, from (6.27), it follows that

$$|T(a) * \phi_t(x)|$$

$$= \left| \int_{\mathbb{R}^n} \phi_t(x - y)T(a)(y) \, dy \right|$$

$$= \left| \frac{1}{t^n} \int_{\mathbb{R}^n} \left[\phi\left(\frac{x - y}{t}\right) - \sum_{\substack{\gamma \in \mathbb{Z}_+^n \\ |\gamma| \leq d}} \frac{\partial^\gamma (\phi(\frac{x-\cdot}{t}))(x_0)}{\gamma!}(y - x_0)^\gamma \right] T(a)(y) \, dy \right|$$

$$\leq \frac{1}{t^n} \left\{ \int_{|y-x_0|<2r_0} \left| \phi\left(\frac{x - y}{t}\right) - \sum_{\substack{\gamma \in \mathbb{Z}_+^n \\ |\gamma| \leq d}} \frac{\partial^\gamma (\phi(\frac{x-\cdot}{t}))(x_0)}{\gamma!}(y - x_0)^\gamma \right| \right.$$

$$\left. \times |T(a)(y)| \, dy + \int_{2r_0 \leq |y-x_0| < \frac{|x-x_0|}{2}} \cdots + \int_{|y-x_0| \geq \frac{|x-x_0|}{2}} \cdots \right\}$$

$$=: \text{VIII}_1 + \text{VIII}_2 + \text{VIII}_3. \tag{6.28}$$

We next estimate VIII_1, VIII_2, and VIII_3, respectively.

First, we deal with VIII_1. Indeed, applying the Taylor remainder theorem, we find that, for any $y \in \mathbb{R}^n$ with $|y - x_0| < 2r_0$, there exists a $t_y \in (0, 1)$ such that

$$
\mathrm{VIII}_1 = \frac{1}{t^n} \int_{|y-x_0|<2r_0} \left| \sum_{\substack{\gamma \in \mathbb{Z}_+^n \\ |\gamma|=d+1}} \frac{\partial^\gamma (\phi(\frac{x-\cdot}{t}))(t_y y + (1-t_y)x_0)}{\gamma!} \right.
$$

$$
\left. \times (y - x_0)^\gamma \right| |T(a)(y)| \, dy
$$

$$
\lesssim \frac{1}{t^{n+d+1}} \int_{|y-x_0|<2r_0} \sum_{\substack{\gamma \in \mathbb{Z}_+^n \\ |\gamma|=d+1}} \left| \partial^\gamma \phi \left(\frac{x - t_y y - (1-t_y)x_0}{t} \right) \right|
$$

$$
\times |y - x_0|^{d+1} |T(a)(y)| \, dy
$$

$$
\lesssim \int_{|y-x_0|<2r_0} \frac{|y - x_0|^{d+1}}{|x - t_y y - (1-t_y)x_0|^{n+d+1}} |T(a)(y)| \, dy. \tag{6.29}
$$

Notice that, for any $y \in \mathbb{R}^n$ with $|y - x_0| < 2r_0$, we have

$$
|y - x_0| < \frac{1}{2}(4r_0) \le \frac{1}{2}|x - x_0|,
$$

which, together with the fact that $t_y \in (0, 1)$, further implies that

$$
\left| x - t_y y - (1-t_y)x_0 \right| \ge |x - x_0| - t_y |y - x_0| > \frac{1}{2}|x - x_0|. \tag{6.30}
$$

From this, (6.29), the Hölder inequality, Lemma 1.5.9 with $p := r$, and Definition 4.3.4(ii), we deduce that

$$
\mathrm{VIII}_1 \lesssim \frac{r_0^{d+1}}{|x - x_0|^{n+d+1}} \left\| T(a) \mathbf{1}_{B(x_0,2r_0)} \right\|_{L^1(\mathbb{R}^n)}
$$

$$
\lesssim \frac{r_0^{d+1}}{|x - x_0|^{n+d+1}} |B(x_0, 2r_0)|^{1-\frac{1}{r}} \left\| T(a) \mathbf{1}_{B(x_0,2r_0)} \right\|_{L^r(\mathbb{R}^n)}
$$

$$
\lesssim \frac{r_0^{d+1}}{|x - x_0|^{n+d+1}} |B(x_0, 2r_0)|^{1-\frac{1}{r}} \left\| a \right\|_{L^r(\mathbb{R}^n)}
$$

$$
\lesssim \frac{1}{\|\mathbf{1}_B\|_X} \frac{r_0^{n+d+1}}{|x - x_0|^{n+d+1}}. \tag{6.31}
$$

Moreover, by the assumption $|x - x_0| \geq 4r_0$, we find that $\frac{r_0}{|x-x_0|} \in (0, \frac{1}{4}]$. This, combined with (6.31) and the assumption $\delta \in (0, 1)$, further implies that

$$\text{VIII}_1 \lesssim \frac{1}{\|\mathbf{1}_B\|_X} \frac{r_0^{n+d+\delta}}{|x - x_0|^{n+d+\delta}}, \tag{6.32}$$

which is the desired estimate of VIII_1.

Now, we estimate VIII_2. For this purpose, we first estimate $T(a)(y)$ with $y \in [B(x_0, 2r_0)]^{\complement}$. Indeed, fix a $y \in [B(x_0, 2r_0)]^{\complement}$. By (1.105) with $f := a$, Definition 4.3.4(iii), and the Taylor remainder theorem, we conclude that, for any $z \in B(x_0, r_0)$, there exists a $\tau_z \in [0, 1]$ such that

$$|T(a)(y)|$$

$$= \left| \int_{\mathbb{R}^n} K(y, z) a(z)\, dz \right|$$

$$= \left| \int_{\mathbb{R}^n} \left[K(y, z) - \sum_{\substack{\beta \in \mathbb{Z}_+^n \\ |\beta| \leq d}} \frac{\partial_{(2)}^\beta K(y, x_0)}{\beta!} (z - x_0)^\beta \right] a(z)\, dz \right|$$

$$= \left| \int_{\mathbb{R}^n} \sum_{\substack{\beta \in \mathbb{Z}_+^n \\ |\beta| = d}} \frac{\partial_{(2)}^\beta K(y, \tau_z z + (1 - \tau_z) x_0) - \partial_{(2)}^\beta K(y, x_0)}{\beta!} \right.$$

$$\left. \times (z - x_0)^\beta a(z)\, dz \right|. \tag{6.33}$$

Observe that, for any $z \in B(x_0, r_0)$,

$$|x_0 - \tau_z z - (1 - \tau_z) x_0| \leq |z - x_0| < r_0 \leq \frac{1}{2}|y - x_0|.$$

Combining this, (6.33), (1.104) with γ, x, y, and z therein replaced, respectively, by β, y, x_0, and $\tau_z z + (1 - \tau_z) x_0$ with $z \in B(x_0, r_0)$, the Hölder inequality, and Definition 4.3.4(ii), we further find that, for any $y \in [B(x_0, 2r_0)]^{\complement}$,

$$|T(a)(y)| \lesssim \int_{|z-x_0|<r_0} \frac{|x_0 - \tau_z z - (1 - \tau_z) x_0|^\delta}{|y - x_0|^{n+d+\delta}} |z - x_0|^d\, |a(z)|\, dz$$

$$\lesssim \frac{r_0^{d+\delta}}{|y - x_0|^{n+d+\delta}} \left\| a \mathbf{1}_{B(x_0, r_0)} \right\|_{L^1(\mathbb{R}^n)}$$

$$\lesssim \frac{r_0^{d+\delta}}{|y-x_0|^{n+d+\delta}} |B(x_0,r_0)|^{1-\frac{1}{r}} \left\| a\mathbf{1}_{B(x_0,r_0)} \right\|_{L^r(\mathbb{R}^n)}$$

$$\lesssim \frac{1}{\|\mathbf{1}_B\|_X} \frac{r_0^{n+d+\delta}}{|y-x_0|^{n+d+\delta}}, \tag{6.34}$$

which is the desired estimate of $T(a)(y)$ with $y \in [B(x_0,2r_0)]^{\complement}$. This, together with the Taylor remainder theorem, (6.30), and the assumption that $\delta \in (0,1)$, further implies that, for any $y \in \mathbb{R}^n$ satisfying $2r_0 \le |y-x_0| < \frac{|x-x_0|}{2}$, there exists a $t_y \in (0,1)$ such that

$$\mathrm{VIII}_2 = \frac{1}{t^n} \int_{2r_0 \le |y-x_0| < \frac{|x-x_0|}{2}} \left| \sum_{\substack{\gamma \in \mathbb{Z}_+^n \\ |\gamma|=d+1}} \frac{\partial^\gamma (\phi(\frac{x-\cdot}{t}))(t_y y + (1-t_y)x_0)}{\gamma!} (y-x_0)^\gamma \right|$$

$$\times |T(a)(y)|\, dy$$

$$\lesssim \frac{1}{t^{n+d+1}} \int_{2r_0 \le |y-x_0| < \frac{|x-x_0|}{2}} \sum_{\substack{\gamma \in \mathbb{Z}_+^n \\ |\gamma|=d+1}} \left| \partial^\gamma \phi\left(\frac{x-t_y y-(1-t_y)x_0}{t} \right) \right|$$

$$\times |y-x_0|^{d+1} \frac{1}{\|\mathbf{1}_B\|_X} \frac{r_0^{n+d+\delta}}{|y-x_0|^{n+d+\delta}}\, dy$$

$$\lesssim \frac{r_0^{n+d+\delta}}{\|\mathbf{1}_B\|_X} \int_{2r_0 \le |y-x_0| < \frac{|x-x_0|}{2}} \frac{1}{|x-t_y y-(1-t_y)x_0|^{n+d+1}}$$

$$\times \frac{1}{|y-x_0|^{n+\delta-1}}\, dy$$

$$\lesssim \frac{1}{\|\mathbf{1}_B\|_X} \frac{r_0^{n+d+\delta}}{|x-x_0|^{n+d+1}} \int_{|y-x_0| < \frac{|x-x_0|}{2}} \frac{1}{|y-x_0|^{n+\delta-1}}\, dy$$

$$\sim \frac{1}{\|\mathbf{1}_B\|_X} \frac{r_0^{n+d+\delta}}{|x-x_0|^{n+d+\delta}}. \tag{6.35}$$

This is the desired estimate of VIII_2.

Finally, we deal with VIII_3. Indeed, for any $y \in \mathbb{R}^n$ satisfying $|y-x_0| \ge \frac{|x-x_0|}{2}$, we have $|y-x_0| \ge 2r_0$. By this, (6.34), and the assumption $\delta \in (0,1)$, we conclude that

$$\mathrm{VIII}_3 \lesssim \int_{|y-x_0| \ge \frac{|x-x_0|}{2}} |\phi_t(x-y)|\, |T(a)(y)|\, dy$$

$$+ \int_{|y-x_0| \ge \frac{|x-x_0|}{2}} \sum_{\substack{\gamma \in \mathbb{Z}_+^n \\ |\gamma| \le d}} \frac{1}{t^{n+|\gamma|}} \left| \partial^\gamma \phi\left(\frac{x-x_0}{t} \right) \right| |y-x_0|^{|\gamma|}\, |T(a)(y)|\, dy$$

$$\lesssim \int_{|y-x_0| \geq \frac{|x-x_0|}{2}} |\phi_t(x-y)| \frac{1}{\|\mathbf{1}_B\|_X} dy \frac{r_0^{n+d+\delta}}{|y-x_0|^{n+d+\delta}} dy$$

$$+ \sum_{\substack{\gamma \in \mathbb{Z}_+ \\ |\gamma| \leq d}} \int_{|y-x_0| \geq \frac{|x-x_0|}{2}} \frac{1}{|x-x_0|^{n+|\gamma|}} |y-x_0|^{|\gamma|}$$

$$\times \frac{1}{\|\mathbf{1}_B\|_X} \frac{r_0^{n+d+\delta}}{|y-x_0|^{n+d+\delta}} dy$$

$$\lesssim \frac{r_0^{n+d+\delta}}{\|\mathbf{1}_B\|_X} \left[\frac{1}{|x-x_0|^{n+d+\delta}} \|\phi\|_{L^1(\mathbb{R}^n)} \right.$$

$$\left. + \sum_{\substack{\gamma \in \mathbb{Z}_+^n \\ |\gamma| \leq d}} \frac{1}{|x-x_0|^{n+|\gamma|}} \int_{|y-x_0| \geq \frac{|x-x_0|}{2}} \frac{1}{|y-x_0|^{n+d+\delta-|\gamma|}} dy \right]$$

$$\sim \frac{1}{\|\mathbf{1}_B\|_X} \frac{r_0^{n+d+\delta}}{|x-x_0|^{n+d+\delta}},$$

which is the desired estimate of VIII_3. Combining this, (6.28), (6.32), and (6.35), we further find that, for any $t \in (0, \infty)$ and $x \in [B(x_0, 4r_0)]^\complement$,

$$|T(a) * \phi_t(x)| \lesssim \frac{1}{\|\mathbf{1}_B\|_X} \frac{r_0^{n+d+\delta}}{|x-x_0|^{n+d+\delta}},$$

which, together with the arbitrariness of t and an argument similar to that used in the estimation of (4.23) with $n+d+1$ therein replaced by $n+d+\delta$, implies that

$$M(T(a), \phi)(x) \lesssim \frac{1}{\|\mathbf{1}_B\|_X} [\mathcal{M}(\mathbf{1}_B)(x)]^{\frac{n+d+\delta}{n}}.$$

This then finishes the proof of Lemma 6.6.5.   □

Applying Lemma 6.6.5, we next show Proposition 6.6.3.

Proof of Proposition 6.6.3 Let all the symbols be as in the present proposition and $r := \max\{2, r_0\}$. Then, by the Hölder inequality, we find that, for any $f \in L^1_{\mathrm{loc}}(\mathbb{R}^n)$, $x \in \mathbb{R}^n$, and $B \in \mathbb{B}$ satisfying that $x \in B$,

$$\left[\frac{1}{|B|} \int_B |f(y)|^{(r/s)'} dy \right]^{\frac{1}{(r/s)'}} \leq \left[\frac{1}{|B|} \int_B |f(y)|^{(r_0/s)'} dy \right]^{\frac{1}{(r_0/s)'}}$$

$$\leq \mathcal{M}^{((r_0/s)')}(f)(x),$$

which further implies that

$$\mathcal{M}^{((r/s)')}(f)(x) \leq \mathcal{M}^{((r_0/s)')}(f)(x).$$

From this, the fact that Assumption 1.2.33 holds true with s and r_0 as in the present proposition, Remark 1.2.18, and Definition 1.2.13(ii), it follows that, for any $f \in L^1_{\mathrm{loc}}(\mathbb{R}^n)$,

$$\left\| \mathcal{M}^{((r/s)')}(f) \right\|_{(X^{1/s})'} \leq \left\| \mathcal{M}^{((r_0/s)')}(f) \right\|_{(X^{1/s})'} \lesssim \|f\|_{(X^{1/s})'}. \tag{6.36}$$

On the other hand, using the assumption $\theta \in (0, \frac{n}{n+d}]$, we find that $d \leq n(\frac{1}{\theta} - 1)$. This further implies that

$$d \leq \left\lfloor n\left(\frac{1}{\theta} - 1\right)\right\rfloor =: d_X. \tag{6.37}$$

Then, combining (6.36), the assumption that X satisfies Assumption 1.2.29 for θ and s as in the present proposition, and Lemma 4.4.3, we conclude that, for any $g \in H^{X,r,d_X,s}_{\mathrm{fin}}(\mathbb{R}^n)$,

$$\|g\|_{H^{X,r,d_X,s}_{\mathrm{fin}}(\mathbb{R}^n)} \sim \|g\|_{H_X(\mathbb{R}^n)} \tag{6.38}$$

with the positive equivalence constants independent of g.

Let $\phi \in \mathcal{S}(\mathbb{R}^n)$ satisfy $\mathrm{supp}\,(\phi) \subset B(\mathbf{0}, 1)$ and $\int_{\mathbb{R}^n} \phi(x)\,dx \neq 0$, $f \in H^{X,r,d_X,s}_{\mathrm{fin}}(\mathbb{R}^n)$, $m \in \mathbb{N}$, $\{\lambda_j\}_{j=1}^m \subset [0, \infty)$, and $\{a_j\}_{j=1}^m$ of (X, r, d_X)-atoms supported, respectively, in the balls $\{B_j\}_{j=1}^m \subset \mathbb{B}$ such that $f = \sum_{j=1}^m \lambda_j a_j$. Using this and the linearity of T, we find that $T(f) = \sum_{j=1}^m \lambda_j T(a_j)$. This implies that

$$M(T(f), \phi) = \sup_{t \in (0,\infty)} |T(f) * \phi_t|$$

$$= \sup_{t \in (0,\infty)} \left| \left[\sum_{j=1}^m \lambda_j T(a_j) \right] * \phi_t \right|$$

$$\leq \sum_{j=1}^m \lambda_j \sup_{t \in (0,\infty)} |T(a_j) * \phi_t| = \sum_{j=1}^m \lambda_j M(T(a_j), \phi).$$

From this, Lemma 6.1.5, and Definition 1.2.13(ii), we deduce that

$$\|M\left(T\left(f\right),\phi\right)\|_{WX}$$

$$\leq \left\|\sum_{j=1}^{m}\lambda_{j}M\left(T\left(a_{j}\right),\phi\right)\right\|_{WX}$$

$$\lesssim \left\|\sum_{j=1}^{m}\lambda_{j}M\left(T\left(a_{j}\right),\phi\right)\mathbf{1}_{4B_{j}}\right\|_{WX} + \left\|\sum_{j=1}^{m}\lambda_{j}M\left(T\left(a_{j}\right),\phi\right)\mathbf{1}_{(4B_{j})^{\complement}}\right\|_{WX}$$

$$=: \mathrm{IX}_{1} + \mathrm{IX}_{2}. \tag{6.39}$$

Next, we first estimate IX_{1}. Indeed, for any $j \in \{1,\ldots,m\}$, applying Lemma 4.3.21 with $\Phi := \|\phi\|_{L^{\infty}(\mathbb{R}^{n})}\mathbf{1}_{B(0,1)}$, $f := T(a_{j})$, the L^{r} boundedness of $\mathcal{M}$, Lemma 1.5.9 with $p := r$, and Definition 4.3.4(ii), we conclude that

$$\left\|M\left(T\left(a_{j}\right),\phi\right)\right\|_{L^{r}(\mathbb{R}^{n})} \lesssim \left\|\mathcal{M}\left(T\left(a_{j}\right)\right)\right\|_{L^{r}(\mathbb{R}^{n})} \lesssim \left\|a_{j}\right\|_{L^{r}(\mathbb{R}^{n})} \lesssim \frac{|B_{j}|^{1/r}}{\|\mathbf{1}_{B_{j}}\|_{X}}.$$

By this, Definition 6.3.3, and an argument similar to that used in the estimation of II_{1} in the proof of Theorem 4.3.18 with $\{\mathcal{M}(a_{j})\mathbf{1}_{2B_{j}}\}_{j\in\mathbb{N}}$ therein replaced by $\{M(T(a_{j}),\phi)\mathbf{1}_{4B_{j}}\}_{j=1}^{m}$, we obtain

$$\mathrm{IX}_{1} \lesssim \left\|\sum_{j=1}^{m}\lambda_{j}M\left(T\left(a_{j}\right),\phi\right)\mathbf{1}_{4B_{j}}\right\|_{X} \lesssim \left\|\left[\sum_{j=1}^{m}\left(\frac{\lambda_{j}}{\|\mathbf{1}_{B_{j}}\|_{X}}\right)^{s}\mathbf{1}_{B_{j}}\right]^{\frac{1}{s}}\right\|_{X},$$
$$\tag{6.40}$$

which is the desired estimate of IX_{1}.

Now, we deal with IX_{2}. Indeed, from (6.37) and Definition 4.3.4, it follows that, for any $j \in \{1,\ldots,m\}$, a_{j} is an (X, r, d)-atom supported in the ball B_{j}. Combining this, Lemma 6.6.5 with $a := a_{j}$ for any $j \in \{1,\ldots,m\}$, Definition 1.2.28(i), (6.26), and Lemma 1.2.11 with r and $\{a_{j}\}_{j\in\mathbb{N}}$ therein replaced, respectively, by s and $\{\frac{\lambda_{j}}{\|\mathbf{1}_{B_{j}}\|_{X}}\mathbf{1}_{B_{j}}\}_{j=1}^{m}$, we further conclude that

$$\mathrm{IX}_{2} \lesssim \left\|\sum_{j=1}^{m}\frac{\lambda_{j}}{\|\mathbf{1}_{B_{j}}\|_{X}}\left[\mathcal{M}\left(\mathbf{1}_{B_{j}}\right)\right]^{\frac{n+d+\delta}{n}}\right\|_{WX}$$

$$\sim \left\|\left\{\sum_{j=1}^{m}\frac{\lambda_{j}}{\|\mathbf{1}_{B_{j}}\|_{X}}\left[\mathcal{M}\left(\mathbf{1}_{B_{j}}\right)\right]^{\frac{n+d+\delta}{n}}\right\}^{\frac{n}{n+d+\delta}}\right\|_{(WX)^{\frac{n+d+\delta}{n}}}^{\frac{n+d+\delta}{n}}$$

$$\lesssim \left\| \left(\sum_{j=1}^{m} \frac{\lambda_j}{\|\mathbf{1}_{B_j}\|_X} \mathbf{1}_{B_j} \right)^{\frac{n}{n+d+\delta}} \right\|_{X^{\frac{n+d+\delta}{n}}}^{\frac{n+d+\delta}{n}}$$

$$\sim \left\| \sum_{j=1}^{m} \frac{\lambda_j}{\|\mathbf{1}_{B_j}\|_X} \mathbf{1}_{B_j} \right\|_X \lesssim \left\| \left[\sum_{j=1}^{m} \left(\frac{\lambda_j}{\|\mathbf{1}_{B_j}\|_X} \right)^s \mathbf{1}_{B_j} \right]^{\frac{1}{s}} \right\|_X , \tag{6.41}$$

which is the desired estimate of IX_2. From this, (6.40), and (6.39), it follows that

$$\|M(T(f),\phi)\|_{WX} \lesssim \left\| \left[\sum_{j=1}^{m} \left(\frac{\lambda_j}{\|\mathbf{1}_{B_j}\|_X} \right)^s \mathbf{1}_{B_j} \right]^{\frac{1}{s}} \right\|_X .$$

This, combined with the fact that there exists an $r_1 \in (0,\infty)$ such that $\mathcal{M}$ is bounded on $(WX)^{1/r_1}$, Lemmas 6.1.5 and 4.1.4(ii), Definition 6.3.3, the choice of $\{\lambda_j\}_{j=1}^{m}$, (4.44), and (6.38) with $g := f$, further implies that

$$\|T(f)\|_{WH_X(\mathbb{R}^n)} \sim \|M(T(f),\phi)\|_{WX}$$

$$\lesssim \|f\|_{H_{\mathrm{fin}}^{X,r,d_X,s}(\mathbb{R}^n)} \sim \|f\|_{H_X(\mathbb{R}^n)} . \tag{6.42}$$

Furthermore, using the assumption that X has an absolutely continuous quasi-norm and [207, Remark 3.12], we conclude that the finite atomic Hardy space $H_{\mathrm{fin}}^{X,r,d_X,s}(\mathbb{R}^n)$ is dense in the Hardy space $H_X(\mathbb{R}^n)$. By this, the fact that (6.42) holds true for any $f \in H_{\mathrm{fin}}^{X,r,d_X,s}(\mathbb{R}^n)$, and a standard density argument, we find that T has a unique extension on $H_X(\mathbb{R}^n)$ and, for any $f \in H_X(\mathbb{R}^n)$,

$$\|T(f)\|_{WH_X(\mathbb{R}^n)} \lesssim \|f\|_{H_X(\mathbb{R}^n)},$$

which then completes the proof of Proposition 6.6.3. □

Now, to show Theorem 6.6.1, we still need the boundedness of Calderón–Zygmund operators from weighted Hardy spaces to weighted weak Hardy spaces. Let $p \in (0,\infty)$ and $\upsilon \in A_\infty(\mathbb{R}^n)$. Recall that the *weighted weak Hardy space* $WH_\upsilon^p(\mathbb{R}^n)$ is defined as in Definition 6.3.3 with $X := L_\upsilon^p(\mathbb{R}^n)$ [see also Remark 6.3.4(ii)]. Then we have the following boundedness of Calderón–Zygmund operators from weighted Hardy spaces to weighted weak Hardy spaces.

Proposition 6.6.6 *Let $s_0 \in (0,1]$, $\upsilon \in A_1(\mathbb{R}^n)$, $d \in \mathbb{Z}_+$, K be a d-order standard kernel with $\delta \in (0,1)$, and T a d-order Calderón–Zygmund operator with kernel K having the vanishing moments up to order d. If $s_0 \in [\frac{n}{n+d+\delta},1]$, then T has a* *unique extension on* $H_v^{s_0}(\mathbb{R}^n)$ *and there exists a positive constant C such that, for any* $f \in H_v^{s_0}(\mathbb{R}^n)$,

$$\|T(f)\|_{WH_v^{s_0}(\mathbb{R}^n)} \le C \|f\|_{H_v^{s_0}(\mathbb{R}^n)}.$$

To prove Proposition 6.6.6, we need some auxiliary lemmas. The following one is the weighted weak vector-valued inequality which was obtained in [5, Theorem 3.1(a)].

Lemma 6.6.7 *Let* $p \in [1, \infty)$, $r \in (1, \infty)$, *and* $v \in A_p(\mathbb{R}^n)$. *Then there exists a positive constant C such that, for any* $\{f_j\}_{j \in \mathbb{N}} \subset L_{\mathrm{loc}}^1(\mathbb{R}^n)$,

$$\left\| \left\{ \sum_{j \in \mathbb{N}} [\mathcal{M}(f_j)]^r \right\}^{\frac{1}{r}} \right\|_{WL_v^p(\mathbb{R}^n)} \le C \left\| \left(\sum_{j \in \mathbb{N}} |f_j|^r \right)^{\frac{1}{r}} \right\|_{L_v^p(\mathbb{R}^n)}.$$

In addition, the following lemma gives the boundedness of the Hardy–Littlewood maximal operator on weighted weak Lebesgue spaces (see, for instance, [151, Corollary 2.6]), which plays a key role in the proof of proof of Proposition 6.6.6.

Lemma 6.6.8 *Let* $p \in (1, \infty)$ *and* $v \in A_p(\mathbb{R}^n)$. *Then there exists a positive constant C such that, for any* $f \in L_{\mathrm{loc}}^1(\mathbb{R}^n)$,

$$\|\mathcal{M}(f)\|_{WL_v^p(\mathbb{R}^n)} \le C \|f\|_{WL_v^p(\mathbb{R}^n)}.$$

Via the above two lemmas, we next prove Proposition 6.6.6.

Proof of Proposition 6.6.6 Let all the symbols be as in the present proposition. We first claim that all the assumptions of Proposition 6.6.3 hold true for the weighted Lebesgue space $L_v^{s_0}(\mathbb{R}^n)$ under consideration. To this end, let $s \in (0, s_0)$ and

$$\theta \in \left(0, \min \left\{ s, \frac{n}{n+d} \right\} \right).$$

Then, by Wang et al. [233, Remarks 2.4(b), 2.7(b), and 3.4(i)], we conclude that the following four statements hold true:

(i) $L_v^{s_0}(\mathbb{R}^n)$ is a BQBF space;
(ii) for any $\{f_j\}_{j \in \mathbb{N}} \subset L_{\mathrm{loc}}^1(\mathbb{R}^n)$,

$$\left\| \left\{ \sum_{j \in \mathbb{N}} \left[\mathcal{M}^{(\theta)}(f_j) \right]^s \right\}^{1/s} \right\|_{L_v^{s_0}(\mathbb{R}^n)} \lesssim \left\| \left(\sum_{j \in \mathbb{N}} |f_j|^s \right)^{1/s} \right\|_{L_v^{s_0}(\mathbb{R}^n)};$$

(iii) $[L_\upsilon^{s_0}(\mathbb{R}^n)]^{1/s}$ is a BBF space and there exists an $r_0 \in (1, \infty)$ such that, for any $f \in L_{\text{loc}}^1(\mathbb{R}^n)$,

$$\left\| \mathcal{M}^{((r_0/s)')}(f) \right\|_{([L_\upsilon^{s_0}(\mathbb{R}^n)]^{1/s})'} \lesssim \|f\|_{([L_\upsilon^{s_0}(\mathbb{R}^n)]^{1/s})'} ;$$

(iv) $L_\upsilon^{s_0}(\mathbb{R}^n)$ has an absolutely continuous quasi-norm.

In addition, let $r_1 \in (0, s_0)$. Then, from the fact that $\upsilon \in A_1(\mathbb{R}^n)$ and Lemma 6.6.8 with $p := \frac{s_0}{r_1}$, it follows that, for any $f \in L_{\text{loc}}^1(\mathbb{R}^n)$,

$$\|\mathcal{M}(f)\|_{WL_\upsilon^{s_0/r_1}(\mathbb{R}^n)} \lesssim \|f\|_{WL_\upsilon^{s_0/r_1}(\mathbb{R}^n)} . \tag{6.43}$$

Combining Definitions 1.2.28(ii) and 2.3.7, we find that

$$WL_\upsilon^{s_0/r_1}(\mathbb{R}^n) = \left[WL_\upsilon^{s_0}(\mathbb{R}^n) \right]^{1/r_1} .$$

This, together with (6.43), further implies that the Hardy–Littlewood maximal operator $\mathcal{M}$ is bounded on $[WL_\upsilon^{s_0}(\mathbb{R}^n)]^{1/r_1}$. Finally, we prove that (6.26) holds true with $X := L_\upsilon^{s_0}(\mathbb{R}^n)$. Indeed, applying Definitions 1.2.28(i) and 2.3.7, and Lemma 6.6.7 with $r := \frac{n+d+\delta}{n}$ and $p := \frac{n+d+\delta}{n}s_0$, we conclude that, for any $\{f_j\}_{j\in\mathbb{N}} \subset L_{\text{loc}}^1(\mathbb{R}^n)$,

$$\left\| \left\{ \sum_{j\in\mathbb{N}} \left[\mathcal{M}(f_j) \right]^{\frac{n+d+\delta}{n}} \right\}^{\frac{n}{n+d+\delta}} \right\|_{[WL_\upsilon^{s_0}(\mathbb{R}^n)]^{\frac{n+d+\delta}{n}}}$$

$$= \left\| \left\{ \sum_{j\in\mathbb{N}} \left[\mathcal{M}(f_j) \right]^{\frac{n+d+\delta}{n}} \right\}^{\frac{n}{n+d+\delta}} \right\|_{WL_\upsilon^{\frac{n+d+\delta}{n}s_0}(\mathbb{R}^n)}$$

$$\lesssim \left\| \left\{ \sum_{j\in\mathbb{N}} \left[\mathcal{M}(f_j) \right]^{\frac{n+d+\delta}{n}} \right\}^{\frac{n}{n+d+\delta}} \right\|_{L_\upsilon^{\frac{n+d+\delta}{n}s_0}(\mathbb{R}^n)}$$

$$\sim \left\| \left(\sum_{j\in\mathbb{N}} |f_j|^{\frac{n}{n+d+\delta}} \right)^{\frac{n}{n+d+\delta}} \right\|_{[L_\upsilon^{s_0}(\mathbb{R}^n)]^{\frac{n+d+\delta}{n}}} ,$$

which implies that (6.26) holds true with $X := L_\upsilon^{s_0}(\mathbb{R}^n)$. This, together with (i) through (iv) in the proof of the present proposition and the fact that $\mathcal{M}$ is bounded on $[WL_\upsilon^{s_0}(\mathbb{R}^n)]^{1/r_1}$, further implies that the weighted Lebesgue space $L_\upsilon^{s_0}(\mathbb{R}^n)$ under consideration satisfies all the assumptions of Proposition 6.6.3. This then finishes the proof of the above claim. Thus, by Proposition 6.6.3 with $X := L_\upsilon^{s_0}(\mathbb{R}^n)$, we conclude that T has a unique extension on $H_\upsilon^{s_0}(\mathbb{R}^n)$ and, for any $f \in H_\upsilon^{s_0}(\mathbb{R}^n)$,

$$\|T(f)\|_{WH_\upsilon^{s_0}(\mathbb{R}^n)} \lesssim \|f\|_{H_\upsilon^{s_0}(\mathbb{R}^n)},$$

which completes the proof of Proposition 6.6.6.   □

Via above preparations, we now show Theorem 6.6.1.

Proof of Theorem 6.6.1 Let all the symbols be as in the present theorem and $f \in H_X(\mathbb{R}^n)$. Then, by the assumption $\theta \in (0, \frac{n}{n+d}]$, we conclude that $d \leq n\left(\frac{1}{\theta} - 1\right)$. This further implies that

$$d \leq \left\lfloor n\left(\frac{1}{\theta} - 1\right) \right\rfloor =: d_X. \tag{6.44}$$

From this, the assumption (i) of the present theorem, and Lemma 4.3.8, it follows that there exists $\{\lambda_j\}_{j\in\mathbb{N}} \subset [0, \infty)$ and $\{a_j\}_{j\in\mathbb{N}}$ of (X, ∞, d_X)-atoms supported, respectively, in the balls $\{B_j\}_{j\in\mathbb{N}} \subset \mathbb{B}$ such that $f = \sum_{j\in\mathbb{N}} \lambda_j a_j$ in $\mathcal{S}'(\mathbb{R}^n)$ and

$$\left\| \left[\sum_{j\in\mathbb{N}} \left(\frac{\lambda_j}{\|1_{B_j}\|_X} \right)^s 1_{B_j} \right]^{\frac{1}{s}} \right\|_X \lesssim \|f\|_{H_X(\mathbb{R}^n)}. \tag{6.45}$$

Combining these and an argument similar to that used in the proof of Theorem 4.8.17 with Remark 4.8.13(ii) therein replaced by Proposition 6.6.6 here, we find that

$$T(f) = \sum_{j\in\mathbb{N}} \lambda_j T(a_j) \tag{6.46}$$

in $\mathcal{S}'(\mathbb{R}^n)$.

Next, we prove that $T(f) \in WH_X(\mathbb{R}^n)$ and

$$\|T(f)\|_{WH_X(\mathbb{R}^n)} \lesssim \|f\|_{H_X(\mathbb{R}^n)}.$$

For this purpose, let $r \in (\max\{2, s\eta'\}, \infty)$ with $\frac{1}{\eta} + \frac{1}{\eta'} = 1$. Then, from (6.44) and Definition 4.3.4, we infer that, for any $j \in \mathbb{N}$, a_j is an (X, r, d)-atom supported in the ball B_j. On the other hand, using the assumption (ii) of the present theorem and an argument similar to that used in the estimation of (4.98) with $(r_0/s)'$ and $(X^{1/s})'$ therein replaced, respectively, by η and Y, we conclude that, for any $f \in L_{loc}^1(\mathbb{R}^n)$,

$$\left\| \mathcal{M}^{((r/s)')}(f) \right\|_Y \lesssim \|f\|_Y. \tag{6.47}$$

In addition, let $\phi \in \mathcal{S}(\mathbb{R}^n)$ satisfy $\mathrm{supp}\,(\phi) \subset B(\mathbf{0}, 1)$ and $\int_{\mathbb{R}^n} \phi(x)\,dx \neq 0$. Then, from (6.46), (6.47), the arguments similar to those used in the estimations of (6.39), (6.40), and (6.41) with $\{\lambda_j\}_{j=1}^m$ and $\{a_j\}_{j=1}^m$ therein replaced, respectively, by $\{\lambda_j\}_{j\in\mathbb{N}}$ and $\{a_j\}_{j\in\mathbb{N}}$, and (6.45), we deduce that

$$\|M\,(T\,(f)\,,\phi)\|_{WX} \lesssim \left\|\sum_{j\in\mathbb{N}}\lambda_j M(T(a_j),\phi)\right\|_{WX}$$

$$\lesssim \left\|\left[\sum_{j\in\mathbb{N}}\left(\frac{\lambda_j}{\|\mathbf{1}_{B_j}\|_X}\right)^s \mathbf{1}_{B_j}\right]^{\frac{1}{s}}\right\|_X \lesssim \|f\|_{H_X(\mathbb{R}^n)},$$

which, combined with the assumption (vi) of the present theorem, Lemmas 6.1.5 and 4.1.4(ii), and Definition 6.3.3, further implies that

$$\|T\,(f)\|_{WH_X(\mathbb{R}^n)} \sim \|M\,(T\,(f)\,,\phi)\|_{WX} \lesssim \|f\|_{H_X(\mathbb{R}^n)}.$$

This then finishes the proof of Theorem 6.6.1.   □

Next, we turn to establish the boundedness of Calderón–Zygmund operators from generalized Herz–Hardy spaces to weak generalized Herz–Hardy spaces. Indeed, the following conclusion gives the boundedness of Calderón–Zygmund operators from $H\dot{\mathcal{K}}_{\omega,0}^{p,q}(\mathbb{R}^n)$ to $WH\dot{\mathcal{K}}_{\omega,0}^{p,q}(\mathbb{R}^n)$.

Theorem 6.6.9 *Let $d \in \mathbb{Z}_+$, $\delta \in (0,1)$, $p \in [\frac{n}{n+d+\delta}, \infty)$, $q \in (0,\infty)$, K be a d-order standard kernel as in Definition 1.5.7, T a d-order Calderón–Zygmund operator with kernel K having the vanishing moments up to order d, and $\omega \in M(\mathbb{R}_+)$ with*

$$-\frac{n}{p} < m_0(\omega) \leq M_0(\omega) < n - \frac{n}{p} + d + \delta$$

and

$$-\frac{n}{p} < m_\infty(\omega) \leq M_\infty(\omega) < n - \frac{n}{p} + d + \delta.$$

Then T has a unique extension on $H\dot{\mathcal{K}}_{\omega,0}^{p,q}(\mathbb{R}^n)$ and there exists a positive constant C such that, for any $f \in H\dot{\mathcal{K}}_{\omega,0}^{p,q}(\mathbb{R}^n)$,

$$\|T(f)\|_{WH\dot{\mathcal{K}}_{\omega,0}^{p,q}(\mathbb{R}^n)} \leq C\|f\|_{H\dot{\mathcal{K}}_{\omega,0}^{p,q}(\mathbb{R}^n)}.$$

To prove this theorem, we need the following weak type inequality about the Hardy–Littlewood maximal operator, which is a part of [90, Theorem 5.6.6].

Lemma 6.6.10 *Let $r \in (1, \infty)$. Then there exists a positive constant C such that, for any $\{f_j\}_{j\in\mathbb{N}} \subset L^1_{\mathrm{loc}}(\mathbb{R}^n)$ and $\lambda \in (0, \infty)$,*

$$\lambda \left\| \mathbf{1}_{\{y\in\mathbb{R}^n:\ \{\sum_{j\in\mathbb{N}}[\mathcal{M}(f_j)(y)]^r\}^{\frac{1}{r}} > \lambda\}} \right\|_{L^1(\mathbb{R}^n)} \leq C \left\| \left(\sum_{j\in\mathbb{N}} |f_j|^r \right)^{\frac{1}{r}} \right\|_{L^1(\mathbb{R}^n)}.$$

Via this estimate, we obtain the weak type vector-valued inequalities on both local and global generalized Herz spaces as follows, which play an important role in the proof of the boundedness of Calderón–Zygmund operators from generalized Herz–Hardy spaces to weak generalized Herz–Hardy spaces.

Proposition 6.6.11 *Let $p \in [1, \infty)$, $q \in (0, \infty)$, $r \in (1, \infty)$, and $\omega \in M(\mathbb{R}_+)$ with*

$$-\frac{n}{p} < m_0(\omega) \leq M_0(\omega) < \frac{n}{p'}$$

and

$$-\frac{n}{p} < m_\infty(\omega) \leq M_\infty(\omega) < \frac{n}{p'},$$

where $\frac{1}{p} + \frac{1}{p'} = 1$. Then there exists a positive constant C such that, for any $\{f_j\}_{j\in\mathbb{N}} \subset L^1_{\mathrm{loc}}(\mathbb{R}^n)$,

$$\left\| \left\{ \sum_{j\in\mathbb{N}} [\mathcal{M}(f_j)]^r \right\}^{\frac{1}{r}} \right\|_{W\dot{\mathcal{K}}^{p,q}_{\omega,0}(\mathbb{R}^n)} \leq C \left\| \left(\sum_{j\in\mathbb{N}} |f_j|^r \right)^{\frac{1}{r}} \right\|_{\dot{\mathcal{K}}^{p,q}_{\omega,0}(\mathbb{R}^n)} \tag{6.48}$$

and

$$\left\| \left\{ \sum_{j\in\mathbb{N}} [\mathcal{M}(f_j)]^r \right\}^{\frac{1}{r}} \right\|_{W\dot{\mathcal{K}}^{p,q}_{\omega}(\mathbb{R}^n)} \leq C \left\| \left(\sum_{j\in\mathbb{N}} |f_j|^r \right)^{\frac{1}{r}} \right\|_{\dot{\mathcal{K}}^{p,q}_{\omega}(\mathbb{R}^n)}. \tag{6.49}$$

Proof Let all the symbols be as in the present proposition and $\{f_j\}_{j\in\mathbb{N}}$ be any given sequence of local integrable functions on $\mathbb{R}^n$. Then we show the present proposition by considering the following two cases on p.

Case (1) $p \in (1, \infty)$. In this case, by Definition 6.0.18 and Theorems 1.6.1 and 1.6.4, we find that

$$\left\| \left\{ \sum_{j\in\mathbb{N}} [\mathcal{M}(f_j)]^r \right\}^{\frac{1}{r}} \right\|_{W\dot{\mathcal{K}}^{p,q}_{\omega,0}(\mathbb{R}^n)} \leq \left\| \left\{ \sum_{j\in\mathbb{N}} [\mathcal{M}(f_j)]^r \right\}^{\frac{1}{r}} \right\|_{\dot{\mathcal{K}}^{p,q}_{\omega,0}(\mathbb{R}^n)}$$

$$\lesssim \left\| \left(\sum_{j\in\mathbb{N}} |f_j|^r \right)^{\frac{1}{r}} \right\|_{\dot{\mathcal{K}}^{p,q}_{\omega,0}(\mathbb{R}^n)}$$

and

$$\left\| \left\{ \sum_{j\in\mathbb{N}} [\mathcal{M}(f_j)]^r \right\}^{\frac{1}{r}} \right\|_{W\dot{\mathcal{K}}^{p,q}_{\omega}(\mathbb{R}^n)} \leq \left\| \left\{ \sum_{j\in\mathbb{N}} [\mathcal{M}(f_j)]^r \right\}^{\frac{1}{r}} \right\|_{\dot{\mathcal{K}}^{p,q}_{\omega}(\mathbb{R}^n)}$$

$$\lesssim \left\| \left(\sum_{j\in\mathbb{N}} |f_j|^r \right)^{\frac{1}{r}} \right\|_{\dot{\mathcal{K}}^{p,q}_{\omega}(\mathbb{R}^n)}.$$

These finish the proof of Proposition 6.6.11 in this case.

Case (2) $p = 1$. In this case, fix a point $\xi \in \mathbb{R}^n$. For any $j \in \mathbb{N}$ and $k \in \mathbb{Z}$, let

$$f_{j,k,1} := f_j \mathbf{1}_{B(\xi, 2^{k-2})},$$

$$f_{j,k,2} := f_j \mathbf{1}_{B(\xi, 2^{k+1})\setminus B(\xi, 2^{k-2})},$$

and

$$f_{j,k,3} := f_j \mathbf{1}_{[B(\xi, 2^{k+1})]^\complement}.$$

Then, obviously, for any $j \in \mathbb{N}$ and $k \in \mathbb{Z}$, $f = f_{j,k,1} + f_{j,k,2} + f_{j,k,3}$. From this, we deduce that, for any $\lambda \in (0, \infty)$,

$$\left\| \mathbf{1}_{\{y\in\mathbb{R}^n: \{\sum_{j\in\mathbb{N}}[\mathcal{M}(f_j)(y)]^r\}^{\frac{1}{r}} > \lambda\}}(\cdot + \xi) \right\|_{\dot{\mathcal{K}}^{1,q}_{\omega,0}(\mathbb{R}^n)}$$

$$\lesssim \left\{ \sum_{k\in\mathbb{Z}} \left[\omega(2^k) \right]^q \right.$$

$$\times \left\| \mathbf{1}_{\{y\in\mathbb{R}^n:\,\{\sum_{j\in\mathbb{N}}[\mathcal{M}(f_{j,k,1})(y)]^r\}^{\frac{1}{r}}>\lambda/3\}} \mathbf{1}_{B(\xi,2^k)\setminus B(\xi,2^{k-1})} \right\|_{L^1(\mathbb{R}^n)}^q \right\}^{\frac{1}{q}}$$

$$+ \left\{ \sum_{k\in\mathbb{Z}} \left[\omega(2^k) \right]^q \right.$$

$$\times \left\| \mathbf{1}_{\{y\in\mathbb{R}^n:\,\{\sum_{j\in\mathbb{N}}[\mathcal{M}(f_{j,k,2})(y)]^r\}^{\frac{1}{r}}>\lambda/3\}} \mathbf{1}_{B(\xi,2^k)\setminus B(\xi,2^{k-1})} \right\|_{L^1(\mathbb{R}^n)}^q \right\}^{\frac{1}{q}}$$

$$+ \left\{ \sum_{k\in\mathbb{Z}} \left[\omega(2^k) \right]^q \right.$$

$$\times \left\| \mathbf{1}_{\{y\in\mathbb{R}^n:\,\{\sum_{j\in\mathbb{N}}[\mathcal{M}(f_{j,k,3})(y)]^r\}^{\frac{1}{r}}>\lambda/3\}} \mathbf{1}_{B(\xi,2^k)\setminus B(\xi,2^{k-1})} \right\|_{L^1(\mathbb{R}^n)}^q \right\}^{\frac{1}{q}}$$

$$=: \mathrm{II}_{\lambda,1} + \mathrm{II}_{\lambda,2} + \mathrm{II}_{\lambda,3}. \tag{6.50}$$

We next estimate $\mathrm{II}_{\lambda,1}$, $\mathrm{II}_{\lambda,2}$, and $\mathrm{II}_{\lambda,3}$, respectively. To achieve this, let

$$\varepsilon \in (0, \{\min\{m_0(\omega), m_\infty(\omega)\} + n, -\max\{M_0(\omega), M_\infty(\omega)\}\})$$

be a fixed positive constant. Then, applying Lemma 1.5.2, we conclude that, for any $0 < t < \tau < \infty$,

$$\frac{\omega(t)}{\omega(\tau)} \lesssim \left(\frac{t}{\tau}\right)^{\min\{m_0(\omega),m_\infty(\omega)\}-\varepsilon} \tag{6.51}$$

and, for any $0 < \tau < t < \infty$,

$$\frac{\omega(t)}{\omega(\tau)} \lesssim \left(\frac{t}{\tau}\right)^{\max\{M_0(\omega),M_\infty(\omega)\}+\varepsilon}. \tag{6.52}$$

For the simplicity of the presentation, let $m := \min\{m_0(\omega), m_\infty(\omega)\} - \varepsilon$ and $M := \max\{M_0(\omega), M_\infty(\omega)\} + \varepsilon$. Thus, we have $m \in (-n, \infty)$ and $M \in (-\infty, 0)$.

Now, we deal with $\mathrm{II}_{\lambda,1}$. Indeed, using the fact that the Hardy–Littlewood maximal operator $\mathcal{M}$ satisfies the size condition (1.91) (see, for instance, [197, Remark 4.4]), together with the Minkowski integral inequality, we find that, for any $k \in \mathbb{Z}$ and $x \in \mathbb{R}^n$ satisfying $2^{k-1} \le |x - \xi| < 2^k$,

$$
\left\{ \sum_{j \in \mathbb{N}} \left[\mathcal{M}(f_{j,k,1})(x) \right]^r \right\}^{\frac{1}{r}} \lesssim \left\{ \sum_{j \in \mathbb{N}} \left[\int_{\mathbb{R}^n} \frac{|f_{j,k,1}(y)|}{|x - y|^n} \, dy \right]^r \right\}^{\frac{1}{r}}
$$

$$
\lesssim \int_{\mathbb{R}^n} \left\{ \sum_{j \in \mathbb{N}} \left[\frac{|f_{j,k,1}(y)|}{|x - y|^n} \right]^r \right\}^{\frac{1}{r}} dy
$$

$$
\sim \int_{B(\xi, 2^{k-2})} \frac{[\sum_{j \in \mathbb{N}} |f_j(y)|^r]^{\frac{1}{r}}}{|x - y|^n} \, dy
$$

$$
\sim \sum_{i=-\infty}^{k-2} \int_{B(\xi, 2^i) \setminus B(\xi, 2^{i-1})} \frac{[\sum_{j \in \mathbb{N}} |f_j(y)|^r]^{\frac{1}{r}}}{|x - y|^n} \, dy.
$$

$$(6.53)$$

On the other hand, for any $k \in \mathbb{Z}$, $i \in \mathbb{Z} \cap (-\infty, k - 2]$, and $x, y \in \mathbb{R}^n$ satisfying $2^{k-1} \le |x - \xi| < 2^k$ and $2^{i-1} \le |y - \xi| < 2^i$, we have

$$
|x - y| \ge |x - \xi| - |y - \xi| > 2^{k-1} - 2^i > 2^{k-2},
$$

which, together with (6.53), implies that, for any $k \in \mathbb{Z}$ and $x \in \mathbb{R}^n$ satisfying $2^{k-1} \le |x - \xi| < 2^k$,

$$
\left\{ \sum_{j \in \mathbb{N}} \left[\mathcal{M}(f_{j,k,1})(x) \right]^r \right\}^{\frac{1}{r}}
$$

$$
\lesssim \sum_{i=-\infty}^{k-2} 2^{-nk} \left\| \left(\sum_{j \in \mathbb{N}} |f_j|^r \right)^{\frac{1}{r}} \mathbf{1}_{B(\xi, 2^i) \setminus B(\xi, 2^{i-1})} \right\|_{L^1(\mathbb{R}^n)}. \qquad (6.54)
$$

In addition, by Lemma 1.1.3 and (6.52), we conclude that, for any $k \in \mathbb{Z}$ and $i \in \mathbb{Z} \cap (-\infty, k - 2]$,

$$
\frac{\omega(2^k)}{\omega(2^i)} \sim \frac{\omega(2^{k-2})}{\omega(2^i)} \lesssim \left(\frac{2^{k-2}}{2^i} \right)^M \sim 2^{(k-i)M}.
$$

From this and (6.54), it follows that, for any $k \in \mathbb{Z}$,

$$
\omega(2^k) \left\| \left\{ \sum_{j \in \mathbb{N}} \left[\mathcal{M}(f_{j,k,1}) \right]^r \right\}^{\frac{1}{r}} \mathbf{1}_{B(\xi,2^k) \setminus B(\xi,2^{k-1})} \right\|_{L^1(\mathbb{R}^n)}
$$

$$
\lesssim \omega(2^k) \sum_{i=-\infty}^{k-2} 2^{-nk} \left\| \mathbf{1}_{B(\xi,2^k) \setminus B(\xi,2^{k-1})} \right\|_{L^1(\mathbb{R}^n)}
$$

$$
\times \left\| \left(\sum_{j \in \mathbb{N}} |f_j|^r \right)^{\frac{1}{r}} \mathbf{1}_{B(\xi,2^i) \setminus B(\xi,2^{i-1})} \right\|_{L^1(\mathbb{R}^n)}
$$

$$
\sim \sum_{i=-\infty}^{k-2} \frac{\omega(2^k)}{\omega(2^i)} \omega(2^i) \left\| \left(\sum_{j \in \mathbb{N}} |f_j|^r \right)^{\frac{1}{r}} \mathbf{1}_{B(\xi,2^i) \setminus B(\xi,2^{i-1})} \right\|_{L^1(\mathbb{R}^n)}
$$

$$
\lesssim \sum_{i=-\infty}^{k-2} 2^{(k-i)M} \omega(2^i) \left\| \left(\sum_{j \in \mathbb{N}} |f_j|^r \right)^{\frac{1}{r}} \mathbf{1}_{B(\xi,2^i) \setminus B(\xi,2^{i-1})} \right\|_{L^1(\mathbb{R}^n)} . \qquad (6.55)
$$

Applying this, the fact that, for any $\lambda \in (0, \infty)$,

$$
\mathbf{1}_{\{y \in \mathbb{R}^n : \{\sum_{j \in \mathbb{N}} [\mathcal{M}(f_{j,k,1})(y)]^r\}^{\frac{1}{r}} > \lambda/3\}} \leq \frac{3}{\lambda} \left\{ \sum_{j \in \mathbb{N}} \left[\mathcal{M}(f_{j,k,1}) \right]^r \right\}^{\frac{1}{r}}, \qquad (6.56)
$$

Lemma 1.2.11, and the assumption $M \in (-\infty, 0)$, we further conclude that, for any $q \in (0, 1]$ and $\lambda \in (0, \infty)$,

$$
\mathrm{II}_{\lambda,1} \lesssim \lambda^{-1} \left(\sum_{k \in \mathbb{Z}} \left[\omega(2^k) \right]^q \left\| \left\{ \sum_{j \in \mathbb{N}} \left[\mathcal{M}(f_{j,k,1}) \right]^r \right\}^{\frac{1}{r}} \mathbf{1}_{B(\xi,2^k) \setminus B(\xi,2^{k-1})} \right\|_{L^1(\mathbb{R}^n)}^q \right)^{\frac{1}{q}}
$$

$$
\lesssim \lambda^{-1} \left\{ \sum_{k \in \mathbb{Z}} \left[\sum_{i=-\infty}^{k-2} 2^{(k-i)M} \omega(2^i) \right. \right.
$$

$$
\times \left. \left. \left\| \left(\sum_{j \in \mathbb{N}} |f_j|^r \right)^{\frac{1}{r}} \mathbf{1}_{B(\xi,2^i) \setminus B(\xi,2^{i-1})} \right\|_{L^1(\mathbb{R}^n)} \right]^q \right\}^{\frac{1}{q}}
$$

$$\lesssim \lambda^{-1} \left\{ \sum_{k\in\mathbb{Z}} \sum_{i=-\infty}^{k-2} 2^{(k-i)Mq} \left[\omega(2^i)\right]^q \right.$$

$$\times \left\| \left(\sum_{j\in\mathbb{N}} |f_j|^r \right)^{\frac{1}{r}} \mathbf{1}_{B(\xi,2^i)\setminus B(\xi,2^{i-1})} \right\|_{L^1(\mathbb{R}^n)}^q \right\}^{\frac{1}{q}}$$

$$\sim \lambda^{-1} \left\{ \sum_{i\in\mathbb{Z}} \left[\omega(2^k)\right]^q \left\| \left(\sum_{j\in\mathbb{N}} |f_j|^r \right)^{\frac{1}{r}} \mathbf{1}_{B(\xi,2^i)\setminus B(\xi,2^{i-1})} \right\|_{L^1(\mathbb{R}^n)}^q \right.$$

$$\times \left. \sum_{k=i+2}^{\infty} 2^{(k-i)Mq} \right\}^{\frac{1}{q}}$$

$$\sim \lambda^{-1} \left\| \left(\sum_{j\in\mathbb{N}} |f_j|^r \right)^{\frac{1}{r}} (\cdot + \xi) \right\|_{\dot{\mathcal{K}}_{\omega,0}^{-1,q}(\mathbb{R}^n)} . \tag{6.57}$$

Moreover, using (6.56), (6.55), the Hölder inequality, and the assumption $M \in (-\infty, 0)$, we find that, for any $q \in (1, \infty)$ and $\lambda \in (0, \infty)$,

$$\mathrm{II}_{\lambda,1} \lesssim \lambda^{-1} \left(\sum_{k\in\mathbb{Z}} \left[\omega(2^k)\right]^q \left\| \left\{ \sum_{j\in\mathbb{N}} \left[\mathcal{M}(f_{j,k,1})\right]^r \right\}^{\frac{1}{r}} \mathbf{1}_{B(\xi,2^k)\setminus B(\xi,2^{k-1})} \right\|_{L^1(\mathbb{R}^n)}^q \right)^{\frac{1}{q}}$$

$$\lesssim \lambda^{-1} \left\{ \sum_{k\in\mathbb{Z}} \left[\sum_{i=-\infty}^{k-2} 2^{(k-i)M}\omega(2^i) \right. \right.$$

$$\times \left. \left. \left\| \left(\sum_{j\in\mathbb{N}} |f_j|^r \right)^{\frac{1}{r}} \mathbf{1}_{B(\xi,2^i)\setminus B(\xi,2^{i-1})} \right\|_{L^1(\mathbb{R}^n)} \right]^q \right\}^{\frac{1}{q}}$$

$$\lesssim \lambda^{-1} \left\{ \sum_{k\in\mathbb{Z}} \left[\sum_{i=-\infty}^{k-2} 2^{\frac{(k-i)Mq'}{2}} \right]^{\frac{q}{q'}} \right.$$

$$\times \left. \sum_{i=-\infty}^{k-2} 2^{\frac{(k-i)Mq}{2}} \left[\omega(2^i)\right]^q \left\| \left(\sum_{j\in\mathbb{N}} |f_j|^r \right)^{\frac{1}{r}} \mathbf{1}_{B(\xi,2^i)\setminus B(\xi,2^{i-1})} \right\|_{L^1(\mathbb{R}^n)}^q \right\}^{\frac{1}{q}}$$

$$\sim \lambda^{-1} \left\{ \sum_{i \in \mathbb{Z}} \left[\omega(2^i) \right]^q \left\| \left(\sum_{j \in \mathbb{N}} |f_j|^r \right)^{\frac{1}{r}} \mathbf{1}_{B(\xi,2^i) \setminus B(\xi,2^{i-1})} \right\|_{L^1(\mathbb{R}^n)}^q \right.$$

$$\left. \times \sum_{k=i+2}^{\infty} 2^{\frac{(k-i)Mq}{2}} \right\}^{\frac{1}{q}}$$

$$\sim \lambda^{-1} \left\| \left(\sum_{j \in \mathbb{N}} |f_j|^r \right)^{\frac{1}{r}} (\cdot + \xi) \right\|_{\dot{\mathcal{K}}_{\omega,0}^{1,q}(\mathbb{R}^n)}, \tag{6.58}$$

which, together with (6.57), then completes the estimate of $\mathrm{II}_{\lambda,1}$.

We next deal with $\mathrm{II}_{\lambda,2}$. Indeed, from Lemma 6.6.10, it follows that, for any $k \in \mathbb{Z}$ and $\lambda \in (0, \infty)$,

$$\left\| \mathbf{1}_{\{y \in \mathbb{R}^n : \, \{\sum_{j \in \mathbb{N}} [\mathcal{M}(f_{j,k,2})(y)]^r\}^{\frac{1}{r}} > \lambda/3\}} \mathbf{1}_{B(\xi,2^k) \setminus B(\xi,2^{k-1})} \right\|_{L^1(\mathbb{R}^n)}$$

$$\leq \left\| \mathbf{1}_{\{y \in \mathbb{R}^n : \, \{\sum_{j \in \mathbb{N}} [\mathcal{M}(f_{j,k,2})(y)]^r\}^{\frac{1}{r}} > \lambda/3\}} \right\|_{L^1(\mathbb{R}^n)}$$

$$\lesssim \lambda^{-1} \left\| \left(\sum_{j \in \mathbb{N}} |f_{j,k,2}|^r \right)^{\frac{1}{r}} \right\|_{L^1(\mathbb{R}^n)}$$

$$\lesssim \lambda^{-1} \left[\left\| \left(\sum_{j \in \mathbb{N}} |f_j|^r \right)^{\frac{1}{r}} \mathbf{1}_{B(\xi,2^{k+1}) \setminus B(\xi,2^k)} \right\|_{L^1(\mathbb{R}^n)} \right.$$

$$+ \left\| \left(\sum_{j \in \mathbb{N}} |f_j|^r \right)^{\frac{1}{r}} \mathbf{1}_{B(\xi,2^k) \setminus B(\xi,2^{k-1})} \right\|_{L^1(\mathbb{R}^n)}$$

$$\left. + \left\| \left(\sum_{j \in \mathbb{N}} |f_j|^r \right)^{\frac{1}{r}} \mathbf{1}_{B(\xi,2^{k-1}) \setminus B(\xi,2^{k-2})} \right\|_{L^1(\mathbb{R}^n)} \right]. \tag{6.59}$$

In addition, by Lemma 1.1.3, we conclude that, for any $k \in \mathbb{Z}$,

$$\omega(2^{k+1}) \sim \omega(2^k) \sim \omega(2^{k-1}).$$

Therefore, using (6.59), we find that, for any $\lambda \in (0, \infty)$,

$$
\begin{aligned}
\mathrm{II}_{\lambda,2} \lesssim \lambda^{-1} \Bigg(& \left\{ \left\| \sum_{k\in\mathbb{Z}} \left[\omega(2^{k+1})\right]^q \left\| \left(\sum_{j\in\mathbb{N}} |f_j|^r\right)^{\frac{1}{r}} \mathbf{1}_{B(\xi,2^k)\backslash B(\xi,2^{k-1})} \right\|^q_{L^1(\mathbb{R}^n)} \right\}^{\frac{1}{q}} \right. \\
& + \left\{ \sum_{k\in\mathbb{Z}} \left[\omega(2^k)\right]^q \left\| \left(\sum_{j\in\mathbb{N}} |f_j|^r\right)^{\frac{1}{r}} \mathbf{1}_{B(\xi,2^k)\backslash B(\xi,2^{k-1})} \right\|^q_{L^1(\mathbb{R}^n)} \right\}^{\frac{1}{q}} \\
& + \left\{ \sum_{k\in\mathbb{Z}} \left[\omega(2^{k-1})\right]^q \left\| \left(\sum_{j\in\mathbb{N}} |f_j|^r\right)^{\frac{1}{r}} \mathbf{1}_{B(\xi,2^k)\backslash B(\xi,2^{k-1})} \right\|^q_{L^1(\mathbb{R}^n)} \right\}^{\frac{1}{q}} \Bigg) \\
\sim \lambda^{-1} & \left\| \left(\sum_{j\in\mathbb{N}} |f_j|^r\right)^{\frac{1}{r}} (\cdot + \xi) \right\|_{\dot{\mathcal{K}}^{1,q}_{\omega,0}(\mathbb{R}^n)},
\end{aligned}
\tag{6.60}
$$

which completes the estimate of $\mathrm{II}_{\lambda,2}$.

Finally, we estimate $\mathrm{II}_{\lambda,3}$. Indeed, applying the fact that the Hardy–Littlewood maximal operator $\mathcal{M}$ satisfies the size condition (1.91) (see, for instance, [197, Remark 4.4]), together with the Minkowski integral inequality, we conclude that, for any $k \in \mathbb{Z}$ and $x \in \mathbb{R}^n$ satisfying $2^{k-1} \leq |x - \xi| < 2^k$,

$$
\begin{aligned}
\left\{ \sum_{j\in\mathbb{N}} \left[\mathcal{M}(f_{j,k,3})(x)\right]^r \right\}^{\frac{1}{r}} &\lesssim \left\{ \sum_{j\in\mathbb{N}} \left[\int_{\mathbb{R}^n} \frac{|f_{j,k,3}(y)|}{|x-y|^n}\, dy\right]^r \right\}^{\frac{1}{r}} \\
&\lesssim \int_{\mathbb{R}^n} \left\{ \sum_{j\in\mathbb{N}} \left[\frac{|f_{j,k,3}(y)|}{|x-y|^n}\right]^r \right\}^{\frac{1}{r}} dy \\
&\sim \int_{[B(\xi,2^{k+1})]^{\complement}} \frac{[\sum_{j\in\mathbb{N}} |f_j(y)|^r]^{\frac{1}{r}}}{|x-y|^n}\, dy \\
&\sim \sum_{i=k+2}^{\infty} \int_{B(\xi,2^i)\backslash B(\xi,2^{i-1})} \frac{[\sum_{j\in\mathbb{N}} |f_j(y)|^r]^{\frac{1}{r}}}{|x-y|^n}\, dy.
\end{aligned}
\tag{6.61}
$$

On the other hand, for any $k \in \mathbb{Z}$, $i \in \mathbb{Z} \cap [k+2, \infty)$, and $x, y \in \mathbb{R}^n$ satisfying $2^{k-1} \leq |x - \xi| < 2^k$ and $2^{i-1} \leq |y - \xi| < 2^i$, we have

$$|x - y| \geq |y - \xi| - |x - \xi| > 2^{i-1} - 2^k \geq 2^{i-2},$$

which, together with (6.61), implies that, for any $k \in \mathbb{Z}$ and $x \in \mathbb{R}^n$ satisfying $2^{k-1} \leq |x - \xi| < 2^k$,

$$\left\{ \sum_{j \in \mathbb{N}} \left[\mathcal{M}(f_{j,k,3})(x) \right]^r \right\}^{\frac{1}{r}}$$
$$\lesssim \sum_{i=k+2}^{\infty} 2^{-ni} \left\| \left(\sum_{j \in \mathbb{N}} |f_j|^r \right)^{\frac{1}{r}} \mathbf{1}_{B(\xi, 2^i) \setminus B(\xi, 2^{i-1})} \right\|_{L^1(\mathbb{R}^n)}. \tag{6.62}$$

In addition, from Lemma 1.1.3 and (6.51), we deduce that, for any $k \in \mathbb{Z}$ and $i \in \mathbb{Z} \cap [k+2, \infty)$,

$$\frac{\omega(2^k)}{\omega(2^i)} \sim \frac{\omega(2^{k+2})}{\omega(2^i)} \lesssim \left(\frac{2^{k+2}}{2^i} \right)^m \sim 2^{(k-i)m}.$$

Applying this and (6.62), we conclude that, for any $k \in \mathbb{Z}$,

$$\omega(2^k) \left\| \left\{ \sum_{j \in \mathbb{N}} \left[\mathcal{M}(f_{j,k,3}) \right]^r \right\}^{\frac{1}{r}} \mathbf{1}_{B(\xi, 2^k) \setminus B(\xi, 2^{k-1})} \right\|_{L^1(\mathbb{R}^n)}$$

$$\lesssim \omega(2^k) \sum_{i=k+2}^{\infty} 2^{-ni} \left\| \mathbf{1}_{B(\xi, 2^k) \setminus B(\xi, 2^{k-1})} \right\|_{L^1(\mathbb{R}^n)}$$

$$\times \left\| \left(\sum_{j \in \mathbb{N}} |f_j|^r \right)^{\frac{1}{r}} \mathbf{1}_{B(\xi, 2^i) \setminus B(\xi, 2^{i-1})} \right\|_{L^1(\mathbb{R}^n)}$$

$$\sim \sum_{i=k+2}^{\infty} 2^{(k-i)n} \frac{\omega(2^k)}{\omega(2^i)} \omega(2^i) \left\| \left(\sum_{j \in \mathbb{N}} |f_j|^r \right)^{\frac{1}{r}} \mathbf{1}_{B(\xi, 2^i) \setminus B(\xi, 2^{i-1})} \right\|_{L^1(\mathbb{R}^n)}$$

$$\lesssim \sum_{i=k+2}^{\infty} 2^{(k-i)(m+n)} \omega(2^i) \left\| \left(\sum_{j \in \mathbb{N}} |f_j|^r \right)^{\frac{1}{r}} \mathbf{1}_{B(\xi, 2^i) \setminus B(\xi, 2^{i-1})} \right\|_{L^1(\mathbb{R}^n)}.$$

From this, the fact that, for any $\lambda \in (0, \infty)$,

$$
\mathbf{1}_{\{y\in\mathbb{R}^n:\, \{\sum_{j\in\mathbb{N}}[\mathcal{M}(f_{j,k,3})(y)]^r\}^{\frac{1}{r}} > \lambda/3\}} \leq \frac{3}{\lambda} \left\{ \sum_{j\in\mathbb{N}} \left[\mathcal{M}(f_{j,k,3})\right]^r \right\}^{\frac{1}{r}}
$$

and an argument similar to that used in the estimations of both (6.57) and (6.58), it follows that, for any $\lambda \in (0, \infty)$,

$$
\mathrm{II}_{\lambda,3} \lesssim \lambda^{-1} \left\| \left(\sum_{j\in\mathbb{N}} |f_j|^r\right)^{\frac{1}{r}} (\cdot + \xi) \right\|_{\dot{\mathcal{K}}_{\omega,0}^{1,q}(\mathbb{R}^n)} . \tag{6.63}
$$

This finishes the estimate of $\mathrm{II}_{\lambda,3}$.

Combining (6.50), (6.57), (6.58), (6.60), and (6.63), we further conclude that, for any $\lambda \in (0, \infty)$,

$$
\lambda \left\| \left\{ \sum_{j\in\mathbb{N}} \left[\mathcal{M}(f_j)\right]^r \right\}^{\frac{1}{r}} (\cdot + \xi) \right\|_{\dot{\mathcal{K}}_{\omega,0}^{1,q}(\mathbb{R}^n)} \lesssim \left\| \left(\sum_{j\in\mathbb{N}} |f_j|^r\right)^{\frac{1}{r}} (\cdot + \xi) \right\|_{\dot{\mathcal{K}}_{\omega,0}^{1,q}(\mathbb{R}^n)} . \tag{6.64}
$$

Thus, from the arbitrariness of ξ, it follows that

$$
\left\| \left\{ \sum_{j\in\mathbb{N}} \left[\mathcal{M}(f_j)\right]^r \right\}^{\frac{1}{r}} \right\|_{W\dot{\mathcal{K}}_{\omega}^{1,q}(\mathbb{R}^n)} \lesssim \left\| \left(\sum_{j\in\mathbb{N}} |f_j|^r\right)^{\frac{1}{r}} \right\|_{\dot{\mathcal{K}}_{\omega}^{1,q}(\mathbb{R}^n)} .
$$

In particular, letting $\xi := \mathbf{0}$ in (6.64), we have

$$
\left\| \left\{ \sum_{j\in\mathbb{N}} \left[\mathcal{M}(f_j)\right]^r \right\}^{\frac{1}{r}} \right\|_{W\dot{\mathcal{K}}_{\omega,0}^{1,q}(\mathbb{R}^n)} \lesssim \left\| \left(\sum_{j\in\mathbb{N}} |f_j|^r\right)^{\frac{1}{r}} \right\|_{\dot{\mathcal{K}}_{\omega,0}^{1,q}(\mathbb{R}^n)} .
$$

Therefore, both (6.48) and (6.49) hold true in this case, and hence we complete the proof of Proposition 6.6.11. $\qquad\square$

In addition, to show Theorem 6.6.9, we also need the following lemma about the convexification of weak local generalized Herz spaces, which is just a simple corollary of both [278, Remark 2.14] and Lemma 1.3.1; we omit the details.

Lemma 6.6.12 *Let* $p, q, s \in (0, \infty)$ *and* $\omega \in M(\mathbb{R}_+)$. *Then*

$$\left[W\dot{\mathcal{K}}_{\omega,0}^{p,q}(\mathbb{R}^n) \right]^{1/s} = W\dot{\mathcal{K}}_{\omega^s,0}^{p/s,q/s}(\mathbb{R}^n)$$

with the same quasi-norms.

Then we show Theorem 6.6.9.

Proof of Theorem 6.6.9 Let all the symbols be as in the present theorem. Then, combining the assumption $m_0(\omega) \in (-\frac{n}{p}, \infty)$ and Theorems 1.2.42 and 1.4.1, we find that the local generalized Herz space $\dot{\mathcal{K}}_{\omega,0}^{p,q}(\mathbb{R}^n)$ under consideration is a BQBF space having an absolutely continuous quasi-norm. Thus, to complete the proof of the present theorem, we only need to show that all the assumptions of Proposition 6.6.3 hold true for $\dot{\mathcal{K}}_{\omega,0}^{p,q}(\mathbb{R}^n)$.

Indeed, let

$$s \in \left(0, \min \left\{ 1, p, q, \frac{n}{\max\{M_0(\omega), M_\infty(\omega)\} + n/p} \right\} \right)$$

and

$$\theta \in \left(0, \min \left\{ s, \frac{n}{n+d} \right\} \right). \tag{6.65}$$

Then, applying Lemma 4.3.25, we conclude that, for any $\{f_j\}_{j\in\mathbb{N}} \subset L_{\mathrm{loc}}^1(\mathbb{R}^n)$,

$$\left\| \left\{ \sum_{j\in\mathbb{N}} \left[\mathcal{M}^{(\theta)}(f_j) \right]^s \right\}^{\frac{1}{s}} \right\|_{\dot{\mathcal{K}}_{\omega,0}^{p,q}(\mathbb{R}^n)} \lesssim \left\| \left(\sum_{j\in\mathbb{N}} |f_j|^s \right)^{\frac{1}{s}} \right\|_{\dot{\mathcal{K}}_{\omega,0}^{p,q}(\mathbb{R}^n)}.$$

This implies that Assumption 1.2.29 holds true with the above θ and s.

Next, we prove that Assumption 1.2.33 holds true with the above s and some $r_0 \in (1, \infty)$. Indeed, let

$$r_0 \in \left(\max \left\{ 1, p, \frac{n}{\min\{m_0(\omega), m_\infty(\omega)\} + n/p} \right\}, \infty \right).$$

Then, from Lemma 1.8.6 with $r := r_0$, we deduce that $[\dot{\mathcal{K}}_{\omega,0}^{p,q}(\mathbb{R}^n)]^{1/s}$ is a BBF space and, for any $f \in L_{\mathrm{loc}}^1(\mathbb{R}^n)$,

$$\left\| \mathcal{M}^{((r_0/s)')}(f) \right\|_{([\dot{\mathcal{K}}_{\omega,0}^{p,q}(\mathbb{R}^n)]^{1/s})'} \lesssim \|f\|_{([\dot{\mathcal{K}}_{\omega,0}^{p,q}(\mathbb{R}^n)]^{1/s})'},$$

which implies that Assumption 1.2.33 holds true.

Let

$$r_1 \in \left(0, \min \left\{ p, \frac{n}{\max\{M_0(\omega), M_\infty(\omega)\} + n/p} \right\} \right).$$

Then, using Lemma 6.1.6 with $r := r_1$, we find that, for any $f \in L^1_{\text{loc}}(\mathbb{R}^n)$,

$$\|\mathcal{M}(f)\|_{[W\dot{\mathcal{K}}^{p,q}_{\omega,0}(\mathbb{R}^n)]^{1/r_1}} \lesssim \|f\|_{[W\dot{\mathcal{K}}^{p,q}_{\omega,0}(\mathbb{R}^n)]^{1/r_1}},$$

which implies that, for the above r_1, the Hardy–Littlewood maximal operator $\mathcal{M}$ is bounded on $[W\dot{\mathcal{K}}^{p,q}_{\omega,0}(\mathbb{R}^n)]^{1/r_1}$.

Finally, we prove that (6.26) holds true with $X := \dot{\mathcal{K}}^{p,q}_{\omega,0}(\mathbb{R}^n)$. Indeed, from the assumptions

$$\min\{m_0(\omega), m_\infty(\omega)\} \in \left(-\frac{n}{p}, \infty \right)$$

and

$$\max\{M_0(\omega), M_\infty(\omega)\} \in \left(-\infty, n - \frac{n}{p} + d + \delta \right),$$

and Lemma 1.1.6, it follows that

$$\min \left\{ m_0 \left(\omega^{\frac{n}{n+d+\delta}} \right), m_\infty \left(\omega^{\frac{n}{n+d+\delta}} \right) \right\}$$
$$= \frac{n}{n+d+\delta} \min\{m_0(\omega), m_\infty(\omega)\} > -\frac{n}{(n+d+\delta)p/n} \tag{6.66}$$

and

$$\max \left\{ M_0 \left(\omega^{\frac{n}{n+d+\delta}} \right), M_\infty \left(\omega^{\frac{n}{n+d+\delta}} \right) \right\}$$
$$= \frac{n}{n+d+\delta} \max\{M_0(\omega), M_\infty(\omega)\}$$
$$< n - \frac{n}{(n+d+\delta)p/n} = \frac{n}{((n+d+\delta)/p)'}. \tag{6.67}$$

These, together with Lemma 6.6.12 with $s := \frac{n}{n+d+\delta}$, the assumption $p \in [\frac{n}{n+d+\delta}, \infty)$, (6.48) with p, q, and ω therein replaced, respectively, by $\frac{n+d+\delta}{n}p$, $\frac{n+d+\delta}{n}q$, and $\omega^{\frac{n}{n+d+\delta}}$, and Lemma 1.3.1 with $s := \frac{n}{n+d+\delta}$, further imply that, for any $\{f_j\}_{j\in\mathbb{N}} \subset L^1_{\mathrm{loc}}(\mathbb{R}^n)$,

$$\left\|\left\{\sum_{j\in\mathbb{N}}[\mathcal{M}(f_j)]^{\frac{n+d+\delta}{n}}\right\}^{\frac{n}{n+d+\delta}}\right\|_{[W\dot{\mathcal{K}}^{p,q}_{\omega,0}(\mathbb{R}^n)]^{\frac{n+d+\delta}{n}}}$$

$$= \left\|\left\{\sum_{j\in\mathbb{N}}[\mathcal{M}(f_j)]^{\frac{n+d+\delta}{n}}\right\}^{\frac{n}{n+d+\delta}}\right\|_{W\dot{\mathcal{K}}^{\frac{n+d+\delta}{n}p,\frac{n+d+\delta}{n}q}_{\omega^{\frac{n}{n+d+\delta}},0}(\mathbb{R}^n)}$$

$$\lesssim \left\|\left(\sum_{j\in\mathbb{N}}|f_j|^{\frac{n+d+\delta}{n}}\right)^{\frac{n}{n+d+\delta}}\right\|_{\dot{\mathcal{K}}^{\frac{n+d+\delta}{n}p,\frac{n+d+\delta}{n}q}_{\omega^{\frac{n}{n+d+\delta}},0}(\mathbb{R}^n)}$$

$$\sim \left\|\left(\sum_{j\in\mathbb{N}}|f_j|^{\frac{n+d+\delta}{n}}\right)^{\frac{n}{n+d+\delta}}\right\|_{[\dot{\mathcal{K}}^{p,q}_{\omega,0}(\mathbb{R}^n)]^{\frac{n+d+\delta}{n}}}.$$

This then implies that (6.26) holds true with $X := \dot{\mathcal{K}}^{p,q}_{\omega,0}(\mathbb{R}^n)$. Combining this, the facts that $\dot{\mathcal{K}}^{p,q}_{\omega,0}(\mathbb{R}^n)$ satisfies both Assumptions 1.2.29 and 1.2.33 with same s and that $\mathcal{M}$ is bounded on $[W\dot{\mathcal{K}}^{p,q}_{\omega,0}(\mathbb{R}^n)]^{1/r_1}$, we find that the local generalized Herz space $\dot{\mathcal{K}}^{p,q}_{\omega,0}(\mathbb{R}^n)$ under consideration satisfies all the assumptions of Proposition 6.6.3. By this, (6.65), and Proposition 6.6.3 with $X := \dot{\mathcal{K}}^{p,q}_{\omega,0}(\mathbb{R}^n)$, we conclude that T has a unique extension on $H\dot{\mathcal{K}}^{p,q}_{\omega,0}(\mathbb{R}^n)$ and, for any $f \in H\dot{\mathcal{K}}^{p,q}_{\omega,0}(\mathbb{R}^n)$,

$$\|T(f)\|_{W H\dot{\mathcal{K}}^{p,q}_{\omega,0}(\mathbb{R}^n)} \lesssim \|f\|_{H\dot{\mathcal{K}}^{p,q}_{\omega,0}(\mathbb{R}^n)},$$

which completes the proof of Theorem 6.6.9.   □

Via Theorem 6.6.9, Remark 4.0.20(ii), and Remark 6.0.25, we immediately obtain the following boundedness of Calderón–Zygmund operators from $HM^{p,q}_{\omega,0}(\mathbb{R}^n)$ to $WHM^{p,q}_{\omega,0}(\mathbb{R}^n)$; we omit the details.

Corollary 6.6.13 *Let $d \in \mathbb{Z}_+$, $p, q \in [1, \infty)$, K be a d-order standard kernel as in Definition 1.5.7 with some $\delta \in (0, 1)$, T a d-order Calderón–Zygmund operator with kernel K having the vanishing moments up to order d, and $\omega \in M(\mathbb{R}_+)$ with*

$$-\frac{n}{p} < m_0(\omega) \le M_0(\omega) < 0$$

and

$$-\frac{n}{p} < m_\infty(\omega) \leq M_\infty(\omega) < 0.$$

Then T has a unique extension on $HM_{\omega,\mathbf{0}}^{p,q}(\mathbb{R}^n)$ and there exists a positive constant C such that, for any $f \in HM_{\omega,\mathbf{0}}^{p,q}(\mathbb{R}^n)$,

$$\|T(f)\|_{WHM_{\omega,\mathbf{0}}^{p,q}(\mathbb{R}^n)} \leq C\|f\|_{HM_{\omega,\mathbf{0}}^{p,q}(\mathbb{R}^n)}.$$

Finally, we turn to show the boundedness of Caldeón–Zygmund operators from the generalized Herz–Hardy space $H\dot{\mathcal{K}}_\omega^{p,q}(\mathbb{R}^n)$ to the weak generalized Herz–Hardy space $WH\dot{\mathcal{K}}_\omega^{p,q}(\mathbb{R}^n)$ as follows.

Theorem 6.6.14 *Let $d \in \mathbb{Z}_+$, $\delta \in (0,1)$, $p \in [\frac{n}{n+d+\delta}, \infty)$, $q \in (0, \infty)$, K be a d-order standard kernel as in Definition 1.5.7, T a d-order Calderón–Zygmund operator with kernel K having the vanishing moments up to order d, and $\omega \in M(\mathbb{R}_+)$ with*

$$-\frac{n}{p} < m_0(\omega) \leq M_0(\omega) < n - \frac{n}{p} + d + \delta$$

and

$$-\frac{n}{p} < m_\infty(\omega) \leq M_\infty(\omega) < 0.$$

Then T is well defined on $H\dot{\mathcal{K}}_\omega^{p,q}(\mathbb{R}^n)$ and there exists a positive constant C such that, for any $f \in H\dot{\mathcal{K}}_\omega^{p,q}(\mathbb{R}^n)$,

$$\|T(f)\|_{WH\dot{\mathcal{K}}_\omega^{p,q}(\mathbb{R}^n)} \leq C\|f\|_{H\dot{\mathcal{K}}_\omega^{p,q}(\mathbb{R}^n)}.$$

To show this theorem, we need the following auxiliary conclusion about the convexification of weak global generalized Herz spaces, which is just a simple consequence of both [278, Remark 2.14] and Lemma 1.3.2; we omit the details.

Lemma 6.6.15 *Let $p, q, s \in (0, \infty)$ and $\omega \in M(\mathbb{R}_+)$. Then*

$$\left[W\dot{\mathcal{K}}_\omega^{p,q}(\mathbb{R}^n)\right]^{1/s} = W\dot{\mathcal{K}}_{\omega^s}^{p/s,q/s}(\mathbb{R}^n)$$

with the same quasi-norms.

Via the above lemma, we now prove Theorem 6.6.14.

Proof of Theorem 6.6.14 Let all the symbols be as in the present theorem. Then, from the assumptions $m_0(\omega) \in (-\frac{n}{p}, \infty)$ and $M_\infty(\omega) \in (-\infty, 0)$, and Theorem 1.2.44, we infer that the global generalized Herz space $\dot{\mathcal{K}}_\omega^{p,q}(\mathbb{R}^n)$ under consideration is a BQBF space. Therefore, in order to finish the proof of the present theorem, it suffices to prove that the assumptions (i) through (vii) hold true for $\dot{\mathcal{K}}_\omega^{p,q}(\mathbb{R}^n)$. For this purpose, let

$$s_0 \in \left(0, \min\left\{1, p, q, \frac{n}{\max\{M_0(\omega), M_\infty(\omega)\} + n/p}\right\}\right),$$

$s \in (0, s_0)$,

$$\theta \in \left(0, \min\left\{s, \frac{n}{n+d}\right\}\right), \tag{6.68}$$

and $\eta \in (1, \infty)$ satisfy

$$\eta < \min\left\{\frac{n}{n(1 - s/p) - s\min\{m_0(\omega), m_\infty(\omega)\}}, \left(\frac{p}{s}\right)'\right\}.$$

Then, repeating the proof of Theorem 4.8.16, we find that the following five statements hold true:

(i) for any $\{f_j\}_{j\in\mathbb{N}} \subset L^1_{\mathrm{loc}}(\mathbb{R}^n)$,

$$\left\|\left\{\sum_{j\in\mathbb{N}}\left[\mathcal{M}^{(\theta)}(f_j)\right]^s\right\}^{1/s}\right\|_{\dot{\mathcal{K}}_\omega^{p,q}(\mathbb{R}^n)} \lesssim \left\|\left(\sum_{j\in\mathbb{N}}|f_j|^s\right)^{1/s}\right\|_{\dot{\mathcal{K}}_\omega^{p,q}(\mathbb{R}^n)};$$

(ii) for any $f \in \mathscr{M}(\mathbb{R}^n)$,

$$\|f\|_{[\dot{\mathcal{K}}_\omega^{p,q}(\mathbb{R}^n)]^{1/s}} \sim \sup\left\{\|fg\|_{L^1(\mathbb{R}^n)}: \|g\|_{\dot{\mathcal{B}}_{1/\omega^s}^{(p/s)',(q/s)'}(\mathbb{R}^n)} = 1\right\}$$

and

$$\|f\|_{[\dot{\mathcal{K}}_\omega^{p,q}(\mathbb{R}^n)]^{1/s_0}} \sim \sup\left\{\|fg\|_{L^1(\mathbb{R}^n)}: \|g\|_{\dot{\mathcal{B}}_{1/\omega^{s_0}}^{(p/s_0)',(q/s_0)'}(\mathbb{R}^n)} = 1\right\}$$

with the positive equivalence constants independent of f;

(iii) both $\|\cdot\|_{\dot{\mathcal{B}}_{1/\omega^s}^{(p/s)',(q/s)'}(\mathbb{R}^n)}$ and $\|\cdot\|_{\dot{\mathcal{B}}_{1/\omega^{s_0}}^{(p/s_0)',(q/s_0)'}(\mathbb{R}^n)}$ satisfy Definition 1.2.13(ii);

(iv) $\mathbf{1}_{B(\mathbf{0},1)} \in \dot{\mathcal{B}}_{1/\omega^{s_0}}^{(p/s_0)',(q/s_0)'}(\mathbb{R}^n)$;

(v) $\mathcal{M}^{(\eta)}$ is bounded on both $\dot{\mathcal{B}}_{1/\omega^s}^{(p/s)',(q/s)'}(\mathbb{R}^n)$ and $\dot{\mathcal{B}}_{1/\omega^{s_0}}^{(p/s_0)',(q/s_0)'}(\mathbb{R}^n)$.

These imply that the assumptions (i) through (v) of Theorem 6.6.1 hold true for $\dot{\mathcal{K}}_\omega^{p,q}(\mathbb{R}^n)$.

Next, we show that Theorem 6.6.1(vi) holds true. Namely, there exists an $r_1 \in (0, \infty)$ such that $\mathcal{M}$ is bounded on $[W\dot{\mathcal{K}}_\omega^{p,q}(\mathbb{R}^n)]^{1/r_1}$. Indeed, let

$$r_1 \in \left(0, \min\left\{p, \frac{n}{\max\{M_0(\omega), M_\infty(\omega)\} + n/p}\right\}\right).$$

From this and Lemma 6.1.11 with $r := r_1$, it follows that, for any $f \in L^1_{\text{loc}}(\mathbb{R}^n)$,

$$\|\mathcal{M}(f)\|_{[W\dot{\mathcal{K}}_\omega^{p,q}(\mathbb{R}^n)]^{1/r_1}} \lesssim \|f\|_{[W\dot{\mathcal{K}}_\omega^{p,q}(\mathbb{R}^n)]^{1/r_1}},$$

which implies that Theorem 6.6.1(iv) holds true for $\dot{\mathcal{K}}_\omega^{p,q}(\mathbb{R}^n)$ with the above r_1.

Finally, we show that the assumption (vii) of Theorem 6.6.1 holds true. Indeed, applying Lemma 6.6.15 with $s := \frac{n}{n+d+\delta}$, (6.66), (6.67), the assumption $p \in [\frac{n}{n+d+\delta}, \infty)$, (6.49) with p, q, and ω therein replaced, respectively, by $\frac{n+d+\delta}{n}p$, $\frac{n+d+\delta}{n}q$, and $\omega^{\frac{n}{n+d+\delta}}$, and Lemma 1.3.2 with $s := \frac{n}{n+d+\delta}$, we find that, for any $\{f_j\}_{j\in\mathbb{N}} \subset L^1_{\text{loc}}(\mathbb{R}^n)$,

$$\left\|\left\{\sum_{j\in\mathbb{N}}\left[\mathcal{M}(f_j)\right]^{\frac{n+d+\delta}{n}}\right\}^{\frac{n}{n+d+\delta}}\right\|_{[W\dot{\mathcal{K}}_\omega^{p,q}(\mathbb{R}^n)]^{\frac{n+d+\delta}{n}}}$$

$$= \left\|\left\{\sum_{j\in\mathbb{N}}\left[\mathcal{M}(f_j)\right]^{\frac{n+d+\delta}{n}}\right\}^{\frac{n}{n+d+\delta}}\right\|_{W\dot{\mathcal{K}}^{\frac{n+d+\delta}{n}p,\frac{n+d+\delta}{n}q}_{\omega^{\frac{n}{n+d+\delta}}}(\mathbb{R}^n)}$$

$$\lesssim \left\|\left(\sum_{j\in\mathbb{N}}|f_j|^{\frac{n+d+\delta}{n}}\right)^{\frac{n}{n+d+\delta}}\right\|_{\dot{\mathcal{K}}^{\frac{n+d+\delta}{n}p,\frac{n+d+\delta}{n}q}_{\omega^{\frac{n}{n+d+\delta}}}(\mathbb{R}^n)}$$

$$\sim \left\|\left(\sum_{j\in\mathbb{N}}|f_j|^{\frac{n+d+\delta}{n}}\right)^{\frac{n}{n+d+\delta}}\right\|_{[\dot{\mathcal{K}}_\omega^{p,q}(\mathbb{R}^n)]^{\frac{n+d+\delta}{n}}}.$$

This further implies that Theorem 6.6.1(vii) holds true with $X := \dot{\mathcal{K}}_\omega^{p,q}(\mathbb{R}^n)$. Thus, the assumptions (i) through (vii) of Theorem 6.6.1 hold true for the global generalized Herz space $\dot{\mathcal{K}}_\omega^{p,q}(\mathbb{R}^n)$ under consideration. Combining this, the fact that $\dot{\mathcal{K}}_\omega^{p,q}(\mathbb{R}^n)$ is a BQBF space, (6.68), and Theorem 6.6.1 with $X := \dot{\mathcal{K}}_\omega^{p,q}(\mathbb{R}^n)$, we conclude that T is well defined on $H\dot{\mathcal{K}}_\omega^{p,q}(\mathbb{R}^n)$ and, for any $f \in H\dot{\mathcal{K}}_\omega^{p,q}(\mathbb{R}^n)$,

$$\|T(f)\|_{WH\dot{\mathcal{K}}_\omega^{p,q}(\mathbb{R}^n)} \lesssim \|f\|_{H\dot{\mathcal{K}}_\omega^{p,q}(\mathbb{R}^n)}.$$

This then finishes the proof of Theorem 6.6.14. $\square$

Via Theorem 6.6.14, Remark 4.0.20(ii), and Remark 6.0.25, we immediately obtain the following boundedness of Calderón–Zygmund operators from the generalized Hardy–Morrey space $HM_\omega^{p,q}(\mathbb{R}^n)$ to the weak generalized Hardy–Morrey space $WHM_\omega^{p,q}(\mathbb{R}^n)$; we omit the details.

Corollary 6.6.16 *Let* d, p, q, ω, K, *and* T *be as in Corollary 6.6.13. Then* T *is well defined on* $HM_\omega^{p,q}(\mathbb{R}^n)$ *and there exists a positive constant* C *such that, for any* $f \in HM_\omega^{p,q}(\mathbb{R}^n)$,

$$\|T(f)\|_{WHM_\omega^{p,q}(\mathbb{R}^n)} \le C\|f\|_{HM_\omega^{p,q}(\mathbb{R}^n)}.$$

6.7 Real Interpolations

As applications, we investigate the real interpolation theorems about generalized Herz–Hardy spaces in this section. Precisely, we show that the real interpolation spaces between the generalized Herz–Hardy spaces and the Lebesgue space $L^\infty(\mathbb{R}^n)$ are just the weak generalized Herz–Hardy spaces.

Let X_0 and X_1 be two quasi-Banach spaces. Recall that the pair (X_0, X_1) is said to be *compatible* if there exists a Hausdorff space $\mathbb{X}$ such that $X_0 \subset \mathbb{X}$ and $X_1 \subset \mathbb{X}$ (see, for instance, [12, p. 20]). For any compatible pair (X_0, X_1) of quasi-Banach spaces, let

$$X_0 + X_1 := \{a \in \mathbb{X} : \exists a_0 \in X_0,\ \exists a_1 \in X_1 \text{ such that } a = a_0 + a_1\}.$$

Moreover, for any $t \in (0, \infty)$, the *Peetre K-functional* $K(t, f; X_0, X_1)$ (see, for instance, [12, p. 38]) is defined by setting, for any $f \in X_0 + X_1$,

$$K(t, f; X_0, X_1) := \inf\left\{\|f_0\|_{X_0} + t\|f_1\|_{X_1} : f = f_0 + f_1,\ f_i \in X_i,\ i \in \{0, 1\}\right\}.$$

We next recall the definition of the real interpolation space between two quasi-Banach spaces X_0 and X_1 as follows (see, for instance, [12, p. 40]).

Definition 6.7.1 Let X_0 and X_1 be two quasi-Banach spaces satisfy that the pair (X_0, X_1) is compatible, $\theta \in (0, 1)$, and $q \in (0, \infty]$. Then the *real interpolation space* $(X_0, X_1)_{\theta,q}$ between X_0 and X_1 is defined to be the set of all the $f \in X_0 + X_1$ such that

$$\|f\|_{\theta,q} < \infty,$$

where, for any $f \in X_0 + X_1$,

$$
\|f\|_{\theta,q} := \begin{cases} \left\{ \int_0^\infty \left[t^{-\theta} K(t, f; X_0, X_1) \right]^q \frac{dt}{t} \right\}^{\frac{1}{q}}, & \text{when } q \in (0, \infty), \\[2mm] \sup_{t \in (0,\infty)} t^{-\theta} K(t, f; X_0, X_1), & \text{when } q = \infty. \end{cases}
$$

Then the following conclusion shows that the real interpolation space between $H\dot{\mathcal{K}}^{p,q}_{\omega,0}(\mathbb{R}^n)$ and $L^\infty(\mathbb{R}^n)$ is just $WH\dot{\mathcal{K}}^{p/(1-\theta),q/(1-\theta)}_{\omega^{1-\theta},0}(\mathbb{R}^n)$.

Theorem 6.7.2 *Let* $p, q \in (0, \infty)$ *and* $\omega \in M(\mathbb{R}_+)$ *satisfy* $m_0(\omega) \in (-\frac{n}{p}, \infty)$ *and* $m_\infty(\omega) \in (-\frac{n}{p}, \infty)$. *Then, for any* $\theta \in (0, 1)$,

$$
\left(H\dot{\mathcal{K}}^{p,q}_{\omega,0}(\mathbb{R}^n), L^\infty(\mathbb{R}^n) \right)_{\theta,\infty} = WH\dot{\mathcal{K}}^{p/(1-\theta),q/(1-\theta)}_{\omega^{1-\theta},0}(\mathbb{R}^n).
$$

To show this real interpolation theorem, we need the following real interpolation result between Hardy spaces associated with ball quasi-Banach function spaces and the Lebesgue space $L^\infty(\mathbb{R}^n)$, which is just [242, Theorem 4.5].

Lemma 6.7.3 *Let* X *be a ball quasi-Banach function space and let* $p_- \in (0, 1)$, $\theta_0 \in (1, \infty)$, *and* $r \in (0, \infty)$ *be such that the following four statements hold true:*

(i) *for any given* $\theta \in (0, p_-)$ *and* $u \in (1, \infty)$, *there exists a positive constant* C *such that, for any* $\{f_j\}_{j\in\mathbb{N}} \subset L^1_{\text{loc}}(\mathbb{R}^n)$,

$$
\left\| \left\{ \sum_{j\in\mathbb{N}} \left[\mathcal{M}(f_j) \right]^u \right\}^{\frac{1}{u}} \right\|_{X^{1/\theta}} \leq C \left\| \left\{ \sum_{j\in\mathbb{N}} |f_j|^u \right\}^{\frac{1}{u}} \right\|_{X^{1/\theta}};
$$

(ii) *X is* θ_0*-concave;*

(iii) *there exists an* $s_0 \in (0, p_-)$, *an* $r_0 \in (s_0, \infty)$, *and a* $C \in (0, \infty)$ *such that* X^{1/s_0} *is a ball Banach function space and, for any* $f \in L^1_{\text{loc}}(\mathbb{R}^n)$,

$$
\left\| \mathcal{M}^{((r_0/s_0)')}(f) \right\|_{(X^{1/s_0})'} \leq C \|f\|_{(X^{1/s_0})'};
$$

(iv) $\mathcal{M}$ *is bounded on* $(WX)^{1/r}$.

Assume that $\theta \in (0, 1)$. *Then*

$$
\left(H_X(\mathbb{R}^n), L^\infty(\mathbb{R}^n) \right)_{\theta,\infty} = WH_{X^{1/(1-\theta)}}(\mathbb{R}^n).
$$

Remark 6.7.4 We should point out that Lemma 6.7.3 has a wide range of applications. Here we present several function spaces to which Lemma 6.7.3 can be applied (see also [242, Section 5]).

(i) Let $p \in (0, \infty)$ and $\upsilon \in A_\infty(\mathbb{R}^n)$. Then, as was mentioned in [242, Subsection 5.4], the weighted Lebesgue space $L_\upsilon^p(\mathbb{R}^n)$ satisfies all the assumptions of Lemma 6.7.3. Therefore, Lemma 6.7.3 with $X := L_\upsilon^p(\mathbb{R}^n)$ holds true. This result coincides with [242, Theorem 5.26].

(ii) Let $\vec{p} := (p_1, \ldots, p_n) \in (0, \infty)^n$. Then, as was pointed out in [242, Subsection 5.2], the mixed-norm Lebesgue space $L^{\vec{p}}(\mathbb{R}^n)$ satisfies all the assumptions of Lemma 6.7.3. This then implies that Lemma 6.7.3 with $X := L^{\vec{p}}(\mathbb{R}^n)$ holds true. This result coincides with [242, Theorem 5.17].

(iii) Let $0 < q \le p < \infty$. Then, as was mentioned in [242, Subsection 5.1], the Morrey space $M_q^p(\mathbb{R}^n)$ satisfies all the assumptions of Lemma 6.7.3. Thus, Lemma 6.7.3 with $X := M_q^p(\mathbb{R}^n)$ holds true. This result coincides with [242, Theorem 5.9].

We now show Theorem 6.7.2.

Proof of Theorem 6.7.2 Let all the symbols be as in the present theorem. Then, by the assumption $m_0(\omega) \in (-\frac{n}{p}, \infty)$ and Theorem 1.2.42, we conclude that the local generalized Herz space $\dot{\mathcal{K}}_{\omega,0}^{p,q}(\mathbb{R}^n)$ under consideration is a BQBF space. Thus, in order to finish the proof of the present theorem, we only need to show that $\dot{\mathcal{K}}_{\omega,0}^{p,q}(\mathbb{R}^n)$ satisfies the assumptions (i) through (iv) of Lemma 6.7.3.

Indeed, let

$$p_- := \min \left\{ 1, p, q, \frac{n}{\max\{M_0(\omega), M_\infty(\omega)\} + n/p} \right\}.$$

Then, for any given $\theta \in (0, p_-)$ and $u \in (1, \infty)$, from Lemma 4.3.10 with $r := \theta$, it follows that, for any $\{f_j\}_{j \in \mathbb{N}} \subset L_{\text{loc}}^1(\mathbb{R}^n)$,

$$\left\| \left\{ \sum_{j \in \mathbb{N}} [\mathcal{M}(f_j)]^u \right\}^{\frac{1}{u}} \right\|_{[\dot{\mathcal{K}}_{\omega,0}^{p,q}(\mathbb{R}^n)]^{1/\theta}} \lesssim \left\| \left(\sum_{j \in \mathbb{N}} |f_j|^u \right)^{\frac{1}{u}} \right\|_{[\dot{\mathcal{K}}_{\omega,0}^{p,q}(\mathbb{R}^n)]^{1/\theta}} . \tag{6.69}$$

This implies that Lemma 6.7.3(i) holds true with the above p_-.

Next, we prove that Lemma 6.7.3(ii) holds true. To this end, let

$$\theta_0 \in (\max\{1, p, q\}, \infty).$$

Then, applying the reverse Minkowski inequality and Lemma 1.3.1 with $s := \theta_0$, we conclude that, for any $\{f_j\}_{j\in\mathbb{N}} \subset \mathscr{M}(\mathbb{R}^n)$,

$$\sum_{j\in\mathbb{N}} \|f_j\|_{[\dot{\mathcal{K}}^{p,q}_{\omega,\mathbf{0}}(\mathbb{R}^n)]^{1/\theta_0}} \le \left\| \sum_{j\in\mathbb{N}} |f_j| \right\|_{[\dot{\mathcal{K}}^{p,q}_{\omega,\mathbf{0}}(\mathbb{R}^n)]^{1/\theta_0}}, \tag{6.70}$$

which then implies that the Herz space $\dot{\mathcal{K}}^{p,q}_{\omega,\mathbf{0}}(\mathbb{R}^n)$ under consideration is θ_0-concave and hence Lemma 6.7.3(ii) holds true.

In addition, let $s_0 \in (0, p_-)$ and

$$r_0 \in \left(\min\left\{ 1, p, \frac{n}{\min\{m_0(\omega), m_\infty(\omega)\} + n/p} \right\}, \infty \right).$$

We now prove that, for this s_0 and r_0, the assumption (iii) of Lemma 6.7.3 holds true. Indeed, by Lemma 1.8.6 with $s := s_0$ and $r := r_0$, we find that $[\dot{\mathcal{K}}^{p,q}_{\omega,\mathbf{0}}(\mathbb{R}^n)]^{1/s_0}$ is a BBF space and, for any $f \in L^1_{\text{loc}}(\mathbb{R}^n)$,

$$\left\| \mathcal{M}^{((r_0/s_0)')}(f) \right\|_{([\dot{\mathcal{K}}^{p,q}_{\omega,\mathbf{0}}(\mathbb{R}^n)]^{1/s_0})'} \lesssim \|f\|_{([\dot{\mathcal{K}}^{p,q}_{\omega,\mathbf{0}}(\mathbb{R}^n)]^{1/s_0})'}. \tag{6.71}$$

This finishes the proof of Lemma 6.7.3(iii).

Finally, we show that Lemma 6.7.3(iv) holds true. Namely, there exists an $r \in (0, \infty)$ such that $\mathcal{M}$ is bounded on $[W\dot{\mathcal{K}}^{p,q}_{\omega,\mathbf{0}}(\mathbb{R}^n)]^{1/r}$. Indeed, let

$$r \in \left(0, \min\left\{ p, \frac{n}{\max\{M_0(\omega), M_\infty(\omega)\} + n/p} \right\} \right).$$

Using this and Lemma 6.1.6, we conclude that, for any $f \in L^1_{\text{loc}}(\mathbb{R}^n)$,

$$\|\mathcal{M}(f)\|_{[W\dot{\mathcal{K}}^{p,q}_{\omega,\mathbf{0}}(\mathbb{R}^n)]^{1/r}} \lesssim \|f\|_{[W\dot{\mathcal{K}}^{p,q}_{\omega,\mathbf{0}}(\mathbb{R}^n)]^{1/r}},$$

which implies that Lemma 6.7.3(iv) holds true for $\dot{\mathcal{K}}^{p,q}_{\omega,\mathbf{0}}(\mathbb{R}^n)$. Combining this, (6.69), (6.70), and (6.71), we further find that the local generalized Herz space $\dot{\mathcal{K}}^{p,q}_{\omega,\mathbf{0}}(\mathbb{R}^n)$ under consideration satisfies the assumptions (i) through (iv) of Lemma 6.7.3. This further implies that, for any $\theta \in (0, 1)$,

$$\left(H\dot{\mathcal{K}}^{p,q}_{\omega,\mathbf{0}}(\mathbb{R}^n), L^\infty(\mathbb{R}^n) \right)_{\theta,\infty} = WH\dot{\mathcal{K}}^{p/(1-\theta),q/(1-\theta)}_{\omega^{1-\theta},\mathbf{0}}(\mathbb{R}^n),$$

which completes the proof of Theorem 6.7.2. $\qquad\square$

Via Theorem 6.7.2, Remark 4.0.20(ii), and Remark 6.0.25, we obtain the following real interpolation theorem which shows that the real interpolation space between $HM_{\omega,0}^{p,q}(\mathbb{R}^n)$ and $L^\infty(\mathbb{R}^n)$ is just the weak space $WHM_{\omega^{1-\theta},0}^{p/(1-\theta),q/(1-\theta)}(\mathbb{R}^n)$; we omit the details.

Corollary 6.7.5 *Let* $p, q \in [1, \infty)$ *and* $\omega \in M(\mathbb{R}_+)$ *with*

$$-\frac{n}{p} < m_0(\omega) \leq M_0(\omega) < 0$$

and

$$-\frac{n}{p} < m_\infty(\omega) \leq M_\infty(\omega) < 0.$$

Then, for any $\theta \in (0, 1)$,

$$\left(HM_{\omega,0}^{p,q}(\mathbb{R}^n), \ L^\infty(\mathbb{R}^n) \right)_{\theta,\infty} = WHM_{\omega^{1-\theta},0}^{p/(1-\theta),q/(1-\theta)}(\mathbb{R}^n).$$

The remainder of this section is devoted to establishing a real interpolation theorem between the generalized Herz–Hardy space $H\dot{\mathcal{K}}_\omega^{p,q}(\mathbb{R}^n)$ and the Lebesgue space $L^\infty(\mathbb{R}^n)$. Indeed, we have the following conclusion which shows that the real interpolation space between $H\dot{\mathcal{K}}_\omega^{p,q}(\mathbb{R}^n)$ and $L^\infty(\mathbb{R}^n)$ is just the weak generalized Herz–Hardy space $WH\dot{\mathcal{K}}_{\omega^{1-\theta}}^{p/(1-\theta),q/(1-\theta)}(\mathbb{R}^n)$.

Theorem 6.7.6 *Let* $p, q \in (0, \infty)$ *and* $\omega \in M(\mathbb{R}_+)$ *satisfy* $m_0(\omega) \in (-\frac{n}{p}, \infty)$ *and*

$$-\frac{n}{p} < m_\infty(\omega) \leq M_\infty(\omega) < 0.$$

Then, for any $\theta \in (0, 1)$,

$$\left(H\dot{\mathcal{K}}_\omega^{p,q}(\mathbb{R}^n), \ L^\infty(\mathbb{R}^n) \right)_{\theta,\infty} = WH\dot{\mathcal{K}}_{\omega^{1-\theta}}^{p/(1-\theta),q/(1-\theta)}(\mathbb{R}^n).$$

To show this theorem, we first establish the following real interpolation theorem about the Hardy spaces associated with ball quasi-Banach function spaces without recourse to associate spaces, which improves the results obtained in [242] (see also Lemma 6.7.3 above).

Proposition 6.7.7 *Let X be a ball quasi-Banach function space and $Y \subset \mathcal{M}(\mathbb{R}^n)$ a linear space equipped with a quasi-seminorm $\|\cdot\|_Y$, and let $p_- \in (0, 1)$, $\theta_0 \in (1, \infty)$, and $r \in (0, \infty)$ be such that the following eight statements hold true:*

(i) *for any given $\theta \in (0, p_-)$ and $u \in (1, \infty)$, there exists a positive constant C such that, for any $\{f_j\}_{j \in \mathbb{N}} \subset L^1_{\mathrm{loc}}(\mathbb{R}^n)$,*

$$\left\| \left\{ \sum_{j \in \mathbb{N}} [\mathcal{M}(f_j)]^u \right\}^{\frac{1}{u}} \right\|_{X^{1/\theta}} \leq C \left\| \left\{ \sum_{j \in \mathbb{N}} |f_j|^u \right\}^{\frac{1}{u}} \right\|_{X^{1/\theta}} ;$$

(ii) *X is θ_0-concave;*
(iii) *$\mathcal{M}$ is bounded on $(WX)^{1/r}$;*
(iv) *for any $s \in (0, p_-)$, $X^{1/s}$ is a ball Banach function space;*
(v) *$\|\cdot\|_Y$ satisfies Definition 1.2.13(ii);*
(vi) *$\mathbf{1}_{B(0,1)} \in Y$;*
(vii) *there exists an $s_0 \in (0, p_-)$ such that, for any $f \in \mathcal{M}(\mathbb{R}^n)$,*

$$\|f\|_{X^{1/s_0}} \sim \sup \left\{ \|fg\|_{L^1(\mathbb{R}^n)} : \|g\|_Y = 1 \right\},$$

where the positive equivalence constants are independent of f;
(viii) *there exists an $r_0 \in (s_0, \infty)$ and a $C \in (0, \infty)$ such that, for any $f \in L^1_{\mathrm{loc}}(\mathbb{R}^n)$,*

$$\left\| \mathcal{M}^{((r_0/s_0)')}(f) \right\|_Y \leq C \|f\|_Y.$$

Then, for any $\theta \in (0, 1)$,

$$\left(H_X(\mathbb{R}^n), L^\infty(\mathbb{R}^n) \right)_{\theta, \infty} = W H_{X^{1/(1-\theta)}}(\mathbb{R}^n).$$

Remark 6.7.8 We should point out that Proposition 6.7.7 is an improved version of the known real interpolation theorem between the Hardy space $H_X(\mathbb{R}^n)$ associated with the ball quasi-Banach function space X and the Lebesgue space $L^\infty(\mathbb{R}^n)$ established in [242, Theorem 4.5]. Indeed, if $Y \equiv (X^{1/s_0})'$ in Proposition 6.7.7, then this proposition goes back to [242, Theorem 4.5].

In order to prove this proposition, we need the following Calderón reproducing formula which was obtained by Calderón [20, p. 219, Section 3] (see also [24, Lemma 4.1] or [267, Lemma 4.6]).

Lemma 6.7.9 *Let $\psi \in \mathcal{S}(\mathbb{R}^n)$ satisfy that $\mathrm{supp}\,(\psi) \subset B(0, 1)$ and $\int_{\mathbb{R}^n} \psi(x)\,dx = 0$. Then there exists a $\phi \in \mathcal{S}(\mathbb{R}^n)$ and $0 < a < b < \infty$ such that*

$$\mathrm{supp}\,(\widehat{\phi}) \subset B(0, b) \setminus B(0, a)$$

and, for any $x \in \mathbb{R}^n \setminus \{0\}$,

$$\int_0^\infty \widehat{\psi}(tx)\widehat{\phi}(tx)\, \frac{dt}{t} = 1.$$

Moreover, let η *be defined by setting* $\widehat{\eta}(0) := 1$ *and, for any* $x \in \mathbb{R}^n \setminus \{0\}$,

$$\widehat{\eta}(x) := \int_1^\infty \widehat{\psi}(tx)\widehat{\phi}(tx)\, \frac{dt}{t}.$$

Then

(i) η *is well defined;*
(ii) $\eta \in C_c^\infty(\mathbb{R}^n)$;
(iii) *there exists a* $\delta \in (0, \infty)$ *such that, for any* $x \in B(0, \delta)$, $\widehat{\eta}(x) = 1$.

We also require the following technical lemma which is just [85, Theorem 1.64].

Lemma 6.7.10 *Let* $\phi \in \mathcal{S}(\mathbb{R}^n)$ *satisfy* $\int_{\mathbb{R}^n} \phi(x)\, dx = 0$ *and, for any* $x \in \mathbb{R}^n \setminus \{0\}$,

$$\int_{\mathbb{R}^n} \widehat{\phi}(tx)\, \frac{dt}{t} = 1.$$

Assume that $f \in \mathcal{S}'(\mathbb{R}^n)$ *vanishes weakly at infinity. Then*

$$f = \int_0^\infty f * \phi_t\, \frac{dt}{t}$$

in $\mathcal{S}'(\mathbb{R}^n)$, *namely,*

$$f = \lim_{\varepsilon \to 0^+,\, A \to \infty} \int_\varepsilon^A f * \phi_t\, \frac{dt}{t}$$

in $\mathcal{S}'(\mathbb{R}^n)$.

Via both Lemmas 6.7.9 and 6.7.10, we next prove Proposition 6.7.7.

Proof of Proposition 6.7.7 Let all the symbols be as in the present proposition and $\theta \in (0, 1)$. Then, by the proof of [242, Theorem 4.5], we find that

$$\left(H_X(\mathbb{R}^n),\, L^\infty(\mathbb{R}^n)\right)_{\theta,\infty} \subset WH_{X^{1/(1-\theta)}}(\mathbb{R}^n). \tag{6.72}$$

Conversely, we next show that

$$WH_{X^{1/(1-\theta)}}(\mathbb{R}^n) \subset \left(H_X(\mathbb{R}^n),\, L^\infty(\mathbb{R}^n)\right)_{\theta,\infty}. \tag{6.73}$$

To this end, let $f \in WH_{\chi^{1/(1-\theta)}}(\mathbb{R}^n)$. Then, from the assumptions (v) through (viii) of the present proposition and Lemma 6.3.13 with X, θ, and s therein replaced, respectively, by $X^{1/(1-\theta)}$, $(r_0/s_0)'$, and $s_0/(1-\theta)$, it follows that f vanishes weakly at infinity. In addition, let $d \geq \lfloor n(1/\min\{\frac{p_-}{\theta_0}, r_0\} - 1)\rfloor$ be a fixed nonnegative integer and $\psi \in \mathcal{S}(\mathbb{R}^n)$ satisfy that $\mathrm{supp}\,(\widehat{\psi}) \subset B(\mathbf{0}, 1)$ and, for any $\gamma \in \mathbb{Z}_+^n$ with $|\gamma| \leq d$,

$$\int_{\mathbb{R}^n} \psi(x) x^\gamma \, dx = 0.$$

Then, using Lemma 6.7.9, we find that there exists a $\phi \in \mathcal{S}(\mathbb{R}^n)$ and $0 < a < b < \infty$ such that

$$\mathrm{supp}\,(\widehat{\phi}) \subset B(\mathbf{0}, b) \setminus B(\mathbf{0}, a)$$

and, for any $x \in \mathbb{R}^n \setminus \{\mathbf{0}\}$,

$$\int_0^\infty \widehat{\psi}(tx)\widehat{\phi}(tx) \, \frac{dt}{t} = 1.$$

These, combined with Lemma 6.7.10 with ϕ therein replaced by $\psi * \phi$ and the fact that f vanishes weakly at infinity, further imply that

$$f = \int_0^\infty f * \psi_t * \phi_t \, \frac{dt}{t} \tag{6.74}$$

in $\mathcal{S}'(\mathbb{R}^n)$.

Let $x_0 := (2, \ldots, 2) \in \mathbb{R}^n$ and η be the same as in Lemma 6.7.9, namely, $\widehat{\eta}(\mathbf{0}) := 1$ and, for any $x \in \mathbb{R}^n \setminus \{\mathbf{0}\}$,

$$\widehat{\eta}(x) := \int_1^\infty \widehat{\psi}(tx)\widehat{\phi}(tx) \, \frac{dt}{t}.$$

By (i), (ii), and (iii) of Lemma 6.7.9, we find that $\eta \in \mathcal{S}(\mathbb{R}^n)$. For any $x \in \mathbb{R}^n$ and $t \in (0, \infty)$, let

$$\widetilde{\phi}(x) := \phi(x - x_0), \quad \widetilde{\psi}(x) := \psi(x + x_0),$$

$$F(x, t) := f * \widetilde{\phi}_t(x), \quad G(x, t) := f * \eta_t(x),$$

and

$$M_\nabla(f)(x) := \sup_{(y,t) \in \Gamma_{3(|x_0|+1)}(x)} \left[|F(y, t)| + |G(y, t)|\right],$$

where, for any $x \in \mathbb{R}^n$, $\Gamma_{3(|x_0|+1)}(x)$ is defined as in (4.5) with $a := 3(|x_0| + 1)$. Then, applying (6.74), we conclude that

$$f(\cdot) = \int_0^\infty f * \tilde{\phi}_t * \tilde{\psi}_t(\cdot)\, \frac{dt}{t} = \int_0^\infty \int_{\mathbb{R}^n} F(y, t)\tilde{\psi}_t(\cdot - y)\, \frac{dy\, dt}{t}$$

in $\mathcal{S}'(\mathbb{R}^n)$. From this and the proof of [242, Lemma 4.4], we deduce that, for any $\alpha \in (0, \infty)$, there exists a $g_\alpha \in L^\infty(\mathbb{R}^n)$ and a $b_\alpha \in \mathcal{S}'(\mathbb{R}^n)$ such that $f = g_\alpha + b_\alpha$ in $\mathcal{S}'(\mathbb{R}^n)$, $\|g_\alpha\|_{L^\infty(\mathbb{R}^n)} \lesssim \alpha$, and

$$\|b_\alpha\|_{H_X(\mathbb{R}^n)} \lesssim \left\| M_\nabla(f)\mathbf{1}_{\{x\in\mathbb{R}^n:\, M_\nabla(f)(x)>\alpha\}} \right\|_X,$$

where the implicit positive constants are independent of both f and α. This, together with the proof of [242, Theorem 4.5], further implies that, for any $t \in (0, \infty)$,

$$t^{-\theta} K\left(t, f; H_X(\mathbb{R}^n), L^\infty(\mathbb{R}^n)\right)$$
$$\lesssim \|M_\nabla(f)\|_{X^{1/(1-\theta)}} \lesssim \|M_\nabla(f)\|_{WX^{1/(1-\theta)}} \sim \|f\|_{WH_{X^{1/(1-\theta)}}(\mathbb{R}^n)}.$$

Using this and Definition 6.7.1, we find that

$$\|f\|_{\theta,\infty} \lesssim \|f\|_{WH_{X^{1/(1-\theta)}}(\mathbb{R}^n)} < \infty.$$

This further implies that $f \in (H_X(\mathbb{R}^n), L^\infty(\mathbb{R}^n))_{\theta,\infty}$ and then finishes the proof of (6.73). Combining this and (6.72), we conclude that

$$\left(H_X(\mathbb{R}^n), L^\infty(\mathbb{R}^n)\right)_{\theta,\infty} = WH_{X^{1/(1-\theta)}}(\mathbb{R}^n)$$

and hence complete the proof of Proposition 6.7.7. □

Via the improved real interpolation result obtained above, we now prove Theorem 6.7.6.

Theorem 6.7.6 Let all the symbols be as in the present theorem. Then, by the assumptions $m_0(\omega) \in (-\frac{n}{p}, \infty)$ and $M_\infty(\omega) \in (-\infty, 0)$, and Theorem 1.2.44, we find that the global generalized Herz space $\dot{\mathcal{K}}_\omega^{p,q}(\mathbb{R}^n)$ under consideration is a BBF space. Therefore, in order to finish the proof of the present theorem, we only need to show that the assumptions (i) through (viii) of Proposition 6.7.7 hold true for $\dot{\mathcal{K}}_\omega^{p,q}(\mathbb{R}^n)$.

Indeed, let

$$p_- := \left\{ 1, p, q, \frac{n}{\max\{M_0(\omega), M_\infty(\omega)\} + n/p} \right\}.$$

Then, for any given $\theta \in (0, p_-)$ and $u \in (1, \infty)$, from Lemma 4.3.10 with $r := \theta$, it follows that, for any $\{f_j\}_{j \in \mathbb{N}} \subset L^1_{\mathrm{loc}}(\mathbb{R}^n)$,

$$\left\| \left\{ \sum_{j \in \mathbb{N}} [\mathcal{M}(f_j)]^u \right\}^{\frac{1}{u}} \right\|_{[\dot{\mathcal{K}}^{p,q}_\omega(\mathbb{R}^n)]^{1/\theta}} \lesssim \left\| \left(\sum_{j \in \mathbb{N}} |f_j|^u \right)^{\frac{1}{u}} \right\|_{[\dot{\mathcal{K}}^{p,q}_\omega(\mathbb{R}^n)]^{1/\theta}}.$$

This implies that Proposition 6.7.7(i) holds true with the above p_-.

Next, we prove that Proposition 6.7.7(ii) holds true with some $\theta_0 \in (1, \infty)$. Indeed, let $\theta_0 \in (\max\{1, p, q\}, \infty)$. Combining this, the reverse Minkowski inequality, and Lemma 1.3.2 with $s := \theta_0$, we conclude that, for any $\{f_j\}_{j \in \mathbb{N}} \subset \mathscr{M}(\mathbb{R}^n)$,

$$\sum_{j \in \mathbb{N}} \|f_j\|_{[\dot{\mathcal{K}}^{p,q}_\omega(\mathbb{R}^n)]^{1/\theta_0}} \leq \left\| \sum_{j \in \mathbb{N}} |f_j| \right\|_{[\dot{\mathcal{K}}^{p,q}_\omega(\mathbb{R}^n)]^{1/\theta_0}},$$

which further implies that the Herz space $\dot{\mathcal{K}}^{p,q}_\omega(\mathbb{R}^n)$ under consideration is θ_0-concave and hence Proposition 6.7.7(ii) holds true.

Let

$$r \in \left(0, \min\left\{ p, \frac{n}{\max\{M_0(\omega), M_\infty(\omega)\} + n/p} \right\} \right).$$

Then, from Lemma 6.1.11, it follows that, for any $f \in L^1_{\mathrm{loc}}(\mathbb{R}^n)$,

$$\|\mathcal{M}(f)\|_{[W\dot{\mathcal{K}}^{p,q}_\omega(\mathbb{R}^n)]^{1/r}} \lesssim \|f\|_{[W\dot{\mathcal{K}}^{p,q}_\omega(\mathbb{R}^n)]^{1/r}}.$$

This implies that the Hardy–Littlewood maximal operator $\mathcal{M}$ is bounded on the weak Herz space $[W\dot{\mathcal{K}}^{p,q}_\omega(\mathbb{R}^n)]^{1/r}$ and hence further implies that, for the above r, Proposition 6.7.7(iii) holds true.

Next, we prove that, for any $s \in (0, p_-)$, $[\dot{\mathcal{K}}^{p,q}_\omega(\mathbb{R}^n)]^{1/s}$ is a BBF space. Indeed, for any $s \in (0, p_-)$, by the assumptions $m_0(\omega) \in (-\frac{n}{p}, \infty)$ and $M_\infty(\omega) \in (-\infty, 0)$, and Lemma 1.1.6, we find that

$$m_0\left(\omega^s\right) = s m_0(\omega) > -\frac{n}{p/s}$$

and

$$M_\infty\left(\omega^s\right) = s M_\infty(\omega) < 0,$$

which, combined with the assumptions $p/s, q/s \in (1, \infty)$, Theorem 1.2.48 with p, q, and ω therein replaced, respectively, by p/s, q/s, and ω^s, and Lemma 1.3.2, further imply that $[\dot{\mathcal{K}}_\omega^{p,q}(\mathbb{R}^n)]^{1/s}$ is a BBF space. This finishes the proof that $\dot{\mathcal{K}}_\omega^{p,q}(\mathbb{R}^n)$ satisfies Proposition 6.7.7(iv).

Finally, we show that there exists a linear space $Y \subset \mathscr{M}(\mathbb{R}^n)$, an $s_0 \in (0, p_-)$, and an $r_0 \in (s_0, \infty)$ such that (v) through (viii) of Proposition 6.7.7 hold true. To this end, let $s_0 \in (0, p_-)$ and

$$r_0 \in \left(\max\left\{ 1, p, \frac{n}{\min\{m_0(\omega), m_\infty(\omega)\} + n/p} \right\}, \infty \right).$$

Then, repeating an argument similar to that used in the proof of Theorem 4.8.16 with η therein replaced by $(r_0/s_0)'$, we conclude that the following four statements hold true:

(i) $\| \cdot \|_{\dot{\mathcal{B}}_{1/\omega^{s_0}}^{(p/s_0)',(q/s_0)'}(\mathbb{R}^n)}$ satisfies Definition 1.2.13(ii);

(ii) $\mathbf{1}_{B(0,1)} \in \dot{\mathcal{B}}_{1/\omega^{s_0}}^{(p/s_0)',(q/s_0)'}(\mathbb{R}^n)$;

(iii) for any $f \in \mathscr{M}(\mathbb{R}^n)$,

$$\|f\|_{[\dot{\mathcal{K}}_\omega^{p,q}(\mathbb{R}^n)]^{1/s_0}} \sim \sup \left\{ \|fg\|_{L^1(\mathbb{R}^n)} : \|g\|_{\dot{\mathcal{B}}_{1/\omega^{s_0}}^{(p/s_0)',(q/s_0)'}(\mathbb{R}^n)} = 1 \right\}$$

with the positive equivalence constants independent of f;

(iv) for any $f \in L^1_{\mathrm{loc}}(\mathbb{R}^n)$,

$$\left\| \mathcal{M}^{((r_0/s_0)')}(f) \right\|_{\dot{\mathcal{B}}_{1/\omega^{s_0}}^{(p/s_0)',(q/s_0)'}(\mathbb{R}^n)} \lesssim \|f\|_{\dot{\mathcal{B}}_{1/\omega^{s_0}}^{(p/s_0)',(q/s_0)'}(\mathbb{R}^n)}.$$

These imply that (v) through (viii) of Proposition 6.7.7 hold true with the above s_0 and r_0, and $Y := \dot{\mathcal{B}}_{1/\omega^{s_0}}^{(p/s_0)',(q/s_0)'}(\mathbb{R}^n)$. Therefore, the assumptions (i) through (viii) of Proposition 6.7.7 hold true for the global generalized Herz space $\dot{\mathcal{K}}_\omega^{p,q}(\mathbb{R}^n)$ under consideration. This then implies that, for any $\theta \in (0, 1)$,

$$\left(H\dot{\mathcal{K}}_\omega^{p,q}(\mathbb{R}^n),\ L^\infty(\mathbb{R}^n) \right)_{\theta,\infty} = WH\dot{\mathcal{K}}_{\omega^{1-\theta}}^{p/(1-\theta),q/(1-\theta)}(\mathbb{R}^n),$$

which completes the proof of Theorem 6.7.6.   □

As an application, by Theorem 6.7.6, Remark 4.0.20(ii), and Remark 6.0.25, we conclude the following real interpolation theorem which shows that the real interpolation space between $HM_\omega^{p,q}(\mathbb{R}^n)$ and $L^\infty(\mathbb{R}^n)$ is just the weak generalized Hardy–Morrey space $WHM_{\omega^{1-\theta}}^{p/(1-\theta),q/(1-\theta)}(\mathbb{R}^n)$; we omit the details.

Corollary 6.7.11 Let p, q, and ω be as in Corollary 6.7.5. Then, for any $\theta \in (0, 1)$,

$$\left(HM_\omega^{p,q}(\mathbb{R}^n),\ L^\infty(\mathbb{R}^n) \right)_{\theta,\infty} = WHM_{\omega^{1-\theta}}^{p/(1-\theta),q/(1-\theta)}(\mathbb{R}^n).$$

Chapter 7
Inhomogeneous Generalized Herz Spaces and Inhomogeneous Block Spaces

The targets of this chapter are threefold. The first one is to introduce the inhomogeneous counterparts of generalized Herz spaces as in Definition 1.2.1 and then find both dual spaces and associate spaces of these inhomogeneous local generalized Herz spaces. The second one is to introduce inhomogeneous block spaces and prove the duality between inhomogeneous global generalized Herz spaces and these inhomogeneous block spaces. Moreover, we also establish the boundedness of some important operators on these block spaces. The last one is, as applications, to investigate the boundedness and the compactness characterizations of commutators on inhomogeneous generalized Herz spaces.

7.1 Inhomogeneous Generalized Herz Spaces

In this section, we first introduce the inhomogeneous counterparts of local and global generalized Herz spaces and investigate their basic properties. Indeed, we obtain the relation between these Herz spaces and the corresponding homogeneous generalized Herz spaces. Then, under some reasonable and sharp assumptions, we show that inhomogeneous generalized Herz spaces are ball (quasi-)Banach function spaces. Furthermore, we show the convexity and the absolutely continuity of quasi-norms of these Herz spaces as well as establish a boundedness criterion of sublinear operators and the Fefferman–Stein vector-valued inequalities on these Herz spaces. Finally, we find both the dual and the associate space of the inhomogeneous local generalized Herz space $\mathcal{K}^{p,q}_{\omega,\mathbf{0}}(\mathbb{R}^n)$, and establish the extrapolation theorems of both the inhomogeneous local and the inhomogeneous global generalized Herz spaces.

We begin this section with the following definitions of inhomogeneous generalized Herz spaces.

© The Author(s), under exclusive license to Springer Nature Singapore Pte Ltd. 2022
Y. Li et al., *Real-Variable Theory of Hardy Spaces Associated with Generalized Herz Spaces of Rafeiro and Samko*, Lecture Notes in Mathematics 2320,
https://doi.org/10.1007/978-981-19-6788-7_7

Definition 7.1.1 Let $p, q \in (0, \infty]$ and $\omega \in M(\mathbb{R}_+)$.

(i) The *inhomogeneous local generalized Herz space* $\mathcal{K}^{p,q}_{\omega,\mathbf{0}}(\mathbb{R}^n)$ is defined to be the set of all the $f \in L^p_{\mathrm{loc}}(\mathbb{R}^n)$ such that

$$\|f\|_{\mathcal{K}^{p,q}_{\omega,\mathbf{0}}(\mathbb{R}^n)} := \left\{ \|f\mathbf{1}_{B(0,1)}\|^q_{L^p(\mathbb{R}^n)} + \sum_{k\in\mathbb{N}} \left[\omega(2^k)\right]^q \|f\mathbf{1}_{B(0,2^k)\setminus B(0,2^{k-1})}\|^q_{L^p(\mathbb{R}^n)} \right\}^{\frac{1}{q}}$$

is finite.

(ii) The *inhomogeneous global generalized Herz space* $\mathcal{K}^{p,q}_{\omega}(\mathbb{R}^n)$ is defined to be the set of all the $f \in L^p_{\mathrm{loc}}(\mathbb{R}^n)$ such that

$$\|f\|_{\mathcal{K}^{p,q}_{\omega}(\mathbb{R}^n)} := \sup_{\xi\in\mathbb{R}^n} \left\{ \|f\mathbf{1}_{B(\xi,1)}\|^q_{L^p(\mathbb{R}^n)} \right.$$
$$\left. + \sum_{k\in\mathbb{N}} \left[\omega(2^k)\right]^q \|f\mathbf{1}_{B(\xi,2^k)\setminus B(\xi,2^{k-1})}\|^q_{L^p(\mathbb{R}^n)} \right\}^{\frac{1}{q}}$$

is finite.

Remark 7.1.2

(i) Obviously, using Definition 7.1.1, we conclude that, for any $f \in \mathcal{M}(\mathbb{R}^n)$,

$$\|f\|_{\mathcal{K}^{p,q}_{\omega}(\mathbb{R}^n)} = \sup_{\xi\in\mathbb{R}^n} \|f(\cdot + \xi)\|_{\mathcal{K}^{p,q}_{\omega,\mathbf{0}}(\mathbb{R}^n)}.$$

(ii) Observe that the inhomogeneous local generalized Herz space $\mathcal{K}^{p,q}_{\omega,\mathbf{0}}(\mathbb{R}^n)$ is always nontrivial. Indeed, by Definition 7.1.1(i), we easily know that $\mathbf{1}_{B(0,1)} \in \mathcal{K}^{p,q}_{\omega,\mathbf{0}}(\mathbb{R}^n)$. However, as is showed in Theorem 7.1.5 later, when $\omega \in M(\mathbb{R}_+)$ satisfying that $m_\infty(\omega) \in (0, \infty)$, then

$$\mathcal{K}^{p,q}_{\omega}(\mathbb{R}^n) = \left\{ f \in \mathcal{M}(\mathbb{R}^n) : \ f = 0 \text{ almost everywhere in } \mathbb{R}^n \right\},$$

and hence the inhomogeneous global generalized Herz space $\mathcal{K}^{p,q}_{\omega}(\mathbb{R}^n)$ is trivial in this case.

(iii) Let $p = q \in (0, \infty]$ and $\omega(t) := 1$ for any $t \in (0, \infty)$. Then, in this case,

$$\mathcal{K}^{p,q}_{\omega,\mathbf{0}}(\mathbb{R}^n) = \mathcal{K}^{p,q}_{\omega}(\mathbb{R}^n) = L^p(\mathbb{R}^n)$$

with the same quasi-norms.

(iv) In Definition 7.1.1(i), for any given $\alpha \in \mathbb{R}$ and for any $t \in (0, \infty)$, let $\omega(t) := t^\alpha$. Then, in this case, the inhomogeneous local generalized Herz space $\mathcal{K}^{p,q}_{\omega,\mathbf{0}}(\mathbb{R}^n)$ coincides with the classical *inhomogeneous Herz space* $K^{\alpha,q}_p(\mathbb{R}^n)$, which was originally introduced in [163, Definition 1.1(b)] (see also

[175, Chapter 1]), with the same quasi-norms. In particular, when $p \in (1, \infty)$, $q = 1$, and $\alpha = n(1 - \frac{1}{p})$, the inhomogeneous local generalized Herz space $\mathcal{K}_{\omega,\mathbf{0}}^{p,q}(\mathbb{R}^n)$ is just the *Beurling algebra* $A^p(\mathbb{R}^n)$ which was originally introduced in [13] (see also [86, Definition 1.1]).

(v) We should point out that, even when $\omega(t) := t^{\alpha}$ for any $t \in (0, \infty)$ and for any given $\alpha \in \mathbb{R}$, the inhomogeneous global generalized Herz space $\mathcal{K}_{\omega}^{p,q}(\mathbb{R}^n)$ is also new.

The following conclusion gives the relation between the local generalized Herz space and its inhomogeneous counterpart.

Theorem 7.1.3 *Let $p, q \in (0, \infty]$ and $\omega \in M(\mathbb{R}_+)$ satisfy $m_0(\omega) \in (0, \infty)$ and $m_\infty(\omega) \in (0, \infty)$. Then*

$$\mathcal{K}_{\omega,\mathbf{0}}^{p,q}(\mathbb{R}^n) = \dot{\mathcal{K}}_{\omega,\mathbf{0}}^{p,q}(\mathbb{R}^n) \cap L^p(\mathbb{R}^n).$$

Moreover, there exist two positive constants C_1 and C_2 such that, for any $f \in \mathcal{K}_{\omega,\mathbf{0}}^{p,q}(\mathbb{R}^n)$,

$$C_1 \|f\|_{\mathcal{K}_{\omega,\mathbf{0}}^{p,q}(\mathbb{R}^n)} \leq \|f\|_{\dot{\mathcal{K}}_{\omega,\mathbf{0}}^{p,q}(\mathbb{R}^n)} + \|f\|_{L^p(\mathbb{R}^n)} \leq C_2 \|f\|_{\mathcal{K}_{\omega,\mathbf{0}}^{p,q}(\mathbb{R}^n)}.$$

Proof Let all the symbols be as in the present theorem. We first show that

$$\mathcal{K}_{\omega,\mathbf{0}}^{p,q}(\mathbb{R}^n) \cap L^p(\mathbb{R}^n) \subset \dot{\mathcal{K}}_{\omega,\mathbf{0}}^{p,q}(\mathbb{R}^n). \tag{7.1}$$

To this end, let $f \in \mathcal{K}_{\omega,\mathbf{0}}^{p,q}(\mathbb{R}^n) \cap L^p(\mathbb{R}^n)$. Then, applying both Definitions 1.2.1 and 7.1.1, we find that

$$\|f\|_{\mathcal{K}_{\omega,\mathbf{0}}^{p,q}(\mathbb{R}^n)}$$

$$\lesssim \left\| f\mathbf{1}_{B(0,1)} \right\|_{L^p(\mathbb{R}^n)} + \left\{ \sum_{k \in \mathbb{N}} \left[\omega(2^k) \right]^q \left\| f\mathbf{1}_{B(0,2^k)\backslash B(0,2^{k-1})} \right\|_{L^p(\mathbb{R}^n)}^q \right\}^{\frac{1}{q}}$$

$$\lesssim \|f\|_{L^p(\mathbb{R}^n)} + \|f\|_{\dot{\mathcal{K}}_{\omega,\mathbf{0}}^{p,q}(\mathbb{R}^n)} < \infty. \tag{7.2}$$

This finishes the proof of (7.1).

Conversely, we prove $\mathcal{K}_{\omega,\mathbf{0}}^{p,q}(\mathbb{R}^n) \subset \dot{\mathcal{K}}_{\omega,\mathbf{0}}^{p,q}(\mathbb{R}^n) \cap L^p(\mathbb{R}^n)$. Indeed, from Lemma 1.1.12, it follows that, for any $k \in \mathbb{N}$,

$$\omega(2^k) \gtrsim 2^{k[m_\infty(\omega)-\varepsilon]}$$

and, for any $k \in \mathbb{Z} \setminus \mathbb{N}$,

$$\omega(2^k) \lesssim 2^{k[m_0(\omega)-\varepsilon]},$$

where $\varepsilon \in (0, \min\{m_0(\omega), m_\infty(\omega)\})$ is a fixed positive constant. We now claim that, for any $f \in \mathcal{K}_{\omega,0}^{p,q}(\mathbb{R}^n)$,

$$\left\| f \mathbf{1}_{[B(0,1)]^\complement} \right\|_{L^p(\mathbb{R}^n)} \lesssim \|f\|_{\mathcal{K}_{\omega,0}^{p,q}(\mathbb{R}^n)}. \tag{7.3}$$

Indeed, by the assumption that, for any $k \in \mathbb{N}$, $\omega(2^k) \gtrsim 2^{k[m_\infty(\omega)-\varepsilon]}$, we conclude that, for any $f \in \mathcal{K}_{\omega,0}^{p,q}(\mathbb{R}^n)$,

$$\left\| f \mathbf{1}_{[B(0,1)]^\complement} \right\|_{L^p(\mathbb{R}^n)}^p$$
$$= \sum_{k\in\mathbb{N}} \left[\omega(2^k)\right]^{-p} \left[\omega(2^k)\right]^p \left\| f \mathbf{1}_{B(0,2^k)\setminus B(0,2^{k-1})} \right\|_{L^p(\mathbb{R}^n)}^p$$
$$\lesssim \sum_{k\in\mathbb{N}} 2^{kp[-m_\infty(\omega)+\varepsilon]} \left[\omega(2^k)\right]^p \left\| f \mathbf{1}_{B(0,2^k)\setminus B(0,2^{k-1})} \right\|_{L^p(\mathbb{R}^n)}^p, \tag{7.4}$$

which, together with Lemma 1.2.11 and $-m_\infty(\omega) + \varepsilon \in (-\infty, 0)$, further implies that, for any $q \in (0, p]$,

$$\left\| f \mathbf{1}_{[B(0,1)]^\complement} \right\|_{L^p(\mathbb{R}^n)}^p$$
$$\lesssim \left\{ \sum_{k\in\mathbb{N}} 2^{kq[-m_\infty(\omega)+\varepsilon]} \left[\omega(2^k)\right]^q \left\| f \mathbf{1}_{B(0,2^k)\setminus B(0,2^{k-1})} \right\|_{L^p(\mathbb{R}^n)}^q \right\}^{\frac{p}{q}}$$
$$\lesssim \left\{ \sum_{k\in\mathbb{N}} \left[\omega(2^k)\right]^q \left\| f \mathbf{1}_{B(0,2^k)\setminus B(0,2^{k-1})} \right\|_{L^p(\mathbb{R}^n)}^q \right\}^{\frac{p}{q}}$$
$$\lesssim \|f\|_{\mathcal{K}_{\omega,0}^{p,q}(\mathbb{R}^n)}^p. \tag{7.5}$$

This finishes the proof of (7.3) when $q \in (0, p]$. On the other hand, applying (7.4), the Hölder inequality, and the assumption $-m_\infty(\omega) + \varepsilon \in (-\infty, 0)$, we find that, for any $q \in (p, \infty)$ and $f \in \mathcal{K}_{\omega,0}^{p,q}(\mathbb{R}^n)$,

$$\left\| f \mathbf{1}_{[B(0,1)]^\complement} \right\|_{L^p(\mathbb{R}^n)}^p$$
$$\lesssim \left\{ \sum_{k\in\mathbb{N}} 2^{kp[-m_\infty(\omega)+\varepsilon](q/p)'} \right\}^{[(q/p)']^{-1}}$$

$$\times \left\{ \sum_{k\in\mathbb{N}} \left[\omega(2^k)\right]^q \left\| f \mathbf{1}_{B(0,2^k)\backslash B(0,2^{k-1})} \right\|_{L^p(\mathbb{R}^n)}^q \right\}^{\frac{p}{q}}$$

$$\sim \left\{ \sum_{k\in\mathbb{N}} \left[\omega(2^k)\right]^q \left\| f \mathbf{1}_{B(0,2^k)\backslash B(0,2^{k-1})} \right\|_{L^p(\mathbb{R}^n)}^q \right\}^{\frac{p}{q}} \lesssim \| f \|_{\mathcal{K}^{p,q}_{\omega,0}(\mathbb{R}^n)}^p.$$

Combining this and (7.5), we further conclude that (7.3) holds true for any $q \in (0,\infty)$ and hence complete the prove of the above claim. Thus, from Definition 7.1.1.1(i), we deduce that, for any $f \in \mathcal{K}^{p,q}_{\omega,0}(\mathbb{R}^n)$,

$$\| f \|_{L^p(\mathbb{R}^n)}^p = \left\| f \mathbf{1}_{B(0,1)} \right\|_{L^p(\mathbb{R}^n)}^p + \left\| f \mathbf{1}_{[B(0,1)]^\complement} \right\|_{L^p(\mathbb{R}^n)}^p \lesssim \| f \|_{\mathcal{K}^{p,q}_{\omega,0}(\mathbb{R}^n)}^p < \infty,$$
(7.6)

which further implies that $\mathcal{K}^{p,q}_{\omega,0}(\mathbb{R}^n) \subset L^p(\mathbb{R}^n)$.

We next show that $\mathcal{K}^{p,q}_{\omega,0}(\mathbb{R}^n) \subset \dot{\mathcal{K}}^{p,q}_{\omega,0}(\mathbb{R}^n)$. To achieve this, let $f \in \mathcal{K}^{p,q}_{\omega,0}(\mathbb{R}^n)$. Then, by the assumption that, for any $k \in \mathbb{Z} \setminus \mathbb{N}$, $\omega(2^k) \lesssim 2^{k[m_0(\omega)-\varepsilon]}$, Lemma 1.2.11, and the assumption $m_0(\omega) - \varepsilon \in (0,\infty)$, we conclude that, for any $q \in [p,\infty)$,

$$\sum_{k\in\mathbb{Z}\backslash\mathbb{N}} \left[\omega(2^k)\right]^q \left\| f \mathbf{1}_{B(0,2^k)\backslash B(0,2^{k-1})} \right\|_{L^p(\mathbb{R}^n)}^q$$

$$\lesssim \sum_{k\in\mathbb{Z}\backslash\mathbb{N}} 2^{kq[m_0(\omega)-\varepsilon]} \left\| f \mathbf{1}_{B(0,2^k)\backslash B(0,2^{k-1})} \right\|_{L^p(\mathbb{R}^n)}^q$$

$$\lesssim \left\{ \sum_{k\in\mathbb{Z}\backslash\mathbb{N}} 2^{kp[m_0(\omega)-\varepsilon]} \left\| f \mathbf{1}_{B(0,2^k)\backslash B(0,2^{k-1})} \right\|_{L^p(\mathbb{R}^n)}^p \right\}^{\frac{q}{p}}$$

$$\lesssim \left[\sum_{k\in\mathbb{Z}\backslash\mathbb{N}} \left\| f \mathbf{1}_{B(0,2^k)\backslash B(0,2^{k-1})} \right\|_{L^p(\mathbb{R}^n)}^p \right]^{\frac{q}{p}} \sim \left\| f \mathbf{1}_{B(0,1)} \right\|_{L^p(\mathbb{R}^n)}^q. \quad (7.7)$$

On the other hand, using the assumption that, for any $k \in \mathbb{Z} \setminus \mathbb{N}$, $\omega(2^k) \lesssim 2^{k[m_0(\omega)-\varepsilon]}$, the Hölder inequality, and the assumption $m_0(\omega) - \varepsilon \in (0,\infty)$ again, we find that, for any $q \in (0,p)$,

$$\sum_{k\in\mathbb{Z}\backslash\mathbb{N}} \left[\omega(2^k)\right]^q \left\| f \mathbf{1}_{B(0,2^k)\backslash B(0,2^{k-1})} \right\|_{L^p(\mathbb{R}^n)}^q$$

$$\lesssim \sum_{k\in\mathbb{Z}\backslash\mathbb{N}} 2^{kq[m_0(\omega)-\varepsilon]} \left\| f \mathbf{1}_{B(0,2^k)\backslash B(0,2^{k-1})} \right\|_{L^p(\mathbb{R}^n)}^q$$

$$
\lesssim \left\{ \sum_{k\in\mathbb{Z}\setminus\mathbb{N}} 2^{kq[m_0(\omega)-\varepsilon](p/q)'} \right\}^{(p/q)'-1} \left[\sum_{k\in\mathbb{Z}\setminus\mathbb{N}} \left\| f\mathbf{1}_{B(0,2^k)\setminus B(0,2^{k-1})} \right\|_{L^p(\mathbb{R}^n)}^p \right]^{\frac{q}{p}}
$$

$$
\sim \left[\sum_{k\in\mathbb{Z}\setminus\mathbb{N}} \left\| f\mathbf{1}_{B(0,2^k)\setminus B(0,2^{k-1})} \right\|_{L^p(\mathbb{R}^n)}^p \right]^{\frac{q}{p}} \sim \left\| f\mathbf{1}_{B(0,1)} \right\|_{L^p(\mathbb{R}^n)}^q ,
$$

which, combined with (7.7), further implies that

$$
\| f \|_{\dot{\mathcal{K}}_{\omega,0}^{p,q}(\mathbb{R}^n)}
$$

$$
= \left\{ \sum_{k\in\mathbb{Z}\setminus\mathbb{N}} \left[\omega(2^k) \right]^q \left\| f\mathbf{1}_{B(0,2^k)\setminus B(0,2^{k-1})} \right\|_{L^p(\mathbb{R}^n)}^q + \sum_{k\in\mathbb{N}} \cdots \right\}^{\frac{1}{q}}
$$

$$
\lesssim \left\{ \left\| f\mathbf{1}_{B(0,1)} \right\|_{L^p(\mathbb{R}^n)}^q + \sum_{k\in\mathbb{N}} \left[\omega(2^k) \right]^q \left\| f\mathbf{1}_{B(0,2^k)\setminus B(0,2^{k-1})} \right\|_{L^p(\mathbb{R}^n)}^q \right\}^{\frac{1}{q}}
$$

$$
\sim \| f \|_{\mathcal{K}_{\omega,0}^{p,q}(\mathbb{R}^n)} < \infty. \tag{7.8}
$$

This implies that $\mathcal{K}_{\omega,0}^{p,q}(\mathbb{R}^n) \subset \dot{\mathcal{K}}_{\omega,0}^{p,q}(\mathbb{R}^n)$. Therefore, we complete the proof that

$$
\mathcal{K}_{\omega,0}^{p,q}(\mathbb{R}^n) = \dot{\mathcal{K}}_{\omega,0}^{p,q}(\mathbb{R}^n) \cap L^p(\mathbb{R}^n).
$$

Moreover, from (7.2), (7.6), and (7.8), it follows that, for any $f \in \mathcal{K}_{\omega,0}^{p,q}(\mathbb{R}^n)$,

$$
\| f \|_{\mathcal{K}_{\omega,0}^{p,q}(\mathbb{R}^n)} \sim \| f \|_{\dot{\mathcal{K}}_{\omega,0}^{p,q}(\mathbb{R}^n)} + \| f \|_{L^p(\mathbb{R}^n)}
$$

with positive equivalence constants independent of f. This finishes the proof of Theorem 7.1.3.   □

Remark 7.1.4 We should point out that, in Theorem 7.1.3, when $\omega(t) := t^\alpha$ for any $t \in (0,\infty)$ and for any given $\alpha \in \mathbb{R}$, Theorem 7.1.3 goes back to [175, Proposition 1.1.2].

For the triviality of inhomogeneous global generalized Herz spaces, we have the following conclusion which shows that, under some assumptions, the Herz space $\mathcal{K}_\omega^{p,q}(\mathbb{R}^n)$ is trivial.

Theorem 7.1.5 *Let* $p, q \in (0,\infty]$ *and* $\omega \in M(\mathbb{R}_+)$ *satisfy* $m_\infty(\omega) \in (0,\infty)$. *Then*

$$
\mathcal{K}_\omega^{p,q}(\mathbb{R}^n) = \left\{ f \in \mathscr{M}(\mathbb{R}^n) : \ f(x) = 0 \ for \ almost \ every \ x \in \mathbb{R}^n \right\}.
$$

Proof Let all the symbols be as in the present theorem and f be a measurable function on $\mathbb{R}^n$ such that

$$\left|\left\{x \in \mathbb{R}^n : |f(x)| > 0\right\}\right| \neq 0.$$

Then, repeating an argument similar to that used in the proof of Theorem 1.2.7 with $\dot{\mathcal{K}}_\omega^{p,q}(\mathbb{R}^n)$ and Definition 1.2.1 therein replaced, respectively, by $\mathcal{K}_\omega^{p,q}(\mathbb{R}^n)$ and Definition 7.1.1, we obtain

$$\|f\|_{\mathcal{K}_\omega^{p,q}(\mathbb{R}^n)} = \infty.$$

This then finishes the proof of Theorem 7.1.5. □

Remark 7.1.6 Let $p, q \in (0, \infty]$ and $\omega \in M(\mathbb{R}_+)$ be such that $m_\infty(\omega) \in (0, \infty)$. Then, combining Theorems 1.2.7 and 7.1.5, we know that, in this case,

$$\dot{\mathcal{K}}_\omega^{p,q}(\mathbb{R}^n) = \mathcal{K}_\omega^{p,q}(\mathbb{R}^n).$$

Now, we show that the inhomogeneous local generalized Herz space $\mathcal{K}_{\omega,\mathbf{0}}^{p,q}(\mathbb{R}^n)$ is a ball quasi-Banach function space.

Theorem 7.1.7 *Let $p, q \in (0, \infty]$ and $\omega \in M(\mathbb{R}_+)$. Then the inhomogeneous local generalized Herz space $\mathcal{K}_{\omega,\mathbf{0}}^{p,q}(\mathbb{R}^n)$ is a ball quasi-Banach function space.*

Proof Let $p, q \in (0, \infty]$ and $\omega \in M(\mathbb{R}_+)$. Obviously, the inhomogeneous local generalized Herz space $\mathcal{K}_{\omega,\mathbf{0}}^{p,q}(\mathbb{R}^n)$ is a quasi-normed linear space satisfying both (i) and (ii) of Definition 1.2.13. Moreover, Definition 1.2.13(iii) is a simple corollary of the monotone convergence theorem. Therefore, using Proposition 1.2.36, we further find that $\mathcal{K}_{\omega,\mathbf{0}}^{p,q}(\mathbb{R}^n)$ is complete and hence $\mathcal{K}_{\omega,\mathbf{0}}^{p,q}(\mathbb{R}^n)$ is a quasi-Banach space.

Thus, to finish the proof of the present theorem, it remains to show that the inhomogeneous local generalized Herz space $\mathcal{K}_{\omega,\mathbf{0}}^{p,q}(\mathbb{R}^n)$ satisfies Definition 1.2.13(iv). To this end, let $B(x_0, r) \in \mathbb{B}$ with $x_0 \in \mathbb{R}^n$ and $r \in (0, \infty)$. Then, for any $k \in \mathbb{N} \cap (\ln(r + |x_0|)/\ln 2 + 1, \infty)$, we have $2^{k-1} > r + |x_0|$. This implies that, for any $x \in B(x_0, r)$,

$$|x| \leq |x - x_0| + |x_0| < r + |x_0| < 2^{k-1}.$$

Therefore, $x \in B(\mathbf{0}, 2^{k-1})$ and hence $B(x_0, r) \subset B(\mathbf{0}, 2^{k-1})$. By this, we further conclude that

$$B(x_0, r) \cap \left[B(\mathbf{0}, 2^k) \setminus B(\mathbf{0}, 2^{k-1})\right] = \emptyset.$$

Combining this and Definition 7.1.1(i), we find that

$$\left\| \mathbf{1}_{B(x_0,r)} \right\|_{\mathcal{K}^{p,q}_{\omega,\mathbf{0}}(\mathbb{R}^n)}^q$$

$$= \left\| \mathbf{1}_{B(x_0,r)} \mathbf{1}_{B(0,1)} \right\|_{L^p(\mathbb{R}^n)}^q$$

$$+ \sum_{k \in \mathbb{N} \cap [1, \frac{\ln(r+|x_0|)}{\ln 2}]} \left[\omega(2^k) \right]^q \left\| \mathbf{1}_{B(x_0,r)} \mathbf{1}_{B(0,2^k) \setminus B(0,2^{k-1})} \right\|_{L^p(\mathbb{R}^n)}^q$$

$$< \infty,$$

which implies that $\mathbf{1}_{B(x_0,r)} \in \mathcal{K}^{p,q}_{\omega,\mathbf{0}}(\mathbb{R}^n)$ and hence finishes the proof of Theorem 7.1.7. □

However, the following example shows that inhomogeneous global generalized Herz spaces may not be ball quasi-Banach function spaces.

Example 7.1.8 Let $p, q \in (0, \infty]$, $\alpha_1 \in \mathbb{R}$, $\alpha_2 \in [0, \infty)$, and

$$\omega(t) := \begin{cases} t^{\alpha_1}(1 - \ln t) & \text{when } t \in (0, 1], \\ t^{\alpha_2}(1 + \ln t) & \text{when } t \in (1, \infty). \end{cases}$$

Then $\mathbf{1}_{B(0,1)} \notin \mathcal{K}^{p,q}_{\omega}(\mathbb{R}^n)$ and hence, in this case, the inhomogeneous global generalized Herz space $\mathcal{K}^{p,q}_{\omega}(\mathbb{R}^n)$ is not a ball quasi-Banach function space.

Proof Let all the symbols be as in the present example and, for any $k \in \mathbb{N}$, $\xi_k \in \mathbb{R}^n$ satisfy $|\xi_k| = 2^k + 1$. Then, by the proof of Example 1.2.41, we find that, for any $k \in \mathbb{N}$,

$$B(0, 1) \subset B(\xi_k, 2^{k+1}) \setminus B(\xi_k, 2^k).$$

From this, Remark 7.1.2(i), and Definition 7.1.1(i), we deduce that

$$\left\| \mathbf{1}_{B(0,1)} \right\|_{\mathcal{K}^{p,q}_{\omega}(\mathbb{R}^n)} \geq \left\| \mathbf{1}_{B(0,1)}(\cdot + \xi_k) \right\|_{\mathcal{K}^{p,q}_{\omega,\mathbf{0}}(\mathbb{R}^n)} = \omega(2^{k+1}) \left\| \mathbf{1}_{B(0,1)} \right\|_{L^p(\mathbb{R}^n)}$$

$$\sim 2^{(k+1)\alpha_2}[1 + (k+1)\ln 2] \to \infty$$

as $k \to \infty$. This further implies that $\mathbf{1}_{B(0,1)} \notin \mathcal{K}^{p,q}_{\omega}(\mathbb{R}^n)$ and hence the inhomogeneous global generalized Herz space $\mathcal{K}^{p,q}_{\omega}(\mathbb{R}^n)$ under consideration is not a BQBF space. This then finishes the proof of Example 7.1.8. □

Under a reasonable and sharp assumption, we next prove that the inhomogeneous global generalized Herz space $\mathcal{K}^{p,q}_{\omega}(\mathbb{R}^n)$ is also a ball quasi-Banach function space. Namely, the following conclusion holds true.

Theorem 7.1.9 *Let* $p, q \in (0, \infty]$ *and* $\omega \in M(\mathbb{R}_+)$ *satisfy* $M_\infty(\omega) \in (-\infty, 0)$. *Then the inhomogeneous global generalized Herz space* $\mathcal{K}_\omega^{p,q}(\mathbb{R}^n)$ *is a ball quasi-Banach function space.*

Proof Let $p, q \in (0, \infty]$ and $\omega \in M(\mathbb{R}_+)$ with $M_\infty(\omega) \in (-\infty, 0)$. Obviously, the inhomogeneous global generalized Herz space $\mathcal{K}_\omega^{p,q}(\mathbb{R}^n)$ is a quasi-normed linear space satisfying both (i) and (ii) of Definition 1.2.13. We now show that Definition 1.2.13(iii) holds true for $\mathcal{K}_\omega^{p,q}(\mathbb{R}^n)$. Indeed, for any given $\{f_m\}_{m \in \mathbb{N}} \subset \mathcal{M}(\mathbb{R}^n)$ and $f \in \mathcal{M}(\mathbb{R}^n)$ satisfying $0 \leq f_m \uparrow f$ almost everywhere as $m \to \infty$, and for any given $\alpha \in (0, \|f\|_{\mathcal{K}_\omega^{p,q}(\mathbb{R}^n)})$, from Remark 7.1.2(i), we deduce that there exists a $\xi \in \mathbb{R}^n$ such that

$$\|f(\cdot + \xi)\|_{\mathcal{K}_{\omega,0}^{p,q}(\mathbb{R}^n)} > \alpha.$$

Using this and the monotone convergence theorem, we conclude that there exists an $N \in \mathbb{N}$ such that, for any $m \in \mathbb{N} \cap (N, \infty)$,

$$\|f_m(\cdot + \xi)\|_{\mathcal{K}_{\omega,0}^{p,q}(\mathbb{R}^n)} > \alpha.$$

By this and Remark 7.1.2(i) again, we find that, for any $m \in \mathbb{N} \cap (N, \infty)$,

$$\alpha < \sup_{\xi \in \mathbb{R}^n} \|f_m(\cdot + \xi)\|_{\mathcal{K}_{\omega,0}^{p,q}(\mathbb{R}^n)} = \|f_m\|_{\mathcal{K}_\omega^{p,q}(\mathbb{R}^n)}.$$

Letting $\alpha \to \|f\|_{\mathcal{K}_\omega^{p,q}(\mathbb{R}^n)}$ and $m \to \infty$, we have

$$\|f\|_{\mathcal{K}_\omega^{p,q}(\mathbb{R}^n)} \leq \lim_{m \to \infty} \|f_m\|_{\mathcal{K}_\omega^{p,q}(\mathbb{R}^n)}.$$

On the other hand, it is easy to show that

$$\lim_{m \to \infty} \|f_m\|_{\mathcal{K}_\omega^{p,q}(\mathbb{R}^n)} \leq \|f\|_{\mathcal{K}_\omega^{p,q}(\mathbb{R}^n)}.$$

Therefore, we conclude that

$$\|f_m\|_{\mathcal{K}_\omega^{p,q}(\mathbb{R}^n)} \uparrow \|f\|_{\mathcal{K}_\omega^{p,q}(\mathbb{R}^n)}$$

as $m \to \infty$ and hence Definition 1.2.13(iii) holds true for $\mathcal{K}_\omega^{p,q}(\mathbb{R}^n)$. From this and Proposition 1.2.36, it follows that the inhomogeneous global generalized Herz space $\mathcal{K}_\omega^{p,q}(\mathbb{R}^n)$ is quasi-Banach space.

Next, we prove that Definition 1.2.13(iv) holds true for $\mathcal{K}_\omega^{p,q}(\mathbb{R}^n)$. Indeed, by Lemma 1.1.12, we find that, for any $k \in \mathbb{N}$,

$$\omega(2^k) \lesssim 2^{k[M_\infty(\omega)+\varepsilon]},$$

where $\varepsilon \in (0, -M_\infty(\omega))$ is a fixed positive constant. Combining this, Definition 7.1.1(i), and the assumption $M_\infty(\omega) + \varepsilon \in (-\infty, 0)$, we further conclude that, for any $B(x_0, r) \in \mathbb{B}$ with $x_0 \in \mathbb{R}^n$ and $r \in (0, \infty)$, and $\xi \in \mathbb{R}^n$,

$$\left\| \mathbf{1}_{B(x_0,r)}(\cdot + \xi) \right\|^q_{\mathcal{K}^{p,q}_{\omega,\mathbf{0}}(\mathbb{R}^n)}$$

$$= \left\| \mathbf{1}_{B(x_0,r)} \mathbf{1}_{B(\xi,1)} \right\|^q_{L^p(\mathbb{R}^n)}$$

$$+ \sum_{k \in \mathbb{N}} \left[\omega(2^k) \right]^q \left\| \mathbf{1}_{B(x_0,r)} \mathbf{1}_{B(\xi,2^k) \setminus B(\xi,2^{k-1})} \right\|^q_{L^p(\mathbb{R}^n)}$$

$$\lesssim \left\| \mathbf{1}_{B(x_0,r)} \right\|^q_{L^p(\mathbb{R}^n)} \left\{ 1 + \sum_{k \in \mathbb{N}} 2^{kq[M_\infty(\omega)+\varepsilon]} \right\} < \infty.$$

This, together with Remark 7.1.2(i), further implies that, for any $x_0 \in \mathbb{R}^n$ and $r \in (0, \infty)$,

$$\left\| \mathbf{1}_{B(x_0,r)} \right\|_{\mathcal{K}^{p,q}_{\omega}(\mathbb{R}^n)} \lesssim 1.$$

Therefore, $\mathbf{1}_{B(x_0,r)} \in \mathcal{K}^{p,q}_{\omega}(\mathbb{R}^n)$ for any $x_0 \in \mathbb{R}^n$ and $r \in (0, \infty)$. This implies that Definition 1.2.13(iv) holds true for $\mathcal{K}^{p,q}_{\omega}(\mathbb{R}^n)$, and hence finishes the proof of Theorem 7.1.9.   □

Remark 7.1.10 By both Examples 1.1.9 and 7.1.8, we find that the assumption $M_\infty(\omega) \in (-\infty, 0)$ in Theorem 7.1.9 is sharp. Indeed, if $M_\infty(\omega) \in [0, \infty)$, then both Examples 1.1.9 and 7.1.8 show that there exists an $\omega \in M(\mathbb{R}_+)$ such that $\mathcal{K}^{p,q}_{\omega}(\mathbb{R}^n)$ is not a ball quasi-Banach function space.

Moreover, the following theorem indicates that the inhomogeneous local generalized Herz space $\mathcal{K}^{p,q}_{\omega,\mathbf{0}}(\mathbb{R}^n)$ is a ball Banach function space when $p, q \in [1, \infty]$.

Theorem 7.1.11 *Let $p, q \in [1, \infty]$ and $\omega \in M(\mathbb{R}_+)$. Then the inhomogeneous local generalized Herz space $\mathcal{K}^{p,q}_{\omega,\mathbf{0}}(\mathbb{R}^n)$ is a ball Banach function space.*

Proof Let $p, q \in [1, \infty]$ and $\omega \in M(\mathbb{R}_+)$. Then, using Theorem 7.1.7, we find that the inhomogeneous local generalized Herz space $\mathcal{K}^{p,q}_{\omega,\mathbf{0}}(\mathbb{R}^n)$ is a BQBF space. Moreover, $\mathcal{K}^{p,q}_{\omega,\mathbf{0}}(\mathbb{R}^n)$ obviously satisfies the triangle inequality due to $p, q \in [1, \infty]$.

Thus, to complete the proof of the present theorem, it remains to show that (1.39) holds true with $X := \mathcal{K}^{p,q}_{\omega,\mathbf{0}}(\mathbb{R}^n)$. To achieve this, we first claim that, for any given $k_0 \in \mathbb{Z}$, (1.39) holds true with $X := \mathcal{K}^{p,q}_{\omega,\mathbf{0}}(\mathbb{R}^n)$ and $B := B(\mathbf{0}, 2^{k_0})$. We show this claim by considering the following two cases on k_0.

Case 1) $k_0 \in \mathbb{Z} \backslash \mathbb{N}$. In this case, we have $B(\mathbf{0}, 2^{k_0}) \subset B(\mathbf{0}, 1)$. By this, the Hölder inequality, and Definition 7.1.1(i), we conclude that, for any $f \in \mathcal{K}_{\omega,\mathbf{0}}^{p,q}(\mathbb{R}^n)$,

$$\int_{B(\mathbf{0},2^{k_0})} |f(y)| \, dy \lesssim \left\| f \mathbf{1}_{B(\mathbf{0},2^{k_0})} \right\|_{L^p(\mathbb{R}^n)} \lesssim \left\| f \mathbf{1}_{B(\mathbf{0},1)} \right\|_{L^p(\mathbb{R}^n)}$$

$$\lesssim \|f\|_{\mathcal{K}_{\omega}^{p,q}(\mathbb{R}^n)}, \tag{7.9}$$

which implies that the above claim holds true in this case.

Case 2) $k_0 \in \mathbb{N}$. In this case, from the Hölder inequality, the fact that, for any given $\alpha \in (0, \infty)$ and $m \in \mathbb{N}$, and for any $\{a_j\}_{j \in \mathbb{N}} \subset \mathbb{C}$,

$$\left(\sum_{j=1}^{m} |a_j| \right)^{\alpha} \leq \max \left\{ 1, m^{\alpha-1} \right\} \sum_{j=1}^{m} |a_j|^{\alpha},$$

and Definition 7.1.1(i), we deduce that, for any $f \in \mathcal{K}_{\omega,\mathbf{0}}^{p,q}(\mathbb{R}^n)$,

$$\int_{B(\mathbf{0},2^{k_0})} |f(y)| \, dy$$

$$\lesssim \left\| f \mathbf{1}_{B(\mathbf{0},2^{k_0})} \right\|_{L^p(\mathbb{R}^n)}$$

$$\sim \left[\left\| f \mathbf{1}_{B(\mathbf{0},1)} \right\|_{L^p(\mathbb{R}^n)}^p + \sum_{k=1}^{k_0} \left\| f \mathbf{1}_{B(\mathbf{0},2^k) \backslash B(\mathbf{0},2^{k-1})} \right\|_{L^p(\mathbb{R}^n)}^p \right]^{\frac{1}{p}}$$

$$\lesssim \left\{ \left\| f \mathbf{1}_{B(\mathbf{0},1)} \right\|_{L^p(\mathbb{R}^n)}^q + \sum_{k=1}^{k_0} \left[\omega(2^k) \right]^q \left\| f \mathbf{1}_{B(\mathbf{0},2^k) \backslash B(\mathbf{0},2^{k-1})} \right\|_{L^p(\mathbb{R}^n)}^q \right\}^{\frac{1}{q}}$$

$$\lesssim \|f\|_{\mathcal{K}_{\omega,\mathbf{0}}^{p,q}(\mathbb{R}^n)}.$$

This, combined with (7.9), then finishes the proof of the above claim.

Applying the above claim, we then show that (1.39) holds true for any $B \in \mathbb{B}$. Notice that, for any $B \in \mathbb{B}$, there exists a $k \in \mathbb{Z}$ such that $B \subset B(\mathbf{0}, 2^k)$. By this and the above claim, we further conclude that, for any $f \in \mathcal{K}_{\omega,\mathbf{0}}^{p,q}(\mathbb{R}^n)$,

$$\int_B |f(y)| \, dy \leq \int_{B(\mathbf{0},2^k)} |f(y)| \, dy \lesssim \|f\|_{\mathcal{K}_{\omega,\mathbf{0}}^{p,q}(\mathbb{R}^n)},$$

which implies that (1.39) holds true with $X := \mathcal{K}_{\omega,\mathbf{0}}^{p,q}(\mathbb{R}^n)$, and hence finishes the proof of Theorem 7.1.9. $\qquad \square$

Similarly, we now show that the inhomogeneous global generalized Herz space $\mathcal{K}_\omega^{p,q}(\mathbb{R}^n)$ is also a ball Banach function space when $p, q \in [1, \infty]$.

Theorem 7.1.12 *Let $p, q \in [1, \infty]$ and $\omega \in M(\mathbb{R}_+)$ satisfy $M_\infty(\omega) \in (-\infty, 0)$. Then the inhomogeneous global generalized Herz space $\mathcal{K}_\omega^{p,q}(\mathbb{R}^n)$ is a ball Banach function space.*

Proof Let all the symbols be as in the present theorem. Then, by the assumption $M_\infty(\omega) \in (-\infty, 0)$ and Theorem 7.1.9, we find that the inhomogeneous global generalized Herz space $\mathcal{K}_\omega^{p,q}(\mathbb{R}^n)$ is a BQBF space. Moreover, notice that $\mathcal{K}_\omega^{p,q}(\mathbb{R}^n)$ satisfies the triangle inequality when $p, q \in [1, \infty]$. Thus, to finish the proof of the present theorem, we only need to show that, for any given $B \in \mathbb{B}$, (1.39) holds true with $X := \mathcal{K}_\omega^{p,q}(\mathbb{R}^n)$. Indeed, from Definition 7.1.1(ii), we deduce that, for any $f \in \mathcal{K}_\omega^{p,q}(\mathbb{R}^n)$,

$$\|f\|_{\mathcal{K}_{\omega,0}^{p,q}(\mathbb{R}^n)} \le \|f\|_{\mathcal{K}_\omega^{p,q}(\mathbb{R}^n)}$$

and hence $f \in \mathcal{K}_{\omega,0}^{p,q}(\mathbb{R}^n)$. Using this, Theorem 7.1.11, and Definition 7.1.1(ii) again, we further conclude that, for any $f \in \mathcal{K}_\omega^{p,q}(\mathbb{R}^n)$,

$$\int_B |f(y)|\, dy \lesssim \|f\|_{\mathcal{K}_{\omega,0}^{p,q}(\mathbb{R}^n)} \lesssim \|f\|_{\mathcal{K}_\omega^{p,q}(\mathbb{R}^n)}.$$

This implies that (1.39) holds true with $X := \mathcal{K}_\omega^{p,q}(\mathbb{R}^n)$, and then finishes the proof of Theorem 7.1.12.   □

7.1.1 Convexities

The main target of this subsection is to show the convexity of the inhomogeneous generalized Herz spaces. For this purpose, we first investigate the relations between these Herz spaces and their convexifications as follows, which are useful in the study of the inhomogeneous generalized Herz–Hardy spaces in next chapters.

Lemma 7.1.13 *Let $p, q, s \in (0, \infty)$ and $\omega \in M(\mathbb{R}_+)$. Then*

$$\left[\mathcal{K}_{\omega,0}^{p,q}(\mathbb{R}^n)\right]^{1/s} = \mathcal{K}_{\omega^s,0}^{p/s,q/s}(\mathbb{R}^n)$$

with the same quasi-norms.

Proof Let all the symbols be as in the present lemma. Then we find that, for any $f \in \mathcal{M}(\mathbb{R}^n)$,

$$\|f\|_{[\mathcal{K}^{p,q}_{\omega,0}(\mathbb{R}^n)]^{1/s}}$$

$$= \left\| |f|^{\frac{1}{s}} \right\|^{s}_{\mathcal{K}^{p,q}_{\omega,0}(\mathbb{R}^n)}$$

$$= \left\{ \left[\int_{|y|<1} |f(y)|^{\frac{p}{s}} \, dy \right]^{\frac{q}{p}} + \sum_{k \in \mathbb{N}} \left[\omega(2^k) \right]^q \left[\int_{2^{k-1} \leq |y| < 2^k} |f(y)|^{\frac{p}{s}} \, dy \right]^{\frac{q}{p}} \right\}^{\frac{s}{q}}$$

$$= \left\{ \left[\int_{|y|<1} |f(y)|^{\frac{p}{s}} \, dy \right]^{\frac{q/s}{p/s}} \right.$$

$$\left. + \sum_{k \in \mathbb{N}} \left[\omega^s(2^k) \right]^{\frac{q}{s}} \left[\int_{2^{k-1} \leq |y| < 2^k} |f(y)|^{\frac{p}{s}} \, dy \right]^{\frac{q/s}{p/s}} \right\}^{\frac{s}{q}}$$

$$= \|f\|_{\mathcal{K}^{p/s,q/s}_{\omega^s,0}(\mathbb{R}^n)}, \tag{7.10}$$

which completes the proof of Lemma 7.1.13. □

Lemma 7.1.14 *Let $p, q, s \in (0, \infty)$ and $\omega \in M(\mathbb{R}_+)$. Then*

$$\left[\mathcal{K}^{p,q}_{\omega}(\mathbb{R}^n) \right]^{1/s} = \mathcal{K}^{p/s,q/s}_{\omega^s}(\mathbb{R}^n)$$

with the same quasi-norms.

Proof Let $p, q, s \in (0, \infty)$ and $\omega \in M(\mathbb{R}_+)$. Then, applying an argument similar to that used in the proof of (7.10), we conclude that, for any $f \in \mathcal{M}(\mathbb{R}^n)$ and $\xi \in \mathbb{R}^n$,

$$\|f(\cdot + \xi)\|_{[\mathcal{K}^{p,q}_{\omega,0}(\mathbb{R}^n)]^{1/s}} = \|f(\cdot + \xi)\|_{\mathcal{K}^{p/s,q/s}_{\omega^s,0}(\mathbb{R}^n)},$$

which, together with Remark 7.1.2(i), further implies that

$$\|f\|_{[\mathcal{K}^{p,q}_{\omega}(\mathbb{R}^n)]^{1/s}} = \|f\|_{\mathcal{K}^{p/s,q/s}_{\omega^s}(\mathbb{R}^n)}.$$

This finishes the proof of Lemma 7.1.14. □

Via both Lemmas 7.1.13 and 7.1.14, we next prove the convexity of inhomogeneous generalized Herz spaces. Namely, we have the following two conclusions.

Theorem 7.1.15 *Let $p, q \in (0, \infty)$, $s \in (0, \min\{p, q\}]$, and $\omega \in M(\mathbb{R}_+)$. Then the inhomogeneous local generalized Herz space $\mathcal{K}^{p,q}_{\omega,0}(\mathbb{R}^n)$ is strictly s-convex.*

Proof Let all the symbols be as in the present theorem. Then, from the assumption $s \in (0, \min\{p, q\}]$, it follows that $p/s, q/s \in [1, \infty)$. This, combined with Lemma 7.1.13 and the Minkowski inequality, implies that, for any sequence $\{f_j\}_{j\in\mathbb{N}}$ of measurable functions in $[\mathcal{K}_{\omega,0}^{p,q}(\mathbb{R}^n)]^{1/s}$, and for any $N \in \mathbb{N}$,

$$\left\| \sum_{j=1}^{N} |f_j| \right\|_{[\mathcal{K}_{\omega,0}^{p,q}(\mathbb{R}^n)]^{1/s}} \leq \sum_{j=1}^{N} \|f_j\|_{[\mathcal{K}_{\omega,0}^{p,q}(\mathbb{R}^n)]^{1/s}} \leq \sum_{j\in\mathbb{N}} \|f_j\|_{[\mathcal{K}_{\omega,0}^{p,q}(\mathbb{R}^n)]^{1/s}}.$$

By this and the monotone convergence theorem, we further find that

$$\left\| \sum_{j\in\mathbb{N}} |f_j| \right\|_{[\mathcal{K}_{\omega,0}^{p,q}(\mathbb{R}^n)]^{1/s}} \leq \sum_{j\in\mathbb{N}} \|f_j\|_{[\mathcal{K}_{\omega,0}^{p,q}(\mathbb{R}^n)]^{1/s}}, \tag{7.11}$$

which completes the proof of Theorem 7.1.15. □

Theorem 7.1.16 *Let $p, q \in (0, \infty)$, $s \in (0, \min\{p, q\}]$, and $\omega \in M(\mathbb{R}_+)$. Then the inhomogeneous global generalized Herz space $\mathcal{K}_{\omega}^{p,q}(\mathbb{R}^n)$ is strictly s-convex.*

Proof Let all the symbols be as in the present theorem. Then, similarly to the estimation of (7.11), using Lemma 7.1.14, we conclude that, for any given sequence $\{f_j\}_{j\in\mathbb{N}}$ of measurable functions in $[\mathcal{K}_{\omega}^{p,q}(\mathbb{R}^n)]^{1/s}$,

$$\left\| \sum_{j\in\mathbb{N}} |f_j| \right\|_{[\mathcal{K}_{\omega}^{p,q}(\mathbb{R}^n)]^{1/s}} \leq \sum_{j\in\mathbb{N}} \|f_j\|_{[\mathcal{K}_{\omega}^{p,q}(\mathbb{R}^n)]^{1/s}}.$$

This finishes the proof of Theorem 7.1.16. □

7.1.2 Absolutely Continuous Quasi-Norms

In this subsection, we investigate the absolutely continuity of the quasi-norms of inhomogeneous generalized Herz spaces. To be precise, we show that the inhomogeneous local generalized Herz space $\mathcal{K}_{\omega,0}^{p,q}(\mathbb{R}^n)$ has an absolutely continuous quasi-norm but find that the inhomogeneous global generalized Herz space may not have an absolutely continuous quasi-norm via a counterexample.

We first prove that $\mathcal{K}_{\omega,0}^{p,q}(\mathbb{R}^n)$ has an absolutely continuous quasi-norm as follows.

Theorem 7.1.17 *Let $p, q \in (0, \infty)$ and $\omega \in M(\mathbb{R}_+)$. Then the inhomogeneous local generalized Herz space $\mathcal{K}_{\omega,0}^{p,q}(\mathbb{R}^n)$ has an absolutely continuous quasi-norm.*

Proof Let $p, q \in (0, \infty)$, $\omega \in M(\mathbb{R}_+)$, and f be a given measurable function in $\mathcal{K}_{\omega,0}^{p,q}(\mathbb{R}^n)$. We now show that f has an absolutely continuous quasi-norm in

$\mathcal{K}^{p,q}_{\omega,\mathbf{0}}(\mathbb{R}^n)$. To this end, let $\{E_i\}_{i\in\mathbb{N}}$ be a sequence of measurable sets satisfying $\mathbf{1}_{E_i} \to 0$ almost everywhere as $i \to \infty$. Then, for any $k, i \in \mathbb{N}$, let

$$a_0 := \left\| f\mathbf{1}_{B(0,1)} \right\|_{L^p(\mathbb{R}^n)},$$

$$a_{0,i} := \left\| f\mathbf{1}_{B(0,1)}\mathbf{1}_{E_i} \right\|_{L^p(\mathbb{R}^n)},$$

$$a_k := \omega(2^k) \left\| f\mathbf{1}_{B(0,2^k)\setminus B(0,2^{k-1})} \right\|_{L^p(\mathbb{R}^n)},$$

and

$$a_{k,i} := \omega(2^k) \left\| f\mathbf{1}_{B(0,2^k)\setminus B(0,2^{k-1})}\mathbf{1}_{E_i} \right\|_{L^p(\mathbb{R}^n)}.$$

From the assumption $f \in \mathcal{K}^{p,q}_{\omega,\mathbf{0}}(\mathbb{R}^n)$, it follows that

$$\left(\sum_{k\in\mathbb{Z}_+} |a_k|^q \right)^{\frac{1}{q}} < \infty. \tag{7.12}$$

This implies that, for any $k \in \mathbb{Z}_+$, $a_k \in [0, \infty)$. Therefore, $f\mathbf{1}_{B(0,1)} \in L^p(\mathbb{R}^n)$ and, for any $k \in \mathbb{N}$,

$$f\mathbf{1}_{B(0,2^k)\setminus B(0,2^{k-1})} \in L^p(\mathbb{R}^n).$$

By this, the facts that, for any $k, i \in \mathbb{N}$,

$$\left| f\mathbf{1}_{B(0,1)}\mathbf{1}_{E_i} \right| \leq \left| f\mathbf{1}_{B(0,1)} \right|$$

and

$$\left| f\mathbf{1}_{B(0,2^k)\setminus B(0,2^{k-1})}\mathbf{1}_{E_i} \right| \leq \left| f\mathbf{1}_{B(0,2^k)\setminus B(0,2^{k-1})} \right|,$$

and the dominated convergence theorem, we conclude that, for any $k \in \mathbb{Z}_+$, $a_{k,i} \to 0$ as $i \to \infty$. Applying this, the fact that, for any $k, i \in \mathbb{N}$, $|a_{k,i}| \leq |a_k|$, (7.12), and the dominated convergence theorem again, we find that

$$\lim_{i\to\infty} \left\| f\mathbf{1}_{E_i} \right\|_{\mathcal{K}^{p,q}_{\omega,\mathbf{0}}(\mathbb{R}^n)} = \lim_{i\to\infty} \left(\sum_{k\in\mathbb{Z}_+} |a_{k,i}|^q \right)^{\frac{1}{q}} = \left(\sum_{k\in\mathbb{Z}_+} \lim_{i\to\infty} |a_{k,i}|^q \right)^{\frac{1}{q}} = 0,$$

which implies that f has an absolutely continuous quasi-norm in $\mathcal{K}^{p,q}_{\omega,\mathbf{0}}(\mathbb{R}^n)$. Thus, by the arbitrariness of f, we further conclude that the inhomogeneous local generalized Herz space $\mathcal{K}^{p,q}_{\omega,\mathbf{0}}(\mathbb{R}^n)$ has an absolutely continuous quasi-norm, and hence complete the proof of Theorem 7.1.17.   □

On the other hand, via the following counterexample, we find that inhomogeneous global generalized Herz spaces may not have absolutely continuous quasi-norms.

Example 7.1.18 Let $p \in (0, \infty)$, $q \in (0, \infty]$, $\alpha \in (-\infty, 0)$, and

$$E := \bigcup_{k \in \mathbb{N}} \left(k - 1 + k^{-\frac{2}{\alpha p}}, k + k^{-\frac{2}{\alpha p}} \right).$$

For any $t \in (0, \infty)$, let $\omega(t) := t^\alpha$. Then the characteristic function $\mathbf{1}_E \in \mathcal{K}^{p,q}_\omega(\mathbb{R})$, but $\mathbf{1}_E$ does not have an absolutely continuous quasi-norm in $\mathcal{K}^{p,q}_\omega(\mathbb{R})$. This implies that the inhomogeneous global generalized Herz space $\mathcal{K}^{p,q}_\omega(\mathbb{R})$ does not have an absolutely continuous quasi-norm.

Proof Let all the symbols be as in the present example. From Definition 7.1.1(i), we deduce that, for any $\xi \in \mathbb{R}$,

$$\left\| \mathbf{1}_E(\cdot + \xi) \right\|_{\mathcal{K}^{p,q}_{\omega,\mathbf{0}}(\mathbb{R})}$$

$$= \left[\left\| \mathbf{1}_E \mathbf{1}_{B(\xi,1)} \right\|^q_{L^p(\mathbb{R})} + \sum_{k \in \mathbb{N}} 2^{k\alpha q} \left\| \mathbf{1}_E \mathbf{1}_{B(\xi,2^k) \setminus B(\xi,2^{k-1})} \right\|^q_{L^p(\mathbb{R})} \right]^{\frac{1}{q}}$$

$$\leq \left[\left\| \mathbf{1}_{B(\xi,1)} \right\|^q_{L^p(\mathbb{R})} + \sum_{k \in \mathbb{N}} 2^{k\alpha q} \left\| \mathbf{1}_E \mathbf{1}_{B(\xi,2^k)} \right\|^q_{L^p(\mathbb{R})} \right]^{\frac{1}{q}}$$

$$\lesssim \left\| \mathbf{1}_{B(\xi,1)} \right\|_{L^p(\mathbb{R})} + \left[\sum_{k \in \mathbb{N}} 2^{k\alpha q} \left\| \mathbf{1}_E \mathbf{1}_{B(\xi,2^k)} \right\|^q_{L^p(\mathbb{R})} \right]^{\frac{1}{q}}. \tag{7.13}$$

Applying this, we find that, for any given $\alpha \in (-\infty, -\frac{1}{p})$ and for any $\xi \in \mathbb{R}$,

$$\left\| \mathbf{1}_E(\cdot + \xi) \right\|_{\mathcal{K}^{p,q}_{\omega,\mathbf{0}}(\mathbb{R})} \lesssim 1 + \left(\sum_{k \in \mathbb{N}} 2^{k\alpha q} 2^{\frac{k\alpha q}{p}} \right)^{\frac{1}{q}} \sim 1. \tag{7.14}$$

On the other hand, by the proof of Example 1.4.4, we conclude that

$$J_{\xi,1} \sim \left[\sum_{k \in \mathbb{N}} 2^{k\alpha q} \left\| \mathbf{1}_E \mathbf{1}_{(\xi,\xi+2^k)} \right\|^q_{L^p(\mathbb{R})} \right]^{\frac{1}{q}} \lesssim 1$$

and

$$J_{\xi,2} \sim \left[\sum_{k \in \mathbb{N}} 2^{k\alpha q} \left\| \mathbf{1}_E \mathbf{1}_{(\xi-2^k,\xi)} \right\|_{L^p(\mathbb{R})}^q \right]^{\frac{1}{q}} \lesssim 1$$

in Example 1.4.4 still hold true for $\alpha = -\frac{1}{p}$. Thus, from Example 1.4.4 again, it follows that, for any given $\alpha \in [-\frac{1}{p}, 0)$ and for any $\xi \in \mathbb{R}$, $J_{\xi,1} \lesssim 1$ and $J_{\xi,2} \lesssim 1$. This, together with (7.13), further implies that, for any given $\alpha \in [-\frac{1}{p}, 0)$ and for any $\xi \in \mathbb{R}$,

$$\left\| \mathbf{1}_E(\cdot + \xi) \right\|_{\mathcal{K}_{\omega,0}^{p,q}(\mathbb{R})}$$

$$\lesssim \left\| \mathbf{1}_{B(\xi,1)} \right\|_{L^p(\mathbb{R})} + \left[\sum_{k \in \mathbb{N}} 2^{k\alpha q} \left\| \mathbf{1}_E \mathbf{1}_{(\xi,\xi+2^k)} \right\|_{L^p(\mathbb{R})}^q \right]^{\frac{1}{q}}$$

$$+ \left[\sum_{k \in \mathbb{N}} 2^{k\alpha q} \left\| \mathbf{1}_E \mathbf{1}_{(\xi-2^k,\xi)} \right\|_{L^p(\mathbb{R})}^q \right]^{\frac{1}{q}}$$

$$\sim \left\| \mathbf{1}_{B(\xi,1)} \right\|_{L^p(\mathbb{R})} + J_{\xi,1} + J_{\xi,2} \lesssim 1.$$

Combining this, (7.14), and Remark 7.1.2(i), we further find that $\left\| \mathbf{1}_E \right\|_{\mathcal{K}_{\omega}^{p,q}(\mathbb{R})} < \infty$, and hence $\mathbf{1}_E \in \mathcal{K}_{\omega}^{p,q}(\mathbb{R})$.

We now show that the characteristic function $\mathbf{1}_E$ does not have an absolutely continuous quasi-norm in the inhomogeneous global generalized Herz space $\mathcal{K}_{\omega}^{p,q}(\mathbb{R})$ under consideration. Indeed, for any $\tilde{k} \in \mathbb{N}$, let

$$F_{\tilde{k}} := \left(\tilde{k}, \infty \right) \text{ and } \xi_{\tilde{k}} := \tilde{k} - 2 + \tilde{k}^{-\frac{2}{\alpha p}}.$$

Then, obviously, for any $x \in \mathbb{R}$, we have $\mathbf{1}_{F_{\tilde{k}}}(x) \to 0$ as $\tilde{k} \to \infty$. In addition, from the assumption that, for any $\tilde{k} \in \mathbb{N}$, $\tilde{k} - 1 + \tilde{k}^{-\frac{2}{\alpha p}} \geq \tilde{k}$, it follows that, for any $\tilde{k} \in \mathbb{N}$,

$$\left(\tilde{k} - 1 + \tilde{k}^{-\frac{2}{\alpha p}}, \tilde{k} + \tilde{k}^{-\frac{2}{\alpha p}} \right) \subset F_{\tilde{k}}.$$

Using this, Remark 7.1.2(i), and Definition 7.1.1(i), we further conclude that, for any $\tilde{k} \in \mathbb{N}$,

$$\left\| \mathbf{1}_E \mathbf{1}_{F_{\tilde{k}}} \right\|_{\mathcal{K}_{\omega}^{p,q}(\mathbb{R})} \geq \left\| \mathbf{1}_{(\tilde{k}-1+\tilde{k}^{-\frac{2}{\alpha p}}, \tilde{k}+\tilde{k}^{-\frac{2}{\alpha p}})} \right\|_{\mathcal{K}_{\omega}^{p,q}(\mathbb{R})}$$

$$\geq \left\| \mathbf{1}_{(\tilde{k}-1+\tilde{k}^{-\frac{2}{\alpha p}}, \tilde{k}+\tilde{k}^{-\frac{2}{\alpha p}})}(\cdot + \xi_{\tilde{k}}) \right\|_{\mathcal{K}_{\omega,0}^{p,q}(\mathbb{R})}$$

$$\geq 2^{\alpha} \left\| \mathbf{1}_{(\widetilde{k}-1+\widetilde{k}^{-\frac{2}{\alpha p}}, \widetilde{k}+\widetilde{k}^{-\frac{2}{\alpha p}})} \mathbf{1}_{(\xi_{\widetilde{k}}+1, \xi_{\widetilde{k}}+2)} \right\|_{L^p(\mathbb{R})}$$

$$= 2^{\alpha} \left| \left(\widetilde{k} - 1 + \widetilde{k}^{-\frac{2}{\alpha p}}, \widetilde{k} + \widetilde{k}^{-\frac{2}{\alpha p}} \right) \right|^{\frac{1}{p}} = 2^{\alpha},$$

which implies that $\mathbf{1}_E$ does not have an absolutely continuous quasi-norm in $\mathcal{K}_{\omega}^{p,q}(\mathbb{R})$, and hence the inhomogeneous global generalized Herz space $\mathcal{K}_{\omega}^{p,q}(\mathbb{R})$ under consideration does not have an absolutely continuous quasi-norm. This finishes the proof of Example 7.1.18.   □

Remark 7.1.19 We should point out that the assumption $\alpha \in (-\infty, 0)$ in Example 7.1.18 is reasonable, which means that, under this assumption, the space $\mathcal{K}_{\omega}^{p,q}(\mathbb{R}^n)$ with $\omega(t) := t^{\alpha}$ for any $t \in (0, \infty)$ is a ball quasi-Banach function space. Indeed, combining Example 1.1.7 and Theorem 7.1.9, we find that, when $\omega(t) := t^{\alpha}$ for any $t \in (0, \infty)$ and for any given $\alpha \in (-\infty, 0)$, the inhomogeneous global generalized Herz space $\mathcal{K}_{\omega}^{p,q}(\mathbb{R})$ is a ball quasi-Banach function space.

7.1.3 Boundedness of Sublinear Operators and Fefferman–Stein Vector-Valued Inequalities

In this subsection, we first establish a boundedness criterion of sublinear operators on inhomogeneous generalized Herz spaces. As applications, we obtain the boundedness of both the Hardy–Littlewood maximal operator and Calderón–Zygmund operators on these spaces immediately. Then we establish the Fefferman–Stein vector-valued inequalities on inhomogeneous generalized Herz spaces. The conclusions obtained in this subsection play important roles in the study of inhomogeneous generalized Herz–Hardy spaces in next chapter.

To begin with, we show the following boundedness criterion of sublinear operators on both inhomogeneous local and inhomogeneous global generalized Herz spaces.

Theorem 7.1.20 *Let* $p \in (1, \infty]$, $q \in (0, \infty]$, *and* $\omega \in M(\mathbb{R}_+)$ *satisfy*

$$-\frac{n}{p} < m_{\infty}(\omega) \leq M_{\infty}(\omega) < \frac{n}{p'},$$

where $\frac{1}{p} + \frac{1}{p'} = 1$. *Assume that* T *is a sublinear operator bounded on* $L^p(\mathbb{R}^n)$ *and there exists a positive constant* $\widetilde{C}$ *such that, for any* $f \in \mathcal{K}_{\omega,\mathbf{0}}^{p,q}(\mathbb{R}^n)$ *and* $x \notin \overline{\mathrm{supp}\,(f)} := \overline{\{x \in \mathbb{R}^n : f(x) \neq 0\}}$,

$$|T(f)(x)| \leq \widetilde{C} \int_{\mathbb{R}^n} \frac{|f(y)|}{|x-y|^n}\, dy. \tag{7.15}$$

Then there exists a positive constant C, independent of f, such that

(i) *for any* $f \in \mathcal{K}_{\omega,0}^{p,q}(\mathbb{R}^n)$,

$$\|T(f)\|_{\mathcal{K}_{\omega,0}^{p,q}(\mathbb{R}^n)} \leq C\left[\tilde{C} + \|T\|_{L^p(\mathbb{R}^n)\to L^p(\mathbb{R}^n)}\right]\|f\|_{\mathcal{K}_{\omega,0}^{p,q}(\mathbb{R}^n)}; \qquad (7.16)$$

(ii) *for any* $f \in \mathcal{K}_\omega^{p,q}(\mathbb{R}^n)$,

$$\|T(f)\|_{\mathcal{K}_\omega^{p,q}(\mathbb{R}^n)} \leq C\left[\tilde{C} + \|T\|_{L^p(\mathbb{R}^n)\to L^p(\mathbb{R}^n)}\right]\|f\|_{\mathcal{K}_\omega^{p,q}(\mathbb{R}^n)}. \qquad (7.17)$$

Proof Let all the symbols be as in the present theorem. We first prove (i). To this end, let $f \in \mathcal{K}_{\omega,0}^{p,q}(\mathbb{R}^n)$. Then, by Definition 7.1.1(i), we find that

$$\|T(f)\|_{\mathcal{K}_{\omega,0}^{p,q}(\mathbb{R}^n)}$$

$$\lesssim \left\|T(f)\mathbf{1}_{B(0,1)}\right\|_{L^p(\mathbb{R}^n)}$$

$$+ \left\{\sum_{k\in\mathbb{N}}\left[\omega(2^k)\right]^q\left\|T(f)\mathbf{1}_{B(0,2^k)\setminus B(0,2^{k-1})}\right\|_{L^p(\mathbb{R}^n)}^q\right\}^{\frac{1}{q}}$$

$$=: A + B. \qquad (7.18)$$

We next deal with A and B, respectively. Indeed, applying both the sublinearity and the $L^p(\mathbb{R}^n)$ boundedness of the operator T and Definition 7.1.1(i), we conclude that

$$A \lesssim \left\|T\left(f\mathbf{1}_{B(0,2)}\right)\mathbf{1}_{B(0,1)}\right\|_{L^p(\mathbb{R}^n)} + \left\|T\left(f\mathbf{1}_{[B(0,2)]^\complement}\right)\mathbf{1}_{B(0,1)}\right\|_{L^p(\mathbb{R}^n)}$$

$$\lesssim \|T\|_{L^p(\mathbb{R}^n)\to L^p(\mathbb{R}^n)}\left\|f\mathbf{1}_{B(0,2)}\right\|_{L^p(\mathbb{R}^n)} + \left\|T\left(f\mathbf{1}_{[B(0,2)]^\complement}\right)\mathbf{1}_{B(0,1)}\right\|_{L^p(\mathbb{R}^n)}$$

$$\lesssim \|T\|_{L^p(\mathbb{R}^n)\to L^p(\mathbb{R}^n)}\left[\left\|f\mathbf{1}_{B(0,1)}\right\|_{L^p(\mathbb{R}^n)} + \left\|f\mathbf{1}_{B(0,2)\setminus B(0,1)}\right\|_{L^p(\mathbb{R}^n)}\right]$$

$$+ \left\|T\left(f\mathbf{1}_{[B(0,2)]^\complement}\right)\mathbf{1}_{B(0,1)}\right\|_{L^p(\mathbb{R}^n)}$$

$$\lesssim \|T\|_{L^p(\mathbb{R}^n)\to L^p(\mathbb{R}^n)}\|f\|_{\mathcal{K}_{\omega,0}^{p,q}(\mathbb{R}^n)} + \left\|T\left(f\mathbf{1}_{[B(0,2)]^\complement}\right)\mathbf{1}_{B(0,1)}\right\|_{L^p(\mathbb{R}^n)}. \qquad (7.19)$$

Notice that, for any $k \in \mathbb{N} \cap [3,\infty)$ and $x, y \in \mathbb{R}^n$ satisfying $|x| < 1$ and $2^{k-1} \leq |y| < 2^k$, we have

$$|x - y| \geq |y| - |x| > 2^{k-1} - 1 = 2^{k-2},$$

which, combined with (7.15) and the Hölder inequality, implies that, for any $x \in \mathbb{R}^n$ satisfying $|x| < 1$,

$$
\left| T\left(f\mathbf{1}_{[B(0,2)]^\complement}\right)(x)\right| \le \widetilde{C}\int_{|y|\ge 2} \frac{|f(y)|}{|x-y|^n}\, dy
$$

$$
\lesssim \widetilde{C}\sum_{k=3}^\infty 2^{-nk}\int_{2^{k-1}\le|y|<2^k}|f(y)|\,dy
$$

$$
\lesssim \widetilde{C}\sum_{k=3}^\infty 2^{-\frac{nk}{p}}\left\| f\mathbf{1}_{B(0,2^k)\setminus B(0,2^{k-1})}\right\|_{L^p(\mathbb{R}^n)}. \tag{7.20}
$$

In addition, from Lemma 1.1.12, it follows that, for any $k \in \mathbb{N}$,

$$
\omega(2^k)\gtrsim 2^{k[m_\infty(\omega)-\varepsilon_1]},
$$

where $\varepsilon_1 \in (0, m_\infty(\omega) + \frac{n}{p})$ is a fixed positive constant. Thus, using (7.20), Lemma 1.2.11, and the assumption $-m_\infty(\omega) - \frac{n}{p} + \varepsilon_1 \in (-\infty, 0)$, we find that, for any $q \in (0, 1]$,

$$
\left\| T\left(f\mathbf{1}_{[B(0,2)]^\complement}\right)\mathbf{1}_{B(0,1)}\right\|_{L^p(\mathbb{R}^n)}
$$

$$
\lesssim \widetilde{C}\sum_{k=3}^\infty 2^{k[-m_\infty(\omega)-\frac{n}{p}+\varepsilon_1]}\omega(2^k)\left\| f\mathbf{1}_{B(0,2^k)\setminus B(0,2^{k-1})}\right\|_{L^p(\mathbb{R}^n)}
$$

$$
\lesssim \widetilde{C}\left\{\sum_{k=3}^\infty 2^{k[-m_\infty(\omega)-\frac{n}{p}+\varepsilon_1]q}\left[\omega(2^k)\right]^q\left\| f\mathbf{1}_{B(0,2^k)\setminus B(0,2^{k-1})}\right\|_{L^p(\mathbb{R}^n)}^q\right\}^{\frac{1}{q}}
$$

$$
\lesssim \widetilde{C}\left\{\sum_{k\in\mathbb{N}}\left[\omega(2^k)\right]^q\left\| f\mathbf{1}_{B(0,2^k)\setminus B(0,2^{k-1})}\right\|_{L^p(\mathbb{R}^n)}^q\right\}^{\frac{1}{q}}\lesssim \widetilde{C}\|f\|_{\mathcal{K}^{p,q}_{\omega,0}(\mathbb{R}^n)}. \tag{7.21}
$$

Moreover, by the assumption that, for any $k \in \mathbb{N}$, $\omega(2^k)\gtrsim 2^{k[m_\infty(\omega)-\varepsilon_1]}$, the Hölder inequality, and the assumption $-m_\infty(\omega) - \frac{n}{p} + \varepsilon_1 \in (-\infty, 0)$ again, we conclude that, for any $q \in (1, \infty)$,

$$
\left\| T\left(f\mathbf{1}_{[B(0,2)]^\complement}\right)\mathbf{1}_{B(0,1)}\right\|_{L^p(\mathbb{R}^n)}
$$

$$
\lesssim \widetilde{C}\sum_{k=3}^\infty 2^{k[-m_\infty(\omega)-\frac{n}{p}+\varepsilon_1]}\omega(2^k)\left\| f\mathbf{1}_{B(0,2^k)\setminus B(0,2^{k-1})}\right\|_{L^p(\mathbb{R}^n)}
$$

$$\lesssim \widetilde{C} \left\{ \sum_{k=3}^{\infty} 2^{k[-m_{\infty}(\omega)-\frac{n}{p}+\varepsilon_1]q'} \right\}^{\frac{1}{q'}}$$

$$\times \left\{ \sum_{k \in \mathbb{N}} \left[\omega(2^k) \right]^q \left\| f \mathbf{1}_{B(0,2^k)\setminus B(0,2^{k-1})} \right\|_{L^p(\mathbb{R}^n)}^q \right\}^{\frac{1}{q}}$$

$$\sim \widetilde{C} \left\{ \sum_{k \in \mathbb{N}} \left[\omega(2^k) \right]^q \left\| f \mathbf{1}_{B(0,2^k)\setminus B(0,2^{k-1})} \right\|_{L^p(\mathbb{R}^n)}^q \right\}^{\frac{1}{q}} \lesssim \widetilde{C} \| f \|_{\mathcal{K}_{\omega,\mathbf{0}}^{p,q}(\mathbb{R}^n)}.$$

Combining this, (7.19), and (7.21), we further obtain

$$A \lesssim \left[\widetilde{C} + \| T \|_{L^p(\mathbb{R}^n) \to L^p(\mathbb{R}^n)} \right] \| f \|_{\mathcal{K}_{\omega,\mathbf{0}}^{p,q}(\mathbb{R}^n)}, \tag{7.22}$$

where the implicit positive constant is independent of both T and f, which completes the estimate of A.

We now deal with B. To achieve this, for any $k \in \mathbb{N}$, let

$$f_{k,0} := f \mathbf{1}_{B(0,2^{-2})},$$

$$f_{k,1} := f \mathbf{1}_{B(0,2^{k-2})\setminus B(0,2^{-2})},$$

$$f_{k,2} := f \mathbf{1}_{B(0,2^{k+1})\setminus B(0,2^{k-2})},$$

and

$$f_{k,3} := f \mathbf{1}_{[B(0,2^{k+1})]^\complement}.$$

Then, from the sublinearity of T, we deduce that

$$B \lesssim \left\{ \sum_{k \in \mathbb{N}} \left[\omega(2^k) \right]^q \left\| T\left(f_{k,0}\right) \mathbf{1}_{B(0,2^k)\setminus B(0,2^{k-1})} \right\|_{L^p(\mathbb{R}^n)}^q \right\}^{\frac{1}{q}}$$

$$+ \left\{ \sum_{k \in \mathbb{N}} \left[\omega(2^k) \right]^q \left\| T\left(f_{k,1}\right) \mathbf{1}_{B(0,2^k)\setminus B(0,2^{k-1})} \right\|_{L^p(\mathbb{R}^n)}^q \right\}^{\frac{1}{q}}$$

$$+ \left\{ \sum_{k \in \mathbb{N}} \left[\omega(2^k) \right]^q \left\| T\left(f_{k,2}\right) \mathbf{1}_{B(0,2^k)\setminus B(0,2^{k-1})} \right\|_{L^p(\mathbb{R}^n)}^q \right\}^{\frac{1}{q}}$$

$$+ \left\{ \sum_{k \in \mathbb{N}} \left[\omega(2^k) \right]^q \left\| T\left(f_{k,3}\right) \mathbf{1}_{B(0,2^k) \setminus B(0,2^{k-1})} \right\|_{L^p(\mathbb{R}^n)}^q \right\}^{\frac{1}{q}}$$

$$=: \widetilde{J}_0 + \widetilde{J}_1 + \widetilde{J}_2 + \widetilde{J}_3. \tag{7.23}$$

Then we estimate $\widetilde{J}_0$, $\widetilde{J}_1$, $\widetilde{J}_2$, and $\widetilde{J}_3$, respectively. Indeed, using Lemma 1.1.12, we find that, for any $k \in \mathbb{N}$,

$$\omega(2^k) \lesssim 2^{k[M_\infty(\omega) + \varepsilon_2]},$$

where $\varepsilon_2 \in (0, -M_\infty(\omega) + \frac{n}{p'})$ is a fixed positive constant. Moreover, for any $k \in \mathbb{N}$ and $x, y \in \mathbb{R}^n$ satisfying $2^{k-1} \leq |x| < 2^k$ and $|y| < 2^{-2}$, we have

$$|x - y| \geq |x| - |y| > 2^{k-1} - 2^{-2} > 2^{k-2},$$

which, together with (7.15) and the Hölder inequality, further implies that, for any $k \in \mathbb{N}$ and $x \in \mathbb{R}^n$ satisfying $2^{k-1} \leq |x| < 2^k$,

$$\left| T\left(f_{k,0}\right)(x) \right| \leq \widetilde{C} \int_{|y| < 2^{-2}} \frac{|f(y)|}{|x - y|^n} \, dy \lesssim \widetilde{C} 2^{-nk} \left\| f \mathbf{1}_{B(0,1)} \right\|_{L^p(\mathbb{R}^n)}.$$

From this, the assumptions that, for any $k \in \mathbb{N}$, $\omega(2^k) \lesssim 2^{k[M_\infty(\omega) + \varepsilon_2]}$ and $M_\infty(\omega) - \frac{n}{p'} + \varepsilon_2 \in (-\infty, 0)$, and Definition 7.1.1(i), it follows that

$$\widetilde{J}_0 \lesssim \widetilde{C} \left\{ \sum_{k \in \mathbb{N}} 2^{k[M_\infty(\omega) + \varepsilon_2]q} 2^{-nkq} \left\| f \mathbf{1}_{B(0,1)} \right\|_{L^p(\mathbb{R}^n)}^q 2^{\frac{nkq}{p}} \right\}^{\frac{1}{q}}$$

$$\sim \widetilde{C} \left\| f \mathbf{1}_{B(0,1)} \right\|_{L^p(\mathbb{R}^n)} \left\{ \sum_{k \in \mathbb{N}} 2^{k[M_\infty(\omega) - \frac{n}{p'} + \varepsilon_2]q} \right\}^{\frac{1}{q}}$$

$$\sim \widetilde{C} \left\| f \mathbf{1}_{B(0,1)} \right\|_{L^p(\mathbb{R}^n)} \lesssim \widetilde{C} \| f \|_{\mathcal{K}_{\omega,0}^{p,q}(\mathbb{R}^n)}. \tag{7.24}$$

This finishes the estimate of $\widetilde{J}_0$.

In addition, by Lemma 1.1.12, we conclude that, for any $0 < t < \tau < \infty$,

$$\left(\frac{\tau}{t} \right)^{m_\infty(\omega) - \varepsilon_3} \lesssim \frac{\omega(\tau)}{\omega(t)} \lesssim \left(\frac{\tau}{t} \right)^{M_\infty(\omega) + \varepsilon_3},$$

where

$$\varepsilon_3 \in \left(0, \min \left\{ m_\infty(\omega) + \frac{n}{p}, -M_\infty(\omega) + \frac{n}{p'} \right\} \right)$$

is a fixed positive constant. From this, repeating an argument similar to that used in the estimations of J_1, J_2, and J_3 in the proof of Theorem 1.5.1 with $\mathbb{Z}$ therein replaced by $\mathbb{N}$, and Definition 7.1.1(i), we deduce that

$$\widetilde{J}_1 \lesssim \widetilde{C} \|f\|_{\mathcal{K}_{\omega,\mathbf{0}}^{p,q}(\mathbb{R}^n)},$$

$$\widetilde{J}_2 \lesssim \|T\|_{L^p(\mathbb{R}^n) \to L^p(\mathbb{R}^n)} \|f\|_{\mathcal{K}_{\omega,\mathbf{0}}^{p,q}(\mathbb{R}^n)},$$

and

$$\widetilde{J}_3 \lesssim \widetilde{C} \|f\|_{\mathcal{K}_{\omega,\mathbf{0}}^{p,q}(\mathbb{R}^n)}.$$

Combining these, (7.23), (7.24), we further find that

$$B \lesssim \left[\widetilde{C} + \|T\|_{L^p(\mathbb{R}^n) \to L^p(\mathbb{R}^n)}\right] \|f\|_{\mathcal{K}_{\omega,\mathbf{0}}^{p,q}(\mathbb{R}^n)}, \tag{7.25}$$

where the implicit positive constant is independent of both T and f. This then finishes the estimate of B. Therefore, applying (7.18), (7.22), and (7.25), we obtain

$$\|T(f)\|_{\mathcal{K}_{\omega,\mathbf{0}}^{p,q}(\mathbb{R}^n)} \lesssim \left[\widetilde{C} + \|T\|_{L^p(\mathbb{R}^n) \to L^p(\mathbb{R}^n)}\right] \|f\|_{\mathcal{K}_{\omega,\mathbf{0}}^{p,q}(\mathbb{R}^n)},$$

where the implicit positive constant is independent of both T and f. This implies that (7.16) holds true.

We next show (7.17). Indeed, for any given $f \in \mathcal{K}_{\omega}^{p,q}(\mathbb{R}^n)$ and for any $\xi \in \mathbb{R}^n$, from Remark 7.1.2(i), it follows that $\|f(\cdot + \xi)\|_{\mathcal{K}_{\omega,\mathbf{0}}^{p,q}(\mathbb{R}^n)} < \infty$ and hence $f(\cdot + \xi) \in \mathcal{K}_{\omega,\mathbf{0}}^{p,q}(\mathbb{R}^n)$. Thus, by (7.16), we conclude that

$$\|T(f)(\cdot + \xi)\|_{\mathcal{K}_{\omega,\mathbf{0}}^{p,q}(\mathbb{R}^n)} \lesssim \left[\widetilde{C} + \|T\|_{L^p(\mathbb{R}^n) \to L^p(\mathbb{R}^n)}\right] \|f(\cdot + \xi)\|_{\mathcal{K}_{\omega,\mathbf{0}}^{p,q}(\mathbb{R}^n)}.$$

By this, the arbitrariness of ξ, and Remark 7.1.2(i) again, we find that

$$\|T(f)\|_{\mathcal{K}_{\omega}^{p,q}(\mathbb{R}^n)} \lesssim \left[\widetilde{C} + \|T\|_{L^p(\mathbb{R}^n) \to L^p(\mathbb{R}^n)}\right] \|f\|_{\mathcal{K}_{\omega}^{p,q}(\mathbb{R}^n)}$$

with the implicit positive constant independent of f. This finishes the proof of (7.17) and hence of Theorem 7.1.20.   □

Remark 7.1.21 We should point out that, in Theorem 7.1.20, if $\omega(t) := t^\alpha$ for any $t \in (0, \infty)$ and for any given $\alpha \in \mathbb{R}$, then Theorem 7.1.20(i) coincides with the result for classical inhomogeneous Herz spaces obtained in [150, Corollary 2.1].

Applying Theorem 7.1.20 and the facts that the Hardy–Littlewood maximal operator satisfies the condition (7.15) (see, for instance, [197, Remark 4.4]) and is bounded on the Lebesgue space $L^p(\mathbb{R}^n)$ when $p \in (1, \infty]$ (see, for instance, [90, Theorem 2.1.6]), we immediately obtain the following two conclusions which give the boundedness of the Hardy–Littlewood maximal operator, respectively, on inhomogeneous local and inhomogeneous global generalized Herz spaces; we omit the details.

Corollary 7.1.22 *Let p, q, and ω be as in Theorem 7.1.20 and $\mathcal{M}$ the Hardy–Littlewood maximal operator as in* (1.54). *Then there exists a positive constant C such that, for any $f \in L^1_{\mathrm{loc}}(\mathbb{R}^n)$,*

$$\|\mathcal{M}(f)\|_{\mathcal{K}^{p,q}_{\omega,0}(\mathbb{R}^n)} \leq C\|f\|_{\mathcal{K}^{p,q}_{\omega,0}(\mathbb{R}^n)}.$$

Remark 7.1.23 We should point out that, in Corollary 7.1.22, if $\omega(t) := t^\alpha$ for any $t \in (0, \infty)$ and for any given $\alpha \in \mathbb{R}$, the conclusion obtained in this corollary goes back to [130, Theorem 4.1].

Corollary 7.1.24 *Let p, q, and ω be as in Theorem 7.1.20 and $\mathcal{M}$ the Hardy–Littlewood maximal operator as in* (1.54). *Then there exists a positive constant C such that, for any $f \in L^1_{\mathrm{loc}}(\mathbb{R}^n)$,*

$$\|\mathcal{M}(f)\|_{\mathcal{K}^{p,q}_{\omega}(\mathbb{R}^n)} \leq C\|f\|_{\mathcal{K}^{p,q}_{\omega}(\mathbb{R}^n)}.$$

Furthermore, let $d \in \mathbb{Z}_+$. Recall that the d-order Calderón–Zygmund operator is defined as in Definition 1.5.8. Then, repeating an argument similar to that used in the proof of Corollary 1.5.10 with Theorem 1.5.1 therein replaced by Theorem 7.1.20, we obtain the following two boundedness criteria of d-order Calderón–Zygmund operators, respectively, on inhomogeneous local and inhomogeneous global generalized Herz spaces; we omit the details.

Corollary 7.1.25 *Let $p \in (1, \infty)$, $q \in (0, \infty)$, and ω be as in Theorem 7.1.20, $d \in \mathbb{Z}_+$, $\delta \in (0, 1]$, K be a d-order standard kernel as in Definition 1.5.7 with the above δ, and T a d-order Calderón–Zygmund operator as in Definition 1.5.8 with kernel K. Then T is well defined on $\mathcal{K}^{p,q}_{\omega,0}(\mathbb{R}^n)$ and there exists a positive constant C such that, for any $f \in \mathcal{K}^{p,q}_{\omega,0}(\mathbb{R}^n)$,*

$$\|T(f)\|_{\mathcal{K}^{p,q}_{\omega,0}(\mathbb{R}^n)} \leq C\|f\|_{\mathcal{K}^{p,q}_{\omega,0}(\mathbb{R}^n)}.$$

Remark 7.1.26 In Corollary 7.1.25, for any given $\alpha \in \mathbb{R}$ and for any $t \in (0, \infty)$, let $\omega(t) := t^\alpha$. Then we point out that, in this case, Corollary 7.1.25 goes back to [175, Remark 5.1.1].

Corollary 7.1.27 *Let $p \in (1, \infty)$, $q \in (0, \infty)$, and ω be as in Theorem 7.1.20, $d \in \mathbb{Z}_+$, $\delta \in (0, 1]$, K be a d-order standard kernel as in Definition 1.5.7 with the above δ, and T a d-order Calderón–Zygmund operator as in Definition 1.5.8 with kernel K. Then T is well defined on $\mathcal{K}^{p,q}_{\omega}(\mathbb{R}^n)$ and there exists a positive constant C such that, for any $f \in \mathcal{K}^{p,q}_{\omega}(\mathbb{R}^n)$,*

$$\|T(f)\|_{\mathcal{K}^{p,q}_{\omega}(\mathbb{R}^n)} \leq C\|f\|_{\mathcal{K}^{p,q}_{\omega}(\mathbb{R}^n)}.$$

Now, we turn to investigate the Fefferman–Stein vector-valued inequality on inhomogeneous generalized Herz spaces. Indeed, repeating an argument similar to that used in the proof of Theorem 1.6.1 via replacing Theorem 1.5.1(i) therein by Theorem 7.1.20(i), we conclude the following Fefferman–Stein vector-valued inequality on inhomogeneous local generalized Herz spaces; we omit the details.

Theorem 7.1.28 *Let* $p, r \in (1, \infty]$, $q \in (0, \infty]$, *and* $\omega \in M(\mathbb{R}_+)$ *satisfy*

$$-\frac{n}{p} < m_\infty(\omega) \leq M_\infty(\omega) < \frac{n}{p'},$$

where $\frac{1}{p} + \frac{1}{p'} = 1$. *Then there exists a positive constant* C *such that, for any* $\{f_j\}_{j\in\mathbb{N}} \subset L^1_{\mathrm{loc}}(\mathbb{R}^n)$,

$$\left\| \left\{ \sum_{j\in\mathbb{N}} [\mathcal{M}(f_j)]^r \right\}^{\frac{1}{r}} \right\|_{\mathcal{K}^{p,q}_{\omega,0}(\mathbb{R}^n)} \leq C \left\| \left(\sum_{j\in\mathbb{N}} |f_j|^r \right)^{\frac{1}{r}} \right\|_{\mathcal{K}^{p,q}_{\omega,0}(\mathbb{R}^n)}.$$

Remark 7.1.29 We should point out that, in Theorem 7.1.28, when $\omega(t) := t^\alpha$ for any $t \in (0, \infty)$ and for any given $\alpha \in \mathbb{R}$, then Theorem 7.1.28 coincides with the known Fefferman–Stein vector-valued inequality on the classical inhomogeneous Herz spaces obtained in [131, Corollary 4.5].

Similarly, we also have the following Fefferman–Stein vector-valued inequality on the inhomogeneous global generalized Herz space $\mathcal{K}^{p,q}_\omega(\mathbb{R}^n)$. The proof of this theorem is just to repeat the proof of Theorem 1.6.1 with Theorem 1.5.1(i) therein replaced by Theorem 7.1.20(ii); we omit the details.

Theorem 7.1.30 *Let* $p, r \in (1, \infty]$, $q \in (0, \infty]$, *and* $\omega \in M(\mathbb{R}_+)$ *satisfy*

$$-\frac{n}{p} < m_\infty(\omega) \leq M_\infty(\omega) < \frac{n}{p'},$$

where $\frac{1}{p} + \frac{1}{p'} = 1$. *Then there exists a positive constant* C *such that, for any* $\{f_j\}_{j\in\mathbb{N}} \subset L^1_{\mathrm{loc}}(\mathbb{R}^n)$,

$$\left\| \left\{ \sum_{j\in\mathbb{N}} [\mathcal{M}(f_j)]^r \right\}^{\frac{1}{r}} \right\|_{\mathcal{K}^{p,q}_\omega(\mathbb{R}^n)} \leq C \left\| \left(\sum_{j\in\mathbb{N}} |f_j|^r \right)^{\frac{1}{r}} \right\|_{\mathcal{K}^{p,q}_\omega(\mathbb{R}^n)}.$$

7.1.4 Dual and Associate Spaces of Inhomogeneous Local Generalized Herz Spaces

The target of this subsection is to investigate dual spaces and associate spaces of inhomogeneous local generalized Herz spaces. To this end, we first introduce the concept of the inhomogeneous local generalized Herz space $\dot{\mathcal{K}}^{p,q}_{\omega,\xi}(\mathbb{R}^n)$ as follows.

Definition 7.1.31 Let $p, q \in (0, \infty]$, $\omega \in M(\mathbb{R}_+)$, and $\xi \in \mathbb{R}^n$. Then the *inhomogeneous local generalized Herz space* $\mathcal{K}^{p,q}_{\omega,\xi}(\mathbb{R}^n)$ is defined to be the set of all the measurable functions f on $\mathbb{R}^n$ such that

$$\|f\|_{\mathcal{K}^{p,q}_{\omega,\xi}(\mathbb{R}^n)} := \left\{ \left\| f \mathbf{1}_{B(\xi,1)} \right\|^q_{L^p(\mathbb{R}^n)} + \sum_{k\in\mathbb{N}} \left[\omega(2^k)\right]^q \left\| f \mathbf{1}_{B(\xi,2^k)\setminus B(\xi,2^{k-1})} \right\|^q_{L^p(\mathbb{R}^n)} \right\}^{\frac{1}{q}}$$

is finite.

Next, we establish the following dual theorem of inhomogeneous local generalized Herz spaces $\dot{\mathcal{K}}^{p,q}_{\omega,\xi}(\mathbb{R}^n)$.

Theorem 7.1.32 *Let* $p \in [1, \infty)$, $q \in (0, \infty)$, $\omega \in M(\mathbb{R}_+)$, *and* $\xi \in \mathbb{R}^n$. *Then the dual space of* $\mathcal{K}^{p,q}_{\omega,\xi}(\mathbb{R}^n)$, *denoted by* $(\mathcal{K}^{p,q}_{\omega,\xi}(\mathbb{R}^n))^*$, *is* $\mathcal{K}^{p',q'}_{1/\omega,\xi}(\mathbb{R}^n)$ *in the following sense:*

(i) *Let* $g \in \mathcal{K}^{p',q'}_{1/\omega,\xi}(\mathbb{R}^n)$. *Then the linear functional*

$$\phi_g : f \mapsto \phi_g(f) := \int_{\mathbb{R}^n} f(y)g(y)\,dy, \qquad (7.26)$$

defined for any $f \in \mathcal{K}^{p,q}_{\omega,\xi}(\mathbb{R}^n)$, *is bounded on* $\mathcal{K}^{p,q}_{\omega,\xi}(\mathbb{R}^n)$.

(ii) *Conversely, any continuous linear functional on* $\mathcal{K}^{p,q}_{\omega,\xi}(\mathbb{R}^n)$ *arises as in* (7.26) *with a unique* $g \in \mathcal{K}^{p',q'}_{1/\omega,\xi}(\mathbb{R}^n)$.

Here $\frac{1}{p} + \frac{1}{p'} = 1$ *and*

$$q' := \begin{cases} \dfrac{q}{q-1} & \text{when } q \in [1, \infty), \\ \infty & \text{when } q \in (0, 1). \end{cases}$$

Moreover, for any $g \in \mathcal{K}^{p',q'}_{1/\omega,\xi}(\mathbb{R}^n)$,

$$\|g\|_{\mathcal{K}^{p',q'}_{1/\omega,\xi}(\mathbb{R}^n)} = \|\phi_g\|_{(\mathcal{K}^{p,q}_{\omega,\xi}(\mathbb{R}^n))^*}.$$

Proof Let all the symbols be as in the present theorem. We first show (i). Indeed, from the Tonelli theorem and the Hölder inequality, we deduce that, for any $f \in \mathcal{K}_{\omega,\xi}^{p,q}(\mathbb{R}^n)$ and $g \in \mathcal{K}_{1/\omega,\xi}^{p',q'}(\mathbb{R}^n)$,

$$
\begin{aligned}
&\|fg\|_{L^1(\mathbb{R}^n)} \\
&= \int_{B(\xi,1)} |f(y)g(y)| \, dy + \sum_{k\in\mathbb{N}} \int_{B(\xi,2^k)\backslash B(\xi,2^{k-1})} |f(y)g(y)| \, dy \\
&\le \left\| f\mathbf{1}_{B(\xi,1)} \right\|_{L^p(\mathbb{R}^n)} \left\| g\mathbf{1}_{B(\xi,1)} \right\|_{L^{p'}(\mathbb{R}^n)} \\
&\quad + \sum_{k\in\mathbb{N}} \left\| f\mathbf{1}_{B(\xi,2^k)\backslash B(\xi,2^{k-1})} \right\|_{L^p(\mathbb{R}^n)} \left\| g\mathbf{1}_{B(\xi,2^k)\backslash B(\xi,2^{k-1})} \right\|_{L^{p'}(\mathbb{R}^n)} \\
&\le \left\{ \left\| f\mathbf{1}_{B(\xi,1)} \right\|_{L^p(\mathbb{R}^n)}^q + \sum_{k\in\mathbb{N}} \left[\omega(2^k) \right]^q \left\| f\mathbf{1}_{B(\xi,2^k)\backslash B(\xi,2^{k-1})} \right\|_{L^p(\mathbb{R}^n)}^q \right\}^{\frac{1}{q}} \\
&\quad \times \left\{ \left\| g\mathbf{1}_{B(\xi,1)} \right\|_{L^{p'}(\mathbb{R}^n)}^{q'} + \sum_{k\in\mathbb{N}} \left[\omega(2^k) \right]^{-q'} \left\| g\mathbf{1}_{B(\xi,2^k)\backslash B(\xi,2^{k-1})} \right\|_{L^{p'}(\mathbb{R}^n)}^{q'} \right\}^{\frac{1}{q'}} \\
&= \|f\|_{\mathcal{K}_{\omega,\xi}^{p,q}(\mathbb{R}^n)} \|g\|_{\mathcal{K}_{1/\omega,\xi}^{p',q'}(\mathbb{R}^n)}. \tag{7.27}
\end{aligned}
$$

This then implies that the linear functional ϕ_g as in (7.26) is bounded on $\mathcal{K}_{\omega,\xi}^{p,q}(\mathbb{R}^n)$ and

$$
\left\| \phi_g \right\|_{(\mathcal{K}_{\omega,\xi}^{p,q}(\mathbb{R}^n))^*} \le \|g\|_{\mathcal{K}_{1/\omega,\xi}^{p',q'}(\mathbb{R}^n)}, \tag{7.28}
$$

which completes the proof of (i).

Conversely, we show (ii). To this end, let $\phi \in (\mathcal{K}_{\omega,\xi}^{p,q}(\mathbb{R}^n))^*$, $A_{\xi,0} := B(\xi, 1)$, and, for any $k \in \mathbb{N}$,

$$
A_{\xi,k} := B(\xi, 2^k) \setminus B(\xi, 2^{k-1}).
$$

Then, by Definition 7.1.31, we find that, for any given $k \in \mathbb{Z}_+$ and for any $f \in L^p(A_{\xi,k})$,

$$
\|f\|_{\mathcal{K}_{\omega,\xi}^{p,q}(\mathbb{R}^n)} \sim \left\| f\mathbf{1}_{A_{\xi,k}} \right\|_{L^p(\mathbb{R}^n)} \sim \|f\|_{L^p(A_{\xi,k})} < \infty,
$$

which further implies that

$$
|\phi(f)| \le \|\phi\|_{(\mathcal{K}_{\omega,\xi}^{p,q}(\mathbb{R}^n))^*} \|f\|_{\mathcal{K}_{\omega,\xi}^{p,q}(\mathbb{R}^n)} \sim \|\phi\|_{(\mathcal{K}_{\omega,\xi}^{p,q}(\mathbb{R}^n))^*} \|f\|_{L^p(A_{\xi,k})}.
$$

Thus, $\phi \in (L^p(A_{\xi,k}))^*$ for any $k \in \mathbb{Z}_+$. Combining this and the Riesz representation theorem (see [17, Theorem 4.11]), we conclude that, for any $k \in \mathbb{Z}_+$, there exists a unique $g_k \in L^{p'}(A_{\xi,k})$ such that, for any $f \in L^p(A_{\xi,k})$,

$$\phi(f) = \int_{A_{\xi,k}} f(y)g_k(y)\,dy. \tag{7.29}$$

Let

$$g := \sum_{k \in \mathbb{Z}_+} g_k.$$

Then, from (7.29) and an argument similar to that used in the proof of Theorem 1.7.3 with $\{g_k\}_{k\in\mathbb{Z}}$ and $\{B(\xi, 2^k) \setminus B(\xi, 2^{k-1})\}_{k\in\mathbb{Z}}$ therein replaced, respectively, by $\{g_k\}_{k\in\mathbb{Z}_+}$ and $\{A_{\xi,k}\}_{k\in\mathbb{Z}_+}$ here, it follows that $g \in \mathcal{K}^{p',q'}_{1/\omega,\xi}(\mathbb{R}^n)$,

$$\|g\|_{\mathcal{K}^{p',q'}_{1/\omega,\xi}(\mathbb{R}^n)} \leq \|\phi\|_{(\mathcal{K}^{p,q}_{\omega,\xi}(\mathbb{R}^n))^*},$$

and, for any $f \in \mathcal{K}^{p,q}_{\omega,\xi}(\mathbb{R}^n)$,

$$\phi(f) = \int_{\mathbb{R}^n} f(y)g(y)\,dy.$$

These, together with (i) and (7.28), further imply that the linear functional ϕ_g as in (7.26) is bounded on $\mathcal{K}^{p,q}_{\omega,\xi}(\mathbb{R}^n)$, $\phi = \phi_g$, and

$$\|\phi\|_{(\mathcal{K}^{p,q}_{\omega,\xi}(\mathbb{R}^n))^*} = \|\phi_g\|_{(\mathcal{K}^{p,q}_{\omega,\xi}(\mathbb{R}^n))^*} \leq \|g\|_{\mathcal{K}^{p',q'}_{1/\omega,\xi}(\mathbb{R}^n)} \leq \|\phi\|_{(\mathcal{K}^{p,q}_{\omega,\xi}(\mathbb{R}^n))^*}.$$

Therefore, we have

$$\|\phi_g\|_{(\mathcal{K}^{p,q}_{\omega,\xi}(\mathbb{R}^n))^*} = \|g\|_{\mathcal{K}^{p,q}_{\omega,\xi}(\mathbb{R}^n)}.$$

Using this and the linearity of ϕ_g about g, we further find that g is unique. This finishes the proof of (ii) and hence Theorem 7.1.32. $\qquad\square$

Using the above dual theorem, we now prove the following conclusion which shows that, when $p, q \in (1, \infty)$, the associate space of the inhomogeneous local generalized Herz space $\mathcal{K}^{p,q}_{\omega,0}(\mathbb{R}^n)$ is just $\mathcal{K}^{p',q'}_{1/\omega,0}(\mathbb{R}^n)$.

Theorem 7.1.33 *Let $p, q \in (1, \infty)$ and $\omega \in M(\mathbb{R}_+)$. Then*

$$\left(\mathcal{K}^{p,q}_{\omega,0}(\mathbb{R}^n)\right)' = \mathcal{K}^{p',q'}_{1/\omega,0}(\mathbb{R}^n)$$

with the same norms, where $(\mathcal{K}^{p,q}_{\omega,0}(\mathbb{R}^n))'$ denotes the associate space of the inhomogeneous local generalized Herz space $\mathcal{K}^{p,q}_{\omega,0}(\mathbb{R}^n)$.

Proof Let all the symbols be as in the present theorem. Then, applying an argument similar to that used in the proof of Theorem 1.7.9 with Theorems 1.2.42 and 1.4.1 therein replaced, respectively, by Theorems 7.1.7 and 7.1.17, we find that

$$\left(\mathcal{K}^{p,q}_{\omega,0}(\mathbb{R}^n)\right)' = \mathcal{K}^{p',q'}_{1/\omega,0}(\mathbb{R}^n)$$

and, for any $f \in (\mathcal{K}^{p,q}_{\omega,0}(\mathbb{R}^n))'$,

$$\|f\|_{(\mathcal{K}^{p,q}_{\omega,0}(\mathbb{R}^n))'} = \|f\|_{\mathcal{K}^{p',q'}_{1/\omega,0}(\mathbb{R}^n)}.$$

This finishes the proof of Theorem 7.1.33. □

Finally, we point out that the dual space of $\mathcal{K}^{p,q}_{\omega,\xi}(\mathbb{R}^n)$ with $p \in (0,1)$, $q \in (0,\infty)$, $\omega \in M(\mathbb{R}_+)$, and $\xi \in \mathbb{R}^n$ is trivial as follows, whose proof is the same as that of Theorem 1.7.5; we omit the details. This theorem plays an important role in the study of inhomogeneous block spaces in the next section and also of independent interest.

Theorem 7.1.34 *Let* $p \in (0,1)$, $q \in (0,\infty)$, $\omega \in M(\mathbb{R}_+)$, *and* $\xi \in \mathbb{R}^n$. *Then*

$$\left(\mathcal{K}^{p,q}_{\omega,\xi}(\mathbb{R}^n)\right)^* = \{0\}.$$

7.1.5 Extrapolation Theorems

In this subsection, we establish the extrapolation theorems of inhomogeneous generalized Herz spaces, which plays an essential role in the study of the boundedness and the compactness characterizations of commutators on these Herz spaces.

We first give the following extrapolation theorem of inhomogeneous local generalized Herz spaces.

Theorem 7.1.35 *Let* $p, q \in (1,\infty)$, $r_0 \in [1,\infty)$, *and* $\omega \in M(\mathbb{R}_+)$ *satisfy*

$$-\frac{n}{p} < m_\infty(\omega) \le M_\infty(\omega) < \frac{n}{p'}.$$

Assume that $\mathcal{F}$ *is a set of all pairs of nonnegative measurable functions* (F, G) *such that, for any given* $\upsilon \in A_{r_0}(\mathbb{R}^n)$,

$$\int_{\mathbb{R}^n} [F(x)]^{r_0} \upsilon(x)\,dx \le C_{(r,[\upsilon]_{A_{r_0}(\mathbb{R}^n)})} \int_{\mathbb{R}^n} [G(x)]^{r_0} \upsilon(x)\,dx,$$

where the positive constant $C_{(r_0, [\upsilon]_{A_{r_0}}(\mathbb{R}^n))}$ is independent of (F, G), but depending on both r_0 and $[\upsilon]_{A_{r_0}}(\mathbb{R}^n)$. Then there exists a positive constant C such that, for any $(F, G) \in \mathcal{F}$ with $\|F\|_{\mathcal{K}^{p,q}_{\omega,0}(\mathbb{R}^n)} < \infty$,

$$\|F\|_{\mathcal{K}^{p,q}_{\omega,0}(\mathbb{R}^n)} \leq C \, \|G\|_{\mathcal{K}^{p,q}_{\omega,0}(\mathbb{R}^n)} \, .$$

To show this theorem, we require two technical lemmas. The following one is related to the boundedness of the Hardy–Littlewood maximal operator on inhomogeneous local generalized Herz spaces.

Lemma 7.1.36 *Let* $p, q \in (0, \infty)$ *and* $\omega \in M(\mathbb{R}_+)$ *satisfy* $m_\infty(\omega) \in (-\frac{n}{p}, \infty)$. *Then, for any given*

$$r \in \left(0, \min\left\{p, \frac{n}{M_\infty(\omega) + n/p}\right\}\right),$$

there exists a positive constant C *such that, for any* $f \in L^1_{\mathrm{loc}}(\mathbb{R}^n)$,

$$\|\mathcal{M}(f)\|_{[\mathcal{K}^{p,q}_{\omega,0}(\mathbb{R}^n)]^{1/r}} \leq C \|f\|_{[\mathcal{K}^{p,q}_{\omega,0}(\mathbb{R}^n)]^{1/r}} \, .$$

Proof Let all the symbols be as in the present lemma and

$$r \in \left(0, \min\left\{p, \frac{n}{M_\infty(\omega) + n/p}\right\}\right).$$

Then, from an argument similar to that used in the proof of Lemma 1.8.5 with Lemma 1.3.1 and Corollary 1.5.4 therein replaced, respectively, by Lemma 7.1.13 and Corollary 7.1.22, it follows that, for any $f \in L^1_{\mathrm{loc}}(\mathbb{R}^n)$,

$$\|\mathcal{M}(f)\|_{[\mathcal{K}^{p,q}_{\omega,0}(\mathbb{R}^n)]^{1/r}} \lesssim \|f\|_{[\mathcal{K}^{p,q}_{\omega,0}(\mathbb{R}^n)]^{1/r}} \, .$$

This then finishes the proof of Lemma 7.1.36.   □

On the other hand, we need the following conclusion about the boundedness of the Hardy–Littlewood maximal operator on associate spaces of inhomogeneous local generalized Herz spaces.

Lemma 7.1.37 *Let* $p, q \in (0, \infty)$ *and* $\omega \in M(\mathbb{R}_+)$ *satisfy* $m_\infty(\omega) \in (-\frac{n}{p}, \infty)$. *Then, for any given*

$$s \in \left(0, \min\left\{p, q, \frac{n}{M_\infty(\omega) + n/p}\right\}\right)$$

and

$$r \in \left(\max \left\{ p, \frac{n}{m_\infty(\omega) + n/p} \right\}, \infty \right],$$

the Herz space $[\mathcal{K}_{\omega,0}^{p,q}(\mathbb{R}^n)]^{1/s}$ *is a ball Banach function space and there exists a positive constant* C *such that, for any* $f \in L_{\mathrm{loc}}^1(\mathbb{R}^n)$,

$$\left\| \mathcal{M}^{((r/s)')}(f) \right\|_{([\mathcal{K}_{\omega,0}^{p,q}(\mathbb{R}^n)]^{1/s})'} \le C \, \|f\|_{([\mathcal{K}_{\omega,0}^{p,q}(\mathbb{R}^n)]^{1/s})'}.$$

Proof Let all the symbols be as in the present lemma,

$$s \in \left(0, \min \left\{ p, q, \frac{n}{M_\infty(\omega) + n/p} \right\} \right),$$

and

$$r \in \left(\max \left\{ p, \frac{n}{m_\infty(\omega) + n/p} \right\}, \infty \right].$$

Then, by the proof of Lemma 1.8.6 via replacing Lemma 1.3.1, Corollary 1.5.4, and Theorem 1.7.9 therein, respectively, by Lemma 7.1.13, Corollary 7.1.22, and Theorem 7.1.33, we conclude that the Herz space $[\mathcal{K}_{\omega,0}^{p,q}(\mathbb{R}^n)]^{1/s}$ is a BBF space and, for any $f \in L_{\mathrm{loc}}^1(\mathbb{R}^n)$,

$$\left\| \mathcal{M}^{((r/s)')}(f) \right\|_{([\mathcal{K}_{\omega,0}^{p,q}(\mathbb{R}^n)]^{1/s})'} \lesssim \|f\|_{([\mathcal{K}_{\omega,0}^{p,q}(\mathbb{R}^n)]^{1/s})'},$$

which completes the proof of Lemma 7.1.37.   □

Via the above two lemmas and the known extrapolation theorem of ball Banach function spaces obtained by Tao et al. [228, Lemma 2.13] (see also Lemma 1.8.3 above), we next prove Theorem 7.1.35.

Proof of Theorem 7.1.35 Let all the symbols be as in the present theorem. Then, repeating the proof of Theorem 1.8.2 with Theorem 1.2.46 and Lemmas 1.8.5 and 1.8.6 used therein replaced, respectively, by Theorem 7.1.11 and Lemmas 7.1.36 and 7.1.37, we find that the inhomogeneous local generalized Herz space $\mathcal{K}_{\omega,0}^{p,q}(\mathbb{R}^n)$ under consideration is a BBF space and the Hardy–Littlewood maximal operator $\mathcal{M}$ as in (1.54) is bounded on both $\mathcal{K}_{\omega,0}^{p,q}(\mathbb{R}^n)$ and $(\mathcal{K}_{\omega,0}^{p,q}(\mathbb{R}^n))'$. From these and

Lemma 1.8.3 with $X := \mathcal{K}_{\omega,0}^{p,q}(\mathbb{R}^n)$, we deduce that, for any $(F, G) \in \mathcal{F}$ and $\|F\|_{\mathcal{K}_{\omega,0}^{p,q}(\mathbb{R}^n)} < \infty$,

$$\|F\|_{\mathcal{K}_{\omega,0}^{p,q}(\mathbb{R}^n)} \lesssim \|G\|_{\mathcal{K}_{\omega,0}^{p,q}(\mathbb{R}^n)}.$$

Thus, the proof of Theorem 7.1.35 is completed.   □

Finally, we establish the extrapolation theorem of inhomogeneous global generalized Herz spaces as follows.

Theorem 7.1.38 *Let* $p, q \in (1, \infty)$, $r_0 \in [1, \infty)$, *and* $\omega \in M(\mathbb{R}_+)$ *satisfy*

$$-\frac{n}{p} < m_\infty(\omega) \leq M_\infty(\omega) < \frac{n}{p'}.$$

Assume that $\mathcal{F}$ *is a set of all pairs of nonnegative measurable functions* (F, G) *such that, for any given* $\upsilon \in A_{r_0}(\mathbb{R}^n)$,

$$\int_{\mathbb{R}^n} [F(x)]^{r_0} \upsilon(x)\, dx \leq C_{(r,[\upsilon]_{A_{r_0}(\mathbb{R}^n)})} \int_{\mathbb{R}^n} [G(x)]^{r_0} \upsilon(x)\, dx,$$

where the positive constant $C_{(r_0,[\upsilon]_{A_{r_0}(\mathbb{R}^n)})}$ *is independent of* (F, G), *but depending on both* r_0 *and* $[\upsilon]_{A_{r_0}(\mathbb{R}^n)}$. *Then there exists a positive constant* C *such that, for any* $(F, G) \in \mathcal{F}$ *with* $\|F\|_{\mathcal{K}_\omega^{p,q}(\mathbb{R}^n)} < \infty$,

$$\|F\|_{\mathcal{K}_\omega^{p,q}(\mathbb{R}^n)} \leq C \|G\|_{\mathcal{K}_\omega^{p,q}(\mathbb{R}^n)}.$$

Proof Let all the symbols be as in the present theorem and $(F, G) \in \mathcal{F}$ with $\|F\|_{\mathcal{K}_\omega^{p,q}(\mathbb{R}^n)} < \infty$. Then, repeating the proof of Theorem 1.8.7 via replacing Remark 1.2.2(ii) and Theorem 1.8.2 therein, respectively, by Remark 7.1.2(i) and Theorem 7.1.35, we obtain

$$\|F\|_{\mathcal{K}_\omega^{p,q}(\mathbb{R}^n)} \lesssim \|G\|_{\mathcal{K}_\omega^{p,q}(\mathbb{R}^n)}.$$

This finishes the proof of Theorem 7.1.38.   □

7.2 Inhomogeneous Block Spaces and Their Applications

Let $p, q \in (0, \infty)$ and $\omega \in M(\mathbb{R}_+)$. In this section, we first introduce the inhomogeneous block spaces $\mathcal{B}_\omega^{p,q}(\mathbb{R}^n)$ and establish the duality between these spaces and inhomogeneous global generalized Herz spaces. As applications, we obtain the boundedness of both sublinear operators and Calderón–Zygmund operators on inhomogeneous block spaces. In particular, the boundedness of powered Hardy–Littlewood maximal operators on inhomogeneous block spaces can be deduced directly from the boundedness of sublinear operators, which plays an important role in the next chapter.

7.2.1 Inhomogeneous Block Spaces

In this subsection, we introduce the inhomogeneous counterparts of block spaces as in Chap. 2. To this end, we first present some necessary concepts. For any $k \in \mathbb{Z}$, we use $\mathcal{D}_k$ as in (2.1) to denote the set of all the standard dyadic cubes on $\mathbb{R}^n$ of level $-k$. Moreover, for any $k \in \mathbb{N}$ and $\xi \in \mathbb{R}^n$, let

$$\mathscr{Q}_k^{(0)} := \mathcal{D}_{k-1} \cap \left[Q(0, 2^{k+1}) \setminus Q(0, 2^k) \right],$$

$$\mathscr{Q}_k^{(\xi)} := \left\{ Q \in \mathcal{Q} : Q - \{\xi\} \in \mathscr{Q}_k^{(0)} \right\},$$

and $\mathscr{Q}_0^{(\xi)} := \{Q(\xi, 2)\}$.

Remark 7.2.1 Obviously, for any $\xi \in \mathbb{R}^n$ and $k \in \mathbb{N}$,

$$\mathscr{Q}_k^{(k)} = \mathcal{Q}_k^{(\xi)},$$

where $\mathcal{Q}_k^{(\xi)}$ is defined as in (2.2). Combining this and Remark 2.1.2, we further find that, for any $\xi \in \mathbb{R}^n$ and $k \in \mathbb{N}$,

$$\sharp \mathscr{Q}_k^{(\xi)} = 2^{2n} - 2^n.$$

Recall that the concept of (ω, p)-blocks is introduced in Definition 2.1.1. Then we give the definition of inhomogeneous block spaces via (ω, p)-blocks as follows.

Definition 7.2.2 Let $p, q \in (0, \infty)$ and $\omega \in M(\mathbb{R}_+)$. Then the *inhomogeneous block space* $\mathcal{B}_\omega^{p,q}(\mathbb{R}^n)$ is defined to be the set of all the measurable functions f on $\mathbb{R}^n$ such that both

$$f = \sum_{l \in \mathbb{N}} \sum_{k \in \mathbb{Z}_+} \sum_{Q \in \mathscr{Q}_k^{(\xi_l)}} \lambda_{\xi_l, k, Q} b_{\xi_l, k, Q} \tag{7.30}$$

almost everywhere in $\mathbb{R}^n$ and

$$\left\{\sum_{l\in\mathbb{N}}\left[\sum_{k\in\mathbb{Z}_+}\sum_{Q\in\mathscr{Q}_k^{(\xi_l)}}\lambda_{\xi_l,k,Q}^q\right]^{\frac{1}{q}}\right\} < \infty,$$

where $\{\xi_l\}_{l\in\mathbb{N}} \subset \mathbb{R}^n$, $\{\lambda_{\xi_l,k,Q}\}_{l\in\mathbb{N},\,k\in\mathbb{Z}_+,\,Q\in\mathscr{Q}_k^{(\xi_l)}} \subset [0,\infty)$, and, for any $l\in\mathbb{N}$, $k\in\mathbb{Z}_+$, and $Q\in\mathscr{Q}_k^{(\xi_l)}$, $b_{\xi_l,k,Q}$ is an $(\omega,\,p)$-block supported in the cube Q. Moreover, for any $f\in\mathscr{M}(\mathbb{R}^n)$,

$$\|f\|_{\mathcal{B}_\omega^{p,q}(\mathbb{R}^n)} := \inf\left\{\sum_{l\in\mathbb{N}}\left[\sum_{k\in\mathbb{Z}_+}\sum_{Q\in\mathscr{Q}_k^{(\xi_l)}}\lambda_{\xi_l,k,Q}^q\right]^{\frac{1}{q}}\right\},$$

where the infimum is taken over all the decompositions of f as above.

We point out that, for any $p, q \in (0, \infty)$ and $\omega \in M(\mathbb{R}_+)$, the inhomogeneous block space $\mathcal{B}_\omega^{p,q}(\mathbb{R}^n)$ is a linear space equipped with the seminorm $\|\cdot\|_{\mathcal{B}_\omega^{p,q}(\mathbb{R}^n)}$. Next, we turn to study the convergence and the embedding properties of inhomogeneous block spaces. Indeed, we have the following result which implies that, under some assumptions, the inhomogeneous block space $\mathcal{B}_\omega^{p,q}(\mathbb{R}^n)$ embeds continuously into the Lebesgue space $L^p(\mathbb{R}^n)$, and hence the series (7.30) converges in $L^p(\mathbb{R}^n)$.

Theorem 7.2.3 *Let* $p \in [1, \infty)$, $q \in (0, \infty)$, *and* $\omega \in M(\mathbb{R}_+)$ *satisfy* $m_\infty(\omega) \in (0, \infty)$. *Then the inhomogeneous block space* $\mathcal{B}_\omega^{p,q}(\mathbb{R}^n)$ *embeds continuously into* $L^p(\mathbb{R}^n)$. *Namely, there exists a positive constant* C *such that, for any* $f \in \mathcal{B}_\omega^{p,q}(\mathbb{R}^n)$,

$$\|f\|_{L^p(\mathbb{R}^n)} \le C \|f\|_{\mathcal{B}_\omega^{p,q}(\mathbb{R}^n)}.$$

Proof Let all the symbols be the same as in the present theorem, and $f \in \mathcal{B}_\omega^{p,q}(\mathbb{R}^n)$. Assume $\{\xi_l\}_{l\in\mathbb{N}} \subset \mathbb{R}^n$, $\{\lambda_{\xi_l,k,Q}\}_{l\in\mathbb{N},\,k\in\mathbb{Z}_+,\,Q\in\mathscr{Q}_k^{(\xi_l)}} \subset [0,\infty)$, and, for any $l\in\mathbb{N}$, $k\in\mathbb{Z}_+$, and $Q\in\mathscr{Q}_k^{(\xi_l)}$, $b_{\xi_l,k,Q}$ is an $(\omega,\,p)$-block supported in the cube Q satisfying that both

$$f = \sum_{l\in\mathbb{N}}\sum_{k\in\mathbb{Z}_+}\sum_{Q\in\mathscr{Q}_k^{(\xi_l)}}\lambda_{\xi_l,k,Q}b_{\xi_l,k,Q} \qquad (7.31)$$

almost everywhere in $\mathbb{R}^n$ and

$$\sum_{l\in\mathbb{N}}\left[\sum_{k\in\mathbb{Z}_+}\sum_{Q\in\mathscr{Q}_k^{(\xi_l)}}\lambda_{\xi_l,k,Q}^q\right]^{\frac{1}{q}}<\infty. \tag{7.32}$$

Then, applying (7.31), the Minkowski inequality, Definition 2.1.1, and Lemma 1.1.3, we find that

$$\|f\|_{L^p(\mathbb{R}^n)}\leq\sum_{l\in\mathbb{N}}\sum_{k\in\mathbb{Z}_+}\sum_{Q\in\mathscr{Q}_k^{(\xi_l)}}\lambda_{\xi_l,k,Q}\left\|b_{\xi_l,k,Q}\right\|_{L^p(\mathbb{R}^n)}$$

$$\lesssim\sum_{l\in\mathbb{N}}\sum_{k\in\mathbb{Z}_+}\sum_{Q\in\mathscr{Q}_k^{(\xi_l)}}\lambda_{\xi_l,k,Q}\left[\omega(2^k)\right]^{-1}$$

$$\sim\sum_{l\in\mathbb{N}}\lambda_{\xi_l,0,Q(\xi_l,2)}+\sum_{l\in\mathbb{N}}\sum_{k\in\mathbb{N}}\sum_{Q\in\mathscr{Q}_k^{(\xi_l)}}\lambda_{\xi_l,k,Q}\left[\omega(2^k)\right]^{-1}.$$

From this, (7.32), and (2.14), it follows that

$$\|f\|_{L^p(\mathbb{R}^n)}\lesssim\sum_{l\in\mathbb{N}}\lambda_{\xi_l,0,Q(\xi_l,2)}+\sum_{l\in\mathbb{N}}\left[\sum_{k\in\mathbb{N}}\sum_{Q\in\mathscr{Q}_k^{(\xi_l)}}\lambda_{\xi_l,k,Q}^q\right]^{\frac{1}{q}}$$

$$\sim\sum_{l\in\mathbb{N}}\left[\sum_{k\in\mathbb{Z}_+}\sum_{Q\in\mathscr{Q}_k^{(\xi_l)}}\lambda_{\xi_l,k,Q}^q\right]^{\frac{1}{q}}<\infty. \tag{7.33}$$

This then implies that $f\in L^p(\mathbb{R}^n)$. Moreover, combining (7.33), the choice of $\{\lambda_{\xi_l,k,Q}\}_{l\in\mathbb{N},\,k\in\mathbb{Z}_+,\,Q\in\mathscr{Q}_k^{(\xi_l)}}$, and Definition 7.2.2, we further obtain

$$\|f\|_{L^p(\mathbb{R}^n)}\lesssim\|f\|_{\dot{\mathcal{B}}_\omega^{p,q}(\mathbb{R}^n)},$$

which completes the proof of Theorem 7.2.3. □

7.2.2 Duality Between Inhomogeneous Block Spaces and Global Generalized Herz Spaces

The main target of this subsection is to establish the duality between inhomogeneous block spaces and inhomogeneous global generalized Herz spaces. Namely, we have the following dual result.

Theorem 7.2.4 *Let* $p \in [1, \infty)$, $q \in (0, \infty)$, *and* $\omega \in M(\mathbb{R}_+)$. *Then the dual space of* $\mathcal{B}_\omega^{p,q}(\mathbb{R}^n)$ *is* $\mathcal{K}_{1/\omega}^{p',q'}(\mathbb{R}^n)$ *in the following sense:*

(i) *Let* $g \in \mathcal{K}_{1/\omega}^{p',q'}(\mathbb{R}^n)$. *Then the linear functional*

$$\phi_g : f \mapsto \phi_g(f) := \int_{\mathbb{R}^n} f(y)g(y)\,dy, \qquad (7.34)$$

defined for any $f \in \mathcal{B}_\omega^{p,q}(\mathbb{R}^n)$, *is bounded on* $\mathcal{B}_\omega^{p,q}(\mathbb{R}^n)$.
(ii) *Conversely, any continuous linear functional on* $\mathcal{B}_\omega^{p,q}(\mathbb{R}^n)$ *arises as in* (7.34) *with a unique* $g \in \mathcal{K}_{1/\omega}^{p',q'}(\mathbb{R}^n)$.

Here $\frac{1}{p} + \frac{1}{p'} = 1$ *and*

$$q' := \begin{cases} \dfrac{q}{q-1} & \text{when } q \in [1, \infty), \\ \infty & \text{when } q \in (0, 1). \end{cases}$$

Moreover, there exist two positive constants C_1 *and* C_2 *such that, for any* $g \in \mathcal{K}_{1/\omega}^{p',q'}(\mathbb{R}^n)$,

$$C_1 \|g\|_{\mathcal{K}_{1/\omega}^{p',q'}(\mathbb{R}^n)} \leq \|\phi_g\|_{(\mathcal{B}_\omega^{p,q}(\mathbb{R}^n))^*} \leq C_2 \|g\|_{\mathcal{K}_{1/\omega}^{p',q'}(\mathbb{R}^n)},$$

where $(\mathcal{B}_\omega^{p,q}(\mathbb{R}^n))^*$ *denotes the dual space of* $\mathcal{B}_\omega^{p,q}(\mathbb{R}^n)$.

To show this dual theorem, we require the following equivalent characterizations of both inhomogeneous local generalized Herz spaces $\mathcal{K}_{\omega,\xi}^{p,q}(\mathbb{R}^n)$ and inhomogeneous block spaces $\mathcal{B}_\omega^{p,q}(\mathbb{R}^n)$.

Lemma 7.2.5 *Let* $p, q \in (0, \infty)$ *and* $\omega \in M(\mathbb{R}_+)$. *Then*

(i) *for any given* $\xi \in \mathbb{R}^n$, *a measurable function* f *belongs to* $\mathcal{K}_{\omega,\xi}^{p,q}(\mathbb{R}^n)$ *if and only if*

$$f = \sum_{k \in \mathbb{Z}_+} \sum_{Q \in \mathscr{Q}_k^{(\xi)}} \lambda_{k,Q} b_{k,Q}$$

almost everywhere in $\mathbb{R}^n$ *and*

$$\left[\sum_{k\in\mathbb{Z}_+}\sum_{Q\in\mathscr{Q}_k^{(\xi)}}\lambda_{k,Q}^q\right]^{\frac{1}{q}} < \infty,$$

where $\{\lambda_{k,Q}\}_{k\in\mathbb{Z}_+,\,Q\in\mathscr{Q}_k^{(\xi)}} \subset [0,\infty)$ *and, for any* $k\in\mathbb{Z}$ *and* $Q\in\mathscr{Q}_k^{(\xi)}$, $b_{k,Q}$ *is an* (ω, p)-*block supported in the cube* Q. *Moreover, for any* $f\in\mathcal{K}_{\omega,\xi}^{p,q}(\mathbb{R}^n)$,

$$\|f\|_{\mathcal{K}_{\omega,\xi}^{p,q}(\mathbb{R}^n)} \sim \left[\sum_{k\in\mathbb{Z}_+}\sum_{Q\in\mathscr{Q}_k^{(\xi)}}\lambda_{k,Q}^q\right]^{\frac{1}{q}},$$

where the positive equivalence constants are independent of both ξ *and* f;
(ii) *a measurable function* f *belongs to* $\mathcal{B}_\omega^{p,q}(\mathbb{R}^n)$ *if and only if*

$$\|f\|_{\mathcal{B}_\omega^{p,q}(\mathbb{R}^n)}^\star := \inf\left\{\sum_{l\in\mathbb{N}}\|f_{\xi_l}\|_{\mathcal{K}_{\omega,\xi_l}^{p,q}(\mathbb{R}^n)}\right\} < \infty,$$

where the infimum is taken over all the sequences $\{\xi_l\}_{l\in\mathbb{N}}\subset\mathbb{R}^n$ *and* $\{f_{\xi_l}\}_{l\in\mathbb{N}}\subset$
$\mathcal{M}(\mathbb{R}^n)$ *such that, for any* $l\in\mathbb{N}$, $f_{\xi_l}\in\mathcal{K}_{\omega,\xi_l}^{p,q}(\mathbb{R}^n)$ *and* $f=\sum_{l\in\mathbb{N}}f_{\xi_l}$ *almost everywhere in* $\mathbb{R}^n$. *Moreover, for any* $f\in\mathcal{B}_\omega^{p,q}(\mathbb{R}^n)$,

$$\|f\|_{\mathcal{B}_\omega^{p,q}(\mathbb{R}^n)} \sim \|f\|_{\mathcal{B}_\omega^{p,q}(\mathbb{R}^n)}^\star,$$

where the positive equivalence constants are independent of f.

Proof Let all the symbols be as in the present lemma. We first prove the necessity of (i). For this purpose, fix a $\xi\in\mathbb{R}^n$ and an $f\in\mathcal{K}_{\omega,\xi}^{p,q}(\mathbb{R}^n)$. For any $k\in\mathbb{Z}_+$ and $Q\in\mathscr{Q}_k^{(\xi)}$, let

$$\lambda_{k,Q} := \omega\left(|Q|^{1/n}\right)\|f\mathbf{1}_Q\|_{L^p(\mathbb{R}^n)}.$$

Moreover, for any $k\in\mathbb{Z}_+$ and $Q\in\mathscr{Q}_k^{(\xi)}$, let $b_{k,Q} := 0$ when $\|f\mathbf{1}_Q\|_{L^p(\mathbb{R}^n)}=0$; otherwise, let

$$b_{k,Q} := \left[\omega\left(|Q|^{\frac{1}{n}}\right)\right]^{-1}\left[\|f\mathbf{1}_Q\|_{L^p(\mathbb{R}^n)}\right]^{-1}f\mathbf{1}_Q.$$

Then, for any $k \in \mathbb{Z}_+$ and $Q \in \mathscr{Q}_k^{(\xi)}$, $b_{k,Q}$ is an (ω, p)-block supported in Q and

$$f = f\mathbf{1}_{Q(\xi,2)} + \sum_{k \in \mathbb{N}} f\mathbf{1}_{Q(\xi,2^{k+1}) \setminus Q(\xi,2^k)}$$

$$= \sum_{k \in \mathbb{Z}_+} \sum_{Q \in \mathscr{Q}_k^{(\xi)}} f\mathbf{1}_Q = \sum_{k \in \mathbb{Z}_+} \sum_{Q \in \mathscr{Q}_k^{(\xi)}} \lambda_{k,Q} b_{k,Q}.$$

Since $Q(\xi, 2) \subset B(\xi, 2)$, from this and Definition 7.1.31, we deduce that

$$\sum_{Q \in \mathscr{Q}_0^{(\xi)}} \lambda_{k,Q}^q = [\omega(2)]^q \left\| f\mathbf{1}_{Q(\xi,2)} \right\|_{L^p(\mathbb{R}^n)}^q$$

$$\lesssim \left\| f\mathbf{1}_{B(\xi,1)} \right\|_{L^p(\mathbb{R}^n)}^q + [\omega(2)]^q \left\| f\mathbf{1}_{B(\xi,2) \setminus B(\xi,1)} \right\|_{L^p(\mathbb{R}^n)}^q$$

$$\lesssim \| f \|_{\mathcal{K}_{\omega,\xi}^{p,q}(\mathbb{R}^n)}^q < \infty. \tag{7.35}$$

On the other hand, by the fact that, for any $k \in \mathbb{N}$,

$$\mathscr{Q}_k^{(\xi)} = \mathcal{Q}_k^{(\xi)},$$

the estimation of (2.17), and Definition 7.1.31 again, we obtain

$$\sum_{k \in \mathbb{N}} \sum_{Q \in \mathscr{Q}_k^{(\xi)}} \lambda_{k,Q}^q \lesssim \sum_{k \in \mathbb{N}} \left[\omega(2^k) \right]^q \left\| f\mathbf{1}_{B(\xi,2^k) \setminus B(\xi,2^{k-1})} \right\|_{L^p(\mathbb{R}^n)}^q \lesssim \| f \|_{\mathcal{K}_{\omega,\xi}^{p,q}(\mathbb{R}^n)}^q.$$

This, combined with (7.35), further implies that

$$\left[\sum_{k \in \mathbb{Z}_+} \sum_{Q \in \mathscr{Q}_k^{(\xi)}} \lambda_{k,Q}^q \right]^{\frac{1}{q}} \lesssim \| f \|_{\mathcal{K}_{\omega,\xi}^{p,q}(\mathbb{R}^n)} < \infty, \tag{7.36}$$

where the implicit positive constant is independent of both ξ and f, which completes the proof of the necessity of (i).

Conversely, we now show the sufficiency of (i). To this end, let $f \in \mathscr{M}(\mathbb{R}^n)$ satisfy that both

$$f = \sum_{k \in \mathbb{Z}_+} \sum_{Q \in \mathscr{Q}_k^{(\xi)}} \lambda_{k,Q} b_{k,Q}$$

almost everywhere in $\mathbb{R}^n$ and

$$\left[\sum_{k\in\mathbb{Z}_+}\sum_{Q\in\mathscr{Q}_k^{(\xi)}}\lambda_{k,Q}^q\right]^{\frac{1}{q}}<\infty,$$

where $\{\lambda_{k,Q}\}_{k\in\mathbb{Z}_+,\,Q\in\mathscr{Q}_k^{(\xi)}}\subset[0,\infty)$ and, for any $k\in\mathbb{Z}_+$ and $Q\in\mathscr{Q}_k^{(\xi)}$, $b_{k,Q}$ is an $(\omega,\,p)$-block supported in the cube Q. Then, using the fact that $B(\xi,2)\subset Q(\xi,4)$, Remark 7.2.1, and Definition 2.1.1, we conclude that

$$\left\|f\mathbf{1}_{B(\xi,2)}\right\|_{L^p(\mathbb{R}^n)}^q\lesssim\left\|\sum_{k=0}^{1}\sum_{Q\in\mathscr{Q}_k^{(\xi)}}\lambda_{k,Q}b_{k,Q}\right\|_{L^p(\mathbb{R}^n)}^q$$

$$\sim\sum_{k=0}^{1}\sum_{Q\in\mathscr{Q}_k^{(\xi)}}\lambda_{k,Q}^q\left\|b_{k,Q}\right\|_{L^p(\mathbb{R}^n)}^q\lesssim\sum_{k=0}^{1}\sum_{Q\in\mathscr{Q}_k^{(\xi)}}\lambda_{k,Q}^q. \qquad (7.37)$$

On the other hand, by the fact that, for any $k\in\mathbb{N}$,

$$\mathscr{Q}_k^{(\xi)}=Q_k^{(\xi)}$$

and the estimation of (2.18), we find that, for any $j\in\mathbb{N}\cap[2,\infty)$,

$$\left\|f\mathbf{1}_{B(\xi,2^j)\setminus B(\xi,2^{j-1})}\right\|_{L^p(\mathbb{R}^n)}^q\lesssim\left[\omega(2^j)\right]^{-q}\sum_{k=j}^{j+1}\sum_{Q\in\mathscr{Q}_k^{(\xi)}}\lambda_{k,Q}^q.$$

Combining this, (7.37), and Definition 7.1.31, we obtain

$$\|f\|_{\mathcal{K}_{\omega,\xi}^{p,q}(\mathbb{R}^n)}\lesssim\left\{\left\|f\mathbf{1}_{B(\xi,2)}\right\|_{L^p(\mathbb{R}^n)}^q+\sum_{j=2}^{\infty}\left[\omega(2^j)\right]^q\left\|f\mathbf{1}_{B(\xi,2^j)\setminus B(\xi,2^{j-1})}\right\|_{L^p(\mathbb{R}^n)}^q\right\}^{\frac{1}{q}}$$

$$\sim\left[\sum_{k\in\mathbb{Z}_+}\sum_{Q\in\mathscr{Q}_k^{(\xi)}}\lambda_{k,Q}^q\right]^{\frac{1}{q}}<\infty,$$

where the implicit positive constants are independent of both ξ and f, which further implies that $f \in \mathcal{K}_{\omega,\xi}^{p,q}(\mathbb{R}^n)$ and hence completes the proof of the sufficiency of (i). Moreover, from this and (7.36), it follows that, for any $f \in \mathcal{K}_{\omega,\xi}^{p,q}(\mathbb{R}^n)$,

$$\|f\|_{\mathcal{K}_{\omega,\xi}^{p,q}(\mathbb{R}^n)} \sim \left[\sum_{k \in \mathbb{Z}_+} \sum_{Q \in \mathscr{Q}_k^{(\xi)}} \lambda_{k,Q}^q\right]^{\frac{1}{q}},$$

where the positive equivalence constants are independent of both ξ and f. This finishes the proof of (i).

Next, repeating the proof of Theorem 2.2.1(ii) via replacing Theorem 2.2.1(i) therein by (i), we find that (ii) holds true, which then completes the proof of Theorem 7.2.4. □

From Lemma 7.2.5, we immediately deduce the following relation between inhomogeneous local generalized Herz spaces and inhomogeneous block spaces, which is useful in the below proof of Theorem 7.2.4; we omit the details.

Lemma 7.2.6 *Let* $p, q \in (0, \infty)$ *and* $\omega \in M(\mathbb{R}_+)$. *Then, for any given* $\xi \in \mathbb{R}^n$, $\mathcal{K}_{\omega,\xi}^{p,q}(\mathbb{R}^n) \subset \mathcal{B}_\omega^{p,q}(\mathbb{R}^n)$. *Moreover, there exists a positive constant* C, *independent of* ξ, *such that, for any* $f \in \mathcal{K}_{\omega,\xi}^{p,q}(\mathbb{R}^n)$,

$$\|f\|_{\mathcal{B}_\omega^{p,q}(\mathbb{R}^n)} \leq C\|f\|_{\mathcal{K}_{\omega,\xi}^{p,q}(\mathbb{R}^n)}.$$

Via the above preparations, we now show Theorem 7.2.4.

Proof of Theorem 7.2.4 Let all the symbols be as in the present theorem. Then, repeating the proof of Theorem 2.2.1 with Lemmas 2.2.2 and 2.2.3 therein replaced, respectively, by Lemmas 7.2.5 and 7.2.6, we find that both (i) and (ii) of the present theorem hold true. This then finishes the proof of Theorem 7.2.4 □

The following conclusion gives the triviality of the dual space of $\mathcal{B}_\omega^{p,q}(\mathbb{R}^n)$ when $p \in (0, 1)$, whose proof is similar to that of Theorem 2.2.4; we omit the details.

Theorem 7.2.7 *Let* $p \in (0, 1)$, $q \in (0, \infty)$, *and* $\omega \in M(\mathbb{R}_+)$. *Then*

$$\left(\mathcal{B}_\omega^{p,q}(\mathbb{R}^n)\right)^* = \{0\}.$$

Applying the above two theorems, we immediately obtain the following two properties of the seminorm $\|\cdot\|_{\mathcal{B}_\omega^{p,q}(\mathbb{R}^n)}$, which shows that inhomogeneous block spaces may be meaningless sometimes; we omit the details.

Theorem 7.2.8 *Let* $p \in [1, \infty)$, $q \in (0, \infty)$, *and* $\omega \in M(\mathbb{R}_+)$ *satisfy* $M_\infty(\omega) \in (-\infty, 0)$. *Then* $f \in \mathcal{B}_\omega^{p,q}(\mathbb{R}^n)$ *if and only if* $f \in \mathscr{M}(\mathbb{R}^n)$ *and* $\|f\|_{\mathcal{B}_\omega^{p,q}(\mathbb{R}^n)} = 0$.

Theorem 7.2.9 *Let $p \in (0, 1)$, $q \in (0, \infty)$, and $\omega \in M(\mathbb{R}_+)$. Then $f \in \mathcal{B}_\omega^{p,q}(\mathbb{R}^n)$ if and only if $f \in \mathscr{M}(\mathbb{R}^n)$ and $\|f\|_{\mathcal{B}_\omega^{p,q}(\mathbb{R}^n)} = 0$.*

7.2.3 Boundedness of Sublinear Operators

The main target of this subsection is to establish the boundedness of both sublinear operators and Calderón–Zygmund operators on inhomogeneous block spaces (see Theorems 7.2.10 and 7.2.15 below). In particular, the boundedness of powered Hardy–Littlewood maximal operators on inhomogeneous block spaces (see Corollary 7.2.14 below) can be concluded directly by the boundedness of sublinear operators on inhomogeneous block spaces, which plays a key role in the study of inhomogeneous generalized Herz–Hardy spaces in the next chapter.

First, we establish the following boundedness criterion of sublinear operators on inhomogeneous block spaces.

Theorem 7.2.10 *Let $p \in (1, \infty)$, $q \in (0, \infty)$, and $\omega \in M(\mathbb{R}_+)$ satisfy*

$$-\frac{n}{p} < m_\infty(\omega) \leq M_\infty(\omega) < \frac{n}{p'},$$

where $\frac{1}{p} + \frac{1}{p'} = 1$. Let T be a bounded sublinear operator on $L^p(\mathbb{R}^n)$ satisfying that there exists a positive constant C such that, for any $f \in \mathcal{K}_{\omega,\mathbf{0}}^{p,q}(\mathbb{R}^n)$ and $x \notin \overline{\mathrm{supp}\,(f)} := \overline{\{x \in \mathbb{R}^n : f(x) \neq 0\}}$,

$$|T(f)(x)| \leq C \int_{\mathbb{R}^n} \frac{|f(y)|}{|x - y|^n}\, dy$$

and, for any $\{f_j\}_{j \in \mathbb{N}} \subset \mathscr{M}(\mathbb{R}^n)$ and almost every $x \in \mathbb{R}^n$,

$$\left| T\left(\sum_{j \in \mathbb{N}} f_j \right)(x) \right| \leq \sum_{j \in \mathbb{N}} |T(f_j)(x)|. \tag{7.38}$$

If T is well defined on $\mathcal{B}_\omega^{p,q}(\mathbb{R}^n)$, then there exists a positive constant C such that, for any $f \in \mathcal{B}_\omega^{p,q}(\mathbb{R}^n)$,

$$\|T(f)\|_{\mathcal{B}_\omega^{p,q}(\mathbb{R}^n)} \leq C \|f\|_{\mathcal{B}_\omega^{p,q}(\mathbb{R}^n)}.$$

In order to prove this theorem, we first establish the following lattice property of inhomogeneous block spaces.

Lemma 7.2.11 *Let $p, q \in (0, \infty)$ and $\omega \in M(\mathbb{R}_+)$. Then a measurable function f on $\mathbb{R}^n$ belongs to the block space $\mathcal{B}_\omega^{p,q}(\mathbb{R}^n)$ if and only if there exists a measurable function $g \in \mathcal{B}_\omega^{p,q}(\mathbb{R}^n)$ such that $|f| \leq g$ almost everywhere in $\mathbb{R}^n$. Moreover, for these f and g,*

$$\|f\|_{\mathcal{B}_\omega^{p,q}(\mathbb{R}^n)} \leq \|g\|_{\mathcal{B}_\omega^{p,q}(\mathbb{R}^n)}.$$

Proof Let all the symbols be as in the present lemma and $f \in \mathcal{M}(\mathbb{R}^n)$. Then, applying an argument similar to that used in the proof of Lemma 2.2.3 with Definition 2.1.3 therein replaced by Definition 7.2.2, we conclude that $f \in \mathcal{B}_\omega^{p,q}(\mathbb{R}^n)$ if and only if there exists a measurable function $g \in \mathcal{B}_\omega^{p,q}(\mathbb{R}^n)$ such that both $|f| \leq g$ almost everywhere in $\mathbb{R}^n$ and

$$\|f\|_{\mathcal{B}_\omega^{p,q}(\mathbb{R}^n)} \leq \|g\|_{\mathcal{B}_\omega^{p,q}(\mathbb{R}^n)},$$

which complete the proof of Lemma 7.2.11.   □

Remark 7.2.12 Let $p, q \in (0, \infty)$, $\omega \in M(\mathbb{R}_+)$, and $f \in \mathcal{M}(\mathbb{R}^n)$. Then, by the argument similar to that used in Remark 2.3.3, we find that $f \in \mathcal{B}_\omega^{p,q}(\mathbb{R}^n)$ if and only if $|f| \in \mathcal{B}_\omega^{p,q}(\mathbb{R}^n)$ and, for any $g \in \mathcal{B}_\omega^{p,q}(\mathbb{R}^n)$,

$$\|g\|_{\mathcal{B}_\omega^{p,q}(\mathbb{R}^n)} = \||g|\|_{\mathcal{B}_\omega^{p,q}(\mathbb{R}^n)}.$$

Then, with the help of the above lemma and the equivalent characterization of inhomogeneous block spaces obtained in Lemma 7.2.5(ii), we obtain the following boundedness criterion of sublinear operators on inhomogeneous block spaces, which plays a key role in the proof of Theorem 7.2.10.

Proposition 7.2.13 *Let $p, q \in (0, \infty)$, $\omega \in M(\mathbb{R}_+)$, and T be a bounded sublinear operator on $\mathcal{K}_{\omega,0}^{p,q}(\mathbb{R}^n)$ such that (7.38) holds true. If T is well defined on $\mathcal{B}_\omega^{p,q}(\mathbb{R}^n)$, then there exists a positive constant C such that, for any $f \in \mathcal{B}_\omega^{p,q}(\mathbb{R}^n)$,*

$$\|T(f)\|_{\mathcal{B}_\omega^{p,q}(\mathbb{R}^n)} \leq C \|f\|_{\mathcal{B}_\omega^{p,q}(\mathbb{R}^n)}.$$

Proof Let all the symbols be as in the present proposition and $f \in \mathcal{B}_\omega^{p,q}(\mathbb{R}^n)$. Then, repeating the proof of Proposition 2.3.4 via replacing Lemmas 2.2.2(ii) and 2.3.2 therein, respectively, by Lemmas 7.2.5(ii) and 7.2.11, we obtain

$$\|T(f)\|_{\mathcal{B}_\omega^{p,q}(\mathbb{R}^n)} \lesssim \|f\|_{\mathcal{B}_\omega^{p,q}(\mathbb{R}^n)},$$

which completes the proof of Proposition 7.2.13.   □

We next show Theorem 7.2.10.

Proof of Theorem 7.2.10 Let all the symbols be as in the present theorem. Then, by both Theorem 7.1.20 and Proposition 7.2.13, we find that, for any $f \in \mathcal{B}_\omega^{p,q}(\mathbb{R}^n)$,

$$\|T(f)\|_{\mathcal{B}_\omega^{p,q}(\mathbb{R}^n)} \lesssim \|f\|_{\mathcal{B}_\omega^{p,q}(\mathbb{R}^n)}.$$

This then finishes the proof of Theorem 7.2.10. □

As an application, we now establish the following boundedness of powered Hardy–Littlewood maximal operators on inhomogeneous block spaces.

Corollary 7.2.14 *Let* $p \in (1, \infty)$, $q \in (0, \infty)$, $r \in [1, p)$, *and* $\omega \in M(\mathbb{R}_+)$ *satisfy*

$$-\frac{n}{p} < m_\infty(\omega) \leq M_\infty(\omega) < n\left(\frac{1}{r} - \frac{1}{p}\right).$$

Then there exists a positive constant C *such that, for any* $f \in L^1_{\mathrm{loc}}(\mathbb{R}^n)$,

$$\left\|\mathcal{M}^{(r)}(f)\right\|_{\mathcal{B}_\omega^{p,q}(\mathbb{R}^n)} \leq C \|f\|_{\mathcal{B}_\omega^{p,q}(\mathbb{R}^n)}.$$

Proof Let all the symbols be as in the present corollary. Then, repeating an argument similar to that used in the proof of Corollary 2.3.5 with Lemma 1.3.1, Corollary 1.5.4, and Proposition 2.3.4 therein replaced, respectively, by Lemma 7.1.13, Corollary 7.1.22, and Proposition 7.2.13, we find that, for any $f \in L^1_{\mathrm{loc}}(\mathbb{R}^n)$,

$$\left\|\mathcal{M}^{(r)}(f)\right\|_{\mathcal{B}_\omega^{p,q}(\mathbb{R}^n)} \lesssim \|f\|_{\mathcal{B}_\omega^{p,q}(\mathbb{R}^n)}.$$

This implies that the powered Hardy–Littlewood maximal operator $\mathcal{M}^{(r)}$ is bounded on $\mathcal{B}_\omega^{p,q}(\mathbb{R}^n)$, and hence finishes the proof of Corollary 7.2.14. □

Finally, we turn to establish the following boundedness of Calderón–Zygmund operators on inhomogeneous block spaces.

Theorem 7.2.15 *Let* $p, q \in (1, \infty)$ *and* $\omega \in M(\mathbb{R}_+)$ *with*

$$-\frac{n}{p} < m_\infty(\omega) \leq M_\infty(\omega) < \frac{n}{p'},$$

where $\frac{1}{p} + \frac{1}{p'} = 1$. *Assume* $d \in \mathbb{Z}_+$ *and* T *is a* d-*order Calderón–Zygmund operator as in Definition 1.5.8. Then* T *is well defined on* $\mathcal{B}_\omega^{p,q}(\mathbb{R}^n)$ *and there exists a positive constant* C *such that, for any* $f \in \mathcal{B}_\omega^{p,q}(\mathbb{R}^n)$,

$$\|T(f)\|_{\mathcal{B}_\omega^{p,q}(\mathbb{R}^n)} \leq C \|f\|_{\mathcal{B}_\omega^{p,q}(\mathbb{R}^n)}.$$

Proof Let all the symbols be as in the present theorem. Then, from an argument similar to that used in the proof of Theorem 2.3.6 with Lemma 2.2.2(ii) and Corollary 1.5.10 therein replaced, respectively, by Lemma 7.2.5(ii) and Corollary 7.1.25, we deduce that T is well defined on $\mathcal{B}_{\omega}^{p,q}(\mathbb{R}^n)$ and, for any $f \in \mathcal{B}_{\omega}^{p,q}(\mathbb{R}^n)$,

$$\|T(f)\|_{\mathcal{B}_{\omega}^{p,q}(\mathbb{R}^n)} \lesssim \|f\|_{\mathcal{B}_{\omega}^{p,q}(\mathbb{R}^n)},$$

which completes the proof of Theorem 7.2.15.   □

7.3 Boundedness and Compactness Characterizations of Commutators

Let $b \in L_{\text{loc}}^1(\mathbb{R}^n)$ and T_Ω be a singular integral operator defined as in (3.4) with homogeneous kernel Ω satisfying (3.2). Recall that the commutator $[b, T_\Omega]$ is defined as in (3.5). In this section, we establish the boundedness and the compactness characterizations of the commutator $[b, T_\Omega]$ on inhomogeneous generalized Herz spaces via the boundedness and the compactness characterizations of commutators on ball Banach function spaces obtained in [228] and Chap. 3.

7.3.1 Boundedness Characterizations

The target of this subsection is to establish the boundedness characterization of commutators on both inhomogeneous local and inhomogeneous global generalized Herz spaces. We first consider commutators on inhomogeneous local generalized Herz spaces. Namely, we have the following conclusion.

Theorem 7.3.1 *Let $p, q \in (1, \infty)$ and $\omega \in M(\mathbb{R}_+)$ satisfy*

$$-\frac{n}{p} < m_\infty(\omega) \leq M_\infty(\omega) < \frac{n}{p'},$$

where $\frac{1}{p} + \frac{1}{p'} = 1$. Assume that Ω is a homogeneous Lipschitz function of degree zero on $\mathbb{S}^{n-1}$ satisfying (3.3), T_Ω a singular integral operator with homogeneous kernel Ω, and $b \in L_{\text{loc}}^1(\mathbb{R}^n)$. Then the commutator $[b, T_\Omega]$ is bounded on the inhomogeneous local generalized Herz space $\mathcal{K}_{\omega,0}^{p,q}(\mathbb{R}^n)$ if and only if $b \in \text{BMO}(\mathbb{R}^n)$. Moreover, there exist two positive constants C_1 and C_2, independent of b, such that

$$C_1\|b\|_{\text{BMO}(\mathbb{R}^n)} \leq \|[b, T_\Omega]\|_{\mathcal{K}_{\omega,0}^{p,q}(\mathbb{R}^n) \to \mathcal{K}_{\omega,0}^{p,q}(\mathbb{R}^n)} \leq C_2\|b\|_{\text{BMO}(\mathbb{R}^n)}.$$

To show this theorem, we first establish the following boundedness of commutators on inhomogeneous local generalized Herz spaces.

Proposition 7.3.2 *Let $p, q \in (1, \infty)$, $\omega \in M(\mathbb{R}_+)$ be as in Theorem 7.3.1, and $r \in (1, \infty]$. Assume that $b \in \mathrm{BMO}(\mathbb{R}^n)$, $\Omega \in L^r(\mathbb{S}^{n-1})$ satisfies both (3.2) and (3.3), and T_Ω is a singular integral operator with homogeneous kernel Ω. Then there exists a positive constant C such that, for any $f \in \mathcal{K}_{\omega,0}^{p,q}(\mathbb{R}^n)$,*

$$\|[b, T_\Omega](f)\|_{\mathcal{K}_{\omega,0}^{p,q}(\mathbb{R}^n)} \leq C \|b\|_{\mathrm{BMO}(\mathbb{R}^n)} \|f\|_{\mathcal{K}_{\omega,0}^{p,q}(\mathbb{R}^n)}.$$

Proof Let all the symbols be as in the present proposition. Then, from Theorem 7.1.11, we deduce that the inhomogeneous local generalized Herz space $\mathcal{K}_{\omega,0}^{p,q}(\mathbb{R}^n)$ under consideration is a BBF space.

Therefore, in order to complete the proof of the present proposition, it suffices to show that $\mathcal{K}_{\omega,0}^{p,q}(\mathbb{R}^n)$ satisfies all the assumptions of Lemma 3.1.3. Indeed, applying Corollary 7.1.22, we conclude that the Hardy–Littlewood maximal operator $\mathcal{M}$ is bounded on $\mathcal{K}_{\omega,0}^{p,q}(\mathbb{R}^n)$, namely, for any $f \in L_{\mathrm{loc}}^1(\mathbb{R}^n)$,

$$\|\mathcal{M}(f)\|_{\mathcal{K}_{\omega,0}^{p,q}(\mathbb{R}^n)} \lesssim \|f\|_{\mathcal{K}_{\omega,0}^{p,q}(\mathbb{R}^n)}. \tag{7.39}$$

On the other hand, from the assumption $M_\infty(\omega) \in (-\frac{n}{p}, \frac{n}{p'})$, it follows that

$$\frac{n}{M_\infty(\omega) + n/p} > \frac{n}{n(1/p' + 1/p)} = 1.$$

Combining this and the assumptions $p, q \in (1, \infty)$, we conclude that

$$\min\left\{ p, q, \frac{n}{M_\infty(\omega) + n/p} \right\} \in (1, \infty).$$

This, together with Lemma 7.1.37, further implies that the Hardy–Littlewood maximal operator $\mathcal{M}$ is bounded on $(\mathcal{K}_{\omega,0}^{p,q}(\mathbb{R}^n))'$, namely, for any $f \in L_{\mathrm{loc}}^1(\mathbb{R}^n)$,

$$\|\mathcal{M}(f)\|_{(\mathcal{K}_{\omega,0}^{p,q}(\mathbb{R}^n))'} \lesssim \|f\|_{(\mathcal{K}_{\omega,0}^{p,q}(\mathbb{R}^n))'}. \tag{7.40}$$

Using this and (7.39), we find that the inhomogeneous local generalized Herz space $\mathcal{K}_{\omega,0}^{p,q}(\mathbb{R}^n)$ under consideration satisfies all the assumptions of Lemma 3.1.3, which completes the proof of Proposition 7.3.2. □

To show Theorem 7.3.1, we also require the following necessity of the boundedness of commutators on inhomogeneous local generalized Herz spaces.

Proposition 7.3.3 *Let $p \in (1, \infty)$, $q \in [1, \infty)$, and $\omega \in M(\mathbb{R}_+)$ be as in Theorem 7.3.1. Assume that $b \in L_{\mathrm{loc}}^1(\mathbb{R}^n)$ and $\Omega \in L^\infty(\mathbb{S}^{n-1})$ satisfies that there exists an open set $\Lambda \subset \mathbb{S}^{n-1}$ such that Ω never vanishes and never changes sign on*

7 Inhomogeneous Generalized Herz Spaces and Inhomogeneous Block Spaces

Λ. *If the commutator $[b, T_\Omega]$ is bounded on the inhomogeneous local generalized Herz space $\mathcal{K}^{p,q}_{\omega,\mathbf{0}}(\mathbb{R}^n)$ and, for any bounded measurable set $F \subset \mathbb{R}^n$ and almost every $x \in \mathbb{R}^n \setminus \overline{F}$,*

$$[b, T_\Omega](\mathbf{1}_F)(x) = \int_F [b(x) - b(y)] \frac{\Omega(x - y)}{|x - y|^n} \, dy, \qquad (7.41)$$

then $b \in \mathrm{BMO}(\mathbb{R}^n)$ and there exists a positive constant C, independent of b, such that

$$\|b\|_{\mathrm{BMO}(\mathbb{R}^n)} \leq C \|[b, T_\Omega]\|_{\mathcal{K}^{p,q}_{\omega,\mathbf{0}}(\mathbb{R}^n) \to \mathcal{K}^{p,q}_{\omega,\mathbf{0}}(\mathbb{R}^n)}.$$

Proof Let all the symbols be as in the present proposition. Then, combining the assumptions $p, q \in [1, \infty)$, and Theorem 7.1.11, we conclude that the inhomogeneous local generalized Herz space $\mathcal{K}^{p,q}_{\omega,\mathbf{0}}(\mathbb{R}^n)$ under consideration is a BBF space. On the other hand, from Corollary 7.1.22, it follows that the Hardy–Littlewood maximal operator $\mathcal{M}$ is bounded on the Herz space $\mathcal{K}^{p,q}_{\omega,\mathbf{0}}(\mathbb{R}^n)$. This implies that $\mathcal{K}^{p,q}_{\omega,\mathbf{0}}(\mathbb{R}^n)$ satisfies all the assumptions of Lemma 3.1.6 and hence finishes the proof of Proposition 7.3.3.   □

We next prove Theorem 7.3.1.

Proof of Theorem 7.3.1 Let all the symbols be as in the present theorem. Then, repeating the proof of Theorem 3.1.1 via replacing Propositions 3.1.2 and 3.1.5 therein, respectively, by Propositions 7.3.2 and 7.3.3, we complete the proof of Theorem 7.3.1.   □

In the remainder of this subsection, we are devoted to establishing the following boundedness characterization of commutators on inhomogeneous global generalized Herz spaces.

Theorem 7.3.4 *Let $p, q \in (1, \infty)$ and $\omega \in M(\mathbb{R}_+)$ satisfy*

$$-\frac{n}{p} < m_\infty(\omega) \leq M_\infty(\omega) < 0,$$

where $\frac{1}{p} + \frac{1}{p'} = 1$. Assume that Ω is a homogeneous function of degree zero on $\mathbb{S}^{n-1}$ satisfying both (3.1) and (3.3), and T_Ω a singular integral operator with homogeneous kernel Ω. Then, for any $b \in L^1_{\mathrm{loc}}(\mathbb{R}^n)$, the commutator $[b, T_\Omega]$ is bounded on the inhomogeneous global generalized Herz space $\mathcal{K}^{p,q}_{\omega}(\mathbb{R}^n)$ if and only if $b \in \mathrm{BMO}(\mathbb{R}^n)$. Moreover, there exist two positive constants C_1 and C_2, independent of b, such that

$$C_1 \|b\|_{\mathrm{BMO}(\mathbb{R}^n)} \leq \|[b, T_\Omega]\|_{\mathcal{K}^{p,q}_{\omega}(\mathbb{R}^n) \to \mathcal{K}^{p,q}_{\omega}(\mathbb{R}^n)} \leq C_2 \|b\|_{\mathrm{BMO}(\mathbb{R}^n)}.$$

To show this boundedness characterization, we first give the following boundedness of commutators on inhomogeneous global generalized Herz spaces.

Proposition 7.3.5 *Let p, q, and ω be as in Theorem 7.3.4 and $r \in (1, \infty]$. Assume that $\Omega \in L^r(\mathbb{S}^{n-1})$ satisfies both (3.2) and (3.3), and T_Ω is a singular integral operator with homogeneous kernel Ω. Then there exists a positive constant C such that, for any $b \in \mathrm{BMO}(\mathbb{R}^n)$ and $f \in \mathcal{K}_\omega^{p,q}(\mathbb{R}^n)$,*

$$\|[b, T_\Omega](f)\|_{\mathcal{K}_\omega^{p,q}(\mathbb{R}^n)} \leq C \|b\|_{\mathrm{BMO}(\mathbb{R}^n)} \|f\|_{\mathcal{K}_\omega^{p,q}(\mathbb{R}^n)}.$$

Proof Let all the symbols be as in the present proposition. Then, repeating the proof of Proposition 3.1.10 with both Definition 1.2.1 and Theorem 1.8.7 therein replaced, respectively, by both Definition 7.1.1 and Theorem 7.1.38, we conclude that, for any $b \in \mathrm{BMO}(\mathbb{R}^n)$ and $f \in \mathcal{K}_\omega^{p,q}(\mathbb{R}^n)$,

$$\|[b, T_\Omega](f)\|_{\mathcal{K}_\omega^{p,q}(\mathbb{R}^n)} \lesssim \|b\|_{\mathrm{BMO}(\mathbb{R}^n)} \|f\|_{\mathcal{K}_\omega^{p,q}(\mathbb{R}^n)}$$

with the implicit positive constant independent of both b and f. This then finishes the proof of Proposition 7.3.2. $\quad\square$

On the other hand, we now establish the following necessity of the boundedness of commutators on inhomogeneous global generalized Herz spaces.

Proposition 7.3.6 *Let $p \in (1, \infty)$, $q \in [1, \infty)$, and $\omega \in M(\mathbb{R}_+)$ be as in Theorem 7.3.4. Assume that $b \in L^1_{\mathrm{loc}}(\mathbb{R}^n)$ and $\Omega \in L^\infty(\mathbb{S}^{n-1})$ satisfies that there exists an open set $\Lambda \subset \mathbb{S}^{n-1}$ such that Ω never vanishes and never changes sign on Λ. If the commutator $[b, T_\Omega]$ is bounded on the inhomogeneous global generalized Herz space $\mathcal{K}_\omega^{p,q}(\mathbb{R}^n)$ and, for any bounded measurable set $F \subset \mathbb{R}^n$ and almost every $x \in \mathbb{R}^n \setminus \overline{F}$,*

$$[b, T_\Omega](\mathbf{1}_F)(x) = \int_F [b(x) - b(y)] \frac{\Omega(x - y)}{|x - y|^n} \, dy,$$

then $b \in \mathrm{BMO}(\mathbb{R}^n)$ and there exists a positive constant C, independent of b, such that

$$\|b\|_{\mathrm{BMO}(\mathbb{R}^n)} \leq C \|[b, T_\Omega]\|_{\mathcal{K}_\omega^{p,q}(\mathbb{R}^n) \to \mathcal{K}_\omega^{p,q}(\mathbb{R}^n)}.$$

Proof Let all the symbols be as in the present proposition. Then, combining the assumption $M_\infty(\omega) \in (-\infty, 0)$, Theorem 7.1.12, and Corollary 7.1.24, we find that the inhomogeneous global generalized Herz space $\mathcal{K}_\omega^{p,q}(\mathbb{R}^n)$ under consideration is a BBF space and the Hardy–Littlewood maximal operator $\mathcal{M}$ is bounded on

$\mathcal{K}_{\omega}^{p,q}(\mathbb{R}^n)$. From this and Lemma 3.1.6, it follows that $b \in \mathrm{BMO}\,(\mathbb{R}^n)$ and

$$\|b\|_{\mathrm{BMO}\,(\mathbb{R}^n)} \lesssim \|[b, T_\Omega]\|_{\mathcal{K}_{\omega}^{p,q}(\mathbb{R}^n) \to \mathcal{K}_{\omega}^{p,q}(\mathbb{R}^n)},$$

which completes the proof of Proposition 7.3.6.   □

Via Propositions 7.3.5 and 7.3.6, we next show Theorem 7.3.4.

Proof of Theorem 7.3.4 Let all the symbols be as in the present theorem. Then, repeating the proof of Theorem 3.1.9 via replacing Propositions 3.1.10 and 3.1.12 therein, respectively, by Propositions 7.3.5 and 7.3.6, we complete the proof of Theorem 7.3.4.   □

7.3.2 Compactness Characterizations

In this subsection, we establish the compactness characterizations of commutators on inhomogeneous generalized Herz spaces. To begin with, we give the compactness characterization of commutators on the inhomogeneous local generalized Herz space $\mathcal{K}_{\omega,0}^{p,q}(\mathbb{R}^n)$ as follows.

Theorem 7.3.7 Let $p, q \in (1, \infty)$ and $\omega \in M(\mathbb{R}_+)$ satisfy

$$-\frac{n}{p} < m_\infty(\omega) \leq M_\infty(\omega) < \frac{n}{p'},$$

where $\frac{1}{p} + \frac{1}{p'} = 1$. Assume that Ω is a homogeneous function satisfying (3.1), (3.2), and (3.3), and T_Ω a singular integral operator with homogeneous kernel Ω. Then, for any $b \in L_{\mathrm{loc}}^1(\mathbb{R}^n)$, the commutator $[b, T_\Omega]$ is compact on the inhomogeneous local generalized Herz space $\mathcal{K}_{\omega,0}^{p,q}(\mathbb{R}^n)$ if and only if $b \in \mathrm{CMO}\,(\mathbb{R}^n)$.

In order to prove this theorem, we first show the following sufficiency of the compactness of commutators on $\mathcal{K}_{\omega,0}^{p,q}(\mathbb{R}^n)$.

Proposition 7.3.8 Let p, q, and ω be as in Theorem 7.3.7. Assume that $b \in L_{\mathrm{loc}}^1(\mathbb{R}^n)$, $\Omega \in L^\infty(\mathbb{S}^{n-1})$ satisfies (3.2), (3.3), and the L^∞-Dini condition, and T_Ω is a singular integral operator with homogeneous kernel Ω. If $b \in \mathrm{CMO}\,(\mathbb{R}^n)$, then the commutator $[b, T_\Omega]$ is compact on the inhomogeneous local generalized Herz space $\mathcal{K}_{\omega,0}^{p,q}(\mathbb{R}^n)$.

Proof Let all the symbols be as in the present proposition and $b \in \mathrm{CMO}\,(\mathbb{R}^n)$. Then, repeating an argument similar to that used in the proof of Proposition 3.2.2 with Theorem 1.2.46, (1.138), and (1.139) therein replaced, respectively, by Theorem 7.1.11, (7.39), and (7.40), we find that the commutator $[b, T_\Omega]$ is compact on the inhomogeneous local generalized Herz space $\mathcal{K}_{\omega,0}^{p,q}(\mathbb{R}^n)$ under consideration. This then finishes the proof of Proposition 7.3.8.   □

Conversely, we establish the following necessity of the compactness of commutators on inhomogeneous local generalized Herz spaces.

Proposition 7.3.9 *Let p, q, and ω be as in Theorem 7.3.7. Assume that $b \in L_{\mathrm{loc}}^1(\mathbb{R}^n)$ and $\Omega \in L^\infty(\mathbb{S}^{n-1})$ satisfies that there exists an open set $\Lambda \subset \mathbb{S}^{n-1}$ such that Ω never vanishes and never changes sign on Λ. If the commutator $[b, T_\Omega]$ is compact on the inhomogeneous local generalized Herz space $\mathcal{K}_{\omega,0}^{p,q}(\mathbb{R}^n)$ and (7.41) holds true, then $b \in \mathrm{CMO}\,(\mathbb{R}^n)$.*

Proof Let all the symbols be as in the present proposition, and let $b \in L_{\mathrm{loc}}^1(\mathbb{R}^n)$ be such that the commutator $[b, T_\Omega]$ is compact on the inhomogeneous local generalized Herz space $\mathcal{K}_{\omega,0}^{p,q}(\mathbb{R}^n)$ under consideration. Then, repeating the proof of Proposition 3.2.5 via replacing Theorem 1.2.46, (1.138), and (1.139) therein, respectively, by Theorem 7.1.11, (7.39), and (7.40), we conclude that $b \in \mathrm{CMO}\,(\mathbb{R}^n)$, and hence complete the proof of Proposition 7.3.9. $\qquad\qquad$ □

Applying Propositions 7.3.8 and 7.3.9, we now give the proof of Theorem 7.3.7.

Proof of Theorem 7.3.7 Let all the symbols be as in the present theorem. Then, from an argument similar to that used in the proof of Theorem 3.2.1 with both Propositions 3.2.2 and 3.2.5 therein replaced, respectively, by both Propositions 7.3.8 and 7.3.9, we deduce that Theorem 7.3.7 holds true, and hence complete the proof of the present theorem. $\qquad\qquad$ □

Next, we are devoted to establishing the compactness characterization of commutators on inhomogeneous global generalized Herz spaces. Indeed, we have the following conclusion.

Theorem 7.3.10 *Let $p, q \in (1, \infty)$ and $\omega \in M(\mathbb{R}_+)$ satisfy*

$$-\frac{n}{p} < m_\infty(\omega) \le M_\infty(\omega) < 0,$$

where $\frac{1}{p} + \frac{1}{p'} = 1$. Assume that Ω is a homogeneous function of degree zero on $\mathbb{S}^{n-1}$ satisfying both (3.1) and (3.3), and T_Ω a singular integral operator with homogeneous kernel Ω. Then, for any $b \in L_{\mathrm{loc}}^1(\mathbb{R}^n)$, the commutator $[b, T_\Omega]$ is compact on the inhomogeneous global generalized Herz space $\mathcal{K}_\omega^{p,q}(\mathbb{R}^n)$ if and only if $b \in \mathrm{CMO}\,(\mathbb{R}^n)$.

To show this theorem, we first give the following sufficiency of the compactness of commutators on the inhomogeneous global generalized Herz space $\mathcal{K}_\omega^{p,q}(\mathbb{R}^n)$.

Proposition 7.3.11 *Let p, q, and ω be as in Theorem 7.3.10. Assume that $b \in L_{\mathrm{loc}}^1(\mathbb{R}^n)$, $\Omega \in L^\infty(\mathbb{S}^{n-1})$ satisfies (3.2), (3.3), and the L^∞-Dini condition, and T_Ω is a singular integral operator with homogeneous kernel Ω. If $b \in \mathrm{CMO}\,(\mathbb{R}^n)$, then the commutator $[b, T_\Omega]$ is compact on $\mathcal{K}_\omega^{p,q}(\mathbb{R}^n)$.*

Proof Let all the symbols be as in the present proposition and $b \in \mathrm{CMO}\,(\mathbb{R}^n)$. Then, from the assumptions $p, q \in (1, \infty)$ and $M_\infty(\omega) \in (-\infty, 0)$, and

Theorem 7.1.12, it follows that the inhomogeneous global generalized Herz space $\mathcal{K}_{\omega}^{p,q}(\mathbb{R}^n)$ under consideration is a BBF space. Thus, to finish the proof of the present proposition, we only need to show that the assumptions (i) through (iii) of Proposition 3.2.11 hold true for $\mathcal{K}_{\omega}^{p,q}(\mathbb{R}^n)$.

First, we show that Proposition 3.2.11(i) holds true for $\mathcal{K}_{\omega}^{p,q}(\mathbb{R}^n)$, namely, both $[b, T_{\Omega}]$ and T_{Ω}^* are well defined on $\mathcal{K}_{\omega}^{p,q}(\mathbb{R}^n)$. Indeed, applying Definition 7.1.1, we find that, for any $f \in \mathcal{K}_{\omega}^{p,q}(\mathbb{R}^n)$,

$$\|f\|_{\mathcal{K}_{\omega,0}^{p,q}(\mathbb{R}^n)} \leq \|f\|_{\mathcal{K}_{\omega}^{p,q}(\mathbb{R}^n)} < \infty,$$

which implies that $f \in \mathcal{K}_{\omega,0}^{p,q}(\mathbb{R}^n)$ and hence

$$\mathcal{K}_{\omega}^{p,q}(\mathbb{R}^n) \subset \mathcal{K}_{\omega,0}^{p,q}(\mathbb{R}^n).$$

Combining this and Proposition 7.3.2, we conclude that $[b, T_{\Omega}](f) \in \mathcal{K}_{\omega,0}^{p,q}(\mathbb{R}^n)$ and hence $[b, T_{\Omega}]$ is well defined on $\mathcal{K}_{\omega}^{p,q}(\mathbb{R}^n)$. Now, we prove that T_{Ω}^* is also well defined on $\mathcal{K}_{\omega}^{p,q}(\mathbb{R}^n)$. Notice that, by the fact that $\mathcal{K}_{\omega}^{p,q}(\mathbb{R}^n) \subset \mathcal{K}_{\omega,0}^{p,q}(\mathbb{R}^n)$, we only need to show that T_{Ω}^* is well defined on $\mathcal{K}_{\omega,0}^{p,q}(\mathbb{R}^n)$. To this end, repeating an argument similar to that used in the proof of Lemma 3.2.12 with Theorem 1.2.46, (1.138), and (1.139) therein replaced, respectively, by Theorem 7.1.11, (7.39), and (7.40), we find that, for any $f \in \mathcal{K}_{\omega,0}^{p,q}(\mathbb{R}^n)$,

$$\left\|T_{\Omega}^*(f)\right\|_{\mathcal{K}_{\omega,0}^{p,q}(\mathbb{R}^n)} \lesssim \|f\|_{\mathcal{K}_{\omega,0}^{p,q}(\mathbb{R}^n)},$$

which then implies that T_{Ω}^* is well defined on $\mathcal{K}_{\omega,0}^{p,q}(\mathbb{R}^n)$ and hence on $\mathcal{K}_{\omega}^{p,q}(\mathbb{R}^n)$. Therefore, Proposition 3.2.11(i) holds true for $\mathcal{K}_{\omega}^{p,q}(\mathbb{R}^n)$.

Next, we prove that $\mathcal{K}_{\omega}^{p,q}(\mathbb{R}^n)$ satisfies Proposition 3.2.11(ii). Indeed, from Corollary 7.1.24, it follows that, for any $f \in L_{\text{loc}}^1(\mathbb{R}^n)$,

$$\|\mathcal{M}(f)\|_{\mathcal{K}_{\omega}^{p,q}(\mathbb{R}^n)} \lesssim \|f\|_{\mathcal{K}_{\omega}^{p,q}(\mathbb{R}^n)},$$

which implies that the Hardy–Littlewood maximal operator $\mathcal{M}$ is bounded on $\mathcal{K}_{\omega}^{p,q}(\mathbb{R}^n)$ and hence Proposition 3.2.11(ii) holds true.

Finally, applying Theorem 7.1.38, we find that the inhomogeneous global generalized Herz space $\mathcal{K}_{\omega}^{p,q}(\mathbb{R}^n)$ under consideration satisfies Proposition 3.2.11(iii). Thus, by the fact that $\mathcal{K}_{\omega}^{p,q}(\mathbb{R}^n)$ is a BBF space and Proposition 3.2.11 with $X := \mathcal{K}_{\omega}^{p,q}(\mathbb{R}^n)$, we find that the commutator $[b, T_{\Omega}]$ is compact on $\mathcal{K}_{\omega}^{p,q}(\mathbb{R}^n)$, which completes the proof of Proposition 7.3.11.   □

To establish the compactness characterization Theorem 7.3.10, we also need the following necessity of the compactness of commutators on $\mathcal{K}_{\omega}^{p,q}(\mathbb{R}^n)$.

Proposition 7.3.12 *Let p, q, and ω be as in Theorem 7.3.10. Assume that $b \in L_{\text{loc}}^1(\mathbb{R}^n)$ and $\Omega \in L^{\infty}(\mathbb{S}^{n-1})$ satisfies that there exists an open set $\Lambda \subset \mathbb{S}^{n-1}$*

such that Ω never vanishes and never changes sign on Λ. If the commutator $[b, T_\Omega]$ is compact on the inhomogeneous global generalized Herz space $\mathcal{K}_\omega^{p,q}(\mathbb{R}^n)$ and satisfies (7.41), then $b \in$ CMO $(\mathbb{R}^n)$.

Proof Let all the symbols be as in the present proposition, and let $b \in L_{\mathrm{loc}}^1(\mathbb{R}^n)$ be such that the commutator $[b, T_\Omega]$ is compact on the inhomogeneous global generalized Herz space $\mathcal{K}_\omega^{p,q}(\mathbb{R}^n)$ under consideration. Then, repeating the proof of Proposition 3.2.13 via Theorems 1.2.48 and 1.8.7 and Corollary 1.5.6 therein replaced, respectively, by Theorems 7.1.12 and 7.1.38 and Corollary 7.1.24, we find that $b \in$ CMO $(\mathbb{R}^n)$, and hence complete the proof of Proposition 7.3.12. □

Via the above two propositions, we now show Theorem 7.3.10.

Proof of Theorem 7.3.10 Let all the symbols be as in the present theorem. Then, repeating an argument similar to that used in the proof of Theorem 3.2.8 with both Propositions 3.2.9 and 3.2.13 therein replaced, respectively, by both Propositions 7.3.11 and 7.3.12, we then complete the proof of Theorem 7.3.10. □

Chapter 8
Hardy Spaces Associated with Inhomogeneous Generalized Herz Spaces

The target of this chapter is devoted to establishing a complete real-variable theory of Hardy spaces associated with inhomogeneous generalized Herz spaces. Precisely, based on the inhomogeneous generalized Herz spaces studied in Chap. 7, we first introduce inhomogeneous generalized Herz–Hardy spaces, inhomogeneous localized generalized Herz–Hardy spaces, and inhomogeneous weak generalized Herz–Hardy spaces in this chapter. Then, via using the known results about Hardy spaces $H_X(\mathbb{R}^n)$ associated with ball quasi-Banach function spaces X as well as some improved characterizations of $H_X(\mathbb{R}^n)$ established in Chaps. 4 through 6, we investigate both the real-variable characterizations and their applications of these Hardy spaces associated with inhomogeneous generalized Herz spaces, which include various maximal function, the atomic, the molecular, and the Littlewood–Paley function characterizations as well as the boundedness of Calderón–Zygmund operators or pseudo-differential operators, duality, the properties of the Fourier transform, and the real interpolation theorems about these Herz–Hardy spaces.

8.1 Inhomogeneous Generalized Herz–Hardy Spaces

In this section, we first introduce the inhomogeneous generalized Herz–Hardy spaces $H\mathcal{K}_{\omega,\mathbf{0}}^{p,q}(\mathbb{R}^n)$ and $H\mathcal{K}_{\omega}^{p,q}(\mathbb{R}^n)$, where $p, q \in (0, \infty)$ and $\omega \in M(\mathbb{R}_+)$. Then we establish some real-variable characterizations and also give some applications of them. To be precise, we show the maximal function, the atomic, the molecular, and the Littlewood–Paley function characterizations of $H\mathcal{K}_{\omega,\mathbf{0}}^{p,q}(\mathbb{R}^n)$ and $H\mathcal{K}_{\omega}^{p,q}(\mathbb{R}^n)$. As applications, we find the dual space of $H\mathcal{K}_{\omega,\mathbf{0}}^{p,q}(\mathbb{R}^n)$ under some reasonable and sharp assumptions and establish the boundedness of the Calderón–Zygmund operators on both $H\mathcal{K}_{\omega,\mathbf{0}}^{p,q}(\mathbb{R}^n)$ and $H\mathcal{K}_{\omega}^{p,q}(\mathbb{R}^n)$ as well as investigate the Fourier transform properties of these generalized Herz–Hardy spaces.

© The Author(s), under exclusive license to Springer Nature Singapore Pte Ltd. 2022
Y. Li et al., *Real-Variable Theory of Hardy Spaces Associated with Generalized Herz Spaces of Rafeiro and Samko*, Lecture Notes in Mathematics 2320,
https://doi.org/10.1007/978-981-19-6788-7_8

To begin with, recall that, for any given $N \in \mathbb{N}$ and for any $f \in \mathcal{S}'(\mathbb{R}^n)$, the non-tangential grand maximal function $\mathcal{M}_N(f)$ of f is defined as in (4.3). Then we introduce the inhomogeneous generalized Herz–Hardy spaces via the operator $\mathcal{M}_N$ as follows.

Definition 8.1.1 Let $p, q \in (0, \infty)$, $\omega \in M(\mathbb{R}_+)$, and $N \in \mathbb{N}$. Then

(i) the *inhomogeneous generalized Herz–Hardy space* $H\mathcal{K}_{\omega,0}^{p,q}(\mathbb{R}^n)$, associated with the inhomogeneous local generalized Herz space $\mathcal{K}_{\omega,0}^{p,q}(\mathbb{R}^n)$, is defined to be the set of all the $f \in \mathcal{S}'(\mathbb{R}^n)$ such that

$$\|f\|_{H\mathcal{K}_{\omega,0}^{p,q}(\mathbb{R}^n)} := \|\mathcal{M}_N(f)\|_{\mathcal{K}_{\omega,0}^{p,q}(\mathbb{R}^n)} < \infty;$$

(ii) the *inhomogeneous generalized Herz–Hardy space* $H\mathcal{K}_\omega^{p,q}(\mathbb{R}^n)$, associated with the inhomogeneous global generalized Herz space $\mathcal{K}_\omega^{p,q}(\mathbb{R}^n)$, is defined to be the set of all the $f \in \mathcal{S}'(\mathbb{R}^n)$ such that

$$\|f\|_{H\mathcal{K}_\omega^{p,q}(\mathbb{R}^n)} := \|\mathcal{M}_N(f)\|_{\mathcal{K}_\omega^{p,q}(\mathbb{R}^n)} < \infty.$$

Remark 8.1.2 We should point out that, in Definition 8.1.1, when $\omega(t) := t^\alpha$ for any $t \in (0, \infty)$ and for any given $\alpha \in \mathbb{R}$, the inhomogeneous generalized Herz–Hardy space $H\mathcal{K}_{\omega,0}^{p,q}(\mathbb{R}^n)$ coincides with the classical *inhomogeneous Herz-type Hardy space* $HK_p^{\alpha,q}(\mathbb{R}^n)$ which was originally introduced in [163, Definition 2.1] (see also [175, Definition 2.1.1]).

8.1.1   Maximal Function Characterizations

The main target of this subsection is to characterize inhomogeneous generalized Herz–Hardy spaces via various maximal functions. Via using the radial and the non-tangential maximal functions presented in Definition 4.1.1, we first show the following maximal function characterizations of the inhomogeneous generalized Herz–Hardy space $H\mathcal{K}_{\omega,0}^{p,q}(\mathbb{R}^n)$.

Theorem 8.1.3 Let $p, q, a, b \in (0, \infty)$, $\omega \in M(\mathbb{R}_+)$, $N \in \mathbb{N}$, and $\phi \in \mathcal{S}(\mathbb{R}^n)$ satisfy $\int_{\mathbb{R}^n} \phi(x)\, dx \neq 0$.

(i) *Let* $N \in \mathbb{N} \cap [\lfloor b+1 \rfloor, \infty)$. *Then, for any* $f \in \mathcal{S}'(\mathbb{R}^n)$,

$$\|M(f, \phi)\|_{\mathcal{K}_{\omega,0}^{p,q}(\mathbb{R}^n)} \lesssim \|M_a^*(f, \phi)\|_{\mathcal{K}_{\omega,0}^{p,q}(\mathbb{R}^n)} \lesssim \|M_b^{**}(f, \phi)\|_{\mathcal{K}_{\omega,0}^{p,q}(\mathbb{R}^n)},$$

$$\|M(f, \phi)\|_{\mathcal{K}_{\omega,0}^{p,q}(\mathbb{R}^n)} \lesssim \|\mathcal{M}_N(f)\|_{\mathcal{K}_{\omega,0}^{p,q}(\mathbb{R}^n)} \lesssim \|\mathcal{M}_{\lfloor b+1 \rfloor}(f)\|_{\mathcal{K}_{\omega,0}^{p,q}(\mathbb{R}^n)}$$

$$\lesssim \|M_b^{**}(f, \phi)\|_{\mathcal{K}_{\omega,0}^{p,q}(\mathbb{R}^n)},$$

and

$$\|M_b^{**}(f,\phi)\|_{\mathcal{K}_{\omega,0}^{p,q}(\mathbb{R}^n)} \sim \|\mathcal{M}_{b,N}^{**}(f)\|_{\mathcal{K}_{\omega,0}^{p,q}(\mathbb{R}^n)},$$

where the implicit positive constants are independent of f.

(ii) *Let $\omega \in M(\mathbb{R}_+)$ satisfy $m_\infty(\omega) \in (-\frac{n}{p}, \infty)$. Assume*

$$b \in \left(\max\left\{ \frac{n}{p}, M_\infty(\omega) + \frac{n}{p} \right\}, \infty \right).$$

Then, for any $f \in \mathcal{S}'(\mathbb{R}^n)$,

$$\|M_b^{**}(f,\phi)\|_{\mathcal{K}_{\omega,0}^{p,q}(\mathbb{R}^n)} \lesssim \|M(f,\phi)\|_{\mathcal{K}_{\omega,0}^{p,q}(\mathbb{R}^n)},$$

where the implicit positive constant is independent of f. In particular, when $N \in \mathbb{N} \cap [\lfloor b+1 \rfloor, \infty)$, if one of the quantities

$$\|M(f,\phi)\|_{\mathcal{K}_{\omega,0}^{p,q}(\mathbb{R}^n)}, \ \|M_a^*(f,\phi)\|_{\mathcal{K}_{\omega,0}^{p,q}(\mathbb{R}^n)}, \ \|\mathcal{M}_N(f)\|_{\mathcal{K}_{\omega,0}^{p,q}(\mathbb{R}^n)},$$

$$\|M_b^{**}(f,\phi)\|_{\mathcal{K}_{\omega,0}^{p,q}(\mathbb{R}^n)}, \ and \ \|\mathcal{M}_{b,N}^{**}(f)\|_{\mathcal{K}_{\omega,0}^{p,q}(\mathbb{R}^n)}$$

is finite, then the others are also finite and mutually equivalent with the positive equivalence constants independent of f.

Proof Let all the symbols be as in the present theorem. Then, repeating the proof of Theorem 4.1.2 with $\max\{M_0(\omega), M_\infty(\omega)\}$, Theorem 1.2.42, and Lemma 1.8.5 therein replaced, respectively, by $M_\infty(\omega)$, Theorem 7.1.7, and Lemma 7.1.36, we find that both (i) and (ii) of the present theorem hold true. This finishes the proof of Theorem 8.1.3. □

Remark 8.1.4

(i) Let $p, q \in (0, \infty)$ and $\omega \in M(\mathbb{R}_+)$. Then the quasi-norm of $H\mathcal{K}_{\omega,0}^{p,q}(\mathbb{R}^n)$ in Definition 8.1.1(i) depends on N. However, using Theorem 8.1.3, we find that the Hardy space $H\mathcal{K}_{\omega,0}^{p,q}(\mathbb{R}^n)$ is independent of the choice of N whenever $\omega \in M(\mathbb{R}_+)$ satisfies $m_\infty(\omega) \in (-\frac{n}{p}, \infty)$, and N satisfies

$$N \in \mathbb{N} \cap \left(1 + \max\left\{ \frac{n}{p}, M_\infty(\omega) + \frac{n}{p} \right\}, \infty \right).$$

(ii) Observe that, if $p = q \in (0, \infty)$ and $\omega(t) := 1$ for any $t \in (0, \infty)$, then, in this case,

$$m_0(\omega) = M_0(\omega) = m_\infty(\omega) = M_\infty(\omega) = 0,$$

$H\mathcal{K}^{p,q}_{\omega,\mathbf{0}}(\mathbb{R}^n)$ coincides with the classical Hardy space $H^p(\mathbb{R}^n)$ in the sense of equivalent quasi-norms, and all the conclusions in Theorem 8.1.3 coincide with the corresponding classical results on the Hardy space $H^p(\mathbb{R}^n)$; see, for instance, [91, 160, 213].

Next, we establish the following maximal function characterizations of the inhomogeneous generalized Herz–Hardy space $H\mathcal{K}^{p,q}_{\omega}(\mathbb{R}^n)$.

Theorem 8.1.5 *Let* $p, q, a, b \in (0, \infty)$, $\omega \in M(\mathbb{R}_+)$, $N \in \mathbb{N}$, *and* $\phi \in \mathcal{S}(\mathbb{R}^n)$ *satisfy* $\int_{\mathbb{R}^n} \phi(x)\, dx \neq 0$.

(i) *Let* $N \in \mathbb{N} \cap [\lfloor b+1 \rfloor, \infty)$ *and* ω *satisfy* $M_\infty(\omega) \in (-\infty, 0)$. *Then, for any* $f \in \mathcal{S}'(\mathbb{R}^n)$,

$$\|M(f, \phi)\|_{\mathcal{K}^{p,q}_{\omega}(\mathbb{R}^n)} \lesssim \|M_a^*(f, \phi)\|_{\mathcal{K}^{p,q}_{\omega}(\mathbb{R}^n)} \lesssim \|M_b^{**}(f, \phi)\|_{\mathcal{K}^{p,q}_{\omega}(\mathbb{R}^n)},$$

$$\|M(f, \phi)\|_{\mathcal{K}^{p,q}_{\omega}(\mathbb{R}^n)} \lesssim \|\mathcal{M}_N(f)\|_{\mathcal{K}^{p,q}_{\omega}(\mathbb{R}^n)} \lesssim \|\mathcal{M}_{\lfloor b+1 \rfloor}(f)\|_{\mathcal{K}^{p,q}_{\omega}(\mathbb{R}^n)}$$

$$\lesssim \|M_b^{**}(f, \phi)\|_{\mathcal{K}^{p,q}_{\omega}(\mathbb{R}^n)},$$

and

$$\|M_b^{**}(f, \phi)\|_{\mathcal{K}^{p,q}_{\omega}(\mathbb{R}^n)} \sim \|\mathcal{M}_{b,N}^{**}(f)\|_{\mathcal{K}^{p,q}_{\omega}(\mathbb{R}^n)},$$

where the implicit positive constants are independent of f.

(ii) *Let* $\omega \in M(\mathbb{R}_+)$ *satisfy*

$$-\frac{n}{p} < m_\infty(\omega) \leq M_\infty(\omega) < 0.$$

Assume $b \in (\frac{n}{p}, \infty)$. *Then, for any* $f \in \mathcal{S}'(\mathbb{R}^n)$,

$$\|M_b^{**}(f, \phi)\|_{\mathcal{K}^{p,q}_{\omega}(\mathbb{R}^n)} \lesssim \|M(f, \phi)\|_{\mathcal{K}^{p,q}_{\omega}(\mathbb{R}^n)},$$

where the implicit positive constant is independent of f. *In particular, when* $N \in \mathbb{N} \cap [\lfloor b+1 \rfloor, \infty)$, *if one of the quantities*

$$\|M(f, \phi)\|_{\mathcal{K}^{p,q}_{\omega}(\mathbb{R}^n)}, \quad \|M_a^*(f, \phi)\|_{\mathcal{K}^{p,q}_{\omega}(\mathbb{R}^n)}, \quad \|\mathcal{M}_N(f)\|_{\mathcal{K}^{p,q}_{\omega}(\mathbb{R}^n)},$$

$$\|M_b^{**}(f, \phi)\|_{\mathcal{K}^{p,q}_{\omega}(\mathbb{R}^n)}, \quad \text{and} \quad \|\mathcal{M}_{b,N}^{**}(f)\|_{\mathcal{K}^{p,q}_{\omega}(\mathbb{R}^n)}$$

is finite, then the others are also finite and mutually equivalent with the positive equivalence constants independent of f.

Remark 8.1.6

(i) Let $p, q \in (0, \infty)$ and $\omega \in M(\mathbb{R}_+)$. Then the quasi-norm of $H\mathcal{K}_\omega^{p,q}(\mathbb{R}^n)$ in Definition 8.1.1(ii) depends on N. However, by Theorem 8.1.5, we conclude that the Hardy space $H\mathcal{K}_\omega^{p,q}(\mathbb{R}^n)$ is independent of the choice of N whenever $\omega \in M(\mathbb{R}_+)$ satisfies

$$-\frac{n}{p} < m_\infty(\omega) \leq M_\infty(\omega) < 0,$$

and N satisfies

$$N \in \mathbb{N} \cap \left(1 + \frac{n}{p}, \infty\right).$$

(ii) Notice that, if $p = q \in (0, \infty)$ and $\omega(t) := 1$ for any $t \in (0, \infty)$, then, in this case,

$$m_0(\omega) = M_0(\omega) = m_\infty(\omega) = M_\infty(\omega) = 0$$

and $H\mathcal{K}_\omega^{p,q}(\mathbb{R}^n)$ coincides with the classical Hardy space $H^p(\mathbb{R}^n)$ in the sense of equivalent quasi-norms. Thus, Theorem 8.1.5 completely excludes the classical Hardy space $H^p(\mathbb{R}^n)$ and, from Remark 7.1.10, we infer that the classical Hardy space $H^p(\mathbb{R}^n)$ is the critical case of $H\mathcal{K}_\omega^{p,q}(\mathbb{R}^n)$ considered in Theorem 8.1.5.

To prove this theorem, we first show the following lemma about the boundedness of the Hardy–Littlewood maximal operator on inhomogeneous generalized Herz spaces.

Lemma 8.1.7 *Let* $p, q \in (0, \infty)$ *and* $\omega \in M(\mathbb{R}_+)$ *satisfy* $m_\infty(\omega) \in (-\frac{n}{p}, \infty)$. *Then, for any given* $r \in (0, \min\{p, \frac{n}{M_\infty(\omega)+n/p}\})$, *there exists a positive constant* C *such that, for any* $f \in L_{\mathrm{loc}}^1(\mathbb{R}^n)$,

$$\left\|\mathcal{M}(f)\right\|_{[\mathcal{K}_\omega^{p,q}(\mathbb{R}^n)]^{1/r}} \leq C\|f\|_{[\mathcal{K}_\omega^{p,q}(\mathbb{R}^n)]^{1/r}}.$$

Proof Let all the symbols be as in the present lemma. Then, for any given $r \in (0, \min\{p, \frac{n}{M_\infty(\omega)+n/p}\})$, using Lemma 1.1.6 and repeating the proof of Lemma 4.1.10 with $\min\{m_0(\omega), m_\infty(\omega)\}$, $\max\{M_0(\omega), M_\infty(\omega)\}$, Lemma 1.3.2, and Corollary 1.5.6 therein replaced, respectively, by $m_\infty(\omega)$, $M_\infty(\omega)$, Lemma 7.1.14, and Corollary 7.1.24, we conclude that, for any $f \in L_{\mathrm{loc}}^1(\mathbb{R}^n)$,

$$\left\|\mathcal{M}(f)\right\|_{[\mathcal{K}_\omega^{p,q}(\mathbb{R}^n)]^{1/r}} \lesssim \|f\|_{[\mathcal{K}_\omega^{p,q}(\mathbb{R}^n)]^{1/r}}.$$

This finishes the proof of Lemma 8.1.7. $\qquad\square$

Via this lemma, we now show Theorem 8.1.5.

Proof of Theorem 8.1.5 Let all the symbols be as in the present theorem. Then, repeating the proof of Theorem 8.1.5 via replacing Theorem 1.2.44 and Lemma 4.1.10 therein, respectively, by Theorem 7.1.9 and Lemma 8.1.7, we find that both (i) and (ii) of the present theorem hold true, and hence complete the proof of Theorem 8.1.5.   □

8.1.2   Relations with Inhomogeneous Generalized Herz Spaces

In this subsection, we investigate the relations between inhomogeneous generalized Herz spaces and associated Hardy spaces. To begin with, the following conclusion shows that, under some reasonable and sharp assumptions, $H\mathcal{K}_{\omega,\mathbf{0}}^{p,q}(\mathbb{R}^n) = \mathcal{K}_{\omega,\mathbf{0}}^{p,q}(\mathbb{R}^n)$ in the sense of equivalent quasi-norms.

Theorem 8.1.8 *Let $p \in (1,\infty)$, $q \in (0,\infty)$, and $\omega \in M(\mathbb{R}_+)$ satisfy*

$$-\frac{n}{p} < m_\infty(\omega) \le M_\infty(\omega) < \frac{n}{p'},$$

where $\frac{1}{p} + \frac{1}{p'} = 1$. Then

(i) $\mathcal{K}_{\omega,\mathbf{0}}^{p,q}(\mathbb{R}^n) \hookrightarrow \mathcal{S}'(\mathbb{R}^n)$.

(ii) *If $f \in \mathcal{K}_{\omega,\mathbf{0}}^{p,q}(\mathbb{R}^n)$, then $f \in H\mathcal{K}_{\omega,\mathbf{0}}^{p,q}(\mathbb{R}^n)$ and there exists a positive constant C, independent of f, such that*

$$\|f\|_{H\mathcal{K}_{\omega,\mathbf{0}}^{p,q}(\mathbb{R}^n)} \le C \|f\|_{\mathcal{K}_{\omega,\mathbf{0}}^{p,q}(\mathbb{R}^n)}.$$

(iii) *If $f \in H\mathcal{K}_{\omega,\mathbf{0}}^{p,q}(\mathbb{R}^n)$, then there exists a locally integrable function $g \in \mathcal{K}_{\omega,\mathbf{0}}^{p,q}(\mathbb{R}^n)$ such that g represents f, which means that $f = g$ in $\mathcal{S}'(\mathbb{R}^n)$,*

$$\|f\|_{H\mathcal{K}_{\omega,\mathbf{0}}^{p,q}(\mathbb{R}^n)} = \|g\|_{H\mathcal{K}_{\omega,\mathbf{0}}^{p,q}(\mathbb{R}^n)},$$

and there exists a positive constant C, independent of f, such that

$$\|g\|_{\mathcal{K}_{\omega,\mathbf{0}}^{p,q}(\mathbb{R}^n)} \le C \|f\|_{H\mathcal{K}_{\omega,\mathbf{0}}^{p,q}(\mathbb{R}^n)}.$$

Proof Let all the symbols be as in the present theorem. Then, repeating an argument similar to that used in the proof of Theorem 4.2.1 with both Theorem 1.2.42 and Lemma 1.8.5 therein replaced, respectively, by both Theorem 7.1.7 and Lemma 7.1.36, we find that (i) through (iii) of the present theorem hold true. This finishes the proof of Theorem 8.1.8.   □

Remark 8.1.9 We should point out that, in Theorem 8.1.8, if $\omega(t) := t^\alpha$ for any $t \in (0, \infty)$ and for any given $\alpha \in \mathbb{R}$, then the conclusion obtained in this theorem goes back to [175, Proposition 2.1.1(1)].

Similarly, the following conclusion shows that, under some reasonable and sharp assumptions, $H\mathcal{K}_\omega^{p,q}(\mathbb{R}^n) = \mathcal{K}_\omega^{p,q}(\mathbb{R}^n)$ with equivalent quasi-norms.

Theorem 8.1.10 *Let $p \in (1, \infty)$, $q \in (0, \infty)$, and $\omega \in M(\mathbb{R}_+)$ satisfy*

$$-\frac{n}{p} < m_\infty(\omega) \leq M_\infty(\omega) < 0.$$

Then

(i) $\mathcal{K}_\omega^{p,q}(\mathbb{R}^n) \hookrightarrow \mathcal{S}'(\mathbb{R}^n)$.

(ii) *If $f \in \mathcal{K}_\omega^{p,q}(\mathbb{R}^n)$, then $f \in H\mathcal{K}_\omega^{p,q}(\mathbb{R}^n)$ and there exists a positive constant C, independent of f, such that*

$$\|f\|_{H\mathcal{K}_\omega^{p,q}(\mathbb{R}^n)} \leq C \|f\|_{\mathcal{K}_\omega^{p,q}(\mathbb{R}^n)}.$$

(iii) *If $f \in H\mathcal{K}_\omega^{p,q}(\mathbb{R}^n)$, then there exists a locally integrable function $g \in \mathcal{K}_\omega^{p,q}(\mathbb{R}^n)$ such that g represents f, which means that $f = g$ in $\mathcal{S}'(\mathbb{R}^n)$,*

$$\|f\|_{H\mathcal{K}_\omega^{p,q}(\mathbb{R}^n)} = \|g\|_{H\mathcal{K}_\omega^{p,q}(\mathbb{R}^n)},$$

and there exists a positive constant C, independent of f, such that

$$\|g\|_{\mathcal{K}_\omega^{p,q}(\mathbb{R}^n)} \leq C \|f\|_{H\mathcal{K}_\omega^{p,q}(\mathbb{R}^n)}.$$

Proof Let all the symbols be as in the present theorem. Then, repeating the proof of Theorem 4.2.10 via replacing both Theorem 1.2.44 and Lemma 4.1.10 therein, respectively, by both Theorem 7.1.9 and Lemma 8.1.7, we conclude that (i), (ii), and (iii) of the present theorem hold true. This then finishes the proof of Theorem 8.1.10. □

8.1.3 Atomic Characterizations

The target of this subsection is to establish the atomic characterization of inhomogeneous generalized Herz–Hardy spaces. To begin with, we show the atomic characterization of the inhomogeneous generalized Herz–Hardy space $H\mathcal{K}_{\omega,\mathbf{0}}^{p,q}(\mathbb{R}^n)$. For this purpose, we first introduce the concepts of both $(\mathcal{K}_{\omega,\mathbf{0}}^{p,q}(\mathbb{R}^n), r, d)$-atoms and the atomic Hardy space $H\mathcal{K}_{\omega,\mathbf{0}}^{p,q,r,s}(\mathbb{R}^n)$ associated with the inhomogeneous local generalized Herz space $\mathcal{K}_{\omega,\mathbf{0}}^{p,q}(\mathbb{R}^n)$ as follows.

Definition 8.1.11 Let $p, q \in (0, \infty)$, $\omega \in M(\mathbb{R}_+)$, $r \in [1, \infty]$, and $d \in \mathbb{Z}_+$. Then a measurable function a on $\mathbb{R}^n$ is called a $(\mathcal{K}_{\omega,0}^{p,q}(\mathbb{R}^n), r, d)$-*atom* if there exists a ball $B \in \mathbb{B}$ such that

(i) $\operatorname{supp}(a) := \{x \in \mathbb{R}^n : a(x) \neq 0\} \subset B$;

(ii) $\|a\|_{L^r(\mathbb{R}^n)} \leq \dfrac{|B|^{1/r}}{\|\mathbf{1}_B\|_{\mathcal{K}_{\omega,0}^{p,q}(\mathbb{R}^n)}}$;

(iii) for any $\alpha \in \mathbb{Z}_+^n$ with $|\alpha| \leq d$,

$$\int_{\mathbb{R}^n} a(x) x^\alpha \, dx = 0.$$

Definition 8.1.12 Let $p, q \in (0, \infty)$, $\omega \in M(\mathbb{R}_+)$ with $m_\infty(\omega) \in (-\frac{n}{p}, \infty)$, $r \in (\max\{1, p, \frac{n}{m_\infty(\omega)+n/p}\}, \infty]$,

$$s \in \left(0, \min\left\{ 1, p, q, \frac{n}{M_\infty(\omega) + n/p} \right\} \right),$$

and $d \geq \lfloor n(1/s - 1) \rfloor$ be a fixed integer. Then the *inhomogeneous generalized atomic Herz–Hardy space* $H\mathcal{K}_{\omega,0}^{p,q,r,d,s}(\mathbb{R}^n)$ is defined to be the set of all the $f \in \mathcal{S}'(\mathbb{R}^n)$ such that there exists $\{\lambda_j\}_{j\in\mathbb{N}} \subset [0, \infty)$ and a sequence $\{a_j\}_{j\in\mathbb{N}}$ of $(\mathcal{K}_{\omega,0}^{p,q}(\mathbb{R}^n), r, d)$-atoms supported, respectively, in the balls $\{B_j\}_{j\in\mathbb{N}} \subset \mathbb{B}$ such that

$$f = \sum_{j\in\mathbb{N}} \lambda_j a_j$$

in $\mathcal{S}'(\mathbb{R}^n)$ and

$$\left\| \left\{ \sum_{j\in\mathbb{N}} \left[\frac{\lambda_j}{\|\mathbf{1}_{B_j}\|_{\mathcal{K}_{\omega,0}^{p,q}(\mathbb{R}^n)}} \right]^s \mathbf{1}_{B_j} \right\}^{\frac{1}{s}} \right\|_{\mathcal{K}_{\omega,0}^{p,q}(\mathbb{R}^n)} < \infty.$$

Moreover, for any $f \in H\mathcal{K}_{\omega,0}^{p,q,r,d,s}(\mathbb{R}^n)$,

$$\|f\|_{H\mathcal{K}_{\omega,0}^{p,q,r,d,s}(\mathbb{R}^n)} := \inf \left\{ \left\| \left\{ \sum_{j\in\mathbb{N}} \left[\frac{\lambda_j}{\|\mathbf{1}_{B_j}\|_{\mathcal{K}_{\omega,0}^{p,q}(\mathbb{R}^n)}} \right]^s \mathbf{1}_{B_j} \right\}^{\frac{1}{s}} \right\|_{\mathcal{K}_{\omega,0}^{p,q}(\mathbb{R}^n)} \right\},$$

where the infimum is taken over all the decompositions of f as above.

Next, we give the atomic characterization of the Hardy space $H\mathcal{K}_{\omega,0}^{p,q}(\mathbb{R}^n)$ as follows.

Theorem 8.1.13 *Let p, q, ω, d, s, and r be as in Definition 8.1.12. Then*

$$H\mathcal{K}^{p,q}_{\omega,0}(\mathbb{R}^n) = H\mathcal{K}^{p,q,r,d,s}_{\omega,0}(\mathbb{R}^n)$$

with equivalent quasi-norms.

To establish this atomic characterization, we need the following two auxiliary lemmas about the Fefferman–Stein vector-valued inequality on inhomogeneous local generalized Herz spaces.

Lemma 8.1.14 *Let $p, q \in (0, \infty)$ and $\omega \in M(\mathbb{R}_+)$ satisfy $m_\infty(\omega) \in (-\frac{n}{p}, \infty)$. Then, for any given $u \in (1, \infty)$ and*

$$r \in \left(0, \min\left\{p, \frac{n}{M_\infty(\omega) + n/p}\right\}\right),$$

there exists a positive constant C such that, for any $\{f_j\}_{j\in\mathbb{N}} \subset L^1_{\mathrm{loc}}(\mathbb{R}^n)$,

$$\left\|\left\{\sum_{j\in\mathbb{N}}[\mathcal{M}(f_j)]^u\right\}^{\frac{1}{u}}\right\|_{[\mathcal{K}^{p,q}_{\omega,0}(\mathbb{R}^n)]^{1/r}} \leq C \left\|\left(\sum_{j\in\mathbb{N}}|f_j|^u\right)^{\frac{1}{u}}\right\|_{[\mathcal{K}^{p,q}_{\omega,0}(\mathbb{R}^n)]^{1/r}}.$$

Proof Let all the symbols be as in the present lemma, $u \in (1, \infty)$, and

$$r \in \left(0, \min\left\{p, \frac{n}{M_\infty(\omega) + n/p}\right\}\right).$$

Then, repeating the proof of Lemma 4.3.10 with both Theorem 1.6.1 and Lemma 1.3.2 therein replaced, respectively, by both Theorem 7.1.28 and Lemma 7.1.14, we find that, for any $\{f_j\}_{j\in\mathbb{N}} \subset L^1_{\mathrm{loc}}(\mathbb{R}^n)$,

$$\left\|\left\{\sum_{j\in\mathbb{N}}[\mathcal{M}(f_j)]^u\right\}^{\frac{1}{u}}\right\|_{[\mathcal{K}^{p,q}_{\omega,0}(\mathbb{R}^n)]^{1/r}} \lesssim \left\|\left(\sum_{j\in\mathbb{N}}|f_j|^u\right)^{\frac{1}{u}}\right\|_{[\mathcal{K}^{p,q}_{\omega,0}(\mathbb{R}^n)]^{1/r}}.$$

This finishes the proof of Lemma 8.1.14. $\qquad\square$

Lemma 8.1.15 *Let $p, q \in (0, \infty)$, $\omega \in M(\mathbb{R}_+)$ satisfy $m_\infty(\omega) \in (-\frac{n}{p}, \infty)$, $s \in (0, \infty)$, and*

$$\theta \in \left(0, \min\left\{s, p, \frac{n}{M_\infty(\omega) + n/p}\right\}\right).$$

Then there exists a positive constant C such that, for any $\{f_j\}_{j\in\mathbb{N}} \subset L^1_{\mathrm{loc}}(\mathbb{R}^n)$,

$$\left\| \left\{ \sum_{j\in\mathbb{N}} \left[\mathcal{M}^{(\theta)}(f_j) \right]^s \right\}^{1/s} \right\|_{\mathcal{K}^{p,q}_{\omega,0}(\mathbb{R}^n)} \leq C \left\| \left(\sum_{j\in\mathbb{N}} |f_j|^s \right)^{1/s} \right\|_{\mathcal{K}^{p,q}_{\omega,0}(\mathbb{R}^n)}.$$

Proof Let all the symbols be as in the present lemma. Then, from Remark 1.2.30(ii) and repeating an argument similar to that used in the proof of Lemma 4.3.11 with Lemma 4.3.10 therein replaced by Lemma 8.1.14, we infer that, for any $\{f_j\}_{j\in\mathbb{N}} \subset L^1_{\mathrm{loc}}(\mathbb{R}^n)$,

$$\left\| \left\{ \sum_{j\in\mathbb{N}} \left[\mathcal{M}^{(\theta)}(f_j) \right]^s \right\}^{1/s} \right\|_{\mathcal{K}^{p,q}_{\omega,0}(\mathbb{R}^n)} \lesssim \left\| \left(\sum_{j\in\mathbb{N}} |f_j|^s \right)^{1/s} \right\|_{\mathcal{K}^{p,q}_{\omega,0}(\mathbb{R}^n)}.$$

This finishes the proof of Lemma 8.1.15.   □

Via both Lemmas 8.1.14 and 8.1.15 above, we now show Theorem 8.1.13.

Proof of Theorem 8.1.13 Let all the symbols be as in the present theorem. Then, applying both Lemmas 4.3.6 and 4.3.8 and repeating the proof of Theorem 4.3.3 with Theorem 1.2.42 and Lemmas 1.8.6 and 4.3.11 therein replaced, respectively, by Theorem 7.1.7 and Lemmas 7.1.37 and 8.1.15, we conclude that

$$H\mathcal{K}^{p,q}_{\omega,0}(\mathbb{R}^n) = H\mathcal{K}^{p,q,r,d,s}_{\omega,0}(\mathbb{R}^n)$$

with equivalent quasi-norms. This finishes the proof of Theorem 8.1.13.   □

Next, we turn to establish the atomic characterization of the inhomogeneous generalized Herz–Hardy space $H\mathcal{K}^{p,q}_{\omega}(\mathbb{R}^n)$. To do this, we first introduce the following definition of $(\mathcal{K}^{p,q}_{\omega}(\mathbb{R}^n), r, d)$-atoms.

Definition 8.1.16 Let $p, q \in (0,\infty)$, $\omega \in M(\mathbb{R}_+)$ with $M_\infty(\omega) \in (-\infty, 0)$, $r \in [1,\infty]$, and $d \in \mathbb{Z}_+$. Then a measurable function a on $\mathbb{R}^n$ is called a $(\mathcal{K}^{p,q}_{\omega}(\mathbb{R}^n), r, d)$-*atom* if there exists a ball $B \in \mathbb{B}$ such that

(i) supp $(a) := \{x \in \mathbb{R}^n : a(x) \neq 0\} \subset B$;

(ii) $\|a\|_{L^r(\mathbb{R}^n)} \leq \dfrac{|B|^{1/r}}{\|\mathbf{1}_B\|_{\mathcal{K}^{p,q}_{\omega}(\mathbb{R}^n)}}$;

(iii) for any $\alpha \in \mathbb{Z}^n_+$ with $|\alpha| \leq d$,

$$\int_{\mathbb{R}^n} a(x)x^\alpha \, dx = 0.$$

Via these atoms, we now give the following concept of the inhomogeneous generalized atomic Herz–Hardy space $H\mathcal{K}_\omega^{p,q,r,d,s}(\mathbb{R}^n)$.

Definition 8.1.17 Let $p, q \in (0, \infty)$, $\omega \in M(\mathbb{R}_+)$ satisfy

$$-\frac{n}{p} < m_\infty(\omega) \leq M_\infty(\omega) < 0,$$

$s \in (0, \min\{1, p, q\})$, $d \geq \lfloor n(1/s - 1) \rfloor$ be a fixed integer, and

$$r \in \left(\max\left\{ 1, \frac{n}{m_\infty(\omega) + n/p} \right\}, \infty \right].$$

Then the *inhomogeneous generalized atomic Herz–Hardy space* $H\mathcal{K}_\omega^{p,q,r,d,s}(\mathbb{R}^n)$ is defined to be the set of all the $f \in \mathcal{S}'(\mathbb{R}^n)$ such that there exists $\{\lambda_j\}_{j\in\mathbb{N}} \subset [0, \infty)$ and a sequence $\{a_j\}_{j\in\mathbb{N}}$ of $(\mathcal{K}_\omega^{p,q}(\mathbb{R}^n), r, d)$-atoms supported, respectively, in the balls $\{B_j\}_{j\in\mathbb{N}} \subset \mathbb{B}$ such that

$$f = \sum_{j\in\mathbb{N}} \lambda_j a_j$$

in $\mathcal{S}'(\mathbb{R}^n)$ and

$$\left\| \left\{ \sum_{j\in\mathbb{N}} \left[\frac{\lambda_j}{\|\mathbf{1}_{B_j}\|_{\mathcal{K}_\omega^{p,q}(\mathbb{R}^n)}} \right]^s \mathbf{1}_{B_j} \right\}^{\frac{1}{s}} \right\|_{\mathcal{K}_\omega^{p,q}(\mathbb{R}^n)} < \infty.$$

Moreover, for any $f \in H\mathcal{K}_\omega^{p,q,r,d,s}(\mathbb{R}^n)$,

$$\|f\|_{H\mathcal{K}_\omega^{p,q,r,d,s}(\mathbb{R}^n)} := \inf \left\{ \left\| \left\{ \sum_{j\in\mathbb{N}} \left[\frac{\lambda_j}{\|\mathbf{1}_{B_j}\|_{\mathcal{K}_\omega^{p,q}(\mathbb{R}^n)}} \right]^s \mathbf{1}_{B_j} \right\}^{\frac{1}{s}} \right\|_{\mathcal{K}_\omega^{p,q}(\mathbb{R}^n)} \right\},$$

where the infimum is taken over all the decompositions of f as above.

Then we have the following atomic characterization of the inhomogeneous generalized Herz–Hardy space $H\mathcal{K}_\omega^{p,q}(\mathbb{R}^n)$.

Theorem 8.1.18 *Let p, q, ω, r, s, and d be as in Definition 8.1.17. Then*

$$H\mathcal{K}_\omega^{p,q}(\mathbb{R}^n) = H\mathcal{K}_\omega^{p,q,r,d,s}(\mathbb{R}^n)$$

with equivalent quasi-norms.

To prove this theorem, we require some preliminary lemmas about both inhomogeneous global generalized Herz spaces and inhomogeneous block spaces. First, the
following one is related to the boundedness of powered Hardy–Littlewood maximal
operators on inhomogeneous block spaces.

Lemma 8.1.19 *Let* $p, q \in (0, \infty)$ *and* $\omega \in M(\mathbb{R}_+)$ *satisfy* $m_\infty(\omega) \in (-\frac{n}{p}, \infty)$.
Then, for any given $s \in (0, \min\{p, q, \frac{n}{M_\infty(\omega)+n/p}\})$ *and*

$$r \in \left(\max\left\{p, \frac{n}{m_\infty(\omega) + n/p}\right\}, \infty\right],$$

there exists a positive constant C *such that, for any* $f \in L^1_{loc}(\mathbb{R}^n)$,

$$\left\|\mathcal{M}^{((r/s)')}(f)\right\|_{\mathcal{B}^{(p/s)',(q/s)'}_{1/\omega^s}(\mathbb{R}^n)} \leq C\|f\|_{\mathcal{B}^{(p/s)',(q/s)'}_{1/\omega^s}(\mathbb{R}^n)}.$$

Proof Let all the symbols be as in the present lemma, $s \in (0, \min\{p, q, \frac{n}{M_\infty(\omega)+n/p}\})$, and

$$r \in \left(\max\left\{p, \frac{n}{m_\infty(\omega) + n/p}\right\}, \infty\right].$$

Then, using Lemma 1.1.6 and repeating an argument similar to that used in the
proof Lemma 4.3.23 with Corollary 2.3.5 therein replaced by Corollary 7.2.14, we
find that, for any $f \in L^1_{loc}(\mathbb{R}^n)$,

$$\left\|\mathcal{M}^{((r/s)')}(f)\right\|_{\mathcal{B}^{(p/s)',(q/s)'}_{1/\omega^s}(\mathbb{R}^n)} \lesssim \|f\|_{\mathcal{B}^{(p/s)',(q/s)'}_{1/\omega^s}(\mathbb{R}^n)}.$$

This finishes the proof of Lemma 8.1.19.   □

We also need the following two technical lemmas about the Fefferman–Stein
vector-valued inequality on inhomogeneous global generalized Herz spaces.

Lemma 8.1.20 *Let* $p, q \in (0, \infty)$ *and* $\omega \in M(\mathbb{R}_+)$ *satisfy* $m_\infty(\omega) \in (-\frac{n}{p}, \infty)$.
Then, for any given $u \in (1, \infty)$ *and*

$$r \in \left(0, \min\left\{p, \frac{n}{M_\infty(\omega) + n/p}\right\}\right),$$

there exists a positive constant C *such that, for any* $\{f_j\}_{j\in\mathbb{N}} \subset L^1_{loc}(\mathbb{R}^n)$,

$$\left\|\left\{\sum_{j\in\mathbb{N}}[\mathcal{M}(f_j)]^u\right\}^{\frac{1}{u}}\right\|_{[\mathcal{K}^{p,q}_\omega(\mathbb{R}^n)]^{1/r}} \leq C\left\|\left(\sum_{j\in\mathbb{N}}|f_j|^u\right)^{\frac{1}{u}}\right\|_{[\mathcal{K}^{p,q}_\omega(\mathbb{R}^n)]^{1/r}}.$$

Proof Let all the symbols be as in the present lemma, $u \in (1, \infty)$, and

$$r \in \left(0, \min\left\{p, \frac{n}{M_\infty(\omega) + n/p}\right\}\right).$$

Then, repeating the proof of Lemma 4.3.24 with both Theorem 1.6.4 and Lemma 1.3.2 therein replaced, respectively, by both Theorem 7.1.30 and Lemma 7.1.14, we conclude that, for any $\{f_j\}_{j\in\mathbb{N}} \subset L^1_{\mathrm{loc}}(\mathbb{R}^n)$,

$$\left\|\left\{\sum_{j\in\mathbb{N}} [\mathcal{M}(f_j)]^u\right\}^{\frac{1}{u}}\right\|_{[\mathcal{K}^{p,q}_\omega(\mathbb{R}^n)]^{1/r}} \lesssim \left\|\left(\sum_{j\in\mathbb{N}} |f_j|^u\right)^{\frac{1}{u}}\right\|_{[\mathcal{K}^{p,q}_\omega(\mathbb{R}^n)]^{1/r}}.$$

This then finishes the proof of Lemma 8.1.20. $\qquad\square$

Lemma 8.1.21 *Let* $p, q \in (0, \infty)$, $\omega \in M(\mathbb{R}_+)$ *satisfy* $m_\infty(\omega) \in (-\frac{n}{p}, \infty)$, $s \in (0, \infty)$, *and*

$$\theta \in \left(0, \min\left\{s, p, \frac{n}{M_\infty(\omega) + n/p}\right\}\right).$$

Then there exists a positive constant C *such that, for any* $\{f_j\}_{j\in\mathbb{N}} \subset L^1_{\mathrm{loc}}(\mathbb{R}^n)$,

$$\left\|\left\{\sum_{j\in\mathbb{N}} \left[\mathcal{M}^{(\theta)}(f_j)\right]^s\right\}^{1/s}\right\|_{\mathcal{K}^{p,q}_\omega(\mathbb{R}^n)} \leq C \left\|\left(\sum_{j\in\mathbb{N}} |f_j|^s\right)^{1/s}\right\|_{\mathcal{K}^{p,q}_\omega(\mathbb{R}^n)}.$$

Proof Let all the symbols be as in the present lemma. Then, applying Remark 1.2.30(ii) and repeating the proof of Lemma 4.3.25 via replacing Lemma 4.3.24 therein by Lemma 8.1.20, we find that, for any $\{f_j\}_{j\subset\mathbb{N}} \subset L^1_{\mathrm{loc}}(\mathbb{R}^n)$,

$$\left\|\left\{\sum_{j\in\mathbb{N}} \left[\mathcal{M}^{(\theta)}(f_j)\right]^s\right\}^{1/s}\right\|_{\mathcal{K}^{p,q}_\omega(\mathbb{R}^n)} \lesssim \left\|\left(\sum_{j\in\mathbb{N}} |f_j|^s\right)^{1/s}\right\|_{\mathcal{K}^{p,q}_\omega(\mathbb{R}^n)}.$$

This finishes the proof of Lemma 8.1.21. $\qquad\square$

Moreover, to show Theorem 8.1.18, we require the following variant of the Hölder inequality of inhomogeneous global generalized Herz spaces.

Lemma 8.1.22 *Let* $p, q \in (1, \infty)$ *and* $\omega \in M(\mathbb{R}_+)$. *Then there exists a positive constant* C *such that, for any* $f, g \in \mathscr{M}(\mathbb{R}^n)$,

$$\|fg\|_{L^1(\mathbb{R}^n)} \leq C \|f\|_{\mathcal{K}^{p,q}_\omega(\mathbb{R}^n)} \|g\|_{\mathcal{B}^{p',q'}_{1/\omega}(\mathbb{R}^n)}.$$

Proof Let all the symbols be as in the present lemma and $f, g \in \mathscr{M}(\mathbb{R}^n)$. Then, repeating an argument similar to that used in the estimation of (2.25) with both (1.111) and Lemma 2.2.2(ii) therein replaced, respectively, by (7.27) and Lemma 7.2.5(ii), we obtain

$$\|fg\|_{L^1(\mathbb{R}^n)} \lesssim \|f\|_{\mathcal{K}_\omega^{p,q}(\mathbb{R}^n)} \|f\|_{\mathcal{B}_{1/\omega}^{p',q'}(\mathbb{R}^n)},$$

which completes the proof of Lemma 8.1.22.   □

In addition, the following equivalent characterization of inhomogeneous global generalized Herz spaces is also an essential tool in the proof of Theorem 8.1.18.

Lemma 8.1.23 *Let $p, q \in (1, \infty)$ and $\omega \in M(\mathbb{R}_+)$ satisfy $M_\infty(\omega) \in (-\infty, 0)$. Then a measurable function f belongs to the inhomogeneous global generalized Herz space $\mathcal{K}_\omega^{p,q}(\mathbb{R}^n)$ if and only if*

$$\|f\|_{\mathcal{K}_\omega^{p,q}(\mathbb{R}^n)}^\star := \sup \left\{ \|fg\|_{L^1(\mathbb{R}^n)} : \|g\|_{\mathcal{B}_{1/\omega}^{p',q'}(\mathbb{R}^n)} = 1 \right\} < \infty.$$

Moreover, there exist two positive constants C_1 and C_2 such that, for any $f \in \mathcal{K}_\omega^{p,q}(\mathbb{R}^n)$,

$$C_1 \|f\|_{\mathcal{K}_\omega^{p,q}(\mathbb{R}^n)} \leq \|f\|_{\mathcal{K}_\omega^{p,q}(\mathbb{R}^n)}^\star \leq C_2 \|f\|_{\mathcal{K}_\omega^{p,q}(\mathbb{R}^n)}.$$

Proof Let all the symbols be as in the present lemma and $f \in \mathscr{M}(\mathbb{R}^n)$. We first show the necessity. For this purpose, assume $f \in \mathcal{K}_\omega^{p,q}(\mathbb{R}^n)$. Then, repeating the estimation of (4.37) with Lemma 4.3.26 therein replaced by Lemma 8.1.22, we obtain

$$\|f\|_{\mathcal{K}_\omega^{p,q}(\mathbb{R}^n)}^\star \lesssim \|f\|_{\mathcal{K}_\omega^{p,q}(\mathbb{R}^n)} < \infty. \tag{8.1}$$

This then finishes the proof of the necessity.

Conversely, we show the sufficiency. To this end, assume

$$\|f\|_{\dot{\mathcal{K}}_\omega^{p,q}(\mathbb{R}^n)}^\star < \infty.$$

Then, repeating an argument similar to that used in the estimation of (4.39) via replacing both Theorems 1.2.44 and 2.2.1 therein, respectively, by both Theorems 7.1.9 and 7.2.4, we find that

$$\|f\|_{\mathcal{K}_\omega^{p,q}(\mathbb{R}^n)} \lesssim \|f\|_{\mathcal{K}_\omega^{p,q}(\mathbb{R}^n)}^\star < \infty. \tag{8.2}$$

This implies that $f \in \dot{\mathcal{K}}_{\omega}^{p,q}(\mathbb{R}^n)$ and hence finishes the proof of the sufficiency. Moreover, combining both (8.1) and (8.2), we conclude that

$$\|f\|_{\mathcal{K}_{\omega}^{p,q}(\mathbb{R}^n)} \sim \|f\|_{\mathcal{K}_{\omega}^{p,q}(\mathbb{R}^n)}^{\star}$$

with the positive equivalence constants independent of f. This finishes the proof of Lemma 8.1.23. □

Via the above lemmas, we next prove Theorem 8.1.18.

Proof of Theorem 8.1.18 Let all the symbols be as in the present theorem. Then, repeating the proof of Theorem 4.3.16 via replacing Theorem 1.2.44 and Lemmas 4.3.23, 4.3.25, and 4.3.27 therein, respectively, by Theorem 7.1.9 and Lemmas 8.1.19, 8.1.21, and 8.1.23, we obtain

$$H\mathcal{K}_{\omega}^{p,q}(\mathbb{R}^n) = H\mathcal{K}_{\omega}^{p,q,r,d,s}(\mathbb{R}^n)$$

with equivalent quasi-norms. This then finishes the proof of Theorem 8.1.18. □

8.1.4 Inhomogeneous Generalized Finite Atomic Herz–Hardy Spaces

In this subsection, we introduce and study finite atomic Hardy spaces associated with inhomogeneous generalized Herz spaces. To this end, we first introduce the following inhomogeneous generalized finite atomic Herz–Hardy space $H\mathcal{K}_{\omega,0,\mathrm{fin}}^{p,q,r,d,s}(\mathbb{R}^n)$.

Definition 8.1.24 Let $p, q \in (0, \infty)$, $\omega \in M(\mathbb{R}_+)$ with $m_\infty(\omega) \in (-\frac{n}{p}, \infty)$,

$$s \in \left(0, \min\left\{1, p, q, \frac{n}{M_\infty(\omega) + n/p}\right\}\right),$$

$d \geq \lfloor n(1/s - 1) \rfloor$ be a fixed integer, and

$$r \in \left(\max\left\{1, p, \frac{n}{m_\infty(\omega) + n/p}\right\}, \infty\right].$$

Then the *inhomogeneous generalized finite atomic Herz–Hardy space* $H\mathcal{K}_{\omega,0,\mathrm{fin}}^{p,q,r,d,s}(\mathbb{R}^n)$, associated with $\mathcal{K}_{\omega,0}^{p,q}(\mathbb{R}^n)$, is defined to be the set of all finite linear combinations of $(\mathcal{K}^{p,q}_{\omega,\mathbf{0}}(\mathbb{R}^n)$, r, $d)$-atoms. Moreover, for any $f \in H\mathcal{K}^{p,q,r,d,s}_{\omega,\mathbf{0},\mathrm{fin}}(\mathbb{R}^n)$,

$$
\|f\|_{H\mathcal{K}^{p,q,r,d,s}_{\omega,\mathbf{0},\mathrm{fin}}(\mathbb{R}^n)} := \left\{ \inf \left\| \left\{ \sum_{j=1}^{N} \left[\frac{\lambda_j}{\|\mathbf{1}_{B_j}\|_{\mathcal{K}^{p,q}_{\omega,\mathbf{0}}(\mathbb{R}^n)}} \right]^s \mathbf{1}_{B_j} \right\}^{\frac{1}{s}} \right\|_{\mathcal{K}^{p,q}_{\omega,\mathbf{0}}(\mathbb{R}^n)} \right\},
$$

where the infimum is taken over all finite linear combinations of f, namely, $N \in \mathbb{N}$, $f = \sum_{j=1}^{N} \lambda_j a_j$, $\{\lambda_j\}_{j=1}^{N} \subset [0,\infty)$, and $\{a_j\}_{j=1}^{N}$ being $(\mathcal{K}^{p,q}_{\omega,\mathbf{0}}(\mathbb{R}^n)$, r, $d)$-atoms supported, respectively, in the balls $\{B_j\}_{j=1}^{N} \subset \mathbb{B}$.

Then, for finite atomic Hardy spaces, we establish the following equivalence on their quasi-norms with those of the corresponding Hardy spaces. Recall that $C(\mathbb{R}^n)$ denotes the set of all continuous functions on $\mathbb{R}^n$.

Theorem 8.1.25 *Let p, q, ω, d, s, and r be as in Definition 8.1.24.*

(i) *If*

$$
r \in \left(\max\left\{ 1, p, \frac{n}{m_\infty(\omega) + n/p} \right\}, \infty \right),
$$

then $\| \cdot \|_{H\mathcal{K}^{p,q,r,d,s}_{\omega,\mathbf{0},\mathrm{fin}}(\mathbb{R}^n)}$ and $\| \cdot \|_{H\mathcal{K}^{p,q}_{\omega,\mathbf{0}}(\mathbb{R}^n)}$ are equivalent quasi-norms on the inhomogeneous generalized finite atomic Herz–Hardy space $H\mathcal{K}^{p,q,r,d,s}_{\omega,\mathbf{0},\mathrm{fin}}(\mathbb{R}^n)$.

(ii) *If $r = \infty$, then $\| \cdot \|_{H\mathcal{K}^{p,q,\infty,d,s}_{\omega,\mathbf{0},\mathrm{fin}}(\mathbb{R}^n)}$ and $\| \cdot \|_{H\mathcal{K}^{p,q}_{\omega,\mathbf{0}}(\mathbb{R}^n)}$ are equivalent quasi-norms on $H\mathcal{K}^{p,q,\infty,d,s}_{\omega,\mathbf{0},\mathrm{fin}}(\mathbb{R}^n) \cap C(\mathbb{R}^n)$.*

Proof Let all the symbols be as in the present theorem. Then, from Lemma 4.4.3 and an argument similar to that used in the proof of Theorem 4.4.2 with both Lemmas 1.8.6 and 4.3.11 therein replaced, respectively, by both Lemmas 7.1.37 and 8.1.15, we infer that both (i) and (ii) of the present theorem hold true, which completes the proof of Theorem 8.1.25. $\square$

Now, we turn to investigate finite atomic Hardy spaces associated with the global Herz space $\mathcal{K}^{p,q}_{\omega}(\mathbb{R}^n)$. We first introduce the following finite atomic Hardy spaces.

Definition 8.1.26 Let p, $q \in (0,\infty)$, $\omega \in M(\mathbb{R}_+)$ with

$$
-\frac{n}{p} < m_\infty(\omega) \leq M_\infty(\omega) < 0,
$$

$s \in (0, \min\{1, p, q\})$, $d \geq \lfloor n(1/s - 1) \rfloor$ be a fixed integer, and

$$
r \in \left(\max\left\{ 1, \frac{n}{m_\infty(\omega) + n/p} \right\}, \infty \right].
$$

Then the *inhomogeneous generalized finite atomic Herz–Hardy space* $H\mathcal{K}_{\omega,\mathrm{fin}}^{p,q,r,d,s}(\mathbb{R}^n)$, associated with $\mathcal{K}_\omega^{p,q}(\mathbb{R}^n)$, is defined to be the set of all finite linear combinations of $(\mathcal{K}_\omega^{p,q}(\mathbb{R}^n),\ r,\ d)$-atoms. Moreover, for any $f \in H\mathcal{K}_{\omega,\mathrm{fin}}^{p,q,r,d,s}(\mathbb{R}^n)$,

$$
\|f\|_{H\mathcal{K}_{\omega,\mathrm{fin}}^{p,q,r,d,s}(\mathbb{R}^n)} := \left\{ \inf \left[\left\| \left\{ \sum_{j=1}^N \left[\frac{\lambda_j}{\|\mathbf{1}_{B_j}\|_{\mathcal{K}_\omega^{p,q}(\mathbb{R}^n)}} \right]^s \mathbf{1}_{B_j} \right\}^{\frac{1}{s}} \right\|_{\mathcal{K}_\omega^{p,q}(\mathbb{R}^n)} \right] \right\},
$$

where the infimum is taken over all finite linear combinations of f, namely, $N \in \mathbb{N}$, $f = \sum_{j=1}^N \lambda_j a_j$, $\{\lambda_j\}_{j=1}^N \subset [0, \infty)$, and $\{a_j\}_{j=1}^N$ being $(\mathcal{K}_\omega^{p,q}(\mathbb{R}^n),\ r,\ d)$-atoms supported, respectively, in the balls $\{B_j\}_{j=1}^N \subset \mathbb{B}$.

Via this concept, we prove the following conclusion which implies that the quasi-norms $\|\cdot\|_{H\mathcal{K}_\omega^{p,q}(\mathbb{R}^n)}$ and $\|\cdot\|_{H\mathcal{K}_{\omega,\mathrm{fin}}^{p,q,r,d,s}(\mathbb{R}^n)}$ are equivalent on $H\mathcal{K}_{\omega,\mathrm{fin}}^{p,q,r,d,s}(\mathbb{R}^n)$.

Theorem 8.1.27 *Let p, q, ω, d, s, and r be as in Definition 8.1.26.*

(i) *If*

$$
r \in \left(\max\left\{ 1, \frac{n}{m_\infty(\omega) + n/p} \right\}, \infty \right),
$$

then $\|\cdot\|_{H\mathcal{K}_{\omega,\mathrm{fin}}^{p,q,r,d,s}(\mathbb{R}^n)}$ and $\|\cdot\|_{H\mathcal{K}_\omega^{p,q}(\mathbb{R}^n)}$ are equivalent quasi-norms on the inhomogeneous generalized finite atomic Herz–Hardy space $H\mathcal{K}_{\omega,\mathrm{fin}}^{p,q,r,d,s}(\mathbb{R}^n)$.

(ii) *If $r = \infty$, then $\|\cdot\|_{H\mathcal{K}_{\omega,\mathrm{fin}}^{p,q,\infty,d,s}(\mathbb{R}^n)}$ and $\|\cdot\|_{H\mathcal{K}_\omega^{p,q}(\mathbb{R}^n)}$ are equivalent quasi-norms on $H\mathcal{K}_{\omega,\mathrm{fin}}^{p,q,\infty,d,s}(\mathbb{R}^n) \cap C(\mathbb{R}^n)$.*

Proof Let all the symbols be as in the present lemma and $f \in H\mathcal{K}_{\omega,\mathrm{fin}}^{p,q,r,d,s}(\mathbb{R}^n)$. Then, using Lemma 4.4.9 and repeating the proof of Theorem 4.4.8 via replacing both Theorem 4.3.16 and Lemma 4.3.25 therein, respectively, by both Theorem 8.1.18 and Lemma 8.1.21, we conclude that, if $r \in (1, \infty)$, then

$$
\|f\|_{H\mathcal{K}_{\omega,\mathrm{fin}}^{p,q,r,d,s}(\mathbb{R}^n)} \sim \|f\|_{H\mathcal{K}_\omega^{p,q}(\mathbb{R}^n)}
$$

and, if $r = \infty$ and $f \in C(\mathbb{R}^n)$, then

$$
\|f\|_{H\mathcal{K}_{\omega,\mathrm{fin}}^{p,q,\infty,d,s}(\mathbb{R}^n)} \sim \|f\|_{H\mathcal{K}_\omega^{p,q}(\mathbb{R}^n)}.
$$

These further imply that both (i) and (ii) of the present theorem hold true, and then finish the proof of Theorem 8.1.27. $\qquad\square$

8.1.5   Molecular Characterizations

This subsection aims to establish the molecular characterizations of inhomogeneous generalized Herz–Hardy spaces. To do this, recall that, for any $j \in \mathbb{N}$, $S_j(B) :=$ $(2^j B) \setminus (2^{j-1} B)$ and $S_0(B) := B$. Then we introduce the $(\mathcal{K}^{p,q}_{\omega,\mathbf{0}}(\mathbb{R}^n), r, d, \tau)$-molecules as follows.

Definition 8.1.28 Let $p, q \in (0, \infty)$, $\omega \in M(\mathbb{R}_+)$, $\tau \in (0, \infty)$, $r \in [1, \infty]$, and $d \in \mathbb{Z}_+$. Then a measurable function m on $\mathbb{R}^n$ is called a $(\mathcal{K}^{p,q}_{\omega,\mathbf{0}}(\mathbb{R}^n), r, d, \tau)$-molecule centered at a ball $B \in \mathbb{B}$ if

(i) for any $j \in \mathbb{Z}_+$,

$$\left\| m \mathbf{1}_{S_j(B)} \right\|_{L^r(\mathbb{R}^n)} \leq 2^{-\tau j} \frac{|B|^{1/r}}{\left\| \mathbf{1}_B \right\|_{\mathcal{K}^{p,q}_{\omega,\mathbf{0}}(\mathbb{R}^n)}};$$

(ii) for any $\alpha \in \mathbb{Z}^n_+$ with $|\alpha| \leq d$,

$$\int_{\mathbb{R}^n} m(x) x^\alpha \, dx = 0.$$

We next show the following molecular characterization of $H\mathcal{K}^{p,q}_{\omega,\mathbf{0}}(\mathbb{R}^n)$.

Theorem 8.1.29 Let $p, q \in (0, \infty)$, $\omega \in M(\mathbb{R}_+)$ with $m_\infty(\omega) \in (-\frac{n}{p}, \infty)$,

$$s \in \left(0, \min \left\{ 1, p, q, \frac{n}{M_\infty(\omega) + n/p} \right\} \right),$$

$d \geq \lfloor n(1/s - 1) \rfloor$ be a fixed integer,

$$r \in \left(\max \left\{ 1, p, \frac{n}{m_\infty(\omega) + n/p} \right\}, \infty \right],$$

and $\tau \in (0, \infty)$ with $\tau > n(1/s - 1/r)$. Then $f \in H\mathcal{K}^{p,q}_{\omega,\mathbf{0}}(\mathbb{R}^n)$ if and only if $f \in \mathcal{S}'(\mathbb{R}^n)$ and there exists a sequence $\{m_j\}_{j \in \mathbb{N}}$ of $(\mathcal{K}^{p,q}_{\omega,\mathbf{0}}(\mathbb{R}^n), r, d, \tau)$-molecules centered, respectively, at the balls $\{B_j\}_{j \in \mathbb{N}} \subset \mathbb{B}$ and a sequence $\{\lambda_j\}_{j \in \mathbb{N}} \subset [0, \infty)$ such that $f = \sum_{j \in \mathbb{N}} \lambda_j m_j$ in $\mathcal{S}'(\mathbb{R}^n)$ and

$$\left\| \left\{ \sum_{j \in \mathbb{N}} \left[\frac{\lambda_j}{\left\| \mathbf{1}_{B_j} \right\|_{\mathcal{K}^{p,q}_{\omega,\mathbf{0}}(\mathbb{R}^n)}} \right]^s \mathbf{1}_{B_j} \right\}^{\frac{1}{s}} \right\|_{\mathcal{K}^{p,q}_{\omega,\mathbf{0}}(\mathbb{R}^n)} < \infty.$$

Moreover, there exist two positive constants C_1 and C_2 such that, for any $f \in HK_{\omega,0}^{p,q}(\mathbb{R}^n)$,

$$C_1 \|f\|_{HK_{\omega,0}^{p,q}(\mathbb{R}^n)} \le \inf \left\{ \left\| \left\{ \sum_{j \in \mathbb{N}} \left[\frac{\lambda_i}{\|\mathbf{1}_{B_j}\|_{K_{\omega,0}^{p,q}(\mathbb{R}^n)}} \right]^s \mathbf{1}_{B_j} \right\}^{\frac{1}{s}} \right\|_{K_{\omega,0}^{p,q}(\mathbb{R}^n)} \right\}$$

$$\le C_2 \|f\|_{HK_{\omega,0}^{p,q}(\mathbb{R}^n)},$$

where the infimum is taken over all the decompositions of f as above.

Proof Let all the symbols be as in the present theorem and $f \in \mathcal{S}'(\mathbb{R}^n)$. Then, applying Lemma 4.5.5 and repeating an argument similar to that used in the proof of Theorem 4.5.2 with Theorem 1.2.42 and Lemmas 1.8.6 and 4.3.11 therein replaced, respectively, by Theorem 7.1.7 and Lemmas 7.1.37 and 8.1.15, we find that $f \in HK_{\omega,0}^{p,q}(\mathbb{R}^n)$ if and only if there exists a sequence $\{m_j\}_{j \in \mathbb{N}}$ of $(K_{\omega,0}^{p,q}(\mathbb{R}^n), r, d, \tau)$-molecules centered, respectively, at the balls $\{B_j\}_{j \in \mathbb{N}} \subset \mathbb{B}$ and a sequence $\{\lambda_j\}_{j \in \mathbb{N}} \subset [0, \infty)$ such that

$$f = \sum_{j \in \mathbb{N}} \lambda_j m_j$$

in $\mathcal{S}'(\mathbb{R}^n)$ and

$$\left\| \left\{ \sum_{j \in \mathbb{N}} \left[\frac{\lambda_j}{\|\mathbf{1}_{B_j}\|_{K_{\omega,0}^{p,q}(\mathbb{R}^n)}} \right]^s \mathbf{1}_{B_j} \right\}^{\frac{1}{s}} \right\|_{K_{\omega,0}^{p,q}(\mathbb{R}^n)} < \infty,$$

and, moreover,

$$\|f\|_{HK_{\omega,0}^{p,q}(\mathbb{R}^n)} \sim \inf \left\{ \left\| \left\{ \sum_{j \in \mathbb{N}} \left[\frac{\lambda_j}{\|\mathbf{1}_{B_j}\|_{K_{\omega,0}^{p,q}(\mathbb{R}^n)}} \right]^s \mathbf{1}_{B_j} \right\}^{\frac{1}{s}} \right\|_{K_{\omega,0}^{p,q}(\mathbb{R}^n)} \right\},$$

where the infimum is taken over all the decompositions of f as in the present theorem. This finishes the proof of Theorem 8.1.29. $\square$

Now, we establish the molecular characterization of the inhomogeneous generalized Herz–Hardy space $HK_{\omega}^{p,q}(\mathbb{R}^n)$. For this purpose, we first introduce the following concept of $(K_{\omega}^{p,q}(\mathbb{R}^n), r, d, \tau)$-molecules.

Definition 8.1.30 Let $p, q \in (0, \infty)$, $\omega \in M(\mathbb{R}_+)$ with $M_\infty(\omega) \in (-\infty, 0)$, $\tau \in (0, \infty)$, $r \in [1, \infty]$, and $d \in \mathbb{Z}_+$. Then a measurable function m on $\mathbb{R}^n$ is called a $(\mathcal{K}_\omega^{p,q}(\mathbb{R}^n), r, d, \tau)$-*molecule* centered at a ball $B \in \mathbb{B}$ if

(i) for any $j \in \mathbb{Z}_+$,

$$\left\| m \mathbf{1}_{S_j(B)} \right\|_{L^r(\mathbb{R}^n)} \leq 2^{-j\tau} \frac{|B|^{1/r}}{\left\| \mathbf{1}_B \right\|_{\mathcal{K}_\omega^{p,q}(\mathbb{R}^n)}};$$

(ii) for any $\alpha \in \mathbb{Z}_+^n$ with $|\alpha| \leq d$,

$$\int_{\mathbb{R}^n} m(x) x^\alpha \, dx = 0.$$

Via these molecules, we characterize the Hardy space $H\mathcal{K}_\omega^{p,q}(\mathbb{R}^n)$ as follows.

Theorem 8.1.31 Let $p, q \in (0, \infty)$, $\omega \in M(\mathbb{R}_+)$ with

$$-\frac{n}{p} < m_\infty(\omega) \leq M_\infty(\omega) < 0,$$

$s \in (0, \min\{1, p, q\})$, $d \geq \lfloor n(1/s - 1) \rfloor$ be a fixed integer,

$$r \in \left(\max\left\{ 1, \frac{n}{m_\infty(\omega) + n/p} \right\}, \infty \right],$$

and $\tau \in (n(\frac{1}{s} - \frac{1}{r}), \infty)$. Then $f \in H\mathcal{K}_\omega^{p,q}(\mathbb{R}^n)$ if and only if $f \in \mathcal{S}'(\mathbb{R}^n)$ and there exists $\{\lambda_j\}_{j \in \mathbb{N}} \subset [0, \infty)$ and a sequence $\{m_j\}_{j \in \mathbb{N}}$ of $(\mathcal{K}_\omega^{p,q}(\mathbb{R}^n), r, d, \tau)$-molecules centered, respectively, at the balls $\{B_j\}_{j \in \mathbb{N}} \subset \mathbb{B}$ such that

$$f = \sum_{j \in \mathbb{N}} \lambda_j m_j$$

in $\mathcal{S}'(\mathbb{R}^n)$ and

$$\left\| \left\{ \sum_{j \in \mathbb{N}} \left[\frac{\lambda_j}{\left\| \mathbf{1}_{B_j} \right\|_{\mathcal{K}_\omega^{p,q}(\mathbb{R}^n)}} \right]^s \mathbf{1}_{B_j} \right\}^{\frac{1}{s}} \right\|_{\mathcal{K}_\omega^{p,q}(\mathbb{R}^n)} < \infty.$$

Moreover, there exist two positive constants C_1 and C_2 such that, for any $f \in HK_\omega^{p,q}(\mathbb{R}^n)$,

$$C_1 \|f\|_{HK_\omega^{p,q}(\mathbb{R}^n)} \leq \inf \left\{ \left\| \left\{ \sum_{j \in \mathbb{N}} \left[\frac{\lambda_i}{\|\mathbf{1}_{B_j}\|_{K_\omega^{p,q}(\mathbb{R}^n)}} \right]^s \mathbf{1}_{B_j} \right\}^{\frac{1}{s}} \right\|_{K_\omega^{p,q}(\mathbb{R}^n)} \right\}$$

$$\leq C_2 \|f\|_{HK_\omega^{p,q}(\mathbb{R}^n)},$$

where the infimum is taken over all the decompositions of f as above.

Proof Let all the symbols be as in the present theorem and $f \in \mathcal{S}'(\mathbb{R}^n)$. Then, using Theorem 4.5.11 and repeating the proof of Theorem 4.3.16 via replacing Theorem 1.2.44 and Lemmas 4.3.23, 4.3.25, and 4.3.27 therein, respectively, by Theorem 7.1.9 and Lemmas 8.1.19, 8.1.21, and 8.1.23, we conclude that $f \in HK_\omega^{p,q}(\mathbb{R}^n)$ if and only if there exists a sequence $\{m_j\}_{j \in \mathbb{N}}$ of $(K_\omega^{p,q}(\mathbb{R}^n), r, d, \tau)$-molecules centered, respectively, at the balls $\{B_j\}_{j \in \mathbb{N}} \subset \mathbb{B}$ and a sequence $\{\lambda_j\}_{j \in \mathbb{N}} \subset [0, \infty)$ such that $f = \sum_{j \in \mathbb{N}} \lambda_j m_j$ in $\mathcal{S}'(\mathbb{R}^n)$ and

$$\left\| \left\{ \sum_{j \in \mathbb{N}} \left[\frac{\lambda_j}{\|\mathbf{1}_{B_j}\|_{K_\omega^{p,q}(\mathbb{R}^n)}} \right]^s \mathbf{1}_{B_j} \right\}^{\frac{1}{s}} \right\|_{K_\omega^{p,q}(\mathbb{R}^n)} < \infty,$$

and, moreover,

$$\|f\|_{HK_\omega^{p,q}(\mathbb{R}^n)} \sim \inf \left\{ \left\| \left\{ \sum_{j \in \mathbb{N}} \left[\frac{\lambda_j}{\|\mathbf{1}_{B_j}\|_{K_\omega^{p,q}(\mathbb{R}^n)}} \right]^s \mathbf{1}_{B_j} \right\}^{\frac{1}{s}} \right\|_{K_\omega^{p,q}(\mathbb{R}^n)} \right\},$$

where the infimum is taken over all the decompositions of f as in the present theorem. This then finishes the proof of Theorem 8.1.31. $\qquad \square$

8.1.6 Littlewood–Paley Function Characterizations

In this subsection, we establish the equivalent characterizations of both the inhomogeneous generalized Herz–Hardy spaces $HK_{\omega,\mathbf{0}}^{p,q}(\mathbb{R}^n)$ and $HK_\omega^{p,q}(\mathbb{R}^n)$ via the Lusin area function, the Littlewood–Paley g-function, and the Littlewood–Paley g_λ^*-function defined as in Definitions 4.6.1 and 4.6.2. First, we show these Littlewood–Paley function characterizations of $HK_{\omega,\mathbf{0}}^{p,q}(\mathbb{R}^n)$ as follows.

Theorem 8.1.32 *Let* $p, q \in (0, \infty)$, $\omega \in M(\mathbb{R}_+)$ *satisfy* $m_\infty(\omega) \in (-\frac{n}{p}, \infty)$,

$$s_0 := \min\left\{1, p, q, \frac{n}{M_\infty(\omega) + n/p}\right\},$$

and $\lambda \in (\max\{1, 2/s_0\}, \infty)$. *Then the following four statements are mutually equivalent:*

(i) $f \in H\mathcal{K}^{p,q}_{\omega,0}(\mathbb{R}^n)$;

(ii) $f \in \mathcal{S}'(\mathbb{R}^n)$, f *vanishes weakly at infinity, and* $S(f) \in \mathcal{K}^{p,q}_{\omega,0}(\mathbb{R}^n)$;

(iii) $f \in \mathcal{S}'(\mathbb{R}^n)$, f *vanishes weakly at infinity, and* $g(f) \in \mathcal{K}^{p,q}_{\omega,0}(\mathbb{R}^n)$;

(iv) $f \in \mathcal{S}'(\mathbb{R}^n)$, f *vanishes weakly at infinity, and* $g^*_\lambda(f) \in \mathcal{K}^{p,q}_{\omega,0}(\mathbb{R}^n)$.

Moreover, for any $f \in H\mathcal{K}^{p,q}_{\omega,0}(\mathbb{R}^n)$,

$$\|f\|_{H\mathcal{K}^{p,q}_{\omega,0}(\mathbb{R}^n)} \sim \|S(f)\|_{\mathcal{K}^{p,q}_{\omega,0}(\mathbb{R}^n)} \sim \|g(f)\|_{\mathcal{K}^{p,q}_{\omega,0}(\mathbb{R}^n)} \sim \|g^*_\lambda(f)\|_{\mathcal{K}^{p,q}_{\omega,0}(\mathbb{R}^n)},$$

where the positive equivalence constants are independent of f.

Proof Let all the symbols be as in the present theorem. Then, applying Lemma 4.6.4 and repeating an argument similar to that used in the proof of Theorem 4.6.3 with $\min\{m_0(\omega), m_\infty(\omega)\}$, $\max\{M_0(\omega), M_\infty(\omega)\}$, Theorem 1.2.42, and Lemmas 1.8.6 and 4.3.11 therein replaced, respectively, by $m_\infty(\omega)$, $M_\infty(\omega)$, Theorem 7.1.7, and Lemmas 7.1.37 and 8.1.15, we find that (i), (ii), (iii), and (iv) of the present theorem are mutually equivalent and, for any $f \in H\mathcal{K}^{p,q}_{\omega,0}(\mathbb{R}^n)$,

$$\|f\|_{H\mathcal{K}^{p,q}_{\omega,0}(\mathbb{R}^n)} \sim \|S(f)\|_{\mathcal{K}^{p,q}_{\omega,0}(\mathbb{R}^n)} \sim \|g(f)\|_{\mathcal{K}^{p,q}_{\omega,0}(\mathbb{R}^n)} \sim \|g^*_\lambda(f)\|_{\mathcal{K}^{p,q}_{\omega,0}(\mathbb{R}^n)}$$

with positive equivalence constants independent of f. This finishes the proof of Theorem 8.1.32.   □

Remark 8.1.33 Let $p \in (1, \infty)$, $q \in [1, \infty)$, and $\omega(t) := t^\alpha$ for any $t \in (0, \infty)$ and for any given $\alpha \in (-\frac{n}{p}, \frac{n}{p'})$ in Theorem 8.1.32. Then, by Remark 8.1.2 and Theorem 8.1.8, we find that, in this case, the inhomogeneous generalized Herz–Hardy space $H\mathcal{K}^{p,q}_{\omega,0}(\mathbb{R}^n)$ coincides with the classical inhomogeneous Herz space $K^{\alpha,q}_p(\mathbb{R}^n)$, and hence Theorem 8.1.32 goes back to [175, Theorem 1.1.1].

In addition, the following conclusion gives various Littlewood–Paley function characterizations of the inhomogeneous generalized Herz–Hardy space $H\mathcal{K}^{p,q}_\omega(\mathbb{R}^n)$.

Theorem 8.1.34 *Let* $p, q \in (0, \infty)$, $\omega \in M(\mathbb{R}_+)$ *satisfy*

$$-\frac{n}{p} < m_\infty(\omega) \leq M_\infty(\omega) < 0,$$

and $\lambda \in (2\max\{1, \frac{1}{p}, \frac{1}{q}\}, \infty)$. Then the following four statements are mutually equivalent:

(i) $f \in H\mathcal{K}_\omega^{p,q}(\mathbb{R}^n)$;
(ii) $f \in \mathcal{S}'(\mathbb{R}^n)$, f vanishes weakly at infinity, and $S(f) \in \mathcal{K}_\omega^{p,q}(\mathbb{R}^n)$;
(iii) $f \in \mathcal{S}'(\mathbb{R}^n)$, f vanishes weakly at infinity, and $g(f) \in \mathcal{K}_\omega^{p,q}(\mathbb{R}^n)$;
(iv) $f \in \mathcal{S}'(\mathbb{R}^n)$, f vanishes weakly at infinity, and $g_\lambda^*(f) \in \mathcal{K}_\omega^{p,q}(\mathbb{R}^n)$.

Moreover, for any $f \in H\mathcal{K}_\omega^{p,q}(\mathbb{R}^n)$,

$$\|f\|_{H\mathcal{K}_\omega^{p,q}(\mathbb{R}^n)} \sim \|S(f)\|_{\mathcal{K}_\omega^{p,q}(\mathbb{R}^n)} \sim \|g(f)\|_{\mathcal{K}_\omega^{p,q}(\mathbb{R}^n)} \sim \left\|g_\lambda^*(f)\right\|_{\mathcal{K}_\omega^{p,q}(\mathbb{R}^n)},$$

where the positive equivalence constants are independent of f.

Proof Let all the symbols be as in the present theorem. Then, using Lemma 4.6.9 and repeating the proof of Theorem 4.6.8 via replacing Theorems 4.1.2 and 4.6.3 therein, respectively, by Theorems 8.1.3 and 8.1.32, we conclude that (i), (ii), (iii), and (iv) of the present theorem are mutually equivalent and, for any $f \in H\mathcal{K}_\omega^{p,q}(\mathbb{R}^n)$,

$$\|f\|_{H\mathcal{K}_\omega^{p,q}(\mathbb{R}^n)} \sim \|S(f)\|_{\mathcal{K}_\omega^{p,q}(\mathbb{R}^n)} \sim \|g(f)\|_{\mathcal{K}_\omega^{p,q}(\mathbb{R}^n)} \sim \left\|g_\lambda^*(f)\right\|_{\mathcal{K}_\omega^{p,q}(\mathbb{R}^n)},$$

where the positive equivalence constants are independent of f. This then finishes the proof of Theorem 4.6.8. □

8.1.7 Dual Space of $H\mathcal{K}_{\omega,0}^{p,q}(\mathbb{R}^n)$

The main target of this subsection is to establish the dual theorem of the inhomogeneous generalized Herz–Hardy space $H\mathcal{K}_{\omega,0}^{p,q}(\mathbb{R}^n)$. To this end, we first introduce the following Campanato-type function space $\mathcal{L}_{\omega,0}^{p,q,r,d,s}(\mathbb{R}^n)$.

Definition 8.1.35 Let $p, q, s \in (0, \infty)$, $r \in [1, \infty)$, $d \in \mathbb{Z}_+$, and $\omega \in M(\mathbb{R}_+)$. Then the *Campanato-type function space* $\mathcal{L}_{\omega,0}^{p,q,r,d,s}(\mathbb{R}^n)$, associated with the inhomogeneous local generalized Herz space $\mathcal{K}_{\omega,0}^{p,q}(\mathbb{R}^n)$, is defined to be the set of all the $f \in L_{\mathrm{loc}}^r(\mathbb{R}^n)$ such that

$$\|f\|_{\mathcal{L}_{\omega,0}^{p,q,r,d,s}(\mathbb{R}^n)}$$

$$:= \sup \left\| \left\{ \sum_{i=1}^m \left[\frac{\lambda_i}{\|\mathbf{1}_{B_i}\|_{\mathcal{K}_{\omega,0}^{p,q}(\mathbb{R}^n)}} \right]^s \mathbf{1}_{B_i} \right\}^{\frac{1}{s}} \right\|_{\mathcal{K}_{\omega,0}^{p,q}(\mathbb{R}^n)}^{-1}$$

$$\times \sum_{j=1}^m \left\{ \frac{\lambda_j |B_j|}{\|\mathbf{1}_{B_j}\|_{\mathcal{K}_{\omega,0}^{p,q}(\mathbb{R}^n)}} \left[\frac{1}{|B_j|} \int_{B_j} \left| f(x) - P_{B_j}^d f(x) \right|^r \, dx \right]^{\frac{1}{r}} \right\}$$

is finite, where the supremum is taken over all $m \in \mathbb{N}$, $\{B_j\}_{j=1}^m \subset \mathbb{B}$, and $\{\lambda_j\}_{j=1}^m \subset [0, \infty)$ with $\sum_{j=1}^m \lambda_j \neq 0$.

Remark 8.1.36 Let all the symbols be as in Definition 8.1.35. Then, obviously, $\mathcal{P}_d(\mathbb{R}^n) \subset \mathcal{L}_{\omega,0}^{p,q,r,d,s}(\mathbb{R}^n)$ and, for any $f \in \mathcal{L}_{\omega,0}^{p,q,r,d,s}(\mathbb{R}^n)$, $\|f\|_{\mathcal{L}_{\omega,0}^{p,q,r,d,s}(\mathbb{R}^n)} = 0$ if and only if $f \in \mathcal{P}_d(\mathbb{R}^n)$. Thus, in what follows, we always identify $f \in \mathcal{L}_{\omega,0}^{p,q,r,d,s}(\mathbb{R}^n)$ with $\{f + P : P \in \mathcal{P}_d(\mathbb{R}^n)\}$.

Then, using [277, Remark 3.3(iii) and Proposition 3.4] with X therein replaced by $\mathcal{K}_{\omega,0}^{p,q}(\mathbb{R}^n)$, we immediately obtain the following equivalent characterizations of the Campanato-type function space $\mathcal{L}_{\omega,0}^{p,q,r,d,s}(\mathbb{R}^n)$; we omit the details.

Proposition 8.1.37 *Let p, q, ω, r, d, and s be as in Definition 8.1.35. Then the following three statements are equivalent:*

(i) $f \in \mathcal{L}_{\omega,0}^{p,q,r,d,s}(\mathbb{R}^n)$;
(ii) $f \in L_{\mathrm{loc}}^r(\mathbb{R}^n)$ and

$$\|f\|_{\mathcal{L}_{\omega,0}^{p,q,r,d,s}(\mathbb{R}^n)}^\star$$

$$:= \sup \inf \left\| \left\{ \sum_{i=1}^m \left[\frac{\lambda_i}{\|\mathbf{1}_{B_i}\|_{\mathcal{K}_{\omega,0}^{p,q}(\mathbb{R}^n)}} \right]^s \mathbf{1}_{B_i} \right\}^{\frac{1}{s}} \right\|_{\mathcal{K}_{\omega,0}^{p,q}(\mathbb{R}^n)}^{-1}$$

$$\times \sum_{j=1}^m \left\{ \frac{\lambda_j |B_j|}{\|\mathbf{1}_{B_j}\|_{\mathcal{K}_{\omega,0}^{p,q}(\mathbb{R}^n)}} \left[\frac{1}{|B_j|} \int_{B_j} |f(x) - P(x)|^r \, dx \right]^{\frac{1}{r}} \right\}$$

is finite, where the supremum is the same as in Definition 8.1.35 and the infimum is taken over all $P \in \mathcal{P}_d(\mathbb{R}^n)$;
(iii) $f \in L_{\mathrm{loc}}^r(\mathbb{R}^n)$ and

$$\|\widetilde{f}\|_{\mathcal{L}_{\omega,0}^{p,q,r,d,s}(\mathbb{R}^n)}$$

$$:= \sup \left\| \left\{ \sum_{i \in \mathbb{N}} \left[\frac{\lambda_i}{\|\mathbf{1}_{B_i}\|_{\mathcal{K}_{\omega,0}^{p,q}(\mathbb{R}^n)}} \right]^s \mathbf{1}_{B_i} \right\}^{\frac{1}{s}} \right\|_{\mathcal{K}_{\omega,0}^{p,q}(\mathbb{R}^n)}^{-1}$$

$$\times \sum_{j \in \mathbb{N}} \left\{ \frac{\lambda_j |B_j|}{\|\mathbf{1}_{B_j}\|_{\mathcal{K}_{\omega,0}^{p,q}(\mathbb{R}^n)}} \left[\frac{1}{|B_j|} \int_{B_j} \left| f(x) - P_{B_j}^d f(x) \right|^r \, dx \right]^{\frac{1}{r}} \right\}$$

is finite, where the supremum is taken over all $\{B_j\}_{j\in\mathbb{N}} \subset \mathbb{B}$ and $\{\lambda_j\}_{j\in\mathbb{N}} \subset [0,\infty)$ satisfying

$$\left\|\left\{\sum_{i\in\mathbb{N}}\left[\frac{\lambda_i}{\|\mathbf{1}_{B_i}\|_{\mathcal{K}^{p,q}_{\omega,0}(\mathbb{R}^n)}}\right]^s \mathbf{1}_{B_i}\right\}^{\frac{1}{s}}\right\|_{\mathcal{K}^{p,q}_{\omega,0}(\mathbb{R}^n)} \in (0,\infty).$$

Moreover, there exist two positive constants C_1 and C_2 such that, for any $f \in L^r_{\mathrm{loc}}(\mathbb{R}^n)$,

$$C_1\|f\|_{\mathcal{L}^{p,q,r,d,s}_{\omega,0}(\mathbb{R}^n)} \leq \|f\|^{\star}_{\mathcal{L}^{p,q,r,d,s}_{\omega,0}(\mathbb{R}^n)} \leq C_2\|f\|_{\mathcal{L}^{p,q,r,d,s}_{\omega,0}(\mathbb{R}^n)}$$

and

$$\|\widetilde{f}\|_{\mathcal{L}^{p,q,r,d,s}_{\omega,0}(\mathbb{R}^n)} = \|f\|_{\mathcal{L}^{p,q,r,d,s}_{\omega,0}(\mathbb{R}^n)}.$$

Next, we give the following dual theorem which shows that the dual space of the inhomogeneous generalized Herz–Hardy space $H\mathcal{K}^{p,q}_{\omega,0}(\mathbb{R}^n)$ is just the Campanato-type function space $\mathcal{L}^{p,q,r',d,s}_{\omega,0}(\mathbb{R}^n)$.

Theorem 8.1.38 *Let $p,q \in (0,\infty)$, $\omega \in M(\mathbb{R}_+)$ with $m_\infty(\omega) \in (-\frac{n}{p},\infty)$,*

$$p_- := \min\left\{1, p, \frac{n}{M_\infty(\omega) + n/p}\right\},$$

$d \geq \lfloor n(1/p_- - 1)\rfloor$ be a fixed integer, $s \in (0, \min\{p_-, q\})$, and

$$r \in \left(\max\left\{1, p, \frac{n}{m_\infty(\omega) + n/p}\right\}, \infty\right].$$

Then $\mathcal{L}^{p,q,r',d,s}_{\omega,0}(\mathbb{R}^n)$ is the dual space of $H\mathcal{K}^{p,q}_{\omega,0}(\mathbb{R}^n)$ in the following sense:

(i) *Let $g \in \mathcal{L}^{p,q,r',d,s}_{\omega,0}(\mathbb{R}^n)$. Then the linear functional*

$$L_g : f \mapsto L_g(f) := \int_{\mathbb{R}^n} f(x)g(x)\,dx, \tag{8.3}$$

initially defined for any $f \in H\mathcal{K}^{p,q,r,d,s}_{\omega,0,\mathrm{fin}}(\mathbb{R}^n)$, has a bounded extension to the inhomogeneous generalized Herz–Hardy space $H\mathcal{K}^{p,q}_{\omega,0}(\mathbb{R}^n)$.

(ii) *Conversely, any continuous linear functional on* $H\mathcal{K}^{p,q}_{\omega,0}(\mathbb{R}^n)$ *arises as in* (8.3) *with a unique* $g \in \mathcal{L}^{p,q,r',d,s}_{\omega,0}(\mathbb{R}^n)$.

Moreover, there exist two positive constants C_1 *and* C_2 *such that, for any* $g \in \mathcal{L}^{p,q,r',d,s}_{\omega,0}(\mathbb{R}^n)$,

$$C_1 \|g\|_{\mathcal{L}^{p,q,r',d,s}_{\omega,0}(\mathbb{R}^n)} \leq \|L_g\|_{(H\mathcal{K}^{p,q}_{\omega,0}(\mathbb{R}^n))^*} \leq C_2 \|g\|_{\mathcal{L}^{p,q,r',d,s}_{\omega,0}(\mathbb{R}^n)},$$

where $(H\mathcal{K}^{p,q}_{\omega,0}(\mathbb{R}^n))^*$ *denotes the dual space of* $H\mathcal{K}^{p,q}_{\omega,0}(\mathbb{R}^n)$.

Proof Let all the symbols be as in the present theorem. Then, from Lemma 4.7.6 and an argument similar to that used in the proof of Theorem 4.7.4 with Theorem 1.2.42 and Lemmas 1.8.6 and 4.3.10 therein replaced, respectively, by Theorem 7.1.7 and Lemmas 7.1.37 and 8.1.14, we infer that both (i) and (ii) of the present theorem hold true. This finishes the proof of Theorem 8.1.38. □

Finally, applying this dual theorem, we immediately obtain the following equivalence of Campanato-type function spaces $\mathcal{L}^{p,q,r,d,s}_{\omega,0}(\mathbb{R}^n)$; we omit the details.

Corollary 8.1.39 *Let* p, q, ω, p_-, d, *and* s *be as in Theorem 8.1.38,*

$$p_+ := \max\left\{1, p, \frac{n}{m_\infty(\omega) + n/p}\right\},$$

$r \in [1, p'_+)$, $d_0 := \lfloor n(1/p_- - 1) \rfloor$, *and* $s_0 \in (0, \min\{p_-, q\})$. *Then*

$$\mathcal{L}^{p,q,r,d,s}_{\omega,0}(\mathbb{R}^n) = \mathcal{L}^{p,q,1,d_0,s_0}_{\omega,0}(\mathbb{R}^n)$$

with equivalent quasi-norms.

8.1.8 Boundedness of Calderón–Zygmund Operators

Let $d \in \mathbb{Z}_+$. Recall that the standard kernel K and the d-order Calderón–Zygmund operator with kernel K are defined, respectively, in Definitions 1.5.7 and 1.5.8. In this subsection, we establish the boundedness of d-order Calderón–Zygmund operators on inhomogeneous generalized Herz–Hardy spaces. First, the following conclusion gives the boundedness of these operators on the inhomogeneous generalized Herz–Hardy space $H\mathcal{K}^{p,q}_{\omega,0}(\mathbb{R}^n)$.

Theorem 8.1.40 *Let* $d \in \mathbb{Z}_+$, $\delta \in (0,1]$, $p,q \in (\frac{n}{n+d+\delta}, \infty)$, K *be a* d-order *standard kernel defined as in Definition 1.5.7,* T *a* d-order *Calderón–Zygmund operator with kernel* K *having the vanishing moments up to order* d, *and* $\omega \in$

$M(\mathbb{R}_+)$ *with*

$$-\frac{n}{p} < m_\infty(\omega) \leq M_\infty(\omega) < n - \frac{n}{p} + d + \delta.$$

Then T has a unique extension on $H\mathcal{K}^{p,q}_{\omega,\mathbf{0}}(\mathbb{R}^n)$ and there exists a positive constant C such that, for any $f \in H\mathcal{K}^{p,q}_{\omega,\mathbf{0}}(\mathbb{R}^n)$,

$$\|T(f)\|_{H\mathcal{K}^{p,q}_{\omega,\mathbf{0}}(\mathbb{R}^n)} \leq C \|f\|_{H\mathcal{K}^{p,q}_{\omega,\mathbf{0}}(\mathbb{R}^n)}.$$

Proof Let all the symbols be as in the present theorem. Then, applying Proposition 4.8.12 and repeating the proof of Theorem 4.8.10 with $\min\{m_0(\omega), m_\infty(\omega)\}$, $\max\{M_0(\omega), M_\infty(\omega)\}$, Theorems 1.2.42 and 1.4.1, and Lemmas 1.8.6 and 4.3.11 therein replaced, respectively, by $m_\infty(\omega)$, $M_\infty(\omega)$, Theorems 7.1.7 and 7.1.17, and Lemmas 7.1.37 and 8.1.15, we find that T has a unique extension on $H\mathcal{K}^{p,q}_{\omega,\mathbf{0}}(\mathbb{R}^n)$ and, for any $f \in H\mathcal{K}^{p,q}_{\omega,\mathbf{0}}(\mathbb{R}^n)$,

$$\|T(f)\|_{H\mathcal{K}^{p,q}_{\omega,\mathbf{0}}(\mathbb{R}^n)} \lesssim \|f\|_{H\mathcal{K}^{p,q}_{\omega,\mathbf{0}}(\mathbb{R}^n)}.$$

This then finishes the proof of Theorem 8.1.40. $\qquad\square$

Remark 8.1.41 We should point out that, in Theorem 8.1.40, when $d = 0$, $p \in (1, \infty)$, and $\omega(t) := t^\alpha$ for any $t \in (0, \infty)$ and for any given $\alpha \in [n(1 - \frac{1}{p}), n(1 - \frac{1}{p}) + \delta)$, then this theorem goes back to [157, Theorem 1].

On the other hand, we prove the following boundedness of Calderón–Zygmund operators on the Hardy space $H\mathcal{K}^{p,q}_\omega(\mathbb{R}^n)$.

Theorem 8.1.42 *Let $d \in \mathbb{Z}_+$, $\delta \in (0, 1]$, $p, q \in (\frac{n}{n+d+\delta}, \infty)$, K be a d-order standard kernel defined as in Definition 1.5.7, T a d-order Calderón–Zygmund operator with kernel K having the vanishing moments up to order d, and $\omega \in M(\mathbb{R}_+)$ satisfy*

$$-\frac{n}{p} < m_\infty(\omega) \leq M_\infty(\omega) < 0.$$

Then T can be extended into a bounded linear operator on $H\mathcal{K}^{p,q}_\omega(\mathbb{R}^n)$, namely, there exists a positive constant C such that, for any $f \in H\mathcal{K}^{p,q}_\omega(\mathbb{R}^n)$,

$$\|T(f)\|_{H\mathcal{K}^{p,q}_\omega(\mathbb{R}^n)} \leq C \|f\|_{H\mathcal{K}^{p,q}_\omega(\mathbb{R}^n)}.$$

Proof Let all the symbols be as in the present theorem. Then, from Theorem 4.8.17 and an argument similar to that used in the proof of Theorem 4.8.16 with $\min\{m_0(\omega), m_\infty(\omega)\}$, $\max\{M_0(\omega), M_\infty(\omega)\}$, Theorem 1.2.44, Lemmas 2.2.3, 2.3.2, 4.3.25, and 4.3.27, and Corollary 2.3.5 therein replaced, respectively, by $m_\infty(\omega)$, $M_\infty(\omega)$, Theorem 7.1.9, Lemmas 7.2.6, 7.2.11, 8.1.21, and 8.1.23, and Corollary 7.2.14, it follows that T is well defined on $H\mathcal{K}_\omega^{p,q}(\mathbb{R}^n)$ and, for any $f \in H\mathcal{K}_\omega^{p,q}(\mathbb{R}^n)$,

$$\|T(f)\|_{H\mathcal{K}_\omega^{p,q}(\mathbb{R}^n)} \lesssim \|f\|_{H\mathcal{K}_\omega^{p,q}(\mathbb{R}^n)}.$$

This finishes the proof of Theorem 8.1.42. □

8.1.9 Fourier Transform

The target of this subsection is to investigate the Fourier transform properties of distributions in the inhomogeneous generalized Herz–Hardy spaces $H\mathcal{K}_{\omega,0}^{p,q}(\mathbb{R}^n)$ and $H\mathcal{K}_\omega^{p,q}(\mathbb{R}^n)$. We first study the Fourier transform in $H\mathcal{K}_{\omega,0}^{p,q}(\mathbb{R}^n)$ as follows.

Theorem 8.1.43 Let $p, q \in (0, 1]$, $\omega \in M(\mathbb{R}_+)$ with $m_\infty(\omega) \in (0, \infty)$, and $p_- \in (0, \frac{n}{M_\infty(\omega)+n/p})$. Then, for any $f \in H\mathcal{K}_{\omega,0}^{p,q}(\mathbb{R}^n)$, there exists a continuous function g on $\mathbb{R}^n$ such that $\widehat{f} = g$ in $\mathcal{S}'(\mathbb{R}^n)$ and

$$\lim_{|x|\to 0^+} \frac{|g(x)|}{|x|^{n(\frac{1}{p_-}-1)}} = 0.$$

Moreover, there exists a positive constant C, independent of both f and g, such that, for any $x \in \mathbb{R}^n$,

$$|g(x)| \le C\|f\|_{H\mathcal{K}_{\omega,0}^{p,q}(\mathbb{R}^n)} \max\left\{1, |x|^{n(\frac{1}{p_-}-1)}\right\}$$

and

$$\int_{\mathbb{R}^n} |g(x)| \min\left\{|x|^{-\frac{n}{p_-}}, |x|^{-n}\right\} dx \le C\|f\|_{H\mathcal{K}_{\omega,0}^{p,q}(\mathbb{R}^n)}.$$

In order to prove this theorem, we first establish the following technical lemma about the quasi-norms of the characteristic function of balls on $\mathcal{K}_{\omega,0}^{p,q}(\mathbb{R}^n)$.

Lemma 8.1.44 Let $p, q \in (0, 1]$, $\omega \in M(\mathbb{R}_+)$ with $m_\infty(\omega) \in (0, \infty)$, and $p_- \in (0, \frac{n}{M_\infty(\omega)+n/p})$. Then there exists a positive constant C such that, for any $B \in \mathbb{B}$,

$$\|\mathbf{1}_B\|_{\mathcal{K}_{\omega,0}^{p,q}(\mathbb{R}^n)} \ge C \min\left\{|B|, |B|^{\frac{1}{p_-}}\right\}.$$

Proof Let all the symbols be as in the present lemma. We first claim that, for any $B(x_0, 2^{k_0}) \in \mathbb{B}$ with $x_0 \in \mathbb{R}^n$ and $k_0 \in \mathbb{Z} \cap (-\infty, 0)$,

$$\left| B(x_0, 2^{k_0}) \right|^{\frac{1}{p_-}} \lesssim \left\| \mathbf{1}_{B(x_0, 2^{k_0})} \right\|_{\mathcal{K}_{\omega,0}^{p,q}(\mathbb{R}^n)}.$$

To show this inequality, we consider the following five cases on x_0.

Case (1) $x_0 = \mathbf{0}$. In this case, notice that $B(\mathbf{0}, 2^{k_0}) \subset B(\mathbf{0}, 1)$. By this and Definition 7.1.1(i), we find that

$$\left| B(\mathbf{0}, 2^{k_0}) \right|^{\frac{1}{p}} = \left\| \mathbf{1}_{B(\mathbf{0}, 2^{k_0})} \right\|_{L^p(\mathbb{R}^n)} \leq \left\| \mathbf{1}_{B(\mathbf{0}, 2^{k_0})} \right\|_{\mathcal{K}_{\omega,0}^{p,q}(\mathbb{R}^n)}. \tag{8.4}$$

On the other hand, from the assumption $m_\infty(\omega) \in (0, \infty)$ and Remark 1.1.5(iii), it follows that

$$p_- < \frac{n}{M_\infty(\omega) + n/p} \leq \frac{n}{m_\infty(\omega) + n/p} < p, \tag{8.5}$$

which, combined with (8.4), further implies that

$$\left| B(\mathbf{0}, 2^{k_0}) \right|^{\frac{1}{p_-}} < \left| B(\mathbf{0}, 2^{k_0}) \right|^{\frac{1}{p}} \leq \left\| \mathbf{1}_{B(\mathbf{0}, 2^{k_0})} \right\|_{\mathcal{K}_{\omega,0}^{p,q}(\mathbb{R}^n)}. \tag{8.6}$$

This finishes the proof of the above claim in this case.

Case (2) $|x_0| \in (0, 3 \cdot 2^{k_0-1})$. In this case, applying both (4.124) and the assumption $k_0 \in \mathbb{Z} \cap (-\infty, 0)$, we find that

$$B \left(\frac{3 \cdot 2^{k_0-1}}{|x_0|} x_0, 2^{k_0-2} \right) \subset \left[B(x_0, 2^{k_0}) \cap \left(B(\mathbf{0}, 2^{k_0}) \setminus B(\mathbf{0}, 2^{k_0-1}) \right) \right]$$

$$\subset \left[B(x_0, 2^{k_0}) \cap B(\mathbf{0}, 1) \right].$$

By this, (8.5), and Definition 7.1.1(i), we conclude that

$$\left| B(x_0, 2^{k_0}) \right|^{\frac{1}{p_-}} < \left| B(x_0, 2^{k_0}) \right|^{\frac{1}{p}} \sim \left| B \left(\frac{3 \cdot 2^{k_0-1}}{|x_0|} x_0, 2^{k_0-2} \right) \right|^{\frac{1}{p}}$$

$$\lesssim \left| B(x_0, 2^{k_0}) \cap B(\mathbf{0}, 1) \right|^{\frac{1}{p}} \lesssim \left\| \mathbf{1}_{B(x_0, 2^{k_0})} \right\|_{\mathcal{K}_{\omega,0}^{p,q}(\mathbb{R}^n)}, \tag{8.7}$$

which implies that the above claim holds true in this case.

Case (3) $|x_0| \in [3 \cdot 2^{k_0-1}, 2^{k_0+1})$. In this case, for any $y \in B(x_0, 2^{k_0})$, we have

$$|y| \leq |y - x_0| + |x_0| < 2^{k_0} + 2^{k_0+1} < 2.$$

This implies that $B(x_0, 2^{k_0}) \subset B(0, 2)$. Therefore, from (8.5) and Definition 7.1.1(i), we deduce that

$$\left| B(x_0, 2^{k_0}) \right|^{\frac{1}{p_-}} < \left| B(x_0, 2^{k_0}) \right|^{\frac{1}{p}}$$

$$\lesssim \left| B(x_0, 2^{k_0}) \cap B(0, 1) \right|^{\frac{1}{p}}$$

$$+ \omega(2) \left| B(x_0, 2^{k_0}) \cap (B(0, 2) \setminus B(0, 1)) \right|^{\frac{1}{p}}$$

$$\lesssim \left\| \mathbf{1}_{B(x_0, 2^{k_0})} \right\|_{\mathcal{K}^{p,q}_{\omega,0}(\mathbb{R}^n)}, \tag{8.8}$$

which completes the proof of the above claim in this case.

Case (4) $|x_0| \in [2^k, 2^{k+1})$ with $k \in \mathbb{Z} \cap [k_0 + 1, 0]$. In this case, for any $y \in B(x_0, 2^{k_0})$, we have

$$|y| \leq |y - x_0| + |x_0| < 2^{k_0} + 2^{k+1} < 2^2,$$

which implies that $B(x_0, 2^{k_0}) \subset B(0, 2^2)$. Thus, using (8.5) and Definition 7.1.1(i), we conclude that

$$\left| B(x_0, 2^{k_0}) \right|^{\frac{1}{p_-}} < \left| B(x_0, 2^{k_0}) \right|^{\frac{1}{p}}$$

$$\lesssim \left| B(x_0, 2^{k_0}) \cap B(0, 1) \right|^{\frac{1}{p}}$$

$$+ \omega(2) \left| B(x_0, 2^{k_0}) \cap (B(0, 2) \setminus B(0, 1)) \right|^{\frac{1}{p}}$$

$$+ \omega(2^2) \left| B(x_0, 2^{k_0}) \cap \left(B(0, 2^2) \setminus B(0, 2) \right) \right|^{\frac{1}{p}}$$

$$\lesssim \left\| \mathbf{1}_{B(x_0, 2^{k_0})} \right\|_{\mathcal{K}^{p,q}_{\omega,0}(\mathbb{R}^n)}. \tag{8.9}$$

This implies that, in this case, the above claim holds true.

Case (5) $|x_0| \in [2^k, 2^{k+1})$ with $k \in \mathbb{N}$. In this case, by the assumption $k_0 \in \mathbb{Z} \cap (-\infty, 0)$ and (4.129), we obtain

$$B(x_0, 2^{k_0}) \subset \left[B(0, 2^{k+2}) \setminus B(0, 2^{k-1}) \right].$$

This, together with (8.5) and Definition 7.1.1(i), further implies that

$$\left| B(x_0, 2^{k_0}) \right|^{\frac{1}{p_-}} < \left| B(x_0, 2^{k_0}) \right|^{\frac{1}{p}}$$

$$\lesssim \omega(2^k) \left| B(x_0, 2^{k_0}) \cap \left(B(0, 2^k) \setminus B(0, 2^{k-1}) \right) \right|^{\frac{1}{p}}$$

$$+ \omega(2^{k+1}) \left| B(x_0, 2^{k_0}) \cap \left(B(0, 2^{k+1}) \setminus B(0, 2^k) \right) \right|^{\frac{1}{p}}$$

$$+ \omega(2^{k+2}) \left| B(x_0, 2^{k+2}) \cap \left(B(0, 2^{k+1}) \setminus B(0, 2) \right) \right|^{\frac{1}{p}}$$

$$\lesssim \left\| \mathbf{1}_{B(x_0, 2^{k_0})} \right\|_{\mathcal{K}_{\omega,0}^{p,q}(\mathbb{R}^n)}.$$

From this, (8.6), (8.7), (8.8), and (8.9), it follows that the above claim holds true.
   In addition, for any $k_0 \in \mathbb{Z}_+$, repeating an argument similar to that used in the estimation of (4.133) with r, $\min\{m_0(\omega), m_\infty(\omega)\}$, and Lemma 1.5.2 therein replaced, respectively, by 2^{k_0}, $m_\infty(\omega)$, and Lemma 1.1.12, we conclude that

$$\left| B(x_0, 2^{k_0}) \right| \lesssim \left\| \mathbf{1}_{B(x_0, 2^{k_0})} \right\|_{\mathcal{K}_{\omega,0}^{p,q}(\mathbb{R}^n)}. \tag{8.10}$$

Assume $B(x_0, r) \in \mathbb{B}$ with $x_0 \in \mathbb{R}^n$ and $r \in (0, \infty)$. Then there exists a $k \in \mathbb{Z}$ such that $r \in [2^k, 2^{k+1})$. Thus, by the above claim and (8.10), we find that

$$\min \left\{ |B(x_0, r)|, |B(x_0, r)|^{\frac{1}{p_-}} \right\}$$

$$\sim \min \left\{ \left| B(x_0, 2^{k_0}) \right|, \left| B(x_0, 2^{k_0}) \right|^{\frac{1}{p_-}} \right\}$$

$$\lesssim \left\| \mathbf{1}_{B(x_0, 2^{k_0})} \right\|_{\mathcal{K}_{\omega,0}^{p,q}(\mathbb{R}^n)} \lesssim \left\| \mathbf{1}_{B(x_0, r)} \right\|_{\mathcal{K}_{\omega,0}^{p,q}(\mathbb{R}^n)}.$$

This finishes the proof of Lemma 8.1.44.   □

   Via the above lemma, we now prove Theorem 8.1.43.

Proof of Theorem 8.1.43   Let all the symbols be as in the present theorem and $f \in H\mathcal{K}_{\omega,0}^{p,q}(\mathbb{R}^n)$. Then, using Lemma 4.9.2 and repeating the proof of Theorem 4.9.1 with $\max\{M_0(\omega), M_\infty(\omega)\}$, Theorem 1.2.42, and Lemmas 4.3.10 and 4.9.4 therein replaced, respectively, by $M_\infty(\omega)$, Theorem 7.1.7, and Lemmas 8.1.14 and 8.1.44, we conclude that there exists a continuous function g on $\mathbb{R}^n$ such that $\widehat{f} = g$ in $\mathcal{S}'(\mathbb{R}^n)$,

$$\lim_{|x| \to 0^+} \frac{|g(x)|}{|x|^{n(\frac{1}{p_-}-1)}} = 0,$$

and, for any $x \in \mathbb{R}^n$,

$$|g(x)| \lesssim \|f\|_{H\mathcal{K}^{p,q}_{\omega,0}(\mathbb{R}^n)} \max\left\{1, |x|^{n(\frac{1}{p_-}-1)}\right\}$$

and

$$\int_{\mathbb{R}^n} |g(x)| \min\left\{|x|^{-\frac{n}{p_-}}, |x|^{-n}\right\} dx \lesssim \|f\|_{H\mathcal{K}^{p,q}_{\omega,0}(\mathbb{R}^n)},$$

where the implicit positive constants are independent of both f and g. This finishes the proof of Theorem 8.1.43.   □

Next, we investigate the Fourier transform of the Hardy space $H\mathcal{K}^{p,q}_\omega(\mathbb{R}^n)$. Indeed, we have the following conclusion.

Theorem 8.1.45 *Let* $p \in (0,1)$, $q \in (0,1]$, $\omega \in M(\mathbb{R}_+)$ *with*

$$n\left(1 - \frac{1}{p}\right) < m_\infty(\omega) \le M_\infty(\omega) < 0,$$

and $p_- \in (0,p)$. *Then, for any* $f \in H\mathcal{K}^{p,q}_\omega(\mathbb{R}^n)$, *there exists a continuous function* g *on* $\mathbb{R}^n$ *such that* $\widehat{f} = g$ *in* $\mathcal{S}'(\mathbb{R}^n)$ *and*

$$\lim_{|x| \to 0^+} \frac{|g(x)|}{|x|^{n(\frac{1}{p_-}-1)}} = 0.$$

Moreover, there exists a positive constant C, *independent of both* f *and* g, *such that, for any* $x \in \mathbb{R}^n$,

$$|g(x)| \le C\|f\|_{H\mathcal{K}^{p,q}_\omega(\mathbb{R}^n)} \max\left\{1, |x|^{n(\frac{1}{p_-}-1)}\right\}$$

and

$$\int_{\mathbb{R}^n} |g(x)| \min\left\{|x|^{-\frac{n}{p_-}}, |x|^{-n}\right\} dx \le C\|f\|_{H\mathcal{K}^{p,q}_\omega(\mathbb{R}^n)}.$$

To show this theorem, we require the following auxiliary estimate for the quasi-norm $\|\cdot\|_{H\mathcal{K}_{\omega}^{p,q}(\mathbb{R}^n)}$ of the characteristic function of balls.

Lemma 8.1.46 *Let* $p \in (0, 1)$, $q \in (0, 1]$, $\omega \in M(\mathbb{R}_+)$ *with*

$$n\left(1 - \frac{1}{p}\right) < m_\infty(\omega) \le M_\infty(\omega) < 0,$$

and $p_- \in (0, p)$. *Then there exists a positive constant* C *such that, for any* $B \in \mathbb{B}$,

$$\|\mathbf{1}_B\|_{\mathcal{K}_{\omega}^{p,q}(\mathbb{R}^n)} \ge C \min\left\{|B|, |B|^{\frac{1}{p_-}}\right\}.$$

Proof Let all the symbols be as in the present lemma and $B(x_0, r) \in \mathbb{B}$ with $x_0 \in \mathbb{R}^n$ and $r \in (0, \infty)$. We show Lemma 8.1.46 by considering the following three cases on r.

Case (1) $r \in (0, 1]$. In this case, we have $B(x_0, r) \subset B(x_0, 1)$. From this, the assumption $p_- \in (0, p)$, Definition 7.1.1(i), and Remark 7.1.2(i), we deduce that

$$|B(x_0, r)|^{\frac{1}{p_-}} \le |B(x_0, r)|^{\frac{1}{p}} \sim \left\|\mathbf{1}_{B(x_0,r)}\right\|_{L^p(\mathbb{R}^n)}$$

$$\lesssim \left\|\mathbf{1}_{B(x_0,r)}(\cdot + x_0)\right\|_{\mathcal{K}_{\omega,0}^{p,q}(\mathbb{R}^n)} \lesssim \left\|\mathbf{1}_{B(x_0,r)}\right\|_{\mathcal{K}_{\omega}^{p,q}(\mathbb{R}^n)}. \tag{8.11}$$

This implies that Lemma 8.1.46 holds true in this case.

Case (2) $r \in (1, 2]$. In this case, we have $B(x_0, 1) \subset B(x_0, r)$. Applying this, the Hölder inequality, Definition 7.1.1(i), and Remark 7.1.2(i), we conclude that

$$|B(x_0, r)| \sim |B(x_0, 1)| \lesssim \left\|\mathbf{1}_{B(x_0,1)}\right\|_{L^p(\mathbb{R}^n)} \lesssim \left\|\mathbf{1}_{B(x_0,1)}(\cdot + x_0)\right\|_{\mathcal{K}_{\omega,0}^{p,q}(\mathbb{R}^n)}$$

$$\lesssim \left\|\mathbf{1}_{B(x_0,1)}\right\|_{\mathcal{K}_{\omega}^{p,q}(\mathbb{R}^n)} \lesssim \left\|\mathbf{1}_{B(x_0,r)}\right\|_{\mathcal{K}_{\omega}^{p,q}(\mathbb{R}^n)}, \tag{8.12}$$

which completes the proof of Lemma 8.1.46 in this case.

Case (3) $r \in (2, \infty)$. In this case, there exists a $k \in \mathbb{N}$ such that $r \in (2^k, 2^{k+1}]$. This implies that

$$\left[B(x_0, 2^k) \setminus B(x_0, 2^{k-1})\right] \subset B(x_0, r). \tag{8.13}$$

In addition, let $\varepsilon \in (0, -n + m_\infty(\omega) + \frac{n}{p})$ be a fixed positive constant. Then, using (8.13), Lemmas 1.1.12 and 1.1.3, Definition 7.1.1(i), and Remark 7.1.2(i), we find that

$$|B(x_0, r)| \sim r^n \lesssim r^{m_\infty(\omega) + \frac{n}{p} - \varepsilon} \lesssim r^{\frac{n}{p}}\omega(r) \sim 2^{\frac{nk}{p}}\omega(2^k)$$

$$\sim \omega(2^k)\left\|\mathbf{1}_{B(x_0,2^k)\setminus B(x_0,2^{k-1})}\right\|_{L^p(\mathbb{R}^n)}$$

$$\lesssim \left\| \mathbf{1}_{B(x_0,r)}(\cdot + x_0) \right\|_{\mathcal{K}^{p,q}_{\omega,0}(\mathbb{R}^n)}$$

$$\lesssim \left\| \mathbf{1}_{B(x_0,r)} \right\|_{\mathcal{K}^{p,q}_{\omega}(\mathbb{R}^n)}.$$

This, together with (8.11) and (8.12), further implies that

$$\min \left\{ |B(x_0,r)|, |B(x_0,r)|^{\frac{1}{p_-}} \right\} \lesssim \left\| \mathbf{1}_{B(x_0,r)} \right\|_{\mathcal{K}^{p,q}_{\omega}(\mathbb{R}^n)},$$

which completes the proof of the present lemma in this case, and hence the whole proof of Lemma 8.1.46.   □

We now prove Theorem 8.1.45 via Lemma 8.1.46.

Proof of Theorem 8.1.45 Let all the symbols be as in the present theorem and $f \in HK^{p,q}_{\omega}(\mathbb{R}^n)$. Then, from Lemma 4.9.2 and an argument similar to that used in the proof of Theorem 4.9.5 via replacing Theorems 1.2.44 and Lemmas 4.3.24 and 4.9.6 therein, respectively, by Theorem 7.1.9 and Lemmas 8.1.20 and 8.1.46, we infer that there exists a continuous function g on $\mathbb{R}^n$ such that $\widehat{f} = g$ in $\mathcal{S}'(\mathbb{R}^n)$,

$$\lim_{|x| \to 0^+} \frac{|g(x)|}{|x|^{n(\frac{1}{p_-} - 1)}} = 0,$$

and, for any $x \in \mathbb{R}^n$,

$$|g(x)| \lesssim \|f\|_{HK^{p,q}_{\omega}(\mathbb{R}^n)} \max \left\{ 1, |x|^{n(\frac{1}{p_-} - 1)} \right\}$$

and

$$\int_{\mathbb{R}^n} |g(x)| \min \left\{ |x|^{-\frac{n}{p_-}}, |x|^{-n} \right\} dx \lesssim \|f\|_{HK^{p,q}_{\omega}(\mathbb{R}^n)}$$

with the implicit positive constants independent of both f and g. This then finishes the proof of Theorem 8.1.45.   □

8.2 Inhomogeneous Localized Generalized Herz–Hardy Spaces

In this section, we introduce inhomogeneous localized generalized Herz–Hardy spaces and then establish their maximal function, atomic, molecular as well as various Littlewood–Paley function characterizations. As applications, we also obtain the boundedness of the pseudo-differential operators on these localized

Hardy spaces as well as investigate the relation between inhomogeneous generalized Herz–Hardy spaces and inhomogeneous localized generalized Herz–Hardy spaces. Recall that, for any given $N \in \mathbb{N}$ and for any $f \in \mathcal{S}'(\mathbb{R}^n)$, the local grand maximal function $m_N(f)$ of f is defined as in (5.1). Then we introduce the inhomogeneous localized generalized Herz–Hardy spaces via this maximal function as follows.

Definition 8.2.1 Let $p, q \in (0, \infty)$, $\omega \in M(\mathbb{R}_+)$, and $N \in \mathbb{N}$. Then

(i) the *inhomogeneous local generalized Herz–Hardy space* $h\mathcal{K}_{\omega,\mathbf{0}}^{p,q}(\mathbb{R}^n)$, associated with the inhomogeneous local generalized Herz space $\mathcal{K}_{\omega,\mathbf{0}}^{p,q}(\mathbb{R}^n)$, is defined to be the set of all the $f \in \mathcal{S}'(\mathbb{R}^n)$ such that

$$\|f\|_{h\mathcal{K}_{\omega,\mathbf{0}}^{p,q}(\mathbb{R}^n)} := \|m_N(f)\|_{\mathcal{K}_{\omega,\mathbf{0}}^{p,q}(\mathbb{R}^n)} < \infty;$$

(ii) the *inhomogeneous local generalized Herz–Hardy space* $h\mathcal{K}_{\omega}^{p,q}(\mathbb{R}^n)$, associated with the inhomogeneous global generalized Herz space $\mathcal{K}_{\omega}^{p,q}(\mathbb{R}^n)$, is defined to be the set of all the $f \in \mathcal{S}'(\mathbb{R}^n)$ such that

$$\|f\|_{h\mathcal{K}_{\omega}^{p,q}(\mathbb{R}^n)} := \|m_N(f)\|_{\mathcal{K}_{\omega}^{p,q}(\mathbb{R}^n)} < \infty.$$

Remark 8.2.2 In Definition 8.2.1, for any given $\alpha \in \mathbb{R}$ and for any $t \in (0, \infty)$, let $\omega(t) := t^{\alpha}$. Then, in this case, the inhomogeneous local generalized Herz–Hardy space $h\mathcal{K}_{\omega,\mathbf{0}}^{p,q}(\mathbb{R}^n)$ goes back to the classical *inhomogeneous local Herz-type Hardy space* $hK_p^{\alpha,q}(\mathbb{R}^n)$ which was originally introduced in [72, Definition 1.2] (see also [175, Section 2.6]).

8.2.1 Maximal Function Characterizations

In this subsection, we characterize the inhomogeneous localized generalized Herz–Hardy spaces via various maximal functions. Recall that several localized radial and localized non-tangential maximal functions are given in Definition 5.1.1. Using these maximal functions, we first establish the following maximal function characterizations of the local Hardy space $h\mathcal{K}_{\omega,\mathbf{0}}^{p,q}(\mathbb{R}^n)$.

Theorem 8.2.3 *Let $p, q, a, b \in (0, \infty)$, $\omega \in M(\mathbb{R}_+)$, $N \in \mathbb{N}$, and $\phi \in \mathcal{S}(\mathbb{R}^n)$ satisfy $\int_{\mathbb{R}^n} \phi(x)\, dx \neq 0$.*

(i) *Let $N \in \mathbb{N} \cap [\lfloor b+1 \rfloor, \infty)$. Then, for any $f \in \mathcal{S}'(\mathbb{R}^n)$,*

$$\|m(f, \phi)\|_{\mathcal{K}^{p,q}_{\omega,0}(\mathbb{R}^n)} \lesssim \|m_a^*(f, \phi)\|_{\mathcal{K}^{p,q}_{\omega,0}(\mathbb{R}^n)} \lesssim \|m_b^{**}(f, \phi)\|_{\mathcal{K}^{p,q}_{\omega,0}(\mathbb{R}^n)},$$

$$\|m(f, \phi)\|_{\mathcal{K}^{p,q}_{\omega,0}(\mathbb{R}^n)} \lesssim \|m_N(f)\|_{\mathcal{K}^{p,q}_{\omega,0}(\mathbb{R}^n)} \lesssim \|m_{\lfloor b+1 \rfloor}(f)\|_{\mathcal{K}^{p,q}_{\omega,0}(\mathbb{R}^n)}$$

$$\lesssim \|m_b^{**}(f, \phi)\|_{\mathcal{K}^{p,q}_{\omega,0}(\mathbb{R}^n)},$$

and

$$\|m_b^{**}(f, \phi)\|_{\mathcal{K}^{p,q}_{\omega,0}(\mathbb{R}^n)} \sim \|m_{b,N}^{**}(f)\|_{\mathcal{K}^{p,q}_{\omega,0}(\mathbb{R}^n)},$$

where the implicit positive constants are independent of f.
(ii) *Let ω satisfy $m_\infty(\omega) \in (-\frac{n}{p}, \infty)$, and*

$$b \in \left(2 \max \left\{ \frac{n}{p}, \frac{n}{q}, M_\infty(\omega) + \frac{n}{p} \right\}, \infty \right).$$

Then, for any $f \in \mathcal{S}'(\mathbb{R}^n)$,

$$\|m_b^{**}(f, \phi)\|_{\mathcal{K}^{p,q}_{\omega,0}(\mathbb{R}^n)} \lesssim \|m(f, \phi)\|_{\mathcal{K}^{p,q}_{\omega,0}(\mathbb{R}^n)},$$

where the implicit positive constant is independent of f. In particular, when $N \in \mathbb{N} \cap [\lfloor b+1 \rfloor, \infty)$, if one of the quantities

$$\|m(f, \phi)\|_{\mathcal{K}^{p,q}_{\omega,0}(\mathbb{R}^n)}, \ \|m_a^*(f, \phi)\|_{\mathcal{K}^{p,q}_{\omega,0}(\mathbb{R}^n)}, \ \|m_N(f)\|_{\mathcal{K}^{p,q}_{\omega,0}(\mathbb{R}^n)},$$

$$\|m_b^{**}(f, \phi)\|_{\mathcal{K}^{p,q}_{\omega,0}(\mathbb{R}^n)}, \ and \ \|m_{b,N}^{**}(f)\|_{\mathcal{K}^{p,q}_{\omega,0}(\mathbb{R}^n)}$$

is finite, then the others are also finite and mutually equivalent with the positive equivalence constants independent of f.

Proof Let all the symbols be as in the present theorem. Then, applying both Lemmas 5.1.4 and 5.1.5 and repeating the proof of Theorem 5.1.2 via replacing $\max\{M_0(\omega), M_\infty(\omega)\}$, Theorems 1.2.42 and 1.3.3, and Lemma 1.8.5 therein, respectively, by $M_\infty(\omega)$, Theorems 7.1.7 and 7.1.15, and Lemma 7.1.36, we find that both (i) and (ii) of the present theorem hold true. This then finishes the proof of Theorem 8.2.3.   □

Remark 8.2.4

(i) Let $p, q \in (0.\infty)$ and $\omega \in M(\mathbb{R}_+)$. Then the quasi-norm of $h\mathcal{K}^{p,q}_{\omega,0}(\mathbb{R}^n)$ in Definition 8.2.1(i) depends on N. However, by Theorem 8.2.3, we conclude that the local Hardy space $h\mathcal{K}^{p,q}_{\omega,0}(\mathbb{R}^n)$ is independent of the choice of N whenever

$\omega \in M(\mathbb{R}_+)$ satisfies $m_\infty(\omega) \in (-\frac{n}{p}, \infty)$, and N satisfies

$$N \in \mathbb{N} \cap \left(1 + 2\max\left\{\frac{n}{p}, \frac{n}{q}, M_\infty(\omega) + \frac{n}{p}\right\}, \infty\right).$$

(ii) Observe that, if $p = q \in (0, \infty)$ and $\omega(t) := 1$ for any $t \in (0, \infty)$, then, in this case,

$$m_0(\omega) = M_0(\omega) = m_\infty(\omega) = M_\infty(\omega) = 0,$$

$h\mathcal{K}_{\omega,0}^{p,q}(\mathbb{R}^n)$ coincides with the classical local Hardy space $h^p(\mathbb{R}^n)$ in the sense of equivalent quasi-norms, and all the conclusions in Theorem 8.2.3 coincide with the corresponding classical results on the local Hardy space $h^p(\mathbb{R}^n)$ established in [89].

Next, we turn to prove the maximal function characterizations of the inhomogeneous local generalized Herz–Hardy space $h\mathcal{K}_\omega^{p,q}(\mathbb{R}^n)$ as follows.

Theorem 8.2.5 *Let* $p, q, a, b \in (0, \infty)$, $\omega \in M(\mathbb{R}_+)$, $N \in \mathbb{N}$, *and* $\phi \in \mathcal{S}(\mathbb{R}^n)$ *satisfy* $\int_{\mathbb{R}^n} \phi(x)\, dx \neq 0$.

(i) *Let* $N \in \mathbb{N} \cap [\lfloor b+1 \rfloor, \infty)$ *and* ω *satisfy* $M_\infty(\omega) \in (-\infty, 0)$. *Then, for any* $f \in \mathcal{S}'(\mathbb{R}^n)$,

$$\|m(f, \phi)\|_{\mathcal{K}_\omega^{p,q}(\mathbb{R}^n)} \lesssim \|m_a^*(f, \phi)\|_{\mathcal{K}_\omega^{p,q}(\mathbb{R}^n)} \lesssim \|m_b^{**}(f, \phi)\|_{\mathcal{K}_\omega^{p,q}(\mathbb{R}^n)},$$

$$\|m(f, \phi)\|_{\mathcal{K}_\omega^{p,q}(\mathbb{R}^n)} \lesssim \|m_N(f)\|_{\mathcal{K}_\omega^{p,q}(\mathbb{R}^n)} \lesssim \|m_{\lfloor b+1 \rfloor}(f)\|_{\mathcal{K}_\omega^{p,q}(\mathbb{R}^n)}$$

$$\lesssim \|m_b^{**}(f, \phi)\|_{\mathcal{K}_\omega^{p,q}(\mathbb{R}^n)},$$

and

$$\|m_b^{**}(f, \phi)\|_{\mathcal{K}_\omega^{p,q}(\mathbb{R}^n)} \sim \|m_{b,N}^{**}(f)\|_{\mathcal{K}_\omega^{p,q}(\mathbb{R}^n)},$$

where the implicit positive constants are independent of f.

(ii) *Let* ω *satisfy*

$$-\frac{n}{p} < m_\infty(\omega) \leq M_\infty(\omega) < 0,$$

and $b \in (2n \max\{\frac{1}{p}, \frac{1}{q}\}, \infty)$. *Then, for any* $f \in \mathcal{S}'(\mathbb{R}^n)$,

$$\|m_b^{**}(f, \phi)\|_{\mathcal{K}_\omega^{p,q}(\mathbb{R}^n)} \lesssim \|m(f, \phi)\|_{\mathcal{K}_\omega^{p,q}(\mathbb{R}^n)},$$

where the implicit positive constant is independent of f. In particular, when
$N \in \mathbb{N} \cap [\lfloor b+1 \rfloor, \infty)$, *if one of the quantities*

$$\|m(f, \phi)\|_{\mathcal{K}_\omega^{p,q}(\mathbb{R}^n)}, \quad \|m_a^*(f, \phi)\|_{\mathcal{K}_\omega^{p,q}(\mathbb{R}^n)}, \quad \|m_N(f)\|_{\mathcal{K}_\omega^{p,q}(\mathbb{R}^n)},$$

$$\|m_b^{**}(f, \phi)\|_{\mathcal{K}_\omega^{p,q}(\mathbb{R}^n)}, \quad and \quad \|m_{b,N}^{**}(f)\|_{\mathcal{K}_\omega^{p,q}(\mathbb{R}^n)}$$

*is finite, then the others are also finite and mutually equivalent with the positive
equivalence constants independent of f.*

Proof Let all the symbols be as in the present theorem. Then, from Lemmas 5.1.4
and 5.1.5 and an argument similar to that used in the proof of Theorem 5.1.10
with $\max\{M_0(\omega), M_\infty(\omega)\}$, Theorems 1.2.44 and 1.3.4, and Lemma 4.1.10 therein
replaced, respectively, by $M_\infty(\omega)$, Theorems 7.1.9 and 7.1.16, and Lemma 8.1.7,
we deduce that both (i) and (ii) of the present theorem hold true. This then finishes
the proof of Theorem 8.2.5.   □

Remark 8.2.6

(i) Let $p, q \in (0.\infty)$ and $\omega \in M(\mathbb{R}_+)$. Then the quasi-norm of $h\mathcal{K}_\omega^{p,q}(\mathbb{R}^n)$ in
Definition 8.2.1(ii) depends on N. However, from Theorem 8.2.5, we deduce
that the local Hardy space $h\mathcal{K}_\omega^{p,q}(\mathbb{R}^n)$ is independent of the choice of N
whenever $\omega \in M(\mathbb{R}_+)$ satisfies

$$-\frac{n}{p} < m_\infty(\omega) \le M_\infty(\omega) < 0,$$

and N satisfies

$$N \in \mathbb{N} \cap \left(1 + 2n \max\left\{\frac{1}{p}, \frac{1}{q}\right\}, \infty\right).$$

(ii) Notice that, if $p = q \in (0, \infty)$ and $\omega(t) := 1$ for any $t \in (0, \infty)$, then, in this
case,

$$m_0(\omega) = M_0(\omega) = m_\infty(\omega) = M_\infty(\omega) = 0$$

and $h\mathcal{K}_\omega^{p,q}(\mathbb{R}^n)$ coincides with the classical local Hardy space $h^p(\mathbb{R}^n)$ in
the sense of equivalent quasi-norms. Therefore, Theorem 8.1.5 completely
excludes the classical local Hardy space $h^p(\mathbb{R}^n)$ and, using Remark 7.1.10,
we find that the classical local Hardy space $h^p(\mathbb{R}^n)$ is the critical case of
$h\mathcal{K}_\omega^{p,q}(\mathbb{R}^n)$ considered in Theorem 8.1.5.

8.2.2 Relations with Inhomogeneous Generalized Herz–Hardy Spaces

The main target of this subsection is to investigate the relation between inhomogeneous generalized Herz–Hardy spaces and inhomogeneous localized generalized Herz–Hardy spaces. First, we give the following relation between $H\mathcal{K}_{\omega,\mathbf{0}}^{p,q}(\mathbb{R}^n)$ and $h\mathcal{K}_{\omega,\mathbf{0}}^{p,q}(\mathbb{R}^n)$ associated with the inhomogeneous local generalized Herz space $\dot{\mathcal{K}}_{\omega,\mathbf{0}}^{p,q}(\mathbb{R}^n)$.

Theorem 8.2.7 *Let* $p, q \in (0, \infty)$, $\omega \in M(\mathbb{R}_+)$ *satisfy* $m_\infty(\omega) \in (-\frac{n}{p}, \infty)$, *and* $\varphi \in \mathcal{S}(\mathbb{R}^n)$ *satisfy*

$$\mathbf{1}_{B(\mathbf{0},1)} \leq \widehat{\varphi} \leq \mathbf{1}_{B(\mathbf{0},2)}.$$

Then there exist two positive constants C_1 *and* C_2 *such that, for any* $f \in \mathcal{S}'(\mathbb{R}^n)$,

$$C_1 \|f\|_{h\mathcal{K}_{\omega,\mathbf{0}}^{p,q}(\mathbb{R}^n)} \leq \|f * \varphi\|_{\mathcal{K}_{\omega,\mathbf{0}}^{p,q}(\mathbb{R}^n)} + \|f - f * \varphi\|_{H\mathcal{K}_{\omega,\mathbf{0}}^{p,q}(\mathbb{R}^n)}$$

$$\leq C_2 \|f\|_{h\mathcal{K}_{\omega,\mathbf{0}}^{p,q}(\mathbb{R}^n)}.$$

Proof Let p, q, ω, and φ be as in the present theorem and $f \in \mathcal{S}'(\mathbb{R}^n)$. Then, using Theorem 5.2.3 and repeating the proof of Theorem 5.2.6 with $\max\{M_0(\omega), M_\infty(\omega)\}$, Theorems 1.2.42 and 1.3.3, and Lemma 1.8.5 therein replaced, respectively, by $M_\infty(\omega)$, Theorems 7.1.7 and 7.1.15, and Lemma 7.1.36, we obtain

$$\|f\|_{h\mathcal{K}_{\omega,\mathbf{0}}^{p,q}(\mathbb{R}^n)} \sim \|f * \varphi\|_{\mathcal{K}_{\omega,\mathbf{0}}^{p,q}(\mathbb{R}^n)} + \|f - f * \varphi\|_{H\mathcal{K}_{\omega,\mathbf{0}}^{p,q}(\mathbb{R}^n)}$$

with positive equivalence constants independent of f. This then finishes the proof of Theorem 8.2.7. □

On the other hand, we obtain the relation between the inhomogeneous generalized Herz–Hardy space $H\mathcal{K}_\omega^{p,q}(\mathbb{R}^n)$ and the inhomogeneous local generalized Herz–Hardy space $h\mathcal{K}_\omega^{p,q}(\mathbb{R}^n)$ as follows.

Theorem 8.2.8 *Let* $p, q \in (0, \infty)$, $\omega \in M(\mathbb{R}_+)$ *satisfy*

$$-\frac{n}{p} < m_\infty(\omega) \leq M_\infty(\omega) < 0,$$

and $\varphi \in \mathcal{S}(\mathbb{R}^n)$ *satisfy*

$$\mathbf{1}_{B(\mathbf{0},1)} \leq \widehat{\varphi} \leq \mathbf{1}_{B(\mathbf{0},2)}.$$

Then there exist two positive constants C_1 and C_2 such that, for any $f \in \mathcal{S}'(\mathbb{R}^n)$,

$$C_1 \|f\|_{h\mathcal{K}^{p,q}_\omega(\mathbb{R}^n)} \leq \|f * \varphi\|_{\mathcal{K}^{p,q}_\omega(\mathbb{R}^n)} + \|f - f * \varphi\|_{H\mathcal{K}^{p,q}_\omega(\mathbb{R}^n)}$$

$$\leq C_2 \|f\|_{h\mathcal{K}^{p,q}_\omega(\mathbb{R}^n)}.$$

Proof Let all the symbols be as in the present theorem. Then, applying Theorem 5.2.3 and repeating an argument similar to that used in the proof of Theorem 5.2.8 with $\max\{M_0(\omega), M_\infty(\omega)\}$, Theorems 1.2.44 and 1.3.4, and Lemma 4.1.10 therein replaced, respectively, by $M_\infty(\omega)$, Theorems 7.1.9 and 7.1.16, and Lemmas 8.1.7, we find that, for any $f \in \mathcal{S}'(\mathbb{R}^n)$,

$$\|f\|_{h\mathcal{K}^{p,q}_\omega(\mathbb{R}^n)} \sim \|f * \varphi\|_{\mathcal{K}^{p,q}_\omega(\mathbb{R}^n)} + \|f - f * \varphi\|_{H\mathcal{K}^{p,q}_\omega(\mathbb{R}^n)}.$$

This then finishes the proof of Theorem 8.2.8. □

8.2.3 Atomic Characterizations

In this subsection, we establish the atomic characterizations of the inhomogeneous local generalized Herz–Hardy spaces $h\mathcal{K}^{p,q}_{\omega,\mathbf{0}}(\mathbb{R}^n)$ and $h\mathcal{K}^{p,q}_\omega(\mathbb{R}^n)$. For this purpose, we first introduce both local-$(\mathcal{K}^{p,q}_{\omega,\mathbf{0}}(\mathbb{R}^n), r, d)$-atoms and local atomic Hardy spaces associated with the inhomogeneous local generalized Herz space $\mathcal{K}^{p,q}_{\omega,\mathbf{0}}(\mathbb{R}^n)$ as follows.

Definition 8.2.9 Let $p, q \in (0, \infty)$, $\omega \in M(\mathbb{R}_+)$, $r \in [1, \infty]$, and $d \in \mathbb{Z}_+$. Then a measurable function a is called a *local-$(\mathcal{K}^{p,q}_{\omega,\mathbf{0}}(\mathbb{R}^n), r, d)$-atom* if

(i) there exists a ball $B(x_0, r_0) \in \mathbb{B}$, with $x_0 \in \mathbb{R}^n$ and $r_0 \in (0, \infty)$, such that

$$\operatorname{supp}(a) := \{x \in \mathbb{R}^n : a(x) \neq 0\} \subset B(x_0, r_0);$$

(ii) $\|a\|_{L^r(\mathbb{R}^n)} \leq \dfrac{|B(x_0,r_0)|^{1/r}}{\|\mathbf{1}_{B(x_0,r_0)}\|_{\mathcal{K}^{p,q}_{\omega,\mathbf{0}}(\mathbb{R}^n)}};$

(iii) when $r_0 \in (0, 1)$, then, for any $\alpha \in \mathbb{Z}^n_+$ such that $|\alpha| \leq d$,

$$\int_{\mathbb{R}^n} a(x) x^\alpha \, dx = 0.$$

Definition 8.2.10 Let $p, q \in (0, \infty)$, $\omega \in M(\mathbb{R}_+)$ with $m_\infty(\omega) \in (-\frac{n}{p}, \infty)$,

$$s \in \left(0, \min\left\{1, p, q, \frac{n}{M_\infty(\omega) + n/p}\right\}\right),$$

$d \geq \lfloor n(1/s - 1) \rfloor$ be a fixed integer, and

$$r \in \left(\max \left\{ 1, p, \frac{n}{m_\infty(\omega) + n/p} \right\}, \infty \right].$$

Then the *inhomogeneous local generalized atomic Herz–Hardy space* $h\mathcal{K}_{\omega,0}^{p,q,r,d,s}(\mathbb{R}^n)$, associated with the inhomogeneous local generalized Herz space $\mathcal{K}_{\omega,0}^{p,q}(\mathbb{R}^n)$, is defined to be the set of all the $f \in \mathcal{S}'(\mathbb{R}^n)$ such that there exists a sequence $\{a_j\}_{j \in \mathbb{N}}$ of local-$(\mathcal{K}_{\omega,0}^{p,q}(\mathbb{R}^n), r, d)$-atoms supported, respectively, in the balls $\{B_j\}_{j \in \mathbb{N}} \subset \mathbb{B}$ and a sequence $\{\lambda_j\}_{j \in \mathbb{N}} \subset [0, \infty)$ such that

$$f = \sum_{j \in \mathbb{N}} \lambda_j a_j$$

in $\mathcal{S}'(\mathbb{R}^n)$ and

$$\left\| \left\{ \sum_{j \in \mathbb{N}} \left[\frac{\lambda_j}{\|\mathbf{1}_{B_j}\|_{\mathcal{K}_{\omega,0}^{p,q}(\mathbb{R}^n)}} \right]^s \mathbf{1}_{B_j} \right\}^{\frac{1}{s}} \right\|_{\mathcal{K}_{\omega,0}^{p,q}(\mathbb{R}^n)} < \infty.$$

Moreover, for any $f \in h\mathcal{K}_{\omega,0}^{p,q,r,d,s}(\mathbb{R}^n)$,

$$\|f\|_{h\mathcal{K}_{\omega,0}^{p,q,r,d,s}(\mathbb{R}^n)} := \inf \left\{ \left\| \left\{ \sum_{j \in \mathbb{N}} \left[\frac{\lambda_j}{\|\mathbf{1}_{B_j}\|_{\mathcal{K}_{\omega,0}^{p,q}(\mathbb{R}^n)}} \right]^s \mathbf{1}_{B_j} \right\}^{\frac{1}{s}} \right\|_{\mathcal{K}_{\omega,0}^{p,q}(\mathbb{R}^n)} \right\},$$

where the infimum is taken over all the decompositions of f as above.

Then we have the following atomic characterization of $h\mathcal{K}_{\omega,0}^{p,q}(\mathbb{R}^n)$.

Theorem 8.2.11 *Let p, q, ω, d, s, and r be as in Definition 8.2.10. Then*

$$h\mathcal{K}_{\omega,0}^{p,q}(\mathbb{R}^n) = h\mathcal{K}_{\omega,0}^{p,q,r,d,s}(\mathbb{R}^n)$$

with equivalent quasi-norms.

Proof Let p, q, ω, d, s, and r be as in Definition 8.2.10. Then, from Lemma 5.3.7 and an argument similar to that used in the proof of Theorem 5.3.3 via replacing Theorem 1.2.42 and Lemmas 1.8.6 and 4.3.11 therein, respectively, by Theorem 7.1.7 and Lemmas 7.1.37 and 8.1.15, we infer that

$$h\mathcal{K}_{\omega,0}^{p,q}(\mathbb{R}^n) = h\mathcal{K}_{\omega,0}^{p,q,r,d,s}(\mathbb{R}^n)$$

with equivalent quasi-norms. This then finishes the proof of Theorem 8.2.11.   □

In the remainder of this subsection, we turn to prove the atomic characterization of the inhomogeneous local generalized Herz–Hardy space $h\mathcal{K}_{\omega}^{p,q}(\mathbb{R}^n)$. To do this, we first introduce the following definition of local-$(\mathcal{K}_{\omega}^{p,q}(\mathbb{R}^n), r, d)$-atoms.

Definition 8.2.12 Let $p, q \in (0, \infty)$, $\omega \in M(\mathbb{R}_+)$ with $M_\infty(\omega) \in (-\infty, 0)$, $r \in [1, \infty]$, and $d \in \mathbb{Z}_+$. Then a measurable function a is called a *local-$(\mathcal{K}_{\omega}^{p,q}(\mathbb{R}^n), r, d)$-atom* if

(i) there exists a ball $B(x_0, r_0) \in \mathbb{B}$, with $x_0 \in \mathbb{R}^n$ and $r_0 \in (0, \infty)$, such that

$$\mathrm{supp}\,(a) := \left\{ x \in \mathbb{R}^n : a(x) \neq 0 \right\} \subset B(x_0, r_0);$$

(ii) $\|a\|_{L^r(\mathbb{R}^n)} \leq \dfrac{|B(x_0, r_0)|^{1/r}}{\|\mathbf{1}_{B(x_0, r_0)}\|_{\mathcal{K}_{\omega}^{p,q}(\mathbb{R}^n)}}$;

(iii) when $r_0 \in (0, 1)$, then, for any $\alpha \in \mathbb{Z}_+^n$ such that $|\alpha| \leq d$,

$$\int_{\mathbb{R}^n} a(x) x^\alpha \, dx = 0.$$

Via these atoms, we now introduce the following atomic Hardy spaces.

Definition 8.2.13 Let $p, q \in (0, \infty)$, $\omega \in M(\mathbb{R}_+)$ with

$$-\frac{n}{p} < m_\infty(\omega) \leq M_\infty(\omega) < 0,$$

$s \in (0, \min\{1, p, q\})$, $d \geq \lfloor n(1/s - 1) \rfloor$ be a fixed integer, and

$$r \in \left(\max\left\{ 1, \frac{n}{m_\infty(\omega) + n/p} \right\}, \infty \right].$$

Then the *inhomogeneous local generalized atomic Herz–Hardy space* $h\mathcal{K}_{\omega}^{p,q,r,d,s}(\mathbb{R}^n)$, associated with the inhomogeneous global generalized Herz space $\mathcal{K}_{\omega}^{p,q}(\mathbb{R}^n)$, is defined to be the set of all the $f \in \mathcal{S}'(\mathbb{R}^n)$ such that there exists a sequence $\{a_j\}_{j \in \mathbb{N}}$ of local-$(\mathcal{K}_{\omega}^{p,q}(\mathbb{R}^n), r, d)$-atoms supported, respectively, in the balls $\{B_j\}_{j \in \mathbb{N}} \subset \mathbb{B}$ and a sequence $\{\lambda_j\}_{j \in \mathbb{N}} \subset [0, \infty)$ such that $f = \sum_{j \in \mathbb{N}} \lambda_j a_j$ in $\mathcal{S}'(\mathbb{R}^n)$ and

$$\left\| \left\{ \sum_{j \in \mathbb{N}} \left[\frac{\lambda_j}{\|\mathbf{1}_{B_j}\|_{\mathcal{K}_{\omega}^{p,q}(\mathbb{R}^n)}} \right]^s \mathbf{1}_{B_j} \right\}^{\frac{1}{s}} \right\|_{\mathcal{K}_{\omega}^{p,q}(\mathbb{R}^n)} < \infty.$$

Moreover, for any $f \in h\mathcal{K}_{\omega}^{p,q,r,d,s}(\mathbb{R}^n)$,

$$\|f\|_{h\mathcal{K}_{\omega}^{p,q,r,d,s}(\mathbb{R}^n)} := \inf\left\{\left\|\left\{\sum_{j\in\mathbb{N}}\left[\frac{\lambda_j}{\|\mathbf{1}_{B_j}\|_{\mathcal{K}_{\omega}^{p,q}(\mathbb{R}^n)}}\right]^s \mathbf{1}_{B_j}\right\}^{\frac{1}{s}}\right\|_{\mathcal{K}_{\omega}^{p,q}(\mathbb{R}^n)}\right\},$$

where the infimum is taken over all the decompositions of f as above.

Then we establish the following atomic characterization of the inhomogeneous local generalized Herz–Hardy space $h\dot{\mathcal{K}}_{\omega}^{p,q}(\mathbb{R}^n)$.

Theorem 8.2.14 *Let* $p, q, \omega, d, s,$ *and* r *be as in Definition 8.2.13. Then*

$$h\mathcal{K}_{\omega}^{p,q}(\mathbb{R}^n) = h\mathcal{K}_{\omega}^{p,q,r,d,s}(\mathbb{R}^n)$$

with equivalent quasi-norms.

Proof Let all the symbols be as in the present theorem. Then, applying Theorems 5.3.14 and 7.1.12 and repeating the proof of Theorem 4.3.16 with Theorem 1.2.44 and Lemmas 4.3.23, 4.3.25, and 4.3.27 therein replaced, respectively, by Theorem 7.1.9 and Lemmas 8.1.19, 8.1.21, and 8.1.23, we conclude that

$$h\mathcal{K}_{\omega}^{p,q}(\mathbb{R}^n) = h\mathcal{K}_{\omega}^{p,q,r,d,s}(\mathbb{R}^n)$$

with equivalent quasi-norms. This then finishes the proof of Theorem 8.2.14. □

8.2.4 Molecular Characterizations

This subsection is devoted to establishing the molecular characterizations of both the inhomogeneous local generalized Herz–Hardy spaces $h\mathcal{K}_{\omega,0}^{p,q}(\mathbb{R}^n)$ and $h\mathcal{K}_{\omega}^{p,q}(\mathbb{R}^n)$. To achieve this, we first show the molecular characterization of $h\mathcal{K}_{\omega,0}^{p,q}(\mathbb{R}^n)$ via introducing the following local-$(\mathcal{K}_{\omega,0}^{p,q}(\mathbb{R}^n), r, d, \tau)$-molecules.

Definition 8.2.15 *Let* $p, q \in (0,\infty)$, $\omega \in M(\mathbb{R}_+)$, $r \in [1,\infty]$, $d \in \mathbb{Z}_+$, *and* $\tau \in (0,\infty)$. *Then a measurable function* m *on* $\mathbb{R}^n$ *is called a* local-$(\mathcal{K}_{\omega,0}^{p,q}(\mathbb{R}^n), r, d, \tau)$-*molecule centered at a ball* $B(x_0, r_0) \in \mathbb{B}$, *with* $x_0 \in \mathbb{R}^n$ *and* $r_0 \in (0,\infty)$, *if*

(i) *for any* $i \in \mathbb{Z}_+$,

$$\left\|m\mathbf{1}_{S_i(B(x_0,r_0))}\right\|_{L^r(\mathbb{R}^n)} \leq 2^{-\tau i}\frac{|B(x_0,r_0)|^{1/r}}{\|\mathbf{1}_{B(x_0,r_0)}\|_{\mathcal{K}_{\omega,0}^{p,q}(\mathbb{R}^n)}};$$

(ii) when $r_0 \in (0, 1)$, then, for any $\alpha \in \mathbb{Z}_+^n$ with $|\alpha| \leq d$,

$$\int_{\mathbb{R}^n} m(x) x^\alpha \, dx = 0.$$

Then we establish the following molecular characterization of the inhomogeneous local generalized Herz–Hardy space $h\mathcal{K}_{\omega,0}^{p,q}(\mathbb{R}^n)$.

Theorem 8.2.16 *Let* $p, q \in (0, \infty)$, $\omega \in M(\mathbb{R}_+)$ *with* $m_\infty(\omega) \in (-\frac{n}{p}, \infty)$,

$$s \in \left(0, \min\left\{1, p, q, \frac{n}{M_\infty(\omega) + n/p}\right\}\right),$$

$d \geq \lfloor n(1/s - 1) \rfloor$ *be a fixed integer,*

$$r \in \left(\max\left\{1, p, \frac{n}{m_\infty(\omega) + n/p}\right\}, \infty\right],$$

and $\tau \in (n(1/s - 1/r), \infty)$. *Then* $f \in h\mathcal{K}_{\omega,0}^{p,q}(\mathbb{R}^n)$ *if and only if* $f \in \mathcal{S}'(\mathbb{R}^n)$ *and there exists a sequence* $\{m_j\}_{j\in\mathbb{N}}$ *of local-*$(\mathcal{K}_{\omega,0}^{p,q}(\mathbb{R}^n), r, d, \tau)$*-molecules centered, respectively, at the balls* $\{B_j\}_{j\in\mathbb{N}} \subset \mathbb{B}$ *and a sequence* $\{\lambda_j\}_{j\in\mathbb{N}} \subset [0, \infty)$ *such that* $f = \sum_{j\in\mathbb{N}} \lambda_j m_j$ *in* $\mathcal{S}'(\mathbb{R}^n)$ *and*

$$\left\| \left\{ \sum_{j\in\mathbb{N}} \left[\frac{\lambda_j}{\|\mathbf{1}_{B_j}\|_{\mathcal{K}_{\omega,0}^{p,q}(\mathbb{R}^n)}} \right]^s \mathbf{1}_{B_j} \right\}^{\frac{1}{s}} \right\|_{\mathcal{K}_{\omega,0}^{p,q}(\mathbb{R}^n)} < \infty.$$

Moreover, there exist two positive constants C_1 *and* C_2 *such that, for any* $f \in h\mathcal{K}_{\omega,0}^{p,q}(\mathbb{R}^n)$,

$$C_1 \|f\|_{h\mathcal{K}_{\omega,0}^{p,q}(\mathbb{R}^n)} \leq \inf\left\{ \left\| \left\{ \sum_{j\in\mathbb{N}} \left[\frac{\lambda_i}{\|\mathbf{1}_{B_j}\|_{\mathcal{K}_{\omega,0}^{p,q}(\mathbb{R}^n)}} \right]^s \mathbf{1}_{B_j} \right\}^{\frac{1}{s}} \right\|_{\mathcal{K}_{\omega,0}^{p,q}(\mathbb{R}^n)} \right\}$$

$$\leq C_2 \|f\|_{h\mathcal{K}_{\omega,0}^{p,q}(\mathbb{R}^n)},$$

where the infimum is taken over all the decompositions of f *as above.*

Proof Let all the symbols be as in the present theorem and $f \in \mathcal{S}'(\mathbb{R}^n)$. Then, from Lemma 5.4.5 and an argument similar to that used in the proof of Theorem 5.4.2 with Theorem 1.2.42 and Lemmas 1.8.6 and 4.3.11 therein replaced, respectively, by Theorem 7.1.7 and Lemmas 7.1.37 and 8.1.15, we infer that $f \in h\mathcal{K}_{\omega,0}^{p,q}(\mathbb{R}^n)$ if and only if there exists a sequence $\{m_j\}_{j\in\mathbb{N}}$ of local-$(\mathcal{K}^{p,q}_{\omega,0}(\mathbb{R}^n),\ r,\ d,\ \tau)$-molecules centered, respectively, at the balls $\{B_j\}_{j\in\mathbb{N}} \subset \mathbb{B}$ and a sequence $\{\lambda_j\}_{j\in\mathbb{N}} \subset [0,\infty)$ such that $f = \sum_{j\in\mathbb{N}} \lambda_j m_j$ in $\mathcal{S}'(\mathbb{R}^n)$ and

$$\left\| \left\{ \sum_{j\in\mathbb{N}} \left[\frac{\lambda_j}{\|\mathbf{1}_{B_j}\|_{\mathcal{K}^{p,q}_{\omega,0}(\mathbb{R}^n)}} \right]^s \mathbf{1}_{B_j} \right\}^{\frac{1}{s}} \right\|_{\mathcal{K}^{p,q}_{\omega,0}(\mathbb{R}^n)} < \infty,$$

and, moreover,

$$\|f\|_{h\mathcal{K}^{p,q}_{\omega,0}(\mathbb{R}^n)} \sim \inf \left\{ \left\| \left\{ \sum_{j\in\mathbb{N}} \left[\frac{\lambda_i}{\|\mathbf{1}_{B_j}\|_{\mathcal{K}^{p,q}_{\omega,0}(\mathbb{R}^n)}} \right]^s \mathbf{1}_{B_j} \right\}^{\frac{1}{s}} \right\|_{\mathcal{K}^{p,q}_{\omega,0}(\mathbb{R}^n)} \right\}$$

with the positive equivalence constants independent of f, where the infimum is taken over all the decompositions of f as in the present theorem. This then finishes the proof of Theorem 8.2.16. □

We next characterize the inhomogeneous local generalized Herz–Hardy space $h\mathcal{K}^{p,q}_{\omega}(\mathbb{R}^n)$ via molecules. To this end, we first introduce the concept of local-$(\mathcal{K}^{p,q}_{\omega}(\mathbb{R}^n),\ r,\ d,\ \tau)$-molecules as follows.

Definition 8.2.17 Let $p, q \in (0,\infty)$, $\omega \in M(\mathbb{R}_+)$ with

$$-\frac{n}{p} < m_\infty(\omega) \leq M_\infty(\omega) < 0,$$

$r \in [1,\infty]$, $d \in \mathbb{Z}_+$, and $\tau \in (0,\infty)$. Then a measurable function m on $\mathbb{R}^n$ is called a *local-*$(\mathcal{K}^{p,q}_{\omega}(\mathbb{R}^n),\ r,\ d,\ \tau)$-*molecule* centered at a ball $B(x_0, r_0) \in \mathbb{B}$, with $x_0 \in \mathbb{R}^n$ and $r_0 \in (0,\infty)$, if

(i) for any $i \in \mathbb{Z}_+$,

$$\left\| m\mathbf{1}_{S_i(B(x_0,r_0))} \right\|_{L^r(\mathbb{R}^n)} \leq 2^{-\tau i} \frac{|B(x_0, r_0)|^{1/r}}{\|\mathbf{1}_{B(x_0,r_0)}\|_{\mathcal{K}^{p,q}_{\omega}(\mathbb{R}^n)}};$$

(ii) when $r_0 \in (0,1)$, then, for any $\alpha \in \mathbb{Z}^n_+$ with $|\alpha| \leq d$,

$$\int_{\mathbb{R}^n} m(x)x^\alpha \, dx = 0.$$

Now, we prove the following molecular characterization of $h\mathcal{K}^{p,q}_{\omega}(\mathbb{R}^n)$.

Theorem 8.2.18 *Let* $p, q \in (0, \infty)$, $\omega \in M(\mathbb{R}_+)$ *with*

$$-\frac{n}{p} < m_\infty(\omega) \leq M_\infty(\omega) < 0,$$

$s \in (0, \min\{1, p, q\})$, $d \geq \lfloor n(1/s - 1) \rfloor$ *be a fixed integer,*

$$r \in \left(\max\left\{ 1, \frac{n}{m_\infty(\omega) + n/p} \right\}, \infty \right],$$

and $\tau \in (0, \infty)$ *with* $\tau > n(1/s - 1/r)$. *Then* $f \in h\mathcal{K}_\omega^{p,q}(\mathbb{R}^n)$ *if and only if* $f \in \mathcal{S}'(\mathbb{R}^n)$ *and there exists a sequence* $\{m_j\}_{j \in \mathbb{N}}$ *of local-*$(\mathcal{K}_\omega^{p,q}(\mathbb{R}^n), r, d, \tau)$-*molecules centered, respectively, at the balls* $\{B_j\}_{j \in \mathbb{N}} \subset \mathbb{B}$ *and a sequence* $\{\lambda_j\}_{j \in \mathbb{N}} \subset [0, \infty)$ *such that* $f = \sum_{j \in \mathbb{N}} \lambda_j m_j$ *in* $\mathcal{S}'(\mathbb{R}^n)$ *and*

$$\left\| \left\{ \sum_{j \in \mathbb{N}} \left[\frac{\lambda_j}{\|\mathbf{1}_{B_j}\|_{\mathcal{K}_\omega^{p,q}(\mathbb{R}^n)}} \right]^s \mathbf{1}_{B_j} \right\}^{\frac{1}{s}} \right\|_{\mathcal{K}_\omega^{p,q}(\mathbb{R}^n)} < \infty.$$

Moreover, there exist two positive constants C_1 *and* C_2 *such that, for any* $f \in h\mathcal{K}_\omega^{p,q}(\mathbb{R}^n)$,

$$C_1 \|f\|_{h\mathcal{K}_\omega^{p,q}(\mathbb{R}^n)} \leq \inf\left\{ \left\| \left\{ \sum_{j \in \mathbb{N}} \left[\frac{\lambda_i}{\|\mathbf{1}_{B_j}\|_{\mathcal{K}_\omega^{p,q}(\mathbb{R}^n)}} \right]^s \mathbf{1}_{B_j} \right\}^{\frac{1}{s}} \right\|_{\dot{\mathcal{K}}_\omega^{p,q}(\mathbb{R}^n)} \right\}$$

$$\leq C_2 \|f\|_{h\mathcal{K}_\omega^{p,q}(\mathbb{R}^n)},$$

where the infimum is taken over all the decompositions of f *as above.*

Proof Let all the symbols be as in the present theorem and $f \in \mathcal{S}'(\mathbb{R}^n)$. Then, using Theorems 5.4.11 and 7.1.12 and repeating the proof of Theorem 4.3.16 with Theorem 1.2.44 and Lemmas 4.3.23, 4.3.25, and 4.3.27 therein replaced, respectively, by Theorem 7.1.9 and Lemmas 8.1.19, 8.1.21, and 8.1.23, we find that $f \in h\mathcal{K}_\omega^{p,q}(\mathbb{R}^n)$ if and only if there exists a sequence $\{m_j\}_{j \in \mathbb{N}}$ of local-$(\mathcal{K}_\omega^{p,q}(\mathbb{R}^n), r, d, \tau)$-molecules centered, respectively, at the balls $\{B_j\}_{j \in \mathbb{N}} \subset \mathbb{B}$ and a sequence $\{\lambda_j\}_{j \in \mathbb{N}} \subset [0, \infty)$ such that $f = \sum_{j \in \mathbb{N}} \lambda_j m_j$ in $\mathcal{S}'(\mathbb{R}^n)$ and

$$\left\| \left\{ \sum_{j \in \mathbb{N}} \left[\frac{\lambda_j}{\|\mathbf{1}_{B_j}\|_{\mathcal{K}_\omega^{p,q}(\mathbb{R}^n)}} \right]^s \mathbf{1}_{B_j} \right\}^{\frac{1}{s}} \right\|_{\mathcal{K}_\omega^{p,q}(\mathbb{R}^n)} < \infty,$$

and, moreover,

$$\|f\|_{h\mathcal{K}_\omega^{p,q}(\mathbb{R}^n)} \sim \inf \left\{ \left\| \left\{ \sum_{j\in\mathbb{N}} \left[\frac{\lambda_i}{\|\mathbf{1}_{B_j}\|_{\mathcal{K}_\omega^{p,q}(\mathbb{R}^n)}} \right]^s \mathbf{1}_{B_j} \right\}^{\frac{1}{s}} \right\|_{\mathcal{K}_\omega^{p,q}(\mathbb{R}^n)} \right\}$$

with the positive equivalence constants independent of f, where the infimum is taken over all the decompositions of f as in the present theorem. This then finishes the proof of Theorem 8.2.18. □

8.2.5 Littlewood–Paley Function Characterizations

The main target of this subsection is to characterize both the inhomogeneous local generalized Herz–Hardy spaces $h\mathcal{K}_{\omega,0}^{p,q}(\mathbb{R}^n)$ and $h\mathcal{K}_\omega^{p,q}(\mathbb{R}^n)$ via the local Lusin area function, the local g-function, and the local g_λ^*-function defined as in Definition 5.1.1. To begin with, we establish the following Littlewood–Paley characterizations of $h\mathcal{K}_{\omega,0}^{p,q}(\mathbb{R}^n)$.

Theorem 8.2.19 *Let* $p,q \in (0,\infty)$, $\omega \in M(\mathbb{R}_+)$ *satisfy* $m_\infty(\omega) \in (-\frac{n}{p}, \infty)$,

$$s_0 := \min \left\{ 1, p, q, \frac{n}{M_\infty(\omega) + n/p} \right\},$$

and $\lambda \in (\max\{1, 2/s_0\}, \infty)$. *Then the following four statements are mutually equivalent:*

(i) $f \in h\mathcal{K}_{\omega,0}^{p,q}(\mathbb{R}^n)$;
(ii) $f \in \mathcal{S}'(\mathbb{R}^n)$ *and* $S_{\mathrm{loc}}(f) \in \mathcal{K}_{\omega,0}^{p,q}(\mathbb{R}^n)$;
(iii) $f \in \mathcal{S}'(\mathbb{R}^n)$ *and* $g_{\mathrm{loc}}(f) \in \mathcal{K}_{\omega,0}^{p,q}(\mathbb{R}^n)$;
(iv) $f \in \mathcal{S}'(\mathbb{R}^n)$ *and* $(g_\lambda^*)_{\mathrm{loc}}(f) \in \mathcal{K}_{\omega,0}^{p,q}(\mathbb{R}^n)$.

Moreover, for any $f \in h\mathcal{K}_{\omega,0}^{p,q}(\mathbb{R}^n)$,

$$\|f\|_{h\mathcal{K}_{\omega,0}^{p,q}(\mathbb{R}^n)} \sim \|S_{\mathrm{loc}}(f)\|_{\mathcal{K}_{\omega,0}^{p,q}(\mathbb{R}^n)} \sim \|g_{\mathrm{loc}}(f)\|_{\mathcal{K}_{\omega,0}^{p,q}(\mathbb{R}^n)}$$

$$\sim \left\| (g_\lambda^*)_{\mathrm{loc}}(f) \right\|_{\mathcal{K}_{\omega,0}^{p,q}(\mathbb{R}^n)},$$

where the positive equivalence constants are independent of f.

Proof Let all the symbols be as in the present theorem. Then, from Theorems 5.5.2, 5.5.11, and 5.5.17 and an argument similar to that used in the proof of Theorem 4.6.3 via replacing Theorem 1.2.42 and Lemmas 1.8.6 and 4.3.11 therein, respectively, by Theorem 7.1.7 and Lemmas 7.1.37 and 8.1.15, we deduce that (i) through (iv) of the present theorem are mutually equivalent and, for any $f \in h\mathcal{K}_{\omega,0}^{p,q}(\mathbb{R}^n)$,

$$\|f\|_{h\mathcal{K}_{\omega,0}^{p,q}(\mathbb{R}^n)} \sim \|S_{\mathrm{loc}}(f)\|_{\mathcal{K}_{\omega,0}^{p,q}(\mathbb{R}^n)} \sim \|g_{\mathrm{loc}}(f)\|_{\mathcal{K}_{\omega,0}^{p,q}(\mathbb{R}^n)}$$

$$\sim \left\|\left(g_\lambda^*\right)_{\mathrm{loc}}(f)\right\|_{\mathcal{K}_{\omega,0}^{p,q}(\mathbb{R}^n)}$$

with the positive equivalence constants independent of f. This then finishes the proof of Theorem 8.2.19. □

Next, we turn to characterize the inhomogeneous local generalized Herz–Hardy space $h\mathcal{K}_\omega^{p,q}(\mathbb{R}^n)$ via various Littlewood–Paley functions as follows.

Theorem 8.2.20 *Let $p, q \in (0, \infty)$, $\omega \in M(\mathbb{R}_+)$ satisfy*

$$-\frac{n}{p} < m_\infty(\omega) \le M_\infty(\omega) < 0,$$

and $\lambda \in (2\max\{1, \frac{1}{p}, \frac{1}{q}\}, \infty)$. Then the following four statements are mutually equivalent:

(i) $f \in h\mathcal{K}_\omega^{p,q}(\mathbb{R}^n)$;
(ii) $f \in \mathcal{S}'(\mathbb{R}^n)$ *and* $S_{\mathrm{loc}}(f) \in \mathcal{K}_\omega^{p,q}(\mathbb{R}^n)$;
(iii) $f \in \mathcal{S}'(\mathbb{R}^n)$ *and* $g_{\mathrm{loc}}(f) \in \mathcal{K}_\omega^{p,q}(\mathbb{R}^n)$;
(iv) $f \in \mathcal{S}'(\mathbb{R}^n)$ *and* $(g_\lambda^*)_{\mathrm{loc}}(f) \in \mathcal{K}_\omega^{p,q}(\mathbb{R}^n)$.

Moreover, for any $f \in h\mathcal{K}_\omega^{p,q}(\mathbb{R}^n)$,

$$\|f\|_{h\mathcal{K}_\omega^{p,q}(\mathbb{R}^n)} \sim \|S_{\mathrm{loc}}(f)\|_{\mathcal{K}_\omega^{p,q}(\mathbb{R}^n)} \sim \|g_{\mathrm{loc}}(f)\|_{\mathcal{K}_\omega^{p,q}(\mathbb{R}^n)}$$

$$\sim \left\|\left(g_\lambda^*\right)_{\mathrm{loc}}(f)\right\|_{\mathcal{K}_\omega^{p,q}(\mathbb{R}^n)},$$

where the positive equivalence constants are independent of f.

Proof Let all the symbols be as in the present theorem. Then, repeating the proof of Theorem 4.6.8 with Theorems 4.1.2, 4.1.8, and 4.6.3, and Lemma 4.6.9(iii) therein replaced, respectively, by Theorems 8.1.3, 8.1.5, and 8.1.32, and Lemma 5.5.24, we find that (i) through (iv) of the present theorem are mutually equivalent and, for any $f \in h\mathcal{K}_\omega^{p,q}(\mathbb{R}^n)$,

$$\|f\|_{h\mathcal{K}_\omega^{p,q}(\mathbb{R}^n)} \sim \|S_{\mathrm{loc}}(f)\|_{\mathcal{K}_\omega^{p,q}(\mathbb{R}^n)} \sim \|g_{\mathrm{loc}}(f)\|_{\mathcal{K}_\omega^{p,q}(\mathbb{R}^n)}$$

$$\sim \left\|\left(g_\lambda^*\right)_{\mathrm{loc}}(f)\right\|_{\mathcal{K}_\omega^{p,q}(\mathbb{R}^n)}$$

with the positive equivalence constants independent of f. This then finishes the proof of Theorem 8.2.20. □

8.2.6 Boundedness of Pseudo-Differential Operators

Recall that the Hörmander class $S_{1,0}^0(\mathbb{R}^n)$ is defined to be the set of all the infinitely differentiable functions σ on $\mathbb{R}^n \times \mathbb{R}^n$ satisfying (5.75). Let $\sigma \in S_{1,0}^0(\mathbb{R}^n)$ and T_σ be a pseudo-differential operator defined as in Definition 5.6.1. In this subsection, we investigate the boundedness of T_σ on inhomogeneous localized generalized Herz–Hardy spaces. First, the following conclusion gives the boundedness of operator T_σ on the inhomogeneous local generalized Herz–Hardy space $h\mathcal{K}_{\omega,\mathbf{0}}^{p,q}(\mathbb{R}^n)$.

Theorem 8.2.21 *Let $p, q \in (0, \infty)$, $\omega \in M(\mathbb{R}_+)$ satisfy $m_\infty(\omega) \in (-\frac{n}{p}, \infty)$, and T_σ be a pseudo-differential operator with the symbol $\sigma \in S_{1,0}^0(\mathbb{R}^n)$. Then T_σ is well defined on $h\mathcal{K}_{\omega,\mathbf{0}}^{p,q}(\mathbb{R}^n)$ and there exists a positive constant C such that, for any $f \in h\mathcal{K}_{\omega,\mathbf{0}}^{p,q}(\mathbb{R}^n)$,*

$$\|T_\sigma(f)\|_{h\mathcal{K}_{\omega,\mathbf{0}}^{p,q}(\mathbb{R}^n)} \leq C\|f\|_{h\mathcal{K}_{\omega,\mathbf{0}}^{p,q}(\mathbb{R}^n)}.$$

Proof Let all the symbols be as in the present theorem. Then, applying Lemma 5.6.4 and repeating an argument similar to that used in the proof of Theorem 5.6.3 with $\max\{M_0(\omega), M_\infty(\omega)\}$, Theorems 1.2.42 and 1.4.1, and Lemmas 1.8.6 and 4.3.11 therein replaced, respectively, by $M_\infty(\omega)$, Theorems 7.1.7 and 7.1.17, and Lemmas 7.1.37 and 8.1.15, we conclude that T_σ is well defined on $h\mathcal{K}_{\omega,\mathbf{0}}^{p,q}(\mathbb{R}^n)$ and, for any $f \in h\mathcal{K}_{\omega,\mathbf{0}}^{p,q}(\mathbb{R}^n)$,

$$\|T_\sigma(f)\|_{h\mathcal{K}_{\omega,\mathbf{0}}^{p,q}(\mathbb{R}^n)} \lesssim \|f\|_{h\mathcal{K}_{\omega,\mathbf{0}}^{p,q}(\mathbb{R}^n)}.$$

This then finishes the proof of Theorem 8.2.21.   □

Remark 8.2.22 We should point out that, in Theorem 8.2.21, when $\omega(t) := t^\alpha$ for any $t \in (0, \infty)$ and for any given $\alpha \in \mathbb{R}$, the result obtained in Theorem 8.2.21 goes back to [233, Corollary 4.14(c)].

In addition, we establish the following boundedness of pseudo-differential operators on the local Hardy space $h\mathcal{K}_\omega^{p,q}(\mathbb{R}^n)$.

Theorem 8.2.23 *Let $p, q \in (0, \infty)$, $\omega \in M(\mathbb{R}_+)$ satisfy*

$$-\frac{n}{p} < m_\infty(\omega) \leq M_\infty(\omega) < 0,$$

and T_σ be a pseudo-differential operator with symbol $\sigma \in S_{1,0}^0(\mathbb{R}^n)$. Then T_σ is well defined on $h\mathcal{K}_\omega^{p,q}(\mathbb{R}^n)$ and there exists a positive constant C such that, for any $f \in h\mathcal{K}_\omega^{p,q}(\mathbb{R}^n)$,

$$\|T_\sigma(f)\|_{h\mathcal{K}_\omega^{p,q}(\mathbb{R}^n)} \leq C\|f\|_{h\mathcal{K}_\omega^{p,q}(\mathbb{R}^n)}.$$

8   Hardy Spaces Associated with Inhomogeneous Generalized Herz Spaces

Proof Let all the symbols be as in the present theorem. Then, using Theorems 4.8.17 and 7.1.12 and repeating the proof of Theorem 4.8.16 with

$$\min\{m_0(\omega), m_\infty(\omega)\}, \ \max\{M_0(\omega), M_\infty(\omega)\},$$

Theorem 1.2.44, Lemmas 2.2.3, 2.3.2, 4.3.25, and 4.3.27, and Corollary 2.3.5 therein replaced, respectively, by $m_\infty(\omega)$, $M_\infty(\omega)$, Theorem 7.1.9, Lemmas 7.2.6, 7.2.11, 8.1.21, and 8.1.23, and Corollary 7.2.14, we conclude that T_σ is well defined on $h\mathcal{K}_\omega^{p,q}(\mathbb{R}^n)$ and, for any $f \in h\mathcal{K}_\omega^{p,q}(\mathbb{R}^n)$,

$$\|T(f)\|_{h\mathcal{K}_\omega^{p,q}(\mathbb{R}^n)} \lesssim \|f\|_{h\mathcal{K}_\omega^{p,q}(\mathbb{R}^n)},$$

which then completes the proof of Theorem 8.2.23.   □

8.3   Inhomogeneous Weak Generalized Herz–Hardy Spaces

In this section, we introduce both inhomogeneous weak generalized Herz spaces and inhomogeneous weak generalized Herz–Hardy spaces. Then we establish the maximal function, the atomic, the molecular, and the Littlewood–Paley function characterizations of these Hardy spaces. As applications, we also obtain the boundedness of Calderón–Zygmund operators from inhomogeneous generalized Herz–Hardy spaces to inhomogeneous weak generalized Herz–Hardy spaces even in the critical case as well as show that the real interpolation spaces between inhomogeneous generalized Herz–Hardy spaces and the Lebesgue space $L^\infty(\mathbb{R}^n)$ are just the new introduced inhomogeneous weak generalized Herz–Hardy spaces.

To begin with, we introduce inhomogeneous weak generalized Herz spaces as follows.

Definition 8.3.1 Let $p, q \in (0, \infty)$ and $\omega \in M(\mathbb{R}_+)$.

(i) The *inhomogeneous weak local generalized Herz space* $W\mathcal{K}_{\omega,\mathbf{0}}^{p,q}(\mathbb{R}^n)$ is defined to be the set of all the measurable functions f on $\mathbb{R}^n$ such that

$$\|f\|_{W\mathcal{K}_{\omega,\mathbf{0}}^{p,q}(\mathbb{R}^n)} := \sup_{\alpha \in (0,\infty)} \left\{ \alpha \left\| \mathbf{1}_{\{x \in \mathbb{R}^n: \, |f(x)| > \alpha\}} \right\|_{\mathcal{K}_{\omega,\mathbf{0}}^{p,q}(\mathbb{R}^n)} \right\} < \infty.$$

(ii) The *inhomogeneous weak global generalized Herz space* $W\mathcal{K}_\omega^{p,q}(\mathbb{R}^n)$ is defined to be the set of all the measurable functions f on $\mathbb{R}^n$ such that

$$\|f\|_{W\mathcal{K}_\omega^{p,q}(\mathbb{R}^n)} := \sup_{\alpha \in (0,\infty)} \left\{ \alpha \left\| \mathbf{1}_{\{x \in \mathbb{R}^n: \, |f(x)| > \alpha\}} \right\|_{\mathcal{K}_\omega^{p,q}(\mathbb{R}^n)} \right\} < \infty.$$

Remark 8.3.2 In Definition 8.3.1, when $\omega(t) := t^\alpha$ for any $t \in (0, \infty)$ and for any given $\alpha \in \mathbb{R}$, the inhomogeneous weak local generalized Herz space $W\mathcal{K}_{\omega,\mathbf{0}}^{p,q}(\mathbb{R}^n)$ coincides with the classical *inhomogeneous weak Herz space* $WK_p^{\alpha,q}(\mathbb{R}^n)$ which was originally introduced in [117] (see also [175, Section 1.4]).

Let $N \in \mathbb{N}$ and the non-tangential grand maximal function $\mathcal{M}_N(f)$ of $f \in \mathcal{S}'(\mathbb{R}^n)$ be as in (4.3). We now introduce the following inhomogeneous weak generalized Herz–Hardy spaces via the operator $\mathcal{M}_N$.

Definition 8.3.3 Let $p, q \in (0, \infty)$, $\omega \in M(\mathbb{R}_+)$, and $N \in \mathbb{N}$.

(i) The *inhomogeneous weak generalized Herz–Hardy space* $WH\mathcal{K}_{\omega,\mathbf{0}}^{p,q}(\mathbb{R}^n)$, associated with the inhomogeneous weak local generalized Herz space $W\mathcal{K}_{\omega,\mathbf{0}}^{p,q}(\mathbb{R}^n)$, is defined to be the set of all the $f \in \mathcal{S}'(\mathbb{R}^n)$ such that

$$\|f\|_{WH\mathcal{K}_{\omega,\mathbf{0}}^{p,q}(\mathbb{R}^n)} := \|\mathcal{M}_N(f)\|_{W\mathcal{K}_{\omega,\mathbf{0}}^{p,q}(\mathbb{R}^n)} < \infty.$$

(ii) The *inhomogeneous weak generalized Herz–Hardy space* $WH\mathcal{K}_\omega^{p,q}(\mathbb{R}^n)$, associated with the inhomogeneous weak global generalized Herz space $W\mathcal{K}_\omega^{p,q}(\mathbb{R}^n)$, is defined to be the set of all the $f \in \mathcal{S}'(\mathbb{R}^n)$ such that

$$\|f\|_{WH\mathcal{K}_\omega^{p,q}(\mathbb{R}^n)} := \|\mathcal{M}_N(f)\|_{W\mathcal{K}_\omega^{p,q}(\mathbb{R}^n)} < \infty.$$

Remark 8.3.4 We point out that, in Definition 8.3.3, if $\omega(t) := t^\alpha$ for any $t \in (0, \infty)$ and for any given $\alpha \in \mathbb{R}$, then the inhomogeneous weak generalized Herz–Hardy space $WH\mathcal{K}_{\omega,\mathbf{0}}^{p,q}(\mathbb{R}^n)$ coincides with the classical *inhomogeneous weak Herz-type Hardy space* $WHK_p^{\alpha,q}(\mathbb{R}^n)$ (see, for instance, [175, Definition 2.1.2]).

8.3.1 Maximal Function Characterizations

The main target of this subsection is to establish the equivalent characterizations of inhomogeneous weak generalized Herz–Hardy spaces via various radial and non-tangential maximal functions given in Definition 4.1.1. To begin with, we prove the following maximal function characterizations of the inhomogeneous weak generalized Herz–Hardy space $WH\mathcal{K}_{\omega,\mathbf{0}}^{p,q}(\mathbb{R}^n)$.

Theorem 8.3.5 *Let* $p, q, a, b \in (0, \infty)$, $\omega \in M(\mathbb{R}_+)$, $N \in \mathbb{N}$, *and* $\phi \in \mathcal{S}(\mathbb{R}^n)$ *satisfy* $\int_{\mathbb{R}^n} \phi(x)\, dx \neq 0$.

(i) *Let $N \in \mathbb{N} \cap [\lfloor b+1 \rfloor, \infty)$. Then, for any $f \in \mathcal{S}'(\mathbb{R}^n)$,*

$$\|M(f,\phi)\|_{W\mathcal{K}^{p,q}_{\omega,0}(\mathbb{R}^n)} \lesssim \|M_a^*(f,\phi)\|_{W\mathcal{K}^{p,q}_{\omega,0}(\mathbb{R}^n)} \lesssim \|M_b^{**}(f,\phi)\|_{W\mathcal{K}^{p,q}_{\omega,0}(\mathbb{R}^n)},$$

$$\|M(f,\phi)\|_{W\mathcal{K}^{p,q}_{\omega,0}(\mathbb{R}^n)} \lesssim \|\mathcal{M}_N(f)\|_{W\mathcal{K}^{p,q}_{\omega,0}(\mathbb{R}^n)} \lesssim \|\mathcal{M}_{\lfloor b+1 \rfloor}(f)\|_{W\mathcal{K}^{p,q}_{\omega,0}(\mathbb{R}^n)}$$

$$\lesssim \|M_b^{**}(f,\phi)\|_{W\mathcal{K}^{p,q}_{\omega,0}(\mathbb{R}^n)},$$

and

$$\|M_b^{**}(f,\phi)\|_{W\mathcal{K}^{p,q}_{\omega,0}(\mathbb{R}^n)} \sim \|\mathcal{M}_{b,N}^{**}(f)\|_{W\mathcal{K}^{p,q}_{\omega,0}(\mathbb{R}^n)},$$

where the implicit positive constants are independent of f.

(ii) *Let $\omega \in M(\mathbb{R}_+)$ satisfy $m_\infty(\omega) \in (-\frac{n}{p}, \infty)$ and*

$$b \in \left(\max \left\{ \frac{n}{p}, M_\infty(\omega) + \frac{n}{p} \right\}, \infty \right).$$

Then, for any $f \in \mathcal{S}'(\mathbb{R}^n)$,

$$\|M_b^{**}(f,\phi)\|_{W\mathcal{K}^{p,q}_{\omega,0}(\mathbb{R}^n)} \lesssim \|M(f,\phi)\|_{W\mathcal{K}^{p,q}_{\omega,0}(\mathbb{R}^n)},$$

where the implicit positive constant is independent of f. In particular, when $N \in \mathbb{N} \cap [\lfloor b+1 \rfloor, \infty)$, if one of the quantities

$$\|M(f,\phi)\|_{W\mathcal{K}^{p,q}_{\omega,0}(\mathbb{R}^n)}, \quad \|M_a^*(f,\phi)\|_{W\mathcal{K}^{p,q}_{\omega,0}(\mathbb{R}^n)}, \quad \|\mathcal{M}_N(f)\|_{W\mathcal{K}^{p,q}_{\omega,0}(\mathbb{R}^n)},$$

$$\|M_b^{**}(f,\phi)\|_{W\mathcal{K}^{p,q}_{\omega,0}(\mathbb{R}^n)}, \quad and \quad \|\mathcal{M}_{b,N}^{**}(f)\|_{W\mathcal{K}^{p,q}_{\omega,0}(\mathbb{R}^n)}$$

is finite, then the others are also finite and mutually equivalent with the positive equivalence constants independent of f.

Remark 8.3.6

(i) Let $p, q \in (0, \infty)$ and $\omega \in M(\mathbb{R}_+)$. Then the quasi-norm of $WH\mathcal{K}^{p,q}_{\omega,0}(\mathbb{R}^n)$ in Definition 8.3.3(i) depends on N. However, from Theorem 8.3.5, it follows that the weak Hardy space $WH\mathcal{K}^{p,q}_{\omega,0}(\mathbb{R}^n)$ is independent of the choice of N whenever $\omega \in M(\mathbb{R}_+)$ satisfies $m_\infty(\omega) \in (-\frac{n}{p}, \infty)$, and N satisfies

$$N \in \mathbb{N} \cap \left(1 + \max \left\{ \frac{n}{p}, M_\infty(\omega) + \frac{n}{p} \right\}, \infty \right).$$

(ii) Observe that, if $p = q \in (0, \infty)$ and $\omega(t) := 1$ for any $t \in (0, \infty)$, then, in this case,

$$m_0(\omega) = M_0(\omega) = m_\infty(\omega) = M_\infty(\omega) = 0,$$

$WH\mathcal{K}^{p,q}_{\omega,\mathbf{0}}(\mathbb{R}^n)$ coincides with the classical weak Hardy space $WH^p(\mathbb{R}^n)$ in the sense of equivalent quasi-norms, and all the conclusions in Theorem 8.3.5 coincide with the corresponding classical results on the weak Hardy space $WH^p(\mathbb{R}^n)$; see, for instance, [153].

To prove this theorem, we first show the following auxiliary lemma about the boundedness of the Hardy–Littlewood maximal operator on inhomogeneous weak local generalized Herz spaces.

Lemma 8.3.7 *Let* $p, q \in (0, \infty)$ *and* $\omega \in M(\mathbb{R}_+)$ *satisfy* $m_\infty(\omega) \in (-\frac{n}{p}, \infty)$. *Then, for any given*

$$r \in \left(0, \min \left\{ p, \frac{n}{M_\infty(\omega) + n/p} \right\} \right),$$

there exists a positive constant C *such that, for any* $f \in L^1_{\mathrm{loc}}(\mathbb{R}^n)$,

$$\|\mathcal{M}(f)\|_{[W\mathcal{K}^{p,q}_{\omega,\mathbf{0}}(\mathbb{R}^n)]^{1/r}} \le C \|f\|_{[W\mathcal{K}^{p,q}_{\omega,\mathbf{0}}(\mathbb{R}^n)]^{1/r}}.$$

Proof Let all the symbols be as in the present lemma and

$$r \in \left(0, \min \left\{ p, \frac{n}{M_\infty(\omega) + n/p} \right\} \right).$$

Then, repeating an argument similar to that used in the proof of Lemma 6.1.6 with both Theorem 1.2.42 and Lemma 4.3.10 therein replaced, respectively, by both Theorem 7.1.7 and Lemma 8.1.14, we find that, for any $f \in L^1_{\mathrm{loc}}(\mathbb{R}^n)$,

$$\|\mathcal{M}(f)\|_{[W\mathcal{K}^{p,q}_{\omega,\mathbf{0}}(\mathbb{R}^n)]^{1/r}} \lesssim \|f\|_{[W\mathcal{K}^{p,q}_{\omega,\mathbf{0}}(\mathbb{R}^n)]^{1/r}}.$$

This then finishes the proof of Lemma 8.3.7. □

With the help of this lemma, we next show Theorem 8.3.5.

Proof of Theorem 8.3.5 Let all the symbols be as in the present theorem. Then, from Lemma 4.1.4 and an argument similar to that used in the proof of Theorem 6.1.1 with $\max\{M_0(\omega), M_\infty(\omega)\}$, Theorem 1.2.42, and Lemma 6.1.6 therein replaced, respectively, by $M_\infty(\omega)$, Theorem 7.1.7, and Lemma 8.3.7, we infer that both (i) and (ii) of the present theorem hold true, and hence complete the proof of Theorem 8.3.5. □

Now, we consider the maximal function characterizations of the inhomogeneous weak generalized Herz–Hardy space $WH\mathcal{K}_\omega^{p,q}(\mathbb{R}^n)$. Indeed, we have the following conclusion.

Theorem 8.3.8 *Let* $p, q, a, b \in (0, \infty)$, $\omega \in M(\mathbb{R}_+)$, $N \in \mathbb{N}$, *and* $\phi \in \mathcal{S}(\mathbb{R}^n)$ *satisfy* $\int_{\mathbb{R}^n} \phi(x)\,dx \neq 0$.

(i) *Let* $N \in \mathbb{N} \cap [\lfloor b + 1 \rfloor, \infty)$ *and* ω *satisfy* $M_\infty(\omega) \in (-\infty, 0)$. *Then, for any* $f \in \mathcal{S}'(\mathbb{R}^n)$,

$$\|M(f, \phi)\|_{W\mathcal{K}_\omega^{p,q}(\mathbb{R}^n)} \lesssim \|M_a^*(f, \phi)\|_{W\mathcal{K}_\omega^{p,q}(\mathbb{R}^n)} \lesssim \|M_b^{**}(f, \phi)\|_{W\mathcal{K}_\omega^{p,q}(\mathbb{R}^n)},$$

$$\|M(f, \phi)\|_{W\mathcal{K}_\omega^{p,q}(\mathbb{R}^n)} \lesssim \|\mathcal{M}_N(f)\|_{W\mathcal{K}_\omega^{p,q}(\mathbb{R}^n)} \lesssim \|\mathcal{M}_{\lfloor b+1 \rfloor}(f)\|_{W\mathcal{K}_\omega^{p,q}(\mathbb{R}^n)}$$

$$\lesssim \|M_b^{**}(f, \phi)\|_{W\mathcal{K}_\omega^{p,q}(\mathbb{R}^n)},$$

and

$$\|M_b^{**}(f, \phi)\|_{W\mathcal{K}_\omega^{p,q}(\mathbb{R}^n)} \sim \|\mathcal{M}_{b,N}^{**}(f)\|_{W\mathcal{K}_\omega^{p,q}(\mathbb{R}^n)},$$

where the implicit positive constants are independent of f.

(ii) *Let* $\omega \in M(\mathbb{R}_+)$ *satisfy* $-\frac{n}{p} < m_\infty(\omega) \leq M_\infty(\omega) < 0$ *and* $b \in (\frac{n}{p}, \infty)$. *Then, for any* $f \in \mathcal{S}'(\mathbb{R}^n)$,

$$\|M_b^{**}(f, \phi)\|_{W\mathcal{K}_\omega^{p,q}(\mathbb{R}^n)} \lesssim \|M(f, \phi)\|_{W\mathcal{K}_\omega^{p,q}(\mathbb{R}^n)},$$

where the implicit positive constant is independent of f. *In particular, when* $N \in \mathbb{N} \cap [\lfloor b + 1 \rfloor, \infty)$, *if one of the quantities*

$$\|M(f, \phi)\|_{W\mathcal{K}_\omega^{p,q}(\mathbb{R}^n)}, \quad \|M_a^*(f, \phi)\|_{W\mathcal{K}_\omega^{p,q}(\mathbb{R}^n)}, \quad \|\mathcal{M}_N(f)\|_{W\mathcal{K}_\omega^{p,q}(\mathbb{R}^n)},$$

$$\|M_b^{**}(f, \phi)\|_{W\mathcal{K}_\omega^{p,q}(\mathbb{R}^n)}, \quad \text{and } \|\mathcal{M}_{b,N}^{**}(f)\|_{W\mathcal{K}_\omega^{p,q}(\mathbb{R}^n)}$$

is finite, then the others are also finite and mutually equivalent with the positive equivalence constants independent of f.

Remark 8.3.9

(i) Let $p, q \in (0, \infty)$ and $\omega \in M(\mathbb{R}_+)$. Then the quasi-norm of $WH\mathcal{K}_\omega^{p,q}(\mathbb{R}^n)$ in Definition 8.3.3(ii) depends on N. However, by Theorem 8.3.8, we find that the weak Hardy space $WH\mathcal{K}_\omega^{p,q}(\mathbb{R}^n)$ is independent of the choice of N whenever $\omega \in M(\mathbb{R}_+)$ satisfies

$$-\frac{n}{p} < m_\infty(\omega) \leq M_\infty(\omega) < 0,$$

and N satisfies

$$N \in \mathbb{N} \cap \left(1 + \frac{n}{p}, \infty\right).$$

(ii) Notice that, if $p = q \in (0, \infty)$ and $\omega(t) := 1$ for any $t \in (0, \infty)$, then, in this case,

$$m_0(\omega) = M_0(\omega) = m_\infty(\omega) = M_\infty(\omega) = 0$$

and $W H \mathcal{K}_\omega^{p,q}(\mathbb{R}^n)$ coincides with the classical weak Hardy space $W H^p(\mathbb{R}^n)$ in the sense of equivalent quasi-norms. Thus, Theorem 8.3.8 completely excludes the classical weak Hardy space $W H^p(\mathbb{R}^n)$ and, based on Remark 7.1.10, we conclude that the classical weak Hardy space $W H^p(\mathbb{R}^n)$ is the critical case of $W H \mathcal{K}_\omega^{p,q}(\mathbb{R}^n)$ considered in Theorem 8.3.8.

To prove this theorem, we first establish the boundedness of the Hardy–Littlewood maximal operator on the convexification of $W \mathcal{K}_\omega^{p,q}(\mathbb{R}^n)$ as follows

Lemma 8.3.10 *Let* $p, q \in (0, \infty)$ *and* $\omega \in M(\mathbb{R}_+)$ *satisfy* $m_\infty(\omega) \in (-\frac{n}{p}, \infty)$. *Then, for any given*

$$r \in \left(0, \min\left\{p, \frac{n}{M_\infty(\omega) + n/p}\right\}\right),$$

there exists a positive constant C *such that, for any* $f \in L^1_{\mathrm{loc}}(\mathbb{R}^n)$,

$$\|\mathcal{M}(f)\|_{[W\mathcal{K}_\omega^{p,q}(\mathbb{R}^n)]^{1/r}} \leq C \|f\|_{[W\mathcal{K}_\omega^{p,q}(\mathbb{R}^n)]^{1/r}}.$$

Proof Let all the symbols be as in the present lemma. Then, using Lemma 6.1.7 and repeating the proof of Lemma 6.1.11 with both Theorem 1.2.44 and Lemma 4.3.24 therein replaced, respectively, by both Theorem 7.1.9 and Lemma 8.1.20, we conclude that, for any given

$$r \in \left(0, \min\left\{p, \frac{n}{M_\infty(\omega) + n/p}\right\}\right)$$

and for any $f \in L^1_{\mathrm{loc}}(\mathbb{R}^n)$,

$$\|\mathcal{M}(f)\|_{[W\mathcal{K}_\omega^{p,q}(\mathbb{R}^n)]^{1/r}} \lesssim \|f\|_{[W\mathcal{K}_\omega^{p,q}(\mathbb{R}^n)]^{1/r}}.$$

This then finishes the proof of Lemma 8.3.10. $\qquad\square$

We next turn to show the maximal function characterizations Theorem 8.3.8

Proof of Theorem 8.3.8 Let all the symbols be as in the present theorem. Then, applying Lemma 4.1.4 and repeating an argument similar to that used in the proof of Theorem 6.1.9 via replacing $\max\{M_0(\omega), M_\infty(\omega)\}$, Theorem 1.2.44, and Lemma 6.1.11 therein, respectively, by $M_\infty(\omega)$, Theorem 7.1.9, and Lemma 8.3.10, we find that both (i) and (ii) of the present theorem hold true. This finishes the proof of Theorem 8.3.8. $\qquad\qquad\qquad\qquad\qquad\qquad\qquad\qquad\qquad\qquad\qquad\qquad\qquad\quad$ $\square$

8.3.2 Relations with Inhomogeneous Weak Generalized Herz Spaces

In this subsection, we investigate the relations between inhomogeneous weak generalized Herz spaces and associated Hardy spaces. We first study the relation between $W\mathcal{K}^{p,q}_{\omega,\mathbf{0}}(\mathbb{R}^n)$ and $WH\mathcal{K}^{p,q}_{\omega,\mathbf{0}}(\mathbb{R}^n)$. Indeed, the following conclusion shows that, under some reasonable and sharp assumptions on the exponents, then

$$WH\mathcal{K}^{p,q}_{\omega,\mathbf{0}}(\mathbb{R}^n) = W\mathcal{K}^{p,q}_{\omega,\mathbf{0}}(\mathbb{R}^n)$$

in the sense of equivalent quasi-norms.

Theorem 8.3.11 *Let* $p \in (1, \infty)$, $q \in (0, \infty)$, *and* $\omega \in M(\mathbb{R}_+)$ *satisfy*

$$-\frac{n}{p} < m_\infty(\omega) \le M_\infty(\omega) < \frac{n}{p'},$$

where $\frac{1}{p} + \frac{1}{p'} = 1$. *Then*

(i) $W\mathcal{K}^{p,q}_{\omega,\mathbf{0}}(\mathbb{R}^n) \hookrightarrow \mathcal{S}'(\mathbb{R}^n)$.

(ii) *If* $f \in W\mathcal{K}^{p,q}_{\omega,\mathbf{0}}(\mathbb{R}^n)$, *then* $f \in WH\mathcal{K}^{p,q}_{\omega,\mathbf{0}}(\mathbb{R}^n)$ *and there exists a positive constant* C, *independent of* f, *such that*

$$\|f\|_{WH\mathcal{K}^{p,q}_{\omega,\mathbf{0}}(\mathbb{R}^n)} \le C \|f\|_{W\mathcal{K}^{p,q}_{\omega,\mathbf{0}}(\mathbb{R}^n)}.$$

(iii) *If* $f \in WH\mathcal{K}^{p,q}_{\omega,\mathbf{0}}(\mathbb{R}^n)$, *then there exists a locally integrable function* g *belonging to* $W\mathcal{K}^{p,q}_{\omega,\mathbf{0}}(\mathbb{R}^n)$ *such that* g *represents* f, *which means that* $f = g$ *in* $\mathcal{S}'(\mathbb{R}^n)$,

$$\|f\|_{WH\mathcal{K}^{p,q}_{\omega,\mathbf{0}}(\mathbb{R}^n)} = \|g\|_{WH\mathcal{K}^{p,q}_{\omega,\mathbf{0}}(\mathbb{R}^n)},$$

and there exists a positive constant C, *independent of* f, *such that*

$$\|g\|_{W\mathcal{K}^{p,q}_{\omega,\mathbf{0}}(\mathbb{R}^n)} \le C \|f\|_{WH\mathcal{K}^{p,q}_{\omega,\mathbf{0}}(\mathbb{R}^n)}.$$

Proof Let all the symbols be as in the present theorem. Then, from Lemma 4.2.6 and an argument similar to that used in the proof of Theorem 6.2.1 with

$$\min\{m_0(\omega), m_\infty(\omega)\}, \quad \max\{M_0(\omega), M_\infty(\omega)\},$$

Theorem 1.2.42, and Lemma 6.1.6 therein replaced, respectively, by $m_\infty(\omega)$, $M_\infty(\omega)$, Theorem 7.1.7, and Lemma 8.3.7, we deduce that (i) through (iii) of the present theorem hold true. This then finishes the proof of Theorem 8.3.11. □

Similarly, we can show that $W H \mathcal{K}_\omega^{p,q}(\mathbb{R}^n) = W \mathcal{K}_\omega^{p,q}(\mathbb{R}^n)$ with equivalent quasi-norms as follows.

Theorem 8.3.12 *Let $p \in (1, \infty)$, $q \in (0, \infty)$, and $\omega \in M(\mathbb{R}_+)$ with*

$$-\frac{n}{p} < m_\infty(\omega) \le M_\infty(\omega) < 0.$$

Then

(i) $W \mathcal{K}_\omega^{p,q}(\mathbb{R}^n) \hookrightarrow \mathcal{S}'(\mathbb{R}^n)$.

(ii) *If $f \in W \mathcal{K}_\omega^{p,q}(\mathbb{R}^n)$, then $f \in W H \mathcal{K}_\omega^{p,q}(\mathbb{R}^n)$ and there exists a positive constant C, independent of f, such that*

$$\|f\|_{W H \mathcal{K}_\omega^{p,q}(\mathbb{R}^n)} \le C \|f\|_{W \mathcal{K}_\omega^{p,q}(\mathbb{R}^n)}.$$

(iii) *If $f \in W H \mathcal{K}_\omega^{p,q}(\mathbb{R}^n)$, then there exists a locally integrable function g belonging to $W \mathcal{K}_\omega^{p,q}(\mathbb{R}^n)$ such that g represents f, which means that $f = g$ in $\mathcal{S}'(\mathbb{R}^n)$,*

$$\|f\|_{W H \mathcal{K}_\omega^{p,q}(\mathbb{R}^n)} = \|g\|_{W H \mathcal{K}_\omega^{p,q}(\mathbb{R}^n)},$$

and there exists a positive constant C, independent of f, such that

$$\|g\|_{W \mathcal{K}_\omega^{p,q}(\mathbb{R}^n)} \le C \|f\|_{W H \mathcal{K}_\omega^{p,q}(\mathbb{R}^n)}.$$

Proof Let all the symbols be as in the present theorem. Then, applying Lemma 4.2.6 and repeating the proof of Theorem 6.2.3 via replacing $\max\{M_0(\omega), M_\infty(\omega)\}$, Theorem 1.2.44, and Lemma 6.1.11 therein, respectively, by $M_\infty(\omega)$, Theorem 7.1.9, and Lemma 8.3.7, we conclude that (i), (ii), and (iii) of the present theorem hold true. This further finishes the proof of Theorem 8.3.12. □

8.3.3 Atomic Characterizations

The main target of this subsection is to characterize both the inhomogeneous weak generalized Herz–Hardy spaces $WH\mathcal{K}^{p,q}_{\omega,\mathbf{0}}(\mathbb{R}^n)$ and $WH\mathcal{K}^{p,q}_{\omega}(\mathbb{R}^n)$ by means of atoms. For this purpose, we first introduce the following inhomogeneous weak generalized atomic Herz–Hardy space $WH\mathcal{K}^{p,q,r,d}_{\omega,\mathbf{0}}(\mathbb{R}^n)$ via $(\mathcal{K}^{p,q}_{\omega,\mathbf{0}}(\mathbb{R}^n), r, d)$-atoms introduced in Definition 8.1.11.

Definition 8.3.13 Let $p, q \in (0, \infty)$, $\omega \in M(\mathbb{R}_+)$ with $m_\infty(\omega) \in (-\frac{n}{p}, \infty)$,

$$p_- \in \left(0, \frac{\min\{p, q, \frac{n}{M_\infty(\omega)+n/p}\}}{\max\{1, p, q\}}\right),$$

$d \geq \lfloor n(1/p_- - 1)\rfloor$ be a fixed integer, and

$$r \in \left(\max\left\{1, p, \frac{n}{m_\infty(\omega) + n/p}\right\}, \infty\right].$$

The *inhomogeneous weak generalized atomic Herz–Hardy space* $WH\mathcal{K}^{p,q,r,d}_{\omega,\mathbf{0}}(\mathbb{R}^n)$, associated with $W\mathcal{K}^{p,q}_{\omega,\mathbf{0}}(\mathbb{R}^n)$, is defined to be the set of all the $f \in \mathcal{S}'(\mathbb{R}^n)$ such that there exists a sequence $\{a_{i,j}\}_{i\in\mathbb{Z},\,j\in\mathbb{N}}$ of $(\mathcal{K}^{p,q}_{\omega,\mathbf{0}}(\mathbb{R}^n), r, d)$-atoms supported, respectively, in the balls $\{B_{i,j}\}_{i\in\mathbb{Z},\,j\in\mathbb{N}} \subset \mathbb{B}$ and three positive constants $c \in (0, 1]$, A, and $\widetilde{A}$, independent of f, satisfying that, for any $i \in \mathbb{Z}$,

$$\sum_{j\in\mathbb{N}} \mathbf{1}_{cB_{i,j}} \leq A,$$

$$f = \sum_{i\in\mathbb{Z}} \sum_{j\in\mathbb{N}} \widetilde{A}2^i \left\|\mathbf{1}_{B_{i,j}}\right\|_{\mathcal{K}^{p,q}_{\omega,\mathbf{0}}(\mathbb{R}^n)} a_{i,j}$$

in $\mathcal{S}'(\mathbb{R}^n)$, and

$$\sup_{i\in\mathbb{Z}} \left\{2^i \left\|\sum_{j\in\mathbb{N}} \mathbf{1}_{B_{i,j}}\right\|_{\mathcal{K}^{p,q}_{\omega,\mathbf{0}}(\mathbb{R}^n)}\right\} < \infty.$$

Moreover, for any $f \in WH\mathcal{K}^{p,q,r,d}_{\omega,\mathbf{0}}(\mathbb{R}^n)$,

$$\|f\|_{WH\mathcal{K}^{p,q,r,d}_{\omega,\mathbf{0}}(\mathbb{R}^n)} := \inf\left\{\sup_{i\in\mathbb{Z}} 2^i \left\|\sum_{j\in\mathbb{N}} \mathbf{1}_{B_{i,j}}\right\|_{\mathcal{K}^{p,q}_{\omega,\mathbf{0}}(\mathbb{R}^n)}\right\},$$

where the infimum is taken over all the decompositions of f as above.

Then we have the following atomic characterization of the inhomogeneous weak generalized Herz–Hardy space $WH\mathcal{K}_{\omega,\mathbf{0}}^{p,q}(\mathbb{R}^n)$.

Theorem 8.3.14 *Let p, q, ω, r, and d be as in Definition 8.3.13. Then*

$$WH\mathcal{K}_{\omega,\mathbf{0}}^{p,q}(\mathbb{R}^n) = WH\mathcal{K}_{\omega,\mathbf{0}}^{p,q,r,d}(\mathbb{R}^n)$$

with equivalent quasi-norms.

Proof Let all the symbols be as in the present theorem. Then, from Lemmas 6.3.5 and 6.3.6 and an argument similar to that used in the proof of Theorem 6.3.2 with $\max\{M_0(\omega), M_\infty(\omega)\}$, Theorem 1.2.42, and Lemmas 1.3.1, 1.8.5, 1.8.6, 4.3.10, and 6.1.6 therein replaced, respectively, by $M_\infty(\omega)$, Theorem 7.1.7, and Lemmas 7.1.13, 7.1.36, 7.1.37, 8.1.14, and 8.3.7, it follows that

$$WH\mathcal{K}_{\omega,\mathbf{0}}^{p,q}(\mathbb{R}^n) = WH\mathcal{K}_{\omega,\mathbf{0}}^{p,q,r,d}(\mathbb{R}^n)$$

with equivalent quasi-norms. This then finishes the proof of Theorem 8.3.14. □

Now, we turn to show the atomic characterization of the inhomogeneous weak generalized Herz–Hardy space $WH\mathcal{K}_{\omega}^{p,q}(\mathbb{R}^n)$. For this purpose, we first introduce the following weak atomic Hardy space $WH\mathcal{K}_{\omega}^{p,q,r,d}(\mathbb{R}^n)$ via $(\mathcal{K}_{\omega}^{p,q}(\mathbb{R}^n),\, r,\, d)$-atoms given in Definition 8.1.16.

Definition 8.3.15 Let $p, q \in (0, \infty)$, $\omega \in M(\mathbb{R}_+)$ with

$$-\frac{n}{p} < m_\infty(\omega) \le M_\infty(\omega) < 0,$$

$p_- \in (0, \min\{p, q\}/\max\{1, p, q\})$, $d \ge \lfloor n(1/p_- - 1)\rfloor$ be a fixed integer, and

$$r \in \left(\max\left\{ 1, \frac{n}{m_\infty(\omega) + n/p} \right\}, \infty \right].$$

The *inhomogeneous weak generalized atomic Herz–Hardy space* $WH\mathcal{K}_{\omega}^{p,q,r,d}(\mathbb{R}^n)$, associated with $W\mathcal{K}_{\omega}^{p,q}(\mathbb{R}^n)$, is defined to be the set of all the $f \in \mathcal{S}'(\mathbb{R}^n)$ such that there exists a sequence $\{a_{i,j}\}_{i\in\mathbb{Z},\, j\in\mathbb{N}}$ of $(\mathcal{K}_{\omega}^{p,q}(\mathbb{R}^n),\, r,\, d)$-atoms supported, respectively, in the balls $\{B_{i,j}\}_{i\in\mathbb{Z},\, j\in\mathbb{N}} \subset \mathbb{B}$ and three positive constants $c \in (0, 1]$, A, and $\widetilde{A}$, independent of f, satisfying that, for any $i \in \mathbb{Z}$,

$$\sum_{j\in\mathbb{N}} \mathbf{1}_{cB_{i,j}} \le A,$$

$$f = \sum_{i\in\mathbb{Z}} \sum_{j\in\mathbb{N}} \widetilde{A}2^i \left\| \mathbf{1}_{B_{i,j}} \right\|_{\mathcal{K}_{\omega}^{p,q}(\mathbb{R}^n)} a_{i,j}$$

612 8 Hardy Spaces Associated with Inhomogeneous Generalized Herz Spaces

in $\mathcal{S}'(\mathbb{R}^n)$, and

$$\sup_{i\in\mathbb{Z}}\left\{2^i\left\|\sum_{j\in\mathbb{N}}\mathbf{1}_{B_{i,j}}\right\|_{\mathcal{K}_\omega^{p,q}(\mathbb{R}^n)}\right\}<\infty,$$

Moreover, for any $f\in WH\mathcal{K}_\omega^{p,q,r,d}(\mathbb{R}^n)$,

$$\|f\|_{WH\mathcal{K}_\omega^{p,q,r,d}(\mathbb{R}^n)}:=\inf\left\{\sup_{i\in\mathbb{Z}}2^i\left\|\sum_{j\in\mathbb{N}}\mathbf{1}_{B_{i,j}}\right\|_{\mathcal{K}_\omega^{p,q}(\mathbb{R}^n)}\right\},$$

where the infimum is taken over all the decompositions of f as above.

Then we have the following atomic characterization of $WH\mathcal{K}_\omega^{p,q}(\mathbb{R}^n)$.

Theorem 8.3.16 *Let p, q, ω, r, and d be as in Definition 8.3.15. Then*

$$WH\mathcal{K}_\omega^{p,q}(\mathbb{R}^n) = WH\mathcal{K}_\omega^{p,q,r,d}(\mathbb{R}^n)$$

with equivalent quasi-norms.

Proof Let all the symbols be as in the present theorem. Then, using Propositions 6.3.11 and 6.3.15 and repeating an argument similar to that used in the proof of Theorem 6.3.10 via replacing $\min\{m_0(\omega), m_\infty(\omega)\}$, $\max\{M_0(\omega), M_\infty(\omega)\}$, Theorem 1.2.44, Lemmas 1.3.2, 2.2.3, 2.3.2, 4.1.10, 4.3.23, 4.3.27, and 6.1.11 therein, respectively, by $m_\infty(\omega)$, $M_\infty(\omega)$, Theorem 7.1.9, Lemmas 7.1.14, 7.2.6, 7.2.11, 8.1.7, 8.1.19, 8.1.23, and 8.3.10, we find that

$$WH\mathcal{K}_\omega^{p,q}(\mathbb{R}^n) = WH\mathcal{K}_\omega^{p,q,r,d}(\mathbb{R}^n)$$

with equivalent quasi-norms. This finishes the proof of Theorem 8.3.16. □

8.3.4 Molecular Characterizations

In this subsection, we establish the molecular characterizations of inhomogeneous weak generalized Herz–Hardy spaces. To this end, via $(\mathcal{K}_{\omega,\mathbf{0}}^{p,q}(\mathbb{R}^n), r, d, \tau)$-molecules introduced in Definition 8.1.28, we first show the following molecular characterization of the inhomogeneous weak generalized Herz–Hardy space $WH\mathcal{K}_{\omega,\mathbf{0}}^{p,q}(\mathbb{R}^n)$.

Theorem 8.3.17 *Let* $p, q \in (0, \infty)$, $\omega \in M(\mathbb{R}_+)$ *satisfy* $m_\infty(\omega) \in (-\frac{n}{p}, \infty)$,

$$p_- \in \left(0, \frac{\min\{p, q, \frac{n}{M_\infty(\omega)+n/p}\}}{\max\{1, p, q\}}\right),$$

$d \geq \lfloor n(1/p_- - 1) \rfloor$ *be a fixed integer,*

$$r \in \left(\max\left\{1, p, \frac{n}{m_\infty(\omega) + n/p}\right\}, \infty\right],$$

and $\tau \in (n(\frac{1}{p_-} - \frac{1}{r}), \infty)$. *Then* f *belongs to the inhomogeneous weak generalized Herz–Hardy space* $WH\mathcal{K}^{p,q}_{\omega,\mathbf{0}}(\mathbb{R}^n)$ *if and only if* $f \in \mathcal{S}'(\mathbb{R}^n)$ *and there exists a sequence* $\{m_{i,j}\}_{i\in\mathbb{Z}, j\in\mathbb{N}}$ *of* $(\mathcal{K}^{p,q}_{\omega,\mathbf{0}}(\mathbb{R}^n)$, r, d, $\tau)$-*molecules centered, respectively, at the balls* $\{B_{i,j}\}_{i\in\mathbb{Z}, j\in\mathbb{N}} \subset \mathbb{B}$ *and three positive constants* $c \in (0, 1]$, A, *and* $\widetilde{A}$, *independent of* f, *such that, for any* $i \in \mathbb{Z}$,

$$\sum_{j\in\mathbb{N}} \mathbf{1}_{cB_{i,j}} \leq A,$$

$$f = \sum_{i\in\mathbb{Z}} \sum_{j\in\mathbb{N}} \widetilde{A} 2^i \left\| \mathbf{1}_{B_{i,j}} \right\|_{\mathcal{K}^{p,q}_{\omega,\mathbf{0}}(\mathbb{R}^n)} m_{i,j}$$

in $\mathcal{S}'(\mathbb{R}^n)$, *and*

$$\sup_{i\in\mathbb{Z}} \left\{ 2^i \left\| \sum_{j\in\mathbb{N}} \mathbf{1}_{B_{i,j}} \right\|_{\mathcal{K}^{p,q}_{\omega,\mathbf{0}}(\mathbb{R}^n)} \right\} < \infty.$$

Moreover, there exist two positive constants C_1 *and* C_2 *such that, for any* $f \in WH\mathcal{K}^{p,q}_{\omega,\mathbf{0}}(\mathbb{R}^n)$,

$$C_1 \|f\|_{WH\mathcal{K}^{p,q}_{\omega,\mathbf{0}}(\mathbb{R}^n)} \leq \inf\left\{ \sup_{i\in\mathbb{Z}} 2^i \left\| \sum_{j\in\mathbb{N}} \mathbf{1}_{B_{i,j}} \right\|_{\mathcal{K}^{p,q}_{\omega,\mathbf{0}}(\mathbb{R}^n)} \right\}$$

$$\leq C_2 \|f\|_{WH\mathcal{K}^{p,q}_{\omega,\mathbf{0}}(\mathbb{R}^n)},$$

where the infimum is taken over all the decompositions of f *as above.*

Proof Let all the symbols be as in the present theorem and $f \in \mathcal{S}'(\mathbb{R}^n)$. Then, from Lemma 6.4.2 and an argument similar to that used in the proof of Theorem 6.4.1 via replacing $\min\{m_0(\omega), m_\infty(\omega)\}$, Definition 6.3.1, Theorems 1.2.42 and 6.3.2, and Lemmas 1.8.6 and 4.3.10 therein, respectively, by $m_\infty(\omega)$, Definition

8.3.13, Theorems 7.1.7 and 8.3.14, and Lemmas 7.1.37 and 8.1.14, it follows that $f \in WH\mathcal{K}_{\omega,0}^{p,q}(\mathbb{R}^n)$ if and only if there exists a sequence $\{m_{i,j}\}_{i\in\mathbb{Z}, j\in\mathbb{N}}$ of $(\mathcal{K}_{\omega,0}^{p,q}(\mathbb{R}^n),\ r,\ d,\ \tau)$-molecules centered, respectively, at the balls $\{B_{i,j}\}_{i\in\mathbb{Z}, j\in\mathbb{N}} \subset \mathbb{B}$ and three positive constants $c \in (0, 1]$, A, and $\widetilde{A}$, independent of f, such that, for any $i \in \mathbb{Z}$,

$$\sum_{j\in\mathbb{N}} \mathbf{1}_{cB_{i,j}} \leq A,$$

$$f = \sum_{i\in\mathbb{Z}} \sum_{j\in\mathbb{N}} \widetilde{A} 2^i \left\| \mathbf{1}_{B_{i,j}} \right\|_{\mathcal{K}_{\omega,0}^{p,q}(\mathbb{R}^n)} m_{i,j}$$

in $\mathcal{S}'(\mathbb{R}^n)$, and

$$\sup_{i\in\mathbb{Z}} \left\{ 2^i \left\| \sum_{j\in\mathbb{N}} \mathbf{1}_{B_{i,j}} \right\|_{\mathcal{K}_{\omega,0}^{p,q}(\mathbb{R}^n)} \right\} < \infty.$$

Moreover, it also holds true that

$$\|f\|_{WH\mathcal{K}_{\omega,0}^{p,q}(\mathbb{R}^n)} \sim \inf \left\{ \sup_{i\in\mathbb{Z}} 2^i \left\| \sum_{j\in\mathbb{N}} \mathbf{1}_{B_{i,j}} \right\|_{\mathcal{K}_{\omega,0}^{p,q}(\mathbb{R}^n)} \right\}$$

with the positive equivalence constants independent of f, where the infimum is taken over all the decompositions of f as in the present theorem. This finishes the proof of Theorem 8.1.28. □

In addition, recall that the concept of $(\mathcal{K}_\omega^{p,q}(\mathbb{R}^n),\ r,\ d,\ \tau)$-molecules is given in Definition 8.1.30. Then, via these molecules, the following conclusion shows the molecular characterization of the inhomogeneous weak generalized Herz–Hardy space $WH\mathcal{K}_\omega^{p,q}(\mathbb{R}^n)$.

Theorem 8.3.18 Let $p, q \in (0, \infty)$, $\omega \in M(\mathbb{R}_+)$ satisfy

$$-\frac{n}{p} < m_\infty(\omega) \leq M_\infty(\omega) < 0,$$

$p_- \in (0, \min\{p, q\}/\max\{1, p, q\})$, $d \geq \lfloor n(1/p_- - 1) \rfloor$ be a fixed integer,

$$r \in \left(\max\left\{ 1, \frac{n}{m_\infty(\omega) + n/p} \right\}, \infty \right],$$

and $\tau \in (n(\frac{1}{p_-} - \frac{1}{r}), \infty)$. Then $f \in WHK_{\omega}^{p,q}(\mathbb{R}^n)$ if and only if $f \in \mathcal{S}'(\mathbb{R}^n)$ and there exists a sequence $\{m_{i,j}\}_{i \in \mathbb{Z}, j \in \mathbb{N}}$ of $(\mathcal{K}_{\omega}^{p,q}(\mathbb{R}^n), r, d, \tau)$-molecules centered, respectively, at the balls $\{B_{i,j}\}_{i \in \mathbb{Z}, j \in \mathbb{N}} \subset \mathbb{B}$ and three positive constants $c \in (0, 1]$, A, and $\tilde{A}$, independent of f, such that, for any $i \in \mathbb{Z}$,

$$\sum_{j \in \mathbb{N}} \mathbf{1}_{cB_{i,j}} \leq A,$$

$$f = \sum_{i \in \mathbb{Z}} \sum_{j \in \mathbb{N}} \tilde{A} 2^i \left\| \mathbf{1}_{B_{i,j}} \right\|_{\mathcal{K}_{\omega}^{p,q}(\mathbb{R}^n)} m_{i,j}$$

in $\mathcal{S}'(\mathbb{R}^n)$, and

$$\sup_{i \in \mathbb{Z}} \left\{ 2^i \left\| \sum_{j \in \mathbb{N}} \mathbf{1}_{B_{i,j}} \right\|_{\mathcal{K}_{\omega}^{p,q}(\mathbb{R}^n)} \right\} < \infty.$$

Moreover, there exist two positive constants C_1 and C_2 such that, for any $f \in WHK_{\omega}^{p,q}(\mathbb{R}^n)$,

$$C_1 \|f\|_{WHK_{\omega}^{p,q}(\mathbb{R}^n)} \leq \inf \left\{ \sup_{i \in \mathbb{Z}} 2^i \left\| \sum_{j \in \mathbb{N}} \mathbf{1}_{B_{i,j}} \right\|_{\mathcal{K}_{\omega}^{p,q}(\mathbb{R}^n)} \right\}$$

$$\leq C_2 \|f\|_{WHK_{\omega}^{p,q}(\mathbb{R}^n)},$$

where the infimum is taken over all the decompositions of f as above.

Proof Let all the symbols be as in the present theorem and $f \in \mathcal{S}'(\mathbb{R}^n)$. Then, using Proposition 6.4.5 and repeating an argument similar to that used in the proof of Theorem 6.4.4 via replacing $\min\{m_0(\omega), m_\infty(\omega)\}$, Definition 6.3.10, Theorems 1.2.44, 1.2.48, and 6.3.10, and Lemmas 4.3.23, 4.3.24, and 4.3.27 therein, respectively, by $m_\infty(\omega)$, Definition 8.3.16, Theorems 7.1.9, 7.1.12, and 8.3.16, and Lemmas 8.1.19, 8.1.20, and 8.1.23, we conclude that $f \in WHK_{\omega}^{p,q}(\mathbb{R}^n)$ if and only if there exists a sequence $\{m_{i,j}\}_{i \in \mathbb{Z}, j \in \mathbb{N}}$ of $(\mathcal{K}_{\omega}^{p,q}(\mathbb{R}^n), r, d, \tau)$-molecules centered, respectively, at the balls $\{B_{i,j}\}_{i \in \mathbb{Z}, j \in \mathbb{N}} \subset \mathbb{B}$ and three positive constants $c \in (0, 1]$, A, and $\tilde{A}$, independent of f, such that, for any $i \in \mathbb{Z}$,

$$\sum_{j \in \mathbb{N}} \mathbf{1}_{cB_{i,j}} \leq A,$$

$$f = \sum_{i \in \mathbb{Z}} \sum_{j \in \mathbb{N}} \tilde{A} 2^i \left\| \mathbf{1}_{B_{i,j}} \right\|_{\mathcal{K}_{\omega}^{p,q}(\mathbb{R}^n)} m_{i,j}$$

in $\mathcal{S}'(\mathbb{R}^n)$, and

$$\sup_{i\in\mathbb{Z}}\left\{2^i\left\|\sum_{j\in\mathbb{N}}\mathbf{1}_{B_{i,j}}\right\|_{\mathcal{K}_\omega^{p,q}(\mathbb{R}^n)}\right\}<\infty.$$

Moreover, we also obtain

$$\|f\|_{WH\mathcal{K}_\omega^{p,q}(\mathbb{R}^n)}\sim\inf\left\{\sup_{i\in\mathbb{Z}}2^i\left\|\sum_{j\in\mathbb{N}}\mathbf{1}_{B_{i,j}}\right\|_{\mathcal{K}_\omega^{p,q}(\mathbb{R}^n)}\right\}$$

with the positive equivalence constants independent of f, where the infimum is taken over all the decompositions of f as in the present theorem. This then finishes the proof of Theorem 8.3.18. □

8.3.5 Littlewood–Paley Function Characterizations

This subsection is devoted to characterizing both the inhomogeneous weak generalized Herz–Hardy spaces $WH\mathcal{K}_{\omega,\mathbf{0}}^{p,q}(\mathbb{R}^n)$ and $WH\mathcal{K}_\omega^{p,q}(\mathbb{R}^n)$ via the Lusin area function, the g-function, and the g_λ^*-function defined as in Definitions 4.6.1 and 4.6.2. First, we give the following Littlewood–Paley function characterization of $WH\mathcal{K}_{\omega,\mathbf{0}}^{p,q}(\mathbb{R}^n)$.

Theorem 8.3.19 Let $p,q\in(0,\infty)$, $\omega\in M(\mathbb{R}_+)$ with $m_\infty(\omega)\in(-\frac{n}{p},\infty)$,

$$p_-:=\min\left\{p,q,\frac{n}{M_\infty(\omega)+n/p}\right\},$$

$$p_+:=\max\left\{p,\frac{n}{m_\infty(\omega)+n/p}\right\},$$

and

$$\lambda\in\left(\max\left\{\frac{2}{\min\{1,p_-\}},1-\frac{2}{\max\{1,p_+\}}+\frac{2}{\min\{1,p_-\}}\right\},\infty\right).$$

Assume that, for any $f\in\mathcal{S}'(\mathbb{R}^n)$, $S(f)$ and $g_\lambda^*(f)$ are as in Definition 4.6.1, and $g(f)$ is as in Definition 4.6.2. Then the following four statements are mutually equivalent:

(i) $f\in WH\mathcal{K}_{\omega,\mathbf{0}}^{p,q}(\mathbb{R}^n)$;

(ii) $f\in\mathcal{S}'(\mathbb{R}^n)$, f vanishes weakly at infinity, and $S(f)\in W\mathcal{K}_{\omega,\mathbf{0}}^{p,q}(\mathbb{R}^n)$;

(iii) $f \in \mathcal{S}'(\mathbb{R}^n)$, f vanishes weakly at infinity, and $g(f) \in W\mathcal{K}_{\omega,0}^{p,q}(\mathbb{R}^n)$;

(iv) $f \in \mathcal{S}'(\mathbb{R}^n)$, f vanishes weakly at infinity, and $g_\lambda^*(f) \in W\mathcal{K}_{\omega,0}^{p,q}(\mathbb{R}^n)$.

Moreover, for any $f \in WH\mathcal{K}_{\omega,0}^{p,q}(\mathbb{R}^n)$,

$$\|f\|_{WH\mathcal{K}_{\omega,0}^{p,q}(\mathbb{R}^n)} \sim \|S(f)\|_{W\mathcal{K}_{\omega,0}^{p,q}(\mathbb{R}^n)} \sim \|g(f)\|_{W\mathcal{K}_{\omega,0}^{p,q}(\mathbb{R}^n)}$$

$$\sim \left\|g_\lambda^*(f)\right\|_{W\mathcal{K}_{\omega,0}^{p,q}(\mathbb{R}^n)},$$

where the positive equivalence constants are independent of f.

To prove this theorem, we need the following auxiliary lemma related to the Fefferman–Stein vector-valued inequality on inhomogeneous weak local generalized Herz spaces.

Lemma 8.3.20 *Let* $p, q \in (0, \infty)$ *and* $\omega \in M(\mathbb{R}_+)$ *satisfy* $m_\infty(\omega) \in (-\frac{n}{p}, \infty)$. *Then, for any given*

$$r \in \left(0, \min\left\{p, \frac{n}{M_\infty(\omega) + n/p}\right\}\right)$$

and $u \in (1, \infty)$, *there exists a positive constant* C *such that, for any* $\{f_j\}_{j\in\mathbb{N}} \subset L_{\mathrm{loc}}^1(\mathbb{R}^n)$,

$$\left\|\left\{\sum_{j\in\mathbb{N}} [\mathcal{M}(f_j)]^u\right\}^{\frac{1}{u}}\right\|_{[W\mathcal{K}_{\omega,0}^{p,q}(\mathbb{R}^n)]^{1/r}} \leq C \left\|\left\{\sum_{j\in\mathbb{N}} |f_j|^u\right\}^{\frac{1}{u}}\right\|_{[W\mathcal{K}_{\omega,0}^{p,q}(\mathbb{R}^n)]^{1/r}}.$$

Proof Let all the symbols be as in the present lemma, $r \in (0, \min\{p, \frac{n}{M_\infty(\omega)+n/p}\})$, and $u \in (1, \infty)$. Then, using Lemma 6.1.7 and repeating the proof of Lemma 6.5.4 with $\max\{M_0(\omega), M_\infty(\omega)\}$, Theorem 1.2.42, and Lemma 4.3.10 therein replaced, respectively, by $M_\infty(\omega)$, Theorem 7.1.7, and Lemma 8.1.14, we find that, for any $\{f_j\}_{j\in\mathbb{N}} \subset L_{\mathrm{loc}}^1(\mathbb{R}^n)$,

$$\left\|\left\{\sum_{j\in\mathbb{N}} [\mathcal{M}(f_j)]^u\right\}^{\frac{1}{u}}\right\|_{[W\mathcal{K}_{\omega,0}^{p,q}(\mathbb{R}^n)]^{1/r}} \lesssim \left\|\left\{\sum_{j\in\mathbb{N}} |f_j|^u\right\}^{\frac{1}{u}}\right\|_{[W\mathcal{K}_{\omega,0}^{p,q}(\mathbb{R}^n)]^{1/r}}.$$

This then finishes the proof of Lemma 8.3.20. □

Via Lemma 8.3.20, we next prove Theorem 8.3.19.

Proof of Theorem 8.3.19 Let all the symbols be as in the present theorem. Then, repeating an argument similar to that used in the proof of Theorem 6.5.1 with Theorem 1.2.42 and Lemmas 1.3.1, 1.8.6, 4.3.10, and 6.5.4 therein replaced, respectively, by Theorem 7.1.7 and Lemmas 7.1.13, 7.1.37, 8.1.14, and 8.3.20, we infer that (i) through (iv) of the present theorem are mutually equivalent. Moreover, we also find that, for any $f \in WH\mathcal{K}_{\omega,0}^{p,q}(\mathbb{R}^n)$,

$$\|f\|_{WH\mathcal{K}_{\omega,0}^{p,q}(\mathbb{R}^n)} \sim \|S(f)\|_{W\mathcal{K}_{\omega,0}^{p,q}(\mathbb{R}^n)} \sim \|g(f)\|_{W\mathcal{K}_{\omega,0}^{p,q}(\mathbb{R}^n)}$$
$$\sim \left\|g_\lambda^*(f)\right\|_{W\mathcal{K}_{\omega,0}^{p,q}(\mathbb{R}^n)},$$

which then completes the proof of Theorem 8.3.19.   □

Now, we turn to establish the following Littlewood–Paley function characterization of the inhomogeneous weak generalized Herz–Hardy space $WH\mathcal{K}_\omega^{p,q}(\mathbb{R}^n)$.

Theorem 8.3.21 *Let $p, q \in (0, \infty)$, $\omega \in M(\mathbb{R}_+)$ satisfy*

$$-\frac{n}{p} < m_\infty(\omega) \leq M_\infty(\omega) < 0,$$

$p_- := \min\{p, q\}$, $p_+ := \frac{n}{m_\infty(\omega)+n/p}$, and

$$\lambda \in \left(\max\left\{\frac{2}{\min\{1, p_-\}}, 1 - \frac{2}{\max\{1, p_+\}} + \frac{2}{\min\{1, p_-\}}\right\}, \infty\right).$$

Assume that, for any $f \in \mathcal{S}'(\mathbb{R}^n)$, $S(f)$ and $g_\lambda^*(f)$ are as in Definition 4.6.1, and $g(f)$ is as in Definition 4.6.2. Then the following four statements are mutually equivalent:

(i) $f \in WH\mathcal{K}_\omega^{p,q}(\mathbb{R}^n)$;
(ii) $f \in \mathcal{S}'(\mathbb{R}^n)$, f vanishes weakly at infinity, and $S(f) \in W\mathcal{K}_\omega^{p,q}(\mathbb{R}^n)$;
(iii) $f \in \mathcal{S}'(\mathbb{R}^n)$, f vanishes weakly at infinity, and $g(f) \in W\mathcal{K}_\omega^{p,q}(\mathbb{R}^n)$;
(iv) $f \in \mathcal{S}'(\mathbb{R}^n)$, f vanishes weakly at infinity, and $g_\lambda^*(f) \in W\mathcal{K}_\omega^{p,q}(\mathbb{R}^n)$.

Moreover, for any $f \in WH\mathcal{K}_\omega^{p,q}(\mathbb{R}^n)$,

$$\|f\|_{WH\mathcal{K}_\omega^{p,q}(\mathbb{R}^n)} \sim \|S(f)\|_{W\mathcal{K}_\omega^{p,q}(\mathbb{R}^n)} \sim \|g(f)\|_{W\mathcal{K}_\omega^{p,q}(\mathbb{R}^n)}$$
$$\sim \left\|g_\lambda^*(f)\right\|_{W\mathcal{K}_\omega^{p,q}(\mathbb{R}^n)},$$

where the positive equivalence constants are independent of f.

To show the above theorem, we first establish the following representation formula of the quasi-norm $\|\cdot\|_{W\mathcal{K}_\omega^{p,q}(\mathbb{R}^n)}$.

Lemma 8.3.22 *Let $p, q \in (0, \infty)$ and $\omega \in M(\mathbb{R}_+)$. Then, for any $f \in \mathscr{M}(\mathbb{R}^n)$,*

$$\|f\|_{W\mathcal{K}_\omega^{p,q}(\mathbb{R}^n)} = \sup_{\xi \in \mathbb{R}^n} \left\|\tau_\xi(f)\right\|_{W\mathcal{K}_{\omega,0}^{p,q}(\mathbb{R}^n)},$$

where, for any $\xi \in \mathbb{R}^n$, the operator τ_ξ is defined as in (4.69).

Proof Let all the symbols be as in the present lemma and $f \in \mathscr{M}(\mathbb{R}^n)$. Then, from Definition 8.3.1(ii), Remark 7.1.2(i), and Definition 8.3.1(i), it follows that

$$\|f\|_{W\mathcal{K}_{\omega}^{p,q}(\mathbb{R}^n)} = \sup_{\lambda \in (0,\infty)} \left[\lambda \left\| \mathbf{1}_{\{y \in \mathbb{R}^n: |f(y)| > \lambda\}} \right\|_{\mathcal{K}_{\omega}^{p,q}(\mathbb{R}^n)} \right]$$

$$= \sup_{\lambda \in (0,\infty)} \left\{ \lambda \sup_{\xi \in \mathbb{R}^n} \left[\left\| \mathbf{1}_{\{y \in \mathbb{R}^n: |f(y)| > \lambda\}} (\cdot - \xi) \right\|_{\mathcal{K}_{\omega,0}^{p,q}(\mathbb{R}^n)} \right] \right\}$$

$$= \sup_{\lambda \in (0,\infty), \, \xi \in \mathbb{R}^n} \left[\lambda \left\| \mathbf{1}_{\{y \in \mathbb{R}^n: |\tau_\xi(f)(y)| > \lambda\}} \right\|_{\mathcal{K}_{\omega,0}^{p,q}(\mathbb{R}^n)} \right]$$

$$= \sup_{\xi \in \mathbb{R}^n} \left\{ \sup_{\lambda \in (0,\infty)} \left[\lambda \left\| \mathbf{1}_{\{y \in \mathbb{R}^n: |\tau_\xi(f)(y)| > \lambda\}} \right\|_{\mathcal{K}_{\omega,0}^{p,q}(\mathbb{R}^n)} \right] \right\}$$

$$= \sup_{\xi \in \mathbb{R}^n} \left\| \tau_\xi(f) \right\|_{\mathcal{K}_{\omega,0}^{p,q}(\mathbb{R}^n)}.$$

This then finishes the proof of Lemma 8.3.22. □

We next give the proof of Theorem 8.3.21 as follows.

Proof of Theorem 8.3.21 Let all the symbols be as in the present theorem. Then, repeating the proof of Theorem 6.5.6 via replacing Theorems 6.1.1(ii), 6.1.9(ii), and 6.5.1, and Lemma 6.5.7 therein, respectively, by Theorems 8.3.5(ii), 8.3.8(ii), and 8.3.19, and Lemma 8.3.22, we infer that (i), (ii), (iii), and (iv) are mutually equivalent and, for any $f \in WH\mathcal{K}_{\omega}^{p,q}(\mathbb{R}^n)$,

$$\|f\|_{WH\mathcal{K}_{\omega}^{p,q}(\mathbb{R}^n)} \sim \|S(f)\|_{W\mathcal{K}_{\omega}^{p,q}(\mathbb{R}^n)} \sim \|g(f)\|_{W\mathcal{K}_{\omega}^{p,q}(\mathbb{R}^n)}$$

$$\sim \left\| g_\lambda^*(f) \right\|_{W\mathcal{K}_{\omega}^{p,q}(\mathbb{R}^n)}.$$

This further finishes the proof of Theorem 8.3.21. □

8.3.6 Boundedness of Calderón–Zygmund Operators

In this subsection, we establish the boundedness of Calderón–Zygmund operators from inhomogeneous generalized Herz–Hardy spaces to inhomogeneous weak generalized Herz–Hardy spaces even in the critical case. Let $d \in \mathbb{Z}_+$. Recall that the standard kernel K and the d-order Calderón–Zygmund operator T with kernel K are defined, respectively, in Definitions 1.5.7 and 1.5.8. To begin with, we show the following boundedness of the operator T from $H\mathcal{K}_{\omega,0}^{p,q}(\mathbb{R}^n)$ to $WH\mathcal{K}_{\omega,0}^{p,q}(\mathbb{R}^n)$.

Theorem 8.3.23 *Let* $d \in \mathbb{Z}_+$, $\delta \in (0, 1)$, $p \in [\frac{n}{n+d+\delta}, \infty)$, $q \in (0, \infty)$, K *be a d-order standard kernel as in Definition 1.5.7, T a d-order Calderón–Zygmund operator with kernel K having the vanishing moments up to order d, and $\omega \in M(\mathbb{R}_+)$ satisfy*

$$-\frac{n}{p} < m_\infty(\omega) \le M_\infty(\omega) < n - \frac{n}{p} + d + \delta.$$

Then T has a unique extension on $H\mathcal{K}^{p,q}_{\omega,0}(\mathbb{R}^n)$ and there exists a positive constant C such that, for any $f \in H\mathcal{K}^{p,q}_{\omega,0}(\mathbb{R}^n)$,

$$\|T(f)\|_{WH\mathcal{K}^{p,q}_{\omega,0}(\mathbb{R}^n)} \le C \|f\|_{H\mathcal{K}^{p,q}_{\omega,0}(\mathbb{R}^n)}.$$

To prove this theorem, we first show the following technical estimate about the Hardy–Littlewood maximal operator.

Lemma 8.3.24 *Let $r \in [1, \infty)$ and $\mathcal{M}$ be the Hardy–Littlewood maximal operator defined as in (1.54). Then there exists a positive constant C such that, for any $\{f_j\}_{j\in\mathbb{N}} \subset \mathcal{M}(\mathbb{R}^n)$ and $x \notin \bigcup_{j\in\mathbb{N}} \mathrm{supp}\,(f_j)$,*

$$\left\{ \sum_{j\in\mathbb{N}} [\mathcal{M}(f_j)(x)]^r \right\}^{\frac{1}{r}} \le C \int_{\mathbb{R}^n} \frac{[\sum_{j\in\mathbb{N}} |f_j(y)|^r]^{\frac{1}{r}}}{|x-y|^n} \, dy,$$

where, for any $j \in \mathbb{N}$, $\mathrm{supp}\,(f_j) := \{x \in \mathbb{R}^n : f_j(x) \ne 0\}$.

Proof Let all the symbols be as in the present lemma, and let $\{f_j\}_{j\in\mathbb{N}} \subset \mathcal{M}(\mathbb{R}^n)$. Then, by the fact that the Hardy–Littlewood maximal operator $\mathcal{M}$ satisfies the size condition (1.91) (see, for instance, [197, Remark 4.4]) and the Minkowski integral inequality, we conclude that, for any $x \notin \bigcup_{j\in\mathbb{N}} \mathrm{supp}\,(f_j)$,

$$\left\{ \sum_{j\in\mathbb{N}} [\mathcal{M}(f_j)(x)]^r \right\}^{\frac{1}{r}} \lesssim \left\{ \sum_{j\in\mathbb{N}} \left[\int_{\mathbb{R}^n} \frac{|f_j(y)|}{|x-y|^n} \, dy \right]^r \right\}^{\frac{1}{r}}$$

$$\lesssim \int_{\mathbb{R}^n} \left[\sum_{j\in\mathbb{N}} \frac{|f_j(y)|^r}{|x-y|^{nr}} \right]^{\frac{1}{r}} dy$$

$$\sim \int_{\mathbb{R}^n} \frac{[\sum_{j\in\mathbb{N}} |f_j(y)|^r]^{\frac{1}{r}}}{|x-y|^n} \, dy.$$

This then finishes the proof of Lemma 7.15. $\square$

Via this lemma, we now establish the following weak type vector-valued inequalities on inhomogeneous generalized Herz spaces, which paly essential roles in the proof of the boundedness of Calderón–Zygmund operators from inhomogeneous generalized Herz–Hardy spaces to inhomogeneous weak generalized Herz–Hardy spaces.

Proposition 8.3.25 *Let $p \in [1, \infty)$, $q \in (0, \infty)$, $r \in (1, \infty)$, and $\omega \in M(\mathbb{R}_+)$ with*

$$-\frac{n}{p} < m_\infty(\omega) \leq M_\infty(\omega) < \frac{n}{p'},$$

where $\frac{1}{p} + \frac{1}{p'} = 1$. Then there exists a positive constant C such that, for any $\{f_j\}_{j \in \mathbb{N}} \subset L^1_{\mathrm{loc}}(\mathbb{R}^n)$,

$$\left\| \left\{ \sum_{j \in \mathbb{N}} [\mathcal{M}(f_j)]^r \right\}^{\frac{1}{r}} \right\|_{W\mathcal{K}^{p,q}_{\omega,0}(\mathbb{R}^n)} \leq C \left\| \left(\sum_{j \in \mathbb{N}} |f_j|^r \right)^{\frac{1}{r}} \right\|_{\mathcal{K}^{p,q}_{\omega,0}(\mathbb{R}^n)} \tag{8.14}$$

and

$$\left\| \left\{ \sum_{j \in \mathbb{N}} [\mathcal{M}(f_j)]^r \right\}^{\frac{1}{r}} \right\|_{W\mathcal{K}^{p,q}_{\omega}(\mathbb{R}^n)} \leq C \left\| \left(\sum_{j \in \mathbb{N}} |f_j|^r \right)^{\frac{1}{r}} \right\|_{\mathcal{K}^{p,q}_{\omega}(\mathbb{R}^n)}. \tag{8.15}$$

Proof Let all the symbols be as in the present proposition, and let $\{f_j\}_{j \in \mathbb{N}} \subset \mathcal{M}(\mathbb{R}^n)$. We next show the present proposition by considering the following two cases on p.

Case (1) $p \in (1, \infty)$. In this case, from Definition 8.3.1 and Theorems 7.1.28 and 7.1.30, it follows that

$$\left\| \left\{ \sum_{j \in \mathbb{N}} [\mathcal{M}(f_j)]^r \right\}^{\frac{1}{r}} \right\|_{W\mathcal{K}^{p,q}_{\omega,0}(\mathbb{R}^n)}$$

$$\leq \left\| \left\{ \sum_{j \in \mathbb{N}} [\mathcal{M}(f_j)]^r \right\}^{\frac{1}{r}} \right\|_{\mathcal{K}^{p,q}_{\omega,0}(\mathbb{R}^n)} \gtrsim \left\| \left(\sum_{j \in \mathbb{N}} |f_j|^r \right)^{\frac{1}{r}} \right\|_{\mathcal{K}^{p,q}_{\omega,0}(\mathbb{R}^n)}$$

and

$$\left\| \left\{ \sum_{j\in\mathbb{N}} [\mathcal{M}(f_j)]^r \right\}^{\frac{1}{r}} \right\|_{W\mathcal{K}_\omega^{p,q}(\mathbb{R}^n)}$$
$$\leq \left\| \left\{ \sum_{j\in\mathbb{N}} [\mathcal{M}(f_j)]^r \right\}^{\frac{1}{r}} \right\|_{\mathcal{K}_\omega^{p,q}(\mathbb{R}^n)} \lesssim \left\| \left(\sum_{j\in\mathbb{N}} |f_j|^r \right)^{\frac{1}{r}} \right\|_{\mathcal{K}_\omega^{p,q}(\mathbb{R}^n)}.$$

These finish the proof of the present proposition in this case.

Case (2) $p = 1$. In this case, fix a $\xi \in \mathbb{R}^n$. Then, using Definition 7.1.1(i), we find that, for any $\lambda \in (0, \infty)$,

$$\left\| \mathbf{1}_{\{y\in\mathbb{R}^n:\ \{\sum_{j\in\mathbb{N}}[\mathcal{M}(f_j)(y)]^r\}^{\frac{1}{r}}>\lambda\}}(\cdot+\xi) \right\|_{\mathcal{K}_{\omega,0}^{1,q}(\mathbb{R}^n)}$$

$$\lesssim \left\| \mathbf{1}_{\{y\in\mathbb{R}^n:\ \{\sum_{j\in\mathbb{N}}[\mathcal{M}(f_j)(y)]^r\}^{\frac{1}{r}}>\lambda\}} \mathbf{1}_{B(\xi,1)} \right\|_{L^1(\mathbb{R}^n)}$$

$$+ \left\{ \sum_{k\in\mathbb{N}} \left[\omega(2^k) \right]^q \right.$$

$$\times \left. \left\| \mathbf{1}_{\{y\in\mathbb{R}^n:\ \{\sum_{j\in\mathbb{N}}[\mathcal{M}(f_j)(y)]^r\}^{\frac{1}{r}}>\lambda\}} \mathbf{1}_{B(\xi,2^k)\setminus B(\xi,2^{k-1})} \right\|_{L^1(\mathbb{R}^n)}^q \right\}^{\frac{1}{q}}$$

$$=: A_\lambda + B_\lambda. \tag{8.16}$$

Now, we deal with A_λ and B_λ, respectively. Indeed, for any $\lambda \in (0, \infty)$, applying the fact that

$$\mathbf{1}_{\{y\in\mathbb{R}^n:\ \{\sum_{j\in\mathbb{N}}[\mathcal{M}(f_j\mathbf{1}_{[B(\xi,2)]^\complement})(y)]^r\}^{\frac{1}{r}}>\lambda\}} \leq \lambda^{-1} \left\{ \sum_{j\in\mathbb{N}} \left[\mathcal{M}\left(f_j\mathbf{1}_{[B(\xi,2)]^\complement} \right) \right]^r \right\}^{\frac{1}{r}}$$

and Lemma 6.6.10 with $\{f_j\}_{j\in\mathbb{N}}$ therein replaced by $\{f_j\mathbf{1}_{B(\xi,2)}\}_{j\in\mathbb{N}}$, we conclude that

$$A_\lambda \lesssim \left\| \mathbf{1}_{\{y\in\mathbb{R}^n:\ \{\sum_{j\in\mathbb{N}}[\mathcal{M}(f_j\mathbf{1}_{B(\xi,2)})(y)]^r\}^{\frac{1}{r}}>\lambda\}} \mathbf{1}_{B(\xi,1)} \right\|_{L^1(\mathbb{R}^n)}$$

$$+ \left\| \mathbf{1}_{\{y\in\mathbb{R}^n:\ \{\sum_{j\in\mathbb{N}}[\mathcal{M}(f_j\mathbf{1}_{[B(\xi,2)]^\complement})(y)]^r\}^{\frac{1}{r}}>\lambda\}} \mathbf{1}_{B(\xi,1)} \right\|_{L^1(\mathbb{R}^n)}$$

$$\lesssim \left\| \mathbf{1}_{\{y \in \mathbb{R}^n : \{\sum_{j \in \mathbb{N}} [\mathcal{M}(f_j \mathbf{1}_{B(\xi,2)})(y)]^r\}^{\frac{1}{r}} > \lambda\}} \right\|_{L^1(\mathbb{R}^n)}$$

$$+ \lambda^{-1} \left\| \left\{ \sum_{j \in \mathbb{N}} \left[\mathcal{M}\left(f_j \mathbf{1}_{[B(\xi,2)]^{\complement}}\right) \right]^r \right\}^{\frac{1}{r}} \right\|_{L^1(\mathbb{R}^n)}$$

$$\lesssim \lambda^{-1} \left[\left\| \left(\sum_{j \in \mathbb{N}} |f_j|^r \right)^{\frac{1}{r}} \mathbf{1}_{B(\xi,2)} \right\|_{L^1(\mathbb{R}^n)} \right.$$

$$\left. + \left\| \left\{ \sum_{j \in \mathbb{N}} \left[\mathcal{M}\left(f_j \mathbf{1}_{[B(\xi,2)]^{\complement}}\right) \right]^r \right\}^{\frac{1}{r}} \right\|_{L^1(\mathbb{R}^n)} \right]. \tag{8.17}$$

In addition, from Lemma 8.3.24 with $\{f_j\}_{j \in \mathbb{N}}$ therein replaced by $\{f_j \mathbf{1}_{[B(\xi,2)]^{\complement}}\}_{j \in \mathbb{N}}$, we infer that, for any $x \in B(\xi, 1)$,

$$\left\{ \sum_{j \in \mathbb{N}} \left[\mathcal{M}\left(f_j \mathbf{1}_{[B(\xi,2)]^{\complement}}\right)(x) \right]^r \right\}^{\frac{1}{r}} \lesssim \int_{|y-\xi| \geq 2} \frac{[\sum_{j \in \mathbb{N}} |f_j(y)|^r]^{\frac{1}{r}}}{|x - y|^n} \, dy. \tag{8.18}$$

Therefore, using (8.17) and repeating an argument similar to that used in the estimations of (7.20), (7.21), and (7.22) with (7.15), $\mathbf{0}$, $|f|$, and $|T(f)|$ therein replaced, respectively, by (8.18), ξ, $(\sum_{j \in \mathbb{N}} |f_j|^r)^{\frac{1}{r}}$, and $\{\sum_{j \in \mathbb{N}} [\mathcal{M}(f_j)]^r\}^{\frac{1}{r}}$, we find that, for any $\lambda \in (0, \infty)$,

$$\mathrm{A}_\lambda \lesssim \lambda^{-1} \left\| \left(\sum_{j \in \mathbb{N}} |f_j|^r \right)^{\frac{1}{r}} (\cdot + \xi) \right\|_{\mathcal{K}^{1,q}_{\omega,0}(\mathbb{R}^n)}, \tag{8.19}$$

which is the desired estimate of A_λ.

Next, we deal with B_λ. For this purpose, for any $k \in \mathbb{N}$, let

$$f_{j,k,0} := f_j \mathbf{1}_{B(\xi,2^{-2})},$$

$$f_{j,k,1} := f_j \mathbf{1}_{B(\xi,2^{k-2}) \setminus B(\xi,2^{-2})},$$

$$f_{j,k,2} := f_j \mathbf{1}_{B(\xi,2^{k+1}) \setminus B(\xi,2^{k-2})},$$

and

$$f_{j,k,3} := f_j \mathbf{1}_{[B(\xi,2^{k+1})]^{\complement}}.$$

Then, for any $\lambda \in (0,\infty)$, we have

$$\begin{aligned}
B_\lambda &\lesssim \left\{ \sum_{k\in\mathbb{N}} \left[\omega(2^k)\right]^q \right. \\
&\qquad \times \left. \left\| \mathbf{1}_{\{y\in\mathbb{R}^n:\, \{\sum_{j\in\mathbb{N}}[\mathcal{M}(f_{j,k,0})(y)]^r\}^{\frac{1}{r}} > \lambda\}} \mathbf{1}_{B(\xi,2^k)\setminus B(\xi,2^{k-1})} \right\|_{L^1(\mathbb{R}^n)}^q \right\}^{\frac{1}{q}} \\
&\quad + \left\{ \sum_{k\in\mathbb{N}} \left[\omega(2^k)\right]^q \right. \\
&\qquad \times \left. \left\| \mathbf{1}_{\{y\in\mathbb{R}^n:\, \{\sum_{j\in\mathbb{N}}[\mathcal{M}(f_{j,k,1})(y)]^r\}^{\frac{1}{r}} > \lambda\}} \mathbf{1}_{B(\xi,2^k)\setminus B(\xi,2^{k-1})} \right\|_{L^1(\mathbb{R}^n)}^q \right\}^{\frac{1}{q}} \\
&\quad + \left\{ \sum_{k\in\mathbb{N}} \left[\omega(2^k)\right]^q \right. \\
&\qquad \times \left. \left\| \mathbf{1}_{\{y\in\mathbb{R}^n:\, \{\sum_{j\in\mathbb{N}}[\mathcal{M}(f_{j,k,2})(y)]^r\}^{\frac{1}{r}} > \lambda\}} \mathbf{1}_{B(\xi,2^k)\setminus B(\xi,2^{k-1})} \right\|_{L^1(\mathbb{R}^n)}^q \right\}^{\frac{1}{q}} \\
&\quad + \left\{ \sum_{k\in\mathbb{N}} \left[\omega(2^k)\right]^q \right. \\
&\qquad \times \left. \left\| \mathbf{1}_{\{y\in\mathbb{R}^n:\, \{\sum_{j\in\mathbb{N}}[\mathcal{M}(f_{j,k,3})(y)]^r\}^{\frac{1}{r}} > \lambda\}} \mathbf{1}_{B(\xi,2^k)\setminus B(\xi,2^{k-1})} \right\|_{L^1(\mathbb{R}^n)}^q \right\}^{\frac{1}{q}} \\
&=: \widetilde{\mathrm{II}}_{\lambda,0} + \widetilde{\mathrm{II}}_{\lambda,1} + \widetilde{\mathrm{II}}_{\lambda,2} + \widetilde{\mathrm{II}}_{\lambda,3}.
\end{aligned} \tag{8.20}$$

Using Lemma 8.3.24 with $\{f_j\}_{j\in\mathbb{N}}$ therein replaced by $\{f_{j,k,0}\}_{j\in\mathbb{N}}$, we conclude that, for any $x \in \mathbb{R}^n$ with $|x| \geq 1$,

$$\left\{ \sum_{j\in\mathbb{N}} \left[\mathcal{M}\left(f_{j,k,0}\right)(x)\right]^r \right\}^{\frac{1}{r}} \lesssim \int_{|y-\xi|<2^{-2}} \frac{[\sum_{j\in\mathbb{N}}|f_j(y)|^r]^{\frac{1}{r}}}{|x-y|^n}\,dy. \tag{8.21}$$

Thus, for any $\lambda \in (0, \infty)$, from the fact that

$$
\mathbf{1}_{\{y \in \mathbb{R}^n : \; \{\sum_{j \in \mathbb{N}} [\mathcal{M}(f_{j,k,0})(y)]^r\}^{\frac{1}{r}} > \lambda\}} \leq \lambda^{-1} \left\{ \sum_{j \in \mathbb{N}} \left[\mathcal{M}\left(f_{j,k,0}\right) \right]^r \right\}^{\frac{1}{r}}
$$

and an argument similar to that used in the estimation of $\widetilde{J}_0$ in the proof of Theorem 7.1.20 with (7.15), $\mathbf{0}$, $|f|$, and $|T(f)|$ therein replaced, respectively, by (8.21), ξ, $(\sum_{j \in \mathbb{N}} |f_j|^r)^{\frac{1}{r}}$, and $\{\sum_{j \in \mathbb{N}} [\mathcal{M}(f_j)]^r\}^{\frac{1}{r}}$, we obtain

$$
\widetilde{\mathrm{II}}_{\lambda,0} \lesssim \lambda^{-1} \left\{ \sum_{j \in \mathbb{N}} \left[\omega(2^k) \right]^q \left\| \left\{ \sum_{j \in \mathbb{N}} \left[\mathcal{M}(f_j) \right]^r \right\}^{\frac{1}{r}} \mathbf{1}_{B(\xi,2^k) \setminus B(\xi,2^{k-1})} \right\|_{L^1(\mathbb{R}^n)}^q \right\}^{\frac{1}{q}}
$$

$$
\lesssim \lambda^{-1} \left\| \left(\sum_{j \in \mathbb{N}} |f_j|^r \right)^{\frac{1}{r}} (\cdot + \xi) \right\|_{\mathcal{K}_{\omega,\mathbf{0}}^{1,q}(\mathbb{R}^n)}, \tag{8.22}
$$

which completes the estimation of $\widetilde{\mathrm{II}}_{\lambda,0}$. On the other hand, repeating the estimations of $\widetilde{\mathrm{II}}_{\lambda,1}$, $\widetilde{\mathrm{II}}_{\lambda,2}$, and $\widetilde{\mathrm{II}}_{\lambda,3}$ in the proof of Proposition 6.6.11 via replacing $\min\{m_0(\omega), m_\infty(\omega)\}$, $\max\{M_0(\omega), M_\infty(\omega)\}$, and $\mathbb{Z}$ therein, respectively, by $m_\infty(\omega)$, $M_\infty(\omega)$, and $\mathbb{N}$, we find that, for any $\lambda \in (0, \infty)$,

$$
\widetilde{\mathrm{II}}_{\lambda,1} \lesssim \lambda^{-1} \left\| \left(\sum_{j \in \mathbb{N}} |f_j|^r \right)^{\frac{1}{r}} (\cdot + \xi) \right\|_{\mathcal{K}_{\omega,\mathbf{0}}^{1,q}(\mathbb{R}^n)},
$$

$$
\widetilde{\mathrm{II}}_{\lambda,2} \lesssim \lambda^{-1} \left\| \left(\sum_{j \in \mathbb{N}} |f_j|^r \right)^{\frac{1}{r}} (\cdot + \xi) \right\|_{\mathcal{K}_{\omega,\mathbf{0}}^{1,q}(\mathbb{R}^n)},
$$

and

$$
\widetilde{\mathrm{II}}_{\lambda,3} \lesssim \lambda^{-1} \left\| \left(\sum_{j \in \mathbb{N}} |f_j|^r \right)^{\frac{1}{r}} (\cdot + \xi) \right\|_{\mathcal{K}_{\omega,\mathbf{0}}^{1,q}(\mathbb{R}^n)}.
$$

These, together with both (8.22) and (8.20), further imply that, for any $\lambda \in (0, \infty)$,

$$B_\lambda \lesssim \lambda^{-1} \left\| \left(\sum_{j \in \mathbb{N}} |f_j|^r \right)^{\frac{1}{r}} (\cdot + \xi) \right\|_{\mathcal{K}_{\omega,0}^{1,q}(\mathbb{R}^n)},$$

which is the desired estimate of B_λ. From this, (8.19), and (8.16), we deduce that, for any $\lambda \in (0, \infty)$,

$$\lambda \left\| \mathbf{1}_{\{y \in \mathbb{R}^n: \; \{\sum_{j \in \mathbb{N}} [\mathcal{M}(f_j)(y)]^r\}^{\frac{1}{r}} > \lambda\}} (\cdot + \xi) \right\|_{\mathcal{K}_{\omega,0}^{1,q}(\mathbb{R}^n)}$$

$$\lesssim \left\| \left(\sum_{j \in \mathbb{N}} |f_j|^r \right)^{\frac{1}{r}} (\cdot + \xi) \right\|_{\mathcal{K}_{\omega,0}^{1,q}(\mathbb{R}^n)}.$$

Combining this and Definition 8.3.1(i), we conclude that

$$\left\| \left\{ \sum_{j \in \mathbb{N}} [\mathcal{M}(f_j)]^r \right\}^{\frac{1}{r}} (\cdot + \xi) \right\|_{W\mathcal{K}_{\omega,0}^{1,q}(\mathbb{R}^n)}$$

$$\lesssim \left\| \left(\sum_{j \in \mathbb{N}} |f_j|^r \right)^{\frac{1}{r}} (\cdot + \xi) \right\|_{\mathcal{K}_{\omega,0}^{1,q}(\mathbb{R}^n)}. \tag{8.23}$$

Therefore, by both Definition 8.3.1(ii) and the arbitrariness of ξ, we have

$$\left\| \left\{ \sum_{j \in \mathbb{N}} [\mathcal{M}(f_j)]^r \right\}^{\frac{1}{r}} \right\|_{W\mathcal{K}_\omega^{1,q}(\mathbb{R}^n)} \lesssim \left\| \left(\sum_{j \in \mathbb{N}} |f_j|^r \right)^{\frac{1}{r}} \right\|_{\mathcal{K}_\omega^{1,q}(\mathbb{R}^n)}.$$

In particular, letting $\xi := \mathbf{0}$ in (8.23), we obtain

$$\left\| \left\{ \sum_{j \in \mathbb{N}} [\mathcal{M}(f_j)]^r \right\}^{\frac{1}{r}} \right\|_{W\mathcal{K}_{\omega,0}^{1,q}(\mathbb{R}^n)} \lesssim \left\| \left(\sum_{j \in \mathbb{N}} |f_j|^r \right)^{\frac{1}{r}} \right\|_{\mathcal{K}_{\omega,0}^{1,q}(\mathbb{R}^n)}.$$

This further implies that both (8.14) and (8.15) hold true in this case, and hence finishes the proof of Proposition 8.3.25. $\quad\square$

Moreover, in order to show Theorem 8.3.23, we also require the following convexification lemma of inhomogeneous weak local generalized Herz spaces, which can be immediately concluded by Lemma 7.1.13 and [278, Remark 2.14]; we omit the details.

Lemma 8.3.26 *Let* $p, q, s \in (0, \infty)$ *and* $\omega \in M(\mathbb{R}_+)$. *Then*

$$\left[W\mathcal{K}_{\omega,\mathbf{0}}^{p,q}(\mathbb{R}^n) \right]^{1/s} = W\mathcal{K}_{\omega^s,\mathbf{0}}^{p/s,q/s}(\mathbb{R}^n)$$

with the same quasi-norms.

Using both Proposition 8.3.25 and Lemma 8.3.26, we now prove Theorem 8.3.23.

Proof of Theorem 8.3.23 Let all the symbols be as in the present theorem. Then, from Proposition 6.6.3 and an argument similar to that used in the proof of Theorem 6.6.9 via replacing (6.48), Theorems 1.2.42 and 1.4.1, and Lemmas 1.3.1, 1.8.6, 4.3.11, 6.1.6, and 6.6.12 therein, respectively, by (8.14), Theorems 7.1.7 and 7.1.17, and Lemmas 7.1.13, 7.1.37, 8.1.15, 8.3.7, and 8.3.26, we infer that T has a unique extension on $H\mathcal{K}_{\omega,\mathbf{0}}^{p,q}(\mathbb{R}^n)$ and, for any $f \in H\mathcal{K}_{\omega,\mathbf{0}}^{p,q}(\mathbb{R}^n)$,

$$\|T(f)\|_{WH\mathcal{K}_{\omega,\mathbf{0}}^{p,q}(\mathbb{R}^n)} \lesssim \|f\|_{H\mathcal{K}_{\omega,\mathbf{0}}^{p,q}(\mathbb{R}^n)}.$$

This then finishes the proof of Theorem 8.3.23. □

Next, we are devoted to establishing the boundedness of Calderón–Zygmund operators from the inhomogeneous generalized Herz–Hardy space $H\mathcal{K}_{\omega}^{p,q}(\mathbb{R}^n)$ to the inhomogeneous weak generalized Herz–Hardy space $WH\mathcal{K}_{\omega}^{p,q}(\mathbb{R}^n)$ as follows.

Theorem 8.3.27 *Let* $d \in \mathbb{Z}_+$, $\delta \in (0, 1)$, $p \in [\frac{n}{n+d+\delta}, \infty)$, $q \in (0, \infty)$, K *be a* d-*order standard kernel as in Definition 1.5.7*, T *a* d-*order Calderón–Zygmund operator with kernel* K *having the vanishing moments up to order* d, *and* $\omega \in M(\mathbb{R}_+)$ *satisfy*

$$-\frac{n}{p} < m_\infty(\omega) \le M_\infty(\omega) < 0.$$

Then T *is well defined on* $H\mathcal{K}_{\omega}^{p,q}(\mathbb{R}^n)$ *and there exists a positive constant* C *such that, for any* $f \in H\mathcal{K}_{\omega}^{p,q}(\mathbb{R}^n)$,

$$\|T(f)\|_{WH\mathcal{K}_{\omega}^{p,q}(\mathbb{R}^n)} \le C\|f\|_{H\mathcal{K}_{\omega}^{p,q}(\mathbb{R}^n)}.$$

To show this theorem, we need a preliminary lemma. Indeed, applying both Lemma 7.1.14 and [278, Remark 2.14], we immediately obtain the following convexification lemma of inhomogeneous weak global generalized Herz spaces, which plays a key role in the proof of Theorem 8.3.27; we omit the details.

Lemma 8.3.28 *Let* $p, q, s \in (0, \infty)$ *and* $\omega \in M(\mathbb{R}_+)$. *Then*

$$\left[W\mathcal{K}_\omega^{p,q}(\mathbb{R}^n) \right]^{1/s} = W\mathcal{K}_{\omega^s}^{p/s,q/s}(\mathbb{R}^n)$$

with the same quasi-norms.

With the help of this lemma, we now turn to prove Theorem 8.3.27.

Proof of Theorem 8.3.27 Let all the symbols be as in the present theorem. Then, using Theorem 6.6.1 and repeating an argument similar to that used in the proof of Theorem 6.6.14 via replacing $\min\{m_0(\omega), m_\infty(\omega)\}$, $\max\{M_0(\omega), M_\infty(\omega)\}$, (6.49), Theorem 1.2.44, and Lemmas 1.3.2, 2.2.3, 2.3.2, 4.3.23, 4.3.25, 6.1.11, and 6.6.15 therein, respectively, by $m_\infty(\omega)$, $M_\infty(\omega)$, (8.15), Theorem 7.1.9, and Lemmas 7.1.14, 7.2.6, 7.2.11, 8.1.19, 8.1.21, 8.3.10, and 8.3.28, we conclude that T is well defined on $H\mathcal{K}_\omega^{p,q}(\mathbb{R}^n)$ and, for any $f \in H\mathcal{K}_\omega^{p,q}(\mathbb{R}^n)$,

$$\|T(f)\|_{W H\mathcal{K}_\omega^{p,q}(\mathbb{R}^n)} \lesssim \|f\|_{H\mathcal{K}_\omega^{p,q}(\mathbb{R}^n)},$$

which completes the proof of Theorem 8.3.27. $\qquad\square$

8.3.7 Real Interpolations

In this subsection, we establish a relation between inhomogeneous generalized Herz–Hardy spaces and inhomogeneous weak generalized Herz–Hardy spaces via the real interpolation. To be precise, we show that the real interpolation spaces between inhomogeneous generalized Herz–Hardy spaces and the Lebesgue space $L^\infty(\mathbb{R}^n)$ are just the inhomogeneous weak generalized Herz–Hardy spaces new introduced in the present section. To begin with, recall that the real interpolation space between two quasi-Banach spaces is defined as in Definition 6.7.1.

We first show the following interpolation theorem about the inhomogeneous generalized Herz–Hardy space $H\mathcal{K}_{\omega,0}^{p,q}(\mathbb{R}^n)$ and the Lebesgue space $L^\infty(\mathbb{R}^n)$.

Theorem 8.3.29 *Let* $p, q \in (0, \infty)$ *and* $\omega \in M(\mathbb{R}_+)$ *satisfy* $m_\infty(\omega) \in (-\frac{n}{p}, \infty)$. *Then, for any* $\theta \in (0, 1)$,

$$\left(H\mathcal{K}_{\omega,0}^{p,q}(\mathbb{R}^n), L^\infty(\mathbb{R}^n) \right)_{\theta,\infty} = W H\mathcal{K}_{\omega^{1-\theta},0}^{p/(1-\theta),q/(1-\theta)}(\mathbb{R}^n).$$

Proof Let all the symbols be as in the present theorem and $\theta \in (0, 1)$. Then, repeating the proof of Theorem 6.7.2 with $\min\{m_0(\omega), m_\infty(\omega)\}$, $\max\{M_0(\omega), M_\infty(\omega)\}$, Theorem 1.2.42, and Lemmas 1.3.1, 1.8.6, 4.3.10, and 6.1.6 therein replaced, respectively, by $m_\infty(\omega)$, $M_\infty(\omega)$, Theorem 7.1.7, and Lemmas 7.1.13, 7.1.37,

8.1.14, and 8.3.7, we find that

$$\left(H\mathcal{K}_{\omega,0}^{p,q}(\mathbb{R}^n),\ L^\infty(\mathbb{R}^n)\right)_{\theta,\infty} = WH\mathcal{K}_{\omega^{1-\theta},0}^{p/(1-\theta),q/(1-\theta)}(\mathbb{R}^n).$$

This finishes the proof of Theorem 8.3.29. □

Finally, we establish the following interpolation theorem which shows that the real interpolation space between $H\mathcal{K}_\omega^{p,q}(\mathbb{R}^n)$ and $L^\infty(\mathbb{R}^n)$ is just the inhomogeneous weak generalized Herz–Hardy space $WH\mathcal{K}_{\omega^{1-\theta}}^{p/(1-\theta),q/(1-\theta)}(\mathbb{R}^n)$.

Theorem 8.3.30 *Let $p, q \in (0, \infty)$ and $\omega \in M(\mathbb{R}_+)$ satisfy*

$$-\frac{n}{p} < m_\infty(\omega) \le M_\infty(\omega) < 0.$$

Then, for any $\theta \in (0, 1)$,

$$\left(H\mathcal{K}_\omega^{p,q}(\mathbb{R}^n),\ L^\infty(\mathbb{R}^n)\right)_{\theta,\infty} = WH\mathcal{K}_{\omega^{1-\theta}}^{p/(1-\theta),q/(1-\theta)}(\mathbb{R}^n).$$

Proof Let all the symbols be as in the present theorem. Then, from Proposition 6.7.7 and an argument similar to that used in the proof Theorem 6.7.6 via replacing $\min\{m_0(\omega), m_\infty(\omega)\}$, $\max\{M_0(\omega), M_\infty(\omega)\}$, Theorems 1.2.44 and 1.2.48, and Lemmas 1.3.2, 2.2.3, 2.3.2, 4.3.23, 4.3.24, 4.3.27, and 6.1.11 therein, respectively, by $m_\infty(\omega)$, $M_\infty(\omega)$, Theorems 7.1.9 and 7.1.12, and Lemmas 7.1.14, 7.2.6, 7.2.11, 8.1.19, 8.1.20, 8.1.23, and 8.3.10, we infer that, for any $\theta \in (0, 1)$,

$$\left(H\mathcal{K}_\omega^{p,q}(\mathbb{R}^n),\ L^\infty(\mathbb{R}^n)\right)_{\theta,\infty} = WH\mathcal{K}_{\omega^{1-\theta}}^{p/(1-\theta),q/(1-\theta)}(\mathbb{R}^n).$$

This then finishes the proof of Theorem 8.3.30. □

Bibliography

1. A. Almeida and D. Drihem, Maximal, potential and singular type operators on Herz spaces with variable exponents, J. Math. Anal. Appl. 394 (2012), 781–795.
2. A. Almeida and S. Samko, Embeddings of local generalized Morrey spaces between weighted Lebesgue spaces, Nonlinear Anal. 164 (2017), 67–76.
3. A. Almeida and S. Samko, Approximation in Morrey spaces, J. Funct. Anal. 272 (2017), 2392–2411.
4. J. Alvarez and M. Milman, H^p continuity properties of Calderón–Zygmund-type operators, J. Math. Anal. Appl. 118 (1986), 63–79.
5. K. F. Andersen and R. T. John, Weighted inequalities for vector-valued maximal functions and singular integrals, Studia Math. 69 (1980), 19–31.
6. T. Aoki, Locally bounded linear topological spaces, Proc. Imp. Acad. Tokyo 18 (1942), 588–594.
7. R. Arai and E. Nakai, Commutators of Calderón–Zygmund and generalized fractional integral operators on generalized Morrey spaces, Rev. Mat. Complut. 31 (2018), 287–331.
8. R. Arai and E. Nakai, An extension of the characterization of CMO and its application to compact commutators on Morrey spaces, J. Math. Soc. Japan 72 (2020), 507–539.
9. A. II. Baernstein and E. T. Sawyer, Embedding and multiplier theorems for $H^p(\mathbb{R}^n)$, Mem. Amer. Math. Soc. 53 (1985), No. 318, iv+82 pp.
10. A. Benedek and R. Panzone, The space L^p, with mixed norm, Duke Math. J. 28 (1961), 301–324.
11. C. Bennett and R. Sharpley, Interpolation of Operators, Pure and Applied Mathematics 129, Academic Press, Inc., Boston, MA, 1988, xiv+469 pp.
12. J. Bergh and J. Löfström, Interpolation Spaces, An Introduction, Grundlehren der Mathematischen Wissenschaften 223, Springer-Verlag, Berlin–New York, 1976, x+207 pp.
13. A. Beurling, Construction and analysis of some convolution algebras, Ann. Inst. Fourier (Grenoble) 14 (1964), 1–32.
14. O. Blasco, A. Ruiz and L. Vega, Non-interpolation in Morrey–Campanato and block spaces, Ann. Scuola Norm. Sup. Pisa Cl. Sci. (4) 28 (1999), 31–40.
15. M. Bownik, Anisotropic Hardy Spaces and Wavelets, Mem. Amer. Math. Soc. 164 (2003), no. 781, vi+122 pp.
16. M. Bownik and L.-A. D. Wang, Fourier transform of anisotropic Hardy spaces, Proc. Amer. Math. Soc. 141 (2013), 2299–2308.
17. H. Brezis, Functional Analysis, Sobolev Spaces and Partial Differential Equations, Universitext, Springer, New York, 2011, xiv+599 pp.

© The Author(s), under exclusive license to Springer Nature Singapore Pte Ltd. 2022
Y. Li et al., *Real-Variable Theory of Hardy Spaces Associated with Generalized Herz Spaces of Rafeiro and Samko*, Lecture Notes in Mathematics 2320,
https://doi.org/10.1007/978-981-19-6788-7

18. H. Q. Bui, Weighted Hardy spaces, Math. Nachr. 103 (1981), 45–62.

19. P. L. Butzer and R. J. Nessel, Fourier Analysis and Approximation, Volume 1: One-Dimensional Theory, Pure and Applied Mathematics 40, Academic Press, New York–London, 1971, xvi+553 pp.

20. A.-P. Calderón, An atomic decomposition of distributions in parabolic H^p spaces, Adv. Math. 25 (1977), 216–225.

21. A.-P. Calderón, Cauchy integrals on Lipschitz curves and related operators, Proc. Nat. Acad. Sci. U.S.A. 74 (1977), 1324–1327.

22. A.-P. Calderón, Commutators, singular integrals on Lipschitz curves and applications, in: Proceedings of the International Congress of Mathematicians (Helsinki, 1978), pp. 85–96, Acad. Sci. Fennica, Helsinki, 1980.

23. A.-P. Calderón, C. P. Calderón, E. Fabes, M. Jodeit and N. M. Rivière, Applications of the Cauchy integral on Lipschitz curves, Bull. Amer. Math. Soc. 84 (1978), 287–290.

24. A.-P. Calderón and A. Torchinsky, Parabolic maximal functions associated with a distribution, Adv. Math. 16 (1975), 1–64.

25. A. P. Calderón and A. Zygmund, On the existence of certain singular integrals, Acta Math. 88 (1952), 85–139.

26. A. P. Calderón and A. Zygmund, Singular integral operators and differential equations, Amer. J. Math. 79 (1957), 901–921.

27. J. Cao, L. D. Ky and D. Yang, Bilinear decompositions of products of local Hardy and Lipschitz or BMO $(\mathbb{R}^n)$ spaces through wavelets, Commun. Contemp. Math. 20 (2018), 1750025, 30 pp.

28. L. Carleson, On convergence and growth of partial sums of Fourier series, Acta Math. 116 (1966), 135–157.

29. D.-C. Chang, S. Wang, D. Yang and Y. Zhang, Littlewood–Paley characterizations of Hardy-type spaces associated with ball quasi-Banach function spaces, Complex Anal. Oper. Theory 14 (2020), Paper No. 40, 33 pp.

30. J. Chen and G. Hu, Compact commutators of rough singular integral operators, Canad. Math. Bull. 58 (2015), 19–29.

31. J. Chen and C. Song, Global stability for the fractional Navier–Stokes equations in the Fourier–Herz space, Math. Methods Appl. Sci. 41 (2018), 3693–3717.

32. W. Chen and D. Yang, Non-convolution type oscillatory singular integral on Hardy space $HK_p(\mathbb{R}^n)$, Approx. Theory Appl. (N.S.) 13(2) (1997), 27–36.

33. Y. Chen, Q. Deng and Y. Ding, Commutators with fractional differentiation for second-order elliptic operators on $\mathbb{R}^n$, Commun. Contemp. Math. 22 (2020), 1950010, 29 pp.

34. Y. Chen and Y. Ding, L^p bounds for the commutators of singular integrals and maximal singular integrals with rough kernels, Trans. Amer. Math. Soc. 367 (2015), 1585–1608.

35. Y. Chen, Y. Ding and G. Hong, Commutators with fractional differentiation and new characterizations of BMO-Sobolev spaces, Anal. PDE 9 (2016), 1497–1522.

36. Y. Chen, Y. Ding and X. Wang, Compactness of commutators for singular integrals on Morrey spaces, Canad. J. Math. 64 (2012), 257–281.

37. Y. Chen, H. Jia and D. Yang, Boundedness of fractional integrals on Hardy spaces associated with ball quasi-Banach function spaces, Tokyo J. Math. (2022) (to appear).

38. Y. Chen, H. Jia and D. Yang, Boundedness of fractional integrals on ball Campanato-type function spaces, Bull. Sci. Math. 182 (2023), Paper No. 103210, 59 pp.

39. Y. Chen, H. Jia and D. Yang, Boundedness of Calderón–Zygmund operators on ball Campanato-type function spaces, Anal. Math. Phys. 12 (2022), Paper No. 118, 35 pp.

40. Y. Z. Chen and K. S. Lau, Some new classes of Hardy spaces, J. Funct. Anal. 84 (1989), 255–278.

41. N. Chikami, On Gagliardo–Nirenberg type inequalities in Fourier–Herz spaces, J. Funct. Anal. 275 (2018), 1138–1172.

42. G. Cleanthous, A. G. Georgiadis and M. Nielsen, Anisotropic mixed-norm Hardy spaces, J. Geom. Anal. 27 (2017), 2758–2787.

43. A. Clop and V. Cruz, Weighted estimates for Beltrami equations, Ann. Acad. Sci. Fenn. Math. 38 (2013), 91–113.

44. R. R. Coifman, A real variable characterization of H^p, Studia Math. 51 (1974), 269–274.

45. R. R. Coifman, Characterization of Fourier transforms of Hardy spaces, Proc. Nat. Acad. Sci. U.S.A. 71 (1974), 4133–4134.

46. R. R. Coifman, P.-L. Lions, Y. Meyer and S. Semmes, Compacité par compensation et espaces de Hardy, C. R. Acad. Sci. Paris Sér. I Math. 309 (1989), 945–949.

47. R. R. Coifman, R. Rochberg and G. Weiss, Factorization theorems for Hardy spaces in several variables, Ann. of Math. (2) 103 (1976), 611–635.

48. L. Colzani, Fourier transform of distributions in Hardy spaces, Boll. Un. Mat. Ital. A (6) 1 (1982), 403–410.

49. D. V. Cruz-Uribe and A. Fiorenza, Variable Lebesgue Spaces, Foundations and Harmonic Analysis, Applied and Numerical Harmonic Analysis, Birkhäuser/Springer, Heidelberg, 2013, x+312 pp.

50. D. V. Cruz-Uribe, A. Fiorenza, J. M. Martell and C. Pérez, The boundedness of classical operators on variable L^p spaces, Ann. Acad. Sci. Fenn. Math. 31 (2006), 239–264.

51. G. Dafni, Nonhomogeneous div-curl lemmas and local Hardy spaces, Adv. Differential Equations 10 (2005), 505–526.

52. G. Dafni, T. Hytönen, R. Korte and H. Yue, The space JN_p: nontriviality and duality, J. Funct. Anal. 275 (2018), 577–603.

53. R. del Campo, A. Fernández, F. Mayoral and F. Naranjo, Orlicz spaces associated to a quasi-Banach function space: applications to vector measures and interpolation, Collect. Math. 72 (2021), 481–499.

54. L. Diening, P. Harjulehto, P. Hästö and M. Růžička, Lebesgue and Sobolev Spaces with Variable Exponents, Lecture Notes in Math. 2017, Springer, Heidelberg, 2011, x+509 pp.

55. L. Diening, P. Hästö and S. Roudenko, Function spaces of variable smoothness and integrability, J. Funct. Anal. 256 (2009), 1731–1768.

56. G. Di Fazio and M. A. Ragusa, Commutators and Morrey spaces, Boll. Un. Mat. Ital. A (7) 5 (1991), 323–332.

57. A. Djeriou and D. Drihem, On the continuity of pseudo-differential operators on multiplier spaces associated to Herz-type Triebel–Lizorkin spaces, Mediterr. J. Math. 16 (2019), Paper No. 153, 25 pp.

58. K. Domelevo, S. Kakaroumpas, S. Petermichl and O. Soler I Gibert, Boundedness of Journé operators with matrix weights, arXiv: 2102.03395.

59. B. Dong and J. Xu, New Herz type Besov and Triebel–Lizorkin spaces with variable exponents, J. Funct. Spaces Appl. 2012, Art. ID 384593, 27 pp.

60. B. Dong and J. Xu, Herz-Morrey type Besov and Triebel–Lizorkin spaces with variable exponents, Banach J. Math. Anal. 9 (2015), 75–101.

61. B. Dong and J. Xu, Variable exponent Herz type Hardy spaces and their applications, Anal. Theory Appl. 31 (2015), 321–353.

62. D. Drihem, Embeddings properties on Herz-type Besov and Triebel–Lizorkin spaces, Math. Inequal. Appl. 16 (2013), 439–460.

63. D. Drihem, Complex interpolation of Herz-type Triebel–Lizorkin spaces, Math. Nachr. 291 (2018), 2008–2023.

64. D. Drihem, Jawerth–Franke embeddings of Herz-type Besov and Triebel–Lizorkin spaces, Funct. Approx. Comment. Math. 61 (2019), 207–226.

65. D. Drihem, Composition operators on Herz-type Triebel–Lizorkin spaces with application to semilinear parabolic equations, Banach J. Math. Anal. 16 (2022), Paper No. 29, 46 pp.

66. D. Drihem, Semilinear parabolic equations in Herz spaces, Appl. Anal. (2022), https://doi.org/10.1080/00036811.2022.2047948.

67. D. Drihem and R. Heraiz, Herz-type Besov spaces of variable smoothness and integrability, Kodai Math. J. 40 (2017), 31–57.

68. D. Drihem and R. Heraiz, Boundedness of some bilinear operators on variable Herz-type Hardy spaces, J. Pseudo-Differ. Oper. Appl. 10 (2019), 601–648.

69. D. Drihem and F. Seghiri, Notes on the Herz-type Hardy spaces of variable smoothness and integrability, Math. Inequal. Appl. 19 (2016), 145–165.

70. J. Duoandikoetxea, Fourier Analysis, Graduate Studies in Mathematics 29, American Mathematical Society, Providence, RI, 2001, xviii+222 pp.

71. L. C. Evans and S. Müller, Hardy spaces and the two-dimensional Euler equations with nonnegative vorticity, J. Amer. Math. Soc. 7 (1994), 199–219.

72. D. Fan and D. Yang, The weighted Herz-type Hardy spaces $h\dot{K}_q^{\alpha,p}(\omega_1, \omega_2)$, Approx. Theory Appl. (N.S.) 13(4) (1997), 19–41.

73. D. Fan and D. Yang, Herz-type Hardy spaces on Vilenkin groups and their applications, Sci. China Ser. A 43 (2000), 481–494.

74. L. C. F. Ferreira and J. E. Pérez-López, Besov-weak-Herz spaces and global solutions for Navier–Stokes equations, Pacific J. Math. 296 (2018), 57–77.

75. C. Fefferman, C., N. M. Rivière and Y. Sagher, Interpolation between H^p spaces: the real method, Trans. Amer. Math. Soc. 191 (1974), 75–81.

76. C. Fefferman and E. M. Stein, Some maximal inequalities, Amer. J. Math. 93 (1971), 107–115.

77. C. Fefferman and E. M. Stein, H^p spaces of several variables, Acta Math. 129 (1972), 137–193.

78. R. Fefferman and F. Soria, The space weak H^1, Studia Math. 85 (1986), 1–16.

79. H. G. Feichtinger, An elementary approach to Wiener's third Tauberian theorem for the Euclidean n-space, in: Symposia Mathematica, Vol. XXIX (Cortona, 1984), pp. 267–301, Sympos. Math., XXIX, Academic Press, New York, 1987.

80. H. G. Feichtinger and F. Weisz, Inversion formulas for the short-time Fourier transform, J. Geom. Anal. 16 (2006), 507–521.

81. H. G. Feichtinger and F. Weisz, The Segal algebra $S_0(\mathbb{R}^d)$ and norm summability of Fourier series and Fourier transforms, Monatsh. Math. 148 (2006), 333–349.

82. H. G. Feichtinger and F. Weisz, Wiener amalgams and pointwise summability of Fourier transform and Fourier series, Math. Proc. Camb. Phil. Soc. 140 (2006), 509–536.

83. H. G. Feichtinger and F. Weisz, Herz spaces and summability of Fourier transforms, Math. Nachr. 281 (2008), 309–324.

84. G. B. Folland, Real Analysis, Modern Techniques and Their Applications, Second edition, Pure Appl. Math. (N. Y.), A Wiley-Interscience Publication, John Wiley & Sons, Inc., New York, 1999, xvi+386 pp.

85. G. B. Folland and E. M. Stein, Hardy Spaces on Homogeneous Groups, Mathematical Notes 28, Princeton University Press, Princeton, N.J.; University of Tokyo Press, Tokyo, 1982, xii+285 pp.

86. J. García-Cuerva, Hardy spaces and Beurling algebras, J. London Math. Soc. (2) 39 (1989), 499–513.

87. J. García-Cuerva and M. J. L. Herrero, A theory of Hardy spaces associated to Herz spaces, Proc. London Math. Soc. (3) 69 (1994), 605–628.

88. A. Gogatishvili and R. Ch. Mustafayev, New pre-dual space of Morrey space, J. Math. Anal. Appl. 397 (2013), 678–692.

89. D. Goldberg, A local version of real Hardy spaces, Duke Math. J. 46 (1979), 27–42.

90. L. Grafakos, Classical Fourier Analysis, third edition, Graduate Texts in Math. 249, Springer, New York, 2014, xviii+638 pp.

91. L. Grafakos, Modern Fourier Analysis, third edition, Graduate Texts in Math. 250, Springer, New York, 2014, xvi+624 pp.

92. L. Grafakos and D. He, Weak Hardy spaces, in: Some Topics in Harmonic Analysis and Applications, Adv. Lect. Math. (ALM) 34, Higher Education Press and International Press, Beijing–Somerville, MA, 2015, pp. 177–202.

93. L. Grafakos, X. Li and D. Yang, Bilinear operators on Herz-type Hardy spaces, Trans. Amer. Math. Soc. 350 (1998), 1249–1275.

94. V. S. Guliev and R. Ch. Mustafaev, Fractional integrals in spaces of functions defined on spaces of homogeneous type, Anal. Math. 24 (1998), 181–200.

95. V. Guliyev, M. Omarova and Y. Sawano, Boundedness of intrinsic square functions and their commutators on generalized weighted Orlicz–Morrey spaces, Banach J. Math. Anal. 9 (2015), 44–62.

96. W. Guo, H. Wu and D. Yang, A revisit on the compactness of commutators, Canad. J. Math. 73 (2021), 1667–1697.

97. G. H. Hardy and J. E. Littlewood, Some properties of fractional integrals. I, Math. Z. 27 (1928), 565–606.

98. G. H. Hardy and J. E. Littlewood, A maximal theorem with function-theoretic applications, Acta Math. 54 (1930), 81–116.

99. J. Hart, R. H. Torres and X. Wu, Smoothing properties of bilinear operators and Leibniz-type rules in Lebesgue and mixed Lebesgue spaces, Trans. Amer. Math. Soc. 370 (2018), 8581–8612.

100. D. He, Square function characterization of weak Hardy spaces, J. Fourier Anal. Appl. 20 (2014), 1083–1110.

101. E. Hernández, G. Weiss and D. Yang, The ϕ-transform and wavelet characterizations of Herz-type spaces, Collect. Math. 47 (1996), 285–320.

102. E. Hernández and D. Yang, Interpolation of Herz-type Hardy spaces, Illinois J. Math. 42 (1998), 564–581.

103. E. Hernández and D. Yang, Interpolation of Herz spaces and applications, Math. Nachr. 205 (1999), 69–87.

104. C. S. Herz, Lipschitz spaces and Bernstein's theorem on absolutely convergent Fourier transforms, J. Math. Mech. 18 (1968/69), 283–323.

105. K.-P. Ho, Atomic decomposition of Hardy spaces and characterization of BMO via Banach function spaces, Anal. Math. 38 (2012), 173–185.

106. K.-P. Ho, Atomic decompositions of weighted Hardy–Morrey spaces, Hokkaido Math. J. 42 (2013), 131–157.

107. K.-P. Ho, Atomic decomposition of Hardy–Morrey spaces with variable exponents, Ann. Acad. Sci. Fenn. Math. 40 (2015), 31–62.

108. K.-P. Ho, Strong maximal operator on mixed-norm spaces, Ann. Univ. Ferrara Sez. VII Sci. Mat. 62 (2016), 275–291.

109. K.-P. Ho, Atomic decompositions and Hardy's inequality on weak Hardy–Morrey spaces, Sci. China Math. 60 (2017), 449–468.

110. K.-P. Ho, Extrapolation, John–Nirenberg inequalities and characterizations of BMO in terms of Morrey type spaces, Rev. Mat. Complut. 30 (2017), 487–505.

111. K.-P. Ho, Young's inequalities and Hausdorff–Young inequalities on Herz spaces, Boll. Unione Mat. Ital. 11 (2018), 469–481.

112. K.-P. Ho, Extrapolation to Herz spaces with variable exponents and applications, Rev. Mat. Complut. 33 (2020), 437–463.

113. K.-P. Ho, Erdélyi-Kober fractional integral operators on ball Banach function spaces, Rend. Semin. Mat. Univ. Padova 145 (2021), 93–106.

114. K.-P. Ho, Spherical maximal function, maximal Bochner–Riesz mean and geometrical maximal function on Herz spaces with variable exponents, Rend. Circ. Mat. Palermo (2) 70 (2021), 559–574.

115. S. Hou, D. Yang and S. Yang, Lusin area function and molecular characterizations of Musielak–Orlicz Hardy spaces and their applications, Commun. Contemp. Math. 15 (2013), 1350029, 37 pp.

116. H. Houamed, Well-posedness and long time behavior for the electron inertial Hall-MHD system in Besov and Kato–Herz spaces, J. Math. Anal. Appl. 501 (2021), 125208, 23 pp.

117. G. Hu, S. Lu and D. Yang, The weak Herz spaces, Beijing Shifan Daxue Xuebao 33 (1997), 27–34.

118. G. Hu, S. Lu and D. Yang, The applications of weak Herz spaces, Adv. in Math. (China) 26 (1997), 417–428.

119. G. Hu, S. Lu and D. Yang, Boundedness of rough singular integral operators on homogeneous Herz spaces, J. Austral. Math. Soc. Ser. A 66 (1999), 201–223.

120. G. Hu and D. Yang, Boundedness of oscillatory singular integrals on HK_p spaces, Approx. Theory Appl. (N.S.) 13(2) (1997), 20–26.

121. L. Huang, D.-C. Chang and D. Yang, Fourier transform of anisotropic mixed-norm Hardy spaces, Front. Math. China. 16 (2021), 119–139.

122. L. Huang, D.-C. Chang and D. Yang, Fourier transform of Hardy spaces associated with ball quasi-Banach function spaces, Appl. Anal. 101 (2022), 3825–3840.

123. L. Huang, J. Liu, D. Yang and W. Yuan, Atomic and Littlewood–Paley characterizations of anisotropic mixed-norm Hardy spaces and their applications, J. Geom. Anal. 29 (2019), 1991–2067.

124. L. Huang, J. Liu, D. Yang and W. Yuan, Dual spaces of anisotropic mixed-norm Hardy spaces, Proc. Amer. Math. Soc. 147 (2019), 1201–1215.

125. L. Huang and X. Wang, Anisotropic variable Campanato-type spaces and their Carleson measure characterizations, Fract. Calc. Appl. Anal. 25 (2022), 1131–1165.

126. L. Huang, F. Weisz, D. Yang and W. Yuan, Summability of Fourier transforms on mixed-norm Lebesgue spaces via associated Herz spaces, Anal. Appl. (Singap.) (2021), https://doi.org/10.1142/S0219530521500135.

127. L. Huang and D. Yang, On function spaces with mixed norms—a survey, J. Math. Study 54 (2021), 262–336.

128. L. Huang, D. Yang and W. Yuan, Anisotropic mixed-norm Campanato-type spaces with applications to duals of anisotropic mixed-norm Hardy spaces, Banach J. Math. Anal. 15 (2021), Paper No. 62, 36 pp.

129. R. A. Hunt, On the convergence of Fourier series, in: Orthogonal Expansions and Their Continuous Analogues, Proc. Conf., Edwardsville, Ill., 1967, pp. 235–255, Southern Illinois Univ. Press, Carbondale, Ill.

130. M. Izuki, Boundedness of sublinear operators on Herz spaces with variable exponent and application to wavelet characterization, Anal. Math. 36 (2010), 33–50.

131. M. Izuki, Vector-valued inequalities on Herz spaces and characterizations of Herz–Sobolev spaces with variable exponent, Glas. Mat. Ser. III 45(65) (2010), 475–503.

132. M. Izuki and T. Noi, Two weighted Herz spaces with variable exponents, Bull. Malays. Math. Sci. Soc. 43 (2020), 169–200.

133. M. Izuki, T. Noi and Y. Sawano, The John–Nirenberg inequality in ball Banach function spaces and application to characterization of BMO, J. Inequal. Appl. 2019, Paper No. 268, 11 pp.

134. M. Izuki and Y. Sawano, The Haar wavelet characterization of weighted Herz spaces and greediness of the Haar wavelet basis, J. Math. Anal. Appl. 362 (2010), 140–155.

135. M. Izuki and Y. Sawano, Characterization of BMO via ball Banach function spaces, Vestn. St.-Peterbg. Univ. Mat. Mekh. Astron. 4(62) (2017), 78–86.

136. H. Jia, J. Tao, D. Yang, W. Yuan and Y. Zhang, Boundedness of fractional integrals on special John–Nirenberg–Campanato and Hardy-type spaces via congruent cubes, Fract. Calc. Appl. Anal. 25 (2022), 2446–2487.

137. H. Jia, J. Tao, D. Yang, W. Yuan and Y. Zhang, Special John–Nirenberg–Campanato spaces via congruent cubes, Sci. China Math. 65 (2022), 359–420.

138. H. Jia, J. Tao, D. Yang, W. Yuan and Y. Zhang, Boundedness of Calderón–Zygmund operators on special John–Nirenberg–Campanato and Hardy-type spaces via congruent cubes, Anal. Math. Phys. 12 (2022), Paper No. 15, 56 pp.

139. H. Jia and H. Wang, Decomposition of Hardy–Morrey spaces, J. Math. Anal. Appl. 354 (2009), 99–110.

140. F. John and L. Nirenberg, On functions of bounded mean oscillation, Comm. Pure Appl. Math. 14 (1961), 415–426.

141. A. Karlovich and A. Lerner, Commutators of singular integrals on generalized L^p spaces with variable exponent, Publ. Mat. 49 (2005), 111–125.

142. A. Y. Karlovich and I. M. Spitkovsky, Pseudodifferential operators on variable Lebesgue spaces, in: Operator Theory, Pseudo-Differential Equations, and Mathematical Physics, pp. 173–183, Oper. Theory Adv. Appl., 228, Birkhäuser/Springer Basel AG, Basel, 2013.

143. C. E. Kenig, H('P) Spaces on Lipschitz Domains, Thesis (Ph.D.)-The University of Chicago, 1978, iii+59 pp.
144. C. E. Kenig, Weighted H^p spaces on Lipschitz domains, Amer. J. Math. 102 (1980), 129–163.
145. V. Kokilashvili, A. Meskhi, H. Rafeiro and S. Samko, Integral Operators in Non-Standard Function Spaces, Vol. 1, Variable Exponent Lebesgue and Amalgam Spaces, Operator Theory: Advances and Applications 248, Birkhäuser/Springer, [Cham], 2016, xx+567 pp.
146. Y. Komori, Notes on singular integrals on some inhomogeneous Herz spaces, Taiwanese J. Math. 8 (2004), 547–556.
147. O. Kováčik and J. Rákosník, On spaces $L^{p(x)}$ and $W^{k,p(x)}$, Czechoslovak Math. J. 41(116) (1991), 592–618.
148. L. D. Ky, New Hardy spaces of Musielak–Orlicz type and boundedness of sublinear operators, Integral Equations Operator Theory 78 (2014), 115–150.
149. A. K. Lerner, S. Ombrosi and I. P. Rivera-Ríos, Commutators of singular integrals revisited, Bull. Lond. Math. Soc. 51 (2019), 107–119.
150. X. Li and D. Yang, Boundedness of some sublinear operators on Herz spaces, Illinois J. Math. 40 (1996), 484–501.
151. Y. Liang, D. Yang and R. Jiang, Weak Musielak–Orlicz Hardy spaces and applications, Math. Nachr. 289 (2016), 634–677.
152. J. E. Littlewood and G. H. Hardy, Some properties of conjugate functions, J. Reine Angew. Math. 167 (1932), 405–423.
153. H. Liu, The weak H^p spaces on homogeneous groups, in: Harmonic Analysis, Tianjin, 1988, Lecture Notes in Math. 1494, Springer, Berlin, 1991, pp. 113–118.
154. J. Liu, Molecular characterizations of variable anisotropic Hardy spaces with applications to boundedness of Calderón–Zygmund operators, Banach J. Math. Anal. 15 (2021), 1–24.
155. J. Liu, L. Huang and C. Yue, Molecular characterizations of anisotropic mixed-norm Hardy spaces and their applications, Mathematics 9 (2021), Art. No. 2216, https://doi.org/10.3390/math9182216.
156. J. Liu, F. Weisz, D. Yang and W. Yuan, Variable anisotropic Hardy spaces and their applications, Taiwanese J. Math. 22 (2018), 1173–1216.
157. W. Liu and S. Lu, Calderón–Zygmund operators on the Hardy spaces of weighted Herz type, Approx. Theory Appl. (N.S.) 13(2) (1997), 1–10.
158. X. Liu, S. Yang, X. Wang and B. Li, Parametrized Littlewood–Paley operators on Herz-type Hardy spaces with variable exponent, Bull. Malays. Math. Sci. Soc. 43 (2020), 4143–4169.
159. Z. Lou, Q. Yang, J. He and K. He, Uniform analytic solutions for fractional Navier–Stokes equations, Appl. Math. Lett. 112 (2021), Paper No. 106784, 7 pp.
160. S.-Z. Lu, Four Lectures on Real H^p Spaces, World Scientific Publishing Co., Inc., River Edge, NJ, 1995, viii+217 pp.
161. S. Lu, Y. Ding and D. Yan, Singular Integrals and Related Topics, World Scientific Publishing Co. Pte. Ltd., Hackensack, NJ, 2007, viii+272 pp.
162. S. Lu and D. Yang, The Littlewood–Paley function and ϕ-transform characterizations of a new Hardy space HK_2 associated with the Herz space, Studia Math. 101 (1992), 285–298.
163. S. Lu and D. Yang, The local versions of $H^p(\mathbb{R}^n)$ spaces at the origin, Studia Math. 116 (1995), 103–131.
164. S. Lu and D. Yang, Oscillatory singular integrals on Hardy spaces associated with Herz spaces, Proc. Amer. Math. Soc. 123 (1995), 1695–1701.
165. S. Lu and D. Yang, The weighted Herz-type Hardy space and its applications, Sci. China Ser. A 38 (1995), 662–673.
166. S. Lu and D. Yang, The decomposition of weighted Herz space on $\mathbb{R}^n$ and its applications, Sci. China Ser. A 38 (1995), 147–158.
167. S. Lu and D. Yang, Embedding theorems for Herz spaces, Chinese Sci. Bull. 40 (1995), 1497–1501.
168. S. Lu and D. Yang, The decomposition of Herz spaces on local fields and its applications, J. Math. Anal. Appl. 196 (1995), 296–313.

169. S. Lu and D. Yang, Hardy–Littlewood–Sobolev theorems of fractional integration on Herz-type spaces and its applications, Canad. J. Math. 48 (1996), 363–380.

170. S. Lu and D. Yang, The continuity of commutators on Herz-type spaces, Michigan Math. J. 44 (1997), 255–281.

171. S. Lu and D. Yang, Herz-type Sobolev and Bessel potential spaces and their applications, Sci. China Ser. A 40 (1997), 113–129.

172. S. Lu and D. Yang, Some characterizations of weighted Herz-type Hardy spaces and their applications, Acta Math. Sinica (N.S.) 13 (1997), 45–58.

173. S. Lu and D. Yang, Multiplier theorems for Herz type Hardy spaces, Proc. Amer. Math. Soc. 126 (1998), 3337–3346.

174. S. Lu and D. Yang, Boundedness of sublinear operators in Herz spaces on Vilenkin groups and its application, Math. Nachr. 191 (1998), 229–246.

175. S. Lu, D. Yang and G. Hu, Herz Type Spaces and Their Applications, Science Press, Beijing, 2008, iv+232 pp.

176. L. Maligranda, Orlicz Spaces and Interpolation, Seminários de Matemática [Seminars in Mathematics] 5, Universidade Estadual de Campinas, Departamento de Matemática, Campinas, 1989, iii+206 pp.

177. W. Matuszewska and W. Orlicz, On certain properties of φ-functions, Bull. Acad. Polon. Sci. Sér. Sci. Math. Astronom. Phys. 8 (1960), 439–443.

178. W. Matuszewska and W. Orlicz, On some classes of functions with regard to their orders of growth, Studia Math. 26 (1965), 11–24.

179. A. Meskhi, H. Rafeiro and M. A. Zaighum, Central Calderón–Zygmund operators on Herz-type Hardy spaces of variable smoothness and integrability, Ann. Funct. Anal. 9 (2018), 310–321.

180. A. Meskhi, H. Rafeiro and M. A. Zaighum, On the boundedness of Marcink-iewicz integrals on continual variable exponent Herz spaces, Georgian Math. J. 26 (2019), 105–116.

181. T. Mizuhara, Notes on embedding theorems for local Hardy spaces, Math. Nachr. 165 (1994), 231–244.

182. C. B. Morrey, On the solutions of quasi-linear elliptic partial differential equations, Trans. Amer. Math. Soc. 43 (1938), 126–166.

183. S. Müller, Hardy space methods for nonlinear partial differential equations, Tatra Mt. Math. Publ. 4 (1994), 159–168.

184. H. Nafis, H. Rafeiro and M. A. Zaighum, A note on the boundedness of sublinear operators on grand variable Herz spaces, J. Inequal. Appl. 2020, Paper No. 1, 13 pp.

185. H. Nafis, H. Rafeiro and M. A. Zaighum, Boundedness of the Marcinkiewicz integral on grand variable Herz spaces, J. Math. Inequal. 15 (2021), 739–753.

186. E. Nakai and Y. Sawano, Hardy spaces with variable exponents and generalized Campanato spaces, J. Funct. Anal. 262 (2012), 3665–3748.

187. E. Nakai and Y. Sawano, Orlicz–Hardy spaces and their duals, Sci. China Math. 57 (2014), 903–962.

188. S. Nakamura and Y. Sawano, The singular integral operator and its commutator on weighted Morrey spaces, Collect. Math. 68 (2017), 145–174.

189. Z. Nan and J. Yuan, Ill-posedness in Fourier–Herz spaces of the 3D Navier–Stokes equations with hyper-dissipation, Colloq. Math. 152 (2018), 123–140.

190. Y. Nie and X. Zheng, Ill-posedness of the 3D incompressible hyperdissipative Navier–Stokes system in critical Fourier-Herz spaces, Nonlinearity 31 (2018), 3115–3150.

191. T. Nogayama, T. Ono, D. Salim and Y. Sawano, Atomic decomposition for mixed Morrey spaces, J. Geom. Anal. 31 (2021), 9338–9365.

192. W. Orlicz, Über konjugierte exponentenfolgen, Studia Math. 3 (1931), 200–211.

193. Y. Pan, Hardy spaces and oscillatory singular integrals, Rev. Mat. Iberoamericana 7 (1991), 55–64.

194. T. S. Quek and D. Yang, Calderón–Zygmund-type operators on weighted weak Hardy spaces over $\mathbb{R}^n$, Acta Math. Sin. (Engl. Ser.) 16 (2000), 141–160.

195. H. Rafeiro and S. Samko, Riesz potential operator in continual variable exponents Herz spaces, Math. Nachr. 288 (2015), 465–475.
196. H. Rafeiro and S. Samko, Maximal operator with rough kernel in variable Musielak–Morrey–Orlicz type spaces, variable Herz spaces and grand variable Lebesgue spaces, Integral Equations Operator Theory 89 (2017), 111–124.
197. H. Rafeiro and S. Samko, Herz spaces meet Morrey type spaces and complementary Morrey type spaces, J. Fourier Anal. Appl. 26 (2020), Paper No. 74, 14 pp.
198. H. Rafeiro and S. Samko, On embeddings of Morrey type spaces between weighted Lebesgue or Stummel spaces with application to Herz spaces, Banach J. Math. Anal. 15 (2021), Paper No. 48, 19 pp.
199. H. Rafeiro and S. Samko, Coincidence of variable exponent Herz spaces with variable exponent Morrey type spaces and boundedness of sublinear operators in these spaces, Potential Anal. 56 (2022), 437–457.
200. S. Rolewicz, On a certain class of linear metric spaces, Bull. Acad. Polon. Sci. Cl. III. 5 (1957), 471–473.
201. N. Samko, Weighted Hardy operators in the local generalized vanishing Morrey spaces, Positivity 17 (2013), 683–706.
202. S. Samko, Variable exponent Herz spaces, Mediterr. J. Math. 10 (2013), 2007–2025.
203. S. Samko, Morrey spaces are closely embedded between vanishing Stummel spaces, Math. Inequal. Appl. 17 (2014), 627–639.
204. S. G. Samko and S. M. Umarkhadzhiev, Grand Morrey type spaces, Vladikavkaz. Mat. Zh. 22 (2020), 104–118.
205. Y. Sawano, A note on Besov–Morrey spaces and Triebel–Lizorkin–Morrey spaces, Acta Math. Sin. (Engl. Ser.) 25 (2009), 1223–1242.
206. Y. Sawano, G. Di Fazio, and D. I. Hakim, Morrey Spaces—Introduction and Applications to Integral Operators and PDE's, Vol. I, Monogr. Res. Notes Math., CRC Press, Boca Raton, FL, 2020, xxi+479 pp.
207. Y. Sawano, K.-P. Ho, D. Yang and S. Yang, Hardy spaces for ball quasi-Banach function spaces, Dissertationes Math. 525 (2017), 1–102.
208. Y. Sawano and S. Shirai, Compact commutators on Morrey spaces with non-doubling measures, Georgian Math. J. 15 (2008), 353–376.
209. Y. Sawano and H. Tanaka, The Fatou property of block spaces, J. Math. Sci. Univ. Tokyo 22 (2015), 663–683.
210. A. Scapellato, Regularity of solutions to elliptic equations on Herz spaces with variable exponents, Bound. Value Probl. 2019, Paper No. 2, 9 pp.
211. S. Semmes, A primer on Hardy spaces, and some remarks on a theorem of Evans and Müller, Comm. Partial Differential Equations 19 (1994), 277–319.
212. C. Shi and J. Xu, Herz type Besov and Triebel–Lizorkin spaces with variable exponent, Front. Math. China 8 (2013), 907–921.
213. E. M. Stein, Harmonic Analysis: Real-Variable Methods, Orthogonality, and Oscillatory Integrals, Princeton Mathematical Series 43, Monographs in Harmonic Analysis III, Princeton University Press, Princeton, NJ, 1993, xiv+695 pp.
214. E. M. Stein and R. Shakarchi, Functional Analysis, Introduction to Further Topics in Analysis, Princeton Lect. Anal. 4, Princeton University Press, Princeton, NJ, 2011, xviii+423 pp.
215. E. M. Stein and G. Weiss, On the theory of harmonic functions of several variables, I: the theory of H^p spaces, Acta Math. 103 (1960), 25–62.
216. J.-O. Strömberg, A modified Franklin system and higher-order spline systems on $\mathbb{R}^n$ as unconditional bases for Hardy spaces, in: Conference on Harmonic Analysis in Honor of Antoni Zygmund, Vol. I, II (Chicago, Ill., 1981), pp. 475–494, Wadsworth Math. Ser., Wadsworth, Belmont, CA, 1983.
217. J.-O. Strömberg and A. Torchinsky, Weights, sharp maximal functions and Hardy spaces, Bull. Amer. Math. Soc. (N.S.) 3 (1980), 1053–1056.
218. J.-O. Strömberg and A. Torchinsky, Weighted Hardy Spaces, Lecture Notes in Math. 1381, Springer-Verlag, Berlin, 1989, vi+193 pp.

219. J. Sun, D. Yang and W. Yuan, Weak Hardy spaces associated with ball quasi-Banach function spaces on spaces of homogeneous type: decompositions, real interpolation, and Calderón–Zygmund operators, J. Geom. Anal. 32 (2022), Paper No. 191, 85 pp.
220. K. Szarvas and F. Weisz, Convergence of multi-dimensional integral operators and applications, Period. Math. Hungar. 74 (2017), 40–66.
221. M. H. Taibleson and G. Weiss, The molecular characterization of certain Hardy spaces, in: Representation Theorems for Hardy Spaces, pp. 67–149, Astérisque, 77, Soc. Math. France, Paris, 1980.
222. J. Tan, Real-variable theory of local variable Hardy spaces, arXiv: 2110.03387.
223. L. Tang, Weighted local Hardy spaces and their applications, Illinois J. Math. 56 (2012), 453–495.
224. L. Tang and J. Xu, Some properties of Morrey type Besov–Triebel spaces, Math. Nachr. 278 (2005), 904–917.
225. J. Tao, Da. Yang and Do. Yang, Beurling–Ahlfors commutators on weighted Morrey spaces and applications to Beltrami equations, Potential Anal. 53 (2020), 1467–1491.
226. J. Tao, D. Yang and W. Yuan, John–Nirenberg–Campanato spaces, Nonlinear Anal. 189 (2019), 111584, 36 pp.
227. J. Tao, D. Yang and W. Yuan, A survey on function spaces of John–Nirenberg type, Mathematics 9 (2021), Art. No. 2264, https://doi.org/10.3390/math9182264.
228. J. Tao, D. Yang, W. Yuan and Y. Zhang, Compactness characterizations of commutators on ball Banach function spaces, Potential Anal. (2021), https://doi.org/10.1007/s11118-021-09953-w.
229. H. Triebel, Theory of Function Spaces, Monographs in Mathematics 78, Birkhäuser Verlag, Basel, 1983, 284 pp.
230. R. M. Trigub and E. S. Belinsky, Fourier Analysis and Approximation of Functions, Kluwer Academic Publishers, Dordrecht, 2004, xiv+585 pp.
231. A. Uchiyama, On the compactness of operators of Hankel type, Tôhoku Math. J. (2) 30 (1978), 163–171.
232. T. Ullrich, Continuous characterizations of Besov–Lizorkin–Triebel spaces and new interpretations as coorbits, J. Funct. Spaces Appl. 2012, Art. ID 163213, 47 pp.
233. F. Wang, D. Yang and S. Yang, Applications of Hardy spaces associated with ball quasi-Banach function spaces, Results Math. 75 (2020), Paper No. 26, 58 pp.
234. H. Wang and Z. Liu, Local Herz-type Hardy spaces with variable exponent, Banach J. Math. Anal. 9 (2015), 359–378.
235. H. Wang and Z. Liu, Boundedness of singular integral operators on weak Herz type spaces with variable exponent, Ann. Funct. Anal. 11 (2020), 1108–1125.
236. H. Wang and Z. Fu, Estimates of commutators on Herz-type spaces with variable exponent and applications, Banach J. Math. Anal. 15 (2021), Paper No. 36, 26 pp.
237. S. Wang and J. Xu, Weighted norm inequality for bilinear Calderón–Zygmund operators on Herz–Morrey spaces with variable exponents, J. Inequal. Appl. 2019, Paper No. 251, 23 pp.
238. S. Wang and J. Xu, Boundedness of vector valued bilinear Calderón–Zygmund operators on products of weighted Herz–Morrey spaces with variable exponents, Chin. Ann. Math. Ser. B 42 (2021), 693–720.
239. S. Wang and J. Xu, Commutators of bilinear Hardy operators on two weighted Herz spaces with variable exponents, Anal. Theory Appl. 37 (2021), 387–403.
240. S. Wang and J. Xu, Bilinear Calderón–Zygmund operators on two weight Herz spaces with variable exponents, Oper. Matrices 15 (2021), 937–958.
241. S. Wang and J. Xu, Boundedness of vector-valued sublinear operators on weighted Herz–Morrey spaces with variable exponents, Open Math. 19 (2021), 412–426.
242. S. Wang, D. Yang, W. Yuan and Y. Zhang, Weak Hardy-type spaces associated with Ball quasi-Banach function spaces II: Littlewood–Paley characterizations and real interpolation, J. Geom. Anal. 31 (2021), 631–696.
243. F. Weisz, Martingale Hardy spaces with continuous time, in: Probability Theory and Applications, pp. 47–75, Math. Appl. 80, Kluwer Acad. Publ., Dordrecht, 1992.

244. F. Weisz, Martingale Hardy spaces and their applications in Fourier analysis, Lecture Notes in Math. 1568, Springer-Verlag, Berlin, 1994, viii+218 pp.
245. F. Weisz, Herz spaces and restricted summability of Fourier transforms and Fourier series, J. Math. Anal. Appl. 344 (2008), 42–54.
246. F. Weisz, Wiener amalgams, Hardy spaces and summability of Fourier series, Math. Proc. Cambridge Philos. Soc. 145 (2008), 419–442.
247. F. Weisz, Pointwise summability of Gabor expansions, J. Fourier Anal. Appl. 15 (2009), 463–487.
248. F. Weisz, Summability of Gabor expansions and Hardy spaces, Appl. Comput. Harmon. Anal. 30 (2011), 288–306.
249. F. Weisz, Herz spaces and pointwise summability of Fourier series, Math. Pannon. 23 (2012), 235–256.
250. F. Weisz, Pointwise convergence in Pringsheim's sense of the summability of Fourier transforms on Wiener amalgam spaces, Monatsh. Math. 175 (2014), 143–160.
251. F. Weisz, Lebesgue points of two-dimensional Fourier transforms and strong summability, J. Fourier Anal. Appl. 21 (2015), 885–914.
252. F. Weisz, Some generalizations of Lebesgue's theorem for two-dimensional functions, Ann. Univ. Sci. Budapest. Sect. Comput. 45 (2016), 277–290.
253. F. Weisz, Lebesgue points and restricted convergence of Fourier transforms and Fourier series, Anal. Appl. (Singap.) 15 (2017), 107–121.
254. F. Weisz, Convergence and Summability of Fourier Transforms and Hardy Spaces, Applied and Numerical Harmonic Analysis, Birkhäuser/Springer, Cham, 2017, xxii+435 pp.
255. F. Weisz, Lebesgue Points and Summability of Higher Dimensional Fourier Series, Birkhäuser/Springer, Cham, 2021, xiii+290 pp.
256. J. Xu, A discrete characterization of Herz-type Triebel–Lizorkin spaces and its applications, Acta Math. Sci. Ser. B (Engl. Ed.) 24 (2004), 412–420.
257. J. Xu, Equivalent norms of Herz-type Besov and Triebel–Lizorkin spaces, J. Funct. Spaces Appl. 3 (2005), 17–31.
258. J. Xu, An admissibility for topological degree of Herz-type Besov and Triebel–Lizorkin spaces, Topol. Methods Nonlinear Anal. 33 (2009), 327–334.
259. J. Xu, Decompositions of non-homogeneous Herz-type Besov and Triebel–Lizorkin spaces, Sci. China Math. 57 (2014), 315–331.
260. J. Xu and D. Yang, Applications of Herz-type Triebel–Lizorkin spaces, Acta Math. Sci. Ser. B (Engl. Ed.) 23 (2003), 328–338.
261. J. Xu and D. Yang, Herz-type Triebel–Lizorkin spaces I, Acta Math. Sin. (Engl. Ser.) 21 (2005), 643–654.
262. J. Xu and X. Yang, Herz–Morrey–Hardy spaces with variable exponents and their applications, J. Funct. Spaces 2015, Art. ID 160635, 19 pp.
263. J. Xu and X. Yang, The molecular decomposition of Herz–Morrey–Hardy spaces with variable exponents and its application, J. Math. Inequal. 10 (2016), 977–1008.
264. J. Xu and X. Yang, Variable exponent Herz type Besov and Triebel–Lizorkin spaces, Georgian Math. J. 25 (2018), 135–148.
265. X. Yan, Z. He, D. Yang and W. Yuan, Hardy spaces associated with ball quasi-Banach function spaces on spaces of homogeneous type: characterizations of maximal functions, decompositions, and dual spaces, Math. Nachr. (2022), https://doi.org/10.1002/mana.202100432.
266. X. Yan, D. Yang and W. Yuan, Intrinsic square function characterizations of Hardy spaces associated with ball quasi-Banach function spaces, Front. Math. China 15 (2020), 769–806.
267. X. Yan, D. Yang, W. Yuan and C. Zhuo, Variable weak Hardy spaces and their applications, J. Funct. Anal. 271 (2016), 2822–2887.
268. D. Yang, Applications of weighted weak Herz-type spaces over Vilenkin groups, Indian J. Math. 41 (1999), 455–479.
269. D. Yang, Y. Liang and L. D. Ky, Real-Variable Theory of Musielak–Orlicz Hardy Spaces, Lecture Notes in Math. 2182, Springer, 2017, xiii+466 pp.

270. Da. Yang, Do. Yang and G. Hu, The Hardy Space H^1 with Non-doubling Measures and Their Applications, Lecture Notes in Math. 2084, Springer-Verlag, Cham, 2013, xiv+653 pp.
271. D. Yang and S. Yang, Weighted local Orlicz Hardy spaces with applications to pseudo-differential operators, Dissertationes Math. 478 (2011), 1–78.
272. D. Yang and S. Yang, Local Hardy spaces of Musielak–Orlicz type and their applications, Sci. China Math. 55 (2012), 1677–1720.
273. D. Yang, W. Yuan and Y. Zhang, Bilinear decomposition and divergence-curl estimates on products related to paces and their dual spaces, J. Funct. Anal. 280 (2021), 108796, 74 pp.
274. Z. Ye and X. Zhao, Global well-posedness of the generalized magnetohydrodynamic equations, Z. Angew. Math. Phys. 69 (2018), Paper No. 126, 26 pp.
275. T.-L. Yee and K.-P. Ho, Hardy's inequalities and integral operators on Herz–Morrey spaces, Open Math. 18 (2020), 106–121.
276. X. Yu and Z. Liu, Boundedness of some integral operators and commutators on homogeneous Herz spaces with three variable exponents, Front. Math. China 16 (2021), 211–237.
277. Y. Zhang, L. Huang, D. Yang and W. Yuan, New ball Campanato-type function spaces and their applications, J. Geom. Anal. 32 (2022), Paper No. 99, 42 pp.
278. Y. Zhang, D. Yang, W. Yuan and S. Wang, Weak Hardy-type spaces associated with ball quasi-Banach function spaces I: Decompositions with applications to boundedness of Calderón–Zygmund operators, Sci. China Math. 64 (2021), 2007–2064.
279. Y. Zhao, D. Yang and Y. Zhang, Mixed-norm Herz spaces and their applications in related Hardy spaces, Anal. Appl. (Singap.) (2022), https://doi.org/10.1142/S0219530522500166.
280. C. Zhuo, Y. Sawano and D. Yang, Hardy spaces with variable exponents on RD-spaces and applications, Dissertationes Math. 520 (2016), 1–74.
281. A. Zygmund, Trigonometric Series, 2nd ed., Vols. I, II, Cambridge University Press, New York, 1959, xii+383 pp., vii+354 pp.

Index

© The Author(s), under exclusive license to Springer Nature Singapore Pte Ltd. 2022
Y. Li et al., *Real-Variable Theory of Hardy Spaces Associated with Generalized Herz Spaces of Rafeiro and Samko*, Lecture Notes in Mathematics 2320,
https://doi.org/10.1007/978-981-19-6788-7

LECTURE NOTES IN MATHEMATICS   🐎 Springer

Editors in Chief: J.-M. Morel, B. Teissier;

Editorial Policy

1. Lecture Notes aim to report new developments in all areas of mathematics and their applications – quickly, informally and at a high level. Mathematical texts analysing new developments in modelling and numerical simulation are welcome.

   Manuscripts should be reasonably self-contained and rounded off. Thus they may, and often will, present not only results of the author but also related work by other people. They may be based on specialised lecture courses. Furthermore, the manuscripts should provide sufficient motivation, examples and applications. This clearly distinguishes Lecture Notes from journal articles or technical reports which normally are very concise. Articles intended for a journal but too long to be accepted by most journals, usually do not have this "lecture notes" character. For similar reasons it is unusual for doctoral theses to be accepted for the Lecture Notes series, though habilitation theses may be appropriate.

2. Besides monographs, multi-author manuscripts resulting from SUMMER SCHOOLS or similar INTENSIVE COURSES are welcome, provided their objective was held to present an active mathematical topic to an audience at the beginning or intermediate graduate level (a list of participants should be provided).

   The resulting manuscript should not be just a collection of course notes, but should require advance planning and coordination among the main lecturers. The subject matter should dictate the structure of the book. This structure should be motivated and explained in a scientific introduction, and the notation, references, index and formulation of results should be, if possible, unified by the editors. Each contribution should have an abstract and an introduction referring to the other contributions. In other words, more preparatory work must go into a multi-authored volume than simply assembling a disparate collection of papers, communicated at the event.

3. Manuscripts should be submitted either online at www.editorialmanager.com/lnm to Springer's mathematics editorial in Heidelberg, or electronically to one of the series editors. Authors should be aware that incomplete or insufficiently close-to-final manuscripts almost always result in longer refereeing times and nevertheless unclear referees' recommendations, making further refereeing of a final draft necessary. The strict minimum amount of material that will be considered should include a detailed outline describing the planned contents of each chapter, a bibliography and several sample chapters. Parallel submission of a manuscript to another publisher while under consideration for LNM is not acceptable and can lead to rejection.

4. In general, **monographs** will be sent out to at least 2 external referees for evaluation.

   A final decision to publish can be made only on the basis of the complete manuscript, however a refereeing process leading to a preliminary decision can be based on a pre-final or incomplete manuscript.

   Volume Editors of **multi-author works** are expected to arrange for the refereeing, to the usual scientific standards, of the individual contributions. If the resulting reports can be forwarded to the LNM Editorial Board, this is very helpful. If no reports are forwarded or if other questions remain unclear in respect of homogeneity etc, the series editors may wish to consult external referees for an overall evaluation of the volume.

5. Manuscripts should in general be submitted in English. Final manuscripts should contain at least 100 pages of mathematical text and should always include

   - a table of contents;
   - an informative introduction, with adequate motivation and perhaps some historical remarks: it should be accessible to a reader not intimately familiar with the topic treated;
   - a subject index: as a rule this is genuinely helpful for the reader.
   - For evaluation purposes, manuscripts should be submitted as pdf files.

6. Careful preparation of the manuscripts will help keep production time short besides ensuring satisfactory appearance of the finished book in print and online. After acceptance of the manuscript authors will be asked to prepare the final LaTeX source files (see LaTeX templates online: https://www.springer.com/gb/authors-editors/book-authors-editors/manuscriptpreparation/5636) plus the corresponding pdf- or zipped ps-file. The LaTeX source files are essential for producing the full-text online version of the book, see http://link.springer.com/bookseries/304 for the existing online volumes of LNM). The technical production of a Lecture Notes volume takes approximately 12 weeks. Additional instructions, if necessary, are available on request from lnm@springer.com.

7. Authors receive a total of 30 free copies of their volume and free access to their book on SpringerLink, but no royalties. They are entitled to a discount of 33.3 % on the price of Springer books purchased for their personal use, if ordering directly from Springer.

8. Commitment to publish is made by a *Publishing Agreement*; contributing authors of multiauthor books are requested to sign a *Consent to Publish form*. Springer-Verlag registers the copyright for each volume. Authors are free to reuse material contained in their LNM volumes in later publications: a brief written (or e-mail) request for formal permission is sufficient.

Addresses:
Professor Jean-Michel Morel, CMLA, École Normale Supérieure de Cachan, France
E-mail: moreljeanmichel@gmail.com

Professor Bernard Teissier, Equipe Géométrie et Dynamique,
Institut de Mathématiques de Jussieu – Paris Rive Gauche, Paris, France
E-mail: bernard.teissier@imj-prg.fr

Springer: Ute McCrory, Mathematics, Heidelberg, Germany,
E-mail: lnm@springer.com

Printed in the United States
by Baker & Taylor Publisher Services